George Wypych

# HANDBOOK OF FILLERS

2nd Edition

Plastics Design Library

Toronto - New York 1999

Published by ChemTec Publishing
38 Earswick Drive, Toronto, Ontario M1E 1C6, Canada

Co-published by Plastics Design Library
a division of William Andrew Inc.
13 Eaton Avenue, Norwich, NY 13815, USA


ISBN 1-895198-19-4



Canadian Cataloguing in Publication Data

Wypych, George
Handbook of Fillers

2nd ed., revised for the second printing
First edition published under title: Fillers
Includes bibliographical references and index

ISBN 1-895198-19-4 (ChemTec Publishing)
ISBN 1-884207-69-3 (William Andrew Inc.)
Library of Congress Catalog Card Number: 98-88518

1. Fillers (Materials). I. Title. II. Title: Fillers

TP1114.W96 1999 668.4'11 C98-901215-8

Printed in Canada by Transcontinental Printing Inc., 505 Consumers Rd.
Toronto, Ontario M2J 4V8

# Table of Contents

# PREFACE

The first edition of this book was written in 1992. At that time it was not obvious that the pace of filler development was accelerating. In the intervening 6 years, much has been done and there are many new filler products on the market and under development. These have opened new and exciting business opportunities which formulators and marketing managers have exploited in a wide range of new products. The new edition of the book covers many of these developments and discusses the potential for future research and development. What was dealt with only as a passing reference in the first edition now requires a chapter to do it justice.

Six years ago there was less pressure than there is now from environment regulations and activists to limit waste, conserve non-renewable resources, deal with fire and explosion risks, shield a wide variety of energy sources, reduce harmful emissions, and recycle scrap. Today, these issues are the basis of stringent requirements. In addition, products must be lighter, stronger, odor free, look good, and be easy to clean. Plastic products are meeting these challenges and, in doing so, are even able to look and feel like natural products.

Fillers have played a major role in meeting these ever more demanding requirements. The introduction of plastic components in automobiles has been rocky. Early attempts to use plastics failed because they lacked strength and weather resistance. Fillers have been responsible for transforming these same plastics to strong durable automotive components. Portable computer have become the truly portable laptop of today due in large part to the lighter, strongly reinforced plastics that are now available. The cases not only look smooth and sleek, they provide shielding from the electromagnetic radiation that used to prevent the use of computers on aircraft in flight. Where filler used to be though of as a means to lower cost of a plastic part they now contribute to the unique properties that sophisticated users demand. In fact, many fillers now cost more than the polymers that they are added to. But such additions make economic sense because of the value that the filler brings to the formulation.

In this book we hope to have dealt with many of these immense opportunities which these new developments have created. We have examined the current technical literature in detail and it is clear that there is almost unlimited future potential to save time, money, and energy while developing new products with unparalleled performance. It would be nice to think that this book would be read cover to cover but we know that most people will skim through it to find the sections that apply to their work or area of interest. We have attempted to structure the book to make it useful both as a textbook and as a series of monographs. It has been categorized in a way that should match the interests of those with specific needs. Where technologies are shared by more than one application we have duplicated the same information in different formats in two or more sections.

The table of contents provides a clear guide to where specific subject material can be found. Wherever possible we have referenced the original source and we encourage researcher to go to these for the additional details that may provide the clarity and depth that their work may need. We would have liked to include more specific examples and explanations but we believe the book should not run to several thousand pages. However, this issue will be addressed by the end of 1999 when we will publish the information on specific grades of fillers on CD-ROM. It will contain much more data on specific fillers and products, data which can be searched and compared electronically.

We deem it a great privilege to have had the opportunity to report on the extensive data from researchers and filler manufacturers and I wish to acknowledge their kind help and many personal efforts to assist me in this project. I am grateful to those who have worked hard and long to generate the data and ideas that have advanced our understanding of filler properties and composite performance. They continue to make this field of technology increasingly more fascinating. I would also like to thank John Paterson who read and corrected much of the manuscript.

*George Wypych*
*Toronto, October 1998*

# ACKNOWLEDGMENT

The author wishes to acknowledge the kind help and many personal efforts from the representatives of the companies manufacturing fillers and equipment. The following companies were kind to share their data and information:
Abrasivos y Maquinaria, S.A., Calle Caspe, 79, 2o, 08013 Barcelona, Spain
Accuratus Ceramic Corporation, 14A Brass Castle Road, Washington, NJ 07882, USA
Ace International Inc., 520 North Gold Street, Centralia, WA 98531-0885, USA
ACuPowder International, LLC, 901 Lehigh Avenue, Union, NJ 07083, USA
AccuRate Bulk Solids Metering, Unit of Schenck AccuRate, 746 East Milwaukee Street, P.O. Box 208, Whitwater, WI 53190, USA
Advanced Ceramics Corporation, 11907 Madison Avenue, Lakewood, OH 44107-5026
Agrashell, Inc., 5934 Keystone Drive, Bath, PA 18014, USA
Akzo Nobel Aramid Products Inc., 801 F Blacklawn Road, Conyers, GA 30207, USA
Albion Kaolin Company, 1 Albion Road, Hephzibah, GA 30815, USA
Alcan Chemicals Europe, Park, Gerrards Cross, Buckinghamshire SL9 0QB, England
American Metal Fibers, Inc., 2889 North Nagel Court, Lake Bluff, IL 60044-1460, USA
American Wood Fibers, 100 Alderson Street, Schofield, WI 54476-0468, USA
AML Industries, Inc., P.O. Box 4110, Warren, OH 44482, USA
Amoco Performance Products, Inc., 4500 McGinnis Ferry Road, Alpharetta, GA 30202, USA
Amspec Chemical Corporation, 751 Water Street, Gloucester City, NJ 08030, USA
Anthracite Industries, Inc., P.O. Box 112, Sunbury, PA 17801, USA
Anval, Inc., 301 Route 17 North, Suite 800, Rutherford, NJ 07070, USA
Applied Carbon Technology, 953 Route 202 North, Somerville, NJ 08876, USA
Asheville Mica Company, 900 Jefferson Avenue, Newport News, VA 23607-6120, USA
Aspect Minerals, Spruce Pine, NC 28777, USA
Ausimont USA Inc., P.O. Box 1838, Morristown, NJ 07962-1838, USA
Barium & Chemicals, Inc., P.O. Box 218, Steubenville, OH 43952-5218, USA
Bel-Tyne Products Ltd., Victoria Works, Brewery Street, Portwood, Stockport SK1 2BQ, England
Bromine Group Dead Sea, P.O. Box 180, Beer Sheva 84101, Israel
Bekaert Corporation, 1395 South Marietta Parkway, Suite 100, Marietta, GA 30067, USA
Buckman Laboratories, Inc., 1256 North McLean Blvd., Memphis, TN 38108, USA
Burgess Pigment, P.O. Box 349, Sandersville, GA 31082
Cabot Corporation, Special Blacks Division, 157 Concord Road, Billerica, MA 01821, USA
Cabot Performance Materials, P.O. Box 1608, County Line Road, Boyertown, PA 19512, USA
Cancarb Ltd., 1702 Brier Park Crescent N.W., Medicine Hat, AB T1A 7G1, Canada
Carborundum Corporation, Boron Nitride Division, 168 Creekside Drive, Amherst, NY 14228-2027, USA
C.E D. Process Minerals Inc., 863 N. Cleveland - Massillon Road, Akron, OH 4433-2167, USA
Celite Corporation (World Minerals, Inc.), Headquarters, P.O. Box 519, Lompoc, CA 93438-0519, USA
Cellulose Filler Factory Corporation, 10200 Worton Road, Chestertown, MD 21620, USA
Charis, Inc., 512 Sweet Briar Drive, Maryville, TN 37804, USA

Charles B. Chrystal Co., Inc., 30 Vesey Street, New York, NY 10007, USA
Chronos Richardson Inc., 15 Gardner Road, Fairfield, NJ 07004, USA
Chemalloy Company, Inc., P.O. Box 350, Bryn Nawr, PA 19010-0350, USA
CIMBAR Performance Minerals, 25 Old River Road S.E., P.O. Box 250, Cartersville, GA 30120, USA
Cleveland Vibrator Company, 2828 Clinton Avenue, Cleveland, OH 44113, USA
Climax Molybdenum Company, Division of Cyprus Amax Company, Centennial Center, Suite 308, P.O. Box 0407, Ypsilanti, MI 48198-0407, USA
Coal Fillers, Inc, P.O. Box 1063, Bluefield, VA 24605, USA
Composite Materials, L L C, 700 Waverly Ave., Mamaroneck, NY 10543, USA
Composite Particles, Inc., 2330 26th Street S.W., Allentown, PA 18103, USA
Columbian Chemicals Company, 600 Parkwood Circle, Suite 400, Atlanta, GA 30339, USA
Cortex Biochem, Inc., 1933 Davis Street, Suite 321, San Leandro, CA 94577, USA
CSM Industries, 21801 Tungsten Road, Cleveland, OH 44117, USA
Degussa AG, Weissfrauenstrasse 9, D-60311 Frankfurt am Main, Germany
Duke Scientific Corporation, 2463 Faber Place, Palo Alto, CA 94303, USA
DUSLO, a.s., Drienova ul. 24, 826 03 Bratislava, Slovak Republic
Eagle Picher Minerals, Inc., 6110 Plumas Street, Reno, NV 89509, USA
ECC International, Ltd., John Keay House, St. Austell, Cornwall PL25 4DJ, England
Electro Abrasives Corporation, 701 Willet Road, Buffalo, NY 14218, USA
EM Corporation, P.O. Box 2400 TR, 2801 Kent Avenue, West Lafayette, IN 47906, USA
Engelhard Corporation, Pigments and Additives Group, 101 Wood Avenue, P.O. Box 770, Iselin, NJ 08830-0770, USA
D. J. Enterprises, Inc., P.O. Box 31366, Cleveland, OH 44131, USA
Evans Clay Company, P.O. Box 595, McIntyre, GA 31054, USA
Expencel, Inc., 2150-H Northmont Parkway, Duluth, GA 30096, USA
Favre & Matthijs SA, Chemin des Fleurettes, 43, CH-1007 Lausanne, Switzerland
Fibertec, 35 Scotland Boulevard, Bridgewater, MA 02324, USA
Fiber Sales & Development Corporation, Checkerboard Sq., St. Louis, MO 6364, USA
Franklin Industrial Minerals, 612 Tenth Avenue, North Nashville, TN 37203, USA
Grefco Minerals, Inc., P.O. Box 637, Lompoc, CA 93438, USA
Halvor Forberg AS, Hegdal, N-3261 Larvik, Norway
Hapman Conveyors, 6002 E. Kilgore Road, P.O. Box 2321, Kalamazoo, MI 49003, USA
Harwick Standard Distribution Corporation, 60 S. Seiberling Street, P.O. Box 9360, Akron, OH 44305-0360, USA
Hitox Corporation of America, P.O. Box 2544, Corpus Christi, TX 78403-2544, USA
Huber, J.M. Corporation, Engineered Minerals Division, One Huber Road, Macon, GA 31298, USA
Hyperion Catalysis International, 38 Smith Place, Cambridge, MA 02138, USA
I H Polymeric Products, Ltd., Meopham Triding Estate, Meopham, Gravesend, Kent DA13 0LT, England
Inco Company, 681 Lawlins Road, Wyckoff, NJ 07481, USA
Interfibe Corporation, 6001 Cochran Road, Solon, OH 44139, USA
JAYGO, Inc., 675 Rahway Avenue, Union, NJ 07083, USA
J.B. Company, 9 Ginter Street, Franklin, NJ, USA
Kentucky-Tennessee Clay Company, 1441 Donelson Pike, Nashville, TN 37217, USA
Keystone Filler & Mfg. Company, 214 Railroad Street, Muncy, PA 17756, USA
Kinetico Inc., 10845 Kinsman Road, P.O. Box 193, Newbury, OH 44065, USA
Kronos Canada, Inc., Suite 206, 45 Sheppard Ave. East, Toronto, Ontario, Canada M2N 5W9
K-Tron, Routes 55 & 553, Pitman, NJ 08071, USA
Lancaster Products, Division of Kercher Industries, Inc., 920 Mechanic Street, Lebanon, PA 17046, USA

Laurel Industries, Inc., 30195 Chagrin Boulevard, Cleveland, OH 44124-5794, USA
Littleford Day, Inc., 7451 Empire Drive, Florence, KY 41042-2985, USA
Luzenac Europe, B.P. 1162, 31036 Toulouse Cedex 1, France
Malvern Minerals Company, 220 Runyon Street, P.O. Box 1238, Hot Springs National Park, AR 71902, USA
Mica-Tek, A Division of Miller and Company, 325 North Center Street, Suite D, Northville, MI 48167-1224, USA
Millennium Inorganic Chemicals, 200 International Circle, Suite 5000, Hunt Valley, MD 21030, USA
MMM Carbon, Avenue Louise 534, B-1050 Brussels, Belgium
Morgan Matroc, Ltd., Bewdley Road Stourport-on Severn, Worcestershire DY13 8QR, England
Nabaltec GmbH, P.O. Box 18 60, D-92409 Schwandorf, Germany
Nanophase Technologies Corporation, 453 Commerce Street, Burr Ridge, IL 60521, USA
Non-Metals, Inc., 1870 West Prince Road, Suite 67, Tucson, AZ 85705, USA
Novamet Specialty Products Corporation, 681 Lawlins Road, Wyckoff, NJ 07481, USA
NOVATEC, 222 E. Thomas Avenue, Baltimore, MD 21225, USA
Nyco Minerals, Inc., 124 Mountain View Drive, Willsboro, NY 12996-0368, USA
Nyacol Products, Inc., Megunco Road, P.O. Box 349, Ashland, MA 01721, USA
Old Hickory Clay Company, P.O. Box 66, Hickory, KY 42051-006, USA
OMG, Inc., World Headquarters, 50 Public Aquare, 3800 Terminal Tower, Cleveland, OH 44113, USA
OMYA/Plüss-Staufer AG, P.O. Box 32, CH-4665 Oftringen, Switzerland
Owens Corning, World Headquarters, One Owens Corning Parkway, Toledo, OH 43659, USA
Pacific Century, Inc., P.O. Box 221016, Chantilly, VA 20153, USA
Palamatic Handling Systems Ltd., Cobnar Wood Close, Chesterfield Trading Estate, Sheepbridge, Chesterfield, Derbyshire S41 9RQ, England
Pierce & Stevens Corporation, 710 Ohio Street, Buffalo, NY 14203, USA
Piqua Minerals, Inc., 1750 West Statler Road, Piqua, OH 45356, USA
Polar Minerals, 1703 Bluff Rd., Mt. Vernon, IN 47620, USA
Plastic Methods Co., Inc., 20 West 37th Street, New York, NY 10018, USA
Polytechs S.A., Zone Industrielle de la Gare, BP 14, 76450 Cany Barville, France
Potters Industries, Inc., Southpoint Corporate Headquarters, P.O. Box 840, Valley Forge, PA 19482-0840, USA
PPG Industries, Inc., One PPG Place, Pittsburgh, PA 15272, USA
PQ Corporation, Corporate Headquarters, P.O. Box 840, Valley Forge, PA 19482-0840, USA
Premier Pneumatics, Inc., 606 North Front St., P O Box 17, Salina, KS 67402-0017, USA
Quarzwerke GmbH, P.O. Box 1780, Kaskadenweg 40, D-50226 Frechen, Germany
ReBase Products, Inc., 70 Collier Street, Barrie, ON L4M 4Z2, Canada
Reheis Ireland, Kilbarrack Road, Dublin 5, Ireland
Piedemont Minerals, Division of RESCO Products, Inc., P.O. Box 7247, Greensboro, NC 27417-0247, USA
Sachtleben Chemie GmbH, Postfach 17 04 54, D-47184 Duisburg, Germany
San Jose Delta Associates, Inc., 482 Sapena Court, Santa Clara, CA 95054, USA
Silberline Manufacturing Co., Inc., Lincoln Drive, P.O. Box B, Tamaqua, PA 18252-0420, USA
Silvered Electronic Mica Co., Inc., P.O. Box 505, 107 Boston Post Road, Willimantic, CT 06226, USA
Solvay S.A. Benelux, rue du Prince Albert 44, B-1050 Bruxelles, Belgium
SOVITEC Iberica S.A., Poligono Industrial, E-Castellbisbal-Barcelona, Spain

Sphere Services, Inc., 1055 Commerce Park Drive, Suite 100, Oak Ridge, TN 37830, USA
Spiroflow-Orthos Systems, Inc., 2806 Gray Fox Road, Monroe, NC 28110, USA
Steward, 1200 East 36th Street, P.O. Box 510, Chattanooga, TN 37401-0510, USA
Strong-Lite Products Corporation, Emmett Sanders Road, P.O. Box 8029, Pine Bluff, AR 71611, USA
Struktol Corporation, 201 East Steels Corners Road, P.O. Box 1649, Stow, OH 44224, USA
Superior Graphite Corporation, 120 South Riverside Plaza, Chicago, IL 60606, USA
Suzorite Mica Products, Inc., 1475 Graham Bell Street, Boucherville, Quebec J4B 6A1, Canada
Syncoglas N.V., Industriepark, Drukkerijstraat 9, B-9240 Zele, Belgium
Synair Corporation, P.O. Box 5269, 2003 Amnicola Highway, Chattanooga, TN 37406, USA
TAM Ceramics, Inc., P.O. Box 67, 4511 Hyde Park Blvd., Niagara Falls, NY 14305-0067, USA
Technic, Inc., Engineered Powders Division, 300 Park East Drive, Woonsocket, RI 02895, USA
Teledyne Advanced Materials, An Allegheny Teledyne Company, 7300 Highway 20 West, Huntsville, AL 35806, USA
TIMCAL, Ltd., Graphites and Technologies, CH-5643 Sins, Switzerland
Tioxide Canada, Inc., 9999 Cavendish Boulevard, Suite 100, Ville Saint-Laurent, Quebec H4M, 2X5, Canada
Transmet Corporation, 4290 Perimeter Drive, Columbus, OH 43228, USA
Toho Rayon Co., Ltd., 3-9 Nihonbashi 3-chome Chuo-ku, Tokyo 103, Japan
3M Chemicals, Specialty Additives, 3M Center building 223-6S-04, St. Paul, MN 55144-1000, USA
United Clays, Inc., 7003 Chadwick Drive, Suite 100, Brentwood, TN 37027, USA
United States Antimony Corporation, P.O. Box 643, Thompson Falls, MT 59873, USA
U.S. Silica Company, P.O. Box 187, Berkeley Springs, WV 25411, USA
Vanderbilt, R.T. Company, Inc., 30 Winfield Street, P.O. Box 5150, Norwalk, CT 06856-5150, USA
Wacker-Chemie GmbH, Hans-Seidel Platz 4, D-81737 München, Germany
Wright Industries, Inc., 225 49th Street, Brooklyn, NY 11220, USA
Zemex Industrial Minerals, 1040 Crown Point Parkway, Suite 270, Atlanta, GA 30338
Zeochem, Chemie Uetikon Subsidiary, P.O. Box 35940, Louisville, KY 40232, USA
Zinc Corporation of America, 300 Frankfort Road, Monaca, PA 15061-2295, USA

# 1

# INTRODUCTION

This introduction:

- Lists the properties of materials which are influenced by fillers
- Lists typical properties of fillers
- Provides definition of the term "filler"
- Defines how fillers function in various applications
- Suggests how fillers may be classified
- Discusses the markets for fillers and the emerging trends in filler use

The introduction will define the scope of the book and provide a brief overview of each chapter.

It is our intention to show how an understanding of the diverse functions of fillers in materials can lead to a well designed material formulation.

## 1.1 EXPECTATIONS FROM FILLERS

What caused fillers to be added to materials in the first place was probably the quest for lower costs. Fillers were inexpensive, thus using them would make the material cheaper. We do not know who the inventor of the idea was but it was probably not one, but many people in many different places. However, as the following discussion shows, cost reduction is no longer the only, or even the most important, consideration for using fillers in formulating composite materials.

These examples which follow list attributes of materials to the formulator's various objectives.

*Cost reduction*[1]

Cost reduction depends on the relative cost of the polymer and the filler. Polymer prices in 1996-7 were approximately:

| | US$/kg | US$/l |
|---|---|---|
| ABS | 1.98 | 2.05 |
| PE | 0.77 | 0.70 |
| PET | 1.65 | 2.67 |
| PP | 0.88 | 0.79 |
| PS | 0.79 | 0.84 |
| PVC | 0.66 | 0.92 |

Filler prices depend greatly on the particle size. In the list below, fillers are divided into *large* particle size materials (up to 100 μm; e.g., ground $CaCO_3$), *medium* particle size (around 10 μm; e.g., clay), *small* particle size (around 1 μm; e.g., $TiO_2$ or precipitated $CaCO_3$), and *very small* particle size (below 0.1 μm; e.g., fumed silica). These are approximate prices:

| | **US$/kg** | **US$/l** |
|---|---|---|
| Large | 0.05 | 0.14 |
| Medium | 0.31 | 0.81 |
| Small | 1.00 | 2.80 |
| Very small | 6.60 | 14.50 |

If we consider only cost, it is the cost per unit volume that must be compared. The table shows that only the use of large particle size fillers (very crude products) can potentially contribute to savings in the manufactured cost of materials made from commodity polymers. At the same time, fillers decrease many mechanical properties of the material so cost reduction is achieved at the expense of performance. Medium particle sized fillers are less attractive economically because costs of processing, inventory and transportation will increase and must be added to the total. This shows that there must be other motives to compound polymers with fillers. These follow.

*Material density*[2]

Fillers can be used either to increase or to decrease the density of a product. Because the density of a filler can be as high as 10 g/cm$^3$ or as low as 0.03 g/cm$^3$, there may be a large difference between the density of the filler and the polymer. Thus a broad range of product densities can be obtained. There are high density products (above 3 g/cm$^3$) such as materials used in appliances or casings for electronic devices. More common are densities below 2 g/cm$^3$, glass fiber filled composites being a typical example. The effective density of the polymer can be decreased by filling a foam with hollow polymer spheres. In this example, the density of a material can be lower than 0.1 $g/cm^3$.

*Optical properties*[3-6]

Optical properties of compounded materials depend on the physical characteristics of the filler and the other major ingredients including the polymer. Most important is the relative refractive index of the two ingredients. Depending on how they match, one can obtain clear or opaque materials. Light absorption by the non-polymer ingredients is essential in preventing UV degradation. Some fillers (e.g., $TiO_2$, ZnO or talc) effectively absorb light. Aluminum trihydroxide in UV curable polyurethanes is noteworthy in that it accelerates the curing process because it is transparent to UV light. Calcinated clay as a filler in greenhouse film at a 10% level drastically reduces infrared absorption during the day and heat loss during the night. This application of physical principles has been an important factor in increasing the productivity of greenhouses.

*Color*

Fillers frequently cause problems in color matching and must be accounted for in product color design. Many fillers have a distinctive color which is useful in material coloring. Recently metal powders have been used in combination with pigments to make the composite appear metal like.

*Surface properties*[7-10]

For hundreds of years sticky surfaces have been dusted with powder (e.g., talc) to keep them separated. Talc is broadly used in cable and profile extrusion to obtain a smooth surface. Similarly, in injection molding, the application of aluminum trihydroxide gives a better surface finish. Talc, $CaCO_3$, and diatomite provide anti-blocking properties. Graphite and other fillers decrease the coefficient of friction of materials. PTFE, graphite and $MoS_2$ allow the production of self-lubricating parts. Here, PTFE, a polymer in powder form, acts as a filler in other polymers. Matte surfaced paint is obtained by the addition of silica fillers.

| | |
|---|---|
| *Product shape*[6,11,12] | Fillers reduce shrinkage of polymer foams. Mica and glass fiber reduce warpage and increase the heat distortion temperature. Intumescent fillers increase in volume rapidly as they degrade thermally expanding the material and blocking fire spread. |
| *Thermal properties*[13] | Fillers may decrease thermal conductivity. The best insulation properties of composites are obtained with hollow spherical particles as a filler. Conversely, metal powders and other thermally conductive materials substantially increase the dissipation of thermal energy. |
| *Electrical properties*[14] | Volume resistivity, static dissipation and other electrical properties can be influenced by the choice of filler. Conductive fillers in powder or fiber form, metal coated plastics and metal coated ceramics will increase the conductivity. Many fillers increase the electric resistivity. These are used in electric cable insulations. Ionic conductivity can be modified by silica fillers. |
| *Magnetic properties*[15] | Ferrites induce ferromagnetic properties and are used to make plastic magnets. |
| *Permeability*[7] | Gas and liquid permeability are influenced by the choice of filler. The platelet structure of mica or talc as a filler in paints and plastics decreases the transmission of gases and liquids. |
| *Mechanical properties*[1,16,17] | All mechanical properties are affected by fillers. Filler combinations may be selected to optimize a variety of mechanical properties. Fillers reinforce and provide abrasion resistance. |
| *Chemical reactivity*[11,18] | Many fillers can be used to influence chemical reactions occurring in their presence. The reaction rate can be decreased or increased. Fillers such as ZnO will react with UV degradation products in PE to limit damage. The pot-life of curing mixtures can be increased. Cure rates can be slowed, exothermic effects can be controlled, incompatible polymers can be blended and molecular mobility reduced. |
| *Rheology*[7,10] | The rheology of many industrial products depends on the filler addition. Examples include sealants, tooth pastes, cosmetics, hotmelts, |

papers, paints, etc. Normally, additions of fillers increase the viscosity and contribute to non-Newtonian flow characteristics, but there are also combinations such as filler mixtures and specially designed glass beads which either reduce the viscosity or do not affect it.

*Morphology*[11,19]

Polymer crystallization and structure are affected by fillers. They may increase or decrease the nucleation rate (and thus the crystallization rate). An increase in the nucleation rate is observed in PET in the presence of mica. Fillers, especially fibers, may also decrease the mechanical properties of filled materials because of their effect on transcrystallinity. The polymer structure at the interface with fillers is different than in the bulk.

*Material durability*[3,18,6,12,20-22]

Fillers which screen radiation and react with degrading molecules contribute to material durability. The opposite effects may also occur where fillers participate in photochemical reactions which decrease photostability. Some fillers are used for their absorption of highly penetrating radiation such as nuclear radiation or filler use in neutron shielding. Thermal degradation can be either decreased and increased by the presence of fillers. Fillers such as borates and montmorillonite also protect materials from biodegradation. The addition of starch generates numerous mechanisms which increase biodegradability by supplying nutrients and also participate to initiate thermal and UV degradation which reduces chain length and allows biological conversions.

*Environmental impact*[23-26]

Fillers contribute to fire retardancy by suppressing fire, increasing autoignition temperature, decreasing smoke formation, increasing char formation, reducing heat transmission rate, preventing dripping, etc. Fillers are used in combinations to balance properties. For example, antimony trioxide increases smoke whereas $Al(OH)_3$ and $Mg(OH)_2$ reduce it. In combination, this allows a balance of properties. It is possible to make paper fire retardant through the proper selection of fillers. Plastics recycling can be improved by incorporating fillers which reduce thermodegradation (stabilize some polymers) complex mixtures of polymer waste are more easily blended if compounded with fillers.

*Performance of other additives* Fillers are instrumental in improving the performance of other additives. Antistatics, blowing agents, catalysts, compatibilizers, coupling agents, organic flame retardants, impact modifiers, rheology modifiers, thermal and UV stabilizers are all influenced by a filler's presence.

*Health & safety* Fillers are probably the least hazardous component among additives. The major exception here is asbestos which is seldom, if ever, used now. Other fillers which may be hazardous are being carefully investigated although disputes still occur when data is incomplete or questionable.

Fillers produced today are manufactured by sophisticated processes. There are numerous examples of surface modification which changes a filler's properties. There are fewer methods of filler synthesis. Preparation of materials for specific medical applications can be carried out using template polymerization.[27] This has become a well established discipline which has contributed to the understanding of polymer structure. Here, the polymer is produced on organic and inorganic (e.g., fillers) templates. By choosing the template structure, polymer properties can be tailored to requirements. Natural biological materials are formed in this manner and synthetic materials can be formed in a similar manner. Filler properties can also be tailored by synthesizing fillers in the presence of other materials. This is used in medical applications where the filler becomes compatible with its surroundings as it forms in body fluids. Artificial bone materials can thus be formed with surface characteristics acceptable to (compatible with) the body's environment. These techniques are at the most advanced levels of engineering and design in filler synthesis.

In summary,

- Fillers usually do not reduce the cost of material manufacturing
- Fillers are not inert materials added to fill space (if they are used in this way, they likely degrade properties of the material)
- Fillers can be modified and tailored to any application
- Fillers modify practically all properties of the material and influence the design, manufacture, and use

- Plastics performance and the performance of other materials are highly influenced by fillers
- The plastics applications base has expanded greatly as the use of fillers has increased

## 1.2 TYPICAL FILLER PROPERTIES

We have outlined the product performance characteristics of fillers. This leads us to an identification of filler properties which allow different fillers to be compared and evaluated. When we go on to develop a definition of fillers in Section 1.3, this list will help to make the definition inclusive yet precise. It will also assist in the classification of fillers discussed in Section 1.4.

| | |
|---|---|
| *Physical state* | All materials discussed are solids but they might be available in a pre-dispersed state |
| *Chemical composition* | May be inorganic or organic and of an established chemical composition. May also be a single element, natural products, mixtures of different materials in unknown proportions (waste and recycled materials), or materials of a proprietary composition |
| *Particle shape*[28] | Spherical, cubical, irregular, block, plate, flake, fiber, mixtures of different shapes |
| *Particle size* | Range from a few nanometers to tens of millimeters (nanocomposites to pavements or textured coatings) |
| *Aspect ratio*[28] | 1 (spherical or cubical) to 1,600 (fibers) |
| *Particle size distribution* | Monodisperse, designed mixture of sizes, Gaussian distribution, irregular distribution |
| *Particle surface area*[29] | From 10 to over 400 $m^2/g$. Depending on the specific surface area particles have different levels of porosity from completely non-porous and smooth to very porous with a range of pore sizes |
| *Particle internal structure* | Hollow to porous to void free solid |
| *Particle-particle association* | Singular, agglomerates, aggregates, flocculated materials |
| *Density* | From 0.03 $g/cm^3$ (expanded polymer beads) to 18.88 $g/cm^3$ (gold) |
| *Refractive index* | Typical range from 1 (air) to 3.2 (iron oxide) |
| *Color* | Full range of colors from colorless and transparent, with increasing opacity through white to black |
| *pH* | From 2 (carbon black) to 12 (calcium hydroxide) |

| | |
|---|---|
| *Moisture* | Traces to 10+% |
| *Oil absorption* | From a few grams to over 1000 g/100 g of filler |
| *Thermal properties* | Thermal expansion coefficient and thermal conductivities vary widely |
| *Electric and magnetic properties* | Wide variations are possible between non-conductive and conductive and between magnetic and non-magnetic |

These and other properties of fillers are used to describe individual products. The potential applications of a filler are determined by its set of properties listed above but, often, other characteristics must be known to select the optimum filler or fillers for specific application. Additional properties are discussed in Chapters 5 to 12.

## 1.3 DEFINITIONS

These different sources define fillers in different ways

| | |
|---|---|
| *Dictionary* | A material used to fill a cavity or increase bulk of something |
| *Technical dictionary*[30] | A material added to a polymer in order to reduce compound cost and/or, to improve processing behavior and/or, to modify product properties |
| *Encyclopedia*[31,32] | Fillers, or extenders as they are called in the coatings industry, are finely divided solids added to polymer systems to improve properties and reduce cost |
| *Handbook*[33] | Fillers are solid additives, different from plastics matrices in composition and structure, which are added to polymers to increase bulk or improve properties |
| *ASTM C 709-91*[34] | In manufactured carbon and graphite product technology, carbonaceous particles comprising the base aggregate in an unbaked green-mix formulation |
| *ASTM C 859*[35] | A general term for a material that is inert under the conditions of use and serves to occupy space and may improve physical properties |
| *ASTM D 123-96*[36] | A relatively inert material added to a plastic to modify its strength, permanence, working properties or other qualities, or to lower costs |
| *ASTM D 1566-95a*[37] | A solid compounding material, usually in a finely divided form, which may be added in relatively large proportions to a polymer for technical or economic reasons |
| *ASTM D 1968-96a*[38] | A material, generally non-fibrous and inorganic, added to the fiber furnish |
| *ASTM D 3878-95c*[39] | A primarily inert solid constituent added to the matrix to modify the composite properties or to lower cost |

These definitions fail in some ways:

- In many instances the filler is regarded as an inert solid used for cost reduction
- Some exclude fibers, some accept fibers as fillers
- None describe conditions under which the filler lowers the cost and/or affects other properties

Although not crucial to the technology itself a more rigorous definition will serve to set boundaries and include all that is vital to filler technology.

The word fill is synonymous with the action of filling, cluttering or dumping as these are very common human activities. It also means saturate, penetrate, infiltrate, impregnate, pack, quench all of which are consistent with what fillers are designed to do. They saturate and pack spaces depending on their shape, particle size distribution, etc. Fillers penetrate and infiltrate materials. But, there are hardly any cases in which the surrounding material penetrates the filler's outside boundary. Their impregnating and quenching activity can be translated into their ability to react or interact with the surrounding material.

Thus, the word "filler" adequately describes the "filler's" potential to perform in multicomponent systems. To follow this simple lead this definition provides the simplicity and precision needed:

> "Filler is a solid material capable of changing the physical and chemical properties of materials by surface interaction or its lack thereof and by its own physical characteristics."

If one compares this definition with the other above, the noticeable differences are as follows:

- It does not attempt to provide an incomplete list of properties. It suggests that a broad scope of properties can be influenced by fillers
- This definition implies the existence of two ways in which a filler performs in a system – through its own properties (e.g., hardness, particle size, particle shape, etc.) and through interactions with the material (the extent of which can vary from strong chemical/physical interaction to almost no interaction at all). This allows us to include all existing fillers (even the degrading fillers which have too large a particle size and too small an interaction to combine with the material in an economical manner)
- The definition does not exclude a material because of its shape, particle size or chemical composition.

We may now judge the definition based on expectations developed in the discussion in Section 1.1. From a cost reduction analysis, it is evident that if a filler has a large particle size and no strong interaction with its surroundings it will decrease the intrinsic mechanical performance of the material. Such fillers are rightly called "degrading fillers". The material density depends not only on the combined densities of the filler and the matrix but also on the interaction with and wetting of (quality of mixing) the filler's surface by the matrix. Optical properties are affected in a similar manner. When transparency is needed, a proper match of refractive indices is required but also the filler must be incorporated with a minimum of voids through good mixing and wetting. Anti-blocking properties and lowering of the coefficient of friction are improved due to crystallization and orientation of the matrix on the filler's surface. Shape retention is affected by interactions on the filler's surface and in intumescent applications, the filler is not only responsible for producing volatiles to expand the material but it also retains the bubbles formed in the process. Thermal, magnetic, and electrical properties depend on both the filler and matrix but also on interactions between the filler and the matrix. Filler particles which are to influence permeability must have shape characteristics which permit close packing and a high affinity for each other and the matrix if permeability is to be maximized or, conversely, little affinity and minimum packing efficiency if minimum permeability is required. Many papers outline reasons for the improvement of mechanical properties. "Interaction" enters into most explanations along with properties such as surface, shape, rigidity, or strength. Chemical reactivity in the presence of a filler can change the probability of a reaction occurring often because the structure of the reactive molecule changes to make reactive groups more accessible. Each chemical reaction requires intimate contact between the chemical groups entering into the reaction. The durability and environmental impact conferred by fillers are caused by similar principles. The effects that fillers have on other components of the formulation are based on the ability of fillers to be UV absorbers, fire retardants, etc. or on mechanisms which cause the filler's surface to interact with additives (e.g., better retention due to absorption, reaction with adhesion promoters, slow release of catalyst, etc.). The rheological and morphological effects of fillers require interactions with surrounding materials.

To further test the definition we should verify that all materials known to be used as fillers can be included in the definition. Organic materials are of concern since other definitions seem to exclude them. This is a serious inconsistency given the fact that wood flour was one of the first fillers used in modern polymers. Today, when many recycled products are used as fillers, their exclusion does not serve any purpose since they do contribute to the improvement of the materials which will use them. They are included in our definition.

Also, fibers are controversial. In one currently used handbook,[33] natural, inorganic fibers such as wollastonite or asbestos have been included among fillers whereas other fibers were included in a separate group with only three materials: glass, aramid, and graphite. But, mixtures of fibrous and particulate materials are found in many composites today and various natural materials having fibrous structures are considered fillers in technical papers. Again our definition includes these examples.

Finally, carbon black and titanium dioxide are frequently classified as colorants, as opposed to fillers.[33] Tire customers "can choose any color as long as it is black". This is economical technology due to the reinforcement and UV protection offered by carbon black. There are other instances in which carbon black is used for these two reasons. It seems wrong to classify it as a colorant. It is rather a filler which bestows several essential benefits due to its properties and its interaction with the matrix. In the paper industry,[40] titanium dioxide is qualified as a filler when it fills the space between fibers and pigments in the surface coating. $CaCO_3$, talc, clay, etc. are also considered pigments in the paper industry. There are very few reasons today to distinguish between fillers and pigments. In the past, it was perhaps simpler because any material which had a particle size below 1 μm was considered a pigment. Today, the majority of fillers fit this criterion. We include titanium dioxide in this discussion because it has physical properties other than color (e.g., very high refractive index, photochemical activity, UV absorption, etc.) which contribute to the performance of the material in which it is compounded. Our definition also eliminates the exclusion, based on chemical composition or particle size, from the group and allows the inclusion of such materials as gold and nanoparticles.

## 1.4 CLASSIFICATION

In the first edition of this handbook,[41] fillers were assigned to groups according to their mineral origin and chemical composition (mineral, glass, carbon black, organic, metal). The group of mineral fillers was further divided according to geological classification. We now prefer not to use the physical origin or the chemical composition of the filler as a division.

Classification by particle size is helpful in classification since particle size will affect performance but, by itself, falls short as a criterion when selecting fillers for applications which require certain levels of conductivity (thermal or electric) or of chemical interaction, etc. In one publication,[33] materials were divided into particulates, fibers, and colorants. These distinctions are not helpful for a material designer. For a classification to be useful in filler applications, it must include the most important properties of fillers which affect the resultant material. The eight most important are as follows:

- Particle size and distribution
- Aspect ratio
- Chemical composition of surface
- Mechanical properties of filler particles
- Electric and thermal conductivity
- Quantitative description of interactions
- Composition of admixtures
- Optical properties

The existing data may allow us to classify materials according to these properties, however, eight major denominators seem too complex to use to apply practically. Thus, we have decided that, of more than 70 groups of fillers in use today, each will be named based on its common use. These are derived from chemical composition (chemical name), method of filler preparation (precipitated, fumed, hydrated, etc.), mineral source, shape of particle, origin (e.g., original waste material from which ground product is manufactured, name of natural organic product, sand, etc), or material structure (e.g., metallized ceramics). This listing of products has some deficiencies but if presented in alphabetical order, fillers are easy to find. However, it is essential to think about fillers in terms of the eight major denominators or the full list of major properties listed in Section 1.1 or Chapter 5. This will provide the greatest benefit in selecting fillers for specific applications.

## 1.5 MARKETS AND TRENDS

Filler market in plastics alone totals over 10,000 tones per year.[42] Calcium carbonate takes about 2/3 of this market. The market is very large but a large segment of it consists of use in products which are not sophisticated technically. The main applications include:

- Plastics
- Construction
- Paper
- Paints and coatings
- Cosmetics and pharmaceuticals
- Fibers
- Food
- Friction materials
- Printing

These applications are covered in detail in Chapter 19.

Four polymers are the largest consumers of fillers (PVC, PP, PA, polyesters). The consumption of each polymer is immediately mirrored by the consumption of the fillers used in this polymer. The most recent changes in PVC consumption were reflected in the consumption of fillers. At the same time, future trends and developments in fillers are more related to the advances of plastics as they replace many traditional materials. For plastics to give the required performance, new filler technology was required. Current developments allow us to predict some future directions in filler markets. These technologies will become more important:

- Nanoparticles
- Conductive fillers
- Surface modification technology
- Filler mixtures
- Non-dusting fillers

- Morphology-specific fillers
- Compatibilizing fillers
- Low-cost reinforcing fillers

Many new applications of plastics (especially in the high technology sector) are becoming possible due to the advances in nanoparticulates and conductive filler technology. The studies in these areas remain closer to laboratory scale than to full production. Surface modification and filler mixtures will be driven by two expectations: increased mechanical properties and to use fillers more as rheology modifiers. Many new products are being tested in this area now and newer products will enter the market in the next few years.

Dust is one of the most troublesome hazards associated with fillers. Thus, compressed (pelletized) fillers will become more important and wetting technology will be more extensively used. New developments in medical applications require compatibility of medical devices with tissues and body liquids. Advances are expected from the synthesis of inorganic materials which will form artificial surfaces which are less intrusive and which meet performance requirements.

The current emphasis on material recycling requires materials to contain additives which will allow the processing of complex mixtures of polymers through compatibilization, increased thermal resistance during reprocessing, allow for filler recovery, and allow the use of ground waste as a filler. All these technologies have high growth potential because of social, regulatory, and economic pressures.

These current developments place an emphasis on the perfection of filler technology. This has resulted in the creation of many very high quality materials which are too expensive to use in most applications. There is a need to develop materials which are substantially more cost-effective but still allow the conservation of matrix materials. This will be driven by environmental concerns. Product life cycle evaluation, an emerging development, will have a strong impact on the choice of future technologies and fillers associated with these technologies.

## REFERENCES


1 Evans L R, Meeting of the Rubber Division, ACS, Montreal, May 5-8, 1996, paper D.
2 Guerrica-Echevarria G, Eguiazabal J I, Nazabal J, *Polym. Degradat. Stabil.*, **53**, No.1, 1996, 1-8.
3 Rabello M S, White J R, *Polym. Composites*, **17**, No.5, 1996, 691-704.
4 Parker A A, Martin E S, Clever T R, *J. Coatings Technol.*, **66**, No.829, 1994, 39-46.
5 Pak S H, Caze C, *J. Appl. Polym. Sci.*, **65**, 1997, 143-53.
6 Dufton P W, **Functional Additives for Plastics**, *Rapra*, Shawbury, 1994.
7 *Int. Polym. Sci. Technol.*, **23**, No.7, 1996, T/1-3.
8 Shin Jen Shiao, Te Zei Wang, *Composites*, **27B**, No.5, 1996, 459-65.
9 *Reinf. Plast.*, **38**, No.11, 1994, 15.
10 Aldcroft D, *Polym. Paint Col. J.*, **184**, No.4366, 1994, 423-5.
11 Alpern V, Shutov F, *Prog. Rubb. Plast. Technol.*, **11**, No.4, 1995, 268-83.
12 AddCon '96, *Rapra*, Shawbury, 1996.
13 Oien H T, Polyurethanes '95. Conference Proceedings, Chicago, Il., 26th-29th Sept.1995, 137-41.
14 Khan S A, Baker G L, Colson S, *Chem. of Mat.*, **6**, No.12, 1994, 2359-63.
15 Anantharaman M R, Kurian P, Banerjee B, Mohamed E M, George M, *Kaut. u. Gummi Kunst.*, **49**, No.6, 1996, 424-6.
16 Oien H T, Polyurethanes '95. Conference Proceedings, Chicago, Il., 26th-29th Sept.1995, 137-41.

17 Donnet J B, Wang T K, *Prog. Rubb. Plast. Technol.*, **11**, No.4, 1995, 261-7.
18 Gordienko V P, Dmitriev Y A, *Polym. Sci., Ser. B*, **37**, Nos.5-6, 1995, 249-50.
19 Okamoto M, Shinoda Y, Okuyama T, Yamaguchi A, Sekura T, *J. Mat. Sci. Lett.*, **15**, No.13, 1996, 1178-9.
20 Sundar K L, Radhakrishnan G, Reddi B R, *Polym. Plast. Technol. Engng.*, **35**, No.4, 1996, 561-6.
21 Ohashi F, Oya A, *J. Mat. Sci.*, **31**, No.13, 1996, 3403-7.
22 Bikiaris D, Prinos J, Panayiotou, *Polym. Degradat. Stabil.*, **56**, 1997, 1-9.
23 Levchik G F, Levchik S V, Lesnikovich A I, *Polym. Degradat. Stabil.*, **54**, Nos 2-3, 1996, 361-3.
24 Nagieb Z A, El-Sakr N S, *Polym. Degradat. Stabil.*, **57**, 1997, 205-9.
25 Baggaley R G, Hornsby P R, Yahya R, Cussak P A, Monk A W, *Fire Mater.*, **21**, 1997, 179-85.
26 Liauw C M, Hurst S J, Lees G C, Rothon R N, Dobson D C, *Prog. Rubb. Plast. Technol.*, **11**, No.2, 1995, 137-53.
27 Polowinski S, **Template Polymerization**, *ChemTec Publishing,* Toronto, 1997.
28 Turner J D, Property Enhancement with Modifiers and Additives. Retec proceedings, New Brunswick, N.J., 18th-19th Oct.1994, 65-87.
29 Allen N S, Edge M, Corrales T, Childs A, Liauw C, Catalina F, Peinado C, Minihan A, *Polym. Degradat. Stabil.*, **56**, 1997, 125-39.
30 Whelan T, **Polymer Technology Dictionary**, *Chapman & Hall*, London, 1994.
31 Kroschwitz J I, **Concise Encyclopedia of Polymer Science and Engineering**, *Wiley*, New York, 1990.
32 Kroschwitz J I, Ed., **Encyclopedia of Polymer Science and Engineering**, 2nd Ed., Vol. 7, *Wiley,* New York, 1987.
33 Schlumpf H P, **Filler and Reinforcements** in **Plastic Additives**, Ed. Gaechter R, Mueller H, *Hanser Verlag,* Munich, 1993.
34 ASTM C 709-91b. Standard Terminology Relating to Manufactured Carbon and Graphite.
35 ASTM C 859-92a. Nuclear Materials.
36 ASTM D 883-96. Standard Terminology Relating to Plastics.
37 ASTM D 1566-95a. Standard Terminology Relating to Rubber.
38 ASTM D 1968-96a. Standard Terminology Relating to Paper and Paper Products.
39 ASTM D 3878-95c. Standard Terminology of High-Modulus Reinforcing Fibers and Their Composites.
40 Hagemeyer R W, Ed., **Pigments for Paper**, *Tappi Press*, Atlanta, 1997.
41 Wypych G, **Fillers**, *ChemTec Publishing*, Toronto, 1993.
42 Hohenberger W, *Kunststoffe Plast Europe*, **86**, 7, 1996, 18-20.

# 2

# Sources of Fillers, Their Chemical Composition, Properties, and Morphology

The information included in this chapter is based on the data selected from the technical information included in the manufacturers literature and research papers. The main goal of this chapter is to provide information on:

- Physical and chemical characteristics of fillers
- Morphology of filler particles
- Sources of fillers
- Manufacturers
- Important commercial grades
- Major applications
- Relevant studies

Data for each filler are presented in the form of a standard table which contains, for a particular filler, only sections for which information was available. The physical characteristics of fillers and other data on characteristic parameters are taken from the manufacturers literature and open literature to show the range of properties rather than values for a particular grade. The information on the characteristics of every grade is extensive and comes from over 150 manufacturers. Large quantity of information gathered is presented as established data in tabular form. A future publication on CD-ROM will present full information on all grades available worldwide.

Commercial information is presented in an abbreviated form in the individual tables. In addition to this information, there is an appendix included at the end of this book which provides references to the manufacturers and distributors of these products worldwide. There is no distinction made in the tables between the manufacturers and distributors.

The text which follows the table for a particular group of fillers discusses manufacturing methods, morphology and explains and amplifies the tabular data.

# 2.1 PARTICULATE FILLERS

## 2.1.1 ALUMINUM FLAKES AND POWDER[1-6]

| | | |
|---|---|---|
| **Names**: aluminum flakes, aluminum pigments, leafing aluminum pigments | | **CAS #**: 7429-90-5 |
| **Chemical formula**: Al | **Functionality**: OH | |
| **Chemical composition**: Al - 95.3-99.97%; oxide content - 1-3%, lubricant content - 0.2-4% | | |
| **Trace elements**: Si - 0.05-.025%, Fe - 0.1-0.4%, other - 0.03-0.05% | | |
| PHYSICAL PROPERTIES | | |
| **Density**, g/cm$^3$: 2.7 | **Mohs hardness**: 2-2.9 | **Melting point**, °C: 660 |
| **Specific heat**, kJ/kg · K: 0.90 | | |
| **Thermal conductivity**, W/K · m: 204 | **Thermal expansion coefficient**, 1/K: 25x10$^{-6}$ | |
| CHEMICAL PROPERTIES | | |
| **Chemical resistance**: excellent corrosion resistance, reacts with alkaline and acidic solutions yielding hydrogen gas | | |
| OPTICAL & ELECTRICAL PROPERTIES | | |
| **Color**: silvery white to chromelike (leafing) metalescent (nonleafing) | | |
| **Resistivity**, Ω-cm: 2.8 x 10$^{-6}$ | | |
| MORPHOLOGY | | |
| **Particle shape**: flat, spherical | **Crystal structure**: cubic | **Particle size**, μm: 10-23 (powder) |
| **Aspect ratio**: 20-100 | **Particle thickness**, μm: 0.1-2 | **Particle length**, μm: 0.5-200 |
| **Sieve analysis**: 0.1-20% retained on 325 (44 μm) sieve | | **Specific surface area**, m$^2$/g: 5-35 |

MANUFACTURER & BRAND NAMES:

Silberline Manufacturing Co., Inc., Tamaqua, PA, USA

manufactures several hundred grades of aluminum powders and flakes. The products are grouped by the particle character (powder, leafing, nonleafing), resistance to acids (non-resistant, resistant), application (general, waterborne, plastics, printing inks, specialty, other (inhibited aluminum pigments, water dispersible aluminum pigments, degradation resistant, sparkle and high series, lenticular series, glitter series, black iron flake, spherical pigments, extra sparkle spheres, metalescent pigments, dedusted flake, colored pigments, resin treated grades)). The following are trade names: Aqua Paste, Aquasil, Aquavex, EternaBrite, Extra Fine, Hydro Paste, Lansford, SilBerCotes, SilBerTones, Silcroma, Sil-O-Wet, Silvar, Silvet, Silvex, Sparkle Silver, Stamford, Super Fine, Tufflake

Transmet, Columbus, OH, USA

Aluminum, copper, brass, and zinc particulate materials manufactured in various shapes of square flake (K-102), rectangular flake (K-101), flat fiber (K-107), flake (K-109), needle (N-101), and tadpole (T-101, T-102, T-103). The symbols in parentheses are the grades numbers for aluminum. If other metal is requested the grade number is derived from the metal number which is the first digit (1 - aluminum, 2 - copper, 3 - zinc, 4 - brass). For example, square flake from brass is K-402. The materials are manufactured by two technologies Melt spin and Spinning cup which are discussed below.

**MAJOR PRODUCT APPLICATIONS**: coatings, inks, roofing, plastics, automotive, powder coatings, containers for sterilizing and storing medical instruments, molding tools, heat sinks for electronic devices, time-delay switch, egg poachers

**MAJOR ADVANTAGES**: heat reflectivity, low emissivity, temperature resistance, moisture and oxygen barrier properties, sealing properties, reinforcement

The technology of production of aluminum powders and flakes dates back to 1930 when a safe process of manufacture was developed by Hall of Columbia University. This method is still used today for most manufactured pigments. The principle of manufacture is based on wet ball milling aluminum in the presence of a lubricant and mineral spirits.

The grinding process depends on the grade to be manufactured and usually takes 5-40 hours. The grade is determined by the particle size and grading is accomplished by filtering the slurry to remove large flakes. Typical leafing grades have 55-65% leafing flakes. The ultraleafing grades have almost 100% leafing flakes. An important difference exists between leafing and nonleafing flakes. Leaving flakes are obtained by the addition of a fatty acid (e.g., stearic acid) lubricant during the milling process. The lubricant coats the surface of flakes which become hydrophobic. There is a large difference in behavior between leafing and nonleafing flakes in coatings. Nonleafing flakes are uniformly distributed through the thickness of coating. They are preferentially oriented parallel to surface but this orientation is not perfect. Leafing flakes are mostly situated close to the paint surface and far from the substrate. Their orientation is much closer to parallel than the orientation of nonleafing flakes. Nonleafing pigments are frequently used with other pigments to obtain colored metallic finish. Leafing flakes give paints a metallic luster and reflectivity. In plastics, a true leafing effect has not yet been accomplished.

Processing of materials containing aluminum flakes must take into account their fragile nature. If flakes are exposed to extensive shearing forces they will degrade. Slow mixing and gradual dilution of flakes normally produces good results.

The commercial products are in most cases in the form of a paste. Standard pastes contain 27-35% mineral spirits. For waterborne applications carrier contains mixture of mineral spirits, nitroethane, and polypropylene glycol. Ink grades contain isopropyl alcohol or ink oil. Plastic grades are dispersed in plasticizer (DOP, DIDP), mineral oil or resin.

Transmet Corporation manufactures flakes by a Rapid Solidification Technology. There are two variations of this method: Melt spin and Spinning cup methods. In the Melt spin method, molten metal of any composition (pure metal or alloy) is driven through an orifice and the shape formed in the orifice (continuous sheet) is rapidly cooled on a chilling block. This metal sheet is cut into segments in the form of flakes (square and rectangular), flat fibers, and ribbons of desired dimensions. Typically, the sheet has thickness of 25 μm and the cut sides (length or width) have a length in the range of 0.5 to 2 mm. In the Spinning cup method, molten metal is driven through an orifice onto a rotating element (spinning cup) which works in manner similar to spray drying equipment. The particles are dispersed in space by tangential forces. In this process, spheres, needles and tadpoles are manufactured. The method can produce a broad range of compositions and shapes. It was determined, based on the rates of chemical reactions, that the shape of particles has a pronounced effect on the reaction rate. The shape of particles and their composition has an effect on their performance in conductive plastics and as reflecting media in coatings. The metal particles produced by this method have found applications in various products which are required to conduct heat and electricity, to shield EMI, and to reflect radiation in roofing materials, in addition to the traditional use of such materials in chemical and metallurgical processes. Figure 19.17 shows the cost of EMI shielding using aluminum flakes in comparison with other materials based on Transmet estimation.

## 2.1.2 ALUMINUM BORATE WHISKERS[7-8]

| | | |
|---|---|---|
| **Name**: aluminum borate whisker | | |
| **Chemical formula**: $(Al_2O_3)_9(B_2O_3)_2$ | | |
| PHYSICAL PROPERTIES | | |
| **Density**, g/cm$^3$: 2.93 | **Thermal expansion coefficient**: 7.4x10$^{-6}$ | |
| **Tensile strength**, GPa: 7.8 | **Tensile modulus**, GPa: 400 | **Compressive strength**, GPa: 3.9 |
| MORPHOLOGY | | |
| **Particle shape**: ribbon or cylinder | **Crystal structure**: single crystal | **Specific surface area**, m$^2$/g: 2.5 |
| **Particle length,** μm: 10-30 | **Particle diameter,** μm: 0.5-1 | **Aspect ratio**: 20-30 |
| MANUFACTURER & BRAND NAME: Shikoku Chemical Corp. - Alborex G | | |
| MAJOR PRODUCT APPLICATIONS: experimental phase as reinforcing filler | | |

## 2.1.3 ALUMINUM OXIDE[9-12]

| | |
|---|---|
| **Names**: anhydrous aluminum oxide, α-, or γ-, or θ-alumina | **CAS #**: 1344-28-1 |
| **Chemical formula**: $Al_2O_3$ | **Functionality**: PBD-coated[10] |
| **Chemical composition**: $Al_2O_3$ - 99.6% | |
| **Trace elements**: $SiO_2$ - 0.02-0.1%, $Fe_2O_3$ - 0.03-0.2%, $TiO_2$ - 0.1%, $Na_2O$ - 0.04-5%, HCl - < 0.5% | |

**PHYSICAL PROPERTIES**

| | | |
|---|---|---|
| **Density**, g/cm$^3$: 3.4-3.9 | **Mohs hardness**: 9 | **Melting point**, °C: 2015-2072 |
| **Thermal conductivity**, W/K · m: 20.5-29.3 | | **Maximum temperature of use**, °C: 1600 |
| **Compressive strength**, MPa: 2000 | | **Surface properties**: hydrophilic |

**CHEMICAL PROPERTIES**

| | | |
|---|---|---|
| **Moisture content**, %: 4-5 | **Adsorbed moisture**, %: 17-27% | **pH of water suspension**: 8-10 |

**OPTICAL & ELECTRICAL PROPERTIES**

| | | |
|---|---|---|
| **Refractive index**: 1.7 | | **Whiteness**: 80-90 |
| **Color**: white through off white to brown | | **Volume resistivity**, Ω-cm: >$10^{14}$ |
| **Dielectric constant**: 9-9.5 | **Dielectric strength**, V/cm: 2560 | **Loss tangent**: 0.0002-0.004 |

**MORPHOLOGY**

| | | |
|---|---|---|
| **Particle shape**: spherical or irregular | | **Pore diameter,** Å: 58-240 |
| **Particle size**, nm: 13-105 | **Crystal structure**: rhombic | **Oil absorption**, g/100 g: 25-225 |
| **Sieve analysis**: 0.05-5% on 45 μm sieve | | **Spec. surface area**, m$^2$/g: 0.3-325 |

**MANUFACTURERS & BRAND NAMES:**

Alcan Chemicals, Gerrards Cross, UK
- Milled grades: RMA, MA, MAFR
- Calcinated alumina: C-70 series, RA (ceramics), Cera (polishing, electrical components), CA, CG, CK (glass, ceramic fibers, etc), Baco (polishing), MA-LS (refractories, ceramics), LS (electrical and engineering components)
- Activated alumina: AA (catalysts, desiccant, fluorine removal from water), Acidsorb (adsorption of HCl from chemical processes), Actibond (refractory binder)

Biotage, Inc.
- Unisphere

Degussa AG, Frankfurt/Main, Germany
- $Al_2O_3$ C

Electro Abrasives Corporation, Buffalo, NY, USA
- Electro-Ox brown aluminum oxide and precision aluminum oxide abrasive

Morgan Matroc, Stourport-on-Seven, UK
- Aluminum oxide

Nanophase Technologies Corporation, Burr Ridge, IL, USA
- NanoTec Aluminum Oxide

The PQ Corporation, Valley Forge, PA, USA
- Nyacol Colloidal Alumina, Nyacol AL20SD

**MAJOR PRODUCT APPLICATIONS**: composites, ceramics, refractories, abrasives, copy toner, electro-optic devices, polishing, electrical and engineering components, acid adsorption, catalyst, nanocomposites

Refractory grades have large particle sizes in the range of 5-25 μm and very low surface area at 0.3-1 $m^2/g$. Their specific gravity is high at 3.95 $g/cm^3$. Calcinated alumina is produced by the Bayer calcination process from aluminum trihydroxide in rotary kilns. During the process, water is removed and stable α-alumina structure is obtained. The particle size of calcinated grades is similar to refractory grades unless they are milled. Smaller particle size grades have a specific surface area of 3-10 $m^2/g$. Activated aluminas have particle sizes in the range of 6-80 μm but very large specific surface areas in the range of 220-325 $m^2/g$. They can readily absorb water to equilibrium at 18-22%.

The grades produced by Nanophase Technologies Corporation are obtained in a synthetic way by evaporation of the metal and its subsequent oxidation. This process produces regular spherical particles as shown in Figure 2.1.[13-14] These materials have properties which cannot be duplicated by conventional grades of alumina obtained from minerals or by chemical synthesis. The nanoparticles are known to enhance mechanical performance of plastic materials (tensile, hardness, wear, etc.). The hardness of compressed ceramics increases as the particle size decreases and it is possible to obtain materials which allow considerable light transmission. These materials are on the market now and they will find many high technology applications.

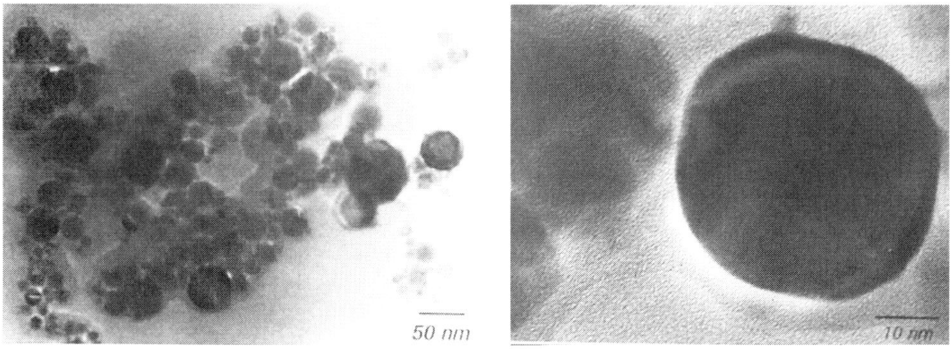


Figure 2.1. TEM of NanoTek aluminum oxide. *Courtesy of Nanophase Technologies Corporation, Burr Ridge, IL, USA.*

## 2.1.4 ALUMINUM TRIHYDROXIDE[15-39]

| | | |
|---|---|---|
| **Names**: aluminum trihydroxide, aluminum hydroxide, hydrated alumina | | **CAS #**: 21645-51-2 |
| **Chemical formula**: $Al(OH)_3$ or $Al_2O_3 \cdot 3H_2O$ | | **Functionality**: OH, methacryl, vinyl, stearic acid, viscosity reducer (Alcan grades S) |
| **Chemical composition**: $Al(OH)_3$ - 94-97%, $Fe_2O_3$ - 0.01%, $SiO_2$ - 0.01-0.03%, $Na_2O$ - 0.2-0.5% | | |
| **Trace elements**: Pb < 0.0005%, As < 0.0002% | | |
| PHYSICAL PROPERTIES | | |
| **Density**, g/cm$^3$: 2.4 | **Mohs hardness**: 2.5-3.5 | **Melting point**, °C: 290 (decomp) |
| **Loss on ignition**, %: 34.5 | | |
| CHEMICAL PROPERTIES | | |
| **Chemical resistance**: amphoteric material | | |
| **Moisture content**, %: 0.1-0.7 | | |
| **pH of water suspension**: 8-10.5 | **Loss on ignition**, %: 34.6% | **Specific conductivity**, μS/cm: 70 |
| OPTICAL & ELECTRICAL PROPERTIES | | |
| **Refractive index**: 1.57-1.59 | **Reflectance**: 89-95 | **Whiteness**: 93 |
| **Color**: bright white (Hunter L = 90-98) | | **Brightness**: 91-98 |
| **Electrical conductivity**, μS/cm: 5 | | **Dielectric constant**: 7 |
| MORPHOLOGY | | |
| **Particle shape**: irregular | **Crystal structure**: gibbsite | |
| **Particle size**, μm: 0.7-55 | **Oil absorption**, g/100 g: 12-41 | **Hegman grind**: 5.5-6 |
| **Sieve analysis**: 325 mesh residue - 0.001-0.15% | | **Spec. surface area**, m$^2$/g: 0.1-12 |

MANUFACTURERS & BRAND NAMES:

Alcan Chemicals, Gerrards Cross, UK
    Alcan AF (toothpaste grade), DH 101 (feedstock grade), FRF (general purpose), FRF LV (particle size optimized to give higher loading), ULV (optimized morphology for high loading and reduced viscosity), CV (modified particle shape improvement of cure time and lower viscosity), Precipitated (rounder particles offer denser particle packing and lower viscosity), Superfine (small particle size 0.5-1.2 μm E grades have much lower ionic impurity for electrical insulation), and Ultrafine (low $Na_2O$ content for application in cables), Flamtard S (zinc stannate), H (zinc hydroxystannate), HB1 (zinc hydroxystannate/zinc borate blend), Z10 & Z15 (zinc borate). Flamtard additives enhance performance of ATH. Cera Hydrate (abrasive)
Amspec Chemical Corporation, Gloucester City, NJ, USA
    Hydromax 100, 109
Charles B. Chrystal Co., Inc., New York, NY, USA
    Aluminum trihydroxide
Franklin Industrial Minerals, Nashville, TN, USA
    DH 35, 55, 80, 100, 200, 280, 500 (number = median particle size x 10)
Hitox Corporation, Corpus Christi, TX, USA
    Haltex 302, 310, 304

*continues on the next page*

**MANUFACTURERS & BRAND NAMES:**
Huber, J.M., Macon, GA, USA
PATH 6, 9, 9HB (optimized as partial replacement of $TiO_2$ in coating applications)
Martinswerk, Bergheim, Germany
Martinal ON-921, OL 104, OL111
Nabaltec GmbH, Schwandorf, Germany
Apyral 1, 2, 3, 4, 8, 15, 16, 24, 22, 40, 60, 90, 120 (number = specific surface x 10)

**MAJOR PRODUCT APPLICATIONS**: carpet backing, coatings, PU-foam, pultrusion, laminates, composites, conveyor belts, cables, flooring, chipboard, tub and shower stalls, coated fabrics, electrical products, polishing, exterior cladding, tiles, synthetic marble, adhesives, coatings and sealants, sheet molding compounds, toothpaste

**MAJOR POLYMER APPLICATIONS**: polyester, epoxy, acrylic, PVC, PP, PE, EVA, polyurethanes, phenolics

The production process for aluminum trihydroxide might be considered a spin off of aluminum metal production where in the first phase, the metallurgical grade of aluminum trihydroxide is produced.[38] At the same time, this grade contains numerous impurities and requires purification. Filler grade production is a separate from the production of the metallurgical grade and yields a pure aluminum trihydroxide. Two properties made aluminum trihydroxide very popular: its flame retarding abilities and its low absorption of UV.

The low absorption of UV makes it a suitable material for applications in UV curable materials. Its flame retarding activity is due to cooling, barrier layer formation, and dilution. The cooling capability of aluminum trihydroxide comes from its ability to release water at elevated temperatures with peak release at around 300°C. The reaction by itself is endothermic and, in addition, water must be evaporated which consumes additional heat energy. Aluminum trihydroxide, after it has been decomposed, forms a barrier which slows the flow of oxygen and formation of gases. Large quantities (e.g., 150 phr) of filler must be used to obtain flame retarding properties (dilution factor). This provides flame retardancy but affects the mechanical and rheological properties of materials. Since the amounts of filler cannot be significantly reduced, additives such as compounds of zinc are used which allow for some reduction in $Al(OH)_3$ concentration. Mechanical properties are improved by the morphology and surface coating of the filler. Grades are available which can be used with many plastics without a fear of degrading their mechanical performance. The problem of rheology of materials during processing and use is addressed by the modification of the morphology of particles and with additives which help to reduce viscosity.

Figures 2.2 and 2.3 show how morphology might be tailored to improve viscosity. Figure 2.2 shows a precipitated grade which is composed of blocky round particles. The careful selection of an appropriate particle size distribution of these morphologically different species resulted in a low viscosity material. Figure 2.3 shows another grade which has platy particles which give a higher viscosity (as might be expected).

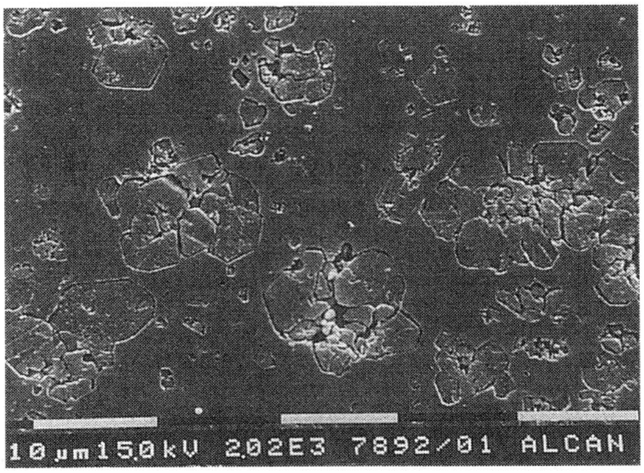


Figure 2.2. SEM of aluminum trihydroxide decreasing viscosity. *Courtesy of Alcan Chemical Europe, Gerrards Cross, UK.*

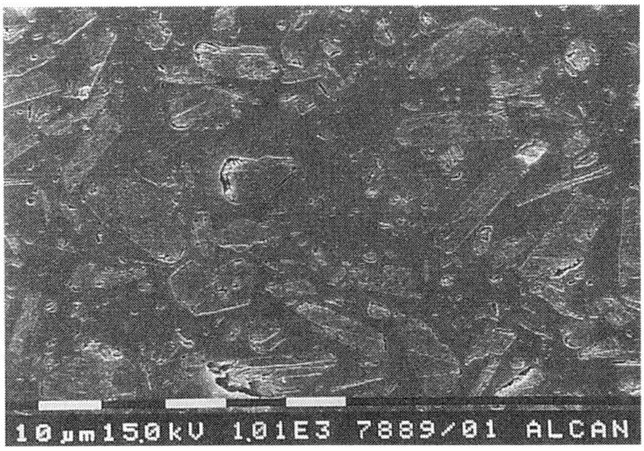


Figure 2.3. SEM of aluminum trihydroxide increasing viscosity. *Courtesy of Alcan Chemical Europe, Gerrards Cross, UK.*

## 2.1.5 ANTHRACITE[43]

| | | |
|---|---|---|
| **Names**: anthracite, semi-anthracite coal, bituminous coal | | **CAS #**: 8029-10-5 |
| **Chemical formula**: C | **Functionality**: OH | |
| **Chemical composition**: carbon - 77%, ash - 6-16% | | |
| **Trace elements**: sulfur - 0.23-1.2%, silica oxide - 2.2-5.4%, alumina - 2%, ferric oxide - 0.4% | | |
| **PHYSICAL PROPERTIES** | | |
| **Density**, g/cm$^3$: 1.31-1.47 | **Mohs hardness**: 2.2 | |
| **CHEMICAL PROPERTIES** | | |
| **Moisture content**, %: 0.5-4 | **pH of water suspension**: 7-7.5 | **Volatiles content**, %: 0.5-20 |
| **ELECTRICAL PROPERTIES** | | |
| **Resistivity**, MΩ-cm: 50 | | |
| **MORPHOLOGY** | | |
| **Sieve analysis**: residue on 325 mesh - traces | | **Particle shape**: irregular |
| **MANUFACTURERS & BRAND NAMES**:<br>Anthracite Industries, Inc., Sunbury, PA, USA<br>4072-C, 505, 7002, 7004, Anthrin Filler, Carbon Filler Oxide<br>Coal Fillers, Inc., Bluefield, VA, USA<br>Austin Black - low specific gravity reinforcing and mineral filler<br>Keystone Filler & Manufacturing Company, Muncy, PA, USA<br>Mineral Black 121 OC, 123, 126, 325BA | | |
| **MAJOR PRODUCT APPLICATIONS**: liner, battery cases | | |
| **MAJOR POLYMER APPLICATIONS**: rubber, EPDM, PP, PE | | |

Anthracite abounds as a mineral and can be cost-effectively mined and ground. It was found[43] that materials containing it have improved strength, stiffness, environmental stress cracking, heat deflection temperature, antistatic properties, weathering resistance, and chemical resistance even if filled with substantial quantities of anthracite (up to 60 wt%). The disadvantages are color, flowability of melt, and increased moisture absorption. One major advantage creates growing interest. Most fillers used today are non-combustible and remain as ash when plastic materials are incinerated at the end of several recycling operations. Anthracite has, by comparison, a very low ash content and provides calorific value.

## 2.1.6 ANTIMONATE OF SODIUM

| | |
|---|---|
| **Name**: sodium antimonate | |
| **Chemical formula**: $NaSbO_3$ | **Functionality**: ONa |
| **Chemical composition**: $Sb_2O_3$ - 70-73%, $Sb_2O_5$ - 80%, $NaSbO_3$ - 95% | |
| **Trace elements**: As - 0.3-0.5%, Pb - 0.6-1%, Fe - 0.004-0.0055%, Cu - 0.004% | |
| **PHYSICAL PROPERTIES** | |
| **Density**, g/cm$^3$: 4.8 | |
| **CHEMICAL PROPERTIES** | |
| **Chemical resistance**: it is soluble in, and reactive with, acids | |
| **Moisture content**, %: 0.5-3 | **Acid soluble matter**, %: 100 |
| **OPTICAL PROPERTIES** | |
| **Refractive index**: 1.75 | **Color**: white to light tan |
| **MORPHOLOGY** | |
| **Sieve analysis**: 325 mesh residue - 12-45% | |
| **MANUFACTURERS & BRAND NAMES**:<br>Laurel Industries, Cleveland, OH, USA<br>Thermogard FR<br>United States Antimony Corporation, Thompson Falls, MT, USA<br>Montana Brand Sodium Antimonate Grade 1 | |
| **MAJOR PRODUCT APPLICATIONS**: chemical intermediate in production of antimony pentoxide; flame retardant in plastics, paints, textiles | |
| **MAJOR POLYMER APPLICATIONS**: PBT, PET, PC, UHDPE, rubber | |

Sodium antimonate must be used with halogen containing compounds for it to act as effective fire retardant. The source of chlorine may come from polymer (e.g., PVC, chlorinated rubber, etc.) or other chlorinated or brominated material. The benefits of using sodium antimonate over antimony oxide include its low tinting strength and the acid scavenging capability. For these reasons, it is used in semi-opaque or dark colored materials and in polymers such as polyesters and polycarbonates which are acid sensitive.

## 2.1.7 ANTIMONY PENTOXIDE

| | | |
|---|---|---|
| **Name**: antimony pentoxide | | **CAS #**: 1314-60-9 |
| **Chemical formula**: $Sb_2O_5$ or $HSb(OH)_6$ in hydrated form | **Functionality**: OH | |
| **Chemical composition**: $Sb_2O_5$ - 92-95% | | |
| PHYSICAL PROPERTIES | | |
| **Density**, g/cm$^3$: 3.8 | | **Melting point**, °C: 380 |
| CHEMICAL PROPERTIES | | |
| **Chemical resistance**: soluble in hot acid | | |
| **Moisture content**, %: 0.2-1% | **pH of water suspension**: 2.5-9 | |
| OPTICAL PROPERTIES | | |
| **Refractive index**: 1.7 | **Tinting strength**: low | **Color**: white to yellow |
| MORPHOLOGY | | |
| **Particle size**, μm: 10-40, 0.025-0.075 (colloidal) | | |
| MANUFACTURER & BRAND NAMES:<br>The PQ Corporation, Valley Forge, PA, USA<br>Nyacol Aqueous Dispersions: A1530, A1540N, A1550 (last two digits give oxide concentration)<br>Nyacol Organic Dispersions: AB40, AP50, APE1540 (last two digits give oxide concentration)<br>BurnEx Powders: Plus A1588LP, Plus A1590, ZTA<br>BurnEx Nano-Dispersible Powders: A1582, ADP480, ADP494 (for dispersions in water, non-polar solvents, and polar solvents, respectively)<br>BurnEx 2000: 10, 20 (dispersed in PP of nano-dispersible grade and organic bromine compound) | | |
| MAJOR PRODUCT APPLICATIONS: textiles, coatings, nonwovens, adhesives, fibers (carpet, draperies, clothing), polyester laminates, wallcoverings, wire insulation, office furniture, automotive interiors, electrical housings, computers, printers, appliances, telecommunication, film, sheet | | |
| MAJOR POLYMER APPLICATIONS: epoxy, polyester, PVC, ABS, HIPS, PP | | |

Antimony pentoxide is an alternative to antimony trioxide. It finds applications in semi-transparent materials and dark colors because of its low tinting strength. As with antimony trioxide, antimony pentoxide must be used together with halogen-containing compounds to function as a flame retardant (see discussion under antimony trioxide). The other advantages of antimony pentoxide include its refractive index which is closer to most materials, its very small particle size, its high specific surface area, and its substantially lower density. Because of its small particle size, its is frequently used in the textile industry since its addition has only a small effect on color or on mechanical properties. Production of fine-denier fibers requires a stable dispersion and a small particle size filler. The flame retardancy of laminates is also improved with antimony pentoxide because small particles are easier to incorporate in the interfiber spaces.

Antimony pentoxide, as an additive for plastic materials such as polyolefins and ABS, is produced in predispersed form containing halogen compounds and a polymeric binder which has a low melting index to aid incorporation.

Incorporation of aqueous dispersions of antimony pentoxide into latex requires a pH adjustment prior to adding it to latex to prevent latex coagulation. Dispersions of antimony pentoxide usually have a pH = 5 which is too low for use in most latex formulations. Adjustment of pH can be made with ammonia but prior to such a pH adjustment it is necessary to dilute the dispersion to a concentration below 40% $Sb_2O_5$.

The use of particulate $Sb_2O_5$ in plastics extrusion requires that some precautions be taken. The extruder temperature setting must be below the level which degrades halogen-containing additive (180-250°C), The vented extruder should be used to remove free moisture. The antimony pentoxide must be kept sealed when not in use to prevent moisture pickup and dust generation should be prevented during handling. If antimony pentoxide is used in materials which do not contain halogen, the formulation should include sufficient halogen-containing additive to provide halogen/antimony mole ratio of 3/1.

## 2.1.8 ANTIMONY TRIOXIDE[39-42]

| | |
|---|---|
| **Name**: antimony trioxide | **CAS #**: 1309-64-4 |
| **Chemical formula**: $Sb_2O_3$ | **Functionality**: none |
| **Chemical composition**: $Sb_2O_3$ - 98-99.5% | |
| **Trace elements**: As - 0.02-0.2%, Pb - 0.04-0.3%, Fe - 0.004-0.01%, Se - 0.005%, $SO_4$ - 0.002-0.05% | |
| **PHYSICAL PROPERTIES** | |
| **Density**, g/cm$^3$: 5.2-5.67 | **Melting point**, $^o$C: 656 |
| **CHEMICAL PROPERTIES** | |
| **Chemical resistance**: reactive with acids and bases | |
| **Moisture content**, %: 0.1 | **Water solubility**, %: 0.001 |
| **pH of water suspension**: 2.0-6.5 | **Acid soluble matter**, %: 100 |
| **OPTICAL PROPERTIES** | |
| **Refractive index**: 2.087 | |
| **Color**: white | **Tinting strength**: high to low |
| **MORPHOLOGY** | |
| **Crystal structure**: cubic or orthorhombic | **Specific surface area**, m$^2$/g: 2-13 |
| **Sieve analysis**: 325 mesh residue - 0.1-0.5% | **Particle size**, μm: 0.2-3 |

**MANUFACTURERS & BRAND NAMES**:

AMSPEC Chemical Corporation, Gloucester City, NJ, USA
: KR (excellent whiteness and tinting strength), KR - Superfine (small particle size for fiber and film), LTS (low tint for darker colors), AMSTAR (utility grade for cost effective applications)

Laurel Industries, Cleveland, OH, USA
: FireShield H (high tint strength), L (low tint strength), HMP (high purity, low trace metals), UltraFine (low particle size, 0.2-0.4 μm gives reduced loss of mechanical properties, and higher tinting strength than H)

United States Antimony Corporation, Thompson Falls, MT, USA
: VF (very fine), MP (micro pure), HT (high tint), LT (low tint), Industrial Grade

**MAJOR PRODUCT APPLICATIONS**: plastics, textiles, paper, paints, rubber, UV resistant pigments

**MAJOR POLYMER APPLICATIONS**: PA, PVC, PP, PE, ABS, HIPS, polyester, polyurethanes, rubber, epoxy

Antimony oxide is usually produced from stibnite (antimony sulfide) or by oxidizing antimony metal.

Many theories attempt to explain the mechanism of flame retardancy. The flame retarding action is thought to take place in the vapor phase above the burning surface. For antimony oxide to work, the halogen and antimony oxide must be found in a vapor phase which will occur at temperatures above 315$^o$C. At these temperatures, antimony halides and oxyhalides are formed and act as flame extinguishing moieties by quenching radicals as they form.

The tinting strength depends on particle size. If particle sizes are below 300 nm they fall below visible range. Above this value, tint strength decreases as the particle size increases. The high tint strength grade usually has particle sizes in a range of 1.1-1.8 μm and the low tint strength grade has particle sizes in a range of 1.8-3 μm. The tint strength can also be affected by crystalline form. The orthorhombic form decreases tint strength.

Different formulations are needed for individual polymers (according to the manufacturer AMSPEC). These concentrations are recommended: PVC: $Sb_2O_3$ - 2-10 phr; PP: $Sb_2O_3$ - 2-4 phr, brominated organic 4-22 phr; ABS: 4:1 organo-Br/$Sb_2O_3$; HIPS: $Sb_2O_3$ - 4 phr, aromatic bromine - 12 phr, polyurethanes: 5-15 phr $Sb_2O_3$ and 5-15 phr halogenated compounds.

The manufacturers offer a wetted grade of antimony oxide to reduce dust. This is made by the addition of 3-4% plasticizer (DIDP, DOP, DINP, or ethylene glycol). Concentrates are produced by manufacturers and specialized companies. United States Antimony Corporation manufacturers concentrates with up to 90% active component. Laurel Industries produce both antimony oxide and organic flame retardants which are sold separately and in ready to use combinations which also include resin carriers. Paraffin is a convenient binder for extrusion and molding applications. Arethon International Plastics Ltd. has a full range of flame retardant masterbatches which are marketed under the brandname Areflam. The active content in these masterbatches is from 50 to 80%. They are prepared with more than 10 carrier resins and have the correct content of halogen-containing material and $Sb_2O_3$ or, in the case of halogen-free masterbatch, appropriate amount of $Al(OH)_3$.

Antimony oxide can be advantageously combined with huntite/hydromagnesite fillers to offer excellent flame retarding properties.[39,42] Also, zinc borate can be used to reduce the amount of antimony trioxide. Other performance enhancing additives include zinc stannate and ammonium octamolybdate.[40]

## 2.1.9 APATITE[44-45]

| | | |
|---|---|---|
| **Names**: apatite, calcium (fluoro, chloro, hydroxyl) phosphate | | |
| **Chemical formula**: $Ca_5(PO_4)_3(OH,F,Cl)$ | **Functionality**: OH, CL, F | |
| PHYSICAL PROPERTIES | | |
| **Density**, g/cm$^3$: 3.1 - 3.2 | **Mohs hardness**: 5 | |
| OPTICAL PROPERTIES | | |
| **Color**: white to yellow | **Brightness**: 58-63 | |
| MORPHOLOGY | | |
| **Particle size**, μm: 43 | **Crystal structure**: hexagonal | **Cleavage**: basal direction |
| MAJOR PRODUCT APPLICATIONS: paper, medical (replacement bones) | | |
| MAJOR POLYMER APPLICATIONS: PMMA | | |

## 2.1.10 ASH, FLY[46-49]

| | | |
|---|---|---|
| **Names**: fly ash | | **CAS #**: 60676-86-0 |
| **Chemical formula**: variable composition | | **Functionality**: variable |
| **Chemical composition**: $SiO_2$ -30-60%, $Al_2O_3$ - 11-19%, $Fe_2O_3$ - 4-11%, MgO - 5-6%, CaO - 2-45% | | |
| **Trace elements**: sodium, boron, potassium, strontium, barium, molybdenum, lithium, vanadium, chromium | | |
| **PHYSICAL PROPERTIES** | | |
| **Density**, g/cm$^3$: 2.1-2.2 | | |
| **CHEMICAL PROPERTIES** | | |
| **Moisture content**, %: 2-20 | | |
| **MORPHOLOGY** | | |
| **Particle shape**: irregular | **Particle size**, μm: 4 | **Porosity**: high |
| **Sieve analysis**: residue on 325 mesh sieve - 5% | | |
| **MAJOR PRODUCT APPLICATIONS**: concrete modification, composite, building materials, polyester mortar | | |
| **MAJOR POLYMER APPLICATIONS**: PP, PE, PU, PET | | |

Fly ash may become more extensively used as a inexpensive filler. It is not used in large quantities at the present time. Research studies[46-49] show that materials can be improved when fly ash is used as a filler. The major hurdle is health and safety since fly ash contains crystalline silica and is, consequently, considered a hazardous material.

## 2.1.11 ATTAPULGITE

| | | |
|---|---|---|
| **Names**: attapulgite, hydrous magnesium aluminum silicate, Fuller's earth, palygorskite, clay | | **CAS #**: 12174-11-7 |
| **Chemical formula**: variable composition | | **Functionality**: OH |
| **Chemical composition**: $SiO_2$ - 50-68%, $Al_2O_3$ - 9-12%, MgO - 3-12%, $Fe_2O_3$ - 3-5% | | |
| **Trace elements**: potassium, sodium, magnesium | | |
| PHYSICAL PROPERTIES | | |
| **Density**, g/cm$^3$: 2.3-2.4 | **Mohs hardness**: 1-2 | **Loss on ignition**, %: 5-23 |
| CHEMICAL PROPERTIES | | |
| **Moisture content**, %: 2-16 | **Adsorbed moisture**, %: 1-6 | **pH of water suspension**: 6.5-9.5 |
| **Volatiles content**, %: 5-15 | | |
| OPTICAL PROPERTIES | | |
| **Color**: buff, tan, cream | | **Refractive index**: 1.57 |
| MORPHOLOGY | | |
| **Particle shape**: irregular, needle | **Crystal structure**: monoclinic | **Oil absorption**, g/100 g: 60-120 |
| **Particle size**, μm: 0.1-20 | **Specific surface area**, m$^2$/g: 120-400 | |
| **Sieve analysis**: residue on 325 mesh sieve - 0.01-8 | | |

MANUFACTURERS & BRAND NAMES:
Milwhite, Inc., Houston TX, USA
Attapulgite A, LMV, RVM, Basco Salt Mud, Econosorb, Fertogel, Gel B, Gel 420-P, Gel 540-P, Gel 601-P, High Yield Attapulgite, Milfines, Milsorb, Milsorb-CG, Supper Gel B
Non-Metals, Inc., Affiliate of The China Non-Metallic Minerals, Tucson, AZ, USA
Attapulgite clay for paint, adsorbent, drilling mud, and fertilizer

MAJOR PRODUCT APPLICATIONS: pesticides, herbicides, fertilizers, absorbents, drilling mud, joint compounds, neutralizers, asphalt thickeners, adhesives, paints, coatings, sealants, environmental remediation materials, antidiarrheal medication, gels

Attapulgite is naturally occurring crystalline hydrated magnesium aluminum silicate. It has a unique three-dimensional chain structure giving unusual colloidal and sorptive properties. Attapulgite is in the range of clay minerals classified as Fuller's earth. The natural mineral is ground, classified, and thermally activated. A high temperature drying produces LVM grade (LVM standing for low volatile matter) and having up to 1% of free moisture and up to 5% of total volatiles. Low temperature drying produces thickeners having up to 12% of free moisture and sorptive products of regular volatile matter, RVM, having 6% free moisture and up to 9% volatiles. Granular grades are manufactured by two basic methods: one includes drying or calcination, followed by grinding and screening to the size; in the other, a raw clay is pugged, extruded, dried or calcinated, followed by grinding and screening. Grades produced by the first method are designed as “A”, whereas extruded grades are "AA". Thus there are four different grades available: AA RVM, A RVM, AA LVM, and A LVM differing in water disintegrability. LVM grades resist disintegration in water whereas RVM grades do not.

There is a wide range of average particle sizes (0.1-20 μm) available. However, most commonly used products are in the range of 0.1-3 μm. Small particle size and high porosity result in a very high BET surface area (120-150 $m^2/g$) and an unusually high oil absorption in a range from 60 to 120%. Attapulgites are unusual in these respects. Also pH, which is in the range of 7.5-9.5, differs from that of kaolins.

Figure 2.4 shows the morphology of attapulgite which reveals the reasons for its high absorptivity.

Figure 2.4. SEM micrograph of Attagel 50. *Courtesy of Rheox, Inc., Hightstown, NJ, USA.*

## 2.1.12 BARIUM METABORATE

| | |
|---|---|
| **Name**: barium metaborate monohydrate | **CAS #**: 13701-59-2 |
| **Chemical formula**: $BaB_2O_4 \cdot H_2O$ | **Functionality**: OH |
| **PHYSICAL PROPERTIES** | |
| **Density**, g/cm$^3$: 3.3 | **Fusion point**, °C: 900-1050 |
| **CHEMICAL PROPERTIES** | |
| **pH of water suspension**: 9.8-10.3 | |
| **OPTICAL PROPERTIES** | |
| **Refractive index**: 1.55-1.60 | |
| **Color**: white | |
| **MORPHOLOGY** | |
| **Oil absorption**, g/100 g: 30 | |
| **MANUFACTURER & BRAND NAME**:<br>Buckman Laboratories, Memphis, TN, USA<br>Busan 11-M1 | |
| **MAJOR PRODUCT APPLICATIONS**: paints, coatings, sealants | |
| **MAJOR POLYMER APPLICATIONS**: alkyd resin, polyurethane, acrylic | |

Barium metaborate is a truly multifunctional additive which inhibits corrosion, increases UV stability, inhibits mold growth, and has flame retarding properties when used in combination with halogenated materials. The commercial product of Buckman Laboratories is a modified product which contains 90% of active ingredient.

## 2.1.13 BARIUM SULFATE[50-57]

| | | |
|---|---|---|
| **Names**: barium sulfate, barite, blanc fixe | | **CAS #**: 7727-43-7 |
| **Chemical formula**: $BaSO_4$ | **Functionality**: none if not surface grafted | |
| **Chemical composition**: $BaSO_4$ - 86-99%, $SrSO_4$ - 1-2%, CaO - 0-10.8%, $Fe_2O_3$ - 0.1-1.4%, $SiO_2$ - 0.9-2.1% | | |
| **Trace elements**: iron, copper, manganese, and lead | | |
| PHYSICAL PROPERTIES | | |
| **Density**, g/cm$^3$: 4.0-4.9 | **Mohs hardness**: 3-3.5 | **Melting point**, $^o$C: 1580 |
| **Linear coefficient of thermal expansion**, $10^{-6}$ 1/K: 10 | | **Loss on ignition**, %: 0.2-2.6 |
| CHEMICAL PROPERTIES | | |
| **Chemical resistance**: resistant to acids and alkalis | | |
| **Moisture content**, %: 0.1-0.3 | **Acid soluble matter**, %: traces | **Volatiles content**, %: 0.1-0.5 |
| **Soluble content**, %: 0.00025-0.4 | **Water solubility**, ppm: 3 | **pH of water suspension**: 6-9.5 |
| OPTICAL & ELECTRICAL PROPERTIES | | |
| **Refractive index**: 1.64 | | **Whiteness**: 94-96 |
| **Color**: white | | **Brightness**: 65-99 |
| **Tinting strength**: medium | | **Reflectance**: 90 |
| **Dielectric constant**: 11.4 | **Resistivity**, Ω: 19.075 | **Conductivity**, μS/cm: 200-300 |
| MORPHOLOGY | | |
| **Particle shape**: depends on grade | **Crystal structure**: orthorhombic | **Oil absorption**, g/100 g: 8-28 |
| **Particle size**, μm: 3-30 (barites and some synthetic grades), 0.7 (blanc fixe), <0.1 (special grades) | | |
| **Sieve analysis**: residues on 325 mesh sieve - 0.01-12%, 0.001% (blanc fixe) | | **Cleavage**: one direction |
| **Specific surface area**, m$^2$/g: 0.4-31 | | **Hegman fineness**: 2.5-7 |

MANUFACTURERS & BRAND NAMES:

Barium and Chemicals, Inc. Steubenville, OH, USA
- Barium Sulfate, 98% Technical Precipitated Grade

CIMBAR, Cartersville, GA, USA
- Bara 2002C, 325C, 200N, 325N, 200M, 325M (industrial grade ground barites)
- Bariace B-30, B-34 ( surface treated barium sulfate with $SiO_2$-$Al_2O_3$ to improve abrasiveness, dispersion, gloss, and hardness; particle size 0.3 μm)
- Barifine, BF-1, BF-10, BF-20, BF21 (ultrafine barium sulfates in particle range of 0.03-0.06 μm, improve dispersion of pigments and prevent flocculation)
- Barimite UF, XF, 22, 200, G-50 (flotation grade barites)
- CIMBAR 325, XF, CF, UF, EX (high purity white barites)
- Polywate (low $BaSO_4$ content materials, filled foam market)

*continued on the next page*

**MANUFACTURERS & BRAND NAMES:**
Hitox Corporation, Corpus Christi, TX, USA
- Bartex 10, 65, 80, OWX - barium sulfate for a broad range of applications, including $TiO_2$ replacement

J.M. Huber Corporation, Macon, GA, USA
- Huberbrite 1, 3, 7, 10, 12 (milled barite, the number refers to median particle size)

Milwhite, Inc., Houston, TX, USA
- Basco Wate (ground barite for drilling fluids)
- Blanca 2, 4, 8 (high quality ground barites; number refers to particle size)
- Marfil 2, 4, 8, 10, 20, 40 (natural ground barite for coatings and plastics, number refers to particle size)

Nippon Chemical Industry Co., Japan
- Barium sulfate AD

Polar Minerals, Mt. Vernon, IN, USA
- 1000 Series includes barites 1075, 1065, 1040 of different particle sizes for paints and coatings
- 2000 Series includes barites 2075, 2065, 2010 of different particle size for plastics, paints, and brake linings
- Blanc Fixe 1090P - precipitated barium sulfate

Sachtleben Chemie GmbH, Duisburg, Germany
- Albaryt and Albaryt Plus (wet processed and chemically bleached grades)
- Barytmehl F, N, G, 901 (natural ground white barites with different particle sizes, F - fine, N - medium, G - coarse)
- Blanc fixe N, F, micro (standard grades)
- Blanc fixe, HXH, HNF (finely precipitated barium sulfate of extremely high purity and brightness)
- Drilling mud grade BS
- EWO (wet processed and chemically bleached grade, slightly coarser than Albaryt)
- Fleur (wet processed and chemically bleached grade slightly coarser than Albaryt and EWO)
- Ground Barites C 101, CH 1177, C 7, C 14, TS (fine powders made by grinding with a lower brightness than Barytmehl but comparable particle sizes)
- K1, K2, K3, K4, M (high purity, synthetic grades having a high brightness (96-98) and high refractive index)
- Sachtoperse HP, HU-N, HU-D (smallest particle size grades from below 0.1 to 0.2 μm, used as nucleating agents and anti-flocculating additives)

ZEMEX Industrial Minerals, Atlanta, GA, USA
- Cherokee 289, 290, 291 (ground barites)

**MAJOR PRODUCT APPLICATIONS**: paints, inks, wood finishes, powder coatings, adhesive, mastics, seals, sealants, coatings, medical, paper, battery products, drilling fluids, brake linings, bowling balls, sound dampening, plastisols, urethane foams, acoustical compounds, insulating materials

**MAJOR POLYMER APPLICATIONS**: PET, PVC, melamine, polyurethanes, alkyd

Barites are the most common barium minerals, found in pure form but also together with many other minerals. The most frequent replacement for barium is that of strontium or radium. Barium sulfate, widely used in industry and in medical applications, originates from natural barites and synthetic materials. The quality of the filler depends on the purity of material used for production and the method of processing (a chemical purification is a complex process which determines the quality of synthetic or reprocessed material). The simplest method of processing includes grinding and dry classification. Finer products are obtained by concentration, wet grinding, bleaching, and classification. The product of highest quality is *blanc fixe* (permanent white). It is produced from the reaction between barium carbonate and sulfuric acid. Since the only other reaction products are water and carbon dioxide, product purity depends on the quality of raw materials used. The particle size distribution depends on process parameters, including the concentration of reactants, the rate of addition, temperature, and efficiency of mixing. These parameters are easily regulated, so particle size distribution. In some applications, the filler must have a narrow range of particle size distribution. The average particle size diameter for natural products is usually in a range from 2 to 30 μm (maximum particle size: 15-75 μm). The price is related to the average particle diameter. *Blanc fixe* being the smallest is most expensive (the average diameter of particles ranges from 0.1 to 4 μm). Oil number depends on particle size, and for *blanc fixe* it is in a range from 12 to 28 g/100 g, whereas for natural products, it is lower, in a range from 7 to 12 g/100 g. Particles are non-porous and of irregular shape in the case of natural product, whereas *blanc fixe* is almost spherical.

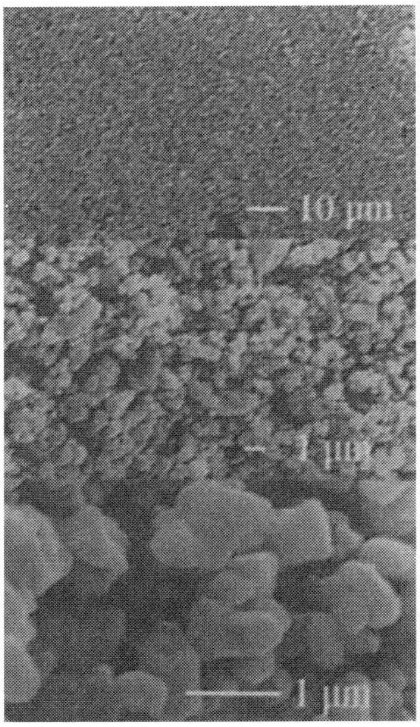


Figure 2.5. SEM micrograph of Blanc fixe micro at different magnifications (upper 1000x, middle 5000x, lower 25,000x). *Courtesy of Sachtleben Chemie, Duisburg, Germany.*

Further information on morphology is discussed below based on electron microscopy data. Figure 2.5. shows morphology of blanc fixe. The particle size of blanc fixe (0.7 μm) is comparable with the particle size of titanium dioxide (0.3 μm). Comparison of blanc fixe with another synthetic grade of barium sulfate, barium sulfate K2, produced by Sachtleben Chemie shows a difference in particle size but the morphological structure is quite similar (Figure 2.6). Figure 2.7 shows a still finer grade developed by Sachtleben Chemie which has particle size similar to titanium dioxide (0.35 μm). This is a quite extraordinary filler which has core made out synthetic barium sulfate (an insulator) coated with a semi-conducting layer of antimony doped with $SnO_2$ (Sacon P401). This material has high brightness, electric conductivity, and light transparency in thin coatings. The material is used to eliminate static charges from plastics and painted surfaces. At approximately 19% PVC material has a percolation threshold and surface resistivity drop rapidly by 8 orders of magnitude. Sachtoperse is still smaller in particle size, from 0.2 μm to below 0.1 μm, depending on grade. This is used as nucleating additive to polymers, such as PET. It decreases cycle time and reduces processing temperature, increases crystallization rate, and prevents flocculation of pigments. Figure 2.8 explains the mechanism by which Sachtoperse prevents pigment flocculation. Pigment particles (lighter particles) adhere to Sachtoperse (smaller darker particles) which act as a spacer. This process results in brighter colors and improved gloss.

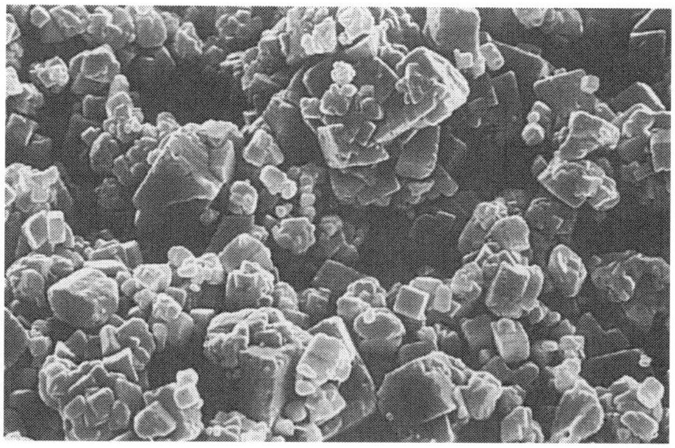

Figure 2.6. SEM micrograph of K2 grade at 2000x magnification. *Courtesy of Sachtleben Chemie, Duisburg, Germany.*

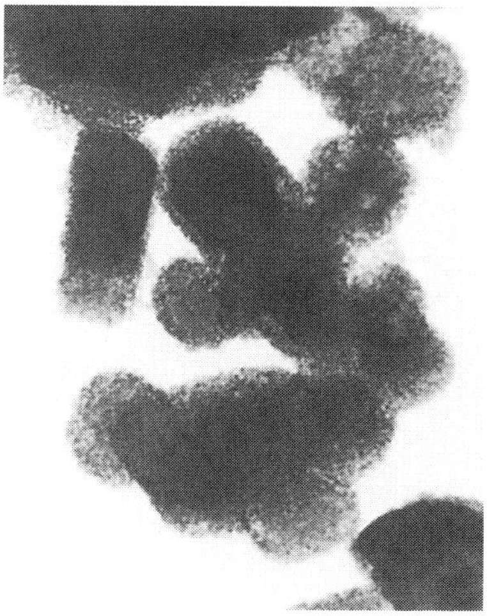

Figure 2.7. TEM micrograph of Sacon P 401 at magnification of 350,000. *Courtesy of Sachtleben Chemie, Duisburg, Germany.*

Figure 2.8. Anti-flocculating action of Sachtoperse HU. *Courtesy of Sachtleben Chemie, Duisburg, Germany.*

When images of synthetic grades are compared with image of ground barites (Figure 2.9), the morphological differences become apparent. These differences are not simply in particle size and distribution but also in the shape of particles.

Figure 2.9. SEM micrograph of milled barite, Huberbrite. *Courtesy of J. M. Huber Corporation, Macon, GA, USA.*

Chemical composition is another factor which determines quality, particularly in chemical and medical applications but also in paints and coatings where it affects brightness. Barium is highly toxic but only in the form of water soluble salt; therefore in every application the water soluble barium must be controlled. Other usual admixtures contain iron, copper, manganese, and lead, and depending on application, their concentration is also restricted. Natural products contain 94-99% $BaSO_4$, whereas *blanc fixe* contains from 97.5 to over 99%.

For some applications, a refractive index is important. A match between the particle size of some barium grades and the refractive index of matrix material allows the formulation of products with desirable optical properties. A series of synthetic barium sulfates is produced by Sachtleben Chemie which have particle sizes between 4 and 10 μm. If the particle size of these barium sulfates is well coordinated with the refractive index of the matrix polymer, semi-opacity combined with translucency results. This permits the formulation of a light disperser in lampshades or in illuminated advertising displays. The correct particle size can be calculated from the equation: $d = (100n - 141)/2$, where n is the refractive index of the resin and d the particle size of barium sulfate.

Barium sulfate has found many applications mainly because of its unique chemical resistance and inertness (for example, it is not affected by acid rain). The other reason for its frequent application is high absorptivity of light and, significantly, X-rays (for use in X-ray detectable materials).

## 2.1.14 BARIUM & STRONTIUM SULFATES

| | | |
|---|---|---|
| **Name**: barium strontium sulfate natural blend | | |
| **Chemical formula**: $BaSO_4$ & $SrSO_4$ | | **Functionality**: none |
| **Chemical composition**: $SrSO_4/BaSO_4$ - 87-95%, $CaCO_3$ - 2.6-5%, CaO - 1.9-2.5%, $Fe_2O_3$ - 0.1-1.7%, $CaSO_4$ - 0.7-3%, $SiO_2$ - 0.1-1% | | |
| PHYSICAL PROPERTIES | | |
| **Density**, g/cm$^3$: 3.8-3.9 | | |
| CHEMICAL PROPERTIES | | |
| **Chemical resistance**: similar to $BaSO_4$ | | |
| **Moisture content**, %: <0.3 | **pH of water suspension**: 7-7.5 | |
| OPTICAL PROPERTIES | | |
| **Color**: white | | **Reflectance**, %: 86-88 |
| MORPHOLOGY | | |
| **Particle size**, μm: 11-20 | **Crystal structure**: rhombic | **Oil absorption,** g/100 g: 9.5-11.5 |
| **Sieve analysis**: retained on 325 mesh sieve - 0.1-2% | | **Hegman fineness**: 3.5 |
| MANUFACTURER & BRAND NAME:<br>Milwhite, Inc., Houston, TX, USA<br>Microwate 10, 20, 40 (natural ground product) | | |
| MAJOR PRODUCT APPLICATIONS: plastics, paints, cellular foams | | |

## 2.1.15 BARIUM TITANATE[58]

| | |
|---|---|
| **Names**: barium titanate | |
| **Chemical formula**: $BaTiO_3$ | **Functionality**: none |
| **Chemical composition**: $BaTiO_3$ - 98.9-99.5% | |
| **Trace elements**: Sr, Ca, Nb, Fe, Si, Al, Mg, Na | |
| **PHYSICAL PROPERTIES** | |
| **Fusion point**, $^o$C: 1250 | **Loss on ignition**, %: 0.8 |
| **CHEMICAL PROPERTIES** | |
| **Moisture content**, %: 0.2 | |
| **OPTICAL & ELECTRICAL PROPERTIES** | |
| **Refractive index**: 2.4 | **Dielectric constant**: 3.8 |
| **MORPHOLOGY** | |
| **Particle size**, μm: 0.07-2.7 | **Specific surface area**, $m^2/g$: 2.4-8.5 |

**MANUFACTURERS & BRAND NAMES**:
Cabot Performance Materials, Boyertown, PA, USA
Hydrothermal Barium Titanate (barium titanate of small particle size obtained by a hydrothermal method)
TAM Ceramics, Niagara Falls, NY, USA
Ticon HPB, HPB-B, TME, F (high purity grades)
Ticon C, P, T (solid state grades)
Ticon 5016 (solid state, high purity grade)
Ticon COF-40, COF-50, COF-70, CN (solid state niobium-doped grades)

**MAJOR PRODUCT APPLICATIONS**: thermistors, capacitors, optics, ferroelectric ceramics, filler for ferroelectric polymers, pyro and piezoelectric composites

**MAJOR POLYMER APPLICATIONS**: poly(vinylidene fluoride)

## 2.1.16 BENTONITE[59-66]

| | |
|---|---|
| **Names**: bentonite, clay, montmorillonite, Na-montmorillonite, Ca-montmorillonite, hydrated sodium calcium aluminum magnesium silicate hydroxide | **CAS #**: 1302-78-9 |
| **Chemical formula**: $(Na, Ca)(Al, Mg)_6(Si_4O_{10})_3(OH)_6 \cdot nH_2O$ | **Functionality**: OH, ONa, OCa |
| **Chemical composition**: $SiO_2$ - 56-72%, $Al_2O_3$ - 13-21%, $Fe_2O_3$ - 0.9-5%, MgO - 1.7-2.4%, CaO - 0.7-2.2%, $Na_2O$ - 0.3-2.7%, $K_2O$ - 0.2-0.3% | |
| **Trace elements**: AS, Ba, Cd, Pb, Se, Hg | |

| | | |
|---|---|---|
| **PHYSICAL PROPERTIES** | | |
| **Density**, g/cm$^3$: 1.6 - 3 | **Mohs hardness**: 1-2 | **Loss on ignition**, %: 8.4-11.9 |
| **CHEMICAL PROPERTIES** | | |
| **Moisture content**, %: 2-14 | **pH of water suspension**: 7-10.6 | **Water solubility**, %: 3 |
| **OPTICAL PROPERTIES** | | |
| **Color**: light cream, buff to tan, light gray, white to off-white | | |
| **MORPHOLOGY** | | |
| **Particle size**, μm: 0.18-1 | **Oil absorption**, g/100 g: 36-52 | |
| **Sieve analysis**: residue on 325 mesh sieve - 2% | | |
| **Specific surface area**, m$^2$/g: 0.8-1.8 | | **Hegman fineness**: 2-7 |

**MANUFACTURERS & BRAND NAMES:**
Charles B. Co., Inc., New York, NY, USA
- Wyoming Granular Bentonite, Bentonite 200, 325 (sodium bentonite)
- Bentonite 34, (silicate of aluminum which swells eight times the volume)
- Cream Bentonite (light color bentonite)
- Bentonite Semi-dried Crude (sodium bentonite)

CIMBAR Performance Minerals, Cartersville, GA USA
- Organotrol 2200, 3300, 3440, 3550, 3660, SA (general purpose thickener and suspension additive)
- Suspengel 16, 30, 200, 325 (high purity bentonite thixotropes)
- Suspengel Ultra, Elite, Micro (high purity bentonite accepted for use in food)

Milwhite, Inc., Houston, TX, USA
- Basco Gel (blended bentonite for viscosity modification)
- Bentonite B (calcium montmorillonite for ceramics and molding)
- Milbond 3 (water treatment and sealant grade)
- Rev-Dust (calcium montmorillonite)

Non-Metals, Inc., Affiliate of The China Non-Metallic Minerals, Tucson, AZ, USA
- HB-Ca, JJ-Ca, JJ-Na, JL-Na, ZL-Na, LL-Ca - Ca and Na bentonites in powder form

**MAJOR PRODUCT APPLICATIONS**: paints, coatings, paper, adhesives, sealants, inks, cosmetics, plastics compounding, , pharmaceuticals, foods, drilling muds, waterproofing

**MAJOR POLYMER APPLICATIONS**: alkyd, polyurethane, butyl resin, PP, PS

Bentonite is a clay derived from the weathering of volcanic ash and composed of the mineral montmorillonite. There are two varieties: sodium bentonite which has high swelling capacity in water and calcium bentonite with negligible swelling capacity. Figures 2.10 and 2.11 show the morphology of ground ore and the

Figure 2.10. Bentonite ground ore. *Courtesy of Rheox, Inc., Highstown, NJ, USA.*

Figure 2.11. Bentonite purified and spray dried. *Courtesy of Rheox, Inc., Hightstown, NJ, USA.*

purified material. The high surface area and a structure which allows water to penetrate mineral layers are responsible for the swelling capabilities of bentonite clays.

In addition to the traditional use in paints as viscosity regulator, bentonite is currently used in the development of new materials with nanocomposite structures.

## 2.1.17 BERYLLIUM OXIDE

| | | |
|---|---|---|
| **Names**: beryllium oxide | | **CAS #**: 1304-56-9 |
| **Chemical formula**: $BeO$ | **Functionality**: none | |
| **Chemical composition**: beryllium oxide - 99.5% | | |
| **Trace elements**: Al, Ca, Mg, Si | | |
| PHYSICAL PROPERTIES | | |
| **Density**, g/cm$^3$: 2.85 | | **Melting point**, °C: 2570 |
| **Thermal conductivity**, W/m·K: 250 | | **Specific heat**, kJ/kg · K: 1.03 |
| **Thermal expansion coefficient**, $10^{-6}$ 1/K: 9 | **Maximum temperature of use**, °C: 1800 | |
| **Tensile modulus**, MPa: 138 | **Poisson ratio:** 0.26 | **Compress. strength**, MPa: 1550 |
| OPTICAL & ELECTRICAL PROPERTIES | | |
| **Color**: white | **Resistivity**, Ω-cm: $10^{17}$ | |
| **Dielectric constant**: 6.8 | **Dielectric strength**, V/cm: 100 | **Loss tangent**: 0.0004 |
| MORPHOLOGY | | |
| **Particle size**, μm: 20 | **Crystal structure**: hexagonal | |
| MANUFACTURERS & BRAND NAMES:<br>Accuratus Ceramic Corporation, Washington, NJ, USA<br>San Jose Delta Associates, Inc., Santa Clara, CA, USA | | |
| MAJOR PRODUCT APPLICATIONS: combination of extremely high thermal conductivity and excellent dielectric properties | | |

## 2.1.18 BORON NITRIDE

| | | |
|---|---|---|
| **Names**: boron nitride | | **CAS #**: 10043-11-5 |
| **Chemical formula**: BN | **Functionality**: none | |
| **Chemical composition**: BN - 95-99.5% | | |
| **Trace elements**: Cu, Al, Mg, Fe, K, Si | | |
| **PHYSICAL PROPERTIES** | | |
| **Density**, g/cm$^3$: 2.25 | **Knoop hardness**, kg/mm$^2$: 11 | **Specific heat**, kJ/kg·K: 794 |
| **Coefficient of expansion**, $10^{-6}$ 1/K: <1 | | |
| **Thermal conductivity**, W/K·m: 250-300 | **Maximum temperature of use**, °C: 985 | |
| **OPTICAL & ELECTRICAL PROPERTIES** | | |
| **Dielectric constant**: 3.9 | **Volume resistivity**, Ω-cm: $10^{15}$ | **Loss tangent**: <0.0002 |
| **MORPHOLOGY** | | |
| **Particle size**, μm: 3-200 | **Crystal structure**: hexagonal | **Spec. surface area**, m$^2$/g: 0.5-25 |
| **MANUFACTURERS & BRAND NAMES**:<br>Accuratus Ceramic Corporation, Washington, NJ, USA<br>Advanced Ceramics Corporation, Lakewood, OH, USA<br>PolarTherm 100 Series (five grades of hexagonal powders of different particle sizes)<br>PolarTherm 300 Series (low density agglomerates)<br>PolarTherm 600 Series (four grades of high density agglomerates)<br>Carborundum Corporation, Amherst, NY, USA<br>CarboTherm (seven grades of different particle sizes for refractory applications)<br>Combat (thirteen grades of different particle sizes for liquid coatings and aerosol sprays)<br>San Jose Delta Associates, Inc., Santa Clara, CA, USA - hot pressed boron nitride shapes | | |
| **MAJOR PRODUCT APPLICATIONS**: rubber pads, liquid encapsulants, underfills, printed circuit boards, adhesives, greases, liquid coatings, aerosol sprays | | |
| **MAJOR POLYMER APPLICATIONS**: silicone, epoxy | | |

Boron nitride filler address the "burning need" of modern electronic industry which is to protect electronic equipment from ever increasing generation of heat by high performance electronic devices. The combination of high electric resistivity with high thermal conductivity gives required performance to electronic adhesives and components.

Figure 2.12 shows SEM micrograph of boron nitride with 8-14 μm particle size.

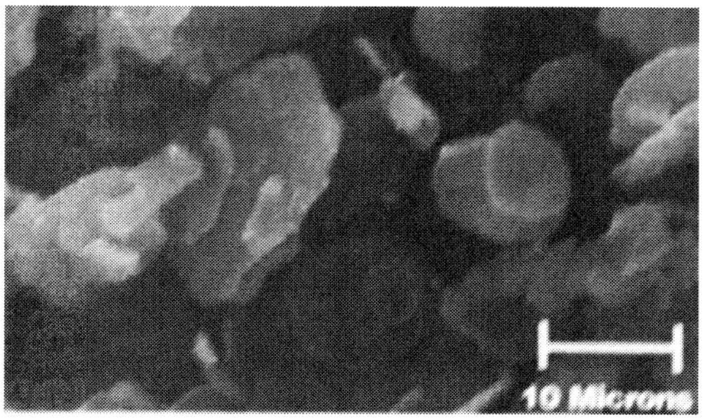


Figure 2.12. PolarTherm PT 120 boron nitride. *Courtesy of Advanced Ceramics Corporation, Lakewood, OH, USA.*

## 2.1.19 CALCIUM CARBONATE[67-138]

| | | |
|---|---|---|
| **Names**: calcium carbonate, limestone, chalk | | **CAS #**: 1317-65-3 |
| **Chemical formula**: $CaCO_3$ | **Functionality**: only from admixtures or surface treatment | |
| **Chemical composition**: $CaCO_3$ - 85-99%, SrO - 0.5%, $MgCO_3$ - 0.4-13%, BaO, MnO, $SiO_2$, $Fe_2O_3$, $Al_2O_3$ | | |
| **Trace elements**: As, Ba, Hg, Pb | | |
| PHYSICAL PROPERTIES | | |
| **Density**, g/cm$^3$: 2.7-2.9 | **Mohs hardness**: 3-4 | **Melting point**, °C: 1339 |
| **Decomposition temp.**, °C: 1150 | **Loss on ignition**, %: 43.5 | **Surface tension**, mJ/m$^2$: 207 |
| **Thermal conductivity**, W/K·m: 2.4-3 | **Linear coefficient of expansion**, 1/K: 4.3-10 x $10^{-6}$ | |
| **Young modulus**, MPa: 35,000 | **Poisson coefficient**: 0.27 | |
| CHEMICAL PROPERTIES | | |
| **Chemical resistance**: reacts with acids | | |
| **Moisture content**, %: 0.01-0.5 | **Water solubility**, %: 0.99 x $10^{-8}$ | **pH of water suspension**: 9-9.5 |
| OPTICAL & ELECTRICAL PROPERTIES | | |
| **Refractive index**: 1.48, 1.65, 1.7 | **Birefringence indices**: 1.48 & 1.65 (calcite) | **Whiteness**: 80-98 |
| **Color**: white to gray | **Reflectance**, %: 86-94 | **Brightness**: 82-94 |
| **Dielectric constant**: 6.1 | **Volume resistivity**, Ω-cm: $10^{10}$ | |
| MORPHOLOGY | | |
| **Particle shape**: irregular | **Crystal structure**: see text | **Hegman fineness**: 2-6.5 |
| **Particle size**, μm: 0.2-30, 0.02-0.4 (precipitated) | | **Oil absorption**, g/100 g: 13-21 |
| **Sieve analysis**: residue on 325 mesh sieve - 0.005-14% | | **Specific surface area**, m$^2$/g: 5-24 |

MANUFACTURERS & BRAND NAMES:

Charles B. Co., Inc., New York, USA

- Calofort U, U70 - small particle size, high specific surface area, precipitated calcium carbonate
- Granulated Oyster Shell - low heavy metals designed for pharmaceutical applications
- Food Grade Calcium Carbonate, FCC Grade - food grades
- Ultrafine Calcium Carbonate - general purpose ground limestone
- Hiflex - surface treated calcium carbonate for easy compounding in water pipes, cables, etc.
- Precipitated USP Grade - 3 grades for pharmaceutical, cosmetic, and food industries
- 402 - surface modified calcium carbonate for PVC plastisols and other plastics

ECC International, Cornwall, UK

- Carbital 110S, 110, 120 - high whiteness grades for PP derived from Italian marble (S stearate coating)
- Polcarb 60 & 90 - PVC extrusion, plastisol and PP sheeting
- Queensfil 25, 240, 300 - footwear, latex, PE masterbatch, PA moldings, all PVC applications
- Polcarb S, SB, 40S, 60S (stearate coated) - cable, extrusion, PE masterbatch and film, all PVC applications, PP molding and sheeting

*continued on the next page*

**MANUFACTURERS & BRAND NAMES:**
J.M. Huber Corporation, Macon, GA, USA
Hubercarb G series (2, 3, 8, 260, 325) milled high brightness grades for paints and coatings
Hubercarb M (6, 4, 3) and S (6, 4) series milled high brightness grades for paints and coatings
Hubercarb Q (325, 6, 4, 3, 2, 1) and W (3, 3N, 4) series milled grades for paints
Milwhite, Inc., Houston, TX, USA
Calfrost MG-NCS dry ground grade for paints, rubber, putties, caulks, adhesives
OMYA/Plüss-Staufer AG, Oftringen, Switzerland
130 companies worldwide producing the large number of grades for different industries under the following brand names:
*paper industry*: Hydrocarb (slurry), Snowcal (slurry), Omyacarb, Setacarb, Omyafil, Covercarb
*paint & coating*: Omyacarb, Durcal, Inducarb, Britomya, Snowcal, Calmote, Granitos, Violette Etikette, Micromya, Omya BSH, Omya BLP, Omyalite, Omya BL, Millicarb, Hydrocarb, Setacarb, Calibrite, Calcigloss, Calcimatt, Calcicoat (slurry), Wical WS
*plastics*: Omyacarb, Millicarb, Omyalite, Omya BRL, Omyalene, Omya EXH 1, Britomya, Snowcal, Omyafoam
*rubber and other industries*
The available grades in one location are given based on the production in Avenza - Carrara/Italy which manufactures grades of high purity for paints and plastics in one of the oldest and world famous location. The grades manufactured in other locations worldwide have similar quality.
The following grades are produced in Carrara:
Omyacarb 1-AV, 1T-AV, 2-AV, 2T-AV, 5-AV, 10-AV, 15-AV, 30-AV. The number signifies (and it is close to) the mean particle size; the letter T stands for the coated grade
Piqua Materials, Inc., Piqua, OH, USA
Piqua Minerals Filler 30, 60, 70, 200, 300, 600, 1800 - dry ground limestone of particle size increasing with grade number
Polar Minerals, Mt. Vernon, IN, USA
Fine Calcium Carbonates 8102, 8103, 8105, 8107 exceptionally pure calcium carbonates of different particle sizes. Also grades are manufactured with the same number symbol followed by letter C which stands for stearate coated grade
Ultrafine Calcium Carbonates 8.14, 8101 particle size 0.2-1.4 μm produced with (C) and without stearate coating
Polishing Marl - a filler designed to replace diatomaceous earth and calcinated kaolin in automotive and household polishing formulations which improves H&S due to the lack of crystalline silica
Solvay Alkali GmbH, Rheinberg, Germany
Rheinberg Plant - Socal P2, P3, N2R, U1R
Giraud, France - Socal 90A, 92E, BO, 31, 311, 312, 322
Angera, Italy - Socal 90AV, 91CV, 92EV, P2V, 312V, 322V
Ebensee, Austria - Socal P2E, N2, NP, $E_2$, $U_1$, $U_1S_1$, $U_1S_2$, $U_3$
the application of grades listed under precipitated grades; pharmaceutical/food grades: P2, U1R, E2, P2V
Suzorite Mica Products, Inc, Boucherville, Canada
Calcium carbonate 80/325 - dry ground limestone

**MAJOR PRODUCT APPLICATIONS:** *milled grades*: plastics, paper, paints and coatings, and numerous other applications difficult to list due to the widespread use
*precipitated grades*: emulsion paints, matt paints, paints containing solvent, printing inks, cigarette paper, fine paper, coated paper, special paper, rigid PVC, rubber, PP, PE, polyester, PVC plastisol, PSF, PU, silicone, polyacrylate, filling materials, pharmaceutical preparations, foodstuffs, beverages, toothpaste, wine deacidification, salt after-treatment, welding electrodes, peroxides

**MAJOR POLYMER APPLICATIONS:** PVC, PE, PP, PS, PA, PSF, PU, silicone, acrylic, rubber, polyester, and many more

Calcium carbonate is the most widely used filler. In the past its use was associated with a substantial cost reduction but today it is the material engineered for the different requirements of modern products. This discussion begins with an introduction to the origins of calcium carbonate which has been given a thorough evaluation in a paper by Bosshard of Omya/Plüss-Staufer AG.[138]

Calcium at 4.8% is the fifth most common elemental constituent of the earth's crust after oxygen, silicon, aluminum, and iron. It is so popular in practical applications because it is found in rocks and minerals which have very high concentration of calcium carbonate. Calcium carbonate is the most common deposit formed in sedimentary rocks. The process of formation of calcium deposits begins with weathering of land surface due to the changes in heat, frost, rain, and the effect of sun. Calcium carbonate is not readily soluble in water but calcium bicarbonate is. The concentration of carbon dioxide in water is thus important for calcium carbonate transportation from the land to the sea since rain water is the carrier. It is estimated that 500,000,000 tons of minerals are carried by rivers to the seas every year out of which about 10-15% of sedimentary rocks containing calcium carbonate are formed.

The soluble form of calcium can be precipitated in the marine environment to form rock by some physical conditions such as warming of the water (carbon dioxide is less soluble in warm water than in cold water and thus calcium carbonate is precipitated), by the use of carbon dioxide by marine plants, or by alterations in the pH of water by ammonia-producing bacteria which also lowers the solubility of calcium carbonate. However, the majority of calcium carbonate deposits are formed from skeletal fragments of organisms living in the marine environment. Some of these organisms inhabit reefs but the majority float free in water. Figure 2.13 shows various shapes of shells formed by Coccolithophorides which can be spherical coccospheres some, such as dicoaster, are star shaped.[138]

These shells not only have spectacular shapes, but they are small and in abundance. They measure 2-25 μm in diameter and there is up to 35,000,000 cells of coccoliths in a liter of sea water. When they die they sink to the sea bed. It is estimated that 68% calcareous mud covers the bottom of Atlantic. By comparison, only 36% of the Pacific is covered with calcareous mud – the difference is believed to be caused by the differences in solubility of carbon dioxide, and thus of calcium carbonate in the two oceans. When shells or a physically formed precipitate reaches the sea bed, a series of other processes occurs preceding formation of rock. The material loosely deposited on the sea bed contains 80-90% water which is gradually expelled by the overlaying sedimentary matter and the process of lithification takes place. The transformation to rock occurs when the residual porosity attains about 30% which requires a pressure of about 300-500 meters of sediment equivalent to about 80 atmospheres. During this slow process, cementation occurs which is based on redissolving of unstable carbonates such as aragonite or vaterite present in sediments and depositing them in pore spaces as calcite or dolomite. The rocks formed in such a manner are then lifted from the sea bottom in geological upheavals and exposed to weathering to continue the cycle.

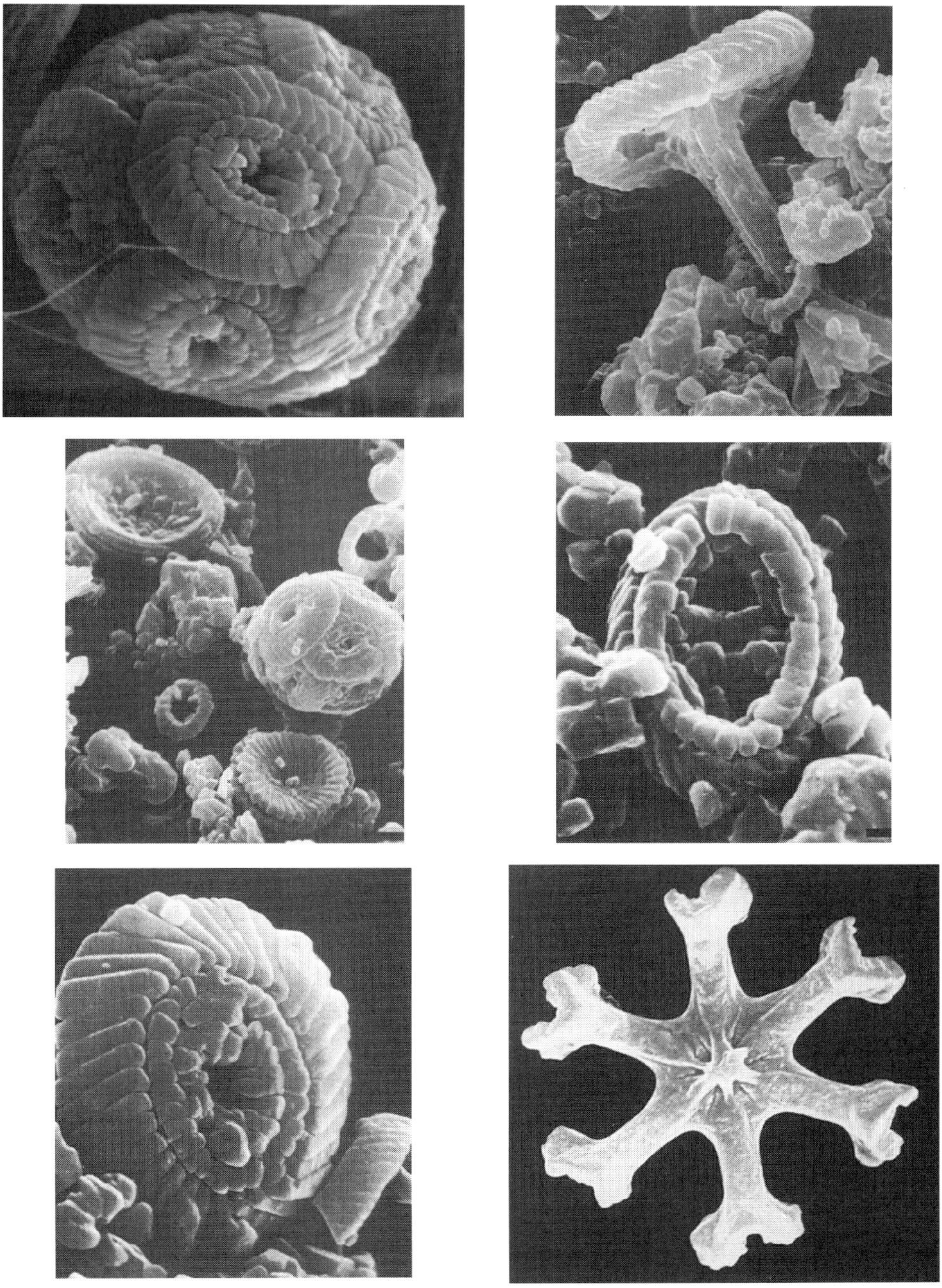

Figure 2.13. Different shapes of coccoliths found in Omya mines. *Courtesy of Omya/Plüss-Staufer AG.*[138] The first micrograph (upper left corner) - *Courtesy of ECC International Ltd., St. Austell, UK.*

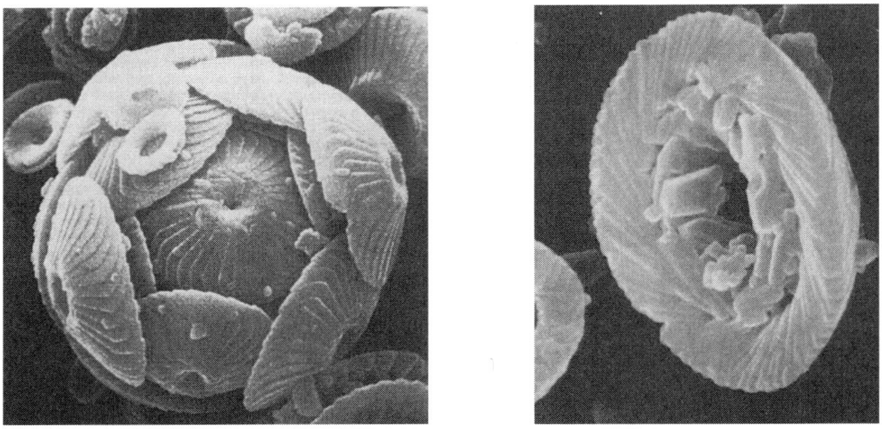

Figure 2.13 (continuation). Different shapes of coccoliths found in Omya mines. *Courtesy of Omya/Plüss-Staufer AG.*[138]

Most of the concerns about global warming has been for land based plants. It can be seen from the proceeding paragraphs that the oceanic conversion of calcium carbonate by microorganisms and of carbon dioxide by plankton are perhaps more important in the regulation of our environment. Incidents such as an underwater volcanic explosion may affect this balance since they alter the temperature of water and the concentration of carbon dioxide in water and, consequently, its internal use and release to the atmosphere.

As was mentioned before, several crystalline forms can be produced. These forms are used to build minerals and rocks. These are defined below. There are three crystalline forms which are mostly used in production of calcium carbonate filler:

| | |
|---|---|
| calcite | a mineral also called calcspar which has trigonal rhombohedral or trigonal scalenohedral form |
| aragonite | orthorhombic crystals |

Figure 2.14 explains differences between these three forms and compares them with morphology of fillers having these crystalline forms as well as with schematic diagrams of the crystals.

During the biological process of formation, each organism produces a specific crystalline form. For example, the mother-of-pearl or pearl itself are aragonite. Here the prismatic layer is formed of calcite. Aragonite is a less stable form and it can be converted by heating to calcite. Both minerals can be easily distinguished by their physical properties such as density (aragonite 2.9 and calcite 2.7), refractive index (aragonite 1.7, calcite with two refractive indices of 1.49 and 1.66 which

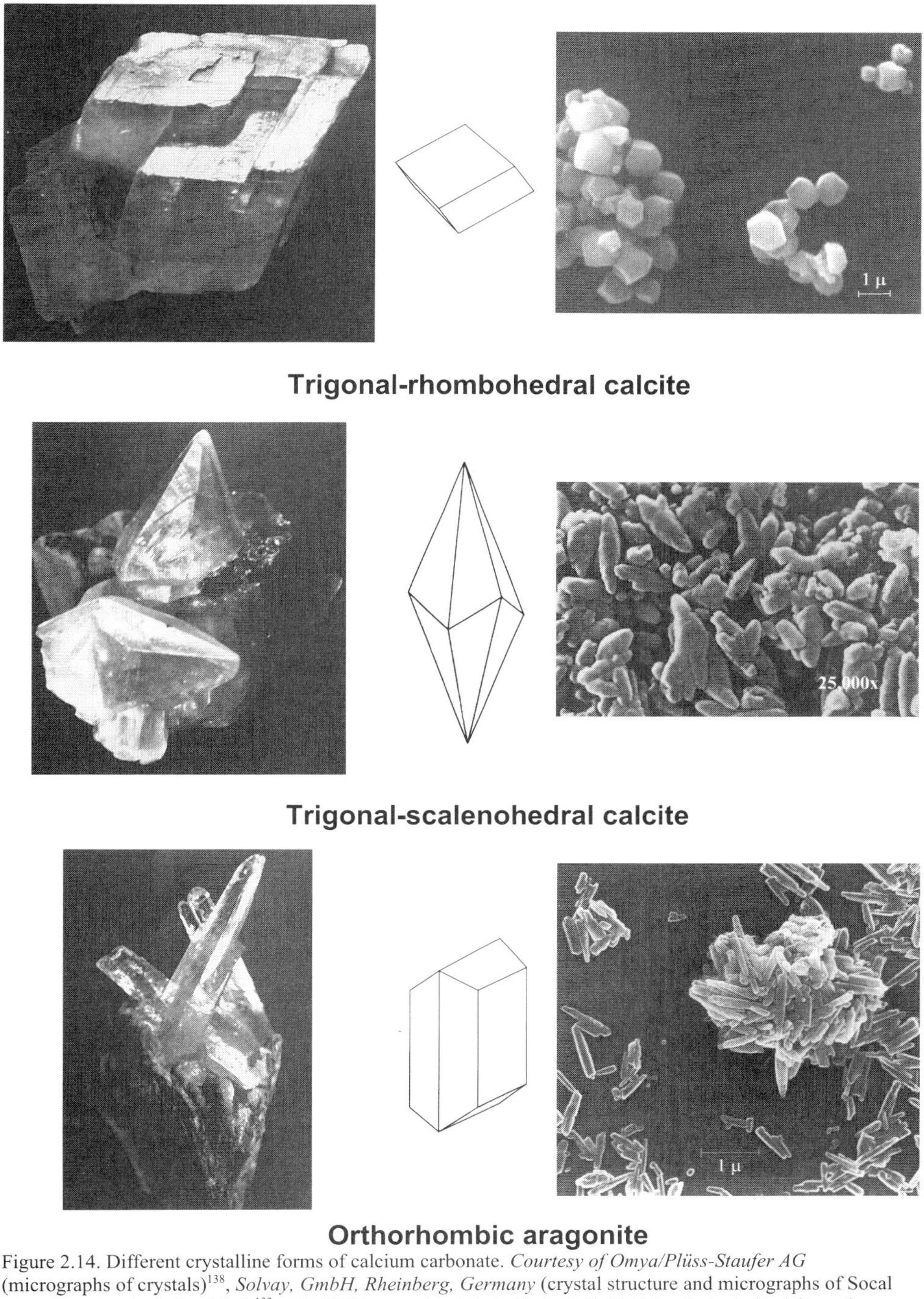


Figure 2.14. Different crystalline forms of calcium carbonate. *Courtesy of Omya/Plüss-Staufer AG* (micrographs of crystals)[138], *Solvay, GmbH, Rheinberg, Germany* (crystal structure and micrographs of Socal trigonal-scalenohedral calcite),[132] and *ECC International Ltd., St. Austell, UK* (rhombohedral calcite and aragonite).

causes a double refraction effect), and hardness (aragonite 3.5-4 and calcite 3). There are several other minerals and rocks associated with calcium carbonate:

| | |
|---|---|
| chalk | a sedimentary rock of soft texture formed from nanofossils |
| dolomite | mineral composed of calcium magnesium carbonate |
| limestone | consolidated sedimentary rock |
| marble | a metamorphic rock originally composed of either calcite, aragonite, or dolomite which was recrystallized to a dense rock under the influence of high pressure and temperature. Its color depends on admixtures (e.g. iron oxide gives yellow to brownish coloration, Carrara marble is white because of high purity) |
| travertine | deposits from spring water in a form of calcite or aragonite which form in caves dripstones (stalactites and stalagmites) |
| vaterite | a hexagonal modification of calcium carbonate which is very unstable and it is readily converted to calcite |

The above review of rock and mineral formation indicates that all calcium carbonates are not the same. Their type and properties depend on their history of formation. In addition to the above processes of formation, the presence of admixtures also determines the process used to extract or refine the filler and its utility. Other minerals such as silicates and clays are formed simultaneously and within calcium carbonate and altogether they form a broad range of mixtures which must be processed. This aspect of the production is underlined in recognition that it is very important for a final product process to use a particular grade of material dependent on the technology of production and the place of origin.

Three major technological processes are used in the production of calcium carbonate filler. These are milling, precipitation, and coating. More than 90% calcium carbonate is processed by milling. Two methods are used: dry and wet. The milling technology was developed for reproducibility and to obtain the required particle size distribution. In addition to general grades, ultrafine grades are also produced by the milling process. If the wet milling process is used, the material is frequently delivered to the customer in the form of a slurry which makes subsequent processes more economical and environmentally friendly. The paper industry uses about 80% of its calcium carbonate in the form of slurry. Also, paints use large quantities of slurried calcium carbonate. Figure 2.15 shows SEM micrograph of milled calcium carbonate. In this process, the crystalline structure of the rock has an important influence on the morphology of the filler.

Figure 2.16 shows a schematic diagram of the production of precipitated calcium carbonate. Such grades are also termed synthetic calcium carbonate since several chemical operations are performed. The first operation is calcination which is performed in a kiln at 900°C. At this stage, calcium carbonate is decomposed to

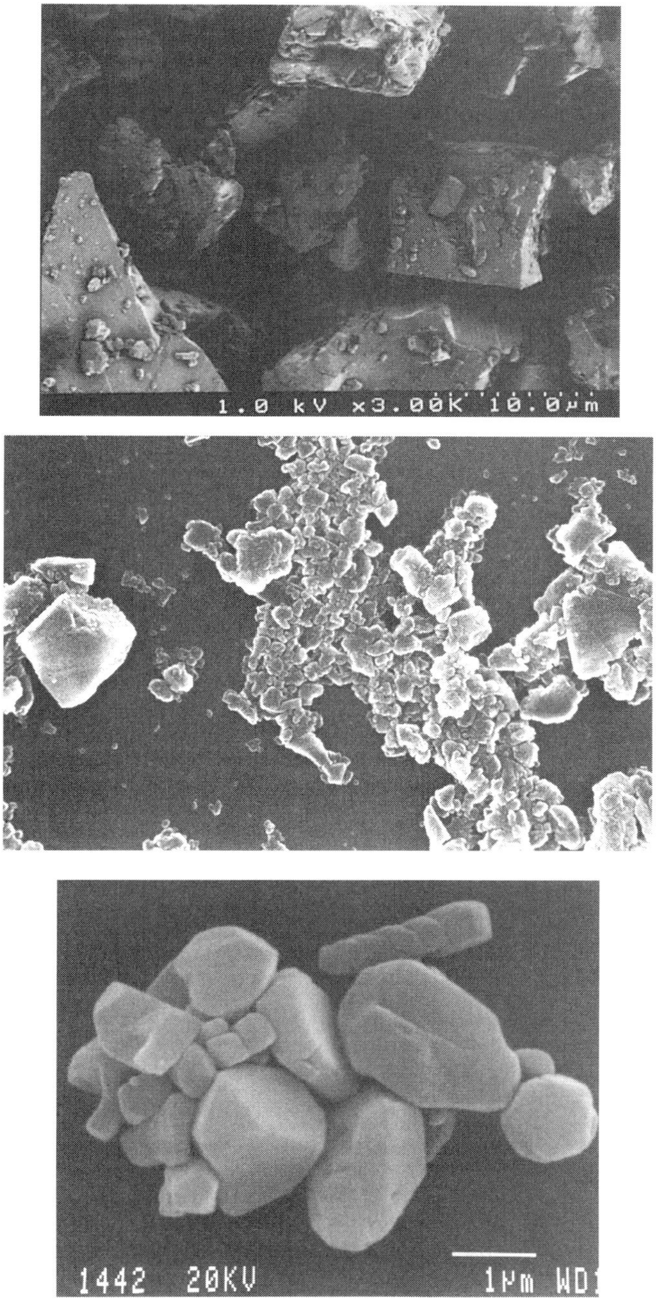


Figure 2.15. SEM of different calcium carbonates. upper - milled calcium carbonate, middle - ultrafine ground calcium carbonate, bottom - chalk. *Courtesy of J.M. Huber Corporation, Macon, GA, USA (upper), and ECC International Ltd., St. Austell, UK (middle and bottom).*

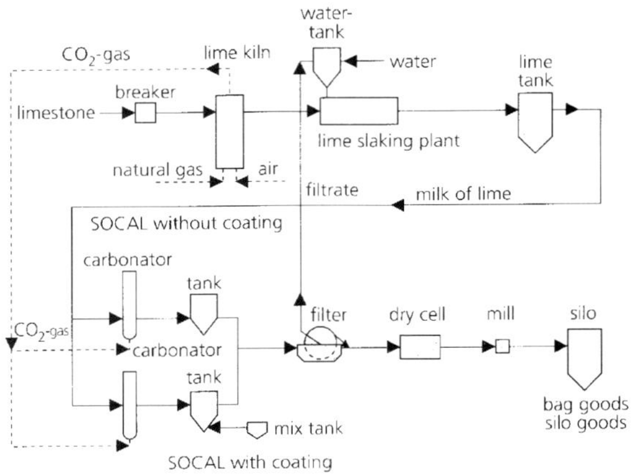


Figure 2.16. Schematic diagram showing the production of precipitated calcium carbonate. *Courtesy of Solvay GmbH, Rheinberg, Germany.*

calcium oxide and carbon dioxide which is used in further step. In the next step, calcium oxide is mixed with water in a process called slaking. This converts calcium oxide to lime and permits a material purification operation to be performed which results in a product of improved purity. In the (sometimes) final operation, the milk of lime is saturated by carbon dioxide which precipitates calcium carbonate. Depending on process parameters such as temperature, degree of purification, and concentration of reagents, different grades are produced which can be distinguished by particle size distribution, or crystalline form, or may be graded for food or pharmaceutical use (Figure 2.14). One additional operation is surface coating during which a 1-3 wt% coating is deposited on the surface of calcium carbonate particles. In most cases, salts of fatty acids are used for coating but titanates and zirconates are also used although less frequently. Grafting various polymers onto the surface is the subject of current research. Rhombohedral calcite is the most likely to be coated. Because of coating its particles do not agglomerate and become hydrophobic. Aragonite or calcite scalenohedral form is likely to be used if calcium carbonate must play the role of a secondary pigment. Here, higher light scattering and brightness are obtained by forming some aggregation. Scanning electron micrographs show that the surface coating, by itself, does not introduce any particular morphological features.

There are also special morphological grades of calcium carbonate which can be used to change the rheological characteristics of materials. One example of such a product is shown in Figure 2.17. The combination of particulates and elongated particles creates special rheological effects. In addition, the elongated particles are

Figure 2.17. SEM micrograph of Viscolite U.

covered by a system of microcracks which contribute to non-Newtonian rheological characteristics which this filler imparts.

## 2.1.20 CALCIUM HYDROXIDE

| | |
|---|---|
| **Names**: calcium hydroxide, carbide lime, lime hydrate, hydrated lime, slaked lime | **CAS #**: 1305-62-0 |
| **Chemical formula**: $Ca(OH)_2$ | **Functionality**: OH |
| **Chemical composition**: $Ca(OH)_2$ - 80-90%, $CaCO_3$ - 10-20% | |
| **PHYSICAL PROPERTIES** | |
| **Density**, g/cm$^3$: 2.2-2.35 | **Melting point**, °C: 272 |
| **CHEMICAL PROPERTIES** | |
| **Chemical resistance**: not resistant to strong acid, phosphorus, maleic anhydride | |
| **Moisture content**, %: 1.5 | **pH of water suspension**: 11.4-12.6 |
| **OPTICAL PROPERTIES** | |
| **Refractive index**: 1.57 | **Color**: gray |
| **MORPHOLOGY** | |
| **Particle shape**: round | **Crystal structure**: hexagonal; **Particle size**, μm: 5 |
| **Specific surface area**, m$^2$/g: 1-6 | |
| **MANUFACTURER & BRAND NAME**: ReBase Products, Inc., Barrie, Canada — White Knight 100 - acetylene production co-product derived from carbide lime | |
| **MAJOR PRODUCT APPLICATIONS**: similar to calcium carbonate | |
| **MAJOR POLYMER APPLICATIONS**: PVC and PE already use the product | |

Calcium hydroxide is a product new to the market. There have been, in past, positive scientific reports of its usefulness. The benefits of calcium hydroxide over calcium carbonate are its functionality, particle shape (more spherical and thus less abrasive to the equipment) (Figure 2.18), its lower density (decreases the density of product and lowers the price), a refractive index closer to many polymers, and its lower cost (approximately half of the price of calcium carbonate). The manufacturing equipment includes an excitement chamber, metered conveying, pneumatic transportation, flash drying, classification, and silo storage. The manufacturer delivers product to customers by its own silo-trucks.

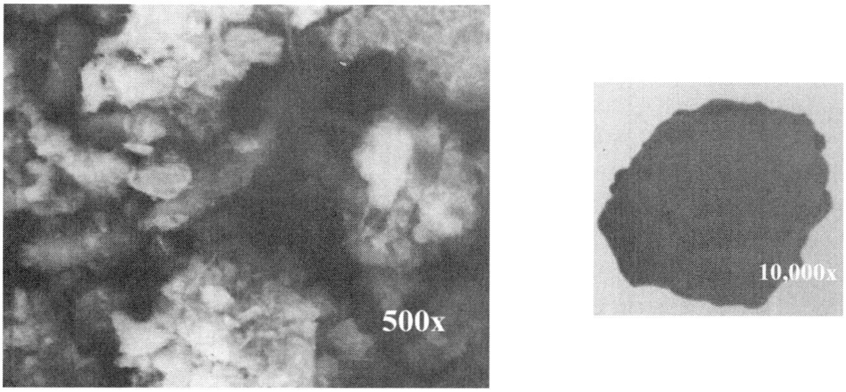


Figure 2.18. SEM micrograph of White Knight 100 calcium hydroxide particle. *Courtesy of ReBase, Barrie, Canada.*

## 2.1.21. CALCIUM SULFATE

| | | |
|---|---|---|
| **Names**: calcium sulfate, gypsum, anhydride | | **CAS #**: 7778-18-9 or 10101-41-4 (dihydrate) |
| **Chemical formula**: $CaSO_4$, $CASO_4 \cdot 2H_2O$ | | **Functionality**: none |
| **Chemical composition**: $CaSO_4$ - 98.7-99%, $SiO_2$ - 0.31% (dihydrate contains $CaSO_4 \cdot 2H_2O$ - 82.3% and $CaCO_3 \cdot MgCO_3$ - 12.2%) | | |
| **Trace elements**: Fe, heavy metals - ppm quantities | | |
| PHYSICAL PROPERTIES | | |
| **Density**, g/cm$^3$: 2.3-3 | **Mohs hardness**: 2 | **Melting point**, °C: 1450 |
| **Decomposition temp.**, °C: 128-63 | **Maximum temperature of use**, °C: 128 | |
| CHEMICAL PROPERTIES | | |
| **Chemical resistance**: reacts with strong mineral acids | | |
| **Moisture content**, %: 0.1 | **pH of water suspension**: 6.8-10.8 | |
| OPTICAL PROPERTIES | | |
| **Refractive index**: 1.52-1.61 | **Color**: white to light gray | |
| MORPHOLOGY | | |
| **Crystal structure**: monoclinic | **Cleavage**: one direction | |
| MANUFACTURERS & BRAND NAMES:<br>Charles B. Chrystal Co., Inc., New York, USA<br>Terra Alba USP Granulated & English - pure forms for pharmaceutical industry<br>NF Grade - calcinated terra alba for food and pharmaceutical industries<br>LP #2 - dihydrate for filling and fire retarding applications<br>204 - anhydrous grade for $TiO_2$ replacement and drying agent | | |
| MAJOR PRODUCT APPLICATIONS: pharmaceutical, food, plastics, paints | | |
| MAJOR POLYMER APPLICATIONS: polyester, PU, PVC | | |

Gypsum shows very little variation in chemical composition, and it is the most common of the sulfate minerals. Its origin is related to a high concentration in sea water (4%) from which it is deposited by sedimentation or evaporation. The last mode of formation may also result in anhydrite formation because both forms are metastable and exist in equilibrium conditions.

The hydrous form of calcium sulfate, *Terra Alba*, contains about 20% water of crystallization. It is processed by fine grinding and air-separation to a selected, white, high purity gypsum. The anhydrous gypsum form is obtained by the same process, the addition of a calcination step in which water is almost entirely removed (only about 0.3% remains). Particles are mostly smaller than 10 μm. Oil absorption is rather high, in the range of 23 to 26 g/100 g. The choice between the hydrous and the anhydrous forms depends on the processing temperature and the moisture sensitivity of the formulation.

Color is another important consideration. Anhydrous forms are brighter than the hydrous ones because of their crystalline form, particle size, and purification during the calcination process. Particle size distribution depends mostly on the grinding process. *Terra Alba*, made by fine grinding and air-separation, has an average particle size of 12 μm, whereas anhydrous calcium sulfate has an average particle size of 7 μm. A fine grinding yields a product with an average particle size equal to 1.4 μm.

## 2.1.22 CARBON BLACK[139-243]

| | | |
|---|---|---|
| **Name**: carbon black | | **CAS #**: 1333-86-4 |
| **Chemical formula**: carbon | **Functionality**: OH, COOH, $SO_3$, ONa | |
| **Chemical composition**: carbon - 95-99% | | |
| **Trace elements**: Zn, Ni, Ba, Si, Fe, Cr, Mg, Al, V, Ca, Sr, Na, K, S | | |
| PHYSICAL PROPERTIES | | |
| **Density**, g/cm$^3$: 1.7-1.9 | | |
| CHEMICAL PROPERTIES | | |
| **Chemical resistance**: reactive with oxidizing agents | | |
| **Moisture content**, %: 0.12-2 | **Volatiles content**, %: 0.1-11 | **Ash content**, %: 0.02-3 |
| **pH of water suspension**: 2-8 | **Water solubility**, %: insoluble | **Total ions**, ppm: 50 |
| OPTICAL PROPERTIES | | |
| **Tinting strength**, %ITRB: 41-164 | | **Jettness index**: 65-99 |
| MORPHOLOGY | | |
| **Particle size**, nm: 14-250 | **Specific surface area**, m$^2$/g: 7-560 | |
| **CTAB surface area**, m$^2$/g: 29-128 | **Iodine number**, g/kg: 19-151 | **Toluene discoloration**, %: 75-85 |
| **Sieve analysis**: 325 mesh residue - 0.002-0.1% | **DBP absorption**, cm$^3$/100 g: 44-192 | |
| **Hegman fineness**: 2-7 | | |

MANUFACTURERS & BRAND NAMES:

Cabot Corporation, Waltham, Mass, USA

manufacturer of a full range of carbon blacks in various grades. The list below includes brand names of products used for different applications:

*plastics*: Monarch, Mogul, Regal, Vulcan, Elftex, Black Pearls (divided by use, such as coloring, electric resistant plastics, conductive plastics, UV protection, good dispersion, blue tone, low cost)

*printing inks*: Mogul, Regal, Elftex, Sterling, Black Pearls (divided by flow, color, news ink, paste ink, liquid ink, printing method, end-use of printed material)

*power cable*: Vulcan

wire and cable: 3000 Series Black Pearls

*solvent coating*: Emperor (surface modified blacks)

*FDA compliant*: Black Pearls

*fine-denier fibers*: Black Pearls

*UV stabilizing blacks*: Elftex, Vulcan, Mogul, Regal, Black Pearls

*high color blacks*: Monarch (fluffy), Black Pearls (pellets)

Cancarb Ltd., Medicine Hat, Alberta, Canada

Thermax, Ultra-Pure N -990, N-991, N-908 all thermal carbon black

Columbian Chemicals Company, Swartz, LA, USA

Performance Furnace carbon blacks - thirty five Raven grades for a full scope of carbon black applications

Conductive Carbon Blacks - Conductex 975 Ultra and SC Ultra

Lampblack Replacements - Raven 22, 16, 14

Specialty Furnace Carbon Blacks - four Raven grades for news-ink, paper and UV protection

*continued on the next page*

**MANUFACTURERS & BRAND NAMES:**
Degussa Corporation, Akron, OH, USA
Corax N110, N220, N234, N299, N326, N330, N339, N347, N550, N650, N660, N754, N762, N774 all furnace blacks
Carbon, Brussels, Belgium, distributed by R.T. Vanderbilt Company, Inc., Norwalk, CT, USA
Ensaco 150, 200, 250 - carbon black produced in form of powder and granules in process similar to furnace black but differing in aerodynamic and thermodynamic conditions. No water quenching is used in the process. The resultant material is closer to acetylene black than furnace black. Unique properties of this carbon black are utilized in dry-cell batteries, paint, plastics, and rubber markets.
Sid Richardson Carbon Black Company, Fort Worth, TX, USA

**MAJOR PRODUCT APPLICATIONS:** tires, plastics, inks, paints, and many other

**MAJOR POLYMER APPLICATIONS:** most polymers and rubbers

Carbon black, initially used as pigment in ink, has the longest history of all the materials discussed in this book. It was produced in China about 3000 B.C. and exported to Japan around 500 A.D. But only in the last 50 years has the technological development in both carbon black production and processing of rubber and polymers resulted in the tremendous variety of products which we know today.

Structurally, carbon black is similar to graphite, composed of large sheets of hexagonal rings formed by carbon atoms separated from each other by a distance of 0.142 nm (i.e., close to the length of the C-C bond in benzene – 0.139 nm). The values of 0.148, 0.134, and 0.12 nm are usually assigned to the single, double, and triple bond distances between two carbons, respectively. This means that the bond length in graphite is between the length of a single and a double bond. The bond length in carbon blacks is also 0.142 nm and hexagonal rings form large sheets, as in the case of graphite. The difference between the graphite and carbon black is in the arrangement of layers. In the case of graphite, the layers are stacked on each other regularly in such manner that each carbon atom has directly above and below it another carbon atom, meaning that the structure has a tri-dimensional order. The distance between carbon atoms in each layer is 0.335 nm. The layers of carbon blacks are also parallel to each other but not arranged in order, usually forming concentric inner layers. Such an arrangement is called a turbostratic structure. The separation distance between parallel layers of carbon blacks varies in the range of 0.350-0.365 nm. The interior of carbon black aggregate is less ordered than its surface and that is why it is chemically more reactive, as confirmed by oxidation studies. Carbon black exposed to high temperature undergoes a graphitization process. Oxygen present in the system reacts with the carbon atoms in the center of particles, resulting in formation of hollow spheres having an increased crystallinity. From what has been said so far, it is not surprising that carbon black has low crystallinity and, in fact, is regarded as amorphous carbon having a degenerated graphitic structure.

The basic reaction of carbon black formation is as follows:

$$C_xH_y \rightarrow xC + \frac{y}{2}H_2 \ \text{-(+)} \ \Delta$$

Therefore, hydrocarbon-containing materials have the potential to be used in carbon black production. Raw materials can be in the form of hydrocarbon gases, such as methane and acetylene, but mostly viscous residual aromatic hydrocarbons are used. Depending on chemical composition, the reaction is exo- or endothermic. Only when carbon black is produced from acetylene the reaction is exothermic and the process demands intensive cooling, whereas in other cases the reaction is endothermic and needs a substantial amount of energy in order to form carbon black.

Several methods can be used for the production of carbon black. The Lampblack Process, the oldest of all, was developed by the Chinese. Initially, vegetable oil was burned in small lamps with tile covers to accumulate the carbon black formed. Later, shallow pans were used in systems with a restricted air supply. Carbon black in this process was recovered from smoke in settling chambers. This method is still used for production of small quantities of carbon black. The Channel Black Process is another method useful in the past and not important for present production. Natural gas is used as a raw material in this process; it is burned in close proximity to steel channels on which carbon black is deposited. Carbon black is removed from the channels by scrapers and falls into hoppers beneath the channels. This process was discontinued in the USA in 1976 because of the price of natural gas, smoke pollution, and low yield. It is still being used in Germany, Eastern Europe, and Japan.

The Thermal Decomposition Process and the Acetylene Black Process are similar in the sense that both processes are conducted in the absence of air and flame, and both use gaseous raw materials. In the Thermal Decomposition Process, natural gas is fed into a generator having a temperature of 1300°C where it undergoes cracking. A stream of product gases, containing carbon black, hydrogen, methane, and other hydrocarbons, is cooled with water sprays and carbon black is removed by bag filters. The process is cyclic in nature because the endothermic reaction requires heating of the generator at 5 minutes intervals. In order to achieve a continuous process, two generators work together in 5 minutes cycles. When one generator is producing, the other is heated, partially by product gases having a high calorific value. A similar process is performed in England with the use of oil, which performs two roles: heating material and raw material for carbon black production. The Acetylene Black Process involves burning the acetylene in a metal retort to attain the process temperature (800-1000°C), then the process is continued in an oxygen-free atmosphere, while heat produced by the exothermic reaction is taken away by a water cooling system. The process gives a product of very low density, which is difficult to compress and resistant to pelletization.

The Oil-Furnace Process is by far the most prevalent method of carbon black production. It is a further development of the Gas Furnace Process. A reactor is fed by liquid hydrocarbon feedstock which is injected, atomized, and mixed with preheated air and auxiliary fuel (usually natural gas). Part of the feedstock is used to maintain the reaction temperature (1450-1800°C) and the remainder is converted to carbon black. The reaction is quenched with water spray and the carbon black is separated from the combustion gases by bag filters and cyclones. The process is completed by pelletizing and drying. An Oil-Furnace Process line is usually equipped with a computer-control system because process conditions greatly affect the product properties.

The installations used in this process are usually very large and they are equipped with energy-saving systems. In the early 1970s, reactor and burner designs were improved, resulting in better mixing and atomization, and lower residence times. A series of new types of carbon blacks was introduced, called "New Technology" or "Improved" carbon blacks. This new development yields products of narrower distribution of aggregate sizes, higher surface activity (higher bound rubber and higher moisture absorption), and more open aggregates (branched, bulky).

The Oil-Furnace Process has superior efficiency and economy. It is also the most versatile process, allowing production of most grades important for industry. Table 2.1 outlines differences between carbon blacks manufactured in five processes.

**Table 2.1. Typical properties of carbon blacks manufactured in different processes.**

| | Furnace | Thermal | Acetylene | Channel | Lamp |
|---|---|---|---|---|---|
| | HAF | MT | | EPC | Lb |
| | N-330 | N-990 | | S300 | |
| Av. particle diameter, nm | 28 | 250 | 40 | 28 | 65 |
| BET surface area, $m^2/g$ | 75 | 7-12 | 65 | 115 | 22 |
| DBP absorption, ml/100 g | 103 | 44 | 250 | 100 | 130 |
| Tinting strength, %SRF | 210 | 35 | 108 | 180 | 90 |
| Toluene extract, % | 0.06 | 0.5 | 0.1 | 0.0 | 0.2 |
| pH | 7.5 | 9-11 | 4.8 | 3.8 | 3.0 |
| Volatile material, % | 1.0 | 0.1 | 0.3 | 5.0 | 1.5 |
| Ash, % | 0.4 | 0.2 | 0.0 | 0.02 | 0.02 |
| Composition, % | | | | | |
| C | 97.9 | 99.6 | 99.7 | 95.6 | 98.0 |
| H | 0.4 | 0.2 | 0.1 | 0.6 | 0.2 |
| S | 0.6 | 0.01 | 0.02 | 0.2 | 0.8 |
| O | 0.7 | 0.1 | 0.2 | 3.5 | 0.8 |
| Raw material | oil or gas | gas | acetylene | gas | coal tar |
| Yield, % theor.carbon | 23-70 | 30-45 | | 1.6-6.0 | |
| Energy use, J/kg | $9.3\text{-}16\times10^7$ | $2.0\text{-}2.8\times10^8$ | | $1.2\text{-}2.3\times10^9$ | |

Acetylene blacks are the purest products manufactured, whereas in the thermal process one can obtain carbon black of the lowest surface area. Channel carbon blacks are surface oxidized as a result of their exposure to air at elevated temperatures. Particles of channel blacks are slightly porous, and the high level of surface oxidation may retard vulcanization rate, but when it is used in polyethylene it improves weathering resistance because the phenol and hydroquinone surface groups have antioxidant properties. A high level of sulfur in oil-furnace blacks depends on the composition of feedstock and can be reduced by its proper choice. It is important in this process that raw materials also contain low levels of alkali metals which affect the size of aggregates. Aromacity of feedstock increases the degree of aggregation, while injection of alkali metal decreases it. One should not be misguided by the results quoted in Table 2.1, which contains data on particular grades but does not reflect their full variety. For example, Oil-Furnace Process blacks have a specific surface area in the range of 25-560 $m^2/g$, particle size from 13 to 75 nm, and carbon content from 90.5 to 98%. Although carbon blacks are produced by various manufacturers according to the standards set by industries, differences exist and the evaluation of products based on a simple comparison of results of their analysis cannot contribute to a reliable technology; therefore their performance should be evaluated during product formulation.

Such a great number of carbon blacks is now manufactured by the industry that without the help of an adequate classification it will be difficult to search for a product that may serve a particular purpose for a carbon black application. Before the Oil-Furnace Process was fully applied, classification was based on both the process of production and the properties of carbon black. Later, the Oil-Furnace Process took some markets from other processes and developed products of similar properties. This practically ruined the former classification (process type became unimportant) and a need for a new classification became apparent. A new classification is based on one letter and three digits (Table 2.2). The letter is N (for normal) and S (for slow), which describes the effect of carbon black on the rate of cure in rubber processing. The first digit refers to the average particle size, as specified in the ASTM Standard. The lower the digit, the smaller particle size; for example, 1 means particle size between 11 and 19 nm, whereas 9 means average particle diameter between 201 and 500 nm. The last two digits are assigned arbitrarily and characterize the set of several properties of carbon blacks, such as iodine adsorption, pour density, etc., which are typical for a particular grade. There is no particular relationship between the last two digits and carbon black properties that can be put in a logical order.

The ASTM classification provides some information about the carbon black type, but the information still can be broadened if one also uses the old classification along with it, and that is why both classifications are frequently used.

The conversion from an old to a new system is not always precise as far as particle size diameter is concerned, but knowing the old designation helps to establish such characteristics as abrasion resistance, reinforcement level, vulcanizate properties, processing properties, typical application, particle size, and electric conductivity. Many designations used in the past were dropped from use because they were related to particular processes which are not used frequently now, such as lampblack (LB), medium flow channel (MFC), etc. When properties are discussed in more detail, we learn that neither classification is sufficient for choosing carbon black arbitrarily, which should be quite obvious, taking into consideration the amount of produced grades and the sophistication of present technology.

**Table 2.2. Carbon black classification**

| ASTM N-type | Designation |
|---|---|
| N100 to N199 | super abrasion furnace, SAF |
| N200 to N299 | high aggregate furnace |
| N300 to N399 | intermediate super abrasion furnace, ISAF |
| N400 to N499 | fine furnace, FF, and conductive furnace, XCF |
| N500 to N599 | fast extrusion furnace, FEF |
| N600 to N699 | high modulus furnace, HMF, general purpose furnace, GPF, and all-purpose furnace, APF |
| N700 to N799 | semi-reinforcing furnace, SFR |
| N800 | fine thermal, FT |
| N907 | medium thermal non-staining, MT-NS |
| N990 | medium thermal, MT |

Let us now discuss the physical properties of carbon blacks currently available in the market. Particle and aggregate size are probably the most important factors characterizing carbon blacks. In order to understand them fully, one should consider the mechanism of carbon black formation. In the Oil-Furnace Process, liquid raw material is atomized in a furnace having a very high temperature. Formation of carbon black is a gradual process in which a few phases can be singled out: droplet vaporization, molecular rearrangement, and decomposition. During molecular rearrangement, large polyaromatic molecules are formed which gradually lose hydrogen and finally become almost pure carbon. It is easy to imagine that such transitions have to be accompanied by a gradual change of state from a liquid to a solid through a viscous state. As long as particles are in the form of liquid droplets, they can easily combine to form larger droplets or disintegrate to smaller ones, depending on the mixing degree and time-temperature relationship of a liquid state. Generally, we can say that during the stage of liquid state, primary particles are formed and their size depends on the process parameters. Viscosity increases with loss of hydrogen and the formation of spherical particles becomes more difficult, but colliding particles may adhere to each other since they are in a viscous state, and partial fusion may occur. This period regulates the size of the aggregates formed, meaning the number and spatial distribution of primary particles forming an agglomerate. As decomposition progresses, aggregates finally reach a stage at which they become solid and can no longer adhere to each other to form durable fusion points. The only way by which the size of an aggregate can increase after this stage is by weak attractive forces that are easy to disrupt during carbon black compounding. When agglomerates are formed, they should not be exposed to the high process temperature since they may undergo crystalline changes known as graphitization. That is why carbon blacks are water-quenched. If water quenching is done too early, it results in carbon black containing an increased amount of tar.

Figures 2.19 and 2.20 show the difference between low structure and high structure carbon blacks.

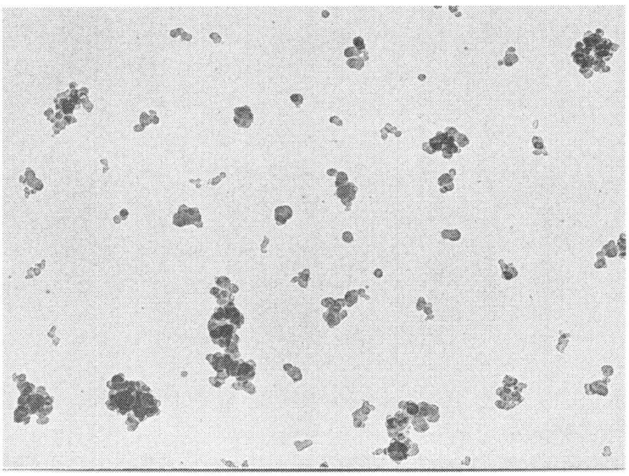

Figure 2.19. TEM micrograph of carbon black N326 (low structure). *Courtesy of Columbian Chemicals Company.*

The above mechanism clearly shows that by varying the process parameters, one can easily regulate the size of the primary particles and the structure of agglomerates. Although the average particle diameter is the basis of the ASTM classification of carbon blacks in processing technology, this factor is usually not used due to technical difficulties with measurement. Particle diameter can be measured by electron microscopy and it is therefore difficult to obtain accurate values for a representative sample size. The use of an image analyzer did not solve this problem. The surface area of carbon black is the most useful parameter relating to particle size and agglomerate size. Several methods are used for this measurement – the simplest, iodine number measurement, is a fast and a precise

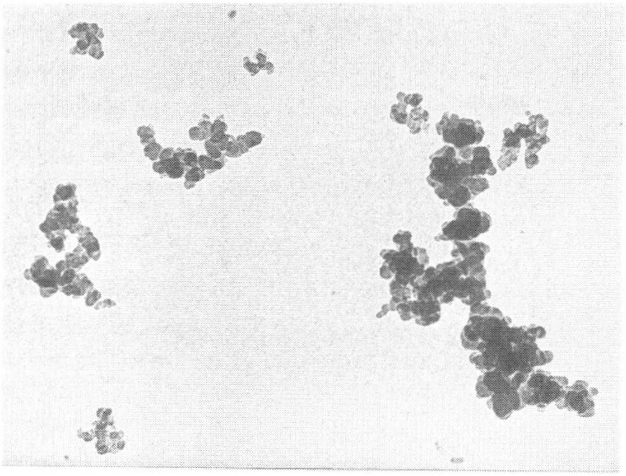

Figure 2.20. TEM micrograph of carbon black N326 (high structure). *Courtesy of Columbian Chemicals Company.*

method, but results are affected by the presence of residual extractable materials and surface oxygen. The BET method is an exact and valuable tool for fast and accurate measurements. The porosity of carbon blacks can be estimated from the difference between the result of the BET method and adsorption of large molecules like cetyltrimethylammonium bromide from aqueous solution. Similar results can be obtained by the so-called '*t*' method, based on the BET principle with the use of controlled conditions of adsorption and a standard sample for comparison. In most grades of carbon black, the porosity is rather low, and a higher porosity usually shows that carbon black was oxidized.

For non-porous particles, the average particle diameter can be closely estimated from the specific surface area since the two values are inversely proportional to one another. Although many efforts have been made to analyze particle size distribution, the size of the particle has only a secondary effect on the size and structure of the aggregate. The reason is that the primary particles are strongly connected in the aggregates and even the most abrasive processing methods do not affect the structure of aggregates. Only up to one fracture per aggregate occurs in rubber processing. This is the primary reason that many efforts have been made to evaluate the structure of aggregates. Three main possibilities exist in the determination of aggregate size and structure:

- Electron microscopy
- Centrifugal sedimentation
- Liquid absorption.

Electron microscopy allows one to analyze the average particle size, the number of particles per agglomerate, and the projected area from which a calculation of the void volume of each aggregate can be done. Centrifugal sedimentation allows direct measurement of the size distribution of aggregates larger than a certain Stokes diameter. The major problem with this method is related to incomplete dispersion and flocculation of aggregates. Finally, the liquid absorption (usually of dibutyl phthalate) gives the void volume in aggregates directly.

Carbon black structure affects the physico-mechanical properties of the material, such as tensile strength, elongation, water absorption, tinting strength, die swell, etc., which are discussed under their respective topics in Chapter 5.

Let us now examine practical examples of carbon blacks chosen from the range of products of the Cabot Corporation, which were selected to show a variety of carbon blacks in respect to their structure and particle size (Table 2.3). Table 2.3 shows that because each carbon black differs in particle size, particle porosity, and aggregate structure, the relationship between parameters cannot have a high correlation.

**Table 2.3. Surface area, particle size and oil absorption of some Cabot grades**

| Type | Surface area, $m^2/g$ | Particle size, nm | Oil absorption, g/100 g |
|---|---|---|---|
| Black Pearls 2000 | 1475 | 15 | 300 |
| Black Pearls 1300 | 560 | 13 | 105 |
| Black Pearls 1100 | 240 | 14 | 50 |
| Vulcan 9 A32 | 140 | 19 | 114 |
| Regal 300 I | 80 | 27 | 72 |
| Sterling SO | 42 | 41 | 120 |
| Sterling NS | 25 | 75 | 70 |

The morphology of carbon black and, in particular, the presence of agglomerates makes it difficult to process. The chemistry of carbon black and, particularly, the chemistry of its surface must be considered in selecting carbon black for a particular application and in determining the best processing method. Heat treatment of carbon black produces both physical and chemical changes in surface activity. Oxygen is usually reacted before the temperature reaches 1000°C, whereas the hydrogen is gradually removed in the temperature range 800 to 1600°C. It is known that oxygen in carbon blacks forms carboxyl, quinone, lactone, and phenolic groups, and they are lost on heating to 950°C. This loss is the volatile content of carbon black. The presence of active groups on the surface of carbon black facilitates wetting, dispersion, and adsorption of moisture. These factors, in turn, increase the reinforcing effect and facilitate dispersability of carbon blacks. Volatile content varies in the range of 0.5 to 11% and the reinforcing types of carbon blacks usually have 2-3% of volatiles. Properties of carbon blacks should also be analyzed for the presence of organic residue, given by the amount extracted by solvents. The organic residue, which is a tar-like product, can migrate to the surface in the compounded product and cause staining.

Research on carbon black continues and the most important topic remain its structure, the effect of functional groups on carbon black properties, the effect of the measured parameters of carbon black on its performance in various systems, and the influence of processing parameters on the product. These and other influences are discussed throughout the book.

## 2.1.23 CERAMIC BEADS[244-247]

| | | |
|---|---|---|
| **Names**: ceramic beads, ceramic spheres, microspheres | | |
| **Chemical formula**: n/a | **Functionality**: OH, silane treatment | |
| **Chemical composition**: silica alumina ceramic, alkali alumino silicate ceramic; $SiO_2$ - 55-65%, $Al_2O_3$ - 25-38%, $Fe_2O_3$ - 0.5-5% | | |
| PHYSICAL PROPERTIES | | |
| **Density**, g/cm$^3$: 0.24-2.5 | **Mohs hardness**: 5-7 | **Softening point**, $^o$C: 980-1400 |
| **Thermal conductivity**, W/K·m: 0.23 | **Compressive strength**, MPa: 1-34 (hollow), 400 (solid) | |
| CHEMICAL PROPERTIES | | |
| **Chemical resistance**: high chemical resistance | | |
| **Moisture content**, %: 0.2-0.5 | **pH of water suspension**: 4-8 | |
| OPTICAL & ELECTRICAL PROPERTIES | | |
| **Color**: white, off-white, gray | **Conductance**, mhos/cm: 200 | **Dielectric constant**: 1.6 |
| MORPHOLOGY | | |
| **Particle shape**: spherical | **Particle size**, μm: 50-350 (hollow), 1-200 (solid) | |
| **Shell thickness**: 10% diameter | **Specific surface area**, m$^2$/g: 0.1-1.1 | |
| **Sieve analysis**: residue on 325 mesh sieve - 0.01-26% | **Hegman fineness**: 3-7 | |

MANUFACTURERS & BRAND NAMES:
Kinetico Incorporated
  Macrolite Ceramic Spheres ML 535, 357, 714, 1430, 3050
PQ Corporation, Valley Forge, PA, USA
  Extendospheres SG standard grade of hollow spheres
  Extendospheres CG medium size hollows spheres
  Extendospheres TG smaller size hollow spheres
  Extendospheres XOL-200 smallest diameter hollow spheres
Sphere Services Inc., The Cenosphere Company, Oak Ridge, TN, USA
  Recyclospheres - ceramic hollow microspheres manufactured from fly ash in three particle size ranges with maximum diameter of 150, 210 and 300 μm
  Bionic Bubble - ceramic hollow microspheres manufactured from fly ash in three particle size ranges with maximum diameter of 75, 100 and 125 μm
Zeelan Industries, Inc., wholly-owned subsidiary of 3M, St. Paul, MN, USA
  Z-light Microspheres G-3400, G-3500, W-1000, W-1012, W-1100, W-1200, W-1300, W-1600 hollow microspheres differing in color (G - gray, W - off-white) and particle size
  Zeeospheres G-200, G-400, G-600, G-800, G-850, W-210, W-410, W-610 solid microspheres differing in color (G - gray, W - white) and particle size

MAJOR PRODUCT APPLICATIONS: *hollow*: bowling balls, cultured marble, plywood patch, roof coatings, refractory materials, grinding wheels, lightweight cement, polymer concrete, exterior insulating finishes, synthetic stucco, asphalt repair compounds, automotive sealants, roofing tiles, carpet backing, chemical resistant coatings, adhesives, sealants, pipe insulation, paint stripper, PVC flooring
*porous*: plastic molds, paints, coatings, sealants, asphalt, rubber, boat construction and repair, lightweight concrete, gypsum wall board, catalyst support, stucco, energy absorbing filler for autobody parts
*solid*: industrial paints, film antiblock, powder coatings, maintenance paints, adhesives, polymer concrete, textured coatings, house paints, low gloss paints, decorative flooring

**MAJOR POLYMER APPLICATIONS**: PP, PE, PS, PA, PVC, PPS, TFE, polyesters, epoxy, polyurethanes, phenolic, silicones

Ceramic spheres are produced from nepheline syenite, aluminum oxide, and bentonite or fly ash. Ceramic spheres have substantially higher densities than glass or polymer beads but are less expensive, more rigid and mechanically resistant due to their thicker walls. They have strength of all spherical materials which give the highest packing density and they improve flow because of the ball-bearing effect. In addition, ceramic spheres reduce dielectric constant, warpage, shrinkage, and improve crack resistance of speckling compounds.

A simple formula allows us to calculate the amount of beads required to replace a filler of higher density: Amount of beads = (density of beads/density of filler) × amount of filler in composition. Considering that beads have a better packing density than the filler they are replacing and produce a lower viscosity in the material, more beads can be added than is calculated from equation and yet maintain the same viscosity in the material. Although ceramic spheres are more rigid than glass spheres, they still require special precautions during handling and mixing. A high shear and prolonged mixing should be avoided during their incorporation. Ceramic beads should be added at the end of the mixing process.

Figure 2.21 shows the morphology of ceramic beads which are composed of a mixture of spherical particles. The unique beads produced by Kinetico have a denser shell to give them more mechanical strength and a porous interior to reduce their density (Figure 2.22).

Figure 2.21. SEM micrograph of Zeeospheres. *Courtesy of 3M, St. Paul, MN, USA.*

Figure 2.22. SEM micrograph of Macrolite – choice of sizes (upper) and cross-section (lower). *Courtesy of Kinetico, Inc., Newbury, OH, USA.*

## 2.1.24 CLAY[248-253]

| | | |
|---|---|---|
| **Names**: clay, ball clay | | **CAS #**: 1332-58-7 |
| **Chemical formula**: composition variable | **Functionality**: OH | |
| **Chemical composition**: $SiO_2$ - 53.3-61.2%, $Al_2O_3$ - 24.3-32.5%, $Fe_2O_3$ - 1.2-1.7%, $TiO_2$, - 1-1.1%, CaO - 0.2-0.3%, MgO - 0.2-0.4%, $K_2O$ - 0.3-1.3%, $Na_2O$ - 0.1-0.3% | | |
| PHYSICAL PROPERTIES | | |
| **Density**, g/cm$^3$: 2.6 | **Mohs hardness**: 2-2.5 | **Loss on ignition**, %: 9.5-12.6 |
| CHEMICAL PROPERTIES | | |
| **Chemical resistance**: reactive with acids and alkalis | | |
| **Moisture content**, %: 3 | **Adsorbed moisture**, %: 5.5-14.5 | **pH of water suspension**: 3.9-9 |
| OPTICAL PROPERTIES | | |
| **Color**: white, tan, gray | | **Brightness**: 60-64 |
| MORPHOLOGY | | |
| **Particle size**, μm: 0.4-5 | **Oil absorption**, g/100 g: 36-40 | |
| **Sieve analysis**: 325 mesh residue 1.6-2.2% | **Specific surface area**, m$^2$/g: 18.9-30.5 | |
| MANUFACTURERS & BRAND NAMES:<br>ECC International, St. Austell, UK<br>Hexafil and Hexafort H - ball clays for plastic and rubber<br>Kentucky-Tennessee Clay Company, Langley, SC, USA<br>#3380, Tenn #6 for rubber compounds, adhesives, plastics and other applications. In addition, the company manufactures a large number of grades for ceramics in Mayfield, KY and Gleason, TN<br>Old Hickory Clay Company, Hickory, KY, USA<br>manufacturer of large number of ball clay grades. No. 5 grade is used as filler in paints and plastics<br>United Clays, Brentwood, TN, USA<br>manufacturer and importer of clays from around the world (China, France, Germany, Indonesia, Thailand, UK, Ukraine) | | |
| MAJOR PRODUCT APPLICATIONS: rubber, adhesives, protective coatings, traffic paint, joint compounds, plastics, cables, belting, footwear, plant lining, tires | | |
| MAJOR POLYMER APPLICATIONS: PVC, rubber, urea formaldehyde, phenol formaldehyde | | |

Popularly-known fillers, such as kaolin clay, China clay, bentonite, Fuller's earth, and vermiculite all are clay minerals. Clay minerals are divided into 5 groups. The kaolinite group includes kaolinite and halloysite; the illite group includes illite; the smectite group includes montmorillonite and hectorite; the palygorskite group includes sepiolite and attapulgite, which, with vermiculite, are precursors of clay fillers. Kaolinites were formed by hydrothermal alteration or weathering of feldspars, and other silicates. Acid conditions favor kaolinite formation, whereas alkaline conditions favor formation of smectites. Both minerals are often accompanied by quartz, iron oxides, mica, and pyrite. The chemical composition of kaolinite is subject to few variations. Illite is more varied. The chemical composition of smectites is similar to pyrophyllite and talc. Montmorillonite is a principal constituent of bentonite clay deposits, which is also the main component of Fuller's earth. Kaolinite is a major component of China clay. Clay fillers are composed of a mixture of various minerals which are found in unique composition in a particular place. The name "clay" implies that particles of the material are very fine.

These fillers are discussed is separate sections, such as attapulgite, bentonite, sepiolite, kaolin, and vermiculite. Here, discussion is limited to ball clay. The name ball clay is derived from the original method of mining this plastic clay in England, where is was cut from the bank in a form of balls weighing 33 lbs. This expression was adopted to a wide range of clay materials which cannot be categorized as kaolins or fire clays.[253] The majority of ball clay is used for production of china and tiles. Only some grades are manufactured for application as fillers. These grades are covered in the table above.

USA ball clays are acidic to neutral and UK ball clay is alkaline which is an important factor in filler reinforcement where acid/base interaction plays a key role.

## 2.1.25 COPPER[254-258]

| | | |
|---|---|---|
| **Names**: copper spheres, copper powder, bronze powder, brass powder | | **CAS #**: 7440-50-8 |
| **Chemical formula**: pure metal or metal alloy | **Functionality**: none | |
| **Chemical composition**: copper powder: Cu - 98.5-99.5%; bronze powder: Cu - 88%, Sn - 10%; brass powder: Cu - 70-90%, Zn - 30-10% | | |
| PHYSICAL PROPERTIES | | |
| **Density**, g/cm$^3$: 8.92 | **Mohs hardness**: 2.5-3 | **Melting point**, °C: 1083 |
| CHEMICAL PROPERTIES | | |
| **Chemical resistance**: reactive with acids, alkalis, and oxygen | | |
| ELECTRICAL PROPERTIES | | |
| **Resistivity**, Ω-cm: 1.6 x $10^{-6}$ | | |
| MORPHOLOGY | | |
| **Particle shape**: dendritic, spherical of spheroidal (water atomized) | | **Aspect ratio**: 1-3 |
| **Particle size**, μm: 1.5-5 | **Sieve analysis**: 325 mesh residue - 0.5% | |
| MANUFACTURERS & BRAND NAMES:<br>AcuPowder, Union, NJ, USA<br>manufacturer of a range of copper, brass, and bronze powders. Ultrafine copper powder 2000, Spherical copper powder A 155 and 500, Bronze powders 5631 and 201 have particle size suitable for thin film applications | | |
| MAJOR PRODUCT APPLICATIONS: conductive plastics and paints | | |
| MAJOR POLYMER APPLICATIONS: epoxy, PP, PA, PE | | |

Copper powder undergoes oxidation when it is contacted with air during cooling process. There are annealed grades available in which the surface oxides are reduced by hydrogen to the pure copper. There are four types of copper powder: electrolytic (irregular porous particles or dendrite shaped aggregates of smaller particles), flake (made by machining), spherical (gas atomized which consists of spherical particles), and spheroidal (water atomized having elongated particles).[258]

## 2.1.26 CRISTOBALITE[259-264]

| | | |
|---|---|---|
| **Names**: cristobalite | | **CAS #**: 14464-46-1 |
| **Chemical formula**: $SiO_2$ | **Functionality**: OH and from silane treatment | |
| **Chemical composition**: $SiO_2$ - 99-99.7%, $Al_2O_3$ - 0.07-0.25%, $Fe_2O_3$ - 0.03-0.05% | | |
| **Trace elements**: Ti, Ca, Na, Mg, K | | |
| **PHYSICAL PROPERTIES** | | |
| **Density**, g/cm$^3$: 2.32 | **Mohs hardness**: 6.5 | **Loss on ignition**, %: 0.15-0.2 |
| **Coefficient of thermal expansion**, 1/K: 54x10$^{-6}$ | | |
| **CHEMICAL PROPERTIES** | | |
| **Chemical resistance**: chemically inert | | |
| **Moisture content**, %: 0.006-0.1 | **pH of water suspension**: 8.5 | |
| **OPTICAL PROPERTIES** | | |
| **Refractive index**: 1.48 | **Brightness**: 91-95 | **Whiteness**: 92-96 |
| **Color**: Y tristimulus value: flour - 90-92, micronized - 95-96 | | |
| **MORPHOLOGY** | | |
| **Crystal structure**: tetragonal | **Oil absorption**, g/100 g: 21-28 | **Hegman fineness**: 5.5-7 |
| **Particle size**, μm: 0-6 (micronized), 0-200 (coarse) | **Specific surface area**, m$^2$/g: 0.4-6.5 | |

**MANUFACTURERS & BRAND NAMES**:

C.E.D. Process Minerals, Akron, OH, USA

Goresil KRS, C-100, C-200, C-325, C-400, 1045, 835, 525, 215, 210 - synthetic cristobalite of varying particle sizes

Quarzwerke, Frechen, Germany

Cristobalite flour M 002, M 006, M 0010, M 3000 - untreated synthetic cristobalite of different particle sizes

Sikron cristobalite flour SF3000, SF4000, SF6000 - micronized untreated cristobalite flours of different particle sizes

Silbond 006 MST, 3000 MST, 3000 RST-M, 4000 MST, 6000 EST, 6000 MST, 6000 RST, 8000 RST, 8000 TST - micronized treated cristobalite flours (EST - epoxysilane, MST - methacrylsilane, RST - trimethylsilane, TST - methyl silane)

**MAJOR PRODUCT APPLICATIONS**: exterior paints, coatings, synthetic plastering compounds, thermoplastic road marking compounds, adhesives, sealants, plastics, abrasives, cables, stucco, kitchen sinks and laminates, dental, military, electronics

**MAJOR POLYMER APPLICATIONS**: epoxy, polyurethane, PMMA, rubber, PVC, unsaturated polyester, silicone, acrylics

Cristobalite is a polymorph of quartz, meaning that it is composed of the same chemistry, $SiO_2$, but has a different structure. Both quartz and cristobalite are polymorphs of quartz group. Cristobalite is not found in sufficient quantities in natural source. For commercial purposes, it is synthetically produced from sand by heating in kiln to 1500°C. The resultant white powder is used as a filler or it is micronized and surface treated. The most important properties of cristobalite are its whiteness and durability on exposure to environmental conditions. Products manufactured by Quarzwerke GmbH are treated with the following silanes: epoxy, methacrylate, trimethyl, and methyl silane.[261-263]

Several essential properties of cristobalite have influence on its applications. They include lower density than quartz (higher volume at the same mass), purity (low catalytic effect on many polymeric systems, excellent properties in exterior coatings due to low level of iron oxide), very low moisture (no need for drying in moisture sensitive systems), pure white color, less abrasive due to filler particle morphology.

## 2.1.27 DIATOMACEOUS EARTH[265-266]

| | | |
|---|---|---|
| **Names**: diatomaceous earth, diatomite | | **CAS #**: 68855-54-9 |
| **Chemical formula**: $SiO_2$ | **Functionality**: OH | |
| **Chemical composition**: $SiO_2$ - 85.5-91.8%, $Al_2O_3$ - 3.2-4.5%, CaO - 0.3-0.6%, $Fe_2O_3$ - 1-1.4%, $K_2O$ - 0.-1.2%, $Na_2O$ - 0.5-3.6%, $TiO_2$ - 0.1-0.2% | | |
| PHYSICAL PROPERTIES | | |
| **Density**, g/cm$^3$: 2-2.5 | **Loss on ignition**, %: 0.1-5 | |
| CHEMICAL PROPERTIES | | |
| **Chemical resistance**: chemically inert | **pH of water suspension**: 6.5-10 | |
| **Moisture content**, %: 0.2-6 | **Adsorbed water**, %: 190-600 | **Water solubility**, %: 0.1-1 |
| OPTICAL PROPERTIES | | |
| **Refractive index**: 1.42-1.48 | **Brightness**: 70-90 | |
| **Color**: white, off white, gray, buff, pink | | |
| MORPHOLOGY | | |
| **Porosity**: 85% (void space), pore size - 1.5-22 μm (in filter aids) | | **Particle size**, μm: 3.7-24.6 |
| **Hegman fineness**: 0-5.5 | **Oil absorption**, g/100 g: 105-190 | |
| **Sieve analysis**: 325 mesh residue - trace to 17.6% | **Specific surface area**, m$^2$/g: 0.7-180 | |
| MANUFACTURERS & BRAND NAMES:<br>Eagle-Picher Minerals, Inc., Reno, NV, USA<br>Celatom Natural Fine Fillers: MN-2, MN-3, MN-4, MN-5, MN-8, LCS-3 natural grades differing in particle size<br>Celatom Flux Fine Fillers: Ultrabloc, Cela-Brite, MW-25, Ultraflat, MW-27, MW-31, MW-32 fillers designed for different applications listed below<br>Celatom line of filtering and polishing media<br>Grefco Minerals, Inc., Torrance, CA, USA<br>Dicalite Natural Diatomite Functional Fillers: 104, CA-3, IG-3, 143, SA-3, 182<br>Dicalite Processed Diatomite Functional Fillers: WF, WFAB, 395, WB-5, L-5, L-10, SP-5, PS, SF-5<br>World Minerals, Inc., Celite, Lompoc, CA, USA<br>Celite 289, 266, 110, 281, 315, 270, 292, 350, White Mist, 499, Super Fine, Super Floss, Snow Floss, HSC - fillers designed for different applications in rubber, paper, paint, polishers, cleaners, catalysts | | |
| MAJOR PRODUCT APPLICATIONS: paints, coatings, rubber, abrasive polishes, cleaning waxes, seed coatings, anticaking agent, antiblock applications, pesticide formulations, asphalt extender, automotive windshields, catalyst support, concrete additive, dental molds, drilling mud, filter papers and pads, specialty papers, paperboard, foundry, waste disposal aids, stucco, battery boxes, plastic film | | |
| MAJOR POLYMER APPLICATIONS: rubber, PE, alkyd, acrylics, silicone | | |

Diatomite is a chalky sedimentary rock composed of skeletal remains of diatomites. Diatomites are single-cell aquatic plants living in the oceans. There is a great variety of diatomites as shown in Figure 2.23. The micrographs show the complicated structure of diatomites which explains their high porosity and thus the effect they have on gelling of liquids and on rheological properties. It is estimated that there are more 25,000 species of diatoms.

Figure 2.23. SEM micrographs of diatomites. *Courtesy of World Minerals, Inc., Lompoc, CA and Grefco Minerals, Inc., Lompoc, CA (figures in the first row).*

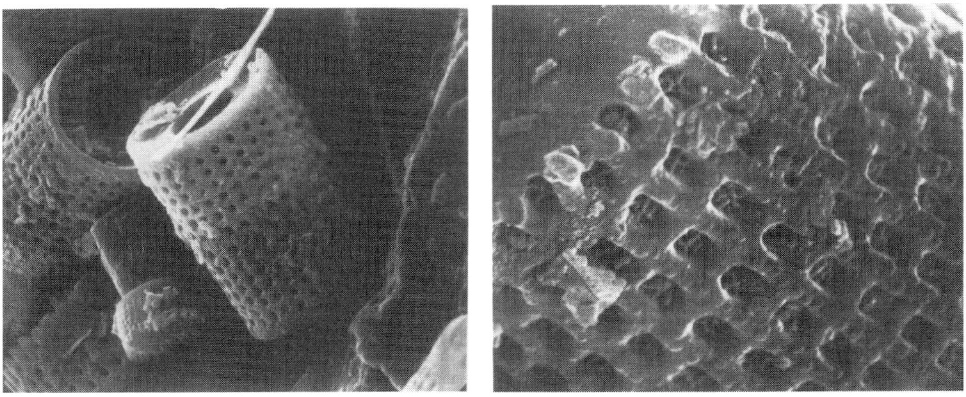

Figure 2.23 (continuation). SEM micrographs of diatomites. *Courtesy of Eagle-Picher, Reno, NV.*

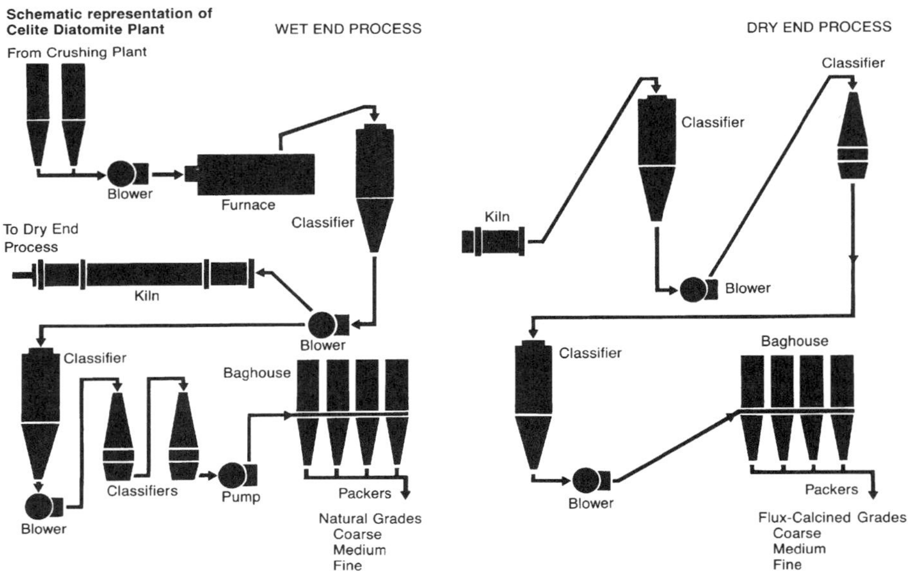


Figure 2.24. Schematic representation of the production process for diatomaceous earth fillers. *Courtesy of World Minerals, Inc., Lompoc, CA, USA.*

Figure 2.24 shows the method of processing of diatomite to different grades of fillers. The natural grades are uncalcinated powders which are crushed and classified according to particle size distribution. In this process moisture is also removed. Natural diatomite contains 40% moisture. In the production of the calcinated and the flux-calcinated products, large kilns are used. The high temperature process causes sintering of the diatom particles to clusters in which the characteristic structures of diatomites are maintained. The process is completed by classification and packaging.[266]

Diatomaceous earth fillers play several roles, such as rheological additives (absorb liquids in pores to increase viscosity on standing and release them on mixing) and flatting agents. They are useful to increase the rate of paint drying (porous filler assists evaporation), to improve sanding properties, to increase mechanical adhesion of coatings, and to reduce the amount of $TiO_2$ needed to produce whiteness or opacity in a material. Due to their chemical inertness, these fillers do not interfere with the other components of the mixture. The grade selected depends on the surface smoothness required, the degree of flatting, and the type of dispersion equipment used. It is also important to chose other co-fillers. It is, for example, known that a combination of diatomaceous earth and talc provides paint with good properties.

## 2.1.28 DOLOMITE[267]

| | |
|---|---|
| **Names**: dolomite | **CAS #**: 16389-88-1 |
| **Chemical formula**: $CaMg(CO_3)_2$ | **Functionality**: none (OH in admixtures) |
| **Chemical composition**: $CaCO_3$ - 55%, $MgCO_3$ - 43%, $SiO_2$ - 0.7%, $Al_2O_3$ - 0.2%, $Fe_2O_3$ - 0.3% | |
| PHYSICAL PROPERTIES | |
| **Density**, g/cm$^3$: 2.85 | **Mohs hardness**: 3.5-4 |
| CHEMICAL PROPERTIES | |
| **Chemical resistance**: reacts with acids | **Moisture content**, %: 0.1 |
| OPTICAL PROPERTIES | |
| **Color**: white, yellow, gray, or brown (if iron is present) | |
| MORPHOLOGY | |
| **Crystal structure**: trigonal | **Cleavage**: three directions forming rhombs |
| MANUFACTURERS & BRAND NAMES:<br>Charles B. Chrystal Co., Inc., New York, USA<br>KF, DF 1000, DF 2000, DF 3000 differing in particle sizes<br>Omya/Plüss-Staufer AG, Oftringen, Switzerland | |
| MAJOR PRODUCT APPLICATIONS: similar to calcium carbonate with exception of food, pharmaceutical and sugar industries | |
| MAJOR POLYMER APPLICATIONS: the same as in calcium carbonate | |

## 2.1.29 FERRITES[268-270]

| | |
|---|---|
| **Names**: ferrites, magnetic fillers | **CAS #**: various |
| **Chemical formula**: Ba/Sr carbonate with ferric oxide, NiZn, MnZn, CuZn, iron silicide, $BaO{\cdot}Fe_2O_3$, $SrO{\cdot}6Fe_2O_3$, $(Mn,Zn)_y(Fe_2O_3)_{2-Zn}$, BaPb, BaSrPb, $Nd_2Fe_{14}B$ | **Functionality**: none |
| **Chemical composition**: variable | |
| PHYSICAL PROPERTIES | |
| **Density**, g/cm$^3$: 2.3-5.1 | |
| CHEMICAL PROPERTIES | |
| **Chemical resistance**: some grades are rust resistant, chemically resistant | |
| ELECTRICAL PROPERTIES | |
| **Dielectric constant**: 8-22 | **Magnetic saturation,** emu/g: 40-109 |
| **Resistivity**, Ω-cm: $10^2$-$10^{10}$ | **Volume resistivity**, Ω-cm: $10^{10}$ |
| MORPHOLOGY | |
| **Particle size**, μm: 0.05-14 | **Oil absorption**, g/100 g: 10.8-14.8 |
| **Aspect ratio**: 1-5 | **Specific surface area**, cm$^2$/g: 210-6000 |

MANUFACTURERS & BRAND NAMES:

Cortex Biochem, San Leandro, CA, USA
- A broad range of biochemical aids used for magnetic separation of biological materials.
- The following lines of products are manufactured: MegaCell (magnetizable cellulose/iron oxide), MagAcrolein (magnetizable polyacrolein/iron oxide), MagaChar (magnetizable charcoal), MagaBeads (magnetizable particles), MagaPhase (ion exchange products)

Steward, Chattanooga, TN, USA
- NiZn Ferrite 72800, 72500
- MnZn Ferrite 73300
- CuZn Ferrite 126800
- Iron silicide Fine, Corse

Wright Industries, Inc., Brooklyn, NY, USA
- Magnetic pigments 5000, 3000, 3006, 4000, 4200, 12672, 112978, 41183

MAJOR PRODUCT APPLICATIONS: plastic magnets, xerographic materials, filters, fibers, energy attenuating powders, microwave absorbing materials

Cortex Biochem has found interesting applications for magnetizable particles in analytical fields. Particles of the analytic aid are prepared from a combination of magnetizable materials (iron oxide) and absorbing material (e.g., charcoal, polyacrolein, ion exchange, cellulose). The particles are dispersed in a biological sample to selectively absorb required compounds. After absorption was accomplished, particles with absorbed substance are removed from solution by a magnetized rod. The materials are used for separation of enzymes, protein, cells or bacteria.

## 2.1.30 FELDSPAR

| | | |
|---|---|---|
| **Names**: feldspar | | **CAS #**: 14808-60-7 |
| **Chemical formula**: (Na or K or Ca)$Al_{1-2}$ $Si_{3-2}O_8$ | **Functionality**: OMe, OH | |
| **Chemical composition**: $SiO_2$ - 68.4-76.8%, $Al_2O_3$ - 14-18.8%, $Fe_2O_3$ - 0.005-0.06%, CaO - 1.1-1.5%, $K_2O$ - 2.8-4.1%, $Na_2O$ - 4.9-6.5% | | |
| **PHYSICAL PROPERTIES** | | |
| **Density**, g/cm$^3$: 2.55-2.76 | **Mohs hardness**: 6-6.5 | |
| **CHEMICAL PROPERTIES** | | |
| **Moisture content**, %: 0.1 | **pH of water suspension**: 8.2-9.3 | |
| **OPTICAL PROPERTIES** | | |
| **Refractive index**: 1.53 | **Brightness**: 90-94 | |
| **Color**: white; L - 96-96.7, a - -0.3 to -0.4, b - 0.5-1.3 | | |
| **MORPHOLOGY** | | |
| **Particle shape**: sub-angular | **Crystal structure**: monoclinic to triclinic | |
| **Particle size**, μm: 3.2-14 | **Oil absorption**, g/100 g: 22-30 | **Hegman fineness**: 0-7 |
| **Sieve analysis**: 325 mesh sieve residue - traces | | **Specific surface area**, m$^2$/g: 0.8-4 |
| **MANUFACTURER & BRAND NAME**:<br>Feldspar Corporation, Atlanta, GA, USA<br>NC-4 - feldspar for ceramic applications<br>Kentucky-Tennessee Clay Company, Mayfield, KY, USA<br>Minspar 3, 4, 7, 10 with particle size decreasing as the grade number increases | | |
| **MAJOR PRODUCT APPLICATIONS**: paints, coatings, plastics, rubber, adhesives, sealants | | |
| **MAJOR POLYMER APPLICATIONS**: alkyd, acrylic, rubber, polyurethanes, epoxy | | |

The feldspar group is a fairly large group with nearly 20 members recognized, but only nine are well known and common. Those few, however, make up the greatest percentage of minerals found in the Earth's crust. The following are some of the more common feldspar minerals: *The plagioclase feldspars*: Albite - sodium aluminum silicate; Oligoclase - sodium calcium aluminum silicate; Andesine - sodium calcium aluminum silicate; Labradorite - calcium sodium aluminum silicate; Bytownite - calcium sodium aluminum silicate; Anorthite - calcium aluminum silicate; *The K-feldspars or alkali feldspars*: Microcline - potassium aluminum silicate; Sanidine - potassium sodium aluminum silicate; Orthoclase - potassium aluminum silicate.

## 2.1.31 GLASS BEADS[271-297]

| | |
|---|---|
| **Names**: glass bead, microspheres, solid beads, hollow microballoons | **CAS #**: 65997-17-3 |
| **Chemical formula**: $SiO_2$ | **Functionality**: OH or depends silane treatment |
| **Chemical composition**: *A-glass*: $SiO_2$ - 72-73%, $Na_2O$ - 13.30-14.3%, $K_2O$ - 0.2-0.6%, CaO - 7.2-9.2%, MgO - 3.5-4%, $Fe_2O_3$ - 0.08-0.2%, $Al_2O_3$ - 0.8-2%; *E-glass*: $SiO_2$ - 52.5%, $Na_2O$ - 0.3%, $K_2O$ - 0.2%, CaO - 22.5%, MgO - 1.2%, $Fe_2O_3$ - 0.2%, $B_2O_3$ - 8.6% | |
| PHYSICAL PROPERTIES | |
| **Density**, g/cm$^3$: 2.46-2.54 (solid), 0.12-1.1 (hollow) | **Mohs hardness**: 6 (A-glass), 6.5 (E-glass) |
| **Softening point**, °C: 704 (A-glass), 846 (E-glass) | **Annealing point**, °C: 548 |
| **Compressive strength**, MPa: up to 70,000 MPa (solid) | **Specific heat**, kJ/kg·K: 1.17 |
| **Young modulus**, GPa: 68.9 (E-glass) | **Poisson ratio**: 0.21 |
| **Coefficient of thermal expansion**: 85x10$^{-7}$ (A-glass), 28x10$^{-7}$ (E-glass) | **Coefficient of friction**: 0.9-1 |
| CHEMICAL PROPERTIES | |
| **Chemical resistance**: resistant to most chemical environments similar to glass | |
| **Silanes used for treatment**: dimethyldiethoxy silane, 3-(methacryloxy) propyltrimethoxy silane, vinyl triethoxy silane, amino silane. Silane coating was estimated to be 0.2 wt%.[277,292] Treatment with epoxy silane has been used[279] followed by PS-maleic anhydride grafting through amine spacer. | |
| **pH of water suspension**: 7-9.4 | |
| OPTICAL & ELECTRICAL PROPERTIES | |
| **Refractive index**: 1.51 (A-glass - soda lime) 1.55 (E-glass - borosilicate) | **Dielectric constant**: 1.2-7.6 |
| **Color**: white or transparent (solid beads) | **Dielectric strength**, V/cm: 4500 |
| **Resistivity**, Ω-cm: 10$^7$ | **Volume resistivity**, Ω-cm: 10$^{12}$-10$^{16}$ |
| MORPHOLOGY | |
| **Particle size**, μm: 7-8 | **Oil absorption**, g/100 g: 17-20; **Wall thickness**, μm: 1-20 |
| **Sieve analysis**: 325 mesh residue - traces to 15% | |
| **DBP absorption**, cm$^3$/100 g: | **Specific surface area**, m$^2$/g: 0.4-0.8 |

MANUFACTURERS & BRAND NAMES:

3M Specialty Additives, St. Paul, MN, USA

Scotchlite Glass Bubbles, General Purpose Series K1, K15, K20, K25, K37, K46, S22, S32, S38, B38, S60 hollow glass bubbles manufactured from soda-lime borosilicate glass (low density beads varying in density in a range from 0.12 to 0.6 g/cm$^3$ which gives a crush strength of 1.7-69 MPa or 250-10,000 psi)

Abrasivos y Maquinara, SA, Barcelona, Spain

Microcel M borosilicate glass spheres in a density range from 0.18 to 0.35 g/cm$^3$ which have a crushing strength of 6-15 MPa or 900-2200 psi

Duke Scientific, Palo Alto, CA USA

Spherical glass material, solid and hollow glass microspheres in ranges of particle size 1-515 μm to be used as standards in scientific studies on sedimentation, separation, insulation, reflection and as spacer pigments

*continued on the next page*

**MANUFACTURERS & BRAND NAMES:**

Grefco Minerals, Inc. Torrance, CA, USA

Dicaperl HP-110, HP-210, HP-510, HP-710, HP-910 two series of these hollow glass bubbles are produced: standard - 10 series and high performance - 20 series which has one of the two proprietary coatings used to increase adhesion (the density is in the range of 0.18-0.25 g/cm$^3$ which indicates that these are light weight bubbles)

JB Company, Franklin, NJ, USA

a specialty products manufacturer produces a variety of solid glass beads, clear and colored, used for decorative purposes in industrial products

MO-SCI Corporation, Rolla, MO, USA

produces a range of specialty products such as Indentisphere (glass microspheres which can be identified by their fluorescent, magnetic or radioactive properties, e.g. identification of explosives); Duraspheres (borosilicate glass spheres for pharmaceutical applications and electronics); Bioactive Glass (restorative purposes in medical and dental applications)

Potter Industries, Inc., Valley Forge, PA, USA

Spheriglass, A-glass 1922, 2024, 2227, 2429, 2530, 2900, 3000, 4000, 5000, 6000 (the higher the number, the smaller particle size in the range of 7-203 μm)

Spheriglass, E-glass 3000E, 4000E, 5000E, 6000E (the higher the number, the smaller particle size in the range of 7-35 μm)

Potter Industries developed the following surface coatings: CP-01, CP-02, CP-03, CP-26.

Glass spheres are offered with a coating for the polymer to which the spheres are to be added.

Spheriglass beads have densities up to 1.08 g/cm$^3$ and they can withstand pressure of 207 MPa (30,000 psi)

Sphericel 110P8 - hollow borosilicate glass spheres developed for paints and thermoplastic molding applications

Sovitec France SA, Florange, France

Micropearl 50, 90, 50100, 1020 - soda-lime glass, solid glass microspheres of different grain sizes in the range of 20-212 μm. Company manufactures the above grades with three surface finishes 215, 216, 217 which are different coupling agents selected for different types of thermoplastic and thermosetting resins. Numerous applications in plastic industry are documented by the results characterizing performance.

The PQ Corporation, Valley Forge, PA, USA

Q-Cel hollow spheres 300, 2116, 2106, 692OL, 636D, 640D, 6717, 7019, 5043 beads differ in density and particle size with the general trend being that larger beads are lighter (low density beads in the density range 0.19-0.48 g/cm$^3$ and working pressures in the range 1.7-21 MPa or 250-3000 psi)

**MAJOR PRODUCT APPLICATIONS**: bowling balls, cast polyester, foam, caulk, explosives, putties, sealants, pipe insulation, potting compounds, speckling compounds, reflective paints, golf balls, pultrusion, aerospace, marine, automotive, composites, and many more

**MAJOR POLYMER APPLICATIONS**: PVC, polyester, polyurethane, epoxy, acrylics, POM, ABS, PA, PC, PE, PI, PMMA, PPO, PP, PS, PSF, melamine, phenoxy, silicone

The processing technology determines the selection of the glass bubbles. To minimize the breakage of bubbles they should be added at the end of the process. Low shear and high flow mixers are required to obtain full benefits. The following mixers are suitable: double planetary, planetary, propeller, flat blade, sigma. The following mixers should not be used for thin wall bubbles: high speed disperser, colloid mill, three roll mill, homogenizers, and impingement mixers. The pumping of materials containing glass bubbles must be carefully controlled. The hydrostatic pressure generated by the pump should be lower than the maximum hydrostatic pressure which glass bubbles can withstand. The clearance between intermeshing gears in gear pumps must be greater than the bubble diameter. The following pumps are suggested by 3M: double diaphragm, piston, progressive cavity and rotary pumps. Similar consideration of pressure should be given to the conditions of extrusion and injection molding. The crush strength of hollow beads manufactured by Potter Industries exceeds 200 MPa (30,000 psi) which is considered sufficient to survive injection molding and high shear mixing equipment. Further discussion of the relationship between crushing strength and density can be found below.

According to Potters Industries, A-glass is suggested for the majority of polymers with the exception of acetal and PTFE where E-glass should be used. The reason for surface modification of glass beads is explicitly illustrated in Figure 2.25. Coated spheres adhere to the matrix but uncoated spheres are easily delaminated from the matrix. Adhesion depends on the selection of coating for a particular matrix polymer.

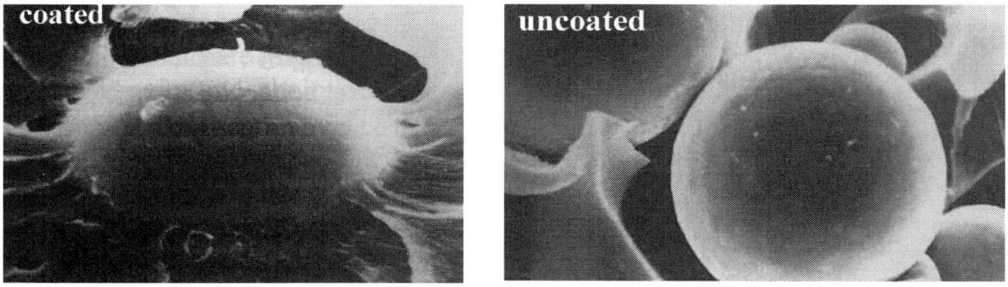


Figure 2.25. Coated and uncoated spheres in polymer matrix. *Courtesy of Potters Industries, Inc., Valley Forge, PA, USA.*

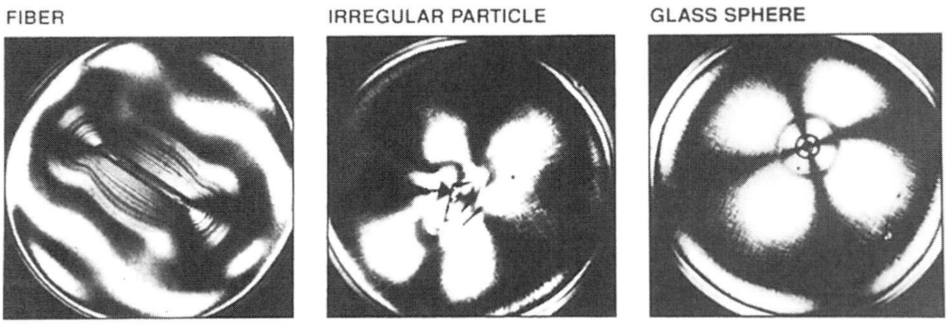


Figure 2.26. Stress distribution around fiber, irregular particle, and glass sphere. *Courtesy of Potters Industries, Inc., Valley Forge, PA, USA.*

Figure 2.26 shows one of the reasons why spherical fillers give good performance in compounded materials. The birefringence patterns show stress distribution in the vicinity of various shapes of inclusions – only with a spherical shape and a good adhesion to the matrix, uniform stress distribution is observed. Stress distribution is an essential element of material design.

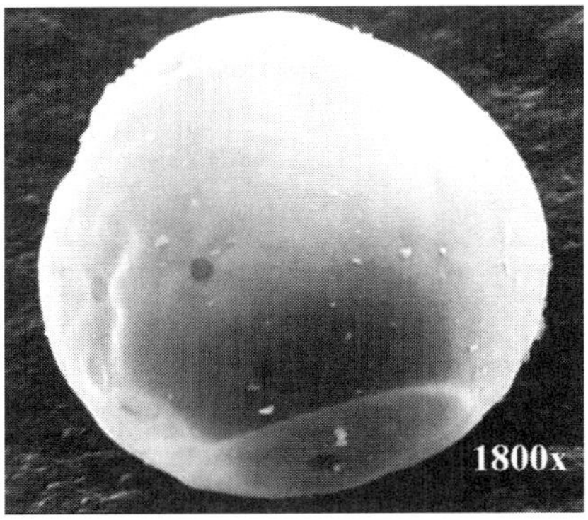


Figure 2.27. Dicaperl HP-510. Magnification 1800x. *Courtesy of Grefco Minerals, Inc., Torrance, CA, USA.*

Glass beads improve or control several properties of materials. These include density reduction, flow properties, viscosity decrease, rheological properties including thickening and non-sag properties, nailing, sanding, shrinkage reduction, impact strength, stiffness, tensile strength, flexural strength, and hardness, explosives performance, and acoustical properties.

The most useful feature of glass beads is their ability to reduce the density of a product. There is a trade off between the mechanical properties of beads and their density. If beads are very light they are also very fragile because their walls are very thin and the types of products and manufacturing methods are limited. But if they can be successfully incorporated they result in a substantial reduction in product density. If the beads are mechanically resistant they have thicker walls and do not reduce density of neat polymers. The density of hollow spheres available in the market varies from 0.12 to 1.1 g/cm$^3$. This means that glass occupies from about 10 to 50% volume of the bead which results in considerable differences in their mechanical performance. The data on the densities and crash strength for individual brands are given in table of manufacturers and brand names.

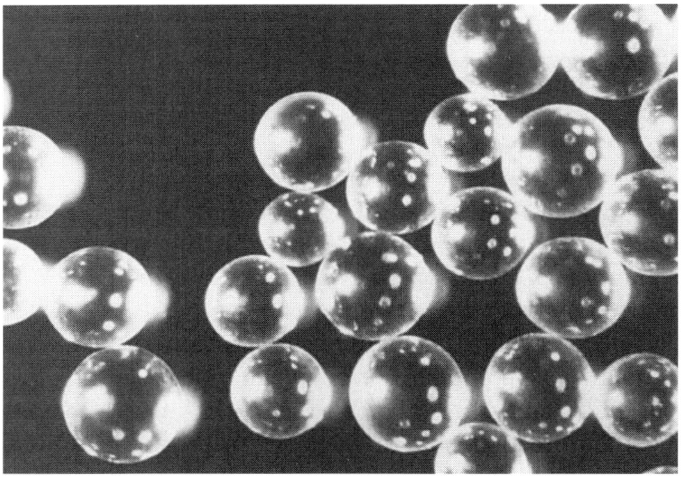

Figure 2.28. Micropearl solid glass beads. *Courtesy of Sovitec France SA, Florange, France.*

Figure 2.27 shows the morphology of a single hollow glass sphere which has a regular spherical shape. Figure 2.28 shows the morphology of solid glass beads.

## 2.1.32 GOLD[298-300]

| | | |
|---|---|---|
| **Names**: gold powder, gold flakes, gold spheres | | **CAS #**: 7400-57-5 |
| **Chemical formula**: Au | **Functionality**: none (possible thiol derivatization)[300] | |
| **Chemical composition**: Au - 99.96% | | |
| **Trace elements**: Cu, Fe, Pd, Ag | | |
| **PHYSICAL PROPERTIES** | | |
| **Density**, g/cm$^3$: 18.8 | **Mohs hardness**: 2.5 - 3 | **Melting point**, °C: 1064 |
| **MORPHOLOGY** | | |
| **Particle size**, μm: 0.8-9 | **Crystal structure**: isometric | **Cleavage**: absent |
| **Specific surface area**, m$^2$/g: 0.05-0.8 | | |
| **MANUFACTURER & BRAND NAMES**:<br>Shoei Chemical, Inc., Japan<br>Technic, Inc., Woonsocket, RI, USA<br>Gold Powder 507, 508, 509, 510 - chemically precipitated, spherical powder for thick conductive inks<br>Gold Flake/Sphere 550 (thick), 555 (thin flake) - chemically precipitated flakes for conductive inks<br>Gold Flake 552, 554, 560 precipitated/mechanically worked flakes for conductive inks and adhesives | | |
| **MAJOR PRODUCT APPLICATIONS**: conductive inks, coatings, and adhesives | | |

Figure 2.29 shows morphology of gold powder and gold flakes.

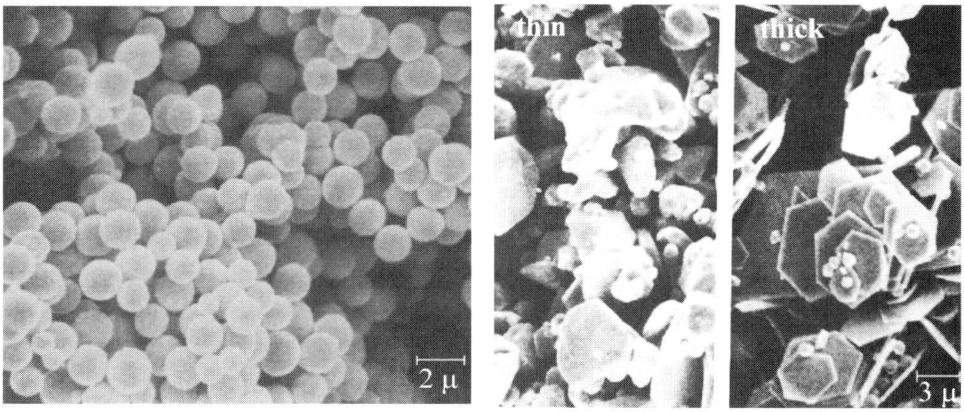


Figure 2.29. Gold powder (magnification 5250x) and thin (550) and thick (555) flakes (magnification 3200x). *Courtesy of Technic, Inc., Woonsocket, RI, USA.*

## 2.1.33 GRAPHITE[301-309]

| | |
|---|---|
| **Names**: graphite, natural graphite | **CAS #**: 7782-42-5 |
| **Chemical formula**: C | **Functionality**: OH |
| **Chemical composition**: carbon 80-99.97% | |
| **Ash content**, %: $SiO_2$ - 48.8, $Al_2O_3$ - 20.8, $Fe_2O_3$ - 22.2, MgO - 2.3, CaO - 1.8, $Na_2O$ - 0.4, $K_2O$ - 2.2, $TiO_2$ - 0.5 | |
| **PHYSICAL PROPERTIES** | |
| **Density**, g/cm$^3$: 2-2.25 | **Mohs hardness**: 1 - 2 |
| **Coefficient of friction**: 0.1-0.6 | |
| **Thermal conductivity**, W/K·m: 110-190 | |
| **CHEMICAL PROPERTIES** | |
| **Moisture content**, %: 0.1-0.5 | **Ash content**, %: 0.03-20 |
| **OPTICAL & ELECTRICAL PROPERTIES** | |
| **Color**: gray | **Resistivity**, mΩ-cm: 0.8-2.5 |
| **MORPHOLOGY** | |
| **Particle size**, μm: 6-96 | **Crystal structure**: hexagonal |
| **Oil absorption**, g/100 g: 75-175 | |
| **Crystallite height**, nm: 60-100 | **Cleavage**: perfect in one direction |
| **Specific surface area**, m$^2$/g: 6.5-20 | **Interlayer distance**, nm: 0.3354-0.336 |

**MANUFACTURERS & BRAND NAMES**:

AML Industries, Inc., Warren, OH, USA
- Natural Graphite Powder - Amlube 611
- High Purity Graphite Powder - Amlube 610, 613

Applied Carbon Technology, Somerville, NJ, USA
- Three types of graphite are manufactured: natural graphite (grades A to H differing in purity and particle size), synthetic graphite (very pure L101 and high ash J101), and amorphous graphite (P100 & P103)

Superior Graphite Co., Chicago, IL, USA
- Several lines of graphite products: amorphous graphite, crystalline flake graphite, crystalline vein graphite, Desulcu, synthetic graphite, ThermoPure. The particle sizes of these graphites are from μm to mm.

Timcal Ltd., Sins, Switzerland
- Timrex KS 6, KS 10, KS 15, KS 25, KS 44, KS 75 graphites of irregular spheroid particle shape used in plastic materials
- Timrex T 44, T 75, T150 angular, flake microporous graphite
- Timrex SG 6, SFG 15, SFG 44, SFG 75 strong anisometric flakes, needles

**MAJOR PRODUCT APPLICATIONS**: extruded profiles, batteries, conductive coatings, brake linings and clutch facings, catalysts, lubricants, self-lubricating parts, pump elements, drive shafts, thrust rings

**MAJOR POLYMER APPLICATIONS**: PA-6, PA-66, PP, PS, LDPE, EPR

Graphite is used in products for the following reasons: conductivity, EMI shielding, lubricating coatings, self-lubricating bearings, lubricants, heat, chemical, and water resistance, flame retardancy, release properties, pigmentation.

Purity, crystalline structure, texture, and particle size are factors which control tribological, thermal, electrical, chemical, and physical properties of products manufactured with graphite.[309-310] Purity can be assessed based on ash content, moisture, and trace elements. For lubricating materials, silicon carbide-free graphite is demanded, because silicon carbide is a highly abrasive material. Such grades are produced by synthetic methods. Superior Graphite Co. patented a high temperature furnace technology which can make graphite having 99.97% carbon. Also Timcal offers grades of similar purity. The following analysis of the effect of graphite is made based on the data from a broad application studies conducted by Timcal.[309-310]

Self-lubricating properties were assessed based on studies of polyamide-6 and polystyrene. The friction coefficient was reduced by 30% with the addition of 30% graphite with only a small increase in wear. The friction coefficient of plastic filled with graphite depends on the purity and the crystallinity of graphite but it also depends on the concentration of graphite. In polystyrene both friction coefficient and the wear decreased as the graphite content was increased up to the peak level of 30%. Further increase in graphite concentration contributed to the increase in both wear and the coefficient of friction. Similar observations for PTFE/graphite system were explained by an increase in the porosity of the composite when it contains more than 30 wt% graphite. It is the porosity that is responsible for an increased wear rate.[304] Glass fiber reinforced SMC and BMC compounds are particularly abrasive. The ratio of graphite to fibers and the overall content must be optimized to achieve a reduction in both wear and friction coefficient.

The mechanical properties of graphite filled plastic can be tailored to meet requirements. The studies on PA-6 show that an addition of graphite increases hardness only slightly (10%). But, the hardness of LDPE can be increased by 25%. If hardness must be increased, a smaller particle size graphite should be selected. The Young's modulus of LDPE can be tripled by the addition of up to 30% graphite. A similar addition to polyamide-6 doubled its Young's modulus. Again smaller particle size graphite is more effective.

Graphite had little influence on the tensile properties of most, but not all grades of LDPE gave the same results. The tensile strength of PA-6 is reduced by the addition of graphite but small particle sized grades have less effect on tensile strength. Elongation of LDPE, similar to other polymers, is reduced as graphite concentration increases but there is more drastic decrease in the case of PA-6 and PA-66. The impact strength of PA-6 and PA-66 is rapidly reduced by an addition of 20-30 wt% graphite. In the case of polypropylene, not only Young's modulus increased by up to 60% by an addition of 30-35 wt% graphite but also its tensile strength was improved. Fine graphite grades improve these properties more rapidly. Impact strength and elongation of PP are decreased in a manner similar to PA. PS is another example of a polymer whose tensile strength is increased by the addition of graphite (~25%) and its Young's modulus is tripled. The elongation of

PS is not significantly reduced but only because PS has very low elongation. Unlike other polymers, the hardness of PS is reduced by the addition of graphite but its impact strength follows the same pattern of being rapidly reduced as the concentration of graphite increases.

The processability of polymers can be improved by addition of graphite. The melt flow index of PS containing graphite gradually decreases as graphite concentration increases even up to 50 wt% graphite. PP gives the same relationship with finer grades decreasing melt flow index more rapidly than the coarse ones. A similar, but less pronounced, effect is observed in LDPE.

The viscosity increase depends on particle size. Smaller particles increase the viscosity of the dispersion more rapidly but there is a big difference between the effect of graphite and carbon black on viscosity. It requires three times as much graphite as carbon black for a similar increase in viscosity.

Antistatic and conductive compounds can be manufactured with graphite. Electrical properties are also very stable. This was determined in a 2 year study during which time the volume resistivity of the graphite containing compound did not change. It is important in formulating these products to consider the effect of other fillers which may be present in the formulation. It was found that the surface resistivity of graphite filled compounds containing large particle sized calcium carbonate or aluminum hydroxide was reduced. The graphite particles should always be kept as small as possible. EPR can be formulated with a high concentration of graphite without losing its flexibility, therefore it is possible to produce flexible electrodes with a surface resistivity of only 1 Ohm.

Graphite helps to improve thermal conductivity and it also helps to process materials faster. Thermal conductivity is improved to a greater extent by graphite of small particle size and high crystallinity.

The addition of graphite to polymeric systems increases their rate of crystallization due to an increased nucleation rate. This increases the molding and extrusion throughput.

Figure 2.30 shows the morphology of graphite which is built up from thin layers of irregularly shaped material.

Figure 2.30. SEM micrograph of Timrex KS 15. *Courtesy of Timcal Ltd., Sins, Switzerland.*

## 2.1.34 HYDROUS CALCIUM SILICATE

| | | |
|---|---|---|
| **Name**: hydrous calcium silicate | | **CAS #**: 14567-73-8 |
| **Chemical formula**: $SiO_2 \cdot CaO \cdot H_2O$ | **Functionality**: OH | |
| **Chemical composition**: $SiO_2$ - 47-49%, CaO - 31-32%, $Al_2O_3$ - 2.3-2.5%, $Fe_2O_3$ - 0.7-0.8%, MgO - 0.6-0.7%, $Na_2O+K_2O$ - 1.2-1.3% | | |
| PHYSICAL PROPERTIES | | |
| **Density**, g/cm$^3$: 2.6 | **Loss on ignition**, %: 14.9-15 | |
| CHEMICAL PROPERTIES | | |
| **Moisture content**, %: 5.5-5.8 | **Adsorbed moisture**, %: 220-550 | **pH of water suspension**: 8.4-9 |
| OPTICAL PROPERTIES | | |
| **Refractive index**: 1.55 | **Color**: gray, white, off-white | **Brightness**: 55-90 |
| MORPHOLOGY | | |
| **Particle size**, μm: 9 | **Oil absorption**, g/100 g: 290 | **Hegman fineness**: 2 |
| **Sieve analysis**: 325 mesh residue 2-8 | **Specific surface area**, m$^2$/g: 95-180 | |
| MANUFACTURER & BRAND NAMES:<br>World Minerals, Inc., Lompoc, CA, USA<br>Micro-Cel A, C, E, T-21, T-26, T-38, T-49, Celkate, Silasorb synthetic fillers obtained from diatomaceous earth and lime in the process discussed below | | |
| MAJOR PRODUCT APPLICATIONS: absorbents, carriers, flatting agents, $TiO_2$ extenders, decolorizers | | |

Hydrous calcium silicate is produced by hydrothermal reaction of diatomaceous earth, hydrated lime, and water. Figure 2.31 gives schematic representation of the process. The product is a material which can absorb 5.5 times of its weight of water.

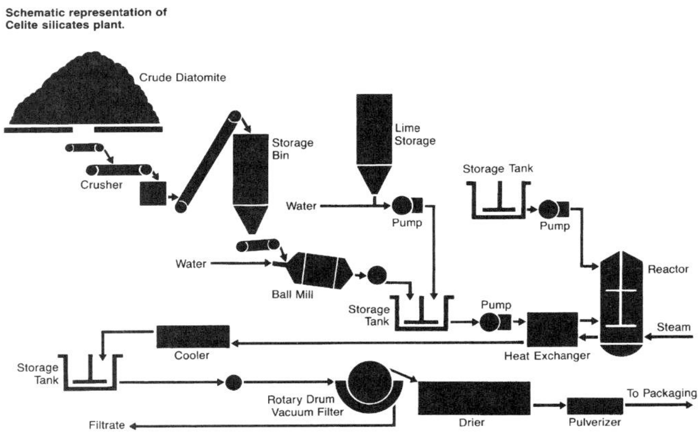


Figure 2.31. Schematic diagram of production of Micro-Cel. *Courtesy of World Minerals, Inc. Lompoc, CA, USA.*

## 2.1.35 IRON OXIDE[310-314]

| | | |
|---|---|---|
| **Names**: iron oxide | | **CAS #**: 1332-37-2 |
| **Chemical formula**: $Fe_2O_3$ | **Functionality**: none | |
| **Chemical composition**: $Fe_2O_3$ - 80-99.5%, $SiO_2$ - 0.03-8%, $CaCO_3$ - 0-5%, $Al_2O_3$ - 0-2%, MgO - 0-2% | | |
| **Trace elements**: Pb, Ni, Cr, Sn | | |
| PHYSICAL PROPERTIES | | |
| **Density**, g/cm$^3$: 4.5-5.8 | **Mohs hardness**: 3.8-5.1 | **Loss on ignition**, %: 3-5 |
| CHEMICAL PROPERTIES | | |
| **Moisture content**, %: 0.1-3 | **pH of water suspension**: 7-9 | |
| OPTICAL PROPERTIES | | |
| **Refractive index**: 2.94-3.22 | **Color**: red, purple, gray, brown (nanosize) | |
| MORPHOLOGY | | |
| **Particle size**, μm: 0.8-10 (26 nm - nanoparticles) | **Specific surface area**, m$^2$/g: 30-60 (nanosize) | |
| **Sieve analysis**: 325 mesh residue from traces to 10% | | **Oil absorption**, g/100 g: 10-35 |
| MANUFACTURERS & BRAND NAMES:<br>Charles B. Chrystal, New York, NY, USA<br>#3 Iron Oxide - coloring pigment for paints and flooring<br>High Purity Iron Oxide - 99% active component, small particle size easy to disperse<br>Miox AS - micaceous iron oxide for primers and coatings<br>Grade W - high purity iron oxide (99% active component) for cement coloring<br>Crocus Martis - polishing grade<br>Nanophase Technologies Corporation, Burr Ridge, IL, USA<br>NanoTec Iron Oxide - nanosize grade | | |
| MAJOR PRODUCT APPLICATIONS: pigment in many materials, coatings, paints, plastics, nanocomposites | | |
| MAJOR POLYMER APPLICATIONS: alkyd, acrylic, polyurethane, epoxy, PP | | |

Iron oxide is a particular example of the wide range of materials which can be obtained from grinding the natural product or synthesis. Figure 2.32 shows the morphology of nanoparticle iron oxide.

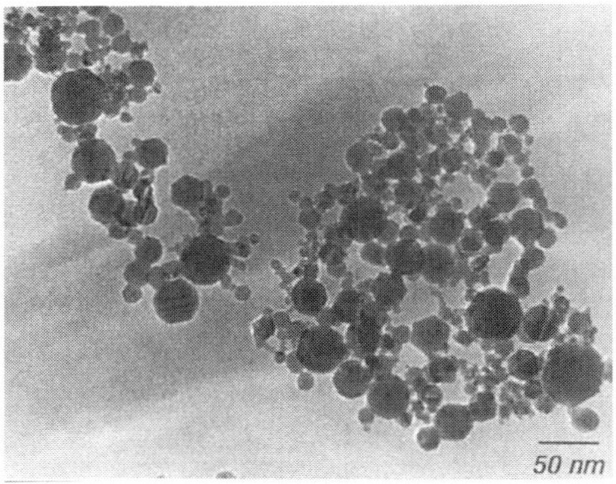

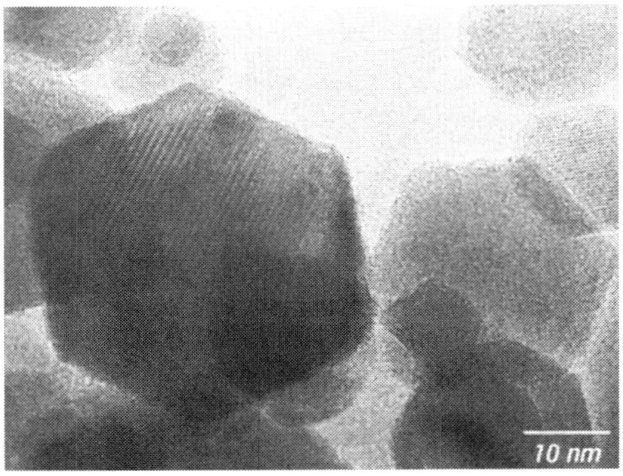


Figure 2.32. TEM micrographs of NanoTec iron oxide. *Courtesy of Nanophase Technologies Corporation, Burr Ridge, IL, USA.*

## 2.1.36 KAOLIN[315-331]

| | |
|---|---|
| **Names**: kaolin - classified, beneficiated, calcinated, aluminum silicate, calcinated silicate, china clay, soft kaolin, hydrated aluminum silicate, kaolinite | **CAS #**: 66402-68-4 |
| **Chemical formula**: $Al_2O_3 \cdot 2SiO_2 \cdot 2H_2O$ | **Functionality**: OH, silane modification |
| **Chemical composition**: $SiO_2$ - 38.5-63%, $Al_2O_3$ - 23-44.5%, $Fe_2O_3$ - 0.2-1%, $TiO_2$ - 0.2-1.9%, $K_2O$ - 0.8-1% | |
| **Trace elements**: Pb, As | |
| PHYSICAL PROPERTIES | |
| **Density**, g/cm$^3$: 2.58-2.62, 2.5-2.63 (calcinated) | |
| **Mohs hardness**: 2, calcinated 4-8 | **Melting point**, °C: 1800 |
| **Loss on ignition**, %: 12.1-14.2, 0.23 (calcinated) | **Specific heat**, kJ/kg·K: 4 |
| CHEMICAL PROPERTIES | |
| **Moisture content**, %: 1-2 (up to 7%), slurry 20-30% | **pH of water suspension**: 3.5-11 |
| OPTICAL & ELECTRICAL PROPERTIES | |
| **Refractive index**: 1.56-1.62 (calcinated 1.62) | **Whiteness**: 88-91 |
| **Color**: white, cream; L* - 95.04-95.70, a* - 0.11-0.30, b* - 5.25-6.4 | **Dielectric constant**: 1.3-2.6 |
| **Brightness**: 69-90 (classified), 85-91 (beneficiated), 84-95 (calcinated) | |
| MORPHOLOGY | |
| **Particle shape**: platy; **Crystal structure**: hexagonal | **Particle size**, μm: 0.2-7.3 |
| **Oil absorption**, g/100 g: 27-48 (classified), 50-60 (beneficiated), 45-120 (calcinated) | |
| **Sieve analysis**: 325 mesh residue - 0.01-2 | **Specific surface area**, m$^2$/g: 8-65 |
| **Hegman fineness**: 3-7 | |

MANUFACTURERS & BRAND NAMES:

Albion Kaolin Co., Hephzibazh, GA, USA
- Albion AP-750 H, AP-750 L, AP-750 M, H-007, S-60, S-75 - adhesives, caulks, sealants, soft rubber products
- Alkoat Plus Slurry, Plus-L Slurry - latex based slurry in high brightness applications
- Britefil 80 Pulverized and Slurry - paper and water-based paints
- Royale Slurry - kaoline dispersed with sodium polyacrylate

Burgess Pigment, Sandersville, GA, USA
- Fine Particle Size #10, #17, #20, #40, #60, Polyclay, Thermo Glace H - hydrous aluminum silicate
- Ultrafine Particle Size #27, #28, #97, #98 - hydrous aluminum silicate. #27 and #28 are spray dried versions for use in water-based systems only
- Air Floated - #80, #86, HC-77 - hydrous aluminum silicate
- Calcinated grades - Icecap K, Iceberg, #30
- Thermo-optic grades - Optiwhite, Optiwhite MX, Optiwhite P, 30 P (see process discussion below)
- Burgess 2212, 2227 - calcinated kaolin surface treated with amino silane

*continued on the next page*

**MANUFACTURERS & BRAND NAMES:**

Charles B. Chrystal Co, Inc., New York, NY, USA
- China Clay Lion - soft kaolin for use in cosmetics, pharmaceuticals, rubber, paint, paper
- Calcinated Kaolin Clay - high brightness
- Plus White Kaolin - paper industry grade
- Electros Kaolin USP - pharmaceutical and cosmetics grade
- Kaolin SIM 90 - exceptional brightness without bleaching

D.J. Enterprises, Inc., Cleveland, OH, USA
- Sillum 200 QP

ECC International, St. Austell, UK
- Supreme, Speswhite, Stockalite, Devolite, Grade B, D, E, Polwhite, GTY - kaolin grades differing in particle size
- PoleStar 200R, 400A, 501 - calcinated kaolins
- Polarite 102A, 103A, 503A - calcinated kaolins, silane coated
- Infilm Clay Range - produced to individual customer requirements

Engelhard Corporation, Iselin, NJ, USA
- ASP 072, 101, 102, 170, 200, 400P, 600, 602, 672, NC, Buca, Catalpo - hydrous aluminum silicate, spray-dried or highly pulverized powders. ASP 101 is stearate coated and ASP NC is delaminated
- Santitone 5, 5HB, Special, SP-33, Whitetex - calcinated kaolins
- Translink 37, 77, 445, 555, HF-900 - calcinated and surface modified: vinyl functionality - 37, 77, amino functionality - remaining grades

Evans Clay Company, McIntyre, GA, USA
- Snofil, Snofil Plus, Hi White - clays for paper industry of different particles sizes
- Snobrite, Snobrite Special, Snobrite PG, Apex, Kaolloid, Hi White R - adhesive, caulk, paint, roofing, rubber grades
- Snobrite slurry - paper, adhesive, paint roofing

J.M. Huber Corporation, Macon, GA, USA
- Polyplate P, P01, 90, HTM - delaminated water washed kaolin grades for water-based coatings. P, 90 and HTM grades are spray dried
- Polygloss 90 - water washed kaolin with ultrafine particles and high brightness
- Huber 35, 35B, 80, 80B, 90, 90B, HG90 - water washed kaolins for water-based systems (all grades) and solvent-based systems (all but with symbol B which means that it disperses only in water). HG means that kaolin was spray-dried.
- Huber 683, 40C, 70C, 90C - structured pigment (683) and calcinated grades (letter C) for water-based and solvent paints and coatings

Kentucky-Tennessee Clay Company, Mayfield, KY, USA
- Suprex, Alumex, Supreme, Rogers - kaolins from two different locations in SC and GA

R.T. Vanderbilt Company, Inc., Norwalk, CT, USA
- Bilt-Plates 145, 156 - primers and paints and unbleached kraft liner board
- Continental Clay - carrier for agricultural chemicals
- Dixie Clay - coatings, primers, crack fillers, caulk
- Langford Clay - low cost reinforcing filler for elastomers
- McNamee Clay - low cost reinforcing filler for elastomers
- Par Clay, Par RG Clay - reinforcing and inert filler for elastomers
- Peerless Clay #2 - crack fillers, traffic and barn paint, floor covering, primer, caulk

Sachtleben Chemie, Duisburg, Germany
- Sachtosil CF, PV - controlled process results in synthetic-like material used as antiblocking additive in films

**MAJOR PRODUCT APPLICATIONS:** cosmetics, pharmaceuticals, rubber, tire, paint, coatings, paper, agriculture, floor covering, crack fillers, primers, films, wire and cable, electrical accessories, can sealants, roofing membranes, syringes, coated fabrics, tennis balls, urethane sealants, foam, gaskets, footwear

**MAJOR POLYMER APPLICATIONS:** alkyd, cellulose, rubber, polyurethanes, PVC, PE, EPDM, EPR, PA, PP

Kaolin is a product of the decomposition of granite and white feldspar. The typical feature of kaolin is extreme fineness. Over the last two centuries, China clay become so popular that it is now the largest export item from the United Kingdom after North Sea oil.

Production of China clay begins with large scale mining. The mined mineral is one part China clay, 3 parts rock, 4 parts sand and one part mica. The other components are separated and either utilized or discarded. After removing of rock material, the remainder is mixed with water and passed through a hydrocyclone to remove fine sand and coarse mica. The further refining process includes thickening of materials obtained from the hydrocyclones through flocculation of particles and their subsequent separation from water in large overflow tanks. In the next stage, fine mica is removed either in hydroseparators or hydrocyclones. To further improve the quality of clay, magnetic separation is applied which removes such minerals as mica, iron oxide, and tourmaline.

From this point the clay becomes suitable for some applications but for others, it must be still refined. Clay classification is one of these stages of refining. For example, paper grades must be very fine and here a centrifugal classifier is usually used to separate finer particles. Some grades are bleached to increase their whiteness. Bleaching can be done by ozone gas or sodium hydrosulfite. Other grades are subjected to grinding. Figure 2.33 shows stacks of china clay before grinding. The grinding process reduces the size and delaminates the stacks resulting in a finer product.

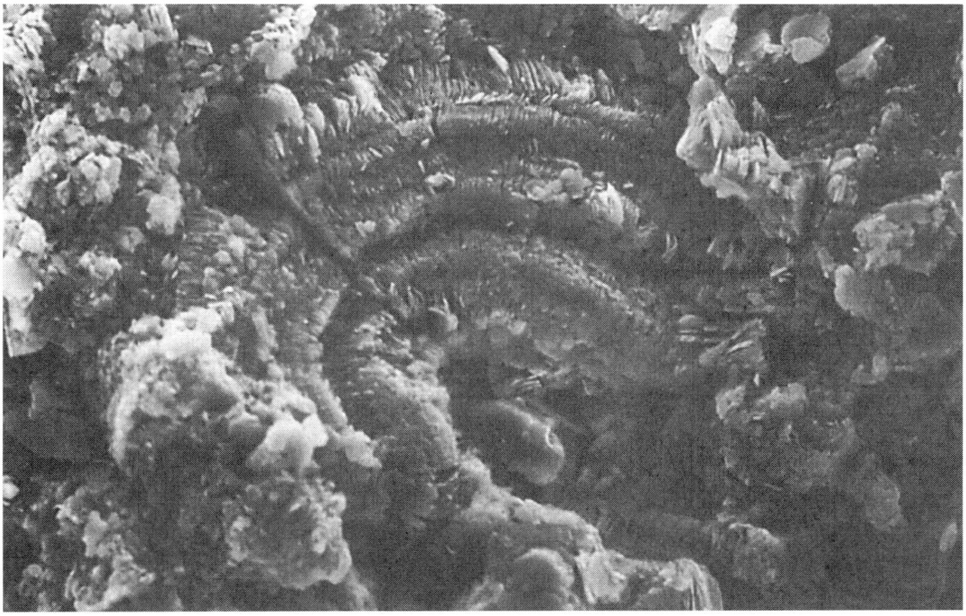

Figure 2.33. SEM micrograph of china clay before processing. *Courtesy of ECC International St. Austell, UK.*

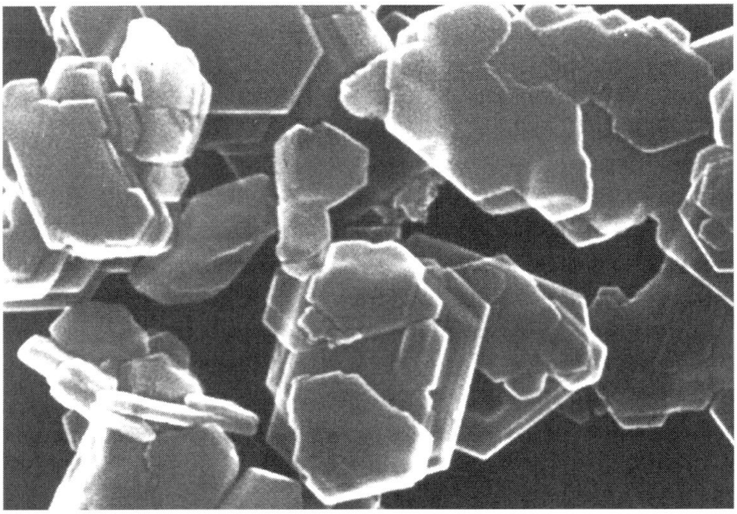

Figure 2.34. SEM micrograph of kaolin. *Courtesy of ECC International St. Austell, UK.*

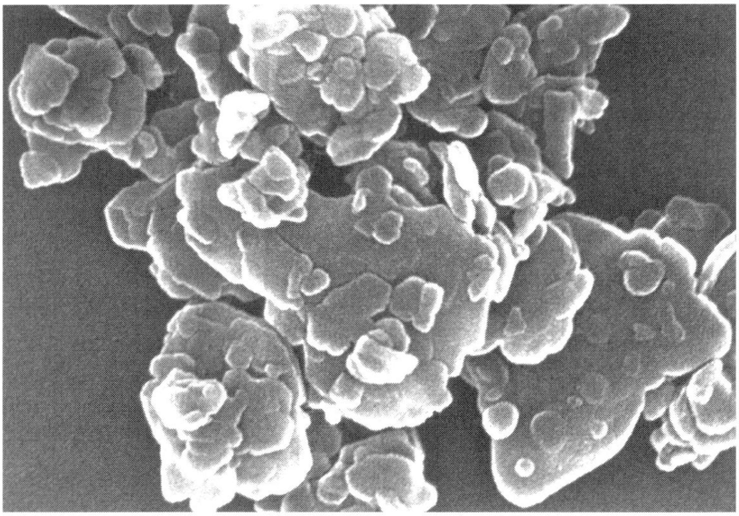

Figure 2.35. SEM micrograph of calcinated kaolin. *Courtesy of ECC International St. Austell, UK.*

Some product is sold in slurry which is a convenient form since it eliminates dust, saves energy, and lowers the cost. The industries which are frequent users of such product are paper and paints. Many other applications require material to be in a powder form, therefore the slurry is flocculated, concentrated (filter presses), and dried. Several dryer types are used such as rotary, tray, fluidized bed process or spray. The clay may be pulverized after some of these drying process depending requirements. Figure 2.34 shows the morphology of kaolin. A typical platy structure is clearly displayed on this photograph.

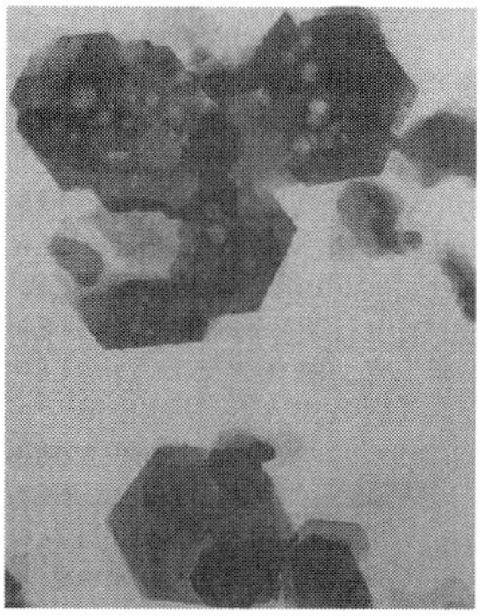

Figure 2.36. SEM micrograph of Optiwhite, thermo-optic grade. *Courtesy of Burges Pigment, Sandersville, GA, USA.*

The process of calcination considerably changes the original properties of the material (see table above). Heating of kaolin above 450°C alters the clay structure and improves electrical resistance and brightness. The process of calcination is conducted in kilns at temperatures between 850 and 1500°C. Figure 2.35 shows calcinated kaolin which differs from dried kaolin by having round edges which is a result of the high temperature treatment.

Burgess Pigment have developed yet another method of kaolin treatment called flash calcination process. The process is conducted by a whirling upward rising stream of hot gas in the form of vortex in which material is dehydrated in a matter of seconds forming the unique morphological structure and a given the grade name "thermo-optic" (Figure 2.36). This material has lower specific gravity and very good hiding power.

Figure 2.37. SEM micrograph of Huber – structured pigment. *Courtesy of J.M. Huber Corporation Macon, GA, USA.*

Huber shows another morphological features of its structured pigment product which is in the form of porous aggregates with high brightness (Figure 2.37). Particles are composed of stacks which form aggregates closer in shape to spherical particles.

## 2.1.37 LITHOPONE

| | | |
|---|---|---|
| **Name**: lithopone | | **CAS #**: 1345-05-7 |
| **Chemical formula**: $ZnS \cdot BaSO_4$ | **Functionality**: none | |
| **Chemical composition**: ZnS - 29-59%, $BaSO_4$ - 70-40%, ZnO - 1% | | |
| PHYSICAL PROPERTIES | | |
| **Density**, g/cm$^3$: 4.2-4.3 | **Mohs hardness**: 3 | |
| CHEMICAL PROPERTIES | | |
| **pH of water suspension**: 7-8 | | |
| OPTICAL & ELECTRICAL PROPERTIES | | |
| **Color**: white | **Conductivity**, mS/cm: 0.3-0.35 | **Brightness**: 98 |
| MORPHOLOGY | | |
| **Particle size**, μm: 0.7 | | |
| **Sieve analysis**: 325 mesh residue - 0.004-0.02% | | **Specific surface area**, m$^2$/g: 3-5 |
| MANUFACTURER & BRAND NAME:<br>Sachtleben Chemie, Duisburg, Germany<br>Lithopone 30 L, 30 D, 30 DS, 60 L - the number is the percentage of ZnS, DS is micronized grade, D is the grade which is easier to disperse | | |
| MAJOR PRODUCT APPLICATIONS: paints (used to replace up to 60% $TiO_2$), coatings, thermoplastics, thermosets and paper | | |
| MAJOR POLYMER APPLICATIONS: melamine resin, polyester, alkyd, acrylic, rubber, PP, ABS, PVC | | |

The advantages of lithopone when used in paints include improved weathering, algae protection, and cost reduction. Up to 60% titanium dioxide can be saved by the use of lithopone due to its excellent hiding power and brightness.

In paint reformulation, several rules must be obeyed to obtain a satisfactory result. One part of titanium dioxide is replaced by 2.5-3 parts of lithopone. The amount of extender pigment should be reduced to compensate for the increased volume of white pigment. requires about 1/3 less wetting agent because it has a lower specific surface area than titanium dioxide. The amount of binder should be reduced in such a manner that total PVC is increased by 2-5 units. The reduction of binder is a logical consequence of the increased packing density. It results in an increase of scattering coefficient. The correct level of binder reduction can be estimated from evaluation of resistance to washing and scrubbing. Finally, the water level should be adjusted to obtain the same pigment/extender/binder solids proportion in the formulation.

## 2.1.38 MAGNESIUM OXIDE[332]

| | |
|---|---|
| **Name**: magnesium oxide | **CAS #**: 1309-48-4 |
| **Chemical formula**: MgO | **Functionality**: OH |
| **Chemical composition**: MgO - 93.72%, $SiO_2$ - 2%, CaO - 3.37% | |
| **PHYSICAL PROPERTIES** | |
| **Density**, g/cm$^3$: 2.4 | **Melting point**, °C: 2852 |
| **Thermal conductivity**, W/mK: 8-32 | **Loss on ignition**, %: 2.72 |
| **Thermal expansion coefficient**, $10^{-6}$/K: 13 | |
| **OPTICAL PROPERTIES** | |
| **Refractive index**: 1.736 | **Color**: white |
| **MORPHOLOGY** | |
| **Sieve analysis**: 325 mesh residue - 3% | **Crystal structure**: cubic |
| **MANUFACTURERS & BRAND NAMES**:<br>Charles B. Chrystal Co., Inc., New York, NY, USA<br>Magnesium oxide - chemical grade for neutralization | |
| **MAJOR PRODUCT APPLICATIONS**: curing agent, acid scavenger | |
| **MAJOR POLYMER APPLICATIONS**: polyester, rubber | |

## 2.1.39 MAGNESIUM HYDROXIDE[333-341]

| | | |
|---|---|---|
| **Name**: magnesium hydroxide | | **CAS #**: 1309-42-8 |
| **Chemical formula**: $Mg(OH)_2$ | **Functionality**: OH and from surface treatment | |
| **Chemical composition**: $Mg(OH)_2$ - 96-98%, possible modifications by silane and fatty acids | | |
| **Trace elements**: Fe, Mn, Cu | | |
| **PHYSICAL PROPERTIES** | | |
| **Density**, g/cm$^3$: 2.4 | **Loss on ignition**, %: 30-30.5 | **Decomposition temp.**, $^o$C: >300 |
| **Decomposition heat.**, kJ/g: 1.1-1.45 | **Decomposition peak**, $^o$C: 320-440 | |
| **CHEMICAL PROPERTIES** | | |
| **Chemical resistance**: reactive with acids | | |
| **Moisture content**, %: 0.2-1 | **Water solubility**, %: traces | **Acid soluble matter**, %: 100 |
| **OPTICAL PROPERTIES** | | |
| **Refractive index**: 1.56-1.58 | **Color**: white | |
| **MORPHOLOGY** | | |
| **Particle size**, μm: 0.5-7.7 | **Crystal structure**: hexagonal | |
| **Specific surface area**, m$^2$/g: 1-30 | **Oil absorption**, g/100 g: 40-50 | |
| **MANUFACTURERS & BRAND NAMES**:<br>Dead Sea Bromine Group, Beer Sheva, Israel<br>Magnesium Hydroxide FR-20<br>Duslo, a.s., Sala, Slovak Republic<br>Duhor N-PL (general use and rubber and PE), C-02 (PP, PS), C-03 (EPDM, EVA), C-041(PA_6) (N-grade is untreated magnesium hydroxide and C grades are surface treated for the use in different polymers as indicated for each grade) | | |
| **MAJOR PRODUCT APPLICATIONS**: cable, building industry | | |
| **MAJOR POLYMER APPLICATIONS**: PA, PVC, PE, PP, EVA, nitrile rubber, HIPS, ABS | | |

Magnesium hydroxide is an emerging filler for fire retardant applications. In this area, it competes with aluminum trihydroxide, antimony oxide, and other fillers based on zinc. Magnesium hydroxide has a different decomposition temperature from aluminum trihydroxide, it is more suitable for polymers with higher decomposition temperature. These aspects and current findings are discussed in detail in Chapter 10.

## 2.1.40 METAL-CONTAINING CONDUCTIVE MATERIALS[342-344]

**Names**: nickel coated carbon fiber, steel fiber, powder, silver coated hollow and solid glass spheres, silver coated mica, silver coated fiber

| | |
|---|---|
| **Chemical formula**: composite materials | **Functionality**: none or derived from coating |

**Chemical composition**: variable composition; silver coatings in Conduct-O-Fil solid glass spheres and fibers - 4-16 wt%, 30% on hollow glass spheres, 65 wt% on mica flakes, 8-19 wt% on copper flakes, 24 wt% on nickel granules, 20 wt% on aluminum particles, nickel coating on Compmat carbon fiber is 24 wt%

**PHYSICAL PROPERTIES**

**Density**, g/cm$^3$: 2.7 (nickel coated carbon fiber Besfight MC), 3.1-3.4 (AgCLAD, silver coated thick-wall spheres), 2.7-2.9 (AgCLAD, fiber coated with silver), 0.6-0.8 (Metalite, silver coated light glass spheres); 2.5-2.8 (solid glass spheres and fibers - Conduct-O-Fil), 1.4-1.65 (hollow glass spheres - Conduct-O-Fil), 9.1-9.2 (silver coated copper products - Conduct-O-Fil), 3.1 (silver coated aluminum powder - Conduct-O-Fil), 4.8 - (silver coated inorganic flake - Conduct-O-Fil), 3.0 (nickel coated Compmat carbon fiber)

**Mohs hardness**: 7 (AgCLAD, thick-walled spheres coated with silver), 5-6 (Metalite, glass light spheres coated with silver)

| | |
|---|---|
| **Tensile strength**, MPa: 3600 (Compmat) | **Tensile modulus**, GPa: 3600 (Compmat) |
| **Elongation**, %: 1.1-1.3 (Compmat) | **Specific heat**, kJ/kg·K: 0.65-1 (Compmat) |

**Compressive strength**, MPa: 345 (AgCLAD, thick-walled spheres coated with silver), 10-20 (Metalite, light glass spheres coated with silver), 70 (Conduct-O-Fil)

**ELECTRICAL PROPERTIES**

**Dry bulk resistivity**, Ω-cm: 0.005-0.008 (silver-coated products of PQ), 0.0017 (silver coated solid and hollow glass spheres - Conduct-O-Fil), 0.004 (silver coated glass fiber - Conduct-O-Fil), 0.0005-0.0006 (silver coated copper powder - Conduct-O-Fil), 0.0012 (silver coated copper flake - Conduct-O-Fil), 0.0007 (silver coated aluminum sphere - Conduct-O-Fil), 0.003 (silver coated inorganic flake - Conduct-O-Fil), 0.006 (silver coated nickel granules - Conduct -O-Fil), 0.0000016 - pure silver

**Specific resistivity**, Ω-cm: 7.5x10$^{-5}$ (Besfight MC), 1.5x10$^{-3}$ (Besfight HTA carbon fiber), 6x10$^{-6}$ (Ni)

**MORPHOLOGY**

**Particle size**, μm: 3 (AgCLAD silver-coated, thick-wall spheres), 45-125 (Metalite, silver-coated, light glass spheres), 50 to 300 mm (Bekinox VS for conductive textiles); Conduct-O-Fil: glass spheres - 12-92, copper flakes - 10-150

| | |
|---|---|
| **Filament diameter**, μm: 6.5-33 | **Aspect ratio**: 15 (silver-coated nickel flakes), 200-1600 (Compmat nickel coated carbon fibers) |

**Thickness of metal coating**, μm: 0.25 (nickel in Besfight MC), silver coating thickness of Conduct-O-Fil S series - 0.05-0.27, 0.4 (nickel in Compmat MCG)

**Specific surface area**, m$^2$/g: 0.6

**Particle thickness**, μm: 1 (silver coated nickel flakes)

**MANUFACTURERS & BRAND NAMES**:

American Metal Fibers, Inc., Lake Bluff, IL, USA

S-207, S-208 (high gauge chopped steel fibers), C-502 (copper fibers), B-401 (brass fibers) - products for brake pad applications

*continued on the next page*

**MANUFACTURERS & BRAND NAMES:**

Anval, Inc., Rutherford, NJ, USA

Anval Metal Powder for Plastic Filler - 304, 316 (non-magnetic stainless steel), 410L, 410, 420 (magnetic stainless steel). Spherical particles obtained by gas atomization of alloys containing different proportions of Cr, Ni, Mo, and Fe. Obtained spherical particles are classified to the required sizes. Materials can be produced to the required size in a range from 15 to 1000 μm.

Bekaert Steel Wire Corporation, Marietta, GA, USA

Bekinox VS (steel fiber), LT (steel mixed with PA), LTW & W (steel mixed with wool), Pes 12/50 (steel mixed with polyester)

Beki-Shield - steel fibers for EMI protection of plastics

Bekitex - metal-containing yarns for conductive textiles

Composite Material L.L.C., Mamaroneck, NY, USA

Compmat - nickel and copper plated graphite fiber roving, chopped fibers of different length in the range of 3 to 25 mm, and prepregs of these fibers for PA, PVA, PP, PE, polyoxazoline, Kynar, PC, ABS, HIPS, PPO, polyvinylpyrrolidone. The fibers are designed for EMI/RFI shielding

Inco Europe Ltd., Swansea, UK

VaporFab - nickel coated carbon fiber by a chemical vapor deposition process for EMI shielding applications

Incoshield - concentrates of nickel-coated carbon fiber in PPS, PC, PMMA, PEI, PA-6, PA-12. Concentrates are used for production of conductive polymers

JB Company, Franklin, NJ, USA

Glass Beads Silver and Gold - metallized beads for decorative applications

MO-SCI Corporation, Rolla, MO, USA

MetaSpheres - glass microspheres coated with Ni, Co, Cu, Ag, Au, Pd, Pt, Rh. Typical coating thickness 2%.

Novamet Specialty Products Corporation, Wyckoff, NJ, USA

Novamet Silver-coated Nickel Flakes - for conductive materials

Novamet Nickel Coated Graphite-60 - graphite powder coated with metal for EMI shielding applications

Plastic Methods Co., Inc., New York, NY, USA

CMC Cathospheres - coppers, nickel, gold, and silver coated glass spheres of diameter sizes from 1.1 to 14 mm, designed for barrel plating which eliminates rack plating

Potters Industries Inc., Affiliate of the PQ Corporation, Valley Forge, PA, USA

Conduct-O-Fil S series - silver coated solid glass spheres. Twelve grades in particle sizes range of 12-92 μm. Materials for conductive adhesives, caulks, coatings, elastomers, greases, inks

Conduct-O-Fil SH - silver coated hollow borosilicate glass spheres containing 30 wt% silver

Conduct-O-Fil SM - silver coated mica flake

Conduct-O-Fil SC - silver coated copper; SC230F8, SC500F20, SC140F19 - flakes, SC325P17 - granules, SC500P18 - powder

Conduct-O-Fil SN - silver coated nickel granules

Conduct-O-Fil SA - silver coated aluminum particles

Conduct-O-Fil PI-1040 - aluminum compatible particles which do not cause galvanic corrosion in gaskets contacted with aluminum

The PQ Corporation, Valley Forge, PA, USA

AgCLAD TW and Filament 32 - a thick-walled spheres and fiber coated with silver, respectively

Metalite SG, CG, SF-20 - light hollow glass spheres coated with silver

Toho Rayon Co., Ltd, Tokyo, Japan

Besfight MC and HTA-CF - carbon fiber nickel-coated with excellent mechanical properties of carbon fiber and good electric conductivity of nickel. The material for conductive plastics.

**MAJOR PRODUCT APPLICATIONS**: adhesives, caulks, sealants, inks, paints, coatings, EMI control, gaskets, decoration, plating, composites, building products, computers, pastes for electronics, stucco, arts and crafts, smoke detectors, covers, printers, copiers

**MAJOR POLYMER APPLICATIONS**: thermosets, silicones, polyurethanes, epoxy, acrylics, PP, PPS, PC, ABS, PEI, PA

This section discusses conductive materials. Although, not consistent with chapter organization, all materials which are composed of two different materials or are in the form of conductive fibers have been included here for easier comparison. Other metallic materials, if they are composed of a single metal, can be found in other sections. Metal coated spheres, flakes, and fibers are manufactured for various applications. Conductive plastics are the most common. Nickel-coated graphite fibers were developed in the 1980s by American Cyanamid. These fibers combine the strength of fiber with the electrical and thermal conductivities of nickel. The choice of nickel is dictated by the fact that it is a relatively inexpensive metal with good corrosion resistance. Typically, 3-5% fibers in material give the static dissipating properties. Toho Rayon, Co. further improved the performance of the material by the use of their technology of carbon fiber manufacturing and a very precise coating of a thin layers of nickel. Figure 2.38 shows the morphology of surface and the cross-section of these fibers from two manufacturers: Toho Rayon, Co. and Composite Materials, L.L.C. The specific resistivity of nickel coated fibers is only one order of magnitude higher than nickel but two orders of magnitude lower than uncoated fiber.

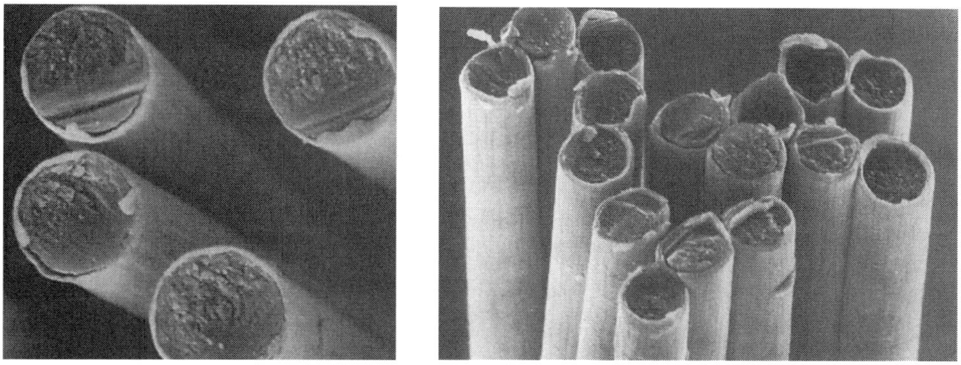

Figure 2.38. SEM micrograph of nickel coated carbon fiber (left - Besfight, Toho Rayon, Co.) (right - Compmat, Composite Materials). *Courtesy of Toho Rayon, Co., Tokyo, Japan and Composite Materials, L.L.C, Mamaroneck, NY, USA.*

Other substrates such as graphite powder and mica are also coated with nickel. Silver is the most conductive metal, being almost 5 times less resistant than nickel. Silver and copper have very similar conductivities but copper is easily oxidized and reacts with acids readily which affects its performance in polymeric systems. The PQ Corporation and Potters Industries, Inc. have developed a whole range of products which are silver coated. Their application is for conductive thermosets, gaskets, sealants, adhesives, paints, coatings, inks, and EMI control applications. Its use in these products saves 1/3 of weight of conductive material.

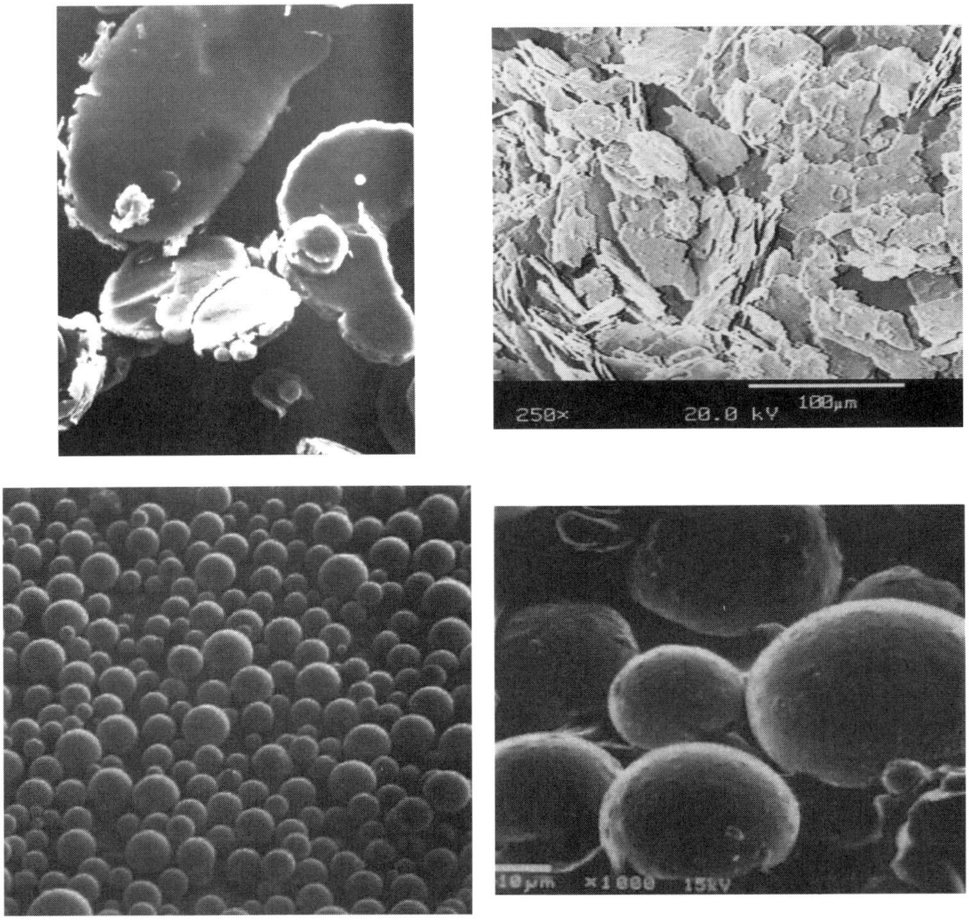


Figure 2.39. SEM micrographs of silver-coated flakes and spheres. Upper left - Novamet' silver coated nickel flakes, upper right - Conduct-O-Fil SC230F8, silver coated copper flakes, bottom - Conduct-O-Fil solid glass spheres coated with silver, left - spheres at 100x magnification, right - spheres in silicon resin. *Courtesy of Novamet Specialty Product Corporation, Wyckoff, NJ, USA and Potters Industries, Inc., Valley Forge, PA, USA.*

Novamet developed a concept to improve the properties of nickel flakes by coating them with 15% silver. The coated flakes have both conductivity and ferromagnetic properties. In addition, because of the differences in density (Ag - 10.5 and Ni - 8.9 g/cm$^3$), it is possible to save 15% in material since conductivity is related to volume rather than to weight and surface conductivity is usually of primary importance. Figure 2.39 shows the morphology of several conductive materials. Metal flakes from silver coated nickel and copper flakes have irregular shapes because they are formed from spherical particles which were first coated with silver and then flattened by mechanical forces. The apparent difference in thickness between the two products is due to the different magnifications. Both products have a similar thicknesses of about 1 μm. Coating consistency is essential since silver must play the role of the corrosion protective metal for copper (nickel is corrosion resistant). The consistency of the silver layer depends, in addition to the conditions of the process, on the properties of the metals involved and on the adhesion between layers. The morphology of spherical particles does not differ from uncoated glass spheres. It can be noted from the micrograph on the right side that spheres have excellent adhesion to silicon resin.

Potters Industries, Inc. developed a new aluminum compatible particles which can be used in gaskets in contact with aluminum. If silver in the gasket were to come in contact with the aluminum of the enclosure, galvanic corrosion may result. The aluminum compatible grade was found to pass 3000 hours in a salt spray chamber without loss of shielding effectiveness.

Plastic Methods Co., Inc. found an interesting application for metal coated glass spheres. The spheres are mixed with a product to be plated or burnished. The balls are light and perfectly round therefore they do not damage the material surface and provide excellent conductivity for even metal distribution in semi-conductor parts and jewelry.

JB Company manufactures glass beads coated with metal for decorative purposes.

## 2.1.41 MICA[345-361]

| | | |
|---|---|---|
| **Name**: mica | | **CAS #**: 1318-94-1, 12001-26-2 |
| **Chemical formula**: $AB_{2-3}(Al,Si)Si_3O_{10}(OH)_2$ A=K, Na, Ca, Ba; B=Al, Fe, Mg, Li; muscovite: $KAl_2(AlSi_3O_{10})(OH)_2$, phlogopite: $KMg_3(AlSi_3O_{10})(OH)_2$ | **Functionality**: OH | |
| **Chemical composition**: muscovite: $SiO_2$ - 44-48%, $Al_2O_3$ - 31-38%, $K_2O$ - 3-11%, $Fe2O_3$ - <1-5.7%; phlogopite: $SiO_2$ - 40-42%, MgO - 21-24%, $Al_2O_3$ - 9-16%, $Fe_2O_3$ - 9-11%, $K_2O$ -10-11% | | |
| **PHYSICAL PROPERTIES** (M) - muscovite mica, (P) - phlogopite mica; most data in this table courtesy of *Polar Minerals*, Mt. Vernon, IN, USA | | |
| **Density**, g/cm$^3$: 2.75-3.2 (M), 2.74-2.95 (P) | **Mohs hardness**: 2.5-4 (M), 2.5-3 (P) | |
| **Decomposition temp.**, $^o$C: 1300 (P) | | **Loss on ignition**, %: 4-9 (M), 2 (P) |
| **Maximum temperature of use**, $^o$C: 500-530 (M), 850-1000 (P) | | **Specific heat**, kJ/kg·K: 0.21 |
| **Linear coefficient of thermal expansion**, 1/$^o$C: 1.5-25x10$^{-6}$ (M), 1-1000x10$^{-6}$ (1 to cleavage); 9-80x10$^{-6}$ (M), 13-14.5x10$^{-6}$ (P) (11 to cleavage) | | |
| **Tensile strength**, MPa: 250-860 | **Tensile modulus**, MPa: 172,000 | **Compressive strength**, MPa: 220 |
| **Coefficient of friction**: 0.1-0.2 (M), 0.2-0.4 (P) | | |
| **CHEMICAL PROPERTIES** | | |
| **Chemical resistance**: very good (M), good (P) | | |
| **Moisture content**, %: 0.3-0.7 | **Water of constitution**, %: 4.5 (M), 3.2 (P) | |
| **pH of water suspension**: 6.5-8.5 (M), 7-8.5 (P) | | |
| **OPTICAL & ELECTRICAL PROPERTIES** | | |
| **Refractive index**: 1.55-1.61 (M), 1.54-1.69 (P) | | **Reflectance**: 87 (M) 42-64 (P) |
| **Color**: white, off-white to beige (M), golden brown to bronze (P) | | **Brightness**: 55-65 |
| **Dissipation factor**: 4.5-8.2x10$^{-2}$ | **Loss tangent**: 0.0013 (M), 0.02-0.04 (P) | |
| **Dielectric constant**: 6.5-9 (M), 5-7 (P) | **Dielectric strength**, V/cm: 7-15 (M), 5-10 (P) | |
| **Specific resistivity**, Ω-cm: 10$^{12}$-10$^{16}$ (M), 10$^{10}$-10$^{13}$ (P) | | **Power factor**: 0.08-0.09 |
| **MORPHOLOGY** | | |
| **Particle shape**: hexagonal | **Crystal structure**: monoclinic | **Cleavage**: basal |
| **Particle size**, μm: 4-70 | **Oil absorption**, g/100 g: 65-72 | |
| **Aspect ratio**: 10-70 | **Particle thickness**, μm: 1.1-2.6 | |
| **Sieve analysis**: 325 mesh residue - 1-45% | | |

**MANUFACTURERS & BRAND NAMES:**

Asheville Mica Company, Newport News, VA, USA

Mica 325, 325FF, 325MF, 325D, AMC, 160 D - dry ground mica for rubber applications.

Aspect Minerals (Zemex), Spruce Pine, NC, USA

AlbaFlex (25, 50, 100, 200, 300, 400), AlbaShield (15, 20, 25, 25-S, 50, 50-S, 1000, 2000) - wet ground muscovite micas

AFlake - ground mica flakes

Cosmetic line for the use in lipstick, face powder, eyeshadow, and nail polish

Surface treated mica

Franklin Industrial Minerals, Kings Mountain, NC, USA

WG-325, HiMod-270, HAR-160, WG-160, H-160 - wet ground muscovite mica

19 grades of dry ground muscovite mica in particle sizes from 17 to 550 μm

Les Produits Mica Suzorite, Inc. (Zemex), Boucherville, PQ, Canada

15-Z, 20-S, 25-Z, 40-S, 40-Z, 50-SD, 60-HK, 60-PE, 60,-PO, 60-PP, 60-S, 60-Z, 80-SF, 150-NY, 150-S, 200-HK, 200-PE, 200-PP, 200-S, 325-HK, 325, PE, 325-PO, 325-PP, 325-S, (SD, Z - purposely not fully delaminated or purified, HK - highly or super-delaminated, SF and S - highly delaminated, PE, PO and PP - surface treated)

Mica-Tek, Northville, MI, USA

Mica-Lyte, Dekorflake, Microfibers, Specular - selected natural, colored, and shaped materials designed as special-effect colorants to impart granite-like, sparkling, and textured appearances to transparent and translucent polymers

Non-Metals, Inc., Affiliate of The China National Non-Metallic Minerals Group, Tucson, AZ, USA

Muscovite Mica Powder - D Series (dry ground), W Series (wet ground)

Polar Minerals, Mt. Vernon, IN, USA

Phlogopite Mica 5200(s), 5100(s), 5040(s), 5010(s) - grades having different particle sizes; (s) means that the product can be supplied with chemical coating

Muscovite Mica 6915, 6912, 6908, 6905 - grades of different particles sizes for plastics and coatings

SG-70, SG-90 - hydrous potassium aluminum silicate produced by patented process which gives high brightness delaminated muscovite mica for joint compounds, adhesives, sealants, coatings

**MAJOR PRODUCT APPLICATIONS:** paints, coatings, composites, plastic parts, sound dampening, foundry coating, lipsticks, face powders, eyeshadows, nail polish, mold release agents, bathwares, housewares, toys, interior decoration, asbestos substitute, filtration aids, asphalt-based compounds and coatings, drilling fluids, insulating heat shields, gaskets, gypsum board, tank linings

**MAJOR POLYMER APPLICATIONS:** ABS, PP, PA, PC, PMP, PE, PET, PBT, PMMA, PS, PVC, rubber

The mica group has about 30 members but only a few are common. Muscovite, phlogopite, and biotite are important representatives of this group. Muscovite is one of the most common of the micas and occurs in a wide variety of geological environments because of its stability. Crystals measuring 2-3 m across are mined in some locations. Muscovite can vary in chemical composition as a result of atomic substitution (Na for K; Mg and F for Al).

Phlogopite is found in metamorphosed magnesium-rich limestones, dolomites, and ultrabasic rocks. Biotite, similar to muscovite, is also widespread. It is usually associated with minerals which were formed under high temperature and pressure. Several elements, other than those included in their typical chemical composition, can be found in these two minerals. These include: Na, Rb, Cs, Ba, F, and Ca. The most important difference between phlogopite and biotite is that biotite contains a substantial amount of iron.

Of the three micas characterized above, muscovite and phlogopite are the most commonly used. Muscovite is almost colorless, phlogopite has a golden brown color, whereas biotite is black. The color influences mica application to a great extent, and practically speaking, muscovite and phlogopite are the only minerals used, with muscovite being the more popular.

Mica fillers are obtained by separation of mica from other minerals which might compose 10-20% of mineral content. Mica is dry or wet milled and classified. The mechanical grinding produces flakes with a low aspect ratio in a range from 20 to 40. The process may include ultrasonic delamination which leads to a high aspect ratio of over 200. Flakes of mica fillers have a thickness in a range from 1 to 3 μm and a width in a range from 10 to 450 μm.

A high aspect ratio contributes greatly to polymer reinforcement, and also allows production of highly-filled polymers. For this reason the aspect ratio should be regarded as the most important single property characterizing the quality of micas. The technology of mica filler manufacture may include surface preparation using silanes, maleated polypropylene wax, and amine acetate. These processes greatly enhance reinforcement. Ultrasonic delamination especially becomes more effective when surface treatment is used. This is related to increased mica wetting, which is usually difficult compared to other fillers. Surface coupling also greatly affects the resistance of the filled polymer to water – one of the most desired mica properties when it is compared with other fillers.

For some applications it is essential to control the concentration of iron which may vary over a broad range, regardless of mineral type. There are muscovite types known to contain up to 5% of $Fe_2O_3$ (although it would be expected to contain none), whereas biotite may contain as little as 2% of $Fe_2O_3$ (it typically contains iron in its chemical formula).

Other reasons are known for mica's frequent use: one, of long standing tradition in the industry, is related to its high resistivity; the other is its effect on thermal expansion. Composites including mica have a low coefficient of thermal expansion comparable to those including glass flakes. In addition, mica is used to reduce shrinkage, warpage, and to improve tensile strength and modulus, high temperature deflection, and permeability.

Mica-Tek has an interesting approach to exploiting the variety of forms and colors of mica. A range of mica-based products have been developed which differ in the color and the shape of particle as well as in their glittering and sparkling effects. These decorative pigments are used in housewares, bathwares, toys, interior decorating, etc.

Figure 2.40 shows SEM micrographs of muscovite and phlogopite mica. The morphological features of both forms are very similar.

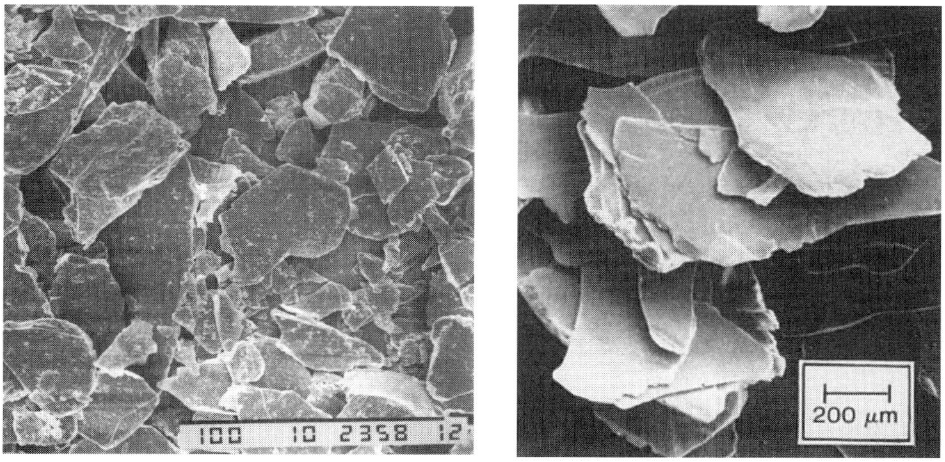


Figure 2.40. The morphology of muscovite (left) and phlogopite (right) mica. *Courtesy of NYCO, Minerals, Inc.,Willsboro, NY, USA (muscovite) and Les Produits Mica Suzorite, Inc., Boucherville, PQ, Canada.*

## 2.1.42 MOLYBDENUM

| | |
|---|---|
| **Name**: molybdenum powder | **CAS #**: 7439-98-7 |
| **Chemical formula**: Mo | **Functionality**: none |
| **Chemical composition**: Mo - 99-99.99% | |
| **Trace elements**: O - 600-1000 ppm | |
| PHYSICAL PROPERTIES | |
| **Density**, g/cm$^3$: 10.2 | **Melting point**, °C: 2610 |
| CHEMICAL PROPERTIES | |
| **Chemical resistance**: soluble in concentrated strong acids | |
| MORPHOLOGY | |
| **Particle size**, μm: 1-50 | **Crystal structure**: cubic |
| MANUFACTURER & BRAND NAMES:<br>CSM Industries, Coldwater, MI, USA<br>OMP - high purity, fine powder obtained from molybdenum trioxide which is hydrogen reduced. It is composed of agglomerated particles<br>MMP highest purity powder produced from ammonium dimolybdate and it is hydrogen reduced and agglomerated, deagglomerated powders are also available<br>SOMP, PDMP - spherically shaped particles produced by spray drying, atomization, and plasma densification are flowable powders | |
| MAJOR PRODUCT APPLICATIONS: electronics, aerospace | |

Figure 2.41 shows an individual particle of molybdenum powder and the agglomerated powder. Spherical particles with a porous structure can be produced from agglomerates (SOMP). PDMP are also spherical particles which have a smooth surface. The agglomerated powder is composed of cubical and elongated particles.

Figure 2.41. SEM micrographs of different grades of molybdenum powder. *Courtesy of CSM Industries, Coldwater, MI, USA.*

## 2.1.43 MOLYBDENUM DISULFIDE[362-366]

| | |
|---|---|
| **Name**: molybdenum disulfide | **CAS #**: 1317-33-5 |
| **Chemical formula**: $MoS_2$ | **Functionality**: S |
| **Chemical composition**: $MoS_2$ - 98% | |
| **Trace elements**: Fe, O | |
| **PHYSICAL PROPERTIES** | |
| **Density**, g/cm$^3$: 4.8-5; **Mohs hardness**: 1 | **Melting point**, °C: 1600 (decomp) |
| **Thermal conductivity**, W/K·m: 0.13-0.19 | **Coefficient of thermal expansion**, 1/°C: 10.7x10$^{-6}$ |
| **Coefficient of friction**: 0.03-0.06 | |
| **CHEMICAL PROPERTIES** | |
| **Acid soluble matter**, %: 95.5 | |
| **MORPHOLOGY** | |
| **Particle size**, μm: 0.4-38 | |
| **MANUFACTURERS & BRAND NAMES**: AML Industries, Inc., Warren, OH, USA: Amlube 510 - technical grade, 511 - fine technical grade; Climax Molybdenum Company, Ypsilanti, MI, USA: Technical, Technical Fine, Super Fine (Suspension) - grades having different particle sizes; EM Corporation, West Lafayette, IN, USA: E-4 - purified molybdenum disulfide powder; Parma-Slik - mixtures of molybdenum disulfide and graphite | |
| **MAJOR PRODUCT APPLICATIONS**: plastic parts (e.g., piston rings, cams, ball bearing retainers, space shuttle bearings, etc.), greases, lubricating aerosols, oil additives, metalworking compounds | |
| **MAJOR POLYMER APPLICATIONS**: PA, PTFE, phenoxy, epoxy, PC, polyarylate | |

The compound occurs as the mineral molybdenite which after refining is also used as lubricating material. The principle of action of molybdenum sulfide is based on the formation of bonds between metal and sulfur. These bonds slip under shear forces and are continuously reformed holding the lubricating film on the surface of the metal.

Figure 2.42 shows morphology of technical grade of molybdenum disulfide.

Figure 2.42. SEM micrograph of molybdenum disulfide. *Courtesy of Climax Molybdenum Company, Ypsilanti, MI, USA.*

## 2.1.44 NICKEL[367-370]

| | | |
|---|---|---|
| **Names**: nickel | | **CAS #**: 7440-02-0 |
| **Chemical formula**: Ni | **Functionality**: none | |
| **Chemical composition**: Ni - >99%, C- 0.1-0.25% | | |
| **Trace elements**: Fe, O | | |
| **PHYSICAL PROPERTIES** | | |
| **Density**, g/cm$^3$: 8.9 | **Specific heat**, kJ/kg·K: 0.44 | **Melting point**, °C: 1455 |
| **Thermal conductivity**, W/K·m: 158 | **Coefficient of thermal expansion**, 1/°C: 13x10$^{-3}$ | |
| **ELECTRICAL PROPERTIES** | | |
| **Resistivity**, Ω-cm: 7.8x10$^{-6}$ | | |
| **MORPHOLOGY** | | |
| **Particle size**, μm: 2.2-9 | **Aspect ratio**: 15-50 | |
| **Particle thickness (flakes)**, μm: 0.4-1.3 | **Specific surface area**, m$^2$/g: 0.6-0.7 | |
| **Sieve analysis**: 325 mesh residue: 1-4% | | |

**MANUFACTURERS & BRAND NAMES:**
INCO Specialty Powder Products, London, UK and AcuPowder International, Union, NJ, USA
- INCO Nickel Powder Type 123 - powder metallurgy
- INCO Filamentary Nickel Powder Types 255, 270, 287 - plastics and electronics

Novamet Specialty Products Corporation, Wyckoff, NJ, USA
- Nickel Flake Powder - leafing and water grade products for protective paints (both grades can be used in solvent-based systems)
- Conductive Nickel Flake Powder HCA-1 - product developed for conductive paints and adhesives which provides EMI shielding when used in surface coatings, inks, and adhesives. The flakes are treated in a controlled atmosphere to give cleaner surface which enhances conductivity
- Conductive Nickel Pigment 525 - dendritic filamentary shape similar to INCO products
- CNS - spherical shape and uniforms size for thick film inks

**MAJOR PRODUCT APPLICATIONS**: EMI/RFI shielding, powder coating, anti-size lubricants, decorative lacquers, waterborne coatings, conductive plastics, non-stick coatings, coatings for cookware, adhesives, inks, sealants

**MAJOR POLYMER APPLICATIONS**: silicone, polyurethanes, epoxy, PE, PP

Nickel in addition to being highly conductive has ferromagnetic properties and it is a relatively inert material. INCO produces nickel powders by thermal decomposition of nickel carbonyl in a process which produces a fine particle metal powder with a spiked or dendritic surface (Figure 2.43). The micrograph on the left hand side shows a singular particle of grade 123. The morphology of Types 255, 270, and 287 is shown in the figure on the right hand side. The dendritic particles are connected to each other to form a chain of a controlled length and porosity.

Figures 2.44 and 2.45 show the morphology of two grades produced by Novamet: flake powder and spherical material.

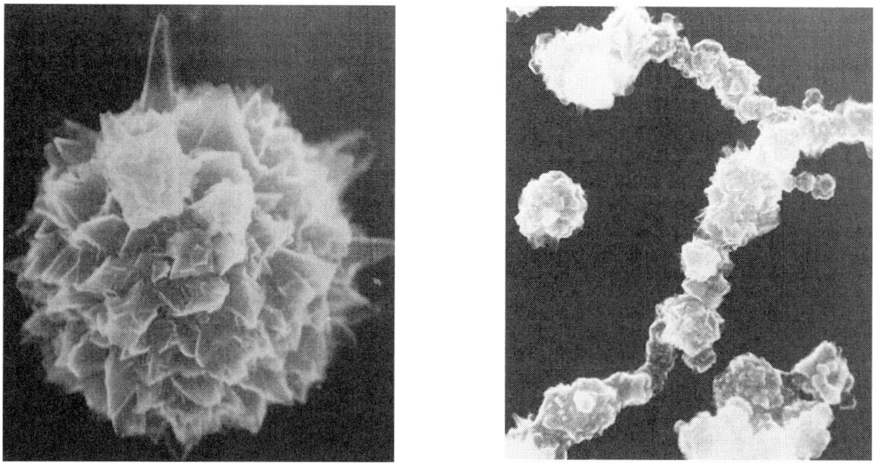

Figure 2.43. INCO nickel powder single particle (left) and chain (right). *Courtesy of INCO Specialty Powder Products, London, UK.*

Figure 2.44. Novamet nickel flakes. *Courtesy of Novamet Specialty Products Corporation, Wyckoff, NJ, USA.*

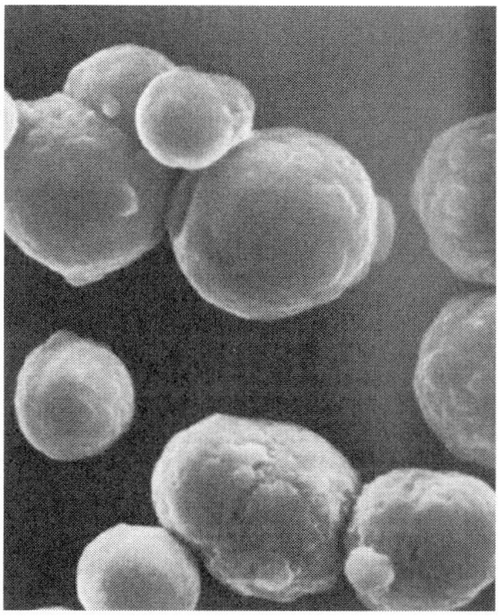

Figure 2.45. Novamet spherical nickel, CNS. *Courtesy of Novamet Specialty Products Corporation, Wyckoff, NJ, USA.*

## 2.1.45 PERLITE[371]

| | | |
|---|---|---|
| **Name**: perlite | | **CAS #**: 93763-70-3 |
| **Chemical formula**: depends on the rock composition | **Functionality**: OH and silane functionality | |
| **Chemical composition**: $SiO_2$ - 71-75%, $Al_2O_3$ - 12-18%, $Na_2O$ - 3-4%, $K_2O$ - 4-5%, $Fe_2O_3$ - 0.5-1.5%, MgO - 0.1-1.5% | | |
| **Trace elements**: Mn, Ti | | |
| PHYSICAL PROPERTIES | | |
| **Density**, $g/cm^3$: 1.2-2.4 | **Mohs hardness**: 5.5 | **Loss on ignition**, %: 1.5 |
| **Specific heat**, kJ/kg · K: 0.88 | **Softening point**, °C: 871 | **Expansion temperature**, °C: 871 |
| CHEMICAL PROPERTIES | | |
| **Chemical resistance**: soluble in hot alkalis and strong acids | | **Water solubility**, %: 1 |
| **Moisture content**, %: 0.5-1 | **pH of water suspension**: 5.5-8.5 | **Acid soluble matter**, %: 3 |
| OPTICAL PROPERTIES | | |
| **Refractive index**: 1.5 | **Brightness**: 74 | |
| **Color**: off-white | | |
| MORPHOLOGY | | |
| **Particle shape**: irregular flake | **Particle size**, μm: 11-37 | **Oil absorption**, g/100 g: 210-240 |
| **Specific surface area**, $m^2/g$: 1.88 | | |
| MANUFACTURERS & BRAND NAMES:<br>Grefco, Inc., Lompoc, CA, USA<br>FF1, FF26, FF36, FF56, FF76 - grades of different particles sizes. FF56 and FF76 have very low effective density of 1.2-1.3 $g/cm^3$. All grades are available with surface modification<br>Strong-Lite Products Corporation, Pine Bluff, AR, USA<br>range of perlite grades mostly for construction and horticulture | | |
| MAJOR PRODUCT APPLICATIONS: construction (thermal insulation, concrete, under-floor insulation), paints, horticulture, filtering, mild abrasives, filler of plastics, caulks, explosives, carrier of agrochemicals | | |
| MAJOR POLYMER APPLICATIONS: PE, PP, PVC | | |

Perlite is a volcanic rock found in many locations. If rapidly heated to 871°C it expands up to 20 times. Figure 2.46 shows the morphology of Perlite FF-56 which is very light filler.

Figure 2.46 SEM micrographs of Perlite FF-56. *Courtesy of Grefco, Inc., Lompoc, CA, USA.*

## 2.1.46 POLYMERIC FILLERS[372-381]

**Names**: plastic microspheres, expandable microspheres, PTFE, PE, PI

**Chemical formula**: this diverse group includes particular materials of different chemical composition which are used as functional fillers

PHYSICAL PROPERTIES

**Density**, g/cm$^3$: Expancel microspheres: unexpanded - 1.05-1.2, expanded - 0.03-0.07; Dualite: 0.065-0.13; PTFE: 2.2; Vistamer HD & UH - 0.94-0.96, Ti - 1.58-2.44

**Melting point**, °C: 329-332 (PTFE)

**Coefficient of friction**: PTFE - <0.1

CHEMICAL PROPERTIES

**Moisture content**, %: Expancel - 1, Dualite - 2

MORPHOLOGY

**Particle size**, μm: Expancel: unexpanded - 6-35, expanded - 15-80; Dualite - 25-140; PTFE: 5-25 (primary particle in Algoflon - 0.15-0.3); Vistamer: 18-290

**Specific surface area**, m$^2$/g: PTFE: 2.5-9

MANUFACTURERS & BRAND NAMES:

AKZO Nobel, Expancel, Inc., Duluth, GA, USA and Sundsvall, Sweden

Unexpanded microspheres 820, 643, 551, 461, 051, 053, 054, 091, 092 hollow particles with thermoplastic shell encapsulating a gas available in wet (WU) and dry (DU) form. The grade numbers signify materials which have different particle diameter, expansion rate, solvent resistance, and temperature of expansion.

Expanded microspheres 551, 461, and 091 expanded hollow particles available in wet (WE) and dry (DE) forms. There also grades of the same type of shell but in different dimensions. The lines differ in particle diameter, density, and solvent resistance

The shell of these microspheres is composed of vinylidene chloride and acrylonitrile copolymer

Ausimont USA, Inc., Montedison Group, Thorofare, NJ, USA

Algoflon L203, L205, L206 - micronized PTFE powders for applications in thermoplastic and thermosetting resins, printing inks, paints, oils and greases, and rubber. The lower the number the smaller the particle size.

Polymist F-5, F-5A, F-5A EX, F-510, XPH-284 free-flowing PTFE powders for applications in thermoplastic and thermosetting resins, printing inks, paints, oils and greases, and rubber. The lower the number the smaller the particle size. The XPH 284 is in compliance with FDA regulation 21 CFR 177.1550 and it is recommended for articles intended for use in contact with food.

Composite Particles, Inc., Allentown, PA, USA

Vistamer HP and UH - HDPE and UHMWPE powders, respectively, with modified surface

Vistamer Ti-911x, Ti-912x surface activated powders of UHMWPE and polyimide, respectively

Pierce & Stevens Corporation, Buffalo, NY, USA

Dualite M6001AE, M6033AE, M6050AE, MS7000 - low density microspheres. Grade M6001A has shell composed of poly(vinylidene chloride) copolymer. All other grades have the shells composed of acrylonitrile copolymer. All grades have calcium carbonate coating. The difference between grades is in particle size, solvent resistance, temperature resistance, and density.

Grade composed of poly(vinylidene chloride) copolymer is less resistant to heat and solvent.

Micropearl F-30, F-50, F-80, F100 - expandable microspheres available in wet and dry forms. These microspheres are marketed for Matsumoto Yushi-Seiyaku, Co., Ltd. Japan

*continued on the next page*

**MANUFACTURERS & BRAND NAMES:**
Sekisui Plastics Co., Ltd., Tokyo, Japan
Techpolymer microspheres manufactured from acrylic and styrenic copolymers in various forms included non-crosslinked, crosslinked, porous and composite. Several manufacturing grades are designed for paints, inks, as resin-modifying agent, delustering, anti-blocking agent, and filler for toiletries and cosmetics
Apamicron beads are of inorganic origin (hydroxyapatite) which have affinity to living organisms and are used in medical applications

**MAJOR PRODUCT APPLICATIONS:** *Expancel*: cultured marble and wood, coatings and sealants, auto and marine fillers, composites, pultruded parts, paints, crack fillers, underbody coatings, elastomer fillers, syntactic foams, cable fillings, explosives, gypsum board, printing inks, paper, paperboard; *Dualite*: boats, automotive components, tub/shower products, automotive underbody coatings, paints, adhesives, sealants, truck caps, side panels, van tops, recreational vehicles, sporting equipment, boats, PVC foam, printing inks, putties, synthetic wood, rubber products, wall papers, non-wovens, molded plastics; *PTFE powders*: broad range of products for thermoplastics and thermosetting resins, paints, coatings, printing inks, oils, greases; *Vistamer grades*: molded parts, adhesives, sealants, paints, coatings, machine parts, pump impellers, valve seats, gears, rings, bearings, liners, wear-plates, guide-rails, cable, steel replacement

**MAJOR POLYMER APPLICATIONS:** *microspheres*: PVC, polyurethanes, polyester, silicone, acrylics, epoxy, rubber; *PTFE powders*: PA, POM, PC, polyesters, PI, PSF, PSO, PPS, polyurethanes, ECTFE, EPDM, SBR, fluorosilicones, NR

Expancel have developed polymeric microspheres which are widely used in various applications. The microsphere's shell is composed of vinylidene chloride and acrylonitrile copolymer and the blowing agent is isobutane. Increasing temperature softens shell and expands gas which at a certain temperature has sufficient pressure to expand the shell. The temperature of expansion is characteristic of the grade but it also depends on the matrix in which Expancel is dispersed. Typically, expansion begins at temperatures form 75 to 135°C and ends between 115 to 195°C depending on the grade of filler. The expansion rate depends on the process conditions. The microspheres can reach up to 50 times of their initial volume. The unexpanded microspheres can be used as a foaming or blowing agents. The expanded microspheres form an ultralow density modifier which does not greatly increase viscosity.

Expanded microspheres maintain their density even after a prolonged heating at temperature range of 140-160°C. Also compression at high pressures (150 bar) does not change the density of the expanded material. In any new formulation, Expancel needs to be checked for the compatibility with the other components of the system. In particular, it should be established whether the microspheres are resistant to the liquids in formulation, such as solvents, plasticizers, curatives, etc. The mixing process is complicated by the fact that microspheres, especially in the pre-expanded form, have a much lower density than the other components of the formulation therefore they float the surface of the mixture which creates difficulties in incorporation and creates the potential for their loss to the surrounding air. Microspheres have good mechanical resistance and can be mixed by high shear mixers. Also, vacuum does not affect microspheres. If mechanical resistance is of

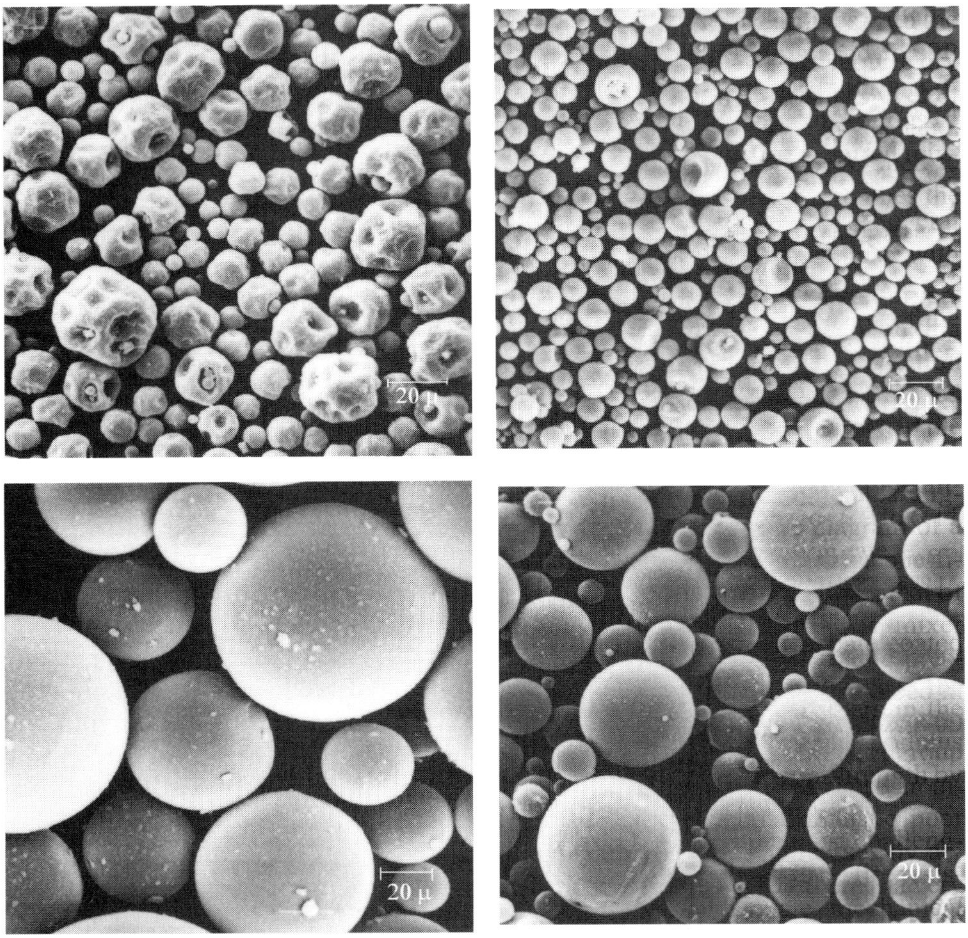


Figure 2.47. Expancel 551 (left) and 091 (right). The top micrograph - unexpanded, the bottom - expanded. *Courtesy of AKZO Nobel, Expancel, Inc., Duluth, GA, USA.*

concern, DU grades should be selected since they are smaller and have thicker walls.

Figure 2.47 shows the morphology of two grades before and after expansion. The 551 grade has more spherical particles before expansion because they have a thicker shell and they will expand to a higher density than the 091 grade. But after expansion both grades form perfectly spherical particles.

A close inspection of micrographs in Figure 2.47 shows that there are small particles attached to the surfaces of microspheres which form surface imperfections. The Figure 2.48 shows a new grade 007 which has very clean surface.

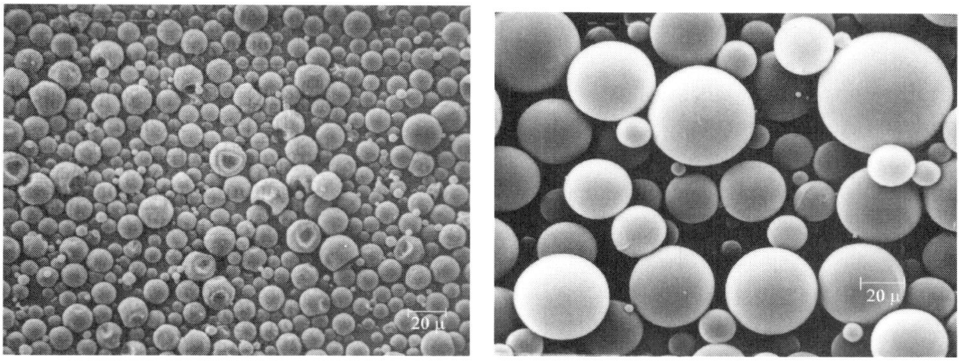


Figure 2.48. Expancel 007 before and after expansion. *Courtesy of AKZO Nobel, Expancel, Inc., Duluth, GA, USA.*

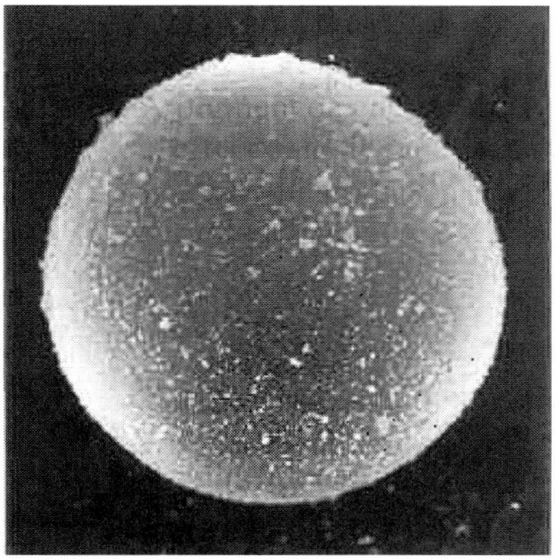

Figure 2.49. SEM micrograph of Dualite microsphere. *Courtesy of Pierce & Stevens, Buffalo, NY, USA.*

Pierce & Stevens Corporation patented the concept of manufacturing polymeric beads with a calcium carbonate coating which is inert and compatible with many materials in which microspheres are dispersed. Figure 2.49 shows individual particle of Dualite which has similar morphological features to other polymeric microspheres in spite of the fact that the surface is coated with calcium carbonate. This shows that the process is capable producing this complex composite with a high degree of precision. Also, particle size distribution curves show a narrow distribution indicating good control over processing. These microspheres resist high shear dispersion, vacuum, pressure, heat and are not affected by methyl ethyl ketone (acrylonitrile shell). The published papers[377-379] give guidelines regarding the application of microspheres in composites, surface finishes, coatings, sealants and adhesives.

Polytetrafluoroethylene powders have found a large number of applications due to their lubricating properties, chemical inertness, improvement to wear characteristics, reduction of the friction coefficient, resistance to UV and weather, effect on non-stick and release properties, increase in rub resistance, improved corrosion resistance, thermal stability, insulating properties, and lack of moisture absorption. Figure 2.50 shows SEM micrographs of two grades of free-flowing powder (Polymist F5 and XPH-284) and one grade of micronized powder

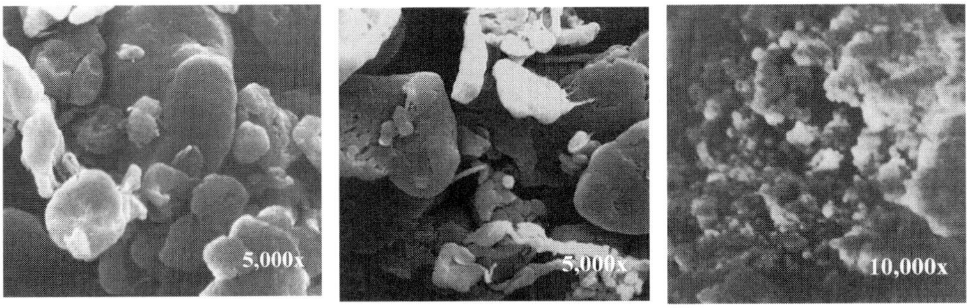


Figure 2.50. SEM micrographs of PTFE powders. Left - Polymist F5 (5000x), center - Polymist XPH-284 (5000x), right - Algoflon L203 (10,000x). *Courtesy of Ausimont USA, Inc., Montedison Group, Thorofare, NJ, USA.*

(Algoflon L203). The micronized grade forms agglomerates of small particles whereas the free-flowing powder is composed of individual particles. Particles have no sharp edges and XPH-284 contains some elongated particles.

Composite Particles, Inc. developed two methods of surface modification of polymeric materials which are used for materials of different shapes and compositions. Here, only the spherical, non-rubber particles are discussed. Further information is included in the section on rubber particles below. One method of surface modification is based on exposing the polymeric powder to a chemically reactive gas atmosphere which oxidizes surface groups to form OH and COOH functionalities. These functionalities are then available for reaction with the components of the matrix into which modified particles are introduced. Vistamer HD and UH are manufactured by this method from polyethylenes of different molecular weights. Two factors can be regulated here: the properties of the core particle and the type and density of functional groups on the surface of these particles. Polyethylene is a material, which without this modification, will not be compatible with most systems. The surface modification allows the incorporation of the material into resins. This improves abrasion resistance, tear strength, and moisture barrier properties and reduces the friction coefficient.

The second method of surface modification permits the formation of a composite particle, the core of which is composed of polymer (UHMWPE or polyimide) and the surface of which is coated with titanium carbide which is hard and abrasion resistant. The composite particles can be incorporated into any suitable matrix resulting in improved abrasion resistance, lowered friction, higher compressive strength, improved creep resistance, etc. This new product is a unique form of raw material which has the potential to improve the properties of many products.

## 2.1.47 PUMICE

| | | |
|---|---|---|
| **Name**: pumice | | |
| **Chemical composition**: $SiO_2$ - 70.9-74.2%, $Al_2O_3$ - 12.5-13.5%, $Fe_2O_3$ - 1.5-2%, CaO - 0.7-1.5%, MgO - 0.2-0.5%, $Na_2O$ - 3.2-4%, $K_2O$ - 3.8-4.5% | | |
| PHYSICAL PROPERTIES | | |
| **Density**, g/cm$^3$: 2.3 | **Mohs hardness**: 5.5 | **Loss on ignition**, %: 3 |
| CHEMICAL PROPERTIES | | |
| **Moisture content**, %: 2 | **Adsorbed moisture**, %: 140 | |
| OPTICAL PROPERTIES | | |
| **Color**: off-white, gray | | |
| MORPHOLOGY | | |
| **Sieve analysis**: 325 mesh residue - 16-22 | **Specific surface area**, m$^2$/g: 0.4-0.6 | |
| MANUFACTURERS & BRAND NAMES:<br>Charles B. Chrystal Co., Inc., New York, NY, USA<br>Chrystal Domestic Pumice - 12 grades differing in particle size<br>Lipari Pumice - high quality Italian pumice<br>Peerless Pumice - the highest quality and uniformity for a broad range of applications | | |
| MAJOR PRODUCT APPLICATIONS: paints (non-skid coatings, textured paints, flatting), chemical carrier, cleaning and polishing liquids, soaps, tooth polishing pastes and powders, cleaning electronic circuit boards | | |

## 2.1.48 PYROPHYLLITE

**Names**: pyrophyllite, aluminum silicate hydroxide

**Chemical formula**: $AlSi_2O_5OH$ | **Functionality**: OH

**Chemical composition**: $SiO_2$ - 68-75%, $Al_2O_3$ - 18-25%, $Fe_2O_3$ - 0.5-0.7%, $TiO_2$ - 0.4%

PHYSICAL PROPERTIES

**Density**, g/cm$^3$: 2.65 - 2.85 | **Mohs hardness**: 1 - 2.5

CHEMICAL PROPERTIES

**Moisture content**, %: 1

OPTICAL PROPERTIES

**Refractive index**: 1.57 | **Brightness**: 66-78

**Color**: white, gray, cream, tan

MORPHOLOGY

**Crystal structure**: monoclinic | **Cleavage**: one direction

**Sieve analysis**: 325 mesh sieve residue - 8.8%, 200 mesh - 1-3%

MANUFACTURERS & BRAND NAMES:
Charles B Chrystal Co., Inc., New York, NY, USA
Pyrophyllite R-200-C
No-Metals, Inc., Affiliate of The China National Non-Metallic Minerals Group, Tucson, AZ, USA
Pyrophyllite
R.T. Vanderbilt Company, Inc., Norwalk, CT, USA
Pyrax A, B, WA

MAJOR PRODUCT APPLICATIONS: paper, rubber, paints, cosmetics

## 2.1.49 RUBBER PARTICLES[382-396]

| | |
|---|---|
| **Names**: rubber particles, rubber filler, ground rubber | |
| **PHYSICAL PROPERTIES** | |
| **Density**, g/cm$^3$: 1.10-1.15 | **Coefficient of friction**: 1.1 |
| **CHEMICAL PROPERTIES** | |
| **Moisture content**, %: 1 | |
| **MORPHOLOGY** | |
| **Particle size**, μm: 75-2000 | |
| **MANUFACTURERS & BRAND NAMES**:<br>Composite Particles, Inc., Allentown, PA, USA<br>Vistamer R 4010, 4030, 4040, 4060, 4100, 4200 - surface activated ground rubber. Grades differ in particle size<br>Vistamer RW 4101, 4014, 4020, 4030, 4040, 4060 - surface activated, cryogenically ground rubber. Grades differ in particle size | |
| **MAJOR PRODUCT APPLICATIONS**: carpet underlay, shoe soles, roof sealant, roller, wheelchair tire, industrial coating, construction panel, industrial enclosures, foam boot-insert, automotive components, marine equipment, slip-resistant coatings, deck coatings, flexible mold, in-line skate wheels | |
| **MAJOR POLYMER APPLICATIONS**: polyurethane, NBR, EVA, PSF, phenoxy, acrylics, epoxy | |

The process developed by Composite Particles, Inc. modifies the surface of ground rubber particles. The modification introduces functional groups such as OH, and COOH which can interact with matrix to form hydrogen and covalent bonding. Numerous research papers presented in this book show that functionalization of the filler is the correct approach to improve the performance of filled materials. If untreated ground rubber is introduced into a polymer matrix the results are usually disappointing. There are two reasons: rubber particles have more affinity to themselves than to the surrounding polymer matrix and this hampers the dispersion which is crucial to the properties. Secondly, rubber particles are defect-causing inclusions, usually of substantial dimensions, which reduce mechanical performance. The situation can be reversed by surface modification of the rubber particles to promote a chemical interaction between filler and matrix. The results reported indicate that there is an improved dispersion of particles, in many matrices including water-based materials. Depending on the matrix, various mechanical properties improved, most notably, tear strength and tensile properties. The coefficient of friction of many materials can be increased by the addition of the surface treated rubber filler, Vistamer, to approach values typical of rubber. The modification method is reported to reduce the odor of ground rubber.

## 2.1.50 SEPIOLITE[397-398]

| | | |
|---|---|---|
| **Names**: sepiolite, hydrated magnesium silicate | | |
| **Chemical formula**: $Mg_4Si_6O_{15}(OH)_2 \cdot 6H_2O$ or $Si_{12}Mg_8O_{30}(OH)_4(H_2O)_4 \cdot 8H_2O$ | **Functionality**: OH | |
| **Chemical composition**: $SiO_2$ - 56.1%, MgO - 24.9%, $Al_2O_3$ - 0.7%, CaO - 1.7% | | |
| PHYSICAL PROPERTIES | | |
| **Density**, g/cm$^3$: 2-2.3 | **Mohs hardness**: 2-2.5 | **Melting point**, $^o$C: 1550 |
| **Loss on ignition**, %: 15 | | |
| CHEMICAL PROPERTIES | | |
| **Moisture content**, %: 8-16 | **pH of water suspension**: 7.5-8.5 | |
| OPTICAL PROPERTIES | | |
| **Color**: white, cream, gray, brown | | |
| MORPHOLOGY | | |
| **Particle size**, μm: 5-7 | **Crystal structure**: orthorhombic | **Micropore volume**, cm$^3$/g: 9.4 |
| **Sieve analysis**: 200 mesh sieve residue - 8% | **Specific surface area**, m$^2$/g: 240-310 | |
| MANUFACTURERS & BRAND NAMES: Non-Metals, Inc. Affiliate of The China National Non-Metallic Minerals Group, Tucson, AZ, USA<br>LH-I, LH-II, LH-III - grades having different sepiolite content | | |
| MAJOR PRODUCT APPLICATIONS: purification agent, asbestos replacement, filler in plastics and rubber, adhesives, blend compatibilizer | | |
| MAJOR POLYMER APPLICATIONS: polyurethane, PS, $PVF_2$, PMMA | | |

The fibrous structure of sepiolite is composed of talc-like ribbons with two sheets of tetrahedral silica units linked by oxygen atoms to a central octahedral sheet of magnesium. It has needle-shaped particles with channels oriented along the fibers which can absorb liquids. Sepiolite has three kinds of water: hygroscopic water, crystallization water, and constitution water. The crystallization water is removed at 500$^o$C and constitution water is removed at 850$^o$C at which point physical properties change brought about by crystal folding of sepiolite.[397]

## 2.1.51 SILICA[399-419]

About a third of all minerals belong to the silicates class, which is divided into five subclasses. Thirty-five other elements participate in the formation of various silicates which form about 95% of the rocky crust of the earth. Most of these (72%) belong to the subclass of tektosilicates called framework silicates. Feldspar and quartz are the most prominent species in this group.

In filler applications, the silicates group of greatest interest is in the subclass of tektosilicates. Four minerals (quartz, tridymite, cristobalite, and opal) belong to the silica group and three of them (quartz, cristobalite, and opal) are used as fillers or materials for their production.

The composition of pure quartz is close to 100% pure $SiO_2$ because the structure of the mineral is so compact and perfect that there is no room for silica replacement by any other element. Also, quartz is insoluble in all acids except HF, which further contributes to its purity. Quartz forms many micro- and cryptocrystalline varieties. Some of them are well-known as semiprecious stones (amethyst, citrine, agate, tiger-eye, etc.).

Unlike quartz, cristobalite has an open structure, allowing some fraction of silicon (2-3%) to be replaced by other elements, such as, Al, Na, or Ca. Still, 95% of the mineral is formed by $SiO_2$. The natural cristobalite does not exist in concentrations that make mining feasible therefore it is produced by synthesis (see separate section on cristobalite). Both minerals are found in volcanic rocks, but quartz, which constitutes 12.5% of the Earth's crust, is found everywhere, since it does not change or erode. Sandstone is one of the sources of quartz.

It should be mentioned here that diatomite or diatomaceous earth, formed from an accumulation of siliceous material of diatoms, is classified as an opal. This mineral is discussed under its commonly accepted name – diatomaceous earth – in the separate section above.

The common availability of silica is not the sole reason for its extensive use. Probably, it is the chemical inertness and durability of silica which determined its popularity. The fillers discussed here include not only natural minerals but also a variety of synthetic products. Natural products can be divided into crystalline and amorphous. Crystalline silica fillers include sands, ground silica (or silica flour), and a form of quartz – tripoli, whereas the amorphous types include diatomaceous earth.

In addition to the natural products, synthetic materials are in common use. Two methods of production are used: pyrogenic or thermal (commonly known as fumed silica grades) and wet process (commonly known as precipitated silica).

This mixture of natural and synthetic materials was taken as a base for creation of the groups below, which are grouped by their common name rather than by their origin.

## 2.1.51.1 FUMED SILICA[420-429]

| | |
|---|---|
| **Names**: fumed silica, pyrogenic silica, thermal silica | **CAS #**: 112945-52-5 for treated differs |
| **Chemical formula**: $SiO_2$ | **Functionality**: OH or modification-dependent |
| **Chemical composition**: $SiO_2$ - 96-99.8%, $Al_2O_3$ - 0.05-1.3%, $Fe_2O_3$ - 0.003-0.06%, $TiO_2$ - 0.03% | |
| **Trace elements**: Al, As, Au, Ba, Ca, Cd, Co, Cr, Cu, Fe, Hg, In, K, Mg, Mn, Mo, Na, Ni, Pb, Sb, Sc, Sn, Th, U, Zn. The trace elements content is below the limits specified by the requirements of major pharmacopoeias | |
| **PHYSICAL PROPERTIES** | |
| **Density**, g/cm$^3$: 2-2.2 | **Decomposition temp.**, °C: >2000 |
| **Loss on ignition**, %: 1-2.5 (hydrophilic), 1-7 (hydrophobic) | |
| **Thermal conductivity**, W/K·m: 0.015 | **Maximum temperature of use**, °C: 850 |
| **CHEMICAL PROPERTIES** | |
| **Chemical resistance**: non-reactive with acids with the exception of HF, unstable in alkalis | |
| **Moisture content**, %: 0.5-2.5 (hydrophilic) 0.5 (hydrophobic) | **Adsorbed moisture**, %: 6 |
| **pH of water suspension**: 3.6-4.5 (hydrophilic), 3.5-11 (hydrophobic) | **Water solubility**, %: 0.015 |
| **OPTICAL & ELECTRICAL PROPERTIES** | |
| **Refractive index**: 1.46 | **Volume resistivity**, Ω-cm: $10^{13}$ |
| **MORPHOLOGY** | |
| **Particle shape**: spherical; **Crystal structure**: amorphous | **Porosity**: non porous |
| **Particle size primary**, nm: 5-40; **Oil absorption**, g/100 g: 100-330 | **Appearance**: fluffy white powder |
| **Density of silanol groups**, 1/nm$^2$: 1.5-4.5 | **Aggregate size**, μm: 0.2-15 |
| **Sieve analysis**: 325 mesh sieve residue - 0.05-1% | **Specific surface area**, m$^2$/g: 50-400 |

**MANUFACTURERS & BRAND NAMES**:

Cabot Corporation, Cab-O-Sil Division, Tuscola, IL, USA

Cab-O-Sil L-90, LM-130, LM-150, M-5, MS-55, H-5, HS-5, EH-5 - hydrophilic grades of fumed silica differing in average primary particle size and BET surface area

Cab-O-Sil TS-720, TS-610, TS-530, TS-500 - hydrophobic grades differing in particle size and BET surface area and treatment chemistry

Cab-O-Sil LM-150D, M-7D, M-75D - densified grades

Degussa AG, Frankfurt/Main, Germany

Aerosil 90, 130, 150, 200, 300, 380, OX50, TT600, MOX80, MOX170 - hydrophilic grades of fumed silica differing in average primary particle size and BET surface area

Aerosil COK 84 - mixture of Aerosil and highly dispersed $Al_2O_3$ in ratio of 5:1 for thickening of aqueous systems

Aerosil R202, R805, R812, R812S, R972, R974, R104, R106, R504, R816 - hydrophobic grades differing in particle size and BET surface area and treatment chemistry

Aerosil K315, K328, K330, K342, DCF784, SATESSA28, SATESSA42 - 30% dispersions

Harwick Standard Distribution Corporation, Akron, OH, USA

Silica S - low cost pyrogenic silica filler for rubber

*continued on the next page*

**MANUFACTURERS & BRAND NAMES:**
Wacker-Chemie GmbH, München, Germany
HDK S13, V15, N20, T30, T40 - hydrophilic grades of fumed silica differing in average primary particle size and BET surface area
HDK H15, H20, H30, H2000, H2000/4, H3004, H2015EP, H2050EP - hydrophobic grades differing in particle size, BET surface area, and treatment chemistry

**MAJOR PRODUCT APPLICATIONS:** paints, coatings, primers, powder coatings, printing inks, pigments, diazo paper, toothpaste, tablets, powders, aerosols, ointments, creams, dry toner, sealants, rubber goods, adhesives, cable and wire, laminates, gel coats, body putties, defoamers, food, insecticides, lubricants, animal feeds, fertilizers, polishes, reproduction papers, waxes

**MAJOR POLYMER APPLICATIONS:** polyurethane, epoxy, silicone, polychloroprene, PSF, acrylics, PVC, polyesters, alkyd, fluoroelastomers, NR, SBR

The product obtained from the vapor process is frequently termed fumed silica because it looks like smoke or fumes. This process was developed by applying carbon black production technology and equipment to silica tetrachloride in an invention by Degussa AG. Fumed silica manufactured is presently based on Degussa's license, which was sold to only a few other corporations. Metallic silicon and gaseous dry HCl are reacted to form silica tetrachloride, which is mixed with hydrogen and air and fed into the burner tube of the reactor where the following reactions occur:

$$2H_2 + O_2 \rightarrow 2H_2O$$
$$SiCl_4 + 2H_2O \rightarrow SiO_2 + 4HCl$$

The reaction temperature is around 1800°C. The HCl formed in the process is recycled. The primary particles of silica leaving the burner are in a molten state; therefore, on collision they are able to coalescence, forming bigger particles. When particles proceed through the reactor, they cool down, and around 1710°C they become solid and are no longer able to recombine. Before this happens, primary particles fuse with one another and form chain-like, branched aggregates. The size of primary grains is usually in the range of 7 to 30 nm, which produces a specific surface area in the final product from 400 to 100 $m^2/g$. Below the melting point of silica (1710°C) particles still collide and form aggregates due to mechanical entanglement or agglomeration. Agglomeration also occurs in the collection process. These mechano-physical aggregates can be disintegrated on mixing during the processing of material formulated with fumed silica. Some trace amounts of HCl (less than 200 ppm) are retained in the product. The process of production of fumed silica sometimes includes compacting, which increases the product density by 2-2.5 times. The manufacturing process can be easily regulated with respect to primary particle size and the size and structure of the aggregate. Figure 2.51 shows the schematic diagram of production process.

Figure 2.52 illustrates the difference between fumed silica and crystalline silica. The diagram for fumed silica does not show absorption peaks whereas the diagram for quartz, which is a crystalline product, does. The amorphous nature of

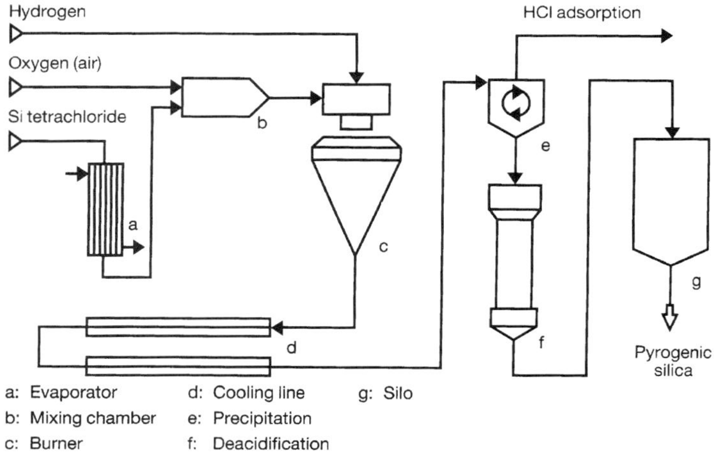


Figure 2.51. Production of Aerosil. *Courtesy of Degussa AG, Frankfurt/Main, Germany.*

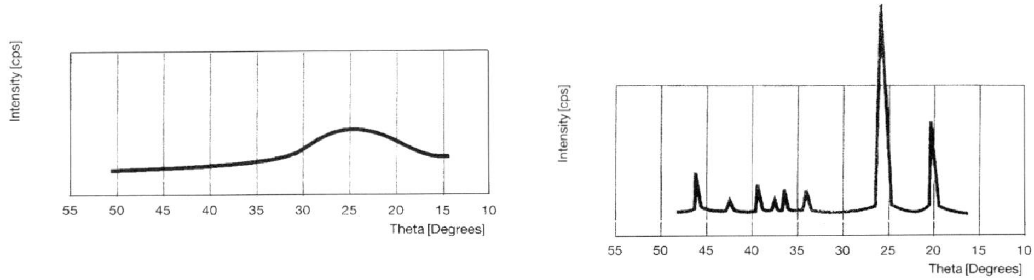


Figure 2.52. X-ray diagram of fumed silica (left) and quartz (right). *Courtesy of Wacker-Chemie GmbH, München, Germany.*

fumed silica is probably caused by the fast cooling process, which takes a few thousandths of a second. This permits the classification of fumed silica as amorphous and is an important benefit for those working with fumed silica that, unlike the crystalline forms of silica, it does not cause silicosis.

Figure 2.53 explains differences between the chemical composition of surfaces of hydrophilic, and silane treated, hydrophobic, fumed silica. The isolated hydroxyl groups and hydrogen-bonded hydroxyl groups are both hydrophilic, whereas the siloxane group is hydrophobic. These chemical groups make the surface of untreated silica hydrophilic and are essential for its properties and applications. Chemical and thermogravimetric analysis indicate that there are approximately 3 to 4.5 hydroxyl groups per square nm of silica surface. On the surface of hydrophobic fumed silica, dimethylsilyl, trimethylsilyl,

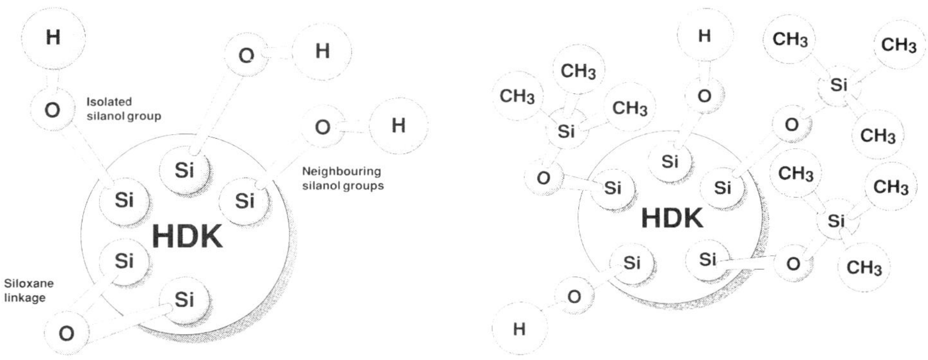


Figure 2.53. Chemical structure of untreated (left) and treated (right) fumed silica surface. *Courtesy of Wacker-Chemie GmbH, München, Germany.*

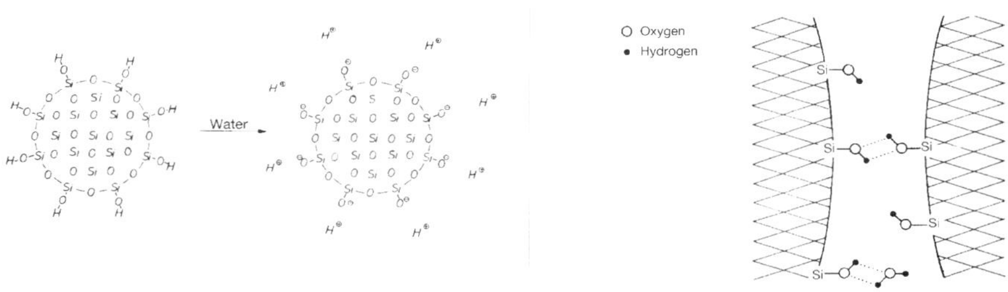


Figure 2.54. The origin of acidic properties of fumed silica (left) and the mechanism of hydrogen bonding (right). *Courtesy of Degussa AG, Frankfurt/Main, Germany.*

dimethylsiloxane, and octyl groups replace some hydroxyl groups. Typically about 1.5 OH groups per square nm remain after treatment. The extent of replacement regulates the hydrophobic properties of fumed silica.

Fumed silica is a weak acid and hydroxyl groups are essential in hydrogen bonding (Figure 2.54).

The mechanism of thickening of liquids by fumed silica is explained by hydrogen bond formation between neighboring aggregates of silica, leading to the formation of a regular network. On the application of shear some of these bonds are broken which reduces viscosity. The initial state is regained when material is left to stand. Hydroxyl groups, needed for this process, are converted to siloxane groups on heating to 110°C, which retards the reaction. Fumed silica, on leaving the factory, has 0.5-2.5% moisture, which is partially needed for the thickening process but, at the same time what water remains is reactive to some of the components in

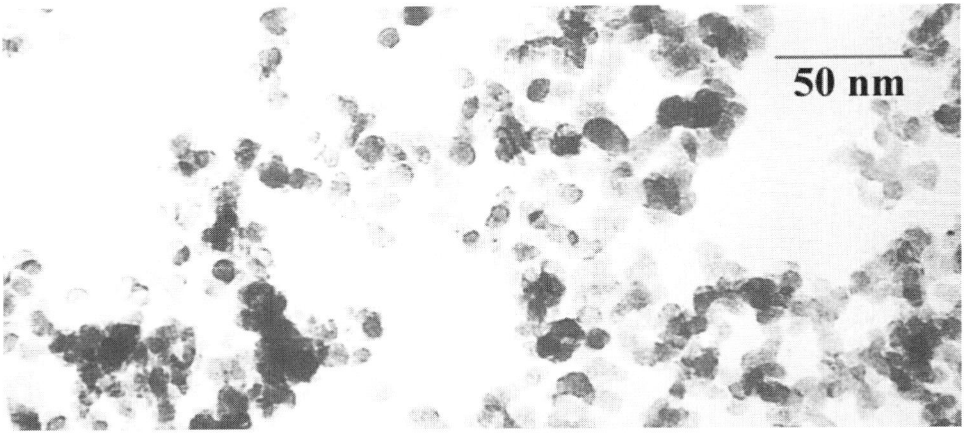


Figure 2.55. SEM micrograph of Wacker HDK N20. Magnification 300,000x. *Courtesy of Wacker-Chemie GmbH, München, Germany.*

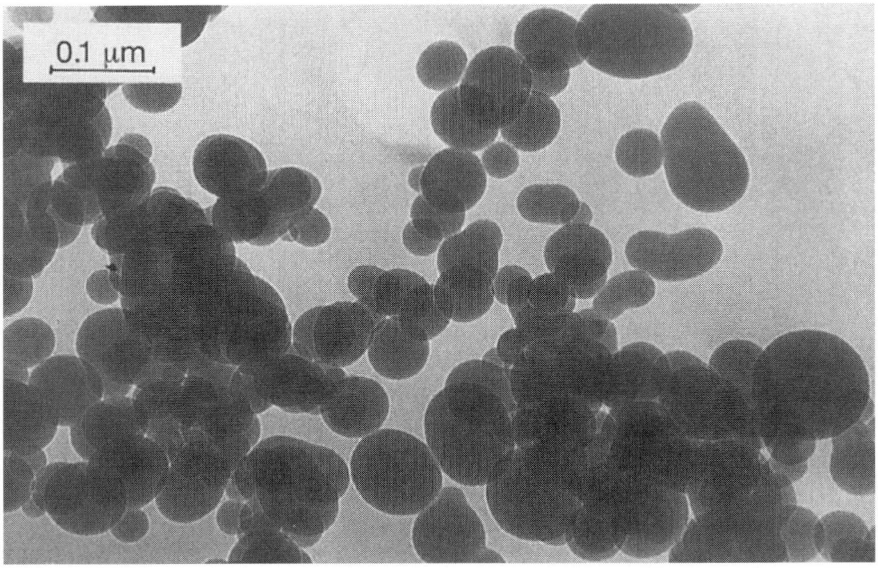


Figure 2.56. TEM micrograph of Aerosil OX50. *Courtesy of Degussa AG, Frankfurt/Main, Germany.*

industrial formulations, such as with ketimines used for polyurethane prepolymer curing.

Figure 2.55 shows the morphology of fumed silica which is composed of grain-like agglomerates. Figure 2.56 shows that particles are spherical. The morphology of primary particles is easier to observe in Aeorosil OX50 which has a larger size of primary particles (40 nm) and TEM display information on the shape of particle in a two dimensional scale. A primary particle of fumed silica is built up of about 10,000 $SiO_2$ units.

The mixing process of fumed silica must be carefully designed to control the degree of thickening. Fumed silica particles are composed of aggregates and agglomerates which are dispersed to form smaller aggregates. Overmixing reduces the size of aggregates too much and aggregates cannot form network of chains interconnected throughout the mixture. Instead, they will form only a partial network. Such overmixing is irreversible process.

In industrial products, the use of fumed silica will confer thixotropy, sag resistance, particle suspension, emulsifiability, reinforcement, gloss reduction, flow enhancement of powders, anti-caking, anti-slip, anti-blocking, etc. Because of its effect on these important properties, fumed silica is widely used in many industries.

## 2.1.51.2 FUSED SILICA

| | | |
|---|---|---|
| **Name**: fused silica | | **CAS #**: 60676-86-0 |
| **Chemical formula**: $SiO_2$ | **Functionality**: none or from silane | |
| **Chemical composition**: $SiO_2$ - 98.5-99%, $Al_2O_3$ - 0.25-1%, $Fe_2O_3$ - 0.05% | | |
| PHYSICAL PROPERTIES | | |
| **Density**, g/cm$^3$: 2.2 | **Mohs hardness**: 7 | **Loss on ignition**, %: 0.1-0.45 |
| **Thermal conductivity**, W/K·m: 1.1 | **Linear coefficient of thermal expansion**, 1/K: 0.5x10$^{-6}$ | |
| CHEMICAL PROPERTIES | | |
| **Chemical resistance**: resistant to acids | **Moisture content**, %: 0.1 | |
| **pH of water suspension**: 9 | | |
| OPTICAL & ELECTRICAL PROPERTIES | | |
| **Color**: white | **Dielectric constant**: 3.78 | **Loss tangent**: <1x10$^{-3}$ |
| **Specific electric conductivity**, S/cm: 10$^{-17}$ -10$^{-18}$ | | |
| MORPHOLOGY | | |
| **Particle size**, µm: 4-28 | **Oil absorption**, g/100 g: 17-27 | |
| **Specific surface area**, m$^2$/g: 0.8-3.5 | | |
| MANUFACTURERS & BRAND NAMES: Denki Kagaku Kogyo Co., Ltd., Ibaraki, Japan — FB-30, FB-35, FB-48, FB-74 - spherical fused amorphous silica; Quarzwerke GmbH, Frechen, Germany — Silbond FW61, FW12, FW100, FW300, FW600 - fused silica flours of different particle sizes. Available with aminosilane (AST grade) and epoxysilane (EST) | | |
| MAJOR PRODUCT APPLICATIONS: encapsulating material for integrated circuits, electric components, conductors | | |
| MAJOR POLYMER APPLICATIONS: epoxy, PPS | | |

Fused silica flour is produced from electrically fused $SiO_2$ by iron free grinding followed by air separation. As an option, it may be coated with silane. Quarzwerke GmbH treats flour with amino and epoxysilanes. Denki Kagaku Kogyo Co., Ltd. manufactures spherical grades of fused amorphous silica.

The properties of this filler can be appreciated when compared with silica sand discussed below in separate section. The comparison shows a very low linear thermal expansion coefficient, thermal conductivity, and very high specific electrical conductivity. These unusual properties, similar to those of the pure quartz crystal, are exploited in applications in electronics.

## 2.1.51.3 PRECIPITATED SILICA[429,432-440]

| | | |
|---|---|---|
| **Name**: precipitated silica | | **CAS #**: 63231-67-4 |
| **Chemical formula**: $SiO_2$ | **Functionality**: OH or from silane | |
| **Chemical composition**: $SiO_2$ - 97.5-99.4%, $Fe_2O_3$ - 0.01-0.1%, $Al_2O_3$ - 0.6%, $TiO_2$ - 0.07%, CaO - 0.5%, MgO - 0.2%, $Na_2SO_4$ - 0.8-1.5% | | |
| **PHYSICAL PROPERTIES** | | |
| **Density**, g/cm$^3$: 1.9-2.1 | **Mohs hardness**: 1 | **Loss on ignition**, %: 3-18 |
| **CHEMICAL PROPERTIES** | | |
| **Moisture content**, %: 3-7 | **Adsorbed moisture**, %: 7-20 | **OH group density**, 1/nm$^2$: 5-12 |
| **pH of water suspension**: 3.5-9 | | |
| **OPTICAL & ELECTRICAL PROPERTIES** | | |
| **Refractive index**: 1.46 | **Dielectric constant**: 1.9-2.8 | **Loss tangent**: 0.00001-0.02 |
| **Color**: white | **Volume resistivity**, Ω-cm: $5.7 \times 10^{11}$-$4.5 \times 10^{14}$ | |
| **MORPHOLOGY** | | |
| **Predominant pore diameter**, nm: 30 | | **Hegman fineness**: 5-7 |
| **Agglomerate size**, μm: 1-40 | **Oil absorption**, g/100 g: 60-320 | **Primary particle size**, nm: 5-100 |
| **Sieve analysis**: 325 mesh sieve residue - 0.002-0.2% | **Specific surface area**, m$^2$/g: 12-800 | |

**MANUFACTURERS & BRAND NAMES**:
Charles B. Chrystal Co., Inc., New York, NY, USA
  Precipitated Silica # 32, #22 - flatting agents, food and pharmaceutical grades
Degussa AG, Frankfurt/Main, Germany
  Ultrasil VN 3, FK 160 FK 300 DS, FK 310 - hydrophilic
  Sipernat D 10, D 17 - hydrophobic
PPG Industries, Pittsburgh, PA, USA
  Lo-Vel 27, 275, 28, 29, 39, 66, HSF, Inhibisil - flatting agents
  Hi-Sil T-600, T-700 - thickeners
Rhône Poulenc, Paris, France
  Zeosil Z91, Z93, Z162, Z172A, Z172B, 175 MP

**MAJOR PRODUCT APPLICATIONS**: tires, sealants, adhesives, coatings, paints, topcoat lacquers, coil coating, micro texture finish, wood finishes, thixotropes, office furniture

**MAJOR POLYMER APPLICATIONS**: nitrocellulose, melamine, polyester, acrylics, silicone, alkyd, epoxy, PVC, EPDM, NR, SBR

Precipitated silica is produced from sodium silicate through its reaction with sulfuric and hydrochloric acids. The following reactions apply:

$$3SiO_2 + Na_2CO_3 \rightarrow 3SiO_2{\cdot}Na_2O + CO_2$$
$$(SiO_2{\cdot}Na_2O)_{aq} + H^+ SO_4^- \rightarrow SiO_2 + Na_2SO_4 + H_2O$$

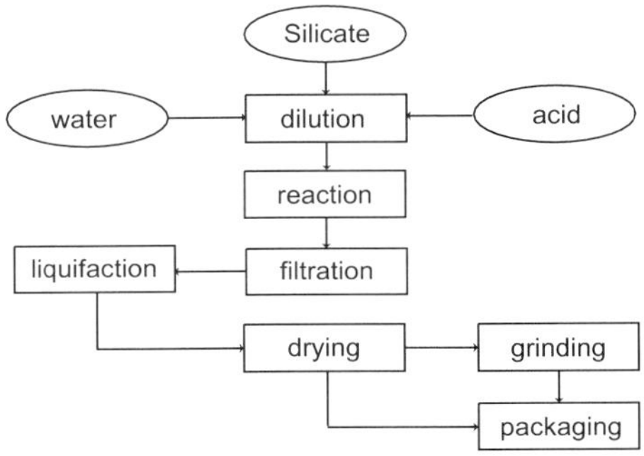


Figure 2.57. Precipitated silica process. After Bomo F, Meeting of the Rubber Division, ACS, Montreal, May 5-8, 1996, paper E.

From these reactions it is quite evident that the concentration of the remaining $Na_2SO_4$ is one of the quality factors. Figure 2.57 shows the schematic diagram of the process.

Concentration of reactants, rates of addition, fraction of theoretical silicate in the reaction, and temperature are the process variables determining the properties of the final product, such as, oil number, specific surface area, porosity, primary particle and agglomerate size and shape, brightness, density, and hardness. After the reaction is complete, the product is separated by filtration, washed, dried, and milled. Final products are sometimes indexed in a manner similar to carbon blacks, which distinguishes the following grades: very high structure (VHS), high structure (HS), medium structure (MS), low structure (LS), and very low structure (VLS).

Moisture concentration in the final product is comparably high (3-7%) and three types of water are available: free water, which can be removed at 105°C; adsorbed water (hydrogen bonded water), which is removed on heating from 105 to 200°C; and constitutional water, which can only be removed in a temperature range from 700 to 900°C. The mechanism of thickening is similar to that of fumed silica and involves bridging between two particles by formation of hydrogen bonding formed by the interaction of silanol and siloxane groups. Precipitated silica has more silanol groups than fumed silica. The product has a lower concentration of silica since it usually contains an admixture of sodium sulfate (approximately up to 1.5%).

Recent advances in the application of precipitated silica in tires will rapidly increase consumption of this filler beyond that which it enjoys in its traditional markets. Regulation of thixotropic properties of industrial products and the flatting of coatings and paints are important applications for these fillers. Figure 2.58 shows the mechanism of flatting. Very good dispersion of precipitated silica facilitates uniform distribution of its agglomerates. The presence of agglomerates close to surface causes surface roughening.

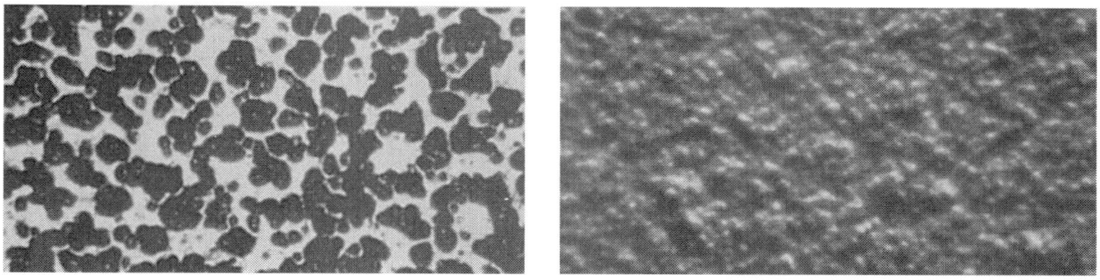

Figure 2.58. Surface flatting mechanism by precipitated silica, Lo-Vel HSF. Left - distribution of agglomerates, right - surface roughness of coating. *Courtesy of PPG Industries, Inc., Pittsburgh, PA, USA.*

## 2.1.51.4 QUARTZ (TRIPOLI)

| | | |
|---|---|---|
| **Names**: microcrystalline silica powder, tripoli, novaculite, quartz silica | | **CAS #**: 14808-60-7 |
| **Chemical formula**: $SiO_2$ | **Functionality**: none or silane modified | |
| **Chemical composition**: $SiO_2$ - 99.1-99.4%, $Fe_2O_3$ - 0.04%, $Al_2O_3$ - 0.1%, $TiO_2$ - 0.02%, CaO - 0.01% | | |
| PHYSICAL PROPERTIES | | |
| **Density**, g/cm$^3$: 2.65 | **Mohs hardness**: 7 | **Loss on ignition**, %: 0.2 |
| **Maximum temperature of use**, $^o$C: 573 | | **Specific heat**, kJ/kg·K: 0.8 |
| CHEMICAL PROPERTIES | | |
| **Moisture content**, %: 0 | **Adsorbed moisture**, %: 8.7 | **pH of water suspension**: 6-7.8 |
| OPTICAL PROPERTIES | | |
| **Refractive index**: 1.55 | **Color**: white | **Brightness**: 80 |
| MORPHOLOGY | | |
| **Particle shape**: platy | **Crystal structure**: trigonal | **Hegman fineness**: 0-7 |
| **Particle size**, μm: 2-19 | **Oil absorption**, g/100 g: 17-20 | |
| **Sieve analysis**: 325 mesh sieve residue - 0.1-1% | | |
| MANUFACTURER & BRAND NAMES:<br>Charles B. Chrystal Co., Inc., New York, NY, USA<br>Silica 3-37 - micronized platy silica<br>Malvern Minerals Company, Hot Springs National Park, AR, USA<br>Novacite 200, 325, 1250, Daper, L-207A, L-337 - grades having different particle size but with the same oil absorption<br>Novakup - silane treated Novacite grades | | |
| MAJOR PRODUCT APPLICATIONS: paints, coatings, corrosion-resistant finishes, casting and potting compounds, powder coatings, grouts, molding articles, electrostatic coatings, pipe linings, silicon rubber articles, abrasive materials | | |
| MAJOR POLYMER APPLICATIONS: polyurethanes, alkyd, acrylics, silicon PVC | | |

The range of materials are produced by Malvern Minerals Company from the high purity mineral – Novaculite - found in Hot Springs, Arkansas. The platy disc shaped particles have many properties important to industrial applications. Novacite has low oil absorption and water sorption, good flatting effect, chemical inertness, and it gives a chalk-free, UV-resistant and non-staining coatings with typical paint binders. Figure 2.59 shows morphological structure of this unique material. The platelet particles combine to form clusters.

Figure 2.59. Novacite morphology. Left - single platelet, middle - distribution of sizes, right - cluster. *Courtesy of Malvern Minerals Company, Hot Springs National Park, AR, USA (micrographs of platelet and cluster).*

## 2.1.51.5 SAND[441-444]

| | | |
|---|---|---|
| **Names**: sand, silica flour, ground silica | | **CAS #**: 14808-60-7 |
| **Chemical formula**: $SiO_2$ | **Functionality**: none or from silane | |
| **Chemical composition**: $SiO_2$ - 97.5-99.8%, $Al_2O_3$ - 0.05-2%, $Fe_2O_3$ - 0.02-0.05% | | |
| PHYSICAL PROPERTIES | | |
| **Density**, g/cm$^3$: 2.65 | **Mohs hardness**: 7 | **Loss on ignition**, %: 0.1-0.55 |
| **Thermal conductivity**, W/K·m: 7.2-13.6 | **Linear thermal expansion coefficient**, 1/K: 14x10$^{-6}$ | |
| CHEMICAL PROPERTIES | | |
| **Moisture content**, %: 0.1 | **pH of water suspension**: 6.8-7.2, 7-9 (silane treated) | |
| OPTICAL & ELECTRICAL PROPERTIES | | |
| **Specific electric conductivity**, S/cm: 10$^{-14}$-10$^{-16}$ | | **Brightness**: 80-88 |
| **Dielectric constant**: 4 | **Reflectance**: 82-90 | |
| MORPHOLOGY | | |
| **Particle size**, μm: 2-90 | **Oil absorption**, g/100 g: 14-28 | **Hegman fineness**: 0-4 |
| **Sieve analysis**: residue on 325 mesh sieve - 0.1-47% | | **Specific surface area**, m$^2$/g: 0.3-6 |
| MANUFACTURERS & BRAND NAMES:<br>Charles B Chrystal Co., Inc., New York, USA<br>High Purity Quartz Type 31/90, Type P, Starsil Spherical Silica - natural silica of different sizes and purity<br>Quarzwerke GmbH, Frechen, Germany<br>Millisil W 3, 4, 6, 8, 10, 12 - iron-free grinding of processed silica sand. Particle size decreases with grade number increasing<br>Sikron SF 300, 500, 600, 800, SH 300, 500 - micronized silica flours<br>Silbond W 6, 12, 100, 600, 800 - silica flours treated with various silanes (AST - amino, EST - epoxy, MST - methylacrylo, RST - trimethyl, TST - methyl, VST - vinyl)<br>US Silica Company, Berkeley Springs, WV, USA<br>Full range of quality sands and silica flours under the following brand names Mystic White, F-series Foundry sands, Penn Sand, Q-Mix, Q-Rok, Sil-Co-Sil, Supersil, Min-U-Sil | | |
| MAJOR PRODUCT APPLICATIONS: high temperature synthesis of wollastonite, synthesis of calcium hydro silicates, sealants, stucco, primers, road marking formulations, resin casting, adhesives, mortars, coatings, paint, lacquers, special papers, construction elements, pin insulators, machine tools, lining for chemical pumps | | |
| MAJOR POLYMER APPLICATIONS: epoxy, polyurethanes, polyesters, PMMA, PVC, PE | | |

The production of sand fillers is simple because it includes, at most, only washing and classification into grades differing in grain size. Because sand has a negligible degree of porosity it has an extremely low specific surface area in the range from 40 to 160 cm$^2$/g. The material usually contains more than 99.7% $SiO_2$, with absorbed water being at a negligible level (0.1%). Ground silica sand is produced in a similar manner, except that pulverizing is included. Ground silica can easily be distinguished under the microscope because it has irregular grains. Grinding considerably increases the surface area into the range from 1000 to 5000 cm$^2$/g, with an average particle size in a range from 16 to 4 μm. High quality grades are produced by grinding sand in iron-free ball mills followed by classification controlled by a laser technique with Cilas-granulometers. Material from this process is stored in moisture-free silo. Figure 2.60 shows the morphology of silica sand.

Figure 2.60. Silica sand (100x). *Courtesy of Quarzwerke GmbH, Frechen, Germany.*

The content of iron is one of important indicators of quality of silica flour for various applications, especially for external coatings. The presence of iron causes formation of rusty streaks which form when the iron oxidizes. The good quality material for these applications should have a $Fe_2O_3$ content below 0.03%.

## 2.1.51.6 SILICA GEL[445-447]

| | | |
|---|---|---|
| **Names**: silica gel, amorphous silica | | **CAS #**: 7699-41-4 |
| **Chemical formula**: $SiO_2$ | | |
| **PHYSICAL PROPERTIES** | | |
| **Density**, g/cm$^3$: 2.2-2.6 | **Mohs hardness**: 6 | |
| **CHEMICAL PROPERTIES** | | |
| **pH of water suspension**: 6.5-7.5 | | |
| **MORPHOLOGY** | | |
| **Particle size**, μm: 2-15 | **Oil absorption**, g/100 g: 80-280 | **Pore radius**, nm: 5-40 |
| **Specific surface area**, m$^2$/g: 40-850 | | |
| **MANUFACTURERS & BRAND NAMES**:<br>Crosfield Group, Warrington, UK<br>Gasil 200 DF, Gasil HP 370<br>Macherey Nägel<br>Nucleosil Nu 100-30, 1000-30 | | |
| **MAJOR PRODUCT APPLICATIONS**: paints, coatings, drying of materials, putties, window spacers | | |
| **MAJOR POLYMER APPLICATIONS**: alkyd, polyurethanes | | |

Silica gel is produced according to the following reaction:

$$Na_2O(SiO_2)_x + H_2SO_4 \rightarrow xSiO_2 + Na_2SO_4 + H_2O$$

The product of reaction contains about 75% water and is subjected to a drying process. Drying takes place in a rotary kiln followed by milling of the material which has been previously washed with hot alkaline water (which reinforces the matrix, decreases shrinkage, and produces larger pores), results in the xerogels. Super-critical drying or replacing water by methanol, before drying, decreases the crushing force and produces aerogels which have up to 94% air space. The average particle size is in a range from 2 to 15 μm. Further changes in the particle size can be accomplished by milling and air classification. The specific surface area is very high due to the high porosity (40-850 m$^2$/g). Hydrogels have a pore radius (7-12 nm) similar to xerogels (5-15 nm), while aerogels have a higher pore radius (10-40 nm). Small particles and high porosity result in high oil absorption, in a range from 80 to 280%.

Silica gels of specific pore size (e.g., Gasil grades) are becoming important in surface matting of paints.

## 2.1.52 SILVER POWDER AND FLAKES[448]

| | | |
|---|---|---|
| **Names**: silver powder, silver flakes, atomized silver powder, silver/palladium powder and flakes | | **CAS #**: 7440-22-4 |
| **Chemical formula**: Ag | | **Functionality**: none |
| **Chemical composition**: Ag - 99.3-99.9%; silver/palladium - all ratios available | | |
| **Trace elements**: heavy metals - 0.02%, Na+K - 0.01-0.02% | | |
| PHYSICAL PROPERTIES | | |
| **Density**, g/cm$^3$: 10.5 | **Mohs hardness**: 2.5-4 | **Melting point**, °C: 962 |
| **Thermal conductivity**, W/K·m: 450 | | **Specific heat**, kJ/kg·K: 0.188 |
| **Tensile strength**, MPa: 290 | | |
| CHEMICAL PROPERTIES | | |
| **Chemical resistance**: soluble in strong acids | | |
| ELECTRICAL PROPERTIES | | |
| **Resistivity**, Ω-cm: $1.59 \times 10^{-6}$ | | |
| MORPHOLOGY | | |
| **Particle shape**: spherical or flake | **Crystal structure**: cubic | **Particle size**, μm: 0.25-25 |
| **Sieve analysis**: 325 mesh residue - traces | | **Specific surface area**, m$^2$/g: 0.15-6 |
| MANUFACTURER & BRAND NAMES:<br>Technic Inc., Woonsocket, RI, USA<br>Silpowder 171, 172, 173, 222, 223, 225, 228, 251, 252, 253, 263, 271, 335, 336, 995 - chemically precipitated powders of different particle sizes for applications listed below<br>Silsphere 514, 517, 519 - chemically precipitated spherical powders<br>Silflakes 131, 132, 134, 135, 138, 235, 237, 239, 241, 242, 282, 285, 299, 255, 450, 556 - mechanically flatted powders to form flakes, mostly for conductive applications<br>Silver/Palladium powders 600 and 700 Series - chemically co-precipitated spherical powders | | |
| MAJOR PRODUCT APPLICATIONS: conductive inks, pastes, coatings, adhesives, thick films, battery plates, electrical contacts, powder metallurgy, capacitor inks | | |
| MAJOR POLYMER APPLICATIONS: epoxy and others | | |

Figure 2.61 shows the morphology of powder (product of chemical precipitation) and flakes made by mechanical flattening of powders.

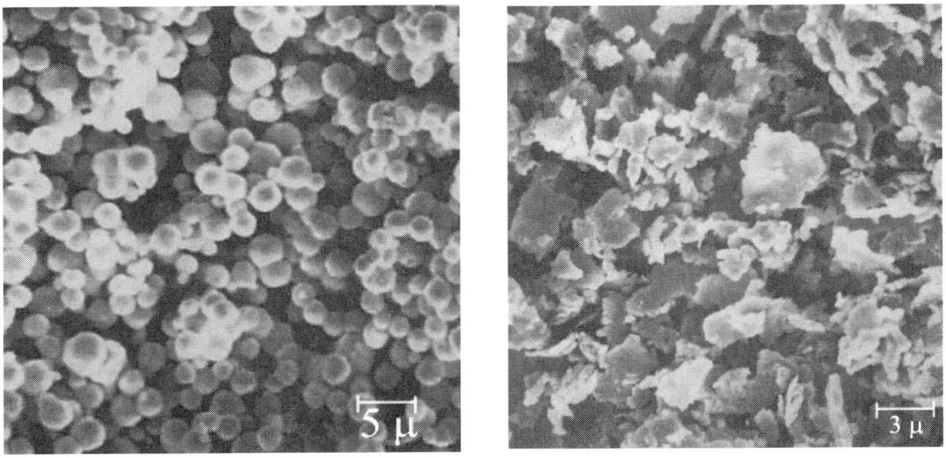


Figure 2.61. Silver powder and flake. *Courtesy of Technic, Inc., Woonsocket, RI, USA.*

## 2.1.53 SLATE FLOUR

| | |
|---|---|
| **Name**: slate flour | **CAS #**: 1335-30-4 |
| **Chemical formula**: variable | **Functionality**: OH |
| **Chemical composition**: $SiO_2$ - 35-62.3%, $Al_2O_3$ - 8.5-20.7%, $Fe_2O_3$ - 2.5-7.65%, CaO - 0.2-2.5%, MgO - 0.4-2%, $Na_2O$ - 0.3-1.2%, $K_2O$ - 2.2-3.6%, carbon - 28.9-29.7% | |
| PHYSICAL PROPERTIES | |
| **Density**, g/cm$^3$: 2.1-2.7 | |
| CHEMICAL PROPERTIES | |
| **Chemical resistance**: reacts with acids and alkalis | |
| **Moisture content**, %: 1 | **pH of water suspension**: 6.5-8.1 |
| OPTICAL PROPERTIES | |
| **Color**: red, light-dark gray | |
| MORPHOLOGY | |
| **Sieve analysis**: residue on 325 mesh sieve - 1% | **Oil absorption**, g/100 g: 22-32 |
| MANUFACTURERS & BRAND NAMES:<br>Keystone Filler & Manufacturing Co., Muncy, PA, USA<br>Light Gray Slate Flour, Dark Gray Slate Flour, Red Slate Flour<br>Charles B. Chrystal, Co., Inc., New York, NY, USA<br>Light Gray Slate Flour, Dark Gray Slate Flour | |
| MAJOR PRODUCT APPLICATIONS: inexpensive filler | |

## 2.1.54 TALC[449-472]

| | | |
|---|---|---|
| **Names**: talc, magnesium silicate hydroxide, phyllosilicate | | **CAS #**: 14807-96-6 |
| **Chemical formula**: $Mg_3Si_4O_{10}(OH)_2$ | | **Functionality**: OH or silane modified |
| **Chemical composition**: $SiO_2$ - 46.4-63.4%, MgO - 24.3-31.9%, CaO - 0.4-13%, $Al_2O_3$ - 0.3-0.8%, $Fe_2O_3$ - 0.1-1.8% | | |
| **Trace elements**: Pb, As, Cd, Zn, Ba, Sb | | |
| PHYSICAL PROPERTIES | | |
| **Density**, g/cm$^3$: 2.7 - 2.85 | **Mohs hardness**: 1-1.5 | **Loss on ignition**, %: 4.8-17 |
| **Thermal conductivity**, W/K·m: 0.02 | | **Maximum temperature of use**, $^o$C: 900 |
| **Thermal expansion coefficient**, 1/K: 8 | | **Specific heat**, kJ/kg·K: 0.82 |
| CHEMICAL PROPERTIES | | |
| **Moisture content**, %: 0.1-0.6 | **pH of water suspension**: 8.7-10.6 | |
| **Water solubility**, %: 0.1 | | **Acid soluble matter**, %: 2 |
| OPTICAL & ELECTRICAL PROPERTIES | | |
| **Refractive index**: 1.57-1.59 | **Brightness**: 78-93 | **Whiteness**: 70-94 |
| **Color**: white | | **Dielectric constant**: 7.5 |
| MORPHOLOGY | | |
| **Particle shape**: platy | **Crystal structure**: monoclinic | **Cleavage**: basal |
| **Particle size**, μm: 1.4-19 | **Oil absorption**, g/100 g: 22-57 | **Hegman fineness**: 0-7 |
| **Aspect ratio**: 5-20 | **Particle thickness**, μm: 0.2-6 | |
| **Sieve analysis**: 325 mesh sieve residue - 0.1-2% | | **Specific surface area**, m$^2$/g: 2.6-35 |

MANUFACTURERS & BRAND NAMES:

Barretts Minerals, Inc., USA
    MP12-50, MP44-26
Canadian Talc Ltd.
    Cantal 45-80
Charles B. Chrystal Co., Inc., New York, NY, USA
    AGSC - talc from South China, meeting CTFA Specification
    Purtalc 6030, 428 - USP Grade
    Talc 523, Delusted, #2 French, #44 - lower price talc
    Micro Talc, 928, Bacteria-free, Sugarloaf grades, Purtalc, Osmanthus, Vertal CO+ - cosmetic grades
    Paper Talc, 9610, 7022, 10-MO - industrial grades

*continued on the next page*

**MANUFACTURERS & BRAND NAMES:**

Luzenac Europe, Toulouse, France

*Paper grades*

Lithicoat P2F, T3F, T4A - grades for matt wood-free art paper which are the mixtures of talc, chlorite and dolomite

Mistron - talc for pitch control

Malusil Naintsch - absorbs interfering anionic substances without impairment of hydrophobic and organophilic character of material due to the activation process which changes its zeta potential

Mistron Vapor C, P2, P5 - microcrystalline talc © - compacted grades with 3% water)

Luzenac 0, 1 - general purpose talc for paper industry

*Plastic grades*

Luzenac 1445, 20M0, 20M00S, 00S - highly lamellar talc with low abrasiveness from French Pyrenees

Steamic 00S, 00S D - micronized and finely ground talc for PP dashboards and bumpers

1N, Extra 5/0-M10, Prever, M8, M10C, M8C, M30 - talc from Val Germanasca mine in Italy for rubber and plastics

*Paint grades*

1N-M20, Prever M10, Extra 5/0 - high purity talc from Val Germanasca mine in Italy

Mistrofil 325, 400 - microcrystalline chlorite

Mistron 705, 754, Monomix, Super-20, Monomix-E, PE-60 - microcrystalline talc

Naintsch E, SE, ASE, - extremely lamellar structure and talcs containing dolomite

Luzenac 00C, 20M0, 10M0, Steabright, Steaopac - various finishes in decorative and industrial paints

Milwhite, Inc., Houston, TX, USA

TDM crude, 85, 92, 95, 98, 300, 325, W-93, W-98, W-286, W-300, W-325, CS-92 - industrial talcs

Westex 60/40, 65/35, 73/27, 80/20 - blended talcs

Westex FF - calcinated talc

Non-Metals, Inc. Affiliate of The China National Non-Metallic Group, Tucson, AZ, USA

talc - a broad range of grades for various applications from four plants located in different parts of China

Pfizer, USA

Microtalc

Polar Minerals, Mt. Vernon, IN, USA

9100 Series (9102, 9103, 9107, 9110) - plastic additives, free of asbestos and high purity

9200 Series (9202, 9202 D, 9205) - rubber, paper, and coatings applications

9300 Series (9305, 9310) 9400 Series (9410) - polypropylene, paint, coatings, polyester, adhesives

9600 Series (9602, 9603, 9607, 9610) - broad range of applications in plastics, rubber, inks, coatings, adhesives and sealants

9800 Series (9810), 9900 Series (9910) - economical grades

Ultra (2000, 3000, 4000, 5000) - cosmetic grades

Gel, Body medium, Fine - polyester talcs

MV Series (310, 305, 610, 607, 603) - talcs for coating industry

Clear Block 80 - anti-blocking additive in LDPE

Surface treated talcs (9603S, 9603Z) - S - silane treated (enhanced interaction), Z - zinc stearate treated (enhanced hydrophobicity)

XX 10, 07, 03, 02 - designed for polyolefins used in automotive and appliance

S.E.T., S.A., Leon, Spain

Specialty Minerals, Easton, PA, USA

PolyTalc AG Series AG

Vanderbilt, R.T. Company, Inc., Norwalk, CT, USA

Nytal 100, 200, 300, 400, 3300, 7700 - paints, coatings, polyolefins

IT FT, 3X, 5X, 325, X -rubber, plastics paints, and coatings

Vantalc 6H, F-2003 - plastics, rubber, paper and coatings applications

*continued on the next page*

**MANUFACTURERS & BRAND NAMES**:
Zemex Industrial Minerals, Atlanta, GA, USA
Benwood 2202, 2203, 2204, 2207, 2210, 2213 - high purity and brightness talc for industrial applications
Pioneer Talc - 767, 1599, 2606, 2620, 2630, 2655, 2661, 2664, 2720, 2871, 2882, 4304, 4306, 4316, 4317, 4319, 4320, 4392, 4404, 4411, 4416, MB-92 - Suzorite talcs for a broad range of applications

**MAJOR PRODUCT APPLICATIONS**: paper, paints, roofing, plastics, ceramics, animal feed, cosmetics, caulking, sound damping, putties, anti-caking agent, sealants, electrical insulation, plaster, lubricant, tile, appliances, garden furniture, food packaging, agricultural film

**MAJOR POLYMER APPLICATIONS**: PP, PE, PC, ABS, PPS, PS, rubber

Talc is the major constituent of rocks known as soapstone or steatite. Its paragenesis is associated with the hydrothermal metamorphism of siliceous dolomites, and thus it might be accompanied by tremolite, which may be of concern for many potential applications.

The composition of talc varies depending on its source. The most important factor is the amount of tremolite present. In the USA, for instance, Montana talcs are considered to be asbestos and tremolite free. The California plate-like talcs contain minor amounts of tremolite (less than 3%), whereas hard talcs contain between 5 to 25% tremolite. Some industrial talcs mined in upper New York State contain 25 to 50% tremolite. The other important component in its composition is water which is chemically combined in the magnesium oxide or brucite layer. Figure 2.62 shows the molecular structure of talc. Talc may lose this water only on heating over 800°C but, if this happens, the plate-like structure is completely lost and talc properties are changed. The planar surfaces of the plate-like structure are held together by very weak van der Waals forces, and therefore talc can be delaminated at relatively low shearing forces, which accounts for the slippery feel of talc, and makes it easy to disperse.

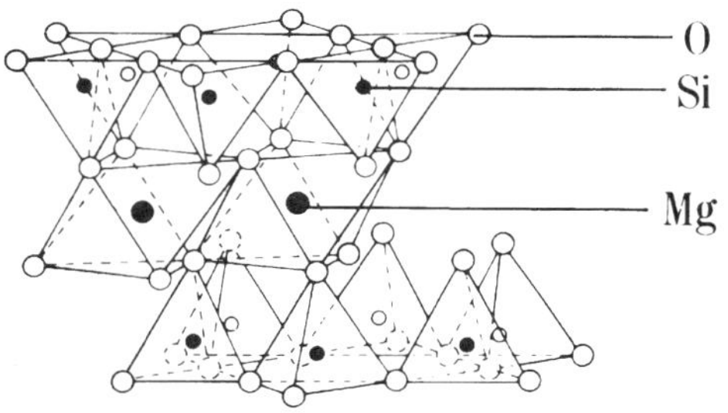

Figure 2.62. Molecular structure of talc. *Courtesy of Luzenac Europe, Toulouse, France.*

Its plate-like structure provides talc-filled materials with important properties, such as, high resistivity and low gas permeability. This comes about because the diffusion path is so complicated. Several other unique properties of talc are structure-related, including its lubricating effect, caused by its easy delamination; its low abrasiveness, because talc is the softest mineral in the Mohs hardness scale; and the hydrophobic properties of its surface. Hydrophobicity can be increased even more by surface coating with zinc stearate. Figure 2.63 shows the plate-like structure of talc.

Figure 2.63. Minstron grade of talc. *Courtesy of Luzenac Europe, Toulouse, France.*

Talc processing is relatively simple. Emphasis is placed on the avoidance of contamination and on a sorting process to sort each talc variety according to mineralogy and color. Frequently manual and optical sorting are employed to obtain a high quality product.[453] There is a wet and a dry process. Dry process begins with selective mining and sorting of the heterogeneous deposit. In the next stage, some ores may be blended and dried but all materials are subjected to grinding. Standard grinding on roller mills results in a coarse material (50 μm). Fine milling in impact mills produces, after classification, finer grades (10-40 μm). The finest grades are obtained by micronization in jet mills (3-10 μm). The wet process separates by flotation those ores which contain a substantial amount of contamination (e.g., with carbonates). This results in materials having a very high concentration of pure talc (97-98%). Before flotation, the material is subjected to primary crushing in impact mill and bag milling which reduce particles to 100 μm. After flotation, the talc is filtered, dried and milled either by impact mills or by jet mill micronization. Some grades have silane surface treatment. The above description of the processes is based on production methods used by Luzenac in various plants worldwide.[453]

In the paper industry, talc was introduced as paper filler by Luzenac in 1905. The widespread use of talc is owed to ability to absorb organic materials, to prevent agglomeration, and to participate in the control of pitch. In recycled papers, talc reduces chemical content in paper manufacture. Talc imparts a smooth texture, reduces porosity and extends the life of machine components due to its lack of abrasiveness. Optimizing ink transfer, talc improves the quality of halftones.

In plastics, the addition of talc improves their heat distortion temperature, dimensional stability, scratch resistance, impact resistance, and reduces the process cycle due to nucleation. Other important properties include high brightness, blocking of infrared in agricultural film, anti-blocking properties, and low absorption of packaged components.

In paints, talcs have a high hiding power, a matting effect and give a satin finish. The morphological structure of talcs gives paints with low moisture permeability. Satin and matt finish in various types of paint is obtained through using talc.

## 2.1.55 TITANIUM DIOXIDE[473-485]

| | |
|---|---|
| **Name**: titanium dioxide | **CAS #**: 13463-67-7 |
| **Chemical formula**: $TiO_2$ | **Functionality**: depends on the surface composition |
| **Chemical composition**: $TiO_2$ - 80-99.5%, $SiO_2$ - 0.15-1.1%, $Al_2O_3$ - 0.3-3.9%, $Fe_2O_3$ - 0.01-2%, $ZrO_2$ - 0.4% | |
| **Trace elements**: Fe, Sn, Nb, Ta, Mg, Mn | |
| PHYSICAL PROPERTIES | |
| **Density**, g/cm$^3$: 3.3-4.25, 4.24 (pure rutile), 3.87 (pure anatase) | **Melting point**, °C: 1825 |
| **Thermal conductivity**, W/K·m: 0.065 | **Loss on ignition**, %: 0.1-2.3 |
| **Coefficient of linear thermal expansion**, $10^{-6}$/K: 8-9.1 | **Mohs hardness**: 6-7 (rutile), 5-6 (anatase) |
| CHEMICAL PROPERTIES | |
| **Chemical resistance**: reacts with acids and alkalis | |
| **Moisture content**, %: 0.2-1.5; **pH of water suspension**: 3.5-10.5 | **Water soluble**, %: 0.3-0.5 |
| OPTICAL & ELECTRICAL PROPERTIES | |
| **Refractive index**: 2.55-2.7; **Tinting strength**: 98-102 | **Brightness**: 99-100 |
| **Relative scattering power**: 64-108 | |
| **Dielectric constant**: 114 (rutile), 48 (anatase) | **Loss angle**: 0.01-0.35 |
| **Brightness**, L*: 93-98; **Undertone**, b*: -6 to -1.5 (gray tints), 1.0-1.9 (white tints) | |
| **Color**: white, buff | **Resistivity**, Ω-cm: 3000-9000 |
| MORPHOLOGY | |
| **Particle shape**: acicular or spherical | **Particle size**, nm: 8-300 |
| **Crystal structure**: tetragonal, orthorhombic, or trigonal in ore and tetragonal in final products | |
| **Hegman fineness**: 6-8 | **Oil absorption**, g/100 g: 10-45 |
| **Sieve analysis**: 325 mesh residue: 0.01% to traces | **Specific surface area**, m$^2$/g: 7-162 |

**MANUFACTURERS & BRAND NAMES**:

Degussa AG, Frankfurt/Main, Germany

P25 - titanium dioxide obtained by pyrogenic process (the same method as used for fumed silica) having particle size of 21 nm and low pH (3.5-4.5)

DuPont, Wilmington, DE, USA

Hitox Corporation, Corpus Christi, TX, USA

Hitox - titanium dioxide obtained by calcination with a small amount (2 wt%) of iron oxide produces buff color. It is economical pigment for coatings, caulks, adhesives, roofing and many plastics

Kemira Pigments, Savannah, GA, USA

*continued on the next page*

**MANUFACTURERS & BRAND NAMES:**

Kronos, Toronto, Ontario, Canada

*Anatase grades*

Kronos 1001, 1002, 1077, E171 (no surface coating), 1014 ($Al_2O_3$ coating), 1015, 1071, 1075 ($Al_2O_3$+$SiO_2$ coating), 1074 ($Al_2O_3$+$MnO_2$+$SiO_2$ coating)

*Plastics*

*Kronos* 2075, 2200, 2210, 2230 ($Al_2O_3$ coating), 2073, 2220, 2222, 2257 ($Al_2O_3$+$SiO_2$ coating)

*Coatings*

Kronos 2059, 2063, 2063 S, 2300 ($Al_2O_3$ coating), 2043, 2044, 2047, 2056, 2057, 2160 ($Al_2O_3$+$SiO_2$ coating), 2190 ($Al_2O_3$+$ZrO_2$ coating), 2065 ($Al_2O_3$+$SiO_2$+ZnO coating), 2310, 2330 ($Al_2O_3$+$SiO_2$+$ZrO_2$ coating)

*Paper laminates*

Kronos 2084, 2088 ($Al_2O_3$ coating), 2081 ($Al_2O_3$+$SiO_2$ coating)

Millennium Inorganic Chemicals, Baltimore, USA

*American Grades*

*Anatase grades*

A-2000, A-3000, A-3100 - products for paper, tire applications, and footwear

*Paper*

Tiona RCS-P (rutile slurry), HSS (anatase slurry)

*Plastics*

Tiona RLC-188 (phosphate and organic coating), RCL-4, RCL-69 ($Al_2O_3$+organic coating), RCL-6 ($Al_2O_3$ and $SiO_2$ coating)

*Coatings - architectural, automotive, coil and powder*

Tiona RCL-9, RCL-535, RCS-9, RCS-535 ($Al_2O_3$+organic coated), RCL-2, RCL-3, RCL-6, RCS-2, RCS-3 ($Al_2O_3$ and $SiO_2$ coating), RCL-628 ($Al_2O_3$ and $ZrO_2$ coating)

*Asia/Pacific Grades*

*Plastics*

Tiona RCL-188 (as above), RCL-69, RCL-128, RCL-181, RCL-575 ($Al_2O_3$+organic coating), RCL-666 ($Al_2O_3$, $SiO_2$, organic coating)

*Coatings - architectural, automotive, coil and powder, low VOC, inks*

RCL-575, RCL-535, RCL-472 ($Al_2O_3$+organic coating), RCL-373, RCL-6 ($Al_2O_3$, $SiO_2$ coated), RCL-666 ($Al_2O_3$, $SiO_2$, organic coating), RCL-628 ($Al_2O_3$, $ZrO_2$, organic coating)

*European Grades*

*Plastics*

Tiona RLC-168 (as above), RCL-4, RCL-69, RCL-535 ($Al_2O_3$+organic coated), RCL-6 ($Al_2O_3$+$SiO_2$ coated), RCL-168 ($Al_2O_3$+$SiO_2$+organic coated)

*Coatings - decorative, industrial, and special purpose*

Tiona RCL-9 ($Al_2O_3$), RCL-472, RCL-535, RCL-552 ($Al_2O_3$+organic), RCL-376, RCL-6 ($Al_2O_3$+$SiO_2$), RCL-388, RCL-666 ($Al_2O_3$+$SiO_2$+organic), RCL-628 ($Al_2O_3$+$ZrO_2$+organic)

Sachtleben Chemie, Duisburg, Germany

Hombitec RM 200, 220, 300, 400 - grades of transparent rutile grades having crystallite size in the range of 10-20 nm coated with $Al_2O_3$ or $ZrO_2$. The product is designed to provide UV protection of coated substrates such as wood, plastics, etc.

TAM Ceramics, Inc., Niagara Falls, NY, USA

Heavy Grade Titanium Dioxide - a product designed for capacitors with a high density of 4.25 $g/cm^3$

*continued on the next page*

| |
|---|
| Tioxide Americas, Inc., Tracy, Canada<br>*Anatase grade*<br>A-HR - uncoated grade for paper, rubber, rubber latex, fibers, road markings, and ceramic systems<br>*Rutile grades*<br>R-Gran 850 - uncoated grade for optical glass and enamel and glaze frits<br>COMET 300, R-BC, R-FC6, R-HD6X, TR23, TR27, TR90 - $Al_2O_3$ coated grades for paper, PE, PP, PVC, ABS, PS, POM, PC, PPO, latex and alkyd paints, appliance enamels, vinyl wall covering, plastic pipe, wood finishes, interior coil coatings, metal decorative and appliance finishes, powder coatings<br>R-XL, TR50, TR60 - $Al_2O_3$ and $SiO_2$ coated for flat latex and alkyd paints, printing inks, colored PVC, exterior coil coatings, automotive finishes, exterior powder coatings<br>TR92 - $Al_2O_3$ and $ZrO_2$ coated is the pigment of choice for a very broad range of applications in paints, enamels, powder coatings and plastics including weather durable materials<br>TR93 - $Al_2O_3$, $SiO_2$, and $ZrO_2$ coated is the most resistant pigment which has high level of opacity, gives excellent gloss and the best UV durability<br>Ultrafine grades - transparent titanium dioxide for UV protection can be used with other pigments at 1% PVC |
| **MAJOR PRODUCT APPLICATIONS**: coatings, plastics, paper, inks, ceramics, capacitors, cosmetics, food, pharmaceuticals, fibers, white concrete, UV stabilizer; the products are not listed considering that most products use titanium dioxide |
| **MAJOR POLYMER APPLICATIONS**: PA, PVC, PE, PP, PPO, POM, PC, PS, ABS, polyester, acrylics, alkyd, polyurethane, melamine, phenoxy |

Titanium dioxide is the most popular pigment used today. The first commercial pigment became available only in 1916 although titanium dioxide was chemically identified first in 1791.[483] Coatings are the largest consumer of titanium dioxide using 57% of the production output, followed by plastics (20%), paper (13%), inks (3%) and ceramics (2%). All other applications accounted for only 5% of global use in 1996. In 1996, a total of 3.3 million tons was produced. Five companies contribute to satisfying 75% of this demand. If the merger between DuPont and Tioxide is approved, DuPont will hold 35% market, followed by Millennium (15%), Kronos (10%), Kerr-Mc-Gee (8%) and Kemira Pigments (7%).

The demand for titanium dioxide is based purely on its physical characteristics. Pigments have two prime functions: to color and to opacify. The coloring characteristics of the pigment depends on its ability to reflect incoming light. Magnesium oxide has the ability to reflect visible light more efficiently than titanium dioxide but it is still an inefficient pigment compared with $TiO_2$ because its capability to opacify is low. Opacifying capability depends on the refractive index and on the absolute difference between the refractive indices of the pigment and the matrix (binder). The most frequently used polymers have refractive indices between 1.45 and 1.6. White powders are considered to be useful as pigments if their refractive index is above 1.7. Titanium dioxide has refractive index between 2.55 (anatase) and 2.7 (rutile). The refractive index of titanium dioxide is higher than any other commercial white pigment. This combined with its reflecting capabilities makes it the most efficient pigment. (It should be noted that air has also very good pigmenting values because its refractive index is 1 which also produces a large difference between it and typical matrices – larger than for zinc oxide, barium sulfate, calcium carbonate).[479,483] The brightness and undertone of pigments depend on their light scattering ability. The brightness is determined by the intensity of reflectance and the undertone by the spectrum of reflected light (ratio of short to long wavelength of reflected light). The difference between anatase and rutile is in their undertone (anatase reflects more short wavelength and has a bluish undertone).

The particle size also has an important influence on the performance of titanium dioxide both as a pigment and as a UV absorber. For the pigment to have maximum opacity, the particle diameter must be equal to half of the wavelength (for a blue/green light to which the eye is most sensitive, the average wavelength is 460 nm, thus a particle diameter of 230 nm gives the maximum opacity). The color of the matrix (binder) has an influence here as well and titanium dioxide must compensate. For this reason, some grades of titanium dioxide are tailored to specific conditions and some are used to eliminate a yellow undertone. This is done by the choice of particle size. For this reason, commercial grades have particle sizes in a range from 200 to 300 nm. The amount of titanium dioxide is also crucial. If too little titanium dioxide is added, the distance between particles is too large and there is no enough opacity. If the amount is too great, it results in lower efficiency due to a particle crowding effect which causes particles to interfere in each other's scattering efficiency. Finally, good dispersion is critical since particles will only give their best performance when they are evenly distributed and separated by binder.

Titanium dioxide is obtained from the following minerals: rutile, anatase, brookite, and ilmenite. The first three minerals contain mostly $TiO_2$, and their structure is octahedral. Both rutile and anatase are tetragonal, the difference being in the mutual arrangement of the octahedra, whereas brookite is orthorhombic. Rutile is the most common mineral, and its geological formation is associated with high temperature. Therefore, it is frequently found in company with other rocks also formed during a secondary high temperature process. Anatase and brookite are found in deposits formed from leaching of gneisses or schists by hydrothermal solutions. Anatase and brookite are converted to rutile upon heating to temperatures above 700°C. Trigonal ilmenite is an earlier constituent of a magma crystallization. By chemical composition, ilmenite is a titanate of ferrous iron. The color of the minerals ranges from yellowish to brownish. Other typical metals, present in small amounts, include Fe, Sn, Nb, Ta, Mg, and Mn.

Figures 2.64-2.66 show the crystalline structures of brookite, rutile, and anatase, respectively.

Most titanium dioxide is produced from ilmenite, which is in abundance. Two processes are used: sulfate and chloride processes. An ilmenite concentrate is reacted with concentrated sulfuric acid in an exothermic reaction. Ferric iron, which is a soluble form under these reaction conditions, is reduced to ferrous. The

Figure 2.64. Brookite. *Courtesy of Tioxide Group PCL, London, UK.*

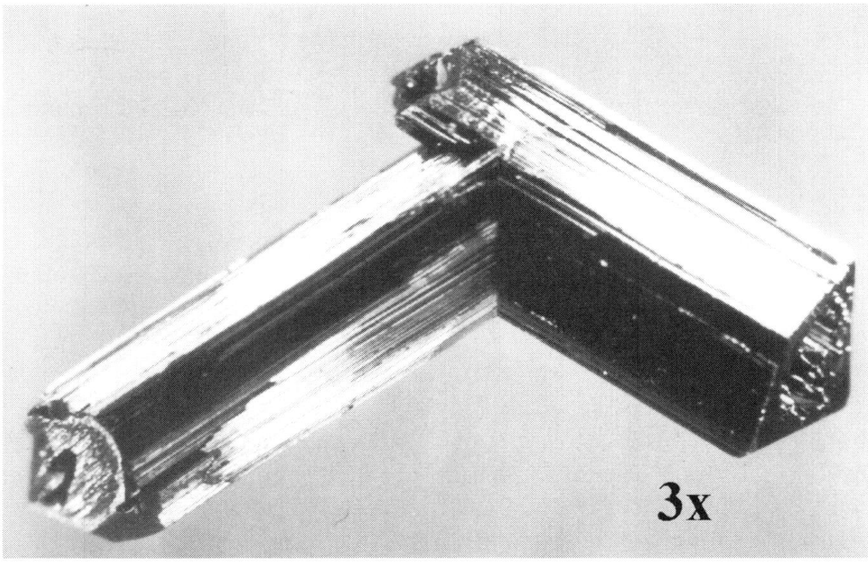


Figure 2.65. Rutile. *Courtesy of Tioxide Group PCL, London, UK.*

undissolved ore and the precipitated iron are removed as contamination. Titanium is precipitated in the form of hydrous titanium oxide after careful nucleation. The precipitate is separated by filtration and washed free of the mother liquor, which removes the traces of iron which would affect color. The washed precipitate is calcinated in a rotary kiln. This process may be followed by the addition of other

Figure 2.66. Anatase. *Courtesy of Tioxide Group PCL, London, UK.*

mineral components to modify properties. Finally, the product is ground and classified. Two processes, nucleation and calcination, determine the crystalline structure formation (e.g., rutile or anatase).

Titanium dioxide is also obtained from the chloride process, which gives an additional option to either hydrolyze titanium tetrachloride with steam or oxidize it with air to the dioxide. In this method, the pigment can be obtained from the gaseous phase. In this method, the feedstock must contain 90% rutile ore. It is not always possible to find such an ore therefore beneficiated feedstock is used which is obtained by various routes. Figure 2.67 compares both manufacturing processes.

The anatase form is manufactured using the sulfate process. The type of crystal (anatase or rutile) produced by the sulfate process depends on the conditions of the process. Generally both crystalline types are produced. The chloride process is used for the production of rutile pigment. New production lines are almost exclusively built for the chloride process because it produces titanium dioxide of higher purity and the operation results in less wastes and produces a smaller quantity of toxic materials. Uncoated rutile is produced in smaller quantities and used in other applications than paints and coatings. Anatase is produced frequently without a coating but Kronos does have a line of coated grades. An inorganic coating is applied in the aqueous slurry by precipitation of one or more hydrated metal oxides and by neutralization of acidic and alkaline compounds. The performance of the inorganic coating depends on the composition of coating ($Al_2O_3$, $SiO_2$, $ZrO_2$, infrequently zinc and tin oxides), the amount of coating (1-15%, typically 5% for paint grades at thickness of 5 nm), the number of deposition stages, and the order and

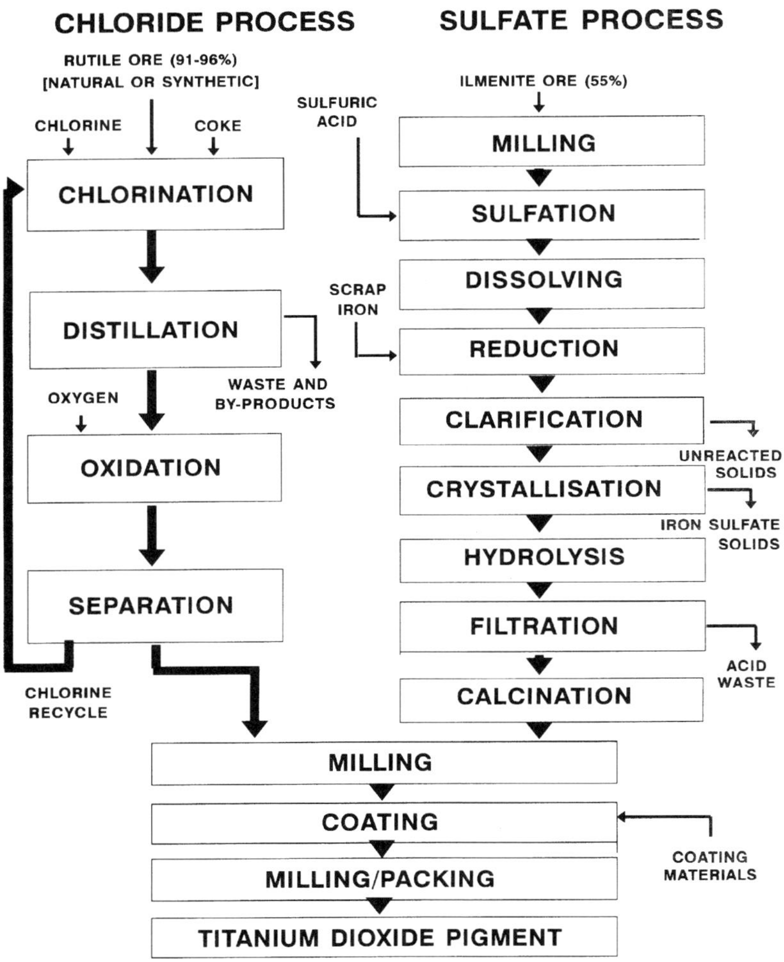


Figure 2.67. Schematic diagrams of chloride and sulfate processes of $TiO_2$ manufacture. *Courtesy of Millennium Inorganic Chemicals, Auburn, Australia.*

rates of deposition of the different coatings. The pH during deposition and after neutralization, the time given to the coating process, the temperature of the process and type of washing aids used all contribute to the performance of the coating and of the coated pigment. Although, zirconium oxide is used as a coating to improve weather stability, the choice of the type of coating used in given application is based on the requirements of the application. An organic treatment is performed to encapsulate particles with a monomolecular layer of a low polarity organic compound, typically trimethylol propane or pentaerythritol (0.3%). This treatment reduces the polarity of $TiO_2$ and improves its ease of dispersion.[483]

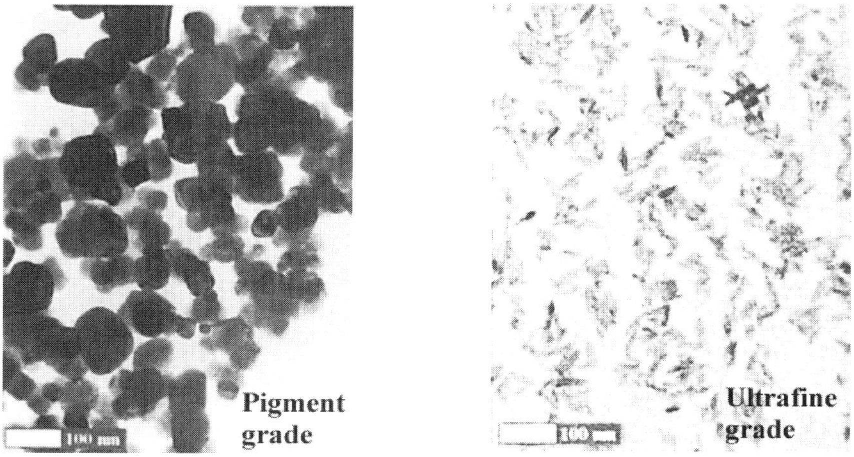


Figure 2.68. Comparison of pigment type and ultrafine titanium dioxide. *Courtesy of Sachtleben Chemie, Duisburg, Germany.*

Because the optimal light scattering of titanium pigments occurs when particle diameter is 0.24 μm, most pigments are manufactured to have the majority of particles closest to that in a range from 0.15 to 0.3 μm, depending on the application and the undertone required. Ultrafine grades are the exception. They typically have particle sizes in a range from 0.015 to 0.035 μm and, because of their small particle size, they are transparent to visible light but absorb in the UV range. Ultralow particle size titanium dioxide is manufactured by Degussa by the same process as fumed silica. $TiCl_4$ is the raw material used in this process. Tioxide manufactures ultrafine $TiO_2$ by the wet process which begins from sodium titanate, $Na_2TiO_3$, which is precipitated from a reaction with hydrochloric acid, neutralized by sodium dioxide, filtered, washed, milled and coated with $SiO_2$, $Al_2O_3$, $ZrO_2$. Additional processes include, filtration and washing after coating, drying, micronizing, and packaging. The control over the process of precipitation affects the crystalline structure of the product. Both anatase and rutile can be obtained in either acicular or spherical morphology. The coating affects the photocatalytic activity of titanium dioxide. The ultrafine, uncoated grades have a high photocatalytic activity of 6.01 mol/g·h. This can be reduced to 0.11 mol/g·h which is similar to that of the coated rutile used for pigment applications (0.07) and much lower than uncoated anatase pigment (0.87).[481] A broad range of properties can be obtained. Typically, surface area, particle size, and oil absorption can all be adjusted but the usual particle sizes are in a range from 7 to 35 nm with surface areas and oil absorptions at the high end of the pigment titanium dioxide. The specific gravity is low at 3.3 g/cm$^3$. The size of the particle can be visualized from the comparison in Figure 2.68.

Figures 2.69-2.71 show the morphology of anatase pigment and rutile with and without coating. The layer of coating can be distinguished on micrograph.

Figure 2.69. Anatase titanium dioxide. *Courtesy of Tioxide Group PCL, London, UK.*

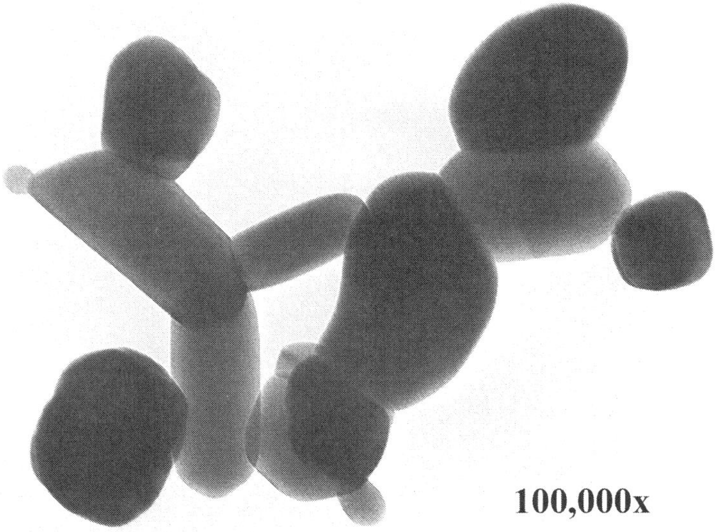


Figure 2.70. Uncoated rutile titanium dioxide. *Courtesy of Tioxide Group PCL, London, UK.*

In addition to the photochemical activity of titanium dioxide, grades have been developed for many other reasons discussed below. Millennium developed its Tiona RCL-188 grade for high performance extrusion. A surface treatment based on phosphate and an undisclosed organic material lowers the energy required for the process, improves the dispersion of pigment even at very high concentrations and without the addition of process aids. When stearates are used in formulation

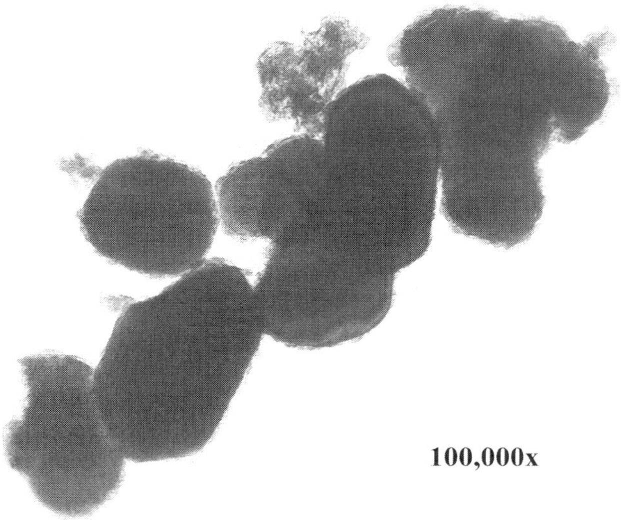


Figure 2.71. Coated rutile titanium dioxide. *Courtesy of Tioxide Group PCL, London, UK.*

there is the potential problem of overlubrication, Tiona RCL-188 does not suffer from this drawback. The surface properties of this grade are compatible with numerous polymers which makes it the material of choice in plastic extrusion applications. The Tiona RLC-4 grade from Millennium coated with a composite organic and $Al_2O_3$ coating is also compatible with numerous polymers. In addition, it is formulated to lower polyethylene yellowing. The product has excellent dispersion characteristics, a low melt flow index, and high tinting strength.

Incorporation of titanium dioxide into paints and coatings depends the grade of $TiO_2$ and on processing conditions. The pigment should be evaluated in the chosen formulation, considering that the final result depends on the quality of dispersion which, in turn, is affected by the pigment, dispersing agent type and amount, and the conditions of mixing. The investigation of this subject is outside the scope of this chapter.

In the paper applications, anatase form has an advantage over rutile in its reflection of light at wavelengths between 380 and 420 nm and on its effect on the abrasion resistance of the paper. The reflection of blue light increases the efficiency of optical brighteners. The scattering efficiency improves as particle size decreases. Tiona A-2000 is a very small particle size grade and, in addition, the slurry containing it has improved calcium resistance. A high concentration of titanium dioxide usually causes the slurry to thicken then gel over time when calcium carbonate is present. Tiona A-2000 is formulated to prevent viscosity changes of the coating slurry when calcium carbonate is added.

## 2.1.56 TUNGSTEN[486]

| | | |
|---|---|---|
| **Name**: tungsten powder | | **CAS #**: 7440-33-7 |
| **Chemical formula**: W | **Functionality**: none | |
| **Chemical composition**: W - 99.5-99.7% | | |
| **Trace elements**: Al, Co, Cr, Cu, Fe, K, Mo, Ni | | |
| PHYSICAL PROPERTIES | | |
| **Density**, g/cm$^3$: 19.35 | **Mohs hardness**: 9 | **Melting point**, °C: 3410 |
| **Thermal conductivity**, W/K · m: 2.35 | **Specific heat**, kJ/kg · K: 0.088 | |
| CHEMICAL PROPERTIES | | |
| **Chemical resistance**: soluble in $HNO_3$ and HF | | |
| OPTICAL & ELECTRICAL PROPERTIES | | |
| **Color**: gray, black | | **Resistivity**, Ω-cm: 5.6x10$^{-6}$ |
| MORPHOLOGY | | |
| **Particle size**, μm: 0.7-18 | **Crystal structure**: cubic | |
| **Sieve analysis**: residue on 325 mesh sieve - traces | | |
| MANUFACTURERS & BRAND NAMES:<br>Teledyne Advanced Materials, Huntsville, AL, USA<br>Tungsten powder C-3, C-5, C-6, C-8, C-10, C-20, C-40, C-60, crystalline - powders of different particle sizes (the higher the number the large the particle) | | |
| MAJOR PRODUCT APPLICATIONS: composites | | |
| MAJOR POLYMER APPLICATIONS: epoxy | | |

## 2.1.57 VERMICULITE

| | | |
|---|---|---|
| **Name**: vermiculite | | **CAS #**: 1318-00-9 |
| **Chemical formula**: $(Mg,Fe^{2+},Al)_3(Al,Si)_4O_{10}(OH)_2 \cdot 4H_2O$ | **Functionality**: OH | |
| **Chemical composition**: $SiO_2$ - 39.4%, MgO - 23.4%, $TiO_2$ - 1.25%, $Al_2O_3$ - 12.1%, $Fe_2O_3$ - 5.5%, FeO - 1.2%, MnO - 0.3%, CaO - 1.5%, $Na_2O$ - 0.8%, $K_2O$ - 2.5% | | |
| **PHYSICAL PROPERTIES** | | |
| **Density**, $g/cm^3$: 2.6 | **Specific heat**, kJ/kg · K: 0.2 | **Melting point**, °C: 1315 |
| **Thermal conductivity**, W/K · m: 0.062-0.065 | **Maximum temperature of use**, °C: 1100 | |
| **Loss on ignition**, %: 5.8 | | |
| **CHEMICAL PROPERTIES** | | |
| **Chemical resistance**: insoluble in water and organic solvents | | |
| **pH of water suspension**: 7 | **Adsorbed water**, %: 240 | |
| **OPTICAL PROPERTIES** | | |
| **Color**: golden-brown | | |
| **MORPHOLOGY** | | |
| **Particle shape**: flakes (after expansion - concertina-shape granules) | **Crystal structure**: monoclinic | |
| **MANUFACTURERS & BRAND NAMES**:<br>Non-Metals, Inc., Affiliate of The China National Non-Metallic Minerals Group, Tucson, AZ, USA<br>Chine Vermiculite Concentrate TG series - golden color, KV series - silver color<br>Strong-Lite, Pine-Bluff, AR, USA<br>expanded and non-expanded vermiculite for various applications | | |
| **MAJOR PRODUCT APPLICATIONS**: insulation, construction, horticulture, paint, packaging, ion-exchange | | |

Vermiculite resembles mica in appearance. In industrial process, vermiculite flakes are rapidly heated at flame temperature approaching 1000°C. Some of the water of hydration is removed and the pressure generated by the water vapor expands (or exfoliates) vermiculite particles which increases in volume by 15 to 20 times. This expansion process must be precisely controlled to achieve the required expansion and to retain its water absorption properties. If the time of heating is extended, vermiculite will no longer absorb water. Thus, different grades may be produced by varying the heating time.

## 2.1.58 WOOD FLOUR AND SIMILAR MATERIALS[487-491]

| | | |
|---|---|---|
| **Names**: wood flour, wood fiber, bark flour, wheat flour | | **CAS #**: 9004-34-6 |
| **Chemical formula**: variable | **Functionality**: OH | |
| **Chemical composition**: protein content up to 15% | | |
| PHYSICAL PROPERTIES | | |
| **Density**, g/cm$^3$: 0.4-1.35 | **Maximum temperature of use**, $^o$C: 200 | |
| CHEMICAL PROPERTIES | | |
| **Moisture content**, %: 2-12 | **Adsorbed moisture**, %: up to 20 | **Ash content**, %: 0.5-0.7 |
| **pH of water suspension**: 5 | | |
| OPTICAL PROPERTIES | | |
| **Color**: buff, tan | | |
| MORPHOLOGY | | |
| **Particle size**, μm: 10-100 | **Oil absorption**, g/100 g: 55-60 | |

MANUFACTURERS & BRAND NAMES:

Ace International Inc., Centralia, WA, USA
- Douglas Fir Wood Flour - A-series (-20/100A, A-100A, A-200A), T-series (T-14, T-70, T-100)
- Alder Bark Flour - Modal regular light, regular dark, spray light, spray dark, superbond - used as glue extender in plywood industry for over forty years
- Wheat Flour - secondary extender in phenolic resin adhesives in plywood industry

Agrashell, Inc., Bath, PA, USA
- Industrial Flour WF-5, WF-7 - nut shell flour

American Wood Fibers, Jessup, MD, USA
- Hardwood grades 2010, 4010, 6010, 8010, 10010, 12010, 14010 - materials of different particle sizes
- Softwood grades 2020, 4020, 6020, 8020, 10020, 12020, 14020 - materials of different particles sizes

MAJOR PRODUCT APPLICATIONS: sheet, pipes, automotive (door panels, air vents, under-dash parts, speaker brackets), toys, flower pots, lawn furniture, cosmetic packaging, garment hangers, brush blocks, paint roller and brush handles, paint pails, tool handles, computer accessories, office organizers, housewares, slats for blinds, speaker housings, vacuum cleaner beater bars, storage crates, toilet seats, pallets, chair supports, adhesives, brake pads, cosmetics

MAJOR POLYMER APPLICATIONS: PP, PE, PVC, PS, polyester, poly(lactic acid), phenoxy, melamine

There are many applications for these fillers because they can improve dimensional stability, increase heat deflection temperature, reduce shrinkage, lower the weight of products, and reduce thermal expansion. Production costs are lowered also, because the wood flour is an inexpensive filler. In some applications, mechanical performance is improved as measured by impact strength and flexural modulus.[488-490] The major drawback of these fillers is their hygroscopic nature which requires a long drying process to remove water prior to production. Their distinct color can be disadvantage but for some products it may be acceptable or even provide a desirable wood-like surface finish reducing the need for additional pigments.

## 2.1.59 WOLLASTONITE[492-494]

| | | |
|---|---|---|
| **Name**: wollastonite | | **CAS #**: 13983-17-0 |
| **Chemical formula**: $CaSiO_3$ | **Functionality**: from silane | |
| **Chemical composition**: CaO - 43-47.5%, $SiO_2$ - 44-52.2%, $Fe_2O_3$ - 0.15-0.4%, $Al_2O_3$ - 0.2-1%, MgO - 0.2-0.8%, MnO - 0.1%, $TiO_2$ - 0.02% | | |
| **PHYSICAL PROPERTIES** | | |
| **Density**, g/cm$^3$: 2.85-2.9 | **Mohs hardness**: 4.5 | **Melting point**, °C: 1540 |
| **Loss on ignition**, %: 0.1-6 | **Coefficient of expansion**: 6.5x10$^{-6}$ | |
| **CHEMICAL PROPERTIES** | | |
| **Moisture content**, %: 0.02-0.6 | **pH of water suspension**: 9.8-10 | **Water solubility**, %: 0.01 |
| **OPTICAL PROPERTIES** | | |
| **Refractive index**: 1.63 | **Color**: white | **Brightness**: 80-94 |
| **MORPHOLOGY** | | |
| **Particle shape**: acicular | **Crystal structure**: monoclinic/anorthic (triclinic) | |
| **Particle length**, μm: 8-650 | **Oil absorption**, g/100 g: 19-47 | **Hegman fineness**: 0-7 |
| **Aspect ratio**: 4-68 | **Particle thickness**, μm: 1-50 | |
| **Sieve analysis**: 325 mesh sieve residue - 0.09-3% | | **Specific surface area**, m$^2$/g: 0.4-5 |

**MANUFACTURERS & BRAND NAMES:**

Fibertec, Bridgewater, MA, USA
- Micronite AP, 1250S, 325, 200S - materials of different particle dimensions and aspect ratios

Non-Metals, Inc. Affiliate of The China National Non-Metallic Minerals Group, Tucson, AZ, USA
- Wollastonite powder LST1, 2, 3, 4, LSP 1, 2 - grades of different brightness and fineness

Nyco Minerals, Willsboro, NY, USA
- Nycor R, Nyad G, 200, 325, 400, 1250 - grades having different particle sizes and aspect ratios
- Wollastocoat 10, 400, Nyad G - surface modified grades
- Nyglos 4, 5, 8 - grades having different particle sizes and aspect ratios

Quarzwerke GmbH, Frechen, Germany
- Tremin 283 - grades 010, 100, 400, 600, 800 (the higher the number the smaller the particle size) with different silane coating (AST - aminosilane, EST - epoxysilane, MST - methacrylsilane, TST - methylsilane, VST - vinylsilane)
- Tremin 939 - grades 100, 300, 600 (the higher the number the smaller the particle size) with different silane coating (AST - aminosilane, EST - epoxysilane, FST - alkylsilane, MST - methacrylsilane, ESST - epoxysilane special, USST - aminosilane special)

Vanderbilt R.T. Company, Inc., Norwalk, CT, USA
- Vansil W-10, W-20, W-30

**MAJOR PRODUCT APPLICATIONS**: coatings, primers, ceramics, adhesives, abrasives, insulating materials, sealants, wallboards

**MAJOR POLYMER APPLICATIONS**: alkyd, acrylics, polyurethanes, epoxy, PP, PA, LCP, PET, SAN, PMMA, fluororubber, phenoxy

Wollastonite is the industrially important mineral of the pyroxene mineral group. It occurs chiefly as a metamorphic mineral in crystalline limestones. Wollastonite has been formed in reaction:

$$CaCO_3 + SiO_2 \rightarrow CaSiO_3 + CO_2$$

For this reaction to occur, a temperature above 450°C is needed to initiate the reaction between calcite and silica. Depending on the composition of minerals in the area where wollastonite was formed, materials with various levels of contamination resulted. The wollastonite mined in New York state can be converted to a high purity product (97-98%) because it contains garnet and diopside as associated minerals. These minerals can be magnetically removed. But calcite, which is a frequent admixture, is very difficult to remove. Wollastonite is the only naturally occurring white mineral which is wholly acicular. The length to diameter ratio (aspect ratio) typically varies from 3:1 to 20:1 but higher aspect ratios are also available. Wollastonite production consists of mining, grinding, separation, classification, and, with some products, treatment with a coupling agent. Commercially available fillers have an aspect ratio similar to the mineral, ranging from 3:1 to 20:1, an average particle diameter of 3.5 μm, and an equivalent spherical diameter distribution in a range from 0.3 to 40 μm. Figure 2.72 shows the morphology of wollastonite filler.

Figure 2.72. SEM micrographs of wollastonite. *Courtesy of NYCO Minerals, Inc. Willsboro, NY, USA (a) and ECC International, St. Austell, UK (b).*

Its specific surface area is very low (0.8-4 $m^2/g$), indicating that the material is not porous. Other characteristic features of wollastonite include a high pH value (9.8), a low coefficient of thermal expansion ($6.5\times10^{-6}/^oC$), and a low moisture content (less than 0.5%). Wollastonite is becoming an increasingly important filler as an asbestos replacement but its most important applications are due to its high brightness, low oil absorption, and reinforcing effect. In latex coatings, its high pH helps in stabilizing pH of the latex which improves the stability and shelf-life of the paint.

In plastics applications, wollastonite reinforces, increases scratch resistance, improves thermal stability, increases welding strength, and decreases warpage and shrinkage. Figure 2.73 demonstrates the effect of surface treatment on reinforcement. In comparative room temperature evaluation of the surface treated and untreated wollastonite as a filler in polypropylene, the surface treated filler was firmly embedded in the matrix whereas the untreated wollastonite delaminated from the matrix. When fractured in liquid nitrogen the samples showed good adhesion between the matrix and surface treated wollastonite whereas untreated wollastonite had small gaps between the matrix and filler.

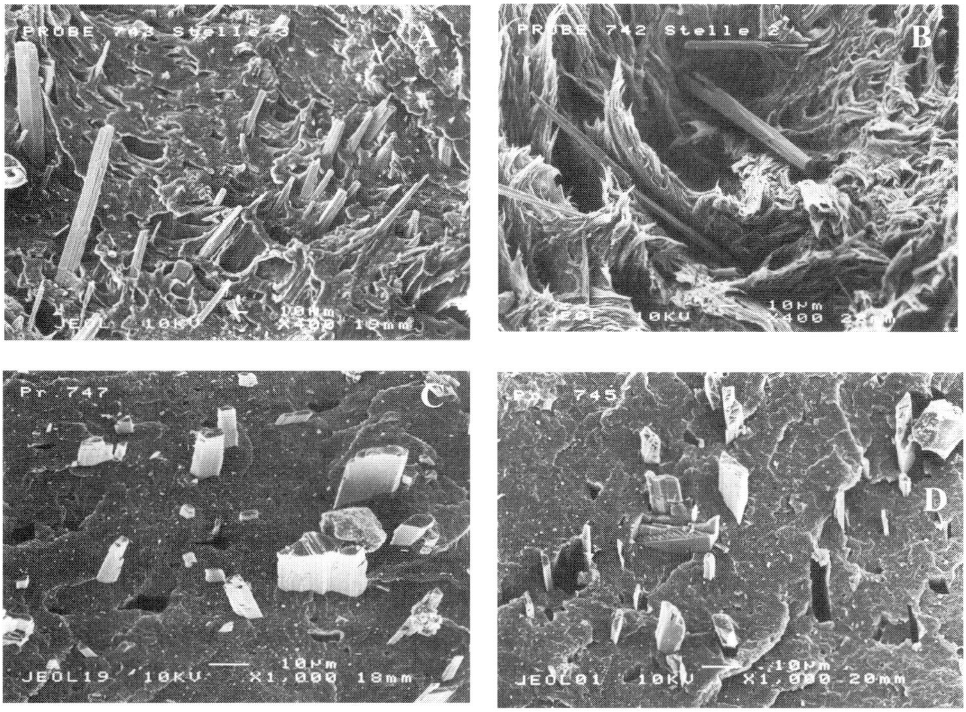


Figure 2.73. SEM micrographs of polypropylene fracture area. Top - fracture at room temperature, bottom - fracture at liquid nitrogen. left - surface treated wollastonite, Tremlin 939, right - untreated wollastonite, Tremlin 939. *Courtesy of D. Skudelny, Quarzwerke GmbH, Frechen, Germany.*

## 2.1.60 ZEOLITES[495-498]

| | | |
|---|---|---|
| **Names**: zeolite, molecular sieves | | **CAS #**: 68989- |
| **Chemical formula**: variable | **Functionality**: OMe | |
| **Chemical composition**: alkali aluminosilicate | | |
| CHEMICAL PROPERTIES | | |
| **Cation type**: K, Na, Ca | | |
| **Moisture content**, %: 1.5 | **Adsorbed moisture**, %: 23-29 | **pH of water suspension**: 10-12 |
| OPTICAL PROPERTIES | | |
| **Color**: white | | |
| MORPHOLOGY | | |
| **Particle size**, μm: 50 | **Oil absorption**, g/100 g: 30-42 | **Pore size**, Å: 3-10 |
| **Hegman fineness**: 5-6 | | |
| MANUFACTURERS & BRAND NAMES:<br>PQ Corporation, Valley Forge, PA, USA<br>PQ Sieves - molecular sieves<br>Valfor - zeolites<br>Zeochem, Louisville, KY, USA<br>Purmol 3A, 3ST, 4A, 5A, 13 - molecular sieves of different pore sizes | | |
| MAJOR PRODUCT APPLICATIONS: plastics, coatings, sealants, caulks, adhesives, pigments, solvents, insulated glass, paper, primers, membranes | | |
| MAJOR POLYMER APPLICATIONS: polyurethanes, polysulfides, PSF, PEI, PPO, PI | | |

Zeolites have found two major applications in polymeric systems: as selective membranes and as *in situ* drying agents. In moisture sensitive systems such as polyurethanes and polysulfides, molecular sieves help to scavenge moisture which extends the shelf-life of moisture cured products manufactured from these polymers. In these applications, 3 Å molecular sieves are safe to use without special precautions because they contain no gas in their pores. Larger pore size sieves should be added under the vacuum to remove gas from the pore volume.

Molecular sieves are also used to scavenge moisture to prevent its condensation in insulated glass units. They are added to adhesive spacers or contained within the spacer which divides the glass panes. The spacer is a barrier to the penetration of the ambient atmosphere into the enclosed space of insulated glass unit.

Molecular sieves can be incorporated in one of two commercial forms: as a powder or as a dispersion in various organic media such as oils or plasticizers.

## 2.1.61 ZINC BORATE[499]

| | | |
|---|---|---|
| **Name**: zinc borate | | **CAS #**: 1332-07-6 |
| **Chemical formula**: $2ZnO_3 \cdot B_2O_3 \cdot 3.5H_2O$ | **Functionality**: OH | |
| **Chemical composition**: ZnO - 37.45%, $B_2O_3$ - 48.05%, $H_2O$ - 14.5% | | |
| PHYSICAL PROPERTIES | | |
| **Density**, g/cm$^3$: 2.8 | | **Melting point**, °C: 980 |
| CHEMICAL PROPERTIES | | |
| **Moisture content**, %: 0.4-0.5 | **pH of water suspension**: 8.1-8.3 | |
| OPTICAL PROPERTIES | | |
| **Refractive index**: 1.59 | **Color**: white | |
| MORPHOLOGY | | |
| **Crystal structure**: triclinic or amorphous | **Specific surface area**, m$^2$/g: 10-15 | |
| **Particle size**, μm: 0.6-1 | **Oil absorption**, g/100 g: 37-44 | |
| MANUFACTURERS & BRAND NAMES:<br>Alcan Chemicals Europe, Gerrards Cross, UK<br>Flamtard Z10 & Z15 - number is equivalent to the specific surface area | | |
| MAJOR PRODUCT APPLICATIONS: flame retarding compositions of polymers listed below | | |
| MAJOR POLYMER APPLICATIONS: PA, PPO, PC, PVC, PE, EVA, EPDM, polychloroprene, polyesters, epoxy | | |

Zinc borate is an inorganic flame retardant which can be used by itself or in combination with aluminum hydroxide or magnesium hydroxide with which it forms synergistic mixtures of high performance flame retardants. It is frequently used as a surface coating on these two fillers. It reduces smoke emission and promotes char formation.

## 2.1.62 ZINC OXIDE[500-502]

| | | |
|---|---|---|
| **Name**: zinc oxide | | **CAS #**: 1314-13-2 |
| **Chemical formula**: ZnO | **Functionality**: none | |
| **Chemical composition**: ZnO - 99.5-99.9% | | |
| **PHYSICAL PROPERTIES** | | |
| **Density**, g/cm$^3$: 5.6 | **Mohs hardness**: 4 | **Melting point**, °C: 1975 |
| **OPTICAL PROPERTIES** | | |
| **Refractive index**: 2.0 | **Color**: white | **Brightness**: 90-94 |
| **MORPHOLOGY** | | |
| **Particle shape**: spherical | **Crystal structure**: hexagonal | **Particle size**, μm: 0.036-3 |
| **Oil absorption**, g/100 g: 10-20 | **Specific surface area**, m$^2$/g: 10-45 | |
| **MANUFACTURERS & BRAND NAMES**:<br>Nanophase Technologies Corporation, Burr Ridge, IL, USA<br>NanoTek zinc oxide - nanoparticle size zinc oxide manufactured by physical vapor synthesis process<br>Societe des Blancs de Zinc de la Mediterranee, Marseille, France<br>Cachet Or - French process zinc oxide<br>Zinc Corporation of America, Monaca, PA, USA<br>Kadox - French process zinc oxide | | |
| **MAJOR PRODUCT APPLICATIONS**: paints, coatings, crosslinker of rubber, sealants | | |
| **MAJOR POLYMER APPLICATIONS**: acrylics, PVC, PC, PE, PP | | |

Zinc oxide is produced either by the French or by the American process. Both processes are pyrometallurgical techniques in which the metal in a vapor state reacts with oxygen, forming zinc oxide. The difference between the methods is in the raw material used for the synthesis. In the French process, pure metal is evaporated, and the final product is as pure as the metal used for its production. In the American process, zinc vapor is obtained directly from an ore by burning it as a mixture with coal or in an electrothermic process where electric current provides the heat. More recently, a new method, somewhat similar to the French process, was introduced by Nanophase Technologies Corporation who patented a physical vapor synthesis process in which zinc metal is vaporized. The vapor is rapidly cooled in the presence of oxygen, causing nucleation and condensation of nanoparticle size zinc oxide. The particles are non-porous and free of contamination.

Figure 2.74 shows the morphology of nanoparticle size zinc oxide which can be compared with zinc oxide obtained in French process (Figure 2.75).

The purest grades of zinc oxide from the French process contain more than 99.99% of ZnO. The purity of zinc oxide is essential in many applications because ZnO is a photochemically active material and impurities may severely affect its properties. Zinc oxide has found many applications due to its photochemical

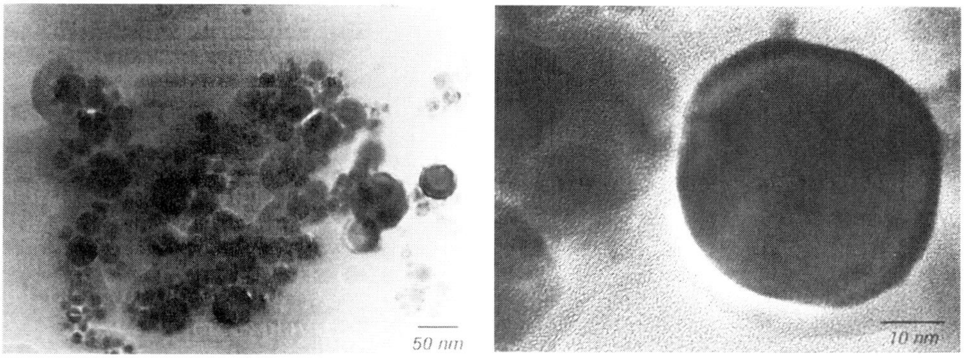


Figure 2.74. TEM micrographs showing NanoTec zinc oxide. *Courtesy of Nanophase Technologies Corporation, Burr Ridge, IL, USA.*

Figure 2.75. SEM micrographs of Kadox 915 manufactured by French process.

properties and chemical reactivity. One of the essential mechanisms of chemical reaction is that in which it forms zinc sulfides, thus preventing product discoloration.

Its particle size is usually in a range from 0.1 to 0.4 μm, and its specific surface area is correspondingly in a range from 20 to 10 $m^2/g$. Nanosize particles have an average particle size of 36 nm and a substantially higher specific surface area at 15-45 $m^2/g$. The high surface area is due to the small particle size, as zinc oxide has little porosity. A product having an average particle size of 0.11 μm has oil absorption as low as 12 g/100 g. Some grades, especially those used in the rubber industry, can be surface modified, usually by the deposition of 0.2-0.4% of stearic acid, propionic acid, or light oil, all of which coatings facilitate mixing.

Several reasons are behind the widespread use of zinc oxide. Zinc oxide is a popular crosslinker for rubber and for various resins. Zinc oxide is also used as an UV stabilizer and as an additive having biocidal activity. It is frequently used in paints. Zinc oxide also has a relatively high refractive index which makes it an efficient white pigment.

## 2.1.63 ZINC STANNATE[503]

| | |
|---|---|
| **Names**: zinc stannate, zinc hydroxystannate | **CAS #**: 12036-37-2 or 12027-96-2 |
| **Chemical formula**: $ZnSnO_3$ and $ZnSn(OH)_6$ | **Functionality**: OH |
| PHYSICAL PROPERTIES | |
| **Density**, g/cm$^3$: 3-3.9 | **Decomposition temp.**, °C: 180-400 |
| CHEMICAL PROPERTIES | |
| **Moisture content**, %: 0.5 | **pH of water suspension**: 9-10 |
| OPTICAL & ELECTRICAL PROPERTIES | |
| **Refractive index**: 1.9 | **Conductivity**, μS/cm: 800; **Color**: white |
| MORPHOLOGY | |
| **Particle size**, μm: 2.5 | |
| MANUFACTURER & BRAND NAME: Alcan Chemicals Europe, Gerrards Cross, UK — Flamtard S (zinc stannate), Flamtard H (zinc hydroxystannate), Flamtard HB1 (zinc hydroxystannate/zinc borate blend) | |
| MAJOR PRODUCT APPLICATIONS: flame retardant in the polymers listed below | |
| MAJOR POLYMER APPLICATIONS: PVC, PE, PA, EVA, EPDM, PC, polyesters, epoxy, polychloroprene | |

Zinc stannate is an inorganic flame retardant which can be used by itself or in combination with aluminum hydroxide or magnesium hydroxide with which it forms synergistic mixtures of high performance flame retardants. It is frequently used as a surface coating on these two fillers. It reduces smoke emission and promotes char formation.

## 2.1.64 ZINC SULFIDE

| | | |
|---|---|---|
| **Names**: zinc sulfide | | **CAS #**: 68611-70-1 |
| **Chemical formula**: ZnS | **Functionality**: | |
| **Chemical composition**: ZnS - 98%, ZnO - 0.2%, $BaSO_4$ - 1% | | |
| PHYSICAL PROPERTIES | | |
| **Density**, g/cm$^3$: 4 | **Mohs hardness**: 3 | **Melting point**, °C: 1700 |
| CHEMICAL PROPERTIES | | |
| **Chemical resistance**: not resistant to strong acids and alkalis | | |
| **Moisture content**, %: 0.3 | **pH of water suspension**: 6-7 | |
| OPTICAL & ELECTRICAL PROPERTIES | | |
| **Refractive index**: 2.37 | **Color**: white | **Brightness**: 98 |
| **Tinting strength**: 55-62% $TiO_2$ | **Conductivity**, mS/cm: 0.2 | |
| MORPHOLOGY | | |
| **Particle size**, μm: 0.3-0.35 | **Oil absorption**, g/100 g: 13-14 | **Specific surface area**, m$^2$/g: 8 |
| **Sieve analysis**: residue on 325 mesh sieve - 0.001-0.01% | | |
| MANUFACTURERS & BRAND NAMES:<br>Sachtleben Chemie GmbH, Duisburg, Germany<br>Sachtolith L (standard paints), HD (high quality paints), HD-S (plastics) | | |
| MAJOR PRODUCT APPLICATIONS: paints, coatings, inks, UV-curable systems, powder coatings, adhesives, insulating and sealing compounds, fibers, paper, sealants, mastics, lubricants | | |
| MAJOR POLYMER APPLICATIONS: alkyd, epoxy, acrylics, PVC, PE, PP, PS, PET, PA | | |

Zinc sulfide is produced by synthetic methods from pure zinc and sulfide obtained as a by-product of barium sulfate synthesis. The precipitated filler has a very small particle size which makes it unsuitable for use as a white pigment. The optimum particle size is obtained by calcination in a continuously operated kiln at 700-800°C. Zinc sulfide crystals grow under these conditions to 0.3 μm which is optimal for white pigment. Depending on the grade, the product of calcination is either deagglomerated or surface treated in a process similar to titanium dioxide. Figure 2.76 shows the morphology of the product obtained by this method.

Zinc sulfide has the next highest refractive index to titanium dioxide and zirconium oxide making it a very efficient pigment. The spectrum of absorption of zinc sulfide resembles more closely anatase than rutile. Because it does not absorb certain UV wavelength, zinc sulfide is useful as a pigment for UV curable materials.

Figure 2.76 implies that zinc sulfide causes low abrasion to the equipment because of its spherical shape and also because of low hardness. Its low oil number

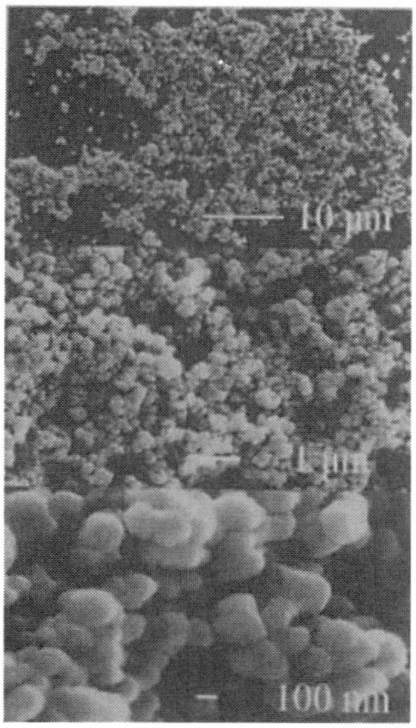


Figure 2.76. SEM micrographs of zinc sulfide, Sachtolith, under three magnifications of 2000x, 10,000x and 150,000x. *Courtesy of Sachtleben Chemie GmbH, Duisburg, Germany.*

means that little binder is needed and minimizes its effect on viscosity of melts and dispersions.

In paint applications, zinc sulfide gives two advantages in addition to its function as a pigment: it gives anti-corrosive properties and acts as efficient algicidal agent. In addition, coatings can be formulated with a reduced level of rheological additives which further improves the anti-corrosive properties of primers.

In plastics applications, zinc sulfide is used for its flame retarding properties. Flame retardant products can be formulated free of antimony and bromine. Zinc sulfide can also be used as a partial replacement of antimony oxide.

## 2.2 FIBERS

### 2.2.1 ARAMID FIBERS[504-513]

| | | |
|---|---|---|
| **Names**: aramid fiber, poly(p-phenylene terephthalamide) | | **CAS #**: 26125-61-1 |
| **Chemical formula**: $(C_{14}H_{10}N_2O_2)_n$ | **Functionality**: NH, COOH, H | |
| PHYSICAL PROPERTIES | | |
| **Density**, g/cm$^3$: 1.44-1.45 | **Decomposition temp.**, °C: 500 | **Melting point**, °C: |
| **Hot air shrinkage**, %: 0.1 | **Loss on ignition**, %: 0.2-0.3 | **Specific heat**, kJ/kg·K: 1.42 |
| **Thermal conductivity**, W/K·m: 0.04-0.05 | **Thermal expansion coefficient**, 1/K: -3.5x10$^{-6}$ | |
| **Tensile strength**, MPa: 2500 | **Residual strength**, 48 h @200°C in %: 90 | |
| **Elongation**, %: 2-3% | | |
| CHEMICAL PROPERTIES | | |
| **Chemical resistance**: low resistance to strong acids and alkalis but substantially better than E-glass[507] | | |
| **Adsorbed moisture**, %: 5-8 | | |
| MORPHOLOGY | | |
| **Fiber length**, mm: 1-6 | **Amount of sizing**, %: 4-6 | **Aspect ratio**: 100-500 |
| **Filament diameter**, μm: 5-18 | **Filament count**, dtex: 1.7 | **Specific surface area**, m$^2$/g: 0.2 |
| MANUFACTURERS & BRAND NAMES:<br>Akzo Nobel Aramid Products, Inc., Conyers, GA, USA<br>Twaron 1010, 1055, 1488 - chopped aramid fiber<br>Twaron 5000, 5010, 5011 - powders with average particle sizes of 450, 110, 55 μm, respectively<br>Composite Particles, Inc., Allentown, PA, USA<br>Vistamer KF - aramid fiber which has surface activated by a patented reactive gas process<br>DuPont, Wilmington, DE, USA<br>Kevlar 29, 49, 149 - Kevlar 149 has lower moisture absorption | | |
| MAJOR PRODUCT APPLICATIONS: composites, wear resistant machine parts, automotive parts, office equipment parts, electrical devices, pumps, brake pads | | |
| MAJOR POLYMER APPLICATIONS: POM, PA, PC, PBT, epoxy, phenoxy, vinyl ester, fluoropolymers | | |

Aramid fiber have been in use for a long time to improve wear resistance of plastic parts. Aramid fiber is superior to other wear resistant additives due to its easier dispersion and minimal effect on mechanical properties of filled materials. Incorporation of fibers increases the impact strength of composites.[506] Further improvements in mechanical properties can be obtained by applying technology developed by Composite Particles, Inc. in which the surface is modified with OH and COOH groups. The presence of these groups was found to increase adhesion to many polymers. The degree of modification should be carefully controlled because the mechanical strength of the fiber and the performance of its composite may be adversely affected.[507]

The high moisture absorption of aramid fibers is their biggest disadvantage. It was reported in the literature that moisture absorption by epoxy laminates degrades their mechanical properties.[504,510] Hygroscopic fibers provide an easy route for moisture ingress. The addition of aramid fibers to epoxy and phenolic composites slightly improves their flame resistance and decreases smoke formation.[505]

## 2.2.2 CARBON FIBERS[514-547]

| | | |
|---|---|---|
| **Names**: carbon fiber, graphite fiber | | **CAS #**: 7440-44-0 |
| **Chemical formula**: C | **Functionality**: OH, COOH, NH | |
| **Chemical composition**: C - 84.3-95.7%, 97-99% (pitch-based), oxygen - 3-7%; sizing agents: epoxy, polyamide (1.3-7%) | | |
| PHYSICAL PROPERTIES | | |
| **Density**, g/cm$^3$: 1.76-1.99, 1.9-2.25 (pitch-based) | | **Mohs hardness**: 0.5-1 |
| **Linear expansion coefficient**, 1/K: -0.1x10$^{-6}$, -1.45x10$^{-6}$ (pitch-based) | | **Specific heat**, kJ/kg·K: 0.71 |
| **Thermal conductivity**, W/K·m: 9-100, 25-1000 (pitch-based fiber), 400 (pure copper), 540 (pitch-based carbon fiber 40/epoxy 60 composite) | | |
| **Maximum temperature of use**, °C: 1300 | | **Young modulus**, GPa: 230-390 |
| **Tensile strength**, MPa: 3000-5500, 1400-3700 (pitch-based) | | **Elongation**, %: 0.4-2 |
| **Tensile modulus**, GPa: 230-500, 160-980 (pitch-based) | | **Coefficient of friction**: 0.1-0.14 |
| OPTICAL & ELECTRICAL PROPERTIES | | |
| **Color**: black | | |
| **Resistivity**, Ω-cm: 3.3x10$^{-2}$-1.5x10$^{-3}$, 10$^{-5}$ (hollow graphite fibrils), 1-3x10$^{-4}$ (pitch-based fiber) | | |
| MORPHOLOGY | | |
| **Fiber length**, μm: 40-160 (milled), 6000 (chopped), 1-10 (hollow graphite fibrils), 3-50,000 (pitch-based) | | |
| **Filament count**: 500-12,000 | **Micropores**, cm$^3$/g: 0.058 | **Pore diameter**, nm: 0.02-0.05 |
| **Filament diameter**, μm: 4-7 (carbon fiber), 0.01 (hollow graphite fibrils), 10-13 (pitch-based) | | |
| **Aspect ratio**: 6-30 (milled); 860 (chopped), 100-1000 (hollow graphite fibrils) | | |
| **Specific surface area**, m$^2$/g: 0.27-0.98, 250-300 (nanofibers[521]), 0.4-0.7 (pitch-based) | | |

MANUFACTURERS & BRAND NAMES:

Amoco Performance Products, Inc., Alpharetta, GA, USA

ThermalGraph DKA X (0.2 mm), CKD X (50 mm), DKE X (0.003-0.005 mm), DKD X (0.2 mm) - pitch-based thermally conductive fibers which have 50% higher longitudinal conductivity than copper. The filament diameter is 10 μm for all fibers and their length is given in parentheses. DKD has higher tensile modulus than DKA.

Thornel VMX-11, VMX-12 - granulated pitch-based fillers for injection molding to enhance electric and thermal conductivity, frictional characteristics and dimensional stability

Thornel K-1100 2K - fiber which has thermal conductivity 2-3 higher than copper and 4-5 times higher than aluminum

T300, T650 - PAN-based carbon fibers

Asahi Chemical Industry, Tokyo, Japan

Courtauld Ltd., UK

Courtelle HM, HT

Hercules Aerospace Espana S.A., Spain

AS

Hyperion Catalysis International, Cambridge, MA, USA

Hollow carbon fibrils

*continued on the next page*

**MANUFACTURERS & BRAND NAMES:**
Toho Rayon Co., Ltd., Tokyo, Japan affiliated with Toho US and Tenax, Germany
Besfight HTA-C6- S, SR, SRS, N, NR, NRS, E - chopped fiber (S, E - epoxy sizing, N - polyamide sizing)
Besfight HTA-CMF- 0040 OH, 0160-E, 0160-OH - milled fiber
Besfight pellet - S-1002C-00, S-1002G-00, L-1002C, L-1002G-00 (PA-66), S-1230C-00, 1230G-00 (POM)
Besfight Prepreg - 100 series (epoxy modified), 300 series (bismaleimide modified)
Toray Industries
Celion G30

**MAJOR PRODUCT APPLICATIONS:** personal computers, aircraft, rockets, satellites, automation equipment, electrical and electronics parts, mechanical parts, medical instruments, fishing rods, golf clubs, tennis rackets, brake pads, composites, mufflers, surface preparation for electrostatic painting

**MAJOR POLYMER APPLICATIONS:** PP, PE, PA, PC, PBT, PEEK, PS, epoxy, polyurethane

The following properties of carbon fibers are exploited in their applications: high tensile strength and modulus, good fatigue resistance and wear lubricity, low density (lower than metal), low linear thermal expansion coefficient, good dimensional stability, heat resistance, electric conductivity, ability to shield electromagnetic waves, x-ray penetrability, good chemical stability and excellent resistance to acids, alkalis, and many solvents. This list shows that carbon fibers have a high potential use in high performance materials. Total world production of carbon fibers is estimated 9,590 tons. North America consumes 40% of total production, Europe and Japan 21% each and the remaining countries 18%. The largest use is in aircraft industry followed by sport and leisure equipment and industrial equipment. Carbon fiber is produced from polyacrylonitrile fiber, rayon or pitch filaments which undergo preoxidation, carbonization and surface treatment. Surface oxidized carbon fibers are also produced to increase adhesion are produced. Also, prepregs are manufactured with various resins (mostly epoxy and bismaleimide) to aid in the incorporation of carbon fibers. Figure 2.77 shows micrograph of the cross-section of carbon fiber which can be compared with Figure 2.38 which shows this fiber coated with nickel.

Figure 2.77. Micrograph of Besfight carbon fiber. *Courtesy of Toho Rayon Co., Ltd., Tokyo, Japan.*

The conditions of carbonization have impact on properties of carbon fibers and their price. The least expensive carbon fibers manufactured from PAN are produced by rapid heating under tension from the initial orientation temperature of 300°C to 1000°C. This process produces low modulus fibers. High strength fibers are heated to 1500°C and the high modulus fiber to 2200°C under argon. These various conditions result in graphite crystals with different structures which affects the mechanical performance of fibers. In the coal-tar or petroleum pitch processes, the initial material is polymerized by heat which helps to remove low molecular weight volatile components. The resultant nematic liquid crystal, or mesophase, is oriented during the spinning operation to form fibers. The third raw material – rayon is used less often because of the environmental impact of the precursor material.[546]

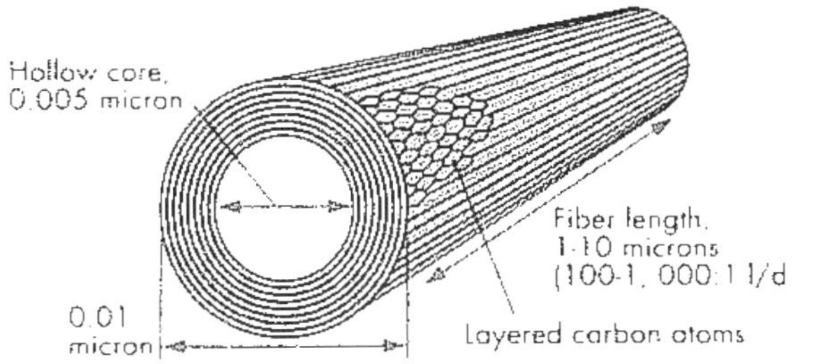


Figure 2.78. A structure of hollow carbon fibers. *Courtesy of Hyperion Catalysis International, Cambridge, MA, USA.*

Hyperion Catalysis International developed a new technology to produce hollow carbon fibrils. The patented technology produces hollow fibrils of very small diameter in a catalytic process using ethylene gas as the raw material. The fibril structure is given in Figure 2.78. The striking feature of these fibrils is their very small diameter. Typically, with these fibrils, seven times less material is required to obtain a conductivity equivalent to products filled with PAN-based carbon fibers and 3 times less than products filled with steel fibers. This performance is due to the high elasticity of these fibers which lowers breakage and allows the fibers to form entangled structures within the body of the plastic material. Efforts are

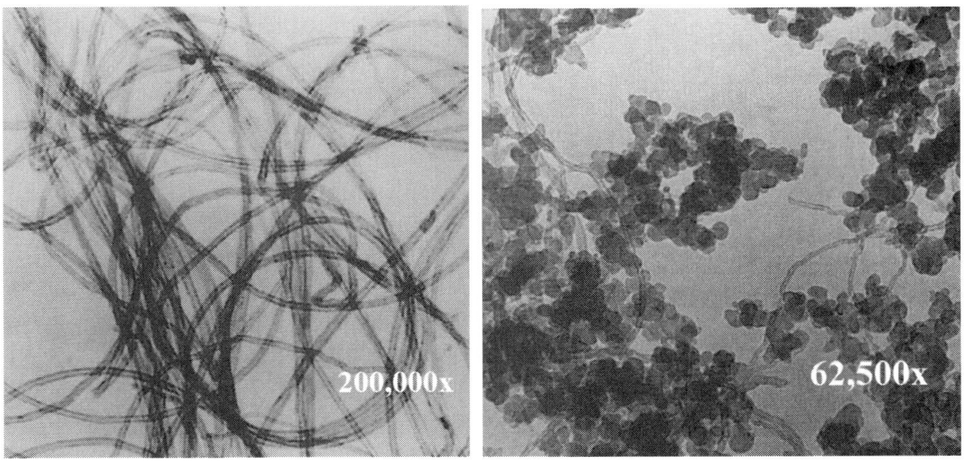


Figure 2.79. Hollow graphite fibrils (left) and fibrils mixed with carbon black (right). *Courtesy of Hyperion Catalysis International, Cambridge, MA, USA.*

being made to simplify the electrostatic painting of parts filled with carbon fibers for automotive and other applications.

Figure 2.79 shows graphite fibrils alone and in comparison with particles of carbon black. Carbon black particles have larger diameter than these hollow tubes.

## 2.2.3 CELLULOSE FIBERS[548-553]

| | | |
|---|---|---|
| **Name**: cellulose fiber | | **CAS #**: 9004-34-6 |
| **Chemical formula**: $(C_6H_{10}O_5)_n$ | | **Functionality**: OH or from modification |
| **Chemical composition**: cellulose content - 45-99.6% | | |
| **Trace elements**: Pb - 10 ppm, As - 1 ppm | | |
| **PHYSICAL PROPERTIES** | | |
| **Density**, g/cm$^3$: 1-1.1 | **Char point**, °C: 290 | **Loss on ignition**, %: 0.3-25 |
| **Maximum temperature of use**, °C: 200 | | |
| **CHEMICAL PROPERTIES** | | |
| **Moisture content**, %: 2-10 | **Adsorbed moisture**, %: 420-1000 | **Ash content**, %: 0.13-0.4 |
| **pH of water suspension**: 4-9 | **Water solubility**, %: 1.5 | |
| **OPTICAL PROPERTIES** | | |
| **Color**: white, gray, brown | | **Brightness**: 86-89 |
| **MORPHOLOGY** | | |
| **Pore size**: 100 Å (only polymers which have molecular weight less than 10,000 can enter pores) | | |
| **Fiber length**, μm: 22-290 | **Oil absorption**, g/100 g: 300-1000 | |
| **Specific surface area**, m$^2$/g: 1 (dry state), 100-200 (accessible to water in wet state) | | |
| **Sieve analysis**: residue on 200 mesh sieve - traces-60% | | **Fiber diameter**, μm: 5-30 |

**MANUFACTURERS & BRAND NAMES**:

Cellulose Filler Factory Corporation, Chestertown, MD, USA affiliate of Cellulose-Füllstoff-Fabrik, Mönchengladbach, Germany
- Technocel 1003/5, 1004, 1004/5, 1004/10, 1004/15, 2004, 202, 40, 90, 150, 180, 200, 300, 750, 2500- recycled and virgin fibers for industrial applications. Fibers differ is color, purity, and particle sizes
- Topcel - products for asphalt reinforcement
- Diacel 40, 90, 150, 200 - pulp for filtration industry (with number increasing particle size increases)
- Sanacel 40, 90, 150, 200, - fibers for cosmetic and pharmaceutical applications
- Qualicel 40, 90, 150, 200, - vegetable fiber for food applications

Fiber Sales & Development Corporation, St. Louis, MO, USA
- Solca-Floc 1016, 10, 20, 40, 60, 100, 200, 300 - fibers of different length manufactured from purified cellulose

Interfibe Corporation, Solon, OH, USA
- White fibers - Gel-Cel W10, W30, W50, 5FT
- Gray fibers - 185, 230, ETF, JMC, JMM, FT, GC66
- Treated fibers - 200, 205, WFP, FTP
- Gel-Cel fibers - 10, 20, 30 - fibers obtained by Jet Process developed to improve uniformity of fibers, modify their morphology, and improve their anti-settling characteristics

**MAJOR PRODUCT APPLICATIONS**: filtration, ceramics, foams, floor tiles, shoe soles, paints, food, building products, welding electrodes, gaskets, stucco, EIFS, asbestos alternative, sealants, roof coatings, athletic surface coatings, crack fillers and sealers, brake pads, clutches, pavement, artificial leather, electrical components, automotive components, household appliances, mastics, putties, patching compounds, grouts

**MAJOR POLYMER APPLICATIONS**: alkyd, polyurethane, acrylic, rubber, melamine resins, phenoxy, polyester, PE, PP, PVC, NBR

Cellulose fibers offer many valuable properties but the most important characteristic is that they are natural in origin. They are safe to use, non-polluting, and energy efficient. These qualities are the major reasons for the growing interest in these fibers. Technical cellulose fibers are produced by recycling of newsprint, magazines, and other paper products. There are also numerous industrial applications for these fibers which exploit their chemical functionality (reactivity) for crosslinking, their ability to retain water and their hydrogen bonding capability for improvement of rheological properties. The shape of fiber helps to prevent cracking, reduce shrinkage, increase green strength, and reinforce materials.

Cellulose content varies. Virgin fibers produced from wood pulp contain 99.6% cellulose and are white. Fibers manufactured from reclaimed materials contain 75% and are gray or brown. Cellulose fibers (especially virgin materials) have a complex morphological structure which facilitates reinforcement (Figure 2.80).

Figure 2.81 shows the fiber surface at a high magnification. The accessability of the fiber surface to interaction with the matrix depends on the differences in fiber morphology relative to the method of their manufacture. The choice of hydrophilic or hydrophobic grades improves their dispersion in different matrices and readily accessible functional groups allow the use fibers to double as reactive crosslinkers.

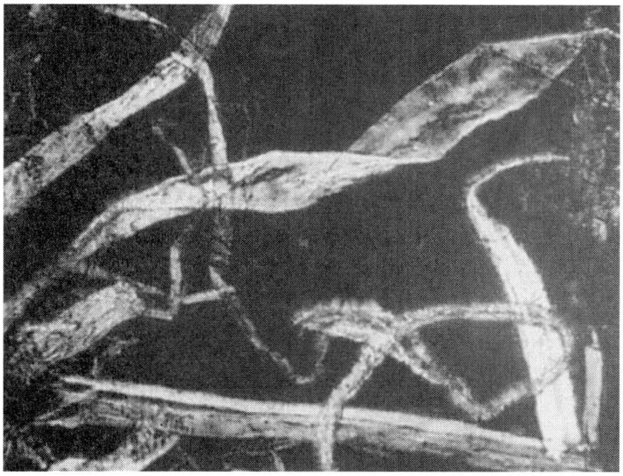

Figure 2.80. The morphology of cellulose fibers. *Courtesy of Cellulose Filler Factory Corporation, Chestertown, MD, USA.*

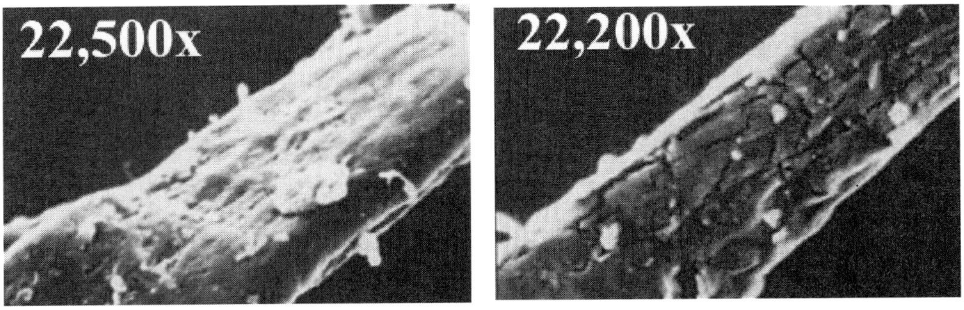


Figure 2.81. SEM micrograph of cellulose fiber, Interfibe WF (left), Interfibe 231 (right). *Courtesy of Interfibe Corporation, Solon, OH, USA.*

## 2.2.4 GLASS FIBERS[554-565]

| | | |
|---|---|---|
| **Name**: glass fibers | | **CAS #**: 65997-17-3 |
| **Chemical formula**: variable | **Functionality**: OH unless modified | |
| **Chemical composition**: $SiO_2$ - 52.5-55.5%, CaO - 21-24%, $Al_2O_3$ - 14-14.5%, $B_2O_3$ - 5-8.6%, sizing 0-3% | | |
| PHYSICAL PROPERTIES | | |
| **Density**, g/cm$^3$: 2.52-2.68 | **Mohs hardness**: 6-6.5 | **Softening point**, °C: 830-920 |
| **Thermal conductivity**, W/K·m: 1 | **Specific heat**, kJ/kg·K: 0.83 | |
| **Young modulus**, MPa: 70,000 | **Poisson ratio**: 0.22 | **Coefficient of friction**: 0.9-1 |
| **Tensile strength**, GPa: 3.1-3.8 | **Elastic modulus**, GPa: 76-81 | **Elongation**, %: 4.5-4.9 |
| CHEMICAL PROPERTIES | | |
| **Moisture content**, %: 0.1-3 | **Adsorbed moisture**, %: 0.3 | **pH of water suspension**: 5-10 |
| OPTICAL & ELECTRICAL PROPERTIES | | |
| **Refractive index**: 1.55-1.56 | **Color**: white | **Loss tangent**: 0.001 |
| **Dielectric constant**: 5.8-6.1 | **Volume resistivity**, Ω-cm: $10^{13}$-$10^{16}$ | |
| MORPHOLOGY | | |
| **Fiber length**, μm: 50-350 (milled grades), 4000-13,000 (chopped grades) | | |
| **Aspect ratio**: 3-800 | **Filament diameter**, μm: 15.8 | |

MANUFACTURERS & BRAND NAMES:

Evans Clay Company, McIntyre, GA, USA

FG 500, 700, 800

Owens Corning, Toledo, OH, USA

Fiberglas - 731 line (cationic size), 737 line (silane) 739 line (no sizing agent) - milled fibers produced in each line in different length sizes but the same filament diameter (15.8 μm) made out of E-glass

Fiberglas 405 - chopped strands made out of E-glass in 1/8, 3/16, 1/4, and ½ lengths for polyester, epoxy and phenolics

Cratec 144A (PP), 408A (PBT, POM, SMA, ABS, SAN, PS, PC, PP), 415A (PE & PC below 15 wt% loading), 489A (products which require FDA approval), 497A (PPS, PPO, PVC, PSF, phenoxy) - chopped glass fiber grades optimized for application in polymers listed in parentheses. All grades have the same fiber length (4 mm) and are produced from E-glass

*continued on the next page*

**MANUFACTURERS & BRAND NAMES:**
PPG Industries, Inc., Fiber Products, Pittsburgh, PA, USA
Chop Vantage 3535 (PA), 3540 (PA, PET, ABS, SAN), 3563 (PET), 3640 (PA-66, PA-46), 3660 (PA), 3763 (PBT, PC), 3793 PBT, ABS, SAN, SMA, PC, PPS, PEI, PES, PEEK), 8016 (chopped strand, mat applications) - chopped fibers with different silane treatment designed for the selected polymers listed in parentheses. Filament diameter is 10 μm and length 3.2 or 4.5 mm.
DeltaChop 3796 (PPS, PEI, PES, PEEK), 8610 (paper, ceramics), 8810 (asbestos replacement in friction applications) - chopped strand of ultrafine fibers with proprietary sizing having filament diameter of 6.5μm and length varying from 3 to 38 mm. The fibers are used in applications listed in parentheses.
MaxiChop 3242 (PP), 3298 (PP), 3617 (PA), 3662 (PA), 3707 (PC), 3762 (PBT, PC), 3790 (PBT, ABS, SAN, SMA, PC), 8018 (non-woven, papers, felts)- chopped strand of fibers with silane sizing having filament diameter of 13 μm (except for 3617 and 3707 which have filament diameter of 17 μm) length for most grades is 3.2 mm except for 3790 (3.2 and 4.8) and 8018 (3 to 38 mm). The fibers are used in applications listed in parentheses.
Type 3075 (bulk molding compounds, BMC), 3156 (thermosets, such as phenoxy, epoxy, polyester, etc.) - chopped strand of 13 and 10 μm silane sized filaments, respectively, cut to the length in a range from 3.2 to 12.8 mm
8239 - wet chopped strand for wet laid mat (diameter - 16 μm, length - 6-32 mm)

**MAJOR PRODUCT APPLICATIONS:** electrical connectors, automotive components, automotive fascia, automotive seals, gaskets and bearings, aerospace components, friction products, putty compounds, adhesives

**MAJOR POLYMER APPLICATIONS:** polyester, epoxy, phenoxy, polyurethanes, PTFE, PP, PE, PBT, POM, SMA, ABS, SAN, PS, PC, PES, PEI, PPS, PPO, PVC, PSF, phenoxy

Glass fibers are produced by two methods, milling and chopping. The milled fibers are milled using a hammer mill which results in a relatively broad (but consistent) length distribution. The diameter depends on the filament diameter manufactured for milling process. The chopped fibers are produced by chopping a bundle of glass filaments to a precise length. The length of chopped fibers is substantially larger than that of the milled fibers. In both cases, fibers may or may not contain sizing or surface modification. If sizing is applied, it is optimized for a certain type or types of polymers. Owen Corning milled fibers are produced with a variety of size coatings for different polymers. Cationic sized milled fiber is suggested for polyester epoxy, phenolic and thermoplastics. Silane modified grades are for urethanes and thermoplastics, and glass fiber without any sizing agent is suggested for use in PTFE and thermoplastics.

Glass fibers are extensively used by industry because of their reinforcing effect, and the improvements they produce in thermal properties such as a reduction in thermal expansion and an increase in heat deflection temperature. The most challenging tasks of fiber application include the incorporation process which must be designed to prevent breakage, improve matrix fiber adhesion, prevent fiber corrosion in some environments, and develop proper fiber orientation.

## 2.2.5 OTHER FIBERS

Numerous fibrous products are used as fillers in plastics materials. Fibers are generally divided into natural and man-made fibers. The natural fibers belong to three groups: vegetable, animal, and mineral fibers. Natural mineral fibers were discussed above in separate sections. The vegetable fibers group is divided into hair fibers (cotton, kapok), bast fibers (flax, hant, jute, ramie) and hard fibers (sisal, hanequen, coir). A typical feature of vegetable fiber is the high cellulose content (65-85%). Other building blocks of vegetable fibers include hemicellulose (5-15%), and lignin (2-15%). In addition to vegetable fibers there is a growing interest in utilization of various waste wood products such as paper and construction wood waste which constitute a significant portion of municipal waste. The properties of wood fibers and cellulose fibers discussed above (Sections 2.1.58 and 2.2.3) show that these materials offer very good properties and are likely to be studied in the future with a growing interest.

Current research indicates that there is a growing interest in natural fibers. Natural fibers from jute were tested in thermosetting and thermoplastic resins.[566-568] Lignin fillers were used in phenol-formaldehyde,[569] SBR, SBS, and SIS[570] and PE[571] with good results. The opportunities for applications of natural fibers in industrial products have been the subject of recent reviews.[572,573] Cellulose whiskers with a high reinforcing value were obtained from wheat straw.[574,575] Wood fibers were found applicable to such diverse materials as polypropylene parts,[576] foams,[577] and polymer blends.[578] The interest in this research is inspired by availability, biological degradability, low cost, and chemical reactivity of these products which can be easily modified by chemical methods. Fibers of animal origin are less important although small amounts are used in adhesives and sealants.

Metal fibers form another group of important materials due to the growing interest in conductive materials.[579-581] Some of these fibers were discussed together with metal powders, flakes and metal coated minerals in Section 2.1.40.

There is also an interest in application of synthetic fibers.[582,583] Two directions are common: surface modification and development of fibers with special morphology. The controlled composition of synthetic fibers gives opportunities to regulate their surface properties to meet specific requirements giving the product formulator new tools to make product improvement. Synthetic fibers can be produced in variety of shapes and sizes which can be tailored to specific applications in new products. Ultra small fibers, some hollow, with a wide variety of surface morphologies can be produced economically to meet specific requirements of a wide variety of high technology products.

## REFERENCES


1 Lei Yang, Schruben D L, *Polym. Engng. Sci.*, **34**, No.14, 1994, 1109-14.
2 Chmutin I A, Ryvkina N G, Ponomarenko A T, Shevchenko V G, *Polym. Sci., Ser. A*, **38**, No.2, 1996, 173-77.
3 Tavman I H, *J. Appl. Polym. Sci.*, **62**, No.12, 1996, 2161-7.
4 **Aluminum-pigmented Coatings for Industrial Maintenance Applications**. *The Aluminum Association*, Washington, 1996.
5 Ferguson R L, **Aluminum Pigments. An Overview**. *Silberline*.
**Leafing Aluminum Pigments**. *Silberline*, Tamaqua.
7 Persson A L, Bertilsson H, *Composite Interfaces*, **3**, No.4, 1996, 321-32.

8 Nichols K, Solc J, Shieu F, Antec '93. Conference Proceedings, New Orleans, La., 9th-13th May 1993, Vol. II, 1938-42.
9 Gordienko V P, Dmitriev Y A, *Polym. Degradat. Stabil.*, **53**, No.1, 1996, 79-87.
10 Garbow J R, Asrar J, Hardiman C J, *Chem. of Mat.*, **5**, No., 1993, 869-75.
11 Grohens Y, Schultz J, *Int. J. Adhesion Adhesives*, **17**, 1997, 163-67.
12 Gallas M R, Rosa A R, Costa T H, da Jornada J A H, *J. Mater. Res.*, **12**, No. 3, 1997, 764-68.
13 Siegel R W, *Sci. American*, **275**, 1996, 74-79.
14 Ford Q, *Ceramic Ind.*, Jan 1998.
15 Weaver A, *Reinf. Plast.*, **40**, No.11, 1996, 52-3.
16 Brown N, Linnert E, *Reinf. Plast.*, **39**, No.11, 1995, 34-7.
17 Miller B, *Plast. World*, **54**, No.1, 1996, 38-43.
18 Levchik G F, Levchik S V, Lesnikovich A I, *Polym. Degradat. Stabil.*, **54**, Nos 2-3, 1996, 361-63.
19 Yeh J T, Yang H M, Huang S S, *Polym. Degradat. Stabil.*, **50**, No.2, 1995, 229-34.
20 Yang Q, Pritchard G, Phipps M A, Rose R G, *Polym. & Polym. Composites*, **4**, No.4, 1996, 239-46.
21 Jang J, Yi J, *Polym. Engng. Sci.*, **35**, No.20, 1995, 1583-91.
22 Dando N R, Kolek P L, Martin E S, Clever T R, *J. Coatings Technol.*, **68**, No.859, 1996, 67-72.
23 Liauw C M, Lees G C, Hurst S J, Rothon R N, Dobson D C, *Plast. Rubb. Comp. Process. Appln.*, **24**, No.4, 1995, 211-9.
24 Rothon R N, *Macromol. Symp.*, **108**, 1996, 221-9.
25 Hong S G, Lin J J,Yuan Ze, *J. Appl. Polym. Sci.*, **59**, No.10, 1996, 1597-605.
26 Jeffs D, *Reinf. Plast.*, **38**, No., 1994, 32-5.
27 McNeill I C, Mohammed M H, *Polym. Degradat. Stabil.*, **48**, No.1, 1995, 189-95.
28 *Reinf. Plast.*, **38**, No.11, 1994, 15.
29 Yang Q, Pritchard G, *Polym. & Polym. Composites*, **2**, No.4, 1994, 233-9.
30 Molesky F, Schultz R, Midgett S, Green D, *J. Vinyl Additive Technol.*, **1**, No.3, 1995, 159-61.
31 Levesque J L, Fraval J T, Antec '93. Conference Proceedings, New Orleans, La., 9th-13th May 1993, Vol. II, 1957-64.
32 Ferm D J, Shen K K, Antec '94. Conference Proceedings, San Francisco, Ca., 1st-5th May 1994, Vol.III, 3522-6.
33 Liptak P, *Int. Polym. Sci. Technol.*, **21**, No.8, 1994, T/50-3.
34 Yamada H, Inagaki S, Okamoto H, Furukawa J, *Int. Polym. Sci. Technol.*, **21**, No.6, 1994, T/29-35.
35 Liptak P, Zelenak P, *Int. Polym. Sci. Technol.*, **20**, No.9, 1993, T/57-9.
36 Parker A A, Martin E S, Clever T R, *J. Coatings Technol.*, **66**, No.829, 1994, 39-46.
37 Lee D H, Condrate R A, Reed J S, *J. Mater. Sci.*, **31,** 1996, 471-8.
38 Dayer A J, Mead NG, **Alcan Superfine ATH in Thermoplastic EVA Cable Compounds**. *Alcan Chemicals*. 1998.
39 Toure B, Lopez Cuesta J M, Gaudon P, Benhassaine A, Crespy A, *Polym. Degradat. Stabil.*, **53**, No.3, 1996, 371-9.
40 Herbert M J, Flame Retardants '96. Conference proceedings, London, 17th-18th Jan.1996, 157-72.
41 Miller B, *Plast. World*, **54**, No.12, 1996, 44-9.
42 Toure B, Lopez Cuesta J-M, Longerey M, Crespy A, *Polym. Degradat. Stabil.*, **54**, Nos 2-3, 1996, 345-52.
43 Kretzschmar B, *Kunststoffe Plast Europe*, **86**, No.4, 1996, 20-2.
44 Nagieb Z A, El-Sakr N S, *Polym. Degradat. Stabil.*, **57**, 1997, 205-9.
45 Weng J, Liu Q, Wolke J G C, Zhang D, De Groot K, *J. Mater. Sci. Lett.*, **16**, 1997, 335-7.
46 Wong K W Y, Truss R W, *Composites Sci. & Technol.*, **52**, No.3, 1994, 361-8.
47 Alkan C, Arslan M, Cici M, Kaya M, Aksoy M, *Resources Conserv. & Recycling*, **13**, Nos.3-4, 1995, 147-54.
48 Sendijarevic A, Sendijarevic V, Wang X, Haidar A, Dutta U, Klempner D, Frisch K C, Polyurethanes '95. Conference Proceedings, Chicago, Il., 26th-29th Sept.1995, 418-26.
49 Rebeiz K S, Rosett J W, Nesbit S M, Craft A P, *J. Mat. Sci. Lett.*, **15**, No.14, 1996, 1273-5.
50 Bijwe J, *Polym. Composites*, **18**, No.3, 1997, 378-96.
51 Leguet X, Ericson M, Chundury D, Baumer G, Antec '97. Conference proceedings, Toronto, April 1997, 2117-34
52 Tsubokawa N, Seno K, *J. Macromol. Sci. A*, **31**, No.9, 1994, 1135-45.
53 Sundar K L, Radhakrishnan G, Reddi B R, *Polym. Plast. Technol. Engng.*, **35**, No.4, 1996, 561-66.
54 Mitsui S, Kihara H, Yoshimi S, Okamoto Y, *Polym. Engng. Sci.*, **36**, No.17, 1996, 2241-6.
55 Erofeev L N, Raevskii A V, Pisarenko T I, Grishin B S, *Int. Polym. Sci. Technol.*, **23**, No.5, 1996,

T/12-4.
56 Hess M, Veeman W, Magusin P, Antec '96. Volume III. Conference proceedings, Indianapolis, 5th-10th May 1996, p.3682-86.
57 Molino L N, Topoleski L D T, *J. Biomed. Mat. Res.*, **31**, No.1, 1996, 131-7.
58 Gregorio R, Cestari M, Bernardino F E, *J. Mat. Sci.*, **31**, No.11, 1996, 2925-30.
59 Visser S A, *J. Appl. Polym. Sci.*, **3**, 1997, 1805-20.
60 Miano F, Rabaioli M R, *Coll. & Surfaces*, **84**, Nos.2/3, 1994, 229-37.
61 Wu J, Lerner M M, *Chem. of Mat.*, **5**, No., 1993, 835-8.
62 Ohashi F, Oya A, *J. Mat. Sci.*, **31**, No.13, 1996, 3403-7.
63 Akelah A, Moet A, *J. Mat. Sci.*, **31**, No.13, 1996, 3589-96.
64 Isayama M, Nomiyama K, Kunitake T, *Adv. Mat.*, **8**, No.8, 1996, 41-4.
65 Le Bras M, Bourbigot S, *Fire & Mat.*, **20**, No.1, 1996, 39-49.
Xavier S F, *Pop. Plast. Packag.*, **41**, No.4, 1996, 61-4.
67 Marentette J M, Norwig J, Stockelmann E, Meyer W H, Wegner G, *Adv. Mater.*, **9**, No.8, 1997, 47-50.
68 Domka L, *Coll. Polym. Sci.*, **272**, No.10, 1994, 1190-202.
69 Wang Y, Lu J, Wang G, *J. Appl. Polym. Sci.*, **64**, 1997, 1275-81.
70 Dolui S K, *J. Appl. Polym. Sci.*, **53**, No.4, 1994, 463-65.
71 Li J X, Silverstein M, Hiltner A, Baer E, *J. Appl. Polym. Sci.*, **52**, No.2, 1994, 255-7.
72 Nakamura S, Kaneko S, Mizutani Y, *J. Appl. Polym. Sci.*, **49**, No.1, 1993, 143-50.
73 Bosse F, Schreiber H P, Eisenberg A, *Macromolecules*, **2**, No.24, 1993, 6447-54.
74 Manolis Sherman L, *Plast. Technol.*, **40**, No.12, 1994, 61-3.
75 Jancar J, DiBenedetto A T, *Sci. & Eng. Composite Materials*, **3**, No. 4, 1994, 217-26.
76 Godard P, Bomal Y, Biebuyck J J, *J. Mat. Sci.*, **28**, No.24, 1993, 605-10.
77 Herzig R, Baker W E, *J. Mat. Sci.*, **28**, No.24, 1993, 6531-9.
78 Jancar J, Dibenedetto A T, *J. Mat. Sci.*, **30**, No.9, 1995, 1601-08.
79 Jancar J, Dibenedetto A T, *J. Mat. Sci.*, **30**, No.9, 1995, 2438-45.
80 Fu Q, Wang G, Liu C, *Polymer*, **36**, No.12, 1995, 2397-401.
81 Jancar J, Dibenedetto A T, Dianselmo A, *Polym. Engng. Sci.*, **33**, No.9, 1993, 559-63.
82 Bataille P, Mahlous M, Schreiber H P, *Polym. Engng. Sci.*, **34**, No.12, 1994, 981-5.
83 Pukanszky B, Belina K, Rockenbauer A, Maurer F H J, *Composites*, **25**, No.3, 1994, 205-14.
84 Chudinova V V, Guzeev V V, Mozzhukhin V B, Pomerantseva E G, Nozrina F D, Zhil'tsov V V, Zubov V P, *Int. Polym. Sci. Technol.*, **21**, No.10, 1994, T/102-4.
85 Ruiz F A, Antec '94. Conference Proceedings, San Francisco, Ca., 1st-5th May 1994, Vol. III, 27-9.
86 Gendron R, Daigneault L E, Tatibouet J, Dumoulin M M, Antec '94. Conference Proceedings, San Francisco, Ca., 1st-5th May 1994, Vol. I, 167-71.
87 Skelhorn D A, Antec '93. Conference Proceedings, New Orleans, La., 9th-13th May 1993, Vol. II, 195-70.
88 Bataille P, Schreiber H P, Mahlous M, Antec '93. Conference Proceedings, New Orleans, La., 9th-13th May 1993, Vol. II, 1757-9.
89 Skelhorn D A, Meeting of the Rubber Division, ACS, Montreal, May 5-8, 1996, paper A.
90 Dave P, Brown K, MacIver W, Chundury D, Draucker C, Lightener L, Antec '97. Conference proceedings, Toronto, April 1997, 514-20.
91 Zaborski M, Slusarski L, *Composite Interfaces*, **3**, No.1, 1995, 9-22.
92 Li Shucai, Peng Weijang, Lu Xiuping, *Int. J. Polym. Mat.*, **29**, Nos.1-2, 1995, 37-42.
93 Ulkem I, Bataille P, Schreiber H P, *J. Macromol. Sci. A*, **31**, No.3, 1994, 291-303.
94 Rockenbauer A, Korecz L, Pukanszky B, *Polym. Bull.*, **33**, No.5, 1994, 585-9.
95 Carraher C E, *Polym. News*, **19**, No.2, 1994, 50-2.
96 Liphard M, Von Rybinski W, Schreck B, *Prog. Coll. & Polym. Sci.*, **95**, 1994, 168-74.
97 *Reinf. Plast.*, **39**, No.4, 1995, 8.
98 Minkova L, Magagnini P L, *Polym. Degradat. Stabil.*, **42**, No.1, 1993, 107-15.
99 Hu X, Xu H, Zhang Z, *Polym. Degradat. Stabil.*, **43**, No.2, 1994, 225-8.
100 Callari J J, *Plast. World*, **52**, No.3, 1994, 22-6.
101 Borden K A, Wei R C, Manganaro C R, *Plast. Compounding*, **1**, No.5, 1993, 51-5.
102 Desai S C, *Pop. Plast. Packag.*, **39**, No.7, 1994, 53-60.
103 Yu Long, Shanks R A, *J. Appl. Polym. Sci.*, **61**, No.11, 1996, 1877-85.
104 Nago S, Mizutani Y, *J. Appl. Polym. Sci.*, **61**, No.1, 1996, 31-5.
105 Khare A, Mitra A, Radhakrishnan S, *J. Mat. Sci.*, **31**, No.21, 1996, 5691-5.
106 Mamat A, Trochu F, Sanschagrin B, *Polym. Engng. Sci.*, **35**, No.19, 1995, 1511-20.

107 Pukanszky B, Maurer F H J, Boode J W, *Polym. Engng. Sci.*, **35**, No.24, 1995, 1962-71.
108 Bomal Y, Godard P, *Polym. Engng. Sci.*, **36**, No.2, 1996, 237-43.
109 Akovali G, Dilsiz N, *Polym. Engng. Sci.*, **36**, No.8, 1996, 1081-86.
110 Gendron R, Daigneault L E, Tatibouet J, Dumoulin M M, *Adv. Polym. Technol.*, **15**, No.2, 1996, 111-25.
111 Kenny J M, Opalicki M, *Composites Part A: Applied Science and Manufacturing*, **27A**, No.3, 1996, 229-40.
112 Kiselev V Y, Vnukova V G, *Int. Polym. Sci. Technol.*, **23**, No.5, 1996, T/88-92.
113 Johnson K C, Antec '96. Volume III. Conference proceedings, Indianapolis, 5th-10th May 1996, 3545-9.
114 Reshadat R, Balke S T, Calidonio F, Dobbin C J B, Antec '96. Volume III. Conference proceedings, Indianapolis, 5th-10th May 1996, 288-92.
115 Cheung T, Tjong S C, Li R K Y,Antec '9. Volume II. Conference proceedings, Indianapolis, 5th-10th May 1996, 2256-59.
116 Wiebking H E, Antec 95. Volume III. Conference proceedings, Boston, Ma., 7th-11th May 1995, 4112-16.
117 Kenny J M, Opalicki M, Molina G, Antec '95. Vol. II. Conference Proceedings, Boston, Ma., 7th-11th May 1995, 2782-89.
118 Gingrich R P, Machado J M, Londa M, Proctor M G, Antec '95. Vol. II. Conference Proceedings, Boston, Ma., 7th-11th May 1995, 2345-50.
119 Cornwell D, *Brit. Plast. Rubb.*, Nov.1996, 47-8.
120 *Rubb. World*, **215**, No.1, 1996, 16-18.
121 Ernst D, *Brit. Plast. Rubb.*, Jan.1997, 4-6.
122 Casoli A, Charmeau J Y, Holl Y, d'Allest J F, *J. Adhesion*, **57**, Nos.1-4, 1996, 133-51.
123 Hohenberger W, *Kunststoffe Plast Europe*, **86**, 7, 1996, 973-77.
124 Hattori K, Morford S, Delany D T, *J. Vinyl and Additive Technol.*, **1**, No.3, 1995. 170-3.
125 Kovacevic V, Lucic S, Hace D, Cerovecki Z, *J. Adhesion Sci. Technol.*, **10**, No.12, 1996, 1273-85.
126 Driscoll S B, Nanavaty P, Stockbower D W, Polymer Additives for Injection Moulding and Extrusion Applications. Retec proceedings, White Haven, Pa., 18th-19th Oct.1995, p.161-85.
127 Oien H T, Polyurethanes '95. Conference Proceedings, Chicago, Il., 2th-29th Sept.1995, 137-41.
128 *Plastics in Canada*, **3**, No.2, 1996, 69.
129 Saad A L G, Younan A F, *Polym. Degradat. Stabil.*, **50**, No.2, 1995, 133-40.
130 Hancock, *Plasticulture*, **79**, 1988, 4-14.
131 Johnson S L, Ahsan T, Polymers, Laminations & Coatings Conference, Conference Proceedings, Atlanta, 1997, 471-477.
132 *Socal. Precipitated Calcium Carbonate. Production, Properties and Applications*. Solvay, 11.96.
133 *Calcium carbonate and dolomite filler and extenders*. Omya/Plüss-Staufer AG. Technical Note # 19.
134 Chan H, *Polym. Paint Colour J.*, Nov. 27, 1991.
135 **The World of Calcium Carbonate**. Omya/Plüss-Staufer Today. Omya/Plüss-Staufer.
136 Huggenberger L, Neubold H B. **Pigments for Paper. Natural Ground Calcium Carbonate as Coating Pigment and Filler**.
137 Schlumpf H-P, *Kunststoffe German Plastics*, **6**, 1983
138 Bosshard A W. **The Origin of Omya Fillers**. Omya/Plüss-Staufer.
139 Ali M H, Abo-Hashem A, *Plast. Rubb. Comp. Process. Appln.*, **24**, No.1, 1995, 47-51.
140 Mandal U K, Tripathy D K, De S K, *Plast. Rubb. Comp. Process. Appln.*, **24**, No.1, 1995, 19-25.
141 Heinrich G, Vilgis T A, *Rubb. Chem. Technol.*, **68**, No.1, 1995, 26-36.
142 Darmstadt H, Roy C, Kaliaguine S, Sahouli B, Blacher S, Pirard R, Brouers F, *Rubb. Chem. Technol.*, **68**, No.2, 1995, 330-41.
143 Meng-Jiao Wang, Wolff S, Freund B, *Rubb. Chem. Technol.*, **67**, No.1, 1994, 27-41.
144 Gruber T C, Zerda T W, Gerspacher M, *Rubb. Chem. Technol.*, **67**, No.2, 1994, 280-7.
145 Coran A Y, Ignatz-Hoover F, Smakula P C, *Rubb. Chem. Technol.*, **67**, No.2, 1994, 237-51.
146 Clarke J, Freakley P K, *Rubb. Chem. Technol.*, **67**, No.4, 1994, 700-15.
147 Wang M-J, Wolff S, Tan E-H, *Rubb. Chem. Technol.*, **66**, No.2, 1993, 178-95.
148 Wolff S, Wang M-J, Tan E-H, *Rubb. Chem. Technol.*, **66**, No.2, 1993, 163-77.
149 Hjelm R P, Wampler W A, Seeger P A, Gerspacher M, *J. Mat. Res.*, **9**, No.12, 1994, 3210-22.
150 Wampler W A, Rajeshwar K, Pethe R G, Hyer R C, Sharma S C, *J. Mat. Res.*, **10**, No.7, 1995, 1811-22.
151 Ming Qiu Zhang, Jia Rui Xu, Han Ming Zeng, Qun Huo, Zhi Yi Zhang, Feng Chun Yun, Friedrich K, *J. Mat. Sci.*, **30**, No.17, 1995, 4226-32.

152 Dick J S, Pawlowski H A, *J. Elastomers Plast.*, **27**, No.1, 1995, 11-38.
153 del Rio C, Acosta J L, *Polymer*, **35**, No.17, 1994, 3752-57.
154 Dutta N K, Choudhury N R, Haidar B, Vidal A, Donnet J B, Delmotte L, Chezeau J M, *Polymer*, **35**, No.20, 1994, 4293-99.
155 Zhu J, Ou Y-C, Feng Y-P, *Polym. Int.*, **37**, No.2, 1995, 105-11.
156 Estrin R I, Pesin O Y, *Int. Polym. Sci. Technol.*, **22**, No.1, 1995, T/12-16.
157 Yu M C, Menashi J, Kaul D J, Antec '94. Conference Proceedings, San Francisco, Ca., 1st-5th May 1994, Vol. III, 2524-28.
158 Okamoto K T, Huang M C T, McGrath P A, Harris R M, Himont USA Inc., Antec '94. Conference Proceedings, San Francisco, Ca., 1st-5th May 1994, Vol. III, 2508-13.
159 Herd C R, Meeting of the Rubber Division, ACS, Montreal, May 5-8, 1996, paper B.
160 Gerspacher M, O'Farrell C P, Yang H H, Wampler W A, Meeting of the Rubber Division, ACS, Montreal, May 5-8, 1996, paper C.
161 Byers J T, Meeting of the Rubber Division, ACS, Cleveland, October 17-20, 1995, paper B.
162 White L, *Eur. Rubb. J.*, **177**, No.3, 1995, 27.
163 Turley R S, Strong A B, *J. Adv. Materials*, **25**, No.3, 1994, 53-9.
164 Foster J K, Sims E S, Venable S W, *Paint & Ink Int.*, **8**, No.3, 1995, 18-21.
165 Vidal A, Wang W, Donnet J B, *Kaut. u. Gummi Kunst.*, **46**, No.10, 1993, 770-78.
166 Donnet J B, Wang W, Vidal A, Wang M J, *Kaut. u. Gummi Kunst.*, **46**, No.11, Nov.1993, 866-71.
167 Wang W D, Haidar B, Vidal A, Donnet J B, *Kaut. u. Gummi Kunst.*, **47**, No.4, 1994, 238-41.
168 Maas S, Gronski W, *Kaut. u. Gummi Kunst.*, **47**, No., 1994, 409-15.
169 Patel A C, *Kaut. u. Gummi Kunst.*, **47**, No.8, 1994, 556-70.
170 Donnet J B, *Kaut. u. Gummi Kunst.*, **47**, No.9, 1994, 628-32.
171 Zaikin A Y, Nigmatullin V A, *Kaut. u. Gummi Kunst.*, **47**, No.10, 1994, 709-14.
172 Cochet P, Bomal Y, *Kaut. u. Gummi Kunst.*, **48**, No.4, 1995, 270-75.
173 Guerbe L, Freakley P K, *Kaut. u. Gummi Kunst.*, **48**, No.4, 1995, 260-69.
174 Probst N, Smet H, *Kaut. u. Gummi Kunst.*, **48**, No.7-8, 1995, 509-11.
175 Soares B G, Gubbels F, Jerome R, Teyssie P, Vanlathem E, Deltour R, *Polym. Bull.*, **35**, No.1/2, 1995, 223-28.
176 Hayashi S, Handa S, Oshibe Y, Yamamoto T, Tsubokawa N, *Polym. J. (Jap.)*, **27**, No.6, 1995, 623-30.
177 Foster J K, Sims E S, Venable S W, *Paint & Ink Int.*, **8**, No.3, 1995, 18-21.
178 Pehlergard P, *Rubb. S. Africa*, **10**, No.5, 1995, 8-12.
179 Monthey S, Duddleston B, Podobnik J, *Rubb. World*, **210**, No.3, 1994, 17-19.
180 Mwila J, Miraftab M, Horrocks A R, *Polym. Degradat. Stabil.*, **44**, No.3, 1994, 351-56.
181 Nasr G M, Badawy M M, Gwaily S E, Shash N M, Hassan H H, *Polym. Degradat. Stabil.*, **48**, No.2, 1995, 237-41.
182 del Rio C, Acosta J L, *J. Appl. Polym. Sci.*, **60**, No.1, 1996, 133-38.
183 Bandyopadhyay S, De P P, Tripathy D K, De S K, *J. Appl. Polym. Sci.*, **58**, No.4, 1995, 719-27.
184 Zhang M, Jia W, Chen X, *J. Appl. Polym. Sci.*, **62**, No.5, 1996, 743-47.
185 Kida N, Ito M, Yatsuyanagi F, Kaido H, *J. Appl. Polym. Sci.*, **61**, No.8, 1996, 1345-50.
186 Achour M E, El Malhi M, Miane J L, Carmona F, *J. Appl. Polym. Sci.*, **61**, No.11, 1996, 2009-13.
187 Kundu P P, Tripathy D K, Gupta B R, *J. Appl. Polym. Sci.*, **61**, No.11, 1996, 1971-83.
188 Bandyopadhyay S, De P P, Tripathy D K, De S K, *J. Appl. Polym. Sci.*, **61**, No.10, 1996, 1813-20.
189 Chiu H-T, Chiu W-M, *J. Appl. Polym. Sci.*, **61**, No.4, 1996, 607-12.
190 Hao Tang, Xinfang Chen, Aoqing Tang, Yunxia Luo, *J. Appl. Polym. Sci.*, **59**, No.3, 1996, 383-87.
191 Donnet J B, Tong Kuan Wang, *Macromol. Symp.*, **108**, 1996, 97-109.
192 Wolff S, *Rubb. Chem. Technol.*, **69**, No.3, 1996, 325-46.
193 Bohin F, Feke D L, Manas-Zloczower I, *Rubb. Chem. Technol.*, **69**, No.1, 1996, 1-7.
194 Qi Li, Feke D L, Manas-Zloczower I, *Rubb. Chem. Technol.*, **68**, No.5, 1995, 836-41.
195 Magee R W, *Rubb. Chem. Technol.*, **68**, No.4, 1995, 590-600.
196 Mori M, Koenig J L, *Rubb. Chem. Technol.*, **68**, No.4, 1995, 551-62.
197 Maas S, Gronski W, *Rubb. Chem. Technol.*, **68**, No.4, 1995, 652-59.
198 Donnet J B, Wang T K, *Prog. Rubb. Plast. Technol.*, **11**, No.4, 1995, 261-67.
199 Modine F A, Duggal A R, Robinson D N, Churnetski E L, Bartkowiak M, Mahan G D, Levinson L M, *J. Mat. Res.*, **11**, No.11, 1996, 2889-94.
200 Sutherland I, Sheng E, Bradley R H, Freakley P K, *J. Mat. Sci.*, **31**, No.21, 1996, 5651-55.
201 Svorcik V, Rybka V, Hnatowicz V, Bacakova L, *J. Mat. Sci. Lett.*, **14**, No.24, 1995, 1723-24.
202 Achour M E, Miane J L, Lahjomri F, El Malhi M, Carmona F, *J. Mat. Sci. Lett.*, **14**, No.20, 1995,

1425-29.
203 Kurian T, De P P, Khastgir D, Tripathy D K, De S K, Peiffer D G, *Polymer*, **36**, No.20, 1995, 3875-84.
204 Unnikrishnan G, Thomas S, Varghese S, *Polymer*, **37**, No.13, 1996, 2687-93.
205 Tan L S, McHugh, *J. Mater. Sci.*, **31**, 1996, 3701-06.
206 Addad J P C, Frebourg P, *Polymer*, **37**, No.19, 1996, 4235-42.
207 Tchoudakov R, Breuer O, Narkis M, Siegmann A, *Polym. Engng. Sci.*, **36**, No.10, 1996, 1336-46.
208 Lawandy S N, Botros S H, Darwish N A, Mounir A, *Polym. Plast. Technol. Engng.*, **34**, No.6, 1995, 861-74.
209 Fu-Yu Hsieh, Bryan C J, Pedley M D, *Fire & Mat.*, **18**, No.6, 1994, 389-91.
210 Nasr G M, Badawy M M, Gwaily S E, Attia G, *Polym. Int.*, **38**, No.3, 1995, 249-55.
211 Ismail H, Freakley P K, Sheng E, *Eur. Polym. J.*, **31**, No.11, 1995, 1049-56.
212 Ryabikova V M, Zigel A N, Sverdlova S I, Vinogradova G A, *Int. Polym. Sci. Technol.*, **23**, No.1, 1996, T/89-90.
213 Khromov M K, Niazashvili G A, *Int. Polym. Sci. Technol.*, **23**, No.5, 1996, T/24-6.
214 Potente H, Flecke J, Antec '96. Vol. I. Conference Proceedings, Indianapolis, 5th-10th May 1996, 178-82.
215 Joo Y L, Lee Y D, Kwack T H, Min T-I, Antec '96. Vol. I. Conference Proceedings, Indianapolis, 5th-10th May 1996, 64-68.
216 Yu M C, Bissell M A, Whitehouse R S, Antec 95. Volume III. Conference proceedings, Boston, Ma., 7th-11th May 1995, 3246-50.
217 Gownder M, Letton A, Hogan H, Antec '95. Vol. II. Conference Proceedings, Boston, Ma., 7th-11th May 1995, 1983-86.
218 Narkis M, Tchoudakov R, Breuer O, Siegmann A, Antec '95. Vol. II. Conference Proceedings, Boston, Ma., 7th-11th May 1995, p.1343-46.
219 Gardiner K, Calvert I A, van Tongeren M J A, Harrington J M, *Ann. Occup. Hyg.*, **40**, No.1, 1996, 65-77.
220 White L, *Eur. Rubb. J.*, **178**, No.8, 1996, 46-52.
221 *Eur. Rubb. J.*, **178**, No.8, 1996, 26.
222 *Eur. Rubb. J.*, **178**, No.7, 1996, 32.
223 White L, *Eur. Rubb. J.*, **178**, No.7, 1996, 30-1.
224 *Eur. Rubb. J.*, **178**, No.5, 1996, 30.
225 *Eur. Rubb. J.*, **178**, No.5, 1996, 28.
226 Nakajima N, *Int. Polym. Processing*, **11**, No.1, 1996, 3-13.
227 Niedermeier W, Raab H, Maier P, Kreitmeier S, Goeritz D, *Kaut. u. Gummi Kunst.*, **48**, No.9, 1995, 611-16.
228 Herd C R, Bomo F, *Kaut. u. Gummi Kunst.*, **48**, No.9, 1995, 588-99.
229 Huybrechts F, *Kaut. u. Gummi Kunst.*, **48**, No.10, 1995, 713-17.
230 Bandyopadhyay S, De P P, Tripathy D K, De S K, *Kaut. u. Gummi Kunst.*, **49**, No.2, 1996, 115-19.
231 de Candia F, Carotenuto M, Gargani L, Guadagno L, Lauretti E, Renzulli A, *Kaut. u. Gummi Kunst.*, **49**, No.2, 1996, 99-101.
232 Leblanc J L, *Kaut. u. Gummi Kunst.*, **49**, No.4, 1996, 258-66.
233 Donnet J B, Custodero E, Wang T K, *Kaut. u. Gummi Kunst.*, **49**, No.4, 1996, 274-79.
234 Hollis R D, Hyche K W, Color and Appearance Retec: Effects in Plastics. Conference proceedings, Oak Brook, Il., 20th-22nd Sept.1994, 94-105.
235 Heuert U, Knorgen M, Menge H, Scheler G, Schneider H, *Polym.Bull.*, **37**, No.4, Oct.1996, 489-96.
236 Karasek L, Meissner B, Asai S, Sumita M, *Polym. J. (Jap.)*, **28**, No.2, 1996, 121-26.
237 Yoshikawa S, Tsubokawa N, *Polym. J. (Jap.)*, **28**, No.4, 1996, 317-22.
238 Forster F, Freund B, *Rubb. S. Africa*, **11**, No.3, 1995, 8-12.
239 Okoroafor M O, Wang A, Bhattacharjee D, Cikut L, Haworth G J, Polyurethanes '95. Conference Proceedings, Chicago, Il., 26th-29th Sept.1995, 303-9.
240 Younan A F, Choneim A M, Tawfik A A A, Abd-El-Nour K N, *Polym. Degradat. Stabil.*, **49**, No.2, 1995, 215-22.
241 Delor F, Lacoste J, Lemaire J, Barrois-Oudin N, Cardinet C, *Polym. Degradat. Stabil.*, **53**, No.3, 1996, 361-69.
242 Tsubokawa N, Hosoya M, Kurumada J, *Reactive & Functional Polym.*, **27**, No.1, 1995, 75-81.
243 Li Q, Manas-Zloczower I, Feke D L, *Rubb. Chem. Technol.*, **69**, No.1, 1996, 8-14.
244 Gabrielson L, Edrisinghe M J, *J. Mat. Sci. Lett.*, **15**, No.13, 1996, 1105-07.
245 Baque T, *Kunststoffe Plast Europe*, **86**, No.8, 1996, 1162-64.

246 Wu S-L, Tung I-C, *Polym. Composites*, **16**, No.3, 1995, 233-39.
247 Schut J H, *Plast. World*, **53**, No.11, 1995, 61-4.
248 Usuki A, Koiwai A, Kojima Y, Kawasumi M, Okada A, Kurauchi T, Kamigaito O, *J. Appl. Polym. Sci.*, **55**, No.1, 1995, 119-23.
249 Lan T, Pinnavaia T J, *Chem. of Mat.*, **6**, No.12, 1994, 2216-19.
250 Al-Turaif H, Unertl W N, Lepoutre P, *J. Adhesion Sci. Technol.*, **9**, No.7, 1995, 801-11.
251 Kurian T, De P P, Tripathy D K, De S K, Peiffer D G, *J. Appl. Polym. Sci.*, **2**, No.10, 1996, 1729-34.
252 Evans L R, Meeting of the Rubber Division, ACS, Montreal, May 5-8, 1996, paper D.
253 **Ball Clay**. *Old Hickory Clay Co*. Hickory, KY, USA.
254 Kouloumbi N, Tsangaris G M, Kyvelidis S T, *J. Coatings Technol.*, **66**, No.839, 1994, 83-8.
255 Suri A, Min K, Antec '97. Conference proceedings, Toronto, April 1997, 1487-91.
256 Larena A, Pinto G, *Polym. Composites*, **16**, No.6, 1995, 536-41.
257 Balashov M M, Makhmudbekova N L, *Int. Polym. Sci. Technol.*, **22**, No.9, 1995, T/84-6.
258 Belanger B, Sanschagrin B, Fisa B, Antec '9. Volume II. Conference proceedings, Indianapolis, 5th-10th May 1996, 1762-67.
259 Van Aken L, *Surface Coatings Int.*, **77**, No.2, 1994, 61-8.
260 Van Aken L, Hoeck R, *Polym. Paint Col. J.*, **183**, No.433, 1993, 442-44.
261 Hofmann F A, Skudelny D, Surface treated mineral fillers - Growing markets for specialty products, 9th Industrial Materials International Congress, March 1990.
262 Skudelny D, *Kunststoffe*, **77**, 1987, 1153-55.
263 Hofman F A, The use of cristobalite as a white extender. The Global Outlook for $TiO_2$ and $TiO_2$ Replacements/Extenders in Coatings, Paper, and Plastics, International Conference, St. Louis, March, 1990.
264 Halvorsen, R, Cristobalite. **A Unique Form of Silica**. *Charis, Inc.*, Maryville.
265 Allen N S, Edge M, Corrales T, Childs A, Liauw C, Catalina F, Peinado C, Minihan A, *Polym. Degradat. Stabil.*, **56**, 1997, 125-39.
266 **Celite. Functional Fillers for Industrial Applications**. *World Minerals, Inc. Celite Corporation*, Lompoc, 11/91.
267 Kovacevic V, Lucic S, Hace D, Glasnovic A, Smit I, Bravar M, *J. Adhesion*, **47**, No.1-3, 1994, 201-15.
268 Anantharaman M R, Kurian P, Banerjee B, Mohamed E M, George M, *Kaut. u. Gummi Kunst.*, **49**, No.6, 199, 424-26.
269 Klapcinski T, Galeski A, Kryszewski M, *J. Appl. Polym. Sci.*, **58**, No.6, 1995, 1007-13.
270 Baranovskii V M, Bondarenko V V, Zadorina E N, Cherenkov A V, Zelenev Y V, *Int. Polym. Sci. Technol.*, **23**, No.6, 199, T/87-9.
271 Asp L E, Sjogren B A, Berglund L A, *Polym. Composites*, **18**, No.1, 1997, 9-15.
272 Sjogren B A, Berglund L A, *Polym. Composites*, **18**, No.1, 1997, 1-8.
273 Meddad A, Fisa B, *J. Mater. Sci.*, **32**, 1997, 1177-85.
274 Roesch J, Barghoorn P, Muelhaupt R, *Makromol. Chem. Rapid Commun.*, **15**, No.9, 1994, 691-96.
275 Benderly D, Siegmann A, Narkis M, *J. Mat. Sci. Lett.*, **14**, No.2, 1995, 132-4.
276 Dufresne A, Lacabanne C, *Polymer*, **34**, No. 15, 1993, 3173-78.
277 Alberola N, Bergeret A, S, *Polym. Composites*, **15**, No.6, 1994, 442-52.
278 Ou Y-C, Yu Z-Z, *Polym. Int.*, **37**, No.2, 1995, 113-17.
279 Friedrich C, Scheuchenpflug W, Neuhaeusler S, Roesch J, *J. Appl. Polym. Sci.*, **57**, No.4, 1995, 499-508.
280 Wang S Q, Inn Y W, *Polym. Int.*, **37**, No.3, 1995, 153-55.
281 Meddad A, Fellahi S, Pinard M, Fisa B, Antec '94. Conference Proceedings, San Francisco, Ca., 1st-5th May 1994, Vol. II, 2284-88.
282 Tamura J, Kawanabe K, Yamamuro T, Nakamura T, Kokubo T, Yoshihara S, Shibuya T, *J. Biomed. Mat. Res.*, **29**, No.5, 1995, 551-59.
283 Palumbo M, Donzella G, Tempesti E, Ferruti P, *J. Appl. Polym. Sci.*, **60**, No.1, 1996, 47-53.
284 Luzinov I, Voronov A, Minko S, Kraus R, Wilke W, Zhuk A, *J. Appl. Polym. Sci.*, **61**, No.7, 1996, 1101-09.
285 Wang J Y, Ploehn H J, *J. Appl. Polym. Sci.*, **59**, No.2, 1996, 345-57.
286 Meddad A, Fisa B, *Macromol. Symp.*, **108**, 1996, 173-82.
287 Lekatou A, Faidi S E, Lyon S B, Newman R C, *J. Mat. Res.*, **11**, No.5, 1996, 1293-304.
288 Hashemi S, Gilbride M T, Hodgkinson J, *J. Mat. Sci.*, **31**, No.19, 1996, 5017-25.
289 Nabi Z U, Hashemi S, *J. Mat. Sci.*, **31**, No.21, 1996, 5593-601.
290 Sinien L, Xiaoguang Z, Zhongneng Q, Huan X, *J. Mat. Sci. Lett.*, **14**, No.20, 1995, 1458-60.

291 Bergeret A, Alberola N, *Polymer*, **37**, No.13, 1996, 2759-65.
293 Kobayashi M, Takahashi T, Takimoto J, Koyama K, *Polymer,* **37**, No.16, 1996, 3745-47.
294 Hashemi S, Din K J, Low P, *Polym. Engng. Sci.*, **36**, No.13, 1996, 1807-20.
295 Schleifstein R A, Property Enhancement with Modifiers and Additives. Retec proceedings, New Brunswick, N.J., 18th-19th Oct.1994, 89-101.
296 Carre A, *J. Adhesion*, **54**, Nos.1-4, 1995, 167-74.
297 Ou Y, Yu Z, Zhu J, Li G, Zhu S, *Chinese J. Polym. Sci.*, **14**, No.2, 1996, 172-82.
298 Kubat J, Kuzel R, Krivka I, Bengtsson P, Prokes J, Stefan O, *Synthetic Metals*, **54**, Nos.1/3, 1993, 187-94.
299 Wool R P, Long J M, *Macromolecules*, **26**, No.19, 1993, 5227-39.
300 Gonsalves K E, Carlson G, Chen X, Kumar J, Aranda F, Perez R, Jose-Yacaman M, *J. Mat. Sci. Lett.*, **15**, No.11, 1996, 948-51.
301 Yang A C M, *Polymer*, **35**, No.15, 1994, 3206-11.
302 Baranovskii V M, Tarara A M, Khomik A A, *Int. Polym. Sci. Technol.*, **20**, No.1, 1993, T/98-9.
303 Knowles J, *Polym. Paint Col. J.*, **185**, No.43, 1995, 26-7.
304 Fengyuan Yan, Wenhua Wang, Qunji Xue, Long Wei, *J. Appl. Polym. Sci.*, **61**, No.7, 1996, 1231-36.
305 Kimura T, Asano Y, Yasuda S, *Polymer*, **37**, No.14, 1996, 2981-87.
306 Fengyuan Yan, Qunji Xue, Shengrong Yang, *J. Appl. Polym. Sci.*, **61**, No.7, 1996, 1223-29.
307 Kuznezov A, Vasnev V, Gribova I, Krasnov A, Gureeva G, Ignatov V, *Int. J. Polym. Mat.*, **32**, Nos.1-4, 1996. 85-91.
308 Vasnev V A, Tarasov A I, Istratov V N, Ignatov V N, Krasnov A P, Kuznetsov A I, Surkova I N, *Reactive & Functional Polym.*, **26**, Nos.1-3, 1995, 177-83.
309 **Timrex. Graphite for carbon brushes and carbon parts**. *Timcal*, Sins, Switzerland, 1996.
310 Kim G M, Michler G H, Gahleitner M, Fiebig J, *J. Appl. Polym. Sci.*, **60**, No.9, 1996, 1391-403.
311 Jyongsik Jang, Jieun Yi, *J. Appl. Polym. Sci.*, **61**, No.12, 1996, 2157-63.
312 Yogo T, Nakamura T, Kikuta K, Sakamoto W, Hirano S, *J. Mat. Res.*, **11**, No.2, 1996, 475-82.
313 Brovkop O O, Sergeeva L M, Slinchenko O A, Fainleib O M, *Polym. Int.*, **40**, No.4, 1996, 299-305.
314 Sergeeva L M, Slinchenko E A, Brovko A A, Fainleib A M, Nedashkovskaya N S, *Polym. Sci., Ser. B*, **38**, Nos.5/6, 199, 225-30.
315 Hindryckx F, Dubois P, Patin M, Jerome R, Teyssie P, Garcia Marti M, *J. Appl. Polym. Sci.*, **56**, No.9, 1995, 1093-105.
316 Helaly F M, El-Sawy S M, Abd El-Ghaffar M A, *J. Elastomers Plast.*, **26**, No.4, 1994, 335-46.
317 Molphy M, Laslett R L, Gunatillake P A, Rizzardo E, Mainwaring D E, *Polym. Int.*, **34**, No.4, 1994, 425-31.
318 Molphy M, Mainwaring D E, Rizzardo E, Gunatillake P A, Laslett R L, *Polym. Int.*, **37**, No.1, 1995, 53-61.
319 Baranovskii V M, Bondarenko S I, Kachanovskaya L D, Zelenev Y V, Makarov V G, Ovcharenko F D, *Int. Polym. Sci. Technol.*, **22**, No.1, 1995, T/91-3.
320 Coussot Ph, Proust S, Ancey Ch, *J. Non-Newtonian Fluid Mech.*, **66**, 1996, 55-70.
321 Zolotnitsky M, Steinmetz J R, *J. Vinyl and Additive Technol.*, **1**, No.2, 1995, 109-13.
322 Balard H, Saada A, Hartmann J, Aouadj O, Papirer E, *Macromol. Symp.*, **108**, 1996, 63-80.
323 Savadori A, Scapin M, Walter R, *Macromol. Symp.*, **108**, 1996, 183-202.
324 Turner J D, Property Enhancement with Modifiers and Additives. Retec proceedings, New Brunswick, N.J., 18th-19th Oct.1994, 65-87.
325 Fajardo W L, **Kaolin in Plastics and Rubber Compounds**. *Engelhard Corporation*, 1993.
326 Washabaugh, F J, *Rubber World*, **10**, 1987.
327 Carr J B, *Plast. Compounding*, **13**, 1990.
328 Sekutowski D, Improved Impact of Mineral/Nylon Systems Through Surface Modification, *Antec '87.*
329 Fajardo W L, Water-Washed Kaolins and Their Use in Rubber Compounds. *Engelhard Corporation.*
330 McGuffog, Clays as Extenders in Decorative Paints. *ECC International.*
331 McGuffog, PoleStar 400A as a Replacement for Heavily Coated Titanium Dioxide. *ECC International.*
332 Itatani K, Yasuda R, Scott Howell F, Kishioka, *J. Mater. Sci.*, **32**, 1997, 2977-84.
333 Rothon R N; Hornsby P R, *Polym. Degradat. Stabil.*, **54**, Nos 2-3, 199, 383-85.
334 Hornsby P R, Wang J, Rothon R, Jackson G, Wilkinson G, Cossick K, *Polym. Degradat. Stabil.*, **51**, No.3, 199, 235-49.
335 Costa L, Camino G, Bertelli G, Borsini G, *Fire & Mat.*, **19**, No.3, 1995, 133-42.
336 Hornsby P R, Watson C L, *J. Mat. Sci.*, **30**, No.21, 1995, 5347-55.
337 Gutman E M, Bobovitch A L, *Eur. Polym. J.*, **32**, No.8, 1996, 979-83.

338 Wang J, Tung J F, Fuad M Y A, Hornsby P R, *J. Appl. Polym. Sci.*, **60**, No.9, 1996, 1425-37.
339 Hornsby P R, Wang J, Cosstick K, Rothon R, Jackson G, Wilkinson G, Flame Retardants '94. Conference proceedings, London, 27th-28th January 1994, 93-108.
340 Hornsby P R, Mthupha A, Antec '93. Conference Proceedings, New Orleans, La., 9th-13th May 1993, Vol. II, 1954-56.
341 Hornsby P R, Wang J, Jackson G, Rothon R N, Wilkinson G, Cosstick K, Antec '94. Conference Proceedings, San Francisco, Ca., 1st-5th May 1994, Vol. III, 2834-39.
342 Rosenov M W K, Bell J A E, Antec '97. Conference proceedings, Toronto, April 1997, 1492-8.
343 Guanghong Lu, Xiaotian Li, Hancheng Jiang, *Composites Sci. & Technol.*, **56**, No.2, 1996, 193-200.
344 Tead S F, Bluem G L, McCormick F B, Plastics for Portable Electronics. Retec Proceedings, Las Vegas, Nv., 5th-th Jan.1995, 114-30.
345 Herzog E, Caseri W, Suter U W, *Coll. Polym. Sci.*, **272**, No.8, 1994, 986-90.
346 Furtado C R G, Nunes R C R, de Siqueira Filho A S, *Polym. Bull.*, **34**, No.5/, 1995, 627-33.
347 Balchandani N, *Pop. Plast. Packag.*, **40**, No.2, 1995, 41-2.
348 Jarvela P A, Shucai L, Jarvela P K, *J. Appl. Polym. Sci.*, **64**, No.10, 1997, 2003-11.
349 Meier L P, Shelden R A, Caseri W R, Suter U W, *Macromolecules*, **27**, No.6, 1994, 1637-42.
350 Watanabe H, Tirrell M, *Macromolecules*, **26**, No.24, 1993, 6455-66.
351 Chiang W Y, Yang W D, Pukanszky B, *Polym. Engng. Sci.*, **34**, No.6, 1994, 485-92.
352 Schott N R, Rahman M, Perez M A, Antec '94. Conference Proceedings, San Francisco, Ca., 1st-5th May 1994, Vol. III, 2846-50.
353 Borden K A, Weil R C, Manganaro C R, Antec '93. Conference Proceedings, New Orleans, La., 9th-13th May 1993, Vol. II, 2167-70.
354 Canova L A, Antec '93. Conference Proceedings, New Orleans, La., 9th--13th May 1993, Vol. II, 1943-49.
355 Canova L A, Fergusson L W, Parrinello L M, Subramanian R, Giles H F, Antec '97. Conference proceedings, Toronto, April 1997, 2112-16.
356 Schott N R, Rahman M, Perez M A, *J. Vinyl and Additive Technol.*, **1**, No.1, 1995, 36-40.
357 Okamoto M, Shinoda Y, Okuyama T, Yamaguchi A, Sekura T, *J. Mat. Sci. Lett.*, **15**, No.13, 1996, 1178-79.
358 Kody R S, Martin D C, *Polym. Engng. Sci.*, **36**, No.2, 1996, 298-304.
359 Hojabr S, Boocock J R B, Antec 95. Volume III. Conference proceedings, Boston, Ma., 7th-11th May 1995, 3620-27.
360 Xanthos M, Grenci J, Dagli S S, Antec '95. Vol.II. Conference Proceedings, Boston, Ma., 7th-11th May 1995, 3194-200.
361 Srivastava V K, Pathak J P, *Polym. & Polym. Composites*, **3**, No.6, 1995, 411-14.
362 Zyuzina G F, Vinogradova N K, Gribova I A, Krasnov A P, *Polym. Sci.*, **36**, No.9, 1994, 1205-08.
363 Shin Jen Shiao, Te Zei Wang, *Composites*, **27B**, No.5, 1996, 459-65.
364 Risdon T J, Loban R D, Effect of Molybdenum Disulfide on the Wear Rates of Polymer Composites. *Climax Molybdenum Company*, Ypsilanti, MI, USA.
365 **Molysulfide? The Unique Solid Lubricant**. *Climax Molybdenum Company*, Ypsilanti, MI, USA
366 **Molysulfide. Particle size analysis.** *Climax Molybdenum Company*, Ypsilanti, MI, USA
367 Fiske T, Gokturk H S, Yazici R, Kalyon D M, *Polym. Eng. Sci.*, **37**, No.5, 1997, 826-37.
368 Mamunya E P, Davidenko V V, Lebedev E V, *Polym. Composites*, **16**, No.4, 1995, 319-24.
369 Fiske T, Gokturk H S, Yazici R, Kalyon D M, Antec '97. Conference proceedings, Toronto, April 1997, 1482-86.
370 Hart A, *Polym. Paint Col. J.*, **18**, No.4378, 199, S3-6.
371 Akin-Oktem G, Tincer T, *J. Appl. Polym. Sci.*, **54**, 1994, 1103-14.
372 Schuster R H, Meeting of the Rubber Division, ACS, Montreal, May 5-8, 1996, paper F.
373 Cha Y J, Choe S, *J. Appl. Polym. Sci.*, **58**, No.1, 1995, 147-57.
374 Bagheri R, Pearson R A, *Polymer*, **36**, No.25, 1995, 4883-85.
375 Sergeeva L M, Skiba S I, Karabanova L V, Polym. Int., **39**, No.4, April 1996, 317-25.
376 Mukha B I, Kolupaev B S, Mukha Y B, *Int. Polym. Sci. Technol.*, **23**, No.6, 199, T/57-9.
377 Undebody Coatings Use Polymeric Microspheres, *Modern Paint & Coatings*, April, 1993.
378 Hollow Filler Reduces Need for Roll-out, *Reinforced Plastics*, **37**, 5, 1993.
379 Clark R F, **The Advantages of Polymeric Microspheres in Adhesive and Sealant**, *Pierce & Stevens, Inc.*
380 Williams M A, Bauman B D, Thomas D A, *Polym. Eng. Sci.*, **31**, 1991, 992-998.

381 Bauman B D, Surface-modified Polymer Particles and Short, Chopped Fiber in Epoxies, SPI Conference, San Francisco, CA 1995.
382 Bagheri R, Williams M A, Pearson R A, *Polym. Eng. Sci.*, **37**, No.2, 1997, 245-51.
383 Hornsby P R, Premphet K, *J. Mater. Sci.*, **32**, 1997, 4767-75.
384 Krysztafkiewicz A, Rager B, Maik M, *Coll. Polym. Sci.*, **272**, No.12, 1994, 1547-59.
385 Becu L, Maazouz A, Sautereau H, Gerard J F, *J. Appl. Polym. Sci.*, **65**, 1997, 2419-31.
386 Mayadunne A, Bhattacharya S N, Kosior E, Boontanjai C, Antec 95. Volume I. Conference proceedings, Boston, Ma., 7th-11th May 1995, 1178-82.
387 Gibala D, Laohapisitpanich K, Thomas D, Hamed G R, *Rubb. Chem. Technol.*, **69**, No.1, 1996, 115-19.
388 Dierkes W, *J. Elastomers Plast.*, **28**, No.3, 1996, 257-78.
389 Pearson R A, Azimi H R, Bagheri R, Qian J Y, Antec '95. Vol. II. Conference Proceedings, Boston, Ma., 7th-11th May 1995, 2656-61.
390 Bauman B D, *Pitture Vern.*, **71**, No.16, 1995, 26-28.
391 Bauman B D, Plastics Finishing: Responding to Tomorrow's Global Requirements. Retec proceedings, Detroit, MI, 21st-22nd March 1995, 62-73.
392 Williams M A, Bagheri R, **The Use of Vistamer Surface Activated Rubber Particles in Adhesives & Floor Coatings**, *Composite Particles, Inc.*
393 Gerace M J, Gerace J M, Williams M A, The Use of Vistamer Surface-Activated Rubber Particles in Structural Adhesives, *Composite Particles, Inc.*
394 Gerace M J, Gerace J M, Williams M A, *Adhesives Age*, December, 1995.
395 Bauman R D, Surface-modified Rubber Particles in Footwear, ACS Meeting Rubber Division, Anaheim, CA, 6-7 May, 1997.
396 Bauman B D, High Value Scrap Ture Recycle, AutoPlas '96, Hong Kong, 29 March, 1996.
397 Torro-Palau A, Fernandez-Garcia J C, Orgiles-Barcelo A C, Pastor-Blas M M, Martin-Martinez J M, *Int. J. Adhesion Adhesives*, **17**, 1997, 111-19.
398 Herrero C R, Morales E, Acosta J L, *J. Appl. Polym. Sci.*, **51**, No.7, 1994, 1189-97.
399 Sunkara H B, Jethmalani J M, Ford W T, *Chem. of Mat.*, **6**, No.4, 1994, 362-64.
400 Krysztafkiewicz A, Rager B, Maik M, Szymanowski J, *Coll. Polym. Sci.*, **272**, No.12, 1994, 1526-35.
401 Li Y, Wang M J, Zhang T, Zhang F, Fu X, *Rubb. Chem. Technol.*, **67**, No.4, 1994, 693-99.
402 Ebengou R H, Cohen-Addad J P, *Polymer*, **35**, No.14, 1994, 2962-69.
403 Janigova I, Chodak I, *Eur. Polym. J.*, **30**, No.10, 1994, 1105-10.
404 Janigova I, Chodak I, *Eur. Polym. J.*, **31**, No.3, 1995, 271-74.
405 Pu Z, Mark J E, Jethmalani J M, Ford W T, *Polym. Bull.*, **37**, No.4, 1996, 545-51.
406 Maeda S, Armes S P, J. *Mat. Chem.*, **4**, No.6, 1994, 935-42.
407 Tino J, Mach P, Hlouskova Z, Chodak I, *J. Macromol. Sci. A*, **A31**, No.10, 1994, 1481-87.
408 Yoshinaga K, Nakanishi K, Hidaka Y, Karakawa H, *Composite Interfaces*, **3**, No.3, 1995, 231-41.
409 Yoshinaga K, Iwasaki M, Teramoto M, Karakawa H, *Polym. & Polym. Composites*, **4**, No.3, 1996, 163-72.
410 Wen J, Mark J E, *J. Appl. Polym. Sci.*, **58**, No.7, 1995, 1135-45.
411 Ou Y C, Yu Z Z, Vidal A, Donnet J B, *J. Appl. Polym. Sci.*, **59**, No.8, 1996, 1321-28.
412 Tertykh V A, *Macromol. Symp.*, **108**, 1996, 55-61.
413 Zaremski M Yu, Chernikova E V, Izmailov L G, Garina E S, Olenin A V, *Macromol. Reports*, **A33**, Suppls.3/4, 1996, 237-42.
414 Roychoudhury A, De P P, Roychoudhury N, Vidal A, *Rubb. Chem. Technol.*, **68**, No.5, 1995, 815-23.
415 Suzuki F, Nakane K, Piao J-S, *J. Mat. Sci.*, **31**, No.5, 1996, 1335-40.
416 Wartewig S, *Polymer*, **37**, No. 21, 1996, 4705-12.
417 Gorl U, Rausch R, Esch H, Kuhlmann R, *Int. Polym. Sci. Technol.*, **23**, No.7, 1996, T/81-7.
418 Torro-Palau A, Fernandez-Garcia J C, Orgiles-Barcelo A C, Martin-Martinez J M, *J. Adhesion*, **57**, Nos.1-4, 1996, 203-25.
419 Ferch H, Toussaint H E, *Kaut. u. Gummi Kunst.*, **49**, No.9, 1996, 589-96.
420 Khan S A, Baker G L, Colson S, *Chem. of Mat.*, **6**, No.12, 1994, 2359-63.
421 Landry C J T, Coltrain B K, Landry M R, Fitzgerald J J, Long V K, *Macromolecules,* **26**, No.14, 5th July 1993, 3702-12.
422 Cochrane H, Lin C S, *Rubb. Chem. Technol.*, **66**, No.1, 1993, 48-60.
423 Shang S W, Williams J W, Soderholm K J M, *J. Mat. Sci.,* **30**, No.17, 1995, 4323-34.
424 Bluemler P, Bluemich B, *Acta Polymerica*, **44**, No.3, 1993, 125-31.
425 Zumbrum M A, *J. Adhesion*, **46**, Nos.1-4, 1994, 181-96.
426 Frounchi M, Burford R P, Chaplin R P, *Polym. & Polym. Composites*, **2**, No.2, 1994, 77-82.

427 **Wacker HDK. The Fumed Silica.**1882-01. *Wacker*, Munich, 1996.
428 **Ca-O-Sil. Untreated Fumed Silica Properties and Functions**. CGEN-8A. *Cabot*, Tuscola, 1993.
429 **Basic Characteristics of Aerosil. Number 11**. *Degussa*, Frankfurt/Main, 1997
430 Skudelny D, **Quartz and Fused silica Fillers for Epoxy Cast Resins**. *Quartzwerke GmbH*, Frechen, Germany.
431 Ohta M, Nakamura Y, Hamada H, Maekawa Z, *Polym. & Polym. Composites*, **2**, No.4, 1994, 215-21.
432 Bomo F, Meeting of the Rubber Division, ACS, Montreal, May 5-8, 1996, paper E.
433 Zaborski M, Slusarski L, Vidal A, *Polim. Tworz. Wielk.*, **38**, No.6, 1993, 263-68.
434 Zaborski M, Vidal A, Papirer E, *Polim. Tworz. Wielk.*, **38**, No.7, 1993, 319-26.
435 Cochet P, Barruel P, Barriquand L, Grobert J, Bomal Y, Prat E, IRC '93/144th Meeting, Fall 1993. Conference Proceedings, Orlando, Fl., 2th-29th Oct.1993, Paper 12.
436 Tabor R L, Polyurethanes '94. Conference proceedings, Boston, Ma., 9th-12th Oct.1994, 288-94.
437 Okel T A, Waddell W H, *Rubb. Chem. Technol.*, **68**, No.1, 1995, 59-76.
438 O'Haver J H, Harwell J H, Evans L R, Waddell W H, *J. Appl. Polym. Sci.*, **59**, No.9, 1996, 1427-35.
439 Kurian T, Khatgir D, De P P, Tripathy D K, De S K, Peiffer D G, *Polymer*, **37**, No.25, 1996, 5597-605.
440 Datta S, Bhattacharya A K, De S K, Kontos E G, Wefer J M, *Polymer*, **37**, No.12, 1996, 2581-85.
441 Sanchez-Solis A, Padilla A, *Polym. Bull.*, **36**, No.6, 1996, 753-58.
442 Sanchez-Solis A, Estrada M R, *Polym. Degradat. Stabil.*, **52**, No.3, 1996, 305-09.
443 Murata N, Nishi S, Hosono S, *J. Adhesion*, **59**, Nos.1-4, 1996, 39-50.
444 Kominar V, Narkis M, Siegmann A, Breuer O, *Sci. & Engng. Composite Materials*, **3**, No.1, 1994, 61-66.
445 Joseph R, Zhang S, Ford W T, *Macromolecules*, **29**, No.4, 1996, 1305-12.
446 Aldcroft D, *Polym. Paint Col. J.*, **184**, No.43, 1994, 423-5.
447 Gailliez-Degremont E, Bacquet M, Laureyns J, Morcellet M, *J. Appl. Polym. Sci.*, **65**, No.5, 1997, 871-82
448 Wei Y, Sancaktar E, *J. Adhesion Sci. Technol.*, **10**, No.11, 1996, 1199-219.
449 Borden K A, Weil R C, Manganaro C R, Antec '94. Conference Proceedings, San Francisco, Ca., 1st-5th May 1994, Vol. III, 2761-65.
450 Seibel S R, Moet A, Bank D H, Nichols K, Antec '93. Conference Proceedings, New Orleans, La., 9th-13th May 1993, Vol. I, 902-5.
451 Sherman L M, *Plast. Technol.*, **43**, No.4, 1997, 26-8.
452 Liu Z, Gilbert M, *J. Appl. Polym. Sci.*, **59**, No.7, 1996, 1087-98.
453 Robert J F, Fragnier P, *Macromol. Symp.*, **108**, 1996, 13-18.
454 Alonso M, Gonzalez A, de Saja J A, *Plast. Rubb. Comp. Process. Appln.*, **24**, No.3, 1995, 131-37.
455 Singhal A, Fina L J, *Polymer*, **37**, No.12, 1996, 2335-43.
456 Rabello M S, White J R, *Polym. Composites*, **17**, No.5, 1996, 691-704.
457 Chang Ho Suh, White J L, *Polym. Engng. Sci.*, **32**, No.11, 1996, 1521-30.
458 Suh C H, White J L, *Polym. Engng. Sci.*, **36**, No.17, 1996, 2188-97.
459 Stricker F, Muelhaupt R, *High Perform. Polym.*, **8**, No.1, 1996, 97-108.
460 *Int. Polym. Sci. Technol.*, **23**, No.7, 199, T/1-3.
461 Gill T S, Xanthos M, Wirsbo Co., Antec '96. Volume II. Conference proceedings, Indianapolis, 5th-10th May 1996, 1757-61.
462 Tiganis B E, Shanks R A, Long Y, Antec '96. Volume II. Conference proceedings, Indianapolis, 5th-10th May 1996, 1744-49.
463 Stockblower D, Antec '96. Volume II. Conference proceedings, Indianapolis, 5th-10th May 1996, 1690-94.
464 Suh C H, White J L, Antec '96. Vol.I. Conference Proceedings, Indianapolis, 5th-10th May 1996, p.958-62.
465 Ishibashi J, Kobayashi A, Yoshikawa T, Shinozaki K, Antec '96. Vol. I. Conference Proceedings, Indianapolis, 5th-10th May 1996, 386-90.
466 Seibel S R, Moet A, Bank D H, Sehanobish K, Antec 95. Volume III. Conference proceedings, Boston, Ma., 7th-11th May 1995, 3966-70.
467 Yu T C, Antec '95. Vol. II. Conference Proceedings, Boston, Ma., 7th-11th May 1995, 2358-68.
468 Chang Ho Suh, White J L, *J. Non-Newtonian Fluid Mechanics*, **62**, Nos.2/3, 1996, 175-206.
469 Kim K J, White J L, *J. of Non-Newtonian Fluid Mechanics*, **66**, Nos.2/3, 1996, 257-70.
470 Wiebking H E, *J. Vinyl and Additive Technol.*, **2**, No.3, 1996, 187-89.
471 Gill T S, Xanthos M, *J. Vinyl and Additive Technol.*, **2**, No.3, 1996, 248-52.
472 Guerrica-Echevarria G, Eguiazabal J I, Nazabal J, *Polym. Degradat. Stabil.*, **53**, No.1, 1996, 1-8.

473 Burke M, Young R J, Stanford J L, *Plast. Rubb. Comp. Process. Appln.*, **20**, No.3, 1993, 121-35.
474 Hedgus C R, Kamel I L, *J. Coatings Technol.*, **65**, No.821, June 1993, 49-61
475 Hegedus C R, Kamel I L, J. *Coatings Technol.*, **65**, No.820, 1993, 23-30.
476 Hegedus C R, Kamel I L, *J. Coatings Technol.*, **65**, No.822, July 1993, 37-43.
477 Hegedus C R, Kamel I L, *J. Coatings Technol.*, **65**, No.820, 1993, 31-43
478 Hansen H, Polymers, Laminations & Coatings Conference, 1995, 53-8.
479 Bergado S, *Plast. Compounding*, **17**, No.7, 1994, 32-8.
480 Qi Wang, Jizhuang Cao, Guangjin Li, Xi Xu, *Polym. Int.*, **41**, No. 3, 1996, 245-49.
481 Robertson D R, Gaw F, AddCon '95, Basel, 1995.
482 **Titanium dioxide pigments**. TD10. *Millenium Inorganic Chemicals.*
483 **The influence of effective pigment dispersion on the economics of paint manufacture**. TD101. *Millenium Inorganic Chemicals.*
484 **Dispersant demand and selection for titanium dioxide pigments**. TD100. *Millenium Inorganic Chemicals.*
485 The durability of Tiona RCL- in high durability media. TD102. *Millenium Inorganic Chemicals.*
486 Nguyen T N, Lethiecq M, Levassort F, Patat F, *Int. J. Polym. Analysis and Characterization*, **1**, No.4, 1995, 277-87.
487 Sain M M, Kokta B V, Antec '93. Conference Proceedings, New Orleans, La., 9th-13th May 1993, Vol. I, 320-24.
488 Berenbrok P A, Liles B E, Antec '97. Conference proceedings, Toronto, April 1997, 2931-2.
489 Zaini M J, Fuad M Y A, Ismail Z, Mansor M S, Mustafah J, *Polym. Int.*, **40**, No.1, 1996, 51-55.
490 Marcovich N E, Reboredo M M, Aranguren M L, *J. Appl. Polym. Sci.*, **61**, No.1, 1996, 119-24.
491 Levit M R, Farrel R E, Gross R A, McCarthy S P, Antec '96. Volume II. Conference proceedings, Indianapolis, 5th-10th May 1996, 1387-91.
492 Zolotnitsky M, Steinmetz J R, Antec '94. Conference Proceedings, San Francisco, Ca., 1st-5th May 1994, Vol. III, 2756-60.
493 Plummer C J G, Wu Y, Gola M M, Kausch H H, *Polym. Bull.*, **30**, No.5, 1993, 587-94.
494 Robinson S M, *Polym. Paint Col. J.*, **184**, No. 4352, 1994, 311.
495 Zhengcai Pu, Mark J E, Beaucage G, Maaref S, Frisch H L, *J. Polym. Sci., Polym. Phys.*, **34**, No.15, 1996, 2657-60.
496 Zhao W, Hasegawa S, Fujita J, Yoshii F, Sasaki T, Makuuchi K, Sun J, Nishimoto S, *Polym. Degradat. Stabil.*, **53**, No.2, 1996, 199-206.
497 Abramova N A, Diikova E U, Lyakhovskii Yu Z, *Polym. Sci.*, **36**, No.9, 1994, 1308-9.
498 Duval J M, Kemperman A J B, Folkers B, Mulder M H V, Desgrandchamps G, Smolders C A, *J. Appl. Polym. Sci.,* **54**, No.4, 1994, 409-18.
499 Benrashid R, Nelson G L, *J. Fire Sci.*, **11**, No.5, 1993, 371-93.
500 Tanaka T, Waki Y, Hamamoto A, Nogami N, Antec '97. Conference proceedings, Toronto, April 1997, 3054-8.
501 Zaborski M, Slusarski L, Donnet J B, Papirer E, *Kaut. u. Gummi Kunst.*, **47**, No.10, 1994, 730-8.
502 Gordienko V P, Dmitriev Y A, *Polym. Sci., Ser. B*, **37**, Nos.5-6, 1995, 249-50.
503 Cusack P, Flame Retardants '96. Conference proceedings, London, 17th-18th Jan.1996, 57-69.
504 Akay M, Mun S K A, Stanley A, *Composites Sci. Technol,* **57**, 1997, 565-71.
505 Hshieh F Y, Beson H D, *Fire Mater.*, **21**, 1997, 41-9.
506 Bhattacharya S K, Bhowmick A K, Singh R K, *J. Mat. Sci.*, **30**, No.1, 1995, 243-7.
507 Hill R, Okoroafor E U, *Composites*, **25**, No.10, 1994, 913-6.
508 Dzenis Y A, Reneker D H, Tsukruk V V, Patil R, *Composite Interfaces*, **2**, No.4, 1994, 307-19.
509 Srinivasan G, Reneker D H, *Polym. Int.*, **36**, No.2, 1995, 195-201.
510 Rebouillat S, Escoubes M, Gauthier R, *J. Adhesion Sci. Technol.*, **10**, No.7, 1996, 635-50.
511 Rebouillat S, Escoubes M, Gauthier R, Vigier A, *J. Appl. Polym. Sci.*, **58**, No.8, 1995, 1305-15.
512 Connor C, Chadwick M M, *J. Mat. Sci.*, **31**, No.14, 1996, 3871-7.
513 Braches E, *Kunststoffe Plast. Europe*, **86**, No.11, 1996, 21-2.
514 Selzer R, Friedrich K, *Composites, Part A*, **28A**, 1997, 595-604.
515 Cantwell W J, Zulkifli R, *J. Mater. Sci. Lett.*, **16**, 1997, 509-11.
516 Agari Y, Ueda A, Nagai S, *J. Appl. Polym. Sci.*, **52**, No.9, 1994, 1223-31.
517 Ibarra L, Macias A, Palma E, *J. Appl. Polym. Sci.*, **57**, No.7, 1995, 831-42.
518 Tiejun Wang, Sherwood P M A, *Chem. of Mat.*, **7**, No.5, 1995, 1031-40.
519 Wang T, Sherwood P M A, *Chem. of Mat.*, **6**, No.6, 1994, 788-95.
520 Katsumata M, Endo M, Ushijima H, Yamanishi H, *J. Mat. Res.*, **9**, No.4, 1994, 841-3.

521 Downs W B, Baker R T K, *J. Mat. Res.*, **10**, No.3, 1995, 625-33.
522 Reis M J, Do Rego A M B, Da Silva J D L, *J. Mat. Sci.*, **30**, No.1, 1995, 118-26.
523 Tsai J S, *J. Mat. Sci.*, **30**, No.8, 1995, 2019-22.
524 Bogoeva-Gaceva G, Burevski D, Dekanski A, Janevski A, *J. Mat. Sci.*, **30**, No.13, 1995, 3543-6.
525 Tsai J S, *Polym. Engng. Sci.*, **34**, No.19, 1994, 1480-4.
526 Byung Suk Jin, Kwang Hee Lee, Chul Rim Choe, *Polym. Int.*, **34**, No.2, 1994, 181-5.
527 Van Beek G A, Pang S S, Lea R H, Antec '93. Conference Proceedings, New Orleans, La., 9th-13th May 1993, Vol. I, 602-4.
528 Rammoorthy M, Muzzy J, Antec '97. Conference proceedings, Toronto, April 1997, 2500-4.
529 Ewing F, Leatherman M, Antec '97. Conference proceedings, Toronto, April 1997, 2924-7.
530 Zihlif A M, Di Liello V, Martuscelli E, Ragosta G, *Int. J. Polym. Mat.*, **29**, Nos.3-4, 1995, 211-20.
531 OgawaT, Ikeda M, *J. Adhesion,* **43**, Nos.1-2, 1993, 69-78.
532 Dong Zhang, *Polym. & Polym. Composites*, **2**, No.3, 1994, 159-64.
533 Tang L-G, Kardos J L, *Polym. Composites*, **18**, No.1, 1997, 100-13.
534 Le Bras M, Bourbigot, Le Tallec Y, Laureyns J., *Polym. Degradat. Stabil.*, **56**, 1997, 11-21.
535 Greso A J, Phillips P J, *Polymer*, **37**, No.14, 1996, 3165-70.
536 Wang P H, Hong K L, Zhu Q R, *J. Appl. Polym. Sci.*, **62**, No.12, 1996, 1987-91.
537 Wang P H, *J. Appl. Polym. Sci.*, **62**, No.10, 1996, 1771-3.
538 Melanitis N, Tetlow P L, Galiotis C, *J. Mat. Sci.*, **31**, No.4, 1996, 851-60.
539 Nofal M M, Zihlif A M, Ragosta G, Martuscelli E, *Polym. Composites*, **17**, No.5, 1996, 705-9.
540 Jin-Shy Tsai, *Polym. Engng. Sci.*, **35**, No.16, 1995, 1313-6.
541 Donnet J B, Wang T K, IRC '95 Kobe International Rubber Conference. Conference proceedings, Kobe, 23rd-27th Oct.1995, 451-4.
542 Dreibelbis G L, Antec 95. Volume III. Conference proceedings, Boston, Ma., 7th-11th May 1995, 4374-6.
543 Ueno H, Tsubokawa N, *Composite Interfaces*, **3**, No.3, 1995, 209-20.
544 Tsutsumi K, Ban K, Shibata K, Okazaki S, Kogoma M, *J. Adhesion*, **57**, Nos.1-4, 1996, 45-53.
545 Ohwaki T, Ishida H, *J. Adhesion*, **52**, Nos.1-4, 1995, 167-86.
546 Peebles L H, *Polym. News*, **21**, No.2, 1966, 55-7.
547 Abdel-Aziz M M, Youssef H A, El Miligy A A, Yoshii F, Makuuchi K, *Polym. & Polym. Composites*, **4**, No.4, 1996, 259-68.
548 de Sena Affonso J E, Nunes R C R, *Polym. Bull.*, **34**, No.5/6, 1995, 669-75.
549 Odberg L, Tanaka H, Glad-Nordmark G, Swerin A, *Coll. & Surfaces*, **86**, 1994, 201-7.
550 Collier J R, Lu M, Fahrurrozi M, Collier B J, *J. Appl. Polym. Sci.*, **61**, No.8, 1996, 1423-30.
551 Nunes R C R, Mano E B, *Polym. Composites*, **16**, No.5, 1995, 421-3.
552 Gassan J, Bledzki A K, Antec '96. Volume II. Conference proceedings, Indianapolis, 5th-10th May 1996, p.2552-7.
553 Vieira A, Nunes R C R, Visconti L L Y, *Polym. Bull.*, **36**, No.6, 1996, 759-66.
554 Horst J J, Spoormaker J L, *J. Mater. Sci.*, **32**, 1997, 3541-51.
555 Thomason J L, Vlug M A, *Composites, Part A*, **28A**, 1997, 277-88.
556 Qiu Q, Kumosa M, *Composites Sci. Technol,* **57**, 1997, 497-507.
557 Averous L, Quantin J C, Crespy A, *Polym. Eng. Sci.*, **37**, No.2, 1997, 329-37.
558 Quintanilla L, Pastor J M, *Polymer*, **35**, No.24, 1994, 5241-6.
559 Turcovsky G, Antec '94. Conference Proceedings, San Francisco, Ca., 1st-5th May 1994, Vol. I, 796-800.
560 Hauser R L, Woods D W, Krause-Singh J, Ferry S R, Antec '93. Conference Proceedings, New Orleans, La., 9th-13th May 1993, Vol. I, 341-6.
561 Murayama H, Min K, Antec '97. Conference proceedings, Toronto, April 1997, 759-65.
562 Wagner A H, Kalyon D M, Yazici R, Fiske T J, Antec '97. Conference proceedings, Toronto, April 1997, 996-1000.
563 Otaigbe J U, Quinn C J, Beall G H, Antec '97. Conference proceedings, Toronto, April 1997, 1826-30.
564 Barbosa S E, Kenny J M, Antec '97. Conference proceedings, Toronto, April 1997, 1855-9.
565 Averous L, Quantin J C, Lafon D, Crespy A, *Int. J. Polym. Analysis and Characterization*, **1**, No.4, 1995, 339-47.
566 Schneider J P, Myers G E, Clemons C M, English B W, *J. Vinyl and Additive Technol.*, **1**, No.2, 1995, 103-8.
567 Gassan J, Bledzki A K, *Polym. Composites*, **18**, No.2, 1997, 179-84.
568 Mannan K M, Robbany Z, *Polymer*, **37**, No.20, 1996, 4639-41.

569 Peng W, Riedl B, *Polymer*, **35**, No.6, 1994, 1280-6.
570 Venencie C, Filliatre C, Leclercq D, Villenave J J, *Pitture Vern.*, **71**, No.15, 1995, 29-39.
571 Casenave S, Ait-Kadi A, Brahimi B, Riedl B, Antec '95. Vol. II. Conference Proceedings, Boston, Ma., 7th-11th May 1995, 1438-42.
572 Burger H, Koine A, Maron R, Mieck K P, *Int. Polym. Sci. Technol.*, **22**, No.8, 1995, T/25-34.
573 Holl M, *Int. Polym. Sci. Technol.*, **21**, No.9, 1994, T/4-9.
574 Hajji P, Cavaille J Y, Favier V, Gauthier C, Vigier G, *Polym. Composites*, **17**, No.4, 1996, 612-9.
575 Helbert W, Cavaille J Y, Dufresne A, *Polym. Composites*, **17**, No.4, 1996, 604-11.
576 Gatenhom P, Hedenberg P, Karlsson J, Felix J, Antec '96. Volume II. Conference proceedings, Indianapolis, 5th-10th May 1996, 2302-4.
577 Malanda L M, Park C B, Balatinecz J J, Antec '96. Volume II. Conference proceedings, Indianapolis, 5th-10th May 1996, p.1900-7.
578 Simonsen J, Rials T G, *J. Thermoplast. Composite Mat.*, **9**, No.3, 1996, 292-302.
579 Gokturk H S, Fiske T J, Kalyon D M, Antec '93. Conference Proceedings, New Orleans, La., 9th-13th May 1993, Vol. I, 605-8.
580 Kumar P, Gawahale A R, Rai B, *Adv. Composite Materials*, **4**, No.4, 1995, 279-85.
581 Topoleski L D T, Ducheyne P, Cuckler J M, *J. Biomed. Mat. Res.*, **29**, No.3, 1995, 299-307.
582 Wawkuschewski A, Cantow H J, Magonov S N, *Polym. Bull.*, **32**, No.2, 1994, 235-40.
583 Nago S, Mizutani Y, *J. Appl. Polym. Sci.*, **53**, No.12, 1994, 1579-87.

# 3

# Transportation, Storage, and Processing of Fillers

## 3.1 FILLER PACKAGING

Fillers are usually packaged in multi-wall paper sacks, pasted valve sacks, or intermediate bulk containers. Sacks are handled in palletised units usually containing 50 sacks of 25 kg each or more. The choice of packaging is made based on consideration of speed of handling, material protection, the flow characteristics of the material, storage volume conservation, suitability for palletising and stacking, purity of product, stability, image projection, cleanliness, environmental safety, and waste disposal.[1] Chronos Richardson has over 100 years of experience with particulate materials. The Company designs equipment for packaging a variety of materials including fillers. The selection of bags includes 20 different designs. The following design criteria must be evaluated and specified:

- Material: paper, polyethylene, polypropylene, polymer metal coated
- Form of material: film, foil, laminate, woven
- Number of plies: 1 to 4
- Material mechanical properties
- Material permeability
- Type of plies: the same material (e.g., 4 layer paper), different layers (e.g., paper with PE in-liner), coating (e.g., 3 layer paper with coating or PE aluminum coated)
- Design: open mouth, cross bottom, pillow type, pinch-bottom, bag with carrier, block bottom
- Valve: external, internal
- Filling level
- Marking and coding

The bag design is also important to the manufacturers of fillers who handle large amounts of material. Bag filling lines are optimized to process specific materials and types of packaging materials. Figure 3.1 shows a fully automatic bagging and palletising line for valve bags developed by Chronos Richardson for a carbon black manufacturer. The carbon black is filled into 25 kg bags at a rate of 700 bags per hour. One of the constraints of the design presented here was that a large amount of material had to be filled with a product at a high temperature. The devel-

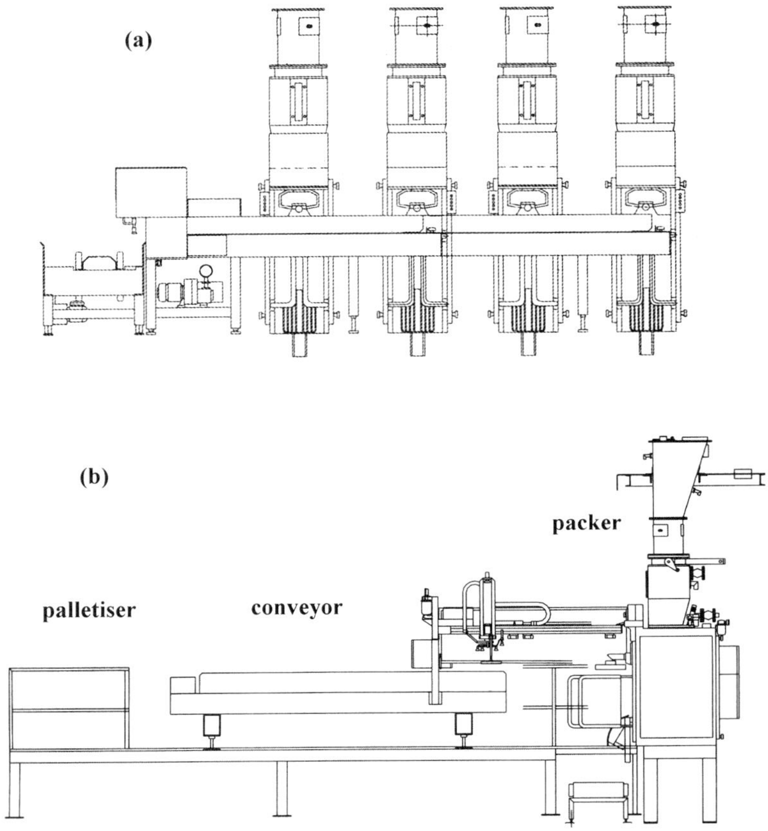


Figure 3.1. Schematic drawing of a fully automatic bagging and palletising line for carbon black packaging. (a) bottom filling packers, (b) side view of packaging line. *Courtesy of Chronos Richardson, Fairfield, New Jersey, USA.*

opment of an automatic line for carbon black is a very challenging task. Carbon black is a very difficult material to convey and it is extremely dusty. The material is charged to a receiving vessel having a special surface treatment. The material is fluidized in the vessel to improve its flow characteristics and to facilitate precise dosing. Automatically filled bags are closed and deposited on a belt conveyor which transfers the bags to a palletising unit.

One cost efficient design is the form-fill-seal line which manufacturers flash cut bags, fills and seals them, places them on pallets, and wraps the pallets in plastic film.[1] The rate and quality of filling can be improved by the use of a spout carousel.

In the three spout carousel design, one bag is filled to the required weight, while the second bag is being air evacuated, and the third is being closed.

Air evacuation is a process to remove air from the material during filling. If air is not removed, the material will normally require a larger bag and bags will be unstable during handling, palletising, and storage. Chronos Richardson developed a unique technology which uses two porous filter lances which plunge into the filled bag and deaerate the product.[2] The filter lances must be designed by the manufacturer for a particular product based on experimental work. Twelve stable layers of bags filled with deaerated product can be put on a pallet.

Intermediate bulk containers vary in size and construction but usually contain about 1000 kg of product. These containers are made from coated or uncoated cloth and are equipped with a lifting collar at the top and a discharge valve at the bottom. Although in a chemical plant environment fillers packed in paper sacks is a common sight only 10% of fillers are transported in packages, the remainder is shipped in bulk. Only industries which are particularly strict about moisture content will prefer material packed in sacks. From the point of view of material handling and exposure to dust, fillers packed in sacks are least safe because they cause the highest emission of dust in the work environment.

## 3.2 EXTERNAL TRANSPORTATION

Fillers are delivered by traditional means of transportation, including rail cars, road vehicles, ships, and barges. Rail cars are used for delivery of bulk powder, paper sacks, and intermediate bulk containers. Rail cars usually have a capacity between 20 and 55 tons. The car for bulk delivery is compartmentalized; usually it has 3 sections, each equipped with release bottom doors, which are usually designed to control the discharge rate. Cars for bulk transportation should be lined with an appropriate coating to avoid contamination of fillers with rust.[1]

Road vehicles are mainly used for delivery of fillers in packed units, but transportation in bulk is also growing. Road tankers for bulk powder transportation can handle up to 50 tons. They are loaded through hatches in the tank roof, and emptied through a pipe (normally 100 mm in diameter) by self-discharging pneumatic conveying equipment which typically can discharge 10 $m^3$ of material per minute. Filler slurries are transported in stainless steel tanks which are also filled from the top, and discharged by positive displacement pumps able to discharge 20 tons in 10 min. The viscosity of slurry depends on the temperature; therefore, tankers used for cold temperature transportation should be insulated.

Transportation of bulk material by ships and barges is more complex because of the need for special equipment for loading and unloading. Loading is usually done by means of fixed or mobile conveyors. With proper equipment and organization, a loading capacity of 1000 tons/hr is realistic. Discharging of fillers is done by means of a variety of cranes and grabs. One crane and grab can usually have a rate of 75-100 tons/h; in larger ports, discharge rates of 300 tons/h are achievable.[3]

## 3.3 FILLER RECEIVING

Fillers can be delivered in bulk by rail or truck. Figure 3.2 shows a vacuum-pressure rail unload system designed by Premier Pneumatics, Inc. The elements of the system are explained on the drawing. Several systems are offered.[4] The dual blower, vacuum pressure system has the highest output at 45,000 kg/h. The system is equipped with PLC controls which include a destination selector and automatic shutdown. A smaller system with a 22,000 kg/h output can be operated by one person. It has a built-in hydraulic system which simplifies the attachment to rail car outlets and variable speed drives which allow operators to control the material feed rate. The company produces all of the required accessories such as Aerolock rotary valve meters with many choices of rotors, diverter valves, piping, couplings, adapters, gates, separators and receivers, and blowers. In fact, every component required for the design and assembly of these systems is offered.

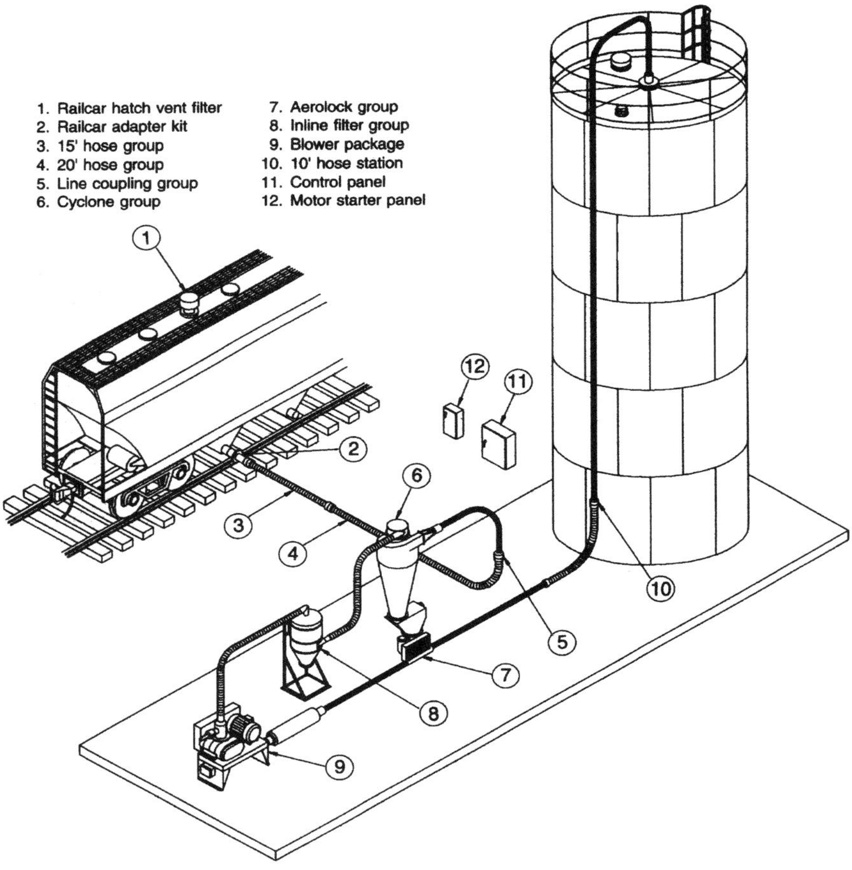


Figure 3.2. Vacuum-pressure unload system. *Courtesy of Premier Pneumatics, Inc., Salina, KS, USA.*

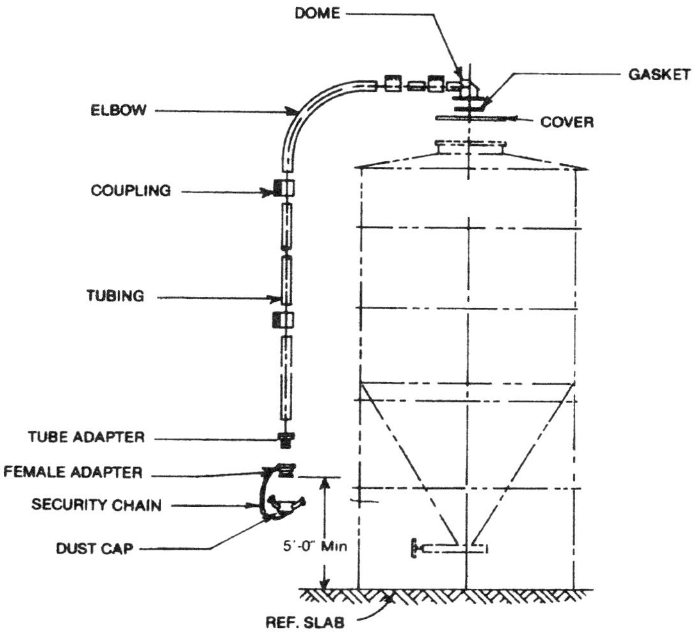


Figure 3.3. Load line assembly. *Courtesy of Premier Pneumatics, Inc., Salina, KS, USA.*

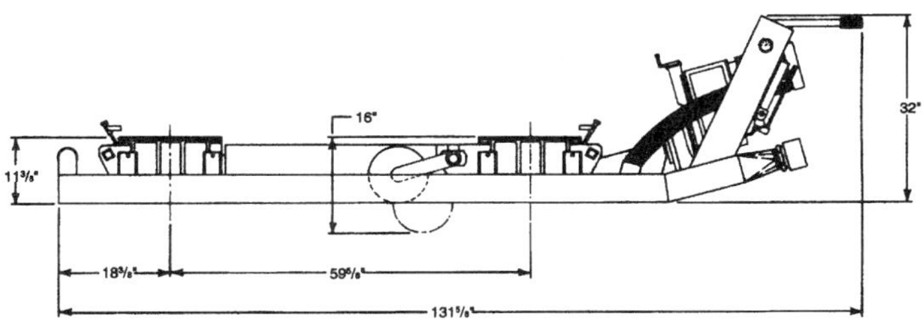


Figure 3.4. Manual railcar unloader. *Courtesy of Premier Pneumatics, Inc., Salina, KS, USA.*

Figure 3.3 shows various elements required to assemble a load line to a silo.

The railcar can also be unloaded by a simpler, manual device (Figure 3.4). This unit can be used for Airslide railcars. The frame serves as a conveying airflow line. It should be noted that an explosion proof design is required for filler unloading.

The truck unloading installations have a selector switch for seven destinations, and an alarm to notify the operator when a storage tank is full. It is equipped with a blower and self-cleaning vent which provide a dust-free exhaust. The bag unloading stations are discussed below as part of in-plant operations.

## 3.4 STORAGE

Storage of fillers is a complex issue, and we will not attempt to discuss all its facets. Fillers are stored in portal frame buildings, general purpose warehouses, storage barges, bunkers, large hoppers, and silos. The choice depends on a large number of circumstances, but mostly on the material properties and the rate of use. Silo storage is very convenient and has these advantages:

- The storage capacity of a silo is several times greater than flat surface storage (floor storage)
- Transportation and packaging costs are reduced
- Equipment cost per unit volume stored is low
- Automatic handling and control is possible
- Controlled conditions of storage (temperature, moisture, etc.) are easily attained
- Quality of stored product is uniform
- The automatic process saves labor costs.

The diameter of a silo is usually 2.5 m or more. They are usually cylindrical with a conical bottom which has a 60° incline to facilitate discharge. There are two types of silos (hoppers): core flow and mass flow. They differ in principle because of the proportion between diameter and height in relation to the rate at which material is disposed. If a silo has a larger ratio of diameter/height and material is disposed in relatively small amounts compared with storage capacity, then material flows in the center (core silos). Mixing of material is minimal. Cohesive material may stop flowing for no reason. The rate of flow is variable and the bulk density of the filler will also vary. The material is not very stable in such a storage vessel and can be suddenly fluidized, leading to a rapid increase in discharge rate which might be hazardous. When the ratio of diameter/height is low (very tall), this becomes a mass flow silo) and the above disadvantages are avoided. Mass flow silos have some disadvantages such as the requirement of a tall building (if the silo must be indoors), high pressure on the side walls (stronger materials needed) and the abrasive action of filler on walls (faster wear). A decision on silo dimensions should be based on the flowability of filler.

The surface coating of the silo also plays an essential role. The exterior coatings are designed for UV stability, corrosion protection, high gloss and color retention. The interior coatings are even more crucial. Coatings must protect against corrosion, be abrasion resistant, and have chemical and thermal resistance. Compatibility with the material being stored contributes to proper discharge.

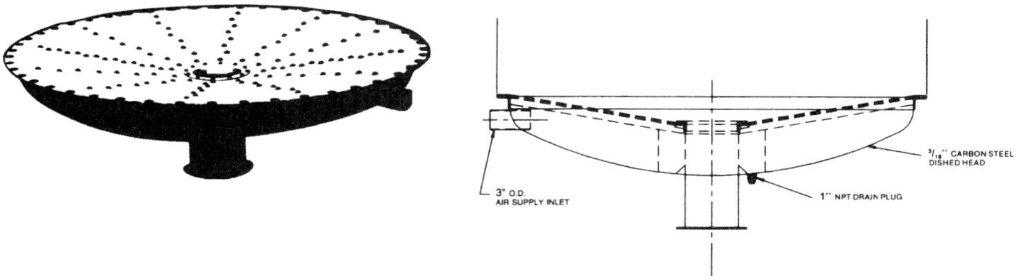


Figure 3.5. Bed activator. *Courtesy of Premier Pneumatics, Inc., Salina, KS, USA.*

Many silos are fitted with a pressure relief disc in the roof which guards against silo damage during filling. Since most fillers have cohesive properties, discharging is aided by air pads or fluidized beds. Some materials gain in cohesion if left undisturbed. Figure 3.5 shows an example of a fluidized bed outlet manufactured by Premier Pneumatics, Inc. By introducing low pressure air into the stored material, the discharge process becomes less restricted and more uniform.

Fillers in the form of an aqueous slurry are stored in concrete or steel tanks. Care must be taken that the slurry temperature does not drop below 10°C. Handling problems develop when the slurry temperature is below 5°C. Overheating (above 35°C) should also be prevented. Materials packed in unitary packages should be stored in conditions specified by the producer. Bags are usually stored in a palletised form, whereas intermediate bulk containers are stacked three high, possibly on pallets or suspended on special metal pallets by means of loops.

A silo is also frequently equipped with a pneumatic conveying system, consisting of a blower, a rotary valve, conveying pipework, and an air/product separator at the discharge point. Some elements of the pipework are given in Figure 3.3. The discharge systems are discussed below. Suction systems can deliver material over a 600 m distance, low-pressure systems over a 1600 m distance, and high-pressure systems 3000 m and more. Conveying pipes have a diameter from 20 to 400 mm and a mass flow rate of up to 400 t/h is typical. A pneumatic conveyor requires more power for operation than a mechanical conveyor. Slurries can be transported by hydraulic conveyors. Conveying distances may be up to 400 km. Pipe diameters are from 60 to 300 mm.

Some silos may be equipped with flow measuring equipment. K-Tron Soder developed a system which is installed directly on the silo (Figure 3.6). This system is suitable only for free flowing material. In most cases the metering of the filler is conducted outside the silo and such solutions are discussed below. There is one exception – the use of a load cell system. K-Tron developed a vibrating wire Smart Force Transducer II which has exceptional performance compared to other sensors.[5] It is a digital sensor designed for process weighing with multiple data regis-

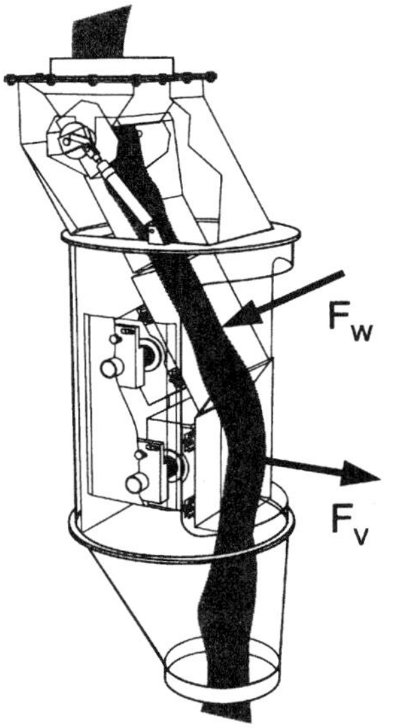


Figure 3.6. Smart flow meter bypass. *Courtesy of K-Tron America, Pitman, NJ, USA.*

ters for data acquisition and advanced digital filtering for highly effective suppression of in-plant vibration. This design overcomes a frequent problem related to vibration in industrial environment which is compensated for by filters and does not affect measurement. There is excellent measurement resolution (1:1,000,000), no need for recalibration, error free data transmission, and data can be sent up to 500 m away. This transducer is also a part of various feeding systems discussed below.

## 3.5 IN-PLANT CONVEYING

When material is stored in a silo, an in-plant conveying system is required to transfer the material from the storage location to the point of use. The main components of such a system include a blower, piping, valve(s), receiver(s), filter(s), and control units.

Blowers are similar in construction to the units used in filler unloading from transporting units but usually reduced in size and capacity. Typical capacities range from 300 to 2000 kg/h. The choice of the right blower is critical for the operation of pneumatic conveying systems. Blowers, according to Premier Pneumatics, Inc. are designed to operate at 75-80% of rated capacity. Premier Pneumatics, Inc. produces three types of blowers: pressure, vacuum, and vacuum/pressure. Mini-Vac™ is the name of a modular system designed by Hapman. The system is suitable for applications where space is limited and high output is required. The system vents through cartridge filters which are easy to replace. This system is used with multiple inlets and receiving points, for container unloading units, and for loading to containers. The schematic drawing shows the system components (Figure 3.7). The choice of a blower depends on the material characteristics and required output.

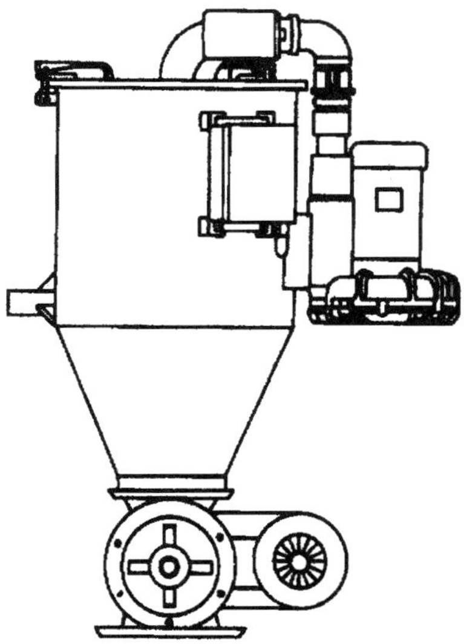

Figure 3.7. Mini-Vac compact blower with rotary valve. *Courtesy of Hapman, Kalamazoo, MI, USA.*

Figure 3.3 shows some elements of piping. The essential elements are pipe diameter, couplings, sight tubes, line branches (tees, wyes), and elbows. Some systems such as conveying systems for carbon black may contain aerators to prevent line plugging. Lines are usually 1.5 to 8" in diameter and are made out of aluminum, steel, or stainless steel. The designer should keep the number of branches, valves, and elbows to a minimum. Each causes obstruction to flow and potential problems in operation.

Several types of valves are used. The tunnel diverter valves allow the use of multiple supply or receiving lines. The A valve diverts the material stream to one of two destinations. The aeropass valve separates air from the material. The slide gate valve opens or closes to control flow. These valves can be either manual or automatic.[6]

Separators and receivers separate material from the conveying air before the point of use. They are either filters or cyclones. Figure 3.8 shows a schematic diagram of part of the line, the filter, receiver, and vacuum blower.

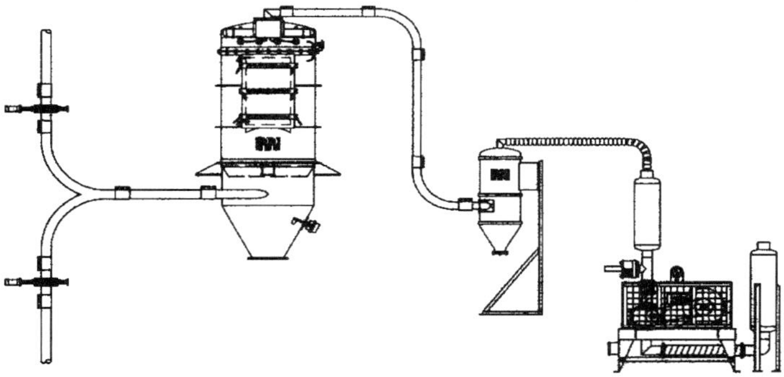

Figure 3.8. Elements of a pneumatic conveying system. *Courtesy of Premier Pneumatics, Inc., Salina, KS, USA.*

Filters are used in conjunction with receiving units and blowers. In applications, where powder dusting is a problem, additional filtering systems are also installed. The whole system is operated from a central controller usually equipped with an alphanumeric backlit LCD display. Various levels of control and automatic operation are available.

Spiroflow-Orthos Systems, Inc. developed several mechanical and aero-mechanical conveying systems which may add to the flexibility of in-plant operations and eliminate unnecessary manual operations and dust. Figure 3.9 shows an aero-mechanical conveyor which can work in vertical, angled, and horizontal arrangement. The wire rope assembly with polyurethane discs moves at a high speed

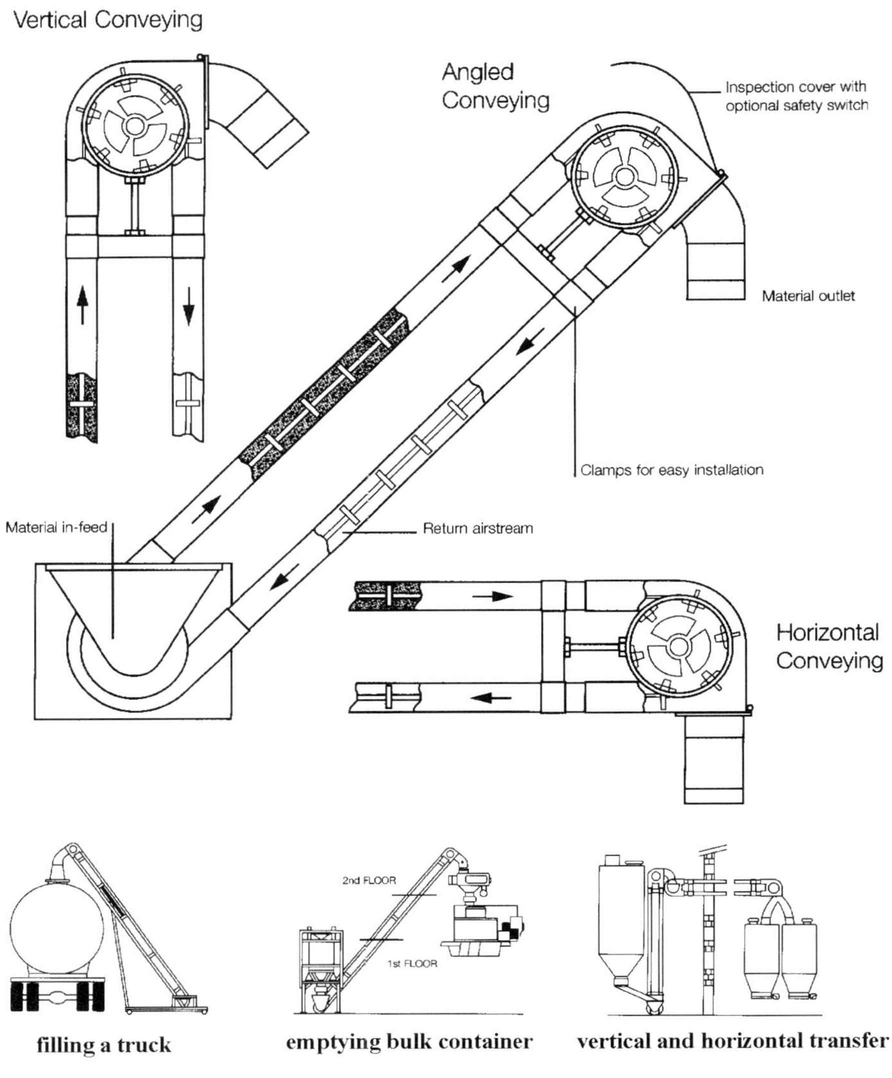


Figure 3.9. Schematic diagram of aero-mechanical conveyor and its applications. *Courtesy of Spiroflow-Orthos Systems, Inc., Monroe, NC, USA.*

transporting the material to its destination. The rate of materials delivery depends on the conveyor size. For example, the 75 mm model can deliver 300 l/min and the 100 mm model 600 l/min. Materials from fine powders to granular particles can be moved by this design. Figure 3.9 shows some typical applications of this conveying system.

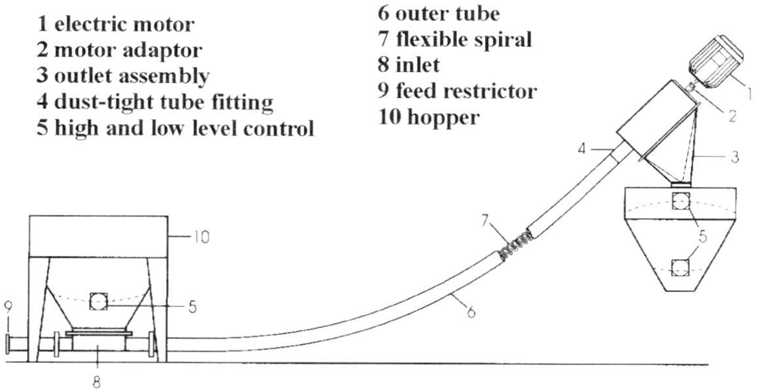

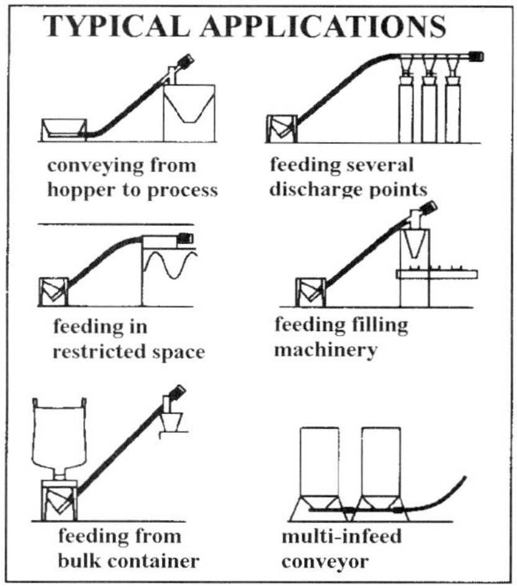


Figure 3.10. Schematic diagram of flexible screw conveyor, spiral design, and examples of application. *Courtesy of Spiroflow-Orthos Systems, Inc., Monroe, NC, USA.*

The flexible screw conveyor is another system with many advantages. Figure 3.10 shows the schematic diagram of a conveyor and examples of its applications. The only moving part of this conveyor is a flexible spiral directly driven by an electric motor and rotating within an outer tube. The system is totally sealed which makes it dust-free and eliminates atmospheric contamination (e.g., humidity). The spiral's gentle action does not degrade the material. This conveyor can convey in any direction and with a variable speed control accurate metering can be achieved. The simple design is easy to dismantle and clean. Different spiral designs can be selected to move different materials. The conveyor has been used for the following fillers: talc, perlite, calcium carbonate, titanium dioxide, bentonite, zinc oxide, carbon black, alumina, silica, diatomaceous earth, quartz sand, and for many foods, pharmaceuticals, and plastics.

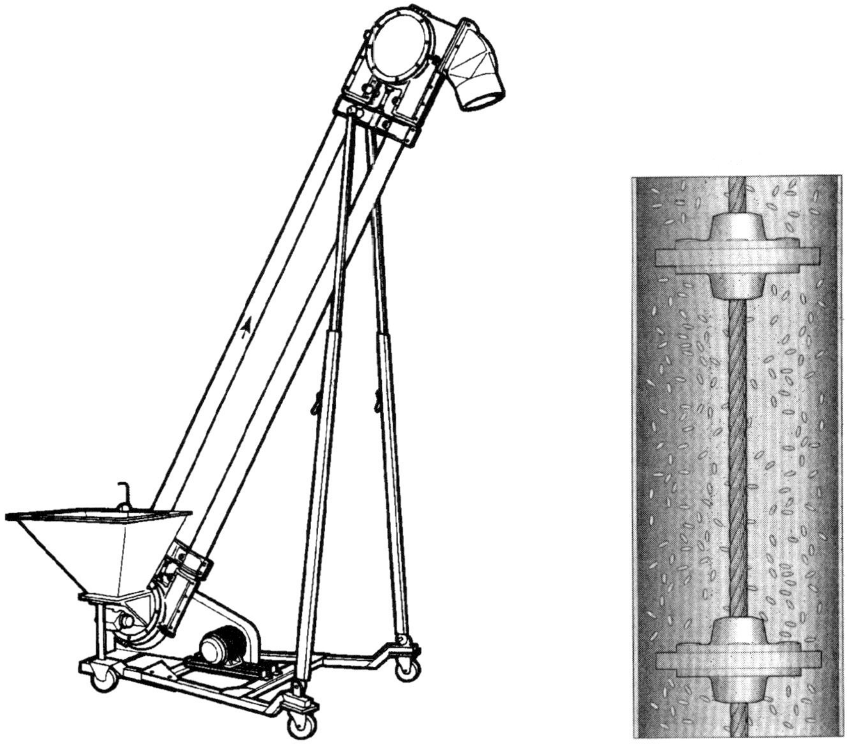

Figure 3.11. AMC aero-mechanical conveyor and the cross-section of conveying element. *Courtesy of Spiroflow-Orthos Systems, Inc., Monroe, NC, USA.*

Figure 3.11 shows an aero-mechanical conveying system which is tubular in construction with a tensioned rope fitted with plastic discs. The discs travel at a high rate creating both air and material displacement. This effect fluidizes the product which limits mechanical damage and material segregation by size. The conveyor works in any arrangement (vertical, horizontal, angled) and can deliver numerous fillers to one or to several destinations.

## 3.6 SEMI-BULK UNLOADING SYSTEMS

Due to the environmental protection many powders are delivered in semi-bulk bags. These heavy units require equipment for unloading. Figure 3.12 shows the system for bulk bag unloading to a bin equipped with a weigh hopper to feed material in a semi-automatic or automatic process.

Figure 3.13 shows Flow-Flexer from K-Tron Soder. During transportation, materials are compacted or lose their fluid properties and will not discharge consistently. Various types of obstruction to the flow occur as illustrated on the left side.

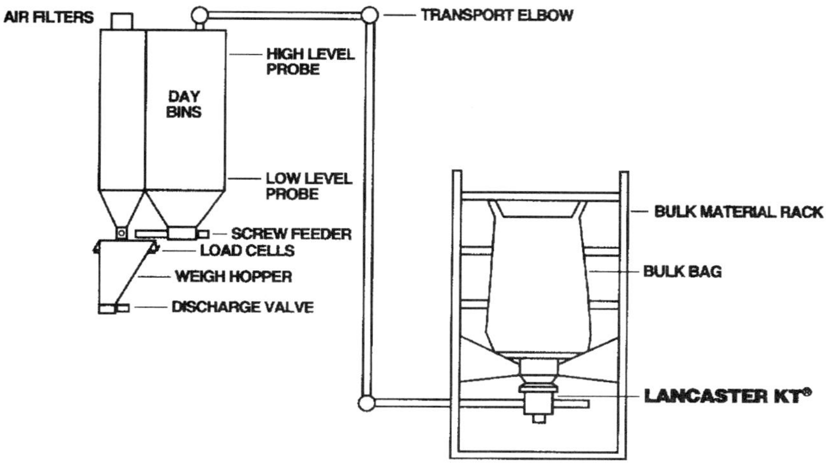


Figure 3.12. Lancaster bulk bag unloader. *Courtesy of Lancaster Products, Lebanon, PA, USA.*

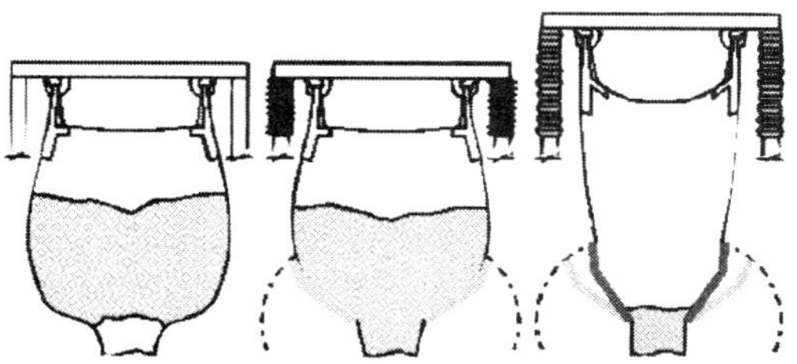

Figure 3.13. Flow-Flexer and Pop-Top bulk bag discharging system. *Courtesy of K-Tron Soder, Pitman, NJ, USA.*

Flow-Flexer bag activators raise and lower the opposite bottom edges of the bag at timed intervals (middle). As the bag becomes lighter, the stroke of the bag activators increases to shape the bottom of the bag into a steep configuration while a Pop-Top bag extension device stretches the bag (right). This device assures complete discharge.

AccuRate developed a combined bulk discharging station and metering unit. This system lifts and positions the bulk bag over a metering unit which can supply material in a known quantity directly to the point of use.

Palamatic, in addition to designing conventional bulk bag discharging systems, has developed another version called the Duo-Pal Dump Station which combines bulk bag and small bag discharging through a dust control unit.

## 3.7 BAG HANDLING EQUIPMENT

There are numerous bag handling systems available for filler users. These range from very simple sack dump stations to complicated lines handling up to 600 sacks per hour. The choice depends on investment and volume.

The UK company, Palamatic, specializes in a full range of solutions. A simple sack discharger requires manual bag discharging but with the use of dust extracting

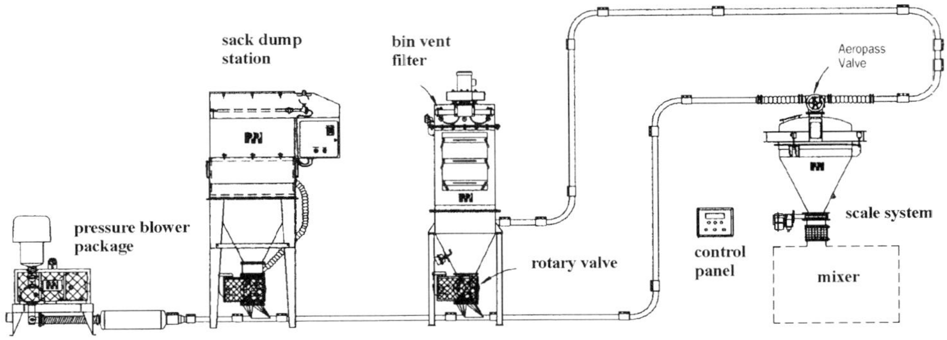


Figure 3.14. Calcium carbonate delivery system from sacks. *Courtesy of Premier Pneumatic, Inc., Salina, KS, USA.*

equipment which protects personnel from exposure to dust. Such a unit does not have any mechanical parts but is equipped with hood, dust vent, and a grating to place the bag on. A large volume sack handling system is composed of several elements, such as a pneumatic bag lifter, a belt conveyor, a photocell to detect the incoming bag, a sack opener, shaker bars to aid content removal, a sack ejector, a dust extraction system, and a bag compactor. The lines are known to perform with carbon black, titanium dioxide, fumed silica, barytes, calcium carbonate, mica, talc, and other fillers. Palamatic also developed a brush unit to remove dust from the surface of bags. The intermediate systems include semi-automatic and automatic sack opener which eliminates dust leakage and risk of injury. Units are available for the safe discharge of dangerous materials.

Bel-Tyne is another company which specializes in bulk handling systems. The range of equipment includes automatic bag slitting machines, manumatic bag slitters, manual bag opening devices, pneumatic bag lifters, bag compactors, and a complete system which can discharge material from bags to storage or production receivers equipped with metering devices. The automatic bag slitting machine is a compact unit with a short belt conveyor which delivers sacks to a bag emptying unit equipped with a dust control system and waste bag disposal system.

Figure 3.14 shows an integrated system offered by Premier Pneumatic, Inc. for feeding mixer from the sack dump station, trough receiver and metering station.

## 3.8 BLENDING

Blending of different fillers is a common operation. It may be conducted in one of the two methods discussed below. Littleford Day manufactures a mixer which is useful in the blending operation and many other applications. Figure 3.15 shows the

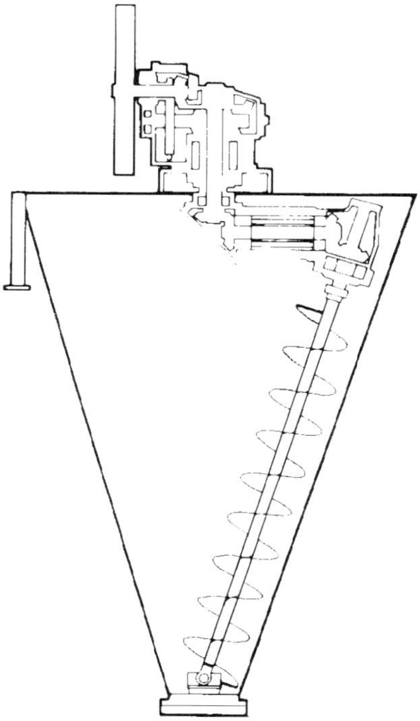

Figure 3.15. Day Mark II mixer. *Courtesy of Day, Inc., Florence, KY, USA.*

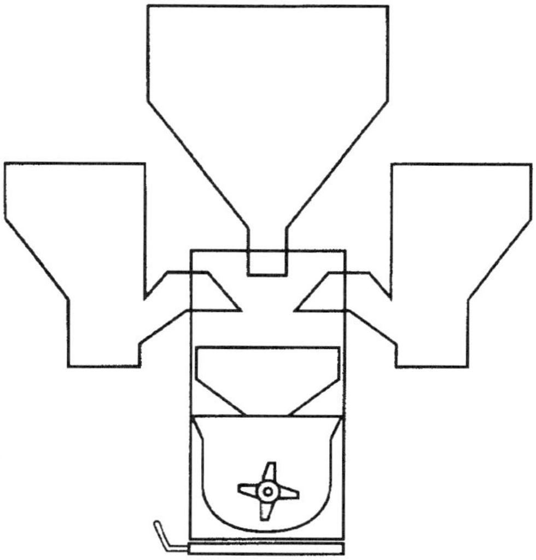

Figure 3.16. NovaBlend design. *Courtesy of Novatec, Baltimore, MD, USA.*

principle of the design. This is an efficient design which can mix up to 4000 lbs of material within 5 min. The screw agitator turns on its axis, producing a lifting action as it spirals the material in an upward flow. The material can be discharged through the bottom or the side. The mixer can be combined with a metering device and used for dosing materials which do not flow well. Other applications include deaeration, vacuum drying, or hot air drying. The unit has a gentle action which does not degrade particles. The mixing action may be increased by the use of a tapered screw design which gives 25% faster mixing.

Novatec developed accurate, sequential metering of up to four materials combined with blending. Figure 3.16 shows the principle of design. The dual load cell under the hopper assures weighing accuracy. Materials are delivered through accurate vibratory feeders. Materials are mixed in the mixing chamber before being delivered to the production unit.

Other designs from Novatec include a gain-in-weight batch blender where 2 to 8 components can be blended together with 0.1% accuracy. During the processing cycle, the operator can view the active filling weight, the actual weight, change formulation, and save formulation. Two hundred formulations can be stored in memory and executed.

## 3.9 FEEDING

Fillers must be fed accurately and consistently. Frequently, a constant feeding rate is required. The characteristic properties of fillers are particle size, friction coefficients (internal and external), flowability, temperature, moisture content, and degree of compression. Several typical feeders are used. A rotary feeder has several constraints, including a volumetric efficiency decrease as the rotor speed increases and feed rate fluctuation. The pressure of the equipment must be below 2 atm. Screw feeders have a variable range due to powder compressibility; the rate of feed is not uniform. A table feeder has a uniform feed rate and a fast response to changes. Its flow pattern is affected by the scraper and the inclination of the hopper. Belt feeders provide a uniform feeding rate except during belt start up. The rate of feeding and the rate of belt movement have been very well correlated. Other typical feeders include vibrating feeders, valves, and dampers.

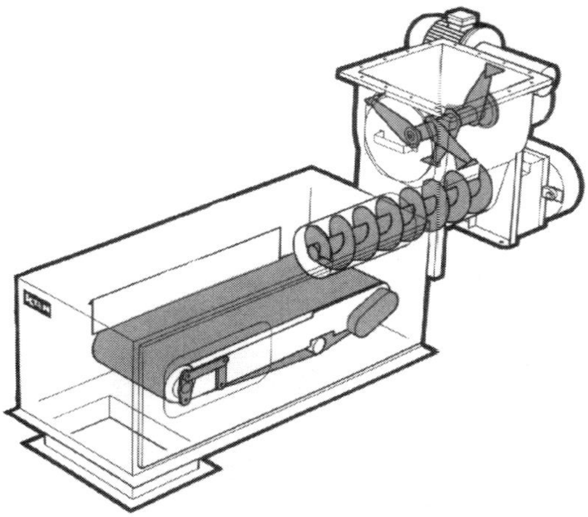

Figure 3.17 Weigh belt feeder. *Courtesy of K-Tron Soder, Pitman, NJ, USA.*

K-Tron Soder specializes in the design and manufacture of feeders useful in dosing and metering of particulate materials. Many solutions are based on their Smart Force Transducer design which gives excellent precision in material weighing. Their range of feeders includes belt, loss-in-weight, and volumetric feeders. Figure 3.17 shows the principle of action of a weigh belt feeder. The feeder schematically shown is designed for poorly flowing bulk materials. The material is delivered to the belt from a hopper or other feed device and is driven through the weight bridge. A computer determines the feed rate based on weight and belt speed. The rate of feed is regulated by belt speed. This type of feeder finds application in feeding glass fibers and glass powders. It delivers material with an accuracy from 0.1 to 1% of batch size.

Figure 3.18 shows the design principle of a loss-in-weight feeder. The hopper rests on weighing modules which sample the weight remaining and adjusts screw rotation accordingly to arrive at the correct feed rate. This feeder can handle a wide range of solid types including free flowing powders, lumpy, moist materials, fibers,

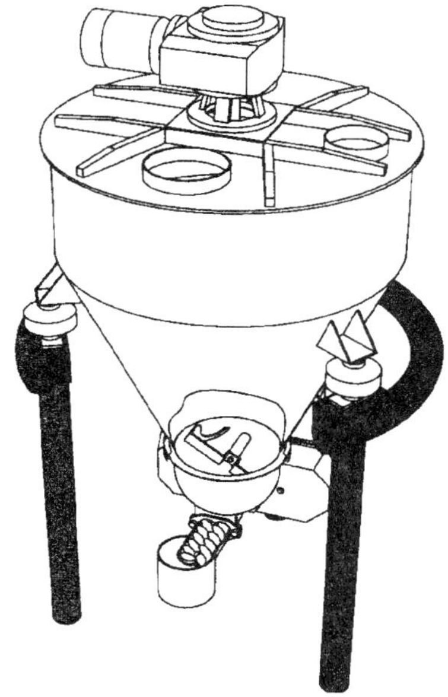

Figure 3.18. Loss-in-weight feeder. *Courtesy of K-Tron Soder, Pitman, NJ, USA.*

and flakes. Figure 3.19 shows the types of screws used to move material. The feeder can move up to 7,000 liters of material per hour. The feeder is microprocessor controlled and can be used in automated designs. It delivers material with an accuracy from 0.1 to 1%. There are many feeder designs which can be used alone or as part of a multiple unit system. All the feeders discussed above are equipped with a feeder control interface, or a feeder line control display, or a multi-line feeder control interface.

AccuRate has a range of weigh belts which are designed for both feeding equipment or dosing material to fill containers. The equipment is microprocessor-controlled and its accuracy is improved due to the application of belt influence compensation. The Company also produces a range of loss-in-weight and volumetric feeders

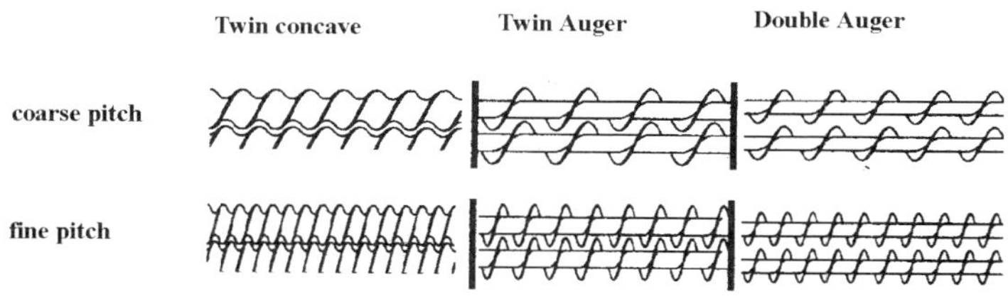


Figure 3.19. Screws used in feeders. *Courtesy of K-Tron Soder, Pitman, NJ, USA.*

which can be used for material ranging from free flowable to difficult to transfer. The materials can be delivered at rates from 15 to 45,000 pounds per hour with a deviation of 0.25% and higher.

Figure 3.20 shows a Multicor Mass Flow Meter which is designed to measure free flowing powders. The material falls on a centrifugal wheel whose rotating guide vanes divert the flow radially outward. The particles moving along guide vanes produce Coriolis forces which generate a measurable torque proportional to the mass flow. These feeders may deliver up 88 tons per hour with metering accuracy and repeatability of 0.5% or better. The feeder is totally enclosed, dust-tight design.

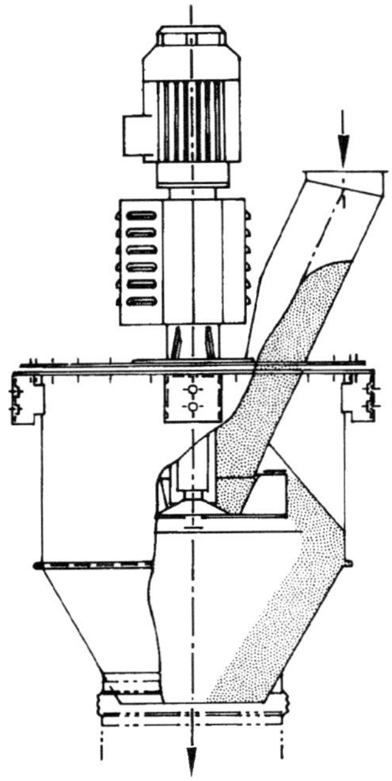

Figure 3.20. Multicor mass flow meter. *Courtesy of AccuRate, Whitewater, WI, USA.*

## 3.10 DRYING

Many technological processes require dry filler. In some cases the moisture level of the filler must be as low as 0.03%. Special drying equipment has been developed to dry fillers. This equipment overcomes the long drying times and ineffectiveness of more conventional drying ovens.

Littleford Day specializes in dryers which use special plow shaped mixing tools which provide a sufficient agitation to filler particles that they form a fluidized bed which is much more accessible to the drying effect of air. In addition, high energy mixing disperses agglomerates. Figure 3.21 shows a schematic diagram of a drying system. The heat transfer coefficient is increased two to four times that of the traditional paddle dryer. The mixer can be operated under atmospheric pressure and vacuum. Mixers are produced in a range of capacities from 300 to 25,000

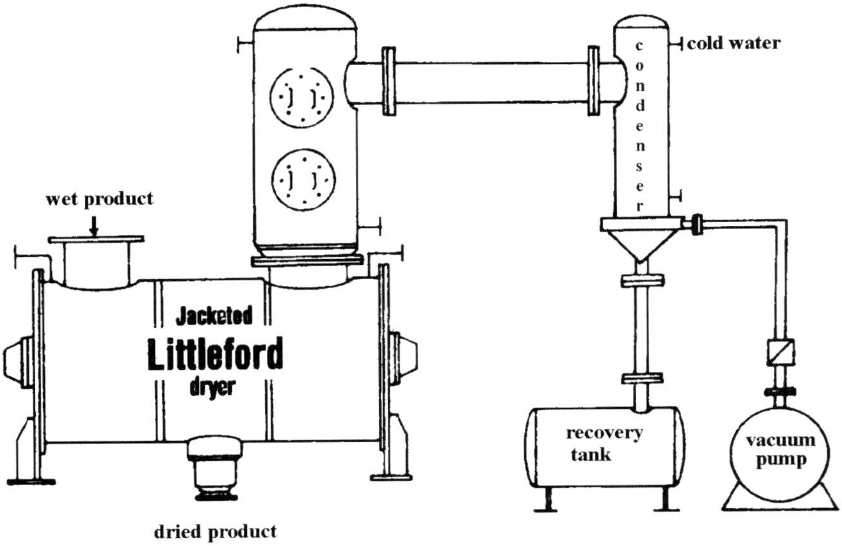


Figure 3.21. Vacuum drying system. *Courtesy of Littleford Day, Florence, KY, USA.*

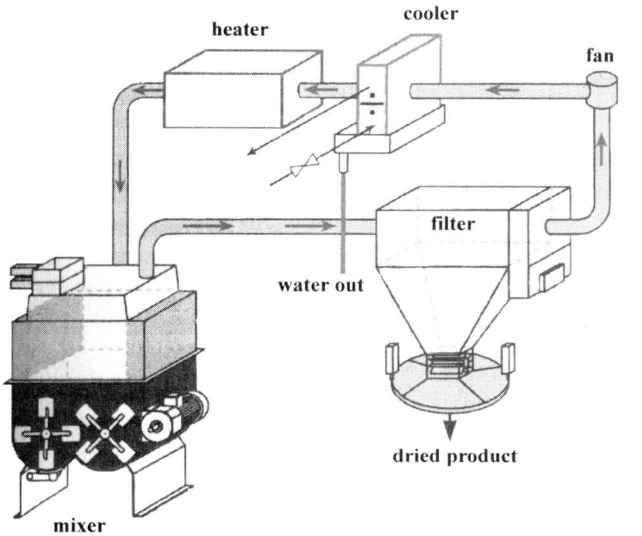


Figure 3.22. Forberg drying system. *Courtesy of Forberg AS, Larvik, Norway.*

liters. The mixer has been used for drying the following fillers: carbon black, carbon fiber, iron oxide, molybdenum disulfide, silicon dioxide, and many other pigments and fillers.

Figure 3.22 shows a concept of a drying system developed in Norway by Forberg AS. Although the shape of the mixer and mixing elements differ from the Littleford design the general idea is very similar. Two rotating shafts, each having 14 paddles, create a fluidization zone which enhances heat exchange. The mixer itself is used for other technological purposes and is known to offer extremely short mixing times, from as little as 10 seconds to 2 minutes. The mixer is very economical both as a mixer and as a drying system. It not only saves energy but processes materials without releasing volatiles to the environment. The mixer is produced in capacities ranging from 20 to 5,000 liters. The following fillers have been known to be processed in this system: bentonite, calcium carbonate, calcium sulfate, chalk, clay, ferrite, fibers, fly ash, glass powder, graphite, metal powders, mica, perlite, silica, sand, talc, vermiculite, and zinc oxide.

Novatec has developed two systems which can be applied to drying fillers. One is an indirect gas fired heater which can be used with any drier to improve the process economy. About 80% of the heating cost can be saved by the use of these heaters. Novatec also offers a portable drying/conveying system which conveys particulate materials through the drier and delivers them directly to the next process step.

Drying efficiency can be evaluated by process monitoring which usually requires that a sample be taken from the drier for testing. Favre & Matthijs SA developed a sampling port which allows sampling without interrupting the process either under vacuum or high pressure (Figure 3.23). When the piston is in the upper position, the sampling bottle can be attached. When the piston is lowered, the sample is taken, then the piston is moved back to the upper position, the pressure equilibrated through the valve, and the sampling bottle detached.

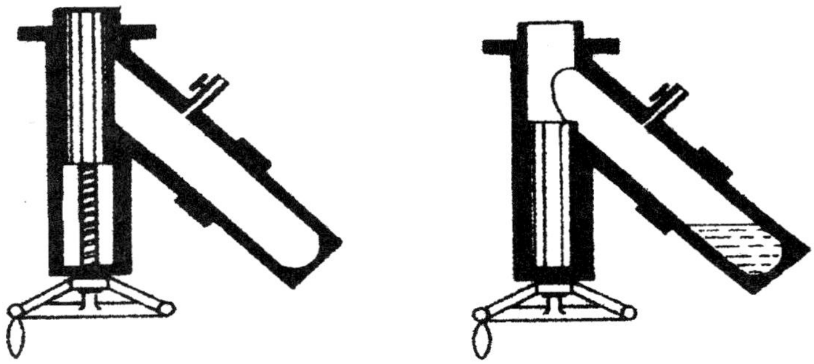

Figure 3.23. Sampling port in closed and opened position. *Courtesy of Favre & Matthijs SA, Lausanne, Switzerland.*

## 3.11 DISPERSION

Selecting dispersion equipment for a specific application is a complex task. Dispersion of the mixture must be complete and the process and equipment must meet economic constraints. But much more is involved.[7] In practice, such simple criteria are complicated by a variety of parameters related to fillers and to the materials in which they are dispersed. These parameters complicate the problem to the degree that it is not easy to formulate general guidelines. In this discussion we will consider the available equipment types most frequently used for filler dispersion and illustrate their applicability with some examples.

A ball mill is an effective means of dispersing solid materials in solids or liquids.[8,9] Ball mills have several advantages which include versatility, low cost of labor and maintenance, the possibility of unsupervised running, no loss of volatiles, and a clean process. The disadvantages are related to discharging viscous and thixotropic mixtures, and considerably lower efficiency when compared with other mixing equipment. The mill base viscosity is usually restricted to about 15-20 Poise, and therefore ball mills are most frequently found in production applications for paints, flexographic, publication gravure, and letterpress news inks, and carbon paper inks which are dispersed at elevated temperatures.

Several general conditions of ball mill operation should be respected:

- The mill should rotate at 50-65% of the theoretical centrifugal speed in order to allow balls to cascade, since the cascading balls grind most effectively and do not cause an excessive loss of ball material
- The ball load should be 40-58% of the total internal mill volume, and the material to be ground should fill only the voids between the balls (a maximum of twice the ball space)

- Viscosity, the order of filler addition, and the quantity of material should be chosen so as not to cause a viscosity increase above the specified range, since the milling efficiency drastically decreases at that point
- The rotation rate of the mill should be chosen giving consideration to millbase viscosity such that the balls should be carried up to a point between 20 to $30^{\circ}$ before the zenith and then cascade down
- The ball diameter should be as small as possible but large enough to permit easy separation from the liquid when discharged
- The wear suffered by the balls generally requires the addition of balls to bring the ball charge up to the correct volume every three months
- If carbon black is to be dispersed, the maximal load of pigment will decrease as particle size decreases because of the effect on millbase viscosity
- The degree of dispersion and jetness achieved when grinding carbon black depends on the wetting properties of the dispersing material and to some degree on the filler form. For instance, pelletized carbon black is easier to disperse than a fluffy type

Sandmills are a logical development of the ball mill idea. In sandmill applications, the following points should be considered:

- The efficiency of a ball mill depends on the number of contact points between the balls
- There is a limit of ball diameter below which centrifuging of mill charge occurs; this limit can only be overcome by a change in the manner of ball movement
- In sandmills, the grinding charge is driven by an impeller. Sand used in such mills has a diameter in ranging from 0.5 to 1 mm; in beadmills, glass beads have a diameter ranging from 1 to 3 mm
- The impeller is mounted centrally in the container and it has several milling discs which rotate at 1,200 to 2,400 rpm
- Advantages include flexibility, ease of operation and maintenance, low contamination, and easy clean-up by solvent washing
- The sand mill has some drawbacks. It is a two stage process (premixing followed by milling). Milling develops high temperatures in the mixture which causes loss of volatiles and requires cooling. If the mill base is high in viscosity or dilatant, the sandmill process may not work at all. Agglomerated or extremely hard pigments are difficult or impossible to disperse
- The practical limit of viscosity is about 20 Poise
- Sand occupies about 50% of the sandmill volume, whereas beads occupy 50-70% of the beadmill volume
- Increasing the volume of the grinding material increases the power requirement and generates more heat; decreasing the volume of grinding material decreases the quality of dispersion

- Dispersion of carbon black is usually done at elevated temperatures in a range from 40 to 150°C
- Inks are generally difficult to feed into a sandmill
- Fluffy carbon blacks can be fed and dispersed without problems, whereas pelleted carbon blacks are difficult to feed
- Some feed problems have been resolved by using a volute type of centrifugal pump and feed tank[3]
- By controlling the ratio of feed to recycle the millbase is kept in constant agitation[8]

Both ball and sand mills operate based on a viscous shear principle, thus the viscosity of the millbase is a critical factor in achieving dispersion. The size of filler particles is critical, especially in sandmills. It was found that the shearing force is inversely proportional to the square of the linear size of filler agglomerate. An agglomerate of diameter of 7 μm attains 100 times the shear stress of an agglomerate of 70 μm in diameter.[7] The difference between the ball mill and the sand mill is in the size and density of the grinding media, which is reflected in their performance. Sandmilling uses small particles of low density, and therefore, there is no noticeable reduction in the size of the sand particle, whereas the balls in ballmills are very much larger and may have a high density (steel), which results in a more complex mechanism of grinding including shattering and impacting which cause this mill to be more effective in disintegrating hard particles and agglomerates containing sintered particles.

There is another mill type called an attritor, which is similar to both the ball mill and the sandmill. In construction, it is similar to a sandmill. It also has a vertical shaft, but in the attritor the agitator bars replace the milling discs of the sandmill. It is also similar to a ball mill because it uses balls, usually ceramic ones 5-15 mm in diameter. Because the motion of the balls is independent of gravity, an attritor can handle thixotropic materials and slightly higher viscosity of millbases, but the principle of action and type of forces operating are similar to those of the ball mill. An attritor applied to pigment dispersion gives several advantages. These include rapid dispersion, the possibility of either a continuous or batch process, low power consumption, small floor space, and easy cleaning and maintenance. Their main disadvantage is high heat generation. Attritors are equipped with a cooling water jacket which can control the heat flow to some extent, but conditions are often too severe for some resins, which may degrade during the process.

Three-roll, one-roll, and stone mills constitute a more mature dispersion technology still in use with medium viscosity millbases. A three-roll mill consists of the feed, center, and apron rolls. In roll mill operation:

- The speeds of feed and apron rolls are adjustable, and each roll rotates with a different speed in order to induce shear in the material at the nip and facilitate the material transfer from one roll to the other

- For mechanical reasons the gap between rolls cannot be less than 10 μm and it is usually ranges from 40 to 50 μm.[7] Small particles will not be affected as they pass through the nip, but agglomerates smaller than the distance between rolls will be disintegrated due to the shear stress imposed on the material
- Shear stress depends on such major factors as the relative speed of the rolls, the viscosity of the millbase, and the tack or adhesion of the millbase to the rolls
- Similarly, the transfer of material from one roll to the other depends on roll speed and the adhesion of the millbase to the rolls
- Mill output depends on the distance between the rolls and the viscosity of the millbase
- The three-roll mill can handle viscosities up to 200 Poise, and therefore can be used for materials not suitable for ball and sandmills
- Due to the introduction of easily dispersed pigments and fillers, three-roll mills have lost some of their importance. This may change in the future when solventless systems of higher viscosities become more common
- The one-roll mill works on a similar principle but the nip is regulated by a pressure bar. Shearing takes place between the roller and the shearing bar. Stone mills have similar principles of operation. The rotor turns on a stator to achieve shearing
- With current raw materials, both the primary particles and agglomerates are very small, and if any positive action can be achieved during the milling process, it can only be done by affecting these small particles. It is thus necessary to operate these machines at very tight gaps which causes abrasion of the mechanical elements, rapid deterioration of equipment, and contamination of the product by the abraded material. This affects the properties of the millbase and the color of the product
- Shattering of the agglomerates can in most cases be achieved during the premixing step which is a necessary step before milling with all except the ball mills

The high-speed impeller or shear mixer is the most common equipment to prepare dispersions of solids in liquid. High speed shear mills and kinetic shear mills have retained their usefulness because of their ability to deagglomerate material that is not adequately dispersed in the premixing step. A high-speed shear mill is composed of two elements - a container and an impeller. These factors are important in the design:

- The ratio of the impeller and tank diameters should be no more than 1:3, 1:2-2.5 is the most common. The smaller the ratio, the higher the shear
- Charge depth should range from 1.5 to 2 diameters of the impeller[2]
- The impeller should be located at 1/3 of the charge depth from the bottom
- Rotor speed and speed range are critical

- Turbulent flow (high rotational speed) gives the best results when applied at the beginning of the process
- Deflocculation and deagglomeration require shearing process which occurs in laminar flow conditions
- The final dilution of the mixed material requires turbulent flow for good mixing
- In the first stage, the viscosity changes from low to high as fillers are incorporated; in the second stage, viscosity remains constantly high because of the disintegration of particles which occurs during the application of the highest shear stress
- Long mixing increases temperature and decreases viscosity. This does not provide the conditions for the best filler dispersion. By extending mixing over, for example, a 15 min period, the degree of dispersion is not improved, but the resin may actually be degraded
- If the quality of dispersion is not satisfactory, the parameters of mixing should be changed. If the expected result cannot be attained, the range of conditions available is not adequate in this particular mill
- In the third stage, the viscosity changes from high to low due to the addition of diluent. The viscosity range which can be handled by high speed mixers is similar to the range of a three-roll mill, i.e., up to about 200 Poise

The range of shear rates available in high-speed mixers is not broad. The flow rate of fluid in motion decreases as viscosity increases and is inversely proportional to the width of the flow passage which, in this case, is the distance between the disperser and the container which is very large in a high speed mixer. It is not so much due to an improvement in mixing equipment that high-speed mixers have become so popular, it is mostly because of the high quality raw materials (pigments, fillers) which are available now. High structure carbon blacks can be more easily dispersed. But with the increased structure, the size of the primary particles decreases, inhibiting dispersion. Because of the interrelation between both parameters, only the medium structure, coarser particles of carbon blacks can be dispersed by high-speed mixers. Other carbon black types demand further treatment. It should be noted that this is only true of a few fillers which are known to possess strongly bonded, small sized particles. In most cases, fillers can be successfully dispersed in high-speed mixers. However, care should be taken that the filler is selected with an appropriate particle size.

High-speed mixers have several important advantages over other existing equipment including the possibility of processing a batch in the same vessel, easy cleaning, and flexibility in color changes. The main disadvantage is that the final dispersion depends greatly on the chosen composition and technology, and these are sometimes limiting factors. Frequently, the proper conditions for quality dispersion cannot be achieved at all.

The basic construction of a high-speed mixer can easily be modified to one's special requirements. For example, a change from impeller to turbine rotor changes both the principle of dispersion and the range of application. The tangential velocities of filler particles can be as high as 500 m/sec. Such particles have a very high kinetic energy, sufficient to cause size reduction. Size reduction is due to particle-particle or particle-wall collisions, and this in turn, is related in efficiency to the relative velocities at the moment of collision. Relative velocity can be increased by decreasing the viscosity of the millbase. The upper limit of millbase viscosity is somewhere around 3 to 4 Poise. It is not viscosity alone which is important but the entire rheological character of the millbase. The best results are obtained when the millbase is nearly Newtonian. For this reason, the dispersion process is best performed in a diluted millbase. As is the case with high-speed mixers, a proper dispersion should be achieved in a matter of 10-20 min. If such is not the case, the conditions of processing should be modified. Once dispersion has been achieved, it should be stabilized, with the mixer continuously running, by the addition of more resin to increase the viscosity in order to prevent sedimentation or flocculation of the pigment.

The other possible modification to such a mixer can be achieved by a substantial lowering of the speed and a change in the motion of the mixing element to planetary. This configuration can process material of a much higher viscosity, up to several thousand Poise. The high speed mixer can be modified in various ways to match its capabilities to the process requirements. Stationary baffles may be added to increase the shear rate. The distance between the rotating and stationary elements can be decreased again increasing the shear rate. The mixer may be designed to work under both pressure and vacuum and with inert gas blanketing which permits deaeration and processing of volatile or moisture sensitive materials.

The other group includes heavy-duty mixers, such as the Banbury mixer and double-arm kneading mixers. The Banbury mixer with a power input of up to 6000 $kW/m^3$ is the strongest and the most powerful mixing unit used by industry. Nearly solid materials are mixed by a rotor which is a heavy shaft with stubby blades rotating at up to 40 rpm. The clearance between the walls and rotor is very small, which induces a very high shear in the material. The high shear generates a great amount of heat which melts the polymer rapidly and allows for quick incorporation of filler. After the filler is incorporated, the dispersion process begins, with rapid distributive mixing along and between two rotors and between the chamber walls and rotor tips. Within 2-3 min, mixing is normally completed and the compound discharged into a pelletizing extruder or a two-roll mill which converts it to a sheet form.[8] Carbon black, which is most frequently processed in a Banbury mixer, is usually placed between two layers of polymeric material in order to reduce dusting.

Double-arm kneading mixers are very popular in some industries. They consist of two counter-rotating blades in a rectangular trough carved at the bottom to form two longitudinal half cylinders and a saddle section. A variety of blade shapes are used, with a clearance between them and the blades and the side walls of up to 1 mm. The most popular blade shapes include: sigma, dispersion, multiwiping overlapping, single-curve, and double-naben blades. It is important for filler dispersion in this mixer that the viscosity of the millbase be kept high enough to create the required shearing force to disperse the material. The strong construction of the mixer and its high power allow one to work with concentrated compositions of pigments which could not be processed by any other method.

High volume production is more and more frequently done by mixing in an extruder.[11,12] This method offers several advantages such as a continuous process, material uniformity, a clean environment, high output, and low labor. The biggest disadvantage of this method is a high investment cost. The twin-screw extruder is the most flexible type of extruder and most appropriate for compounding. Their screw design can be varied as can the method of dosing and the output rate. The abrasiveness of the filler may affect the life-span of the equipment, and particle size and its distribution may influence the quality of filler dispersion and material uniformity. But in general, there is adequate machinery available for almost all requirements. For instance, glass-fiber reinforced materials can be produced by this technique with little change to the initial structure and dimensions of the glass fibers, which shows the versatility of the technology. The production rate of this method is comparable to the Banbury mixer, and an additional advantage comes from the fact that the material can be completely processed in one pass through the machinery.

Finally, one should mention the press mixer, which is a recent development. A press mixer resembles, in its general principle, the high-speed mixer. It has been developed to deal with the high viscosities and heat generated by the mixing process. The mixer has two shafts: one powering the mixing element, called a mixing tool, the other moving one of the container bottoms. The mixing tool is a very strong mixing element occupying approximately 2-3% of the entire mixer volume. This tool can rotate and can be moved with high speed between both bottoms, creating rapid mixing in the whole volume. The bottom moves axially and, because it is well sealed against the side walls, it exerts pressure on the mixed material, increasing the mixing efficiency because the mixing is done on a compressed material. This mixer is suitable for both liquid and solid materials. It is equipped with a method of removal of heat generated during the mixing process. Both the container sides and bottoms and the mixing tool have refrigerant flowing through them which can cool solid rubber by 50°C in a matter of a few minutes. The order of component addition, which is important in other mixers, is less important. The mixer is simply loaded with all components and content is rapidly mixed to the utmost uniformity by the powerful tools provided. The press mixer may even influence the material selection process because it affects the particle size of the filler.

The importance of the proper dispersion of fillers and the complexity of techniques for measuring the degree of dispersion are reflected in numerous publications. Further information on the mixing of fillers is included in Sections 18.5 (dispersion) and 18.10 (mixing).

## REFERENCES


1 **Weighing, Filling, Bagging**. *Chronos Richardson*, AD/ASM e 7006.
2 **Luftentzug aus fluidisierten Producten bringt**. *Chronos Richardson*. 9/93/2.5.
3 **Handling, Storage, Distribution of China Clay**. Techn. Bull. 5M/1/84. *ECC International Ltd.*, St. Austell, England, 1984.
4 **Bulk Unloading & Storage Systems**. Brochure n. 405. *Premier Pneumatics, Inc.*, 12/96.
5 Foley J, **Smart Force Transducer II**. *K-Tron America*, 4/98
6 **Diverter Valves**. Brochure no. 352, *Premier Pneumatic, Inc.*, 7/96.
7 **Dispersion of Tioxide Pigments in Non-aqueous Media. Techn.** Bull. 875. *Tioxide Int.*, London, 1976.
8 **Dispersion of Carbon Black for Plastics, Inks, Coatings, and Other Special Applications**. Techn. Rep. S-31. *Cabot Corp.*, Boston, 1977.
9 Funt J M, *Rubber World*, **193** (5), 21 (1986).
10 Hess K-M, *Kunststoffe*, **73**, 282 (1983).
11 Jakopin S, *Adv. Chem. Ser.*, **134**, 114 (1974).

# 4

# QUALITY CONTROL OF FILLERS

This chapter contains discussion of analytical methods which are used to determine properties of fillers discussed in Chapter 2. Only a general principle of each method is given. The details of the method can be found in the referenced standards. The goals of the chapter are to:

- Provide background on the data included in the Chapter 2
- Clarify situations where differences exist between standards which may create discrepancies in data presented by different suppliers

The methods are generally very simple. The information gleaned from these tests gives a good indication of the filler's properties but is usually not sufficient to use as a set of data for screening fillers for potential applications. Substantially more information is required to assess quality of particular product.

## 4.1 ABSORPTION COEFFICIENT[1]

The spectrophotometric method measures the amount of light transmitted through a film of ethylene polymer containing carbon black. The absorption of the sample is compared with a standard to evaluate carbon black dispersion and the amount of carbon black.

## 4.2 ACIDITY OR ALKALINITY OF WATER EXTRACT[2-3]

Part 4 of ISO 787 specifies the method of determination and Part 3 specifies how the extract should be prepared. The material for testing is extracted in boiling water for 5 minutes and filtered to obtain a clear filtrate. An aliquot of filtered extract is titrated either with hydrochloric acid or sodium or potassium hydroxide solution in the presence of an indicator or evaluated by potentiometric determination.

The ASTM method differs from ISO in extract preparation which is obtained by a 5 minute extraction at room temperature. The method of determination is based on titration in the presence of an indicator. The acid used for titration is sulfuric acid.

## 4.3 ASH CONTENT[3-4]

The sample of filler or pigment is dried at 105°C to remove water and then ashed at 900-1000°C for a total 30 minutes.[3] This method is mostly used for mineral fillers.

The ash in carbon black is determined after drying at 125°C in a 550°C muffle furnace.[4] The duration in the furnace is up to 16 hours depending on crucible type.

The furnace treatment is continued until a constant weight is obtained unlike in the previous standard where constant (short) ashing time is used. The method permits the use of a microwave furnace which typically shortens the time to 2 to 6 hours.

When the instruments are available, the above methods can be replaced by thermogravimetric analysis which is more informative and simpler to conduct.

## 4.4 BRIGHTNESS[5]

Brightness is the term for a numerical value of reflectance of blue light (400-500 nm) from a sample under 45° illumination. The method is used to compare materials in paper and other industries. The specimen of material is compared in a brightness tester with standard specimens made from paper or opal glass which should be replaced monthly. The method gives results which measure the effectiveness of the bleaching process and accounts for the amount and the type of optical brighteners used. The result is a measure of paper quality and gives an indication of its price.

## 4.5 COARSE PARTICLES[6]

The method is used for determination of the amount of coarse particles in a particulate material or their dispersion. The particles are considered coarse if they do not pass through a 45 μm sieve. The process of sieving is conducted with wet material and it is aided by water flushing and brushing. The material retained on the sieve is determined gravimetrically after drying.

## 4.6 COLOR[2,7]

The Part 1 of ISO 787 gives a color comparison method for pigments and extenders. The specimen and the standard pigment are dispersed in a specific binder under controlled conditions. The resultant pastes of pigments are spread on a substrate and visually compared.

The Part 25 of ISO 787 specifies a colorimetric method of comparison. The similar method of dispersion is used but a more precise definition of binder is given. In addition, fumed silica is used as an ingredient in the dispersion. The results of testing give relative hue and lightness differences for a broad range of materials from white to black.

The ASTM standard specifies details of method which is in principle a method of using color computer to determine CIE tristimulus values and other parameters of color which can be calculated.

Each method discussed in this section has different precision of determination and results are not comparable. In evaluation of this data it is essential to take note of the method used.

## 4.7 CTAB SURFACE AREA[8]

This method gives a specific surface area contained in micropores of carbon black. The micropores cannot be penetrated by hexadecyltrimethylammonium bromide

(CTAB). The method is used to characterize rubber grades of carbon black. A sample of carbon black, previously dried at 125°C is treated with a standard solution of hexadecyltrimethylammonium bromide and mixed to aid its adsorption. The excess of hexadecyltrimethylammonium bromide is determined by titration of the filtrate.

## 4.8 DBP ABSORPTION NUMBER[9]

The method is based on the measurement of the torque required to mix carbon black with n-dibutyl phthalate (DBP). n-dibutyl phthalate is added from a constant rate burette to a powdery sample of carbon black. The end of the titration is detected by reaching a predetermined torque level. The test helps in determining and controlling the quality of carbon black and relating values to its structure. It also helps to predict formulation that will give good processing characteristics. A simplified procedure uses manual mixing of fillers (see oil absorption below) but the results are not comparable.

## 4.9 DENSITY[2,10,11]

Part 10 of ISO 787 gives a pycnometer method of density determination.[2] Two methods are suggested. One method uses simple wet pycnometer in which the sample displaces water or some other liquid and the result is determined by a gravimetric method. The other method uses vacuum to remove air from the sample followed by the introduction of a portion of the liquid under vacuum. There is an inevitable difference in the results and the precision of each method. The differences in the determined values may also come from the choice of liquid used for displacement.

Part 23 of ISO 787 contains a description of an alternative method which allows to remove air entrained in the sample of a powdered material. The powder is placed in a special tube, mixed with an excess of the displacement liquid more than sufficient to cover its surface, and placed in centrifuge to remove air.

The change in a material's density caused by a filler addition can be measured by a method which relies on the change of weight of the material when immersed in a liquid (either water or other liquid). The method discussed here[10] is fast and precise and it is suitable for the determination of density of filled materials.

A Scott volumeter is suggested as being suitable for measuring the density of metal powders.[11] The method gives a bulk density of the metal powder and results can be related to the measurement of tamped volume (see below). The Scott volumeter is more complex and precise than the ISO method. The result is given as apparent density.

## 4.10 ELECTRICAL PROPERTIES[12,13]

Methods of testing conductive materials are used to evaluate specimens containing conductive fillers. Two ASTM standards contain details of specimen testing for resistance[12] and EMI shielding effectiveness.[13] The details of specimen preparation are given.

## 4.11 EXTRACTABLES[14]

The method employs the fact that toluene discolors as it dissolves extractable substances in carbon black.[14] A previously dried sample of carbon black at 125°C is extracted in toluene for 60 seconds, filtered, and its color intensity is measured in a spectrophotometer at 425 nm. The change in transmission of solution of extractables is recorded.

## 4.12 FINES CONTENT[15]

This method determines fines present in pelleted carbon black. Material passing through a 125 μm sieve is considered fines. The material remaining on the sieve is weighed to determine the percent fines.

## 4.13 HEATING LOSS[16]

Heating loss is used to determine moisture content in carbon black. The drying is performed at 125°C for 30 min. Under these conditions moisture is removed but some other volatile materials may also be lost. The automatic equipment such as drying balances is also used (note that carbon black does not absorb infrared rapidly therefore, other sources of heat are normally used). This method gives precise readings because it avoids errors due to reabsorption of moisture.

## 4.14 HEAT STABILITY[2]

Heat stability is determined according to the Part 21 of ISO 787. The specimen is dispersed in a binder and tested in the form of a film having a wet thickness of 75 to 120 μm. The temperature of exposure in a ventilated oven is selected based on the anticipated exposure of the material in its intended application.

## 4.15 HEGMAN FINENESS[17]

This method is used to determine the fineness of grind of a pigment in a vehicle. It uses a gage with a wedge shaped depression which has depth starting at zero and going to 100 μm. The paste material is spread with the use of metal spreader and result read from a scale of 0 to 8 (0 means depth of 100 μm, 3 - 65 μm, 6 - 25 μm, and 8 - 0 μm). The point of termination of the speckled pattern on the surface of the sample is the measure of the fineness of grind.

## 4.16 HIDING POWER[18]

Hiding power of pigment in paint can be measured by reflectometry without the use of standard. It is calculated from the determined values of reflectivity and the scattering coefficient.

## 4.17 IODINE ABSORPTION NUMBER[19]

A carbon black sample is treated with excess of iodine. The excess iodine is then titrated with a sodium thiosulfate solution. The result is expressed as adsorbed iodine per unit of mass of the sample. The iodine number depends on amount of volatiles, surface porosity, and extractables. The iodine number correlates with the nitrogen specific surface area. It is a simple method used to evaluate the quality of carbon black.

## 4.18 LIGHTENING POWER OF WHITE PIGMENTS[2]

Two alternate methods are proposed in the Part 27 of ISO 787. In both cases a standard blue paste is dispersed with the white pigment to be tested in an automatic muller or by hand using a hand muller or a palette knife. In the first method, two sample pastes containing the same amounts of the test and the standard pigment are dispersed. The amount of pigment added is normalized for the frequently used pigments such as zinc oxide, lithopone, and titanium dioxide. The mulled samples are compared for intensity of color. In the second method, the sample is compared with a set (usually five) of standard pigments at different concentrations. From a visual comparison, the match closest to the standard sample is selected and that value is used to calculate the hiding power of pigment which is expressed as a ratio of the weight of pigment in the test to that of the standard sample.

## 4.19 LOSS ON IGNITION[3]

This method of determination is identical to that described above for the method of ash determination.

## 4.20 MECHANICAL AND RELATED PROPERTIES[20-27]

The mechanical properties of filled materials are evaluated using standard methods developed for specific matrix materials. Carbon black is usually evaluated in natural rubber. There is a standard method of sample preparation and tensile strength, modulus, and elongation of the prepared samples are determined.[20] A similar standard was developed for styrene-butadiene rubber.[21] Other materials are tested according to a general standard for plastic materials which gives procedures of testing shrinkage,[22] flexural properties,[23] deflection temperature under load,[24] tensile properties,[25] impact resistance,[26] and compressive strength.[27]

## 4.21 OIL ABSORPTION[2,28]

The Part 5 of ISO 787 gives a method for determining the oil absorption of pigments and extenders.[2] A refined linseed oil is dispersed in small portions from a burette and mixed with powder using palette knife until smooth consistency is obtained. Different amounts of powder are taken depending on the expected oil absorption. Oil absorption is expressed as a percent of the mass of powder.[2]

A simple spatula method is also given by the ASTM standard which is essentially similar to that described above. The only difference is in the method of endpoint detection which in the ASTM standard is a very stiff, putty-like paste. The result is expressed as the amount of oil absorbed by 100 g of powder.

## 4.22 PARTICLE SIZE[29,30]

The average particle size of metal powders is determined by the Fisher sub-sieve sizer. The method uses air permeability to determine particle size. The method is designed for coarser metal powders having particle sizes in the range of 0.2 to 50 μm. The method should not be used for flakes or fibers.

The most frequently used method for particle size distribution is based on an optical particle counter.[30] Determination of monosize particles, flakes, and fibers is not accurate. In these cases either electron or optical microscopy are the most suitable techniques.

## 4.23 PELLET STRENGTH[31]

The automated pellet hardness tester is computer controlled and transports pellets to a measuring gage. The result is given as the force required to crush a pellet of a measured diameter.[31]

## 4.24 pH[2,3,32]

According to the Part 9 of ISO 787, a 10% suspension of filler is made up in freshly distilled water at room temperature and pH measurement of suspension is made.[2]

In an ASTM standard method,[3] a suspension is made with warm water and cooled to room temperature for measurement. An alternative method allows one to use colorimetric indicators in the measurement.

The method developed for carbon black uses either a boiling slurry or a sonically dispersed slurry of carbon black in water.[32]

## 4.25 RESISTANCE TO LIGHT[2]

Resistance to light is determined for pigments dispersed in the material in which they to be used. Two methods of exposure are used: under glass outdoors or in an artificial weathering unit equipped with a xenon arc as a source of radiation. The result of exposure is compared with a standard exposed to the same conditions. The evaluation is based on the color differences between the exposed and shadowed parts of the specimens.

## 4.26 RESISTIVITY OF AQUEOUS EXTRACT[2]

The Part 14 of ISO 787 gives details of the method.[2] A sample is prepared in boiling water. If the filler is hydrophobic some methanol is added to increase its wettability. The extract is filtered, cooled to room temperature, and measured in a conductivity cell. The result is expressed as resistivity.

## 4.27 SIEVE RESIDUE[2,33,34]

Two methods of determining of sieve residue are given in ISO 787. The Part 7 describes manual procedure.[2] A suspension of powder in water is prepared with the aid of a dispersion agent. The suspension is poured onto the sieve and washed with water containing the dispersing agent. The amount of residue is determined by a gravimetric method. The result is given as a percentage of the total mass of the tested powder. The Part 18 describes a mechanical flashing procedure. A system of rotating jets is used for flushing. Other details of the methods are similar.

When determining carbon black residue on a sieve, the method uses water to transfer carbon black to sieve through funnel. The sieve is then flushed with water from rubber hose. The residue is dried at 125°C and the results presented in ppm.[33] A similar method of determination is described for lime and limestone.[34]

## 4.28 SOLUBLE MATTER[2,3]

ISO 787 specifies two methods of determination of matter soluble in water. The Part 3 gives the hot extraction method. The material is boiled in water for 5 min, cooled to room temperature, filtered, extract is evaporated, and soluble matter determined gravimetrically. In Part 8, the cold extraction method is specified. Extraction is done at room temperature for 1 h. The next steps are the same as in hot extraction method.

The ASTM method is the same as hot extraction method in ISO procedure.[3]

## 4.29 SPECIFIC SURFACE AREA[35,36]

Details of several different methods for determining the specific surface area of carbon black are described in ASTM D 3037. The different types of equipment used and procedures are included in separate sections. Another standard[36] gives full details of procedure of conventional Brunauer, Emmett, and Teller (BET) method based on multilayer gas adsorption. The results of determination are in both cases given in the surface area in square meters per gram of substance.

## 4.30 SULFUR CONTENT[37]

Several methods of sulfur determination are used for carbon black. They include oxygen bomb calorimetry, high-temperature combustion with an iodometric detection procedure and an infrared detection procedure.[37] The results are given as percentage of sulfur.

## 4.31 TAMPED VOLUME[2]

The tamped volume or apparent density is determined according to Part 11 of ISO 787. The material is passed through a sieve to disperse agglomerates and placed in tarred graduated measuring cylinder. The cylinder is then placed in a tamping volumeter and tamped for 250 revolutions. The volume read from the cylinder is divided by the mass of powder and given as a percent.

## 4.32 TINTING STRENGTH[2,38-40]

ISO 787 gives a choice of two methods of determination of tinting strength. The visual comparison method is given in Part 16. A standard white paste is prepared either with a mechanical muller or spatula mixing. In a similar method, tinting pastes of a standard pigment and the test pigment are prepared. The pastes are mixed in the right proportions with white pigment paste and their tinting strength and undertones compared visually. Part 24 describes a photometric method. In essence, the method is the same but in place of a visual comparison, tristimulus values are measured or samples are measured at 550 nm.

For printing ink dispersions, either a visual comparison is made or the tinting strength is calculated according to the equation from spectrophotometric data.[38] The specimen is prepared by mixing tinting paste with base and comparing the result with a standard tinting paste mixed in the same proportions.

A carbon black test sample is obtained by mixing carbon black and zinc oxide with epoxidized soybean oil. The mixture is milled in a mechanical muller with frequent scraping. The specimen is prepared by film drawdown, roller spreader or by the glass slide method. Reflectometer readings are obtained. The result is a comparison of the tint strength of standard with the test sample expressed in tint units.[39]

White pigments are measured in compositions containing a black letdown vehicle using a reflectance measurement. The test pigment is compared with a standard sample.[40]

There is much compositional freedom in these methods which makes a comparison of results from different sources very difficult and unreliable.

## 4.33 VOLATILE MATTER[2]

The volatile matter according to the Part 2 of ISO 787 is determined gravimetrically by weighing the sample to a constant mass after a series of drying intervals at 105°C.[2]

## 4.34 WATER CONTENT[3]

The water content is determined by azeotropic distillation in the Dean-Stark apparatus.[3]

## 4.35 WATER-SOLUBLE SULFATES, CHLORIDES AND NITRATES[2]

Part 13 of ISO 787 determines water-soluble sulfates, chlorides and nitrates. The sample extract can be prepared by either cold or hot extraction method described in Section 4.28. The sulfates in the extract are determined by precipitation with barium chloride, the chlorides are determined by titration with silver nitrate, and the nitrates are determined by a colorimetric method using Nessler reagent.[2] Part 19 gives an alternative method of determination of nitrates by a salicylic acid method.

## REFERENCES


1 ASTM D 3349-93. Absorption coefficient of ethylene polymer materials pigmented with carbon black.
2 ISO 787. General methods of test for pigments and extenders.
3 ASTM D 1208-96. Common properties of certain pigments.
4 ASTM D 1506-95. Carbon black – ash content.
5 ASTM D 985-93. Brightness of pulp, paper, and paper board (directional reflectance at 457 nm).
6 ASTM D 185-95. Coarse particles in pigments, pastes, and paints.
7 ASTM E 308-96. Computing the colors of objects by using CIE system.
8 ASTM D 3765-96. Carbon black – CTAB (cetyltrimethylammonium bromide) surface area.
9 ASTM D 2414-96. Carbon black – n-dibutyl phthalate absorption number.
10 ASTM D 792-91. Density and specific gravity (relative density) of plastics by displacement.
11 ASTM B 329-95. Apparent density of metal powders and compounds using the Scott volumeter.
12 ASTM D 257-93. DC resistance or conductance of insulating materials.
13 ASTM D 4935-94. Measuring the electromagnetic shielding effectiveness of planar materials.
14 ASTM D 1618-97. Carbon black extractables – toluene discoloration.
15 ASTM D 1508-93. Carbon black, pelleted – fines content.
16 ASTM D 1509-95. Carbon black – heating loss.
17 ASTM D 1210-96. Fineness of dispersion of pigment-vehicle systems by Hegman-type gage.
18 ASTM D 2805-96. Hiding power of paints by reflectometry.
19 ASTM D 1510-96. Carbon black - iodine absorption number.
20 ASTM D 3192-96. Carbon black evaluation in NR (natural rubber).
21 ASTM D 3191-96. Carbon black in SBR (styrene-butadiene rubber) - recipe and evaluation procedure.
22 ASTM D 955-96. Measuring shrinkage from mold dimensions of molded plastics.
23 ASTM D 790-96. Flexural properties of unreinforced and reinforced plastics and electrical insulating materials.
24 ASTM D 648-96. Deflection temperature of plastics under flexural loaf.
25 ASTM D 638-96. Tensile properties of plastics.
26 ASTM D 256-93. Determining the pendulum impact resistance of notched specimens of plastics.
27 ASTM C 695-95. Compressive strength of carbon and graphite.
28 ASTM D 281-95. Oil absorption of pigments by spatula rub-out.
29 ASTM B 330-93. Average particle size of powders of refractory metals and their compounds by the Fisher sub-sieve sizer.
30 ASTM F 661-92. Particle count and size distribution measurement in batch samples of filter evaluation using an optical particle counter.
31 ASTM D 5230-96. Carbon black – automated individual pellet crush strength.
32 ASTM D 1512-95. Carbon black – pH value.
33 ASTM D 1514-95. Carbon black – sieve residue.
34 ASTM C 110-96. Physical testing of quicklime, hydrated lime and limestone.
35 ASTM D 3037-93. Carbon black – surface area by nitrogen adsorption.
36 ASTM D 4820-96. Carbon black – surface area by multipoint BET nitrogen adsorption.
37 ASTM D 1619-94. Carbon black – sulfur content.
38 ASTM D 2066-97. Relative tinting strength of paste-type printing ink dispersions.
39 ASTM D 3265-96. Carbon black – tint strength.
40 ASTM D 2745-93. Relative tinting strength of white pigments by reflectance measurements.

# 5

# PHYSICAL PROPERTIES OF FILLERS AND FILLED MATERIALS

The following information is analyzed in this chapter:

- Physical properties of fillers
- The effect of physical properties of fillers on the properties of filled materials
- The universal principles governing the relationships between the properties of fillers and effect of fillers on the properties of filled materials.

Some examples are given to illustrate the nature of these relationships and the effects obtained.

## 5.1 DENSITY[1-15]

The data in Table 5.1 show that the range of densities of fillers is very wide ranging from 0.03 to 19.36 g/cm$^3$. If we allow that air can also be considered a filler and platinum may be potentially applied in conductive materials, fillers occupy the full spectrum of density of known materials. But it is apparent from the table that most fillers have densities in a range from 2 to 3 g/cm$^3$.

The effect of filler density on the density of filled product can be closely approximated by the additivity rule. If a more precise method of density estimation is required or filler/matrix mixtures are far from being perfect, several corrections are necessary. System density becomes nonlinear close to the critical volume concentration (CVC). The critical volume concentration determines the amount of conductive filler which rapidly increases the conductivity of the composite. Figure 5.1 shows that at, or close to the critical volume concentration, density decreases. This density difference can be detected either after the CVC (polyethylene), before (polystyrene) or the two depressions are observed – one before and one after the CVC (polymethylmethacrylate) is reached.[15] This density depression is due to filler-matrix interaction.

## Table 5.1. Density of fillers

| **Density range**, g/cm$^3$ | **Fillers** (filler density is given in parentheses) |
|---|---|
| 0.1-0.39 | expanded polymeric microspheres (0.03-0.13), hollow glass beads (0.12-1.1), thin-wall, hollow ceramic spheres (0.24) |
| 0.4-0.69 | wood flour (0.4-1.35), porous ceramic spheres (0.6-1.05), silver coated glass beads (0.6-0.8) |
| 0.7-0.99 | thicker wall, hollow ceramic spheres (0.7-0.8), polyethylene fibers and particles (0.9-0.96) |
| 1-1.99 | cellulose fibers (1-1.1), unexpanded polymeric spheres (1.05-1.2), rubber particles (1.1-1.15), expanded perlite (1.2), anthracite (1.31-1.47), aramid fibers (1.44-1.45), carbon black (1.7-1.9), PAN-based carbon fibers (1.76-1.99), precipitated silica (1.9-2.1), pitch-based carbon fibers (1.9-2.25) |
| 2-2.99 | fumed and fused silica (2-2.2), graphite (2-2.25), sepiolite (2-2.3), diatomaceous earth (2-2.5), fly ash (2.1-2.2), slate flour (2.1-2.7), PTFE (2.2), calcium hydroxide (2.2-2.35), silica gel (2.2-2.6), boron nitride (2.25), pumice (2.3), attapulgite (2.3-2.4), calcium sulfate (2.3-3), ferrites (2.3-5.1), cristobalite (2.32), aluminum trihydroxide (2.4), magnesium oxide and hydroxide (2.4), unexpanded perlite (2.4), solid ceramic spheres (2.4-2.5), solid glass beads (2.46-2.54), kaolin and calcinated kaolin (2.5-2.63), silver coated glass spheres and fibers (2.5-2.8), glass fibers (2.52-2.68), feldspar (2.55-2.76), clay (2.6), hydrous calcium silicate (2.6), vermiculite (2.6), quartz ans sand (2.65), pyrophyllite (2.65-2.85), aluminum powders and flakes (2.7), talc (2.7-2.85), nickel coated carbon fiber (2.7-3), calcium carbonate (2.7-2.9), mica (2.74-3.2), zinc borate (2.8), beryllium oxide (2.85), dolomite (2.85), wollastonite (2.85-2.9), aluminum borate whiskers (2.93) |
| 3-4.99 | zinc stannate and hydroxystannate (3-3.9), silver coated aluminum powder (3.1), apatite (3.1-3.2), barium metaborate (3.3), titanium dioxide 3.3-4.25), antimony pentoxide (3.8), zinc sulfide (4), barium sulfate and barite (4-4.9), lithopone (4.2-4.3), iron oxides (4.5-5.8), sodium antimonate (4.8), silver coated inorganic flakes (4.8), molybdenum disulfide (4.8-5) |
| 5-6.99 | antimony trioxide (5.2-5.67), zinc oxide (5.6) |
| 7-8.99 | nickel powder and flakes (8.9), copper powder (8.92) |
| 9 and above | silver coated copper powders and flakes (9.1-9.2), molybdenum powder (10.2), silver powder and flakes (10.5), gold powder (18.8), tungsten powder (19.35) |

Figure 5.2 shows the influence of filler concentration on the density of polymer calculated from the following equation:

$$d_{p,p} = \frac{d_c - d_{MF}V_{MF}}{1 - V_{MF}} \qquad [5.1]$$

where
$d_{p,p}$ density of polymer
$d_c$ density of composite
$d_{MF}$ density of filler
$V_{MF}$ volume fraction of filler

Below the critical concentration of filler some polymer is converted to the interphase layer where the polymer has a higher density because of closer packing,

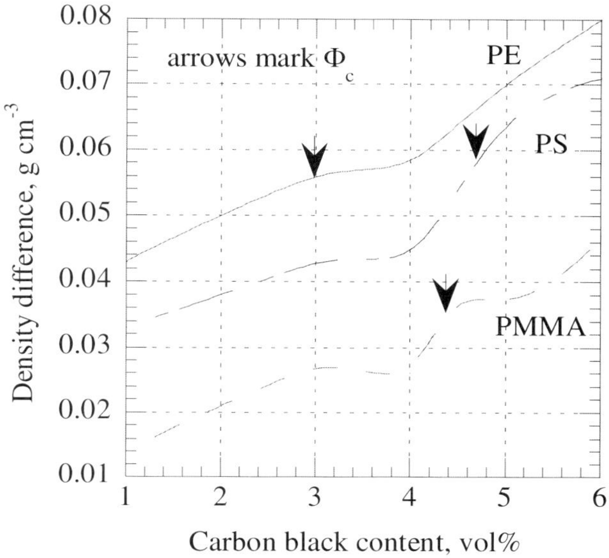


Figure 5.1. Density of composite vs. concentration of carbon black around the CVC. [Data from Weeling B, Electrical Conductivity in Heterogeneous Polymer Systems. Conductive Polymers, Conference Proceedings, 1992, Bristol, UK.]

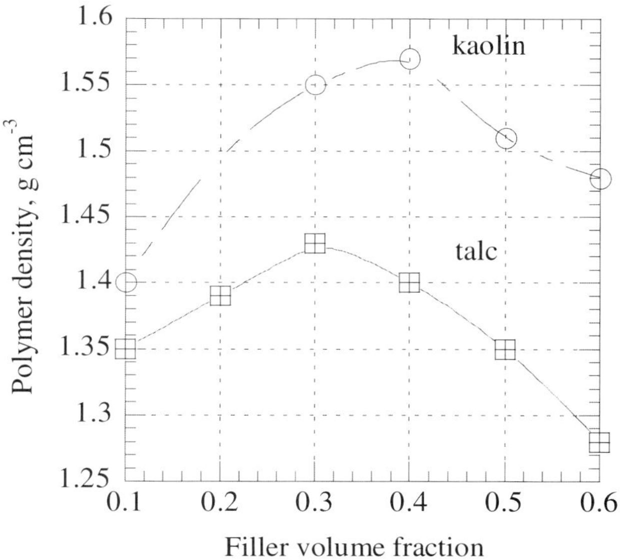


Figure 5.2. Polymer density vs. volume fraction of filler. [Adapted, by permission, from Magrupov M A, Umarov A V, Saidkhodzhaeva K S, Kasimov G A, *Int. Polym. Sci. Technol.*, **23**, No.1, 1996, T/77-9.]

therefore the density of the polymer increases. Above the critical concentration of filler, there is not enough polymer to cover the surface which increases the free volume and the density of the composite decreases.[10]

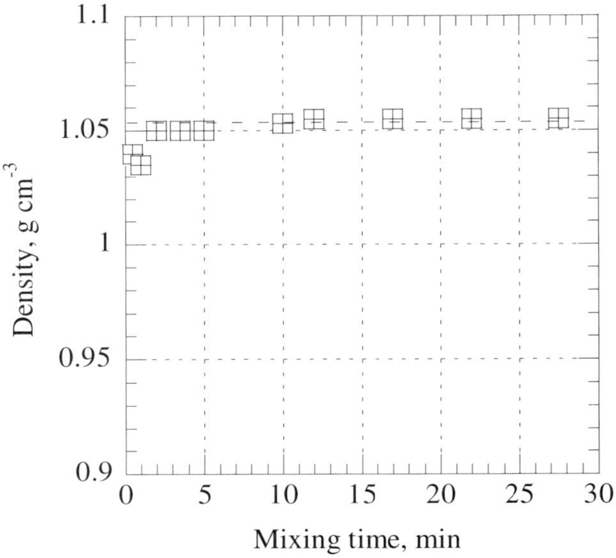


Figure 5.3. Density of SBR containing 30 phr carbon black vs. mixing time. [Adapted, by permission, from Clarke J, Freakley P K, *Rubb. Chem. Technol.*, **67**, No.4, 1994, 700-15.]

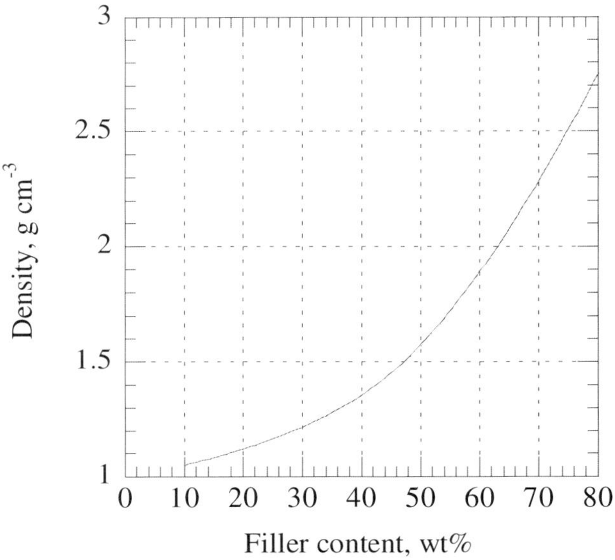


Figure 5.4. Density of copper/polyamide composite vs. filler content. [Data from Larena A, Pinto G, *Polym. Composites*, **16**, No.6, 1995, 536-41.]

Figure 5.3 shows the effect of mixing on the density of composite. The line gives the theoretical density of the composite calculated by the additivity rule. The density of composite at 0 mixing time was calculated assuming that the DBPA value for carbon black was equivalent to the air content of the carbon black pellets. The graph shows that the ultimate density is approached at a very early stage of the mixing process.

Composite density can be expected to vary because of the uneven distribution of filler particles in the manufactured product. This is very typical of the injection molding process where filler is distributed in a complex pattern of flow. In glass reinforced polystyrene parts, manufactured by injection molding, the density varied between 0.9 and 1.4 g/cm$^3$ depending on the process conditions and locations from which the sample was taken.[7]

The other reason for variable density is traced to air voids in the material, related to the method of filler incorporation. Figure 5.4 shows the relationship of recorded densities for copper particles of different sizes in polyamide. The particle size did not have an influence. The variations were related to incorporation methods and filler content. The lines show calculated densities at different void volume contents. The void volume content varied between 10 and 20%.[8]

## 5.2 PARTICLE SIZE[16-46]

According to the data in Table 5.2, only primary particles of fumed and precipitated silica and ultrafine titanium dioxide are produced in sizes lower than 10 nm. The next group includes nanoparticles which are manufactured by chemical methods and metal evaporation techniques combined with oxidation. Mineral fillers of the smallest particle sizes belong to the group of particles with a size above 100 nm. All pigments also belong to the same group (0.1-0.5 μm) together with some synthetic fillers. Metal powders have still larger particles above 0.5 μm. The fillers used in the largest quantities have particles in the range of 1-10 μm. The largest particles are produced for materials used either for decoration (e.g., sand in stucco), as an inexpensive products (e.g., sand in unsaturated polyester composites), or are composed of materials difficult to pulverize (rubber particles).

It is apparent from the data that particles of a few nanometers in size can only be made on industrial scale by synthetic methods. On the other hand, these particles are either intentionally or unintentionally aggregated and agglomerated in their powder forms. Thus, for the dispersion of fillers, agglomerate and aggregate size is usually as relevant as the primary particle size. Fillers, which are obtained by various milling and classification processes, can also be obtained in the form of small particles, but usually not below 100 nm.

The most difficult part of particle size estimation is related to the determination methods themselves. Particle size determination is complicated by size distribution, the presence of particle associations, and the shape of particles. If particles are not spherical, more than one parameter is needed to describe them and if the shape of the particle is irregular, numerous parameters are needed to express their dimensions. The method used for particle size determination (sieving, light scattering, microscopy, etc.) determines what dimensional aspects are measured. In addition, different methods are more useful than the others for the determination of particles in certain size ranges. All these procedural difficulties make it difficult to find a precise method. The more precise analysis can only be done within the scope of a well controlled experiment aimed at understanding a certain property. Particle size is, however, the one property of a filler that influences every aspect of its use and the success of many applications. In view of the fact that there is no general way of dimensioning filler particles we will deal with the particle size of specific fillers throughout the book and make no attempt here to deal with specifics.

**Table 5.2. The average particle size of different fillers**

| **Particle size range**, μm | **Filler** (the range of the average particle sizes for a particular filler is given in parentheses) |
|---|---|
| below 0.01 | primary particles of fumed silica (0.005-0.04), primary particles of precipitated silica (0.005-0.1), ultrafine titanium dioxide (0.008-0.04) |
| 0.011-0.03 | aluminum oxide (0.013-0.1), carbon black (0.14-0.25), precipitated calcium carbonate (0.02-0.4), colloidal antimony pentoxide (0.025-0.075), iron oxide nanoparticles (0.026) |
| 0.031-0.06 | zinc oxide (0.036-3), ferrites (0.05-14) |
| 0.061-0.1 | barium titanate (0.07-2.7) |
| 0.1-0.5 | blanc fixe (0.1-0.7), attapulgite (0.1-20), bentonite (0.18-1), titanium dioxide pigment (0.19-0.3), antimony trioxide (0.2-3), kaolin (0.2-7.3), aggregates of fumed silica (0.2-15), calcium carbonate (0.2-22), silver powders and flakes (0.25-25), zinc sulfide (0.3-0.35), ball clay (0.4-5), molybdenum disulfide (0.4-38), magnesium hydroxide (0.5-7.7) |
| 0.6-1 | zinc borate (0.6-1), lithopone (0.7), aluminum trihydroxide (0.7-55), tungsten powder (0.7-18), gold powder (0.8-9), iron oxide (0.8-10) |
| 1-5 | precipitated silica agglomerates (1-40), ceramic beads (1-50), talc (1.4-19), copper powder (1.5-5), silica gel (2-15), quartz (tripoli) (2-19), sand (2-3000), nickel powder (2.2-9), zinc stannate (2.5), barites and synthetic barium sulfates (3-30), feldspar (3.2-14), diatomaceous earth (3.7-24.6), fly ash (4), fused silica (4-28), mica (4-70), calcium hydroxide (5), sepiolite (5-7), PTFE (5-25) |
| 6-10 | unexpanded polymeric spheres (6-35), graphite (6-96), glass beads (7-8) |
| 10-100 | aluminum powder (10-23), antimony pentoxide (10-40), wood flour (10-100), perlite (11-37), expanded polymeric spheres (15-140), beryllium oxide (20), apatite (43) |
| above 100 | porous ceramic beads (100-350), rubber particles (100-2000), coarse sand (500-3000) |

## 5.3 PARTICLE SIZE DISTRIBUTION[17,28,30,33,35,45,47-56]

Figure 5.5 compares two grades of kaolin manufactured in a form of slurry. A medium particle size kaolin (Britefil 80 Slurry) is used in the paper industry where small particle size is not critical. Another grade of kaolin (Royal Slurry) is used in

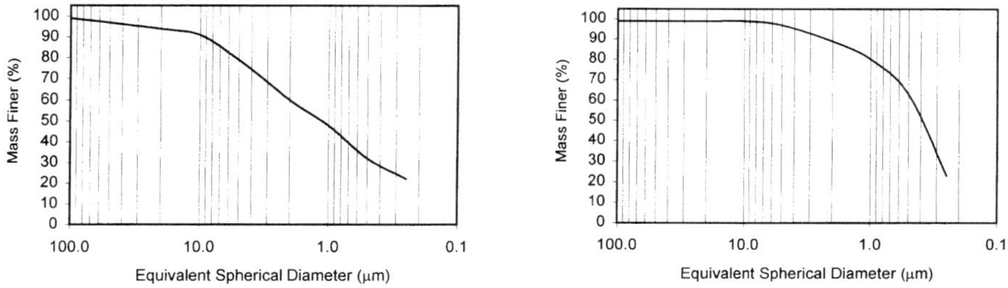


Figure 5.5. Particle size distribution of Britefil 80 Slurry (left) and Royal Slurry (right). *Courtesy of Albion Kaolin Co., Hephzibah, GA, USA.*

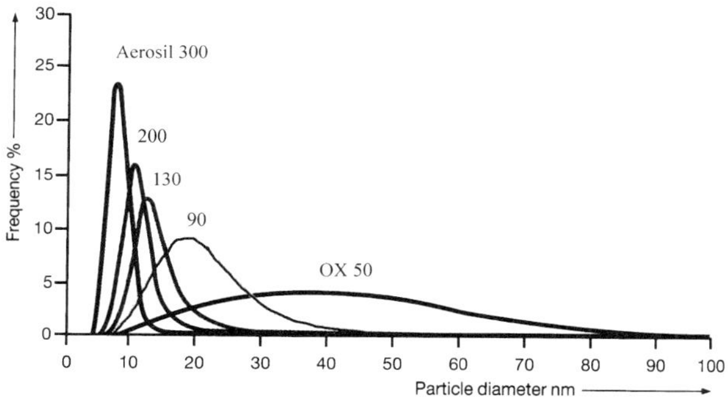


Figure 5.6. Particle size distribution of different grades of Aerosil. *Courtesy of Degussa AG, Frankfurt/Main, Germany.*

specialty applications where fine grade is needed. This grade is milled to a smaller particle size and stabilized with a dispersant. This example shows that milling technology is capable of tailoring particle size distribution to requirements.

Figure 5.6 shows that pyrogenic manufacturing gives excellent control over particle size distribution and median particle size. These grades of fumed silica differ in properties and require a different technological approaches to their dispersion since small particle size filler is more difficult to disperse. At the same time, smaller particle sizes give more transparent products and better reinforcement.

Figure 5.7 shows particle size distribution of synthetic barium sulfate. The characteristic feature of these curves is their steepness which denotes a very narrow particle size distribution which was obtained by controlling the conditions of precipitation. The development of this kind of particle size distribution in a small particle sized filler allows for substantial improvement in the gloss of coatings.

Similar benefits can be shown with the talcs presented in Figure 5.8. The following are the properties of these talcs related to their particle size distribution:

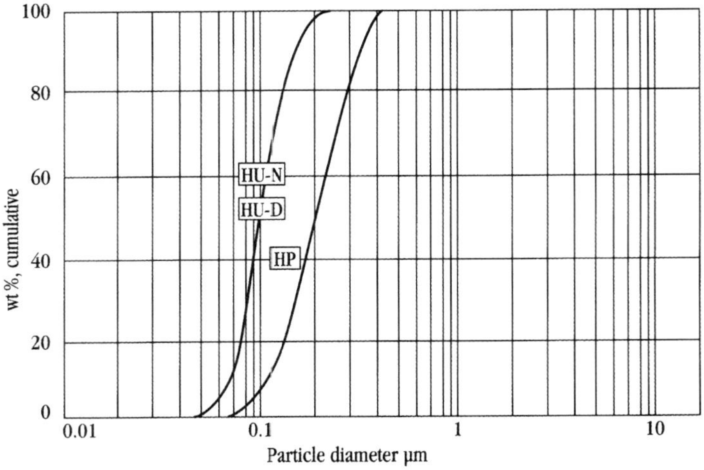


Figure 5.7. Particle size distribution of Sachtoperse. *Courtesy of Sachtleben Chemie GmbH, Duisburg, Germany.*

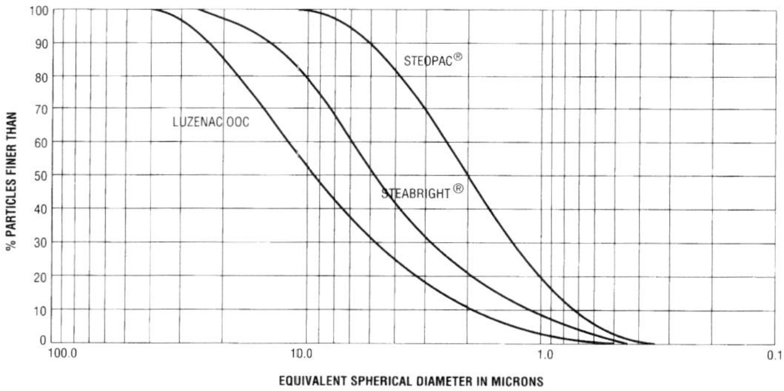


Figure 5.8. Particle size distribution of different talcs. *Courtesy of Luzenac Group, Toulouse, France.*

| | *Luzenac 00C* | *Steabright* | *Steopac* |
|---|---|---|---|
| Whiteness, % | 84.2 | 87.7 | 88.9 |
| Oil absorption, g/100 g | 35 | 50 | 62 |
| Opacity | 0.99 | 0.992 | 0.995 |
| Matting (85° sheen) | 1.4 | 1.8 | 2.8 |

The three talcs have the same composition (talc: 40-41%, chlorite: 57-59%). The differences in properties can be attributed to the way in which they were processed.

A general conclusion from this is that industry can manufacture a variety of particle size distributions tailored to the requirements of the application. Particle size distribution is controlled by the technological parameters of filler production and the methods of classification as well as blending.

Graphing does not always provide the best means of comparing particle size distribution unless the materials are very divergent (as the selected examples). A mathematical form of data presentation is sometimes more convenient. Granulometry in number and in weight is calculated from the following equations:[54]

$$L_n = \frac{\sum_i d_i \times n_i}{\sum_i n_i} \qquad L_w = \frac{\sum_i d_i^2 \times n_i}{\sum_i d_i \times n_i} \quad [5.2]$$

where:
$d_i$ particle diameter
$n_i$ number of particles

The results are either expressed as a ratio - $L_w/L_n$ or a dispersity factor is calculated:

$$D = \frac{L_w - L_n}{L_n} \quad [5.3]$$

In a study of the synthesis of a monodisperse colloidal silica, it was possible to control the particle size distribution.[45] A range of products was obtained with ratios $L_w/L_n$=1.03-33. This again shows that it is possible to tailor particle size to the requirements. We now need to determine what the ratio should be and why.

In plastic products, the particle size distribution of the filler has influence on viscosity and on the amount of filler which can be incorporated. The obvious benefits of mixing particles of different sizes are discussed below. This inevitably leads to a discussion of packing density and critical pigment volume concentration. In some plastics, a certain stress distribution is required and, in such cases, monodisperse, spherical particles are best.

Fillers may also play the role of a pigment and when they do, the particle size distribution is important for several reasons. Figure 5.9 shows that the tint strength and opacity depend particle size. This graph is based on the following relationship developed from the scattering theory of Mie:

$$d_{opt} \sim \frac{\lambda}{1.6(n_p - n_B)} [nm] \quad [5.4]$$

where:
$d_{opt}$ optimum particle diameter
$\lambda$ wavelength of the incident light
$n_p$ refractive index of pigment
$n_B$ refractive index of matrix

According to this relationship there is a direct interdependence between scattering power and particle diameter. This equation suggests that pigment having different particle size distributions may have different scattering properties not only in terms

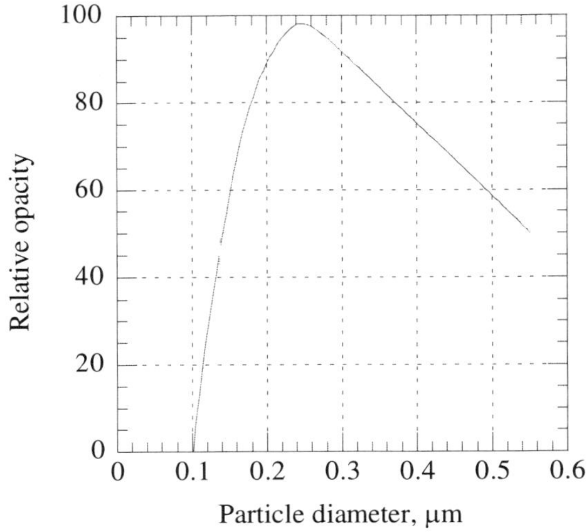


Figure 5.9. Relative opacity vs. particle diameter.

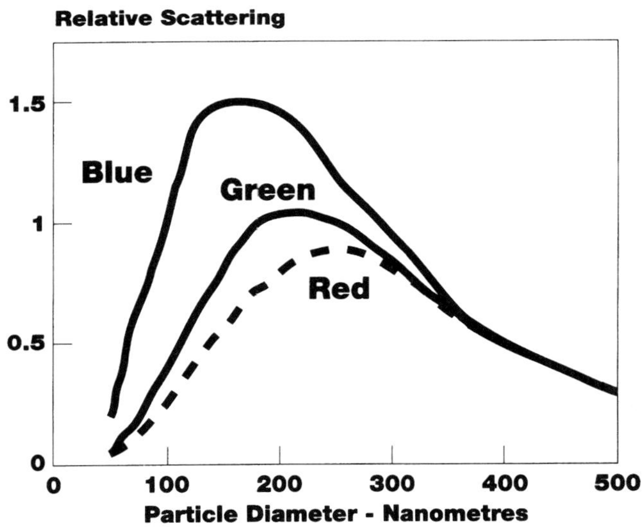


Figure 5.10. Scattering of rutile titanium dioxide. *Courtesy of Millennium Inorganic Chemicals, Auburn, Australia.*

of hiding and opacity but also may influence the color of reflected light. Figure 5.10 shows the effect of particle diameter on scattering of blue, green and red light. Changes to the particle size distribution will change the undertone of the pigment allowing a system to be tailored to the requirements. Certain grades may be capable of providing optical brightening or of masking the yellow color.

## 5.4 PARTICLE SHAPE[23,45,57-59]

The morphology of filler particles can be compared using the SEM and TEM micrographs included in Chapter 2. Here, only summary is included in the form of table (Table 5.3).

### Table 5.3. Typical shapes of fillers particles

| Shape | Filler examples |
|---|---|
| spherical | aluminum powder, aluminum oxide, carbon black, ceramic beads, copper powder, fumed silica, glass beads, silver powder, titanium dioxide, zinc oxide |
| cubic | calcium hydroxide, calcium hydroxide, feldspar |
| tabular | barite, feldspar, sand |
| dendritic | copper powder, nickel powder |
| flake | aluminum flake, graphite, kaolin, mica, perlite, tripoli, sliver flake, talc, vermiculite |
| elongated | aluminum borate whisker (ribbons or cylinders), aramid (fiber), attapulgite (needle), carbon fiber, cellulose fiber , glass fiber, titanium dioxide (acicular), wood flour (fiber), wollastonite (acicular) |
| irregular | aluminum oxide, aluminum hydroxide, anthracite, attapulgite, barite, calcium carbonate, clay, dolomite, fly ash, magnesium hydroxide, perlite, precipitated silica |

Each particle shape brings with it certain advantages. Spherical particles give the highest packing density, a uniform distribution of stress, increase melt flow and powder flow, and lower viscosity. Cubic and tabular shapes give good reinforcement and packing density. Dendritic particles have a very large surface area available for interaction. Flakes have large reflecting surfaces, facilitate orientation, and lower the permeability of liquids, gases and vapors. Elongated particles give superior reinforcement, reduce shrinkage and thermal expansion and facilitate thixotropic properties. Irregular particles may not possess special advantages but they are generally easier to make and are thus inexpensive fillers. These properties are discussed in other chapters of this book.

## 5.5 PARTICLE SURFACE MORPHOLOGY AND ROUGHNESS[23,58,60-68]

The particle surface of mineral fillers can be estimated from a knowledge of the crystal structure, since the milling process cleaves the crystals according to a typical pattern of cleavage for a particular mineral. Many crystals, particularly these of mineral origin, cleave in only one direction and form plate like particles. Table 5.4 summarizes the crystal structure and cleavage pattern of some fillers of mineral origin.

The information in the table shows that the shape of filler particles is determined by their crystal structure and cleavage. The surface area of crystal is increased by milling but it retains the original features of the mineral. This information can be compared with micrographs in Chapter 2. The interactions that may occur on such a filler surface depend on the crystal structure which dictates a defined pattern of chemical organization and on the functional groups which are available on the surface for the eventual reaction with the matrix.

### Table 5.4. Crystal structure and cleavage pattern of selected mineral fillers

| **Crystal structure** | **Fillers** (typical cleavage is given in parentheses) |
|---|---|
| hexagonal | apatite (indistinct in one basal direction), graphite (perfect in one direction), kaolin |
| monoclinic | aluminum trihydroxide (one direction), attapulgite, bentonite (perfect in one direction), calcium sulfate (one direction and distinct in two others), feldspar (good in 2 directions forming nearly right angled prisms), mica (perfect in one direction producing thin sheets or flakes), pyrophyllite (perfect in one direction), talc (perfect in one direction, basal), vermiculite |
| orthorhombic | barite (perfect in one direction, less so in another direction), calcium carbonate - aragonite (one direction), sepiolite |
| tetragonal | cristobalite (absent) |
| triclinic | feldspar (perfect in one and good in another direction forming nearly right angled prisms), wollastonite (perfect in two directions at near 90 degrees forming prisms with a rectangular cross-sections) |
| trigonal | calcium carbonate - calcite (perfect in three directions, forming rhombohedrons), dolomite (perfect in three directions forming rhombi), quartz |

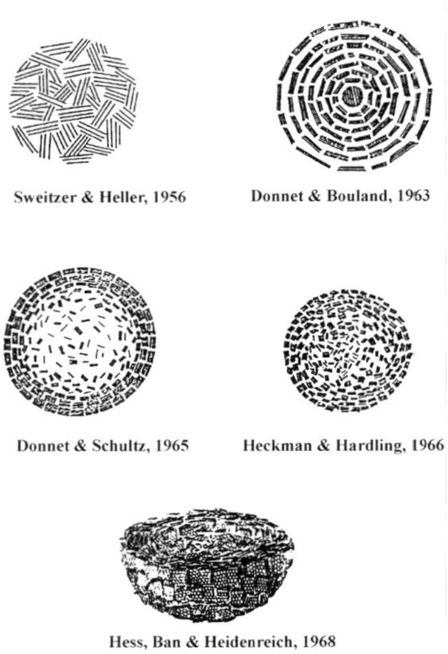


Figure 5.11. The models of carbon black particles. [Adapted, by permission, from Donnet J B, *Kaut. u. Gummi Kunst.*, **47**, No.9, 1994, 628-32.]

In synthetic materials, the surface organization also depends on the internal structure of particles. Carbon black is good example. Figure 5.11 shows the models of carbon black primary particles. The most recent model developed by Hess, Ban and Heidenreich is commonly accepted as being characteristic of carbon black particles. The particle is composed of small elements which are interconnected to form quasi-spherical particles.[58] Recent studies[62] indicate that the core of the particle is less dense and filled with voids but organized in such a way that graphitic scales form on the surface which makes surface rough and accommodating to polymer chains. In a different process where carbon and aramid fiber are formed there are also numerous imperfections on the surface.[63,65] With the advent of atomic force microscopy these imperfections can now be observed and surface roughness can be estimated in numerical form. This surface roughness is important in the development of adhesive forces between the filler and matrix.

The surface roughness of filled materials is obviously not related to filler surface imperfections but it is very much determined by the shape of filler particles.[60,64] The effect of glass fibers in plastics and flatting agents are specific examples of the influence of specific shaped particles on surface roughness.

## 5.6 SPECIFIC SURFACE AREA[69-80]

Specific surface area is a convenient method of characterizing fillers. The results can be correlated to many performance characteristics and to the properties of filled systems. Table 5.5 gives a summary of the specific surface area of some fillers.

**Table 5.5. Specific surface area of some fillers**

| **Specific surface area range**, $m^2/g$ | **Filler** (the range of specific surface area for the filler is given in parentheses) |
|---|---|
| 0-0.49 | aluminum oxide (0.3-1), aluminum trihydroxide (0.1-12), aramid fibers (0.2), barium sulfate (0.4-31), carbon fibers (0.2-1), ceramic beads (0.1-1), glass beads (0.4-0.8), gold powder and flakes (0.05-0.8), pumice (0.4-0.6), sand (0.3-6), silver powder and flakes (0.15-6), wollastonite (0.4-5) |
| 0.5-0.99 | bentonite (0.8-1.8), boron nitride (0.5-25), cristobalite (0.4-7), diatomaceous earth (0.7-180), feldspar (0.8-4), nickel powder and flake (0.6-0.7), fused silica (0.8-3.5) |
| 1-4.99 | aluminum borate whisker (2.5), antimony trioxide (2-13), barium titanate (2.4-8.5), calcium hydroxide (1-6), lithopone (3-5), magnesium hydroxide (1-30), talc (2.6-35), cellulose fibers (1) |
| 5-9.99 | aluminum powder and flakes (5-35), calcium carbonate (5-24), graphite (6-20), kaolin (8-65), titanium dioxide (7-162), zinc sulfide (8) |
| 10-49.99 | clay (18-30), nanosize iron oxide (30-60), precipitated silica (12-800), silica gel (40-850), thermal and lamp carbon blacks (10-30), zinc oxide (10-45) |
| 50-99.99 | acetylene carbon blacks (65-80), furnace carbon black (50-1475), fumed silica (50-400) |
| above 100 | activated alumina (220-325), attapulgite (120-400), ferrites (210-6000), hydrous calcium silicate (100-180), sepiolite (240-310) |

Larger, non-porous particles, such as metals, particles fused by heat, glass spheres, have the lowest specific surface areas. These are followed by mineral particles especially from minerals which cleave to the smooth surfaces of crystals. Fillers which have small particles but are not very porous occupy the middle range of specific surface area. Very small particles, formation of aggregates, and minerals of high porosity give fillers having the highest specific surface areas.

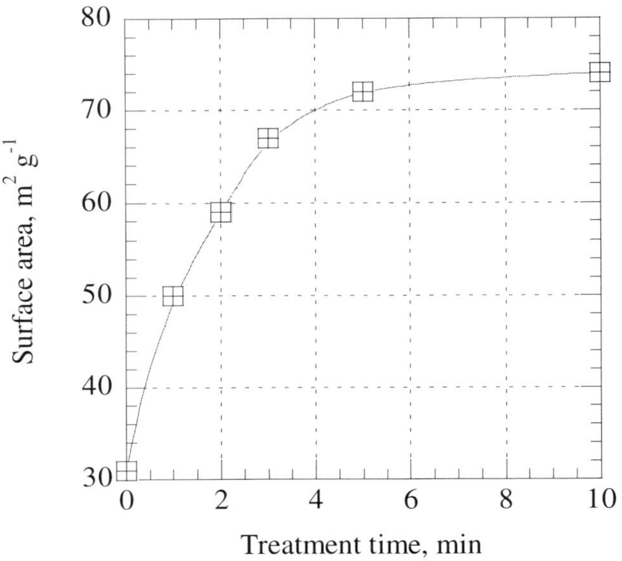


Figure 5.12. Specific surface area of carbon fibers vs. treatment time in oxygen plasma. [Adapted, by permission, from Byung Suk Jin, Kwang Hee Lee, Chul Rim Choe, *Polym. Int.*, **34**, No.2, 1994, 181-5.]

From this short analysis, it is evident that specific surface area comprises the total surface of particles including its pores and includes at least part of the free volume in aggregates. For non-porous particles it is useful for calculation of the average particle size. It is also used to calculate the average particle size of materials (such as for example carbon black) which are porous but for which particle size cannot be more precisely determined because of the effect of its structure.

Specific surface area, related to the particle size is a very important parameter. As with particle size, it is useful in helping us to understand how the properties of filled materials are so strongly influenced by fillers.

The specific surface area depends on filler treatment. The treatment of carbon based materials is one of such examples (Figure 5.12).[72] Surface oxidation increases the specific surface area of carbon fibers.

## 5.7 POROSITY[24,39,69,81-87]

The two extreme cases are zeolite (the smallest pore size) and diatomaceous earth (the largest volume of pores). Zeolites are manufactured with predesigned pore sizes to match the sizes of molecules which can fit into these pores and become absorbed into the pore area. Applications for zeolites include moisture scavenging and selective absorption of various chemical components of mixtures. Diatomaceous earth at the other end of the scale is not selective at all. The large number of pores allows it to absorb 190-600% of its own mass. Applications include absorption of liquids and regulation of rheological properties. The mechanism of rheological control is simple. When the liquid and diatomaceous earth is mixed and left to stand, the liquid flows into the pores and the viscosity of mixture increases. But when it is mixed again the liquid flows out of the pores and the viscosity drops.

**Table 5.6. Pore volume and size of some fillers**

| Filler | Pore volume, $cm^3/g$ | Pore diameter, nm |
|---|---|---|
| Aluminum oxide | | 5.8-24 |
| Aluminum oxide[82] | | 5.8 |
| Calcite[24] | 0.0026-0.0136 | |
| Calcium carbonate (ultrafine)[84] | 0.1-0.8 (increasing with particle size decreasing) | |
| Carbon fiber | 0.058 | 0.02-0.05 |
| Carbon fiber[87] | | 0.017-0.052 |
| Diatomaceous earth | 85% of total particle volume | |
| Microporous polypropylene fibers[83] | | 230 |
| Precipitated silica[39] | 0.2-0.45 | 2-60 (aggregates) |
| Precipitated silica[69] | 0.1-4.2 | 7.4-152 |
| Quartz[24] | 0.0193-0.0676 | |
| Sepiolite | 9.4 | |
| Silica gel | | 5-40 |
| Zeolites | | 0.3-1 |

Many effects can be produced by the pores in filler particles. One is that pores in silica reinforce rubber.[39] During mixing, rubber chains migrate into the pores which increase the adhesion between the phases. The selective absorption of low molecular weight components affects the performance of paints and other materials. Microporous membranes and fibers are produced to clean water and selectively absorb certain solutes.

## 5.8 PARTICLE-PARTICLE INTERACTION AND SPACING[36,70,71,88-90]

Figure 5.13 shows the potential energy between two neighboring particles. The London-van der Waals forces are attractive and Coulombic forces are repulsive. Their relative magnitudes determine if particles are attracted by each other or repelled. Two methods can be used to overcome the barrier if there is a need to form an agglomeration of particles. One is to reduce distance by using shear forces (mixing) which force particles to come into contact by overcoming the barrier of repulsion. The second method is to increase the ionic concentration which increases attractive forces. Figure 5.14 shows the effect of both methods. The results indicate

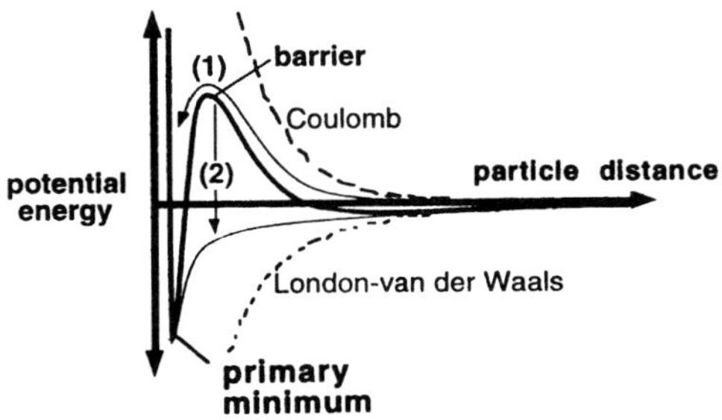


Figure 5.13. Potential energy curve for two colloidal particles. [Adapted, by permission, from Schueler R, Petermann J, Schulte K, Wentzel H P, *Macromol. Symp.*, **104**, 1996, 261-8.]

that by increasing ionic concentration with copper chloride, the contact between particles causes a decrease in resistivity at a lower concentration of carbon black than was possible by applying shear.

Some fillers have a natural tendency to agglomerate (or flocculate) as can be seen from Figure 5.15. Clay particles have a different charge on their crystal face from their crystal edge. Depending on pH these particles are either in a deflocculated state (alkaline environment) or flocculated state (acid environment) as shown in Figure 5.15.

These effects are exploited in commercial applications. In one, conductive particles are expected to come to close contact with each other in conductive plastics. In another, the flocculated state is required in regulating rheological properties of coatings. But in many other cases, the opposite effect is required – the filler is incorporated to form a homogeneous well dispersed mixture.

The two terms: agglomeration and flocculation require some clarification. Agglomeration is defined as a gathering of smaller particles into larger size units.

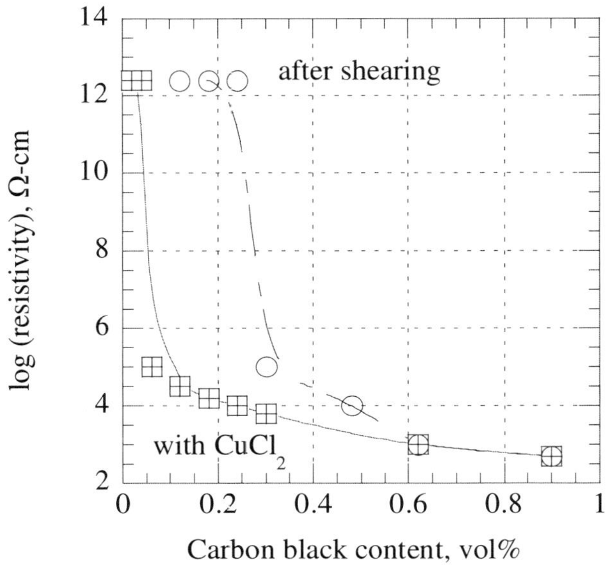


Figure 5.14. Resistivity of epoxy resin vs. carbon black concentration. [Data from Schueler R, Petermann J, Schulte K, Wentzel H P, *Macromol. Symp.*, **104**, 1996, 261-8.]

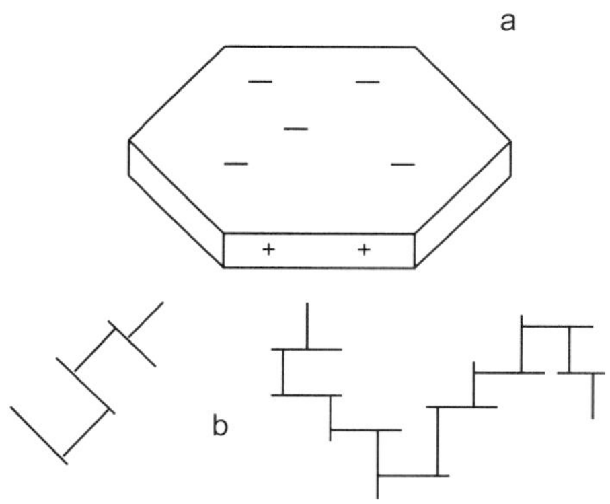


Figure 5.15. Positive and negative charges on clay particles (a). Flocculated state (b).

This phenomenon occurs in fillers during storage. As the storage time gets longer, the agglomeration of particles increases to the extent that stored filler requires substantially higher dispersion forces than does freshly manufactured filler. The mechanical forces to which the filler is exposed during transportation and the compaction that occurs as a result of storing several layers of bags or layers of filler in silo increase agglomeration. The word agglomeration is used to describe changes in the particulate materials in their solid state. Many industrial compaction methods are based on agglomeration. Flocculation is a similar process but usually occurs in a liquid medium. The name is derived from the word "flock" which describes the appearance of flocculated particles. The flocculation process is often associated with the coagulation of particles in water treatment with flocculants. It is also occurs in paints but this is usually undesirable. More information on this subject is included in the separate sections below.

The mean particle spacing can be calculated using the following equation:

$$s = (k\phi^{-1/3} - 1)d \qquad [5.5]$$

where:

| | |
|---|---|
| s | interparticle spacing |
| d | particle diameter |
| $\phi$ | volume fraction |

In this equation, the coefficient k depends on particle arrangement. For face-centered particles in their closest arrangement, the value for k is 0.906.

## 5.9 AGGLOMERATES[3,29,39,77,89,91-95]

Both agglomeration and flocculation lead to a similar result, in the sense that two or more particles join together to form a bigger one. Filler particles are mostly composed of primary particles but some are pre-formed aggregates (carbon blacks). Agglomeration and flocculation adversely affects the dispersion stability of fillers. But there are many technological advantages of agglomeration.

Van der Waals forces are primarily responsible for agglomeration of fillers during production and storage. These forces are especially important during the dispersion of fillers. For agglomeration to occur the sum of all environmental forces (gravity, inertia, drag, etc.) must be smaller than the forces between the adhering partners:

$$T_a = \frac{\sum_i B_i}{\sum_j E_j} > 1 \qquad [5.5]$$

where:
$T_a$ tendency to adhere
$B_i$ binding forces
$E_j$ environmental forces

This equation shows the forces that cause agglomeration and deagglomeration. The forces causing adhesion between particles can be grouped as follows:

- **Bridging**: sintering, melting, the effect binders, chemical reaction
- **Adhesion and cohesion**: the effect of viscous binders and adsorption layers
- **Attraction forces**: van der Waals, hydrogen bonding, electrostatic and magnetic
- **Interfacial forces**: liquid bridges ($H_2O$ – hydrogen bonding), capillary.

The agglomeration forces can be measured by determining the tensile strength of compacted fillers. Tensile strength depends on the packing density and the type of filler. Tensile strength and, therefore, agglomeration also depends on the type of mechanical processes used for filler dispersion. Pelletized carbon black does not return to its former agglomeration after grinding, and the intensity of grinding determines the resultant packing density and the tensile strength. Organic treatment of the titanium dioxide surface may decrease agglomeration as manifested by a lower tensile strength of similarly compacted material at the same packing densities. Agglomeration of titanium dioxide particles has been found to be due to water adsorption through liquid bridging, rather than by van der Waals forces, which usually prevail with carbon blacks.

Agglomeration has an effect on fillers used in various industrial processes. Dispersion of carbon black, especially that having very fine particles, is difficult. On the other hand, the agglomeration process is broadly used in the pharmaceutical industry to pelletize various ingredients where the mechanical strength of pellets is important. It is well-known that carbon black is not composed of individual primary particles but of primary particles joined together into aggregates. Even a prolonged effort to grind materials containing carbon black does not result in a change of their aggregates' size. Forces holding individual particles together are sufficiently strong to resist even very intensive grinding or mixing. Other agglomeration processes are based on the formation of hydrogen bonds. Individual particles such as fumed silica form networks of aggregates.

From the above discussion, one can see that agglomeration, depending on the type of mechanism, leads to formation of aggregates which can be weakly bonded or have very strong bonds, resisting even extensive grinding. Apart from the mechanism of bonding and type of bonding forces utilized, the differences in agglomeration are related to particle size, type of surface, chemical groups available on the surface, moisture level, effect of surface treatment, method of filler production, etc. Agglomeration processes are complex in nature and, if they are to be either prevented or enhanced, the nature of agglomeration must be carefully studied.

Several processes benefit from agglomeration. They include: wet mixing, suspending, rheological modification, drying, fluidized-bed processes, clarification, briquetting, tableting, pelletizing, and sintering. Processes negatively affected by agglomeration include: dispersion, dry grinding, screening, dry mixing, conveying, silos storage, etc.

## 5.10 AGGREGATES AND STRUCTURE[23,39,50,52,56,62,70,91,96-107]

Aggregates and structure are very important morphological features of carbon black and to a lesser extent of silica fillers. The aggregate of carbon black is a cluster of primary particles which are fused together and can be separated only by extensive mechanical forces which seldom exist in typical mixing operations. The aggregates of silica are formed by chemical and physical-chemical interactions which cause the formation of an assembly of particles which are the smallest units not subdivided by mixing.[39] The aggregate can be quantified by the size of the primary particles, the number of primary particles in the aggregate, and their geometrical arrangement in the aggregate. The term "structure" encompasses all these three parameters to give a general measure of the aggregate. A low structure carbon black contains less particles and limited branching. It is perceived as spherical assemblage of particles. A high structure carbon black is represented more by a grape-like structure with numerous branches.

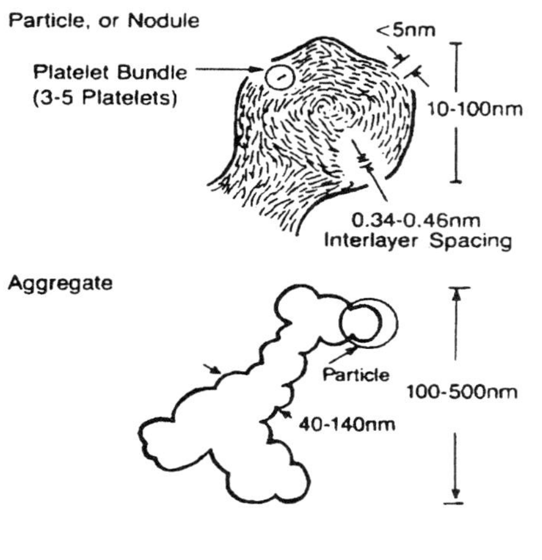


Figure 5.16. Structure of carbon black primary particle and aggregate. [Adapted, by permission, from Byers J T, Meeting of the Rubber Division, ACS, Cleveland, October 17-20, 1995, paper B.]

Figure 5.16 gives a schematic diagram which compares various dimensions in carbon black particles and aggregates. Compared with the small dimensions of voids within particle and the particle itself, the aggregate is a fairly large object of irregular morphological structure. As much as the application of carbon black is related to its morphology, its structure relates to vehicle (or binder) demand.

Scientists are continuing to make a major effort to determine the structure of carbon black and to apply this knowledge to its manufacture and application.

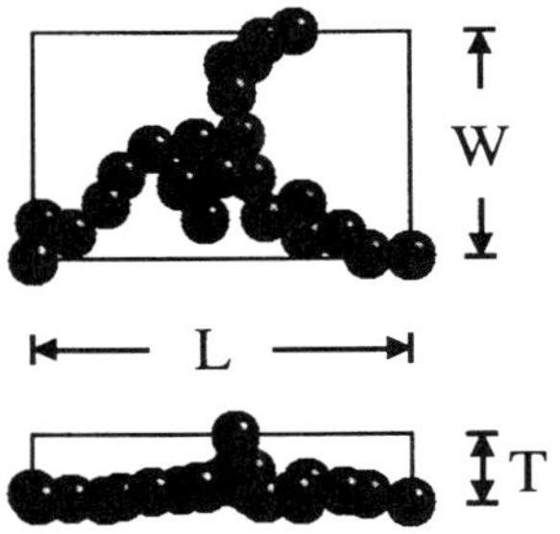


Figure 5.17. Two views of N220 aggregate model obtained by 90° rotation. [Adapted, by permission, from Gruber T C, Zerda T W, Gerspacher M, *Rubb. Chem. Technol.*, **67**, No.2, 1994, 280-7.]

Several methods are used including oil absorption, transmission electron microscopy, compression, and thermoporometry. The analytical results must be further analyzed by various algorithms to transform the results to a form which can be used for the prediction of properties of the compounded materials. Various forms of microscopy are applied in research studies and the findings have contributed to the further understanding of this complex subject. Figure 5.17 illustrates the essential problem related to microscopy. Because of the very small size of primary particles, only TEM gives sufficient resolution to elucidate morphological features. But, TEM can produce only two dimensional micrographs which do not display the spatial distribution of primary particles in the aggregate. In addition, the image projected depends on the viewing angle. Figure 5.17 shows the same aggregate displayed from angular views which differ by 90°.[96] The aim of this study[96] was to develop a technique for three dimensional analysis of carbon black aggregates. The results indicate that tread-grades of carbon black are planar and highly branched similar to the aggregates displayed in Figure 5.17.

High surface area carbon black was studied using small angle neutron scattering and contrast variation. It was found that aggregates are built out of 4-6 primary particles which can be represented by a prolate ellipsoid with semi-axes at 14.5 and 76.4 nm. This method can determine the average number of particles forming the aggregate.

In the case of carbon black, the aggregates are distributed in the matrix rather than individual particles, it is therefore important in some applications (e.g., conductive plastics) to evaluate the distance between these aggregates. It is now possible to measure these distances by atomic force microscopy coupled with straining device.[106] There is a linear relationship between the parallel distance between aggregates dispersed in SBR and strain value. For 10 phr of N 234, the mean distance between aggregates varied in a range from 1.85 to 3.42 μm. For practical purposes, a modified equation [5.4] is used to determine the interaggregate distance:

$$s = [k(\beta\phi)^{-1/3} - 1]d_{St} \quad [5.6]$$

where:

| | |
|---|---|
| s | interparticle spacing |
| k | coefficient of spatial arrangement |
| $\beta$ | $=1 + (0.7325 \times DBPA - 15.75) \times 10^{-2}$, Medalia's coefficient based on DBP absorption |
| $d_{St}$ | Stokes particle diameter |
| $\phi$ | volume fraction |

This is a complex area of investigations and far from being complete. Until mathematical criteria characterizing the structure are developed, the available quality control and research data is the only source of information that can be used to select carbon black for specific application.

## 5.11 FLOCCULATION AND SEDIMENTATION[89,108-112]

Flocculation of pigment is a mechanism exploited to facilitate a higher retention of pigment in the paper manufacture. Heteroflocculation is induced by the addition of cationic polyacrylamide to the pulp and clay mixture. The retention of clay is dramatically improved and clay distribution becomes more even. This is an example of how a controlled flocculation process may help to achieve certain technological goals. In paint production, too, the addition of flocculants not only inhibits phase separation but also allows the reversal of separation by preventing sediment compaction. On the other hand, a good dispersion of pigment can be completely reversed by the addition of auxiliary agents which eliminate particle charge (decreasing $\zeta$-potentials – for more information see separate section below). Such an addition affects not only the durability of the product but also its brightness, color, and opacity. Flocculation also depends on the pigment concentration. The higher the flocculation gradient, the more the pigment flocculates.

Figure 5.18 shows a schematic representation of montmorillonite particles in dispersions. This diagram helps us to distinguish between different types of flocculation. Figure 5.18a depicts internal mutual flocculation which is described in Figure 5.15. As a result of electrostatic and van der Waals forces between the edges and faces of particles, a house-of-cards structure is formed (the pH of the dispersion or its ionic strength influence this effect). Under shearing conditions, the orientation of particles changes (5.18b) which affects viscosity. Figure 5.18c shows face-to-face flocculation or heteroflocculation. Heteroflocculation requires a second component such as polyvalent cation used in paper manufacturing. The polyvalent cation reverses the surface charge and changes the electrokinetic potential, resulting in the collapse of a voluminous gel structure into compact face-to-face packing.[108]

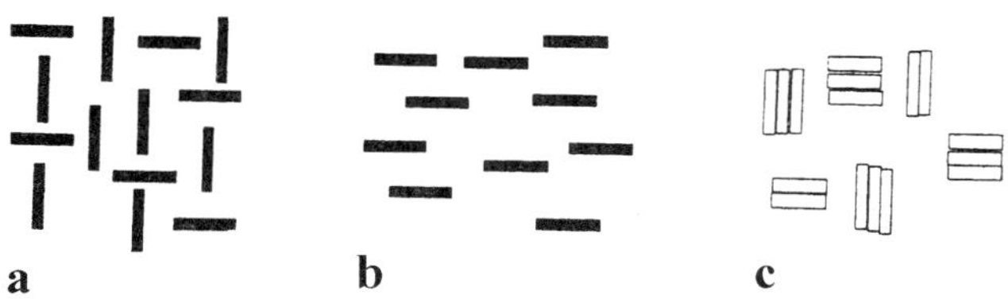


Figure 5.18, A schematic representation of montmorillonite particles in dispersion. [Adapted, by permission, from Miano F, Rabaioli M R, *Coll. & Surfaces*, **84**, Nos.2/3, 1994, 229-37.]

Flocculation affects filler packing and therefore it also affects surface roughness and gloss. The composition of fillers (pigments) can be changed by coflocculation. Special additives are used to promote this effect because coflocculation is seen as one of the mechanisms which can be used to overcome flooding and floating. Co-flocculating agents, by bridging two different particles, restrict their movement which contributes to a better color development in the material or a more uniform composition in the case of filled material. Excessive co-flocculation detracts from gloss and changes rheological properties.

The rheology of the suspension is affected through the particle interaction coefficient:

$$\sigma = \sigma_S + \sigma_P \qquad [5.7]$$

where

| | |
|---|---|
| $\sigma_S$ | contribution of solvent, flocculating agents, etc. |
| $\sigma_P =$ | $\sigma_{PC} / D_1$, summation of all individual particle contributions to the particle interaction coefficient |
| $\sigma_{PC}$ | particle contribution constant |
| $D_1$ | number average particle size |

This equation has been confirmed by experimental results.[109] These have shown that the interaction parameter increases as the particle size decreases. The particle interaction coefficient, σ, in the following equation is required to describe the viscosity-concentration relationship of suspensions:

$$\ln \frac{\eta}{\eta_0} = \frac{[\eta]\varphi_n}{\sigma - 1}\left[\left(\frac{\varphi_n - \varphi}{\varphi_n}\right)^{1-\sigma} - 1\right] \qquad [5.8]$$

where

| | |
|---|---|
| $\sigma$ | particle interaction coefficient |
| $\eta$ | suspension viscosity |
| $\eta_0$ | suspending medium viscosity |
| $[\eta]$ | intrinsic viscosity |
| $\varphi_n$ | particle packing fraction |
| $\varphi$ | suspension particle volume concentration |

Filler particles can be modified to decrease flocculation. Kaolin particles modified by a graft of poly(ethylene oxide) showed an increase in the upper critical flocculation temperature. Stabilization of particle dispersion was due to an enhanced steric stabilization.[112]

In rubber systems containing carbon black, flocculation may cause substantial changes in mechanical properties. Flocculation in these systems counteracts filler dispersion. Carbon black flocculation occurs in filled rubber stock during storage or during vulcanization in the absence of shear.[111] Temperature is the important kinetic factor which affects the flocculation rate (Figure 5.19). In addition to temperature and time, flocculation depends on the type of carbon black and its concentration.

Sedimentation occurs readily in suspensions in low viscosity liquids. The sedimentation coefficient is given by the equation:

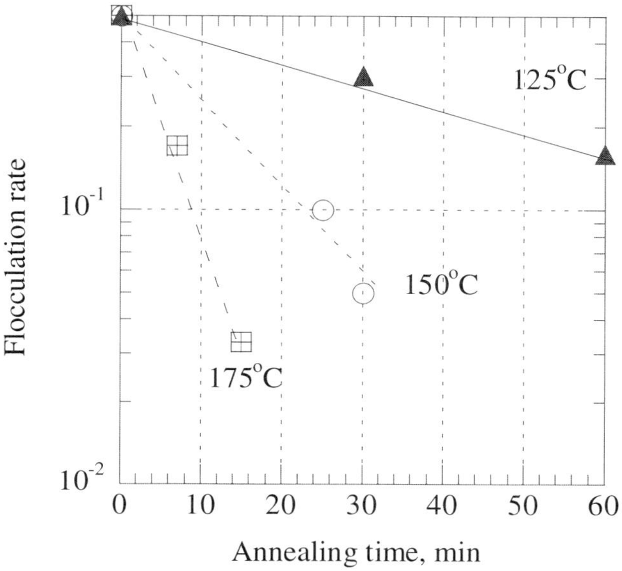


Figure 5.19. Rate of carbon black flocculation at different temperatures. [Adapted, by permission, from Boehm G G A, Nguyen M N, *J. Appl. Polym. Sci.*, **55**, No.7, 1995, 1041-50.]

$$s_0 = \frac{4/3\pi R^3(\rho_f - \rho_p)}{6\pi\eta_0 R} \qquad [5.9]$$

where
R particle radius
$\rho_f$ density of the fluid
$\rho_p$ density of a particle
$\eta_0$ viscosity of fluid medium

Since particles absorb components of the system to form adlayers (or bound polymer layers) the radius of the particle has to be corrected as follows:[110]

$$R_e = R + \Delta r = \frac{(\phi V \rho_f)^{1/3}}{W} \qquad [5.10]$$

where
$R_e$ effective radius of particle
$\Delta r$ thickness of adlayer
$\phi$ packing factor
V bulk sediment volume
W weight of particles

## 5.12 ASPECT RATIO[113-117]

Aspect ratio is the length of a particle divided by its diameter. Table 5.7 provides information on aspect ratios of some fillers.

## Table 5.7. Aspect ratio of some fillers

| **Aspect ratio range** | **Filler** (actual aspect ratios are given in parentheses) |
|---|---|
| 1-3 | ferrites (1-5); majority of particulate fillers |
| 3-10 | milled carbon fiber (6-30), milled glass fiber (3-25), talc (5-20), wollastonite (4-68) |
| 10-20 | silver-coated nickel flakes (15), nickel flakes (15-50) |
| 20-100 | aluminum flakes (20-100), mica (10-70) |
| above 100 | aramid fibers (100-500), chopped carbon fibers (860), chopped glass fibers (250-800), hollow graphite fibrils (100-1000), nickel-coated carbon fibers (200-1600) |

The majority of fillers fall into a group of low aspect ratio fillers (below 10). Reinforcing elongated particles of mineral origin have an aspect ratio between 10 and 70. Fibers (except for milled fibers) have aspect ratios well above 100. The aspect ratio of fibers is a critical parameter in composites[115,117] and in providing electrical and shielding properties.[116] For reinforcement, high aspect ratios are more effective. Also, in electrical applications high aspect ratio fillers give good performance at substantially lower concentrations and a typical aspect ratio is in a range from 20 to 100. The initial aspect ratio of filler is not necessarily retained in the final product because of degradation of fiber length during processing.

## 5.13 PACKING VOLUME[1,3,9,17,20,90,109,113,118-128]

The maximum packing volume of a filler can be calculated for different geometrical arrangements, determined after the filler is dispersed in a liquid media (e.g. oil). It is calculated by dividing the tamped bulk density by specific gravity of filler. Table 5.8 compares the data obtained from calculation for monodispersed spheres in different arrangements with determined values.

The data in the Table 5.8 show that a high packing volume can be obtained in real systems as compared with theoretical calculation results. A particle size decrease results in a decrease in the maximum volume packing fraction. A surface coating can increase the maximum volume packing fraction by reducing the thickness of the bound polymer. The above data shows that a higher packing was obtained in experimental systems than was predicted for monodispersed spheres. This is a result of the mixture of particle sizes which fill voids more efficiently.

In polymeric systems, particle size has to be corrected for the thickness of the occluded polymer layer. This can be done by the use of the volume coefficient of separation, α, given by the following equation:

$$\alpha = (1 + h / d)^3 \qquad [5.11]$$

where
h the thickness of the matrix interlayer
d particle diameter

**Table 5.8. Maximum packing volume calculated for monodispersed spheres and determined for some fillers[9,90]**

| Spatial configuration or fillers in different media | Maximum packing volume fraction |
|---|---|
| *Theoretical calculations* | |
| Hexagonal or pyramidal arrangement (maximum packing) | 0.74 |
| Double staggered layout | 0.70 |
| Random close packing | 0.64 |
| Random loose packing (simple staggered) | 0.60 |
| Cubic | 0.52 |
| *Experimental results* | |
| Glass beads in polyethylene | 0.68 |
| Ground calcium carbonate (10 μm) in polyethylene | 0.52 |
| Precipitated calcium carbonate (2 μm) in polyethylene | 0.44 |
| Ground calcium carbonate (1 μm) in mineral oil | 0.55 |
| Ground calcium carbonate (3 μm) in mineral oil | 0.59 |
| Precipitated calcium carbonate (0.6 μm) in mineral oil | 0.30 |
| Surface treated ground calcium carbonate (1 μm) in mineral oil | 0.77 |
| Surface treated ground calcium carbonate (3 μm) in mineral oil | 0.76 |

The maximum volume packing fraction can also be estimated but with much lower precision by dividing bulk density by specific density. The lower precision results from the fact that particle packing depends on an arrangement of loosely packed particles which is not ideal for measuring bulk density. Table 5.9 gives data calculated for a large number of grades of different fillers using this method. Tamped density was taken as the bulk density which gives more realistic values.

The values in Table 5.9 are far from the theoretical values presented in the Table 5.8. Only a few fillers included in the last row come close to the values from theoretical calculations. In most cases, fillers are manufactured to offer a broad range of packing densities so that one can be selected according to the requirements which may not always be maximum packing. The information on maximum packing volume is important to realize cost savings and to maximize mechanical properties. If cost savings is an important consideration then the filler or fillers combination which offer the most efficient packing and thus the highest level of filler incorporation should be selected. Otherwise, maximum packing density and a correction for bound polymer should be always evaluated to ensure that fillers are not used in excessive amounts. Mechanical properties decrease rapidly as maximum volume packing is approached.

**Table 5.9. Maximum packing volume fraction, $\phi_M$, of some fillers calculated by dividing tamped density by specific density of filler**

| $\phi_M$ range | **Filler** (the range of $\phi_M$ for a given filler is given in parentheses) |
|---|---|
| 0.01-0.099 | aluminum flakes (0.07-0.17), fumed silica (0.02-0.06), graphite (0.09-0.46), milled glass fiber (0.07-0.43), nickel powder (0.9-0.33) |
| 0.1-0.19 | calcium carbonate (0.18-0.53), carbon black (0.15-0.28), kaolin (0.11-0.34), PTFE powder (0.12-0.15), talc (0.16-0.42), silver flake (0.17-0.4), silver powder (0.13-0.52), silver spheres (0.1-0.48), titanium dioxide (0.19-0.3), wollastonite (0.13-0.47), zinc sulfide (0.17-0.23) |
| 0.2-0.29 | aluminum trihydroxide (0.2-0.55), chopped glass fiber (0.21-0.28), cristobalite (0.26-0.36), gold spheres (0.21-0.48), mica (0.22-0.42) |
| 0.3-0.39 | barite (0.35-0.50) |
| above 0.4 | aluminum needles and tadpoles (0.47-0.6), hollow glass beads (0.53-0.66), polymeric beads (0.4), silica flour (0.5-0.65), stainless steel powder (0.63) |

Packing density must be understood when lowering the viscosity of system, increasing thermal conductivity and heat dissipation, increasing electric conductivity, designing electronic devices which are protected from overloading, designing materials of high and low specific densities, etc.

The data in Figure 5.20 demonstrates another aspect of packing density. Nanoparticle size $Al_2O_3$ was slurried in water and compressed in a die. The results show that the density of pellet is very close to the specific density of the material if sufficient pressure is applied. In real applications, high pressures result from various forces operating in the system such as equipment conditions, crystallization, shrinkage, and chemical bonding. All these factors influence the potential maximum loading in a real systems.

Three factors associated with particle packing are common use: critical volume fraction (or loading), effective volume fraction, and critical pigment volume concentration. The effective volume fraction of a filler includes the filler and the elastomer immobilized within the aggregates. This is given by the equation:[3]

$$\phi_e = (\phi_a \times \alpha) + \phi_t \qquad [5.12]$$

where

$\phi_a$ volume fraction of agglomerates
$\alpha$ volume fraction of immobilized rubber
$\phi_t$ volume fraction of carbon black

The coefficient $\alpha$ which corrects for the incorporated polymer layer in a manner similar to equation [5.11] is obtained for carbon black from the oil absorption and calculated from the equation:

$$\alpha = DBPA / [DBPA + (100 / \rho)] \qquad [5.13]$$

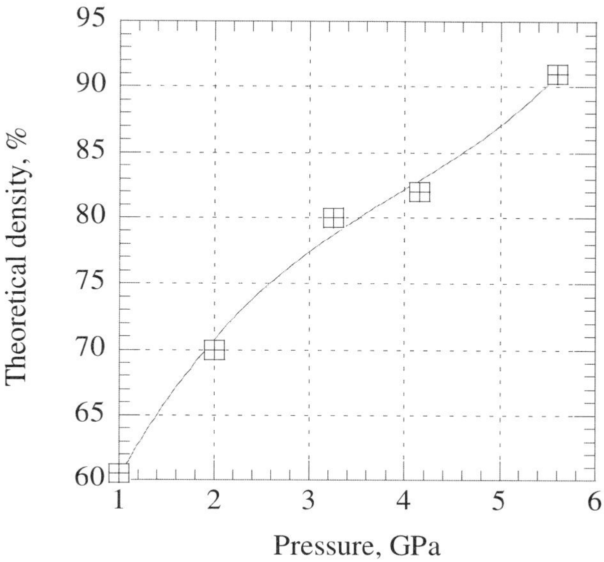


Figure 5.20. Density of $Al_2O_3$ samples vs. compression pressure. [Adapted, by permission, from Gallas M R, Rosa A R, Costa T H, da Jornada J A H, *J. Mater. Res.*, **12**, No. 3, 1997, 764-8.]

where
DBPA dibutyl phthalate absorption
$\rho$ carbon black density

The critical volume fraction of filler is the volume of filler above which a property change occurs or above which the rate of change of that property is increased. Figure 5.21 illustrates the meaning of this critical value in the studies of carbon black flocculation. The critical volume fraction of N347 carbon black used in this study is at 13 vol%. At 20 phr (10 vol%), there is no change in the excess storage modulus because the carbon black aggregates are too far apart and unable to migrate far enough to flocculate. At 30 phr (14 vol%), the composition is just above the critical volume fraction of filler and small changes occur. If still more carbon black is added (50 phr) flocculation occurs rapidly.

Figure 5.22 shows that the critical filler volume fraction depends on the structure of carbon black which is here characterized by DBP absorption.

The critical volume fraction of the filler has a different application in the case of conductive materials. As the amount of conductive filler is increased, the material reaches a percolation threshold. Below the percolation threshold concentration, the electric conductivity is similar to that of matrix. Above the percolation threshold conductivity rapidly increases. Above the critical volume fraction of filler which is, in turn, a concentration above the percolation threshold, there is a rapid increase in conductivity.[94] The critical volume fraction depends on the type of filler and its particles size. For example, for silver powder, it ranges from 5 to 20 vol% for

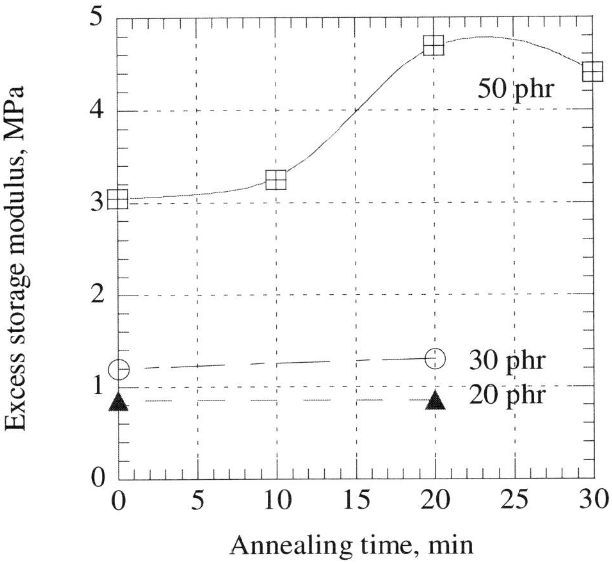


Figure 5.21. Excess storage modulus of carbon black filled polybutadiene vs. annealing time. [Adapted, by permission, from Boehm G G A, Nguyen M N, *J. Appl. Polym. Sci.*, **55**, No.7, 1995, 1041-50.]

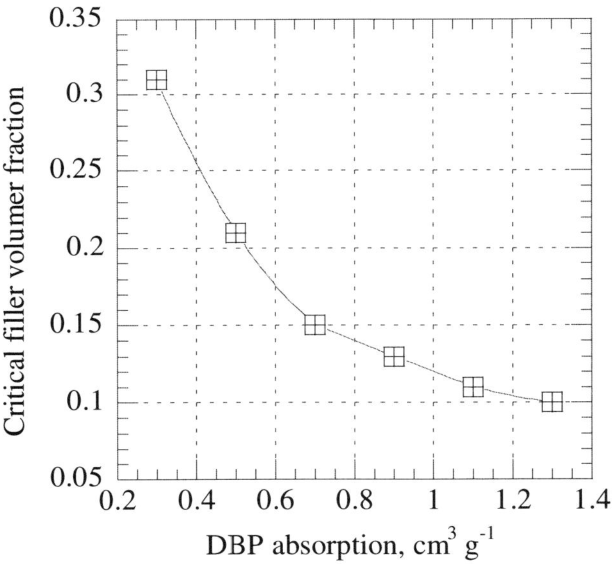


Figure 5.22. Critical volume fraction of carbon black vs. DBP absorption. [Adapted, by permission, from Boehm G G A, Nguyen M N, *J. Appl. Polym. Sci.*, **55**, No.7, 1995, 1041-50.]

particle sizes in the range of 0.5 to 9 μm (the smaller the particles size the smaller the critical volume fraction).

The concept of critical pigment volume concentration was introduced about 50 years ago by Asbeck and van Loo to explain the sudden change in paint properties around a certain concentration of pigment. Above this concentration, gloss rapidly decreases, porosity and water permeability increases, and the film becomes brittle. This is caused by the fact that there is not enough binder to fill the voids between particles and encapsulate them. Solvent-based paints are usually formulated well below the critical pigment concentration. The critical pigment concentration is calculated from the equation:[128]

$$CPVC = \frac{V_{pigment} + V_{filler}}{V_{pigment} + V_{filler} + bV_{vehicle}} \qquad [5.14]$$

where

$V_{pigment}$ volume of pigment
$V_{filler}$ volume of filler
$V_{vehicle}$ volume of resin
b constant, $b = 1$ for solvent paints and $b > 1$ for latex paints

## 5.14 pH[129-130]

### Table 5.10 pH of filler slurry

| pH range | **Filler** (the range of pH for a filler is given in parentheses) |
|---|---|
| 1-2.9 | antimony pentoxide (2.5-9), antimony trioxide (2-6.5), carbon black (2-8) |
| 3-4.9 | ceramic beads (4-8), cellulose fibers (4-9), clay (3.9-9), kaolin (3.5-11) fumed silica - hydrophilic (3.6-4.5), fumed silica - hydrophobic (3.5-11), precipitated silica (3.5-9), titanium dioxide (3.5-10.5) |
| 5-6.9 | attapulgite (6.5-9.5), barium sulfate (6-9.5), calcium sulfate (6.8-10.8), diatomaceous earth (6.5-10), glass fibers (5-10), muscovite mica (6.5-8.5), perlite (5.5-8.5), quartz (6-7.8), sand (6.8-7.2), silica gel (6.5-7.5), slate flour (6.5-8.1), wood flour (5), zinc sulfide (6-7) |
| 7-8.9 | aluminum oxide (8-10), aluminum trihydroxide (8-10.5), anthracite (7-7.5), barium and strontium sulfate (7-7.5), bentonite (7-10.6), cristobalite (8.5), feldspar (8.2-9.3), glass beads (7-9.4), hydrous calcium silicate (8.4-9), iron oxide (7-9), lithopone (7-8), phlogopite mica (7-8.5), sepiolite (7.5-8.5), talc (8.7-10.6), vermiculite (7), zinc borate (8.1-8.3) |
| 9-10.9 | barium metaborate (9.8-10.3), calcium carbonate (9-9.5), fused silica (9), wollastonite (9.8-10), zeolites (10-12), zinc stannate (9-10) |
| 11 and above | calcium hydroxide (11.4-12.6) |

The majority of fillers have a pH close to neutral. But many fillers have a broad range of pH which is either due to their origin, manufacturing technology, or surface treatment. The pH of filler may strongly affect interaction with other components of the mixture, so it is possible to chose fillers for specific application. While this gives additional methods of influencing properties of materials, it requires care in selecting an appropriate filler.

Fillers can be degraded either by too high or too low a pH[130] or modified by polymer conformation. The modifications causes a change in the surface coating of the filler.[129]

## 5.15 ζ-POTENTIAL[108,131-134]

The electric charge distribution in the plane of shear (or in the plane perpendicular to the surface) is referred to as ζ-potential (zeta-potential). The surface charges on the pigment or filler particles are formed as a result of dissociation of functional surface groups or adsorption of countercharges from the liquid phase. The development of surface charge on the particle is accompanied by the formation of countercharge in the surrounding medium which results in the electrochemical double layer. This double layer plays an essential role in stabilization of colloids and suspensions. Stabilization occurs when the liquid phase has a high dielectric constant, thus the stabilization effect is more pronounced in water rather than in solvent media.

The ζ-potential depends on the pH of the liquid phase. The pH at which the ζ-potential is zero is called the isoelectric point. The isoelectric point of each filler depends on its surface structure. In the case of titanium dioxide, the isoelectric point depends on the surface coating. A $SiO_2$ coating decreases the isoelectric point whereas $Al_2O_3$ increases it.[134]

Also electrolytes and polyelectrolytes affect the ζ-potential. Studies on montmorillonite clays showed that an excess of $Na^+$ ions in solution does not produce changes in ζ-potential although it is known that $Na^+$ ions react with the edges of clay. Thus, only interaction with the face of crystal affects ζ-potential. $Ca^{2+}$ ions can replace sodium counterions on the montmorillonite face and this replacement causes a shift towards negative values of ζ-potential. When $Ca^{2+}$ ions replace sodium counterions on the montmorillonite face they cause deflocculation and an increase in viscosity.[108]

The measurement of ζ-potential was used to control the flotation recovery of kaolin and calcium carbonate from waste paper.[131] The addition of a cationic polymer changes its usually negative values of ζ-potential of kaolin and calcium carbonate (-60 and -40 mV, respectively). The ζ-potential becomes positive when the concentration of polyelectrolyte reaches $5\times10^{-4}$ g/l then gradually increases until a plateau is reached at about $1\times10^{-3}$ g/l. The final ζ-potential is higher for kaolin than calcium carbonate. This interaction with the polyelectrolyte results in large particles which are more readily separated and recovered.[131]

The ζ-potential of colloidal silica surface treated by acrylate copolymers is affected by pH. The ζ-potential of untreated colloidal silica at a pH of 4 is -7 mV and it decreases to -32 mV at a pH of 7.[132] Modification of the surface of colloidal silica changes its surface properties and behavior. In another study on filler modification,[133] hydroxyapatite was modified for medical applications with several different silanes. The ζ-potential depended to a large extent on silane composition and the pH of surrounding liquid.

## 5.16 SURFACE ENERGY[6,20,23,66,72,74,84,90,104,112,135-159]

The following subjects, which are related to surface energy, are included in this discussion: wettability, acid-base interaction, and work of adhesion. The interrelation is well illustrated by the set of equations.

Particles in a matrix are either spontaneously wetted or remain unwetted by polymer depending on the relative magnitudes of their solid/vapor surface energy, $\gamma_{SV}$, and liquid/vapor surface energy, $\gamma_{LV}$. The following equations may be used to calculate these energies:[90]

equation of state

$$\cos\theta = 1 + b\ln(\gamma_c / \gamma_{LV}) \qquad [5.15]$$

solid/vapor surface energy

$$\gamma_{SV} = [b\exp(1/b - 1)]\gamma_c \qquad [5.16]$$

liquid/solid surface energy

$$\gamma_{LS} = \gamma_{SV} + \gamma_{LV} - \gamma_{LV}\left[1 + b\exp\left(1 - \frac{1}{b}\right) + b\exp\left(1 - \frac{1}{b}\right)\ln\frac{\gamma_{SV}}{b\gamma_{LV}}\right] \qquad [5.17]$$

where

| | |
|---|---|
| $\theta$ | contact angle of filler wetted by a liquid |
| b | Lee interaction parameter |
| $\gamma_c$ | critical surface tension |
| $\gamma_{LV}$ | liquid/vapor surface energy |
| $\gamma_{SV}$ | solid/vapor surface energy |
| $\gamma_{LS}$ | liquid/solid surface energy |

Both surface tension energies can be determined from contact angle measurement and $\bar{b}$ can be obtained as a geometrical mean between the b values of the constituents. Plotting the surface energy ratio between filler and polymer vs. extent of interaction, $\bar{b}$, it is possible to obtain the matrix shown in Figure 5.23. The results of similar determinations for any given system can be plotted on this matrix to establish in which zone the actual system resides. The lines separating various zones on the matrix were plotted based on the following relationships:

equilibrium work of adhesion

$$W_{LS} = \gamma_{LV}[b\ln(\gamma_{SV} / \gamma_{LV}) + b + 1 - b\ln b] \qquad [5.18]$$

Harkins spreading coefficient

$$\lambda_{LS} = \gamma_{LV} b[\exp(1 - 1/b)]\{1 + \ln[\gamma_{SV} b\gamma_{LV}]\} - 1 \qquad [5.19]$$

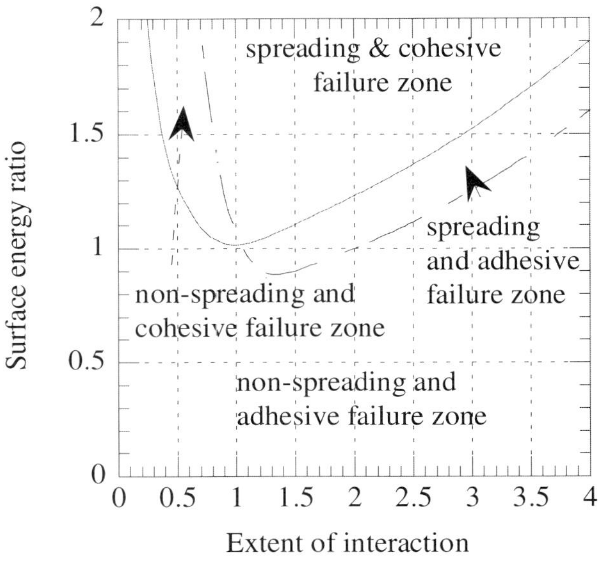


Figure 5.23. Spreading and failure characteristics predicted from the theory of adhesion. [Adapted, by permission, from Bomal Y, Godard P, *Polym. Engng. Sci.*, 36, No.2, 1996, 237-43.]

The method of determination is given elsewhere.[158,159] For our purposes, the above discussion shows that both wetting of fillers and the adhesion between filler and the matrix is governed by the principles of the theory of adhesion based on the surface energy properties of the filler and the matrix.

This method allows one to evaluate an unknown system. The following discussion concentrates on the surface properties of different fillers. The current level of understanding has been developed from principles proposed by Fowkes who indicated that the work of adhesion has two components:

$$W_a = W^d + W^{sp} \qquad [5.20]$$

where
$W^d$ contribution of dispersive, non-specific or London-type forces
$W^{sp}$ contribution of specific interactions such as dipole-dipole, H-bonding, acid-base, etc.

Accordingly, the surface free energy of a solid can be expressed as a sum of dispersive and specific components:

$$\gamma_S = \gamma_S^d + \gamma_S^{sp} \qquad [5.21]$$

where
$\gamma_S^d$ dispersive component of surface free energy
$\gamma_S^{sp}$ specific component of surface free energy

The dispersive component is associated with polymer-filler interaction and the specific component is associated with filler networking and agglomeration. The dispersive component of different fillers is more conveniently measured by inverse gas chromatography although it can also be measured by contact angle methods. The work of adhesion is given by the following equation, which has been modified to account for Fowkes theory,

$$W_a = 2Na[(\gamma_1^d \gamma_2^d)^{0.5} + (\gamma_1^p \gamma_2^p)^{0.5}] \quad [5.22]$$

where

| | |
|---|---|
| N | Avogadro number |
| a | surface area of adsorbed molecule |
| 1,2 | subscripts denoting filler and polymer or pigment and liquid |
| d,p | superscripts denoting dispersive and polar components |

The work of adhesion increases as the dispersive component of surface free energy increases. Table 5.11 gives the values of the dispersive component available in the literature for different fillers.

**Table 5.11. Dispersive components of different fillers**

| Filler | $\gamma^d$, mJ/m$^2$ | Reference |
|---|---|---|
| γ-aluminum oxide | 92 | 136 |
| calcium carbonate (Socal Solvay, Milicarb Omya, Albacar 5970)<br>calcium carbonate precipitated & maleated | 52/48/53<br>64.3 & 32.8 | 136<br>139 |
| carbon black (range for numerous grades)<br>carbon black oxidized and unoxidized<br>carbon black | 40-120<br>41.9-43.4<br>51 | 23<br>145<br>149 |
| carbon fiber treated by plasma in different concentration of $CF_4/O_2$ | 17.7-36.9 | 143 |
| fumed silica | 80 | 136 |
| magnesium oxide | 95 | 136 |
| muscovite mica | 70 | 136 |
| silica<br>silica precipitated (Zeosil 175)<br>silica precipitated (Zeosil 175), esterified with alcohols $C_{16}$-$C_1$<br>silica precipitated (Zeosil 175), methacryl and vinyl silane modified | 49.8<br>105<br>46-87<br>84 & 84 | 20<br>137<br>137<br>137 |
| talc | 130 | 136 |
| titanium dioxide<br>non-coated<br>$Al_2O_3$ and $SiO_2$ coated | 76<br>50.3<br>104.3 & 124.8 | 136<br>20<br>20 |
| zinc oxide | 52 | 136 |

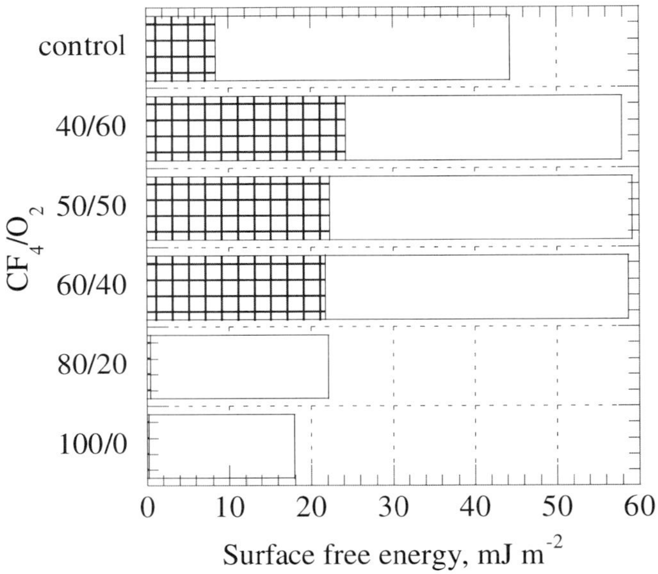


Figure 5.24. Surface energy components of carbon fibers treated with plasma in the presence of different gas composition. Open bars - $\gamma_S^d$, shaded bars - $\gamma_S^p$. [Data from Tsutsumi K, Ban K, Shibata K, Okazaki S, Kogoma M, *J. Adhesion*, **57**, Nos.1-4, 1996, 45-53.]

Various surface modifying operations such as silane coating, maleation, oxidation, surface coating have a noticeable effect on surface energy.

Figure 5.24 shows the effect of oxidation on dispersive and polar components of surface free energy. Carbon fibers were exposed to plasma treatment in the presence of various ratios of $CF_4$ and $O_2$. The untreated sample and the samples exposed to a substantial concentrations of oxygen show increase in the polar component. High concentrations of $CF_4$ gas reduced its dispersive component and converted the surface to a PTFE-like material as confirmed by XPS studies.[143]

Acid-base interaction which results from polar interaction can be predicted from the inverse gas chromatography data. The basic relationship used in this type of studies is:[148]

$$\Delta H_{ab} = Ka\,DN + Kd\,AN \qquad [5.23]$$

where

$\Delta H_{ab}$ enthalpy of absorption

Ka, Kd the solids' acid-base interaction parameters

AN DN literature values of vapors' acid-base interaction

The values of Ka and Kd can be measured from the plots of $\Delta H_{ab}$/AN vs. DN/AN. The methods of determination and result interpretations are discussed elsewhere.[66,136148,157]

## 5.17 MOISTURE[160-170]

It is usually important to know how much moisture is present in a filler and whether or not the filler is hygroscopic. Table 5.12 gives an overview of typical moisture concentration in some fillers (the fillers are qualified to a particular group based on their lower limiting value of the moisture concentration range). The information in the table is based on data for a large number of grades which vary in moisture content.

**Table 5.12. Moisture in fillers**

| **Moisture range**, % | **Filler** (the range of moisture concentration for a filler is given in parentheses) |
|---|---|
| below 0.1 | calcium carbonate (0.01-0.5), cristobalite (0.006-0.1), quartz (tripoli) (traces), wollastonite (0.02-0.6) |
| 0.1-0.19 | aluminum trihydroxide (0.1-0.7), barium sulfate (0.1-0.3), calcium sulfate (0.1), carbon black (0.12-2), glass fiber (0.1-3), graphite (0.1-0.5), iron oxide (0.1-3), fused silica (0.1), sand (0.1), talc (0.1-0.6) |
| 0.2-0.39 | antimony pentoxide (0.2-1), antimony trioxide (0.1), barium titanate (0.2), ceramic beads (0.2-0.5), diatomaceous earth (0.2-6 ), magnesium hydroxide (0.2-1), mica (0.3-0.7), titanium dioxide (0.2-1.5), zinc sulfide (0.3) |
| 0.4-0.99 | anthracite (0.5-4), perlite (0.5-1), fumed silica hydrophobic (0.5), fumed silica hydrophilic (0.5-2.5), sodium antimonate (0.5-3), zinc borate (0.4-0.5) |
| 1-4.99 | aluminum oxide (4-5), aramid fiber (1-8), attapulgite (2-16), bentonite (2-14), ball clay (3), calcium hydroxide (1.5), cellulose fiber (2-10), fly ash (2-20), kaolin (1-2), pumice (2), pyrophyllite (1), rubber particles (1), precipitated silica (3-7), slate flour (1), wood flour (2-12), zeolite (1.5) |
| 5-9.99 | hydrous calcium silicate (5.5-5.8), sepiolite (8-16) |
| above 10 | calcium carbonate slurry (10-30), kaolin slurry (20-30), titanium dioxide slurry (10-20) |

The presence of water in a filler is not usually beneficial. However, in paper manufacture and in water based paints, where aqueous slurries can be used, moisture level is of no major concern. Four benefits of using slurry are: lower cost, better dispersion, elimination of dust, and easier handling. The cost is reduced because the process of manufacture does not require drying which is an expensive step and packaging and handling is simpler with a slurry. Better dispersion contributes to improved quality in the final product due to the fact that slurries are usually stabilized to limit agglomeration. Whereas, when fillers are dried, the drying process results in the production of agglomerates. Environmental impact is reduced due to the fact that there is less waste and no packaging materials are involved. Drying processes burn large amounts of fuels and there are generally less environmentally friendly.

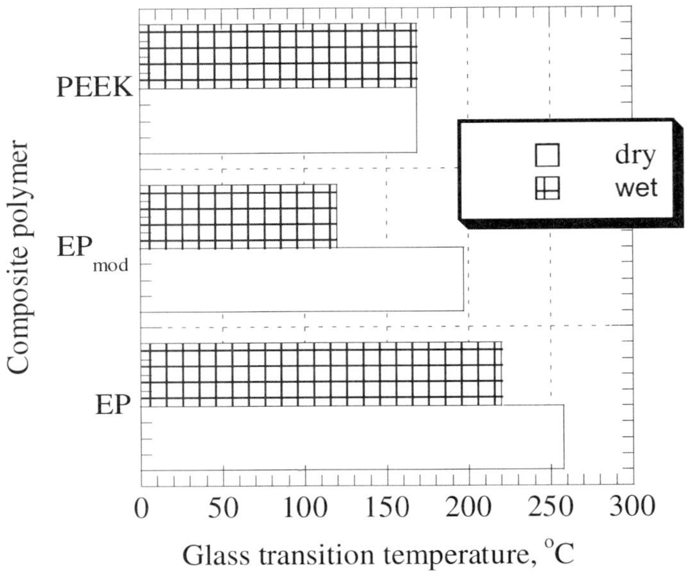


Figure 5.25. Glass transition of composites containing carbon fiber under dry and wet conditions. [Adapted, by permission from Selzer R, Friedrich K, *Composites, Part A*, **28A**, 1997, 595-604.]

In most other processes, the presence of moisture in filler either requires a process correction in the amount of the active ingredient or the moisture must be removed. In the case of hygroscopic fillers (which are very important to industry), the surface of the filler must be treated to lower moisture uptake. Montmorillonite,[167] glass beads and fibers,[165] silica,[164] titanium dioxide,[163] aramid fiber,[161] rubber particles,[169] and carbon fiber were studied to improve their moisture absorption and impart the hydrophobic properties.[160]

Figure 5.25 shows that the glass transition of composites containing carbon fibers may be affected by water uptake. The glass transition of carbon fiber/PEEK composite remains the same under dry and wet conditions. But carbon fiber/epoxy composites may experience a decrease in $T_g$ as high as 77°C depending on the properties of the matrix resin.[160]

Composites containing aramid fibers rapidly regain moisture which results in a lowering their initial mechanical properties.[168] Figures 11.14 and 16.15 show the kinetics of moisture absorption by different fibers.[166] Figure 8.26 shows how moisture content affects compressive strength of aramid/epoxy laminates. Figure 5.26 shows the effect of moisture content on the interlaminar strength of epoxy/aramid laminates. Different fibers and epoxy resins were used in this study but the results follow a relationship of a linear decrease of adhesion as the moisture content decreases.

Figure 5.27 shows that a substantial amount of moisture is absorbed by glass beads/epoxy composites. The addition of glass beads increases the moisture uptake

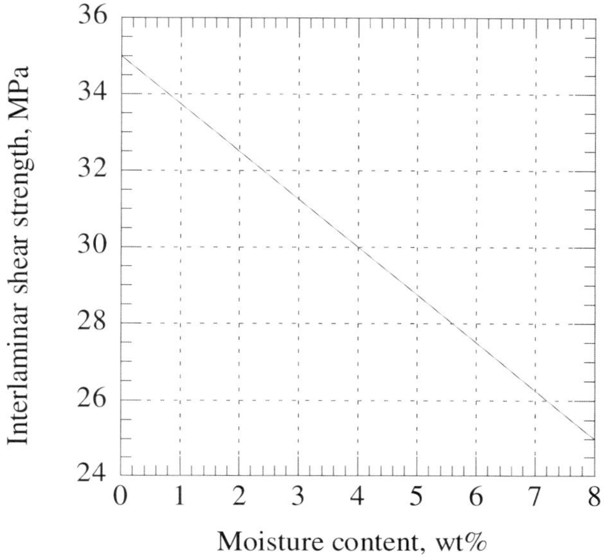


Figure 5.26. Interlaminar strength vs. moisture content in epoxy/aramid fiber laminates. [Data from Akay M, Mun S K A, Stanley A, *Composites Sci. Technol,* **57**, 1997, 565-71.]

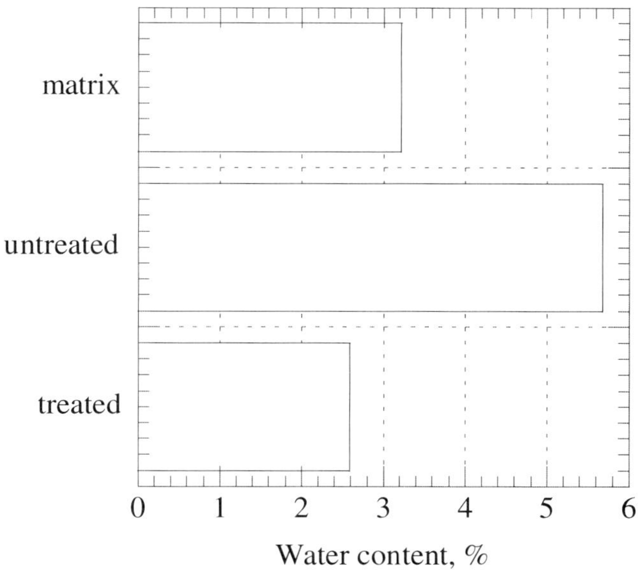


Figure 5.27. Water content in epoxy/glass beads composites. [Data from Wang J Y, Ploehn H J, *J. Appl. Polym. Sci.*, **59**, No.2, 1996, 345-57.]

over that of the plain matrix but a surface treatment of glass beads with silane decreases the water uptake to a value below the plain matrix.

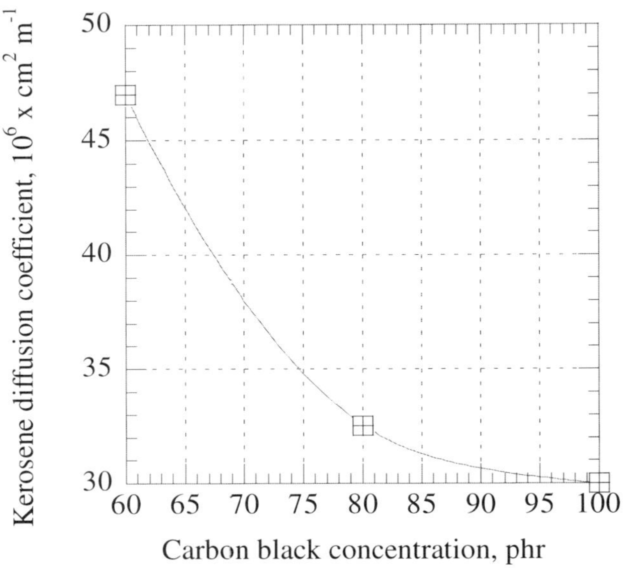


Figure 5.28. Kerosene diffusion coefficient in SBR rubber vs. carbon black concentration. [Adapted, by permission, from Nasr G M, Badawy M M, *Polym. Test.*, **15**, No.5, 1996, 477-84.]

In the rubber industry, moisture absorbed on the surface of silicate, impacts the rate and extent of cure and results in sponge-like textures. In moisture cured systems such as polyurethanes, polysulfides and silicones, moisture causes a premature reduction in shelf-life. In extrusion and injection molding the moisture absorbed on fillers contributes to various defects and a strict regime must be followed regarding the drying time and the conditions prior to processing. Lacing, a less well known phenomenon, is caused by the absorption of moisture on the surface of titanium dioxide.[163]

## 5.18 ABSORPTION OF LIQUIDS AND SWELLING[171-189]

Information on absorption of liquids and gases by filled materials remains limited even though it is very important to two areas of applications: filled reactive systems and solvent resistant materials.

In the study of precipitated silica grades (Zeosil), the absorption of four different amines was studied. The effect of the amine type on absorption was generally stronger than the silica grade but the absorption of all grades of silica increased when the concentration of functional groups (OH) on the surface was increased. Two grades had higher absorption levels because they had a pH below seven. Extraction by water removed a large part of the absorbed amine but 10-20% of the initial amine concentration always remained absorbed. This study may explain some reasons for retarded and incomplete cures in systems which contain fillers.

A mathematical model was proposed for evaluating the diffusion of a material which can react with the filler (e.g., acid). The proposed method permits the study of process kinetics for different concentrations of penetrant and filler.[179]

SBR filled with intercalated montmorillonite had substantially lower toluene uptake compared with the same rubber filled with carbon black (see Figure 15.42). Figure 5.28 shows that the diffusion coefficient of kerosene, which defines penetration rate, decreases when the concentration of carbon black in SBR vulcanizates is increased.[176] Figure 15.33 compares the uptake rate of benzene by unfilled rubber and by silica and carbon black filled rubber. Both fillers reduce the solvent uptake but carbon black is more effective.

Similarly, swelling of polyethylene filled with 35 and 50 wt% calcite was reduced. Table 5.13 gives equilibrium swelling of polyethylene in different solvents.

**Table 5.13. Equilibrium swelling of calcite-filled polyethylene. Data from Ref.[178]**

| Solvent | HDPE | HDPE+35 wt% calcite | HDPE+50 wt% calcite |
|---|---|---|---|
| heptane | 4.0 | 3.8 | 2.7 |
| o-xylene | 6.7 | 5.3 | 4.7 |
| tetrachloroethylene | 14.0 | 10.8 | 6.7 |

The swelling rate of polybutadiene/carbon black mixtures was reduced when the mixture was swollen, dried and swollen again.[182] This experiment, together with other studies conducted by NMR, explains the reasons for the reduction in swelling polymer/filler composites. As discussed in Chapter 7, the addition of filler and its interaction with polymer results in a bound fraction of polymer on the filler surface. During mixing, the interaction between the polymer and the filler surface is a chaotic process which causes the surface of filler to be incompletely covered by interacted chain segments. Swelling increases chain mobility and allows the chains to rearrange themselves to provide a more perfect coverage which increases the amount of bound polymer. The bound polymer fraction is then more difficult to swell which reduces the rate of solvent diffusion.

An increase in concentration of carbon fiber in SBR reduced the swelling rate but increased swelling anisotropy. Longer fibers (6 and 1 mm long fibers were studied) were more effective in the reduction of swelling in length direction but have almost no influence on swelling in the width direction. Increased anisotropy of swelling with fiber loading is explained by the increased fiber orientation with loading which thus only affects swelling behavior in the direction of orientation.

## 5.19 PERMEABILITY AND BARRIER PROPERTIES[190-197]

Plate like particles act as a barrier to gas diffusion by increasing the tortuosity of the diffusion pathway according to the following equation:

$$\frac{P_c}{P_p} = \frac{\phi_f}{1 + (W/2T)\phi_p} \qquad [5.24]$$

where

| | |
|---|---|
| $P_c$ | permeability of composite |
| $P_p$ | permeability of unfilled polymer |
| $\phi_p$ | volume fraction of polymer |
| $\phi_f$ | volume fraction of filler |
| W | particle width |
| T | particle thickness |

Figure 15.22 shows the effect of changes in the volume fraction of clay on $CO_2$ permeability. Permeability decreases most dramatically when the aspect ratio (particle width divided by particle thickness, W/T) is increased.[191] Figure 19.20 gives an example of the effect of talc loading on the oxygen permeability of HDPE film.[195,197] The practical application of mica in corrosion resistant coatings is widespread. The same principles apply to both liquids and gases. Section 5.12 gives the ranges of aspect ratios of available fillers.

Limiting the diffusion of oxygen improves the weather stability of materials due to reduced photooxidation.[194] This subject is discussed in Chapter 11.

There is still another aspect of permeability which has an influence on the durability of coatings. This is partially related to critical pigment volume concentration, CPVC (see Section 5.13 in this chapter) but it is also related to pigment-filler interaction relative to surface energy. A study on the effect of titanium dioxide on durability of coatings, containing different grades of titanium dioxide with different PVCs, shows that an increase in PVC decreases the resistance of the coating to salt spray but durability was also related to the grade of titanium dioxide used.[190] If the titanium dioxide did not have any surface coating, specimens of coatings cracked at very low concentration of pigment (PVC=6.4) well below the CPVC. By comparison, coatings containing titanium dioxide coated with $Al_2O_3$ and $SiO_2$ did not crack at PVC=17 which is slightly above the CPVC. This shows that permeability is also governed by pigment-filler interactions and the effect that a pigment has on the durability of a binder.

Fillers influence the performance of semi-permeable membranes. Semi-permeable membranes were obtained by stretching a highly filled film.[192] In another application, zeolites were used to obtain polymer membranes used in gas separation.[193]

## 5.20 OIL ABSORPTION

Oil absorption is a widely used parameter to characterize the effect of filler on rheological properties of filled materials. If oil absorption is low, the filler has little effect on the viscosity. The effect of particle shape on rheology should be considered since it is known that spherical particles aid flow due to their ball bearing effects. Fillers which have medium oil absorption are useful as co-thickeners. Filler having a very high oil absorption are used as thickeners and absorbents. Particle morphology (see Chapter 2 to view different morphological structures) may contribute to high oil absorptions (several hundred times the mass of filler) if the particles have exceptionally high porosities. Oil absorption must also be considered in applications which need filler for reinforcement.

The reinforcement by fillers increases as the filler concentration increases since the reinforcing mechanism is related to the presence of active sites on the filler surface which are available for reaction or interaction with matrix polymer. But this increase is limited by the effect a filler has on the rheological properties of a mixed material. There is a certain filler concentration above which the reinforcing effect of the dispersed filler is lost. Carbon black can serve as a simple example. Acetylene black has many useful properties but it cannot be used effectively for reinforcement because its structure does not permit high loadings whereas some furnace blacks can be loaded to high concentrations.

Table 5.14 gives an overview of oil absorptions. The oil absorptions are based on various grades to show the available variety.

**Table 5.14. Oil absorption of fillers**

| **Oil absorption range**, g/100 g | **Filler** (the range for a particular filler group is given in parentheses) |
|---|---|
| below 10 | barium sulfate (8-28), barium & strontium sulfates (9.5-11.5) |
| 10-19.9 | aluminum trihydroxide (12-41), calcium carbonate (13-21), ferrites (10.8-14.8), glass beads (17-20), iron oxide (10-35), fused silica (17-27), quartz, tripoli (17-20), sand (14-28), titanium dioxide (10-45), wollastonite (19-47), zinc sulfide (13-14) |
| 20-29.9 | aluminum oxide (25-225), cristobalite (21-28), feldspar (22-30), kaolin (27-48), slate flour (22-32), talc (22-57) |
| 30-49.9 | barium metaborate (30), ball clay (36-40), bentonite (36-52), carbon black (44-300), magnesium hydroxide (40-50) |
| 50-100 | attapulgite (60-120), graphite (75-175), kaolin beneficiated (50-60) kaolin calcinated (50-120), mica (65-72), precipitated silica (60-320), silica gel (80-280), wood flour (55-60) |
| over 100 | cellulose fiber (300-1000), diatomaceous earth (105-190), fumed silica (100-330), hydrous calcium silicate (290), perlite (210-240) |

## 5.21 HYDROPHILIC/HYDROPHOBIC PROPERTIES[147,198-199]

In water-based systems, it is important that the filler is compatible with water, usually, filler dispersion occurs in an aqueous medium before a polymer emulsion is added. The manufacturers of fillers for water-based systems frequently provide a simple demonstration of the change in the filler's hydrophobicity by comparing the

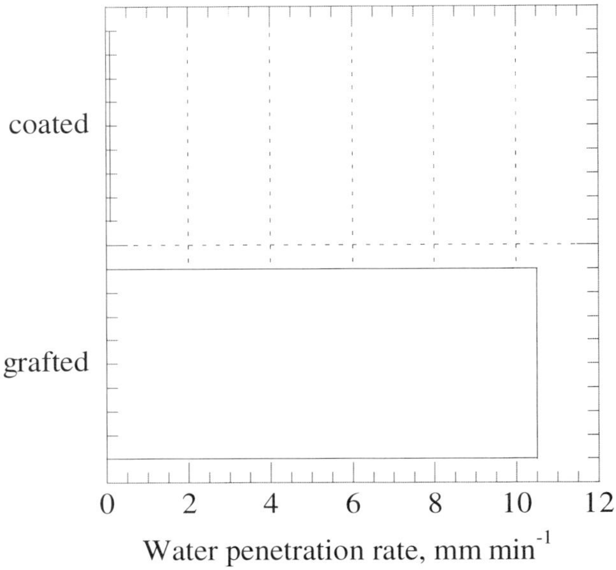


Figure 5.29. Penetration rate of water through column packed with grafted and stearate coated barium sulfate. [Adapted, by permission, from Tsubokawa N, Seno K, *J. Macromol. Sci. A*, **31**, No.9, 1994, 1135-45.]

unmodified filler which floats on water with the modified filler which mixes readily with water. There are numerous methods of increasing hydrophilic properties of fillers. These include grafting, surface coating, oxidation, etc. Figure 5.29 demonstrates the results of acrylamide grafted on the barium sulfate in comparison with stearate coated barium sulfate. These two products display a spectacular difference in behavior since the stearate coated barium sulfate floats on water in spite of the fact that its density is four times higher than that of water while the acrylamide grafted product readily sinks into water and mixes without difficulties. The penetration rate of water through a column packed with filler provides a method of quantifying these observations. Surface grafting with a hydrophilic polymer gives a substantial improvement in the compatibility of the filler and water.[147]

However, the hydrophilic surface of fillers is often a serious disadvantages, considering that the majority of polymers are hydrophobic. This single feature frequently diminishes the economic advantage gained from the use of relatively inexpensive filler because the cost of its dispersion outbalances the reduced cost of the material. Two research groups in Poland[198,199] contributed data which shows the broad spectrum of possibilities of filler modifications. The results of filler modification were quantified by using the degree of hydrophobicity calculated from the following equation:

$$N = 100(mH_i^B - nH_i^B) / mH_i^B \qquad [5.25]$$

where
$mH_i^B$ heat of immersion in benzene of modified filler
$nH_i^B$ heat of immersion in benzene of unmodified filler

The heats of immersions were measured in a differential calorimeter. Table 5.15 gives data for 2 wt% coating of the filler surface. More detailed information can be found in the original papers which, in addition to the full calorimetric data for 1, 2, and 3 wt% coatings, gives a set of mechanical properties of rubber vulcanizates and polyurethanes containing these modified fillers. Also, results for proprietary coatings are given which demonstrate the further improvement of quality in such fillers.

**Table 5.15. Degree of hydrophobicity of various fillers. Data from refs. 198 and 199**

| Surface modifier | Chalk | Precipitated $CaCO_3$ | Kaolin | Precipitated $SiO_2$ |
|---|---|---|---|---|
| stearic acid | 20.1 | 21.7 | 9.2 | |
| magnesium stearate | 21.1 | 20.1 | | |
| calcium stearate | 20.7 | 20.1 | 8.3 | |
| oleic acid | 21.1 | 23.0 | | |
| tall oil | 22.7 | 21.7 | | |
| tetrabutylammonium chloride | 23.7 | 22.6 | 26.2 | |
| sodium dodecylsulfate | 11.9 | 21.7 | 14.4 | |
| sodium glutamate | 13.8 | 23.6 | | |
| polyethylene glycol (10,000) | 9.8 | 8.4 | 15.4 | |
| mecaptosilane (A-189) | 8.2 | 4.6 | 27.8 | 24.6 |
| aminosilane (A-1100) | 5.6 | 3.4 | 18.5 | |
| isostearoil titanate (KRTTS) | 26.6 | 29.1 | 31.8 | |
| 9-butyl-3,6-dioxa-azatridecanol | | | | 27.3 |
| 3.6-dioxa-9-thiaheptadecanol | | | | 25.9 |
| 7,10,13,16-tetrathiadocosane | | | | 23.8 |

The fatty acid derivatives give a very good performance on calcium carbonate but are inferior on kaolin. The results of mechanical testing show that the ease of dispersion and mechanical properties of fillers are governed by interactions with the matrix polymer. Thus, mechanical testing of the filled material must be carried out before the best coating can be selected for a given polymer.

## 5.22 OPTICAL PROPERTIES[200-208]

Gloss and brightness are the most important indicators of paints and paper quality. Fillers and pigments influence both properties. The gloss of paper depends on the amount of pigment (relative to the coat weight) and the amount of thickener. The surface gloss of paints depends primarily on the film-forming properties of the resin although fillers may also influence gloss if they cause surface roughening (see Sections 5.4 and 5.13). Since gloss is the result of surface smoothness, the degree of pigment dispersion has an impact.[200] The paint formulation should be designed to assist dispersion of fillers and pigments but it should also include consideration of the processes occurring during drying. Two stages of paint drying are distinguished.[201] The first, wetter, stage involves removal of the majority of solvent. Surface tension dominates this stage. The surface of the drying film remains smooth. Surface tension remains constant throughout the drying process, and the compressive strength of the structure during the first stage is lower than the surface tension. In the second stage, the compressive strength and the yield stress increase until they exceed the surface tension. The yield stress of high gloss paint increases slower than that of low gloss paint. If the film shrinks, surface roughness develops.

As $TiO_2$ concentration increases, gloss increases because increasing the concentration of this pigment increases the refractive index. But, conversely gloss decreases as the amount of pigment increases because particulates roughen the surface. So the two mechanisms compete. In flocculated systems, the structure develops earlier during the drying process than in paints containing well-dispersed pigments. An increase in gloss and color strength follows the dispersion of coarse agglomerates.[202] During the service life of a paint, the gloss changes because of chalking – a phenomenon related to binder degradation. A degraded surface contains a particulate deposit which affects light reflection.[204] In black ink formulations, gloss mostly depends on particle size and the structure of carbon black. Smaller particle diameter and low structure of carbon black help to give the high gloss to black inks.

Brightness can be affected by fillers. The tables for individual fillers in Chapter 2 contain information on brightness. In white coatings, a yellowish undertone may be caused by binder or by fillers. This undertone can be eliminated by the addition of small quantities of a blue or violet pigment, carbon black with bluish tinge or fine particles of aluminum powder.[134] But such correction usually causes a loss of brightness. If optical brighteners are used, a loss of brightness can be avoided.

The hiding power of the pigmented material is a measure of its ability to hide a colored substrate or differences in the substrate color. The hiding power of the film is determined from the following equation:[134]

$$CR = L^*_B / L^*_w \qquad [5.26]$$

where
CR contrast ratio

$L^*_B$ brightness over a black substrate
$L^*_w$ brightness of a white substrate

According to DIN 53 778, the coating is considered fully opacifying if the contrast ratio, CR≥0.98. In inks and paints, hiding power or tinting strength is the most important factor characterizing the quality of the pigment. Particle size distribution is the major factor affecting the tinting strength of a filler. The type of filler, in conjunction with other components of the composition, determines the processes occurring during storage. Flocculation of pigment is usually responsible for a change of the initial hiding power.

In printing inks, which are pigmented with carbon black, the tone can be corrected by the choice of carbon black type. Since the tinting strength increases when the particle size and structure of carbon black decrease, the natural tendency is to use material of a very fine particle size. Black inks are usually required to have a blue tone, which is contrary to the choice of carbon black based on the particle size, because as the particle size decreases, the brown tone becomes more pronounced.

## 5.23 REFRACTIVE INDEX

Refractive index influences light scattering in fillers and pigments. A correct choice in the refractive index of the particulate material and binder permits a formulation of transparent materials containing fillers (for further information on light scattering see Chapter 2, especially section on titanium dioxide and Section 5.3). Table 5.16 gives an overview of refractive indices of various fillers.

**Table 5.16 Refractive indices of fillers**

| Refractive index range | Filler (refractive index of a particular group of fillers is given in parentheses) |
|---|---|
| 1 | air (1) |
| 1.3-1.49 | calcium carbonate - calcite (birefringence: 1.48 & 1.65), cristobalite (1.48), diatomaceous earth (1.42-1.48), fumed silica (1.46), precipitated silica (1.46) |
| 1.5-1.69 | aluminum trihydroxide (1.57-1.59), attapulgite (1.57), barium metaborate (1.55-1.6), barium sulfate (1.64), calcium hydroxide (1.57), calcium sulfate (1.52-1.61), feldspar (1.53), glass beads, flakes and fibers (1.51 A-glass and 1.55 E-glass), hydrous calcium silicate (1.55), kaolin (1.56-1.62), magnesium hydroxide (1.56-1.58), mica (1.55-1.69), perlite (1.5), pyrophyllite (1.57), quartz (1.56), talc (1.57-1.59), wollastonite (1.63), zinc borate (1.59) |
| 1.7-1.99 | aluminum oxide (1.7), antimony pentoxide (1.7), calcium carbonate - aragonite (1.7), magnesium oxide (1.736), sodium antimonate (1.75), zinc stannate (1.9) |
| 2-2.19 | antimony trioxide (2.087), zinc oxide (2) |
| 2.2 and above | barium titanate (2.4), iron oxide (2.94-3.22), titanium dioxide (2.55-2.7), zinc sulfide (2.37) |

The fillers in the Table 5.16 are divided into six groups. The first group includes air which is a good “pigment” because the difference in refractive index between air and most binders is in the range of 0.4-0.6 therefore it has a scattering power comparable to zinc oxide. The second group consists of fillers which are the most suitable materials for transparent products since their refractive indices fall into a range similar to many polymers. The third group consists of typical fillers. Even if they have a white color, their contribution to coloring is very small because of the small difference between their refractive index and that of binder. It is considered that material is a pigment if its refractive index is above 1.7 which is the case of the last three groups. The last group contains the most important white pigments. Because of its very high refractive index, titanium dioxide has the highest scattering power of white pigments.

## 5.24 FRICTION PROPERTIES[209-214]

Fillers are available with a range of frictional properties from self-lubricating through severely abrasive which permits applications which range from slide bearings to brake pads.

Polytetrafluoroethylene, molybdenum disulfide, graphite, and aramid fibers reduce the frictional coefficient. These may be used as single friction additive, in combination with other fillers, and in combination with silicone oil. Table 5.17 illustrates effect of PTFE on the frictional properties of different polymers.

**Table 5.17. Wear factor and dynamic coefficient of friction of different polymers containing PTFE Polymist.** *Courtesy of Ausimont USA, Inc.*

| Polymer | PTFE, % | Wear factor | | Dynamic coefficient of friction | |
|---|---|---|---|---|---|
| | | Unmodified | Modified | Unmodified | Modified |
| POM | 20 | 65 | 15 | 0.21 | 0.15 |
| ECTFE | 10 | 1000 | 27 | 0.29 | 0.11 |
| PA-6 | 20 | 200 | 15 | 0.26 | 0.19 |
| PA-66 | 20 | 200 | 12 | 0.28 | 0.18 |
| PC | 20 | 2500 | 70 | 0.38 | 0.14 |
| PBT | 20 | 210 | 15 | 0.25 | 0.17 |
| PPS | 20 | 540 | 55 | 0.24 | 0.10 |
| PU | 15 | 340 | 60 | 0.37 | 0.32 |

The coefficient of friction and wear are substantially reduced by the incorporation of PTFE powder. Molybdenum disulfide has an even broader range of application temperatures than PTFE (-150 to 300°C, PTFE up to 260°C) and provides even better performance under high load. For this reason it is used either in combination with PTFE or alone. Aramid fibers give additionally reinforcement therefore are frequently found in combinations with other fillers.

Many fillers play a prominent role in brake pads and clutch linings. These include fibers such as aramid, glass, carbon, steel, and cellulose; low cost fillers such as barites, calcium carbonate and clay; frictional modifiers such as alumina, metallic flakes and powders. The combination of these materials with binders gives a broad range of brake pad materials.

Numerous other materials are used as a components of proprietary polishing and abrasive materials with a variety of uses.

## 5.25 HARDNESS[8,215,216]

Hardness of fillers is summarized in Table 5.18.

**Table 5.18. Hardness of fillers**

| Mohs hardness | **Filler** (the range for a particular filler is given in parentheses) |
|---|---|
| 1 | attapulgite (1-2), bentonite (1-2), carbon fibers (0.5-1), graphite (1-2), molybdenum disulfide (1), precipitated silica (1), pyrophyllite (1-2.5), talc (1-1.5) |
| 2 | aluminum flakes and powder (2-2.9), aluminum trihydroxide (2.5-3.5), anthracite (2.2), calcium sulfate (2), clay (2-2.5), copper (2.5-3), gold (2.5-3), kaolin (2), mica (2.5-4), sepiolite (2-2.5), silver (2.5-4) |
| 3 | barium sulfate (3-3.5), calcium carbonate (3-4), dolomite (3.5-4), iron oxide (3.8-5.1), lithopone (3), zinc sulfide (3) |
| 4 | calcinated kaolin (4-8), wollastonite (4.5), zinc oxide (4) |
| 5 | apatite (5), ceramic beads (5-7) perlite (5.5), pumice (5.5), silver-coated, light glass spheres (5-6), titanium dioxide - anatase (5-6) |
| 6 | cristobalite (6.5), feldspar (6-6.5), glass beads, flakes and fibers (6 for A-glass and 6.5 for E-glass), silica gel (6), titanium dioxide - rutile (6-7) |
| 7 | fused silica (7), silver-coated, thick-wall glass spheres (7), quartz (7), sand (7) |
| 9 | aluminum oxide (9), carborundum (9-10), tungsten powder (9) |

The most popular fillers are soft materials in the hardness range of 1-3. Silica fillers are hard and frequently abrasive. Most grades of silicas have a hardness in the range 6-7.

The effect which fillers have on the hardness of filled materials is detailed in the data in Figure 5.30. Graphite is a soft material but still it may either increase or decrease the hardness of a polymer depending on its interaction and particle size. In polyamide-66, small particle size graphite increases hardness while coarse particles have little influence on the hardness of the composite. In polypropylene, all grades of graphite substantially increase hardness. But with polystyrene (not shown here), hardness is decreased by all grades of graphite. The effect depends on the interaction between polymer and filler.

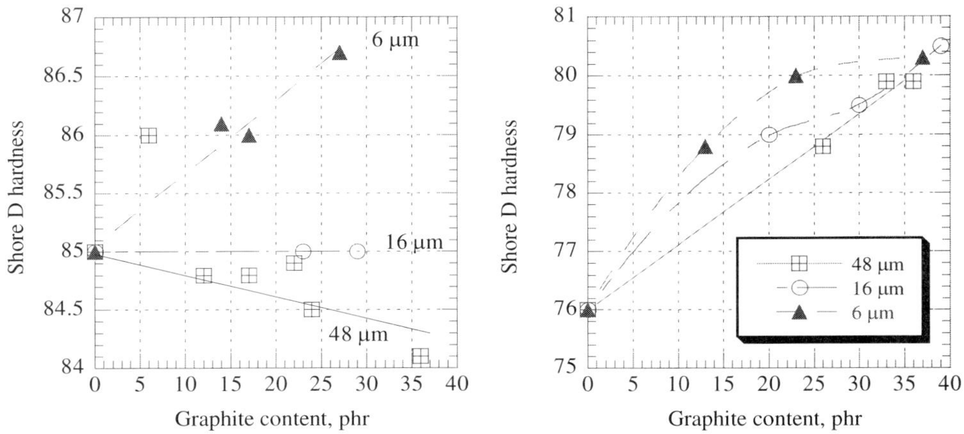


Figure 5.30. Hardness of composite vs. graphite concentration and type. Left -PA-66, right - PP. *Courtesy of Timcal Ltd., Sins, Switzerland.*

The general trend in filled material is that fillers increase hardness as the filler concentration is increased. In highly filled materials, especially those filled with silica flour, the hardness of the composite approaches the hardness of the filler. Several different fillers were found to induce a softening effect in aged PDMS.[215] While a freshly prepared composite increased in hardness as the filler concentration was increased, the aged material reached a minimum hardness around 20 wt% filler. The hardness then increased gradually as the spaces between particles were taken up by the filler.[215]

## 5.26 INTUMESCENT PROPERTIES[217-220]

Natural graphite brings intumescence to products used in construction and other applications where fire retardancy is important. The growing interest in intumescent products stems from findings that the most effective method of decreasing the combustibility of plastics is to use additives which cause carbonization of the organic materials. The material should also retain the formed gases and expand to built an insulating layer.

A combination of materials must be used to regulate the kinetics of such processes as degradation, gas formation, char formation, and foam growth. The major components include a carbonization catalyst, a carbonization agent, and a blowing agent. These components are designed to form gaseous products which cause expansion of the product (e.g., coating or sealant). The design of the product must also include mechanisms which allow it to retain these formed gases. With foams the pressure in the bubble must be balanced by the surface tension and mechanical properties of the bubble wall for the gas to be retained. For this reasons, it is important to design composition which changes its properties under increasing heat to re-

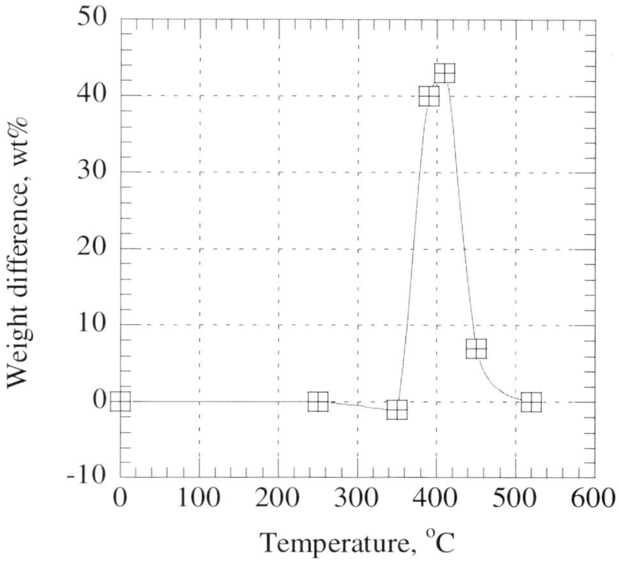


Figure 5.31. Weight difference of intumescent formulation based on LDPE vs. temperature. [Adapted, by permission, from Le Bras M, Bourbigot, Le Tallec Y, Laureyns J., *Polym. Degradat. Stabil.*, **56**, 1997, 11-21.]

tain sufficient mechanical properties such that gas is retained. Both the resin and the filler play a part in this process.

The success of graphite in this applications shows that filler with plate like structures should be considered when intumescent materials are being formulated. Recent developments in intumescent paints[219] show that performance can be improved if a layer of organic material is inserted between the layers of the plate like filler. The degradation of this material in the enclosed space increases the expansion rate and the retention of gas inside the degrading material. Based on this principle any plate like filler has the potential to be useful in an intumescent application. The composition of filler is also important. When clay was used as a filler in fire retardant applications, it was found that some of its components interfere with the action of carbonization catalysts and detract from the overall performance of the system in terms of limiting oxygen index.[218]

Figure 5.31 shows a curve typical of the performance of intumescent material. The degradation process should occur rapidly which generates an insulation layer in a short period of time and keeps the temperature of adjacent layers sufficiently low to prevent their degradation. In the graph, the height of peak is important since it shows the amount of retained material.

## 5.27 THERMAL CONDUCTIVITY[78,126,189,221-230]

Table 5.19 gives an overview of the thermal conductivity of various fillers. The data in the table is skewed towards thermal insulators at one and at thermal conductors at the other range since data for other fillers are seldom available because they are not intended for heat insulating or conducting applications. In most cases, nonmetallic fillers are thermal insulators but pitch-based carbon fiber is the exception. It has a higher thermal conductivity than any metal.

**Table 5.19. Thermal conductivity of fillers**

| **Thermal conductivity range**, W/K·m | **Filler** (thermal conductivity given in parentheses) |
|---|---|
| below 10 | aramid fiber (0.04-0.05), calcium carbonate (2.4-3), ceramic beads (0.23), glass fiber (1), magnesium oxide (8-32), fumed silica (0.015), fused silica (1.1), molybdenum disulfide (0.13-0.19), PAN-based carbon fiber (9-100), sand (7.2-13.6), talc (0.02), titanium dioxide (0.065), tungsten (2.35), vermiculite (0.062-0.065) |
| 10-29 | aluminum oxide (20.5-29.3), pitch-based carbon fiber (25-1000) |
| 100-199 | graphite (110-190), nickel (158) |
| above 200 | aluminum flakes and powder (204), beryllium oxide (250), boron nitride (250-300), copper (483), gold (345), silver (450) |

Figure 15.17 shows that high aspect ratio carbon fibers are used to make materials electrically conductive. Figure 15.19 shows that thermal conductivity depends only on the amount of carbon fibers, not on their length or aspect ratio.[126] Mathematical modelling which shows that high aspect ratio fibers should increase thermal conductivity but some practical experiments disprove this.[221] Several other models analyzed in a review paper[224] are in agreement with the experimental data and this analysis confirms that the thermal conductivity of filler and its concentration are the main parameters determining the thermal conductivity of composite.[224] A composite based on epoxy resin (60 parts) and pitch-based carbon fiber (40 parts) had a thermal conductivity of 540 W/K·m which is higher than the thermal conductivity of metal. In another study,[225] the thermal conductivity of HDPE filled with aluminum particles was found to be 3.5 W/K·m.

In modern electronic devices there is a need to manufacture materials which have high thermal conductivity and a high electrical resistance. The data in the Table 5.19 show that such a requirement can be easily fulfilled using boron nitride or beryllium oxide. Both fillers have excellent thermal conductivity and they are electrical insulators.

Some of the insulating fillers found in the first row of Table 5.19 are used in foams and adhesives designed for insulation in modern appliances.[229,230]

## 5.28 THERMAL EXPANSION COEFFICIENT[231-234]

Table 5.20 contains data on the linear thermal expansion coefficient of various fillers. The data indicate that most fillers, especially these used for reinforcement, have much lower coefficient of thermal expansion than metals and plastics. This is an important fact which should be considered in formulating plastics exposed to wide temperature swings since one of the requirements of filler addition is to reduce thermal expansion and improve dimensional stability of plastics. This data also shows that it is preferable to use mineral fillers for thermally conductive plastics because they have low thermal expansion coefficient.

**Table 5.20. The linear thermal expansion coefficient, α, of different fillers in temperature range of 20-200°C**

| α **range**, $10^{-6}$ $K^{-1}$ | **Fillers** (the value of α for a particular group of fillers given in parentheses) |
|---|---|
| below 5 | aramid fiber (-3.5), boron oxide (<1), calcium carbonate (4.3-10), calcinated kaolin (4.9), carbon fiber (-0.1 to -1.45), fused silica (0.5), glass beads and fiber from E-glass (2.8), pyrophyllite (3.5) |
| 5-9.9 | beryllium oxide (9), glass beads and fiber from A-glass (8.5), mica (7.1-14.5), talc (8), titanium dioxide (8-9.1), wollastonite (6.5) |
| 10-14.9 | barium sulfate (10-17.8), dolomite (10.3), magnesium oxide (13), molybdenum disulfide (10.7), quartz (14), sand (14) |
| 15-19.9 | feldspar (19) |
| 20-29.9 | aluminum flakes and powder (25) |
| 30-100 | cristobalite (56) |

Thermal expansion can be used as simple method of verifying the adhesion between the filler and the matrix. If the adhesion is poor the composite will have high thermal expansion.[232]

## 5.29 MELTING TEMPERATURE

Melting temperatures of fillers are given in the tables for individual fillers in Chapter 2. These temperatures are usually so high that they do not have much relevance to filler choice. The only area when the melting or decomposition temperature of the filler may become relevant is in the processes of filler recovery from waste plastics. Such studies were not found in the literature.

Fillers such as magnesium hydroxide and aluminum trihydroxide are used as flame retardants because their decomposition product – water – is an active ingredient in flame retardancy. These fillers are discussed in detail in Chapter 12.

## 5.30 ELECTRICAL PROPERTIES[4,8,52,75,78,89,102,126-7,177,185-6,189,204,224,234-272]

One single property of filler – electric conductivity – affects many properties of the final products. These properties include electric insulation, conductivity, superconductivity, EMI shielding, ESD protection, dirt pickup, static decay, antistatic properties, electrocatalysis, ionic conductivity, photoconductivity, electromechanical properties, thermo-electric conductivity, electric heating, paintability, biocompatibility, etc. Possession of one of these properties in a polymer can make it useful in industry and everyday use. Examples are given in Chapter 19. Here, the electrical properties of fillers are summarized and the general effect of a filler's conductivity on the properties of filled materials is analyzed.

## Table 5.21. Electrical properties of fillers

| Filler | Resistivity 'Ω-cm | Dielectric constant | Dielectric strength V/cm | Loss tangent |
|---|---|---|---|---|
| Aluminum | $2.8 \times 10^{-6}$ | | | |
| Aluminum oxide | $10^{14}$-$10^{22}$ | 9-9.5 | 2560 | 0.0002-0.004 |
| Aluminum trihydroxide | | 7 | | |
| Anthracite | 50 | | | |
| Barium sulfate | 19.075 | 11.4 | | |
| Barium titanate | | 3.8 | | |
| Beryllium oxide | $10^{17}$ | 6.8 | 100 | 0.0004 |
| Boron nitride | $10^{15}$ | 3.9 | | <0.0002 |
| Calcium carbonate | $10^{10}$ | 6.1-8.5 | | |
| Carbon fiber | $10^{-2}$-$10^{-5}$ | | | |
| Carbon fiber, Ni-coated | $6 \times 10^{-6}$ | | | |
| Ceramic beads | | 1.6 | | |
| Copper | $1.6 \times 10^{-6}$ | | | |
| Ferrites | $10^{2}$-$10^{10}$ | 8-22 | | |
| Fumed silica | $10^{13}$ | | | |
| Fused silica | $10^{17}$-$10^{18}$ | 3.78 | | 0.001 |
| Glass beads | $10^{7}$ | 1.2-7.6 | 4500 | 0.015-0.058 |
| Glass fibers | $10^{13}$-$10^{16}$ | 5.8-6.1 | | 0.001 |
| Gold | $2.1 \times 10^{-6}$ | | | |
| Graphite | 0.8-2.5 | | | |
| Kaolin | | 1.3-2.6 | | |
| Mica | | | | 0.0013-0.04 |
| Nickel | $7.8 \times 10^{-6}$ | | | |
| Precipitated silica | $10^{11}$-$10^{14}$ | 1.9-2.8 | | 0.00001-0.02 |
| Sand | $10^{14}$-$10^{16}$ | 4 | | 0.0002 |
| Silver | $1.6 \times 10^{-6}$ | | | |
| Steel | $72 \times 10^{-6}$ | | | |
| Talc | | 7.5 | | |
| Titanium dioxide | $3\text{-}9 \times 10^{3}$ | 48 (anatase)<br>114 (rutile) | | 0.01-0.35 |
| Tungsten | $5.6 \times 10^{-6}$ | | | |

Table 5.22 gives resistivity and dielectric constants of selected polymers.

**Table 5.22. Resistivity and dielectric constants of some polymers**

| Polymer | Resistivity, Ω-cm | Dielectric constant |
|---|---|---|
| Epoxy resin | $10^{12}$-$10^{14}$ | 3.5-6 |
| Polyethylene | $>10^{15}$ | 2.3 |
| Polypropylene | $>10^{15}$ | 2.2-2.6 |
| Polystyrene | $>10^{16}$ | 2.5-2.65 |
| Polytetrafluoroethylene | $10^{18}$ | 2 |
| Polyvinyl chloride | $10^{12}$-$10^{16}$ | 3.2-4 |
| Silicone | $>10^{12}$ | 3.5 |

When the two tables are compared it is evident that there is a wide choice in fillers which either enhance or retain dielectric properties of polymers. It is more difficult to formulate conductive polymers where consideration must be given to how the filler can change properties of the polymer. Electrically conductive polymers can be divided into three groups:[255]

| | **resistivity range, Ω-cm** | **applications** |
|---|---|---|
| low conductors | $10^6$-$10^{11}$ | antistatic protection |
| semi-conductors | $10^2$-$10^6$ | EMI shielding, ESD dissipation |
| conductors | below $10^2$ | heaters, sensors, elastic conductors |

Comparing the range of conductivity of low conductors with resistivity of some fillers in Table 5.21 shows that the task of their formulation is not difficult.

For EMI shielding applications, numerous processes are used, some require conductive fillers. These applications include parts molded with conductive filler and conductive paints. Conductive fillers used in commercial applications include aluminum, silver, nickel, and copper flakes and powders, stainless steel fibers, and fibers and flakes coated by nickel and silver. Thermoplastic compounds can provide up to 65-70 dB of electromagnetic noise attenuation but obtaining values over 45 dB is difficult. Static dissipative compounds (ESD) are mostly produced with carbon black which accounts for approximately 90% of the market but many other fillers are also used.

Several general principles determine the amount of filler which must be incorporated. Figure 5.32 shows a typical relationship between the concentration and conductivity. The initial addition of conductive fillers does very little to the change of conductivity until a threshold concentration or percolation threshold is attained.

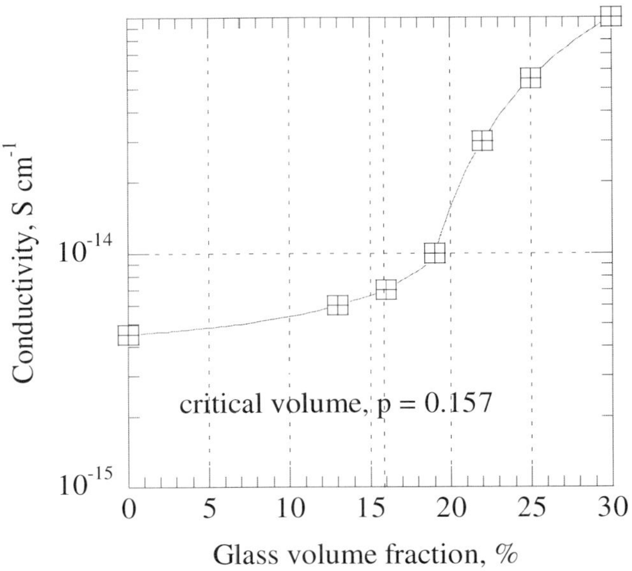


Figure 5.32. Conductivity of epoxy resin filled with silver coated glass beads vs. volume concentration. [Adapted, by permission, from Lekatou A, Faidi S E, Lyon S B, Newman R C, *J. Mat. Res.*, **11**, No.5, 1996, 1293-304.]

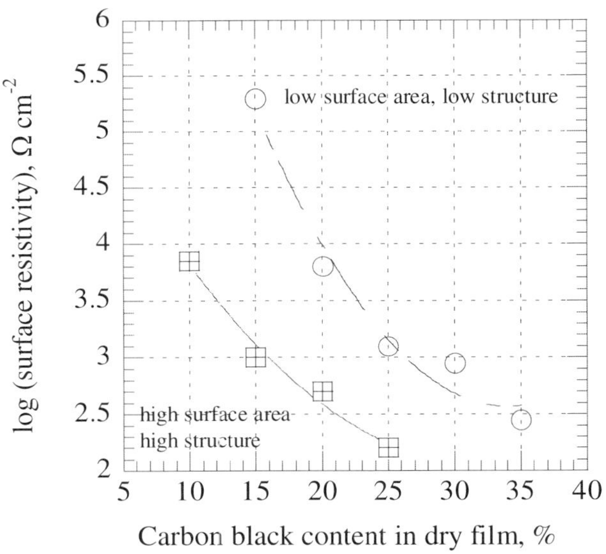


Figure 5.33. Resistivity of acrylic resin vs. concentration of carbon black. [Adapted, by permission, from Foster J K, Sims E S, Venable S W, *Paint & Ink Int.*, **8**, No.3, 1995, 18-21.]

The amount of filler required to reach this threshold value depends on the conductivity of the particular filler, its particle shape, and its interaction with matrix. After percolation threshold, conductivity increases rapidly. The steepness of the increase

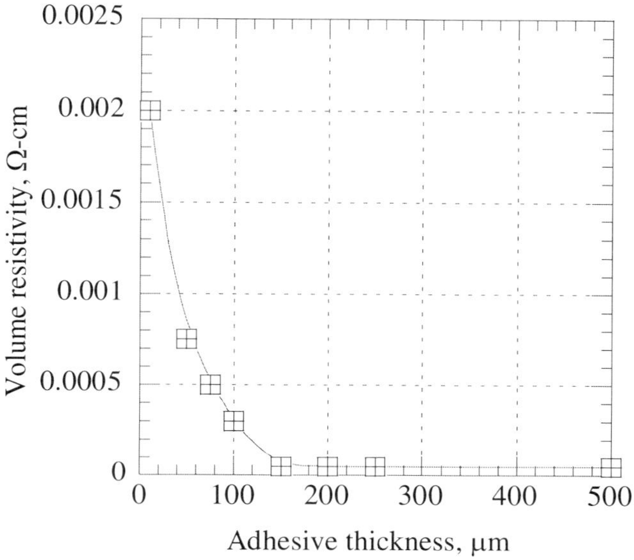


Figure 5.34. Resistivity vs. adhesive layer thickness. [Adapted, by permission, from Wei Y, Sancaktar E, *J. Adhesion Sci. Technol.*, **10**, No.11, 1996, 1199-219.]

is controlled mostly by the particle shape and the intrinsic conductivity of the filler. Finally, the conductivity reaches a plateau the value of which depends both on the conductivities of the filler and the matrix.

Particle size, and in the case of carbon black, its structure, and the amount used determine the properties of the filled composite (Figure 5.33). The smaller the particle and the higher the structure, the less carbon black is required. The same holds true for particulate materials (see Figures 15.10 and 15.37).

A third important filler parameter is related to its shape. Figure 15.17 shows that the aspect ratio of carbon fiber affects conductivity. If the fiber is milled to almost spherical particles, its percolation threshold concentration is substantially increased.

In very thin conductive layers such as adhesives, paints or inks, the layer thickness plays a big part (Figure 5.34).[267] The graph shows that a certain thickness is required before a full conductivity effect is obtained.

## 5.31 MAGNETIC PROPERTIES[273-279]

Two other sections are devoted to the magnetic properties of fillers. Filler materials are discussed in Section 2.1.29 and some examples of such products are included in Section 19.23. Figure 5.35 shows that fiber orientation strongly influences magnetic properties. Figure 5.36 shows that the shape of the manufactured article may determine how the filler particles are oriented. This, in turn may determine if the filler is being used effectively. In addition to orientation, aspect ratio, particle size

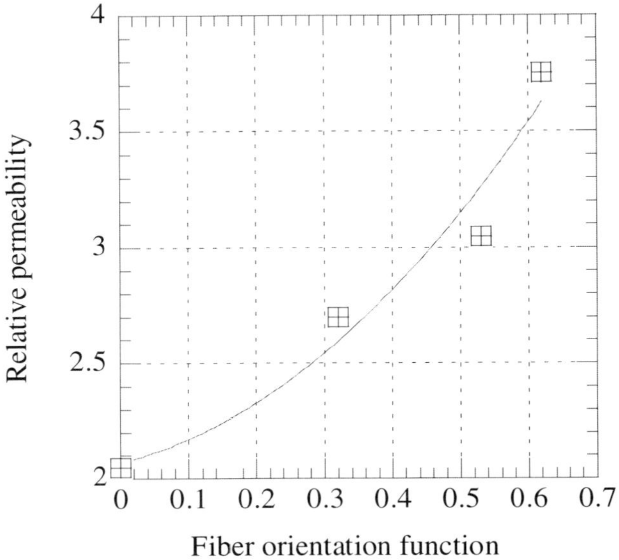


Figure 5.35. Relative permeability vs. nickel fiber orientation in HDPE matrix. [Adapted, by permission, from Fiske T, Gokturk H S, Yazici R, Kalyon D M, *Polym. Eng. Sci.*, **37**, No.5, 1997, 826-37.]

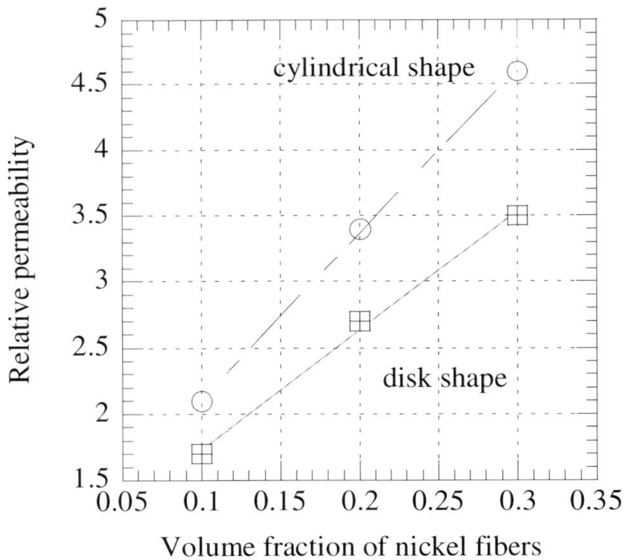


Figure 5.36. Relative permeability vs. volume fraction of nickel fibers in HDPE depending on article shape. [Adapted, by permission, from Fiske T, Gokturk H S, Yazici R, Kalyon D M, *Polym. Eng. Sci.*, **37**, No.5, 1997, 826-37.]

and method of processing affect properties of manufactured materials with magnetic properties.

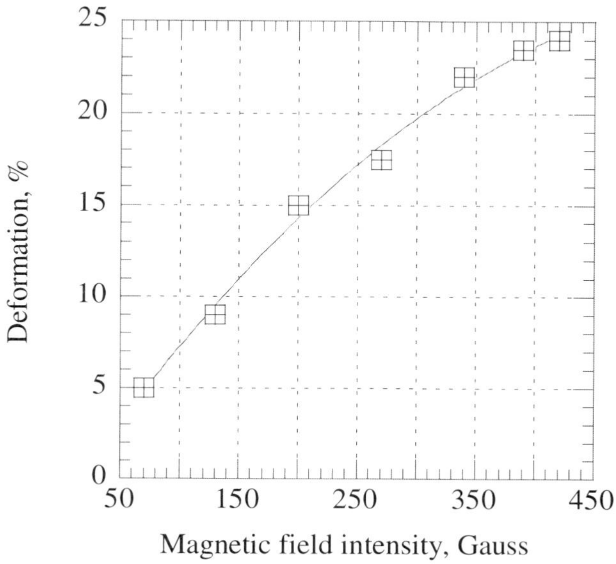


Figure 5.37. Deformation vs. magnetic field in polyacryloamide gel containing ferrite. [Adapted, by permission, from Klapcinski T, Galeski A, Kryszewski M, *J. Appl. Polym. Sci.*, **58**, No.6, 1995, 1007-13.]

Figure 5.37 gives an interesting example of a magneto-mechanical device. Polyacrylamide gel containing ferrite was magnetized in compressed stage. Application of magnetic field causes deformation of gel depending on magnetic field intensity and *vice versa.*

## REFERENCES


1 Gallas M R, Rosa A R, Costa T H, da Jornada J A H, *J. Mater. Res.*, **12**, No. 3, 1997, 764-8.
2 Coran A Y, Ignatz-Hoover F, Smakula P C, *Rubb. Chem. Technol.*, **67**, No.2, 1994, 237-51.
3 Clarke J, Freakley P K, *Rubb. Chem. Technol.*, **67**, No.4, 1994, 700-15.
4 Wampler W A, Rajeshwar K, Pethe R G, Hyer R C, Sharma S C, *J. Mat. Res.*, **10**, No.7, 1995, 1811-22.
5 Schneider J P, Myers G E, Clemons C M, English B W, *J. Vinyl and Additive Technol.*, **1**, No.2, 1995, 103-8.
6 Wang W D, Haidar B, Vidal A, Donnet J B, *Kaut. u. Gummi Kunst.*, **47**, No.4, 1994, 238-41.
7 Ogadhoh S O, Papathanasiou T D, *Composites Part A: Applied Science and Manufacturing*, **27A**, No.1, 1996, 57-63.
8 Larena A, Pinto G, *Polym. Composites*, **16**, No.6, 1995, 536-41.
9 Golubev A I, *Int. Polym. Sci. Technol.*, **22**, No.11, 1995, T/67-9.
10 Magrupov M A, Umarov A V, Saidkhodzhaeva K S, Kasimov G A, *Int. Polym. Sci. Technol.*, **23**, No.1, 1996, T/77-9.
11 Kiselev V Y, Vnukova V G, *Int. Polym. Sci. Technol.*, **23**, No.5, 1996, T/88-92.
12 Mushack R, Luttich R, Bachmann W, *Eur. Rubb. J.*, **178**, No.7, 1996, 24-9.
13 Jesionowski T, Krysztafkiewicz A, *Pigment Resin Technol.*, **25**, No.3, 1996, 4-14.
14 Jones D W, Rizkalla A S, *J. Biomedical Materials Research (Applied Biomaterials)*, **33**, No.2, 1996, 89-100.
15 Weeling B, Electrical Conductivity in Heterogeneous Polymer Systems. Conductive Polymers, Conference Proceedings, 1992, Bristol, UK.
16 Itatani K, Yasuda R, Scott Howell F, Kishioka, *J. Mater. Sci.*, **32**, 1997, 2977-84.
17 Kauly T, Keren B, Siegmann A, Narkis M, *J. Mater. Sci.*, **32**, 1997, 693-9.

18 Tsagaropoulos G, Eisenberg A, *Macromolecules*, **28**, No.1, 1995, 396-8.
19 Maeda S, Armes S P, *Synthetic Metals*, **73**, No.2, 1995, 151-5.
20 Hegedus C R, Kamel I L, *J. Coatings Technol.*, **65**, No.820, 1993, 31-43
21 Vratsanos L A, Farris R J, *Polym. Engng. Sci.*, **33**, No.22, 1993, 1458-65.
22 Skelhorn D A, Antec '93. Conference Proceedings, New Orleans, La., 9th-13th May 1993, Vol. II, 1965-70.
23 Byers J T, Meeting of the Rubber Division, ACS, Cleveland, October 17-20, 1995, paper B.
24 Kovacevic V, Lucic S, Hace D, Glasnovic A, Smit I, Bravar M, *J. Adhesion*, **47**, No.1-3, 1994, 201-15.
25 Hornsby P R, Wang J, Cosstick K, Rothon R, Jackson G, Wilkinson G, Flame Retardants '94. Conference proceedings, London, 27th-28th January 1994, 93-108.
26 Maeda S, Armes S P, J. *Mat. Chem.*, **4**, No.6, 1994, 935-42.
27 Zaborskaya L V, Dovgualo V A, Yurkevich O R, *J. Adhesion Sci. Technol.*, **9**, No.1, 1995, 61-71.
28 Yoshinaga K, Hidaka Y, *Polym. J. (Jap.)*, **26**, No.9, 1994, 1070-9.
29 Gerspacher M, O'Farrel C P, Wampler W A, *Rubb. World*, **212**, No.3, 1995, 26-9.
30 Cha Y J, Choe S, *J. Appl. Polym. Sci.*, **58**, No.1, 1995, 147-57.
31 Yu Long, Shanks R A, *J. Appl. Polym. Sci.*, **61**, No.11, 1996, 1877-85.
32 Magee R W, *Rubb. Chem. Technol.*, **68**, No.4, 1995, 590-600.
33 Tan L S, McHugh, *J. Mater. Sci.*, **31**, 1996, 3701-6.
34 Pukanszky B, Voros G, *Polym.Composites*, **17**, No.3, 1996, 384-92.
35 Kovacevic V, Lucic S, Hace D, Glasnovic A, *Polym. Engng. Sci.*, **36**, No.8, 1996, 1134-9.
36 Hashemi S, Din K J, Low P, *Polym. Engng. Sci.*, **36**, No.13, 1996, 1807-20.
37 Mitsui S, Kihara H, Yoshimi S, Okamoto Y, *Polym. Engng. Sci.*, **36**, No.17, 1996, 2241-6.
38 Lawandy S N, Botros S H, Darwish N A, Mounir A, *Polym. Plast. Technol. Engng.*, **34**, No.6, 1995, 861-74.
39 Gorl U, Rausch R, Esch H, Kuhlmann R, *Int. Polym. Sci. Technol.*, **23**, No.7, 1996, T/81-7.
40 Johnson K C, Antec '96. Volume III. Conference proceedings, Indianapolis, 5th-10th May 1996, 3545-9.
41 Stockblower D, Antec '96. Volume II. Conference proceedings, Indianapolis, 5th-10th May 1996, 1690-4.
42 Ishibashi J, Kobayashi A, Yoshikawa T, Shinozaki K, Antec '96. Vol. I. Conference Proceedings, Indianapolis, 5th-10th May 1996, 386-90.
43 Nguyen T N, Lethiecq M, Levassort F, Patat F, *Int. J. Polym. Analysis and Characterization*, **1**, No.4, 1995, 277-87.
44 Chen L, Liu K, Yang C Z, *Polym. Bull.*, **37**, No.3, 1996, 377-83.
45 Yoshinaga K, Iwasaki M, Teramoto M, Karakawa H, *Polym. & Polym. Composites*, **4**, No.3, 1996, 163-72.
46 Sanchez-Solis A, Estrada M R, *Polym. Degradat. Stabil.*, **52**, No.3, 1996, 305-9.
47 Minkova L, *Coll. Polym. Sci.*, **272**, No.2, 1994, 115-20.
48 Svistkov A L, *Polym. Sci.*, **36**, No.3, 1994, 337-41.
49 Canova L A, Antec '93. Conference Proceedings, New Orleans, La., 9th-13th May 1993, Vol. II, 1943-9.
50 Herd C R, Meeting of the Rubber Division, ACS, Montreal, May 5-8, 1996, paper B.
51 Bergado S, *Plast. Compounding*, **17**, No.7, 1994, 32-8.
52 Garncarek Z, Piasecki R, Borecki J, Maj A, Sudol M, *J. Phys. D*, **29**, No.5, 1996, 1360-6.
53 Scaffaro R, Pedretti U, La Mantia F P, *Eur. Polym. J.*, **32**, No.7, 1996, 869-75.
54 Averous L, Quantin J C, Lafon D, Crespy A, *Int. J. Polym. Analysis and Characterization*, **1**, No.4, 1995, 339-47.
55 Kretzschmar B, *Kunststoffe Plast Europe*, **86**, No.4, 1996, 20-2.
56 Donnet J B, Custodero E, Wang T K, *Kaut. u. Gummi Kunst.*, **49**, No.4, 1996, 274-9.
57 Meijerink J I, Eguchi S, Ogata M, Ishii T, Amagi S, Numata S, Sashima H, *Polymer*, **35**, No.1, 1994, 179-86.
58 Donnet J B, *Kaut. u. Gummi Kunst.*, **47**, No.9, 1994, 628-32.
59 Jarvela P A, Jarvela P K, *J. Mat. Sci.*, **31**, No.14, 1996, 3853-60.
60 Arakawa K, Iwami E, Kimura K, Nomaguchi K, *J. Reinf. Plast. Comp.*, **13**, No.12, 1994, 1100-15.
61 Gent A N, Lai S-M, *Rubb. Chem. Technol.*, **68**, No.1, 1995, 13-25.
62 Hjelm R P, Wampler W A, Seeger P A, Gerspacher M, *J. Mat. Res.*, **9**, No.12, 1994, 3210-22.
63 Srinivasan G, Reneker D H, *Polym. Int.*, **36**, No.2, 1995, 195-201.
64 Wada N, Uchiyama Y, *Int. Polym. Sci. Technol.*, **21**, No.10, 1994, T/23-34.

65 Dzenis Y A, Reneker D H, Tsukruk V V, Patil R, *Composite Interfaces*, **2**, No.4, 1994, 307-19.
66 Al-Turaif H, Unertl W N, Lepoutre P, *J. Adhesion Sci. Technol.*, **9**, No.7, 1995, 801-11.
67 Donnet J B, Tong Kuan Wang, *Macromol. Symp.*, **108**, 1996, 97-109.
68 Donnet J B, Wang T K, *Prog. Rubb. Plast. Technol.*, **11**, No.4, 1995, 261-7.
69 Okel T A, Waddell W H, *Rubb. Chem. Technol.*, **68**, No.1, 1995, 59-76.
70 Wang M-J, Wolff S, Tan E-H, *Rubb. Chem. Technol.*, **66**, No.2, 1993, 178-95.
71 Wolff S, Wang M-J, Tan E-H, *Rubb. Chem. Technol.*, **66**, No.2, 1993, 163-77.
72 Byung Suk Jin, Kwang Hee Lee, Chul Rim Choe, *Polym. Int.*, **34**, No.2, 1994, 181-5.
73 Molesky F, Schultz R, Midgett S, Green D, *J. Vinyl Additive Technol.*, **1**, No.3, 1995, 159-61.
74 Vidal A, Wang W, Donnet J B, *Kaut. u. Gummi Kunst.*, **46**, No.10, 1993, 770-8.
75 Probst N, Smet H, *Kaut. u. Gummi Kunst.*, **48**, No.7-8, 1995, 509-11.
76 Agarwal S, Campbell G A, Antec 95. Volume I. Conference proceedings, Boston, Ma., 7th-11th May 1995, 839-42.
77 Cochet P, Barruel P, Barriquand L, Grobert J, Bomal Y, Prat E, IRC '93/144th Meeting, Fall 1993. Conference Proceedings, Orlando, Fl., 26th-29th Oct.1993, Paper 162.
78 Nasr G M, Badawy M M, Gwaily S E, Shash N M, Hassan H H, *Polym. Degradat. Stabil.*, **48**, No.2, 1995, 237-41.
79 Allen N S, Edge M, Corrales T, Childs A, Liauw C, Catalina F, Peinado C, Minihan A, *Polym. Degradat. Stabil.*, **56**, 1997, 125-39.
80 Wolff S, *Rubb.Chem.Technol.*, **69**, No.3, 1996, 325-46.
81 Papathanasiou T D, Lee P D, *Polym. Composites*, **18**, No.2, 1997, 242-53.
82 Grohens Y, Schultz J, *Int. J. Adhesion Adhesives*, **17**, 1997, 163-7.
83 Nago S, Mizutani Y, *J. Appl. Polym. Sci.*, **53**, No.12, 1994, 1579-87.
84 Shiroki Y, *Int. Polym. Sci. Technol.*, **20**, No.6, 1993, T/12-21.
85 Luzinov I, Voronov A, Minko S, Kraus R, Wilke W, Zhuk A, *J. Appl. Polym. Sci.*, **61**, No.7, 1996, 1101-9.
86 Nago S, Mizutani Y, *J. Appl. Polym. Sci.*, **61**, No.1, 1996, 31-5.
87 Jin-Shy Tsai, *Polym. Engng. Sci.*, **35**, No.16, 1995, 1313-6.
88 Wu J, Lerner M M, *Chem. of Mat.*, **5**, No.6, 1993, 835-8.
89 Schueler R, Petermann J, Schulte K, Wentzel H P, *Macromol. Symp.*, **104**, 1996, 261-8.
90 Bomal Y, Godard P, *Polym. Engng. Sci.*, **36**, No.2, 1996, 237-43.
91 Li Y, Wang M J, Zhang T, Zhang F, Fu X, *Rubb. Chem. Technol.*, **67**, No.4, 1994, 693-9.
92 Baranovskii V M, Tarara A M, Khomik A A, *Int. Polym. Sci. Technol.*, **20**, No.1, 1993, T/98-9.
93 Liauw C M, Lees G C, Hurst S J, Rothon R N, Dobson D C, *Plast. Rubb. Comp. Process. Appln.*, **24**, No.4, 1995, 211-9.
94 Vogeler T, Kreuzer M, Tschudi T, Simoni F, *Molecular Crystals & Liquid Crystals*, **282**, 1996, 419-28.
95 Kovacevic V, Lucic S, Hace D, Cerovecki Z, *J. Adhesion Sci. Technol.*, **10**, No.12, 1996, 1273-85.
96 Gruber T C, Zerda T W, Gerspacher M, *Rubb. Chem. Technol.*, **67**, No.2, 1994, 280-7.
97 Maas S, Gronski W, *Kaut. u. Gummi Kunst.*, **47**, No.6, 1994, 409-15.
98 Li Q, Manas-Zloczower I, Feke D L, *Rubb. Chem. Technol.*, **69**, No.1, 1996, 8-14.
99 Qi Li, Feke D L, Manas-Zloczower I, *Rubb. Chem. Technol.*, **68**, No.5, 1995, 836-41.
100 Potente H, Flecke J, Antec '96. Vol. I. Conference Proceedings, Indianapolis, 5th-10th May 1996, 178-82.
101 Eisenbach C D, Ribbe A, Goeldel A, *Kaut. u. Gummi Kunst.*, **49**, No.6, 1996, 406-10.
102 Foster J K, Sims E S, Venable S W, *Paint & Ink Int.*, **8**, No.3, 1995, 18-21.
103 Long Y, Shanks R A, Antec '96. Volume III. Conference proceedings, Indianapolis, 5th-10th May 996, 3673-5
104 Wolff S, Wang M J, Tan E H, *Kaut. u. Gummi Kunst.*, **47**, No.12, 1994, 873-84.
105 Tokita N, Shieh C H, Ouyang G B, Patterson W J, *Kaut. u. Gummi Kunst.*, **47**, No.6, 1994, 416-20.
106 Maas S, Gronski W, *Rubb. Chem. Technol.*, **68**, No.4, 1995, 652-9.
107 Lin C R, Lee Y D, *Macromol. Theory & Simulations*, **5**, No.6, 1996, 1075-104.
108 Miano F, Rabaioli M R, *Coll. & Surfaces*, **84**, Nos.2/3, 1994, 229-37.
109 Sudduth R D, *J. Appl. Polym. Sci.*, **54**, No.9, 1994, 1243-62.
110 Hedgus C R, Kamel I L, *J. Coatings Technol.*, **65**, No.821, June 1993, 49-61
111 Boehm G G A, Nguyen M N, *J. Appl. Polym. Sci.*, **55**, No.7, 1995, 1041-50.
112 Molphy M, Mainwaring D E, Rizzardo E, Gunatillake P A, Laslett R L, *Polym. Int.*, **37**, No.1, 1995, 53-61.
113 Chaturvedi M, Antec '97. Conference proceedings, Toronto, April 1997, 1865-8.

114 Rockenbauer A, Korecz L, Pukanszky B, *Polym. Bull.*, **33**, No.5, 1994, 585-9.
115 Zhong-Fu Li, Grubb D T, Phoenix S L, *Composites Sci. & Technol.*, **54**, No.3, 1995, 251-66.
116 Guanghong Lu, Xiaotian Li, Hancheng Jiang, *Composites Sci. & Technol.*, **56**, No.2, 1996, 193-200.
117 Hashemi S, Gilbride M T, Hodgkinson J, *J. Mat. Sci.*, **31**, No.19, 1996, 5017-25.
118 Mayadunne A, Bhattacharya S N, Kosior E, Boontanjai C, Antec 95. Volume I. Conference proceedings, Boston, Ma., 7th-11th May 1995, 1178-82.
119 Godard P, Bomal Y, Biebuyck J J, *J. Mat. Sci.*, **28**, No.24, 1993, 6605-10.
120 Ye B S, Svenson A L, Bank L C, *Composites*, **26**, No.10, 1995, 725-31.
121 Shiga T, Okada A, Kurauchi T, *J. Appl. Polym. Sci.*, **58**, No.4, 1995, 787-92.
122 Jancar J, *Macromol. Symp.*, **108**, 1996, 163-72.
123 Golubev A I, *Int. Polym. Sci. Technol.*, **22**, No.10, 1995, T/40-1.
124 Golubev A I, *Int. Polym. Sci. Technol.*, **22**, No.10, 1995, T/42-3.
125 Konstantellos B, Sideridis E, *Coll. Polym. Sci.*, **273**, No.4, 1995, 307-16.
126 Agari Y, Ueda A, Nagai S, *J. Appl. Polym. Sci.*, **52**, No.9, 1994, 1223-31.
127 Ming Qiu Zhang, Jia Rui Xu, Han Ming Zeng, Qun Huo, Zhi Yi Zhang, Feng Chun Yun, Friedrich K, *J. Mat. Sci.*, **30**, No.17, 1995, 4226-32.
128 Hoy L L, *J. Coatings Technol.*, **68**, No.853, 1996, 33-9.
129 Qiu Q, Kumosa M, *Composites Sci. Technol,* **57**, 1997, 497-507.
130 Lee D H, Condrate R A, Reed J S, *J. Mater. Sci.*, **32**, 1997, 471-8.
131 Liphard M, Von Rybinski W, Schreck B, *Prog. Coll. & Polym. Sci.*, **95**, 1994, 168-74.
132 Yoshinaga K, Nakanishi K, Hidaka Y, Karakawa H, *Composite Interfaces*, **3**, No.3, 1995, 231-41.
133 Dupraz A M P, de Wijn J R, v. d. Meer S A T, de Groot K, *J. Biomed. Mat. Res.*, **30**, No.2, 1996, 231-8.
134 **Kronos. Titanium Dioxide Pigments in Industrial Coatings**. *Kronos*, Leverkusen, 1996, DS 1805E/6965E.
135 Tang L-G, Kardos J L, *Polym. Composites*, **18**, No.1, 1997, 100-13.
136 Balard H, Papirer E, *Prog. Org. Coatings*, **22**, No.1-4, 1993, 1-17.
137 Zaborski M, Slusarski L, Vidal A, *Int. Polym. Sci. Technol.*, **20**, No.11, 1993, T/99-104.
138 Pukanszky B, Belina K, Rockenbauer A, Maurer F H J, *Composites*, **25**, No.3, 1994, 205-14.
139 Zaborski M, Slusarski L, *Composite Interfaces*, **3**, No.1, 1995, 9-22.
140 Zumbrum M A, *J. Adhesion*, **46**, Nos.1-4, 1994, 181-96.
141 Donnet J B, Wang W, Vidal A, Wang M J, *Kaut. u. Gummi Kunst.*, **46**, No.11, Nov.1993, 866-71.
142 Wang W, Vidal A, Donnet J-B, Wang M-J, *Kaut. u. Gummi Kunst.*, **46**, No.12, Dec.1993, 933-40.
143 Tsutsumi K, Ban K, Shibata K, Okazaki S, Kogoma M, *J. Adhesion*, **57**, Nos.1-4, 1996, 45-53.
144 Torro-Palau A, Fernandez-Garcia J C, Orgiles-Barcelo A C, Martin-Martinez J M, *J. Adhesion*, **57**, Nos.1-4, 1996, 203-25.
145 Asai S, Sumita M, *J. Macromol. Sci. B,* **34**, No.3, 1995, 283-94.
146 Rammoorthy M, Muzzy J, Antec '97. Conference proceedings, Toronto, April 1997, 2500-4.
147 Tsubokawa N, Seno K, *J. Macromol. Sci. A*, **31**, No.9, 1994, 1135-45.
148 Schreiber H P, Antec '95. Vol. II. Conference Proceedings, Boston, Ma., 7th-11th May 1995, 2446-51.
149 Ou Y C, Zhu J, Feng Y P, *J. Appl. Polym. Sci.*, **59**, No.2, 1996, 287-94.
150 Mele P, Alberola N D, *Composites Sci. & Technol.*, **56**, No.7, 1996, 849-53.
151 Shang S W, Williams J W, Soderholm K J M, *J. Mat. Sci.,* **30**, No.17, 1995, 4323-34.
152 Pak S H, Caze C, *J. Appl. Polym. Sci.*, **65**, 1997, 143-53.
153 Bosse F, Schreiber H P, Eisenberg A, *Macromolecules*, **26**, No.24, 1993, 6447-54.
154 Hegedus C R, Kamel I L, J. *Coatings Technol.*, **65**, No.820, 1993, 23-30.
155 Bataille P, Mahlous M, Schreiber H P, *Polym. Engng. Sci.*, **34**, No.12, 1994, 981-5.
156 Ulkem I, Bataille P, Schreiber H P, *J. Macromol. Sci. A*, **31**, No.3, 1994, 291-303.
157 Persson A L, Bertilsson H, *Composite Interfaces*, **3**, No.4, 1996, 321-32.
158 Lee M C H. **Adhesive Chemistry - Development and Trends**. *Plenum*, New York, 1985.
159 Lee M C H, *J. Appl. Polym. Sci.*, **33**, 1987, 2479.
160 Selzer R, Friedrich K, *Composites, Part A*, **28A**, 1997, 595-604.
161 Akay M, Mun S K A, Stanley A, *Composites Sci. Technol,* **57**, 1997, 565-71.
162 Kouloumbi N, Tsangaris G M, Kyvelidis S T, *J. Coatings Technol.*, **66**, No.839, 1994, 83-8.
163 Hansen H, Polymers, Laminations & Coatings Conference, 1995, 653-8.
164 Evans L R, Meeting of the Rubber Division, ACS, Montreal, May 5-8, 1996, paper D.
165 Wang J Y, Ploehn H J, *J. Appl. Polym. Sci.*, **59**, No.2, 1996, 345-57.
166 Connor C, Chadwick M M, *J. Mat. Sci.*, **31**, No.14, 1996, 3871-7.

167 Akelah A, Moet A, *J. Mat. Sci.*, **31**, No.13, 1996, 3589-96.
168 Rebouillat S, Escoubes M, Gauthier R, *J. Adhesion Sci. Technol.*, **10**, No.7, 1996, 635-50.
169 Bauman B D, Plastics Finishing: Responding to Tomorrow's Global Requirements. Retec proceedings, Detroit, Mi, 21st-22nd March 1995, 62-73.
170 Gassan J, Bledzki A K, *Polym. Composites*, **18**, No.2, 1997, 179-84.
171 Darmstadt H. Roy C, Kaliaguine S, Sahouli B, Blacher S, Pirard R, Brouers F, *Rubb. Chem. Technol.*, **68**, No.2, 1995, 330-41.
172 Bogoeva-Gaceva G, Burevski D, Dekanski A, Janevski A, *J. Mat. Sci.*, **30**, No.13, 1995, 3543-6.
173 Zaborski M, Slusarski L, Vidal A, *Polim. Tworz. Wielk.*, **38**, No.6, 1993, 263-8.
174 Zaborski M, Vidal A, Papirer E, *Polim. Tworz. Wielk.*, **38**, No.7, 1993, 319-26.
175 Burnside S D, Giannelis E P, *Chem. of Mat.*, **7**, No.9, 1995, 1597-600.
176 Nasr G M, Badawy M M, *Polym. Test.*, **15**, No.5, 1996, 477-84.
177 Lekatou A, Faidi S E, Lyon S B, Newman R C, *J. Mat. Res.*, **11**, No.5, 1996, 1293-304.
178 Fatoev I I, Sitamov S, Manin V N, *Int. Polym. Sci. Technol.*, **22**, No.9, 1995, T/26-8.
179 Budtov V P, Gandel'sman M I, Mulin Yu A, Suleimanov I E, Shterenzon A L, *Polym. Sci., Ser. A*, **37**, No.10, 1995, 1064-70.
180 Unnikrishnan G, Thomas S, Varghese S, *Polymer*, **37**, No.13, 1996, 2687-93.
181 Ou Y C, Yu Z Z, Vidal A, Donnet J B, *J. Appl. Polym. Sci.*, **59**, No.8, 1996, 1321-8.
182 Addad J P C, Frebourg P, *Polymer*, **37**, No.19, 1996, 4235-42.
183 Xu P, Mark J E, *Eur. Polym. J.*, **31**, No.12, 1995, 1191-5.
184 Ismail H, Freakley P K, Sutherland I, Sheng E, *Eur. Polym. J.*, **31**, No.11, 1995, 1109-17.
185 Abdel-Aziz M M, Youssef H A, El Miligy A A, Yoshii F, Makuuchi K, *Polym. & Polym. Composites*, **4**, No.4, 1996, 259-68.
186 Younan A F, Choneim A M, Tawfik A A A, Abd-El-Nour K N, *Polym. Degradat. Stabil.*, **49**, No.2, 1995, 215-22.
187 Strauss M, Kilian H G, Freund B, Wolff S, *Coll. Polym. Sci.*, **272**, No.10, 1994, 1208-17.
188 de Sena Affonso J E, Nunes R C R, *Polym. Bull.*, **34**, No.5/6, 1995, 669-75.
189 Abdel-Aziz M M, Gwaily S E, *Polym. Degradat. Stabil.*, **55**, 1997, 269-74.
190 Hegedus C R, Kamel I L, *J. Coatings Technol.*, **65**, No.822, July 1993, 37-43.
191 Lan T, Kaviratna D, Pinnavaia T J, *Chem. of Mat.*, **6**, No.5, 1994, 573-5.
192 Nakamura S, Kaneko S, Mizutani Y, *J. Appl. Polym. Sci.*, **49**, No.1, 1993, 143-50.
193 Duval J M, Kemperman A J B, Folkers B, Mulder M H V, Desgrandchamps G, Smolders C A, *J. Appl. Polym. Sci.,* **54**, No.4, 1994, 409-18.
194 Rabello M S, White J R, *Polym. Composites*, **17**, No.5, 1996, 691-704.
195 Gill T S, Xanthos M, Wirsbo Co., Antec '96. Volume II. Conference proceedings, Indianapolis, 5th-10th May 1996, 1757-61.
196 Xanthos M, Grenci J, Dagli S S, Antec '95. Vol.II. Conference Proceedings, Boston, Ma., 7th-11th May 1995, 3194-200.
197 Gill T S, Xanthos M, *J. Vinyl and Additive Technol.*, **2**, No.3, 1996, 248-52.
198 Domka L, *Coll. Polym. Sci.*, **272**, No.10, 1994, 1190-202.
199 Krysztafkiewicz A, Rager B, Maik M, Szymanowski J, *Coll. Polym. Sci.*, **272**, No.12, 1994, 1526-35.
200 Hemenway C P, Latimer J J, *Tappi J.*, **68** (5), 102 (1985).
201 Braun J H, *J. Coat. Technol.*, **63** (8), 1991, 43.
202 Schröder J, *Prog. Org. Coat.*, **15**, 1988, 337.
203 Völz H G, *Prog. Org. Coat.*, **18**, 1990, 289.
204 Braun J H, *J. Coat. Technol.*, **62** (785), 1990, 39.
205 Parker A A, Martin E S, Clever T R, *J. Coatings Technol.*, **66**, No.829, 1994, 39-46.
206 Godovsky D Yu, *Adv. Polym. Sci.*, **119**, 1995, 81-122.
207 Athey R D, Kirkland T, Lindblom G, Swoboda J, *Eur. Coatings J.*, No.11, 1995, 793-8.
208 Murata N, Nishi S, Hosono S, *J. Adhesion*, **59**, Nos.1-4, 1996, 39-50.
209 Zhang C-Z, Liu W-M, Xue Q-J, Shen W-C, *J. Appl. Polym. Sci.*, **66**, 1997, 85-93.
210 Lee I, *J. Mat. Sci.*, **30**, No.23, 1995, 6019-22.
211 Shin Jen Shiao, Te Zei Wang, *Composites*, **27B**, No.5, 1996, 459-65.
212 Srivastava V K, Pathak J P, *Polym. & Polym. Composites*, **3**, No.6, 1995, 411-4.
213 Kuznezov A, Vasnev V, Gribova I, Krasnov A, Gureeva G, Ignatov V, *Int. J. Polym. Mat.*, **32**, Nos.1-4, 1996. 85-91.
214 Bijwe J, *Polym. Composites*, **18**, No.3, 1997, 378-96.
215 White L, *Eur. Rubb. J.,* **177**, No.3, 1995, 27.

216 Yang A C M, *Polymer*, **35**, No.15, 1994, 3206-11.
217 Le Bras M, Bourbigot, Le Tallec Y, Laureyns J., *Polym. Degradat. Stabil.*, **56**, 1997, 11-21.
218 Le Bras M, Bourbigot S, *Fire & Mat.*, **20**, No.1, 1996, 39-49.
219 Miller B, *Plast. World*, **54**, No.12, 1996, 44-9.
220 Gibov K M, Mamleev V S, *J. Appl. Polym. Sci.*, **66**, 1997, 329-38.
221 Ramani K, Vaidyanathan A, *J. Composite Mat.*, **29**, No.13, 1995, 1725-40.
222 Baranovskii V M, Bondarenko S I, Kachanovskaya L D, Zelenev Y V, Makarov V G, Ovcharenko F D, *Int. Polym. Sci. Technol.*, **22**, No.1, 1995, T/91-3.
223 Ruiz F A, Polymers, Laminations & Coatings Conference, 1995, 647-51.
224 Bigg D M, **Thermal and Electrical Conductivity of Polymers Materials**, Eds. Godovsky Y K, Privalko V P, *Springer*, Berlin 1995.
225 Tavman I H, *J. Appl. Polym. Sci.*, **62**, No.12, 1996, 2161-7.
226 Nyilas A, Rehme R, Wyrwich C, Springer H, Hinrichsen G, *J. Mat. Sci. Lett.*, **15**, No.16, 1996, 1457-9.
227 Nasr G M, Badawy M M, Gwaily S E, Attia G, *Polym. Int.*, **38**, No.3, 1995, 249-55.
228 Priss L S, *Int. Polym. Sci. Technol.*, **23**, No.7, 1996, T53-6.
229 Oien H T, Polyurethanes '95. Conference Proceedings, Chicago, Il., 26th-29th Sept.1995, 137-41.
230 Okoroafor M O, Wang A, Bhattacharjee D, Cikut L, Haworth G J, Polyurethanes '95. Conference Proceedings, Chicago, Il., 26th-29th Sept.1995, 303-9.
231 Nichols K, Solc J, Shieu F, Antec '93. Conference Proceedings, New Orleans, La., 9th-13th May 1993, Vol. II, 1938-42.
232 McCabe J F, Wassell R W, *J. Mat. Sci. Mat. In Med.*, **6**, No.11, 1995, 624-9.
233 Baranovskii V M, Bondarenko V V, Zadorina E N, Cherenkov A V, Zelenev Y V, *Int. Polym. Sci. Technol.*, **23**, No.6, 1996, T/87-9.
234 Strumpler R, Maidorn G, Garbin A, Ritzer L, Greuter F, *Polym. & Polym. Composites*, **4**, No.5, 1996, 299-304.
235 Mamunya E P, Shumskii V F, Lebedev E V, *Polym. Sci.*, **36**, No.6, 1994, 835-8.
236 Mateev M M, Totev D L, *Kaut. u. Gummi Kunst.*, **49**, No.6, 1996, 427-31.
237 Chan C M, *Polym. Engng. Sci.*, **36**, No.4, 1996, 495-500.
238 Achour M E, El Malhi M, Miane J L, Carmona F, *J. Appl. Polym. Sci.*, **61**, No.11, 1996, 2009-13.
239 Acosta J L, Jurado J R, *J. Appl. Polym. Sci.*, **57**, No.4, 1995, 431-7.
240 Khan S A, Baker G L, Colson S, *Chem of Mat.*, **6**, No.12, 1994, 2359-63.
241 Ali M H, Abo-Hashem A, *Plast. Rubb. Comp. Process. Appln.*, **24**, No.1, 1995, 47-51.
242 Bowen C P, Bhalla A S, Newnham R E, Randall C A, *J. Mat. Res.*, **9**, No.3, 1994, 781-8.
243 Abramova N A, Diikova E U, Lyakhovskii Yu Z, *Polym. Sci.*, **36**, No.9, 1994, 1308-9.
244 Mamunya E P, Davidenko V V, Lebedev E V, *Polym. Composites*, **16**, No.4, 1995, 319-24.
245 Lei Yang, Schruben D L, *Polym. Engng. Sci.*, **34**, No.14, 1994, 1109-14.
246 Chellappa V, Chiou Z W, Jang B Z, Antec '94. Conference Proceedings, San Francisco, Ca., 1st-5th May 1994, Vol. II, p.2371-4.
247 Fiske T J, Gokturk H S, Kalyon D M, Antec '93. Conference Proceedings, New Orleans, La., 9th-13th May 1993, Vol. I, 614-7.
248 Gokturk H S, Fiske T J, Kalyon D M, Antec '93. Conference Proceedings, New Orleans, La., 9th-13th May 1993, Vol. I, 605-8.
249 Suri A, Min K, Antec '97. Conference proceedings, Toronto, April 1997, 1487-91.
250 Zaikin A Y, Nigmatullin V A, *Kaut. u. Gummi Kunst.*, **47**, No.10, 1994, 709-14.
251 Soares B G, Gubbels F, Jerome R, Teyssie P, Vanlathem E, Deltour R, *Polym. Bull.*, **35**, No.1/2, 1995, 223-8.
252 Svorcik V, Micek I, Jankovskij O, Rybka V, Hnatowicz V, Wang L, Angert N, *Polym. Degradat. Stabil.*, **55**, 1997, 115-21.
253 Hao Tang, Xinfang Chen, Aoqing Tang, Yunxia Luo, *J. Appl. Polym. Sci.*, **59**, No.3, 1996, 383-7.
254 Porter J R, Groseth C K, Little D W, Polyurethanes '95. Conference Proceedings, Chicago, Il., 26th-29th Sept.1995, 532-7.
255 Klason C, McQueen D H, Kubat J, *Macromol. Symp.*, **108**, 1996, 247-60.
256 Trotignon J P, Tcharkhtchi A, *Macromol. Symp.*, **108**, 1996, 231-45.
257 Modine F A, Duggal A R, Robinson D N, Churnetski E L, Bartkowiak M, Mahan G D, Levinson L M, *J. Mat. Res.*, **11**, No.11, 1996, 2889-94.
258 Gregorio R, Cestari M, Bernardino F E, *J. Mat. Sci.*, **31**, No.11, 1996, 2925-30.
259 Svorcik V, Rybka V, Hnatowicz V, Bacakova L, *J. Mat. Sci. Lett.*, **14**, No.24, 1995, 1723-4.
260 Dyrda V I, Meshchaninov S K, *Int. Polym. Sci. Technol.*, **22**, No.12, 1995, T/14-6.

261 Chmutin I A, Ryvkina N G, Ponomarenko A T, Shevchenko V G, *Polym. Sci., Ser. A*, **38**, No.2, 1996, 173-7.
262 Kimura T, Asano Y, Yasuda S, *Polymer*, **37**, No.14, 1996, 2981-7.
263 Tang H, Chen X, Luo Y, *Eur. Polym. J.*, **32**, No.8, 1996, 963-6.
264 Roth S, Mair H J, *Int. Polym. Sci. Technol.*, **22**, No.11, 1995, T/1-6.
265 Hedvig P, *Int. Polym. Sci. Technol.*, **23**, No.3, 1996, T/23-8.
266 Stukhlyak P D, Shkodzinskii O K, Mytnik N M, Shovkun A P, Kovalyuk B P, *Int. Polym. Sci. Technol.*, **23**, No.6, 1996, T/81-4.
267 Wei Y, Sancaktar E, *J. Adhesion Sci. Technol.*, **10**, No.11, 1996, 1199-219.
268 Sethi R S, Goosey M T, London, *Chapman & Hall*, 1995, p.1-36. **Special Polymers for Electronics and Optoelectronics**, Chilton J A, Goosey M T, Ed.
269 Karasek L, Meissner B, Asai S, Sumita M, *Polym. J. (Jap.)*, **28**, No.2, 1996, 121-6.
270 Saad A L G, Younan A F, *Polym. Degradat. Stabil.*, **50**, No.2, 1995, 133-40.
271 Vasnev V A, Tarasov A I, Istratov V N, Ignatov V N, Krasnov A P, Kuznetsov A I, Surkova I N, *Reactive & Functional Polym.*, **26**, Nos.1-3, 1995, 177-83.
272 Narkis M, Tchoudakov R, Breuer O, Siegmann A, Antec '95. Vol. II. Conference Proceedings, Boston, Ma., 7th-11th May 1995, p.1343-6.
273 Fiske T, Gokturk H S, Yazici R, Kalyon D M, *Polym. Eng. Sci.*, **37**, No.5, 1997, 826-37.
274 Fiske T J, Gokturk H, Kalyon D M, *J. Appl. Polym. Sci.*, **65**, 1997, 1371-7.
275 Fiske T, Gokturk H S, Yazici R, Kalyon D M, Antec '97. Conference proceedings, Toronto, April 1997, 1482-6.
276 Klapcinski T, Galeski A, Kryszewski M, *J. Appl. Polym. Sci.*, **58**, No.6, 1995, 1007-13.
277 Yogo T, Nakamura T, Kikuta K, Sakamoto W, Hirano S, *J. Mat. Res.*, **11**, No.2, 1996, 475-82.
278 Stassen S, Cloots R, Vanderbemden P, Godelaine P A, Bougrine H, Rulmont A, Ausloos M, *J. Mat. Res.*, **11**, No.5, 1996, 1082-5.
279 Fiske T J, Gokturk H, Kalyon D M, Antec '96. Volume II. Conference proceedings, Indianapolis, 5th-10th May 1996, 1768-72.

# 6

# CHEMICAL PROPERTIES OF FILLERS AND FILLED MATERIALS

## 6.1 REACTIVITY

The properties of filled materials are critically dependent on the interphase between the filler and the matrix polymer. The type of interphase depends on the character of the interaction which may be either a physical force or a chemical reaction. Both types of interaction contribute to the reinforcement of polymeric materials. Formation of chemical bonds in filled materials generates much of their physical properties. An interfacial bond improves interlaminar adhesion, delamination resistance, fatigue resistance, and corrosion resistance. These properties must be considered in the design of filled materials, composites, and in tailoring the properties of the final product. Other consequences of filler reactivity can be explained based on the properties of monodisperse inorganic materials having small particle sizes. The controlled shape, size and functional group distribution of these materials develop a controlled, ordered structure in the material. The filler surface acts as a template for interface formation which allows the reactivity of the filler surface to come into play. Here are examples:

The first example refers to the creation of functional groups on the filler's surface during controlled synthesis of the filler.[1] Silica-gel, prepared from tetraethyl orthosilicate under acidic conditions, has OH groups on its surface. A similar synthesis under basic conditions deposits alkoxy groups ($OCH_2CH_3$) due to an incomplete hydrolysis of the substrate. This simple example shows the numerous possibilities that may deposit different groups on the filler surface. More examples of formation of functional groups are given in Section 6.2.

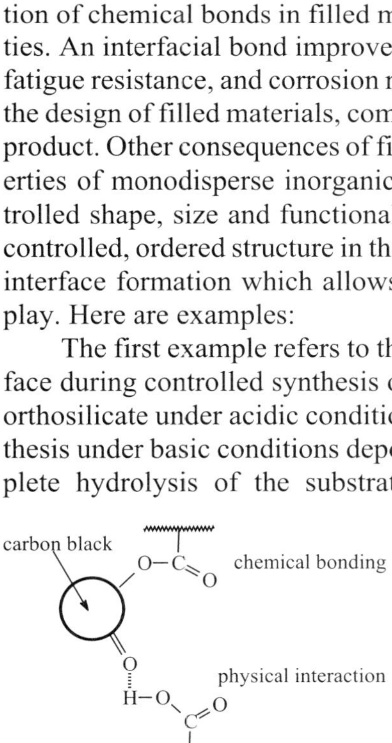

Figure 6.1. Interaction between XNBR and carbon black.[2]

Figure 6.1 shows the difference between physical interaction and chemical bonding,[2,3] although in both cases chemical bonding is involved. In the case of covalent bond formation, the link is more permanent (requires a higher energy to disrupt it ) and therefore it is considered a chemical bond. Hydrogen bonding can be disso-

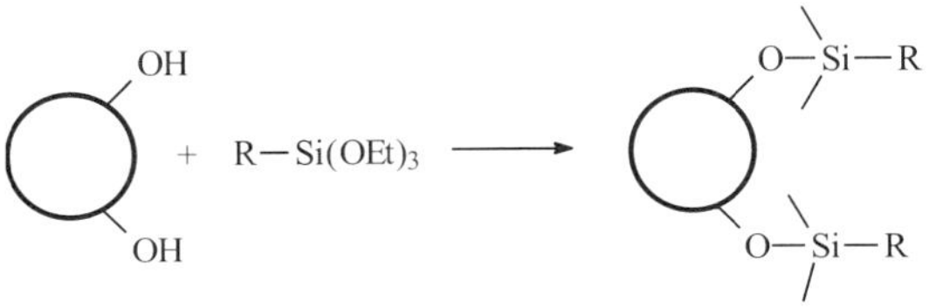


Figure 6.2. Reaction between monodispersed silica and silane.[4]

ciated by thermal energy (or a low level of energy) and then reformed again. Thus, hydrogen bonding is considered a physical interaction. It gives some flexibility to the system since the density of this crosslinking can be altered by changes in energy conditions. Other physical interactions, such as van der Waals forces, have been omitted from this discussion since they are not chemical reactions. In this example, oxidized carbon black (OH and O groups on its surface) reacts with carboxylated nitrile rubber to form the product shown in Figure 6.1. This product is believed to be formed in a two-stage process. During mixing, a carboxyl group from XNBR forms hydrogen bonds with a neighboring OH group. In the second step, a molecule of water is released and the covalent bond forms later during molding. Molding assists in covalent bond formation because the process supplies sufficient energy for the reaction to proceed. This reaction is a spontaneous reaction which occurs without any special interference simply because both the correct reactants and the proper conditions are available.

Many well-controlled reactions occur when a filler and a reactant (e.g., polymer) are selected to take advantage of the chemical interaction. Reactions with silanes are typical examples of such reactions. Figure 6.2 shows an example of such a reaction.[4-6] Silane reactions provide a simple means of a conversion of functional groups. There may be two advantages: the possibility of forming a new functionality and a means to regulate coating thickness. Depending on the choice of R we can introduce unlimited numbers and kinds of functional groups. Depending on the effective length of the molecule R we can regulate the thickness of the monolayer coating. In one experiment,[5] the length of the molecule was varied from 1.92 to 13.0 nm. The length of the organic group and its concentration affects the thickness of the coating. By choosing the appropriate concentration, one can obtain monomolecular or multimolecular layers which can have further technological implications for the property of the material.

The chemical reaction can be discontinued when the silane coating has been applied or it may be continued by adding other reagents in a following step. The example described is a multistage process in which the reaction with silane is followed by many other reactions. The aim being to coat particles with *in situ* formed polymer. These reactions, called grafting reactions, may use all common organic polymerization reactions, such as radical, ring-opening, addition, etc.

Barium sulfate was first modified with 12-hydroxystearate and the product used for further grafting with acrylamide (Figure 6.3).[7] This experiment demonstrates that 12-hydroxystearate can be used as an initiation site for further polymerization. Also, polymer chains grow on the initiation sites formed by functional

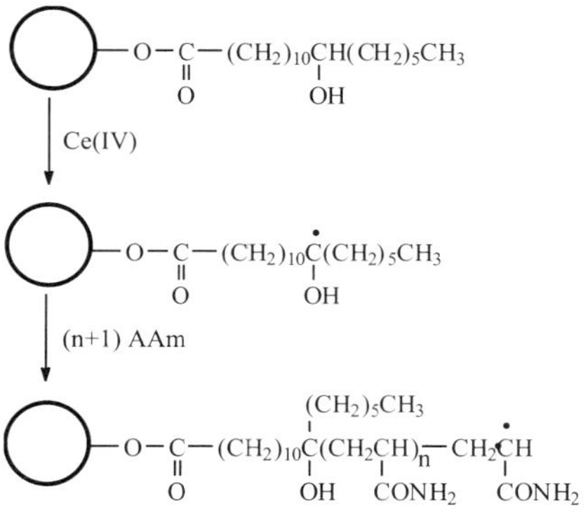


Figure 6.3. Acrylamide grafting on the 12-hydroxystearate previously reacted with the surface of $BaSO_4$.[7]

groups introduced onto the surface of fillers. Ceric ion alone can initiate polymerization of acrylamide but the conversion is low. If acrylamide is mixed with $BaSO_4$ (without modification with 12-hydroxystearate) the conversion of acrylamide increases by about 100% but because there are no active sites available for grafting, no grafting of polymer occurs on the surface of $BaSO_4$. Conversion, in the presence of $BaSO_4$ modified with 12-hydroxystearate, increases by about 220% and about 10% of the polymer formed is grafted onto the surface of the filler. The reaction proceeds according to first order kinetics (see Figures 6.9 to 6.11 in Section 6.3). Its rate is linearly increased by increasing the concentration of acrylamide, the concentration of catalyst, and the concentration of the modified filler. In summary, these reactions do not show any exceptional characteristics in comparison with the reactions without filler. But the filler's presence, and especially the presence of a modified filler, increases the initiation rate and thus the overall reaction rate. More examples of surface modification are given in Section 6.3.

Chemical reaction depends on the presence of reactive substrates and on the probability of their encounters. Thus, the possibilities of reactions can be numerous. The literature describes reactions of OH groups on the surface of kaolin with isocyanates,[8] vulcanization of nitrile rubber by ZnO,[9] reactions of carboxyl groups on the filler surface with amines and epoxy groups,[10] reactions of carboxyl groups with diols,[11] and many others.[12-15] The presence of a reactant on the surface of a material particle increases the probability of chemical reaction. Other factors include statistical probabilities, surface barriers which affect contact, dilution factors, molecular mobility, and viscosity changes in the system. These are discussed in other sections of this book.

There is one particular factor which affects reactivity in systems containing fillers. This is exemplified by the work on the restriction of spin probe motions.[16] Nitroxyl radicals were studied in polyethylene filled with various fillers. Because of chemisorption of these radicals, their activity in the system was restricted. This phenomenon may affect the chemical reaction but in this context it is an essential mechanism which explains the somewhat disappointing performance of some UV stabilizers. UV stabilization is discussed in Chapter 11. This example is given now to show that the fillers present in a system may change performance characteristics by restricting the reactivity or the availability of system components. They may also enhance retention of components by slowing down their physical loss during processing or in subsequent exposure of the material to environmental forces.

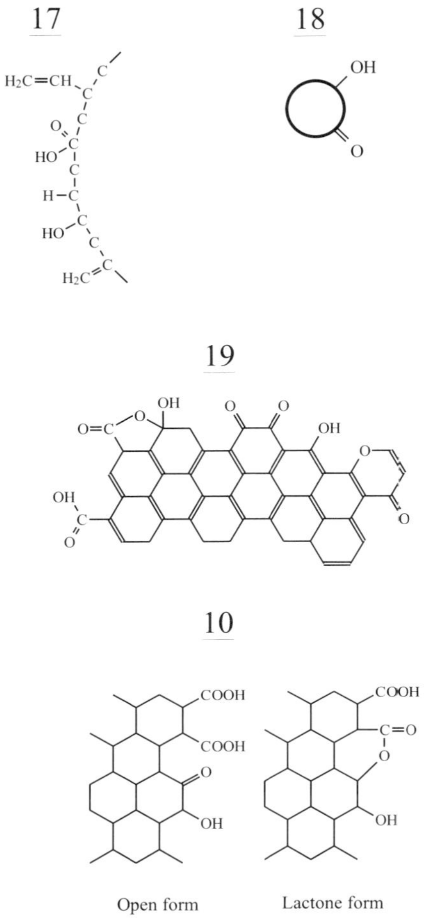


Figure 6.4. Various groups on CB surface. (Numbers are reference numbers).

## 6.2 CHEMICAL GROUPS ON THE FILLER SURFACE

Carbon black, because it is used so extensively, is one of the most frequently investigated fillers. However, findings have controversial elements in some of the details which attribute importance to surface groups.

Surface groups range from simple to complex structures (Figure 6.4). All agree on the presence of hydroxyl groups or oxygen on the surface but other groups such as carboxyl, lactone and unsaturations do not show on all formulas. The differences relate to the type of material tested. Some structures are typical of carbon black and some are typical of carbon fibers which differ not only in surface chemistry but also in surface morphology.

These differences in the chemical structure of the surface depend not only on the process of manufacture but also on additional treatments or processing conditions. In oxidized carbon fibers, the concentration of carbonyl and, more particularly carboxyl groups, is substantially increased at the expense of hydroxyl groups.[10] In the treatment of carbon fibers, several methods of oxidation are used. Liquid phase oxidation is carried out by the electrochemical and chemical methods whereas gaseous oxidation is carried out in air, oxygen or in the presence of catalysts. Plasma treatment is also used for the surface oxidation of formed fibers. Different methods of oxidation produce different surface characteristics. For example, interlaminar strength is improved by a factor of 10 by electrolytic oxidation over crude oxidation in air.

XPS data show that adequate treatment time is needed to obtain the required concentration of functional groups on the surface of carbon black. As oxygen plasma treatment continues, the concentration of C-C bonds gradually decreases.[20] C-O bonds increase only during the early stages of the process, whereas both C=O and O-C=O continue to increase throughout the process. This confirms previously referred studies.[10] The formation of surface groups improves interfacial adhesion which contributes to reinforcement. But reinforcement also requires a strong fiber.

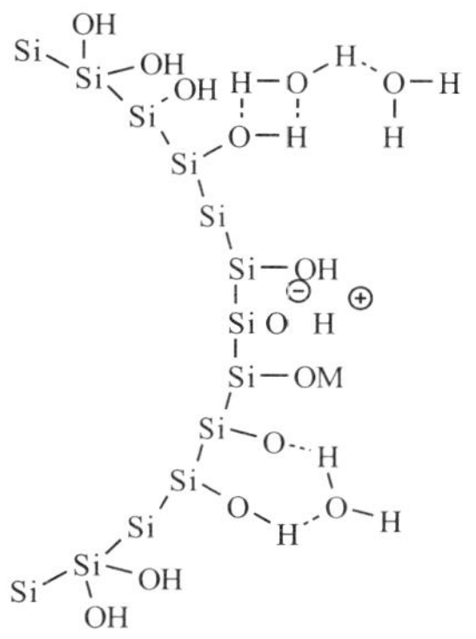

Figure 6.5. Silica, clay and talc particle.[25]

Fiber strength cannot be maintained if the oxidation process goes too far. It is therefore important to find a balance between fiber properties and the ability of the fiber's surface to interact with the surrounding materials.[21] Moderate oxidation generally gives the best performance.

IR studies give some insight into the type of chemical groups to which the hydroxyl group becomes attached. It is speculated that hydroxyl groups are parts of substituted phenols, phenols, alcohols and enols.[22] IR analysis also indicates that lactones, dicarbonyl compounds, carboxylic groups and carbonyl groups are present. Some of these groups are engaged in hydrogen bonding.

ESCA has been used in the surface analysis of carbon black oxidized by various methods. Again, oxidation in air contributes to the most substantial loss of C-C bonds. Keto-enol groups were detected only in the samples which were oxidized in air. When other oxidative processes were employed, the groups detected were OH, C=O, and COOH.[23] All other analytic methods provided similar information.

The groups present on the carbon black surface may also come from chemical treatments. In one report,[24] peroxide groups were introduced by radical trapping and then used for radical graft polymerization. In such a method, the entrapped radical plays the role of an initiator.

Figure 6.5 shows various functional groups which may be detected on silica, talc, and clay surfaces.[25] The surface character of carbon black differs in that it is mostly nonpolar whereas the surface of silica is polar. Thus carbon black is more compatible with hydrocarbon polymers which are also nonpolar. Silica and other similar fillers (talc, clay) have more affinity to each other than to nonpolar polymers. This is a major factor in the inferior performance in rubber applications where interfacial adhesion is reduced.

Figure 6.6 shows the distribution of hydroxyl groups on the surface of silicates such as aluminum, calcium, magnesium, and magnesium-aluminum silicates.[26] The surfaces of these fillers are dominated by silicate groups which occupy space as isolated, vicinal, geminal hydroxyl groups and sometimes form siloxane groups on the surface.

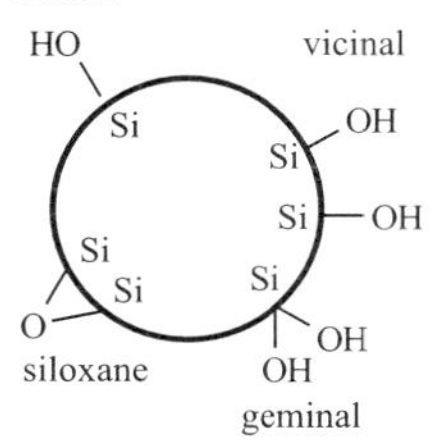


Figure 6.6. Silicate surface groups.[26]

The pH of the material surrounding kaolin will determine whether or not its surface will have OH groups. When the pH is above 7, the deprotonation of hydroxyl groups occurs which eliminates the active functional groups from the surface. The chemical changes are consistent with the ability of kaolin to flocculate in suspensions.[27]

**Table 6.1: Filler modification**

| Modification | Reason | Typical fillers | Refs. |
|---|---|---|---|
| *Physical treatment methods* | | | |
| thermal (800-1050°C) | improved dispersion | talc | 33 |
| thermal | interaction with CSPE | silica | 40 |
| oxygen plasma | reinforcement | carbon fibers | 20 |
| surface oxidation (various) | interfacial adhesion | carbon fibers | 10 |
| microwave plasma | water resistance | aramid fibers | 48 |
| acetylene gas, plasma | reinforcement | $CaCO_3$, carbon fibers | 30 |
| *Acid treatment* | | | |
| hydrochloric | rubber crosslinking | ZnO | 6 |
| stearic | reinforcement | $CaCO_3$ | 29 |
| stearic | surface hydrophobization | clay, $CaCO_3$ | 35 |
| stearic | dispersion | $Al(OH)_3$, $Mg(OH)_2$ | 57 |
| fatty metal soaps | dispersion | $Al(OH)_3$, $Mg(OH)_2$ | 57 |
| maleic derivatives | interaction with $H(CH_2)_nH$ | $CaCO_3$ | 32 |
| *Isocyanates* | | | |
| isocyanate | reinforcement | hydroxyapatite | 49 |
| isocyanate | colloidal behavior | kaolin | 27 |
| polyethylene glycol, isocyanate | resistance to solvents | kaolin | 8 |
| *Other low-molecular* | | | |
| dimeric aluminates | reinforcement | $CaCO_3$ | 34 |
| oxyethylenes with N and S | surface hydrophobization | silica | 45 |
| hexadecanol | interaction with matrix | silica | 40,41 |
| dicarboxylic acid anhydride | sedimentation | $Al(OH)_3$ | 51 |
| doping and coating | weather resistance | $TiO_2$ | 59 |
| *Grafting and resin coating* | | | |
| radical trapping | grafting initiation | carbon black | 31 |
| polymerization | improved dispersion | carbon black | 31 |
| various polymers | colloidal dispersion | carbon black, silica | 15,43 |
| polyethers | dispersability | carbon whisker | 38 |
| acrylamide | toughening, reinforcement | $CaCO_3$ | 37 |
| acrylamide | wettability, hydrophilic | $BaSO_4$ | 7 |
| maleic anhydride PP | coupling | mica, talc | 52-6 |
| functionalized polymers | dispersion, adhesion | $Al(OH)_3$, $Mg(OH)_2$ | 57 |
| polybutadiene coating | chromatographic media | $Al_2O_3$ | 58 |
| resin coating | water resistance | aramid fiber | 48 |

**Table 6.1: continuation**

| Modification | Reason | Typical fillers | Refs. |
|---|---|---|---|
| *Silane or titanate treatment* | | | |
| silanes | coupling | clay | 35 |
| silanes | controlled coating thickness | silica | 5 |
| silanes | increase/decrease adhesion | silica | 42 |
| silanes | understanding surface | fumed silica | 44 |
| silanes | fire retardant improvement | $Mg(OH)_2$ | 46 |
| silanes | reinforcement | wollastonite | 47 |
| silanes | adhesion to matrix | basalt, sludge | 50 |
| silanes | whisker orientation | AlB whisker | 53 |
| silanes | coupling and adhesion | kaoline, talc, mica | 54 |
| silanes | coupling | $Al(OH)_3$, $Mg(OH)_2$ | 57 |
| silanes | coupling | GF, silica, quartz | 60 |
| silanes | ion exchange | hydroxyapatite | 6 |
| silanes | reinforcement | natural fibers | 61 |
| silanes | nanoparticle synthesis | ceramic, metal | 62 |
| silanes | dense covering | silica | 63 |
| silanes | matrix-mineral adhesion | silica, steel, plastic | 64 |
| silanes | reinforcement | wollastonite | 65 |
| polymeric silanes | nanoparticles | silica | 39 |
| titanates | dispersion, coupling | $Al(OH)_3$, $Mg(OH)_2$ | 57 |
| titanates | coupling, reinforcement | kaolin, silicate, $CaCO_3$ | 36 |

Moisture is also a factor in controlling the concentration of functional groups on the filler surface.[28] Hydrated silicic acid has many times more OH groups than anhydrous silicic acid. The number of functional groups can also be maximized by a dispersion or particle size. For example, talc has numerous groups on its crystal side faces, therefore the number of OH groups is substantially increased with size reduction (delamination). This is consistent with the observation that fine talc gives better reinforcement of rubber than a coarse grade.

Calcium carbonate does not have functional groups (its surface is inert), therefore interaction can only be improved by chemical modification. Some hydroxyl groups can be found from admixtures such as $Ca(OH)_2$ but these admixtures may limit the ways in which calcium carbonate can be used because these admixtures increase the amount of absorbed moisture. Functional groups are frequently hydrophilic thus they attract water molecules. In many applications, moisture can either cause product instability, reduce cure rate, or reduce reinforcement. Caution is needed in selecting surface treatment to generate functional groups.

A similar analysis of functional groups in organic fillers is not feasible. These materials may be very complex mixtures (natural products) differing in chemical composition and surface organization or very diverse (man-made organic fillers).

## 6.3 FILLER SURFACE MODIFICATION

The following subjects are discussed in this section:

- Modification methods
- Reasons for employing specific method
- Examples of different fillers
- Examples of chemical reactions
- Reaction yield
- Modified material properties

Table 6.1 summarizes modification methods and reasons why such modification methods are used with the common fillers.

Table 6.1 shows that:

- Silanes are by far the most popular materials used for filler modification
- Silanes are also the most versatile (useful in modifying many types of fillers)
- Reinforcement and improvement of interface adhesion are the most frequent reasons for filler modification

Several approaches are used to modify fillers. One approach aims at increasing the number of active sites on the filler surface. This is usually done either by physical treatment or by acid treatment.

A second (and largest) group of methods includes the reaction of existing active groups to change their chemical composition. Acetic treatment, isocyanation, grafting, addition of other low molecular weight substances, and silane modifications fall into this category.

When modified by one of these methods, fillers become reactive with other chemical groups (a change in functionality) or change in surface character from hydrophobic to hydrophilic (and *vice versa*). Fillers are usually hydrophilic and do not easily combine with most polymeric materials which are usually hydrophobic. Such modification not only contributes to reinforcement but is also very useful in increasing the interaction of particles to impart rheological properties, prevent sedimentation, aid dispersion, or prevent agglomeration. These reactions deposit different coating densities. Coatings can be monomolecular or consist of numerous interacting layers. Coating thickness can also be varied by the length of the grafted polymer chain.

If functional groups are not available on the filler's surface, the filler's surface cannot be modified. This is the case with calcium carbonate. It may be coated with a layer of stearates, reacted with carboxylic acid, or exposed to a pyrrolytic process (acetylene gas, plasma treatment) which forms reactive surface groups. Other fillers may be coated with a layer of polymer or a low molecular weight substance. This method is used frequently with fibers to either protect them against damage during processing (carbon fibers and glass fibers are fragile) or to assure that they will be wetted by the polymer matrix.

Calcium carbonate is a useful filler for the reinforcement of poly(vinyl acetate). Unlike other fillers, calcium carbonate can react with the carboxylic groups of poly(vinyl acetate). Stearic acid treatment is similar. Stearic acid is bound to the surface molecules to form insoluble calcium stearate. It is estimated that 3.2% of stearic acid covers only 40% of the surface. For 100% coverage, 8% stearic acid is needed.[29] The acid used for this reaction must be chosen with care. Best results are obtained from acids with hydroxyl group (hydroxyundecanoic acid).[33] This occurs due to hydrogen bonding between neighboring groups. The grafting ratio of hydroxyundecanoic acid (4.3 $mol/nm^2$) is better than that of stearic acid (3.5 $mol/nm^2$). Note that the concentration of acid in this experiment[33] was a magnitude lower than in the previously discussed experiment.[29] The density of the surface coverage is essential for orientation of the matrix chains on the filler's surface. If there is no coating on the calcium carbonate surface, it will have a high surface energy. This attracts some segments of the polymer chain to cover the filler surface. Most polymer structures do not interact with the filler surfaces. If a moderate coverage of fatty acid is applied, a polymer interacts with the surface of filler in between fatty acid chains. This gives a more dense and a more uniform interaction and better orientation of the polymer chains on the filler's surface. A further increase in coverage by fatty acid (more than 5%) does not leave enough space for the polymer to interact with the filler's surface which reduces the reinforcing effect of the filler.[35]

Several interesting and unusual methods are used to treat carbon black. Polymer radicals formed during the decomposition of peroxide polymers may be trapped on the surface of carbon black.[31] Such radicals reacted on the surface to form natural sites for further polymerization. This is because trapped radicals function as an initiator to initiate further polymerization reaction of methyl methacrylate on these sites. Polymerization of isobutyl vinyl ether gave very good grafting yields (23.5% and 16.2) when hydroxyl and carbonyl groups were first converted to sodium phenolates or carboxylated or by amine groups, respectively.[15] Methyl-2-oxazoline gave even higher grafting yields (24.8 to 32.9%).

Properties of carbon fibers were modified by oxygen plasma. Figure 6.7 shows changes in the O/C ratio versus the duration of the plasma treatment.[20] The amount of oxygen increases rapidly during the first minute of treatment. This is accompanied by a very rapid increase in carboxyl group concentration which later becomes stable. Other functional groups such as hydroxyl and carbonyl are stable throughout the first 3 minutes of the treatment. Only 10% of the original tensile strength is lost during the first 3 minutes of the treatment.

This discussion refers to modification of carbon whisker.[38] Ring opening polymerization of cyclic ethers was used to modify the whisker surface. To increase the number of functional groups, the whisker was also pretreated with $HNO_3$ which increased the concentration of hydroxyl and carboxyl groups by about 50%. Figure 6.8 shows that increasing the grafting temperature decreased the grafting yield.

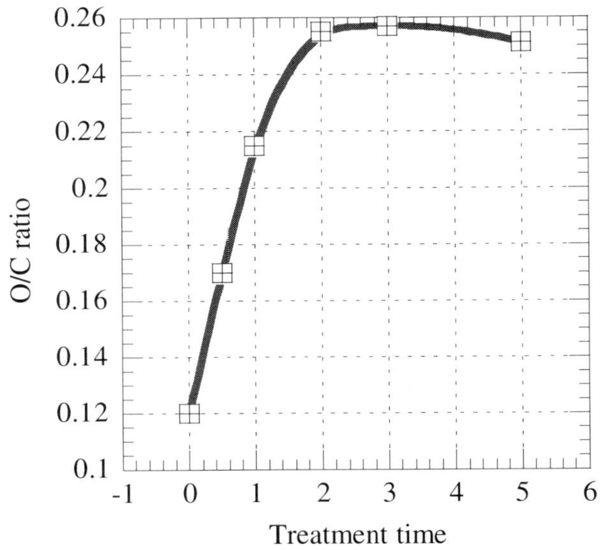


Figure 6.7. Variation of O/C elemental ratio as a function of treatment time. [Adapted, by permission, from Byung Suk Jin, Kwang Hee Lee, Chul Rim Choe, *Polym. Int.*, **34**, No.2, 1994, 181-5.]

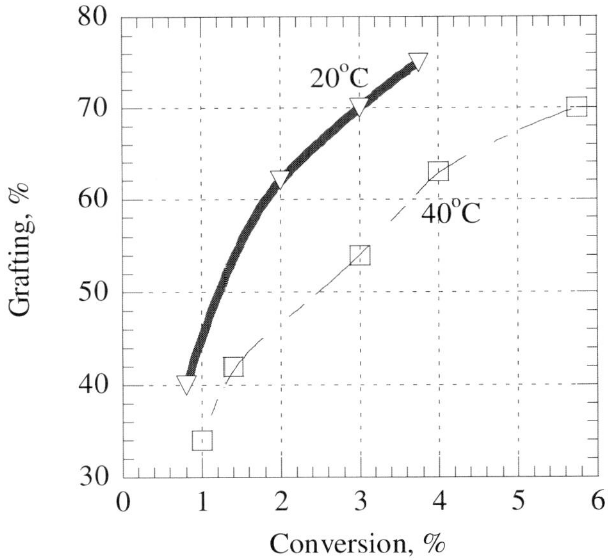


Figure 6.8. Effect of temperature on the grafting of poly(THF) onto carbon whisker. [Adapted, by permission from, Tsubokawa N, Yoshihara T, *Coll. & Surfaces*, **81**, 1993, 195-201.]

More ungrafted polymer is formed at higher temperatures because the higher temperatures facilitate chain transfer in the growing polymer cation.

Surface modification of silica is much easier than that of calcium carbonate because silica has numerous surface groups. It is not the greater reactivity of the silica which makes it easier but also the way in which silanes or titanates orient on the surface. Calcium carbonate, which does not have active sites, is coated with a random layer of modifying compounds. In silica, the modifier molecules are oriented perpendicular to the filler's surface.[36] There is a growing interest in methods of preparation of monodispersed particles of colloidal silica with grafted silane. These become sites for further polymerization. Alternately, polymeric silane is used for modification.[5,39] The nanoparticles had sizes of 11 or 42 nm and they were reacted with polymeric silanes, such as trimethoxysilyl-terminated poly(maleic anhydride-styrene) (PM-ST), PMMA or PS. On small particles, 8.6 molecules of PM-ST were bound to a particle compared with 590 molecules bound to larger particles.[39] The polymeric modifier (PM-ST) is a relatively rigid molecule and can be stretched to a molecular length of 15 nm which is relatively large compared to the smaller particle size (11 nm). This suggests that polymer chains are in a sterically crowded condition on the surface of small particles and the modifier encounters strong steric repulsion forces resulting in lower coverage. A broader study included 6 different low molecular weight silanes, polymeric silanes and further polymerization on the silane initiated sites.[5] The aim of the study was to control the thickness of the coating. For much larger particles of silica (with a diameter of 450 nm), the thickness of the coating varied from 0.6 to 73.1 nm depending on the modifier type and its concentration. It was not possible to control coating thickness by consecutive radical polymerization over previously initiated sites. But the use of some low molecular coupling agents enables the layer thickness to be controlled by concentration. Synthesis is simple, involving a simple mixing of reagents in solution with the filler. The temperature (from 0°C to reflux) and the time of reaction depend on the reactivity of the reagents. The yield of the reaction is also influenced by these conditions. Treatment by silanes is often conducted in bulk in which all the ingredients (including polymer) are present. This is a convenient method but results are not nearly as precise as those discussed above where silica is modified under controlled conditions.

The modification of a filler surface with isocyanates is a simple process which involves the reaction of hydroxyl groups on the filler surface with monomeric isocyanate. 2,4-toluene diisocyanate or hexamethylene diisocyanate are commonly used.[8,27,49] Since isocyanates are bifunctional they can be further reacted with polyols to form a coating on the surface or they can be used for the reinforcement of polyurethane. A strong covalent bonding can be verified by controlled extraction with the solvent. Bound material will not be removed from the filler's surface.

Mica, because of its platelet structure is a very useful filler. Its performance is improved by increasing the compatibility between filler and polymer. Silane modification is one simple and frequently used method. An alternative method involves a polymeric modifier which, in the case of polypropylene formulations, is polypropylene modified by maleic anhydride.[52,56] Such modifiers act more as compatibilizers. They are added in small amounts to a system containing both mica and polypropylene. The polymeric component of the compatibilizer mixes with the polymer interphase and reacts with the filler. Similar technology is used for PP filled with talc.[55] These types of reactive compatibilizers (or polymeric modifiers of the filler surface) are of growing interest, considering that rather small additions of inexpensive material (relative to the cost of silane treatment) give the required reinforcement.

Wollastonite and kaolin are most frequently modified by silanes.[35,47,54] Several water-borne products are now available for this purpose, including those with functional amines, diamines, and vinyl-amines.[54]

Surface grafting of barium sulfate is interesting from the point of view of the kinetics of such reactions.[7] Barium sulfate like calcium carbonate, is an inert filler. So it is necessary to modify its surface. First, barium chloride is reacted with sodium sulfate in the presence of a small amount of sodium 12-hydroxystearate. This introduces a controlled number of hydroxyl stearate sites onto the barium sulfate surface. The reaction is followed by a redox graft polymerization of acrylamide initiated by the hydroxyl stearate groups and ceric ion as a catalyst. Figures 6.9 to 6.11 show the effect of reaction substrates concentrations on polymerization rate.

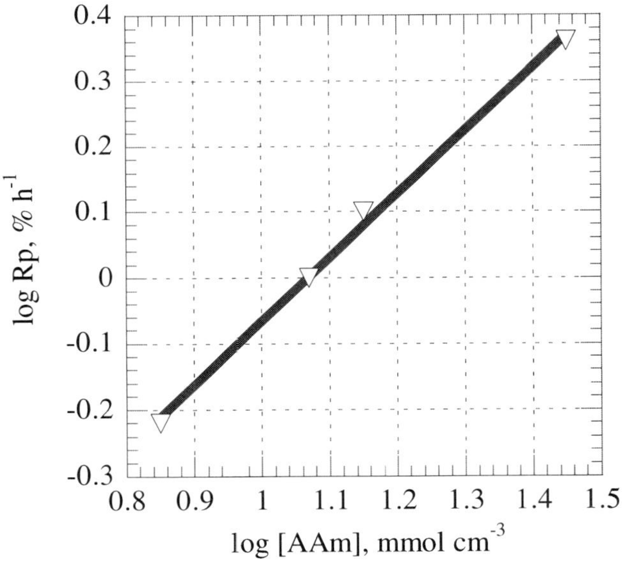


Figure 6.9. Polymerization rate, $R_p$, versus acrylamide concentration. [Adapted, by permission, from Tsubokawa N, Seno K, *J. Macromol. Sci. A*, **31**, No.9, 1994, 1135-45.]

In addition to the data included in Figures 6.9 to 6.11, it should be mentioned that grafting only occurs when all three of the ingredients are present (active site, monomer, and catalyst). The grafting reaction constitutes about 12% of the total

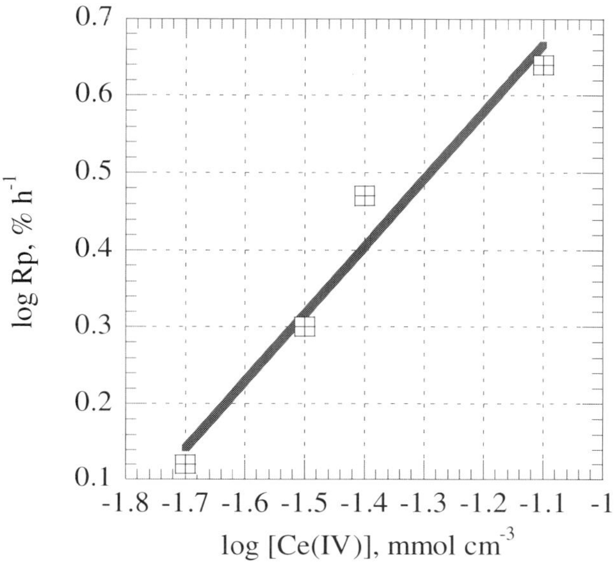


Figure 6.10. Polymerization rate, $R_p$, versus ceric ion (catalyst) concentration. [Adapted, by permission, from Tsubokawa N, Seno K, *J. Macromol. Sci. A*, **31**, No.9, 1994, 1135-45.]

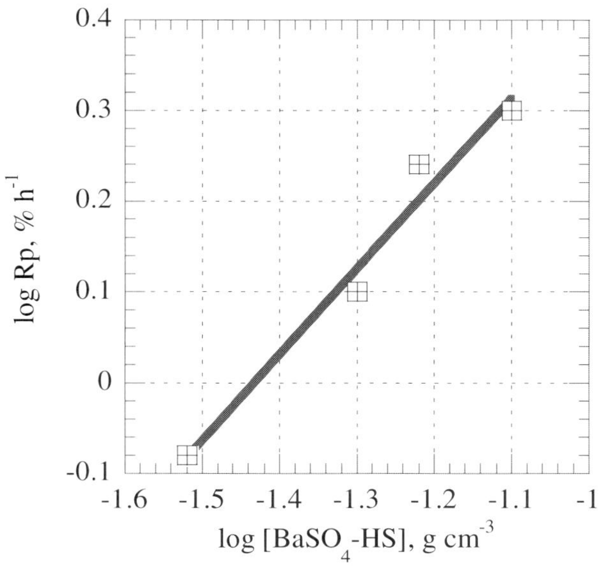


Figure 6.11. Polymerization rate, $R_p$, versus $BaSO_4$ 12-hydroxystearate modified concentration. [Adapted, by permission, from Tsubokawa N, Seno K, *J. Macromol. Sci. A*, **31**, No.9, 1994, 1135-45.]

polymer produced. The remaining polymer is not attached to the filler particles. The graphs show that the following kinetic equation is valid:

$$R_p = k[AAm][Ce(IV)][BaSO_4 - HS] \qquad [6.1]$$

The modification of aluminum hydroxide by dicarboxylic anhydride has similar kinetics (Figure 6.12).[51] Figure 6.13 shows that the linkages formed are durable since they withstand of 30 h Soxlet extraction with n-hexane. Only when more than 1% of dicarboxylic anhydride is used does it becomes associated with the filler through physical forces. In this condition it can be removed by extraction. The concentration of reactive functional groups on the filler surface has a strong influence on the modification processes.

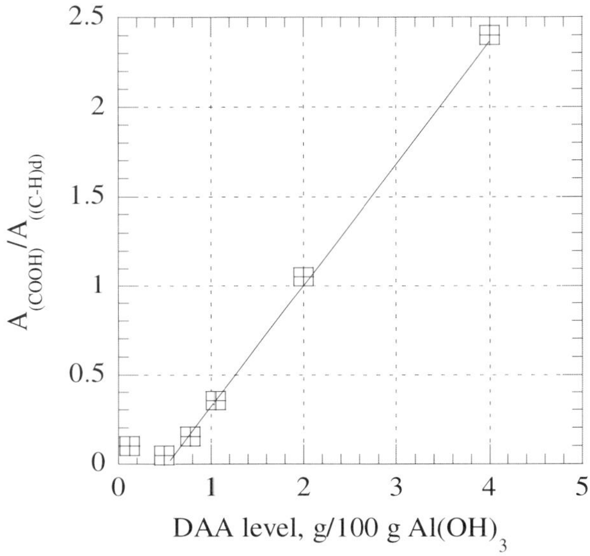


Figure 6.12. Carboxyl group IR absorption versus amount of dicarboxylic anhydride used for modification. [Adapted, by permission, from Liauw C M, Lees G C, Hurst S J, Rothon R N, Dobson D C, *Plast. Rubb. Comp. Process. Appln.*, **24**, No.4, 1995, 211-9.]

Titanium dioxide can be improved by doping with metals. Titanium dioxide participates in photochemical processes. Its mechanism involves the formation of positive holes in the valence band and electron promotion to the conductive band irradiated by UV. Both electrons and holes react with the surrounding material. By doping $TiO_2$ crystals with various metals, electron and hole recombination centers are formed. Also, the crystal is coated with a layer of hydrous oxides which decompose hydroxyl radicals. This is applied to various grades of $TiO_2$ which gives them a unique performance in applications where UV durability is required.

Silanes play an important role in the modification of fillers. Silanes are coupling agents. Coupling, in technical terms, means a device for connecting things. Here, coupling means chemical or physical bridging of two different chemical materials which otherwise would have had a weaker association. Coupling depends on:

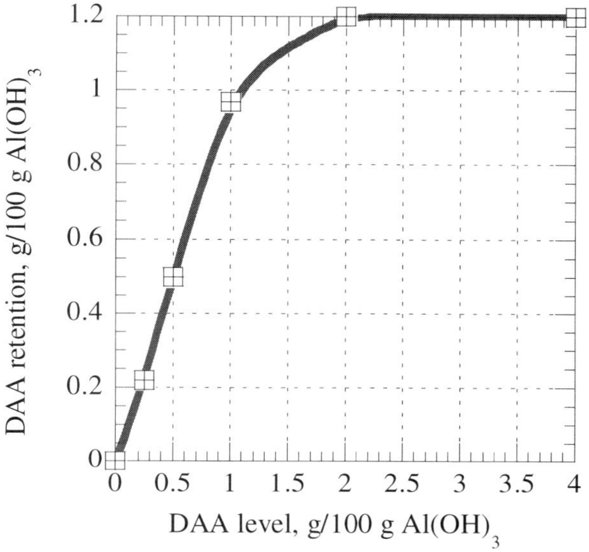


Figure 6.13. Retention of dicarboxylic anhydride after Soxlet extraction of modified $Al(OH)_3$ versus amount of modifier. [Adapted, by permission, from Liauw C M, Lees G C, Hurst S J, Rothon R N, Dobson D C, *Plast. Rubb. Comp. Process. Appln.*, **24**, No.4, 1995, 211-9.]

- The chemical structure and mechanism of action of the coupling agent
- The character and chemical composition of the filler surface
- The chemical composition and, thus, the reactivity of the polymeric material
- The surface tension of the material being coupled
- The effect of system rheology
- The mechanism of physical and chemical adsorption of the coupling agent on the filler surface
- The molecular coverage and molecular orientation of the coupling agent
- The mobility of the coupling agent and other components of the system
- The effect of pH, solvent, etc. on adsorption
- The effect of filler surface preparation on adsorption and bond stability
- The reactivity of the organic part of the coupling agent with the polymer

From this list, the complexity of the coupling phenomenon can be estimated. It has taken forty years of practical experiments to develop our current understanding of these processes.

The chemical formula of the coupling agent can be written as follows:

$$R_nAX_{4-n} \qquad [6.2]$$

where:

*R* a group responsible for polymer binding
*X* a group which combines with a filler
*A* a four-valent central atom connecting both groups in one chemical moiety.

Silicon, titanium, and zirconium, members of the IVth group of the periodic table, are the elements used as a central atom of the coupling compound. They are able to form four-valent compounds. The above structure is sometimes written in the more detailed form of a chemical compound able to perform six functions:[84]

$$(Y - R_1 - Z - O)_n AX_{4-n} \qquad [6.3]$$

where:

*Y* provides bonding reactivity with the polymer
$R_1$ provides van der Waals attraction and entanglement via long carbon chains
*Z* provides antioxidant effect, acid resistance, and corrosion protection via chemical groups involved (alkyl, carboxyl, sulfonyl, phenolic phosphate, etc.)
*X* hydrolyzable portion of molecule able to combine with filler
*A* provides transesterification and transalkylation catalytic activities, as well as affects other processes simultaneously performed in the system (curing, foaming, etc.)
*n* controls functionality of each substrate involved in the coupling reaction (filler and polymer).

**Table 6.2: Physical properties of some organofunctional silanes**

| Formula | Mol. weight | Density, g/cm$^3$ | $n_D^{25}$ | Boiling point, °C |
|---|---|---|---|---|
| $CH_2=CHSi(OCH_2CH_3)_3$ | 190.3 | 0.894 | 1.397 | 161 |
| $CH_2=C(CH_3)C(O)O(CH_2)_3Si(OCH_3)_3$ | 248.1 | 1.045 | 1.429 | 255 |
| $EpoxyCH_2O(CH_2)_3Si(OCH_3)_3$ | 236.1 | 1.069 | 1.427 | 290 |
| $HS(CH_2)_3Si(OCH_3)_3$ | 238.3 | 1.072 | 1.440 | 212 |
| $H_2N(CH_2)_3Si(OCH_2CH_3)_3$ | 221.3 | 0.942 | 1.420 | 217 |

This chemical formula describes the functions of coupling agents. The functional groups in available compounds, containing titanium, zirconium, and silicon, can be identified in corresponding catalogs of these products.[66,67] Several hundred compounds are available and discussed in these catalogues[66,67] which list prospective coupling agents. Table 6.2 outlines the basic properties of organofunctional silanes which have found broad application in industry. The chemical structure of these compounds influences the mechanism in which they are involved. The first involves hydrolysis, according to the following equation:

$$RAX_3 + 3H_2O \rightarrow RA(OH)_3 + 3HX \qquad [6.4]$$

Because the X group is usually either alkoxy or chlorine, HX denotes alcohol or hydrogen chloride. In the next stage, the hydrolyzed silane undergoes a condensation according to the following reaction:

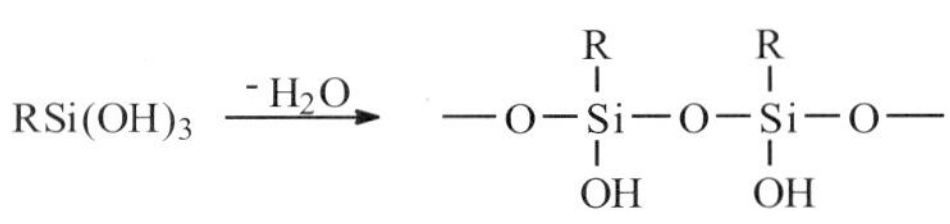

This stage is probably the most important contributor to coupling success. If the reaction occurs just as the silane is added or, later during material storage, it may retard silane mobility in the system (as the molecular weight of the silane increases, its rate of migration to the surface is slowed). This results in a less efficient silane use or in system instability. Here, the silane is used for the production of homopolymer entangled in the polymer chains, rather than for forming an interface between the filler and the polymer.

It has long been known that the rate of silane homopolymerization is increased by pH or metal salt catalysis and decreased by increased concentration and higher temperature. Most silanes are hydrolyzed most rapidly at pH between 3 and 5. Solution stability depends on the rate of homopolymerization to siloxane polymer. This is affected by pH, the presence of soluble salts of lead, zinc, iron, etc., and silane concentration. A pH in the range of 4 to 5 generally favors the monomeric form and retards polymerization. The formation of homopolymer can be detected as silane loses solubility and forms a gel which is not active in the coupling process. It is, then, desirable to retain silane in the monomeric or dimeric form. In the next two steps a bond is formed with the substrate (e.g., filler).

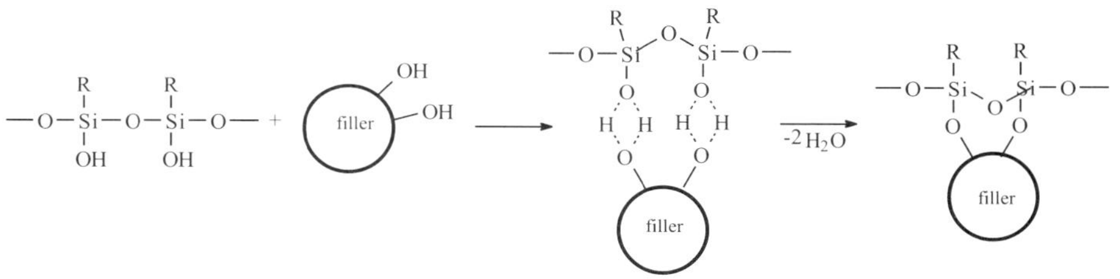


The first step is hydrogen bonding followed by bond formation and then by release of water which hydrolyzes other molecules should there be a shortage of water in the system.

The character of the deposition of γ-(methacryloxy)-propyl-trimethoxysilane (MPS) on the surface of clay and calcium carbonate was studied.[68] While most of MPS resists tetrahydrofuran washing when deposited on clay, MPS is removed by THF from calcium carbonate. Physico-sorbed layers can be removed by a solvent, whereas chemisorbed layers cannot be. Calcium carbonate does not contain hydroxyl groups (only some are available in admixtures), unlike clay which has a surface composed of Si-OH and Al-OH functionalities capable of covalent bond formation with silanol. Clay retains 66% of the silane applied, whereas only 19% of silane remained on the surface of calcium carbonate.

Further experiments have been done to determine the molecular weight of silane oligomers formed on the surfaces of various fillers and the amount of silane retained after THF washing. Table 6.3 shows how the retention of silane compares with the pH of the filler slurry. Neutral pH favors retention of silane, whereas a ba-

**Table 6.3: Silane retention percentage on various fillers after THF washing[69]**

| Acidic | Neutral | Basic |
|---|---|---|
| Iron(III) oxide (pH=1.9) 82%<br>Zirconium oxide (3.1) 78%<br>Aluminum oxide (3.3) 100%<br>Clay (4.1) 77%<br>Tin (IV) oxide (4.1) 77%<br>Tungsten oxide (4.8) 45%<br>Tin oxide (5.3) 100<br>Iron(II) oxide (5.7) 100<br>Copper(II) oxide (6.1) 18% | Aluminum silicate (6.6) 87%<br>Titanium oxide (6.6) 84%<br>Amorphous silica (6.9) 83%<br>Nickel oxide (7.0) 96%<br>Kaolin (7.1) 96%<br>Zinc oxide (7.6) 100% | Calcium hydroxide (12.3) 57%<br>Magnesium oxide (11.1) 98%<br>Glass spheres (10.7) 32%<br>Barium hydroxide (10.5) 46%<br>Lead oxide (10.0) 80%<br>Wollastonite (9.9) 21%<br>E-glass (9.5) 50%<br>Calcium carbonate (9.4) 19%<br>Calcium metasilicate (9.4) 34%<br>Mica (9.3) 55% |
| Average = 74% | Average = 89% | Average = 49% |

sic species does not favor retention. A pH of 4-5 is needed to stabilize a monomeric form.

At neutral pH, high molecular weight silane structures are obtained which provide more protection against chemical and water attack at the interface, but they also reduce interpenetration of resin into the silane layers (a similar effect was discussed above for calcium carbonate coating density).

The reversibility of the reaction is another important feature of coupling by silanes, titanates, and zirconates. The bond formed in the second stage (see chemical reaction above) is not a permanent bond but is an equilibrium reaction which depends on the amount of water in the system. This is the most important concept in the coupling mechanism. Bonds can form, break, and reform. Water immersion affects the interface, causing bond breakage. Bonds can be reformed again if the internal stress in the polymer matrix does not cause permanent delamination which separates the surfaces.

The third stage involves the reaction of the organic part of the silane molecule with a polymer, if such mechanism is available; e.g., when the organic part contains groups which can react with the polymer in question. The reactive group of the organic part of silane must react with the polymer. But, it is also very essential when this reaction occurs. If the rate of this reaction is too high, the polymer binds silane before it can reach the filler surface, thus silane mobility is retarded. This silane molecule will never participate in interface formation but will remain entrapped in the system, forming an inefficient bond.

In summary, depending on polymer type, amino, mercapto, epoxy, or vinyl are the most common functional groups which react with the polymer. Alkoxy or chlorine groups are often used to react with the filler surface, chlorine being less popular because it produces hydrogen chloride, a corrosive material.

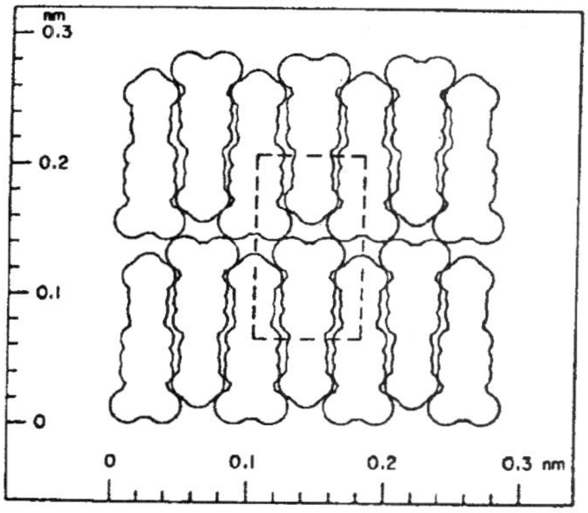


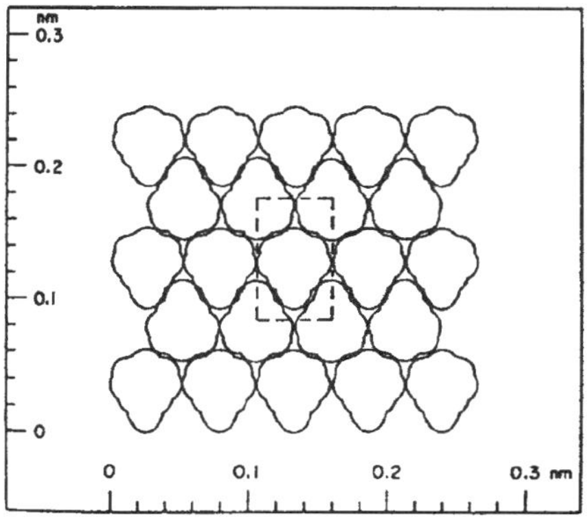


Figure 6.14. Closely-packed molecular configurations for silanetriol oriented parallel (left) and perpendicular (right) to the substrate surface. [Adapted, by permission, from Miller J D, Ishida H, *Surface Sci.*, **148**, 1973, 601.]

Efficiency is determined by the wetting characteristics of coupling agents, the surface area of the filler occupied by the coupling agent, the molecular mobility of coupling agents, their effect on viscosity, the effect of solvents on adsorption, the molecular orientation of coupling agents on the filler surface, their configuration, and molecular packing. The effect of coupling depends on the density of bonds formed on both sides of the interface, which primarily depends on the availability of the coupling agent at the interface. Critical surface tensions of most silanes are generally much lower than those of the surfaces (glass, fillers, metals) which they wet, indicating that surface wettability does not create a barrier.

The molecular mobility of the coupling agent is probably the single most important factor which contributes to the efficiency of its action. Two aspects of the mobility are equally essential: the reactivity of chemical groups in the coupling agent with chemical groups on the system components, and the mobility of the coupling agent, which depends on its physical interaction with the mixture components. The coupling agent must undergo a chemical reaction with the reactive group of the polymer or any other component of the mixture. But such a reaction should occur only when the reactive molecule is delivered to the interface which must be improved. To accomplish this requires planning of the mechanisms such that the rate of migration of the coupling agent is much higher than the rate of its reaction with the organic component. If this mechanism is not provided, more of this expensive component must be added and polymer properties will be modified. In this situation, the pot-life of the reactive system may affect the adhesion. Better adhesion will be obtained in a freshly prepared mixture than in the same system after storage, as a result of the partial reaction of silane.

Orientation of γ-methacryloxypropyltrimethoxysilane (MPS) was studied in detail.[70] Two approaches were adapted: one was experimental, using FTIR with a hemispherical diffuse reflectance attachment, the other involved molecular model-

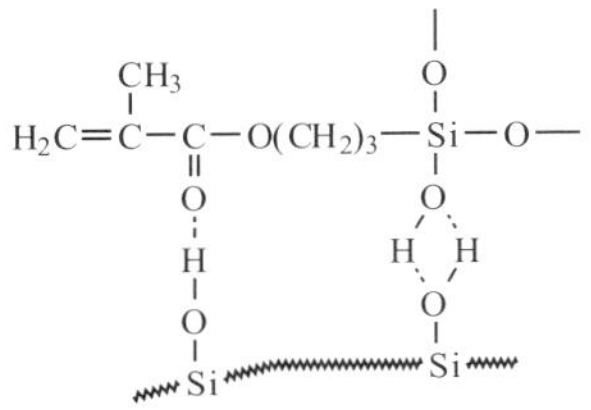


Figure 6.15. Silane arrangement on substrate surface.[70]

ling. Two molecular projections were analyzed, as displayed in Figure 6.14. Calculations showed that at a parallel orientation to the substrate, the MPS occupies a surface area of 0.55 $nm^2$. At a perpendicular orientation to the substrate surface one molecule of MPS occupies a surface area of 0.24 $nm^2$. From spectroscopic data it was calculated that the MPS molecule occupies 0.6 $nm^2$ on clay and 0.59 $nm^2$ on lead oxide which indicates that the molecules are in a parallel orientation (Figure 6.15).

These experiments show that the silane molecule must be held on the surface by hydrogen bonding at two centers of the molecule. The hydrogen bonding interaction with a clay surface is stronger than that with a lead oxide surface, as suggested by the 9 $cm^{-1}$ difference in the frequency shift. The hydrogen bonding appears to be related to the ability of the surface to donate protons.

The close similarity of the data from modelling and experiments suggests that silane, if allowed to migrate to the surface of the filler, forms a closely packed monomolecular layer. The surface area occupied is related to the orientation of the molecule and its size, rather than to the type of substrate (filler) on which it is deposited. According to other studies,[71,72] silane molecules can also be oriented horizontally, and their actual orientation probably depends on the method of application (concentration, type of solvent, etc.).

The data in Table 6.4 have been developed from many years of practical experience. It provides information on the suitable types of coupling agents differentiated by the organic portion of the molecule. In polymers which are not reactive, such as polyethylene, polypropylene, etc., adhesion is built up by hydrogen bond formation. Methacrylosilanes provide this effect with these materials. Experimental work is always recommended to evaluate each combination of coupling agent, substrate, and polymer. So many diverse factors are involved that theoretical predictions are not always reliable.

**Table 6.4: Preferred silanes for certain resins**

| Resin | Silane functional group |
|---|---|
| Epoxy | Epoxy, amine |
| Melamine | Amine |
| Polyamide | Epoxy, amine |
| Phenolic | Epoxy, amine |
| Polybutadiene | Vinyl, methacryl, mercapto |
| Polyester | Vinyl, methacryl |
| Polyethylene | Vinyl, methacryl |
| Polypropylene | Methacryl |
| Polyvinylchloride | Mercapto, amine |
| Urethane | Methacryl, mercapto, amine |

Several new developments in silanes have been reported recently.[64] One deficiency of existing silanes was that they were not applicable to high-temperature polymers such as polyimides, either because they were not thermally stable or because they did not facilitate adhesion. A new aromatic imide silane was synthesized with two silane groups attached to the terminal phenyls in the molecule. A

**Table 6.5: Changes in material properties caused by use of modified filler**

| Property change | Filler | Modification | Refs. |
|---|---|---|---|
| Reinforcement | glass beads | silane | 73 |
| Increased impact resistance | glass beads<br>graphite fiber | compatibilizer<br>polymerization | 74<br>75 |
| Increase in tensile strength | clay<br>silica | silane | 26<br>42,76 |
| Decrease in elongation | talc | phosphate coating<br>hexadecanol, silica | 77<br>76 |
| Increase in interlaminar shear strength | carbon fiber | thermal & epoxy<br>oxidation | 78<br>10 |
| Increase in tear strength | clay<br>$CaCO_3$ | silane | 26<br>28 |
| Increase in abrasion resistance | clay | silane | 26 |
| Increase in flexural modulus | talc | phosphate coating | 77 |
| Improvement in shear stress transfer | cellulose fibers | maleic anhydride | 79 |
| Decrease in Mullins' effect | silica | hexadecanol, silanes | 76 |
| Decrease in compression set | clay | silane | 26 |
| Increase in heat resistance | clay | silane | 26 |
| Decrease in sedimentation | $Al(OH)_3$ | dicarboxylic anhydride | 51 |
| Improvement in colloidal stability | kaolin<br>carbon black | polyurethane coating<br>grafting | 8<br>11 |
| Improvement in dispersion | carbon whisker | grafting | 38 |
| Decrease in viscosity | clay<br>$Al(OH)_3$ | silanes<br>dicarboxylic anhydride | 26<br>51 |
| Increase in melt flow | $CaCO_3$ | stearate | 80 |
| Effect on rheological properties | sepiolite | thermal treatment | 81 |
| Control of particle size and coating | colloidal silica | grafting | 5 |
| Improvement in whiteness | $CaCO_3$ | stearate | 80 |
| Lower electric conductivity | graphite | polymerization | 72 |
| Increase in polymer MW | graphite | polymerization | 75 |
| Transcrystallinity occurs | cellulose fiber | maleic anhydride | 79 |
| Increase in rubber-filler bonding | carbon black | oxidation<br>silane | 82<br>83 |
| Converts hydrophilic to hydrophobic | montmorillonite | grafting | 84 |

**Table 6.5: continuation**

| Property change | Filler | Modification | Refs. |
|---|---|---|---|
| Reduction in water uptake | montmorillonite | grafting | 84 |
| Increase in crosslink density | silica<br>$CaCO_3$ | hexadecanol, silanes<br>maleic derivatives | 76<br>32 |
| Decrease in filler-filler interaction | silica | hexadecanol, silanes | 76 |
| Increase in solvent resistance | kaolin | polyurethane coating | 8 |
| Decrease in specific interaction | fumed silica | silanes | 44 |

new polymeric silane was also developed. This has a polyethyleneimine backbone. It has film forming properties and can be used for the reinforcement of the interphase. Silanes have recently been used in the fabrication of integrated circuits where it adheres to polyimide, silicon and the layered metal patterns. Other area of developments in silane technology involve various methods of polymerization. Two relatively new methods are: photograft polymerization (photosensitive silane) and plasma polymerization of organosilanes. There are many new applications which use mixed silanes which allow various properties to be optimized.

## 6.4 EFFECT OF FILLER MODIFICATION ON MATERIAL PROPERTIES

Modification of fillers alters their properties. This section discusses:

- The types of changes which can be expected from modifications
- The extent of such changes

Table 6.5 lists the properties affected by various filler modifiers. Table 6.5 is not given as a comprehensive list of applications of these fillers. Fillers and specific properties of filled materials are dealt in detail in other parts of the book. The table summarizes the types of changes which can be expected from modifications.

The extent of changes caused by modifications is illustrated by the specific cases discussed below. Figure 6.16 shows the influence of modification of glass beads with 3-aminopropyltrimethylsilane by maleic anhydride grafted polypropylene (Exxelor PO2011) on the notched Charpy impact of polypropylene containing a compatibilizer and a non-functionalized rubber (EPM). The addition of glass beads to PP caused a reduction of Charpy impact from 6 to 4 $kJ/m^2$. Yield stress was reduced from 32 to 18 MPa. This is caused by poor interfacial adhesion between PP and amine functional beads. The addition of a maleic grafted PP compatibilizer increased the yield stress to 31-32 MPa in the concentration range of compatibilizer. The Charpy impact improved at the lowest level of compatibilizer and remained constant. Although glass beads are coated with only 0.02% 3-aminopropyltrimethoxysilane, this is sufficient to react with the smallest amount of compatibilizer (no more reactive sites left). If EPM is dispersed as a separate

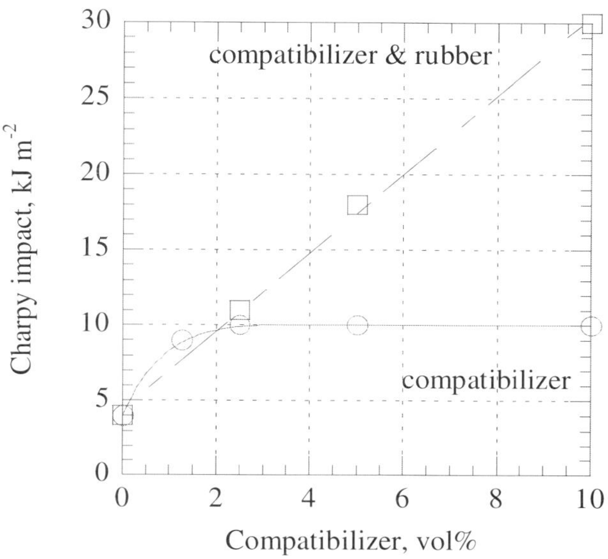


Figure 6.16. Charpy impact of PP containing 30% glass beads, compatibilizer (PP-g-MA) and rubber (EPM). [Data from ref. 74.]

microphase, its Charpy impact can be dramatically improved with a slight loss of yield stress. This shows that it is not just simple surface modification but complex interaction of the several components of the formulation and the reactive processing. The results of a good composition design can surpass the performance of the matrix polymer.

Figure 6.17 shows that crystalline dimensions affect interlaminar shear strength. Crystalline dimensions on the surface of carbon fibers can be measured by Raman spectroscopy. The ratio of intensity of two bands (1355 and 1575 cm$^{-1}$) is proportional to the crystalline dimensions on the surface. The crystalline width increases considerably from 3 to over 12.5 nm when the temperature of manufacture of carbon fibers is increased from 1000 to 3000°C. The increase in band ratio correlates with an increase in interlaminar shear strength.

The amount of carboxylic anhydride used for modification of $Al(OH)_3$ determines the sedimentation rate of filler particles and the increase in water-based slurry viscosity.[51] A limiting value of viscosity is attained at relatively low levels of modifier. This amount of modifier is sufficient to react with the available sites on the $Al(OH)_3$ providing conditions for the breakdown of the aggregates and a separation of individual particles. Since the reaction decreases particle-particle interaction these processes are likely to occur. The sedimentation volume curve can be explained in the same way.

At low levels of addition, aggregates breakdown with simple shaking (before measurement) and later reagglomerate as they settle. These data show the effect of

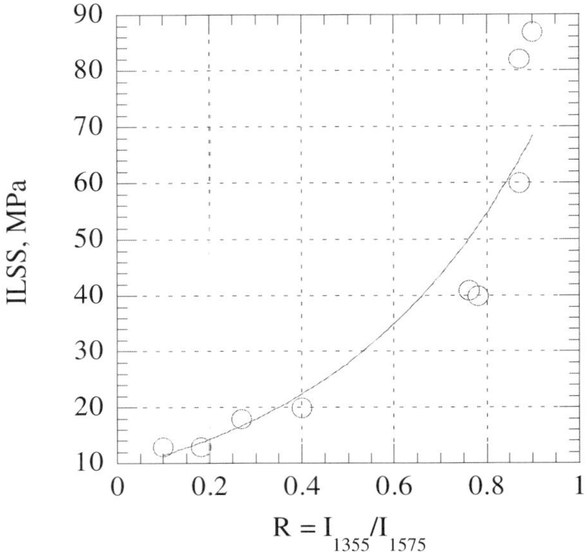


Figure 6.17. Ratio of Raman peak intensities at 1355 and 1575 $cm^{-1}$ vs. interlaminar shear strength of composites containing carbon fibers of different origin. [Data from Tang L-G, Kardos J L, *Polym. Composites*, **18**, No.1, 1997, 100-13.]

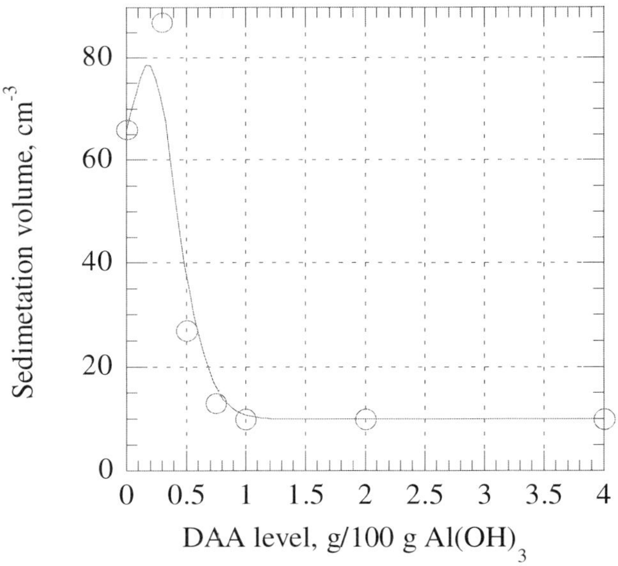


Figure 6.18. Sedimentation volume of $Al(OH)_3$ vs. amount of dicarboxylic anhydride used for its modification. [Adapted, by permission, from Liauw C M, Lees G C, Hurst S J, Rothon R N, Dobson D C, *Plast. Rubb. Comp. Process. Appln.*, **24**, No.4, 1995, 211-9.]

modification on particle-particle interactions. Not only rheological properties but other properties such mechanical properties, Mullins' effect, etc. are affected.

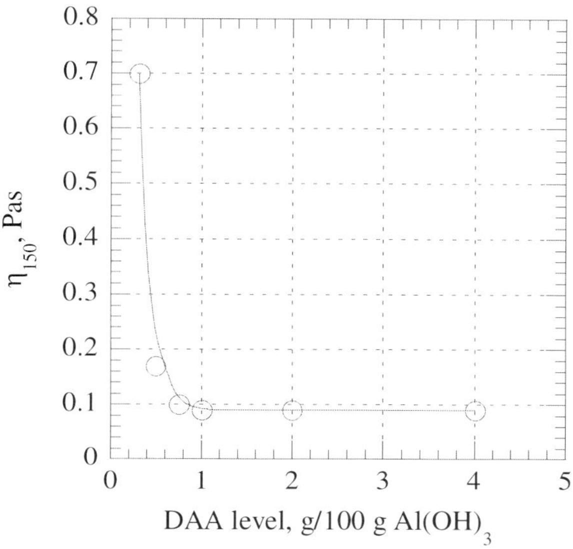


Figure 6.19. Slurry viscosity vs. amount of dicarboxylic anhydride used for $Al(OH)_3$ modification. [Adapted, by permission, from Liauw C M, Lees G C, Hurst S J, Rothon R N, Dobson D C, *Plast. Rubb. Comp. Process. Appln.*, **24**, No.4, 1995, 211-9.]

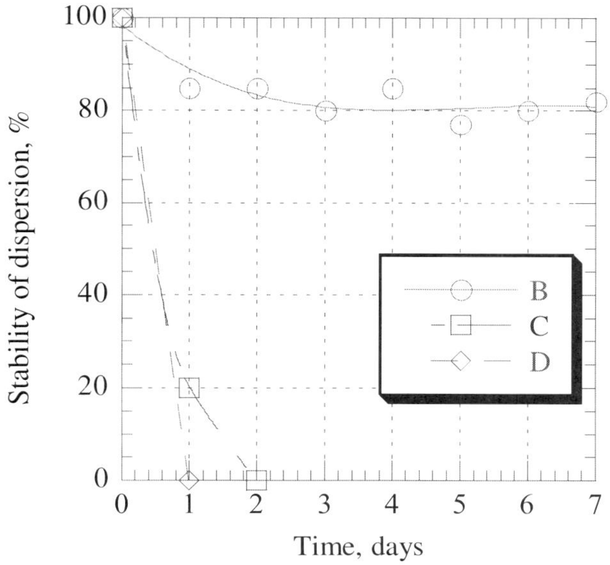


Figure 6.20. Stability of grafted carbon black (A), carbon black with physically absorbed polymer (B) and untreated carbon black (C). [Adapted, by permission, from Tsubokawa N, Hosoya M, Kurumada J, *Reactive & Functional Polym.*, **27**, No.1, 1995, 75-81.]

Carbon black grafting has a similar effect on dispersion stability (Figure 6.20).[11] Modified carbon black is substantially improved. The graphs show that there is a substantial difference between the physical dispersion of the polymer and grafting. The absorption of polymer on the surface does not give improvement over untreated carbon black. Direct condensation of carboxyl groups on the surface of carbon black with N,N'-dicyclohexylcarbodiimide, followed by the reaction with the polyol gives substantial improvement. Grafted polymer chains inhibit aggregation and contribute to the formation of a stable colloidal dispersion.

The same results were obtained when polytetrahydrofuran was grafted onto carbon whisker.[38]

## 6.5 RESISTANCE TO VARIOUS CHEMICAL MATERIALS

The literature on the subject of chemical resistance of fillers and filled materials remains sparse but some essential data can be reported. A comparison was made between E-glass and Kevlar with respect to their resistance to acids and bases.[85] E-glass is severely attacked by most acids. The most aggressive is nitric acid. Two weeks immersion causes a loss of more than 90% of its original strength. E-glass is resistant to acetic and phosphoric acids. Only 39% of its original strength is retained in ammonia and 64% in NaOH. In all these cases, the samples were immersed for 2 weeks in 2M solutions. Kevlar in most cases resists acids very well with a loss of only 10% of its original strength. The exceptions are HCl ( a loss of 4%) and $HNO_3$ (loss of 60%). Other immersions, such as in $H_2SO_4$, $H_3PO_4$, and NaOH cause losses of about 30%.

Morphological studies explain the mechanisms of E-glass corrosion.[86] According to these studies, acid corrosion of E-glass is caused by calcium and aluminum depletion which varies depending on the acid type, fiber type, and acid concentration. Oxalic and sulfuric acids are more corrosive than nitric and hydrochloric acids. This difference is due to the fact that, in oxalic acid, precipitated products are formed which decrease the concentration of leachates in solution. In addition to the loss of mineral content, fibers develop axial and spiral cracks. Crack formation depends on the rate of material depletion.

$CaCO_3$ is not resistant to the attacks of $H_2SO_4$ and HCl because the products of reaction are water soluble. Polyolefins filled with $CaCO_3$ experience a rapid loss of weight on acid immersion which depends on the concentration of filler. This loss of mass causes increased porosity of the filled materials.[87] The effect of $TiO_2$ on corrosion resistance was also evaluated. Different grades varied in their degree of interaction with the binder. If the interaction decreases, the corrosion resistance also decreases. The corrosion damage in salt spray was proportional to the concentration of $TiO_2$.[88] Filled epoxy resins used in food contact applications were evaluated by SEM. Introduction, even in small concentrations (10%), of filler causes inhomogeneities in the fracture zone but the surface remains similar to unfilled material. But, under high magnification, all surfaces had small (about 9 nm) microcracks which might permit reagents to diffuse. Larger particle sized fillers (barium sulfate and iron oxide) caused more inhomogeneity in the filled material than did smaller particles. Small particles were well coated with resin.[89]

Solvents produce different effects than do corrosive chemicals. Both silica and carbon black filled natural rubbers were more resistant to solvents than unfilled rubber.[90] Also, the cure time was important, indicating that the bound rubber plays a role in the reduction of a solvent sorption. The diffusion coefficient of solvents into rubbers decreases with longer cure times and higher fillers loadings. Polychloroprene rubber swollen with solvent has a lower compression set when it is filled with carbon black.[91]

## 6.6 CURE IN FILLER'S PRESENCE

This section contains information on the cure response of UV-curable and thermosetting polymers in the presence of fillers. The discussion includes:

- Advantages and disadvantages of the use of fillers
- How fillers interfere with cure

The kinetics of reaction is discussed in Section 6.10 and polymerization reactions in Section 6.7. Grafting is discussed in Section 6.8, crosslink density in Section 6.9, and bound rubber in Chapter 7. Here, UV-curable materials, epoxy resins, polyurethanes, rubbers, polyesters, and phenolic resins are discussed.

The application of fillers in light-curable resins is considered a sensitive issue because filler particles are known to reflect and absorb light radiation which may potentially affect curing rates. This makes it an interesting subject to evaluate. Several fillers were studied in this context.[92-95] Silica was used to fill 2,2'-bis[4-(-methacryloxy-2-hydroxy-propoxy)-phenyl]-propane – a material used for restorative purposes in dental applications. Figure 6.21 shows the degree of conversion vs. silica content. The increase in conversion rate depends on the path length of UV radiation through the material. Multiple radiation scattering by the filler causes a better use of light energy due to the complex path that the light beam is forced to take and by better light distribution throughout the material. In order to take full advantage of scattering, the size of particles must be half of the wavelength of the activating light. Silica particles are very small (0.04 μm) but they form agglomerates which are larger (0.2 μm). The size of these agglomerates is close to half of the wavelength of the activating light (0.4-0.5 μm).[92]

Printing inks and wood fillers also take advantage of light curing.[93] $Al(OH)_3$ is the usual choice in this application. Materials were cured by a mercury lamp and results evaluated by differential photocalorimetry and UV-visible spectroscopy. The sample weight and amount of filler were the essential variables. Addition of filler increased UV cure rate. At a low filler concentration (13.2%) the rate was only ~10% more than with no filler. At a medium filler load (19.7%) the rate was doubled. At very high loads (56.5%) and in larger specimens used for the evaluation of wood fillers, the curing rates tripled over products filled with the more conventional $CaCO_3$ and clay. These increased rates of cure are due to a filler more

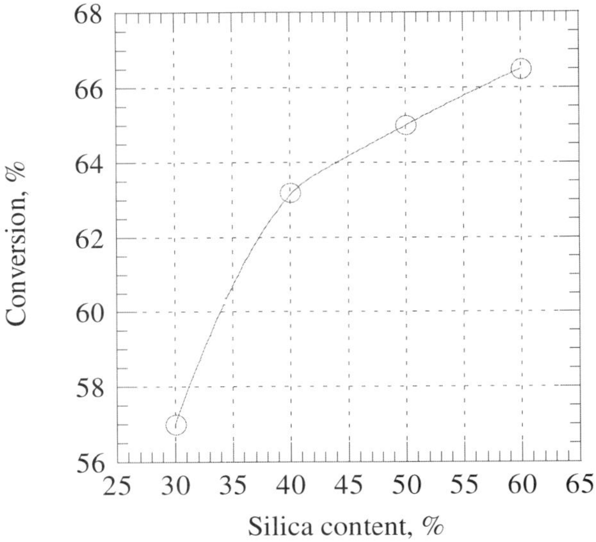


Figure 6.21. Degree of conversion as a function of silica content. [Adapted, by permission, from Kim S, Jang J, *Polym. Test.*, **15**, No.6, 1996, 559-71.]

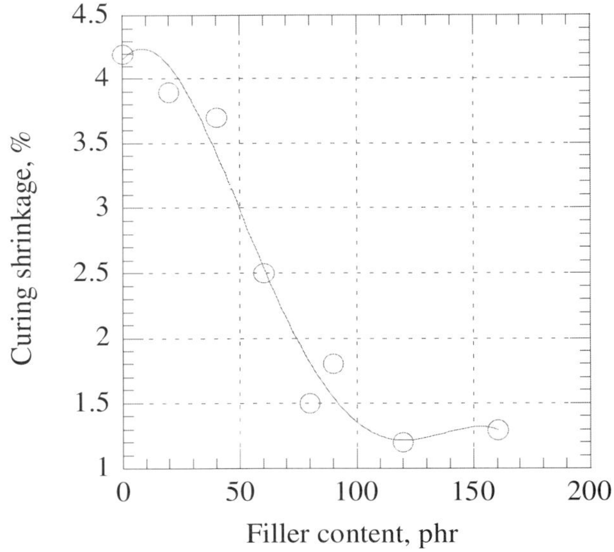


Figure 6.22. Curing shrinkage of UV curable adhesive vs. filler content. [Adapted, by permission, from Murata N, Nishi S, Hosono S, *J. Adhesion*, **59**, Nos.1-4, 1996, 39-50.]

transparent to UV light than is the resin which causes the increase in UV flux and penetration depth. This conclusion is further reinforced by other data presented.[94] The evaluation of light transmittance through a sample of a filled polymer (UV-

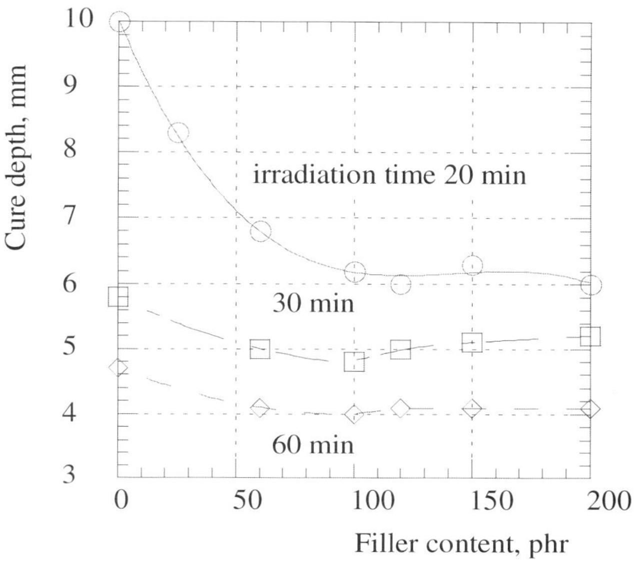


Figure 6.23. The effect of quartz filler on the depth of cure at three irradiation times. [Adapted, by permission, from Murata N, Nishi S, Hosono S, *J. Adhesion*, **59**, Nos.1-4, 1996, 39-50.]

cured polyurethane) shows that the unfilled polymer has the same transmittance as a polymer filled with $Al(OH)_3$ but the filled polymer has substantially improved diffuse transmittance over the neat polymer. Thus, more light is used in a process which accelerates the cure. Particle size does not seem to have any influence on transmittance nor on cure.

Quartz was used as a filler in the manufacture of optical devices from epoxy in a UV-curable system.[95] Figure 6.22 shows that addition of a filler can substantially reduce curing shrinkage which is highly desirable in the precise manufacture of these materials. Reduced shrinkage is the result of a low thermal expansion coefficient of quartz in comparison with the resin.

The presence of a filler does not significantly affect cure depth (Figure 6.23). Small quantities of a filler in a system with very long cure times have some influence. But this is irrelevant considering that the best optical and mechanical properties are obtained at filler loadings of 100-150 phr. Several metal oxides ($Fe_2O_3$, $Al(OH)_3$, and ZnO) were tested in brominated epoxy resins cured at elevated temperatures. $Fe_2O_3$ was found to have a catalytic effect on the cure. This is due to the reaction of terminal epoxy groups with surface hydroxyl groups on the filler. Adding $Al(OH)_3$ also caused an increase in the rate of cure. This is attributed to an accelerated homopolymerization of epoxy on the alumina surface. ZnO gave the greatest acceleration of cure rate but no explanation is yet known. All three fillers affect reaction rates, reaction orders, activation energies, and reaction exotherms.

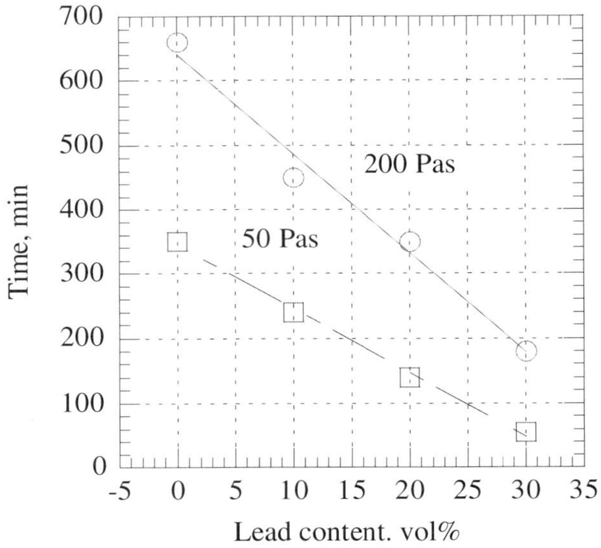


Figure 6.24. Polyurethane formation in the presence of lead powder. Reaction time to reach certain viscosity. [Adapted, by permission, from Caillaud J L, Deguillaume S, Vincent M, Giannotta J C, Widmaier J M, *Polym. Int.*, **40**, No.1, 1996, 1-7. ]

Lead powder was used as a filler in polyurethane.[97] It is clear from the graph (Figure 6.24) that the addition of metal catalyzes the curing process. The reasons for this catalytic effect are unknown. It is suspected that surface impurities may act as a system catalyst but which impurities is not known.

Densified polyurethane foam was used as a filler in rubber in an attempt to recycle this material.[98] Small additions (up to 30%) did not much affect the cure rate but as the quantity was increased the rate of vulcanization slowed probably due to the effect of dilution and increasing viscosity.

In interpenetrating polymer networks, chemical crosslinking and phase separation and their timing affect properties. Fumed silica, alumina, and carbon fiber were used in a network developed from polyurethane and polyesteracrylate.[99] The presence of fillers affected many properties. Conversion rates were higher in the presence of fillers. Also, microphase separation was affected. As a result of these two changes the filled material was unrecognizable from the unfilled material.

In rubber, fillers play a role in the vulcanization process. These fillers are not considered here. The effect of carbon black on the vulcanization rate is still a matter of some dispute. Older papers presented data indicating that carbon black slows down the vulcanization rate. A more recent study[100] shows that the vulcanization rate of rubber actually speeds up with the addition of carbon black. The proposed explanation suggests that changes in the mechanism of vulcanization occur in the presence of carbon black. Addition of ground rubber together with carbon black did not affect the vulcanization rate.[101] In ferromagnetic applications, ferrites were

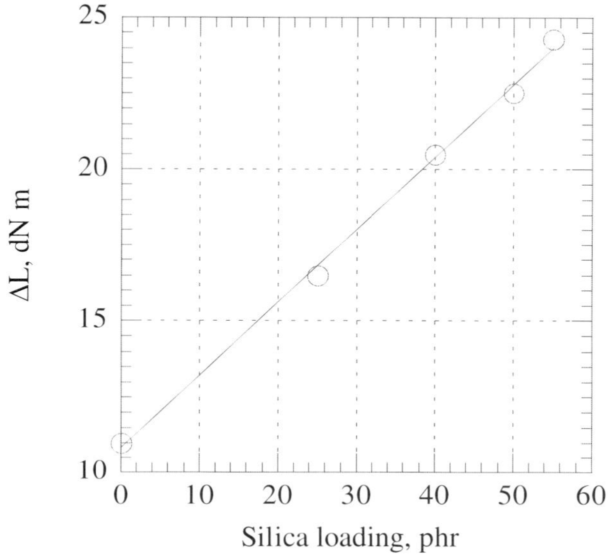


Figure 6.25. Torque difference vs. silica loading. [Adapted, by permission, from Cochrane H, Lin C S, *Rubb. Chem. Technol.*, **66**, No.1, 1993, 48-60.]

added to rubber.[102] Cure rates were substantially faster especially in the presence of barium ferrite.

Silicone polymers are unique in that they require a filler to improve their properties. The filler discussed here (silica) is similar in functionality to the polymer.[103] Figure 6.25 shows the difference between the initial and the final torque which is a measure of crosslink density. The reaction rate is proportional to the silica loading. This indicates that both the filler and polymer contribute to the measured property. Also, properties depend on the number of hydroxyl groups on the filler's surface since they participate in the reaction.

$Al(OH)_3$ inhibits the curing reaction of allylester resin.[104] $CaCO_3$ and glass fiber exert a similar effect on the cure of unsaturated polyesters.[105,106] Both the reaction rate constant and the activation energy are higher in the presence of a filler than in the neat resin. Many papers have been published dealing with the rate of reaction in the presence of these fillers. There is no consensus. Some report acceleration, some no effect, others rate reduction. The reasons are also inconclusive. It is difficult to say if cure inhibitions reported in recent papers are a special case or if the results reported depend on factors which are still to be determined.

Lignin fillers decreased the cure rate of phenol-formaldehyde resin.[107] Here, the filler acts as a diluent and does not have the ability to affect the reaction kinetics by interaction with the polymer. Glass fibers also decreased the rate of cure of a phenolic resin in another study.[108]

In conclusion, the effects of particulate fillers differ from those of high aspect ratio fillers. Particulate fillers seem to increase cure rate in most cases, especially if they contain active groups on their surface which either may react with the resin or change reaction mechanism. High aspect ratio fillers seem to decrease the reaction rate due perhaps to their more localized influence. Questions still remain as to how accurate these findings are. Reactive systems have been studied by rheological and physical methods and little is known about the mechanisms of reaction or kinetics determined by following the concentrations of substrates during the reaction. Physical and rheological methods give information on the entire system as determined by changes in viscoelastic properties or in thermodynamic properties. These depend not only on the chemical reaction but also on the association, crystallization, and orientation which are properties unrelated to cure.

## 6.7 POLYMERIZATION IN FILLER'S PRESENCE

Catalytic polymerization on solid surfaces is becoming a more attractive method for the production of polymer filler composites.[109-111] The process involves three major stages: preparation of filler, surface activation of filler, and polymerization on the filler surface. Fillers usually tested for this application include kaolin, tufa, dolomite, perlite, $Al_2O_3$, $SiO_2$, $CaSO_4$, mica and wollastonite. The best results were obtained with $CaSO_4$, wollastonite, $Al_2O_3$, dolomite, and kaolin. Before the filler can be used, it must be dried because the deposition of catalyst requires moisture-free conditions. The catalyst is a combination of Al/Ti/Mg in different proportions each of which gives a different efficiency. The deposition of catalyst is a simple process involving the treatment of filler particles with solvent solutions of organometallic compounds. The polymerization of ethylene in a slurry is a highly efficient process. This is sometimes called polymerization filling. There is no elaborate process mechanism nor is the morphology or chemistry well understood, but the results of this synthesis surpass all conventional mixing techniques. The products from this process have better elongation than unfilled polymers and show improvements in almost all mechanical properties.

Montmorillonite is an effective complex with the initiator in the polymerization of methyl methacrylate.[112] It not only accelerates the polymerization but also improves such mechanical properties as hardness and compression strength. Pentabromobenzyl acrylate was mechano-polymerized in the presence of $Mg(OH)_2$.[113] The polymerization occurred at a reduced temperature and a flame retarded product was produced.

Figure 6.26 gives information on the effect of carbon black loading on the polymerization efficiency of pyrrole. The polymerization rate was reduced as the loading of carbon black was increased. The reduced rate is caused by the oligomer coupling on the surface of the carbon black and by the absorption of the chemical oxidant needed for polymerization.[114]

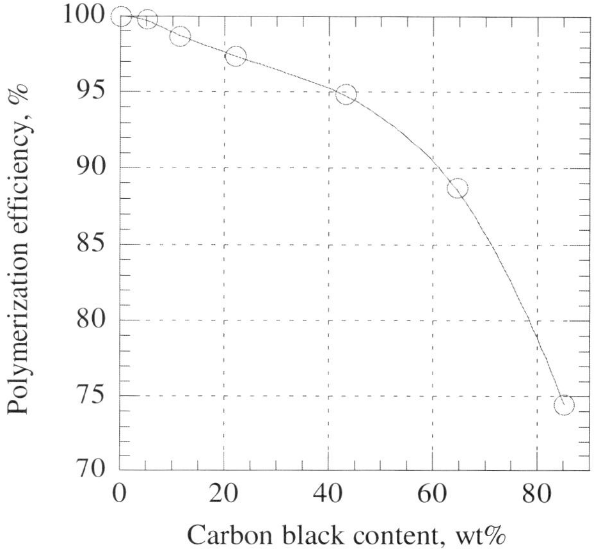


Figure 6.26. Effect of carbon black on polymerization efficiency of pyrrole. [Data from Wampler W A, Rajeshwar K, Pethe R G, Hyer R C, Sharma S C, *J. Mat. Res.*, **10**, No.7, 1995, 1811-22.]

The free radical polymerization of styrene initialized by iniferter is influenced by chemical binding of iniferter on the surface of the silica.[115] This reaction is used for grafting the polymer onto the surface of the silica. A similar approach is used when carbon whisker is incorporated during the graft-polymerization of methyl methacrylate.[116] Depending on how the whisker is prepared, surface conversion can be increased up to twelve times compared to a polymerization with no whisker present. The addition of graphite to the polyesterification reaction doubles the molecular weight of the polymer.[117]

The presence of a filler in a polymerization reaction can often produce an improved material. Now, the challenge is to take advantage of these new findings and develop cost effective commercial processes.

## 6.8 GRAFTING

Grafting has already been discussed in this chapter. These additional remarks highlight the technology. Three general types of grafting have evolved. Grafting to specific, well defined substrates, grafting to natural products and grafting during another primary or secondary process such as during mixing. Most of the processes discussed above fall into the first category.[7,15,31,37,38,43] A well-controlled process must begin with a filler with appropriate functional groups which permit further synthesis. But the conditions of grafting must be very strictly met because the polymerization of organic monomers requires tight control. In the majority of cases discussed in the literature, the technology is developed for laboratory conditions. Grafting is a simple one- or two-step process. The first step is the conversion of functional groups to form initiating sites for polymerization. The second step involves chain extension. Typically, grafting is performed in solvent slurries which is an expensive process. The results of these syntheses show that products meet requirements and that the process can be controlled to obtain the designed thickness and distribution of coverage. Now, such synthesis is limited in practice to very expensive products because at this time processes are technologically too complex, too expensive, and frequently environmentally undesirable. This research, has helped significantly in improving our understanding of the range of properties offered by grafted fillers but has made little contribution to the mainstream processing of plastics for which the associated costs are too high.[41,84,118,119]

The second type is a group of processes which involve grafting onto natural materials or waste products.[120,121] Here, the goal is to utilize these materials in a simple and economical fashion. The conditions of grafting are not well controlled because of the complex nature of the substrates. This affects results. Products of these grafting processes are useful to merely fill polymers without detracting from their properties. This work will prove useful, particularly in the recovery of materials from recycling streams. At the same time, a substantial effort will be required to develop these processes with an economical application in mind.

The third group is the most promising because grafting during material processing adds only the cost of the raw materials. Two options are available for the development of these processes: a third additive is used to react with the filler and interact with the polymer (e.g., reactive compatibilizer)[32,52-8] or the filler surface is modified by a simple process (e.g., silanization) to allow reactive grafting during the manufacturing process.[74] Both routes are already in use and new applications and research will contribute to the further improvement of materials.

## 6.9 CROSSLINK DENSITY

In some cases, the crosslink density of a polymer can be affected by the filler. These include:

- The filler particle contains several functional groups which react with different polymer chains
- The filler surface is modified to contain a group which can react with polymer chains
- The modification of the filler surface reacts with a similar group on another filler particle

All of these mechanisms which affect crosslink density were confirmed by experimental studies. The classic case of a reactive particle filler is silica filled polysiloxane (Figure 6.25).[103] Silica particles have numerous OH groups which react with the crosslinking component of polysiloxane. Modification of silica by silanes reduces reinforcement.

Modification of the silica surface with mercaptosilane makes it reactive with rubber, resulting in an improvement in mechanical properties.[122] Modified, precipitated cellulose can reinforce butadiene-acrylonitrile copolymer by forming covalent bonds.[123] Maleic derivatives of EPM react with $CaCO_3$ to increase crosslink density.[32] There are other examples where a variety of functions can be utilized to modify the crosslink density.

One of the reactions which occurs on the surface of filler particles is that involving silanes. Vinyl silanes and mercapto silanes being typical examples. Kaolin modified with an isocyanate can react with polyols.[8] Magnetic resonance spectroscopy was used to identify various crosslinks involving the filler[124] and this shows that crosslinked rubber chains were attached to the surface of the carbon black.

Zinc oxide is a reactive filler commonly used in rubber vulcanization. The crosslink density of rubber can be doubled by reaction of ZnO with HCl.[9] Only a few specific fillers have the catalytic activity to promote crosslinking but fillers can take part directly in crosslinking processes initiated by an external source such as γ-radiation.[125] Generally, fillers reduce the effect of radiation. But γ-rays are not screened by the filler so the protection given by fillers comes from reduction in chain mobility which lessens the probability of photoconversion.

In summary, fillers have a very limited effect on matrix crosslinking except when they are used as crosslinkers or when the effect is caused by the physical properties of the filler (e.g., $Al(OH)_3$ in UV crosslinked systems).

## 6.10 REACTION KINETICS

The kinetics of reactions which occur when fillers are present depend on the reaction type and on the analytical methods used to follow the kinetics of such reactions. A few examples of kinetic modelling are given below.

Figures 6.9–6.11 provide a data set of a grafting reaction of acrylamide onto the surface of barium sulfate which had been previously reacted with 12-hydroxystearate.[7] The steps of this reaction are given below:

Primary radical formation:

$$BaSO_4{-}HS + Ce(IV) \overset{K}{\rightleftharpoons} [Complex] \xrightarrow{k_d} Ba\dot{S}O_4 + Ce(III) + H^{\oplus}$$

Initiation:

$$BaSO_4 + M \xrightarrow{k_i} BaSO_4{-}M^{\bullet}$$

Propagation:

$$BaSO_4{-}M^{\bullet} + (n\text{-}1)M \xrightarrow{k_p} BaSO_4{-}M^{\bullet}_n$$

Unimolecular termination:

$$BaSO_4{-}M^{\bullet}_n \xrightarrow{k_t} BaSO_4{-}M_n$$

The following equation applies according to the steady-state principle as applied to active intermediates:

$$R_p = K\frac{k_p k_i}{k_t}[AAm][Ce(IV)][BaSO_4 - HS] \quad [6.5]$$

where:

| | |
|---|---|
| $R_p$ | reaction rate |
| AAm | acrylamide concentration |
| Ce(IV) | catalyst concentration |
| $BaSO_4$–HS | reactive site concentration |

This equation is typical of most bimolecular reactions studied by the analysis of substrates and reaction products.

The equation is a simple case of a mechanistic model. Models such as this may give better predictions but may not always apply because of the complexity of the reactions. Phenomenological models are expressed by simple rate equations which ignore the details of the reaction. Phenomenological models are typically used to follow cure rates in polymeric systems which are difficult to follow by chemical analysis. This is because reaction products become insoluble during the course of the reaction and, consequently, are not detected in an analysis of the solution.

A model which can be applied to reactions involving fillers is expressed in the simplest form given as an equation of the nth order:[126]

$$\frac{d\alpha}{dt} = k(1-\alpha)^n \quad [6.6]$$

where:

| | |
|---|---|
| $\alpha$ | cumulative conversion at given time t |
| $k$ | rate constant which obeys Arrhenius dependence (Eq 6.7) |
| $n$ | reaction exponent |

The Arrhenius temperature dependence is:

$$k = k_o \exp(-E/RT) \quad [6.7]$$

where:

| | |
|---|---|
| $E$ | activation energy, kJ/mol |
| $R$ | universal gas constant |
| $T$ | temperature |

These equations can be used to calculate the reaction rate and activation energy of the process. The equations are simple and cannot account for the complexity in each reaction stage nor for all of the physical processes such as phase separation, gelation, or vitrification which determine the outcome of reaction. Several other variations of these equations have been developed to deal better with these complexities.

The order of the reaction can be calculated with more precision using the following equation:

$$\ln\frac{\dot{\alpha}}{(1-\alpha_i)^n} = -\frac{E_\alpha}{RT_i} + \ln(k_o) \quad [6.8]$$

where:

$\dot{\alpha}$ conversion rate
$\alpha$ conversion
$E_\alpha$ activation energy
$k_o$ Arrhenius frequency factor

If fillers are involved, the expression is changed to:

$$\frac{d\alpha}{dt} = (k_1^* + k_2^*\alpha^m)(1-\alpha)^n \quad [6.9]$$

where:
$k_i^*$ functions of filler content
m, n reaction exponents

The filler presence affects constants which can be used in comparison with the unfilled system.

There are other equations derived from these which have a close relationship to them but deal with other aspects of the complexities.

## 6.11 MOLECULAR MOBILITY

Chain mobility should be considered from both a chemical and a physical standpoint. Chemical reactions require reagents to be physically in contact. The morphology of the interphase organization restricts chain motions which might be considered either a chemical or a physical phenomenon.

There are three types of chains or chain segments which may be involved in filler-polymer interactions:

- A chain segment which is restricted in its motion because it has been adsorbed on the surface of the filler and has possibly reacted with it
- Adjacent segments because they are restricted by a proximity of the bound segment
- Chains which belong to the polymer bulk since they behave as they were in the unfilled polymer.

The first two types cannot be dissolved in a good solvent whereas the third type can. The first type can be designated to be in a tight region and is called a rigid segment. The adjacent segments form elastic loops and they belong to the loose region and are called elastic segments. From the point of view of chemical reactivity, the rigid segment is the one which has either undergone one of the chemical reactions discussed above or it has been absorbed on the surface by very strong physical interactions. This chain segment can participate in other chemical reactions if the energy level is sufficient to rupture its connection to the filler surface or if the action of binding to the surface of the filler changes its configuration. Elastic segments have the ability to participate in reactions but the probability of such reactions are less than that of the unrestricted polymer chain.

Several analytic techniques can contribute information on molecular mobility with NMR being the most useful. Two spin-spin relaxation times are observed: short (tight region) and long (loose region). The values and ratios for materials of different compositions can give an insight into the behavior of these two segmental types. Modification of the surface of silica with hexadecanol decreases the ratio of loosely bound to tightly bound signals when it is dispersed in natural rubber. This shows first that rigid segments are formed in the reaction of rubber with silanol groups. Second, it shows that the compatibility between rubber and modified silica was greater than between rubber and unmodified silica. The signal of loose segments is three times stronger.[41] Similar studies on many different systems confirm the validity of observations using a spin-spin relaxation time, $T_2$.[124]

A study of SBR rubber showed a difference in the behavior over the temperature range of 20-70°C between the unfilled rubber and rubber filled with carbon black.[127] $T_2$ changes by a few tens percent for filled rubber in this temperature range and it almost doubles in the unfilled rubber. The increased temperature contributes to the increased molecular mobility but this effect is retarded by the bound segments in the filled rubber.[127]

Monte Carlo simulations show the differences between chains on elongation.[128] Restricted chains (chains attached to the filler's surface) have a modulus similar to free chains when the elongation is small whereas a substantially higher modulus is observed when restricted chains are subjected to large deformations. These simulations produce results which are bourne out by experimental work. NMR studies suggest that fixation (attachment) of one monomeric unit to the filler's surface hinders random motions (a characteristic of free chains) of approximately 4 monomeric units on both sides of the contact point.[129] On the other hand, the diffusion of free chains is progressively reduced as the number of the adsorbed segments is increased. If the filler has a low potential for bonding (e.g., $CaCO_3$) then the system is not affected by the concentration of interacting components because a sufficient number of functional groups does not exist to make any observable difference.

Dynamic mechanical analysis of filled systems confirms analytical observations.[130] The thickness of the restricted mobility region in carbon filled rubber is proportional to the activity of the carbon black.

The mobility of low molecular weight additives in the presence of fillers is important for the same reasons. Recent studies show that UV stabilizers are immobilized on the surfaces of filler particles. Nitroxyl radicals were used as spin probes in silica filled polymers.[131] Experimental work confirms that absorption occurs on the OH groups of silica but it was shown that a certain minimum concentration of filler is required to trigger this absorption effect.

The forces which come into play in a filled system are not restricted to affecting the mobility of chains. They also influence the filler particle distribution. The migration of filler particles has been modeled for an injection molding process.[132] A spectacular effect was observed when jute fiber was used as a filler.[133] As the moisture level of jute was increased, the fibers rotated about their axes. This changes the distribution and orientation of the jute fibers and has an effect on the properties of the composite.

In summary, the molecular mobility of high molecular weight substances in the presence of fillers is very different from the mobility of low molecular weight materials (most especially, in liquid systems). The effect of these phenomena on chemical processes and reactions is limited. The most pronounced effect on the chemistry is the lack of reaction homogeneity. Molecules which have interacted with surfaces preferentially undergo localized chemical conversions with their nearest neighbors. The reaction mechanisms of these adsorbed segments are different because their conformation and configuration are affected by this act of interaction which causes a shift in the preferred reaction in the unfilled systems. The largest influences of molecular mobility are on the organization of the interface, the effect on mechanical and rheological properties, and on morphology. These subjects are discussed in the following chapters.

## REFERENCES


1 Joseph R, Zhang S, W T, *Macromolecules*, **29**, No.4, 1996, 1305-12.
2 Bandyopadhyay S, De P P, Tripathy D K, De S K, *Kaut. u. Gummi Kunst.*, **49**, No.2, 1996, 115-9.
3 Bandyopadhyay S, De P P, Tripathy D K, De S K, *J. Appl. Polym. Sci.*, **58**, No.4, 1995, 719-27.
4 Yoshinaga K, Hidaka Y, *Polym. J. (Jap.)*, **26**, No.9, 1994, 1070-9.
5 Yoshinaga K, Iwasaki M, Teramoto M, Karakawa H, *Polym. & Polym. Composites*, **4**, No.3, 1996, 163-72.
6 Dupraz A M P, de Wijn J R, v. d. Meer S A T, de Groot K, *J. Biomed. Mat. Res.*, **30**, No.2, 1996, 231-8.
7 Tsubokawa N, Seno K, *J. Macromol. Sci. A*, **31**, No.9, 1994, 1135-45.
8 Molphy M, Laslett R L, Gunatillake P A, Rizzardo E, Mainwaring D E, *Polym. Int.*, **34**, No.4, 1994, 425-31.
9 Zaborski M, Slusarski L, Donnet J B, Papirer E, *Kaut. u. Gummi Kunst.*, **47**, No.10, 1994, 730-8.
10 Tang L-G, Kardos J L, *Polym. Composites*, **18**, No.1, 1997, 100-13.
11 Tsubokawa N, Hosoya M, Kurumada J, *Reactive & Functional Polym.*, **27**, No.1, 1995, 75-81.
12 Kurian T, Khatgir D, De P P, Tripathy D K, De S K, Peiffer D G, *Polymer*, **37**, No.25, 1996, 5597-605.
13 Jesionowski T, Krysztafkiewicz A, *Pigment Resin Technol.*, **25**, No.3, 1996, 4-14.
14 Gordienko V P, Dmitriev Y A, *Polym. Sci., Ser. B*, **37**, Nos.5-6, 1995, 249-50.
15 Yoshikawa S, Tsubokawa N, *Polym. J. (Jap.)*, **28**, No.4, 1996, 317-22.
16 Tino J, Mach P, Hlouskova Z, Chodak I, *J. Macromol. Sci. A*, **A31**, No.10, 1994, 1481-7.
17 Byers J T, Meeting of the Rubber Division, ACS, Cleveland, October 17-20, 1995, paper B.
18 Bandyopadhyay S, De P P, Tripathy D K, De S K, *Polymer*, **37**, No.2, 1996, 353-7.
19 Wolff S, *Rubb.Chem.Technol.*, **69**, No.3, 1996, 325-46.
20 Byung Suk Jin, Kwang Hee Lee, Chul Rim Choe, *Polym. Int.*, **34**, No.2, 1994, 181-5.
21 Bogoeva-Gaceva G, Burevski D, Dekanski A, Janevski A, *J. Mat. Sci.*, **30**, No.13, 1995, 3543-6.
22 Datta S, De S K, Kontos E G, Wefer J M, Wagner P, Vidal A, *Polymer*, **37**, No.15, 1996, 3431-5.
23 Wang W, Vidal A, Donnet J-B, Wang M-J, *Kaut. u. Gummi Kunst.*, **46**, No.12, Dec.1993, 933-40.
24 Hayashi S, Handa S, Tsubokawa N, *J. Polym. Sci., Polym. Chem.*, **34**, No.8, 1996, 1589-95.
25 Byers J T, Meeting of the Rubber Division, ACS, Cleveland, October 17-20, 1995, paper B.
26 Evans L R, Meeting of the Rubber Division, ACS, Montreal, May 5-8, 1996, paper D.
27 Molphy M, Mainwaring D E, Rizzardo E, Gunatillake P A, Laslett R L, *Polym. Int.*, **37**, No.1, 1995, 53-61.
28 Shiroki Y, *Int. Polym. Sci. Technol.*, **20**, No.6, 1993, T/12-21.
29 Kovacevic V, Lucic S, Hace D, Cerovecki Z, *J. Adhesion Sci. Technol.*, **10**, No.12, 1996, 1273-85.
30 Akovali G, Dilsiz N, *Polym. Engng. Sci.*, **36**, No.8, 1996, 1081-6.
31 Hayashi S, Handa S, Oshibe Y, Yamamoto T, Tsubokawa N, *Polym. J. (Jap.)*, **27**, No.6, 1995, 623-30.
32 Zaborski M, Slusarski L, *Composite Interfaces*, **3**, No.1, 1995, 9-22.
33 Balard H, Papirer E, *Prog. Org. Coatings*, **22**, No.1-4, 1993, 1-17.
34 Liao K, Chen X, Zheng C, *J. Appl. Polym. Sci.*, **57**, No.10, 1995, 1245-50.
35 Domka L, *Coll. Polym. Sci.*, **272**, No.10, 1994, 1190-202.

36 Krysztafkiewicz A, Rager B, Maik M, *Coll. Polym. Sci.*, **272**, No.12, 1994, 1547-59.
37 Wang Y, Lu J, Wang G, *J. Appl. Polym. Sci.*, **64**, 1997, 1275-81.
38 Tsubokawa N, Yoshihara T, *Coll. & Surfaces*, **81**, 1993, 195-201.
39 Yoshinaga K, Sueishi K, Karakawa H, *Polym. Adv. Technol.*, **7**, No.1, 1996, 53-6.
40 Roychoudhury A, De P P, Roychoudhury N, Vidal A, *Rubb. Chem. Technol.*, **68**, No.5, 1995, 815-23.
41 Ou Y C, Yu Z Z, Vidal A, Donnet J B, *J. Appl. Polym. Sci.*, **59**, No.8, 1996, 1321-8.
42 Ohta M, Nakamura Y, Hamada H, Maekawa Z, *Polym. & Polym. Composites*, **2**, No.4, 1994, 215-21.
43 Yoshinaga K, Hidaka Y, *Polym. J. (Jap.)*, **26**, No.9, 1994, 1070-9.
44 Zumbrum M A, *J. Adhesion*, **46**, Nos.1-4, 1994, 181-96.
45 Krysztafkiewicz A, Rager B, Maik M, Szymanowski J, *Coll. Polym. Sci.*, **272**, No.12, 1994, 1526-35.
46 Rothon R N; Hornsby P R, *Polym. Degradat. Stabil.*, **54**, Nos 2-3, 1996, 383-5.
47 Turner J D, Property Enhancement with Modifiers and Additives. Retec proceedings, New Brunswick. N.J., 18th-19th Oct.1994, 65-87.
48 Rebouillat S, Escoubes M, Gauthier R, *J. Adhesion Sci. Technol.*, **10**, No.7, 1996, 635-50.
49 Bos M, Van Dam G W, Jongsma T, Bruin P, Pennings A J, *Composite Interfaces*, **3**, No.2, 1995, 169-76.
50 Pritykin L M, Razumova O V, Sokolova Y A, Antonov S M, Bolshakov V I, *Int. Polym. Sci. Technol.*, **23**, No.3, 1996, T/80-1.
51 Liauw C M, Lees G C, Hurst S J, Rothon R N, Dobson D C, *Plast. Rubb. Comp. Process. Appln.*, **24**, No.4, 1995, 211-9.
52 Borden K A, Wei R C, Manganaro C R, *Plast. Compounding*, **16**, No.5, 1993, 51-5.
53 Nichols K, Solc J, Shieu F, Antec '93. Conference Proceedings, New Orleans, La., 9th-13th May 1993, Vol. II, 1938-42.
54 Zolotnitsky M, Steinmetz J R, Antec '94. Conference Proceedings, San Francisco, Ca., 1st--5th May 1994, Vol. III, 2756-60.
55 Borden K A, Weil R C, Manganaro C R, Antec '94. Conference Proceedings, San Francisco, Ca., 1st-5th May 1994, Vol. III, 2761-5.
56 Chiang W Y, Yang W D, Pukanszky B, *Polym. Engng. Sci.*, **34**, No.6, 1994, 485-92.
57 Liauw C M, Hurst S J, Lees G C, Rothon R N, Dobson D C, *Prog. Rubb. Plast. Technol.*, **11**, No.2, 1995, 137-53.
58 Garbow J R, Asrar J, Hardiman C J, *Chem. of Mat.*, **5**, No.6, 1993, 869-75.
59 Spriet C in **Enhancing Polymers Using Additives and Modifiers II**, *Rapra*, Shawbury, 1996.
60 Pape P G, Property Enhancement with Modifiers and Additives. Retec proceedings, New Brunswick, N.J., 18th-19th Oct.1994, 201-9.
61 Gassan J, Bledzki A K, Antec '96. Volume II. Conference proceedings, Indianapolis, 5th-10th May 1996, p.2552-7.
62 Schmidt H K, *Macromol. Symp.*, **101**, 1996, 333-42.
63 Tertykh V A, *Macromol. Symp.*, **108**, 1996, 55-61.
64 Suzuki N, Ishida H, *Macromol. Symp.*, **108**, 1996, 19-53.
65 Enhancing Polymers Using Additives and Modifiers, *Rapra*, Shawbury, 1993.
66 Monte S J, Sugerman G, Ken-React Reference Manual - Titanate and Zirconate Coupling Agents, Kenrich Pertochemical Inc., Bayonne, 1985.
67 Silicon Compounds, Register and Review, Pertrarch Systems, Inc., Bristol, 1984.
68 Ishida H, Miller D J, *Macromolecules*, **17**, 1984, 1659.
69 Milewska-Duda J, *Polimery*, **29**, 1984 19.
70 Miller D J, Ishida H, *Surface Sci.*, **148**, 1984, 601.
71 Favis B D, Blanchard L P, Leonard J, Prud'homme, *Polym. Compos.*, **5**, 1984, 11.
72 Favis B D, Blanchard L P, Leonard J, Prud'homme, *J. Appl. Polym. Sci.*, **28**, 1983, 1235.
73 Bergeret A, Alberola N, *Polymer*, **37**, No.13, 1996, 2759-65.
74 Roesch J, Barghoorn P, Muelhaupt R, *Makromol. Chem. Rapid Commun.*, **15**, No.9, 1994, 691-6.
75 Vasnev V A, Tarasov A I, Istratov V N, Ignatov V N, Krasnov A P, Kuznetsov A I, Surkova I N, *Reactive & Functional Polym.*, **26**, Nos.1-3, 1995, 177-83.
76 Zaborski M, Slusarski L, Vidal A, *Int. Polym. Sci. Technol.*, **20**, No.11, 1993, T/99-104.
77 Liu Z, Gilbert M, *J. Appl. Polym. Sci.*, **59**, No.7, 1996, 1087-98.
78 Bogoeva--Gaceva G, Burevski D, Dekanski A, Janevski A, *J. Mat. Sci.*, **30**, No.13, 1995, 3543-6.
79 Gatenhom P, Hedenberg P, Karlsson J, Felix J, Antec '96. Volume II. Conference proceedings, Indianapolis, 5th-10th May 1996, 2302-4.
80 Skelhorn D A, Antec '93. Conference Proceedings, New Orleans, La., 9th-13th May 1993, Vol. II,

1965-70.
81 Torro-Palau A, Fernandez-Garcia J C, Orgiles-Barcelo A C, Pastor-Blas M M, Martin-Martinez J M, *Int. J. Adhesion Adhesives*, **17**, 1997, 111-9.
82 Bandyopadhyay S, De P P, Tripathy D K, De S K, *Polymer*, **37**, No.2, 1996, 353-7.
83 Bandyopadhyay S, De P P, Tripathy D K, De S K, *J. Appl. Polym. Sci.*, **61**, No.10, 1996, 1813-20.
84 Akelah A, Moet A, *J. Mat. Sci.*, **31**, No.13, 1996, 3589-96.
85 Hill R, Okoroafor E U, *Composites*, **25**, No.10, 1994, 913-6.
86 Qiu Q, Kumosa M, *Composites Sci. Technol*, **57**, 1997, 497-507.
87 Fatoev I I, Sitamov S, Manin V N, *Int. Polym. Sci. Technol.*, **22**, No.10, 1995, T/88-90.
88 Hegedus C R, Kamel I L, *J. Coatings Technol.*, **65**, No.822, July 1993, 37-43.
89 Lambert C, Chauvet H, Larroque M, *Polymer*, **37**, No.15, 1996, 3441-5.
90 Unnikrishnan G, Thomas S, Varghese S, *Polymer*, **37**, No.13, 1996, 2687-93.
91 Lawandy S N, Botros S H, Darwish N A, Mounir A, *Polym. Plast. Technol. Engng.*, **34**, No.6, 1995, 861-74.
92 Kim S, Jang J, *Polym. Test.*, **15**, No.6, 1996, 559-71.
93 Dando N R, Kolek P L, Martin E S, Clever T R, *J. Coatings Technol.*, **68**, No.859, 1996, 67-72.
94 Parker A A, Martin E S, Clever T R, *J. Coatings Technol.*, **66**, No.829, 1994, 39-46.
95 Murata N, Nishi S, Hosono S, *J. Adhesion*, **59**, Nos.1-4, 1996, 39-50.
96 Hong S G, Lin J J,Yuan Ze, *J. Appl. Polym. Sci.*, **59**, No.10, 1996, 1597-605.
97 Caillaud J L, Deguillaume S, Vincent M, Giannotta J C, Widmaier J M, *Polym. Int.*, **40**, No.1, 1996, 1-7.
98 Sims G L A, Sombatsompop N, *Cell. Polym.*, **15**, No.2, 1996, 90-104.
99 Sergeeva L M, Skiba S I, Karabanova L V, *Polym. Int.*, **39**, No.4, April 1996, 317-25.
100 Mori M, Koenig J L, *Rubb. Chem. Technol.*, **68**, No.4, 1995, 551-62.
101 Gibala D, Laohapisitpanich K, Thomas D, Hamed G R, *Rubb. Chem. Technol.*, **69**, No.1, 1996, 115-9.
102 Anantharaman M R, Kurian P, Banerjee B, Mohamed E M, George M, *Kaut. u. Gummi Kunst.*, **49**, No.6, 1996, 424-6.
103 Cochrane H, Lin C S, *Rubb. Chem. Technol.*, **66**, No.1, 1993, 48-60.
104 Jang J, Yi J, *Polym. Engng. Sci.*, **35**, No.20, 1995, 1583-91.
105 Kenny J M, Opalicki M, Molina G, Antec '95. Vol. II. Conference Proceedings, Boston, Ma., 7th-11th May 1995, 2782-9.
106 Kenny J M, Opalicki M, *Composites Part A: Applied Science and Manufacturing*, **27A**, No.3, 1996, 229-40.
107 Peng W, Riedl B, *Polymer*, **35**, No.6, 1994, 1280-6.
108 Murayama H, Min K, Antec '97. Conference proceedings, Toronto, April 1997, 759-65.
109 Novokshonova L A, Meshkova I N, *Polym. Sci.*, **36**, No.4, 1994, 517-25.
110 Hindryckx F, Dubois P, Jerome R, Teyssie P, Marti M G, *J. Appl. Polym. Sci.*, **64**, 1997, 423-38.
111 Hindryckx F, Dubois P, Jerome R, Teyssie P, Marti M G, *J. Appl. Polym. Sci.*, **64**, 1997, 439-54.
112 Al-Esaimi M M, *J. Appl. Polym. Sci.*, **64**, 1997, 367-72.
113 Gutman E M, Bobovitch A L, *Eur. Polym. J.*, **32**, No.8, 1996, 979-83.
114 Wampler W A, Rajeshwar K, Pethe R G, Hyer R C, Sharma S C, *J. Mat. Res.*, **10**, No.7, 1995, 1811-22.
115 Zaremski M Yu, Chernikova E V, Izmailov L G, Garina E S, Olenin A V, *Macromol. Reports*, **A33**, Suppls.3/4, 1996, 237-42.
116 Ueno H, Tsubokawa N, *Composite Interfaces*, **3**, No.3, 1995, 209-20.
117 Kuznezov A, Vasnev V, Gribova I, Krasnov A, Gureeva G, Ignatov V, *Int. J. Polym. Mat.*, **32**, Nos.1-4, 1996,
118 Friedrich C, Scheuchenpflug W, Neuhaeusler S, Roesch J, *J. Appl. Polym. Sci.*, **57**, No.4, 1995, 499-508.
119 Luzinov I, Voronov A, Minko S, Kraus R, Wilke W, Zhuk A, *J. Appl. Polym. Sci.*, **61**, No.7, 1996, 1101-9.
120 Casenave S, Ait-Kadi A, Brahimi B, Riedl B, Antec '95. Vol. II. Conference Proceedings, Boston, Ma., 7th-11th May 1995, 1438-42.
121 Sain M M, Kokta B V, Antec '93. Conference Proceedings, New Orleans, La., 9th-13th May 1993, Vol. I, 320-4.
122 Bomo F, Meeting of the Rubber Division, ACS, Montreal, May 5-8, 1996, paper E.
123 de Sena Affonso J E, Nunes R C R, *Polym. Bull.*, **34**, No.5/6, 1995, 669-75.
124 Legrand A P, *Macromol. Symp.*, **108**, 1996, 81-96.
125 Bataille P, Mahlous M, Schreiber H P, *Polym. Engng. Sci.*, **34**, No.12, 1994, 981-5.

126 Yousefi A, Lafleur P G, Gauvin R, *Polym. Composites*, **18**, No.2, 1997, 157-68.
127 de Candia F, Carotenuto M, Gargani L, Guadagno L, Lauretti E, Renzulli A, *Kaut. u. Gummi Kunst.*, **49**, No.2, 1996, 99-101.
128 Mark J E, *Macromol. Symp.*, **101**, 1996, 423-33.
129 Cohen Addad J P, Euradh '94. Conference Proceedings, Mulhouse, 12th-15th Sept.1994, 25-30.
130 Mandal U K, Tripathy D K, De S K, *Plast. Rubb. Comp. Process. Appln.*, **24**, No.1, 1995, 19-25.
131 Tino J, Mach P, Hlouskova Z, Chodak I, *J. Macromol. Sci. A*, **A31**, No.10, 1994, 1481-7.
132 Bormashenko E Y, Zagoskin A M, *Int. Polym. Sci. Technol.*, **23**, No.1, 1996, T/80-1.
133 Mannan K M, Robbany Z, *Polymer*, **37**, No.20, 1996, 4639-41.

# 7

# Organization of Interface and Matrix Containing Fillers

This chapter analyzes how a filler is distributed in materials and what interaction occurs between the filler and the matrix. These two factors make a major contribution to reinforcement of the filled materials. We will outline the principles governing filler distribution and interaction and explain the relevance of reported studies. Chapters 5, 6, and 10 contain discussion of other related phenomena such as particle size of fillers, chemical reactivity in filled systems, and morphology, respectively. Chapter 8 shows impact of organization and filler presence on mechanical properties of filled systems. The information included in the above chapters helps us to understand how to use fillers to improve the performance of a material.

## 7.1 PARTICLE DISTRIBUTION IN MATRIX

Idealized distribution of filler particles in a matrix can be predicted by various models as discussed in Chapter 5. Here, an attempt is made to examine empirical data on filler distribution and to determine factors in actual filler which cause that distribution differs from an ideal model used to predict packing density of the filler.

Filler particles generated *in situ* can be perceived as ideally distributed within the matrix. Experimental studies show that the situation is more complex.[1] Poly(dimethyl siloxane) network was swollen to equilibrium in tetraethylorthosilicate which was then hydrolyzed to produce an *in situ* filler. Such an experiment gives the almost ideal conditions of uniform distribution because both matrix and the filler precursor are chemically similar. There are numerous factors which affect how uniformly a filler is distributed. These include:

- The choice of hydrolysis catalyst
- The hydrolysis time
- The sample thickness

The most uniform distribution was obtained when the hydrolysis time was long, sample was thin, and the catalyst basic. If conditions were reversed (short hydrolysis time, bulky sample, and acidic catalyst), filler was preferentially formed on the peripheries of the sample. What is the force which drives the precursor out of its initial equilibrium? The most likely scenario is that a fast process leads to the formation of a silica skin which inhibits the transport of the water required to hydrolyze the inside layers of the precursor. Under such circumstances, tetraethylorthosilicate, even though it is a larger molecule than water, diffuses to the surface to equilibrate its concentration to replace the already converted portions.

In another similar example nanocomposite was formed in a polyurethane matrix.[2] Solvent soluble polyurethane had pyridine groups attached which formed complexes with metal salts. Films were then formed and subjected to a reducing agent in order to produce particulate metal filler. In this case the distribution of the filler which was formed was not uniform because the filler had tendency to aggregate (even though it was chemically attached to the matrix prior to the reduction). The following were factors controlling size and shape of these metal particles:

- The concentration of metal iron
- The morphology of polyurethane
- The polarity of the matrix

Polymeric segments could prevent excessive aggregation. These experiments show that there is no reasons to expect a polymer-filler system to have a homogeneous distribution. Even under such ideal conditions as described in the above two cases (the chemical affinity of substrates and the anchoring of the filler precursor) it did not occur. The general conclusion from these experiments is that a homogeneous distribution of filler in the matrix is rather the exception than the rule.

Studies on randomness of filler distribution in polymethylacrylate nanocomposite are interesting.[3] In this experiment, silica particles were formed both before and after matrix polymerization. The results indicated that the concentration of silica was a controlling factor in the stress-strain relationship rather than the uniformity of particle distribution. Also, there was no anisotropy of mechanical properties regardless of the sequence of filler formation. This outcome cannot be expected to be duplicated in all other systems. For example, when nickel coated fibers were used in an EMI shielding application.[4] When compounded with polycarbonate resin, fibers had a much worse performance than when a dry blend was prepared first and then incorporated into the polymer (Figure 7.1). In this case, pre-blending protected the fiber from breakage.

Calcium carbonate treated with stearic acid gave improved performance to poly(vinyl acetate) composites but only if the filler particles were sufficiently small.[5] Smaller particles tend to agglomerate if they are not coated. Coating prevents agglomeration and improves their interaction with the matrix. Large particles do not interact with the matrix but form defects in the composite. All three examples show that

- Uniform distribution of filler particles in a matrix does not guarantee improved performance
- At least two factors, filler surface availability and potential for interaction, contribute to improved filler distribution

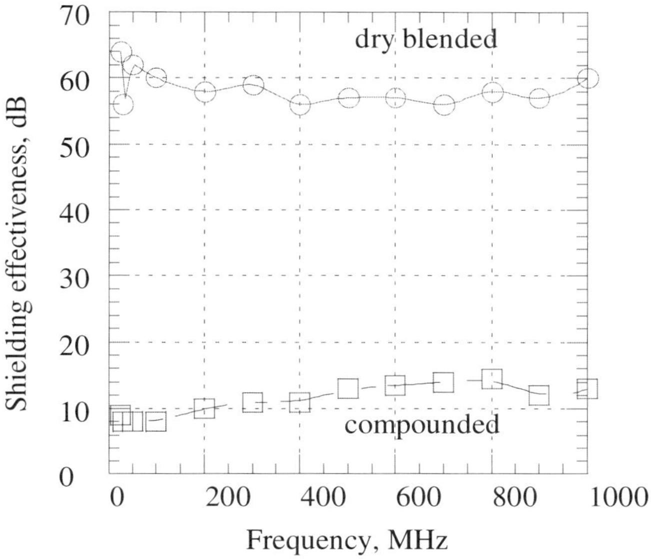


Figure 7.1. Shielding effectiveness. [Adapted, by permission, from Rosenov M W K, Bell J A E, Antec '97. Conference proceedings, Toronto, April 1997, 1492-8.]

Usually, a uniform distribution of the filler will give the most available surface for interaction. However, the nature of this surface has a strong influence on the properties of the filled material. In some applications, where perhaps thermal and electric conductivity improvements are sought, a uniform distribution will not necessary improve properties.

In thermoplastic melts, filler particles migrate due to a temperature gradient in the article during cooling. This produces an interphase tension at the particle-melt boundary. These forces cause particle movements from the cold regions into the melt.[6] Pressure sensitive adhesive containing fumed silica particles has a much lower tack on its surface than it has at the bottom of a cast film.[7] XPS analysis shows that the surface contains about 8 times more silicon than the bottom inferring that silica particles preferentially migrate to the surface.

The above systems are fairly simple, homogeneous systems since they contain only one polymer in the matrix. Blending polymers makes the behavior more complex. In polypropylene/polycarbonate blends, carbon black is preferentially located in the polycarbonate phase.[8] A blend which is better mixed is less conductive than a blend in which carbon black predominantly resides in the polycarbonate phase where it can form a conductive network. These are properties which control morphology (and related electric conductivity):

- Polarity
- Crystallinity
- Viscosity

These properties determine how carbon black will be distributed within the blend. These properties are not those of the filler but are the essential properties of the matrix. The matrix thus has strong influence on particle distribution. SEM studies showed that high vinyl polybutadiene and styrene-butadiene copolymers had morphologically identical carbon black distribution.[9] However, their mechanical properties were very different. NMR analysis indicated that the difference in mechanical behavior is related to the interaction and more precisely to the molecular motions in rubbery matrix.

The initial form of the filler is another complicating factor. Good and consistent dispersion of filler contributes to product properties. There two goals when filler is to be mixed:

- To reduce the size of filler particles (the intensity factor)
- To obtain a uniform distance between particles (the extensity factor)

The first determines property development, the second uniformity of these properties. Under the same conditions of mixing, the initial form of the filler plays dominant role. Consider carbon black. Pellets lose their initial shape during mixing process but are more difficult to disperse than non-pelletized blacks. Mixing can reduce the size of agglomerates but has much less influence on aggregates, and primary particles are not affected by mixing process. Filler form can affect product performance just as much as the intensity of processing (mixing).[10]

Filler particle distribution can be further complicated by the processing method since mixing is seldom last step of the process. Some classical examples are connected with the molding processes.[11-13] In the injection molding process, particles concentration around the gate axis increases as the mix passes the gate. Stream of particles is then diverted from the front surfaces towards the sides of the mold.[11] These phenomena cause the surfaces of injection molded parts to contain a lower concentration of filler particles than do their centers. The advancing surface contains increasingly more filler particles when the particle size is increased. This phenomenon produces two gradients of particle concentration. One in a plane perpendicular to the material flow (core-skin structure) with as a skin depleted of particles.[12] The other is along the flow direction with a higher concentration of particles close to the advancing front and a lower near the gate. Other effects are related to particle orientation in the matrix discussed below.[13]

We have outlined factors which affect particle distribution in a matrix. This distribution depends partly on filler properties but predominantly on the combination of properties of the pair filler-matrix. Filler distribution in a matrix depends on intended application. Some, such as applications which use fillers for reinforcement, require a homogeneous distribution of particles. In others, such as mentioned above electrical conductive materials, adhesives), a uniform distribution of filler particles may decrease their effectiveness.

## 7.2 ORIENTATION OF FILLER PARTICLE IN A MATRIX

Three aspects of orientation are considered here:

- How can orientation of filler particle be achieved?
- What kind of results can be expected?
- How does the orientation of filler particles affect material properties?

The simplest method of particle orientation involves material compression.[14] In an experiment, ferrite powders were dispersed in linear polyacrylamide, the gel was crosslinked, and the swollen gel compressed. This process resulted in particle orientation. Industrial processes, such as extrusion, injection, compression, and blow molding, fiber spinning, and thermoforming induce orientation due to flow.[15,16] Typical parameters which control fiber orientation in these processes include:

- Particle shape
- Filler concentration
- Viscosity of the matrix
- Rate of flow
- Shape and length of the die
- Length of the flow path in the cavity
- Thickness of the wall

Many other parameters may be involved depending on the method of processing. Such orientation not only occurs when processing from a solution or a melt but may also occur by inducing strain in the material.[15] Glass fiber reinforced polyamide-6 was subjected to such a strain. Heated specimens were extended under controlled strain and cooled under extension. Hencky strain was calculated from the following equation:

$$\varepsilon = \int d\varepsilon = \int_{L_o}^{L_f} \frac{dL}{L} = \ln\left(\frac{L_f}{L_o}\right) \qquad [7.1]$$

where:

$\varepsilon$ Hencky strain
$L_o$ initial length
$L_f$ length after extension

Fiber orientation was determined by microradiography. The images of microradiographs were digitized and their orientational distribution determined by image processing software. The fiber orientation function was calculated from the following equation:

$$J = 2\int_{-\pi/2}^{\pi/2} \cos^2\theta\, q(\theta) d(\theta) - 1 \qquad [7.2]$$

where:

$\theta$ the fiber orientation angle
$q(\theta)$ the distribution of fiber orientation angles

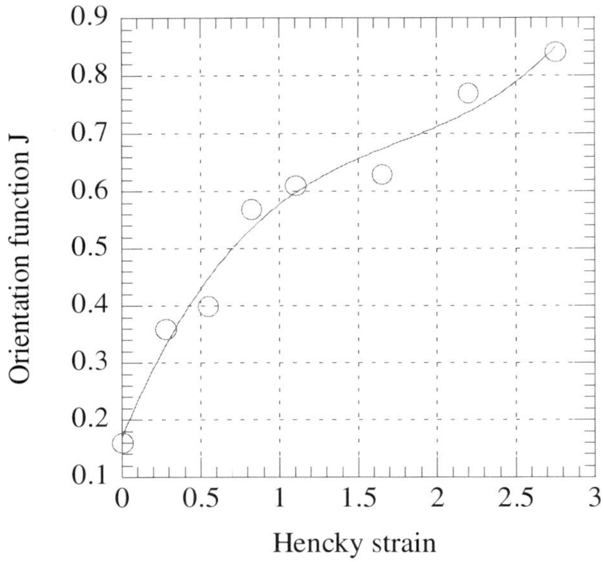


Figure 7.2. Orientation distribution function, J, vs. Hencky strain. [Adapted, by permission, from Wagner A H, Kalyon D M, Yazici R, Fiske T J, Antec '97. Conference proceedings, Toronto, April 1997, 996-1000.]

The fiber orientation function, J, equals 0 for random distribution and 1 for unidirectional distribution. Figure 7.2 shows the results obtained. This simple experiment shows that extension by ~1.5% causes a very high orientation of fibers (J = 0.84). The experiments show that a fairly large rate of flow is needed in the industrial processes to induce fiber orientation. For example, when the injection rate was increased from $6\times10^{-8}$ to $1.7\times10^{-7}$ $m^3/s$, the J value increased from 0.53 to 0.63 for HDPE filled with 20% glass fibers.[17] For the same conditions, the J value decreased from 0.53 to 0.32 when the gate diameter decreased from 3.2 mm to 1.7 mm. The maximum injection rate is limited by product requirement. For example, high speed injection of carbon fiber filled resin reduces electrical conductivity but improves appearance. It is necessary to find a compromise between appearance and conductivity.[18]

Shear controlled orientation technology was developed to optimize plastic properties by orientation of filler particles.[19] In this patented technology, the single feed is split into a plurality of feeds which can supply pressure to the mold cavity independent of the feed channel. Figure 7.3 shows feed arrangements. The shear is applied by a controlled movement of pistons which imposes microscopic shear. A perfect alignment of fibers can be obtained.

Fiber orientation can be induced by simultaneous shearing and application of electric fields.[20] Such conditions were simulated in a plate rheometer in which the plates were also inducing an electric field. Dielectric particles of filler were oriented in the same direction as that of the electric field. The time to reach an equili-

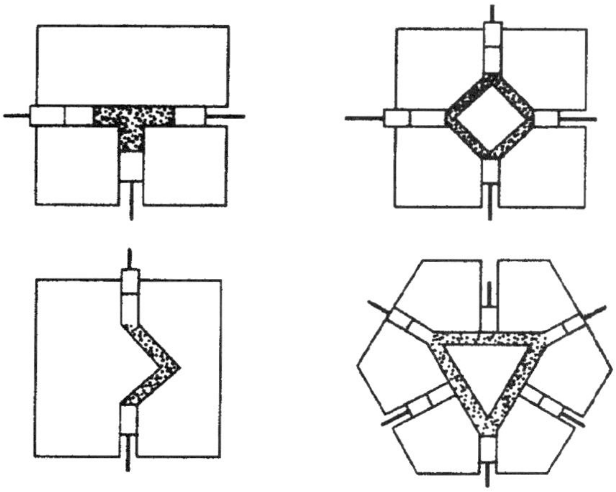

Figure 7.3. Feed arrangements to produce orientation of fibers. [Adapted, by permission, from Allan P S, Bevis M J, *Materials World*, **2**, No.1, 1994, 7-9.]

brated orientation was also measured. The 90% of all fibers aligned within 100 s. Exposure to magnetic field also produces orientation.[14]

These processes attempt to order particles in a predictable manner. How much orientation can be achieved? Figure 7.4 shows the effect of orientation of nickel fibers.[17] The graph shows that fibers are mostly aligned in the flow direction. The orientation was enhanced by an increased rate of flow but only for shorter fi-

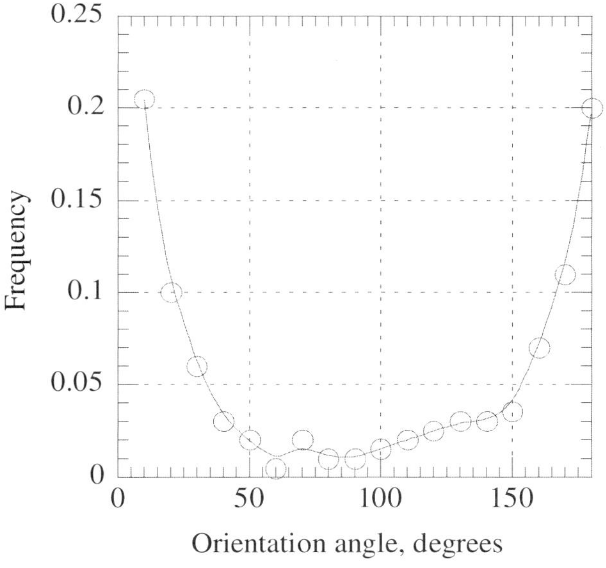


Figure 7.4. Fiber orientation distribution. [Adapted, by permission, from Fiske T, Gokturk H S, Yazici R, Kalyon D M, Antec '97. Conference proceedings, Toronto, April 1997, 1482-6.]

bers. The rate of flow did not have any effect on longer fibers (aspect ratio of 50). The distribution of fibers along a cross-section of cylindrical parts can take one of several forms.[16] Combinations of radial structure and onion-like structures can be distinguished. The core and skin have different structures. Increasing the concentration of filler increases the influence of the extrusion rate on the orientation of talc

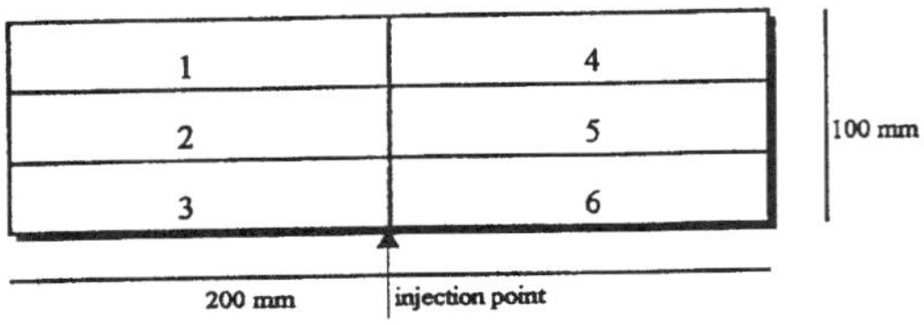


Figure 7.5. Longitudinal selection of test plate segments. [Adapted, by permission, from Barbosa S E, Kenny J M, Antec '97. Conference proceedings, Toronto, April 1997, 1855-9.]

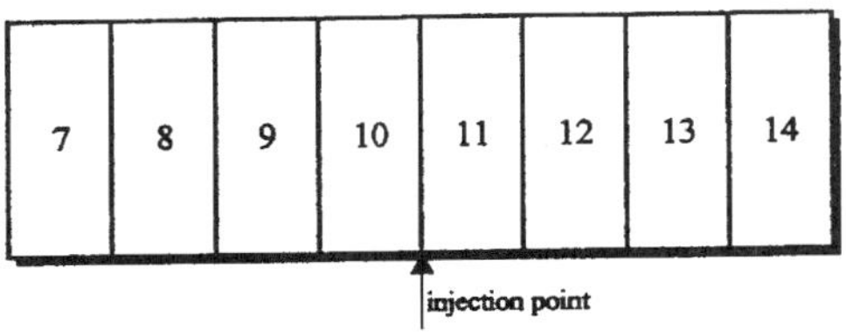


Figure 7.6. Widthwise selection of test plate segments. [Adapted, by permission, from Barbosa S E, Kenny J M, Antec '97. Conference proceedings, Toronto, April 1997, 1855-9.]

which as the concentration becomes greater than 20%, becomes more radially oriented with extrusion rate increasing.[16]

Also, the proportion of fibers in the skin (or shell) and those in the core depends on the rates of flow.[21] Low injection rates and low temperatures causes an expansion of the shell (skin) region.[22] These relationships also affect the orientation of polymer chains in filled and unfilled polymers during processing.[22] Orientation of fiber in blow molding of bottles filled with fibers caused anisotropy of properties. Tensile strength was increased in the machine direction.[23] At the same time, talc filled bottles had more uniform tensile properties than unfilled bottles.[24]

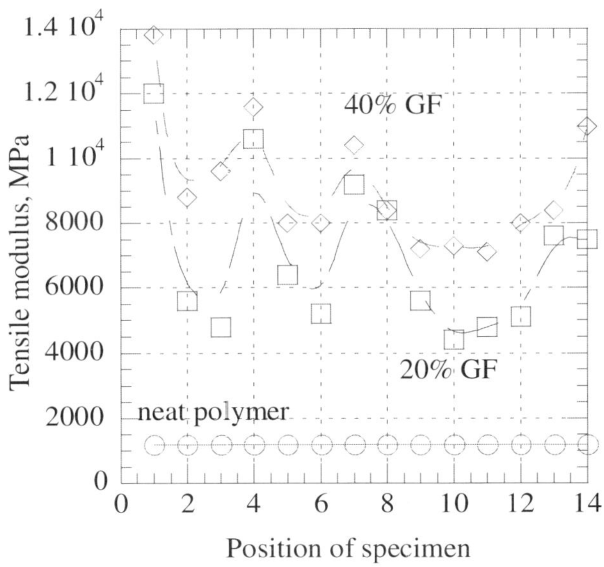


Figure 7.7. Tensile modulus of glass fiber reinforced polypropylene vs. position of sample. [Adapted, by permission, from Barbosa S E, Kenny J M, Antec '97. Conference proceedings, Toronto, April 1997, 1855-9.]

The distribution of fiber in an injection molded plate is shown in Figures 7.5 to 7.7.[25] Specimens for tensile testing were selected from injection molded plate as shown in Figures 7.5 and 7.6. Neat polypropylene has a low but consistent modulus throughout the width and length of the plate. The addition of fiber seems to increase the tensile modulus although readings were less uniform than expected. Fibers certainly travel preferentially with the front of injected material because the most distant segments have always the highest modulus. The lowest readings are from specimens close to the injection point. Increased concentration of filler adds to the uniformity of readings (more uniform readings for samples containing 40% fibers than for these containing 20% fibers). Blow molding of talc filled plastic bottles[13] and thermoformed talc filled thermoplastics produced materials which yielded similar test results.[26]

The relative magnetic permeability depends on the fiber orientation function, J (see Eq 7.2):

$$\mu' = \mu_o + 4J^2 \qquad [7.3]$$

where:

$\mu_o$ the relative magnetic permeability of a matrix filled with spherical particles of nickel

Figure 7.8 shows relationship between the relative magnetic permeability and the fiber orientation function.[17] The results came from an experiment previously discussed (see Figure 7.4).

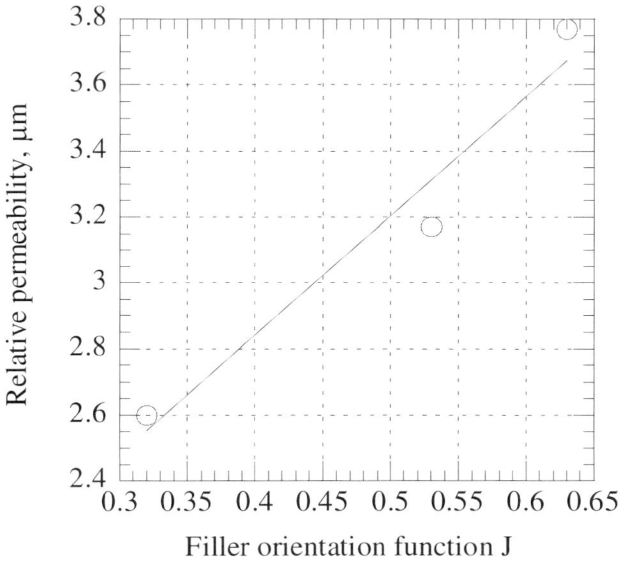


Figure 7.8. Relative magnetic permeability vs. fiber orientation function. [Adapted, by permission, from Fiske T, Gokturk H S, Yazici R, Kalyon D M, Antec '97. Conference proceedings, Toronto, April 1997, 1482-6.]

A polyamide gel filled with ferrite powder forms a ferroelastic material.[14] When the material is in a compressed state or in a magnetic field, filler particles are oriented. When the stress is released or the magnetic field is removed, particles become disoriented. When no stress is applied, material has no magnetic properties. Application of stress gives a measurable magnetic field (1-30 gauss) which depends on the extent of compression. When conditions are reversed, i.e. if material is put into a magnetic field it deforms to orient the filler. This is a good example describing memory of particles distribution and energy conversion (magnetic to mechanical and *vice versa*).

Fillers are known to affect UV stability of materials but little is known as to how the orientation of filler particles affects UV stability. Coatings and films designed with corrosion protecting barrier properties may contain dispersed mica or talc. The plates orientation prevents penetration of diffusing materials. To protect against UV degradation such a barrier would also be useful to exclude oxygen because photochemical changes are accelerated by oxygen. Until recently, there was little evidence that orientation of particles contributed to UV stabilization. A recent paper[27] seems to give experimental evidence that it does. Talc filled polypropylene test bars, prepared by injection molding, were exposed to UV radiation. Initial degradation of the filled material was faster than the unfilled control specimens containing filler but in longer exposure filled specimens were better protected and did not go through a further change on continued exposure. This may be explained by the fact that the material surface (the skin) did not contain many oriented particles, but internal core did. Therefore, the interior was protected by less a permeable barrier formed by the oriented particles of talc which did not allow oxygen to penetrate.

Figure 7.9 shows that orientation of fiber affects the wear resistance of a material.[28] The lowest wear occurs when fibers are perpendicular to the matting surface. If both matting surfaces are made out of fiber filled materials, the wear properties can be further optimized by the choice of fiber orientation in respect to both surfaces.

We have shown that orientation of filler particles can affect many properties (sometimes in unexpected ways). The best orientation depends on the property which is to be optimized and on the materials in the application. Extensive use is being made of these means of improving properties. Many materials can be further improved by application of these principles.

## 7.3 VOIDS

The term "void" may mean different things in relationship to filling and fillers. But all the meanings have one common denominator – they play a role in material reinforcement. A void may be

- An air filled space created around the filler particle by incomplete wetting or debonding

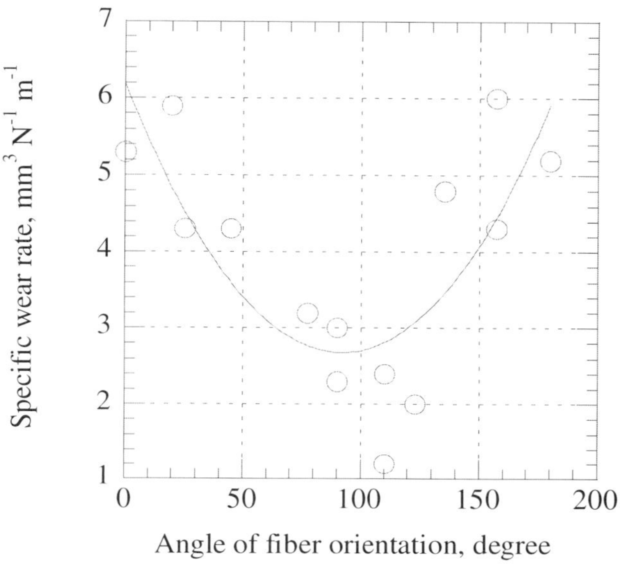


Figure 7.9. Specific wear rate vs. angle of fiber orientation. [Adapted, by permission, from Wada N, Uchiyama Y, Hosokawa M, *Int. Polym. Sci. Technol.*, **21**, No.3, 1994, T/53-63.]

- An air bubble created in a filled material for the purpose other than toughening
- A very small air bubble (microvoid) created to toughen polymer
- The free space around filler such as carbon black or fumed silica which are characteristic of the structure of those materials

All these terms are used to explain various phenomena related to fillers.

A model was developed to estimate properties of polymer composites which have voids of various sizes (large and small).[29] Such voids are typically found between fiber tows (macrovoids) and inside the fiber tows (microvoids) in composites produced by liquid molding. The presence of these voids does prevent the matrix from adhering to the fiber which reduces the composite's mechanical performance. Larger voids do not seem to affect performance as much as smaller voids do. In practice, the volume of voids in normal production is within 5% of the total volume of the composite. At 5% void volume, the mechanical strength of composite can be reduced by as much as 30%. This is considered a substantial imperfection but it is found in practice. The model developed predicts values of mechanical properties which correlate with void volume. In another application, magnetic resonance imaging helped to determine voids in solid rocket propellants and liners similar to those used in space shuttle.[30] The voids were found to be in close proximity to the filler particles.

In the production of microporous sheets used as separation membranes, voids are internally created to make material permeable.[31] Polypropylene was highly filled with calcium carbonate (~200% phr) and biaxially stretched. The stretching separates the matrix from the filler particles. This creates a soft membrane with gas and water vapor permeability (but liquid water does not penetrate the membrane). Changing filler content, particle size, and degree of stretching one can modify membrane properties. The degree of stretching is important because it determines mechanical properties in both directions of stretching. These properties can be optimized to give balanced tensile properties in both directions which is important in the practical applications of a membrane.

Void creation was simulated by mixing polystyrene with rigid particles.[32] This study was intended to develop an understanding of how impact resistance could be improved by incorporating voids rather than incorporating rubber. Addition of 1% rigid particles performed better than the impact modified resin. The optimum size of particles was ~2 μm. This effect is due to improvement in crack growth resistance due to shear deformation or crazing. An epoxy system was improved in a similar experiment but non-adhering organic particles were used to create microvoids.[33] Regardless of the particle type, at least a twofold improvement in fracture toughness was obtained.

Void volume is one of the main parameters used to characterize the structure of carbon black.[34-6] Void volume enters into the equation used to characterize interaggregate distance, IAD, as follows:

$$IAD = \frac{(1 - VV)}{(N_2SA \times \phi)} \qquad [7.4]$$

where:
IAD interaggregate distance, nm
VV void volume, $cm^3/g$
$N_2SA$ nitrogen surface area, $m^2/g$
$\phi$ volume fraction

Void volume can be calculated using this equation after determining the surface area by nitrogen absorption (BET method) or by filling the voids with liquid. Typical values of void volume are 0.6-0.7 $cm^3/g$ for carbon black and 0.5-0.97 $cm^3/g$ for silica.

## 7.4 MATRIX-FILLER INTERACTION

Studies on filled systems all find that the experimental observations can be explained by matrix-filler interaction. This interaction is a complex process involving:

- A chemical reaction between the filler and matrix materials (Chapters 6 and 7)
- Physical interaction (van der Waals forces and hydrogen bonding)
- Changes in morphology of interacting components
- Mechanical interlocking

These processes modify surface layers of both interacting materials (filler and matrix) and form an interphase which differs in properties from the bulk matrix. The formation of the interphase is responsible for changes in the physical and mechanical properties of filled materials and usually improves material reinforcement.[37-47]

In ABS/glass beads system, thermo-physical measurements show that interaction between the ABS and the glass causes a decrease in heat capacity and increase in thermal conductivity due to restriction in molecular motions of the resin surrounding glass beads.[37] Molecular transformations in the vicinity of filler particles is a local phenomenon dependent on the concentration of filler particles.[39] At low loadings (10% and below), the particles are surrounded by a tightly bound polymer covered by a layer of loosely bound chains. As the loading increases, the areas of loosely bound polymer begin to overlap, consisting the entire matrix to be influenced by the filler. If filler loading is high, there is little space left for loosely bound polymer and the participation of the layers of tightly bound polymer increases. At different filler loadings, a change in the interaction mechanism is likely to occur.

Particle-particle interaction affects maximum particle packing and it is influenced, in turn, by surface coating of the filler.[41] Packing density of the filler is not a simple geometrical phenomenon. It depends on interaction with the matrix. This interaction changes the morphology of matrix.[43] Adhered layers of polymer have a different conformation from polymer in the crystalline structures formed due to the polymer crystallization. The character of filler determines the conformation of the surrounding polymer and therefore strongly influences mechanical and chemical properties. In many filled systems, a second glass transition temperature can be detected due to the presence of adsorbed layers on the surface of filler.[44]

These and other phenomena are the subject of discussion in the following sections of this chapter.

## 7.5 CHEMICAL INTERACTIONS

Chapter 6 discussed chemical reactions which occur on the surface of the filler and with the filler surface. Now we focus on how the interphase is created and on how interface chemistry affects the formation of interphase.

Figures 7.10 to 7.12 show three models of interaction between the surface of the filler and the matrix.[35,39,48] Each model was developed to examine interaction in different system. They complement each other and show the complexity of interaction. The models also help us to distinguish between chemical and physical interactions.

This model was mentioned in the previous section.[39] An increase in the amount of filler decreases the average particle distance. When a relatively small number of particles is present (7.10A), particles influence the surrounding matrix but there is enough available bulk which is not subject to interactions with the filler surface (line-shaded areas are particles of filler, black areas correspond to the

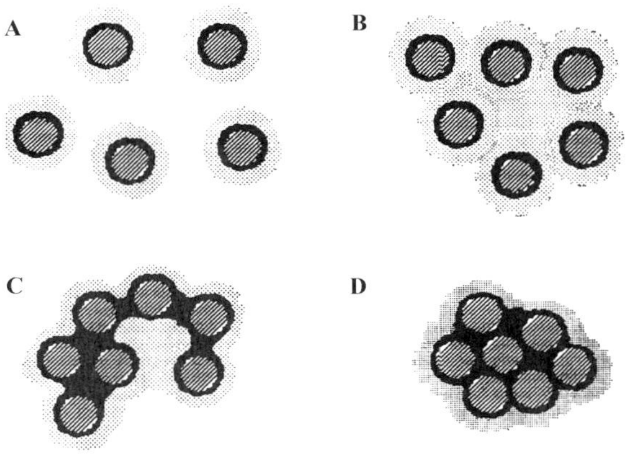


Figure 7.10. Schematic model of morphological transformations in filled polymers. A - silica content less than 10 wt% ($d>d_{cr}$), B - silica content ~10 wt% ($d=d_{cr}$), C - silica content ~20 wt% ($d<d_{cr}$), D - silica content over 50 wt%. [Adapted, by permission, from Tsagaropoulos G, Eisenberg A, *Macromolecules*, **28**, No.18, 1995, 6067-77.]

tightly bound resin, gray areas to loosely bound resin). There is only one glass transition. This indicates that the mobility of polymer next to the immobilized layer is not significantly affected. The magnitude of the first glass transition does decrease indicating that some polymer is involved in the formation of tightly bound layers. When the concentration of particles becomes close to critical, $d_{cr}$, the mobility of all chains in the composite becomes affected (7.10B). At this point, the second glass transition temperature can be detected, signifying formation of a substantial amount of loosely bound polymer. As more filler particles are incorporated (7.10C) more of the polymer becomes tightly bound pretty much in amount proportional to the increase in filler concentration. The second glass transition temperature becomes less easy to detect. At high concentration of filler (7.10D), most of the polymer is immobilized. The conclusions from this model are:

- Tightly and loosely bound polymer are two distinct and different physical materials
- Filler-matrix interaction affects chain mobility
- Chain mobility is controlled by the concentration of filler

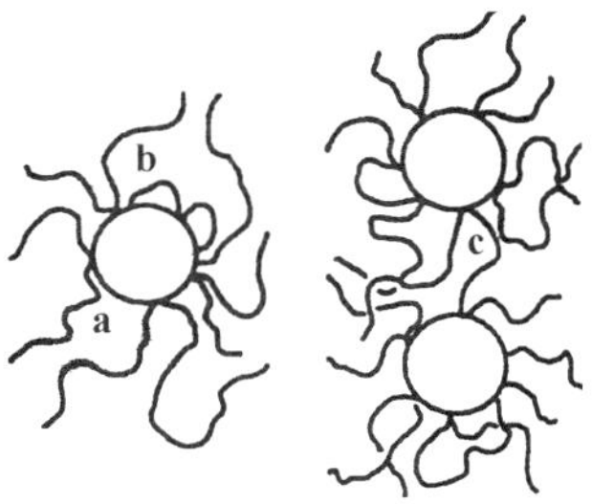


Figure 7.11. The concept of segmental interaction with a carbon black surface. [Adapted, by permission, from Wolff S, *Rubb.Chem.Technol.*, **69**, No.3, 1996, 325-46.]

The second model shows us how chains can get attached to the surface of carbon black (Figure 7.11).[35] The chain can be attached at single point (a), at more than one point (b), or the chain can bridge two or more filler particles (c). This model does not indicate if a chemical reaction is involved but it can be anticipated that, since reactive functional groups can be positioned in the middle of chain or at its ends, the reactions may involve chain segment or terminal group. It might be thought that both models complement each other by showing that the probability of particles bridging increases with filler concentration. In rubber, a gel is formed

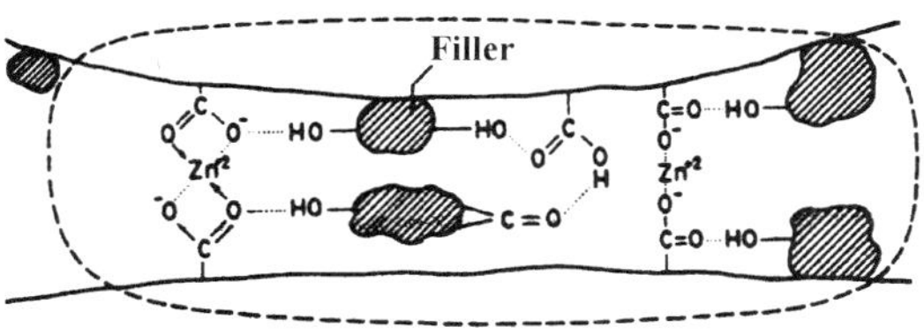


Figure 7.12. A mechanism of interaction between filler and ionic groups in the restricted mobility region in EPDM. [Adapted, by permission, from Datta S, De S K, Kontos E G, Wefer J M, Wagner P, Vidal A, *Polymer*, **37**, No.15, 1996, 3431-5.]

but only when configuration (c) occurs (two or more particles are connected).

Figure 7.12 depicts a model of chemical interactions proposed for a system of carbon black and maleated EPDM.[48] There are two types of links. Hydrogen bonding and covalent bonding which is characterized by the attachment of filler particles to the chain. Because the rubber is crosslinked there is more opportunity for it to form interaction.

This system follows the Guth and Gold equation:

$$\frac{E'_f}{E'_g} = 1 + 1.3\phi + 14.3\phi^2 \qquad [7.5]$$

where:

$E'_f$ storage modulus of filled system
$E'_g$ storage modulus of unfilled rubber
$\phi$ volume fraction of filler

The Eq 7.5 shows that the degree of reinforcement (which correlates to the ratio of the storage moduli) increases with increasing filler concentration (similar to the first model (Figure 7.10)). A second glass transition temperature is not detected but tanδ decreases as the concentration of filler increases indicating that the number of crosslinks increases. It is interesting to analyze the numerical values of coefficients at the right side of Eq 7.5. The Guth and Gold equation such as Eq 7.5 has the following general form:

$$E'_f = E'_o(1 + \alpha\phi + \beta\phi^2) \qquad [7.6]$$

$\alpha$ coefficient which depends on filler dispersion
$\beta$ coefficient which depends on molecular interaction

For van der Waals interaction, $\alpha = 2.5$ and $\beta = 14.1$. For the system containing carboxylated acrylonitrile rubber and ISAF carbon black, $\alpha = 4$ and $\beta = 42$. If ISAF carbon black is oxidized, $\alpha$ remains the same and $\beta$ increases to 53 which is consistent with the fact that oxidized carbon black has more reactive sites and therefore molecular interaction should increase.[49] When the system is vulcanized, $\beta$ further increases to 62 for ISAF carbon black and to 68 for oxidized carbon black, meaning that additional interactions occur. Additional mixing also increases the value of $\alpha$. This is what mixing is intended to do. Because mixing increases the value of α beyond what is considered typical of van der Waals forces, it plays an essential role in promoting chemical reactions.

Ayala *et al.*[50] proposed the following equation to describe rubber-filler interaction:

$$I = \sigma / \eta \quad [7.7]$$

σ the slope of stress-strain curve in the relatively linear region
η filler-filler networking parameter calculated from ratio of storage modulus at high and low strains

This equation implies that reduction of filler-filler interaction increases rubber-filler interaction which is what mixing does. Figure 7.13 confirms the usefulness of equation 7.7.

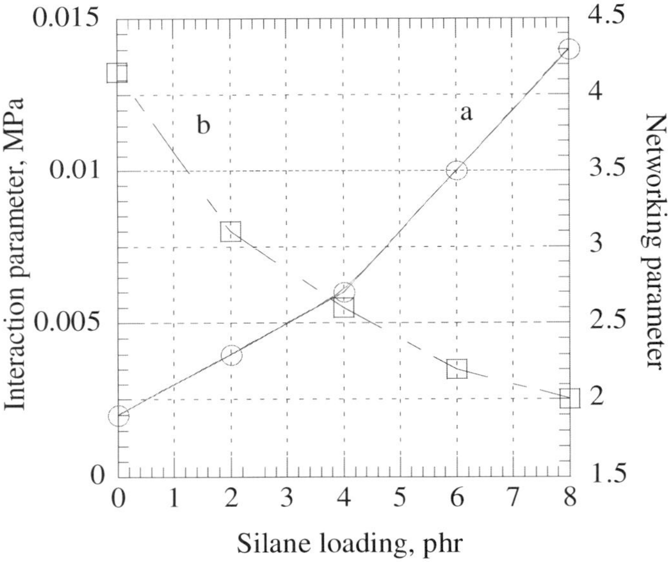


Figure 7.13. Interaction parameter (a), I, and networking parameter (b), η, vs. concentration of 3-aminopropyltriethoxysilane in NBR-carbon black system. [Data from Bandyopadhyay S, De P P, Tripathy D K, De S K, *J. Appl. Polym. Sci.*, **61**, No.10, 1996, 1813-20.]

Silane is added to increase bonding between rubber and filler. The experimental results show that the physical interaction between filler particles is decreased and chemical bonding is increased. Other papers by the same research group show similar trends for other systems.[52,53] Several research methods are used to identify the exact nature of chemical interactions.[54-57] $^1$H magic-angle spinning NMR, IR, Raman, and ESCA are the most successful techniques used for these purposes.

This left us conclude that:

- The effect of chemical sites on the filler surface should be interpreted in the context of the sphere of influence of chemical interactions

- This sphere of influence includes the nature of chemical bonding, properties of the formed structure, and the concentration of filler particles
- An increase in filler concentration alters the mechanism of interaction and the interaction influences the properties of the materials
- The methods of measurement can distinguish between chemical bonding and physical interaction.

## 7.6 OTHER INTERACTIONS

The interactions other than the formation of covalent bonds include:

- Van der Waals forces
- Ionic interactions
- Hydrogen bonding
- Acid-base interaction
- Mechanical interlocking
- Other interactions

Van der Waals forces are made up of:

- Dispersion forces (London)
- Orientation forces (Keesom)
- Induction forces (Debye)

The dispersion force is the major factor in non-chemical interaction. London derived an equation characterizing dispersion energy for attracting two spherical molecules:

$$E_d = -\frac{3}{4}\left[\frac{1}{(4\pi\varepsilon_o)^2}\right]\left(\frac{I\alpha^2}{r^6}\right) \qquad [7.8]$$

where:

$\varepsilon_o$ permittivity in free space,
I ionization constant,
$\alpha$ electronic polarizability of molecule,
r separation distance.

The London theory was later modified to account for retardation effects occurring at greater separation distances:

$$E_r = -\left[\frac{1}{(4\pi\varepsilon_o)^2}\right]\left(\frac{\alpha^2}{r^6}\right)\left(\frac{hc}{r}\right) \qquad [7.9]$$

where:

h Planck constant,
c light velocity.

Dispersion forces act over long separation distances (from interatomic distance to 10 nm and above) and are affected by nearby bodies. These forces align and orient molecules. Powerful as they are, dispersion forces do decrease rapidly as the separation distance between two interacting bodies increases. The energy is inversely proportional to the sixth or seventh power of separation distance. Electrons traveling around the nucleus form an asymmetric charge distribution which produces a dipole which generates short-lived electric fields which, in turn, induce dipoles in the neighborhood. Dipoles are attracted by each other and this is what generates the force of dispersion.

Keesom analyzed the effect of the orientation of dipoles on the energy of interaction between the molecules:

$$E_o - \frac{2}{3}\left[\frac{1}{(4\pi\varepsilon_o)^2}\right]\left(\frac{\mu_1^2\mu_2^2}{kTr^6}\right) \qquad [7.10]$$

where:

μ dipole moment,
k Boltzmann constant,
T absolute temperature

Note that temperature is a parameter of the equation. As the material temperature rises during processing, the value of orientation energy becomes negligible. In a typical system conflicting dipole fields are created which significantly reduce dipole-dipole net interaction. Keesom forces, unlike London forces, do not apply to nonpolar substances because both dipoles, which participate in the interaction, must be permanent dipoles (London forces do not require the presence of permanent dipoles).

Debye modified the Keesom equation to account for experimental observations. He showed that the energy of interaction was not as greatly reduced by temperature as was predicted by the Keesom equation:

$$E_i = \left[\frac{1}{(4\pi\varepsilon_o)^2}\right]\left(\frac{\alpha_1\mu_2^2 + \alpha_2\mu_1^2}{r^6}\right) \qquad [7.11]$$

This equation gives the induction energy for dipole-induced dipole interaction; for dipole-nonpolar, $\alpha_1\mu_2^2$ is neglected.

London dispersion forces account for more than 75% (up to as much as 100%) of total interaction energy. Very polar small molecules such as water are an exception. These owe most of their interaction energy to hydrogen bonding (only 24% of the attractive forces are contributed by dispersion interaction). Particle size plays a very important role. Over large distances, the attractive forces between particles, since it is inversely proportional to the seventh power of the separation distance, become negligible. Van der Waals forces are at least a few hundred times lower than that of covalent bonds but are strong enough to cause difficulties in the dispersion of some grades of carbon black so that the desired increase in tensile strength, due to the reinforcing effect, is not achieved. The type of interacting material is also important. Molecules prefer to interact with molecules of their own kind and the hydrophobic-hydrophilic effect is significant.

The equation for van der Waals forces as applied to particle interactions was developed by Hamaker:[5]

$$F = \frac{A_1}{6} d_2 (d + t)\left(\frac{1}{a^2} + \frac{1}{b^2} + \frac{2}{ab}\right) \qquad [7.12]$$

where:

| | |
|---|---|
| $A_1$ | Hamaker coefficient, |
| d | diameter of two equal spheres, |
| t | separation distance between two spheres, |
| a | $= t^2 + 2dt$, |
| b | $= a + d^2$. |

The Hamaker coefficient is the sum of three terms: London, Keesom, and Debye. His equation is an integration of the forces acting between a pair of particles across the phase boundary. Hartley[58] found that Hamaker's constant for carbon blacks is in the range of $0.6\text{-}5.8\times10^{-19}$ J, which is in agreement with the literature and theory. Van der Waals forces play a significant role in carbon black dispersions, but not in the dispersion of titanium dioxide. Titanium dioxide has an adsorbed layer of moisture, which not only reduces van der Waals forces but causes a liquid bridging force that dominates flocculation. Thus, the cohesiveness of both carbon black and $TiO_2$ depends, on entirely different principles.

This theoretical work was utilized by Good and Fowkes who developed theories relating van der Waals forces to surface tension and to the work of adhesion. These concepts are discussed in Chapters 5 & 14.

Ionic interaction is believed to play a role in the reinforcement of EPDM crosslinked by ZnO with modified silica particles.[59] A restricted mobility region is formed by ionic aggregates. In other work, muscovite mica was modified by various cations.[60] Polymers with crown ethers were absorbed on such modified mica. It was discovered that the ionic radius was an important parameter in the absorption process. Radii in the range of 130-150 pm ($K^+$, $Rb^+$, and $Ba^{2+}$) were optimum for absorption. Ionic forces are equivalent to covalent bonding forces. The highest energy is attained when two interacting ions are in close proximity, i.e., separated by the length of typical bond. If the distance of separation is larger than the bond length, the energy of interaction rapidly decreases.

Covalent forces binding atoms in molecules range from 200 to 900 kJ/mol. The energy of hydrogen bonding is in the range of 8 to 42 kJ/mol which is small compared with covalent bonding force but considerably higher than that attributed to van der Waals interactions. Because of the relatively low energy required for bond formation and breaking, hydrogen bonding plays an important role at room temperature. It has an essential effect on interaction between surfaces of inorganic materials which contain hydroxyl groups on their surface and organic molecules present in their proximity. Silicone rubber reinforcement is an example of hydrogen bonding which has an increased apparent crosslink density.[61] Figure 6.25 shows how silica loading increases $\Delta L$ (apparent crosslink density). Here, two forces, hydrogen bonding and polymer absorption on the surface of silica particles, are responsible.

The acidity or basicity of a solid surface is determined by its isoelectric point, $I_s$. Water is basic on an acidic surface and acidic on a basic surface. The measure of bond energy is given by the equation:

$$\Delta H \cong \frac{DN \times AN}{100} \qquad [7.13]$$

where:
DN donor number
AN acceptor number

Acid-base interaction affects the mutual interaction between a solid (e.g., filler) and a liquid (e.g., solvent, polymer, etc.), as well as between a liquid and a liquid. This type of interaction may also affect the conformation of the polymer molecule when it is in contact with another acceptor/donor. More information on acid-base interaction is included in Chapters 5 and 14.

Mechanical interlocking is commonly thought of a macroscopic phenomenon in the adhesion between a substrate and an adhesive. But the interaction between polymer and filler plays a role and is elegantly exemplified in the rubber-silica system.[62] The authors[62] investigated the size of voids in fumed silica in its original form and after compounding with rubber. Pore size in the original silica was determined by a mercury porosimeter. After the silica was compounded with rubber the silica pores were cleaned by pyrolizing the rubber at 480°C. The pore sizes were measured again using the mercury porosimeter. It was also determined that the pyrolysis conditions do not affect the pore size of silica. It was found that the silica grade which caused the most reinforcement of rubber had widened pores after it was compounded. The increase in the size of pores depended on the conditions of mixing and on the formulation of rubber. Reinforcement required the initial size of silica pores to be large enough to allow penetration by rubber chains.

There are other interactions. The surface of carbon black has a tri-dimensional structure dependent on the conditions of its preparation. The rubber chains are thought to fit into imperfections in the surface and produce reinforcing effect.[63] In another paper, the reinforcement effect was correlated with surface roughness.[64] This paper postulated that the chains align themselves over the uneven surface to cause reinforcement.

Another interaction is responsible for the recovery of the material after it is subjected to stress.[65] Rubber bridging the neighboring particles of filler is an example. Some particles are connected through several rubber chains which makes their association more permanent and assures filler-filler contact. These filler-filler contacts are responsible for the recovery since, unlike chain-filler contacts, they store the strain energy which is then used in the recovery process. Chain-filler contacts can easily debond or rearrange in different location and this process does not result in recovery of the initial shape.

A study of paint technology reveals other interactions.[66] The layer of paint in immediate contact with the surface of the substrate is depleted of filler. The next layer is enriched with filler. Between the last layer and the bulk of paint there is still polymer-rich layer. This effect is attributed to the affinity of the polymer with the substrate. This affinity leads to polymer migration. It also causes binder orientation which leads to the increased interaction with filler. This in turn, causes migration of filler particles from adjacent layers.

Polymers filled with ultrafine metal particles form periodic stripes.[67] These stripes are thought to be caused by an inhomogeneous electric field which induces electrostatic interactions among the polarized polymer chains. The phenomenon is known as mutual dielectrophoresis.

Thus, many complex phenomena can affect the organization of the interface and this, in turn, affects how fillers contribute to the reinforcement.

## 7.7 INTERPHASE ORGANIZATION

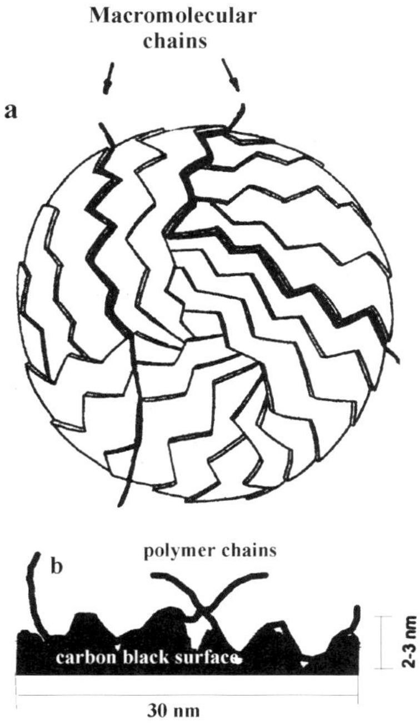


Figure 7.14. Conformation of a chain on the surface of carbon black. (a) direct view of surface, (b) cross-section through the surface. [After refs. 63, 64.]

Carbon black research[63-65,68,69] focused on a study of the structure at the interface and attempt to explain reinforcement. More recently, other fillers have been investigated.[66,70-75] A variety of surface structures have been postulated for carbon black to explain the organization at the interface.[63] Figure 7.14 gives examples of how the surface of a filler can contribute to interface organization.[63,64] The attachment of chain to the surface of the filler is accomplished through a process called "wetting" and its removal through a process of "dewetting". Chains which are removed from the surface by strain can become attached again which is consistent with some mechanisms of reinforcement (e.g., molecular slippage).

Another concept of the structure at the interface has been proposed to overcome some of difficulties associated with the previous model (Figure 7.15).[65] In this model, a distinction is made between two types of contacts. FF is filler-filler contact and FM is filler-matrix contact. Special emphasis is given to the contact FF which is believed to explain energy storage during strain. The

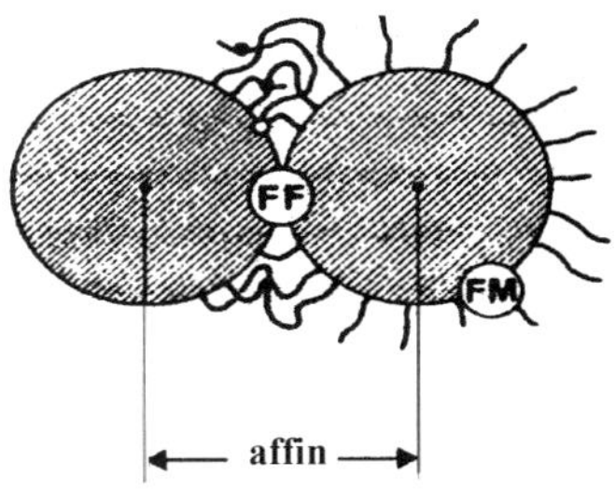


Figure 7.15. The basic reinforcing component. [Adapted, by permission, from Strauss M, Pieper T, Peng W, Kilian H G, *Makromol. Chem., Macromol. Symp.*, **76**, 1993, 131-6.]

contact FF is well protected by surrounding bonds. The separation of filler particles is limited to a certain distance when the strain is less than a critical. This allows to store energy due to the action of the van der Waals and the elastic forces of the protecting chains.

The polymer chains are not the only components of the mixtures which are capable of interacting with the filler surface. Other additives can be adsorbed on the surface to create a situation in which monolayer or multilayer coverage competes to form an association with the surface.[69] Such coverages contribute to the organization of interface. The first layer of adsorbed components in the formulation has an impact on the entire organization of the interphase because it affects configuration of adsorbed chains and the crystallization processes around the adsorbed layer.

Investigations of polymer blends has developed an increased understanding of interphase organization. In blends two interfaces exists: the interface between two matrix types and distribution of filler and its interfaces with this matrices. The interphase of carbon black in blends of natural rubber and EPDM depends on the character of carbon black (surface groups available for interaction), the viscosity, the molecular weight, and on the order of mixing.[68] These organizations determine the mechanical properties of rubber for tires.

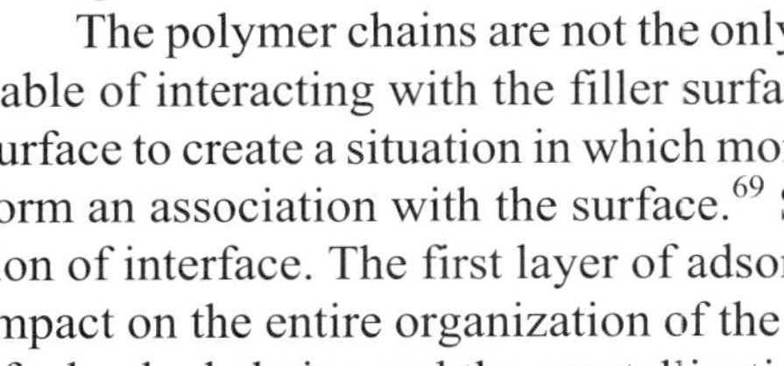


Figure 7.16. Interphase between paint and substrate. [Adapted, by permission, from Roche A A, Dole P, Bouzziri M, *J. Adhesion Sci. Technol.*, **8**, No.6, 1994, 587-609. ]

Figure 7.16 shows interface formation with painted substrate.[66] The mechanism of organization was discussed in the previous section. The alignment in the polymer layer plays a large part in polymer-filler interaction in the adjacent layers. The way in which polymer is configured on the substrate surface determines if polymer chains are readily available for interaction with filler. This example shows that it is not only the filler and the matrix which play a role in the interphase organization.

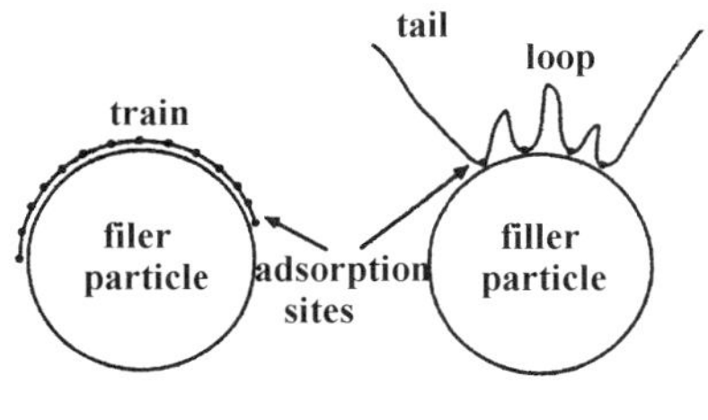


Figure 7.17. Polymer chain absorption onto a solid surface. [Adapted, by permission, from Hedgus C R, Kamel I L,US, J. Coatings Technol., 65, No.821, June 1993, 49-61.]

In Figure 7.17, we have two chain adsorption methods on the surfaces of silica or titanium dioxide.[70] On larger particles, the chain assumes a flat coverage of the surface (train). On smaller particles, the curvature of surface does not allow for train configuration. Instead, loops and tails are formed. In the first case, the interphase is thinner than in the second and

Figure 7.18. Optical micrograph with crossed polars of bamboo fiber in maleated polypropylene. [Adapted, by permission, from Mi Y, Chen X, Guo Q, *J. Appl. Polym. Sci.*, **64**, 1997, 1267-73.]

fewer chains participate in the formation of the interface. On the other hand chain configuration is different in both cases.

Figure 7.18 shows how crystalline structure is affected by the presence of fiber. Here, bamboo fiber was used for polypropylene reinforcement.[71] A nucleation occurs on the surfaces of fiber. Spherulites grow from the fiber surface. Such growth results in transcrystallinity. The maleation of polypropylene increases interaction because of reactivity with OH groups on the fiber surface. This organization contributes to the reinforcement.

Glass fibers sized with polyurethane and polyvinyl acetate formed different interfaces. This was due to the differences in reactivity and miscibility. Polyurethane forms a stronger interface because it is reactive and miscible with epoxy resin.[74] Surface tension of glass surface in a molten state correlates with the interface formation with polymer.[75] The diffusion at interface contributes to a complex structure controlling properties of the interphase. The analysis of the diffusion at the interphase has helped to develop an understanding of the formation of metal-polymer interfaces and plastic welding.

In summary, numerous effects influence interphase formation. The most important influences depend on the type of active groups on particle surface, particle size, surface shape, and interaction with the matrix. The interphase can be modified by mixing process, the order of addition, filler concentration, and the orientation of the chains on the surfaces among other possible causes of interphase modification.

## 7.8 INTERFACIAL ADHESION

Interfacial adhesion can be predicted from available models or from data on the mechanical performance of filled systems.[5,76-9] The following equation describes the reversible work of adhesion:

$$W_{AB} = \gamma_A + \gamma_B - \gamma_{AB} \quad [7.14]$$

where:

$\gamma_A$ and $\gamma_B$ surface free energies of adhering substances
$\gamma_{AB}$ interfacial energy

The interaction depends on the morphology and the chemical structure of both the filler and the matrix. The condition which outlines the limit of the stress, which the bonding can withstand, is determined from the following equation:

$$\sigma^D = -C_1\sigma^T + \sqrt{\frac{C_2 W_{mf}}{R}} \quad [7.15]$$

where:

| | |
|---|---|
| $\sigma^D$ | dewetting stress |
| $\sigma^T$ | thermal stress |
| $C_1, C_2$ | constants |
| $W_{mf}$ | work of adhesion between matrix and filler |
| R | average radius of filler particles |

The dewetting model is useful in predicting critical stress from a knowledge of tensile yield stress.[5] The results of tensile testing can be used to predict adhesion of polymer to filler particles of different sizes. The following model is useful for this purpose:

$$\sigma_c = \sigma_p(1 - a\Phi_f^b + c\Phi_f^d) \quad [7.16]$$

where:

| | |
|---|---|
| $\sigma_c$ | tensile strength of composite |
| $\sigma_p$ | tensile strength of polymer |
| a, b, c, d | coefficients |
| $\Phi_f$ | volume fraction of filler |

Coefficient "a" is related to stress concentration. Coefficients "c" and "d" are related to the adhesion of the polymer matrix to filler. In an experiment involving different sizes of calcium carbonate in poly(vinyl acetate), small and medium particles had a much larger values of coefficients "c" and "d" than did large particles. This is in agreement with an experiment which shows that large particles decrease the mechanical properties of composites.

Interfacial adhesion can also be estimated from the Suetsugu-Sakairi equation:[80]

$$\sigma_c = K\Phi\chi V_f + \sigma_m(1 - V_f) \quad [7.17]$$

where:

| | |
|---|---|
| $\sigma_c$ | composite strength |
| K | coefficient reflecting the orientation and the length distribution of glass fiber |
| $\sigma_m$ | matrix strength |
| $\Phi$ | interfacial adhesion parameter |
| $\chi$ | glass fiber aspect ratio |
| $V_f$ | volume fraction of glass fiber |

This equation was used to estimate the interfacial adhesion in comparison with the acid-base properties of glass fibers in LDPE.[79] The effect of surface treatment of glass beads on their interfacial adhesion to PET was also estimated from a mechanical property measurement.[78] A mathematical model describing the adsorption of polymers on filler surfaces related coupling density to the average area available for coupling between rubber and filler surface.[76]

## 7.9 INTERPHASE THICKNESS

Several methods are used to determine the thickness of the interphase.[70] Table 7.1 lists the most important methods and the results of the thickness of an interphase obtained from several sources. The equation below gives the correction parameter, B, in relationship to the matrix-filler interphase:[81,82]

$$B = (1 + \Delta R / R)^3 \quad [7.18]$$

where:
$\Delta R$ effective thickness of interphase
R average radius of filler particles

**Table 7.1. Interphase thickness**

| System | ΔR, nm | R, nm | ΔR/R | Method of determination | Refs. |
|---|---|---|---|---|---|
| Natural rubber/silica | 5.2 | 8.7 | 0.60 | DMA | 83 |
| SBR/silica | 4.0 | 7.9 | 0.51 | DMA | 83 |
| NBR/silica | 3.7 | 8.0 | 0.46 | DMA | 83 |
| Immobilized layer | 0.5-2 | | | general range | 39 |
| Restricted mobility | 2.5-9 | | | general range | 39 |
| Rubber/carbon black (CB) | 10 | | | | 84 |
| Immobilized layer | 0.4-1.3 | | | range for rubber/CB | 45 |
| Restricted mobility | 3-6.6 | | | range for rubber/CB | 45 |
| PMMA/$TiO_2$ | 70 | 150 | 0.47 | viscosity | 70 |
| PMMA/$TiO_2$ | 65 | 96 | 0.67 | viscosity | 70 |
| PMMA/$TiO_2$ | 51 | 73 | 0.63 | viscosity | 70 |
| PMMA/silica | 17 | 8 | 2.18 | viscosity | 70 |
| PMMA/glass | 1.4 | 18 | 0.08 | viscosity | 70 |
| PMMA/mica | 0.11 | 0.75 | 0.15 | viscosity | 70 |
| PS/glass | 1 | 20 | 0.05 | viscosity | 70 |
| PS/mica | 0.06 | 0.66 | 0.09 | viscosity | 70 |
| PVA/PS particles | 0.24 | 3 | 0.08 | ultracentrifuge | 85 |
| Elastomer/carbon black | 0.13 | 2.2 | 0.06 | bound rubber | 86 |

This equation includes the parameters used in Table 7.1 to characterize the interphase thickness. The results presented are much affected by the method of measurement. The methods of measurement are indirect therefore it is quite difficult to estimate what the potential error of measurement may be. There are some data in the literature (not included here) which show thicknesses several orders of magnitude higher than the results presented in the Table 7.1. There is a need for further studies to give credible values of the interphase thickness which are necessary to establish other related properties of filled materials such as effective filler volume, bound rubber, reinforcement, etc.

The thickness of the interphase depends on the reactivity of the filler surface with the matrix material. It also depends on their physical affinity.[87] Increased acid-base interaction between chlorinated polyethylene and titanium dioxide increases the thickness of the adsorbed layer. There is a maximum of thickness of interphase which depends on the properties of polymer bulk. The acid-base interaction is more dependent on how the filler is modified than on the matrix properties themselves. Both filler and matrix are responsible for the formation of an equilibrium, although each contributes in a different way.

## 7.10 FILLER-CHAIN LINKS

Here, a distinction is made regarding the "permanence" of filler chain bonds. This subject has evolved throughout this chapter and it is an important factor in understanding the mechanism of reinforcement.

Chains arrive at the filler's surface at different time scales. The early arriving chains can select any part of the free surface and have an increased probability of forming consecutive links with other points on the surface after forming an initial contact point. This process continues until most of the available sites are occupied. The chains which arrive first have a high probability of forming strong links because they can be either attached at several segments along their length or form a "train" configuration which involves many neighboring segments of the same chain. Latecomers find the filler surface mostly occupied by existing links. The probability of their forming stable links is severely reduced because only a few isolated sites are available. These chains can be removed from the surface more easily than chains with more permanent linkages. In crosslinked systems, the total network developed can be expressed by the following equation:[88]

$$N = N_c + N_{st} + N_{un} \tag{7.19}$$

where:
$N_c$ chemical network density
$N_{st}$ network formed by stable links
$N_{un}$ network formed by unstable links

This equation gives a quantitative description of the storage modulus of a filled material:

$$G'(\gamma) = (N_c + N_{st} + N_{un}(\gamma))kT \tag{7.20}$$

where:
$G'(\gamma)$ storage modulus dependent on $\gamma$

$\gamma$ deformation amplitude
k Boltzmann constant
T temperature

Equation 7.20 shows that the changes in mechanical properties are first affected by temperature and then by the network of unstable links. Only after the unstable links are consumed stable links take the impact of changes occurring in the material. A similar logic can be applied to show that filler concentration also plays an essential role, considering that with small addition of filler most chains will form weak bonds because of very high competition for free surfaces on filler particles and the effect of reinforcement will be diminished. This is expressed by a simple equation:[89]

$$\gamma_{n,B} = 1 + \gamma_n \qquad [7.21]$$

where:
$\gamma_{n,B}$ number of adsorbed segments per chain
$\gamma_n$ adsorption index

When filler concentration is low, $\gamma_{n,B} \approx 1$. Each filler is bound only once. Carbon black filled rubber does not form gel if only small amounts of carbon black are used. The molecular weight of polymer in the matrix affects the fraction of bound polymer according to the equation:[90]

$$\phi_B = \phi_M (1 - \phi_M / 4) \qquad [7.22]$$

where:
$\phi_B$ fraction of bound polymer
$\phi_M$ $= M_n^{1/2}$maximum fraction of polymer which can be bound

As molecular weight increases, $\phi_B$ increases as does the probability of multiple connections.

## 7.11 CHAIN DYNAMICS

Section 6.11 is devoted to molecular mobility and, in it, the properties of macromolecular chains in filled systems are discussed. This section includes a brief evaluation of chain dynamics in relationship to "weak" and "strong" bondings of polymer chains which were introduced as a concept in the previous paragraphs.

Based on NMR studies, which play a prominent role in clarifying the mechanisms of interaction, monomeric units (or interacting segments) can be divided into three groups:[90]

- Those fixed on the surface – magnetic interactions of protons attached to these monomeric units are strong and the relaxation process is characterized by a high relaxation rate, $\sigma_B$. These units behave in a manner similar to the units of polymer in a glassy state.

- Forming loops and tails – these monomeric units have the freedom of a random rotation. Conformational fluctuations are restricted by fixed points on the filler's surface. The relaxation rate of these units, $\sigma_L$, is reduced according to the equation: $\sigma_L = \sigma_B / <n>$; where <n> is the mean number of skeletal bonds in one loop. With this relationship the relaxation rate decreases with the number of units in the loop. These units behave in a manner similar to polymer gels.
- Free chains – these units have the freedom of motion typical of an unfilled matrix. Their relaxation rates, $\sigma_F$, are given by the following equation: $\sigma_F = \sigma_B(\sigma_B \tau_c)$; where $\sigma_B(\sigma_B \tau_c)$ is a reduction factor related to $\tau_c$ which is the mean correlation time of random motions involved in the dynamics of the chain.

The importance of this classification is in characterizing the dynamics of diffusional processes and the strength of topological constraints to which the monomeric units (chain segments) are exposed.

NMR determines two types of spin-spin relaxation times: short, $T_{2s}$, and long, $T_{2l}$, which are for tightly and loosely bound polymer, respectively.[83] From modification studies of silica particles, it has been found that silanol groups are the main factor in increasing $T_{2s}$. Any reduction in silanol group concentration results in an increase of $T_{2l}$ and a decrease in $T_{2s}$. This is in accordance with the logical prediction of the behavior of such system. Computer simulations of networks in conjunction with experimental studies gives further insight into the chain dynamics in filled systems.[91]

## 7.12 BOUND RUBBER

Bound rubber is the fraction of polymer which is not extracted by a good solvent from a rubber-filler mix. It is a measure of rubber reinforcement as well as of filler activity towards the rubber. This concept was introduced in 1925 by Twiss.[92] Although, the traditional term "bound rubber" is commonly used for rubber compounds, the concept can also be applied to other macromolecular materials. The amount of bound rubber is given by the following equations:[89]

$$B = 1 - \int_0^\infty w(y)\exp(-qy)dy, \quad q = cPM_0 / A_0N_A = cPM_0D / N_A, \quad y = M / M_0 \qquad [7.23]$$

where:

| | |
|---|---|
| w(y)dy | molar mass distribution |
| q | fraction of adsorbed segments |
| y | number of segments per polymer chain (degree of polymerization) |
| c | filler loading (filler to polymer mass ratio) |
| P | specific surface area of filler |
| $M_0$ | molar mass of segment |
| M | molar mass of polymer |
| $A_0$ | filler surface area per one active site |
| $N_A$ | Avogadro number |

D number of active sites per unit filler surface area

Eq 7.23 can be converted to the following form:

$$B = \gamma \frac{4+\gamma}{(2+\gamma)^2}, \quad \gamma = cP\overline{M}_w D / N_A \qquad [7.24]$$

where:

$\gamma$ number of adsorbed segments per primary mass average molecule (the so-called adsorption index)
$\overline{M}_w$ mass average molar mass

Specific bound rubber, L, is another factor, frequently used in comparative studies:

$$L = \lim_{cP \to 0} (B / cP) = D\overline{M}_w / N_A \qquad [7.25]$$

It is a very convenient factor because it allows the amount of bound rubber and the available active surface to be related.

In laboratory practice, a small sample of rubber is extracted with solvent (usually toluene) at room temperature for a specified period of time (1 week) and the percentage of bound rubber, $R_B$, is calculated from the equation:[46]

$$R_B = \frac{W_{fg} - W[m_f / (m_f + m_p)]}{W[m_p / (m_f + m_p)]} \times 100 \qquad [7.26]$$

where:

$W_{fg}$ weight of carbon black and gel
W weight of specimen of rubber taken for extraction
$m_f$ weight of filler in composition
$m_p$ weight of polymer in compound

In order to establish the nature of the bonds, the specimen is also treated with ammonia. Under these conditions only chemically bound rubber remains absorbed on the filler's surface and physically bound polymer is extracted. Silica-rubber gels contain mostly physical bonding.

The temperature of extraction has an effect on the result of the determination (Figure 7.19). At moderate temperatures, there is very little change in the amount of bound rubber determined. At temperatures above 70°C there is substantial increase in the amount of extracted rubber. This data shows that most carbon black is adsorbed by physical forces.

The amount of bound rubber depends on carbon black loading (Figure 7.20).[57] Experimental studies[35,46] show that small additions of carbon black (below 40 phr) obey different relationship than larger additions. The cross-section of both relationships gives a critical coherent loading. This data also shows that bound rubber increases rapidly above 30 phr. Above 80 phr, the bound rubber content begins to level off. NMR studies show a very restricted chain mobility above 80 phr.[57]

Figure 7.21 shows that bound rubber increases as the surface area of carbon black increases.[35,46] This is a classical experiment which shows that the amount of bound rubber depends on the surface area of the filler. High structure carbon blacks

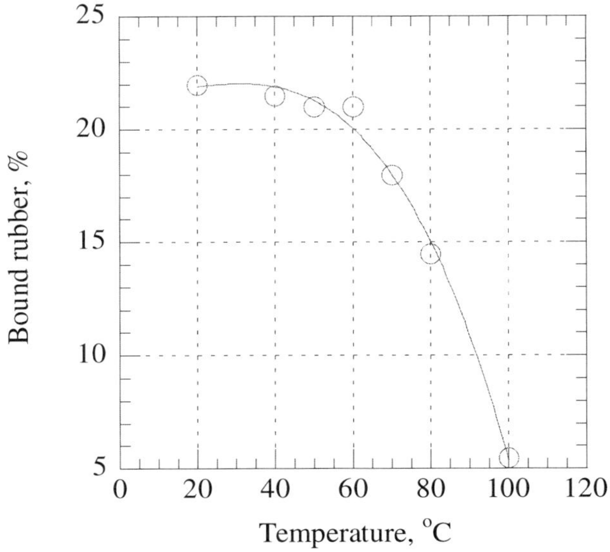


Figure 7.19. Bound rubber as a function of extraction temperature for N330. [Adapted, by permission, from Wolff S, Wang M-J, Tan E-H, *Rubb. Chem. Technol.*, **66**, No.2, 1993, 163-77.]

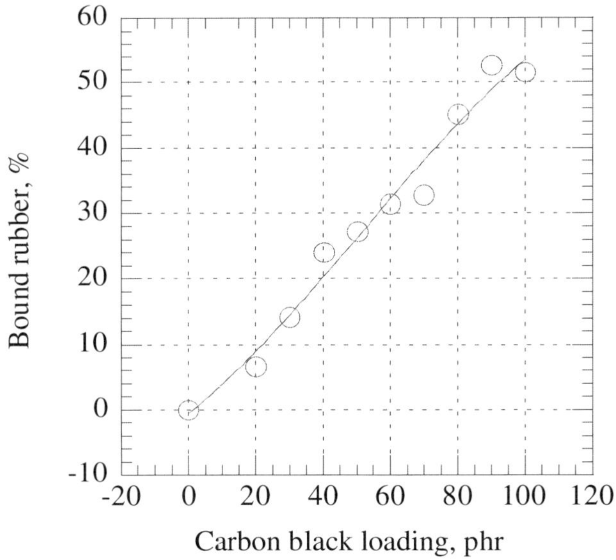


Figure 7.20. Bound SBR vs. carbon black (N110) loading. [Adapted, by permission, from Datta N K, Choudhury N R, Haidar B, Vidal A, Donnet J B, Delmotte L, Chezeau J M, *Polymer*, **35**, No.20, 1994, 4293-9.]

adsorb more rubber than do low structure carbon blacks, having the same surface area, because of the increased probability of multiple adsorption, less graphitization, and a higher tendency to break aggregates during mixing.[89]

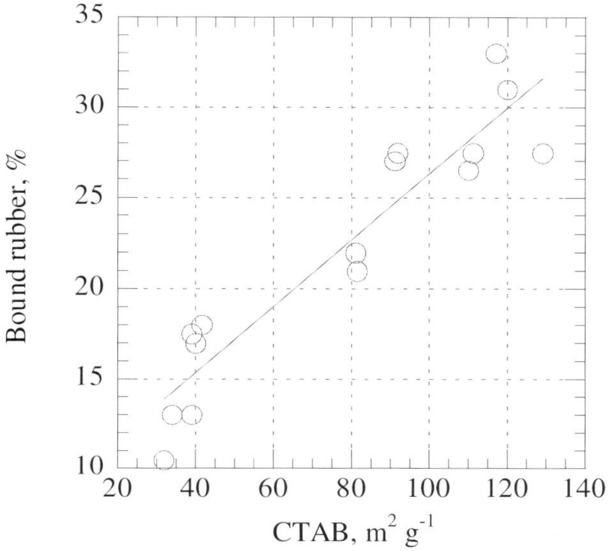


Figure 7.21. Bound rubber vs. CTAB surface area for various carbon blacks at 50 phr loading in SBR. [Adapted, by permission, from Wolff S, Wang M-J, Tan E-H, *Rubb. Chem. Technol.*, **66**, No.2, 1993, 163-77.]

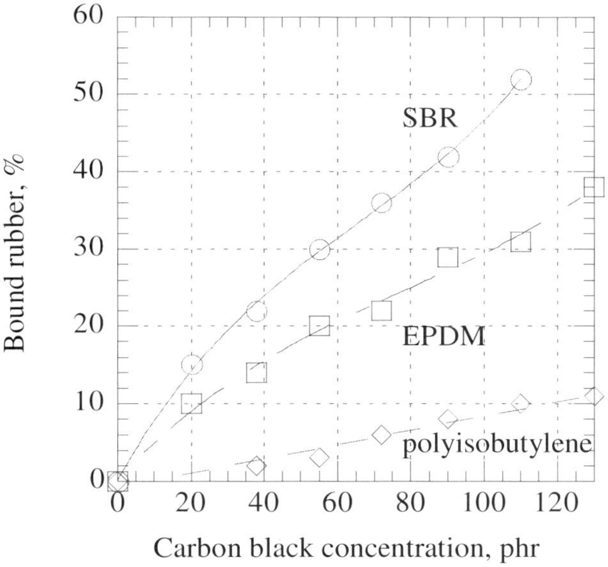


Figure 7.22. Bound rubber in various systems. [Adapted, by permission, from Karasek L, Sumita M, *J. Mat. Sci.*, **31**, No.2, 1996, 281-9.]

The type of rubber also has an influence on the amount of bound rubber (Figure 7.22).[84] It depends on the chemical structure of the rubber, unsaturations, and on the thermal, thermo-mechanical, and oxidative stability of the rubber.

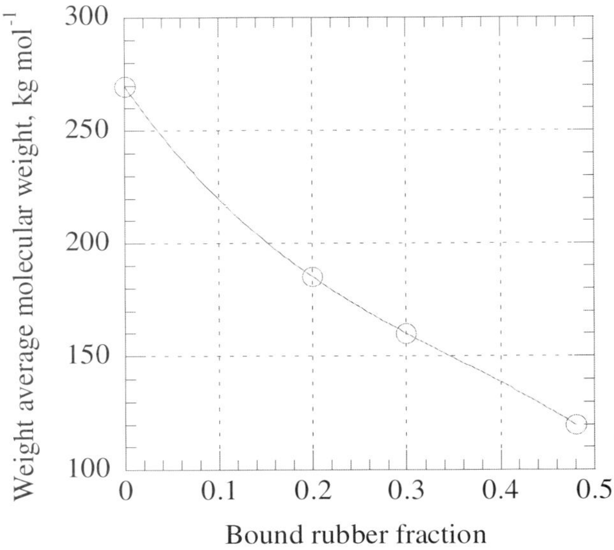


Figure 7.23. Molecular weight of extracted rubber vs. amount of bound rubber. [Adapted, by permission, from Karasek L, Sumita M, *J. Mat. Sci.*, **31**, No.2, 1996, 281-9.]

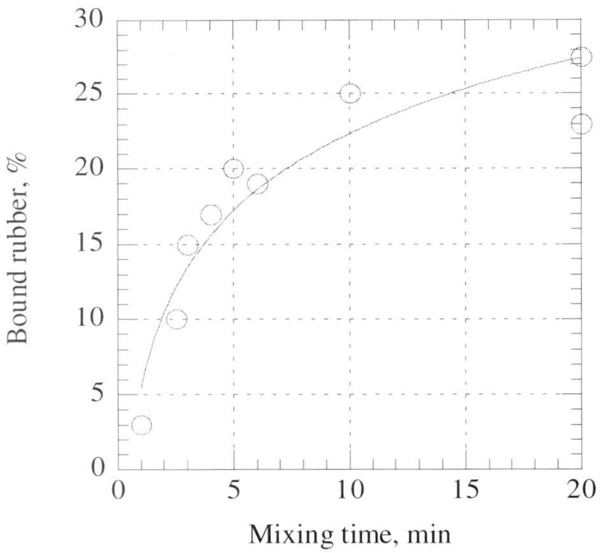


Figure 7.24. Bound rubber formation during mixing. [Adapted, by permission, from Leblanc J L, *Prog. Rubb. Plast. Technol.*, **10**, No.2, 1994, 112-29.]

Longer polymer chains are preferentially absorbed by carbon black. Figure 7.23 shows the molecular weight of extracted rubber vs. the amount of bound rubber. Because of the preferential adsorption of longer chains, the molecular weight of extracted rubber decreases as the amount of bound rubber increases.

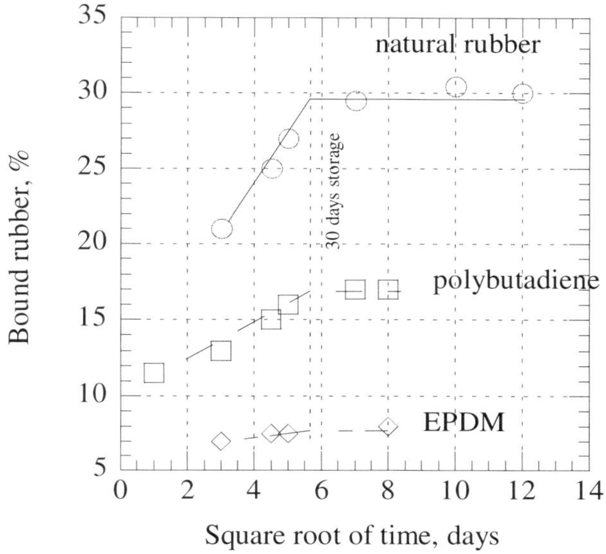


Figure 7.25. Bound rubber vs. storage maturation. [Adapted, by permission, from Leblanc J L, *Prog. Rubb. Plast. Technol.*, **10**, No.2, 1994, 112-29.]

Mixing energy, mixing time, and the processing temperature are the parameters affecting the amount of the bound rubber. Figure 7.24 shows the effect of mixing time.[45] There is a certain effective mixing time which is required to attain an equilibrium state. Extended mixing beyond this point causes much smaller changes in bound rubber. There is also a period at the beginning of mixing during which the wetting and the breakdown of aggregates occur. During this induction period, only a small gain in bound rubber was observed.

Not only mixing increases bound rubber but also the storage time. This is called storage maturation (Figure 7.25).[45] This maturation process is very long because it involves a diffusion which is very slow process with macromolecular materials.

Chemical modification of filler surface reduces the surface area available for interaction. This reduces bound rubber (Figure 7.26).[69,93] The quantity of adsorbing additives on the filler surface must be strictly controlled because these additives compete with the reinforcing effect of the bound rubber. Thermal treatment of rubber increased the quantity of bound rubber but only when rubber was added prior to the addition of low molecular processing additives.[94] This shows that there was competition between the low molecular additive and the rubber for adsorption sites.

When the behavior of carbon black and silica is compared in compounded rubber, it is evident that silica adsorbs less rubber than carbon black. In addition to the differences in the chemical compositions of the surfaces this difference is caused by the differences in the dispersive components of surface energies of each filler. Car-

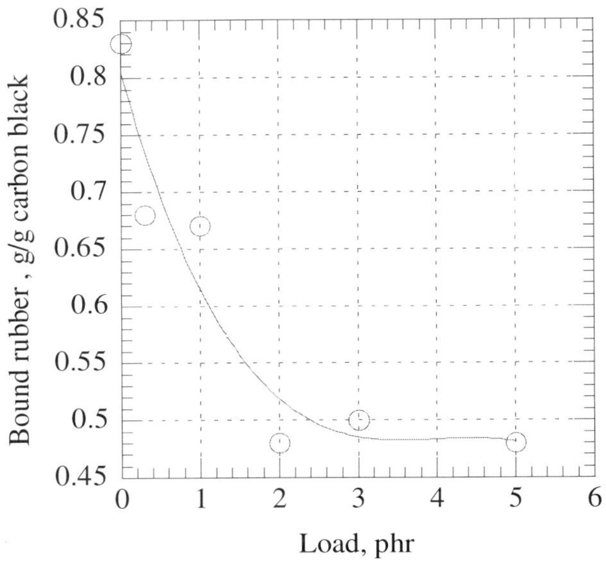


Figure 7.26. Effect of multifunctional additive on bound rubber. [Adapted, by permission, from Ismail H, Freakley P K, Sheng E, *Eur. Polym. J.*, **31**, No.11, 1995, 1049-56.]

bon black has higher dispersive component of surface energy than silica which is the reason for its better dispersion and interaction.[95,96]

The atomic force microscope is used to observe bound rubber on the filler surface.[97] The highest concentration of bound rubber was found in the regions between carbon black particles. A further review of the theory of gel formation can be found in the literature.[98] In the case of some polymers, an uncertainty exists as to whether the determined values are correct because of their solubility (or its lack).[77] Polyethylene is an example. In addition to the complication of solubility, polyethylene modified by maleic anhydride can form covalent bonds with the filler which substantially increases the amount of filler bound polymer.[77] Solvents which interact poorly with silica do not affect the polymer-filler linkages and they give high readings.[99] Also, treatment with ammonia may give a confusing result in the presence of a filler which has been treated previously with low molecular weight substances. Ammonia treatment either removes low molecular substances or reacts with the polymer, which increases the amount of gel formed.[99]

## 7.13 DEBONDING

Debonding (also called dewetting) is one mechanism of the failure of filler reinforced composites which are subjected to either continuous stress or fluctuating stresses. Debonding may also be used as a method of production for some of the materials discussed in Section 7.3.

Eq 7.15 gives a simple description of the stress acting on an isolated particle. In reality, more particles are involved in the dissipation of local stresses in filled materials. The interacting stress fields of neighboring particles modify Eq 7.15:[100,101]

$$\sigma^D = \left( -\frac{\sigma^T}{2} + C\sqrt{\frac{W_{mf}}{R}} \right)(1 + m\phi^{1/3}) \qquad [7.27]$$

where:

$\sigma_D$ debonding stress
$\sigma_T$ thermal stress
C constant
$W_{mf}$ reversible work of adhesion
R radius of inclusion (filler)
m constant
$\phi$ fraction of inclusion (filler)

This equation shows that debonding stress increases with adhesion and filler fraction and decreases with particle size. Figure 7.27 shows the effect of particle size on prediction of yield stress based on the debonding simulated by an equation derived from Eq 7.27. Decreasing particle size increases the stress required for debonding.

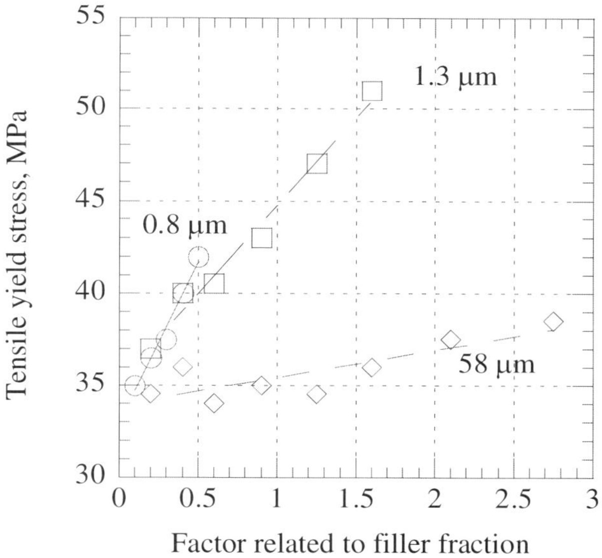


Figure 7.27. Yield stress prediction for different particle sizes. [Adapted, by permission, from Pukanszky B, Voros G, *Polym.Composites*, **17**, No.3, 1996, 384-92.]

Figure 7.28 shows that the tensile strength (reduced to account for the volume fraction of the filler and its interaction) increases with the volume fraction of the filler.[101]

Figure 7.29 shows that coefficient of interaction increases as adhesion increases. Calcium carbonate is treated to increase filler-polymer interaction. Fig-

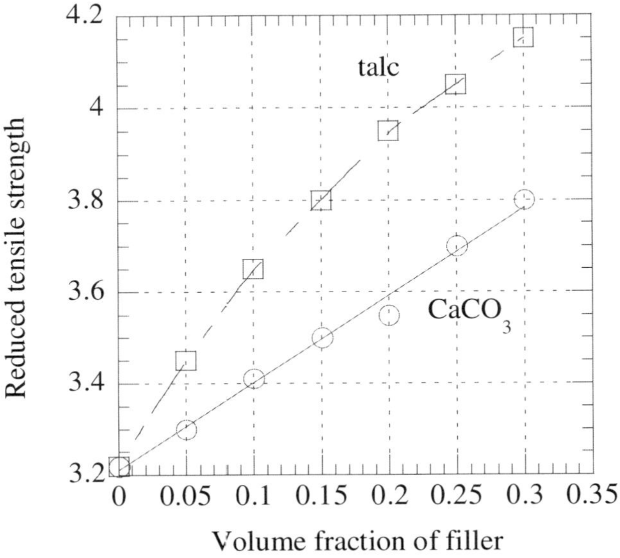


Figure 7.28. Reduced tensile strength vs. volume fraction of filler. [Adapted, by permission, from Pukanszky B, Belina K, Rockenbauer A, Maurer F H J, *Composites*, **25**, No.3, 1994, 205-14.]

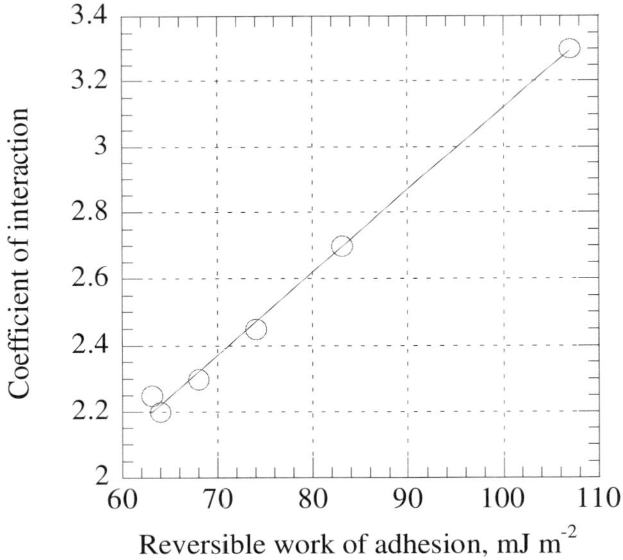


Figure 7.29. Interaction between surface treated $CaCO_3$ and PP vs. work of adhesion. [Adapted, by permission, from Pukanszky B, Belina K, Rockenbauer A, Maurer F H J, *Composites*, **25**, No.3, 1994, 205-14.]

ure 7.29 demonstrates that the effect was achieved. Figures 7.27-7.29 give experimental evidence that Eq 7.27 is generally correct.

The typical stress-strain behavior of filled composites has three stages: elastic, debonding, and crazing or shear yielding. These stages are related to the state of filler-matrix bond.[102-5] Initially all filler particles, having a volume fraction, $\phi$, are bonded to the matrix (bonded filler fraction, $\phi_b = \phi$). Under stress, particles gradually debond and new fraction of debonded particles, $\phi_d$, is formed ($\phi_b = \phi - \phi_d$). The fraction of completely debonded material is now $\phi_d = \phi$ and $\phi_b = 0$. This has been used in a practical way to obtain a permeable membrane from highly filled material after a stretching process.[31] Two important principles can be derived from this analysis of filler fractions. One is the rate of debonding and the other is the volume increase due to the debonding. The rate of debonding is expressed as

$$\frac{d\phi_d}{dt} = (\phi - \phi_d)K\sigma \exp(B\overline{\sigma}) \quad [7.28]$$

where:

| | |
|---|---|
| t | time |
| K | debonding rate constant |
| $\sigma$ | nominal stress |
| $\overline{\sigma}$ | effective stress |
| B | debonding rate constant |

The rate of debonding decreases as the number of debonded particles increases and as the stress increases. The debonding constants characterize the interaction and the influence of neighboring particles. Their values depend on the filler concentration and on the adhesion of the filler to the matrix. The volume increase due to debonding is given by the equation:

$$\varsigma_d = \int_0^\infty \phi_d d\varepsilon \quad [7.29]$$

where:

| | |
|---|---|
| $\varepsilon$ | strain |

The volume increase depends on the filler fraction and on the applied strain. This is confirmed in practice.[31] Debonding correlates with loss of stiffness. The first part of the stress-strain curve (elastic stage) is related to the strains beyond which debonding occurs. In glass bead filled polypropylene, this strain was 0.7%.[106]

In mixtures of particles, the stress of debonding is not uniform. Higher stress is needed to debond from smaller particles.[107] Adhesion is inversely proportional to the cube root of the diameter of the particles.[107] Experiments confirmed that large particle sized filler decreased the tensile strength of composites.[5] The filler concentration effect is not linear. Up to a certain concentration, filler did increase the tensile properties but beyond certain level there is a reverse effect.[108] This may relate to the interactions described in previous sections where the quality of bonding (weak or strong) depended on filler concentration.

A simple equation is derived from the first law of thermodynamics:[109]

$$\delta U = \delta U_{strain} + \delta U_{surface} = \delta W + \delta Q \quad [7.30]$$

where:

U energy

W work

Q heat

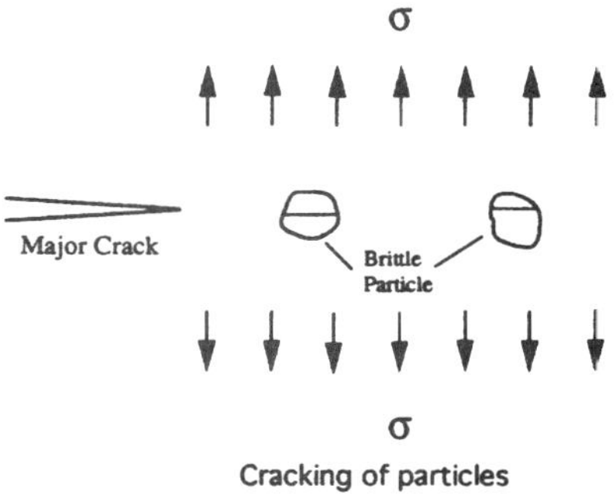


Figure 7.30. Possible crack growth mechanism. [Adapted, by permission, from Xu X X, Crocrombe A D, Smith P A, *Int. J. Fatigue*, **16**, No.7, 1994, 469-77.]

Figure 7.31. Particle splitting. [Adapted, by permission, from Li J X, Silverstein M, Hiltner A, Baer E, *J. Appl. Polym. Sci.*, **52**, No.2, 1994, 255-67.]

This energy balance depends on the energy of the applied strain and the energy of surface. We have consistently assumed that the energy input is lower than the cohesive energy of filler particles. But this is not always true.[110-112]

The mechanical strength of the filler particle may be lower than the adhesive bond strength between the filler and the matrix. This effect is illustrated in Figure 7.30. Concentrated stress causes particle cracking. An SEM micrograph of this event is illustrated in Figure 7.31.

## 7.14 MECHANISMS OF REINFORCEMENT

Einstein developed the concept of hydrodynamic reinforcement which is expressed by the equation:

$$f = \frac{\eta}{\eta_0} = 1 + 2.5\phi \quad [7.31]$$

where:

f hydrodynamic reinforcement factor

$\eta$ viscosity of suspension

$\eta_0$ viscosity of solvent

$\phi$ filler volume fraction

This simple model was later extended by Guth and Gold to include interparticular disturbances. One form of this model is given by Eq 7.6. This model modified by Thomas fits some experimental data:

$$f = 1 + 2.5\phi + 10.05\phi^2 + A\exp(B\phi) \qquad [7.32]$$

where:
A coefficient = 0.00273
B coefficient = 16.6

Figure 7.32 shows that Guth & Gold equation fits data for lower filler volume fractions but Thomas model gives a good prediction of experimental results throughout a very broad range of filler concentrations.[113] This model is fairly universal and it is one of the popular models used for interpretation of experimental data. At the same time, it is clearly visible that the model does not consider most factors, discussed throughout this chapter, which are thought to influence reinforcement of polymers. In one recent review[114] on polymer reinforcement, it is stressed that no consistent model exists (except for the above equations derived from Einstein's concept) which may be used to follow polymer reinforcement. Because of the lack of phenomenological model there are numerous publications which deal with the subject of experimental data by proposing empirical relationships or microscopic models which can explain observed results.[34,35,38,42,59,64,65,115-20] Some findings are discussed below together with much earlier proposal which still remains valid.

One earlier model was developed by Dannenberg to explain observations of behavior of compounded rubber.[121] Figure 7.33 shows how this model works. Polymer chains are connected with filler particles. Depending on strain, chains remain relaxed, are fully extended, slip, or matrix undergoes structural changes. It is im-

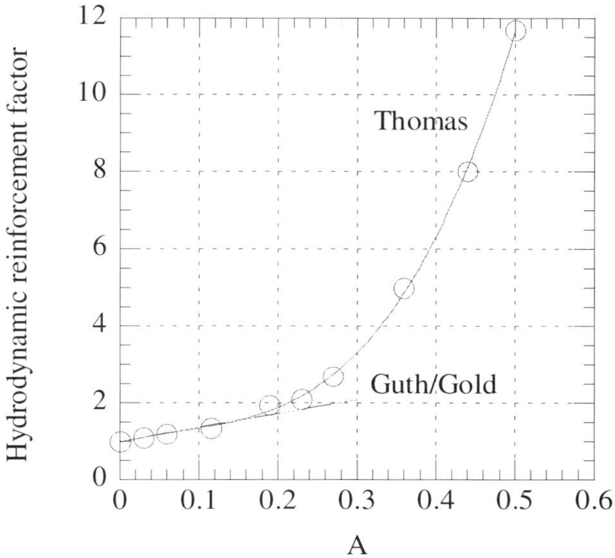


Figure 7.32. Hydrodynamic reinforcement factor vs. filler volume fraction. [Adapted, by permission, from Eggers H, Schummer P, *Rubb. Chem. Technol.*, **69**, No.2, 1996, 253-65.]

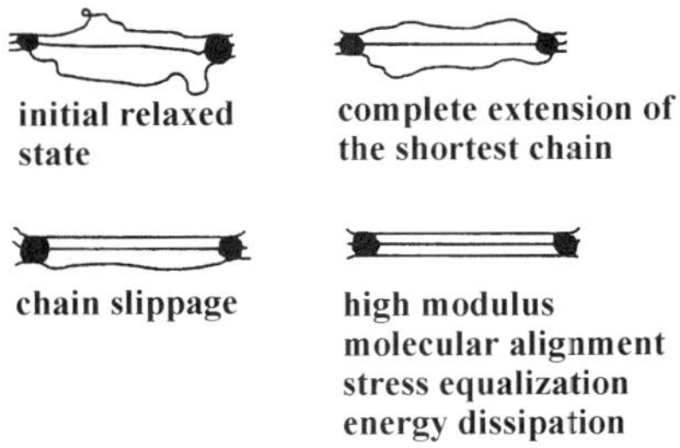


Figure 7.33. Molecular slippage model. [Adapted, by permission, from Dannenberg E M, *Rubber Chem. Technol.*, **48**, 1975, 410.]

portant model which explains why certain stress is fully relaxed and larger stresses cause changes in the material but, at the same time it is only descriptive model – not useful in interpretation of experimental data.

Model previously developed by Kraus[122] has got additional interpretation in recent works.[113] Kraus gave simple equation:

$$\phi_{eff} = \beta\phi \qquad [7.33]$$

where:

$\phi_{eff}$ effective concentration of filler
$\beta$ effectiveness factor
$\phi$ filler volume fraction

On surface it is very simple model but effective concentration of filler includes observation that some layer of polymer is bound to the surface of filler and the mechanisms of this bonding is mathematically expressed by effectiveness factor. The recent model assumes that filler particles are spheres which might be connected to form chain-like agglomerates. Each particle is surface coated with matrix polymer. The elastomeric layer is considered immobilized. The effective filler volume is higher than filler volume fraction by the amount of adsorbed polymer. The effectiveness factors is given by equation:

$$\beta = \frac{V_{sphere} + V_{layer} + n\Delta V}{V_{sphere}} = 1 + 6\frac{h}{d_p} + 12\left(1 - \frac{n}{4}\right)\left(\frac{h}{d_p}\right) + 8\left(1 - \frac{n}{2}\right)\left(\frac{h}{d_p}\right) \qquad [7.34]$$

where:

V volume
n mean number of adjacent particles
$d_p$ mean particle diameter

Figure 7.34 shows that the model fits experimental data for carbon black and silica particles.

Several performance characteristics of rubber such as abrasion resistance, pendulum rebound, Mooney viscosity, modulus, Taber die swell, and rheological properties can be modeled by Eq 7.34.[34] A complex mathematical model, called “links-nodes-blobs” was also developed and experimentally tested to express the properties of a filled rubber network system.[42] Blobs are the filler aggregates, nodes are crosslinks and links are interconnecting chains. The model not only allows for

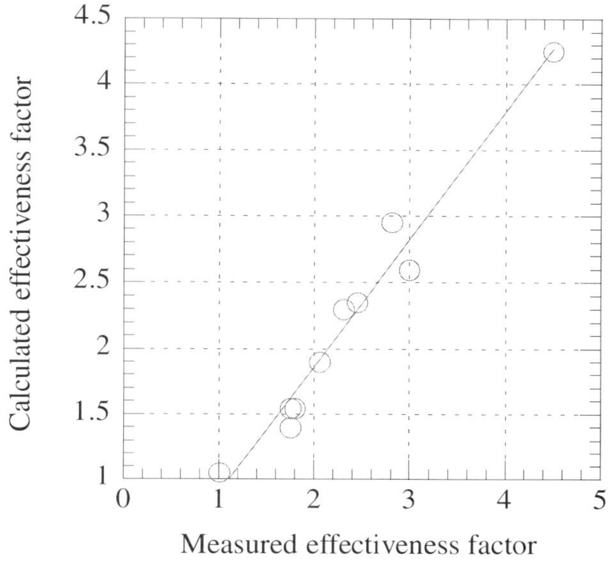


Figure 7.34. Modeling of effectiveness factor. [Adapted, by permission, from Eggers H, Schummer P, *Rubb. Chem. Technol.*, **69**, No.2, 1996, 253-65.]

positional changes but assumes the fracture of links. Ten different rubbers were tested and simulated according to the model with good correlation. The success of this percolation model for inelastic filler network indicates that computerized predictions will soon be able to give much closer approximations of experimental results.

Figure 7.15 shows pictorial elements of another model which has been proposed.[38,65] This model was examined by WAXS analysis. An assumption was made that the composite consists of rubber matrix, filler particle, and boundary layer, which diffract waves without interfering with each other. The total radial density difference function, $\sigma_{total}$, was calculated from the following equation:

$$\sigma_{total} = (1 - v_F - v_B)\sigma_M + kv_F\sigma_F + cv_B\sigma_B \qquad [7.35]$$

where:

$v_F$ volume fraction of filler
$v_B$ volume fraction of boundary layer
$\sigma_M$ radial density difference function of rubber matrix
$\sigma_F$ radial density difference function of filler particle
$\sigma_B$ radial density difference function of boundary layer
k, c normalization constants due to different scattering power of carbon black and adhesion layer

This model characterizes surface contacts, deals with agglomerates, and explains rubber swelling. It was further developed to characterize reinforcement as a non-Gaussian phenomenon.[38] The model deals with intra-cluster forces and the stress-strain cycle. It is used in the experimental part on uniaxial compression to explain observed anisotropy of filler-rubber contacts. This is another example of the progress being made in the fundamental treatment of reinforcement.

Rheological tests can also be used to determine the reinforcing potential of silica.[35] The following equation can be used:

$$\frac{D_{max} - D_{min}}{D^0_{max} - D^0_{min}} - 1 = \alpha_F \frac{m_F}{m_P} \qquad [7.36]$$

where:

| | |
|---|---|
| $D_{max} - D_{min}$ | torque difference of filled system |
| $D^0_{max} - D^0_{min}$ | torque difference of the gum |
| $m_F/m_P$ | filler loading |
| $\alpha_F$ | filler constant characterizing morphology of filler |

Eq 7.36 was used to evaluate the effect of filler type and loading on rebound, modulus, compression. It also permits a comparison with the parameters which characterize morphology.

The distance between aggregates, $\delta_{aa}$, can be obtained from the following equation:[120]

$$\delta_{aa} = \frac{6000}{\rho S}(k\phi^{-1/3}\beta^{-1/3} - 1)\beta^{1.43} \qquad [7.37]$$

where:

| | |
|---|---|
| $\rho$ | density of filler |
| S | specific surface area of filler |
| k | constant based on filler packing |
| $\beta$ | expansion factor (or ratio of effective filler volume fraction to filler volume fraction) |

The distance between aggregates is a value which correlates with many properties of filled rubber. Figure 7.35 gives an example of the correlation with tanδ.[120] Other applications were made with these properties: ball rebound, effect of graphitization on properties of carbon black and parameters of carbon black which characterize structure.

Some results of experimental studies have been interpreted based on the Anderson-Farris model.[123,124] This model is based on assumptions from a modified first law of thermodynamics:[119]

$$\delta Q + \delta W = \delta U + G_c \delta A \qquad [7.38]$$

where:

| | |
|---|---|
| $\delta Q$ | net heat transferred into the system |
| $\delta W$ | net external work done on the system |
| $\delta U$ | net internal energy in the system |
| $G_c\delta A$ | the surface energy dissipated |

This equation is based on a model assuming that the work energy put into the system is either stored as internal strain energy or is used to form a new surface area through debonding. Based on these assumptions, several functional relationships were developed to characterize energy released, uniaxial tensile, change in surface

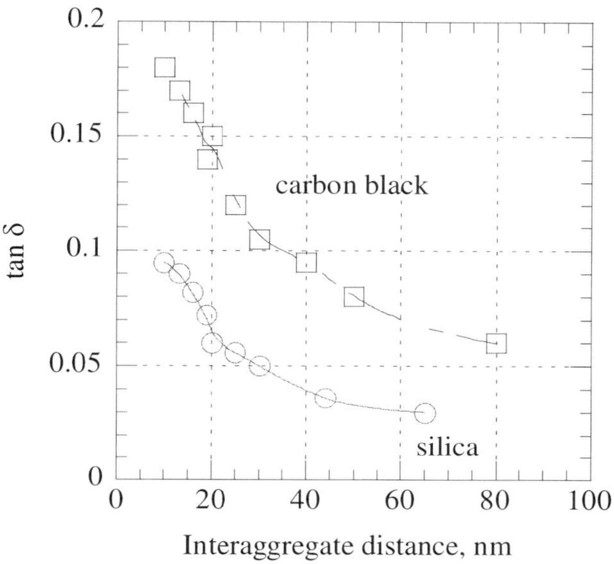


Figure 7.35. tanδ of natural rubber filled with carbon black and silica vs. interaggregate distance, $\delta_{aa}$. [Adapted, by permission, from Wang M-J, Wolff S, Tan E-H, *Rubb. Chem. Technol.*, **66**, No.2, 1993, 178-95.]

area, changes in modulus, and Poisson ratio. From relationships, the model can be applied to predict many properties of filled materials. Ten samples of HDPE containing glass beads were used to verify the model.[119] The results show that the model gave a very good prediction of the stress-strain curve. The model predicts nonlinearity due to the particles debonding.

Several other models were proposed based on a series of studies.[115,116,118] These models address specific cases related to work done on experimental materials. A broad discussion of the mechanisms of reinforcement can be found in the specialized monograph by one of the experts in the field.[125] There has been a concerted effort to analyze materials in many different ways and we have tried to present much of this work in this chapter. Many successful attempts have been made to develop universal relationships which explain the reasons for reinforcement and material behavior under external stresses.

## 7.15 BENEFITS OF ORGANIZATION ON MOLECULAR LEVEL

This section is not intended as a list of all the benefits of interphase formation. They are the subject of this book and the properties of materials are discussed in detail in the individual chapters. It is not appropriate to identify a single property of a material as the most significant. Here are some concluding remarks and examples of other benefits. This will perhaps show that interfacial interactions are not used only for reinforcement.

A recent paper[126] brings several points of interest for this discussion. Examples of two materials (abalone shell and spider web fiber) are examined. These natural materials benefit from the molecular organization to the extent still not conquered by the scientific discoveries. Abalone shell is composed of calcium carbonate and polysaccharides and proteins as binders. The impact resistance of this material is remarkable. Calcium carbonate is not known in our applications as reinforcing material. Natural material differs in structural organization and interaction with the binder from man made materials. Similarly, the strength of fibers produced by spiders is achieved through the morphology of the natural polymer. Again, no man made polymer has been able to duplicate this level of performance. Although these examples show that the current technology has been unable to achieve the remarkable performance of these natural materials, recent developments provide evidence that rapid progress is being made towards better performing materials.

Natural products are highly compatible with other surrounding materials particularly, growing tissues. In the future, materials used for medical applications may have the ability to influence one's body to deposit layers of material which is compatible with the body. By crystallization of filler-like materials in the presence of body fluids, surfaces have been artificially synthesized to be similar to natural materials. It may be possible to induce grafting of a surface through the natural processes occurring in the organism. A goal of such work would be to develop highly specific interfaces which will be recognized by many organisms and ultimately would be specifically compatible with one.

The third essential point of the cited publication[126] also makes us realize that nature uses a very small number of compounds as building blocks (as demonstrated by the widespread presence of silica or calcium carbonate). But the natural design of these structures builds products of very diverse properties. The design differences are generally not chemical but structural. The important lesson for designers is that it is not the number of available monomers but their sequence and structural organization which imparts their unique properties.

Figure 7.36 shows that natural graphite from Siberia can be used to synthesize copolymers with different properties.[127] An increase in the specific surface area results in the formation of copolymers with shorter blocks. By varying the structure of the filler and its concentration, one is able to tailor copolymers to a desired structure.

The amount of carbon black, its particle size and structure, the filler-matrix interaction, and the processing technique determine the electrical properties of a product. At a certain concentration of filler, the conductivity of the material increases dramatically. This concentration is known as the percolation threshold and the conductivity of the material is expressed by equation:

$$\sigma = \sigma_0 (X - X_c)^s \qquad [7.39]$$

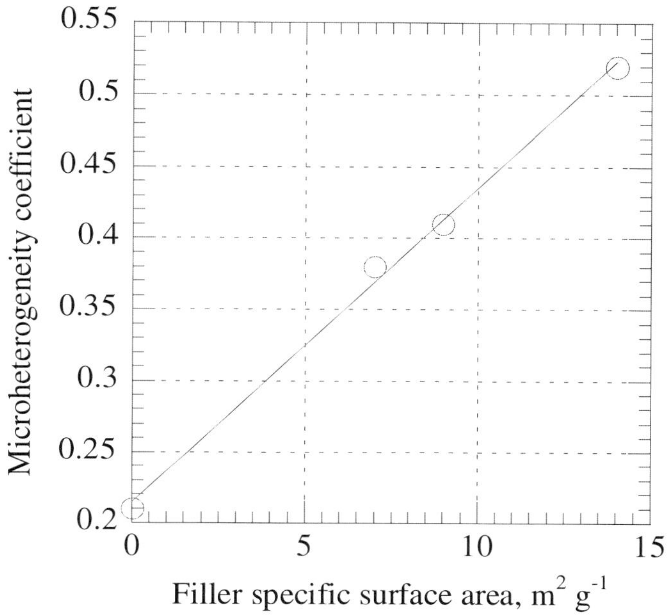


Figure 7.36. Influence of filler's surface on microheterogeneity coefficient of copolymers. [Adapted, by permission, from Vasnev V A, Tarasov A I, Istratov V N, Ignatov V N, Krasnov A P, Kuznetsov A I, Surkova I N, *Reactive & Functional Polym.*, **26**, Nos.1-3, 1995, 177-83.]

where:

| | |
|---|---|
| $\sigma_0$ | conductivity of filler particles |
| X | volume fraction of filler |
| $X_c$ | volume fraction of filler at percolation threshold |
| s | a quantity determining the power of the conductivity increasing above $X_c$ |

Figure 7.37 shows the effect of the percolation threshold on a material's conductivity.[128] In this example the material has s = 7.75 which is a very high value compared with other data found in the literature. The value of s depends on the structure and surface area of the filler used for production of the material. The filler properties determine the interface formation which permit the electron tunneling mechanism to occur.

Figure 3.38 shows that reaction between $Al(OH)_3$ and dicarboxylic acid anhydride affects the sedimentation volume of filler.[129] The limiting value of sedimentation was obtained by modifying the filler surface with a monolayer of a suitable modifier. A similar modification affects the performance of this filler in polymer-filler composites. Thus, different properties were affected by the surface coverage of filler and by the filler-matrix interactions.

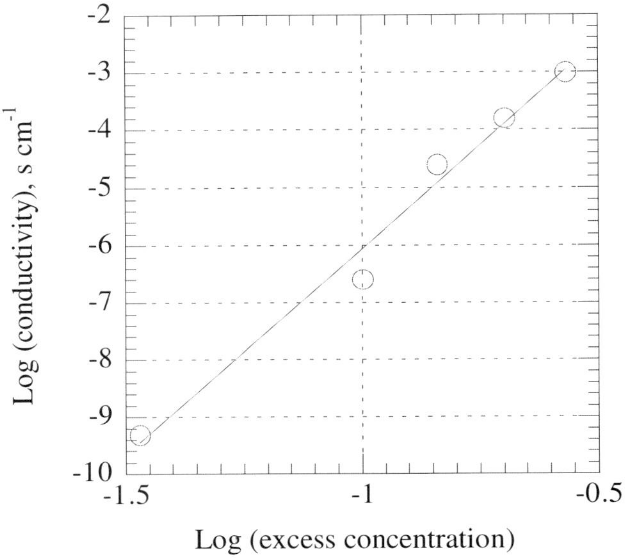


Figure 7.37. Conductivity of SBR-carbon black vs. excess concentration. [Adapted, by permission, from Karasek L, Meissner B, Asai S, Sumita M, *Polym. J. (Jap.)*, **28**, No.2, 1996, 121-6.]

## REFERENCES


1 Shuhong Wang, Mark J E, *Macromol. Reports*, **A31**, Nos.3 & 4, 1994, 253-6.
2 Chen L, Liu K, Yang C Z, *Polym. Bull.*, **37**, No.3, 1996, 377-83.
3 Pu Z, Mark J E, Jethmalani J M, Ford W T, *Polym. Bull.*, **37**, No.4, 1996, 545-51.
4 Rosenov M W K, Bell J A E, Antec '97. Conference proceedings, Toronto, April 1997, 1492-8.
5 Kovacevic V, Lucic S, Hace D, Cerovecki Z, *J. Adhesion Sci. Technol.*, **10**, No.12, 1996, 1273-85.
6 Bormashenko E Y, Zagoskin A M, *Int. Polym. Sci. Technol.*, **23**, No.1, 1996, T/80-1.
7 Kano Y, Akiyama S, Yui H, Kasemura T, *Polym. Networks Blends*, **7**, 117-20 (1997).
8 Tchoudakov R, Breuer O, Narkis M, Siegmann A, *Polym. Networks Blends*, **6**, 1-8 (1996).
9 de Candia F, Carotenuto M, Gargani L, Guadagno L, Lauretti E, Renzulli A, *Kaut. u. Gummi Kunst.*, **49**, No.2, 1996, 99-101.
10 Guerbe L, Freakley P K, *Kaut. u. Gummi Kunst.*, **48**, No.4, 1995, 260-9.
11 Papathanasiou T D, *Int. Polym. Processing*, **11**, No.3, Sept.1996, 275-83.
12 Ogadhoh S O, Papathanasiou T D, *Composites Part A: Applied Science and Manufacturing*, **27A**, No.1, 1996, 57-63.
13 Chang Ho Suh, White J L, *Polym. Engng. Sci.*, **36**, No.11, 1996, 1521-30.
14 Klapcinski T, Galeski A, Kryszewski M, *J. Appl. Polym. Sci.*, **58**, No.6, 1995, 1007-13.
15 Wagner A H, Kalyon D M, Yazici R, Fiske T J, Antec '97. Conference proceedings, Toronto, April 1997, 996-1000.
16 Kim K J, White J L, *J. of Non-Newtonian Fluid Mechanics*, **66**, Nos.2/3, 1996, 257-70.
17 Fiske T, Gokturk H S, Yazici R, Kalyon D M, Antec '97. Conference proceedings, Toronto, April 1997, 1482-6.
18 Dreibelbis G L, Antec 95. Volume III. Conference proceedings, Boston, Ma., 7th-11th May 1995, 4374-6.
19 Allan P S, Bevis M J, *Materials World*, **2**, No.1, 1994, 7-9.
20 Barbosa S E, Kenny J M, Ercoli D R, Bibbo M A, Antec '97. Conference proceedings, Toronto, April 1997, 2189-93.
21 Lee S C, Yang D Y, Youn J R, Antec '97. Conference proceedings, Toronto, April 1997, 642-8.
22 Gerard P, Raine J, Pabiot J, Antec '97. Conference proceedings, Toronto, April 1997, 526-31.
23 Suh C H, White J L, Antec '96. Vol.I. Conference Proceedings, Indianapolis, 5th--10th May 1996,

p.958-62.
24 Wagner A H, Kalyon D M, Tan V, Antec '96. Volume II. Conference proceedings, Indianapolis, 5th-10th May 1996, 2195-9.
25 Barbosa S E, Kenny J M, Antec '97. Conference proceedings, Toronto, April 1997, 1855-9.
26 Suh C H, White J L, *Polym. Engng. Sci.*, **36**, No.17, 1996 2188-97.
27 Rabello M S, White J R, *Polym. Composites*, **17**, No.5, 1996, 691-704.
28 Wada N, Uchiyama Y, Hosokawa M, *Int. Polym. Sci. Technol.*, **21**, No.3, 1994, T/53-63.
29 Wu Y, Shivpuri R, Lee L J, Antec '97. Conference proceedings, Toronto, April 1997, 1849-54.
30 Merwin L H, Nissan R A, Stephens T S, Wallner A S, *J. Appl. Polym. Sci.*, **62**, No.2, 1996, 341-8.
31 Nakamura S, Kaneko S, Mizutani Y, *J. Appl. Polym. Sci.*, **49**, No.1, 1993, 143-50.
32 Mitsui S, Kihara H, Yoshimi S, Okamoto Y, *Polym. Engng. Sci.*, **36**, No.17, 1996, 2241-6.
33 Bagheri R, Pearson R A, *Polymer*, **36**, No.25, 1995, 4883-5.
34 Patel A C, *Kaut. u. Gummi Kunst.*, **47**, No.8, 1994, 556-70.
35 Wolff S, *Rubb.Chem.Technol.*, **69**, No.3, 1996, 325-46.
36 Li Q, Manas-Zloczower I, Feke D L, *Rubb. Chem. Technol.*, **69**, No.1, 1996, 8-14.
37 Tsukuda R, Sumimoto S, Ozawa T, *J. Appl. Polym. Sci.*, **59**, No.6, 1996, 1043-6.
38 Kilian H G, Strauss M, Hamm W, *Rubb. Chem. Technol.*, **67**, No.1, 1994, 1-16.
39 Tsagaropoulos G, Eisenberg A, *Macromolecules*, **28**, No.18, 1995, 6067-77.
40 Mandal U K, Tripathy D K, De S K, *Plast. Rubb. Comp. Process. Appln.*, **24**, No.1, 1995, 19-25.
41 Bomal Y, Godard P, *Polym. Engng. Sci.*, **36**, No.2, 1996, 237-43.
42 Lin C R, Lee Y D, *Macromol. Theory & Simulations*, **5**, No.6, 1996, 1075-104.
43 Wu J, Lerner M M, *Chem. of Mat.*, **5**, No.6, 1993, 835-8.
44 Tsagaropoulos G, Eisenberg A, *Macromolecules*, **28**, No.1, 1995, 396-8.
45 Leblanc J L, *Prog. Rubb. Plast. Technol.*, **10**, No.2, 1994, 112-29.
46 Wolff S, Wang M-J, Tan E-H, *Rubb. Chem. Technol.*, **66**, No.2, 1993, 163-77.
47 Reis M J, Do Rego A M B, Da Silva J D L, *J. Mat. Sci.*, **30**, No.1, 1995, 118-26.
48 Datta S, De S K, Kontos E G, Wefer J M, Wagner P, Vidal A, *Polymer*, **37**, No.15, 1996, 3431-5.
49 Bandyopadhyay S, De P P, Tripathy D K, De S K, *Kaut. u. Gummi Kunst.*, **49**, No.2, 1996, 115-9.
50 Ayala J A, Hess W M, Joyce G A, Kistler F D, *Rubb. Chem. Technol.*, **66**, 1993, 772.
51 Bandyopadhyay S, De P P, Tripathy D K, De S K, *J. Appl. Polym. Sci.*, **61**, No.10, 1996, 1813-20.
52 Datta S, Bhattacharya A K, De S K, Kontos E G, Wefer J M, *Polymer*, **37**, No.12, 1996, 2581-5.
53 Kurian T, De P P, Khastgir D, Tripathy D K, De S K, Peiffer D G, *Polymer*, **36**, No.20, 1995, 3875-84.
54 Sain M M, Kokta B V, Antec '93. Conference Proceedings, New Orleans, La., 9th-13th May 1993, Vol. I, 320-4.
55 Tang L-G, Kardos J L, *Polym. Composites*, **18**, No.1, 1997, 100-13.
56 Yoshinaga K, Nakanishi K, Hidaka Y, Karakawa H, *Composite Interfaces*, **3**, No.3, 1995, 231-41.
57 Tsutsumi K, Ban K, Shibata K, Okazaki S, Kogoma M, *J. Adhesion*, **57**, 1996, 45-53.
58 Hartley P A, Parfitt G P, Pollack L B, *Powder Technol.*, **42**, 1985, 35.
59 Kurian T, Khatgir D, De P P, Tripathy D K, De S K, Peiffer D G, *Polymer*, **37**, No.25, 1996, 5597-605.
60 Herzog E, Caseri W, Suter U W, *Coll. Polym. Sci.*, **272**, No.8, 1994, 986-90.
61 Cochrane H, Lin C S, *Rubb. Chem. Technol.*, **66**, No.1, 1993, 48-60.
62 Gorl U, Rausch R, Esch H, Kuhlmann R, *Int. Polym. Sci. Technol.*, **23**, No.7, 1996, T/81-7.
63 Donnet J B, *Kaut. u. Gummi Kunst.*, **47**, No.9, 1994, 628-32.
64 Donnet J B, Wang T K, *Prog. Rubb. Plast. Technol.*, **11**, No.4, 1995, 261-7.
65 Strauss M, Pieper T, Peng W, Kilian H G, *Makromol. Chem., Macromol. Symp.*, **76**, 1993, 131-6.
66 Roche A A, Dole P, Bouzziri M, *J. Adhesion Sci. Technol.*, **8**, No.6, 1994, 587-609.
67 El-Shall M S, Slack W, *Macromolecules,* **28**, No.24, 1995, 8456-8.
68 Herd C R, Bomo F, *Kaut. u. Gummi Kunst.*, **48**, No.9, 1995, 588-99.
69 Ismail H, Freakley P K, Sheng E, *Eur. Polym. J.*, **31**, No.11, 1995, 1049-56.
70 Hedgus C R, Kamel I L, *J. Coatings Technol.*, **65**, No.821, June 1993, 49-61.
71 Mi Y, Chen X, Guo Q, *J. Appl. Polym. Sci.*, **64**, 1997, 1267-73.
72 Dufresne A, Lacabanne C, *Polymer*, **34**, No. 15, 1993, 3173-8.
73 Wool R P, Long J M, *Macromolecules*, **26**, No.19, 1993, 5227-39.
74 Gerard J F, Chabert B, *Macromol. Symp.*, **108**, 1996, 137-46.
75 Carre A, *J. Adhesion*, **54**, Nos.1-4, 1995, 167-74.
76 Heinrich G, Vilgis T A, *Rubb. Chem. Technol.*, **68**, No.1, 1995, 26-36.
77 Hindryckx F, Dubois P, Patin M, Jerome R, Teyssie P, Garcia Marti M, *J. Appl. Polym. Sci.*, **56**, No.9, 1995, 1093-105.

78 Ou Y, Yu Z, Zhu J, Li G, Zhu S, *Chinese J. Polym. Sci.*, **14**, No.2, 1996, 172-82.
79 Pak S H, Caze C, *J. Appl. Polym. Sci.*, **65**, 1997, 143-53.
80 Suetsugu K, Sakairi T, *Kobunshi Ronbunshu*, **44**, 1987, 369.
81 Ziegel K D, J. *Colloid Interface Sci.*, **29**, 1969, 72.
82 Shenoy A V, Saini D R, *Polym. Compos.*, **7**, 1986, 96.
83 Ou Y C, Yu Z Z, Vidal A, Donnet J B, *J. Appl. Polym. Sci.*, **59**, No.8, 1996, 1321-8.
84 Karasek L, Sumita M, *J. Mat. Sci.*, **31**, No.2, 1996, 281-9.
85 Garvey M J, Tadros T F, Vincent B, *J. Colloid Inter. Sci.*, **49**, 1974, 57.
86 Pliskin I, Tokita N, *J. Appl. Polym. Sci.*, **16**, 1972, 473.
87 Boluk M Y, Schreiber H P, *Polym. Compos.*, **7**, 1986, 295.
88 Maier P G, Goeritz D, *Kaut. u. Gummi Kunst.*, **49**, No.1, 1996, 18-21.
89 Meissner B, *Rubb. Chem. Technol.*, **68**, No.2, 1995, 297-310.
90 Cohen Addad J P, Euradh '94. Conference Proceedings, Mulhouse, 12th-15th Sept.1994, 25-30.
91 Mark J E, *Macromol. Symp.*, **101**, 1996, 423-33.
92 Twiss D F, *J. Chem. Soc.*, **44**, 1925, 1067.
93 Ismail H, Freakley P K, Sutherland I, Sheng E, *Eur. Polym. J.*, **31**, No.11, 1995, 1109-17.
94 Donnet J B, Wang W, Vidal A, Wang M J, *Kaut. u. Gummi Kunst.*, **46**, No.11, Nov.1993, 866-71.
95 Karasek L, *Int. Polym. Sci. Technol.*, **21**, No.10, 1994, T/35-40.
96 Wolff S, Wang M J, Tan E H, *Kaut. u. Gummi Kunst.*, **47**, No.12, 1994, 873-84.
97 Niedermeier W, Raab H, Maier P, Kreitmeier S, Goeritz D, *Kaut. u. Gummi Kunst.*, **48**, No.9, 1995, 611-6.
98 Karasek L, Meissner B, *J. Appl. Polym. Sci.*, **52**, No.13, 1994, 1925-31.
99 Roychoudhury A, De P P, Roychoudhury N, Vidal A, *Rubb. Chem. Technol.*, **68**, No.5, 1995, 815-23.
100 Pukanszky B, Voros G, *Polym.Composites*, **17**, No.3, 1996, 384-92.
101 Pukanszky B, Belina K, Rockenbauer A, Maurer F H J, *Composites*, **25**, No.3, 1994, 205-14.
102 Meddad A, Fisa B, *J. Mater. Sci.*, **32**, 1997, 1177-85.
103 Meddad A, Fisa B, *J. Appl. Polym. Sci.*, **65**, 1997, 2013-24.
104 Meddad A, Fellahi S, Pinard M, Fisa B, Antec '94. Conference Proceedings, San Francisco, Ca., 1st-5th May 1994, Vol. II, 2284-8.
105 Meddad A, Fisa B, *Macromol. Symp.*, **108**, 1996, 173-82.
106 Sjogren B A, Berglund L A, *Polym. Composites*, **18**, No.1, 1997, 1-8.
107 Babich V F, Lipatov Yu S, Todosijchuk T T, *J. Adhesion*, **55**, Nos.3-4, 1996, 317-27.
108 Dubnikova I L, Gorokhova E V, Gorenberg A Y, Topolkaraev V A, *Polym. Sci., Ser. A*, **37**, No.9, 1995, 951-8.
109 Vratsanos L A, Farris R J, *Polym. Engng. Sci.*, **33**, No.22, 1993, 1458-65.
110 Xu X X, Crocrombe A D, Smith P A, *Int. J. Fatigue*, **16**, No.7, 1994, 469-77.
111 Xu X X, Crocombe A D, Smith P A, *Int. J. Fatigue*, **17**, No.4, 1995, 279-86.
112 Li J X, Silverstein M, Hiltner A, Baer E, *J. Appl. Polym. Sci.*, **52**, No.2, 1994, 255-67.
113 Eggers H, Schummer P, *Rubb. Chem. Technol.*, **69**, No.2, 1996, 253-65.
114 Vilgis T A, Heinrich G, *Macromolecules*, **27**, No.26, 1994, 7846-54.
115 Jancar J, *Macromol. Symp.*, **108**, 1996, 163-72.
116 Wang Z, *J. Appl. Polym. Sci.*, **60**, No.12, 1996, 2239-43.
117 Alpern V, Shutov F, *Prog. Rubb. Plast. Technol.*, **11**, No.4, 1995, 268-83.
118 Mele P, Alberola N D, *Composites Sci. & Technol.*, **56**, No.7, 1996, 849-53.
119 Wong F C, Ait-Kadi A, *J. Appl. Polym. Sci.*, **55**, No.2, 1995, 263-78.
120 Wang M-J, Wolff S, Tan E-H, *Rubb. Chem. Technol.*, **66**, No.2, 1993, 178-95.
121 Dannenberg E M, *Rubber Chem. Technol.*, **48**, 1975, 410.
122 Kraus G, *Rubber Chem. Technol.*, **44**, 1971, 199.
123 Anderson L L, Farris R J, *Polym. Eng. Sci.*, **28**, 1988, 522.
124 Anderson L L, PhD Thesis. University of Massachussetts, 1989.
125 Lipatov Y S, **Polymer Reinforcement**, *ChemTec Publishing*, Toronto, 1995.
126 Carraher C E, *Polym. News*, **19**, No.2, 1994, 50-2.
127 Vasnev V A, Tarasov A I, Istratov V N, Ignatov V N, Krasnov A P, Kuznetsov A I, Surkova I N, *Reactive & Functional Polym.*, **26**, Nos.1-3, 1995, 177-83.
128 Karasek L, Meissner B, Asai S, Sumita M, *Polym. J. (Jap.)*, **28**, No.2, 1996, 121-6.
129 Liauw C M, Hurst S J, Lees G C, Rothon R N, Dobson D C, *Prog. Rubb. Plast. Technol.*, **11**, No.2, 1995, 137-53.

# 8

# The Effect of Fillers on the Mechanical Properties of Filled Materials

## 8.1 TENSILE STRENGTH AND ELONGATION

Tensile strength testing is by far the most popular method of evaluating of filled materials. This can be seen from the numerous publications which analyze the subject.[1-56] The information in this section is organized to provide the following information:

- Generalized models describing tensile properties of filled materials
- The effects of different fillers on tensile properties
- Methods of improving of tensile properties

A general equation describes the effect of the volume fraction of a filler on tensile strength:

$$\sigma_c = \sigma_p(1 - a\phi_f^b + c\phi_f^d) \qquad [8.1]$$

where:

$\sigma_c$ tensile strength of composite
$\sigma_b$ tensile strength of polymer matrix
$\phi_f$ volume fraction of filler
a, b, c, d constants

Without knowing the values of these coefficients, it is not possible to predict if tensile strength of the composite increases or decreases as the volume fraction of the filler increases. It is also obvious from the form of the equation that constants can be selected to describe certain features of the filler's behavior. For example, constant "a" is usually related to stress concentration. In composites, in which the filler has very poor adhesion, a = 1.21 or a = 1.23 for non-spherical particles.[1] The constant "b" is usually assigned the arbitrary value of 0.67. Constants "c" and "d" relate to the effect of particle size. The smaller the particle size, the larger are the values of these constants. When the values of these four constants are known or approximated, it makes it possible to predict the tensile strength of various composites. Since the last term in Eq 8.1 is positive, a decrease in the particle size of the filler should result in an increase in tensile strength. Many modifications of the above equation or its parameters (constants) are used to explain experimental data.

For low concentrations of filler, the Einstein equation usually fits experimental data:

$$\sigma_c = \sigma_p (1 + a\phi_\phi^b) \quad [8.2]$$

In the Einstein equation, b = 1 for spherical particles at low concentration and "a" depends on the adhesion between the matrix and the filler. This equation predicts that the addition of filler increases tensile strength which was found to be not always the case, so this equation has been modified by various researchers. The Nicolais and Narkis equation[57] is a common modification in which a=1.21 and b=2/3.[3,4,8,11]

A modified Nielsen model[58] is another frequently used equation,[1,3,9,10] especially in the form proposed by Nicolais and Narkis:[57]

$$\sigma_c = \sigma_p \frac{(1-\phi_f)}{1+2.5\phi_f} \exp(B\phi_f) \quad [8.3]$$

In this equation "B" is a parameter characterizing the interaction.

Some other equations are also in use. One is the Piggott and Leinder equation:[59]

$$\sigma_c = \lambda\sigma_p - \chi\phi_f \quad [8.4]$$

where:

$\lambda$ stress concentration factor

$\chi$ constant dependent on particle-matrix adhesion

which correlates well with experimental measurements made on polymer composites.

Neither of the above equations considers the filler particle as the potential week point in the composite. Instead, the above equations assume that either the matrix fails or loss of adhesion between the filler and the matrix is responsible for failure. The equation below gives the balance of stress in a composite:

$$\phi_f k\sigma^e + (1-\phi_f) < \sigma^m = \sigma^e \quad [8.5]$$

where:

k proportionality constant for stress transfer

$\sigma^e$ external load

$<\sigma^m>$ average stress in the matrix

$\phi_f k\sigma^e$ load carried by the filler

Properties of filler can be compared with the stress applied to the filler particle.[5]

In fiber-filled composites, the Kelly and Tyson equation[60] can be used to estimate the effect of properties of fiber on the load bearing properties of a composite:

$$\sigma_c = \eta_o \frac{\sigma_f L_f}{2L_c} \phi_f + (1 - \phi_f)\sigma_p \qquad [8.6]$$

where:

| | |
|---|---|
| $\eta_o$ | fiber orientation efficiency factor |
| $\sigma_f$ | tensile strength of fiber |
| $L_f$ | mean fiber length |
| $L_c$ | critical fiber length |

In this equation, the mechanical properties, length, and orientation of the fiber are accounted for. In fiber-filled composites, mechanical properties depend also on fiber-fiber proximity:

$$N = A\phi_f L_f / d \qquad [8.7]$$

where:

| | |
|---|---|
| N | the average number of virtual touches per fiber |
| A | coefficient ($=8/\pi^2$ for random in-plane orientation) |
| d | fiber diameter |

The results of tensile testing are frequently presented in the form of stress-strain curves or are related to the tensile modulus as given by equation:

$$E = \frac{\sigma}{\varepsilon} = \frac{F / A}{(l_1 - l_o) / l_o} \qquad [8.8]$$

where:

| | |
|---|---|
| $\sigma$ | tensile stress |
| $\varepsilon$ | tensile strain |
| F | tensile force |
| A | original cross-sectional area |
| $l_o$ | original length |
| $l_1$ | final length |

The results of experimental studies summarized in the Table 8.1 show the potential effect of different fillers on tensile properties of filled materials. The first column gives a list of pairs of polymer and filler for which data on the tensile properties are available in the literature. For each pair, the actual concentration of filler used in the system is given in column 2. Either the specific concentrations are given (e.g., 10 & 20) or the concentration range (e.g., 5–50) if more than two concentrations of filler were tested. The concentration is given in weight percent unless otherwise specified. For the concentration of filler given in the second column, the respective changes of the tensile strength are given in the third column. The values in the third column are percentage of increase (plus sign) or decrease (minus sign) of the tensile strength of the filled material relative to the unfilled polymer. In the last column, short comments are given either to indicate what might have caused the observed changes (e.g., interaction, particle size, modification, etc.) or to give data on relative change of elongation of these samples.

**Table 8.1. Effect of fillers on tensile properties of filled materials**

| Filler/polymer | Conc. range, wt% | Tensile Strength (+) decrease (-), % | Refs. | Comments |
|---|---|---|---|---|
| PARTICULATE, INORGANIC FILLERS | | | | |
| *Alumino-silicate* | | | | |
| PVAc | 10.5% | 0−+35 | 45 | decreases with interaction increasing |
| *Aluminum hydroxide* | | | | |
| chloroprene | 10 & 20 | +11 & +13 | 61 | |
| epichlorohydrin | 10 & 20 | -37 & -41 | 61 | |
| epoxy | 5−50 | -15−-36 | 17 | elongation change: -9−-82 |
| *Antimony trioxide* | | | | |
| EEA | 4 & 8 | no effect | 62 | |
| EVA | 4 & 8 | no effect | 62 | |
| PE | 4 & 8 | no effect | 62 | |
| *Barium ferrite* | | | | |
| natural rubber | 5 & 10 | -8 & -1 | 40 | |
| *Calcite* | | | | |
| PVAc | 10.5% | +2−+13 | 45 | depending on particle size |
| *Calcium carbonate* | | | | |
| PE | 2−25 vol% | +50−+10 | 10 | phosphate modified |
| PE | 2−10 vol% | -5−-50 | 10 | not modified |
| PVAc | 5−20 | +70−+75 | 1 | size 3.6 μm, elongation decreases |
| PVAc | 5−20 | +50−+58 | 1 | size 5.2 μm, elongation decreases |
| PVAc | 5−20 | +40−0 | 1 | size 16.8 μm, elongation decreases |
| PVAc | 5−20 | +55−+72 | 1 | size 3.6 μm, stearic acid coated |
| PP | 10−40 | -5−-21 | 28 | size 18 μm, elong. const for 10-20% |
| PP | 5−30 vol% | -30−-45 | 38 | compression molding, no orientation |
| PP | 5−30 vol% | 0−+20 | 38 | injection molding, particles oriented |
| PP | 5−30 vol% | -30−-40 | 53 | |
| *Clay* | | | | |
| EPDM | 10−35 | -1−-10 | 34 | |
| *Copper* | | | | |
| PA11 | 5−55 | -10−+7 | 26 | |
| *Hydroxyapatite* | | | | |
| polyurethane | 57 | -45/+25 | 21 | untreated/treated with isocyanate |
| *Glass beads* | | | | |
| epoxy | 10−40 vol% | -25−-60 | 5 | no adhesion |
| epoxy | 20−40 vol% | no effect | 5 | good adhesion |
| PA | 5−40 vol% | -15−+22 | 8 | increase only at 40 vol% |
| POM | 10−30 | -15−-40 | 6 | |
| POM | 24 vol% | -43−-47 | 4 | particle size in range 7÷36 μm |
| PP | 10−50 vol% | -11−-46 | 8 | debonding stress; no treatment |
| PS | 5−25 vol% | -5−-15 | 5 | poor adhesion |
| PS | 3−10 vol% | +5−+15 | 5 | good adhesion |
| *Magnesium carbonate* | | | | |
| LCP | 20−60 | +7−+30 | 29 | elongation rapidly reduced |
| *Magnesium hydroxide* | | | | |
| PEK | 65 | -15 | 63 | |
| PP | 10−50 | -1−-23 | 64 | |
| *Mica* | | | | |
| PA66 | 20−40 | -13−+27 | 50 | increases as particle size decreases |
| PBT | 15−40 | +12−+65 | 50 | increases with concentration |
| PP | 5−22 vol% | +14−-7 | 9 | no surface treatment, elongation decr. |
| PP | 5−22 vol% | +18−+14 | 9 | 8 wt% acrylic acid treatment |
| *Miconite* | | | | |
| PP | 10−60 | +100−+150 | 43 | hydrated K-Mg aluminosilicate (3 μm) |
| *Nanoparticles* | | | | |
| epoxy | 2−24 | +60−+1800 | 56 | montmorillonite; layered composite |

| Filler/polymer | Conc. range, wt% | Tensile Strength (+) decrease (-), % | Refs. | Comments |
|---|---|---|---|---|
| *Silica, fumed* | | | | |
| PDMS | 30−50 | +5−+40 | 65 | increases as particle size decreases |
| *Silica, precipitated* | | | | |
| EPM | 50 | +500−+700 | 37 | depending on surface treatment |
| *Talc* | | | | |
| PE | 2−10 | +15−+80 | 19,25 | |
| PP | 40 | +25−+44 | 33 | depending on phosphate coating |
| PP | 5−30 vol% | -20−-25 | 38 | compression molding, no orientation |
| PP | 5−30 vol % | 0−+80 | 38 | injection molding, filler oriented |
| PP | 5−30 vol% | -25−-36 | 53 | |
| *Wollastonite* | | | | |
| LCP | 20−60 | +5−+15 | 29 | elongation rapidly reduced |
| PA66 | 15−35 | -19−-25 | 13 | used in combination with glass fiber |
| FIBROUS FILLERS | | | | |
| *Aramid fiber* | | | | |
| fluoroelastomer | 10 | +260 | 66 | |
| *Carbon fiber* | | | | |
| PP | 60 | +4−+23 | 31 | depending on surface treatment |
| *Glass fiber* | | | | |
| ABS | 30 | +40 | 12 | |
| LCP | 20÷60 | +15−+40 | 29 | elongation rapidly reduced |
| PA6 | 50 | +100 | 12 | |
| PA66 | 30 | +100 | 12 | |
| PAI | 30 | +54 | 12 | |
| PBT | 30 | +75 | 12 | |
| PE | 30 | +60−+185 | 12 | |
| PEK | 2−7 | +105 | 23 | |
| PEK | 10−22 vol% | +50−+90 | 7 | long glass fiber |
| PEEK | 30 | +75 | 12 | |
| PES | 30 | +55 | 12 | |
| POM | 10−30 | +25−+75 | 6 | |
| PP | 2−7 | +30−+100 | 7 | long glass fiber |
| PP | 30 | +50 | 12 | |
| PP | 30 | +90 | 24 | |
| PSU | 30 | +67 | 12 | |
| *Polyamide fiber* | | | | |
| natural rubber | 5−15 | -40−-64 | 15 | elongation decreases -23 to -86 |
| **ORGANIC & RECYCLED FILLERS** | | | | |
| *Carbon black* | | | | |
| EPDM | 10−60 | +60−+370 | 16 | elongation change: 0 to -22 |
| fluoroelastomer | 20 | +200 | 66 | elongation increase by 100% |
| natural rubber | 20−100 | +40−+100 | 2 | elongation change: -30 to -70% |
| *Cellulose* | | | | |
| natural rubber | 5−25 | +35−+55 | 18 | |
| *Fly ash* | | | | |
| PE | 10−50 | +50−+150 | 35 | small particles |
| PE | 10−40 | -15−+20 | 35 | large particles |
| *Lignin* | | | | |
| PE | 22−72 vol% | -60−-93 | 22 | elongation also rapidly decreases |
| *PU foam, ground* | | | | |
| natural rubber | 20−80 | -20−+60 | 2,30 | increase peak around 30 phr |
| *Wood flour* | | | | |
| PP | 20−50 | -2−+10 | 3 | elongation rapidly decreasing |

The data in Table 8.1 shows how the tensile strength of composite can be improved. The following factors contribute to the improvement of tensile strength:

- Particle size (nanoparticles, carbon black, and fumed silica are examples of small particles which typically contribute to an increase in tensile strength; compare the effect of particle size on PVAc adhesive properties where different sizes of calcium carbonate were used)
- Particle shape (an aspect ratio increase in a certain range improves tensile properties; see examples for fibrous fillers and mica)
- Interaction with the matrix (untreated calcium carbonate in PE decreases tensile strength but after phosphate modification tensile strength is increased; glass beads may decrease or increase tensile strength depending on their interfacial adhesion; mica and talc give a similar effect in PP; polyamide fiber does not reinforce natural rubber because of its lack of interaction)
- Concentration (the relationship of tensile strength is not a linear function of concentration; there is a certain critical concentration above which a further increase in filler's concentration decreases tensile strength)
- Proper choice of pair filler-matrix (there should be interaction between the filler and the matrix; some combinations produce adverse results; there are cases (see alumino-silicate with PVAc) where an increased interaction reduces tensile strength due to increasing material stiffness)

Figure 8.1 illustrates the effect that the shape of a particle has on tensile properties.[6] Both relationships are linear with volume fraction of filler but they point out at different directions. The experimental data for glass beads fit Einstein's model, Eq 8.2, with a=-1.72 and b=1. The negative value of coefficient "a" indicates that the presence of glass beads has a weakening effect on the composite due to debonding. Weak adhesion and debonding reduce the volume fraction of the composite which can carry the applied load. The glass fiber data follow Kelly and Tyson model, Eq 8.6. It was calculated from the model that the fiber orientation efficiency factor is 0.3. This factor is larger than the value of 0.2 which is generally used for randomly oriented fibers. The higher value is a result of the test specimens being prepared by injection molding which tends to orient the fibers.

Figure 8.2 shows the effect of particle spacing on the tensile properties of a glass bead filled composite. Glass beads addition typically decreases the tensile strength properties of a composite. An increase in interparticle spacing contributes to the increased tensile strength of the composite.[4]

The elongation is usually inversely proportional to tensile strength which means that increasing the tensile strength of filled material usually contributes to a decrease in elongation. Table 8.1 reports two cases (EPDM and fluoropolymer reinforced with carbon black) which are different. In the first case (EPDM), elongation remains constant over a certain range of carbon black. At the second case (fluoropolymer) both tensile and elongation are increased when fillers are added.

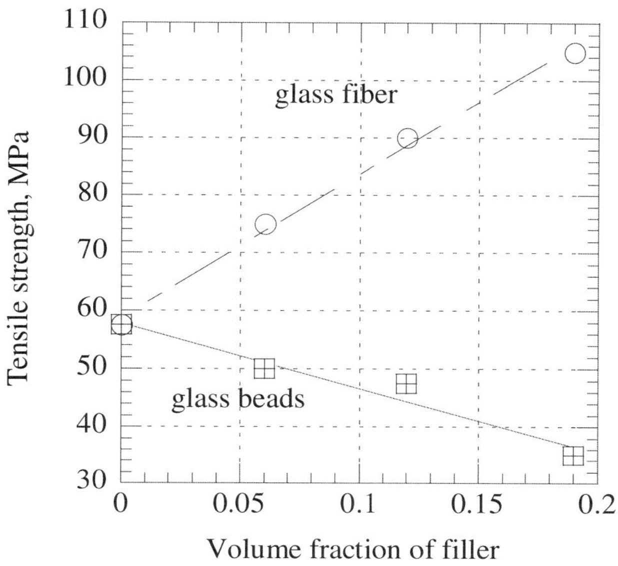


Figure 8.1. Tensile strength of POM filled with glass fibers and glass beads. [Adapted, by permission, from Hashemi S, Gilbride M T, Hodgkinson J, *J. Mat. Sci.*, **31**, No.19, 1996, 5017-25.]

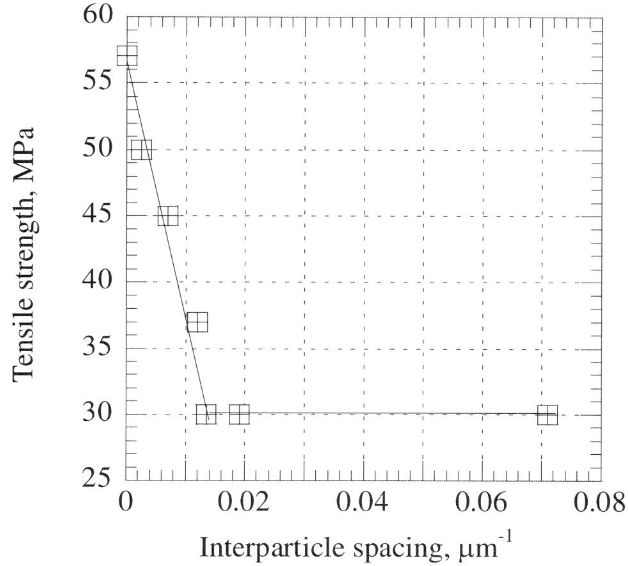


Figure 8.2. The effect of reciprocal interparticle spacing on the tensile strength of POM filled with glass beads. [Adapted, by permission, from Hashemi S, Din K J, Low P, *Polym. Engng. Sci.*, **36**, No.13, 1996, 1807-20.]

Such properties can be obtained with small, interacting particles which contribute to a physical crosslinking of a relatively weak matrix. But in most cases, a reduction of elongation is an expected result of reinforcement.

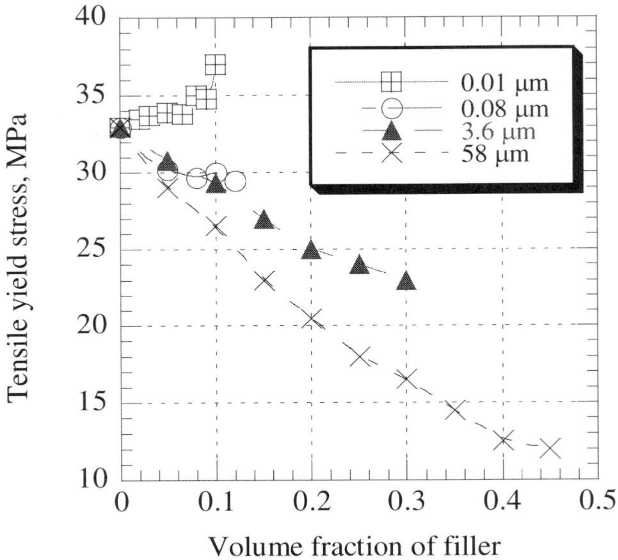


Figure 8.3. Tensile yield stress of particulate filled PP vs. filler content. [Adapted, by permission, from Voros G, Pukanszky, *J. Mat. Sci.*, **30**, No.16, 1995, 4171-8.]

## 8.2 TENSILE YIELD STRESS

Tensile yield stress gives additional information on filler-matrix interactions and consequently it is one of the preferred methods of composite testing.[5,33,53,67-77] Figure 8.3 shows that the particle size affects yield stress of PP composites.[67] Only when filler particles become very small does the yield stress value increase as the concentration increases. The smaller the particle size the higher the value of tensile yield stress. The three largest particles are $CaCO_3$ and the smallest one is silica. Thus, yield stress behavior not only depends on particle size but also on the interaction with the matrix. If the matrix is deficient in the smallest particles of $CaCO_3$ the yield stress decreases. The stress which initiates yielding can be expressed by the equation:

$$\sigma_y = \sigma_{y0}[1 - \phi_f / (1 - \phi_f < \sigma^\infty >_f / \sigma^e)] \qquad [8.9]$$

where:

$\sigma_y$ external stress initiating yielding
$\sigma_{y0}$ yield stress of matrix
$\phi_f$ volume fraction of filler
$<\sigma^\infty>_f$ stress inside filler particle placed into infinite matrix
$\sigma^e$ external stress
$<\sigma^\infty>_f/\sigma^e$ = k, dimensionless quantity

This equation can be rearranged into:

$$\frac{1-\phi_f}{\sigma_y} = \frac{1}{\sigma_{y0}} - \frac{k}{\sigma_{y0}}\phi_f \qquad [8.10]$$

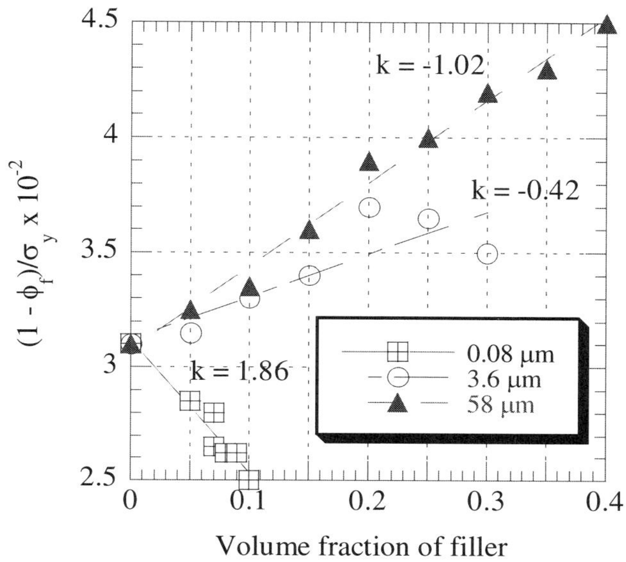


Figure 8.4. Plot of Eq 8.10. [Adapted, by permission, from Voros G, Pukanszky, *J. Mat. Sci.*, **30**, No.16, 1995, 4171-8.]

Plotting $(1 - \phi_f)\sigma_y$ versus $\phi_f$ should give straight lines with an intercept at $1/\sigma_{y0}$ and a slope of $k/\sigma_{y0}$. Figure 8.4 shows this relationship for PP/$CaCO_3$ composites from Figure 8.3. The three lines show the strong dependence of factor k on particle size. The studies were conducted for PP, PVC and LDPE.[67] Factor k depends also on the polymer used with the same fillers indicating further that the value of factor k (and yield stress) depends on polymer-filler interaction.

Figure 8.5 compares tensile yield stress for PP with two fillers.[53] In both cases, tensile yield stress decreases significantly as filler concentration increases. At higher concentrations of talc (values above 0.15 are not plotted on Figure 8.5), the composite breaks without yielding. The difference is explained by the crystallization behavior of polypropylene on the filler surface which changes the mechanical properties of composite. This shows that an additional parameter (the orientation of the polymer) may play a role in tensile yield stress behavior.

If there is perfect adhesion (no debonding), tensile yield stress increases as the concentration of the filler increases (Figure 8.6).[5] Filler particle size is also important. As the particle size of the filler decreases, the curves become more steep and the yield stress increases along with concentration increasing.

Figures 8.7 and 8.8 show applications in which various fillers increase tensile yield stress as their concentration increases. There is a linear increase in tensile yield stress (Figure 8.7) due to a strong interfacial bonding between the carbon fiber and the matrix.[75] The presence of a coupling agent increases adhesion and this is re-

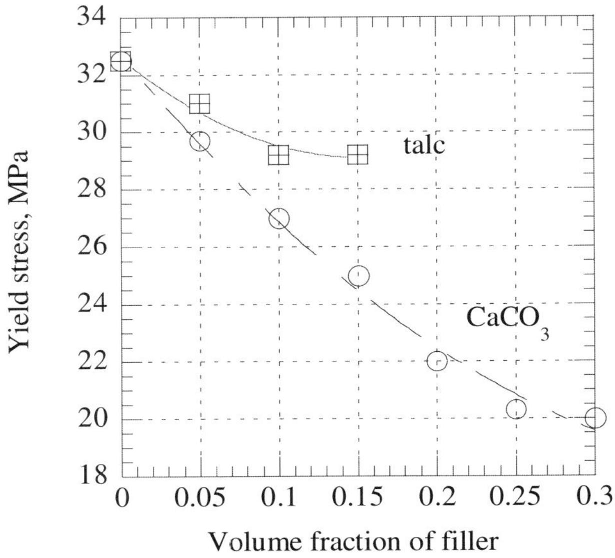


Figure 8.5. Tensile yield stress versus volume fraction of calcium carbonate and talc. [Adapted, by permission, from Pukanszky B, Belina K, Rockenbauer A, Maurer F H J, *Composites*, **25**, No.3, 1994, 205-14.]

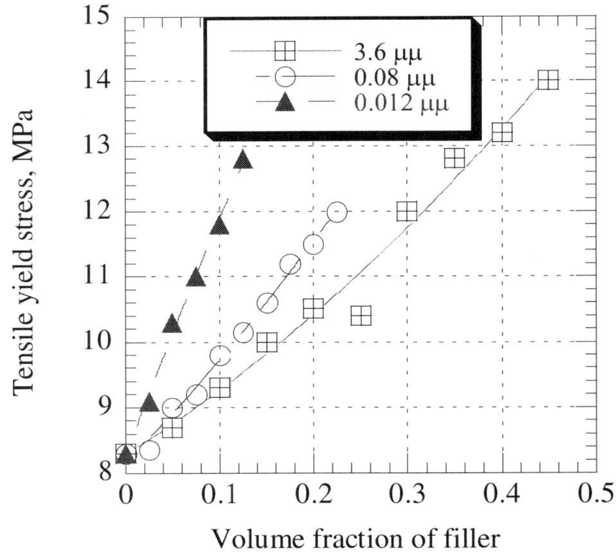


Figure 8.6. Tensile yield stress of PE composites in the case of perfect adhesion. [Adapted, by permission, from Pukanszky B, Voros G, *Polym. Composites*, **17**, No.3, 1996, 384-92.]

sponsible for the behavior presented in Figure 8.8.[73] Without a coupling agent the same composite has a tensile strength substantially lower than its yield stress.

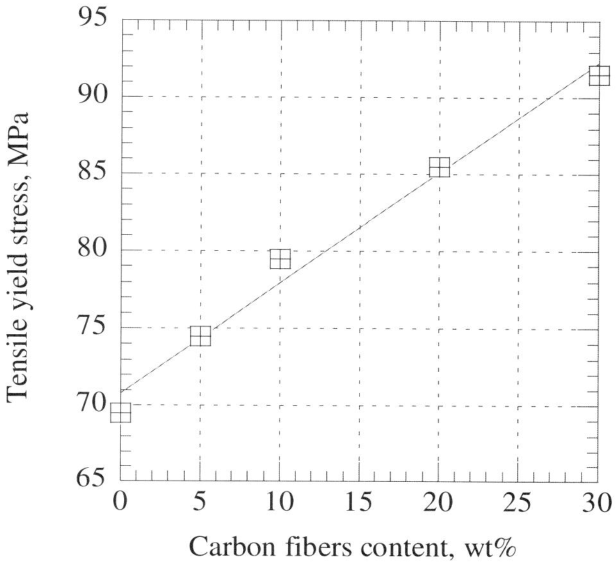


Figure 8.7. Tensile yield stress as a function of carbon fiber concentration in polycarbonate. [Adapted, by permission, from Zihlif A M, Di Liello V, Martuscelli E, Ragosta G, *Int. J. Polym. Mat.*, **29**, Nos.3-4, 1995, 211-20.]

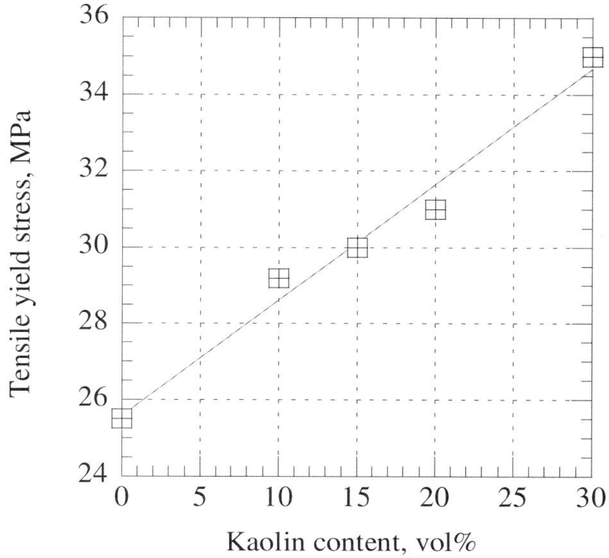


Figure 8.8. Tensile yield stress of filled HDPE with a coupling agent as a function of kaolin concentration. [Adapted, by permission, from Savadori A, Scapin M, Walter R, *Macromol. Symp.*, **108**, 1996, 183-202.]

Calcium carbonate which is the most frequently used filler in PVC, decreases tensile yield stress (Figure 8.9).[71] There is good interaction and adhesion between

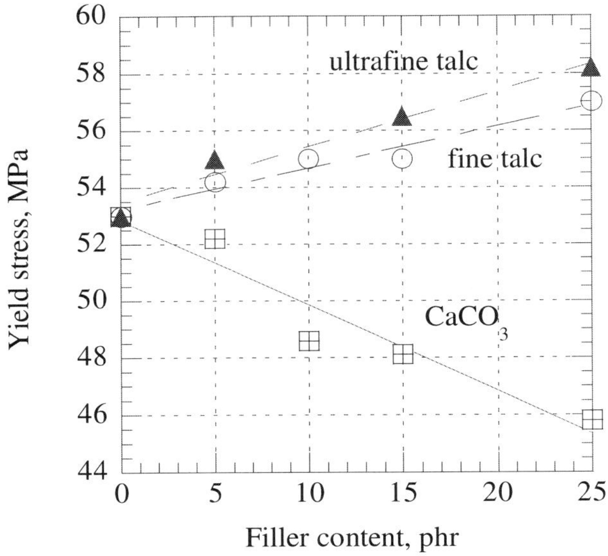


Figure 8.9. Tensile yield stress of PVC vs. filler loading. [Adapted, by permission, from Wiebking H E, Antec 95. Volume III. Conference proceedings, Boston, Ma., 7th-11th May 1995, 4112-6.]

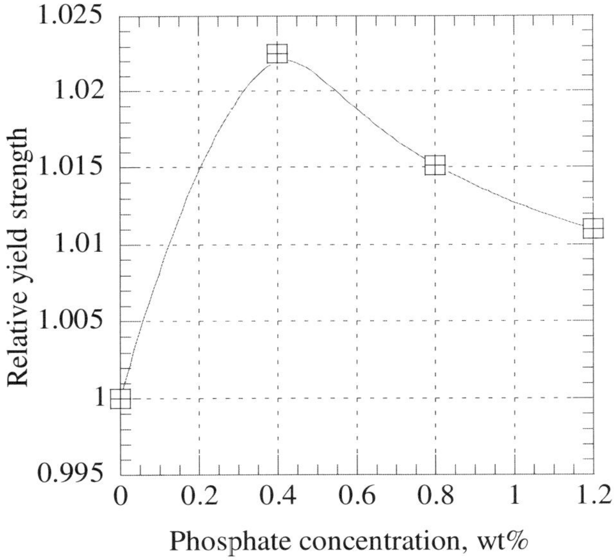


Figure 8.10. Effect of phosphate concentration on the tensile yield stress of talc filled polypropylene. [Adapted, by permission, from Liu Z, Gilbert M, *J. Appl. Polym. Sci.*, **59**, No.7, 1996, 1087-98.]

talc and PVC and the composite has a high tensile yield stress. Particle size is a less important factor.

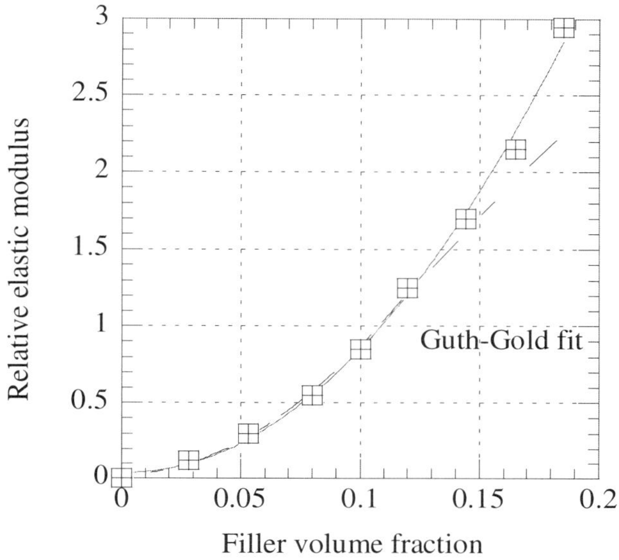


Figure 8.11. Comparison of the prediction of the Guth-Gold equation with experimental data for N330-filled SBR. [Adapted, by permission, from Wang M-J, Wolff S, Tan E-H, *Rubb. Chem. Technol.*, **66**, No.2, 1993, 178-95.]

Tensile yield strength can be improved by surface treatment of filler (Figure 8.10).[33] The coating influenced the crystallinity by contributing to nucleation. This, in turn, changes the mechanical properties of the composite. At smaller additions of phosphate, polypropylene has much higher crystallinity. A concentration of phosphate below 0.5 wt% gives the greatest tensile yield stress improvement.

In summary, tensile yield stress depends on filler particle size, concentration and on the interaction between the matrix and the filler. There are various means of improving tensile yield stress through the proper selection of filler for a particular polymers and through the surface modification of filler.

## 8.3 ELASTIC MODULUS

Elastic modulus or Young modulus are frequently used to characterize filled systems.[22,33,53,72,75,77-90] Einstein's viscosity equation modified by Guth and Gold predicts:

$$E = E_o(1 + 2.5\phi + 14.1\phi^2) \qquad [8.11]$$

predicts that elastic modulus, E, increases as the filler concentration, $\phi$, increases. Its prediction is quite precise at low concentrations (Figure 8.11).[78] At high filler concentrations the rate of change of elastic modulus deviates from that predicted by the equation.

Materials filled with rigid particles follow closely the predicted growth in elastic modulus as filler concentration increases. Many examples can be found in the

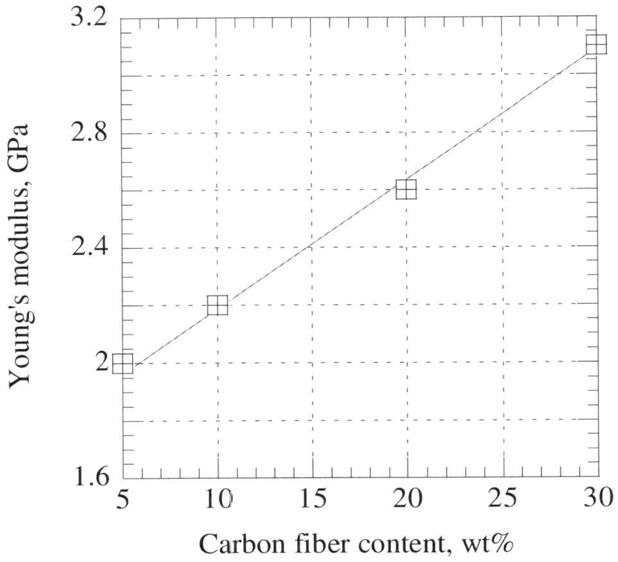


Figure 8.12. Tensile Young's modulus of a copolycarbonate composite as a function of carbon fiber concentration. [Adapted, by permission, from Zihlif A M, Di Liello V, Martuscelli E, Ragosta G, *Int. J. Polym. Mat.*, **29**, Nos.3-4, 1995, 211-20.]

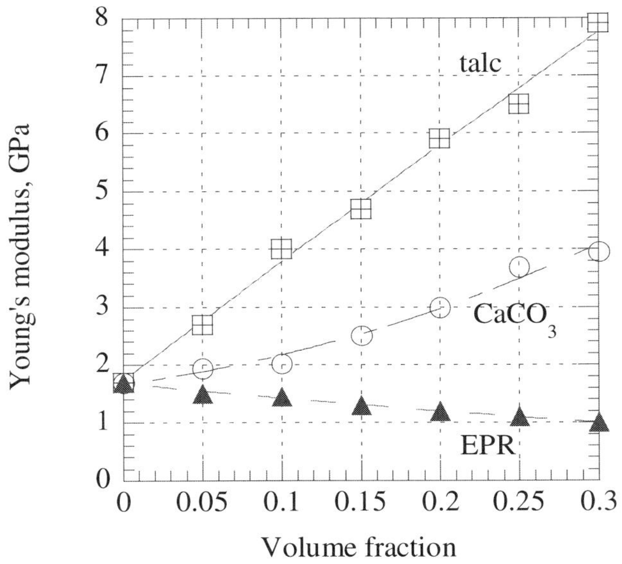


Figure 8.13. Young's modulus vs. volume fraction of filler. [Adapted, by permission, from Pukanszky B, Maurer F H J, Boode J W, *Polym. Engng. Sci.*, **35**, No.24, 1995, 1962-71.]

literature to confirm this prediction. Figure 8.12 shows the relationship for polycarbonate filled with carbon fibers.[75] The stiffness of the material increases linearly as

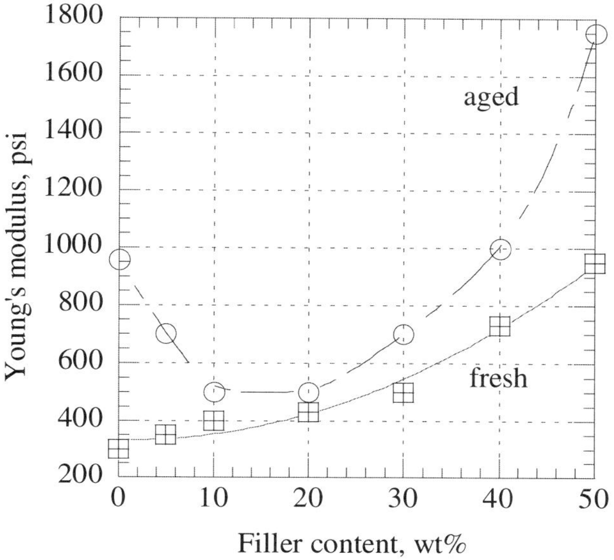


Figure 8.14. Young's moduli for fresh and aged silicone elastomer containing ZnO. [Adapted, by permission, from Yang A C M, *Polymer*, **35**, No.15, 1994, 3206-11.]

carbon fiber concentration increases and the material becomes increasingly brittle due to the nature of the fibers.

Figure 8.13 shows relationships for 3 materials.[72] Both calcium carbonate and talc generate an increased modulus whereas the addition of an elastic material such as EPR slightly reduces the value of Young's modulus. The theory predicts this because filler is composed of rigid particles, for example, calcium carbonate or talc. Better adhesion and plate like structure of talc are instrumental in rapid increase of Young's modulus.

In most of the experimental cases,[9,15,32,53,80-82,84-90] Young's modulus increases as predicted by Eq 8.11. The decrease of Young's modulus was noted when EPR[72] and lignin[22] were added.

There are other applications of elastic modulus. One can be to determine the adhesion between a filler and the matrix. To do this, elastic modulus is measured twice: once on the fresh sample and again on a sample which has been prestressed to specific strain. The decrease in Young's modulus is a measure of debonding.[83,91] Other means of composite degradation, such as those caused by UV, thermal, or water immersion, also cause Young's modulus to decrease (Figures 8.14 and 8.15).[32,88] Thermal aging (Figure 8.14) causes a drop in Young's modulus at lower concentrations of filler followed by an increase. Similar effects were produced by other fillers, such as iron oxide and graphite.

Samples of epoxy resin filled with glass microspheres have a reduced elastic modulus after water immersion. The loss of elastic modulus is more pronounced as

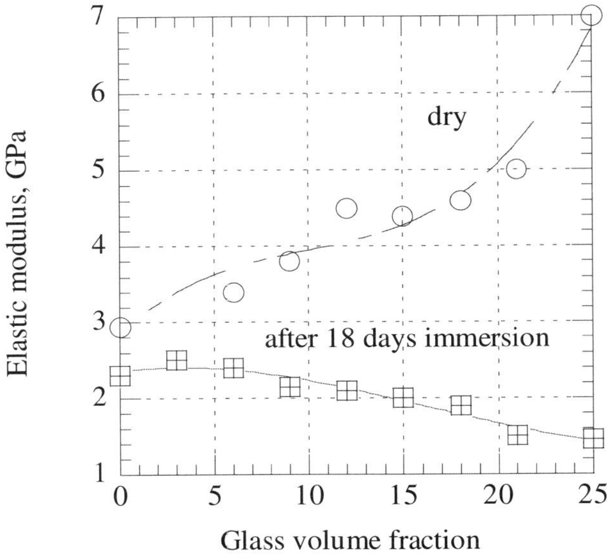


Figure 8.15. Young's modulus of epoxy reinforced with silane-coated glass microspheres vs. volume fraction of filler. [Adapted, by permission, from Lekatou A, Faidi S E, Lyon S B, Newman R C, *J. Mat. Res.*, **11**, No.5, 1996, 1293-304.]

the concentration of microspheres is increased. But, without water immersion, the elastic modulus of the composite increases as the concentration of microspheres is increased (Figure 8.15).

## 8.4 FLEXURAL STRENGTH AND MODULUS

Flexural modulus is a convenient measure of composite stiffness. Fillers can contribute significantly to a stiffness increase.[3,6,20,23,24,27,28,31,33,41,42,50,64,70,71,74,92-101] The simple Einstein equation, Eq 8.2 permits a fit of experimental data as shown in Figure 8.16.[6] A different coefficient is needed for glass beads (a=-1.30) than for glass fiber (a=1.71). Flexural strength is about 1.5 to 2 times higher than tensile strength. The Einstein equation does not consider shape and particle size. It is known[50,94] that the flexural modulus depends on particle size. Larger particles of wood flour increase flexural modulus.[3]

The aspect ratio of the filler has significant impact on flexural modulus (Figure 8.17).[23] The values of coefficient, a, given on the graph are the values of coefficient from Einstein equation, Eq 8.2. This coefficient varies in proportion to aspect ratio of filler. The higher the aspect ratio the higher the steepness of graph.

Attempts to improve flexural strength by surface treatment of fillers have not, to date, been successful. A variety of silanes, titanates, and fatty acids and their derivatives have been used to coat magnesium hydroxide for use as a filler in polypropylene.[64] Almost all composites had inferior flexural properties. In the few cases where some improvement was seen, it was 10% more then the unfilled material.

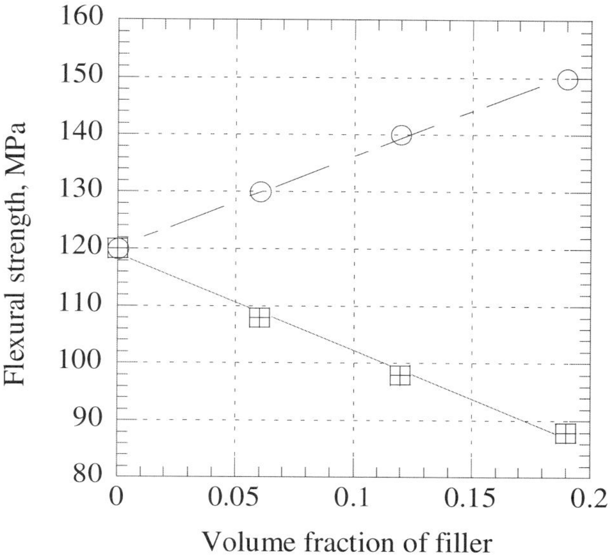


Figure 8.16. Flexural strength of POM vs. volume fraction of glass beads and glass fibers. [Adapted, by permission, from Hashemi S, Gilbride M T, Hodgkinson J, *J. Mat. Sci.*, **31**, No.19, 1996, 5017-25.]

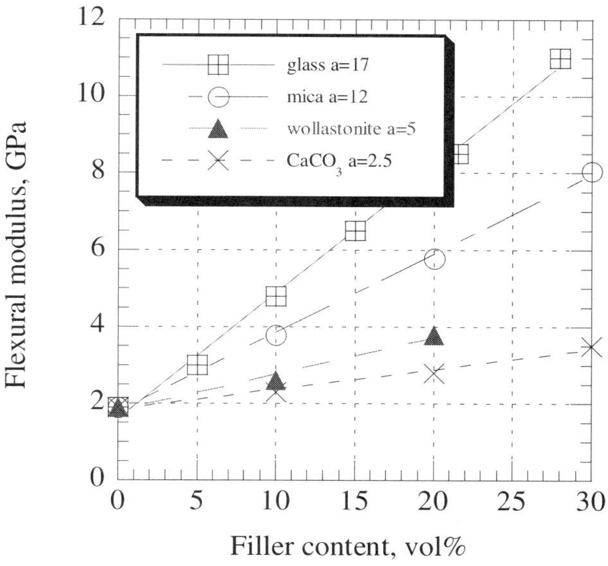


Figure 8.17. Flexural modulus of polyketone with different fillers. [Adapted, by permission, from Gingrich R P, Machado J M, Londa M, Proctor M G, Antec '95. Vol. II. Conference Proceedings, Boston, Ma., 7th-11th May 1995, 2345-50.]

Treatment of ultrafine talc with an acrylic modifier for use as a filler in rigid PVC always resulted in a gradual decrease of flexural modulus as the modifier concen-

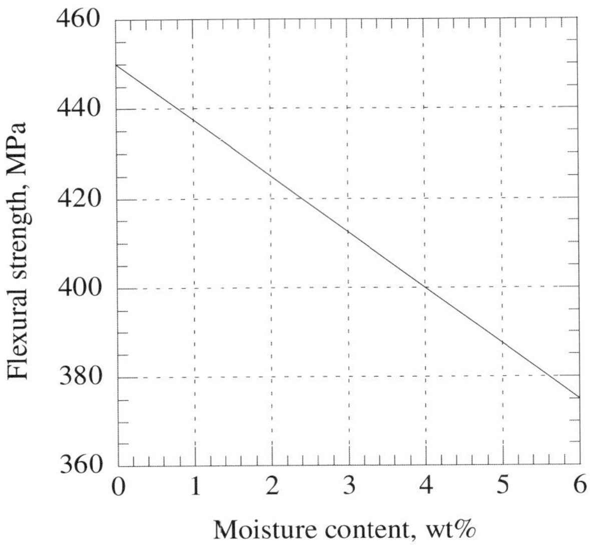


Figure 8.18. Flexural strength vs. moisture content in Kevlar reinforced epoxy. [Adapted, by permission, from Akay M, Mun S K A, Stanley A, *Composites Sci. Technol,* **57**, 1997, 565-71.]

tration was increased.[70] Similar results were obtained both with phosphate coated talc[33] and modified carbon black.[96]

Better mixing methods and processing techniques which align fibers in the composite seem to be the most promising avenues to improve flexural modulus with filler additions.[98-9] Mixing speed choice allows to increase flexural strength by 25%.

Finding a way to balance often conflicting requirements is the most challenging aspect of product development work. If a material is formulated to be fire resistant, UV stable, or moisture resistant, it may have inadequate mechanical properties. But with a careful and imaginative use of filler and filler coatings, a balance can be found.[92,100,101]

Figure 8.18 shows the effect of moisture content on the flexural strength of a composite.[92] The hygroscopic nature of both the fiber and the matrix contribute to the deterioration of the composite.

## 8.5 IMPACT RESISTANCE

Fillers improve the impact strength of the filled materials.[3-4,7,14,17,20,23-4,41,43-6,50,64,69-75,81,87,89,96,97,101-119] A model analysis of impact strength improvement is discussed below in the section on fracture and toughness.

The results of experimental studies which are summarized in Table 8.2 show the potential effect of different fillers on impact properties of filled materials. The information in Table 8.2 is presented in the same format as explained in introduction to Table 8.1.

**Table 8.2. Effect of fillers on impact properties of filled materials**

| Filler/polymer | Conc. range, wt% | Impact strength increase (+) decrease (-), % | Refs. | Comments |
|---|---|---|---|---|
| PARTICULATE, INORGANIC FILLERS | | | | |
| *Calcium carbonate* | | | | |
| PEK | 10–30 | -55–-70 | 23 | |
| PP | 2 | +45 | 112 | improvement due to nucleation |
| PP | 10–60 | -40–+50 | 107 | increase with stearate coating |
| PP | constant | 0–+50 | 103 | increases with particle % > 1 μm |
| PP | 5–30 vol% | +20–+65 | 105 | no adhesion; peak at 10 vol% |
| PP | 5–30 vol% | +10–+25 | 105 | perfect adhesion; peak at 10 vol% |
| PVC | 2–45 vol% | +40–+150 | 104 | peak value at 15 vol% |
| PVC | 5–25 | +20–+80 | 70 | |
| *Fly ash* | | | | |
| PP | 5–20 | -50–-65 | 87 | |
| *Glass beads* | | | | |
| PP | 30 vol % | -35 | 89 | |
| *Kaolin* | | | | |
| PE | 10–30 | +140–+360 | 73 | |
| *Magnesium hydroxide* | | | | |
| PP | 10–60 | -27–-70 | 64 | metal stearate coating can double IS |
| *Mica, muscovite* | | | | |
| PA66 | 20–40 | -30–-15 | 50 | varies with particle size and amount |
| PBT | 15–40 | +15–-35 | 50 | varies with particle size and amount |
| PEK | 10–30 | -50–-70 | 23 | |
| PP | 2–20 vol% | +5–+26 | 105 | used in combination with $CaCO_3$ |
| *Mica, phlogopite* | | | | |
| PP | 2–20 vol% | 0–+15 | 105 | used in combination with $CaCO_3$ |
| *Micaceous* | | | | |
| PP | 10–60 | 0÷-20 | 43 | hydrated K-Mg aluminosilicate |
| *Talc* | | | | |
| PP/EP blend | 1.2 | +24 | 112 | improvement due to nucleation |
| PP | 10 | -15 | 69 | |
| PVC | 5–20 | +13–-15 | 71 | |
| PVC | 5–25 | -13–-50 | 70 | |
| *Titanium dioxide* | | | | |
| PS | 10–35 | +9–+35 | 111 | dispersion improves IS |
| *Various* | | | | |
| PS | 0.1 | +15–+220 | 110 | particle size determines IS |
| *Wollastonite* | | | | |
| PA66 | 15–35 | no change | 13 | |
| PEK | 10–20 | -35–-40 | 23 | |
| FIBROUS FILLERS | | | | |
| *Carbon fiber* | | | | |
| PC | 5–30 | +40–+120 | 75 | |
| *Glass fiber* | | | | |
| PE | 2–7 | 0–+7 | 7 | |
| PEK | 10–20 | +10–+40 | 23 | |
| PP | 10–60 | +300–+2750 | 102 | constant fiber length at 6 mm |
| PP | 25 | +650–+1050 | 102 | variable fiber length: 3–12 mm |
| PP | 2–7 | +100–+250 | 7 | |
| *Organic fillers* | | | | |
| *Carbon black* | | | | |
| ABS | 3 | -20 | 114 | varies depending on size and mixing |
| PP | 15–60 | no change | 74 | |
| PP copolymer | 15–60 | -30–+200 | 74 | peak value at 30% |

The data compiled in Table 8.2 show how impact strength can be improved. The following factors contribute to the improvement:

- Particle size (in many cases a certain range of particle size substantially increases impact strength)
- Particle shape (aspect ratio is the most important parameter; the use of fibers is the most certain method of impact strength improvement)
- Particle rigidity (hollow particles and fillers which have low hardness substantially decrease the impact strength)
- Interaction with the matrix (does not help in most cases; see for example calcium carbonate; in some cases (magnesium hydroxide) surface coating by metal stearate rapidly improves impact strength). Interaction with matrix is relevant in fiber reinforcements (see section on fracture resistance below)
- Concentration has a mixed influence (in fillers which improve impact strength, increased concentration increases impact strength)
- Nucleation (the presence of fillers or fillers with nucleating agents contribute to changes of crystallinity which increases impact strength)

Figure 8.19 shows the effect of the particle size of particulate fillers on impact resistance.[110] Several fillers were used in this study, including calcium phosphate, barium sulfate, calcium carbonate, and white carbon. The average particle diameters of these fillers ranged from 0.8 to 30 μm. The impact strength of composites containing different fillers is plotted against the average particle diameter. Particle size had a pronounced effect on impact strength whereas the influence of chemical composition was negligible. Maximum reinforcement was obtained with particles having diameter of 2 μm.

Figure 8.20 shows the effect of adhesion between the filler and the matrix on impact strength.[105] The maximum performance is obtained at low filler concentrations because interparticle distances and the formation of agglomerates are the controlling factors in impact strength improvement. If particles have perfect adhesion, the matrix is constrained to a greater degree because of matrix interaction with filler surface, leading to additional embrittlement of the material.

## 8.6 HARDNESS

Literature contains little data on the effect of fillers on hardness but what is available indicates that the addition of fillers increases hardness.[18,34,40,65,120-3]

Figure 8.21 explains that the reasons for this increase of hardness are more complicated than the increase caused by adding a harder material.[120] Fillers which have relatively large particle size do not interact and therefore their effect on hardness is due to their higher hardness. But the gain in hardness is very small because these particles are surrounded by an elastic matrix which moderates the effect of their hardness. Much larger gains are observed with semi-reinforcing grades due to the formation of an interlayer with mechanical properties more similar to the filler than to the matrix. In this case, the actual size of the particle is increased by the

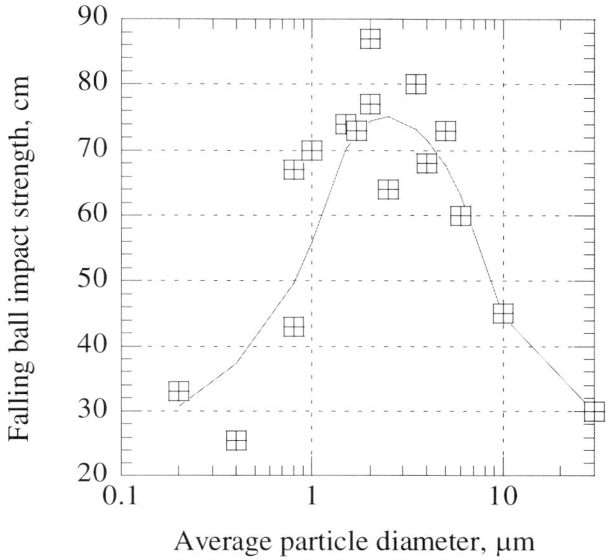


Figure 8.19. Falling ball impact strength vs. impact strength of filled polystyrene. [Adapted, by permission, from Mitsui S, Kihara H, Yoshimi S, Okamoto Y, *Polym. Engng. Sci.*, **36**, No.17, 1996, 2241-6.]

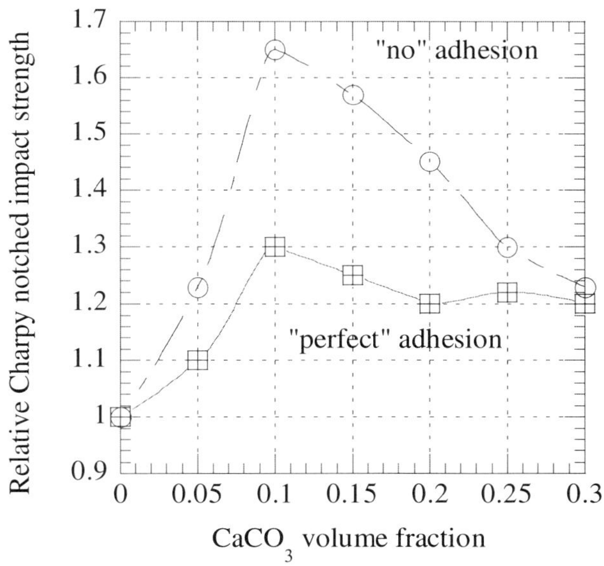


Figure 8.20. Charpy notched impact strength of calcium carbonate filled polypropylene. [Adapted, by permission, from Jancar J, Dibenedetto A T, Dianselmo A, *Polym. Engng. Sci.*, **33**, No.9, 1993, 559-63.]

thickness of its adsorbed layer, therefore even small particles occupy a substantial space in composites. Reinforcing fillers introduce another variable related to formation of physical crosslinks which can be very numerous because of the small size

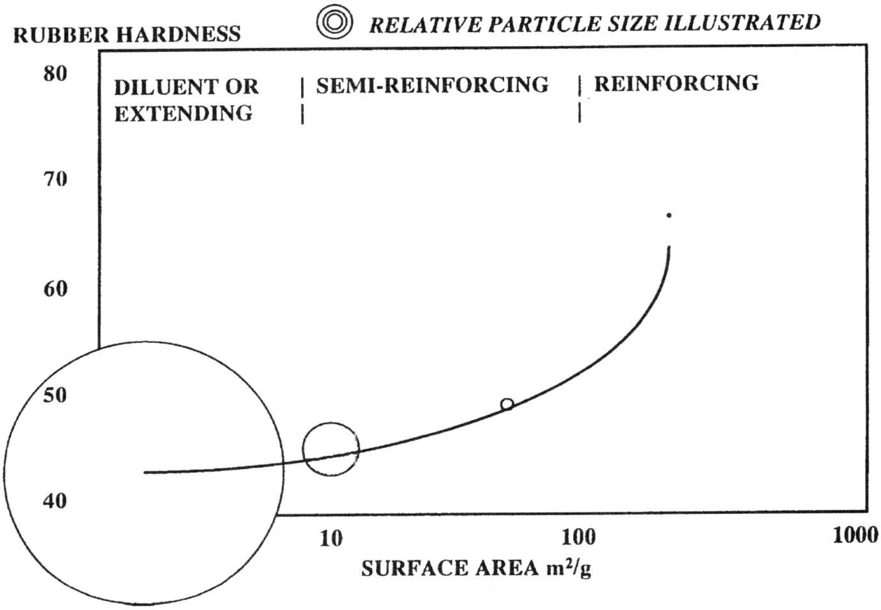


Figure 8.21. Rubber hardness vs. surface area of silica filler. [Adapted, by permission, from Evans L R, Meeting of the Rubber Division, ACS, Montreal, May 5-8, 1996, paper D.]

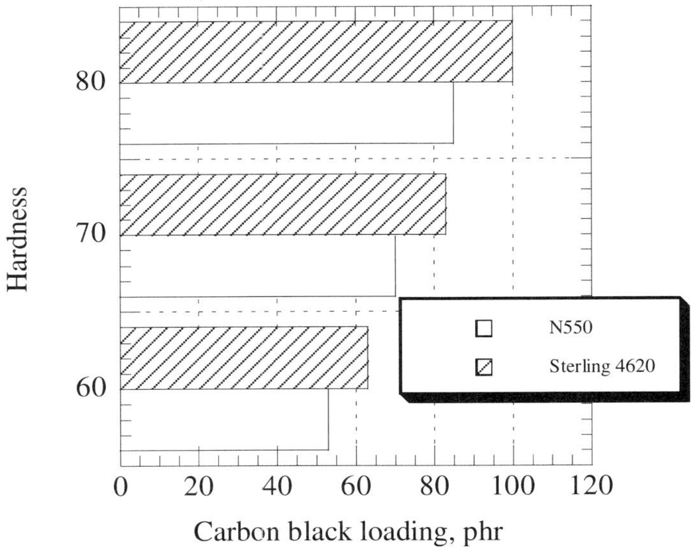


Figure 8.22. Carbon black loading for the same hardness. [Adapted, by permission, from Monthey S, Duddleston B, Podobnik J, *Rubb. World*, **210**, No.3, 1994, 17-9.]

of the particles. These physical crosslinks further reinforce the rubber resulting in its increased hardness.

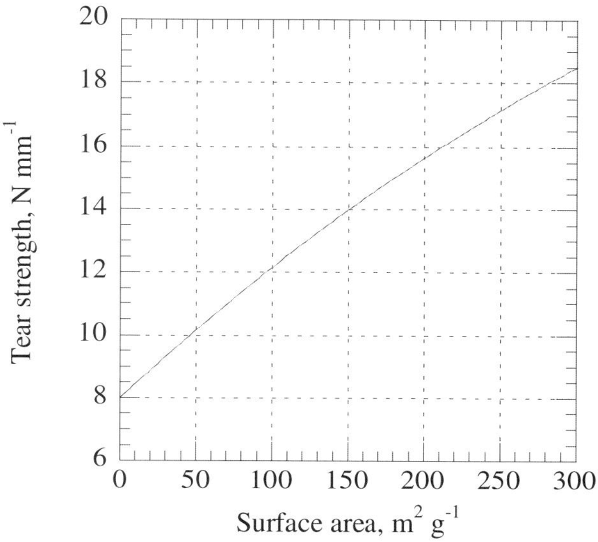


Figure 8.23. Silicone rubber tear strength vs. surface area of silica. [Data from Okel T A, Waddell W H, *Rubb. Chem. Technol.*, **68**, No.1, 1995, 59-76.]

Figure 8.22 shows that the same hardness can be obtained by a wide range of levels of carbon black.[121] It is possible to add more carbon black and retain the same hardness. So, other properties such as compression set, can be improved without sacrificing the elasticity of the product.

Small additions of non-interacting fillers cause only small changes in product hardness. Usually, such additions produce a 10-20% increase in the hardness compared to unfilled material.[18,34,40,65,120,123]

## 8.7 TEAR STRENGTH

Tear strength data are very limited.[18,34,66,123-7] Tearing energy is given by the following equation:

$$G_c = 2kcW \quad [8.12]$$

where:

| | |
|---|---|
| k | function of extension ratio |
| c | crack length |
| W | strain energy density on the crack path |

This equation indicates that tear strength can be increased by modulation of strain energy. A straight tear line with a smooth surface of tear path is indicative of low tear strength. Filler can improve tear strength in two ways. It may either form obstacles in the tear path which disrupts the smooth tear surface and changes the crack direction or by interacting with the matrix and adhering to it so that stress is transferred to the matrix over larger surface area.

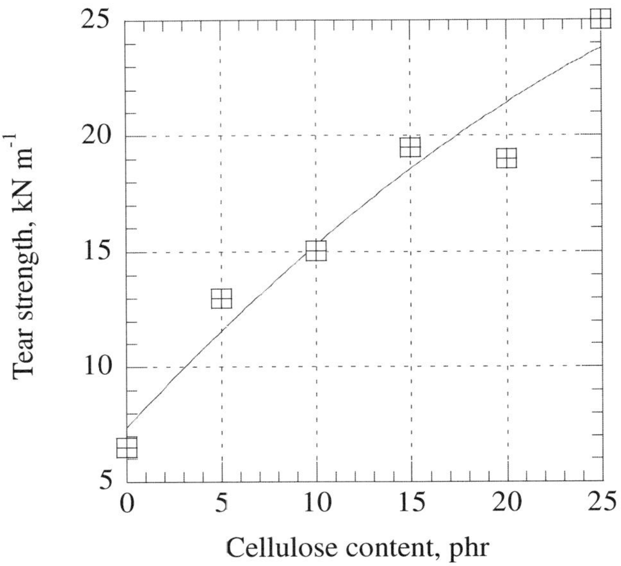


Figure 8.24. Tear strength of NR/BR vs. the amount of cellulose. [Data from Vieira A, Nunes R C R, Visconti L L Y, *Polym. Bull.*, **36**, No.6, 1996, 759-66.]

A filler with a high surface area increases the interaction with the matrix and thus increases tear strength (Figure 8.23).[125] When rubber is filled with silica the large surface area of the silica interacts with the rubber and adheres to it. This adhesive interaction allows energy to be stored or dissipated.

Fibers, due to their high aspect ratio, are the most efficient method of improving tear strength (Figure 8.24).[18] Even such weak fibers as cellulose fibers can increase tear strength by a factor of 6. Fibers form large obstacles in the path of crack growth. Fibers with better adhesion to matrix are more efficient.

The quantities of multifunctional additives used must be selected with care as each type and grade of carbon black requires a specific but different amount to achieve optimum performance. To achieve optimum adhesion, the concentration of additive should coat the surface of carbon black with a monomolecular layer. Building additional layers on the carbon black surface reduces adhesion which also reduces tear strength.

## 8.8 COMPRESSIVE STRENGTH

Compressive strength depends on the stiffness of the material, thus, all of the parameters which affect stiffness, including the effect of fillers, influence compressive strength.[17,79,92,95,128-9] The following equation associates compressive strength with other mechanical properties:

$$\sigma_{cc} = KG_R = K\left[\frac{E_\tau}{2(1+\nu)}\right] \qquad [8.13]$$

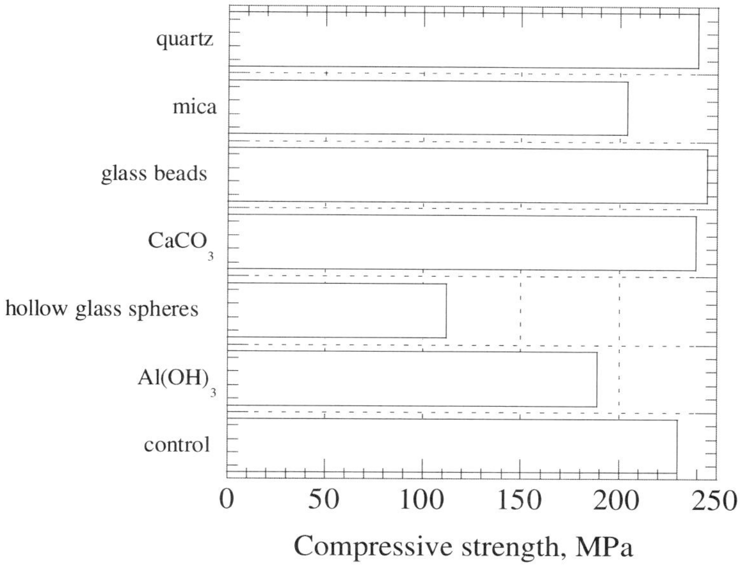


Figure 8.25. The effect of particulate fillers on the compressive strength of epoxy resin. [Adapted, by permission, from Yang Q, Pritchard G, *Polym. & Polym. Composites*, **2**, No.4, 1994, 233-9.]

where

| | |
|---|---|
| $\sigma_{cc}$ | compressive strength |
| K | constant |
| $G_R$ | matrix shear modulus |
| $E_\tau$ | Young's modulus |
| $\nu$ | Poisson's ratio |

This equation shows that the parameters which affect Young's modulus may affect the compressive strength. Figure 8.25 shows results for several fillers.[95] Aluminum trihydrate, hollow glass beads, and mica decrease compressive strength. The first two fillers are found to decrease Young's modulus as well. It is not clear why mica caused a decrease in compressive modulus. It would be expected to increase it. Remaining fillers increased compressive strength.

It was previously reported (Figure 8.18) that moisture reduces flexural strength of Kevlar filled epoxy. Figure 8.26 shows that moisture reduces its compressive strength.[92]

## 8.9 FRACTURE RESISTANCE

The fracture resistance of a material depends on all of the properties which have been discussed including tensile strength, yield stress, elastic modulus, flexural strength, and impact resistance, all of which depend, in part, on fillers. Fillers, consequently, are important determinants of fracture resistance.[4,6,10,26,32,56,73,87-8,104-5,109,116,125,130-45] Only those phenomena which are related

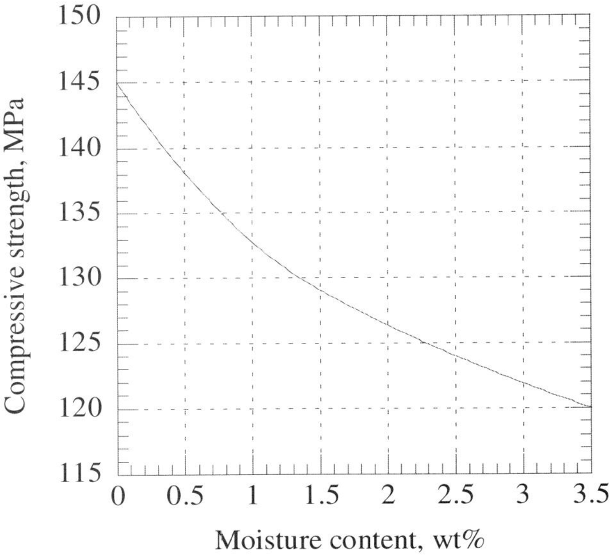


Figure 8.26. Compressive strength of Kevlar reinforced epoxy vs. moisture content. [Adapted, by permission, from Akay M, Mun S K A, Stanley A, *Composites Sci. Technol,* **57**, 1997, 565-71.]

to impact, flexural, or tensile stresses which can cause materials to fail. All phenomena related to cyclic forces are discussed under fatigue in a separate section.

This discussion includes the following:

- Modes of fracture
- Mechanism of fracture
- Microstructure of filler inclusions
- Changes in matrix due to impact
- Material toughening
- Methods of fracture prediction and modelling

Five fracture modes were observed during tensile experiments (Figure 8.27).[133] The most ductile compositions fracture during the strain-hardening (mode A) or the neck propagation (mode B). Modes C and D are typical of quasi-brittle fracture. A thinned region is formed at the neck formation (mode C). In this case, stress drops to the draw stress. In mode D, specimens fracture through macro-shearbanding. These bands are formed across the specimen and fracture occurs after a yield maximum is exceeded. Mode E is a brittle failure perpendicular to the loading direction. The fracture occurs before the yield point. The description of the mechanisms of these modes of fracture which follows is based on SEM observations of the fracture of calcium terephthalate and calcium carbonate filled thermoplastic polyester.[134]

The mechanism of fracture according to mode A is given by Figure 8.28.[134] The fracture surface had a rough region where a crack was propagated by a ductile

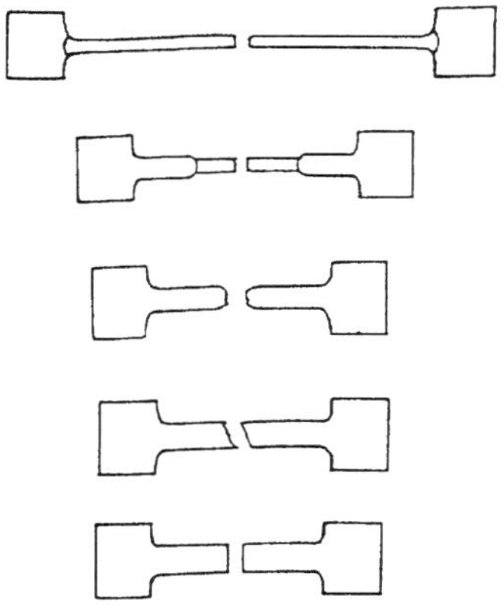


Figure 8.27. Schematic representation of the five tensile fracture modes. [Adapted, by permission, from Li J X, Silverstein M, Hiltner A, Baer E, *J. Appl. Polym. Sci.*, **52**, No.2, 1994, 255-67.]

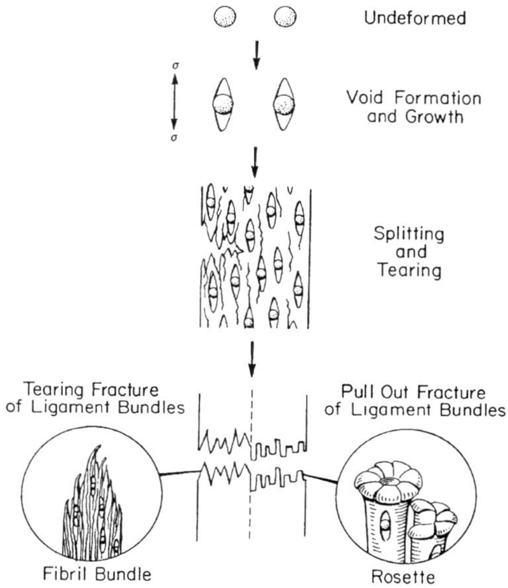


Figure 8.28. Schematic representation of mode A fracture. [Adapted, by permission, from Li J X, Hiltner A, Baer E, *J. Appl. Polym. Sci.*, **52**, No.2, 1994, 269-83.]

tearing. Strain hardening enables the polymer to sustain these loads. At a low concentration of filler (typical of mode A behavior), there is enough polymer matrix to withstand an external load without fracture. The surface has a pullout region (the initial fracture site) and a quasi-cleavage rosette region (where continuation of tearing occurs). The morphological features include debonded particles and elongated voids. Fracture occurs when the local strain in the ligaments reaches the fracture strain of the matrix. The formation of a rosette pattern requires a certain amount of polymer to be present and thus it will occur only at very low filler loadings.

Figure 8.29 shows the mechanism of fracture according to mode B.[134] Void formation and growth is similar to the previous mechanisms. The differences include the coalescence of voids and a singular tearing fracture initiated by a critical size of void created from coalescence. The fracture initiates from either the side or the center (mode A from the side only) and there is no rosette region. Fiber bundles are short and small in diameter.

Figure 8.30 shows the mechanism of fracture by mode C.[134] There was also a void formation and coalescence occurred in this specimen but only in lateral direction. Fracture was initiated from the center and had numerous secondary cracks. This mode of failure is typical of critical volume fraction of filler where the

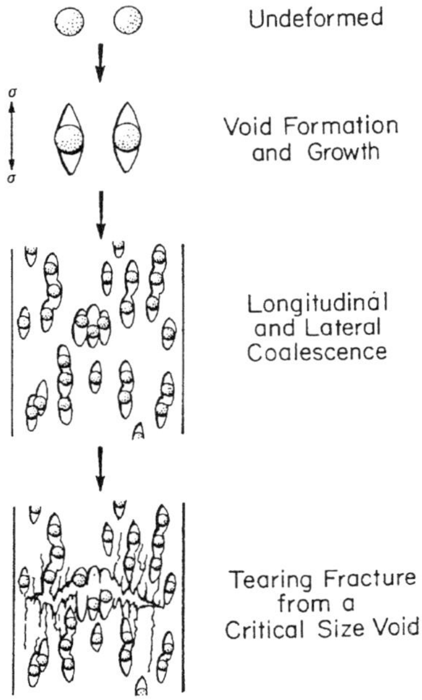


Figure 8.29. Schematic representation of mode B fracture. [Adapted, by permission, from Li J X, Hiltner A, Baer E, *J. Appl. Polym. Sci.*, **52**, No.2, 1994, 269-83.]

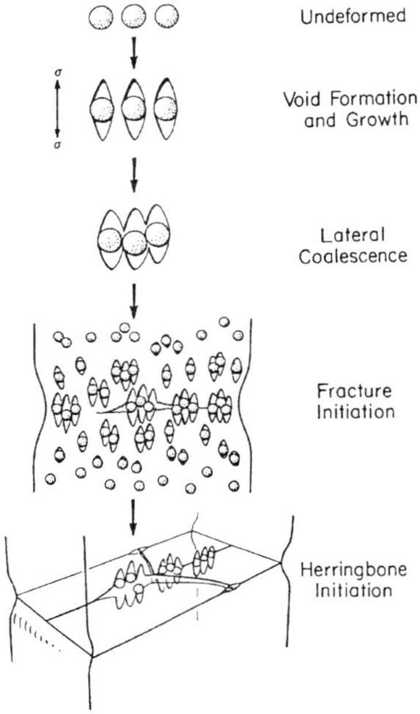


Figure 8.30. Schematic representation of mode C fracture. [Adapted, by permission, from Li J X, Hiltner A, Baer E, *J. Appl. Polym. Sci.*, **52**, No.2, 1994, 269-83.]

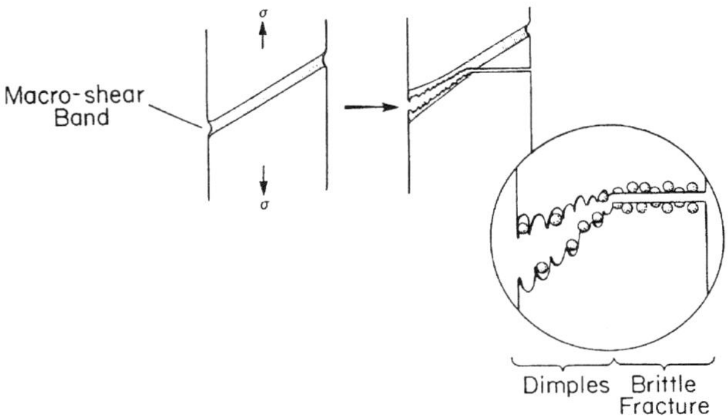


Figure 8.31. Schematic representation of mode D fracture. [Adapted, by permission, from Li J X, Hiltner A, Baer E, *J. Appl. Polym. Sci.*, **52**, No.2, 1994, 269-83.]

particles are sufficiently close to diminish the sizes of the ligaments which are needed to resist fracture.

Figure 8.31 shows the mechanism of fracture by mode D.[134] Fracture was initiated at one side and propagated at an angle to the other side. As the rate of crack development increased, the fracture path deviated to become almost perpendicular to the direction of loading. This part did not exhibit the stress-whitening effect which indicates brittle fracture. Particles in this specimen were not debonded but some were cracked (see Figure 7.31). In fractures

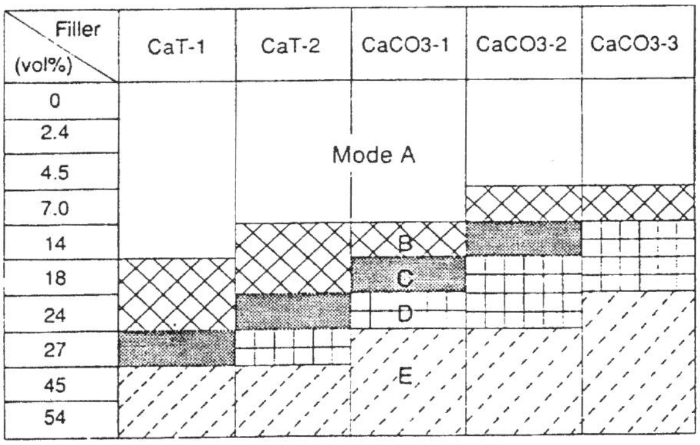


Figure 8.32. Map of fracture modes for various fillers and their concentrations. [Adapted, by permission, from Li J X, Silverstein M, Hiltner A, Baer E, *J. Appl. Polym. Sci.*, **52**, No.2, 1994, 255-67.]

by mode E, there was no indication of debonding but the particles themselves fractured along with the undeformed matrix.

Figure 8.32 shows the effect of filler type and concentration on the mode of fracture.[133] Several factors are responsible for the behavior. CaT (calcium terephthalate) fillers have good adhesion to the matrix and have an elongated shape. $CaCO_3$ (1) and (2) are both untreated fillers of smaller particle sizes (2.2 and 4.1 μm, respectively). $CaCO_3$ (3) is a stearate coated filler (better dispersion, but poor adhesion to matrix) with a particle size of 6.1 μm. The mode of fracture depends on filler concentration, the degree of adhesion to the matrix, and particle size.

While we have a comprehensive picture of tensile fracture modes, a similar picture of impact fractures is yet to be developed.[104] Composite behavior during impact is more complex than the behavior during tensile stress. Attempts at mathematical modelling are being made. The steps including crack initiation, propagation, yield, crazing, voiding and debonding (essentially similar to that discussed above) are being analyzed.

The way in which a filler is incorporated has an effect on fracture resistance. Figure 8.33 shows a schematic representation of the microstructure of fillers.[141] The rubber particles are generalized as rubber particles added in a toughening process (a), rubber or polymer coating in core-shell microstructure, bound polymer, or surface coating (b).

Figure 8.34 shows the results of impact on composites containing such particles.[141] Fillers without a coating allow the formation of numerous microcracks which weaken the material. Rubber particles lower the effect of impact and microcracking. Fillers coated by rubber (polymer) are more effective due to formation of yielding zone. This explanation also forms immediate question regarding the thickness of such a coating. A thick coating lowers toughness due to increased elastomeric behavior. A discussion of this and other structural phenomena follows.

Several characteristics of the matrix and filler-matrix interphase are involved in material toughening. These include: the particle size of filler, interfacial adhesion, filler concentration (already discussed), filler surface composition, the crystallization of the matrix, shell thickness, stress whitening, and strain hardening.

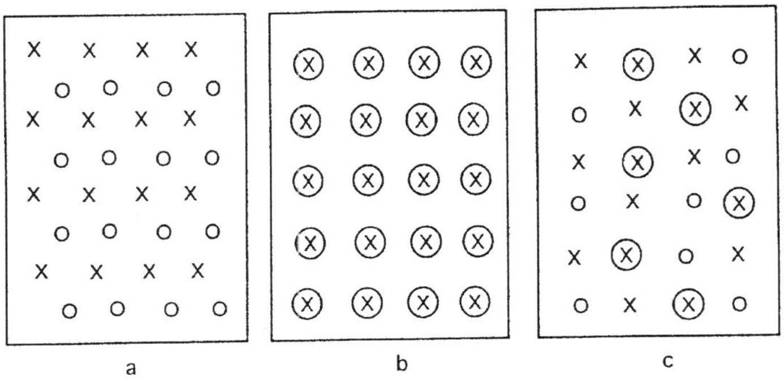


Figure 8.33. Schematic representation of three representations of microstructure of fillers (×) and rubber particles (O) in polymer matrix. [Adapted, by permission, from Yu Long, Shanks R A, *J. Appl. Polym. Sci.*, **61**, No.11, 1996, 1877-85.]

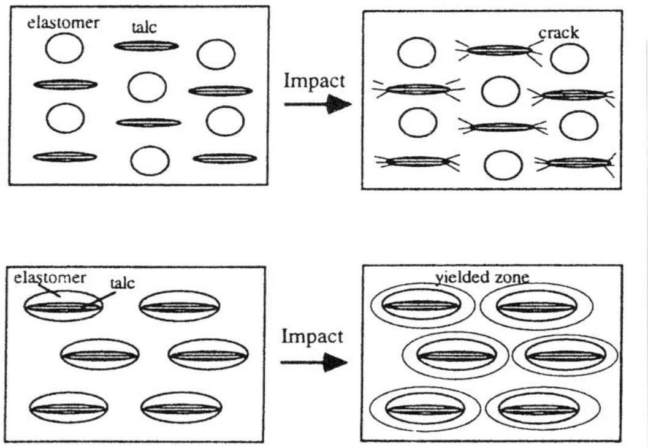


Figure 8.34. Morphological model of impact fracture of composites. [Adapted, by permission, from Li J X, Silverstein M, Hiltner A, Baer E, *J. Appl. Polym. Sci.*, **52**, No.2, 1994, 255-67.]

The effect of filler surface composition can be exemplified by a simple copper filler.[26] Two types of copper spheres were used in polyamide-11 composites. Both grades were produced by atomization but one filler had an oxidized surface whereas the other was reduced to pure copper. The mechanical properties of the copper composite were improved when oxidized particles were used because it had rougher surface which gave better adhesion. SEM micrographs showed that fracture surfaces were different in each case. The fracture path avoided the oxidized copper particles and failure occurred in the matrix whereas large debonded areas were seen on the reduced copper. This is an example of good mechanical adhesion being developed on a rough surface. The improved adhesion contributed to a change in fracture mode from adhesive to cohesive failure. Various methods used to enhance adhesion are discussed in Chapter 6. The organization of the interphase can also increase adhesion as discussed in Chapter 7. The thickness of shell (or interphase) is critical for materials toughness. To obtain a higher modulus than that of the matrix, a very thin shell is required along with very good adhesion. Increasing shell thickness rapidly lowers the rigidity of the material.

The addition of small particles of rubber to brittle polymers is the frequently used method to toughen materials. The improvement is due to increased crack growth resistance due to cavitation of rubber particles followed by deformation and crazing.[143] The role of such particles is to redirect the stress and distribute it onto a larger surface area. With that in mind, a theory was put forward that perhaps rubber particles are not necessary at all. The concept was tested by comparing the effect of rubber particles and holes created by the incorporation of thin wall latex particles. Figure 8.35 shows a fracture surface of epoxy modified with 10 vol% of such holes.[143] The particle sizes of latex particles were 0.4 and 1 μm and their walls were very thin. Control experiments were conducted with rubber particles having particles size of 0.2 and 0.55 μm. The fracture toughness of epoxy filled with holes was 2.30 and 1.95 MPa $m^{1/2}$ and of the rubber containing samples 2.05 and 2.2. The incorporation of particles may have affected crystallization, so DSC analyses were conducted which indicated no difference between filled and unfilled material. This example shows the importance of the matrix in the mechanism of toughening (since holes by themselves could not contribute to toughening).

Three phenomena are responsible for the response of a material to stress: the strain hardening, crazing, and stress whitening. The first two phenomena are completely beyond the scope of this book. Each increase mechanical performance of the matrix by orientation then crystallization of the matrix to counterbalance crack propagation. Stress whitening is a change of appearance around the stressed area which is thought to originate from void formation by separation of the polymer-filler interface. It is known from studies on silica in silicones that stress whitening can be decreased by reducing the pH of the silica and by lowering the $Na_2O$ content.[125] Both factors improve surface wetting and improve adhesion.

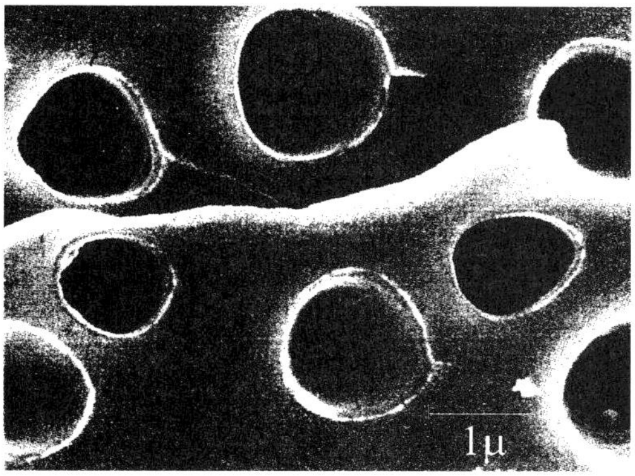

Figure 8.35. SEM micrograph of the fracture surface of an epoxy resin modified with 10 vol% latex particles. [Adapted, by permission, from Bagheri R, Pearson R A, *Polymer*, **36**, No.25, 1995, 4883-5.]

Many mathematical methods have been developed to interpret data. Some, more common models are given below. Fracture strength can be calculated from modified Einstein equation:

$$\sigma_c = \sigma_m(1 - 1.21\phi^{2/3}) \quad [8.14]$$

where:

$\sigma_c$ fracture strength of the composite

$\sigma_m$ fracture strength of the matrix
$\phi$ filler volume fraction

This equation applies to systems in which there is no adhesion between the filler and the matrix. The equation predicts that the fracture strength of the composite is reduced as the filler concentration increases.

For random-packed monodisperse spheres where the concentration range is $0.56>\phi>0$, the following equation applies:

$$\sigma_c = \sigma_m(1 + 1.06\phi^2) \quad [8.15]$$

This equation predicts that there will be some small gains in fracture strength as filler concentration increases. For fiber-filled composite the following equation was found to be supported by experimental data:[6]

$$\sigma_c = \sigma_m(1 + 1.64\phi) \quad [8.16]$$

This equation explains why fiber reinforcement is so important in increasing fracture resistance. The above equations are important for basic classification of fillers in terms of their influence on fracture resistance but they are deficient in describing the effect of the matrix and the interaction with filler.

One method used is the energy rate interpretation of the J-integral. It is assumed that J has two constant values one at the crack initiation point, $J_c$, and the other at the failure point, $J_R$, given by equations:[109]

$$J_c = -\frac{1}{B}\frac{\Delta U_c}{\Delta a} \qquad J_R = -\frac{1}{B}\frac{\Delta U_T}{\Delta a} \quad [8.17]$$

where:
B specimen thickness
$U_c$ energy of initiation
a crack length
$U_T$ total energy of fracture

The values of these two energies can be found by plotting U per unit thickness (U/B) vs. a. J-values can be determined from the slopes of the lines. Plotting J-values against the volume fraction of filler permits an estimate of the effect of filler on crack initiation and on the energy of fracture.

Other methods of data interpretation include the calculation of critical stress intensity (fracture toughness),[4] crack growth rate,[142] fracture energy,[104] and fracture resistance.[104] These models predict stress distribution, and rheological behavior of the matrix around the particle.

## 8.10 WEAR

The abrasive wear of plastics occurs as a result of strong adhesive interaction, fatigue, macroshearing, abrasive action, thermal and thermooxidative interaction, corrosion, cavitation, etc. Fillers are involved in these processes because mineral fillers are abrasive and cause wear of the mating surfaces, other fillers are used to reduce wear.[146-50]

The wear volume of plastic material is given by equation:

$$W_s = K\frac{\mu P}{E}\frac{DW}{I_s} \quad [8.18]$$

where:

| | |
|---|---|
| K | proportionality constant |
| μ | coefficient of friction |
| P | force |
| E | modulus |
| D | sliding distance |
| W | load |
| $I_s$ | interlaminar shear strength |

A filler used as a wear decreasing additive should not lower the permissible strain, μP/E.

Both the matrix and the filler contribute to wear resistance. Typical polymers used in these applications include polyamide, polyacetal, polybutene terephthalate, and polycarbonate.[150] These polymers have the right balance of required properties, such as low friction coefficient, good mechanical properties, impact strength, and dimensional stability. Filler selection depends on the value of its friction coefficient, its having a minimal influence on the mechanical properties of the matrix polymer, and its good adhesion to the matrix. Frequently, it is difficult to meet these criteria because fillers which have a low friction coefficient do not combine well with other materials. Polytetrafluoroethylene, frequently used as anti-wear additive, is such an example. The addition of 20% PTFE to polyamide-66 reduces its tensile strength by 40%. In order to compensate for this effect, PTFE is frequently used in combination with glass fibers. Glass fibers increase the abrasiveness of the material but also reinforce it and thus balance the losses due to the use of PTFE.

Typical fillers used for reduction of wear include PTFE, silicone, graphite powder, molybdenum disulfide, and aramid fibers. Good results were also reported with mica and zirconia combination.[151] Figure 8.36 shows the effect of mica and mica in combination with zirconia on the wear resistance of an epoxy resin.[151]

Figure 8.37 shows that additions of graphite and $MoS_2$ are capable of making PTFE even more wear resistant.[146] This effect is limited to a relatively small range of filler concentration because PTFE does not interact with fillers and its mechanical properties deteriorate rapidly as filler concentration increases. The addition of 40% filler (graphite or $MoS_2$) reduces the elongation from 240% to close to zero. Addition of only 20% filler reduces the elongation to about half of the unfilled value.[146]

Figure 7.9 illustrates the effect of fiber orientation on wear rate.[153-4] The surface roughness of the mating surfaces influences wear. Adhesion of fibers to the matrix and their dispersion are other essential parameters of the aramid fibers performance.[150] Large surface area of the aramid fibers caused mechanical inter-

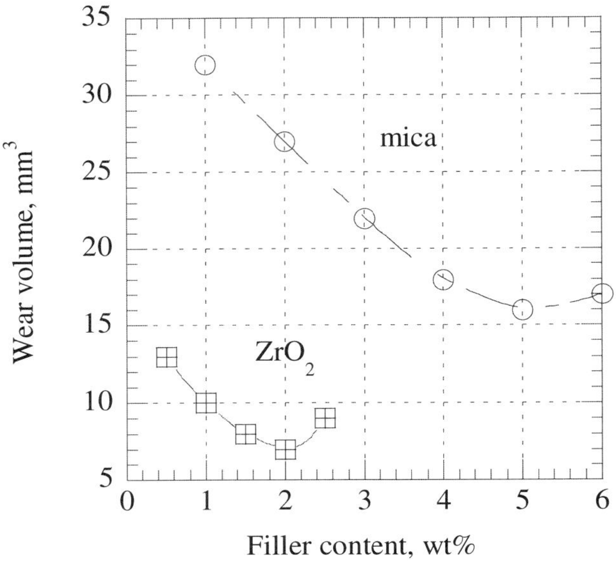


Figure 8.36. Wear volume vs. amount of filler. [Data from Srivastava V K, Pathak J P, *Polym. & Polym. Composites*, **3**, No.6, 1995, 411-4.]

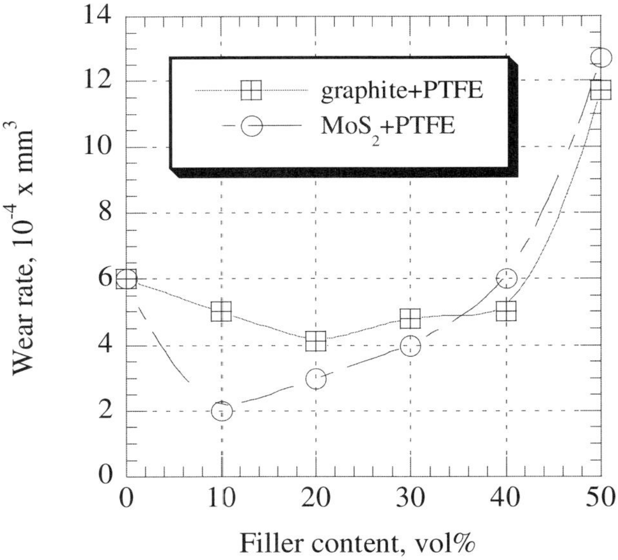


Figure 8.37. Wear rate vs. filler content. [Adapted, by permission, from Fengyuan Yan, Qunji Xue, Shengrong Yang, *J. Appl. Polym. Sci.*, **61**, No.7, 1996, 1223-9.]

locking and limited their dispersion in the matrix. An application of a sizing agent increased adhesion but also caused the fiber to form bundles which were difficult to separate after cutting process. A special process oil was used to prevent bundling.

## 8.11 FRICTION

**Table 8.3. Friction coefficient of some plastics**

| Polymer | Filler | Amount, wt% | Dynamic coefficient of friction |
|---|---|---|---|
| Polycarbonate | PTFE | 10 | 0.12 |
| Polycarbonate | Aramid | 15 | 0.15 |
| Polycarbonate | PTFE | 15<br>15 | 0.10 |
| Polyoxymethylene | PTFE<br>Aramid<br>Silicone | 10<br>5<br>3 | 0.05 |
| Polybutyleneterephthalate | PTFE | 15 | 0.20 |
| Polyethyleneterephthalate | PTFE | 10 | 0.15 |
| Polyamide-6 | PTFE | 15 | 0.23 |
| Polyamide-6 | PTFE<br>Silicone | 18<br>2 | 0.11 |
| Polyamide-6,6 | PTFE<br>Glass fiber | 15<br>30 | 0.26 |
| Polyamide-6,6 | PTFE<br>Glass fiber<br>Silicone | 13<br>30<br>2 | 0.14 |
| Polyamide-6,6 | PTFE<br>Carbon fiber | 15<br>30 | 0.11 |
| Polyamide-11 | PTFE | 15 | 0.31 |
| Polyamide-12 | PTFE | 20 | 0.20 |
| Polyamide, amorphous | PTFE | 20 | 0.22 |
| Polypropylene | PTFE<br>Glass fiber | 15<br>20 | 0.09 |
| Polyurethane, thermoplastic | PTFE | 15 | 0.32 |

This section adds information on the influence of filler on the wear behavior of plastics.[149,152-4] The friction coefficient of a material depends on the applied load according to the following equation:

$$\mu = KW^{0.3-0.4} \tag{8.19}$$

where:

K proportionality coefficient

W load

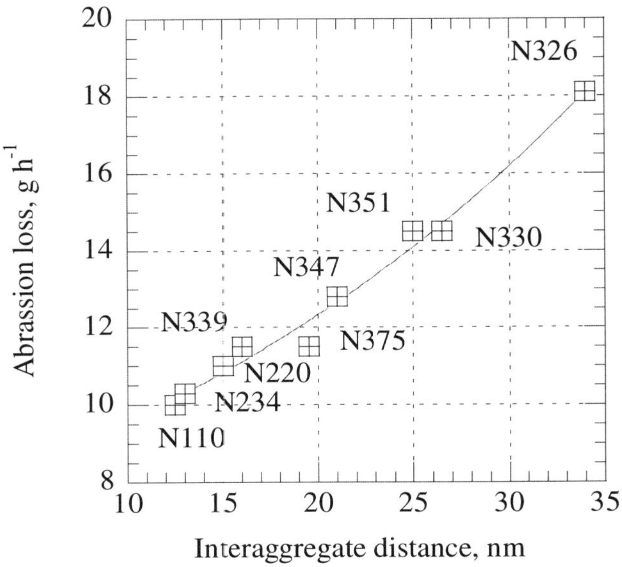


Figure 8.38. Abrasion loss of SBR containing different types of carbon black vs. interaggregate distance. [Adapted, by permission, from Patel A C, *Kaut. u. Gummi Kunst.*, **47**, No.8, 1994, 556-70. ]

Table 8.3 gives friction coefficients of plastics containing fillers intended as wear reduction additives.

Generally, the friction coefficient decreases as the load of filler increases but there is a critical quantity above which the friction coefficient decreases. The correct amount of anti-wear additive for a particular material and a particular applied load can be determined by simple morphological observation of the surface. The expected wear pattern forms surface debris whereas if the part is not wearing well its surface will melt and become shiny.

## 8.12 ABRASION

Two aspects of abrasion will be discussed, namely the abrasion resistance of filled materials and the use of fillers in the friction materials.[34,123,126,156-7] Each has its own specificity and differs from other two.

Rubber, due to its elastomeric properties, usually, has a low abrasion resistance. Fillers such as carbon black and silica can be added to impart abrasion resistance. Figure 8.38 shows the extent to which different grades of carbon black are abraded.[156] As interaggregate distance decreases the abrasion loss decreases as well.

Figure 8.39 shows that increasing the filler concentration decreases abrasion loss. Because of chemical interaction between the hydroxyl groups on clay and the ionic crosslinker in EPDM, clay reduces the abrasion loss more effectively than carbon black. Carbon black forms a weak interaction with the backbone

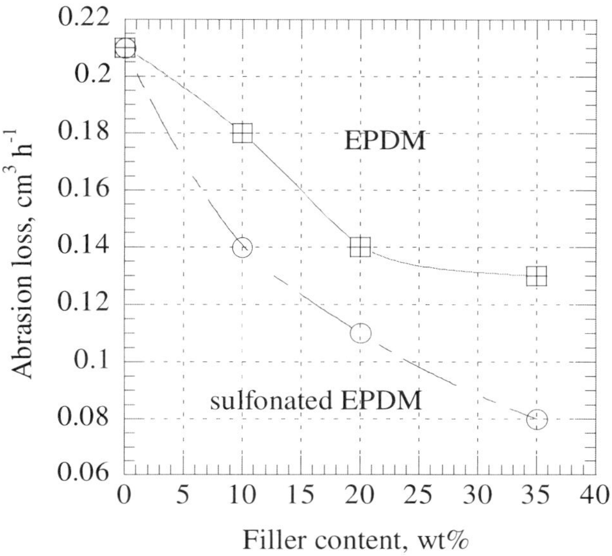


Figure 8.39. Abrasion loss of EPDM vs. filler content. [Data from Kurian T, De P P, Khastgir D, Tripathy D K, De S K, Peiffer D G, *Polymer*, **36**, No.20, 1995, 3875-84.]

chains.[34,123] Increased adhesion between the filler and the matrix contributes to an increase in abrasion resistance.[126] Studies on natural rubber filled with silica demonstrated this performance improvement. The adhesion between the filler and the matrix was increased by a multifunctional additive. As the concentration of the additive was increased the abrasion loss decreased.

Typical friction materials used in the automotive industry are brake pads and clutches.[157] In the past, asbestos was the filler of choice but after 1983, the use of asbestos was gradually phased out and alternative replacements were found after extensive tests on over 1200 different materials. There are three main technologies used for brake pad production: metallic brake linings, carbon-carbon composites, and resin-bonded friction materials. Resin bonded friction materials will be discussed here. The friction material must fulfill several requirements which makes the choice of filler a complicated process. In addition to a sufficient friction coefficient (typically 0.4-0.5), brakes should not suffer a substantial and long-lasting loss of friction during repeated cycles of overheating and returning to normal temperature. Brakes should recover quickly from friction loss due to exposure to water, they should have high tolerance to different magnitudes of load and wear. Meeting all these requirements is challenging. Fillers are seldom used alone but rather in combination. These may include a conductor to facilitate heat dissipation (metal powder, most frequently copper or brass), a particulate filler for surface filling and cost reduction (typical candidates are barytes, calcium carbonate, and clays) and reinforcement (fibers which can withstand heat and maintain mechanical and fric-

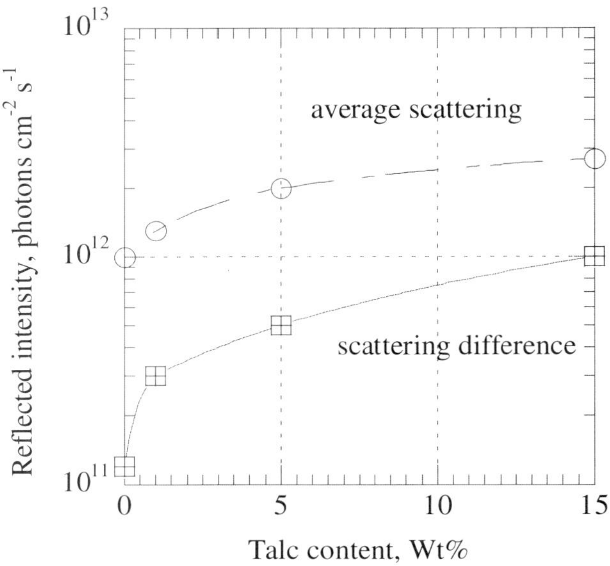


Figure 8.40. Average scattering and scattering difference vs. talc load in polypropylene. [Adapted, by permission, from Kody R S, Martin D C, *Polym. Engng. Sci.*, 36, No.2, 1996, 298-304.]

tional properties, such as aramid, glass, carbon, steel, cellulosic). All fibers used today are still no match for asbestos which had thermal stability, high mechanical properties, and sufficient elasticity to prevent fracture (typical of glass and carbon fibers). The biggest challenge of new technologies is to overcome existing relationship between coefficient of friction and wear rate which makes more durable brakes less efficient.

## 8.13 SCRATCH RESISTANCE

In spite of the fact that scratch resistance is very important requirement only a limited number of credible studies have been done. This research is very difficult to conduct.[158-9] Recently, image analysis was employed to overcome some of the technical difficulties associated with the interpretation of observations. In addition to the damage of the scratch itself, the surrounding areas show stress whitening which adds to the perceived damage. The use of plastics in the automotive industry, especially the wide use of polypropylene makes these studies very important in the development of high quality finished products.

The method adapted in instrumental studies includes several steps: surface scratching using a controlled testing apparatus, analyzing the scratch area by reflected polarized light in an optical microscope, image acquisition, and analysis of the digital image.[158] Figure 8.40 shows of the data obtained when the scratch surface of talc filled propylene was analyzed. As the amount of talc was increased, scratch resistance decreased. The average scattering of reflected polarized light

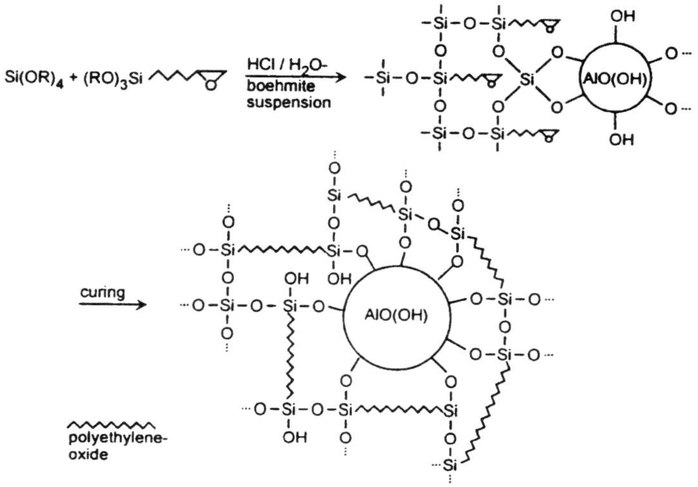


Figure 8.41. Reaction scheme and structural model of boehmite-based nanocomposite. [Adapted, by permission, from Schmidt H K, *Macromol. Symp.*, **101**, 1996, 333-42.]

correlates with void formation due to the impact-related separation of filler from the matrix. The scattering difference is related to changes in chain orientation resulting from scratch damage. Results can be improved by increasing the adhesion between filler and matrix only an incremental improvement is possible because of an inherent deficiency in talc (low hardness).

Figure 8.41 shows the structure of a novel nanocomposite and how it was formed.[159] This nanocomposite has high scratch resistance. It is currently used for an optical lens coating. Its scratch resistance was increased by factor of 3 in the diamond scratch test when compared with conventional hard coatings and by factor of 2 in lens testing where lenses are rotated in drum of abrasive material. These excellent properties are due to a structure which makes both the filler and the matrix a singular, ordered material.

## 8.14 FATIGUE

Fillers contribute to the elastomeric properties of materials and thus affect the fatigue resistance of compounded products.[73,124,132,140,144,160-177] The discussion below includes:

- The principles governing fatigue resistance
- The mechanisms of fatigue deterioration of filled systems
- The effect of fillers on improvement of fatigue resistance

Paris' law relates crack propagation rate to stress intensity:

$$\frac{da}{dN} = C(\Delta K)^m \qquad [8.20]$$

where:
a crack length
N number of cycles
C, m material dependent constants
$\Delta K$ $= K_{max} - K_{min}$, stress intensity factor range

The values of constants C and m are not associated with any particular physical meaning. $\Delta K$ increases when filler load increases. This indicates that the filler increases fatigue resistance. Crack growth data can be conveniently obtained by interfacing tensile testing machine operating in a cyclic mode with a software/hardware combination capable of determining crack length at one cycle intervals.[140] The following equation is used to calculate results:

$$\left(\frac{da}{dN}\right)_n = \frac{a_{n+1} - a_{n-1}}{N_{n+1} - N_{n-1}} \qquad [8.21]$$

where:
n iteration number

Figure 8.42 shows one outcome of such studies.[140] The addition of surface treated glass beads requires substantially higher stress to support the same growth rate of cracks compared with neat epoxy. Good adhesion and stress distribution are responsible for improvement.

The other factor which is important in fatigue resistance studies is related to dissipation of energy. The energy accumulated in the material due to extension must be dissipated. Energy dissipation is given by the equation:

$$W_d = \int_0^{2\pi} \sigma(t)\frac{d\gamma}{dt}dt = \pi G''\gamma_0^2 \qquad [8.22]$$

where:
$W_d$ energy dissipated in a complete strain cycle per unit volume
$\sigma$ stress response
$\gamma$ input strain
t time
G″ loss modulus ($G'' = G' \tan\delta$)

Figure 8.43 shows difference in behavior between filled and unfilled materials.[174] There is a large difference in the energy stored by the two systems due to the reinforcing action of carbon black. Also, filled systems differ in their reaction to continuous stress. In a filled system, nonlinear behavior is observed due in part to the level of energy involved in cycling. This affects efficiency of energy dissipation and formed crosslinks.

Stress softening is a phenomenon related to filler reinforcement. When a material is extended to a certain strain, returned to zero strain and stretched again, the second stress-strain curve lies below the first one. There are several reasons for this phenomenon, including incomplete elastic recovery, conversion of the hard phase

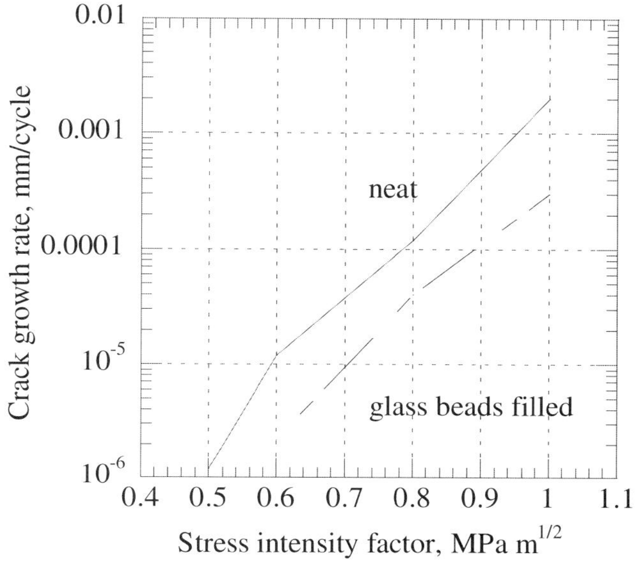


Figure 8.42. Crack growth rate of epoxy with and without glass beads vs. stress intensity factor. [Adapted, by permission, from Azimi H R, Pearson R A, Hertzberg R W, *J. Appl. Polym. Sci.*, **58**, No.2, 1995, 449-63.]

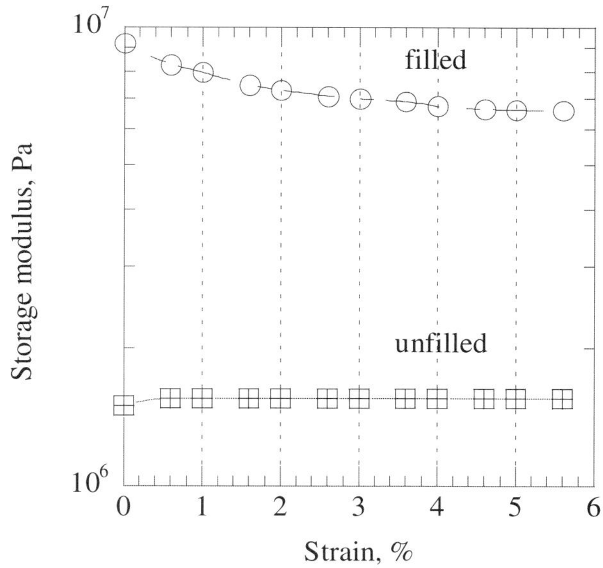


Figure 8.43. Storage modulus of nitrile rubber with and without carbon black. [Adapted, by permission, from Gownder M, Letton A, Hogan H, Antec '95. Vol. II. Conference Proceedings, Boston, Ma., 7th-11th May 1995, 1983-6.]

to soft phase, breaks in the network, and chain slippage from detachment points.[160]

The equation for loss of deformation energy gives a mathematical interpretation of this phenomenon:

$$\Delta W = \left[1 - \int_0^{\varepsilon} \sigma d\varepsilon \int_0^{\varepsilon} \sigma_0 d\varepsilon\right] 100 \qquad [8.23]$$

where:

| | |
|---|---|
| $\Delta W$ | loss of deformation energy |
| $\sigma$ | stress value of second cycle (deformation) |
| $\varepsilon$ | maximal strain |
| $\sigma_0$ | stress value of the first cycle |

The Mooney-Rivlin equation is frequently used to interpret experimental results related to stress softening:

$$g(\varepsilon) = 1 + \frac{C_2}{C_1}\frac{1}{\lambda} \qquad [8.24]$$

where:

| | |
|---|---|
| $g(\varepsilon)$ | damping function (based on the fact that the material's response does not remain proportional to the deformation) |
| $C_1$ | elastic constant (a measure of the connectivity of network strands) |
| $C_2$ | stress relaxation constant (contribution of trapped entanglements to the equilibrium modulus) |
| $\lambda$ | extension ratio |

Failure occurs in two steps: crack initiation and crack propagation. The examples above and some examples presented below show that the addition of filler may help in interfering with the propagation step. But the filler introduces inhomogeneity to the matrix and thus may contribute to crack initiation. In this respect the properties of the matrix are very important. If the matrix has a brittle character, its fatigue resistance depends on crack initiation. Once a crack is initiated, its propagation is a very fast process. In ductile matrix there is a larger resistance to crack propagation, therefore the introduction of a crack initiation site (e.g., fillers) is not as detrimental to overall fatigue resistance. Matrix-filler adhesion and the matrix character determine the influence of the filler on crack initiation.

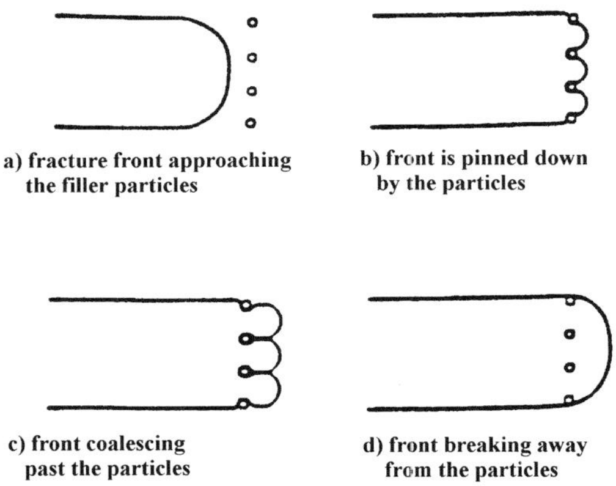


Figure 8.44. Crack front propagation slowed down by a pinning mechanism. [Adapted, by permission, from Azimi H R, Pearson R A, Hertzberg R W, *J. Appl. Polym. Sci.*, **58**, No.2, 1995, 449-63.]

Crack propagation is affected by the properties of the filler and its interaction with

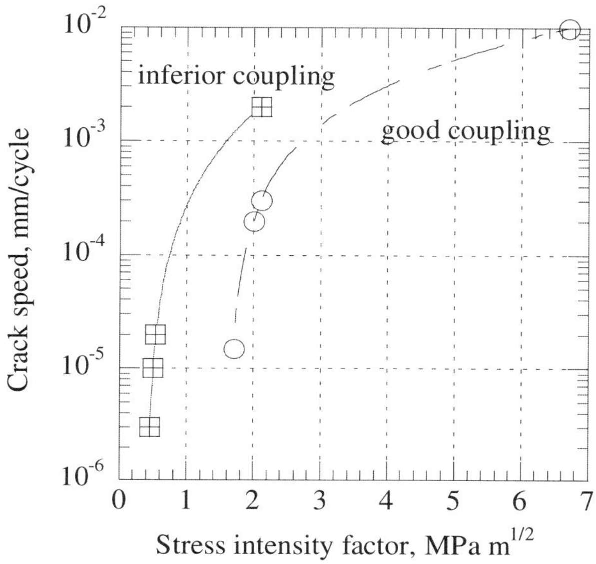


Figure 8.45. Effect of coupling on crack rate development in kaolin-filled HDPE. [Adapted, by permission, from Savadori A, Scapin M, Walter R, *Macromol. Symp.*, **108**, 1996, 183-202.]

the matrix. Figure 8.44 shows the mechanism by which filler particles may slow crack propagation.[140] A crack front approaches filler particles which have good adhesion to the matrix. The front of the crack is slowed by filler particles because of their interaction with the tip stress field. Cavitation and coalescence of voids is followed by the matrix breaking away from particles and the crack front progressing to the next obstacle.[140]

Figure 8.45 shows experimental data from a different source[73] which confirms the pinning mechanism. Much higher stress intensity (ΔK from Eq 8.20) is required to support the same crack growth rate in material with good adhesion as compared with material with poor adhesion.

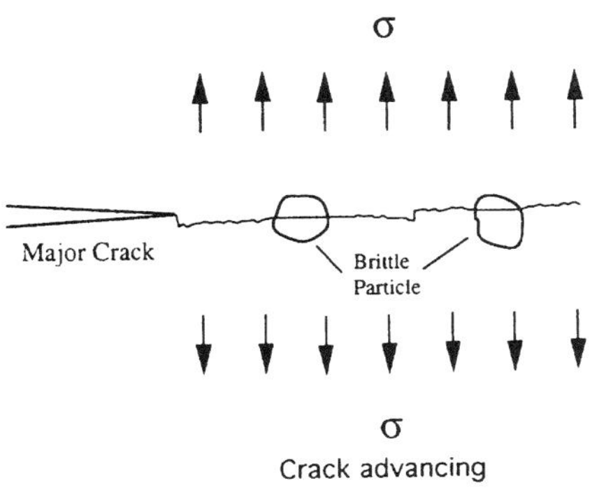


Figure 8.46. Crack growth mechanism. [Adapted, by permission, from Xu X X, Crocrombe A D, Smith P A, *Int. J. Fatigue*, **16**, No.7, 1994, 469-77.]

Figure 8.46 shows the effect of crack advancement on integrity of filler particle.[164] The advancing crack can be delayed by the pinning mechanism if stress is lower than adhesion and filler particle cohesion. If stress is higher than particle cohesion then particles may break. The

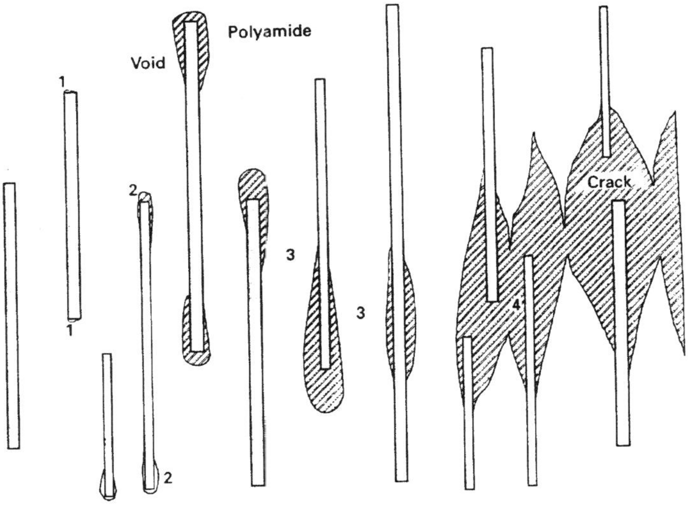


Figure 8.47. Failure mechanism in fiber reinforced system. [Adapted, by permission from Horst J J, Spoormaker J L, *J. Mater. Sci.*, **32**, 1997, 3641-51.]

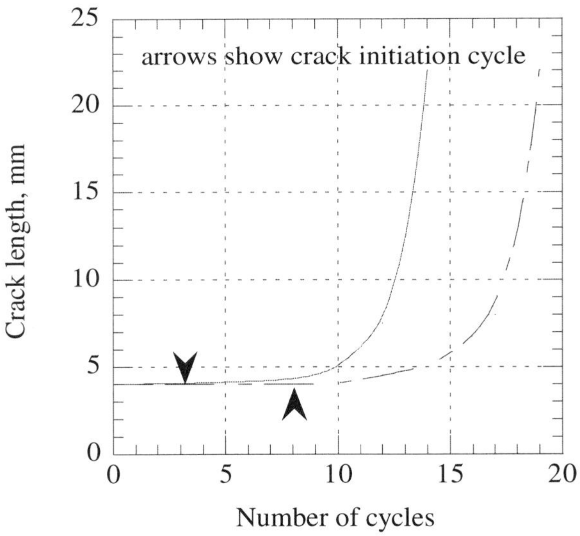


Figure 8.48. Crack length vs. number of cycles for talc filled PC/ABS blend. [Adapted, by permission, from Seibel S R, Moet A, Bank D H, Sehanobish K, Antec 95. Volume III. Conference proceedings, Boston, Ma., 7th-11th May 1995, 3966-70.]

experimental evidence of the operation of this mechanism is given in Figure 7.31.

The fiber debonding mechanism is given in Figure 8.47.[162] Several stages are involved, including debonding, void formation, coalescence, and finally crack formation. Because of a higher stress around the fiber ends, debonding is more likely

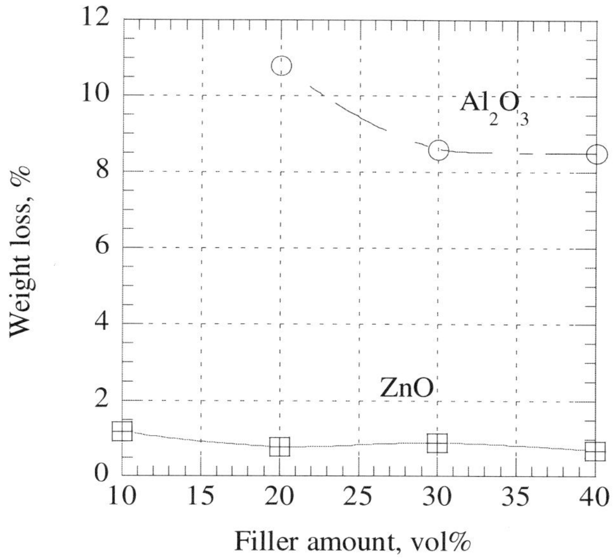


Figure 8.49. Weight loss vs. filler amount in PDMS. [Data from Visser S A, *J. Appl. Polym. Sci.*, **64**, 1997, 1499-1509.]

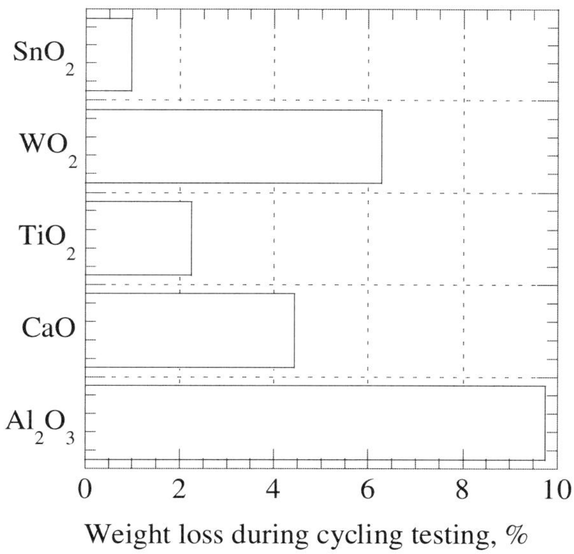


Figure 8.50. Weight loss vs. filler type. [Data from Visser S A, *J. Appl. Polym. Sci.*, **63**, 1997, 1805-20.]

to occur in these areas. There is substantial evidence from SEM studies which confirms that such mechanisms operate in composites.

Several examples below illustrate the effect of fillers on fatigue resistance. Figure 8.48 shows that crack length depends on the number of cycles.[144,166] Talc increases the time to failure by about 50%. Both crack initiation and crack propagation were improved by the addition of talc.

Figure 8.49 shows the effect of filler concentration on fatigue resistance.[170] The weight loss during the cycling stress experiment is caused by the reaction of ionic fragments with polymer chains. As a result of these reactions, volatile species are lost. In this experiment each sample was subjected to 60 cycles.

Figure 8.50 shows the results of similar studies for different fillers in PDMS.[169] In most cases the increased Young modulus of filled composite corresponds to less weight loss. This points again to the effect of interaction on failure resistance.

## 8.15 FAILURE

These remarks evaluate the effect of filler-related phenomena on failure of plastic materials.[80,90,175-182] Several reasons for the failure of plastics are filler related. They include delamination of laminated composite materials, debonding in particulate filled materials, stress cracking of filler particles, yielding, cavitation, and corrosion.

Failure analysis of composite pipe[182] shows that the majority of problems are related to bundling of fibers which prevents the binder from wetting the individual fibers, the porosity of the resin which allows penetration of the composite by corroding liquids, and a lack of adhesion between the matrix and a sand filler. All these failures can be substantially reduced and the useful life of pipes extended by application of proper technological practices. A high porosity in the binder layer increased corrosion rate which, in turn, reduced the original fiber strength to 1/3 of its initial value over a period of 10 years. With good fiber protection, mechanical strength was reduced by only 30%.[182]

Figures 8.51 and 8.52 show differences between the behavior of polypropylene filled with glass beads at different temperatures.[175] In both cases, the debonding between filler and the matrix requires the lowest level of energy and confirms that this is the most likely mode of failure. The volume fraction of filler has little effect on debonding, cavitation, and yielding at 0°C. At -60°C, yielding is improved by increasing concentration of filler. Debonding is initiated at the poles and begins plastic yielding in the matrix which ultimately leads to failure.[90] Strain required to initiate failure is reduced when the filler concentration is increased.[90]

The adhesion between the matrix and the filler has an important influence on the mechanism of failure. In polyester filled with quartz, uncoupled (low adhesion) quartz was delaminated from the matrix if the quartz particles were in the path of the crack growth. Silane coupled quartz particles showed many instances of particle cracking on the pathway of crack growth. Apparently, adhesive forces were higher than the cohesion of filler material.[178]

In polymer blends, the distribution of filler between the two or more component polymers of blends influenced tensile, tear, and fatigue failures. For example,

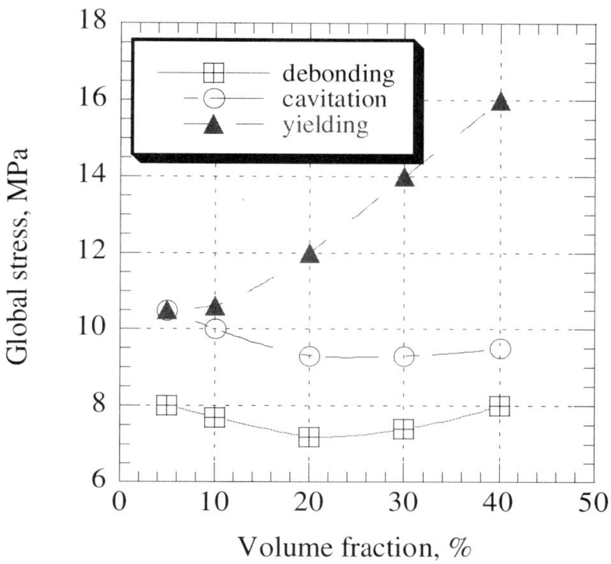


Figure 8.51. Global stress vs. volume fraction of glass beads in polypropylene at -60°C. [Adapted, by permission, from Asp L E, Sjogren B A, Berglund L A, *Polym. Composites*, **18**, No.1, 1997, 9-15.]

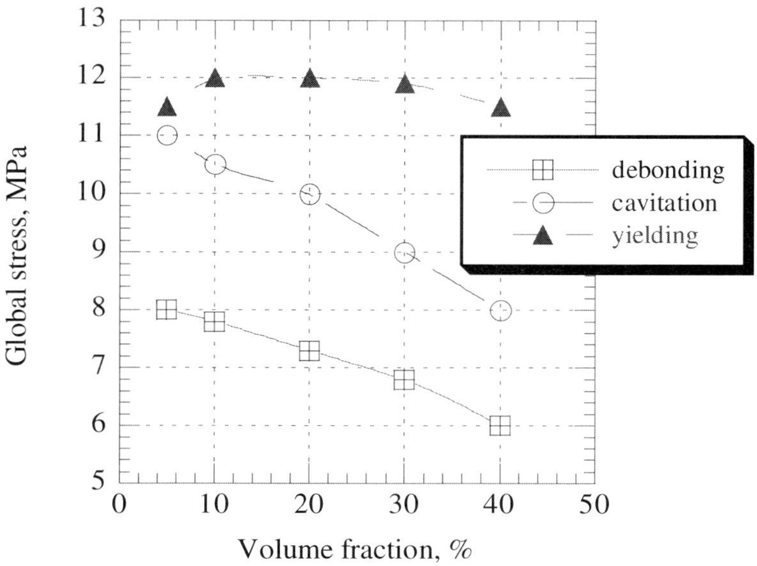


Figure 8.52. Global stress vs. volume fraction of glass beads in polypropylene at 0°C. [Adapted, by permission, from Asp L E, Sjogren B A, Berglund L A, *Polym. Composites*, **18**, No.1, 1997, 9-15.]

when carbon black was distributed equally in two phases the fatigue life was 30% better than when all of the carbon black resided in only one phase. Also, uniform distribution of carbon black increased failure resistance.

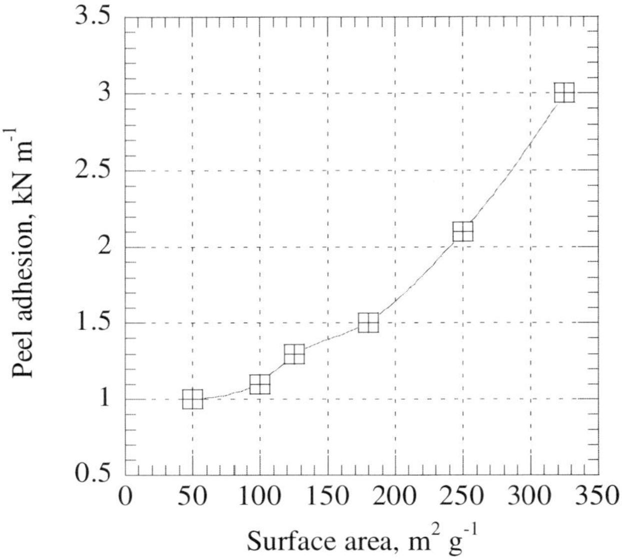


Figure 8.53. Effect of fumed silica surface area on peel adhesion of silicone sealant. [Data from Cochrane H, Lin C-S, *Rubber World*, 1985.]

## 8.16 ADHESION

This section presents the available data on adhesion of plastic to other substrates in relationship to the concentration of filler.[1,45,77,105,164-5,183-200] Adhesion between filler and the matrix has been discussed in other sections. Figure 8.53 shows the effect of fumed silica on adhesion of silicone sealant. This might be application specific since the adhesion of a silicone sealants is improved by silanes. The increased surface area of fumed silica is associated with an increased concentration of functional groups on its surface, so the improvement is related to the effect of the silane rather than that of the filler.[200]

In another specific application that of hot melt adhesives, the choice of filler has an effect on adhesion (Figure 8.54).[199] Four types of spheres (K-20, S-22, Z-light, and ML 3050) increase adhesion compared with unfilled adhesive. The first two (K-20 and S-22) are hollow microspheres of very low density. The remaining two have a lower density than glass but have thicker walls. Three fillers ($CaCO_3$, Zeospheres, and aluminum) are solid products. All fillers which decrease adhesion have substantially higher thermal conductivity than the fillers which increase adhesion. From the studies of crystallization profiles, it is apparent that longer crystallization times allow the network to orient itself on the substrate, resulting in better adhesion.[199]

In $TiO_2$ filled coatings either using different amounts of $TiO_2$ or using $TiO_2$ with a variety of surface treatments no substantial difference in adhesion was

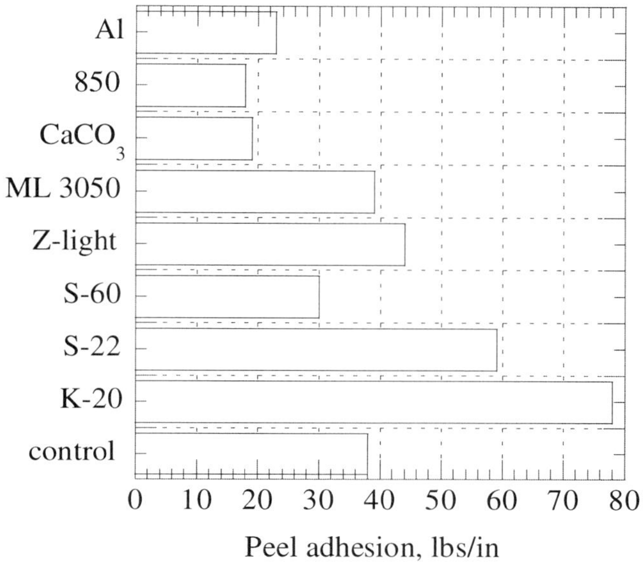


Figure 8.54. Effect of filler type on adhesion of polyurethane adhesive. [Adapted, by permission, from Oien H T, Polyurethanes '95. Conference Proceedings, Chicago, Il., 26th-29th Sept.1995, 137-41.]

found.[185,191] Figure 7.16 indicates that the orientation of polymer on the substrate's surface is responsible for excluding most of the filler from the layer adjacent to the substrate.

The adhesion of PVAc adhesive was improved by the addition of calcite, alumino-silicate and starch but gains were rather small and the mode of failure was altered from cohesive failure in the unfilled sample to adhesive failures in the filled samples.[45]

In PET films containing $CaCO_3$, the adhesion of the film was less than that of unfilled film. This is due to the replacement of the adhesion promoting surface by filler which has no adhesive properties.[197] In UV cured adhesives, quartz filler contributed to a faster development of adhesion due to its transparency to the UV radiation used for curing.[198] In polymer blends filler was accumulated at the interphase between polymers. Adhesion depended on the interaction of filler with both polymers and on the particle size of filler.[188]

In general, fillers may increase the adhesion of plastics to other materials if they can become involved in the mechanism of adhesion (e.g., participate in coupling), change the properties or the orientation of the polymer (e.g., influence curing, elastomeric properties, orientation, etc.). But in many instances fillers may decrease adhesion if they are concentrated at the interface or increase the modulus (higher stress at the interface) of materials or increase viscosity (retarded wetting).

The addition of filler to the material which is combined with other substrates by intermediate layer (adhesive), usually improves adhesion. The adhesion of

LDPE coating to polyester film, clay-coated liner board, and aluminum foil was improved by an addition of calcium carbonate. Adhesion increases as the amount of calcium carbonate is increased and at a 30% level it is doubled.[189] The adhesion of SBR to a polyurethane adhesive was substantially improved (400%) by the addition of silica but only when the surface was roughened. The amount of silica did not affect adhesion of unroughened SBR.[195] The use of MgO with silane coupling increased adhesion of bromobutyl liner by a factor of four.[190] Generally, in the reported data, surface roughening of the filled composite contributes to a better adhesion to other substrates joined by adhesives.

## 8.17 THERMAL DEFORMATION

Heat distortion temperature is one important property of plastic materials which can be improved by the incorporation of filler.[12-3,20,23-4,43,74] Table 8.4 shows what can be obtained with various systems.

Table 8.4 shows that substantial gains can be obtained by filling crystalline polymers but amorphous polymers are not much affected by reinforcement. Also, particulate fillers are substantially less effective than fibrous fillers. Glass fiber is the most useful filler in this application. Figure 8.55 shows the effect of two grades of particulate fillers on the heat deflection temperature of polypropylene.[43] Small changes are observed at smaller additions followed by a rapid increase in HDT above a 30% filler content. The particle size has only small difference.

## 8.18 SHRINKAGE

There are several reports on the influence of fillers on shrinkage.[9,115,193,198,201-4] Figure 8.56 shows mold shrinkage vs. concentration of filler.[9] Mold shrinkage can be reduced to half of the value for unfilled resin by the incorporation of mica. Additional reduction of shrinkage is possible if the interaction between filler and the matrix can be increased. This can be achieved by reacting polypropylene with maleic anhydride.[9]

Mold shrinkage is even more efficiently reduced by glass fiber or combination of mica and glass fiber. Glass fiber alone reduces mold shrinkage more effectively than mica. For example, mold shrinkage of in this experiment[201] was 0.28% for 40% glass fiber in polypropylene and 0.96% for 40% mica. If combination of mold shrinkage with warpage reduction is required combination of mica with glass fiber gives better results than the use of glass fiber alone but in such compositions shrinkage increases with mica concentration increasing.[201]

Figure 8.57 shows the anisotropy of mold shrinkage.[202] Unfilled polypropylene (RTP100) shows very big difference between longitudinal and transverse shrinkage. Polypropylene containing 40% calcium carbonate has very similar shrinkage in both directions. In addition, the shrinkage of filled material is consistently lower and less dependent on cycle time. Lower holding times can be used with filled material.[202]

**Table 8.4. Heat deflection temperature of various systems**

| Polymer | Filler | Amount, % | HDT, °C @ 18.6 | Ref. |
|---|---|---|---|---|
| ABS | -<br>glass fiber | 0<br>30 | 172<br>358 | 12 |
| Polyamide-6 | -<br>glass fiber | 0<br>30 | 75<br>212 | 12 |
| Polyamide-66 | -<br>glass fiber<br>wollastonite | 0<br>30<br>20 | 95<br>248<br>247 | <br>12<br>13 |
| Polyketone | -<br>glass fiber<br>mica<br>wollastonite<br>$CaCO_3$ | 0<br>20<br>20<br>20<br>20 | 105<br>210<br>165<br>120<br>95 | 23 |
| Polyetheretherketone | -<br>glass fiber | 0<br>30 | 155<br>315 | 12 |
| Polybutyleneterephthalate | -<br>glass fiber | 0<br>30 | 65<br>210 | 12 |
| Polypropylene | -<br>glass fiber<br>carbon black<br>talc<br>$CaCO_3$<br>aluminosilicate | 0<br>30<br>30<br>40<br>40<br>50 | 65<br>148<br>130<br>135<br>116<br>120 | <br>12<br>74<br>20<br>20<br>43 |
| Polyethersulfone | -<br>glass fiber | 0<br>30 | 201<br>216 | 12 |
| Polyphenylenesulfone | -<br>glass fiber | 0<br>30 | 172<br>181 | 12 |
| Polyarylate | -<br>glass fiber | 0<br>30 | 175<br>180 | 12 |
| Polyimide | -<br>glass fiber | 0<br>30 | 275<br>348 | 12 |
| Polyamide-imide | -<br>glass fiber | 0<br>30 | 270<br>275 | 12 |
| Polyetherimide | -<br>glass fiber | 0<br>30 | 200<br>210 | 12 |

Proper choice of filler allows to modify shrinkage of UV curable transparent adhesives.[198] The shrinkage of UV curable adhesive was reduced by factor 2-3 by incorporation of quartz which has better UV light transparency than conventionally

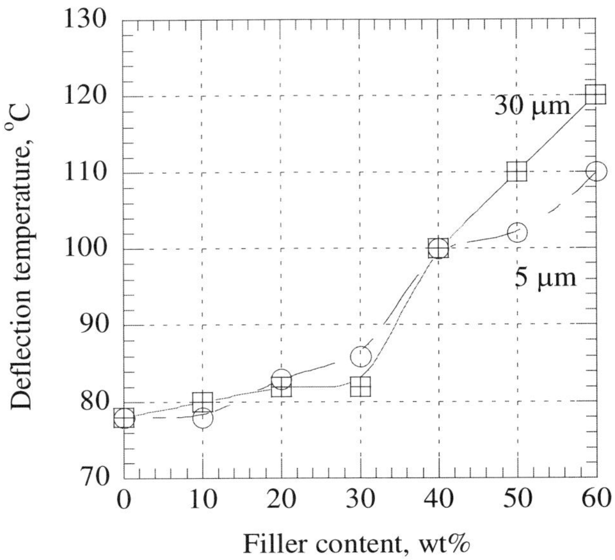


Figure 8.55. Heat deflection temperature of polypropylene containing hydrated K-Mg aluminosilicate. [Adapted, by permission, from Schott N R, Rahman M, Perez M A, *J. Vinyl and Additive Technol.*, **1**, No.1, 1995, 36-40.]

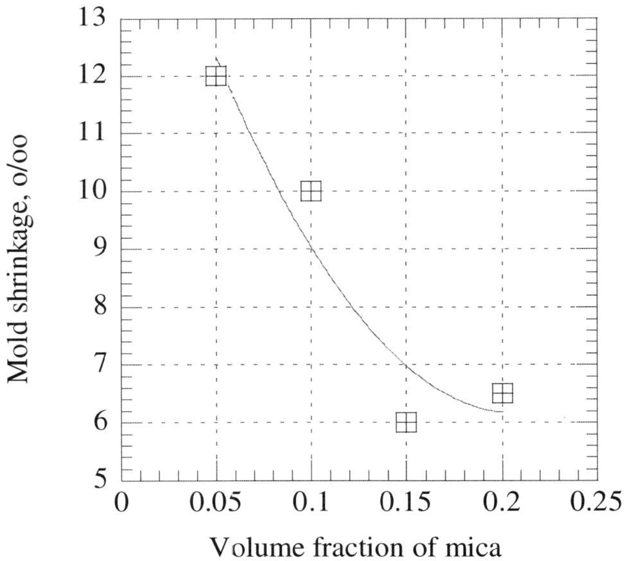


Figure 8.56. Mold shrinkage of mica/polypropylene composites vs. concentration of filler. [Data from Chiang W Y, Yang W D, Pukanszky B, *Polym. Engng. Sci.*, **34**, No.6, 1994, 485-92.]

used fillers. Due to more uniform curing and development of network less shrinkage was observed.

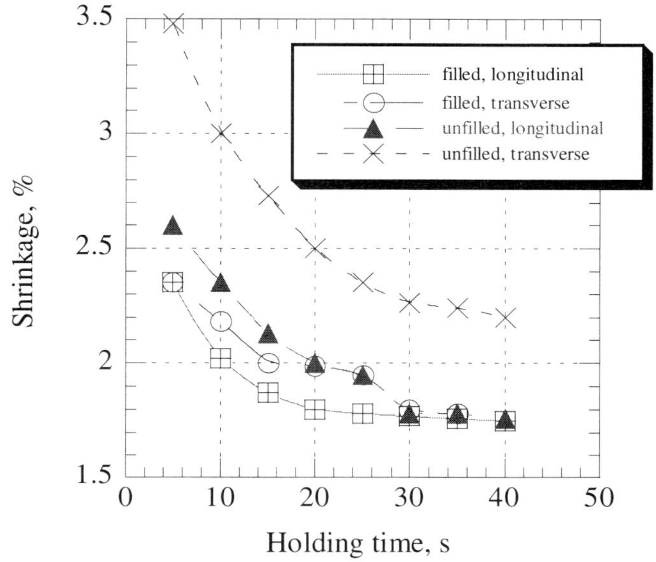


Figure 8.57. Mold shrinkage vs. holding time. [Data from Mamat A, Trochu F, Sanschagrin B, *Polym. Engng. Sci.*, **35**, No.19, 1995, 1511-20.]

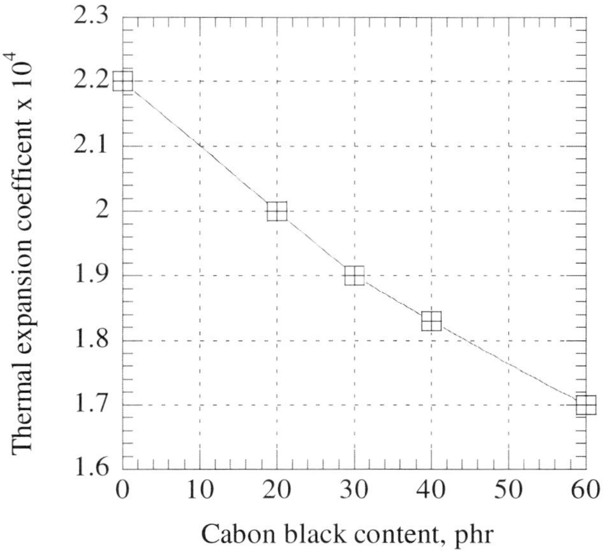


Figure 8.58. Thermal expansion coefficient of polyisoprene rubber vs. carbon black amount. [Data from Priss L S, *Int. Polym. Sci. Technol.*, **23**, No.7, 1996, T53-6.]

Figure 8.58 shows the effect of carbon black on thermal expansion coefficient of rubber.[204] Increasing rubber content contributes to a decrease of thermal expansion.

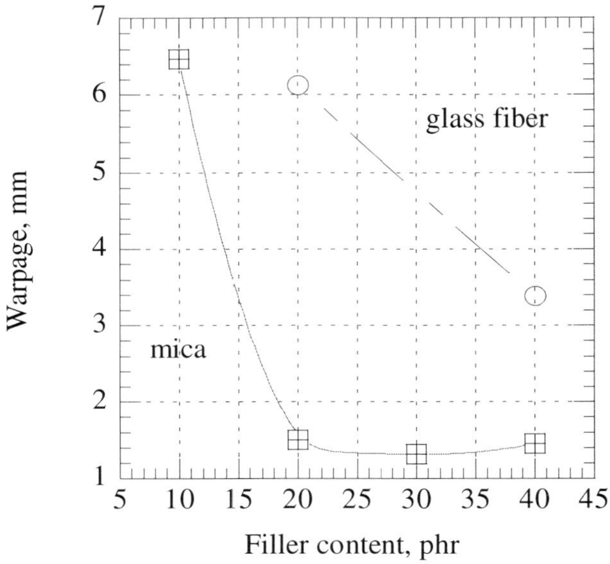


Figure 8.59. Warpage vs. filler content in polypropylene. [Data from Canova L A, Fergusson L W, Parrinello L M, Subramanian R, Giles H F, Antec '97. Conference proceedings, Toronto, April 1997, 2112-6.]

## 8.19 WARPAGE

Warpage results from residual stresses within a molded part. The ejected part is not constrained by the mold and, if its residual stresses are higher than the modulus of the material, material distortion occurs.[201,205-6] Some fillers can help to decrease warpage.

Figure 8.59 shows the difference in warpage between mica filled and glass fiber filled polypropylene.[201] An increase of mica concentration to 20% is sufficient for a substantial reduction in warpage. Glass fiber can also reduce warpage but it is much less effective than mica. Glass fiber is very efficient in shrinkage reduction (see previous section), so combinations of mica and glass fiber usually give a good balance of properties. In this experiment, the best compromise between shrinkage and warpage was attained when the polypropylene was filled with 20% mica and 20% glass fiber.[201]

Warpage is a function of stress distribution within the material. Stress distribution depends, in turn, on the distribution of filler particles. If filler distribution is not uniform (see section 7.2) stress distribution will vary in different sections of the part. Typically, warpage close to the edges has a different direction of deflection than in the center of the part. Also, there is a higher negative deflection in the region of the part remote from the injection nozzle because these regions are filler deficient.

The type of filler plays an important role. Fillers which have a platelike structure are more efficient than fibers, and fibers are more efficient then particulates.

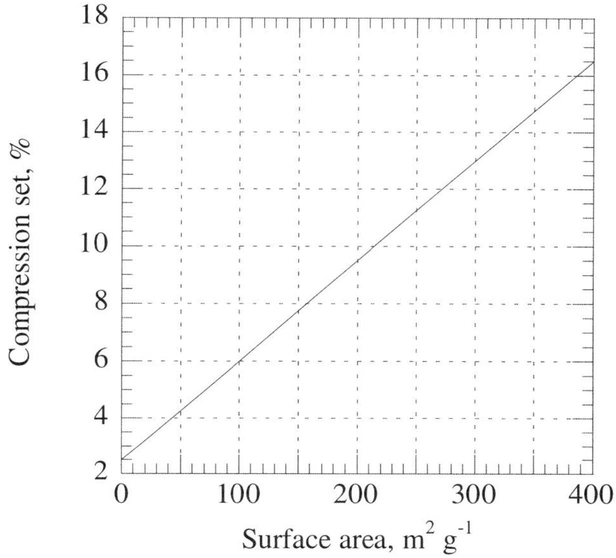


Figure 8.60. Compression set of vulcanized silicone rubber vs. surface area of silica. [Data from Okel T A, Waddell W H, *Rubb. Chem. Technol.*, **68**, No.1, 1995, 59-76.]

Orientation of filler particles is also important therefore warpage is influenced by processing conditions which tend to orient the filler such as rate of flow, temperature, ejection temperature, and material crystallization conditions. The current literature does not provide much information on this subject.

## 8.20 COMPRESSION SET

Compression set is an important property of elastomers which is affected by the choice of filler.[18,121,125,207-9] Studies were conducted on silica in silicon rubber vulcanizates. Figure 8.60 shows the relationship between the surface area of silica and compression set.[125] As the surface area increases compression set increases. The increase surface area contributes to an increase in the number of functional groups on the surface of silica. These groups can potentially react with siloxane. When they do, there is a good interaction of filler with matrix which contributes to reduction of compression set (Figure 8.61).[125]

Acidic silica has more active hydroxyl groups on its surface which increase reactivity. The amount of moisture in silica also increases compression set. This is because promoting its water is necessary to initiate reaction with siloxane and thus its condensation with hydroxyl groups on the surface of silica.

Figure 8.62 shows the compression set of rubber with different fillers at 1:1 proportion to rubber.[207] Fillers, such as precipitated calcium carbonate, whiting, calcinated clay, each of which have limited interaction with the matrix give a substantially lower compression set. As the interaction between filler and the matrix

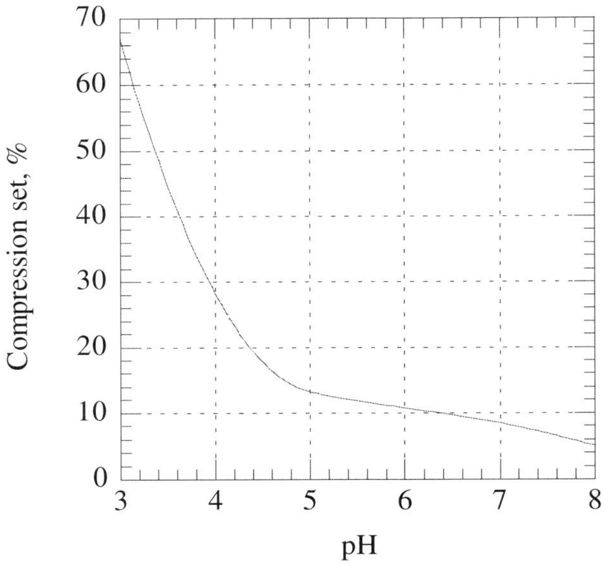


Figure 8.61. Compression stress of vulcanized silicone rubber vs. pH of silica. [Adapted, by permission, from Okel T A, Waddell W H, *Rubb. Chem. Technol.*, **68**, No.1, 1995, 59-76.]

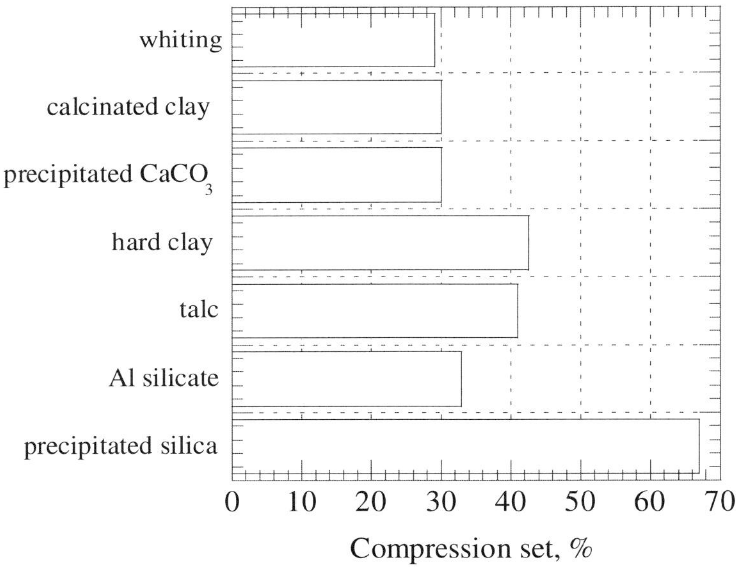


Figure 8.62. Compression set of EPDM compounded with different fillers. [Adapted, by permission, from Mushack R, Luttich R, Bachmann W, *Eur. Rubb. J.*, **178**, No.7, 1996, 24-9.]

increases, compression set increases. This is also found with different grades of silica.[125]

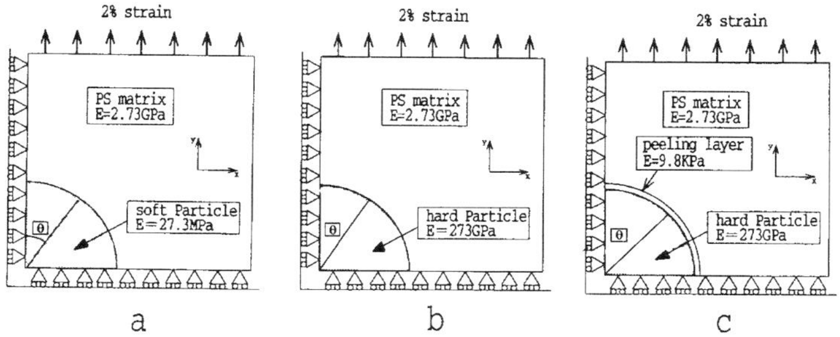


Figure 8.63. Finite element method stress analysis around the particulate in polystyrene. (a) the system with dispersed softer particles, (b) the system with dispersed harder particles, (c) the system with dispersed particles having a peeling layer (adsorbed polymer). [Adapted, by permission, from Mitsui S, Kihara H, Yoshimi S, Okamoto Y, *Polym. Engng. Sci.*, **36**, No.17, 1996, 2241-6.]

Cellulose fibers in NBR were found to increase compression set as the concentration of fiber was increased which is consistent with the fact that cellulose interacts with the matrix and increases tensile strength, modulus and abrasion resistance.[18] A comparison of two grades of carbon black shows that one grade decreases compression set in 5 different rubbers whereas the other grade does not change compression set in a broad range of concentrations.[121] The reasons for this are not explained.

## 8.21 LOAD TRANSFER

This section discusses the structural contribution that fillers make by their participation in stress transfer in filled systems.[5,67,73,90,110,175,204,210-213] The intensity of stress at high concentration of particles (interacting stress fields) is given by:

$$\sigma^* = \frac{\sigma^e}{1 + m\phi^{1/3}} \qquad [8.25]$$

where:

$\sigma^e$ external load
m proportionality constant
$\phi$ filler volume concentration

An increased concentration of filler decreases stress around individual particles. Figure 8.63 shows the diagrams used for stress analysis by the finite element method.[110] Figure 8.64 shows the stress concentration factor, C(θ) vs. the angle of the maximum stress concentration point, θ.[110] Soft particles and particles with a peeling layer have a maximum stress concentration at $\theta = 90^o$. This explains why delamination occurs at the poles, perpendicular to the applied strain. The peeling layer forms in the strain direction and crazes propagate from the peeling layer perpendicular to the applied strain.[110]

Stress distribution and load transfer can be conveniently measured by two methods: Raman spectroscopy[210] and NMR.[212] Figure 8.65 shows that the Raman frequency correlates well with the applied strain.[210] The Raman absorption peak

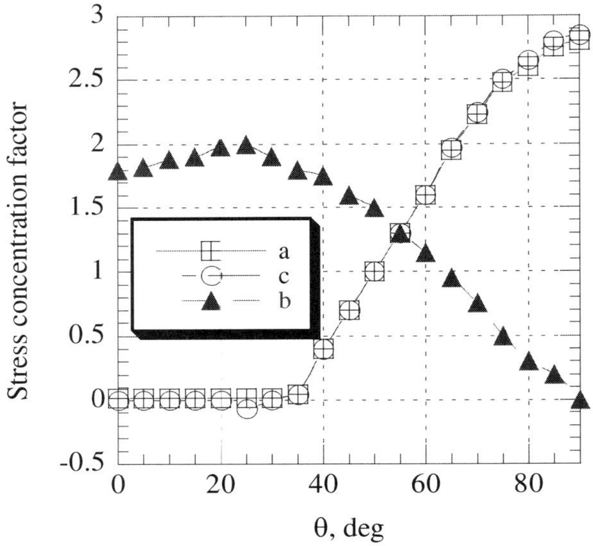


Figure 8.64. Concentration coefficient of the maximum principal stress vs. the highest stress concentration point. Particles types are labeled as in Figure 8.63. [Adapted, by permission, from Mitsui S, Kihara H, Yoshimi S, Okamoto Y, *Polym. Engng. Sci.*, **36**, No.17, 1996, 2241-6.]

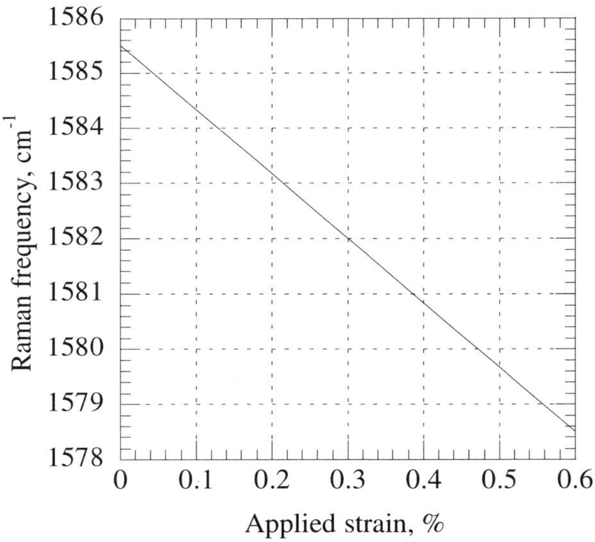


Figure 8.65. Raman frequency shift vs. tensile strain applied to carbon fibers. [Adapted, by permission, from Leveque D, Auvray M H, *Composites Sci. & Technol.*, **56**, No.7, 1996, 749-54.]

shifts relative to the applied tensile strain. From this data it is possible to obtain information on elastic load transfer and debonded part. NMR imaging,[212] can be used to calculate the strain map based on measurements of the relaxation time, $T_2$.

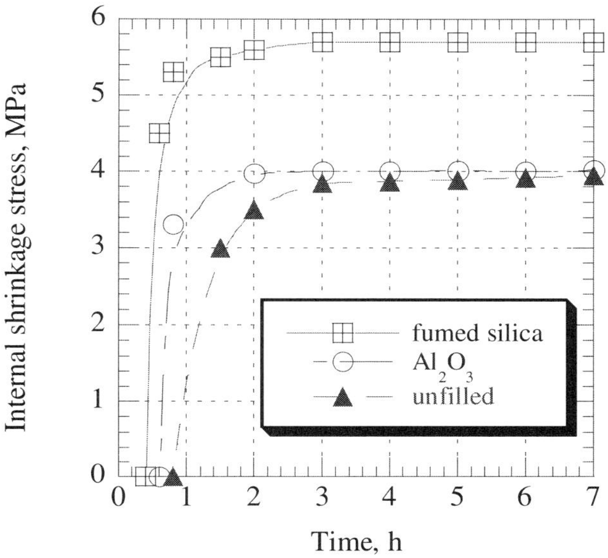


Figure 8.66. Internal shrinkage stress vs. interpenetrating network formation time for PU/PEA=10/90 network. (1) unfilled network, (2) alumina-filled, (3) fumed silica-filled. [Data from Sergeeva L M, Skiba S I, Karabanova L V, Polym. Int., 39, No.4, April 1996, 317-25.]

## 8.22 RESIDUAL STRESS

The residual stress is associated with changes of volume during processing (cooling, evaporation, crystallization) and during the useful lifetime of the finished products (temperature differences combined with differences in the thermal expansion coefficients of the various materials in formulation). The effect of filler can be predicted.[193,203,210,214] The residual thermal strain was determined in carbon fiber composites by Raman spectroscopy.[210] This stress was not large (-0.1%) but it was concentrated at the fiber ends.

In sheet molded laminates, residual stress decreased as the fiber fraction increased and as the angle of bundle orientation decreased.[214] An increased load can reduce stress as the stresses are distributed throughout the laminate (more interacting surfaces). When the stress is high enough to equal matrix strength, it can cause damage to the laminate. Balancing of stress is a convenient method of improving properties of fiber laminates.

Figure 8.66 shows the effect of different fillers on the residual stress in an interpenetrating network.[203] The stress is determined by the chemical nature of filler. Filler which has a good interaction with the network (such as fumed silica) increases shrinkage stress.

In paints, shrinkage stress influences paint quality. Paint may undergo mud cracking rather than forming a uniform film. Cracking is associated with critical volume packing. If the concentration of filler is higher than critical volume pack-

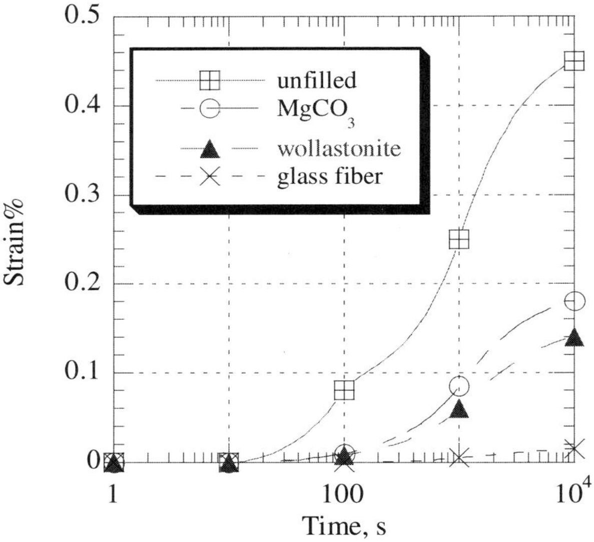


Figure 8.67. Creep strain of LCP vs. time for different fillers. [Adapted, by permission, from Scaffaro R, Pedretti U, La Mantia F P, *Eur. Polym. J.*, **32**, No.7, 1996, 869-75.]

ing, filler particles cannot move closer together and cracking is very likely to occur. The film forming properties of binder also influence shrinkage stress and cracking.[193]

## 8.23 CREEP

The creep resistance of materials depends on filler-matrix interaction and, therefore, is very much related to fillers use.[7,29,215-7] A simple equation shows creep strain:

$$\frac{\varepsilon_c(t)}{\varepsilon_m(t)} = \frac{E_m}{E_c} \qquad [8.26]$$

where

| | |
|---|---|
| $\varepsilon_c$ | strain of filled polymer |
| $\varepsilon_m$ | strain of matrix (unfilled polymer) |
| t | time |
| $E_m$ | Young's modulus of matrix |
| $E_c$ | Young's modulus of filled polymer |

This equation shows that if the creep strain is lower than that predicted by the equation, filler particles induce stress in the surrounding matrix (if no debonding occurs). Figure 8.67 shows the effect of different fillers on creep and Figure 8.68 shows the effect of filler concentration.[29] Creep strain decreases when the filler interaction with matrix increases. This is demonstrated in experiments which show that magnesium carbonate gives a higher strain than does wollastonite and glass fiber. The increased concentration of filler decreases strain. Both results are consistent with the effect of fillers on Young's modulus.

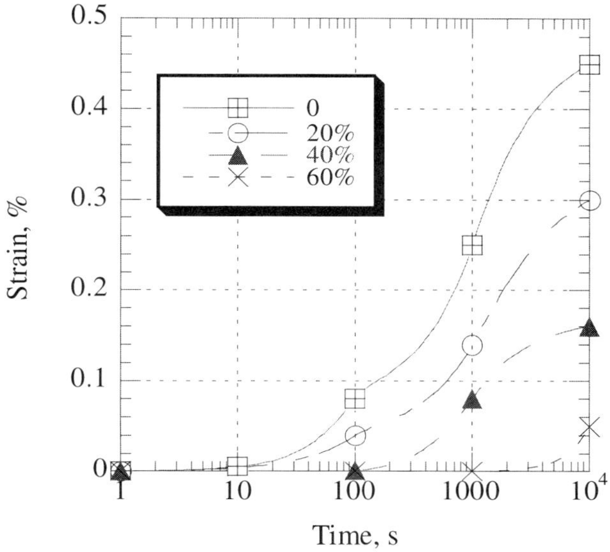


Figure 8.68. Creep strain of LCP vs. time for different concentrations of magnesium carbonate. [Adapted, by permission, from Scaffaro R, Pedretti U, La Mantia F P, *Eur. Polym. J.*, **32**, No.7, 1996, 869-75.]

It was demonstrated that the effect of filler can be fully utilized when stress has sufficient value. At low stress, all glass bead filled composites had the same creep. When stress was increased, it became apparent that surface treatment with different coupling agents produced materials with different creep characteristics. The best results were obtained when glass beads were treated with a combination of silane and bismaleimide. It was also discovered that the creep damage areas appear in the polar regions, in a manner similar to short-term stress-induced damage. After initial debonding microcracks occur at the end of the debonding zone and lead to failure.[215]

The addition of a small amount of glass fiber (7%) to recycled PE and PP drastically reduced creep. This combination was found useful for production of packaging materials from recycled milk bottles.

## REFERENCES


1 Kovacevic V, Lucic S, Hace D, Cerovecki Z, *J. Adhesion Sci. Technol.*, **10**, No.12, 1996, 1273-85.
2 Sombatsompop N, Sims G L A, *Cell. Polym.*, **15**, No.5, 1996, 317-34.
3 Zaini M J, Fuad M Y A, Ismail Z, Mansor M S, Mustafah J, *Polym. Int.*, **40**, No.1, 1996, 51-5.
4 Hashemi S, Din K J, Low P, *Polym. Engng. Sci.*, **36**, No.13, 1996, 1807-20.
5 Pukanszky B, Voros G, *Polym.Composites*, **17**, No.3, 1996, 384-92.
6 Hashemi S, Gilbride M T, Hodgkinson J, *J. Mat. Sci.*, **31**, No.19, 1996, 5017-25.
7 Bush S F, Tonkin J D, Antec '97. Conference proceedings, Toronto, April 1997, 3081-85.
8 Meddad A, Fellahi S, Pinard M, Fisa B, Antec '94. Conference Proceedings, San Francisco, Ca., 1st-5th May 1994, Vol. II, 2284-8.
9 Chiang W Y, Yang W D, Pukanszky B, *Polym. Engng. Sci.*, **34**, No.6, 1994, 485-92.
10 Fu Q, Wang G, Liu C, *Polymer*, **36**, No.12, 1995, 2397-401.

11 Godard P, Bomal Y, Biebuyck J J, *J. Mat. Sci.*, **28**, No.24, 1993, 6605-10.
12 Maxwell J, **Plastics in High Temperature Applications**, *Rapra*, Shawbury, 1992.
13 Enhancing Polymers Using Additives and Modifiers, *Rapra*, Shawbury, 1993.
14 Ou Y, Yu Z, Zhu J, Li G, Zhu S, *Chinese J. Polym. Sci.*, **14**, No.2, 1996, 172-82.
15 Saad A L G, Younan A F, *Polym. Degradat. Stabil.*, **50**, No.2, 1995, 133-40.
16 Younan A F, Choneim A M, Tawfik A A A, Abd-El-Nour K N, *Polym. Degradat. Stabil.*, **49**, No.2, 1995, 215-22.
17 Yang Q, Pritchard G, Phipps M A, Rose R G, *Polym. & Polym. Composites*, **4**, No.4, 1996, 239-46.
18 Vieira A, Nunes R C R, Visconti L L Y, *Polym. Bull.*, **36**, No.6, 1996, 759-66.
19 Gill T S, Xanthos M, *J. Vinyl and Additive Technol.*, **2**, No.3, 1996, 248-52.
20 Hattori K, Morford S, Delany D T, *J. Vinyl and Additive Technol.*, **1**, No.3, 1995. 170-3.
21 Bos M, Van Dam G W, Jongsma T, Bruin P, Pennings A J, *Composite Interfaces*, **3**, No.2, 1995, 169-76.
22 Casenave S, Ait-Kadi A, Brahimi B, Riedl B, Antec '95. Vol. II. Conference Proceedings, Boston, Ma., 7th-11th May 1995, 1438-42.
23 Gingrich R P, Machado J M, Londa M, Proctor M G, Antec '95. Vol. II. Conference Proceedings, Boston, Ma., 7th-11th May 1995, 2345-50.
24 Hojabr S, Boocock J R B, Antec 95. Volume III. Conference proceedings, Boston, Ma., 7th-11th May 1995, 3620-7.
25 Gill T S, Xanthos M, Wirsbo Co., Antec '96. Volume II. Conference proceedings, Indianapolis, 5th-10th May 1996, 1757-61.
26 Belanger B, Sanschagrin B, Fisa B, Antec '96. Volume II. Conference proceedings, Indianapolis, 5th-10th May 1996, 1762-7.
27 Montsinger L V, Antec '96. Volume II. Conference proceedings, Indianapolis, 5th-10th May 1996, 2546-9.
28 Johnson K C, Antec '96. Volume III. Conference proceedings, Indianapolis, 5th-10th May 1996, 3545-9.
29 Scaffaro R, Pedretti U, La Mantia F P, *Eur. Polym. J.*, **32**, No.7, 1996, 869-75.
30 Sims G L A, Sombatsompop N, *Cell. Polym.*, **15**, No.2, 1996, 90-104.
31 Akovali G, Dilsiz N, *Polym. Engng. Sci.*, **36**, No.8, 1996, 1081-6.
32 Lekatou A, Faidi S E, Lyon S B, Newman R C, *J. Mat. Res.*, **11**, No.5, 1996, 1293-304.
33 Liu Z, Gilbert M, *J. Appl. Polym. Sci.*, **59**, No.7, 1996, 1087-98.
34 Kurian T, De P P, Tripathy D K, De S K, Peiffer D G, *J. Appl. Polym. Sci.*, **62**, No.10, 1996, 1729-34.
35 Alkan C, Arslan M, Cici M, Kaya M, Aksoy M, *Resources Conserv. & Recycling*, **13**, Nos.3-4, 1995, 147-54.
36 Cochet P, Barruel P, Barriquand L, Grobert J, Bomal Y, Prat E, IRC '93/144th Meeting, Fall 1993. Conference Proceedings, Orlando, Fl., 26th-29th Oct.1993, Paper 162.
37 Zaborski M, Slusarski L, Vidal A, *Polim. Tworz. Wielk.*, **38**, No.6, 1993, 263-8.
38 Rockenbauer A, Korecz L, Pukanszky B, *Polym. Bull.*, **33**, No.5, 1994, 585-9.
39 Ulkem I, Bataille P, Schreiber H P, *J. Macromol. Sci. A*, **31**, No.3, 1994, 291-303.
40 Anantharaman M R, Kurian P, Banerjee B, Mohamed E M, George M, *Kaut. u. Gummi Kunst.*, **49**, No.6, 1996, 424-6.
41 Schneider J P, Myers G E, Clemons C M, English B W, *J. Vinyl and Additive Technol.*, **1**, No.2, 1995, 103-8.
42 Zolotnitsky M, Steinmetz J R, *J. Vinyl and Additive Technol.*, **1**, No.2, 1995, 109-13.
43 Schott N R, Rahman M, Perez M A, *J. Vinyl and Additive Technol.*, **1**, No.1, 1995, 36-40.
44 Clemens M L, Doyle M D, Lees G C, Briggs C C, Day R C, Flame Retardants '94. Conference proceedings, London, 27th-28th January 1994, 193-202.
45 Kovacevic V, Lucic S, Hace D, Glasnovic A, Smit I, Bravar M, *J. Adhesion*, **47**, No.1-3, 1994, 201-15.
46 Pechulis M, Vautour D, Antec '97. Conference proceedings, Toronto, April 1997, 1860-4.
47 Barbosa S E, Kenny J M, Antec '97. Conference proceedings, Toronto, April 1997, 1855-9.
48 Davis R M, Gardner S H, Marand E, Laot C, Reifsnider K, DeSmidt H, McGrath J, Tan B, Antec '97. Conference proceedings, Toronto, April 1997, 2494-9.
49 Kumar P, Gawahale A R, Rai B, *Adv. Composite Materials*, **4**, No.4, 1995, 279-85.
50 Canova L A, Antec '93. Conference Proceedings, New Orleans, La., 9th-13th May 1993, Vol. II, 1943-9.
51 Liptak P, *Int. Polym. Sci. Technol.*, **21**, No.8, 1994, T/50-3.
52 Sato T, *Int. Polym. Sci. Technol.*, **21**, No.3, 1994, T/45-52.

53 Pukanszky B, Belina K, Rockenbauer A, Maurer F H J, *Composites*, **25**, No.3, 1994, 205-14.
54 Byung Suk Jin, Kwang Hee Lee, Chul Rim Choe, *Polym. Int.*, **34**, No.2, 1994, 181-5.
55 Bataille P, Mahlous M, Schreiber H P, *Polym. Engng. Sci.*, **34**, No.12, 1994, 981-5.
56 Lan T, Pinnavaia T J, *Chem. of Mat.*, **6**, No.12, 1994, 2216-9.
57 Nicolais L, Narkis M, *Polym. Eng. Sci.*, **11**, 1971, 194.
58 Nielsen L E, *J. Appl. Phys.*, **4**, 1970, 4626.
59 Piggott M R, Leidner J, *J. Appl. Polym. Sci.*, **18**, 1974, 1619.
60 Kelly A, Tyson W R, *J. Mech. Phys. Solids*, **6**, 1965, 13.
61 Yamada H, Inagaki S, Okamoto H, Furukawa J, *Int. Polym. Sci. Technol.*, **21**, No.6, 1994, T/29-35.
62 Enhancing Polymers Using Additives and Modifiers II, *Rapra*, Shawbury, 1996, paper 5.
63 Miller B, *Plast. World*, **54**, No.1, 1996, 38-43.
64 Hornsby P R, Watson C L, *J. Mat. Sci.*, **30**, No.21, 1995, 5347-55.
65 Cochrane H, Lin C S, *Rubb. Chem. Technol.*, **66**, No.1, 1993, 48-60.
66 Bhattacharya S K, Bhowmick A K, Singh R K, *J. Mat. Sci.*, **30**, No.1, 1995, 243-7.
67 Voros G, Pukanszky, *J. Mat. Sci.*, **30**, No.16, 1995, 4171-8.
68 Cheung T, Tjong S C, Li R K Y,Antec '96. Volume II. Conference proceedings, Indianapolis, 5th-10th May 1996, 2256-9.
69 Liu Z, Gilbert M, *J. Appl. Polym. Sci.*, **59**, No.7, 1996, 1087-98.
70 Wiebking H E, *J. Vinyl and Additive Technol.*, **2**, No.3, 1996, 187-9.
71 Wiebking H E, Antec 95. Volume III. Conference proceedings, Boston, Ma., 7th-11th May 1995, 4112-6.
72 Pukanszky B, Maurer F H J, Boode J W, *Polym. Engng. Sci.*, **35**, No.24, 1995, 1962-71.
73 Savadori A, Scapin M, Walter R, *Macromol. Symp.*, **108**, 1996, 183-202.
74 Chiu H-T, Chiu W-M, *J. Appl. Polym. Sci.*, **61**, No.4, 1996, 607-12.
75 Zihlif A M, Di Liello V, Martuscelli E, Ragosta G, *Int. J. Polym. Mat.*, **29**, Nos.3-4, 1995, 211-20.
76 Jancar J, Dibenedetto A T, *J. Mat. Sci.*, **30**, No.6, 1995, 1601-8.
77 Jancar J, DiBenedetto A T, *Sci. & Eng. Composite Materials*, **3**, No. 4, 1994, 217-26.
78 Wang M-J, Wolff S, Tan E-H, *Rubb. Chem. Technol.*, **66**, No.2, 1993, 178-95.
79 Palumbo M, Donzella G, Tempesti E, Ferruti P, *J. Appl. Polym. Sci.*, **60**, No.1, 1996, 47-53.
80 Jancar J, *Macromol. Symp.*, **108**, 1996, 163-72.
81 Averous L, Quantin J C, Lafon D, Crespy A, *Int. J. Polym. Analysis and Characterization*, **1**, No.4, 1995, 339-47.
82 Beloshenko V A, Kozlov G V, Slobodina V G, Prut E V, Grinev V G, *Polym. Sci., Ser. B*, **37**, Nos.5-6, 1995, 316-8.
83 Babich V F, Lipatov Yu S, Todosijchuk T T, *J. Adhesion*, **55**, Nos.3-4, 1996, 317-27.
84 Guerrica-Echevarria G, Eguiazabal J I, Nazabal J, *Polym. Degradat. Stabil.*, **53**, No.1, 1996, 1-8.
85 Jones D W, Rizkalla A S, *J. Biomedical Materials Research (Applied Biomaterials)*, **33**, No.2, 1996, 89-100.
86 Owen A J, Koller I, *Polymer*, **37**, No.3, 1996, 527-30.
87 Wong K W Y, Truss R W, *Composites Sci. & Technol.*, **52**, No.3, 1994, 361-8.
88 Yang A C M, *Polymer*, **35**, No.15, 1994, 3206-11.
89 Roesch J, Barghoorn P, Muelhaupt R, *Makromol. Chem. Rapid Commun.*, **15**, No.9, 1994, 691-6.
90 Sjogren B A, Berglund L A, *Polym. Composites*, **18**, No.1, 1997, 1-8.
91 Y. S. Lipatov, **Polymer Reinforcement**, *ChemTec Publishing*, Toronto, 1995.
92 Akay M, Mun S K A, Stanley A, *Composites Sci. Technol,* **57**, 1997, 565-71.
93 Leguet X, Ericson M, Chundury D, Baumer G, Antec '97. Conference proceedings, Toronto, April 1997, 2117-34.
94 Ohta M, Nakamura Y, Hamada H, Maekawa Z, *Polym. & Polym. Composites*, **2**, No.4, 1994, 215-21.
95 Yang Q, Pritchard G, *Polym. & Polym. Composites*, **2**, No.4, 1994, 233-9.
96 Ou Y C, Zhu J, Feng Y P, *J. Appl. Polym. Sci.*, **59**, No.2, 1996, 287-94.
97 Gahleitner M, Bernreitner K, Knogler B, Neissl W, *Macromol. Symp.*, **108**, 1996, 127-36.
98 Tan L S, McHugh A J, Gulgun M A, Kriven W M, *J. Mat. Res.*, **11**, No.7, 1996, 1739-47.
99 Tan L S, McHugh A J, *J. Mater. Sci.*, **31**, 1996, 3701-6.
100 Sanchez-Solis A, Padilla A, *Polym. Bull.*, **36**, No.6, 1996, 753-58.
101 Toure B, Lopez Cuesta J-M, Longerey M, Crespy A, *Polym. Degradat. Stabil.*, **54**, Nos 2-3, 1996, 345-52.
102 Thomason J L, Vlug M A, *Composites, Part A*, **28A**, 1997, 277-88.
103 Herzig R, Baker W E, *J. Mat. Sci.*, **28**, No.24, 1993, 6531-9.

104 Pukanszky B, Maurer F H J, *Polymer*, **36**, No.8, 1995, 1617-25.
105 Jancar J, Dibenedetto A T, Dianselmo A, *Polym. Engng. Sci.*, **33**, No.9, 1993, 559-63.
106 Jancar J, DiBenedetto A T, Antec '94. Conference Proceedings, San Francisco, Ca., 1st-5th May 1994, Vol. II, 1710-2.
107 Skelhorn D A, Antec '93. Conference Proceedings, New Orleans, La., 9th-13th May 1993, Vol. II, 1965-70.
108 Plummer C J G, Wu Y, Gola M M, Kausch H H, *Polym. Bull.*, **30**, No.5, 1993, 587-94.
109 Nabi Z U, Hashemi S, *J. Mat. Sci.*, **31**, No.21, 1996, 5593-601.
110 Mitsui S, Kihara H, Yoshimi S, Okamoto Y, *Polym. Engng. Sci.*, **36**, No.17, 1996, 2241-6.
111 Qi Wang, Jizhuang Cao, Guangjin Li, Xi Xu, *Polym. Int.*, **41**, No. 3, 1996, 245-9.
112 Tiganis B E, Shanks R A, Long Y, Antec '96. Volume II. Conference proceedings, Indianapolis, 5th-10th May 1996, 1744-9.
113 McIlvaine J, Antec 95. Volume III. Conference proceedings, Boston, Ma., 7th-11th May 1995, 3346-9.
114 Yu M C, Bissell M A, Whitehouse R S. Antec 95. Volume III. Conference proceedings, Boston, Ma., 7th-11th May 1995, 3246-50.
115 Yu T C, Antec '95. Vol. II. Conference Proceedings, Boston, Ma., 7th--11th May 1995, 2358-68.
116 Wang K J, Sue H J, Antec '95. Vol. II. Conference Proceedings, Boston, Ma., 7th-11th May 1995, 1758-64.
117 Yeh J T, Yang H M, Huang S S, *Polym. Degradat .Stabil.*, **50**, No.2, 1995, 229-34.
118 Toure B, Lopez Cuesta J M, Gaudon P, Benhassaine A, Crespy A, *Polym. Degradat. Stabil.*, **53**, No.3, 1996, 371-9.
119 Vasnev V A, Tarasov A I, Istratov V N, Ignatov V N, Krasnov A P, Kuznetsov A I, Surkova I N,Russia, *Reactive & Functional Polym.*, **26**, Nos.1-3, 1995, 177-83.
120 Evans L R, Meeting of the Rubber Division, ACS, Montreal, May 5-8, 1996, paper D.
121 Monthey S, Duddleston B, Podobnik J, *Rubb. World*, **210**, No.3, 1994, 17-9.
122 Abdel-Aziz M M, Gwaily S E, *Polym. Degradat. Stabil.*, **55**, 1997, 269-74.
123 Kurian T, De P P, Khastgir D, Tripathy D K, De S K, Peiffer D G, *Polymer*, **36**, No.20, 1995, 3875-84.
124 Kim S G, Lee S H, *Rubb. Chem. Technol.*, **67**, No.4, 1994, 649-61.
125 Okel T A, Waddell W H, *Rubb. Chem. Technol.*, **68**, No.1, 1995, 59-76.
126 Ismail H, Freakley P K, Sutherland I, Sheng E, *Eur. Polym. J.*, **31**, No.11, 1995, 1109-17.
127 Ismail H, Freakley P K, Sheng E, *Eur. Polym. J.*, **31**, No.11, 1995, 1049-56.
128 Tamura J, Kawanabe K, Yamamuro T, Nakamura T, Kokubo T, Yoshihara S, Shibuya T, *J. Biomed. Mat. Res.*, **29**, No.5, 1995, 551-9.
129 Nofal M M, Zihlif A M, Ragosta G, Martuscelli E, *Polym. Composites*, **17**, No.5, 1996, 705-9.
130 Hornsby P R, Premphet K, *J. Mater. Sci.*, **32**, 1997, 4767-75.
131 Qiu Q, Kumosa M, *Composites Sci. Technol*, **57**, 1997, 497-507.
132 Becu L, Maazouz A, Sautereau H, Gerard J F, *J. Appl. Polym. Sci.*, **65**, 1997, 2419-31.
133 Li J X, Silverstein M, Hiltner A, Baer E, *J. Appl. Polym. Sci.*, **52**, No.2, 1994, 255-67.
134 Li J X, Hiltner A, Baer E, *J. Appl. Polym. Sci.*, **52**, No.2, 1994, 269-83.
135 Ping Gu, Ping Xie, Beaudoin J J, *Cement & Concrete Comp.*, **15**, No.3, 1993, 173-80.
136 Dose T M, Antec '97. Conference proceedings, Toronto, April 1997, 3314-8.
137 Okiyokota M, Hamada H, Hiragushi M, Hasegawa T, Antec '97. Conference proceedings, Toronto, April 1997, 3319-21.
138 Donnet J B, *Kaut. u. Gummi Kunst.*, **47**, No.9, 1994, 628-32.
139 Khan S A, Baker G L, Colson S, *Chem. of Mat.*, **6**, No.12, 1994, 2359-63.
140 Azimi H R, Pearson R A, Hertzberg R W, *J. Appl. Polym. Sci.*, **58**, No.2, 1995, 449-63.
141 Yu Long, Shanks R A, *J. Appl. Polym. Sci.*, **61**, No.11, 1996, 1877-85.
142 Liu C T, Ravi-Chandar K, J. *Reinf. Plast. Comp.*,**15**, No.2, 1996, 196-207.
143 Bagheri R, Pearson R A, *Polymer*, **36**, No.25, 1995, 4883-5.
144 Seibel S R, Moet A, Bank D H, Sehanobish K, Antec 95. Volume III. Conference proceedings, Boston, Ma., 7th-11th May 1995, 3966-70.
145 Conductive Polymers, *Rapra*, Shawbury, 1992.
146 Fengyuan Yan, Qunji Xue, Shengrong Yang, *J. Appl. Polym. Sci.*, **61**, No.7, 1996, 1223-9.
147 Fengyuan Yan, Wenhua Wang, Qunji Xue, Long Wei, *J. Appl. Polym. Sci.*, **61**, No.7, 1996, 1231-6.
148 Shiqi He, Qianchu Liu, Yiu-Wing Mai, Bryant R, *J. Mat. Sci. Mat. In Med.*, **7**, No.10, 1996, 611-6.
149 Shin Jen Shiao, Te Zei Wang, *Composites*, **27B**, No.5, 1996, 459-65.
150 Braches E, *Kunststoffe Plast. Europe*, **86**, No.11, 1996, 21-2.
151 Srivastava V K, Pathak J P, *Polym. & Polym. Composites*, **3**, No.6, 1995, 411-4.

152 Zhang C-Z, Liu W-M, Xue Q-J, Shen W-C, *J. Appl. Polym. Sci.*, **66**, 1997, 85-93.
153 Wada N, Uchiyama Y, Hosokawa M, *Int. Polym. Sci. Technol.*, **21**, No.3, 1994, T/53-63.
154 Wada N, Uchiyama Y, *Int. Polym. Sci. Technol.*, **21**, No.10, 1994, T/23-34.
155 Knowles J, *Polym. Paint Col. J.*, **185**, No.4366, 1995, 26-7.
156 Patel A C, *Kaut. u. Gummi Kunst.*, **47**, No.8, 1994, 556-70.
157 Bijwe J, *Polym. Composites*, **18**, No.3, 1997, 378-96.
158 Kody R S, Martin D C, *Polym. Engng. Sci.*, **36**, No.2, 1996, 298-304.
159 Schmidt H K, *Macromol. Symp.*, **101**, 1996, 333-42.
160 Schuster R H, Meeting of the Rubber Division, ACS, Montreal, May 5-8, 1996, paper F.
161 Kontou E, *J. Reinf. Plast. Comp.*, **13**, No.8, 1994, 756-66.
162 Horst J J, Spoormaker J L, *J. Mater. Sci.*, **32**, 1997, 3641-51.
163 Zhou J, Li G, Li B, He T, *J. Appl. Polym. Sci.*, **65**, 1997, 1857-64.
164 Xu X X, Crocrombe A D, Smith P A, *Int. J. Fatigue*, **16**, No.7, 1994, 469-77.
165 Xu X X, Crocombe A D, Smith P A, *Int. J. Fatigue*, **17**, No.4, 1995, 279-86.
166 Seibel S R, Moet A, Bank D H, Nichols K, Antec '93. Conference Proceedings, New Orleans, La., 9th-13th May 1993, Vol. I, 902-5.
167 Jia N, Kagan V A, Antec '97. Conference proceedings, Toronto, April 1997, 1844-8.
168 Topoleski L D T, Ducheyne P, Cuckler J M, *J. Biomed. Mat. Res.*, **29**, No.3, 1995, 299-307.
169 Visser S A, *J. Appl. Polym. Sci.*, **63**, 1997, 1805-20.
170 Visser S A, *J. Appl. Polym. Sci.*, **64**, 1997, 1499-1509.
171 Trotignon J P, Tcharkhtchi A, *Macromol. Symp.*, **108**, 1996, 231-45.
172 Dyrda V I, Meshchaninov S K, *Int. Polym. Sci. Technol.*, **22**, No.12, 1995, T/14-6.
173 Visser S A, Hewitt C E, Binga T D, *J. Polym. Sci., Polym. Phys.*, **34**, No.9, 1996, 1679-89.
174 Gownder M, Letton A, Hogan H,Antec '95. Vol. II. Conference Proceedings, Boston, Ma., 7th-11th May 1995, 1983-6.
175 Asp L E, Sjogren B A, Berglund L A, *Polym. Composites*, **18**, No.1, 1997, 9-15.
176 Cantwell W J, Zulkifli R, *J. Mater. Sci. Lett.*, **16**, 1997, 509-11.
177 Wang S Q, Inn Y W, *Rheol. Acta*, **33**, No.2, 1994, 108-16.
178 Kominar V, Narkis M, Siegmann A, Breuer O, *Sci. & Engng. Composite Materials*, **3**, No.1, 1994, 61-6.
179 Jancar J, Dibenedetto A T, *J. Mat. Sci.*, **30**, No.9, 1995, 2438-45.
180 Rodrigues F, Antec '93. Conference Proceedings, New Orleans, La., 9th-13th May 1993, Vol. I, 332-7.
181 Herd C R, Bomo F, *Kaut. u. Gummi Kunst.*, **48**, No.9, 1995, 588-99.
182 Hauser R L, Woods D W, Krause-Singh J, Ferry S R, Antec '93. Conference Proceedings, New Orleans, La., 9th-13th May 1993, Vol. I, 341-6.
183 Vilgis T A, Heinrich G, *Macromolecules*, **27**, No.26, 1994, 7846-54.
184 Gent A N, Lai S-M, *Rubb. Chem. Technol.*, **68**, No.1, 1995, 13-25.
185 Hegedus C R, Kamel I L, *J. Coatings Technol.*, **65**, No.822, July 1993, 37-43.
186 Zhuk A V, Knunyants N N, Oshmyan V G, *Polym. Sci.*, **36**, No.4, 1994, 572-5.
187 Vratsanos L A, Farris R J, *Polym. Engng. Sci.*, **33**, No.22, 1993, 1458-65.
188 Kiselev V Ya, *Int. Polym. Sci. Technol.*, **21**, No.7, 1994, T/52-5.
189 Ruiz F A, Polymers, Laminations & Coatings Conference, 1995, 647-51.
190 Cochet P, Bomal Y, *Kaut. u. Gummi Kunst.*, **48**, No.4, 1995, 270-5.
191 Roche A A, Dole P, Bouzziri M, *J. Adhesion Sci. Technol.*, **8**, No.6, 1994, 587-609.
192 Lee I, *J. Mat. Sci.*, **30**, No.23, 1995, 6019-22.
193 Hoy L L, *J. Coatings Technol.*, **68**, No.853, 1996, 33-9.
194 Kiselev V Y, Vnukova V G, *Int. Polym. Sci. Technol.*, **23**, No.5, 1996, T/88-92.
195 Torro-Palau A, Fernandez-Garcia J C, Orgiles-Barcelo A C, Martin-Martinez J M, *J. Adhesion*, **57**, Nos.1-4, 1996, 203-25.
196 Tsutsumi K, Ban K, Shibata K, Okazaki S, Kogoma M, *J. Adhesion*, **57**, Nos.1-4, 1996, 45-53.
197 Casoli A, Charmeau J Y, Holl Y, d'Allest J F, *J. Adhesion*, **57**, Nos.1-4, 1996, 133-51.
198 Murata N, Nishi S, Hosono S, *J. Adhesion*, **59**, Nos.1-4, 1996, 39-50.
199 Oien H T, Polyurethanes '95. Conference Proceedings, Chicago, Il., 26th-29th Sept.1995, 137-41.
200 Cochrane H, Lin C-S, *Rubber World*, 1985.
201 Canova L A, Fergusson L W, Parrinello L M, Subramanian R, Giles H F, Antec '97. Conference proceedings, Toronto, April 1997, 2112-6.
202 Mamat A, Trochu F, Sanschagrin B, *Polym. Engng. Sci.*, **35**, No.19, 1995, 1511-20.
203 Sergeeva L M, Skiba S I, Karabanova L V, *Polym. Int.*, **39**, No.4, April 1996, 317-25.

204 Priss L S, *Int. Polym. Sci. Technol.*, **23**, No.7, 1996, T53-6.
205 Schultz R, Antec '96. Vol. I. Conference Proceedings, Indianapolis, 5th-10th May 1996, 519-24.
206 Heberlein D E, Antec '94. Conference Proceedings, San Francisco, Ca., 1st-5th May 1994, Vol. I, 791-5.
207 Mushack R, Luttich R, Bachmann W, *Eur. Rubb. J.*, **178**, No.7, 1996, 24-9.
208 Lawandy S N, Botros S H, Darwish N A, Mounir A, *Polym. Plast. Technol. Engng.*, **34**, No.6, 1995, 861-74.
209 Wang W D, Haidar B, Vidal A, Donnet J B, *Kaut. u. Gummi Kunst.*, **47**, No.4, 1994, 238-41.
210 Leveque D, Auvray M H, *Composites Sci. & Technol.*, **56**, No.7, 1996, 749-54.
211 Kim G M, Michler G H, Gahleitner M, Fiebig J, *J. Appl. Polym. Sci.*, **60**, No.9, 1996, 1391-403.
212 Bluemler P, Bluemich B, *Acta Polymerica*, **44**, No.3, 1993, 125-31.
213 Sinien L, Yongli W, Zhifa D, Yuchun O, Xiaoping F, Antec '96. Volume II. Conference proceedings, Indianapolis, 5th-10th May 1996, 2298-300.
214 Kabelka J, Hoffmann L, Ehrenstein G W, *J. Appl. Polym. Sci.*, **62**, No.1, 1996, 181-98.
215 Sinien L, Xiaoguang Z, Zhongneng Q, Huan X, *J. Mat. Sci. Lett.*, **14**, No.20, 1995, 1458-60.
216 Simhambhatla M, Leonov A I, *Rheol. Acta*, **34**, No.4, 1995, 329-38.
217 Dufresne A, Lacabanne C, *Polymer*, **34**, No. 15, 1993, 3173-8.

# 9

# The Effect of Fillers on Rheological Properties of Filled Materials

## 9.1 VISCOSITY

Viscosity determinations are the simplest means of the rheological characterization of materials.[1-18] Viscosity measurements lack sufficient precision for the generalization purposes because filled materials are most frequently non-Newtonian liquids and a singular numerical parameter cannot adequately describe complex properties. In spite of this deficiency, many important conclusions can be drawn from viscosity data.

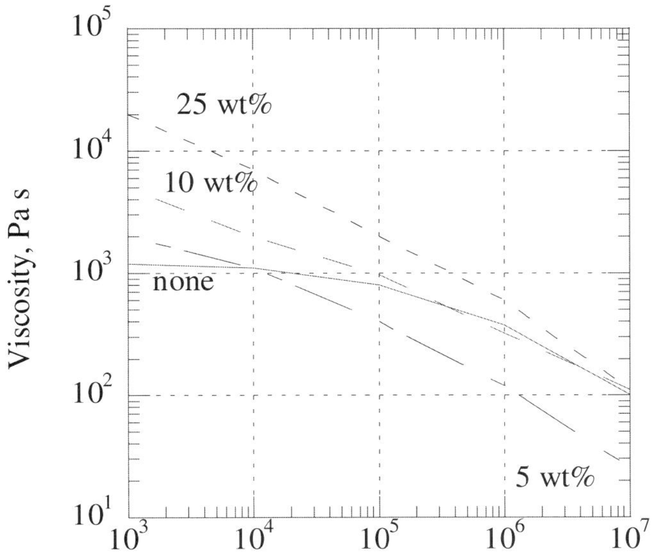


Figure 9.1. The effect of carbon black on the viscosity of polycarbonate. [Adapted, by permission, from Joo Y L, Lee Y D, Kwack T H, Min T I, Antec '96. Vol. I. Conference Proceedings, Indianapolis, 5th-10th May 1996, 64-8.]

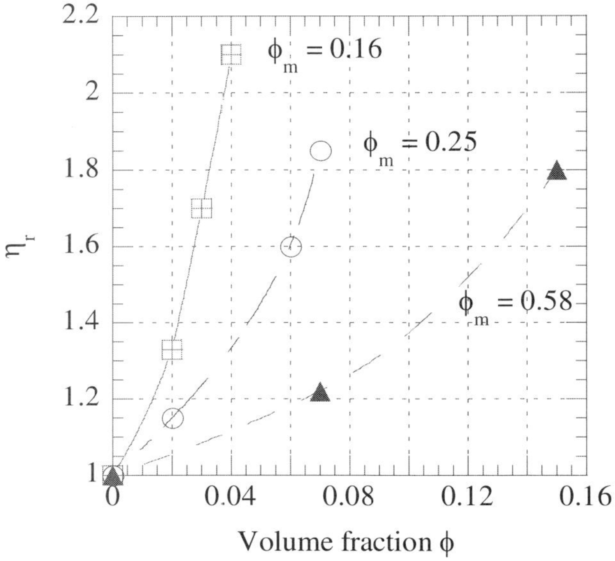


Figure 9.2. Reduced viscosity of polypropylene filled with different grades of calcium carbonate as a function of volume fraction. [Adapted, by permission from Gendron R, Daigneault L E, Tatibouet J, Dumoulin M M, *Adv. Polym. Technol.*, **15**, No.2, 1996, 111-25.]

Increasing the amount of filler in a filled material increases shear stress and viscosity (Figure 9.1).[6] In addition to an increase in viscosity with increasing additions of carbon black, zero-shear viscosity and shear-thinning increase more rapidly when the amount of carbon black is increased. This data shows that there is always a certain region in which viscosity does not increase as rapidly (in this case, in the range of shear stress of $10^5$ -$10^6$). It is this region which is best suited for processing.

Note that the viscosity increase with filler addition depends on the properties of filler such as its maximum packing fraction (Figure 9.2).[9] With an increased volume fraction, viscosity increases but this increase would not be the same for example, for three different grades of calcium carbonate. What causes this is difference in their maximum packing fraction. The rate of viscosity increase depends on the ratio $\phi/\phi_m$ where $\phi$ is volume fraction of filler added and $\phi_m$ is the maximum packing fraction.

Viscosity measurements helped to understand the reaction rate in the formation of polyurethane in the presence of lead powder (Figure 9.3).[11] A smaller addition of lead powder (10%), does not accelerate the reaction rate but larger amounts of lead powder increase the reaction rate rapidly. At 30% lead powder, the reaction rate is increased by a factor of 3.

As investigation of the process of mixing based on viscosity data shows (Figure 9.4) that viscosity increases at the beginning of mixing but continues falling shortly afterwards.[16] This rapid decrease of viscosity occurs only at the beginning

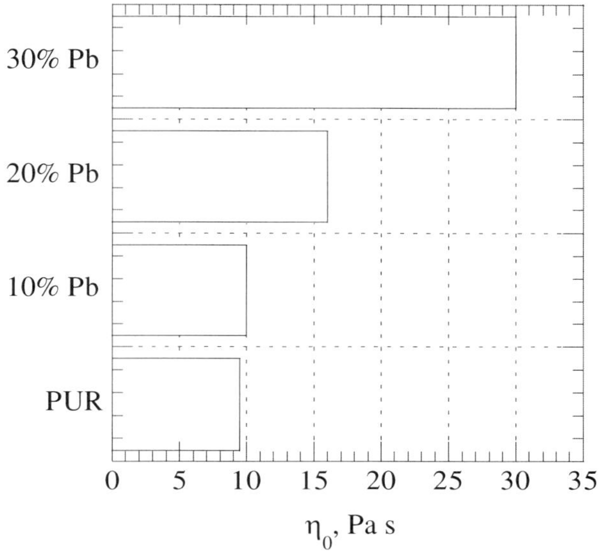


Figure 9.3. Viscosity of formation of polyurethane in the presence of lead powder. [Data from Caillaud J L, Deguillaume S, Vincent M, Giannotta J C, Widmaier J M, *Polym. Int.*, **40**, No.1, 1996, 1-7.]

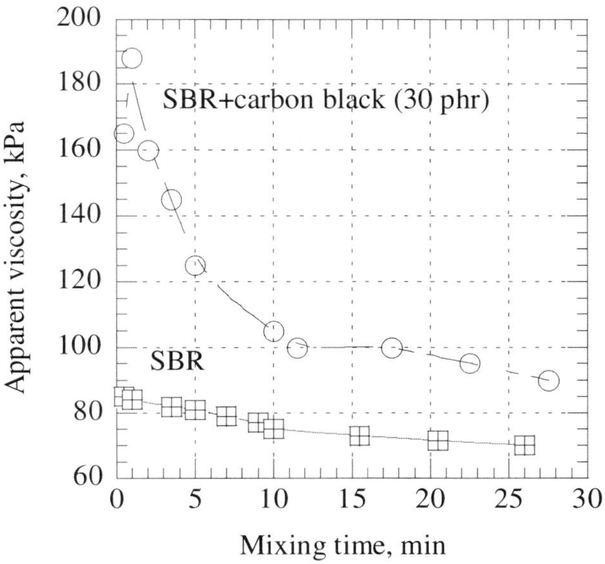


Figure 9.4. Viscosity of SBR filled with carbon black vs. mixing time. [Adapted, by permission, from Clarke J, Freakley P K, *Rubb. Chem. Technol.*, **67**, No.4, 1994, 700-15.]

of the mixing process. Further decreases in viscosity are more difficult to obtain. Figure 9.5 gives information on the effect of amount of added filler and quality of mixing.[16] A poorly mixed compound increases in relative viscosity less rapidly be-

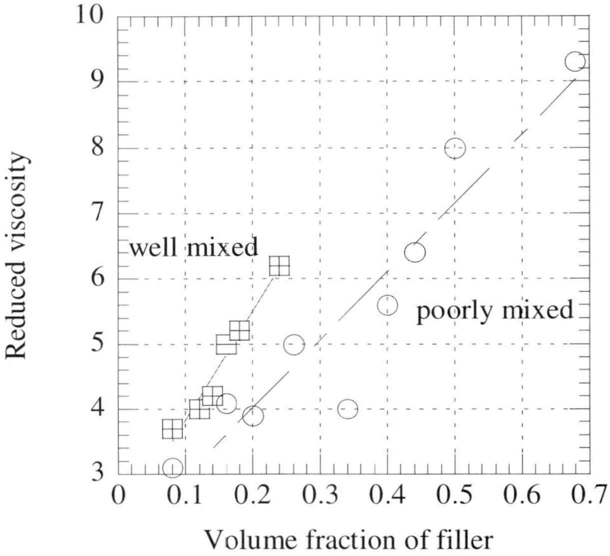


Figure 9.5. Reduced viscosity of SBR vs. fraction of carbon black in relationship to quality of mixing. [Adapted, by permission, from Clarke J, Freakley P K, *Rubb. Chem. Technol.*, **67**, No.4, 1994, 700-15.]

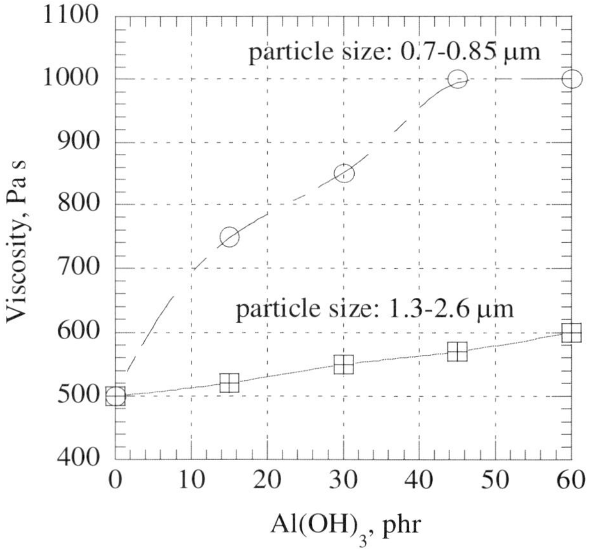


Figure 9.6. Viscosity of plasticized PVC vs. fraction of $Al(OH)_3$. [Data from Liptak P, Zelenak P, *Int. Polym. Sci. Technol.*, **20**, No.9, 1993, T/57-9.]

cause undispersed aggregates behave in a manner similar to spherical particles which create less resistance to mixing.

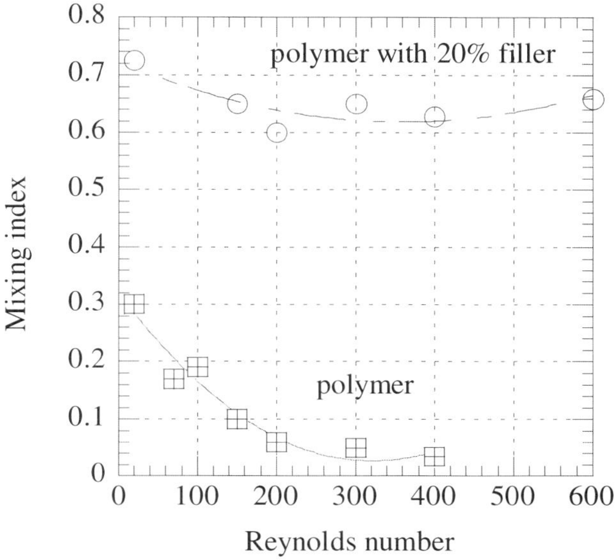


Figure 9.7. Mixing index as a function of volume fraction of filler and Reynolds number. [Data from Agarwal S, Campbell G A, Antec 95. Volume I. Conference proceedings, Boston, Ma., 7th-11th May 1995, 839-42.]

The quality of mixing depends both on filler properties and on the dispersing medium or its rate of flow during the mixing process. If a filler has a low oil absorption, it does not affect viscosity over a certain range of loading (Figure 9.6). But a filler having a high oil absorption increases viscosity as the load of filler increases.[14] Selectivity is a measure of filler particle segregation. The closer to zero the selectivity is the better the dispersion of the filler. Increase Reynolds number on mixing increases the effectiveness of mixing (Figure 9.7).[5] So the same filler will give better results if the mixing medium and the mixing process are matched.

## 9.2 FLOW

Many industrial processes are affected by the influence of particulate materials on the flow properties of material. Flow properties of materials can be adjusted by fillers to meet the requirements. Flow properties can also be adversely affected by numerous phenomena related to the presence of filler in formulations.[13,19-24] One common example is related to the flow of industrial slurries which contain concentrated suspensions of small particles.[19] Such suspensions are usually non-Newtonian fluids with a yield stress which is formed through strong interactions between particles. During flow, these interactions are continuously broken and rebuilt. A solid deposit formed on the slopes and walls is an adverse effect of this property.

Materials with a yield value have a specific stress value below which the material is deformed as an elastic solid and above which it flows. Concrete mixes are

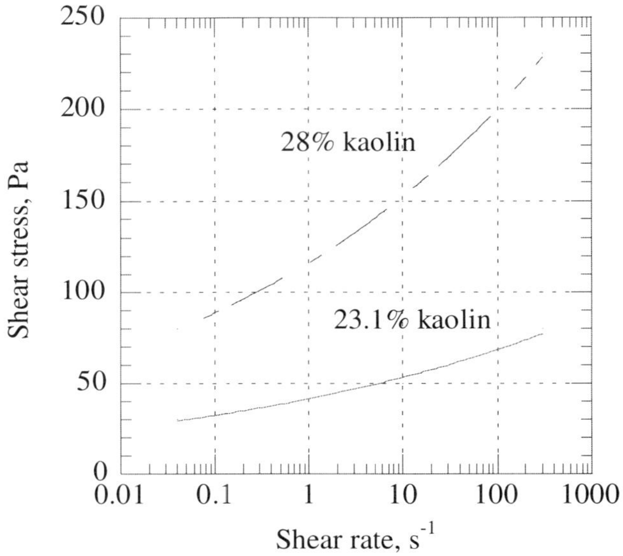


Figure 9.8. Water clay mixtures. [Adapted, by permission, from Coussot Ph., Proust S, Ancey Ch, *J. Non-Newtonian Fluid Mech.*, **66**, 1996, 55-70.]

tested by the "slump test" in which a truncated cone is filled with a concrete mix and slump and flow is measured. Slump stoppage depends on the rheological properties of mix which cause flow to stop when shear stress falls below yield value. Another demonstration of this effect is observed in sealants during the "sag test". This test determines the thickness of a layer formed on the wall of testing device at which thickness flow stops (the stress formed by gravity falls below yield stress). These properties can be more precisely measured by rheometry as Figure 9.8 shows. A small difference in the concentration of a clay filler substantially increases the yield value. In spite of the much higher precision of rheometric measurements, rheological measurements cannot replace practical tests because they do not capture the complex relationships of interplaying factors in practical applications.

One difficulty in flow measurement in suspended systems is caused by particle distribution close to walls. A slip layer forms on the surface of the walls which consists of particle-free binder. The thickness of this layer is given by equation:[21]

$$\delta = u_s \eta_s / \tau_R \qquad [9.1]$$

where:

| | |
|---|---|
| $\delta$ | layer thickness |
| $u_s$ | slip velocity at the wall |
| $\eta_s$ | Newtonian shear viscosity |
| $\tau_R$ | corrected shear stress |

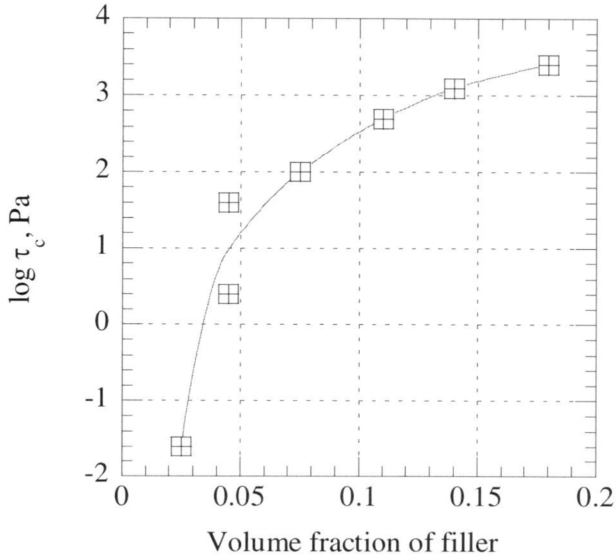


Figure 9.9. Flow limits vs. filler content. [Adapted, by permission, from Mamunya E P, Shumskii V F, Lebedev E V, *Polym. Sci.*, **36**, 1994, 835-8.]

Wall slip affects flow properties in tubes.[20] Concentration of particles in the cross-sectional areas of a tube changes due to the radial migration of particles. Such process affects also filtration.

Flow restrictions are also observed in flow of melts.[23] Figure 9.9 shows flow limits of carbon black filled polyolefins. Data show that yield stress appears at low concentrations of carbon black and that the type of the matrix does not affect flow characteristics which are caused by the presence and properties of fillers. The viscosity of these systems are well described by Fedor's equation:

$$\frac{G}{G_p} = \left(\frac{1.25\phi}{F - \phi}\right)^2 \qquad [9.2]$$

where:

| | |
|---|---|
| G | viscosity of filled system |
| $G_p$ | viscosity of matrix |
| $\phi$ | filler fraction |
| F | packing factor |

Viscosity increase and, therefore, flow decrease depend on filler concentration and the packing factor which is related to filler volume. Polyethylene filled with metal particulates behaves in a similar way.[22] Flow was decreasing linearly with the concentration of metal particles.

## 9.3 FLOW INDUCED FILLER ORIENTATION

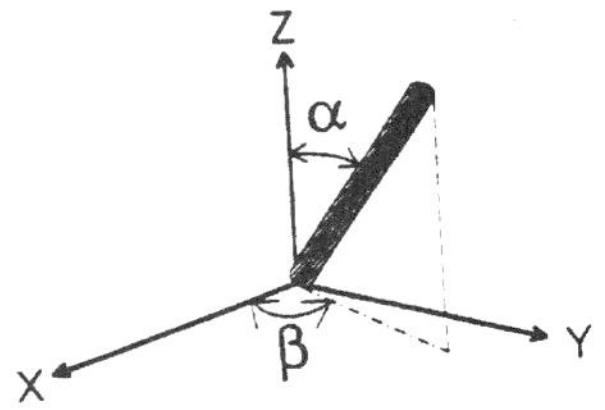


Figure 9.10. Coordinate system used to determine fiber orientation. Z is a direction of elongational flow. [Adapted, by permission, from Kobayashi M, Takahashi T, Takimoto J, Koyama K, *Polymer*, **36**, No.20, 1995, 3927-33.]

To make practical use of fillers, a knowledge of filler orientation during the flow is needed.[25-31] The best description of orientation principles can be obtained from modelling.[25] Figure 9.10 gives the coordinate system used to describe orientation of fibers.

The elongational flow field is determined from equation:

$$\tan\alpha = \exp\left(-\frac{3}{2}\varepsilon\right)\tan\alpha_0 \quad \beta = \beta_0 \qquad [9.3]$$

where:

$\alpha,\beta$ angles given in Figure 9.10
$\varepsilon$ elongational strain rate

The rotary motion is given by the following equation:

$$\frac{d\alpha}{dt} = -\frac{3}{4}\varepsilon R\sin(2\alpha) \quad \frac{d\beta}{dt} = 0 \qquad [9.4]$$

$$R = \frac{r_p^2 - 1}{r_p^2 + 1} \quad \tan\alpha = \exp\left(-\frac{3}{2}\varepsilon Rt\right)\tan\alpha_0$$

where:

R shape factor
t time
$r_p$ aspect ratio
$\alpha_0$ initial orientation angle

From the above equations it can be seen that orientation increases with aspect ratio and elongational strain rate.

Figure 9.11 shows that the average angle α of whiskers decreases with the elongational strain rate, ε.[25] Figure 9.12 shows the effect of the aspect ratio of ferrite on the apparent permeability of a composite.[30] Both sets of experimental data are consistent with the model. The orientation of fiber increases with the elongational strain rate and fiber aspect ratio.

It should be noted from Figure 9.12 that the magnetic permeability increases as the effective aspect ratio increases. Conductive thermoplastics, which contain carbon fiber, have a maximum injection rate above which electric conductivity will not increase. These composites also depend on fiber orientation which increases the

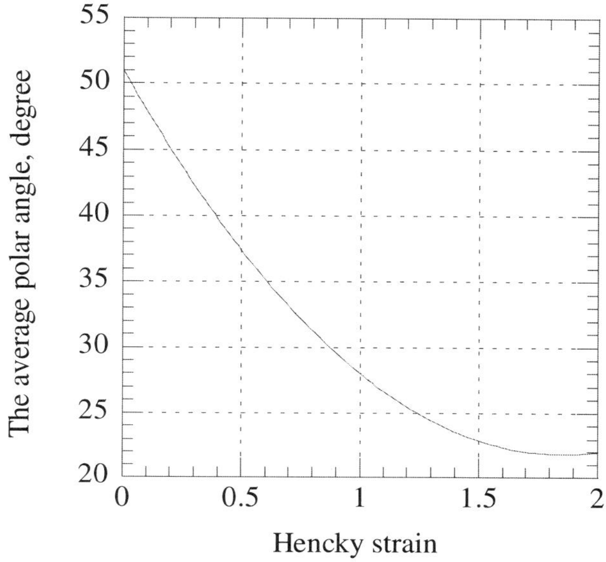


Figure 9.11. The average polar angle, α vs. elongational strain rate, ε. [Data from Kobayashi M, Takahashi T, Takimoto J, Koyama K, *Polymer*, **36**, No.20, 1995, 3927-33.]

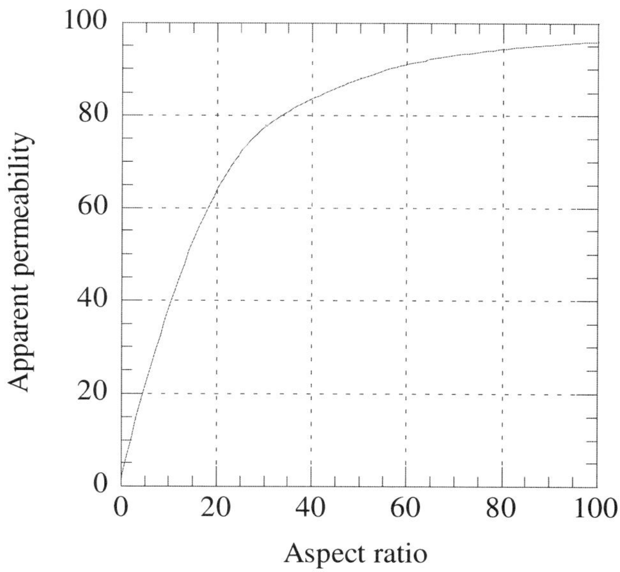


Figure 9.12. Apparent magnetic permeability vs. effective aspect ratio. [Adapted, by permission, from Fiske T, Gokturk H S, Yazici R, Kalyon D M, *Polym. Eng. Sci.*, **37**, No.5, 1997, 826-37.]

number of contacts and thus the conductivity. Higher injection rates increase the number of fiber breakage which reduces conductivity.[29]

Talc filled thermoplastic materials were studied in rheometers of different geometries (elongational, capillary, parallel plate). Geometry of the testing method and the flow paths had an important influence on the orientation of talc particles.[27] In addition to flow decrease, an increased concentration of filler had a pronounced effect on both flow and orientation. The cross-sectional distribution of particles depends also on the size of particulate material.[28] The concentration of particles on the free surface of advancing material increases as particle size increases.

## 9.4 TORQUE

Torque increases as viscosity increases which is typically a result of an increase in filler loading.[32-4] The viscosity of mixtures containing fillers depends on the nature and concentration of filler, its shape, size and interaction with matrix. Figure 9.13 shows how various fillers cause torque increase.[32] A magnesium carbonate addition causes a relatively small increase in torque. Glass fiber creates an extreme effect, increasing torque very rapidly. Silicon dioxide causes a torque increase between these two extremes. The high aspect ratio of glass fiber is responsible for its extreme effect and the low interaction of magnesium carbonate with the matrix causes the torque increase to be slight.

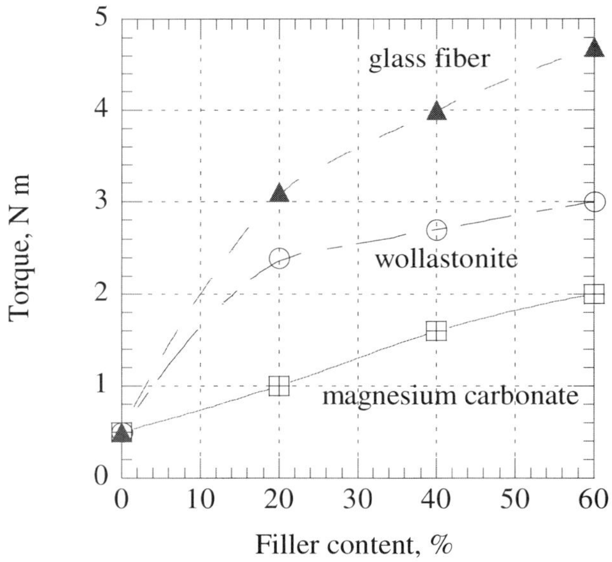


Figure 9.13. Torque vs. filler content. [Adapted, by permission, from Scaffaro R, Pedretti U, La Mantia F P, *Eur. Polym. J.*, **32**, No.7, 1996, 869-75.]

Measurement of torque also provides data on the effect of fillers on the curing rates of reactive systems.[33-4]

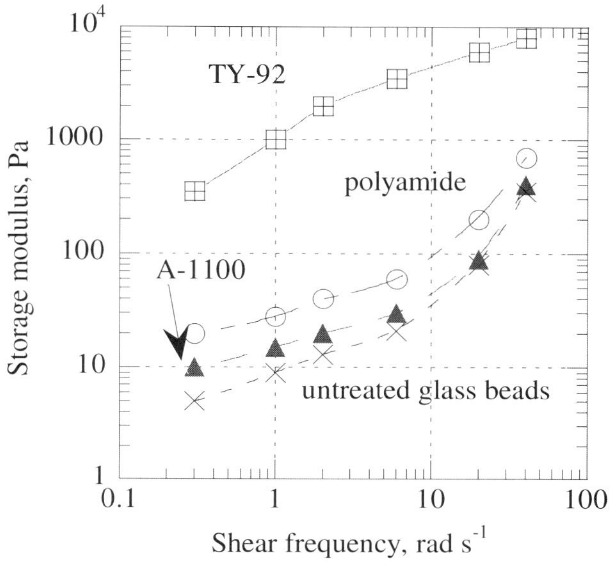


Figure 9.14. Storage modulus of glass bead filled polyamide-6 vs. shear frequency. [Adapted, by permission, from Ou Y-C, Yu Z-Z, *Polym. Int.*, **37**, No.2, 1995, 113-7.]

## 9.5 VISCOELASTICITY

Rheological properties of filled systems are complex and formulation specific, largely dependent on fillers and other materials, especially materials which form a matrix.[5,27,35-40] Flow through tubes demonstrates the unusual properties of filled system. Plug flow is typical of filled systems much different from the characteristics of unfilled system.[5] This phenomenon is frequently observed with highly filled systems which behave in a manner similar to both solids and liquids.

Figure 9.14 compares untreated glass beads with glass beads treated with γ-aminopropyltriethoxysilane. The untreated beads decrease storage modulus in all probability because the surface treatment decreases hydrogen bonding between polymer chains. Introduction of surface treatment (TY-92), which improves surface adhesion, contributes to a considerable increase in storage modulus. The loss modulus of a system containing TY-92 decreases as Figure 9.15 shows.[35]

The storage modulus of a polymer filled with polymeric particles is less dependent on frequency when these particles do not interact with the matrix polymer. They form clusters during storage which contribute to the non-Newtonian behavior of the filled polymer.[40]

Carbon black interacts strongly with polymer (HDPE) to produce a large increase in storage modulus in a manner similar to the surface treated glass beads.[36] The storage modulus is less sensitive to frequency. The storage modulus increase is explained by the effect of modifier on crosslinking.

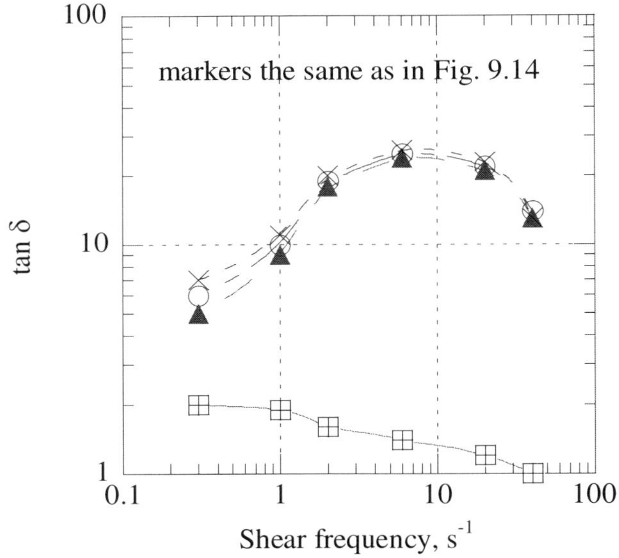


Figure 9.15. Loss modulus of glass beads filled polyamide-6 vs. shear frequency. [Adapted, by permission, from Ou Y-C, Yu Z-Z, *Polym. Int.*, **37**, No.2, 1995, 113-7.]

Both the storage and the loss moduli have linear relationship with filler concentration (iron particles) when the measurements of compounded gels are done in a magnetic field.[39]

## 9.6 DYNAMIC MECHANICAL BEHAVIOR

The addition of a filler which interacts with the matrix restricts molecular mobility which can be measured using the dynamic mechanical analysis.[12,41-49] Figure 9.16 shows the effect of particle size of glass beads on tan δ. Smaller beads with higher surface area available for interaction restrict molecular mobility.[42] Increasing the concentration of filler causes a decrease in the magnitude of main relaxation (related to $T_g$). In some cases the concentration of filler does not influence $T_g$. Changes in dynamic mechanical properties are related to matrix-filler interaction and chemical coupling.

In Chapter 7, the Figure 7.10 models tightly and loosely bound polymer and shows gradual changes which occur when filler concentration increases. Figure 9.17 complements this model with experimental data which show changes in tan δ peak positions associated with tightly and loosely bound polymer. The second tan δ peak changes when a loosely bound polymer transits to a tightly bound polymer.[43]

Figure 9.18 gives one more example of how interaction affects dynamic mechanical properties. Two polymers were tested with various concentrations of alumina. Polystyrene was almost unaffected by various concentrations of filler. Sulfonated polystyrene interacts more strongly with its filler than polystyrene which contributes to increase in $T_g$.[49]

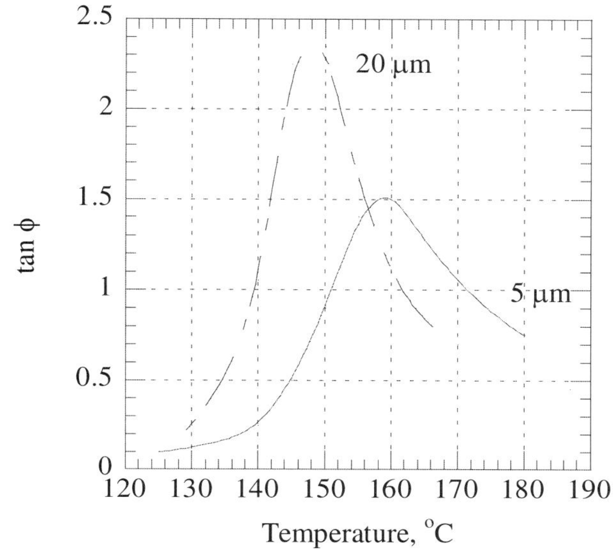


Figure 9.16. Tan δ vs. temperature for styrene-methacrylic acid copolymer filled with glass beads of different diameter. [Data from Bergeret A, Alberola N, *Polymer*, **37**, No.13, 1996, 2759-65.]

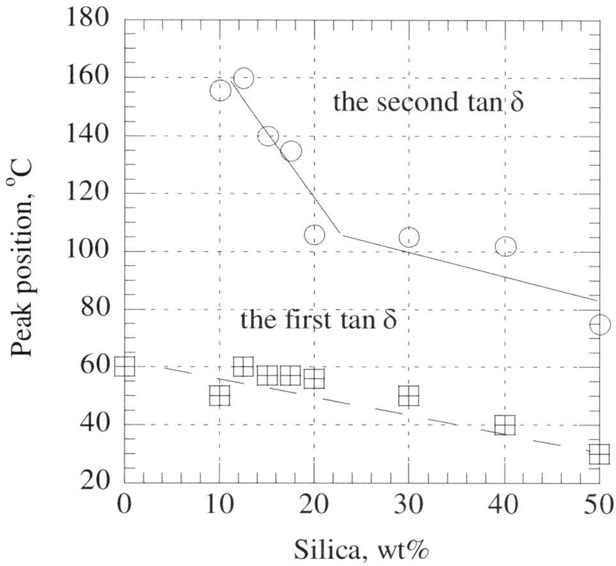


Figure 9.17. Peak positions of the first and second tan δ peaks vs. silica contents in PVAc composites. [Adapted, by permission, from Tsagaropoulos G, Eisenberg A, *Macromolecules*, **28**, No.18, 1995, 6067-77.]

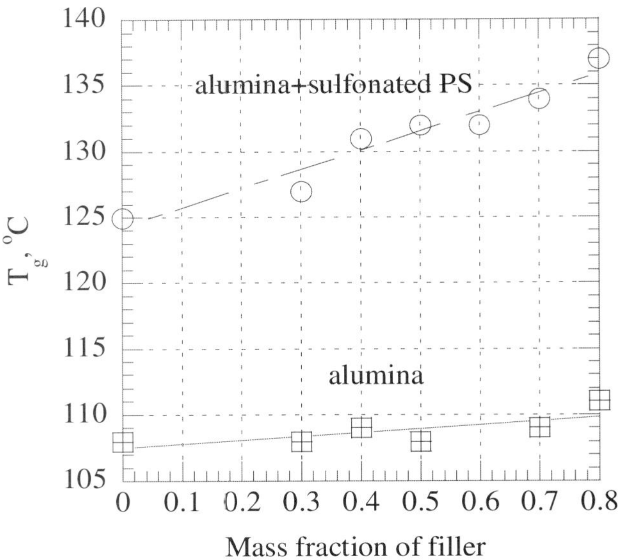


Figure 9.18. $T_g$ versus concentration of alumina. [Data from Cousin P, Smith P, *J. Polym. Sci., Polym. Phys.*, **32**, No.3, 1994, 459-68.]

## 9.7 COMPLEX VISCOSITY

Complex viscosity data also shows how fillers interact with the matrix.[27,50-58] Figures 9.19 and 9.20 show the effect of filler loading on the complex viscosity of two polymers (PP and PPS).[27] Two conclusions can be drawn from these figures: polymer type affects the rate of viscosity increase and the increase of viscosity is not proportional to the concentration of filler. Figure 9.21 shows that viscosities of polypropylene loaded with 10 and 20% calcium carbonate are almost identical whereas a large increase in viscosity is observed on addition of 40% calcium carbonate.[54]

The process of mixing has an interesting influence on complex viscosity (Figure 9.22).[58] The first mixing was performed in a Brabender, the second by a hand held spatula for a prolonged time (5 hr). The viscosity of the mixture was considerably decreased and, in addition, the shear-thinning properties of the material were reduced. It should be noted that, for a proper evaluation of fillers, the mixing regime is a very important consideration since viscosity may change by as much as 1000 times depending upon how the compound was mixed. By proper mixing, it is possible to change material properties so as to be able to process it rather than by adding complex combinations of additives.

Figures 9.23 and 9.24 show that complex viscosity decays with time of mixing. The magnitude of this decay depends on the surface treatment of the filler as well as on its concentration.[57] Again the conditions of the experiment influence the

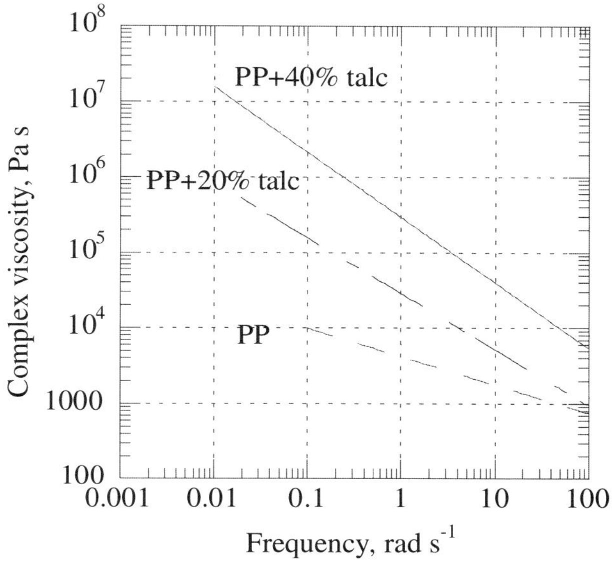


Figure 9.19. Complex viscosity of talc filled PP vs. frequency. [Data from Chang Ho Suh, White J L, *J. Non-Newtonian Fluid Mechanics*, **62**, Nos.2/3, 1996, 175-206.]

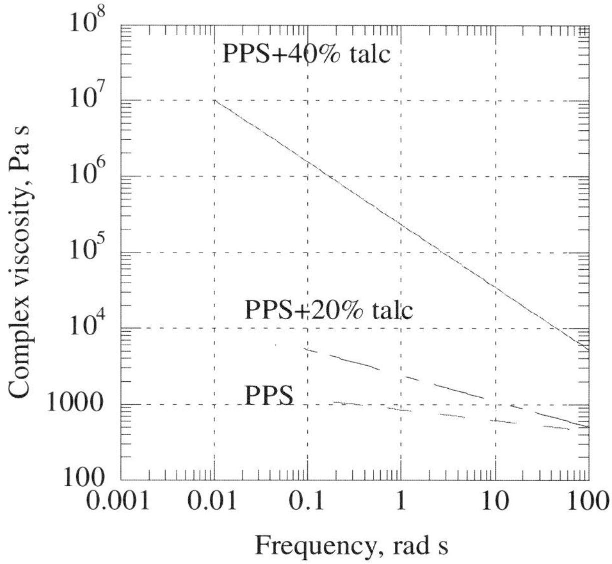


Figure 9.20. Complex viscosity of talc filled PPS vs. frequency. [Adapted, by permission, from Chang Ho Suh, White J L, *J. Non-Newtonian Fluid Mechanics*, **62**, Nos.2/3, 1996, 175-206.]

result. The slower decay in the complex viscosity of a mixture of particles of glass beads treated with fluorosilane is explained by a reduction in the interfacial adhesion followed by a reduction in dynamic slip.

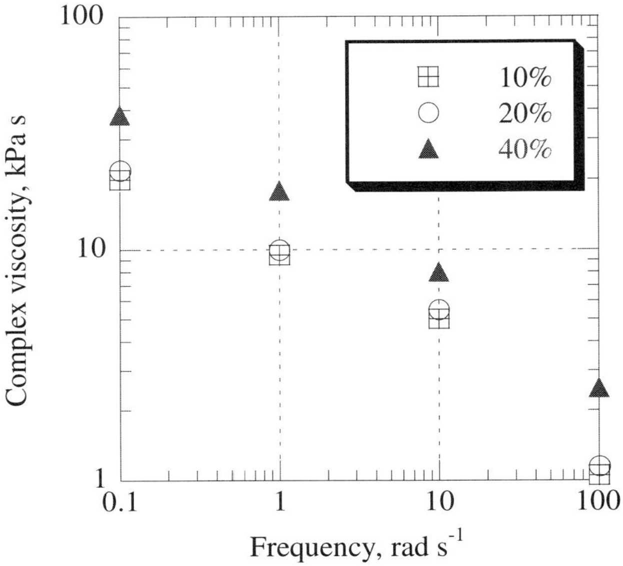


Figure 9.21. Complex viscosity of calcium carbonate filled PP vs. frequency. [Adapted, by permission, from Johnson K C, Antec '96. Volume III. Conference proceedings, Indianapolis, 5th-10th May 1996, 3545-9.]

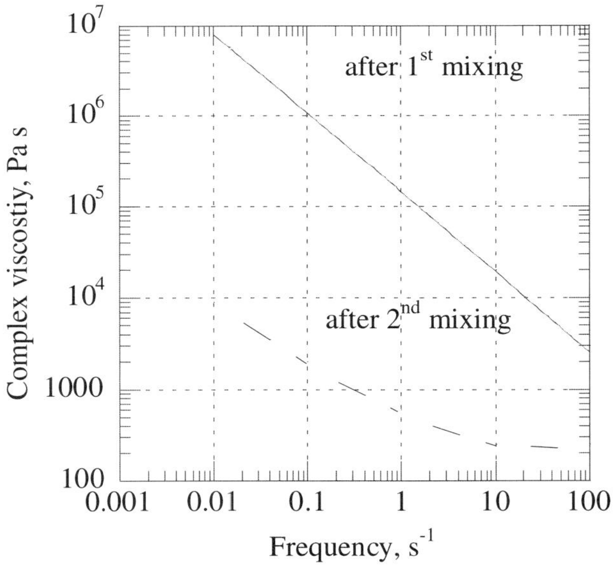


Figure 9.22. Complex viscosity of 31% $TiO_2$ in polybutene vs. frequency. [Data from Carreau P J, Lavoie P A, Bagassi M, *Macromol. Symp.*, **108**, 1996, 111-26.]

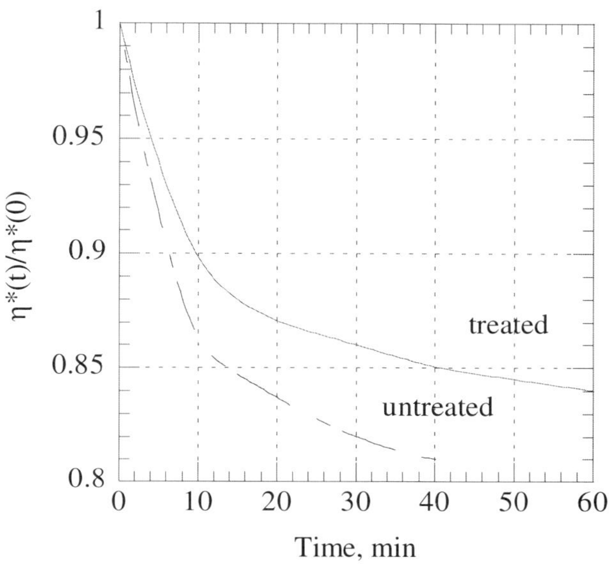


Figure 9.23. Decay of reduced complex viscosity of PDMS filled with glass beads vs. determination time. [Adapted, by permission, from Wang S Q, Inn Y W, *Rheol. Acta*, **33**, No.2, 1994, 108-16.]

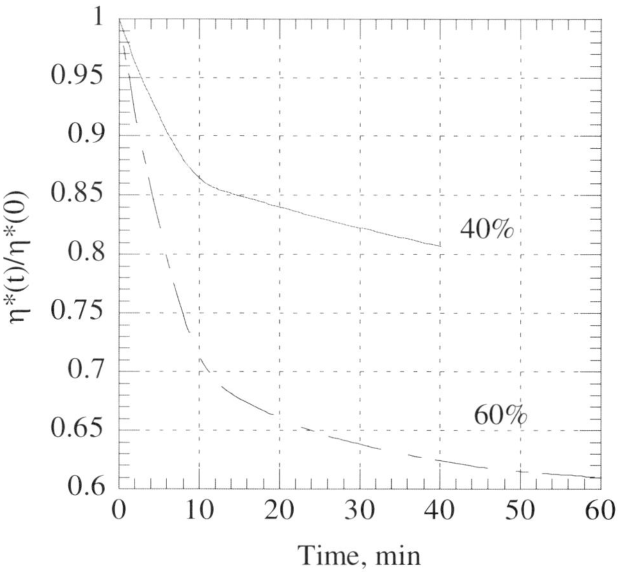


Figure 9.24. Effect of glass beads concentration on the decrease of reduced complex viscosity of PDMS vs. determination time. [Adapted, by permission, from Wang S Q, Inn Y W, *Rheol. Acta*, **33**, No.2, 1994, 108-16.]

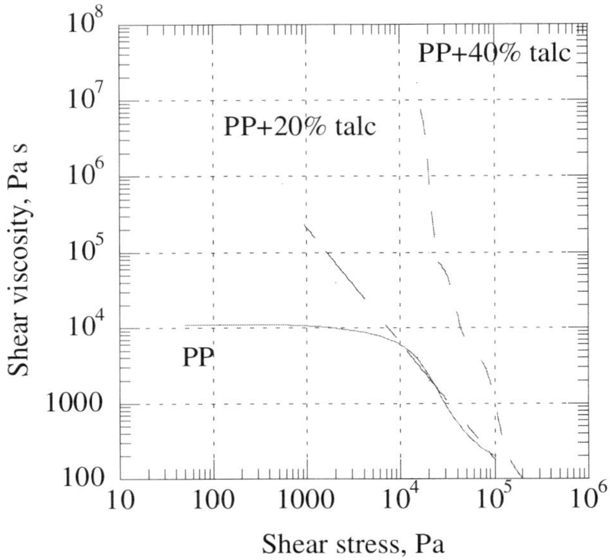


Figure 9.25. Viscosity of talc filled PP vs. shear stress. [Data from Chang Ho Suh, White J L, *J. Non-Newtonian Fluid Mechanics*, **62**, Nos.2/3, 1996, 175-206.]

## 9.8 SHEAR VISCOSITY

A filled system's rheology depends on conditions of shearing.[6,11,19,21,27,40,59-61] Figures 9.25 and 9.26 show the effect of shear stress on two polymers filled with talc.[27] The reaction of both systems differs in the rates of viscosity change and in the character of non-Newtonian properties but the responses of both systems are similar in their reactions to high shear rates where viscosities of the filled and the neat polymer are almost the same.

Figure 9.27 shows dependence of viscosity on filler concentration. Note that, as with previous data, viscosity increases more rapidly at lower shear rates.[61]

## 9.9 ELONGATIONAL VISCOSITY

The mode of deformation induced in elongational viscosity studies differs from other methods of measurement.[27,62-3] Figure 9.28 shows that elongation to break decreases when a filler content increases but, at the same time, the relationship with rate is linear and only slightly affected by changes in elongation rate.[27]

In fiber filled systems, the elongation of a material correlates with the orientation of fibers (Figure 9.29). This phenomenon is frequently exploited in industrial processes to increase reinforcement and other properties which depend on fiber orientation. Hencky strain shown in Figure 9.29 is a parameter entering in the equation of elongational viscosity.[62]

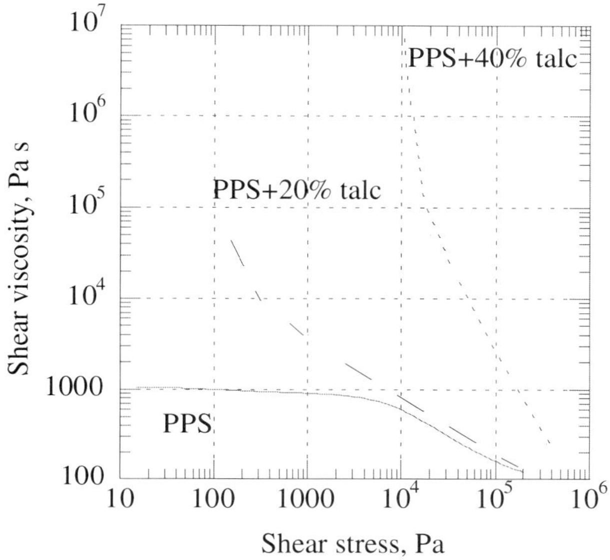


Figure 9.26. Viscosity of talc filled PPS vs. shear rate. [Data from Chang Ho Suh, White J L, *J. Non-Newtonian Fluid Mechanics*, **62**, Nos.2/3, 1996, 175-206.]

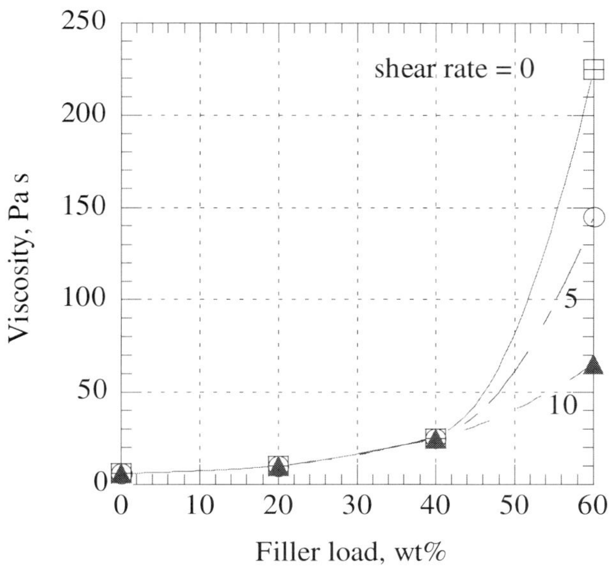


Figure 9.27. Shear viscosity vs. filler concentration. [Adapted, by permission, from, Cheng J, Bigio D I, Briber R M, Antec '97. Conference proceedings, Toronto, April 1997, 162-7.]

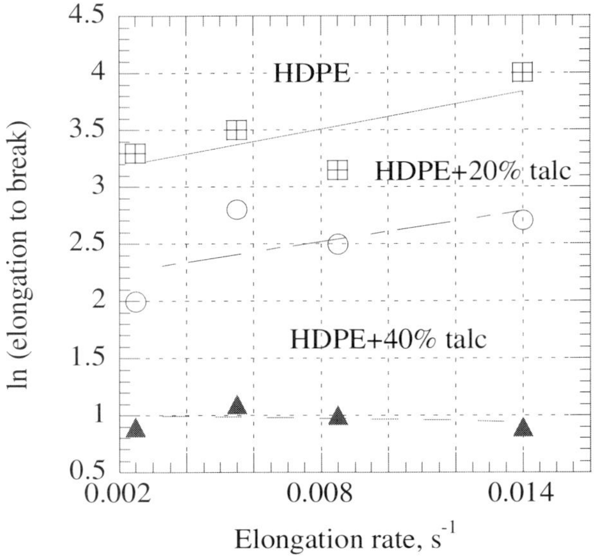


Figure 9.28. Elongation to break of HDPE filled with talc vs. elongation rate. [Adapted, by permission, from Chang Ho Suh, White J L, *J. Non-Newtonian Fluid Mechanics*, **62**, Nos.2/3, 1996, 175-206.]

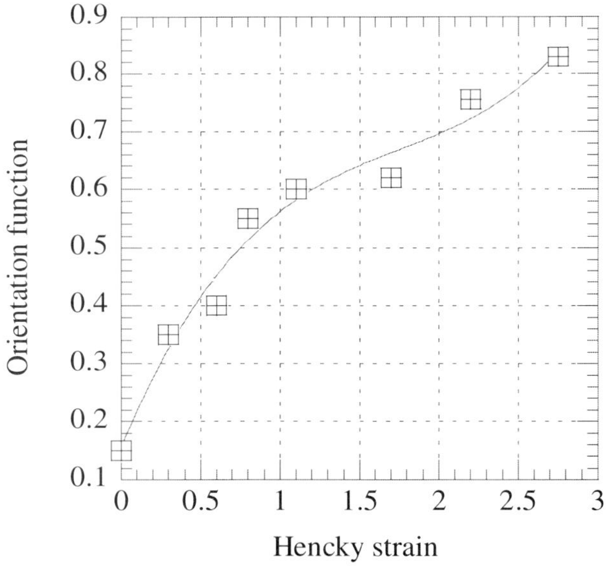


Figure 9.29. Fiber orientation function vs. Hencky strain for PA-6 filled with glass fiber. [Adapted, by permission, from Wagner A H, Kalyon D M, Yazici R, Fiske T J, Antec '97. Conference proceedings, Toronto, April 1997, 996-1000.]

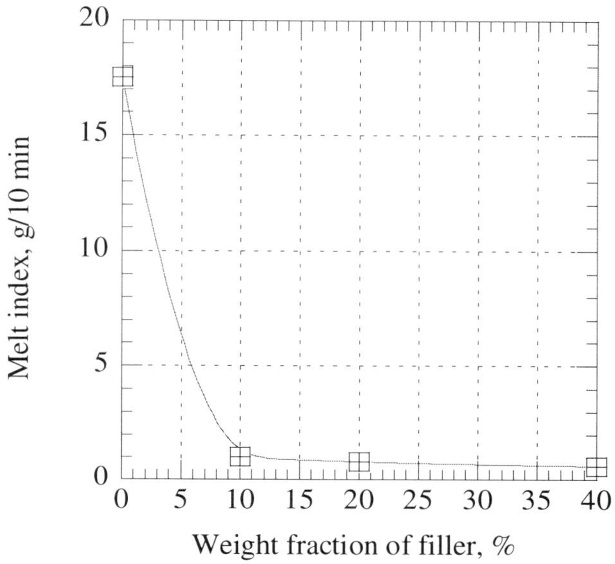


Figure 9.30. Melt flow index vs. weight fraction of calcium carbonate in PP. [Adapted, by permission, from Johnson K C, Antec '96. Volume III. Conference proceedings, Indianapolis, 5th-10th May 1996, 3545-9.]

## 9.10. MELT RHEOLOGY

The effect of filler on polymer melts is discussed in several papers.[10,12,52,64-72] Figure 9.30 shows that the melt flow index decreases rapidly on calcium carbonate addition. The changes as concentration increases are much less pronounced.[52] This study was conducted for a large particle sized (50-400 μm) grade of calcium carbonate.

Figure 9.31 shows the effect of coating of calcium carbonate with stearic acid on the relative melt viscosity.[70] The surface coating reduces the melt viscosity by a factor of 3 at higher concentrations of filler. The maximum packing depends on filler type and particle size. For 10 μm calcium carbonate, the maximum packing was 0.52, for precipitated calcium carbonate (2 μm) it was 0.44 and for glass beads 0.68. The surface treated calcium carbonate had maximum packing of 0.77.

The melt flow index of filled PP changes during reprocessing (Figure 9.32).[72] The magnitude of the change depends on the amount of talc added. Any level of talc addition improved the polymer's resistance to reprocessing but the optimum conditions were at low concentrations of filler.

## 9.11 YIELD VALUE

The yield value of a filled system depends on its filler and the matrix.[23,27,58,72-3] On the filler side, yield value depends on particle size and on the functional groups which interact with the matrix to form easily recoverable bonds. For carbon black

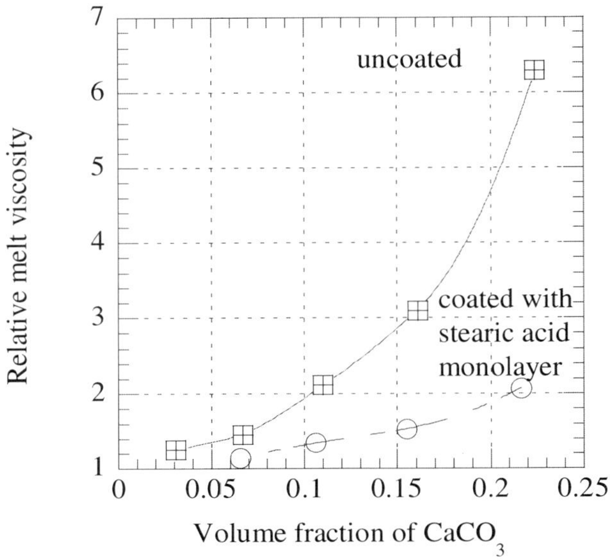


Figure 9.31. Relative melt viscosity vs. fraction of calcium carbonate in LDPE. [Data from Bomal Y, Godard P, *Polym. Engng. Sci.*, **36**, No.2, 1996, 237-43.]

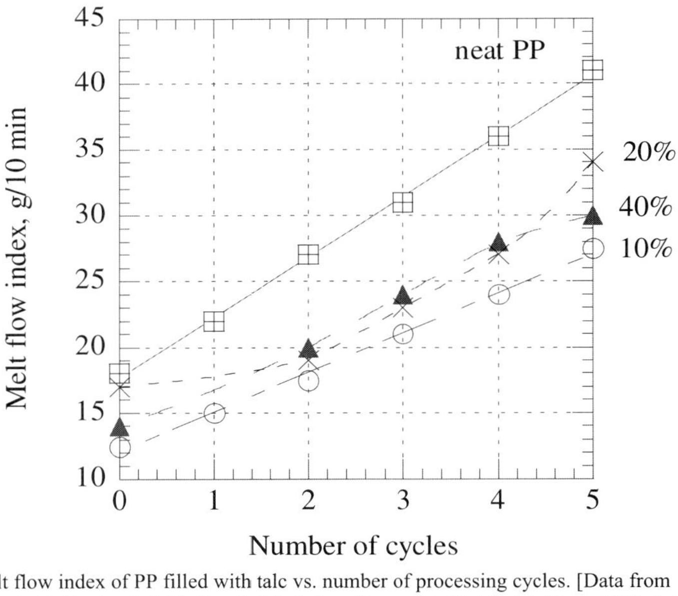


Figure 9.32. Melt flow index of PP filled with talc vs. number of processing cycles. [Data from Guerrica-Echevarria G, Eguiazabal J I, Nazabal J, *Polym. Degradat. Stabil.*, **53**, No.1, 1996, 1-8.]

in PPS, yield values range between 15 and 90 kPa. Smaller values were found for calcium carbonate in PPS (1.5 to 40 kPa). Fillers dispersed in PE usually resulted in low yield values.

Figure 9.9 shows that yield stress increases with a filler volume fraction.[23] Polymer type does not play a significant role here since neither polymer interacts with filler. Small additions of carbon black increase yield stress rapidly. Several equations are used to describe yield stress. The Casson equation is the most frequently used. In this equation the yield stress of a filled system depends on the yield stress and the viscosity of the matrix and on the applied shear rate.[58]

## REFERENCES


1 Brown N, Linnert E, *Reinf. Plast.*, **39**, No.11, 1995, 34-7.
2 Elfving K, Soderberg B, *Reinf. Plast.*, **40**, No.6, 1996, 64-5.
3 Cochet P, Barruel P, Barriquand L, Grobert J, Bomal Y, Prat E, IRC '93/144th Meeting, Fall 1993. Conference Proceedings, Orlando, Fl., 26th-29th Oct.1993, Paper 162.
4 Yu M C, Menashi J, Kaul D J, Antec '94. Conference Proceedings, San Francisco, Ca., 1st-5th May 1994, Vol. III, 2524-8.
5 Agarwal S, Campbell G A, Antec 95. Volume I. Conference proceedings, Boston, Ma., 7th-11th May 1995, 839-42.
6 Joo Y L, Lee Y D, Kwack T H, Min T I, Antec '96. Vol. I. Conference Proceedings, Indianapolis, 5th-10th May 1996, 64-8.
7 Ishibashi J, Kobayashi A, Yoshikawa T, Shinozaki K, Antec '96. Vol. I. Conference Proceedings, Indianapolis, 5th-10th May 1996, 386-90.
8 Kiselev V Y, Vnukova V G, *Int. Polym. Sci. Technol.*, **23**, No.5, 1996, T/88-92.
9 Gendron R, Daigneault L E, Tatibouet J, Dumoulin M M, *Adv. Polym. Technol.*, **15**, No.2, 1996, 111-25.
10 Qi Wang, Jizhuang Cao, Guangjin Li, Xi Xu, *Polym. Int.*, **41**, No. 3, 1996, 245-9.
11 Caillaud J L, Deguillaume S, Vincent M, Giannotta J C, Widmaier J M, *Polym. Int.*, **40**, No.1, 1996, 1-7.
12 Kurian T, Khatgir D, De P P, Tripathy D K, De S K, Peiffer D G, *Polymer*, **37**, No.25, 1996, 5597-605.
13 Kerber M L, Ponomarev I N, Lapshova O A, Grinenko E S, Sabsai O Y, Dubinskii M B, Burtseva I V, *Polym. Sci. Ser. A*, **38**, No.8, 1996, 867-74.
14 Liptak P, Zelenak P, *Int. Polym. Sci. Technol.*, **20**, No.9, 1993, T/57-9.
15 Hedgus C R, Kamel I L, *J. Coatings Technol.*, **65**, No.821, June 1993, 49-61
16 Clarke J, Freakley P K, *Rubb. Chem. Technol.*, **67**, No.4, 1994, 700-15.
17 Khan S A, Baker G L, Colson S, *Chem. of Mat.*, **6**, No.12, 1994, 2359-63.
18 Gahleitner M, Bernreitner K, Neissl W, *J. Appl. Polym. Sci.*, **53**, No.3, 1994, 283-9.
19 Coussot Ph, Proust S, Ancey Ch, *J. Non-Newtonian Fluid Mech.*, **66**, 1996, 55-70.
20 Yaras P, Yilmazer U, Kalyon D M, Antec '93. Conference Proceedings, New Orleans, La., 9th-13th May 1993, Vol. III, 2604-6.
21 Aral B, Kalyon D M, Antec '93. Conference Proceedings, New Orleans, La., 9th--13th May 1993, Vol. III, 2607-10.
22 Balashov M M, Makhmudbekova N L, *Int. Polym. Sci. Technol.*, **22**, No.9, 1995, T/84-6.
23 Mamunya E P, Shumskii V F, Lebedev E V, *Polym. Sci.*, **36**, No.6, 1994, 835-8.
24 Simhambhatla M, Leonov A I, *Rheol. Acta*, **34**, No.4, 1995, 329-38.
25 Kobayashi M, Takahashi T, Takimoto J, Koyama K, *Polymer*, **36**, No.20, 1995, 3927-33.
26 Plummer C J G, Wu Y, Gola M M, Kausch H H, *Polym. Bull.*, **30**, No.5, 1993, 587-94.
27 Chang Ho Suh, White J L, *J. Non-Newtonian Fluid Mechanics*, **62**, Nos.2/3, 1996, 175-206.
28 Papathanasiou T D, *Int. Polym. Processing*, **11**, No.3, Sept.1996, 275-83.
29 Dreibelbis G L, Antec 95. Volume III. Conference proceedings, Boston, Ma., 7th-11th May 1995, 4374-6.
30 Fiske T, Gokturk H S, Yazici R, Kalyon D M, *Polym. Eng. Sci.*, **37**, No.5, 1997, 826-37.
31 Averous L, Quantin J C, Crespy A, *Polym. Eng. Sci.*, **37**, No.2, 1997, 329-37.
32 Scaffaro R, Pedretti U, La Mantia F P, *Eur. Polym. J.*, **32**, No.7, 1996, 869-75.
33 de Sena Affonso J E, Nunes R C R, *Polym. Bull.*, **34**, No.5/6, 1995, 669-75.
34 Tan L S, McHugh A J, *J. Mater. Sci.*, **31**, 1996, 3701-6.
35 Ou Y-C, Yu Z-Z, *Polym. Int.*, **37**, No.2, 1995, 113-7.
36 Zhu J, Ou Y-C, Feng Y-P, *Polym. Int.*, **37**, No.2, 1995, 105-11.

37 Fuelber C, Bluemich B, Unseld K, Herrmann V, *Kaut. u. Gummi Kunst.*, **48**, No.4, 1995, 254-9.
38 Leblanc J L, *Kaut. u. Gummi Kunst.*, **49**, No.4, 1996, 258-66.
39 Shiga T, Okada A, Kurauchi T, *J. Appl. Polym. Sci.*, **58**, No.4, 1995, 787-92.
40 Liqing Sun, Aklonis J J, Salovey R, *Polym. Engng. Sci.*, **33**, No.20, 1993, 1308-19.
41 Datta S, De S K, Kontos E G, Wefer J M, Wagner P, Vidal A, *Polymer*, **37**, No.15, 1996, 3431-5.
42 Bergeret A, Alberola N, *Polymer*, **37**, No.13, 1996, 2759-65.
43 Tsagaropoulos G, Eisenberg A, *Macromolecules*, **28**, No.18, 1995, 6067-77.
44 Vaidyanathan J, Vaidyanathan T K, *J. Mat. Sci. Mat. In Med.*, **6**, No.11, 1995, 670-4.
45 Ou Y C, Yu Z Z, Vidal A, Donnet J B, *J. Appl. Polym. Sci.*, **59**, No.8, 1996, 1321-8.
46 Peng W, Riedl B, *Polymer*, **35**, No.6, 1994, 1280-6.
47 Meijerink J I, Eguchi S, Ogata M, Ishii T, Amagi S, Numata S, Sashima H, *Polymer*, **35**, No.1, 1994, 179-86.
48 Mandal U K, Tripathy D K, De S K, *Plast. Rubb. Comp. Process. Appln.*, **24**, No.1, 1995, 19-25.
49 Cousin P, Smith P, *J. Polym. Sci., Polym. Phys.*, **32**, No.3, 1994, 459-68.
50 Hornsby P R, Wang J, Cosstick K, Rothon R, Jackson G, Wilkinson G, Flame Retardants '94. Conference proceedings, London, 27th-28th January 1994, 93-108.
51 Otaigbe J U, Quinn C J, Beall G H, Antec '97. Conference proceedings, Toronto, April 1997, 1826-30.
52 Kenny J M, Opalicki M, Molina G, Antec '95. Vol. II. Conference Proceedings, Boston, Ma., 7th-11th May 1995, 2782-9.
53 Stockblower D, Antec '96. Volume II. Conference proceedings, Indianapolis, 5th-10th May 1996, 1690-4.
54 Johnson K C, Antec '96. Volume III. Conference proceedings, Indianapolis, 5th-10th May 1996, 3545-9.
55 Kenny J M, Opalicki M, *Composites Part A: Applied Science and Manufacturing*, **27A**, No.3, 1996, 229-40.
56 Wang S Q, Inn Y W, *Polym. Int.*, **37**, No.3, 1995, 153-5.
57 Wang S Q, Inn Y W, *Rheol. Acta*, **33**, No.2, 1994, 108-16.
58 Carreau P J, Lavoie P A, Bagassi M, *Macromol. Symp.*, **108**, 1996, 111-26.
59 Chang Ho Suh, White J L, *Polym. Engng. Sci.*, **36**, No.11, 1996, 1521-30.
60 Persson A L, Bertilsson H, *Composite Interfaces*, **3**, No.4, 1996, 321-32.
61 Cheng J, Bigio D I, Briber R M, Antec '97. Conference proceedings, Toronto, April 1997, 162-7.
62 Wagner A H, Kalyon D M, Yazici R, Fiske T J, Antec '97. Conference proceedings, Toronto, April 1997, 996-1000.
63 Kobayashi M, Takahashi T, Takimoto J, Koyama K, *Polymer,* **37**, No.16, 1996, 3745-7.
64 Murayama H, Min K, Antec '97. Conference proceedings, Toronto, April 1997, 759-65.
65 Clemens M L, Doyle M D, Lees G C, Briggs C C, Day R C, Flame Retardants '94. Conference proceedings, London, 27th-28th January 1994, 193-202.
66 Asai S, Sumita M, *J. Macromol. Sci. B,* **34**, No.3, 1995, 283-94.
67 Locati G, Poggio S, Rathenow J, *Polym. Test.*, **15**, No.5, 1996, 443-54.
68 Jeyaseelan R S, Giacomin A J, *Polym. Gels & Networks*, **3**, No.2, 1995, 117-33.
69 Yu T C, Antec '95. Vol. II. Conference Proceedings, Boston, Ma., 7th-11th May 1995, 2358-68.
70 Bomal Y, Godard P, *Polym. Engng. Sci.*, **36**, No.2, 1996, 237-43.
71 Friedrich C, Scheuchenpflug W, Neuhaeusler S, Roesch J, *J. Appl. Polym. Sci.*, **57**, No.4, 1995, 499-508.
72 Guerrica-Echevarria G, Eguiazabal J I, Nazabal J, *Polym. Degradat. Stabil.*, **53**, No.1, 1996, 1-8.
73 Rockenbauer A, Korecz L, Pukanszky B, *Polym. Bull.*, **33**, No.5, 1994, 585-9.

# 10

# MORPHOLOGY OF FILLED SYSTEMS

## 10.1 CRYSTALLINITY

The crystalline structure of composite materials can be highly varied. The measurements of crystallinity show how the combined interference of the various components of the composite influences the structure. Filled material is composed of crystalline and amorphous regions separated by an interphase which is a diffuse boundary between these two states. The crystallinity of the binder material depends on the fraction of crystalline structures and on their size. Filler may affect both the fraction and the size of crystallites. But, those two measures of crystalline structure are often insufficient and the measurement of crystallinity may give confusing information if the results are taken without further analysis of the fine structure of the material. Table 10.1 gives examples of the effect of fillers on material crystallinity from the current literature.[1-13]

**Table 10.1. Effect of fillers on crystallinity of polymers**

| Polymer | Filler (%) | Processing method | Polymer crystallinity, % | Composite crystallinity, % | Reference |
|---|---|---|---|---|---|
| UHMWPE | bauxite (45) | extrusion | 52 | 28 | 1 |
| PTFE | ferrite (14) | hot pressing | 60 | 57 | 2 |
| LDPE | talc (11) | film | 51 | 61 | 3 |
| PP | kaolin (0.3) | hot pressing | 63 | 46 | 9 |
| PP | kaolin (7) | hot pressing | 63 | 58 | 9 |
| PP | $CaCO_3$ (30) | compression | 67 | 68 | 10 |
| PP | talc (30) | compression | 67 | 78 | 10 |
| PA-66 | GF (80) | compression | 40-45* | 36-45* | 11 |
| PP | $TiO_2$ (30) | injection | 45-47** | 46 | 13 |
| HDPE | $TiO_2$ (30) | injection | 62-65** | 63 | 13 |

*depends on annealing temperature in a range from 20 to 300°C
**depends on specimen taken either from skin or core (filled material uniform)

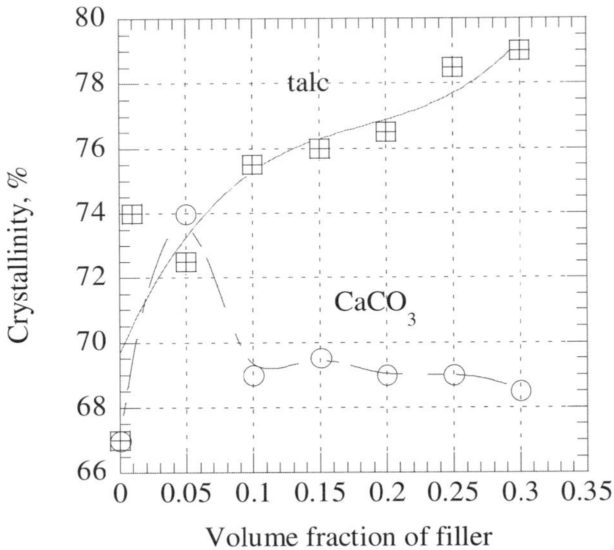


Figure 10.1. Crystallinity of PP composites as a function of $CaCO_3$ and talc concentration. [Adapted, by permission, from Pukanszky B, Belina K, Rockenbauer A, Maurer F H J, *Composites*, **25**, No.3, 1994, 205-14.]

It is difficult to conclude from the data in Table 10.1 whether the filler addition increases or decreases crystallinity. The lack of a clear pattern in the results is caused by differences in filler treatment and processing which cause the development of different structures as the material crystallizes. These points will be further discussed in the next paragraphs.

Figure 10.1 shows the effect of the addition of fillers to polypropylene on its crystallinity.[10] This study was conducted under the same conditions for all specimens tested. There is a difference in the effect of $CaCO_3$ and talc. Calcium carbonate lacks surface functional groups so it tends to have a very small influence on crystallinity and the crystallization behavior. Talc has interacting functional groups on its surface which cause the increase in crystallinity along with the concentration increase.

Several studies[2,3,9] have shown that small additions of filler cause substantial changes in crystallinity (either a large increase or decrease). Whether it was an increase or decrease in crystallinity, these small additions caused a substantial increase in tensile strength and a reduction in elongation. This indicates that the crystalline structure is formed by a nucleation process (see below) which is capable of producing reinforcement.

Surface treatment of a filler may also affect crystallinity. Phosphate coating on talc increased the crystallinity at a low concentration of coating (up to 0.5%). But there was a decrease in crystallinity when the talc was coated with higher concentrations of phosphate.[6]

Other fillers also produce variable degrees of crystallinity depending on their surface treatment. Carbon black is such an example. Graphitization of carbon black may increase surface crystallinity by 30%.[4,5]

## 10.2 CRYSTALLIZATION BEHAVIOR

Crystallization rate, nucleation, size of crystalline units, crystalline structure, crystal modification, transcrystallinity, and crystal orientation are the most relevant characteristics of crystallization behavior in the presence of fillers.[7,10,14-34] Here the discussion is focused on crystallization rate. The other topics are discussed in the following sub-chapters.

Crystallization kinetics is estimated from the Avrami equation:[20]

$$\tau_c(\theta) = \tau_c(\infty)[1 - \exp(-K\theta^n)] \qquad [10.1]$$

where:

| | |
|---|---|
| $\tau_c(\theta)$ | amount of crystalline material at time $\theta$ |
| $\tau_c(\infty)$ | maximum amount of crystallinity reached after the completion of the primary crystallization (the Avrami equation does not account for secondary crystallization) |
| K | temperature dependent factor |
| $\theta$ | time |
| n | exponent related to the dimensionality of crystallites |

This equation deals with the temperature-dependence and crystallite-size- dependence of crystallinity. Frequently, the crystallization half-time, $t_{1/2}$, is reported in the research data. The time to reach one half of the total crystallization is $t_{1/2}$. The time to achieve maximum crystallization is $\tau_c(\infty)$. Figure 10.2 shows the relationship between $t_{1/2}$ and temperature, T, for silica-filled PDMS. The value for $\tau_c(\infty)$ is higher for the filled system than for the unfilled system. The value increases as the temperature increases. Only a small difference was noted for two different filler loadings . The temperature shift shows that less supercooling is required with the filled than with the unfilled system. Fillers produce a nucleation effect which initiates the crystallization process.[20]

Figure 10.3 shows the effect of silica concentration on crystallization rate. This behavior is independent of temperature but the absolute value of the crystallization rate is temperature dependent.[20] Two mechanisms must operate to give such behavior. For lower volume fractions of filler (below 0.26) the crystallization rate is high because more nucleation sites are available. Adsorption of polymer on the surface of silica organizes the adsorbed layer which causes a more ordered structure to develop as the material cools down. When the concentration of silica increases, the silica particles form obstacles to the free movement of crystallizing chains and crystallization is stopped at a filler volume fraction of 0.45. Before this happens, the rate of crystallization gradually decreases. At higher concentrations of filler, there is an insufficient number of polymer molecules adsorbed and crystalline structures do not form because of conformational constraints. The maximum rate of crystallization is determined by two competing processes: nucleation and impingement.

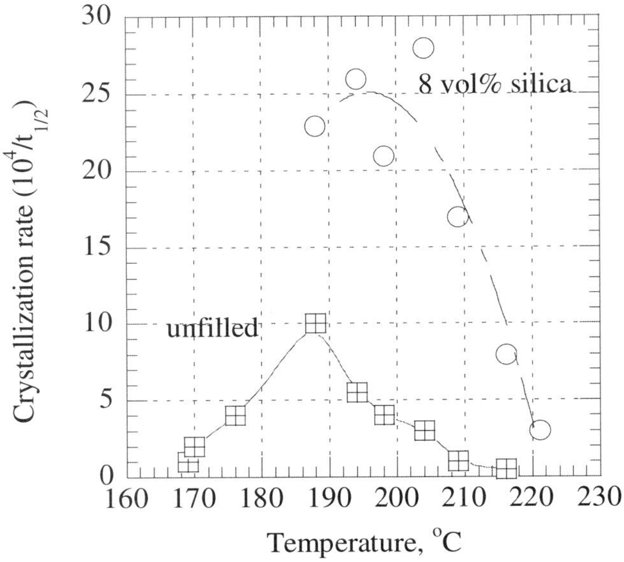


Figure 10.2. Crystallization rate of PDMS filled with silica. [Adapted, by permission, from Ebengou R H, Cohen-Addad J P, *Polymer*, **35**, No.14, 1994, 2962-9.]

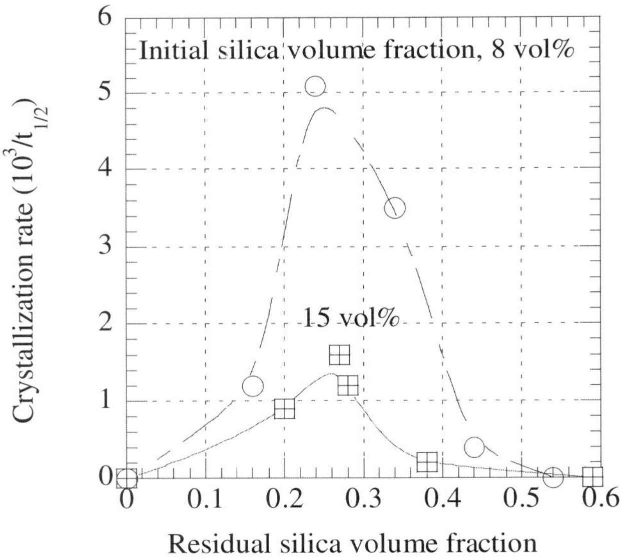


Figure 10.3. Crystallization rate as a function of silica volume fraction in PDMS. [Adapted, by permission, from Ebengou R H, Cohen-Addad J P, *Polymer*, **35**, No.14, 1994, 2962-9.]

Figure 10.4 shows the differences in the Avrami exponent for PVDF filled with carbon black and copper. In the case of carbon black, the rate of crystallization

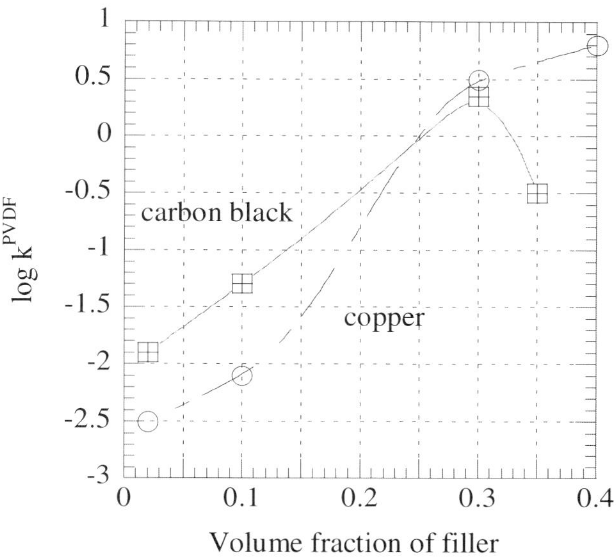


Figure 10.4. Avrami exponent vs. filler concentration for PVDF filled with carbon black and copper. [Data from del Rio C, Acosta J L, *Polymer*, **35**, No.17, 1994, 3752-7.]

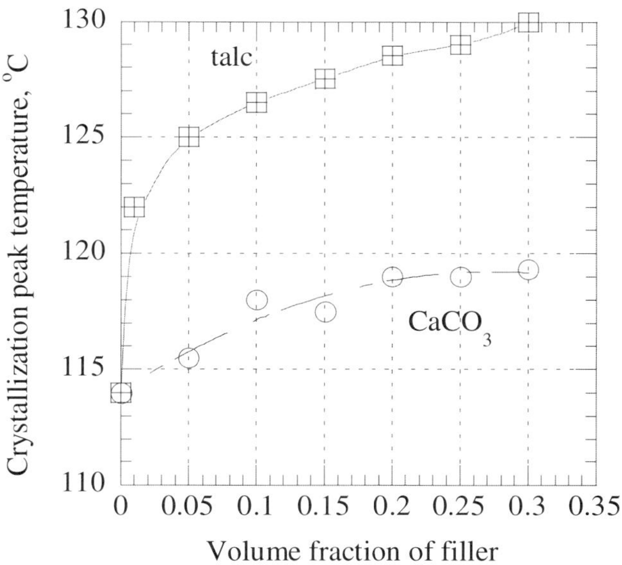


Figure 10.5. Crystallization peak temperature vs. volume fraction of filler. [Adapted, by permission, from Pukanszky B, Belina K, Rockenbauer A, Maurer F H J, *Composites*, **25**, No.3, 1994, 205-14.]

decreases above a concentration of 30% unlike in the copper-filled system. The difference is due to the differences in surface activities of the two fillers.[19]

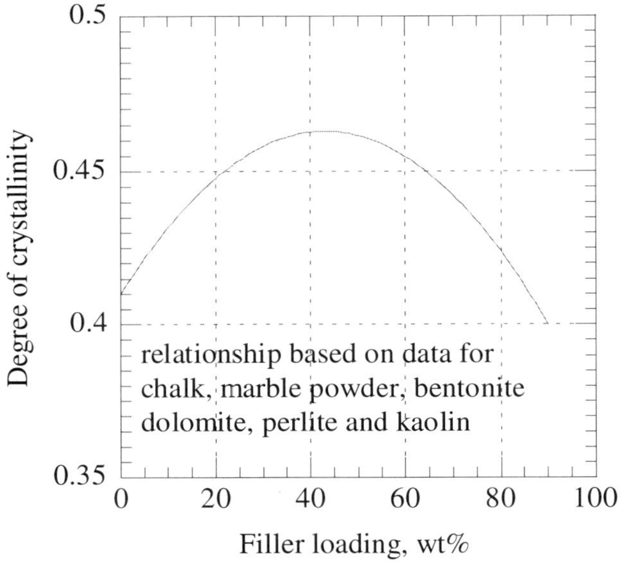


Figure 10.6. Degree of crystallinity vs. concentration of different fillers. [Adapted, by permission, from Minkova L, *Coll. Polym. Sci.*, **272**, No.2, 1994, 115-20.]

The above observations are similar to those obtained for peroxide-crosslinked polyethylene. An addition of filler, such as silica, results in an increased crystallization rate and a decrease in the crystallization half-time, $t_{1/2}$.[21,22]

Figure 10.5 shows the effect of fillers on crystallization peak temperature.[10] The effect of $CaCO_3$ is much less pronounced than that of talc.

Figure 10.6 gives a summary of data on different fillers in UHMWPE. The total degree of crystallinity, as determined by the enthalpy of crystallization, increases with filler concentrations up to 40-50% and then gradually decreases.[16] This decrease is caused by filler aggregation which decreases its nucleation ability.

## 10.3 NUCLEATION

Many papers have been presented on the nucleation process.[7,10,20,23,31,32,35-39] However, the mechanism involved is still disputed.[38] The most important properties of a nucleating agent are its free surface and its ability to organize molecules in conformation which facilitates rapid crystallization. Nucleating agents include various organic materials and inorganic materials, including fillers. Modification of the filler surface may enhance its nucleating abilities. Nucleation provides the benefit of faster processes and better mechanical properties in the final products.

One essential principle of crystallization is expressed by the Lauritzen and Hoffman theory:[20]

$$K = K_0 N \exp\left(-\frac{n\Delta G_\eta}{k_B T}\right)\exp\left(-\frac{nK_g}{T\Delta T}\right) \qquad [10.2]$$

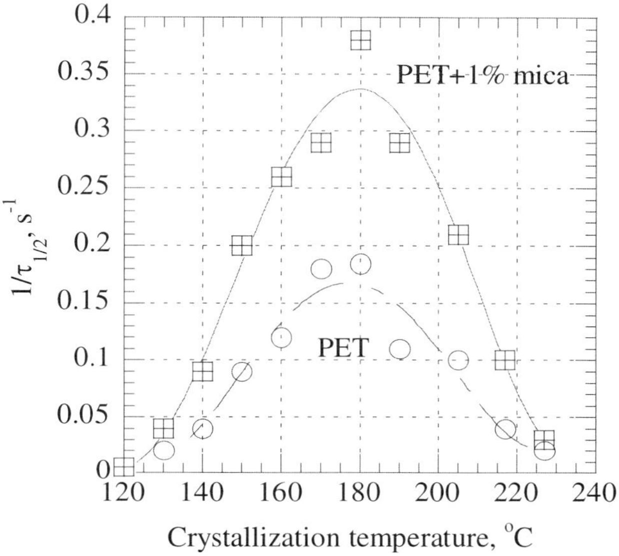


Figure 10.7. Crystallization rate vs. temperature. [Data from Okamoto M, Shinoda Y, Okuyama T, Yamaguchi A, Sekura T, *J. Mat. Sci. Lett.*, **15**, No.13, 1996, 1178-9.]

$$K_g = \frac{4b_0\gamma\gamma_e T_m^0}{k_B \Delta h_f}$$

where:

| | |
|---|---|
| K | coefficient of Avrami equation see Eq 10.1 |
| $K_0$ | constant accounting for geometric parameters of polymer chain and crystalline lamella |
| N | concentration of germ nuclei per unit volume |
| n | coefficient of Avrami equation see Eq 10.1 |
| $\Delta G_\eta$ | free enthalpy of activation governing short distance diffusion of crystallizing elements |
| $k_B$ | Boltzmann constant |
| T | isothermal crystallization temperature |
| $\Delta T$ | $= T_m^0 - T$, degree of supercooling |
| $T_m^0$ | equilibrium melting temperature |
| $b_0$ | parameter of unit cell |
| $\gamma$ | interfacial free energy for lateral surface |
| $\gamma_e$ | interfacial free energy for folding surface |
| $\Delta h_f$ | enthalpy of fusion per unit volume of structural unit |

This equation stresses the importance that diffusion of the crystallizing species has on the crystallization rate, concentration of nuclei, the molecular arrangement, and the required degree of supercooling. Figure 10.2 shows that the rate of crystallization is increased in the presence of silica. This is an effect of nucleation. The filler surface also lowers the free enthalpy barrier which promotes the formation of nuclei.

Figure 10.7 shows how nucleation can be applied to industrial processes. An addition of 1 % mica to PET increases its crystallization rate by a factor of 2. Similar results were obtained with small additions of talc.[37]

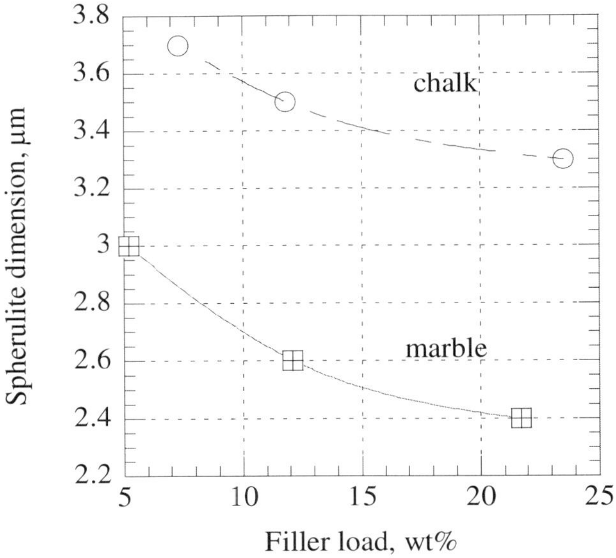


Figure 10.8. Effect of chalk and marble on the size of spherulites in HDPE. [Data from Minkova L, Magagnini P L, *Polym. Degradat. Stabil.*, **42**, No.1, 1993, 107-15.]

Attempts to add fillers to polymer blends produced interesting results.[32] Carbon black was added to a polymer blend containing polycarbonate and polypropylene. Carbon black is known to act as a nucleating agent in polypropylene, however, no increase in the temperature of crystallization was observed. Morphological studies showed that carbon black was preferentially located in the polycarbonate phase therefore it did not affect the nucleation of polypropylene.

Nucleating agents not only shorten the time of crystallization but also improve the mechanical properties of materials. Polypropylene processed with a nucleating agent (2% $CaCO_3$) had its impact strength and modulus increased by 50%.[31]

## 10.4 CRYSTAL SIZE

Figure 10.8 shows the effect of fillers on the dimensions of spherulites. The size of HDPE spherulites crystallized without a filler was 3.9 μm. As filler was added and increased in concentration, the size of the spherulites became progressively smaller.[7]

Figure 10.9 shows the kinetics of spherulite growth in polypropylene containing different amounts of $CaCO_3$. Polypropylene with no filler grew spherulites of a large size over a long period of time. The addition of $CaCO_3$ reduced the ultimate size of the spherulite and also shortened the time to reach an equilibrium size.[25]

Temperature also has an effect on crystallite size. PVDF containing carbon black had crystallites with mean dimensions of 22.4, 20.1, and 16.2 μm when specimens were respectively, slowly cooled, air cooled, and quenched.[24]

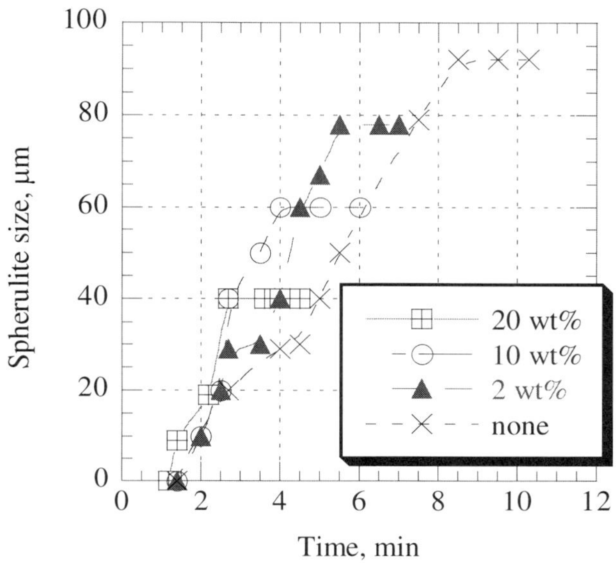


Figure 10.9. Isothermal crystallization of PP containing different concentrations of $CaCO_3$. [Adapted, by permission, from Khare A, Mitra A, Radhakrishnan S, *J. Mat. Sci.*, **31**, No.21, 1996, 5691-5.]

In experiments conducted to obtain controlled sizes of filler particles formed in a matrix, several polymers were used as the matrix.[15] Copolymers were synthesized from polyethylene oxide (does not interact with $CaCO_3$) and poly(methacrylic acid) (reacts with *in situ* crystallizing $CaCO_3$). In the presence of polyethylene oxide, crystals grew to similar sizes as without any polymer. The presence of the poly(methacrylic acid) crystal size of $CaCO_3$ was reduced by a factor 5 to 10 depending on the concentration of the filler precursor.

## 10.5 SPHERULITES

Figure 10.10 illustrates the kinetics of spherulite formation with and without fillers.[10] The left half of each photograph shows spherulite growth without a filler. Two attributes of this growth are evident:

- The process of crystallization is slower in filled than unfilled system (the right halves of photographs)
- Spherulites are larger when no filler is present

This helps to confirm that nucleation, crystallization rate, and spherulite size are strongly influenced by the presence of fillers. It is still uncertain what role a filler plays in the mechanism of nucleation.

In one publication,[13] an extensive morphological study was conducted on the effect of $TiO_2$ on the morphology of crystallized PP and HDPE. The authors did not find any evidence of a modified morphology around the particles and concluded that spherulites grew until they were stopped by the surface of the filler unless the

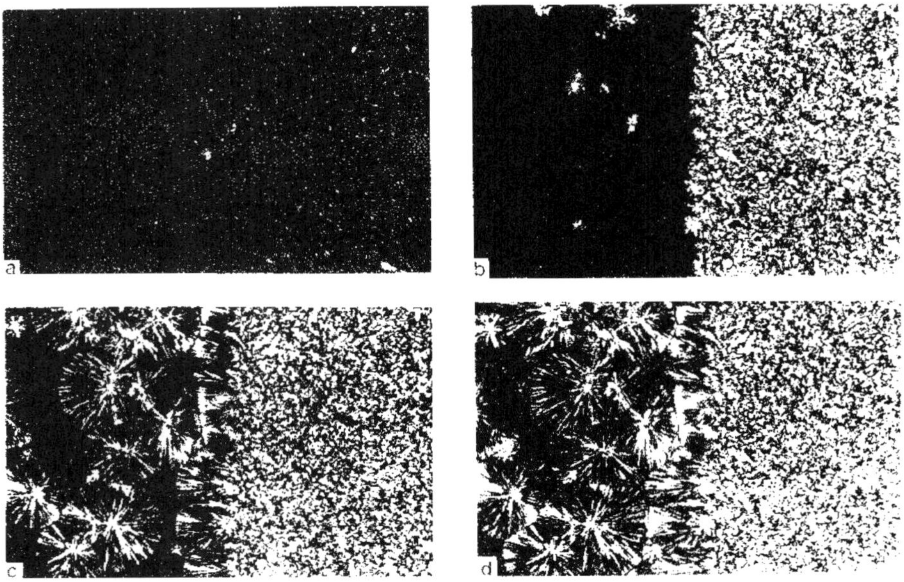

Figure 10.10. Polarization micrographs of PP/talc showing the nucleating effect of talc taken at different times of crystallization. (a) 0.5 min, (b) 7.5, (c) 31, and (d) 38. The left half of photograph is without filler and the right half of the photograph is with 0.5% talc. [Adapted, by permission, from Pukanszky B, Belina K, Rockenbauer A, Maurer F H J, *Composites*, **25**, No.3, 1994, 205-14.]

filler particle became embedded into spherulites and was not ejected by the forces in the crystallizing material. The results of image analysis are given in Figure 10.11. The majority of clusters contain only one crystal when the polypropylene contained 10% $TiO_2$. The number of multicrystal clusters increased only when more filler (40%) was added. Dispersion is never perfect, so it can be assumed that, with a lower filler content, spherulites are growing at the surface of the filler. The fact that no special morphological feature was detected does not preclude the possibility that the growth of the spherulite is initiated on the crystal surface rather than stopped by it as the authors have indicated. The addition of a larger amount of filler contributes to the formation of agglomerates.[16]

Figure 10.12 proposes a mechanism by which crystalline structure in LDPE is formed. The filler particle is close to the face of the crystal structure.[3] If this mechanism of spherulite formation is accepted, there should be no unusual morphological structures in the material since alignment occurs on a molecular level. The filler acts as a template on which the chain is aligned and this makes further folding much easier. The adhesion between the filler surface and the matrix is not high because of weak hydrogen bonding. This makes detachment easy as observed in many filled materials. In order to form a stronger bonding the material must form additional structures (see transcrystallinity below).

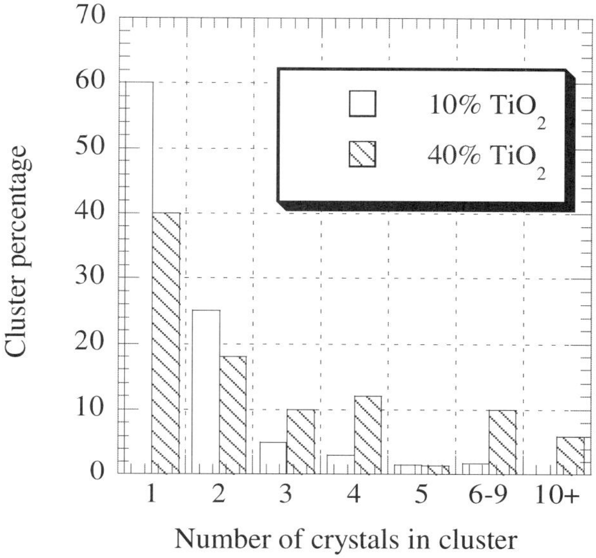


Figure 10.11. Cluster size distribution for $TiO_2$ particles in injection molded PP. [Data from Burke M, Young R J, Stanford J L, *Plast. Rubb. Comp. Process. Appln.*, **20**, No.3, 1993, 121-35.]

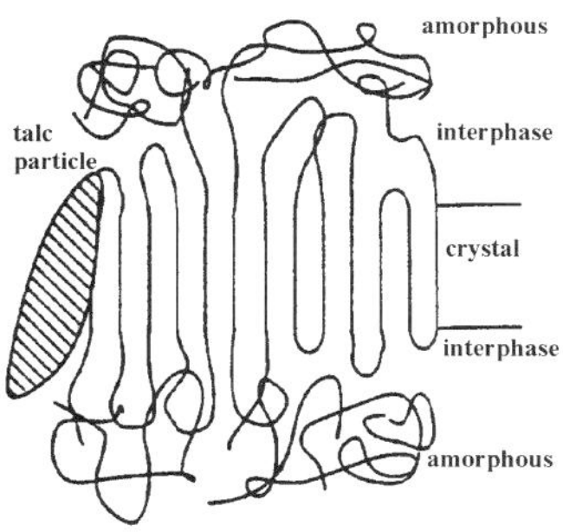


Figure 10.12. Schematic structure of filled and crystallized LDPE. [Adapted, by permission, from Singhal A, Fina L J, *Polymer*, **37**, No.12, 1996, 2335-43.]

The filler affects spherulite size only if cooling rates are low.[16] At a high cooling rate (e.g., 20°C/min), the nucleating role of the filler becomes much less significant.

While the presence of a filler affects the way a matrix crystallizes, the opposite is also true. In studies of *in situ* formation of calcium carbonate in different copolymers, different crystalline forms of calcium carbonate were found.[15] Calcium carbonate crystallized without a polymer had a rhombohedral morphology. When crystallized in the presence of polyethylene oxide its morphology remained rhombohedral because the polymer does not interact with the crystal of calcium carbonate as it forms. The calcium carbonate crystals which formed in the presence of methacrylic acid copolymer had an elongated structure not found in the other two cases.

## 10.6 TRANSCRYSTALLINITY

The cooling of polymer melt in the presence of a foreign surface which can nucleate crystalline growth inhibits the lateral growth of spherulites. Crystallization occurs in a direction normal to the surface.[38] This is called transcrystallinity. It can im-

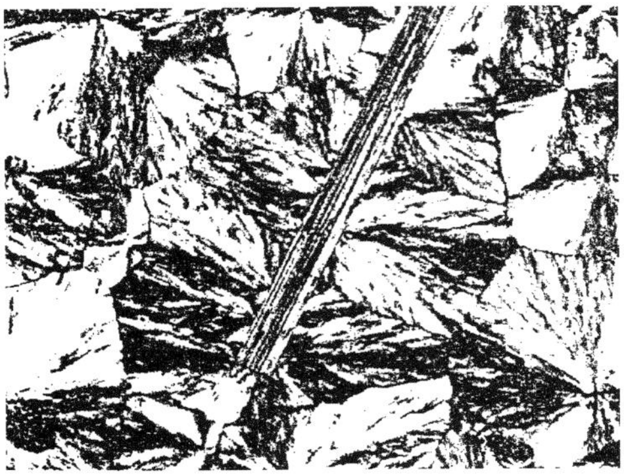

Figure 10.13. Optical micrograph with crossed polars of bamboo fiber in PP matrix (x100). [Adapted, by permission, from Mi Y, Chen X, Guo Q, *J. Appl. Polym. Sci.*, **64**, 1997, 1267-73.]

prove the adhesion and the mechanical properties of composites. Figure 10.13 shows spherulite formation on the surface of a bamboo fiber. There is a nucleation phenomenon on the surface of fibers but the normal three-dimensional growth is hindered. Figure 7.18 shows transcrystallinity which occurs due to changes in the conditions of the process.

The strength of the material and the adhesion between fiber and matrix depend on the thickness of the transcrystalline layer. Figure 10.14 shows the effect of some process conditions (in this case temperature) on the thickness of the transcrystalline layer.[30]

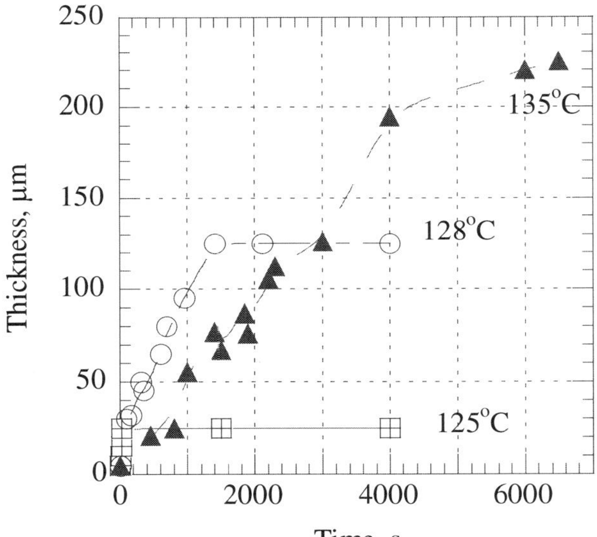


Figure 10.14. The thickness of transcrystalline layer versus time at different melt temperatures in cellulose reinforced PP. [Adapted, by permission, from Gatenhom P, Hedenberg P, Karlsson J, Felix J, Antec '96. Volume II. Conference proceedings, Indianapolis, 5th-10th May 1996, 2302-4.]

Many process parameters are responsible for transcrystallinity.[40-42] The most extensive transcrystallinity is observed under rapid pulling of fibers and high cooling rates. Other parameters include the viscosity of the polymer melt, the rate of shear, the fiber/matrix wettability, and the temperature gradient between matrix and fiber.[40] Figure 10.15 shows the effect of interlayer thickness on the fracture en-

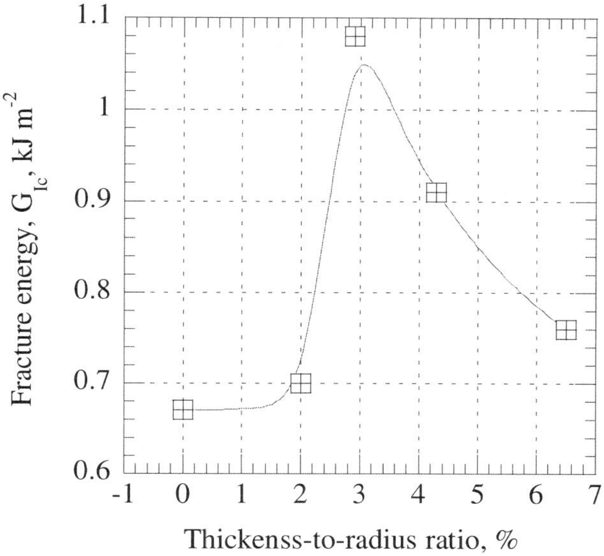


Figure 10.15. Fracture energy vs. interlayer thickness in glass beads-reinforced epoxy. [Adapted, by permission, from Gerard J F, Chabert B, *Macromol. Symp.*, **108**, 1996, 137-46.]

ergy of epoxy reinforced with glass beads. Large gains in mechanical properties can be obtained by tailoring the properties of the interlayer.[41]

Formation of the transcrystalline structure also depends on the geometry of the chain and the fiber surface. Carbon fibers and polyamides are a good match. This makes the chain arrangement on the surface of the fiber very precise and thus the resultant composite is very strong.[42]

## 10.7 ORIENTATION

Three processes of orientation occur simultaneously during the processing of filled materials. These are: filler particle orientation (see Chapter 7), chain orientation (or conformation change) as related to filler particle, and the direction of crystallite growth.[25,27,43-46] Often orientation is detrimental to the material produced. These processes are very difficult to study. Some information is available but more is needed.

Talc is always an attractive subject of such studies due to its platelet structure. In thermoforming and compression molding processes of three resins (PP, HDPE, and PPS), each containing 20% talc, the talc particles were always parallel to the specimen surface, regardless of the resin used.[27] Crystallites grew in a direction normal to the surface of talc particles and thus were perpendicular to the specimen surface. But in the case of unfilled HDPE, crystallites grew parallel to the specimen surface. There was no difference in crystallite growth direction in the case of polypropylene with and without talc.

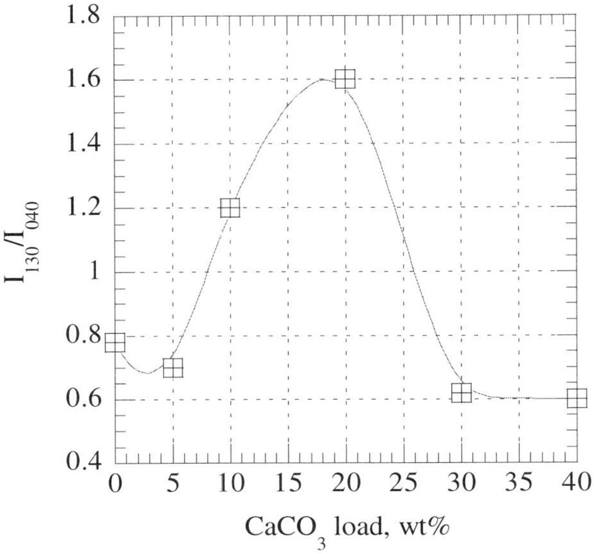


Figure 10.16. The ratio of intensities of the 130 and 040 reflections as a function of $CaCO_3$ concentration. [Adapted, by permission, from Khare A, Mitra A, Radhakrishnan S, *J. Mat. Sci.*, **31**, No.21, 1996, 5691-5.]

The orientation of PMMA chains on the surface of alumina was found to be affected by acid-base interactions. Due to these interactions, the *trans* conformation was more common at the interface than the *gauche* conformation which was prevalent in bulk.[43]

In injection molded composites of polypropylene containing short glass fibers, the fiber orientation depended on the flow pattern (which, in turn, is related to mold thickness, the position of the gate, and flow rate).[45] Substantial variation was detected along the thickness of the sample. Crystallites followed a pattern of fiber distribution but they grew in a direction perpendicular to the direction of the fiber and specimen surface. The direction of spherulite growth was different in neat resin where crystallites grew parallel to the surface of the mold (specimen).

Figure 10.16 gives data on orientation of crystallites in polypropylene containing various amounts of $CaCO_3$. Maximum orientation of crystallites is obtained when the concentration of calcium carbonate is in the range of 15-20%.[25]

## REFERENCES

1 Beloshenko V A, Kozlov G V, Slobodina V G, Prut E V, Grinev V G, *Polym. Sci., Ser. B*, **37**, Nos.5-6, 1995, 316-8.
2 Baranovskii V M, Bondarenko V V, Zadorina E N, Cherenkov A V, Zelenev Y V, *Int. Polym. Sci. Technol.*, **23**, No.6, 1996, T/87-9.
3 Singhal A, Fina L J, *Polymer*, **37**, No.12, 1996, 2335-43.
4 Donnet J B, Wang T K, *Prog. Rubb. Plast. Technol.*, **11**, No.4, 1995, 261-7.
5 Donnet J B, Tong Kuan Wang, *Macromol. Symp.*, **108**, 1996, 97-109.
6 Liu Z, Gilbert M, *J. Appl. Polym. Sci.*, **59**, No.7, 1996, 1087-98.
7 Minkova L, Magagnini P L, *Polym. Degradat. Stabil.*, **42**, No.1, 1993, 107-15.

8 Suri A, Min K, Antec '97. Conference proceedings, Toronto, April 1997, 1487-91.
9 Baranovskii V M, Bondarenko S I, Kachanovskaya L D, Zelenev Y V, Makarov V G, Ovcharenko F D, *Int. Polym. Sci. Technol.*, **22**, No.1, 1995, T/91-3.
10 Pukanszky B, Belina K, Rockenbauer A, Maurer F H J, *Composites*, **25**, No.3, 1994, 205-14.
11 Quintanilla L, Pastor J M, *Polymer*, **35**, No.24, 1994, 5241-6.
12 Abramova N A, Diikova E U, Lyakhovskii Yu Z, *Polym. Sci.*, **36**, No.9, 1994, 1308-9.
13 Burke M, Young R J, Stanford J L, *Plast. Rubb. Comp. Process. Appln.*, **20**, No.3, 1993, 121-35.
14 Weng J, Liu Q, Wolke J G C, Zhang D, De Groot K, *J. Mater. Sci. Lett.*, **16**, 1997, 335-7.
15 Marentette J M, Norwig J, Stockelmann E, Meyer W H, Wegner G, *Adv. Mater.*, **9**, No.8, 1997, 647-50.
16 Minkova L, *Coll. Polym. Sci.*, **272**, No.2, 1994, 115-20.
17 Mi Y, Chen X, Guo Q, *J. Appl. Polym. Sci.*, **64**, 1997, 1267-73.
18 Baranovskii V M, Tarara A M, Khomik A A, *Int. Polym. Sci. Technol.*, **20**, No.1, 1993, T/98-9.
19 del Rio C, Acosta J L, *Polymer*, **35**, No.17, 1994, 3752-7.
20 Ebengou R H, Cohen-Addad J P, *Polymer*, **35**, No.14, 1994, 2962-9.
21 Janigova I, Chodak I, *Eur. Polym. J.*, **31**, No.3, 1995, 271-4.
22 Janigova I, Chodak I, *Eur. Polym. J.*, **30**, No.10, 1994, 1105-10.
23 Chunmin Ye, Jingjiang Liu, Zhishen Mo, Gongben Tang, Xiabin Jing, *J. Appl. Polym. Sci.*, **60**, No.11, 1996, 1877-81.
24 Zhang M, Jia W, Chen X,Jilin, *J. Appl. Polym. Sci.*, **62**, No.5, 1996, 743-7.
25 Khare A, Mitra A, Radhakrishnan S, *J. Mat. Sci.*, **31**, No.21, 1996, 5691-5.
26 Kerber M L, Ponomarev I N, Lapshova O A, Grinenko E S, Sabsai O Y, Dubinskii M B, Burtseva I V, *Polym. Sci. Ser. A*, **38**, No.8, 1996, 867-74.
27 Suh C H, White J L, *Polym. Engng. Sci.*, **36**, No.17, 1996 2188-97.
28 Stricker F, Muelhaupt R, *High Perform. Polym.*, **8**, No.1, 1996, 97-108.
29 Xu P, Mark J E, *Eur. Polym. J.*, **31**, No.12, 1995, 1191-5.
30 Gatenhom P, Hedenberg P, Karlsson J, Felix J, Antec '96. Volume II. Conference proceedings, Indianapolis, 5th-10th May 1996, 2302-4.
31 Tiganis B E, Shanks R A, Long Y, Antec '96. Volume II. Conference proceedings, Indianapolis, 5th-10th May 1996, 1744-9.
32 Benderly D, Siegmann A, Narkis M, *J. Mat. Sci. Lett.*, **15**, No.15, 1996, 1349-52.
33 Ou Y, Yu Z, Zhu J, Li G, Zhu S, *Chinese J. Polym. Sci.*, **14**, No.2, 1996, 172-82.
34 Gordienko V P, Dmitriev Y A, *Polym. Degradat. Stabil.*, **53**, No.1, 1996, 79-87.
35 Chiang W Y, Yang W D, Pukanszky B, *Polym. Engng. Sci.*, **34**, No.6, 1994, 485-92.
36 Zhang Xiaomin, Li Jingshu, Yin Zhihui, Yin Jinghua, *J. Appl. Polym. Sci.*, **62**, No.2, 1996, 313-8.
37 Okamoto M, Shinoda Y, Okuyama T, Yamaguchi A, Sekura T, *J. Mat. Sci. Lett.*, **15**, No.13, 1996, 1178-9.
38 Xavier S F, *Pop. Plast. Packag.*, **41**, No.4, 1996, 61-64.
39 Liauw C M, Hurst S J, Lees G C, Rothon R N, Dobson D C, *Prog. Rubb. Plast. Technol.*, **11**, No.2, 1995, 137-53.
40 Cai Y, Petermann J, Wittich H, *J. Appl. Polym. Sci.*, **65**, 1997, 67-75.
41 Gerard J F, Chabert B, *Macromol. Symp.*, **108**, 1996, 137-46.
42 Greso A J, Phillips P J, *Polymer*, **37**, No.14, 1996, 3165-70.
43 Grohens Y, Schultz J, *Int. J. Adhesion Adhesives*, **17**, 1997, 163-7.
44 Vasnev V A, Tarasov A I, Istratov V N, Ignatov V N, Krasnov A P, Kuznetsov A I, Surkova I N, *Reactive & Functional Polym.*, **26**, Nos.1-3, 1995, 177-83.
45 Gerard P, Raine J, Pabiot J, Antec '97. Conference proceedings, Toronto, April 1997, 526-31.
46 Haynes A R, Coates P D, *J. Mat. Sci.*, **31**, No.7, 1996, 1843-55.

# 11

# Effect of Fillers on Degradative Processes

## 11.1 IRRADIATION

Many filled systems are exposed to irradiation during processing or use. Such processes include radiation crosslinking and vulcanization, development of antistatic properties, production of γ-radiation shields, and sterilization.[1-11] The effect of fillers in these applications is studied.

Several studies look at the crosslinking of PVC compounds containing $CaCO_3$.[1-3] Exposure of a PVC compound to γ-radiation will change its properties. Properties affected include tensile strength and Young modulus which are increased and elongation which is decreased. Figure 11.1 shows that the presence of calcium carbonate had minimal influence on crosslink density. Similarly, calcium carbonate did not influence the performance of the crosslinker (trimethylol propane trimethacrylate).

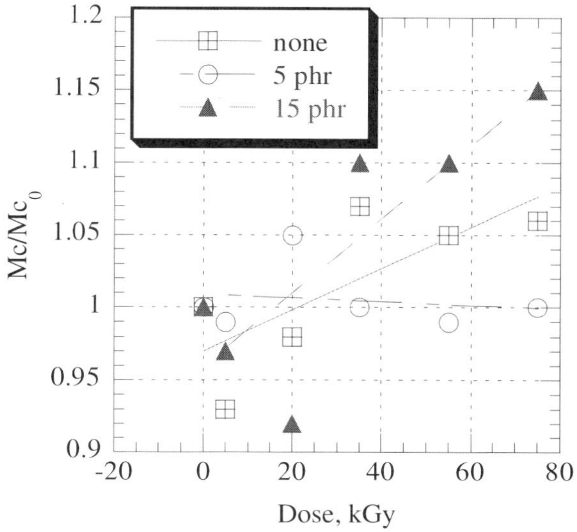


Figure 11.1. Molecular weight between crosslinks vs. dose of γ-radiation at different levels of calcium carbonate. [Adapted, by permission, from Bataille P, Mahlous M, Schreiber H P, *Polym. Engng. Sci.*, **34**, No.12, 1994, 981-5.]

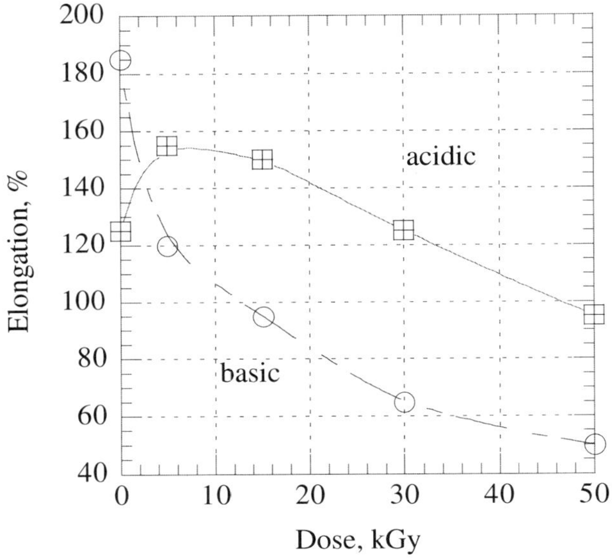


Figure 11.2. Elongation vs. γ-rays dose for PVC filled with 15 phr $CaCO_3$. [Adapted, by permission, from Ulkem I, Bataille P, Schreiber H P, *J. Macromol. Sci. A*, **31**, No.3, 1994, 291-303.]

Figure 11.2 shows that the use of acidic filler helps to preserve elongation when the filled material is exposed to γ-radiation.[3] Both acidic and basic calcium carbonate were evaluated. The acidic version was obtained through surface treatment of normally basic calcium carbonate. Due to lower acid/base interaction, Young modulus and yield stress of the compounds containing the acidic filler were lower but elongation was less affected.

Elongation was better retained also by the addition of 5% soot to LDPE. The material underwent a rapid crosslinking at 50-60 kGy which improved its elongation by a factor of 4. At the same time, its tensile strength was decreased by 30%.[4] Polyethylene containing carbon black was found to be resistant to ionizing radiation.[5-6] The impact strength of carbon black filled HDPE and HDPE/EPDM was improved after exposure to γ-radiation.[6]

Figure 11.3 gives data on the radical decay in PE filled with silica. The increased addition of filler gradually decreases radical decay. This data has significance in two areas. The data shows first that polymer chains are gradually immobilized as the amount of filler is increased. Second, a lower rate of radical decay signifies an increasing probability that chemical conversions and localized reactions are occurring.[7]

Radiation vulcanization of carbon fiber reinforced styrene-butadiene rubber causes a substantial increase in crosslink density (Figure 11.4) and tensile strength (Figure 11.5).[8] This magnitude of change is possible only when the interaction between the filler and the matrix is improved. When irradiated in the presence of air,

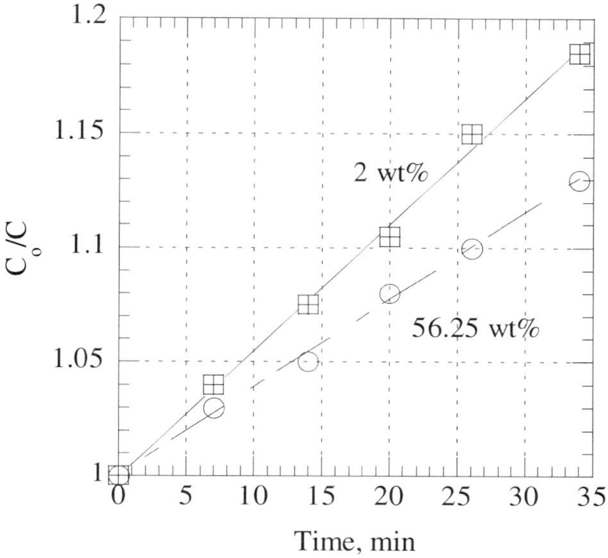


Figure 11.3. Free radical decay in silica filled PE. [Adapted, by permission, from Szocs F, Klimova M, Chodak I, Chorvath I, *Eur. Polym. J.*, **32**, No.3, 1996, 401-2.]

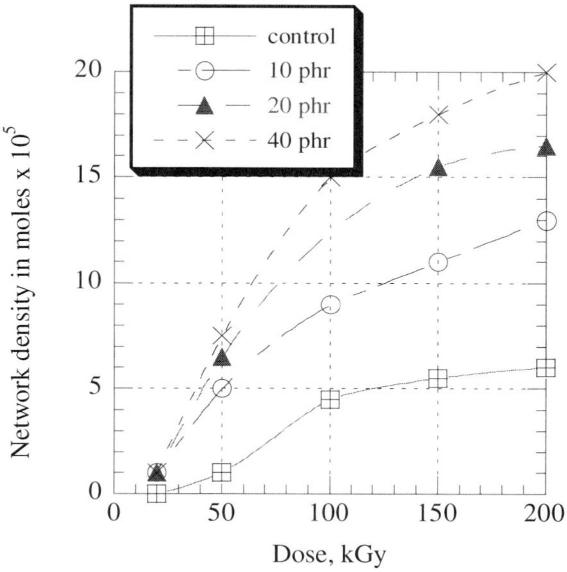


Figure 11.4. Effect of irradiation on crosslink density of SBR filled with carbon fiber. [Adapted, by permission, from Abdel-Aziz M M, Youssef H A, El Miligy A A, Yoshii F, Makuuchi K, *Polym. & Polym. Composites*, **4**, No.4, 1996, 259-68.]

carbon fibers gain functionality which substantially increases their adhesion resulting in a spectacular improvement in properties. SEM studies show that as the dose of radiation increases, the adhesion of the matrix to the fiber increases. Here, a

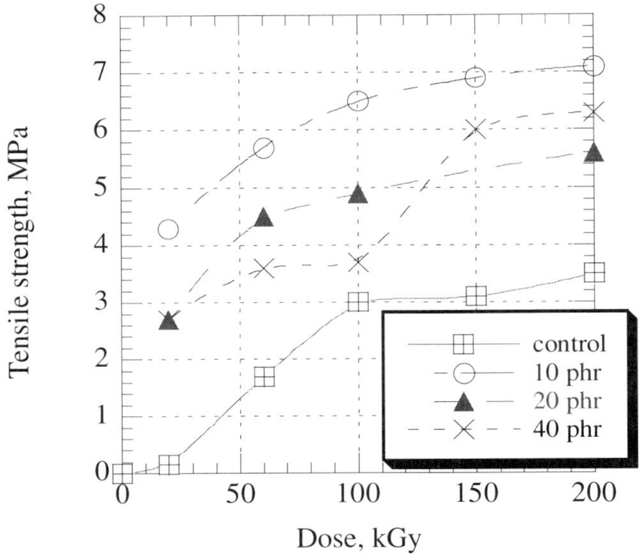


Figure 11.5. Effect of irradiation on tensile strength of SBR filled with carbon fibers. [Adapted, by permission, from Abdel-Aziz M M, Youssef H A, El Miligy A A, Yoshii F, Makuuchi K, *Polym. & Polym. Composites*, **4**, No.4, 1996, 259-68.]

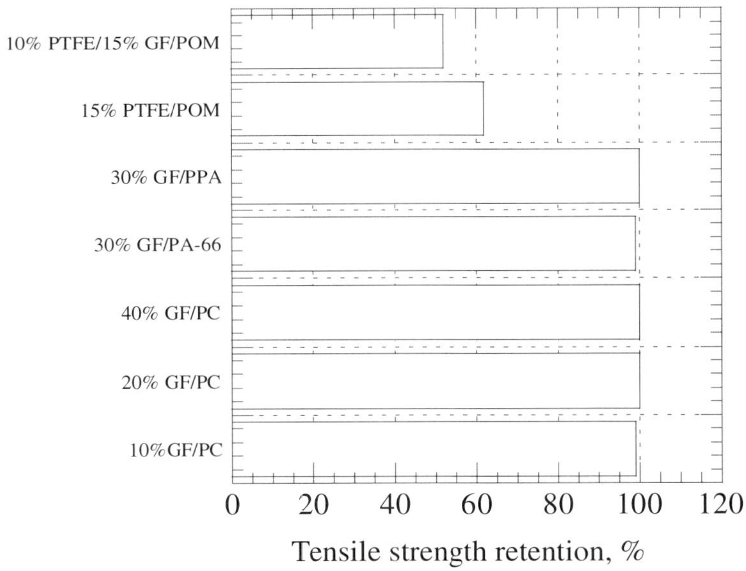


Figure 11.6. Tensile strength retention after exposure to γ-radiation at 3.5 MRad. [Adapted, by permission, from McIlvaine J, Antec 95. Volume III. Conference proceedings, Boston, Ma., 7th-11th May 1995, 3346-9.]

study of the morphology of the fracture surface of the fibers shows they have a matrix deposit on their surface.

Styrene-butadiene rubber loaded with lead oxide was studied to determine its effectiveness as shield for $\gamma$-radiation. The material did have the required performance but it gradually hardened on exposure to radiation.[9]

The effect of radiation sterilization on several plastics was studied (Figure 11.6). With exception of acetal, none were affected by radiation. Six month after exposure, no effect of radiation sterilization was found on impact and tensile strength.[10]

## 11.2 UV RADIATION

Fillers commonly constitute more than 50% of the total composition of processed polymers. Although their effect on weathering resistance has either been demonstrated in service or predicted based on theoretical assumptions, the number of weathering studies is rather small.[12-42] This is perhaps because of the more pressing need to study polymers, which are the components most responsible for the physical properties and durability the composite materials.

Titanium dioxide, the white pigment used in many products, is probably the most extensively studied filler. When titanium dioxide is irradiated with a radiation wavelength of less than 405 nm, the absorbed energy is sufficient to promote electrons from the valence band to the conduction band. Positive holes are formed in the valence band with both holes and electrons able to move within the crystal lattice. During such movement, some holes and electrons will recombine, but these available on the crystal surface can initiate chemical reactions. Electrons may combine with oxygen forming radicals, whereas positive holes combine with hydroxyl groups, forming hydroxyl radicals. These radicals may then react with organic matter or water, initiating a radical reaction chain. In order to limit such reactions, producers of titanium dioxide pigments have developed methods to promote the recombination reactions. This is done either by using admixtures of transition metals (zinc or aluminum) or by coating the $TiO_2$ particles with alumina or silica. Transition metals act as a recombination center for both electrons and holes. Coating helps to destroy the hydroxyl radicals by facilitating their recombination to water and oxygen. Such measures improve the quality of titanium dioxide but they do not completely eliminate radical formation.

Considering the fact that titanium dioxide plays the dual role of stabilizer and sensitizer, one may anticipate that its effect depends not only on the type and quality of titanium dioxide but also on the properties of the binder (polymer). Figures 11.7 and 11.8 show the effect of titanium dioxide on mass loss during the weathering of a durable and a non-durable binder. The durable binder is sufficiently stable to withstand weathering without the need for UV stabilization. The addition of titanium dioxide causes formation of free radicals in the vicinity of its crystals in the binder, which triggers rapidly accelerating degradative changes leading to the decreased weather stability of the material. A non-durable binder (Figure 11.8) is also the subject of radical formation in the vicinity of titanium dioxide particles but the

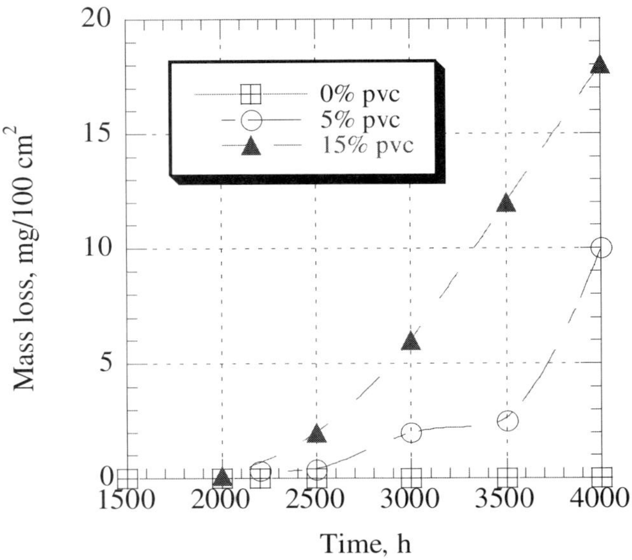


Figure 11.7. Mass loss of a durable binder vs. exposure time relative to pigment load. [Adapted, by permission, from Simpson L A, *Austral. OCCA Proc. News*, **20**, 1983, 6.]

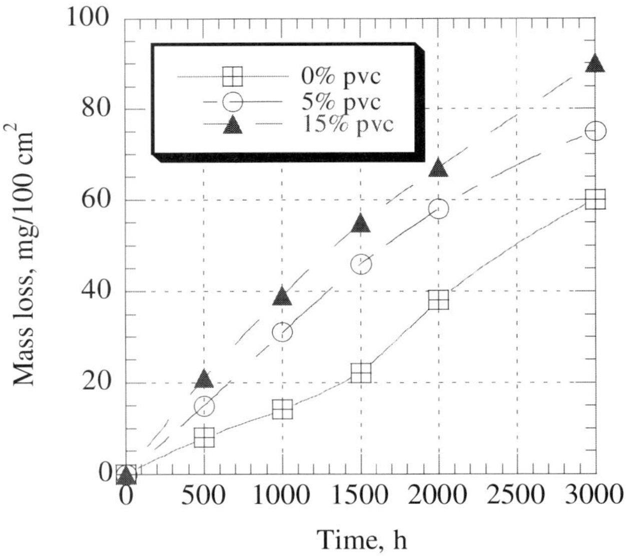


Figure 11.8. Mass loss of a non-durable binder vs. exposure time relative to pigment load. [Adapted, by permission, from Simpson L A, *Austral. OCCA Proc. News*, **20**, 1983, 6.]

protective effect of the pigment is sufficient to offset the negative effect of radical process, resulting in a net improvement in weather stability of the material. Since

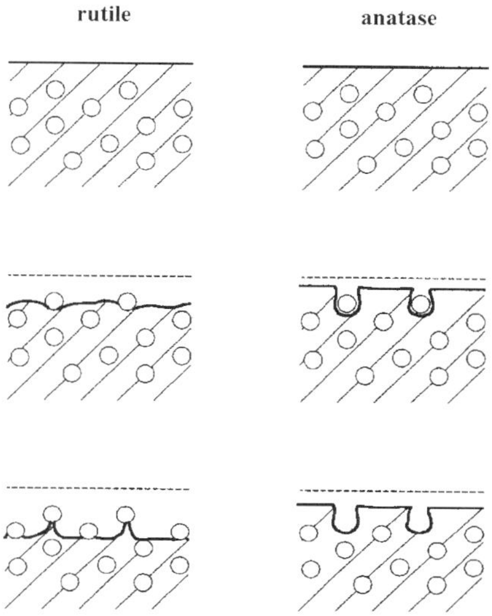


Figure 11.9. Model of binder degradation. [After Braun J H, *Prog. Org. Coat.*, **15**, 1987, 249.]

most of the changes in the material occur around titanium dioxide particles, eventually the binder is sufficiently eroded that pigment and binder separation or chalking occur (Figure 11.9).

Different grades of titanium dioxide produce different effects. The mechanism of chalking described by Figure 11.9 results in changes in material gloss. Both durable and non-durable binder respond to an increasing concentration of titanium dioxide in similar ways, i.e., gloss decreases more rapidly with a higher concentration of $TiO_2$.

Other parameters affect how titanium dioxide participates in degradative processes. In the non-durable binder, titanium dioxide acts to screen the binder from UV radiation. The efficiency of screening depends on the degree to which $TiO_2$ is dispersed. Flocculated pigment has a lower screening power. Less flocculated pigments inhibit both gloss deterioration and mass loss of the material.

The temperature at which degradation occurs is also an important factor. Gloss is better retained when samples are irradiated at lower temperatures. This suggests that the rate of formation of free radicals is controlled by two processes: one being photon absorption; the other being the reaction of the excited species followed by radical formation. The quantum efficiency of radical formation is reduced when radicals recombine. Because radicals are formed within a rigid matrix, it is difficult for them to escape (the cage escape efficiency is reduced). At low temperatures, the matrix is more rigid than at high temperatures and recombination is the more probable outcome. Around $T_g$, the cage escape efficiency is rapidly reduced.

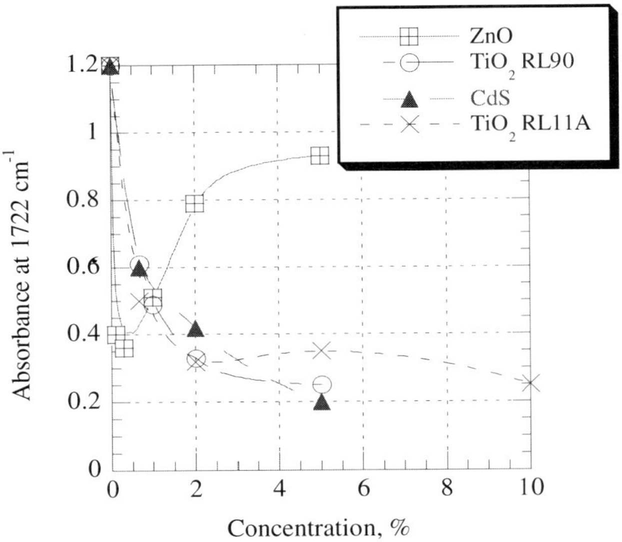


Figure 11.10. Carbonyl content in photooxidized EP copolymer film vs. filler concentration. [Adapted, by permission, from Lacoste J, Singh R P, Boussand J, Arnaud R, *J. Polym. Sci., Polym. Chem.*, **25**, 1987, 2799.]

The effect of pigment properties on photochemical activity is shown in Figure 11.10. Surface passivated titanium dioxide (RL90) and CdS both decrease the amount of carbonyl group formation as the concentration of pigment increases. ZnO and untreated titanium dioxide also contribute to a decrease in carbonyl group formation but only at low concentrations. Above a certain level, each cause an increase in the formation of carbonyl groups. Treated $TiO_2$ and CdS have poor photocatalytic activity and they participate in photodegradative processes by actively screening radiation. In the first part of the curve (Figure 11.10), there is a very low quantum yield of photocatalysis due to electron-hole recombination. ZnO and untreated $TiO_2$ provide screening at low concentrations. At higher concentrations, their photocatalytic effect becomes predominant and carbonyl concentration increases. On the other hand, ZnO has long been known to stabilize some polymers. Formation of zinc caboxylates was found to contribute to the stabilizing efficiency of ZnO in HDPE.[39] From a comparison of the effect of several metal oxides including ZnO and $TiO_2$ on the stability of LDPE, it was postulated that the stabilizing activity of a filler depends on its ability to induce crystallinity to the matrix.[40]

Small additions of fine particle $TiO_2$ are currently used to preserve the vitamin content of milk packaged in plastic films. As small as 0.5-1% concentrations of $TiO_2$ are sufficient to retain more than 90% of vitamin A in milk exposed to UV for 3 days.[36] This example shows that fillers may not only protect materials against degradation but they also protect the contents. A 2% addition of fine particle $TiO_2$ (spherical particles of 15-30 nm diameter) absorbs all UV light below 350 nm (the most degrading part of UV) at film thickness of 70 μm.

A recent study[35] shows another possible application of $TiO_2$. Combination of peroxides and $TiO_2$ or ZnO was used for controlled degradation of PVC on exposure to sun light. It is possible to degrade this PVC in only one month.

The tensile strength of the material and other mechanical properties were better preserved when pigment, regardless of the color, was added. Studies of fourteen iron oxide pigments in a PVC matrix showed that pigments with a large inherent ESR spectrum strength, have poor weatherability.[16] PVC plates with an ESR spectrum strength lower than 1 exhibited very good weatherability, equivalent to five years of outdoor exposure. The type of pigment used in PVC greatly influenced the UV stability of the polymer.

An extensive studies[26] were conducted on the effect of pigments and $TiO_2$ on the degradation and stabilization properties of polymer matrices. These properties are important: dispersibility, light absorbing characteristics, semi-conductor properties, metal content, influence on polymer matrix, surface properties, composition of products of degradation. This list could be expanded to include: pigment surface area, absorption of components of matrix (e.g., stabilizers), wavelength of emitted radiation by pigment on energy absorption, generation of singlet oxygen, hydrogen abstraction, effect on polymer morphology (some pigments interfere in crystallization), interaction with polymer, etc. The way in which a pigment interacts with the polymer network is known to have an effect on the UV stability of the material but this effect can vary widely. For example, the stability of a composition to radiation at 375 nm can be increased by increasing pigment concentration. But there are also exceptions. Ultramarine blue increases durability of PP by 75% although it does not absorb UV. $TiO_2$ absorbs of UV as much as does channel black but has only a fraction of its stabilizing activity. Some types of dyes, such as, for example, azo condensation yellow, red and orange are known to decrease the stability of some polymers (e.g., PP fiber). In some polymers (e.g., PVC) most pigments (especially inorganic and carbon black) considerably increase durability. Phthalocyanine blue is a relatively good stabilizer for PP fiber but a poor pigment for PVC. Some dyes behave differently in low relative humidity than they do in moisture close to saturation. This shows that there is a substantial amount of work to be done to explore the very wide range of pigment, dye, and polymer combinations.

The addition of $CaCO_3$ to PP causes a slight reduction in carbonyl formation.[24] The efficiency of some antioxidants, such as Irganox 1010, was found to be reduced by the presence of $CaCO_3$. In another study,[31] PP stability was increased by the addition of $CaCO_3$ especially in combination with small addition of $TiO_2$ (0.5%) or HALS. In polyurethanes, $CaCO_3$ acts as a heat sink.[32]

The addition of talc to PP increased the absorption of UV light somewhat due to the opacity of the filler but the absorption of UV was negligible compared to $TiO_2$. This is related to the relatively large particle size of talc. No substantial difference was detected in stability of filled and unfilled PP exposed to UV radiation.[38]

Silica, in the concentrations in which it is typically used, does not affect radical decay during the degradation of PMMA by UV, nor is the radical composition affected.[28] Large additions (above 50%) modify the material structure due to matrix absorption on the silica surface which also causes an increase in the radical decay rate. These data are contradictory to the data presented in Figure 11.3.[7] Sand was added to PVC[34] and PE[33] and their photodegradation was monitored by molecular weight determination and measurements of changes in flexural strength and insoluble matter. Both sand containing polymers were more stable.

Carbon black is the best UV screening compound and provides long-term protection. Carbon black not only screens out UV but also inhibits photooxidation through a complex series of autooxidative mechanisms. Not only is the particle size of carbon black important (the best performance is in the range of 15-25 nm), but also the chemical composition of its surface. It was proven experimentally that the best results were obtained when Channel Black was used. Channel Black is no longer manufactured by the channel process but by more modern techniques which are able to simulate the channel process and produce a surface composition similar to the original. Carbon black is widely used for the production of weather-resistant materials. It effectively protects polymers used as durable binders. In polymers which are less UV stable, carbon black affects thermal stability more than UV stability.[37] The performance of carbon black filled materials is concentration dependent. At very low concentrations (below 1%) carbon black may reduce the photolytic stability of a material. The photolytic stability is increased at higher loadings and at least 2.5% is needed for minimum protection.

This shows that there is still inadequate information to assess the effect of fillers on photodegradation of filled materials. The processes occurring during photodegradation are complex in nature and as such require extensive studies. Specialized methods and equipment are needed to investigate changes in materials.[43]

## 11.3 TEMPERATURE

Many fillers were studied in relationship to their effect on the thermal stability of polymers.[3,11,37,44-71] Table 11.1 gives a summary of the findings. It shows that the role of the filler ranges from causing a decrease in the degradation rate through no effect to causing an increase in the degradation rate. Its behavior in any system is influenced by the presence of impurities and the potential reactivity of all system components. The examples of the results of experimental studies characterize the range of effects.

Figure 11.11 shows that talcs from different sources behave differently in polypropylene. The thermal stability of compounds depends on type and amount of impurities which are different depending on the origin of mineral and on the method of processing.

**Table 11.1 Effect of fillers on the thermal stability of polymers**

| Filler | Polymer | Testing Method | Findings | Refs. |
|---|---|---|---|---|
| $Al(OH)_3$ | EEA<br>PVC<br>PBMA<br>PVB | TG, TVA, IR<br>HCl absorption<br>TG, GC-MS<br>TG, GC-MS | Retards thermal degradation (endothermic decomposition)<br>Higher degradation rate for uncrosslinked<br>Inhibits monomer evolution, promotes ester decomposition<br>Reduces degradation temperature | 61<br>44<br>65<br>64 |
| $CaCO_3$ | LPDE, EEA | TG, DSC, TVA | Reduces degradation rate measured by volatiles | 63 |
| Carbon black | PE, PB | Peroxides | Antioxidant, radical scavenger, peroxide decomposition | 59 |
| Carbon fiber | Phenoxy | XPS | Fiber protected against oxidation by coating | 47 |
| Chalk | PE | TG | No change in degradation rate | 58 |
| Glass fiber | PES, PEEK | Tensile strength | Improved retention of tensile strength | 71 |
| Graphite | Silicone | Young modulus | Increased Young modulus indicates degradation | 51 |
| Iron oxide | Silicone | Young modulus | Increased Young modulus indicates degradation | 51 |
| $Mg(OH)_2$ | PA<br>PBB<br>PA<br>EEA | TGA<br>Weight loss<br>TGA<br>TG | Chains scission due to hydrolysis<br>Substantially lower weight loss up to 20% filler<br>Reduced thermal stability<br>Improved thermal stability | 52<br>66<br>70<br>61 |
| Marble | PE | TG | No change in degradation rate | 58 |
| Silica | PBMA<br>PVB<br>PP | TG, GC-MS<br>TG, GC-MS<br>Carbonyl | Improved thermal stability<br>Decreased thermal stability<br>Stability reduced (absorption of thermal stabilizers) | 65<br>64.<br>62 |
| Talc | PP | Heat aging | Stability depends on source of filler | 45 |
| Zeolites | PP | TG, GC-MS | Some zeolites can be used as catalysts of degradation | 11 |

Calcium carbonate, especially the coated grades as well as some grades of silica are the most inert fillers and they do not much affect thermal degradation rates of many polymers. But, there are examples in which the thermal stability of some polymers can be improved. Figure 11.12 gives data on the dehydrochlorination rate of PVC with and without coated calcium carbonate.[72] The compound containing filler has improved thermal stability. This is due to the participation of calcium carbonate and its stearate coating in a reaction with hydrogen chloride which is an autocatalyst for the thermal degradation of PVC.

Carbon black can increase the thermal stability of many polymers because of its properties. Phenoxyl and quinoid groups on the surface of carbon black function as antioxidants.[59] These groups also participate in the catalytic decomposition of peroxides which contributes to a reduction in degradation rate. Quinone, polynuclear structures, polyconjugated double bonds, and carbonyl groups all scavenge radicals. Many polymers and rubbers benefit from these properties of carbon black.

Molybdenum disulfide is known to stabilize polyarylate. It was postulated that two mechanisms may be responsible for this process: the formation of coordination

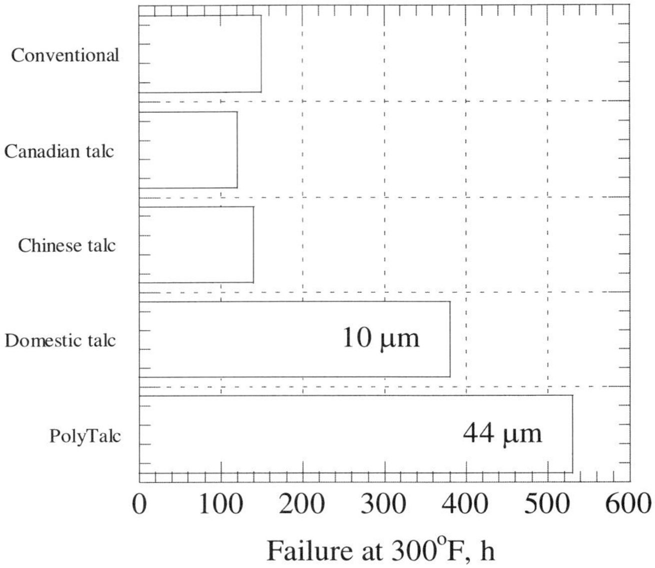


Figure 11.11. Heat aging of 40% talc in PP at 150°C. [Adapted, by permission, from Sherman L M, *Plast. Technol.*, **43**, No.4, 1997, 26-8.]

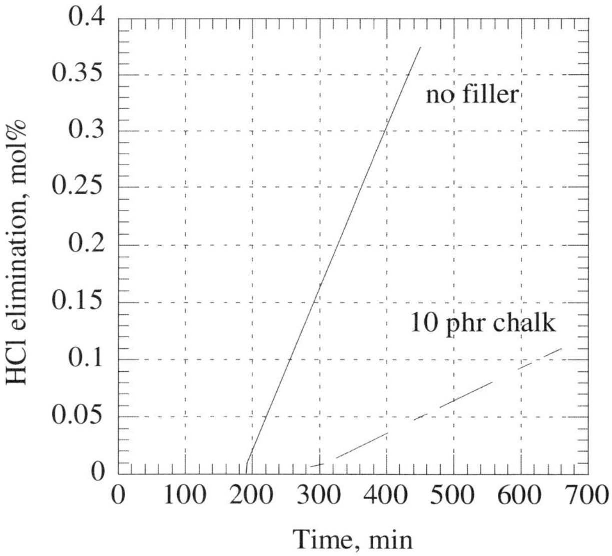


Figure 11.12. Dehydrochlorination rate of PVC with and without 10% calcium carbonate. [Adapted, by permission, from Braun D, Kraemer K, **Recycling of PVC & Mixed Plastic Waste**, La Manta F P, Ed., *ChemTec Publishing*, Toronto, 1996.]

complexes between carboxyl groups and molybdenum disulfide and the reaction with oxygen (antioxidative effect).

The physical properties of some fillers play a role in their function as stabilizers. $Al(OH)_3$ undergoes endothermic decomposition which lowers temperature of material. Loss of water from $Mg(OH)_2$ may increase stability in some cases. In others, it may cause degradation. This is discussed below. The platelet structure of some fillers (e.g., talc or mica) contributes to an increased thermal stability because the degradation rate is increased as oxygen concentration increases. The structure formed by the platelets reduces the diffusion rate of oxygen.

There are many examples which show that a filler may reduce thermal stability of a polymer. Impurities in the form of metal salts such as formed by Co, Cd, Fe, Zn, etc. provide classic cases where, in their presence, thermal degradation is adversely affected. Water formed from the decomposition of fillers such as $Al(OH)_3$ or $Mg(OH)_2$ can hydrolyze the backbone of polyamide and polyester which degrades the polymer.[52,61]

Silica and other fillers affect thermal stability indirectly by adsorbing thermal stabilizers which prevents them from acting as stabilizers.[62] Some zeolites were used to catalyze the degradation of polypropylene during waste processing. The type of cation was essential in decreasing the degradation temperature (e.g., $Na^+$).[11]

## 11.4 LIQUIDS AND VAPORS

Resistance to water is the most important property of composites.[73-81]

Figure 11.13 shows that in a jute filled epoxy resin, water intake increases with time of immersion and with the amount of fiber.[73] This jute fibers readily absorb water. A surface treatment of the jute with epoxy silane reduces the water intake. Tensile properties of a composite containing surface treated fiber remain constant up to a moisture content of 5%. Water absorption by natural fibers, such as cellulose causes the formation of internal forces capable of rotating fibers around their axis. This twisting motion introduces stress into the structure of composite.[79]

Aramid fiber rapidly absorbs water on immersion (Figure 11.14). A NMR study of water absorption shows that water is absorbed into the voids of the fiber. These contain sodium salts (mostly sodium carbonate). This explains why water is so rapidly absorbed in the beginning of immersion.[78] Aramid fibers are resistant to bases and are fairly resistant to acids except for $HNO_3$.[77] E-glass can withstand immersion in $H_3PO_4$ and acetic acid but it is not resistant to bases and strong acids. E-glass fiber is severely affected by oxalic acid which extracts about 25% of its weight.[75] Acid corrosion of E-glass fibers is attributed to calcium and aluminum depletion. The severity of this depletion depends on the acid type and on the method of fabrication of the fiber.

Carbon fiber in polycarbonate composites absorbs water.[80] Figure 11.15 shows compressive yield stress of composite vs. aging time in boiling water. Compressive yield stress increases with fiber addition as well as with the aging time. Aging in boiling water also enhances Young modulus of the composite.

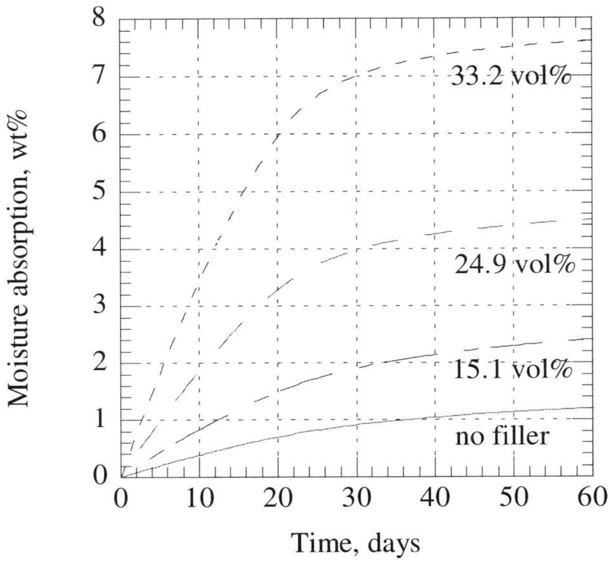


Figure 11.13. Moisture absorption of epoxy containing jute fiber vs. exposure time. [Adapted, by permission, from Gassan J, Bledzki A K, *Polym. Composites*, **18**, No.2, 1997, 179-84.]

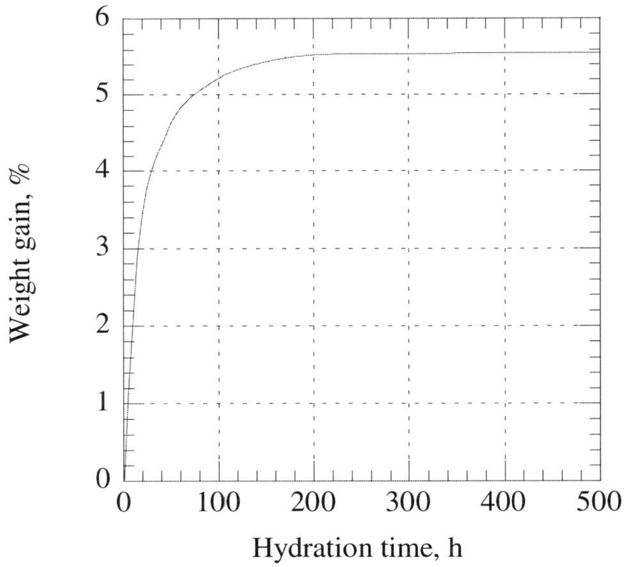


Figure 11.14. Weight increase vs. time of exposure of aramid fibers to 100% relative humidity at room temperature. [Adapted, by permission, from Connor C, Chadwick M M, *J. Mat. Sci.,* **31**, No.14, 1996, 3871-7.]

Orientation of fiber is detrimental to performance of composites which contain carbon fibers (Figures 11.16 and 11.17).[74] Tensile strength of composites containing fibers oriented in a direction parallel to the surface is not affected by

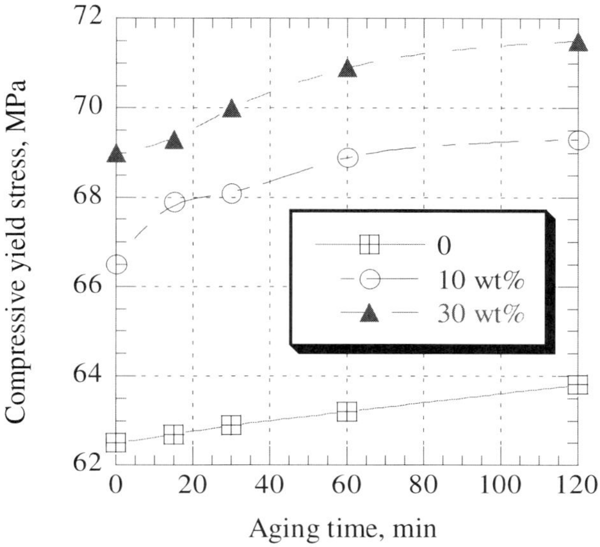


Figure 11.15. Yield stress of carbon fiber/polycarbonate composite vs. aging time in boiling water. [Adapted, by permission, from Nofal M M, Zihlif A M, Ragosta G, Martuscelli E, *Polym. Composites*, **17**, No.5, 1996, 705-9.]

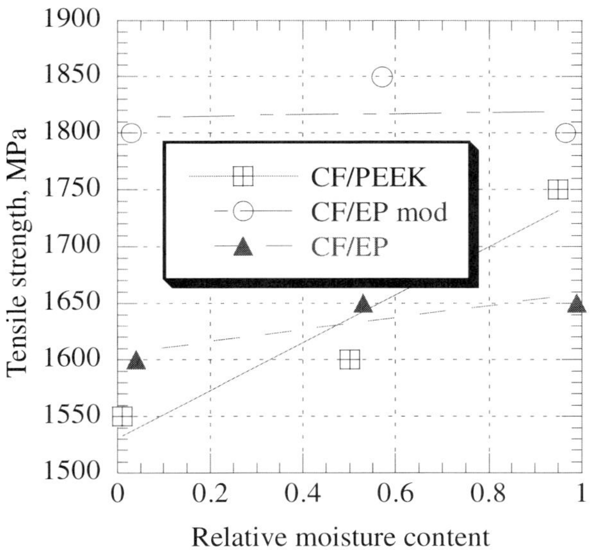


Figure 11.16. Tensile strength of laminates with fibers oriented parallel to the surface. [Adapted, by permission, from Selzer R, Friedrich K, *Composites, Part A*, **28A**, 1997, 595-604.]

moisture content. Composites which have fibers oriented in a direction perpendicular to the surface, lose tensile strength as moisture increases (exception – carbon fiber/PEEK composite). Similar effects on tensile modulus, compression modulus and elongation have been observed.

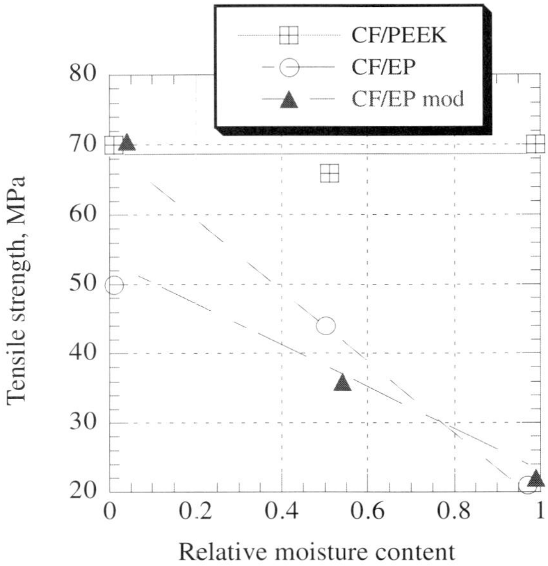


Figure 11.17. Tensile strength of laminates with fibers oriented perpendicular to the surface. [Adapted, by permission, from Selzer R, Friedrich K, *Composites, Part A*, **28A**, 1997, 595-604.]

## 11.5 STABILIZATION

Pigments and filler in combination with UV stabilizers may influence the stabilizing of the stabilizer.[41,59,62,69,82-86] A typical stabilizing system used today consists of UV absorber and HALS. In most cases, formulation-specific solutions have been developed which give optimal performance. Still, these systems have many inherent deficiencies related to the properties of both types of stabilizers. UV absorbers perform only when an adequate surface concentration of stabilizer is available. Because of the nature of the mechanism of adsorption, UV absorbers cannot protect surface (to a depth of about 10 μm). In addition, UV stabilizers are not permanent and their concentration is gradually reduced during exposure to UV. This loss of stabilizer is enhanced when the surface layers are degraded. Here, fillers can help. Ultrafine grades of $TiO_2$ have improved absorption when their particle size is close to 20 nm. The pigment has a relatively low opacity (it does not absorb visible light) but it absorbs UV radiation very strongly.[85] Most $TiO_2$ is produced to achieve high opacity which is at a maximum when the particle size is in a range from 150 to 300 nm. New grades absorb UV light as well as UV stabilizers. The concentration of 0.2% ultrafine $TiO_2$ was found to give better performance in PP than 0.1% UV absorber. Three advantages can be attributed to the use of pigment over an organic UV absorber: it is permanent, its performance is not lost when surface layers are degraded and it is less expensive. These attributes ensure a bright future for $TiO_2$ in this application.

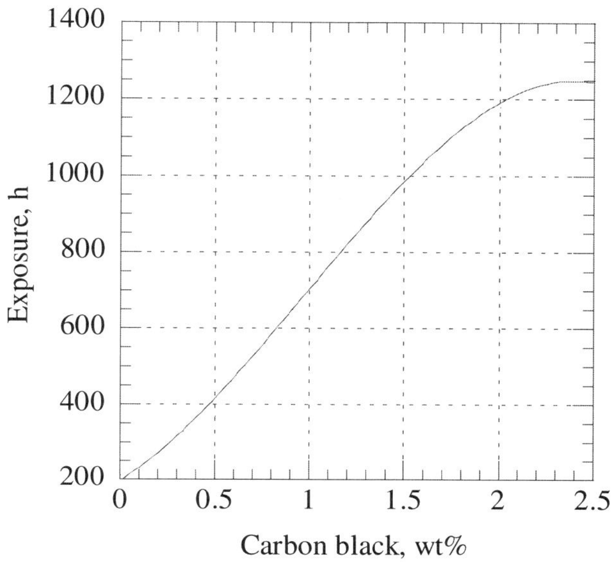


Figure 11.18. Exposure of PE stabilized by carbon black vs. concentration. [Adapted, by permission, from Turley R S, Strong A B, *J. Adv. Materials*, **25**, No.3, 1994, 53-9.]

HALS has also several limitations. Since it is volatile it evaporates slowly. Polymeric HALS are not volatile and will be retained much longer. But, they are less mobile and thus less reactive. HALS can react with radicals on the surface of the material which gives it advantage over UV absorber. But if the volatile HALS is used more of the surface material will be lost to evaporation. Carbon black is able to perform some functions of HALS (see Section 11.3).[59] Figure 11.18 gives the concentration of carbon black required to obtain a predetermined length of service. Properties of the carbon black chosen are also important. The best results are obtained with carbon blacks which have the correct balance of particle size and backscattering efficiency.[86]

The stabilizing effect of fillers extends to their interaction with UV stabilizers. HALS is readily adsorbed on the surface of fillers such as silica. The adsorption mechanism is by hydrogen bonding which immobilizes HALS.[82] Selection of the appropriate HALS is important. Tertiary HALS is not as strongly adsorbed by fillers as secondary HALS.[69] The adsorption of HALS is frequently thought of to be a disadvantage but it can enhance the stabilizing activity of HALS when the filler acts as controlled release agent. Systems can be formulated which enhance HALS performance based on this principle.[83]

## 11.6 DEGRADABLE MATERIALS

Starch and cellulosic materials are frequently used as fillers in degradable materials.[76,87-93] The addition of starch to LDPE in combination with a pro-oxidant increases the photooxidation rate and the formation of hydroperoxides and carbonyl

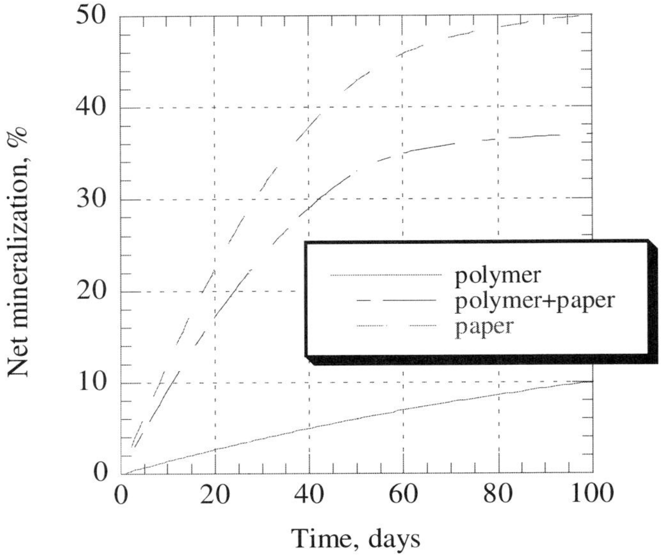


Figure 11.19. Kinetics of degradation in soil aerobic test. [Adapted, by permission, from Levit M R, Farrel R E, Gross R A, McCarthy S P, Antec '96. Volume II. Conference proceedings, Indianapolis, 5th-10th May 1996, 1387-91.]

groups. Starch alone does not increase the photooxidation rate.[93] The addition of starch to LDPE increases its stability in 80°C water.[76] Slower degradation in water is due to leaching out of the pro-oxidant. The addition of starch causes biodegradation process under soil burial conditions.[92] Further increase in the degradation rate can be achieved by preheating polyethylene filled with starch.[91]

Cellulosic materials such as wood flour, paper, and rayon improve biodegradation of poly(lactic acid) in aerobic soil (Figure 11.19). The polymer degrades at a rate similar to paper.[90]

Hydroxyapatite and magnesium oxide improve biodegradation of polylactides. [88,89] The degradation occurs in the bulk material whereas, in unfilled material, degradation occurs by surface erosion.

## REFERENCES

1 Bataille P, Mahlous M, Schreiber H P, *Polym. Engng. Sci.*, **34**, No.12, 1994, 981-5.
2 Bataille P, Schreiber H P, Mahlous M, Antec '93. Conference Proceedings, New Orleans, La., 9th-13th May 1993, Vol. II, 1757-9.
3 Ulkem I, Bataille P, Schreiber H P, *J. Macromol. Sci. A*, **31**, No.3, 1994, 291-303.
4 Mateev M M, Totev D L, *Kaut. u. Gummi Kunst.*, **49**, No.6, 1996, 427-31.
5 Svorcik V, Micek I, Jankovskij O, Rybka V, Hnatowicz V, Wang L, Angert N, *Polym. Degradat. Stabil.*, **55**, 1997, 115-21.
6 Ou Y C, Zhu J, Feng Y P, *J. Appl. Polym. Sci.*, **59**, No.2, 1996, 287-94.
7 Szocs F, Klimova M, Chodak I, Chorvath I, *Eur. Polym. J.*, **32**, No.3, 1996, 401-2.
8 Abdel-Aziz M M, Youssef H A, El Miligy A A, Yoshii F, Makuuchi K, *Polym. & Polym. Composites*, **4**, No.4, 1996, 259-68.
9 Abdel-Aziz M M, Gwaily S E, *Polym. Degradat. Stabil.*, **55**, 1997, 269-74.
10 McIlvaine J, Antec 95. Volume III. Conference proceedings, Boston, Ma., 7th-11th May 1995, 3346-9.

11 Zhao W, Hasegawa S, Fujita J, Yoshii F, Sasaki T, Makuuchi K, Sun J, Nishimoto S, *Polym. Degradat. Stabil.*, **53**, No.2, 1996, 199-206.
12 Braun J H, *Prog. Org. Coat.*, **15**, 1987, 249.
13 Skledar S, *Mater. Plast.*, **21**, 1984, 29.
14 Simpson L A, *Austral. OCCA Proc. News*, **20**, 1983, 6.
15 Koyama M, Tanaka A, Ichijima M, Oki Y, *Kinzoku Hyomen Gijutsu,* **37**, 1986, 25.
16 Strelkova L D, Fedoseeva G F, Potepalova S N, Batueva L I, Lebedev W P, *Plast. Massy*, 1986, 5, 46.
17 Dolezel B, Adamirova L, *Plasty Kauc.*, **20**, 1983, 18.
18 Charlesby A, *J. Radioanal. Nucl. Chem.*, **101**, 1986, 401.
19 Bowley H J, Gerrard D L, Williams K J P, Biggin I S, *J. Vinyl Technol.,* **8**, 1986, 176.
20 Andrady A L, Shultz A R, *J. Appl. Polym. Sci.*, **33**, 1987, 1389.
21 Pickett J E, *J. Appl. Polym. Sci.*, **33**, 1987, 525.
22 Skledar S, *Angew. Makromol. Chem.*, **137**, 1985, 149.
23 Gupta B D, Verdu J, *J. Polym. Eng.*, **8**, 1988, 81.
24 Pan J, Xu H, Qi J, Ce J, Ma Z, *Polym. Degradat. Stabil.*, **33**, 1991, 67.
25 Genova-Dimitrova P, *Polym. Degradat. Stabil.*, **33**, 1991, 355.
26 Klemchuk P P, *Polym. Photochem.*, **3**, 1983, 1.
27 Saad A L G, Aziz A W, *Polym. Degradat. Stabil.*, **41**, 1993, 31.
28 Davydov E Ya, Pustoschniy V P, Vorotnikov A P, Pariyskiy G B, *Intern. J. Polym. Mater.*, **16**, 1992, 295.
29 Imhof J, Stern P, Egger A, *Angew. Makromol. Chem.*, **176/177**, 1990, 185.
30 Gardette J-L, Lemaire J, *Polym Degradat. Stabil.*, **33**, 1991, 77.
31 Rysavy D, Tkadleckova H, *Polym. Degradat. Stabil.*, **37**, 1992, 19.
32 Dolui, *J. Appl. Polym. Sci.*, **53**, 1994, 463.
33 Sanchez-Solis A, Estrada M R, *Polym. Degradat. Stabil.*, **52**, No.3, 1996, 305-9.
34 Sanchez-Solis A, Padilla A, *Polym. Bull.*, **36**, No.6, 1996, 753-58.
35 Hidaka H, Suzuki Y, Nohara K, Horikoshi S, Hisamatsu Y, Pelizzetti E, Serpone N, *J. Polym. Sci., Polym. Chem.*, **34**, No.7, 1996, 1311-6.
36 Gaw F, Enhancing Polymers with Additives and Modifiers, Rapra, 1993,
37 Delor F, Lacoste J, Lemaire J, Barrois--Oudin N, Cardinet C, *Polym. Degradat. Stabil.*, **53**, No.3, 1996, 361-9.
38 Rabello M S, White J R, *Polym. Composites*, **17**, No.5, 1996, 691-704.
39 Gordienko V P, Dmitriev Y A, *Polym. Sci., Ser. B*, **37**, Nos.5-6, 1995, 249-50.
40 Gordienko V P, Dmitriev Y A, *Polym. Degradat. Stabil.*, **53**, No.1, 1996, 79-87.
41 Turley R S, Strong A B, *J. Adv. Materials*, **25**, No.3, 1994, 53-9.
42 Thomas R W, Ancelet C R, Brzuskiewicz J E, Antec '93. Conference Proceedings, New Orleans, La., 9th-13th May 1993, Vol. II, 2005-8.
43 Wypych G, **Handbook of Material Weathering**, ChemTec Publishing, Toronto, 1995.
44 Liptak P, *Int. Polym. Sci. Technol.*, **21**, No.8, 1994, T/50-3.
45 Sherman L M, *Plast. Technol.*, **43**, No.4, 1997, 26-8.
46 Dolui S K, *J. Appl. Polym. Sci.*, **53**, No.4, 1994, 463-5.
47 Wang T, Sherwood P M A, *Chem. of Mat.*, **6**, No.6, 1994, 788-95.
48 Nichols M E, Pett R A, *Rubb. Chem. Technol.*, **67**, No.4, 1994, 619-28.
49 Kim S G, Lee S H, *Rubb. Chem. Technol.*, **67**, No.4, 1994, 649-61.
50 Zyuzina G F, Vinogradova N K, Gribova I A, Krasnov A P, *Polym. Sci.*, **36**, No.9, 1994, 1205-8.
51 Yang A C M, *Polymer*, **35**, No.15, 1994, 3206-11.
52 Hornsby P R, Wang J, Jackson G, Rothon R N, Wilkinson G, Cosstick K, Antec '94. Conference Proceedings, San Francisco, Ca., 1st-5th May 1994, Vol. III, 2834-9.
53 Jia N, Kagan V A, Antec '97. Conference proceedings, Toronto, April 1997, 1844-8.
54 Zumbrum M A, *J. Adhesion*, **46**, Nos.1-4, 1994, 181-96.
55 Fuelber C, Bluemich B, Unseld K, Herrmann V, *Kaut. u. Gummi Kunst.*, **48**, No.4, 1995, 254-9.
56 Cochet P, Bomal Y, *Kaut. u. Gummi Kunst.*, **48**, No.4, 1995, 270-5.
57 Minkova L, *Coll. Polym. Sci.*, **272**, No.2, 1994, 115-20.
58 Minkova L, Magagnini P L, *Polym. Degradat. Stabil.*, **42**, No.1, 1993, 107-15.
59 Mwila J, Miraftab M, Horrocks A R, *Polym. Degradat. Stabil.*, **44**, No.3, 1994, 351-6.
60 Dole P, Chauchard J, *Polym. Degradat. Stabil.*, **47**, No.3, 1995, 441-8.
61 McNeill I C, Mohammed M H, *Polym. Degradat. Stabil.*, **48**, No.1, 1995, 189-95.

62 Allen N S, Edge M, Corrales T, Childs A, Liauw C, Catalina F, Peinado C, Minihan A, *Polym. Degradat. Stabil.*, **56**, 1997, 125-39.
63 McNeill I C, Mohammed M H, *Polym. Degradat. Stabil.*, **49**, No.2, 1995, 263-73.
64 Nair A, White R L, *J. Appl. Polym. Sci.*, **60**, No.11, 1996, 1901-9.
65 Nair A, White R L, *J. Appl. Polym. Sci.*, **60**, No.11, 1996, 1911-20.
66 Gutman E M, Bobovitch A L, *Eur. Polym. J.*, **32**, No.8, 1996, 979-83.
67 Magee R W, *Rubb. Chem. Technol.*, **68**, No.4, 1995, 590-600.
68 Dyrda V I, Meshchaninov S K, *Int. Polym. Sci. Technol.*, **22**, No.12, 1995, T/14-6.
69 Kikkawa K, *Polym. Degradat. Stabil.*, **49**, No.1, 1995, 135-43.
70 Hornsby P R, Wang J, Rothon R, Jackson G, Wilkinson G, Cossick K, *Polym. Degradat. Stabil.*, **51**, No.3, 1996, 235-49.
71 Maxwell J, **Plastics in High Temperature Applications**, Pegamon Press, Oxford, 1992.
72 Braun D, Kraemer K, **Recycling of PVC & Mixed Plastic Waste**, *ChemTec Publishing*, Toronto, 1996.
73 Gassan J, Bledzki A K, *Polym. Composites*, **18**, No.2, 1997, 179-84.
74 Selzer R, Friedrich K, *Composites, Part A*, **28A**, 1997, 595-604.
75 Qiu Q, Kumosa M, *Composites Sci. Technol*, **57**, 1997, 497-507.
76 Hakkarainen M, Albertsson A-C, Kalsson S, *J. Appl. Polym. Sci.*, **66**, 1997, 959-67.Hill R, Okoroafor E U, *Composites*, **25**, No.10, 1994, 913-6.
77 Hill R, Okoroafor E U, *Composites*, **25**, No.10, 1994, 913-6.
78 Connor C, Chadwick M M, *J. Mat. Sci.*, **31**, No.14, 1996, 3871-7.
79 Mannan K M, Robbany Z, *Polymer*, **37**, No.20, 1996, 4639-41.
80 Nofal M M, Zihlif A M, Ragosta G, Martuscelli E, *Polym. Composites*, **17**, No.5, 1996, 705-9.
81 Sutherland I, Sheng E, Bradley R H, Freakley P K, *J. Mat. Sci.*, **31**, No.21, 1996, 5651-5.
82 Tino J, Mach P, Hlouskova Z, Chodak I, *J. Macromol. Sci. A*, **A31**, No.10, 1994, 1481-7.
83 Hu X, Xu H, Zhang Z, *Polym. Degradat. Stabil.*, **43**, No.2, 1994, 225-8.
84 Bergado S, *Plast. Compounding*, **17**, No.7, 1994, 32-8.
85 Robertson D R, Gaw F, AddCon '95, Basel, 1995.
86 Mathews C, Enhancing Polymers Using Additives and Modifiers II, Rapra, 1996.
87 Svorcik V, Rybka V, Hnatowicz V, Bacakova L, *J. Mat. Sci. Lett.*, **14**, No.24, 1995, 1723-4.
88 Van der Meer S A T, de Wijn J R, Wolke J G C, *J. Mat. Sci. Mat. In Med.*, **7**, No.6, 1996, 359-61.
89 Liu Q, De Wijn J R, Bakker D, Van Blitterswijk C A, *J. Mat. Sci. Mat. In Med.*, **7**, No.9, 1996, 551-7.
90 Levit M R, Farrel R E, Gross R A, McCarthy S P, Antec '96. Volume II. Conference proceedings, Indianapolis, 5th-10th May 1996, 1387-91.
91 Albertsson A-C, Barenstedt C, Karlsson S, *J. Environmental Polym. Degradat.*, **1**, No.4, 1993, 241-5.
92 Griffin G J L, *Polym. Degradat. Stabil.*, **45**, No.2, 1994, 241-7.
93 Albertsson A C, Griffin G J L, Karlson S, Nishimoto K, Watanabe Y, *Polym. Degradat. Stabil.*, **45**, No.2, 1994, 173-8.

# 12

# Environmental Impact of Fillers

This chapter contains information on the flammability and fire resistance of filled materials. It also covers the recycling of filled materials.[1-21]

## 12.1 DEFINITIONS

The terms listed in Table 12.1 are commonly used in the evaluation of materials flammability and to describe the impact of fire.

**Table 12.1 Definitions**

| Term | Definition | Unit |
|---|---|---|
| Char yield | Total residue of combustion (carbonaceous char + inorganic material) | wt% |
| Ignition time | The period required for the entire surface to burn with a sustained, luminous flame | s |
| Peak rate of heat release | Maximum heat released during a fire (peak value of heat release). This value expresses maximum intensity of fire | $kW/m^2$ |
| Rate of heat release | Average value of heat release rate during a specified period of time. This value correlates with heat release in a room burn situation where all of the material is not ignited at the same time | $kW/m^2$ |
| Specific extinction area | The measure of smoke obscuration averaged over the whole test period | $m^2/kg$ |
| Fire performance index | The ratio of ignition time to peak rate of heat release. This parameter relates to the time to flashover in a full scale fire | $m^2s/kW$ |
| Smoke parameter | The product of the average specific extinction area and the peak rate of heat release. This parameter indicates the amount of smoke generated | MW/kg |
| Limiting oxygen index | The minimum concentration of oxygen in an oxygen-nitrogen atmosphere required to initiate and support a flame for more than 3 min | % $O_2$ |
| Propensity of flashover | The ratio of ignition time to peak rate of heat release. It is the same as fire performance index | $m^2s/kW$ |
| Smoke production rate | A product of the average mass loss rate and the average specific extinction area | $m^2/s$ |
| CO yield | Carbon monoxide yield per unit surface area | $g/m^2$ |

Fillers can play an important role in limiting the flammability of materials and in reducing the damage and injuries caused by fires.

## 12.2 LIMITING OXYGEN INDEX

**Table 12.2 Limiting oxygen index (LOI) for different materials**

| Filler | | Polymer | LOI | | Refs. |
|---|---|---|---|---|---|
| Type | Conc. wt% | | control | with filler | |
| $Al(OH)_3$ | 20/30/40/50 | Flexible PVC | 25.7 | 28.5/29.9/30.3/33.5 | 15 |
| $Al(OH)_3$ | 33/50/60 | PMMA | | 19.8/22.1/27.5 | 27 |
| $Al(OH)_3$ | 55 | EVA | 18.5 | 30.5 | 27 |
| Anthracite | 40/50/60 | HDPE | 18.7 | 20.2/21.7/22.5 | 12 |
| Apatite | 1-4 | Wood pulp | 17 | 23 | 2 |
| $Ca_2B_6O_{11}$ | 18.4/42.1/48.8 | EPM | 18.3 | 19.0/20.9/21.9 | 9 |
| $CaCO_3$ | 55 | EVA | 18.3 | 29.0 | 27 |
| Glass fiber | 30 | PEI | 47 | 32 | 18 |
| Glass fiber | 30 | PEEK | 19 | 43 | 18 |
| Glass fiber | 30 | PES | 38 | 41 | 18 |
| $Mg(OH)_2$ | 40/50/60 | EVA | 17.5 | 22.0/24.0/42.5 | 10 |
| $Mg(OH)_2$ | 55 | EVA | 18.5 | 38.5 | 27 |
| $Mg(OH)_2$ | 20/30/40/50 | Flexible PVC | 25.7 | 28.6/30.4/31.3/30.3 | 15 |
| $Mg(OH)_2$ | 60 | PA-6 | 24.1 | 51.3-70.0* | 7,25 |
| $Mg(OH)_2$ | 60 | PA-66 | 26.5 | 45.9-57.4* | 7,25 |
| $Mg(OH)_2$ | 10/30/60 | PP | 17.5 | 18/20/27 | 29 |
| $Sb_2O_3$ | 2.5/5 | Polyester | 26.8 | 39.2/42.9 | 14 |
| Talc | 10/30/60 | PP | 17.5 | 17.5/21/20.5 | 29 |

*depending on particle size (the smaller the particle the higher the LOI)

Limiting oxygen index (LOI) is the parameter most frequently used to characterize the improvements in fire retardancy.[1-3,6-7,9-16,18-19,22-30] Table 12.2 gives a summary of data obtained for various fillers. The data in Table 12.2 show that even the addition of very common and inexpensive fillers such as calcium carbonate or talc increases the LOI value. From the data presented, $Sb_2O_3$ and $Mg(OH)_2$ are the most efficient in increasing LOI.

Figure 12.1 shows the effect of increasing the concentration of different fillers on the LOI value.[7] $Mg(OH)_2$ produces a much larger effect than the other fillers listed but high concentrations are required to obtain a substantial effect on LOI.

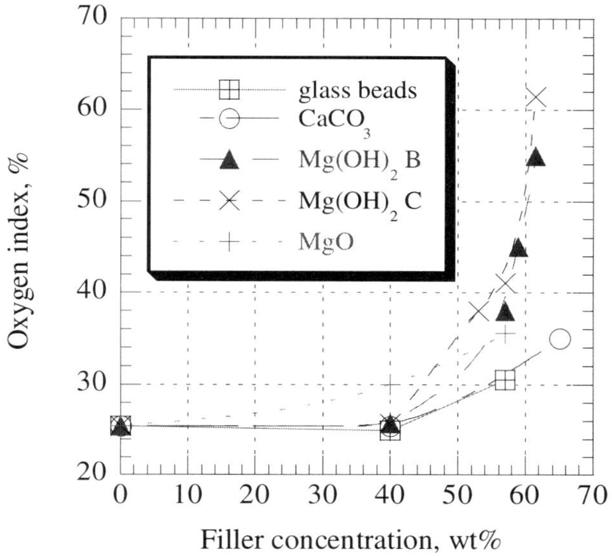


Figure 12.1. Limiting oxygen index vs. filler concentration in PA-66. [Adapted, by permission, from Hornsby P R, Wang J, Jackson G, Rothon R N, Wilkinson G, Cosstick K, Antec '94. Conference Proceedings, San Francisco, Ca., 1st-5th May 1994, Vol. III, 2834-9.]

The performance of fillers can be improved by the use of combinations of organic fire retardants and mineral fillers.[11,14,22] Substantially better results can be obtained by the surface coating of fillers. When zinc hydroxystannate was used to coat $Al(OH)_3$ and $Mg(OH)_2$, the LOI value was improved by 18 to 35% for $Mg(OH)_2$ and 28 to 36% for $Al(OH)_3$.[15]

## 12.3 IGNITION AND FLAME SPREAD RATE

Autoignition temperature, ignition time, and flame spread rate are influenced by fillers.[8,9,11,13,14,15,27,29,31] The ignition times of various filled systems are given in Table 12.3. The data in Table 12.3 show that the results are not very precise and depend on the method used. $Al(OH)_3$ performed best.

Autoignition tests are performed either at constant temperature (the specimen is held at 430°C and the time to ignition is measured),[29] or at varying temperatures (the specimen is placed in a 100% oxygen environment and the temperature at which ignition occurs is recorded),[31] or in varying oxygen concentrations.[8] Figure 12.2 shows the effect of carbon black on the autoignition temperature of a fluoroelastomer. Carbon black increases the autoignition temperature because it forms stable oxides on the charred surface.[31]

Autoignition of epoxy and phenolic composites with glass fiber, aramid and graphite was affected by the oxygen concentration to a limited degree.[8] But with epoxy composites, neither fiber type nor concentration of oxygen had an effect on the autoignition time (~50 s). In phenolic composites, the fiber type affected

**Table 12.3. Ignition time**

| Filler, wt% | | Polymer | Ignition time, s | | Refs. |
|---|---|---|---|---|---|
| Type | Concentration | | control | with filler | |
| $Al(OH)_3$ | 55 | EVA | <2 | 20 | 27 |
| $Al(OH)_3$ | 20/30/40/50 | PVC flexible | 16 | 22/31/78/43 | 15 |
| Glass beads | 55 | EVA | <2 | <2 | 27 |
| $Mg(OH)_2$ | 20/30/40/50 | PVC flexible | 16 | 16/11/27/21 | 15 |
| $Mg(OH)_2$ | 55 | EVA | <2 | 20 | 27 |
| Precipitated $CaCO_3$ | 55 | EVA | <2 | <2 | 27 |
| $Sb_2O_3$ | 2.5/5 | Polyester | 25 | 22/24 | 14 |

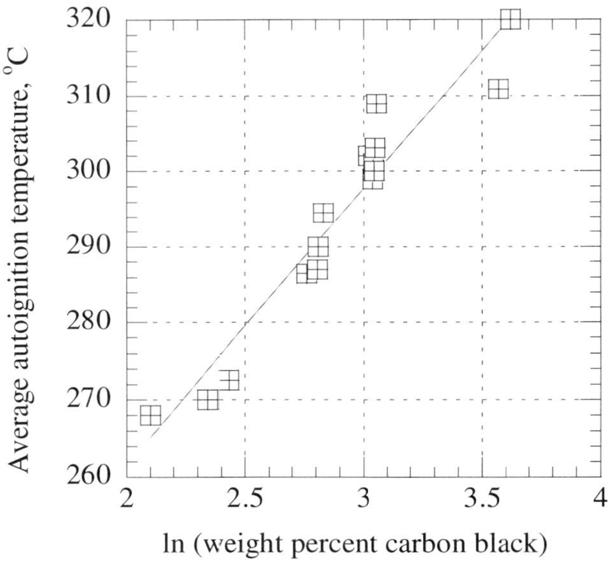


Figure 12.2. Autoignition temperature of fluoroelastomer vs. carbon black concentration. [Adapted, by permission, from Fu-Yu Hsieh, Bryan C J, Pedley M D, *Fire & Mat.*, **18**, No.6, 1994, 389-91.]

autoignition time more than oxygen concentration. Graphite-reinforced phenolic composites were the most stable.

Figures 12.3 and 12.4 show the effect of talc and $Mg(OH)_2$ in polypropylene on autoignition time and temperature. $Mg(OH)_2$ improved autoignition time more than did talc.[29] $Mg(OH)_2$ had some effect on autoignition temperature.

Flame propagation is expressed by length (cm),[8] rate (mm/s),[8,9,11,13] and flame persistence (s).[9,11,13] Figure 12.5 shows data on the rate of flame spread for a PP-PE

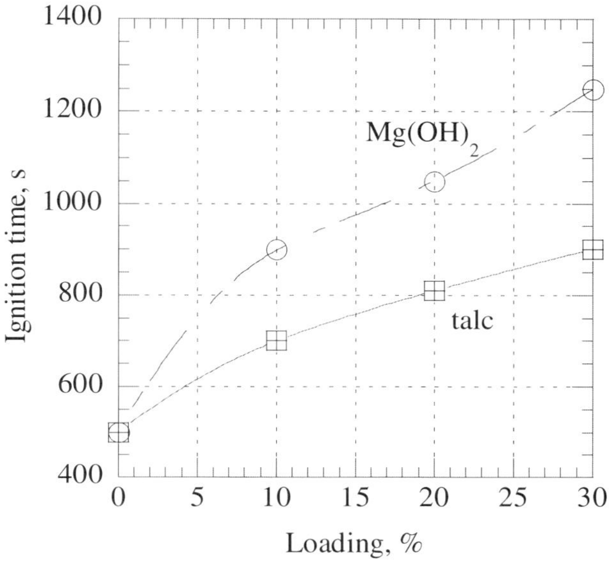


Figure 12.3. Autoignition time measured @ 430°C for PP containing talc and $Mg(OH)_2$. [Data from Costa L, Camino G, Bertelli G, Borsini G, *Fire & Mat.*, **19**, No.3, 1995, 133-42.]

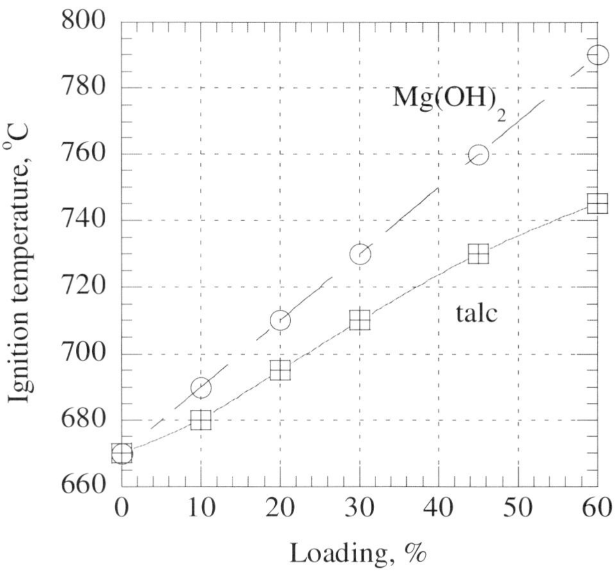


Figure 12.4. Autoignition temperature measured by glow wire test for PP containing talc and $Mg(OH)_2$. [Data from Costa L, Camino G, Bertelli G, Borsini G, *Fire & Mat.*, **19**, No.3, 1995, 133-42.]

copolymer filled with Portacarb (a mixture of hydromagnesite and huntite).[9,11,13] The addition of filler has reduced the rate of flame spread in an effective manner.

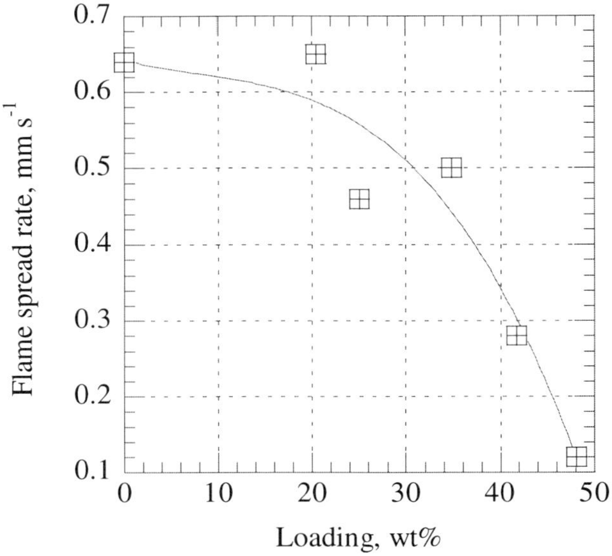


Figure 12.5. Rate of flame spread vs. Portacarb concentration. [Data from Toure B, Lopez Cuesta J M, Gaudon P, Benhassaine A, Crespy A, *Polym. Degradat. Stabil.*, **53**, No.3, 1996, 371-9.]

## Table 12.4. Heat transmission rate

| Filler, wt% | | Polymer | Heat transmission rate, kW/m$^2$ | | Refs. |
|---|---|---|---|---|---|
| Type | Concentration | | control | with filler | |
| $Al(OH)_3$* | 30/40/50 | PVC flexible | 290 | 173/120/104 | 15 |
| $Al(OH)_3$ | 30/40/50 | PVC flexible | 232 | 133/98/80 | 15 |
| fumed silica** | 5/8 | ABS | 1500 | 900/800 | 23 |
| fumed silica** | 3/5 | PET | 1600 | 400/320 | 23 |
| fumed silica** | 5/8 | PA-66 | 2000 | 700/500 | 23 |
| Glass beads* | 60 | PP | 550 | 240 | 10 |
| $Mg(OH)_2$* | 60 | PP | 550 | 100 | 10 |
| $Mg(OH)_2$ | 30/60 | PP | 450 | 280/130 | 29 |
| $Mg(OH)_2$* | 30/60 | PP | 700 | 380/180 | 29 |
| $Mg(OH)_2$ | 30/40/50 | PVC flexible | 232 | 161/161/156 | 15 |
| $Mg(OH)_2$* | 30/40/50 | PVC flexible | 290 | 208/214/203 | 15 |
| $Sb_2O_3$* | 2.5/5 | Polyester | 188 | 124/117 | 14 |
| Talc | 30/60 | PP | 450 | 380/310 | 29 |
| Talc* | 30/60 | PP | 700 | 480/460 | 29 |

*peak value (other values for average rate of heat release)
**fumed silica compounded with PDMS (the values given are peak values from cone calorimeter data)

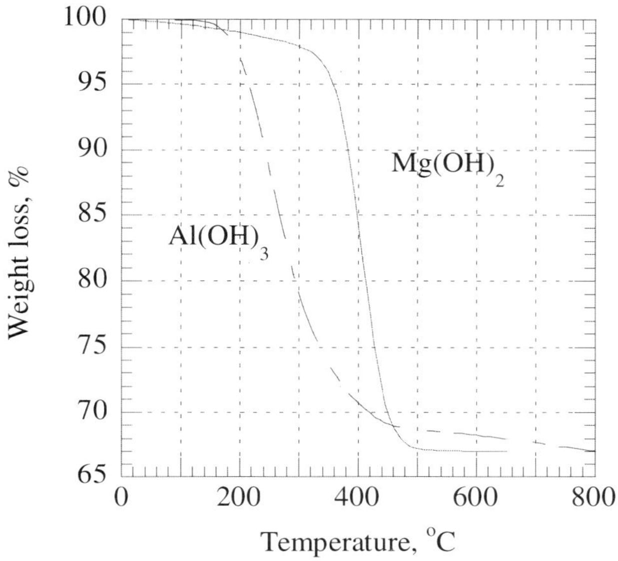


Figure 12.6. Thermogravimetric analysis of $Al(OH)_3$ and $Mg(OH)_2$. [Adapted, by permission, from Yeh J T, Yang H M, Huang S S, *Polym. Degradat. Stabil.*, **50**, No.2, 1995, 229-34.]

Flame persistence is reduced to 0 s by the addition of 48 wt% $Sb_2O_3$ containing decabromodiphenyl oxide.

## 12.4 HEAT TRANSMISSION RATE

Heat transmission rate, peak of heat transmission rate, propensity of flashover (or fire performance index) are the parameters which characterize heat generation during a fire. These parameters are affected by fillers.[7,8,10,14,15,23,25,29] Table 12.4 reviews the data from the literature. The addition of certain fillers such as $Al(OH)_3$, $Mg(OH)_2$, and $Sb_2O_3$ substantially reduces the heat transmission rate. Fillers (with the exception of $Sb_2O_3$) must be used at high concentrations (e.g., 60%) to give the best performance. The modification of fillers by zinc hydroxystannate further reduces the heat transmission rate.[15]

The addition of glass fiber, aramid, or graphite has little effect on the peak heat release rates of epoxy and phenolic resins.[8]

## 12.5 DECOMPOSITION AND COMBUSTION

The decomposition of some fillers and their effect on decomposition and combustion of polymers is discussed below.[7,8,17,25,29,30,32-36]

Hydroxides of metals, such as aluminum and magnesium and borates containing crystalline water, decompose at specific temperatures and liberate water (Figures 12.6 and 12.7). Magnesium hydroxide is more thermally stable since it begins to decompose at 320°C as compared with 220°C for aluminum hydroxide. The amount of water produced from $Mg(OH)_2$ is slightly lower than that produced from

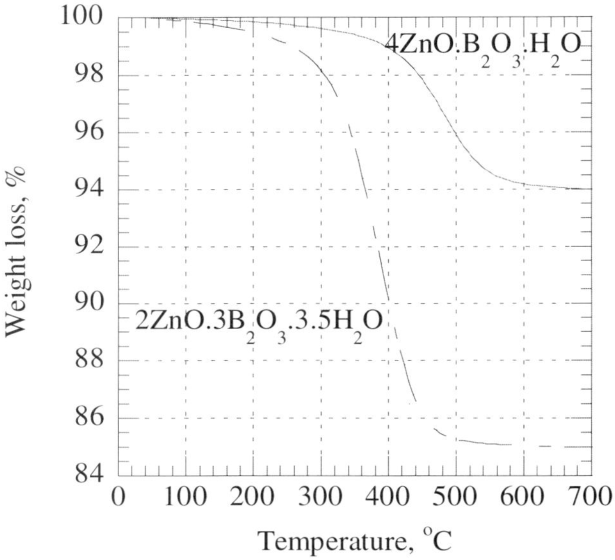


Figure 12.7. Thermogravimetric analysis of zinc borate. [Data from Shen K K, O'Connor R, Addcon '96, Brussels, 1996.]

$Al(OH)_3$.[30] In both cases, the reaction is endothermic which contributes a cooling effect to the degrading polymer. Borates degrade at different temperatures depending on their composition (ammonium - 120°C, barium metaborate - 200°C, zinc borate - 290°C).[21]

Since the decomposition reaction occurs at a specific temperature, the performance of these fillers depends on the properties of the polymers in which they are used. For example, $Mg(OH)_2$ performs better in polyethylene than $Al(OH)_3$ because it remains stable during compounding and decomposes at a temperature closer to the decomposition of PE (300-400°C). In unsaturated polyesters, $Al(OH)_3$ starts to release water at 200°C.[17] The major endothermic peak occurs at 300°C with a heat of decomposition of 300 kJ/mol. About 90% of the water is released between 200 and 400°C. A considerable amount of heat is absorbed before the polymer is affected. The water also dilutes combustible gases and hinders the access of oxygen to the polymer surface. Figure 12.8 shows the difference between talc and a fire retardant filler in PP.[29] Talc causes an increase in the combustion rate as its concentration increases, whereas $Mg(OH)_2$, used at a sufficient concentration (above 20%), decreases the rate of combustion.

It should be noted that the particular grade of filler is important (Figure 12.9). In this study,[35] four different grades of $Mg(OH)_2$ (A to D) were used. These differed in particle size and surface area. Only one filler (A) with the largest particle size (7.7 μm) and the largest surface area (18.9 $m^2/g$) slowed PP decomposition. $Mg(OH)_2$ which performs well, has a small, aggregated structure with irregular

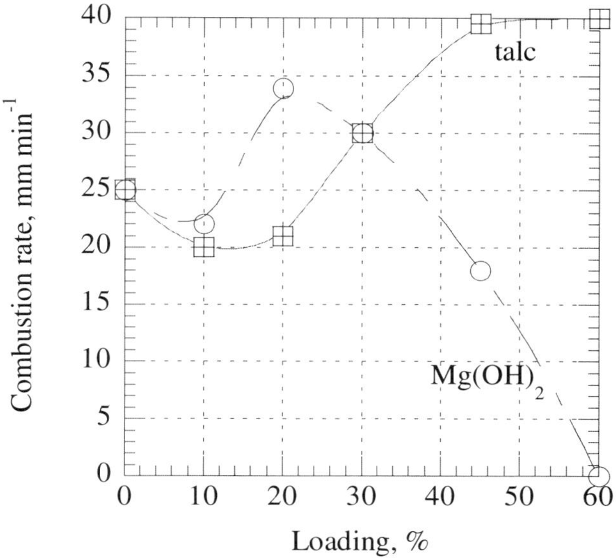


Figure 12.8. Rate of combustion of filled PP. [Data, from Costa L, Camino G, Bertelli G, Borsini G, *Fire & Mat.*, **19**, No.3, 1995, 133-42.]

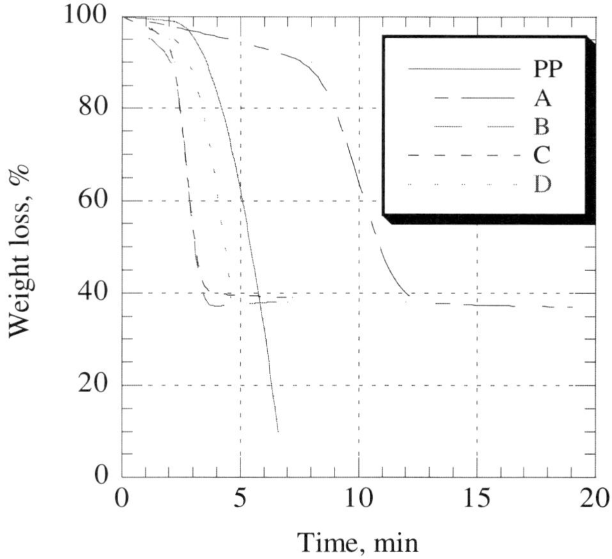


Figure 12.9. Decomposition of $Mg(OH)_2$ filled PP. [Data from Hornsby P R, Mthupha A, Antec '93. Conference Proceedings, New Orleans, La., 9th-13th May 1993, Vol. II, 1954-6.]

crystallites. Other materials were hexagonal and were composed of much larger crystallites.

Studies of polyamides[7,25] also illustrate mechanisms involving $Mg(OH)_2$. $Mg(OH)_2$ decomposes before the decomposition of PA-6 and after the decomposition of PA-66. The water formed by $Mg(OH)_2$ before the decomposition of PA-6 gives a fire retardant effect. However, it is possible that the water thus formed may hasten the degradation of the polymer by hydrolyzing it.

Polypropylene was more thermally stable when filled with $Mg(OH)_2$.[35] Calcium carbonate did not stabilize LDPE but the emission of low molecular weight hydrocarbons was reduced due to dilution of the polymer.[33] Calcium carbonate[37] stabilizes PVC due to its reaction with HCl. A similar mechanism is found in the stabilization of poly(ethylene acrylate) and ethylene ethyl acrylate copolymer with calcium carbonate where less acids are produced and the interaction with the polymer changes the mechanism of degradation.[33] In this case, the type of calcium carbonate, its coating, and its particle size all have an important impact on performance. Similar behavior was found in the case of $Mg(OH)_2$ and $Al(OH)_3$.[34] Formation of acetic and propionic acid was decreased during the degradation of ethylene ethyl acrylate copolymer. The endothermic effect was a stabilizing factor if the filler degraded before the polymer. When the degradation temperature of the polymer was higher than the decomposition temperature of the filler, the filler reacted with functional groups in the polymer and affected the mechanism of degradation. Carbon monoxide emission was substantially reduced when zinc borate was used in combination with $Al(OH)_3$.[36]

## 12.6 EMISSION OF GASEOUS COMPONENTS AND HEAVY METALS

There is not much data available on the effect that fillers have on what is emitted during the decomposition of filled products. But some work has been done.[33-36,38-42]

Decomposition of PA-6 and PA-66 produces $NH_3$, $H_2O$, CO, $CO_2$, and hydrocarbons.[39] The addition of magnesium hydroxide decreases the amounts of volatiles produced but the chemical components and their proportions are very similar to unfilled polymers. The addition of zeolites to polypropylene changes the mechanism of degradation depending on the zeolite type, its morphology, and dispersion but, in the investigation, the composition of the decomposition products was not determined.[41]

The rate of emission of CO and $CO_2$ from flexible PVC depends on the type of filler added.[36] Additions of zinc borate to $Al(OH)_3$ and $Mg(OH)_2$ substantially reduced the emission of carbon monoxide. The presence of antimony oxide increased emissions of both CO and $CO_2$. The data do not explain why the behavior is different.

The addition of $Al(OH)_3$, $Mg(OH)_2$, MgO, and $TiO_2$ to ethylene ethyl acrylate copolymer did not change the chemical components of the volatiles.[34] The only observable difference was related to the concentration of ethylene (higher with $Al(OH)_3$ than in the presence of the other fillers) and $CO_2$ (lower with $Al(OH)_3$ than with $Mg(OH)_2$ and MgO). There was no detectable difference in the degradation products from LDPE, poly(ethyl acrylate), and ethylene ethyl acrylate copolymer either unfilled or filled with calcium carbonate.[33]

A study of twelve silica fillers in rubber compounds showed that all of the fillers contributed to the formation of nitrosamines. There was a substantial difference in the amounts of nitrosamines detected in the presence of different fillers (from $1.1 \times 10^{-4}$ to $14.8 \times 10^{-4}$ mol/kg) but this difference could not be correlated with properties such as their structure or surface area.[38]

## 12.7 SMOKE

Fire retardant fillers affect smoke formation.[6,8,10,14,18,26,29,36,43-46] Table 12.5 gives some data on the specific extinction area. The data show that, with the exceptions of $Al(OH)_3$ and $Mg(OH)_2$, fillers have a small effect on smoke suppression.

The effect of fillers on smoke suppression depends on the particle size and crystalline structure of the filler. A new fire retardant, based on a hydrated potassium-magnesium aluminosilicate, in two grades – coarse and fine is available. The fine grade is twice as effective as the coarse grade.[6,26]

## 12.8 CHAR

The effect of fillers on char formation was considered in several studies[14,15,22,35,46] and the data is summarized in Table 12.6.

High efficiency in char formation is one of the reasons for the strong performance of $Mg(OH)_2$ and $Al(OH)_3$ as fire retarding additives. This performance can still be enhanced by surface coating the fillers with zinc hydroxystannate. $Sb_2O_3$ is volatized from the material and therefore does not affect char formation. Formation of char is an effective method of increasing the fire resistance of materials. The material which forms carbonaceous char has a reduced ability to supply the gaseous fuels required to fuel the fire.

## 12.9 RECYCLING

Fillers play various roles in the recycling of materials. These include: the use of waste materials as a substitute for fillers (filler replacement), the effect that fillers have on recycling methods and waste production (filler impact on quantity of wastes), recovery of fillers from scrap (filler recovery), and the application of fillers to make materials recyclable (material improvement).[47-68] These subjects are discussed below.

A growing volume of waste materials, especially vulcanized rubbers and crosslinked polymers are proving difficult to recycle. As an alternative to their disposal in landfills, there have been many attempts to grind these materials and use the products as a substitute for fillers in composite materials.[48,49,51,54,59-64] Other non-plastic materials such as glass, paper, natural fibrous materials, and fly ash are also used for filler replacement. There is extensive literature on the use of ground tires as filler replacements.[59,66] This is a specialized topic with only a minor relationship to fillers.

**Table 12.5. Specific extinction area**

| Filler, wt% | | Polymer | Specific extinction area, $m^2/kg$ | | Refs. |
|---|---|---|---|---|---|
| Type | Concentration | | control | with filler | |
| $Al(OH)_3$ | 30/40/50 | PVC flexible | 963 | 622/186/192 | 15 |
| $Al(OH)_3 + Sb_2O_3$ | 18+9 | PVC flexible | | 678 | 36 |
| $Al(OH)_3$ + ZB | 18+9 | PVC flexible | | 928 | 36 |
| $Al(OH)_3 + Mg(OH)_2$ | 30+30 | PVC flexible | | 493 | 36 |
| Glass beads | 60 | PP | 610 | 600 | 10 |
| $Mg(OH)_2$ | 30/40/50 | PVC flexible | 963 | 398/379/309 | 15 |
| $Mg(OH)_2$ | 60 | PP | 610 | 250 | 10 |
| $Mg(OH)_2$ | 30/60 | PP | 600 | 610/320 | 29 |
| $Sb_2O_3$ | 2.5/5 | Polyester | 720 | 663/671 | 14 |
| $Sb_2O_3$ | 10 | PVC flexible | 777 | 538 | 45 |
| Talc | 30/60 | PP | 600 | 610/530 | 29 |
| Zinc borate (ZB) | 10 | PVC flexible | 777 | 719 | 45 |

**Table 12.6 Char yield**

| Filler, wt% | | Polymer | Char yield, % | | Refs. |
|---|---|---|---|---|---|
| Type | Concentration | | control | with filler | |
| $Al(OH)_3$ | 20/30/40/50 | PVC flexible | 3 | 15/20/30/28 | 15 |
| $Al(OH)_3$ | 20/50 | PVC flexible | 3 | 15/28 | 14 |
| $Mg(OH)_2$ | 20/30/40/50 | PVC flexible | 3 | 25/32/36/38 | 15 |
| $Mg(OH)_2$ | 20/50 | PVC flexible | 7 | 19/35 | 14 |
| $Sb_2O_3$ | 2.5/5 | Polyester | 20 | 20/20 | 14 |

Polyurethane is a very common crosslinked polymer and many materials produced from it end up as waste. Given the quantity of material, recycling is a major problem. Cryogenic pulverization systems have been developed which can process PU foam to particles smaller than 1 mm (preferably <100 μm).[63] These particles are homogenized with polyol and then reacted with isocyanates to produce foam. This foam with 5% pulverized PU foam has a density equivalent to a similar foam produced without the recycled material. A further increase in filler content causes a density increase.[63] Pulverized PU foam particles were also tried as a filler in natural rubber vulcanizates with good results.[60,61] Figure 12.10 shows the effect of PU

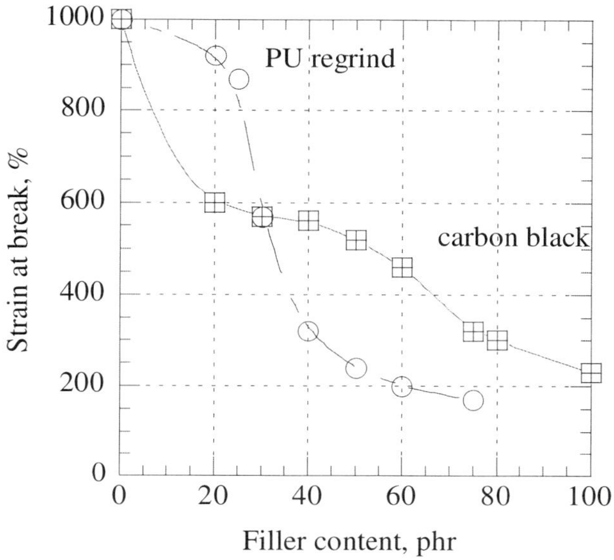


Figure 12.10. Ultimate strain of natural rubber vs. filler content. [Adapted, by permission, from Sombatsompop N, Sims G L A, *Cell. Polym.*, **15**, No.5, 1996, 317-34.]

regrind on the ultimate strain of vulcanized rubber compared to the effect of carbon black. Lower concentrations of regrind can produce better results than rubber filled with carbon black. The ultimate stress is lower than with carbon black filled vulcanized rubber.

Small quantities of epoxy regrind (5-10%) can be added to epoxy resins without affecting the tensile ord flexural strength, tensile or flexural modulus, or dielectric properties.[51] The biggest hurdle in recycling these epoxy materials is the cost of regrinding which at the moment is prohibitive. Molding compounds for automotive applications are used with 15% recycled material.[48] The regrind replaces the same percentage of calcium carbonate which makes parts up to 7% lighter.

Another area of recycling activity involves the use of alternative materials to replace fillers. Fly ash has been successfully used in polyethylene, providing the composition of fly ash chosen was suitable. Fly ash containing a large amount of CaO was not suitable.[54] The use of natural or recovered fibers (wood pulp, paper, plant fibers, etc.) is a growing market due to their low cost, biodegradability, and natural occurrence in renewable resources. Treated flax fiber in PP gives a reinforcement comparable to E-glass.[62] In a more unusual application, material recovered from ground glass bottles was used in polyurethane and acrylic paints. In acrylic paints, this filler had good flattening characteristics.[64] Plastics recovered from municipal wastes perform better as a filler when cellulose recovered from paper is added to them.[66]

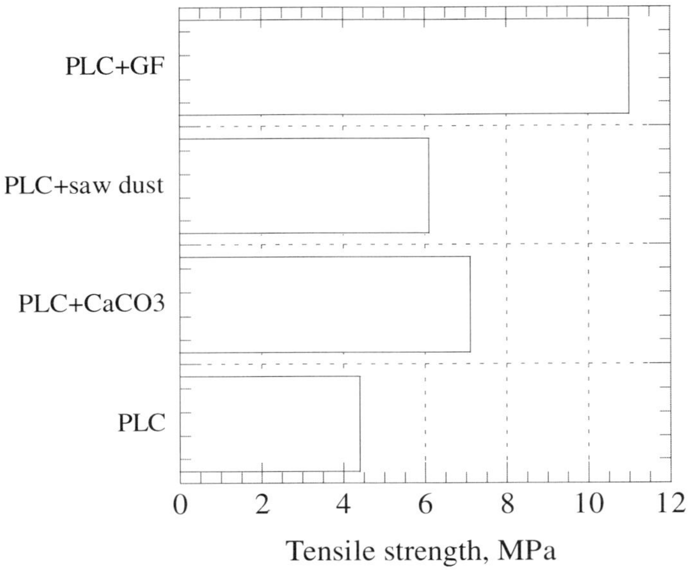


Figure 12.11. Tensile strength of post-consumer plastic containers reprocessed with different fillers. [Adapted, by permission, from La Mantia F P, **Recycling of PVC & Mixed Plastic Waste**, *ChemTec Publishing*, Toronto, 1996.]

Fillers have a noticeable impact on waste reprocessing methods. Expired products contain a large proportion of fillers which must be dealt with.[65] A plant in Bernau, Germany processes circuit boards using incineration. The material contains 49% glass wool which remains after incineration.[67] Solid wastes from electronic scrap contain 19% glass fiber. To date, no use has been found for this material. Trials are under way to remelt it into glass. Composting of materials containing fillers is difficult because the filler causes them to degrade more slowly. An addition of a small amount of calcium carbonate (7%) to starch slows down its composting process rate by half.[67]

Filler recovery from expired materials is still in its infancy. Glass fibers from recycled sheet-molding compounds processed by chemical means have the potential to replace up to half of the glass fibers used in bulk-molding compounds.[49] Clay is used as a decolorant in the lubricating oil industry. This clay can be reused with good results without any prior treatment as a neutral filler in vulcanized rubber.[56] Paper recycling requires that calcium carbonate and kaolin be removed prior to recycling. Removal by flotation was studied.[53]

Fillers have the potential to be used to improve recycled materials.[52] Two major problems in plastic recycling can potentially be addressed by fillers. Recycled plastics are frequently processed as a mixture of different polymers with various concentrations of each. This is especially true in the case of municipal wastes. There is a problem of compatibility between polymer components of these wastes

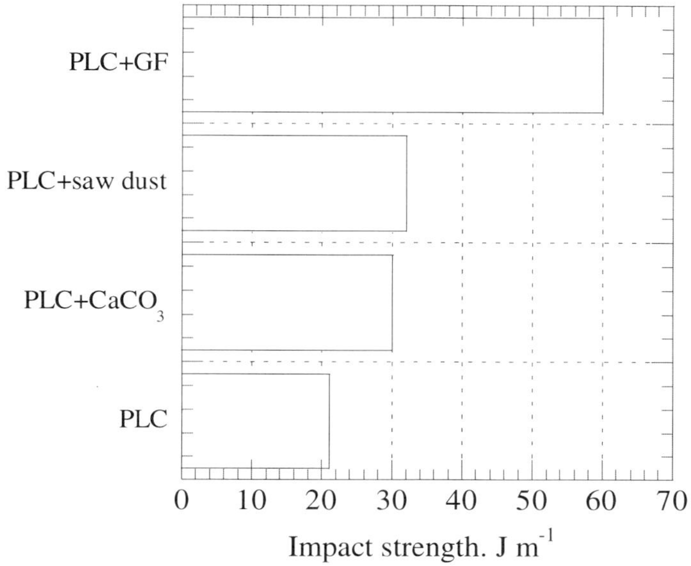


Figure 12.12. Impact strength of post-consumer plastic containers reprocessed with different fillers. [Adapted, by permission, from La Mantia F P, **Recycling of PVC & Mixed Plastic Waste**, *ChemTec Publishing*, Toronto, 1996.]

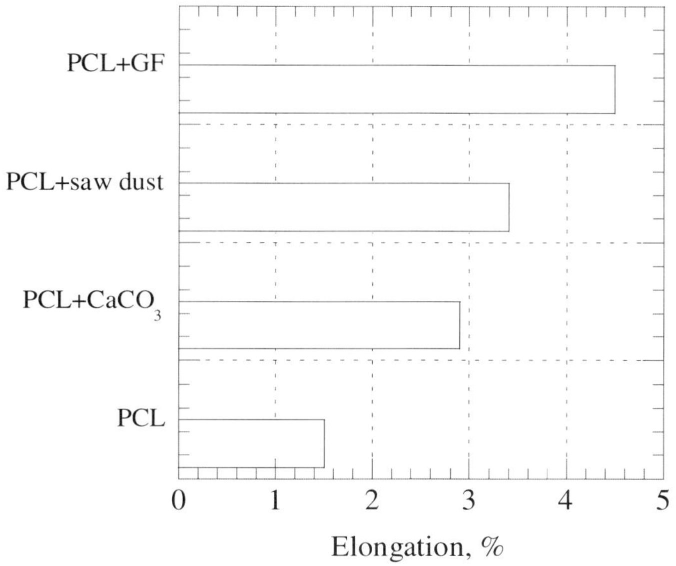


Figure 12.13. Elongation of post-consumer plastic containers reprocessed with different fillers. [Adapted, by permission, from La Mantia F P, **Recycling of PVC & Mixed Plastic Waste**, *ChemTec Publishing*, Toronto, 1996.]

which results in products having inferior mechanical characteristics. The mechanical characteristics of mixed plastic blends can be improved by the addition of cal-

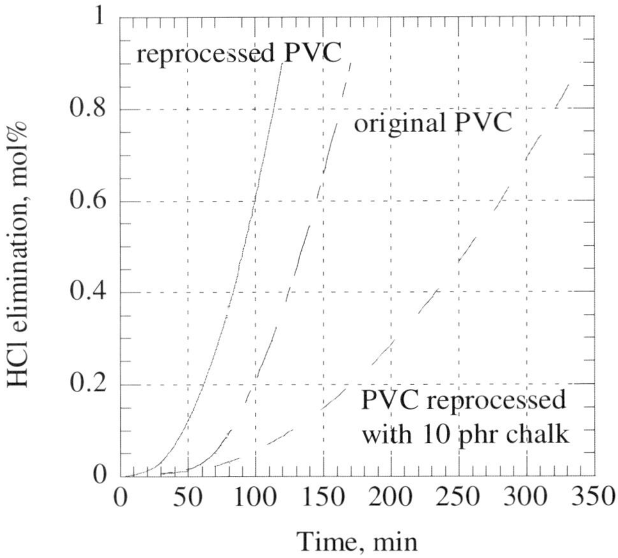


Figure 12.14. Dehydrochlorination rate of PVC. [Adapted, by permission, from Braun D, Kraemer K, **Recycling of PVC & Mixed Plastic Waste**, La Mantia F P, Ed. *ChemTec Publishing*, Toronto, 1996.]

cium carbonate, talc, or glass fibers.[66] A recycled PE/PS blend was considerably improved by the addition of treated wood fiber.[57] Figures 12.11 to 12.13 show the improvement of properties in post-consumer plastic containers by adding various fillers. Tensile and impact strength and elongation were improved.[68]

Figure 12.14 shows the effect of calcium carbonate on PVC reprocessing.[68] An addition of 10% chalk improves the thermal stability of the material to the extent that it performs better than the material before reprocessing.

## REFERENCES

1 Le Bras M, Bourbigot, Le Tallec Y, Laureyns J., *Polym. Degradat. Stabil.*, **56**, 1997, 11-21.
2 Nagieb Z A, El-Sakr N S, *Polym. Degradat. Stabil.*, **57**, 1997, 205-9.
3 Benrashid R, Nelson G L, *J. Fire Sci.*, **11**, No.5, 1993, 371-93.
4 Krisher J A, Marshall S S, Antec '97. Conference proceedings, Toronto, April 1997, 2928-30.
5 Nichols K, Solc J, Shieu F, Antec '93. Conference Proceedings, New Orleans, La., 9th-13th May 1993, Vol. II, 1938-42.
6 Schott N R, Rahman M, Perez M A, Antec '94. Conference Proceedings, San Francisco, Ca., 1st-5th May 1994, Vol. III, 2846-50.
7 Hornsby P R, Wang J, Jackson G, Rothon R N, Wilkinson G, Cosstick K, Antec '94. Conference Proceedings, San Francisco, Ca., 1st-5th May 1994, Vol. III, 2834-9.
8 Hshieh F Y, Beson H D, *Fire Mater.*, **21**, 1997, 41-9.
9 Toure B, Lopez Cuesta J-M, Longerey M, Crespy A, *Polym. Degradat. Stabil.*, **54**, Nos 2-3, 1996, 345-52.
10 Rothon R N; Hornsby P R, *Polym. Degradat. Stabil.*, **54**, Nos 2-3, 1996, 383-5.
11 Toure B, Lopez Cuesta J M, Gaudon P, Benhassaine A, Crespy A, *Polym. Degradat. Stabil.*, **53**, No.3, 1996, 371-9.
12 Kretzschmar B, *Kunststoffe Plast Europe*, **86**, No.4, 1996, 20-2.

13 Toure B, Lopez-Cuesta J, Benhassaine A, Crespy A, *Int. J. Polym. Analysis and Characterization*, **2**, No.3, 1996, 193-202.
14 Cusack P, Flame Retardants '96. Conference proceedings, London, 17th-18th Jan.1996, 57-69.
15 Baggaley R G, Hornsby P R, Yahya R, Cussak P A, Monk A W, *Fire Mater.*, **21**, 1997, 179-85.
16 Elfving K, Soderberg B, *Reinf. Plast.*, **40**, No.6, 1996, 64-5.
17 Brown N, Linnert E, *Reinf. Plast.*, **39**, No.11, 1995, 34-7.
18 Maxwell J, **Plastics in High Temperature Applications**, Pergamon Press, Oxford, 1992.
19 Smith R, Enhancing Polymers Using Additives and Modifiers II, Shawbury, 1996.
20 Martin P A, Buszad D L, Papez M, Addcon `96, Brussels, 1996.
21 Shen K K, O'Connor R, Addcon `96, Brussels, 1996.
22 Levchik G F, Levchik S V, Lesnikovich A I, *Polym. Degradat. Stabil.*, **54**, Nos 2-3, 1996, 361-3.
23 Pape P G, Romenesko D J, Antec '97. Conference proceedings, Toronto, April 1997, 2941-52.
24 Clemens M L, Doyle M D, Lees G C, Briggs C C, Day R C, Flame Retardants '94. Conference proceedings, London, 27th-28th January 1994, 193-202.
25 Hornsby P R, Wang J, Cosstick K, Rothon R, Jackson G, Wilkinson G, Flame Retardants '94. Conference proceedings, London, 27th-28th January 1994, 93-108.
26 Schott N R, Rahman M, Perez M A, *J. Vinyl and Additive Technol.*, **1**, No.1, 1995, 36-40.
27 Rothon R N, *Macromol. Symp.*, **108**, 1996, 221-9.
28 Le Bras M, Bourbigot S, *Fire & Mat.*, **20**, No.1, 1996, 39-49.
29 Costa L, Camino G, Bertelli G, Borsini G, *Fire & Mat.*, **19**, No.3, 1995, 133-42.
30 Yeh J T, Yang H M, Huang S S, *Polym. Degradat. Stabil.*, **50**, No.2, 1995, 229-34.
31 Fu-Yu Hsieh, Bryan C J, Pedley M D, *Fire & Mat.*, **18**, No.6, 1994, 389-91.
32 Innes J D, Addcon `96, Brussels, 1996.
33 McNeill I C, Mohammed M H, *Polym. Degradat. Stabil.*, **49**, No.2, 1995, 263-73.
34 McNeill I C, Mohammed M H, *Polym. Degradat. Stabil.*, **48**, No.1, 1995, 189-95.
35 Hornsby P R, Mthupha A, Antec '93. Conference Proceedings, New Orleans, La., 9th-13th May 1993, Vol. II, 1954-6.
36 Ferm D J, Shen K K, Antec '94. Conference Proceedings, San Francisco, Ca., 1st-5th May 1994, Vol.III, 3522-6.
37 Braun D, Kraemer K, **Recycling of PVC & Mixed Plastics**, Editor La Mantia F P, *ChemTec Publishing*, Toronto, 1996.
38 Gorl U, De Kok J J, Bomal Y, Cochet P, Mueller H, *Kaut. u. Gummi Kunst.*, **47**, No.6, June 1994, 430-4.
39 Hornsby P R, Wang J, Rothon R, Jackson G, Wilkinson G, Cossick K, *Polym. Degradat. Stabil.*, **51**, No.3, 1996, 235-49.
40 Kettrup A A, Lenoir D, Thumm W, Kampke-Thiel K, Beck B, *Polym. Degradat. Stabil.*, **54**, Nos 2-3, 1996, 175-80.
41 Zhao W, Hasegawa S, Fujita J, Yoshii F, Sasaki T, Makuuchi K, Sun J, Nishimoto S, *Polym. Degradat. Stabil.*, **53**, No.2, 1996, 199-206.
42 *Reinf. Plast.*, **40**, No.10, 1996, 66-70.
43 Modesti M, Simioni F, Albertin P, *Cell. Polym.*, **13**, No.2, 1994, 113-24.
44 Molesky F, Schultz R, Midgett S, Green D, *J. Vinyl Additive Technol.*, **1**, No.3, 1995, 159-61.
45 Herbert M J, Flame Retardants '96. Conference proceedings, London, 17th-18th Jan.1996, 157-72.
46 Miller B, *Plast. World*, **54**, No.12, 1996, 44-9.
47 Smock D, *Plast. World*, **54**, No.12, 1996, 35
48 *Reinf. Plast.*, **39**, No.4, 1995, 8.-9.
49 Winter H, Mostert H A M, Smeets P J H M, Paas G, *J. Appl. Polym. Sci.*, **57**, No.11, 1995, 1409-17.
50 Mayadunne A, Bhattacharya S N, Kosior E, Boontanjai C, Antec 95. Volume I. Conference proceedings, Boston, Ma., 7th-11th May 1995, 1178-82.
51 Whalen J P, Poston D D, Antec '94. Conference Proceedings, San Francisco, Ca., 1st-5th May 1994, Vol. I, 924-8.
52 Jacobson K E, Eller R, Polyolefins VIII. Conference Proceedings, Houston,Tx., 21st-24th Feb.1993, 494-509.
53 Liphard M, Von Rybinski W, Schreck B, *Prog. Coll. & Polym. Sci.*, **95**, 1994, 168-74.
54 Alkan C, Arslan M, Cici M, Kaya M, Aksoy M, *Resources Conserv. & Recycling*, **13**, Nos.3-4, 1995, 147-54.
55 Long Y, Tiganis B E, Shanks R A, *J. Appl. Polym. Sci.*, **58**, No.3, 1995, 527-35.
56 Chen X, Zhang S, Wang X, Yao X, Chen J, Zhou C, *J. Appl. Polym. Sci.*, **58**, No.8, 1995, 1401-5.

57 Simonsen J, Rials T G, *J. Thermoplast. Composite Mat.*, **9**, No.3, 1996, 292-302.
58 Graham J, Hendra P J, Mucci P, *Plast. Rubb. Comp. Process. Appln.*, **24**, No.2, 1995, 55-67.
59 Dierkes W, *J. Elastomers Plast.*, **28**, No.3, 1996, 257--78.
60 Sombatsompop N, Sims G L A, *Cell. Polym.*, **15**, No.5, 1996, 317-34.
61 Sims G L A, Sombatsompop N, *Cell. Polym.*, **15**, No.2, 1996, 90-104.
62 Burger H, Koine A, Maron R, Mieck K P, *Int. Polym. Sci. Technol.*, **22**, No.8, 1995, T/25-34.
63 Sims G L A, Angus M W, Crosley I, Antec '95. Vol. II. Conference Proceedings, Boston, Ma., 7th-11th May 1995, 2166-70.
64 Athey R D, Kirkland T, Lindblom G, Swoboda J, *Eur. Coatings J.*, No.11, 1995, 793-8.
65 Hohenberger W, *Kunststoffe Plast Europe*, **86**, 7, 1996, 18-20.
66 La Mantia F P, **Recycling of Plastic Materials**, *ChemTec Publishing*, Toronto, 1993.
67 Krause H H, Penninger J M L, **Conversion of Polymer Wastes & Energetics**, *ChemTec Publishing*, Toronto, 1994.
68 La Mantia F P, **Recycling of PVC & Mixed Plastic Waste**, *ChemTec Publishing*, Toronto, 1996.

# 13

# Influence of Fillers on Performance of Other Additives and Vice Versa

Product formulations are complex, consisting of numerous additives designed to perform certain functions. The performance of these additives depends on other components of a mixture. Similarly, fillers are added to perform certain tasks, and their performance might be enhanced or retarded by other components of the mixture. This chapter reviews the current understanding of these interactions, in order to highlight potential improvements or potential risks related to the application of fillers in complex formulations, which contain components that may interact due to physical or chemical forces.

## 13.1 ADHESION PROMOTERS

There are two groups of chemical materials which have similar chemical composition. These are adhesion promoters and coupling agents. In this section, the main object of the discussion is the performance of the entire system (product) in the relationship to the surrounding materials (frequently called substrates) with which the material must form durable bonding. Other aspects of physical and chemical interaction are discussed in one of the following sections entitled "Coupling Agents".

Adhesion promoters are most frequently used in coatings, adhesives, and sealants, but they also find application in other processes where adhesion between two layers of materials is of importance, namely in coated fabric, pultrusion, and lamination of several layers (e.g., calendering). It is surprising to find that in spite of so many applications, the literature on the subject almost does not exist.[1-4] Interfacial adhesion between layers of bromobutyl-based tire inner liner was found to be a problem.[1] Adhesion increased when 3.75 phr MgO and 3 phr 3-aminopropyltriethoxysilane was used. It should be noted that very large amount of silane had to be used to promote adhesion. In another study,[2] the effect of drying temperature on the penetration of silane into polymer (PVC) layer was investigated. At lower drying temperatures silane was able to diffuse more effectively into the polymer layer. Sulfur containing silanes were found to promote adhesion of organic materials to metals and inorganic substrates.[3] Salt spray resistance and adhesion of epoxy coatings were improved by the addition of silanes.[4]

It is noticeable from the above short review of publications that the effect of fillers is not evaluated. It is therefore necessary to use a fundamental approach to assess the problem. Three groups of chemical materials are mainly used to enhance adhesion of organic material to various substrates. These include silanes, polyurethanes, and tackifying resins. In most cases of the above mentioned products, adhesion promoters will work in formulations containing various fillers. Considering that fillers contain similar reactive groups to many substrates which are joined together, it is quite easy to anticipate that there must be a competition between functional groups on the filler surface and similar groups on the substrate surface for reaction with an adhesion promoter.

In the case of silanes, hydroxyl groups are found on both filler and substrate, and these groups compete for reaction with silane. It well-known in the practice of adhesives that primer gives superior adhesion to the same silane added into the mass of adhesive. The reason is related to the mechanism of cure of silanes which occurs in the presence of moisture. Silane must migrate to the substrate surface, and react with moisture, and then free hydroxyl groups on the substrate surface in order to form bonding, affecting overall adhesion. If hydrolysis of methoxy or ethoxy groups (the first step of silane reactions) occurs in the bulk of adhesive, it is very likely that the next step of reaction (reaction between two hydroxyl groups) also occurs in the bulk, producing a small molecular weight oligomer or bonding with filler, which will further immobilize the silane molecule. It is thus simple to understand that silane added to the adhesive bulk cannot be as effective as when applied as a primer because it is partially used for ineffective reactions. Extending this explanation to answer our question regarding the effect of fillers on silanes, it is easy to predict that fillers and water present in the formulation will retard the performance of adhesion promoter. This phenomenon was confirmed in many technological attempts. For example, fumed silica is an excellent rheological additive, but it cannot be effectively used in some polyurethane coatings or adhesives because it contains both reactive hydroxyl groups and water which react with adhesion promoter (silane), and such materials do not have adhesion to substrates. Silanized fumed silica does not consume adhesion promoter but has a much lower efficiency in changing the rheology of these systems.

A similar mechanism acts in the case of reactive adhesion promoters such as polyurethanes. Here water is a factor, as well as surface functional groups of fillers. The adhesion promoter is exhausted within the bulk of organic material (coating, adhesive, etc.) and unable to perform the task.

Tackifying materials are solutions of resins having high green strength. These, in turn, are easily absorbed by some fillers, and thus fillers may affect adhesion.

It is thus always important to evaluate fillers from the point of view of their effect on adhesion of the final product. On the other hand, fillers may also improve mechanical adhesion due to the fact that they increase surface roughness.

## 13.2 ANTISTATICS

Conductive fillers, intrinsically conductive polymers, and organic additives are used as antistatics. There is no common product available which has a combination of the above. The only known combinations are particulate and fibrous conductive fillers, which are claimed to produce a better effect.

There is a potential interference between fillers and organic antistatics. It should be mentioned that there are no general rules here, as there is no universal antistatic agent. Antistatic agents may affect the dispersion of fillers and their interaction with the matrix. The dispersion of fillers can be affected in both directions – it can be improved (some antistatics are titanates and zirconates which are also known to be dispersing agents) and it can be more difficult (surface charges on filler particles aid the mixing process, which antistatics eliminate, making dispersion of carbon black or graphite very difficult). Interaction between filler and matrix depends on adjacent functional groups on the surfaces of both interacting materials. Organic antistatics form a layer on the surface of fillers by which they change the character of its surface as well as isolate filler from the matrix polymer. The order of addition of both antistatic agent and filler should always be considered as an important factor.

Also, fillers may affect the performance of antistatic agents. Organic antistatic agents work well when they are on the surface and when the surface has a certain level of adsorbed moisture. Since antistatic agents are cationic, anionic, and nonionic surface active agents, they absorb very well on most fillers, which has to be compensated by larger additions in formulations containing fillers. When antistatics are mixed with fillers, they should be first predispersed in other organic materials in order to minimize the effect of their absorption on filler particles. Some fillers have a tendency to absorb water, which is another factor which may reduce the effectiveness of antistatic agents.

## 13.3 BLOWING AGENTS

Fillers may affect foaming in many ways.[5-8] The addition of filler increases the viscosity of the system and may affect cure rate of reactive systems. Considering that bubble size depends on the balance of internal pressure of gas inside the bubble and viscosity of the material in the surroundings, size of the bubble depends on both curing rate and viscosity of the system. With viscosity increasing, the same pressure of gas results in lower expansion of the bubble. At lower viscosity, bubbles may coalesce, forming larger bubbles that are less uniformly distributed throughout the material. Balancing viscosity and the rate of gas formation is an important task

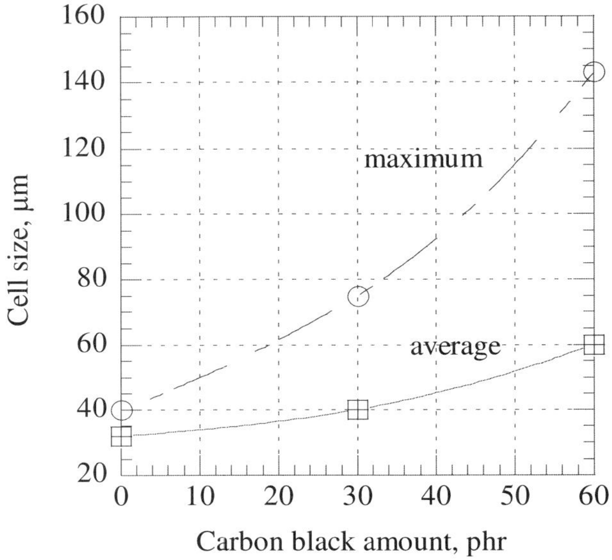


Figure 13.1. Cell size of vulcanized EPDM vs. concentration of carbon black. [Data from Guriya K C, Tripathy D K, *J. Appl. Polym. Sci.*, **62**, No.1, 1996, 117-27.]

of foaming. In addition, fillers have substantially higher density than the matrix, which makes systems containing fillers more difficult to foam.

The density of foam and its mechanical and thermal performances depend on the cell size and distribution. Fillers may affect cell sizes and distribution in several ways. On one hand, fillers have a nucleating effect, and they will participate in nucleation together with nucleating agents added to a formulation. Also, fillers are known to induce air pockets into the body of material, especially if they are not properly wetted. These air pockets will then be enlarged by a gas diffusing from the surroundings. The formation of foam depends on the properties of the matrix polymer, and fillers are known to change properties of the polymer due to interaction and adsorption.

From the above list of influences, it is pertinent that many properties of filler affect the foaming process and thus the performance of foaming agent. These properties include particle size and size distribution, surface area, shape, concentration, formation of agglomerates, distribution in the matrix, surface properties, and functional groups. It can also be predicted that these numerous parameters affect foaming in a complex way. This complexity eliminated fillers from foams in 1960s and 1970s; later, many regulations (fire protection, toxicity), cost, and increasing performance expectations brought fillers back to foams. Figure 13.1 shows the effect of filler on cell size. Increased concentration of carbon black contributes to the increase of average cell size and maximum cell size.[8]

## 13.4 CATALYSTS

The effect of fillers on the performance of catalysts attracted attention in current works.[9-13] Fillers are used as active supports of catalysts in the molecular design of polymers. The works on metallocene catalysts are far beyond the scope of this publication, and their review can be found in a specialized publication.[13] Several fillers are used for catalyst support. The most prominent members of this group include kaolin, tufa, dolomite, perlite, $Al_2O_3$, $SiO_2$, $CaSO_4$, and wollastonite.[11,12] These fillers participate in single-site catalytic systems but they can also be used for other purposes and should be considered in applications when catalyst is used in filled systems. One common application of a filler/catalyst system is in filler elastomer technology where monomers are polymerized in the presence of catalyst absorbed on filler, resulting in reactor-grade filled elastomers. These elastomers have properties which cannot be achieved by simple mixing. Typically, 25-50% active groups on the filler surface are utilized to attach catalyst, and these catalysts have efficiency increased by a factor of 10 to 1000. In addition to changes in catalytic activity, such reactive systems change polymer morphology, forming very regular polymers, designed for the structure needed. Although the above processes are highly specialized technologies applied on the industrial scale to produce special grades of polymers, the general principles are also applied in simple filled systems, which are the subject of our discussion. Therefore, the observations gathered from the synthesis of metallocene catalyzed polymers have relevance to these simple systems.

The role of filler in catalyst-containing systems is not restricted to morphological changes and reaction rate increases. Fillers may also retard the chemical processes of catalyzed reactions. One example is given in Figure 13.2. The silica surface contains reactive hydroxyl groups which contribute to the absorption of common accelerators used in rubber processing. Compounds containing silica cure with reduced rate, since accelerator is depleted. Typical methods used to overcome this problem include either increased addition of accelerator or, more efficiently, the process of filler surface precoating with, for example, polyethylene glycol.[9] Surface groups are neutralized by the coating and cannot further absorb accelerator. Such a precoating uses a controlled order of addition, based on the physical principle that substances adsorbed first on the surface are very difficult to replace.

The other interesting case of interaction between filler and catalyst was reported in the literature.[12] A new filler was developed for automotive bulk molding and sheet molding compounds. This filler is composed of calcium carbonate coated on the surface by a thermoplastic polymer. Because of the presence of coating, catalyst cannot combine with the surface groups of filler which results in a higher degree of styrene-based polymer crosslinking.

It is thus pertinent from this short review that catalysts and fillers have a very high affinity, and their interaction may increase or decrease polymerization rate as well as change the molecular structure of the resultant polymer.

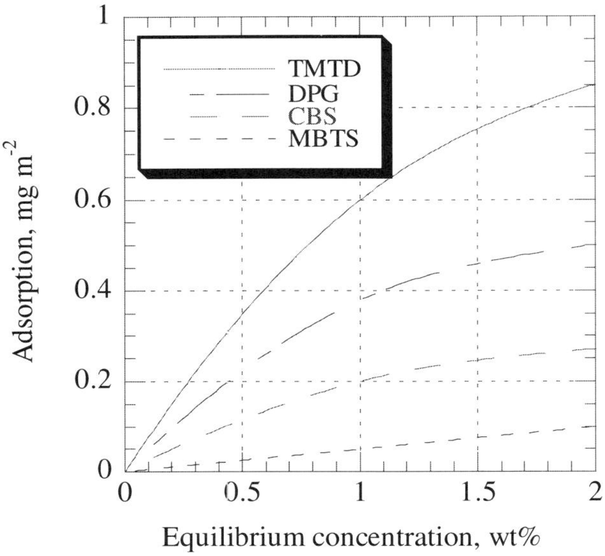


Figure 13.2. Adsorption of rubber accelerators on a silica surface. [Data from Bomo F, Meeting of the Rubber Division, ACS, Montreal, May 5-8, 1996, paper E.]

## 13.5 COMPATIBILIZERS

The most frequent compatibilizers of polymer blends are composed of block copolymers which have a chemical similarity to polymer components of the blend. Their role is to provide a vehicle for increasing the compatibility of normally incompatible polymers. It is usually quite a simple task if blend is only composed of polymers. In more complex systems, compatibilizer may more readily interact with other components (e.g., filler) than with the polymer which should be compatibilized. This is one example of the different effects of fillers found in studies of polymer blends.[14-20] A blend of PP/PA-6 was filled with glass beads.[15] Ionomers and PP functionalized by reaction with maleic anhydride were used as compatibilizers. Addition of compatibilizer in the presence of filler may be used for encapsulation of filler – quite typical of studied systems. Filler encapsulation was a much more rapid process than blend compatibilization, which is typically a slow process. Thus, if compatibilizer was added in the presence of filler, filler was preferentially encapsulated. If the blend was compatibilized first and then filler added, best results were obtained. If high strength and high toughness are to be obtained simultaneously, it is beneficial to encapsulate filler in rubber toughened thermoplastic. Such a goal was set in studies of HDPE/EPDM/carbon black blends. Figure 13.3 shows the effect of interfacial modifier on the first normal stress difference, $N_1$. The addition of rubber significantly increased melt elasticity due to dispersion of filler and formation of a core-shell dispersed phase. In addition, the viscosity and dynamic storage modulus increase, and dynamic loss modulus decreases.[18]

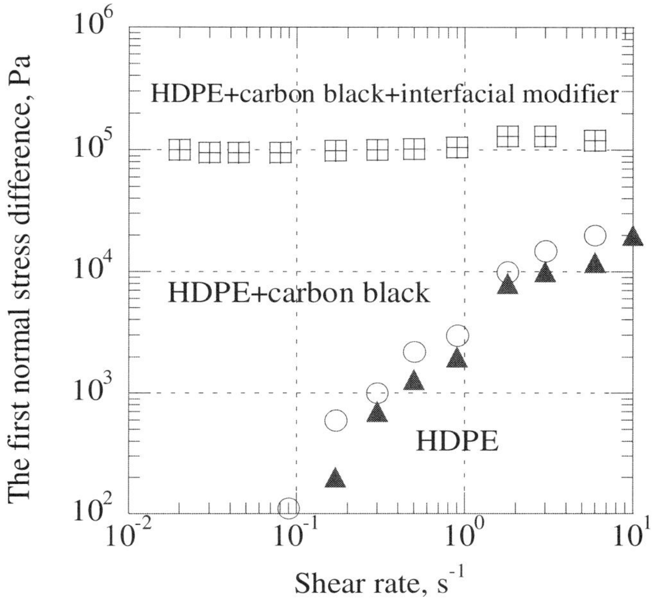


Figure 13.3. $N_1$ vs. $\dot{\gamma}$ for HDPE/carbon black composites. [Adapted, by permission, from Zhu J, Ou Y-C, Feng Y-P, *Polym. Int.*, **37**, No.2, 1995, 105-11.]

In another experiment,[14] acrylic acid was absorbed on the surface of mica, forming an interlayer in a form of a coating, having thickness proportional to the amount of absorbed monomer. The rigidity of PP/mica composites is inversely proportional to the thickness of the interlayer. In PP filled with glass beads and compatibilized with EPM, the core-shell structure is less efficient in obtaining a strength/toughness balance than the presence of an EPM microphase.[17]

The incompatible system containing poly(vinylidene fluoride) and polystyrene was compatibilized by sepiolite addition. In this system, filler played the role of compatibilizer, giving the best performance at 10% concentration.

## 13.6 COUPLING AGENTS

One function of a coupling agent is to interact with filler. It is thus proven *a priori* that coupling agents affect filler properties, and modified fillers affect the system. Numerous works use this modification.[21-53] In this discussion, we summarize the effects obtained. Table 13.1 contains this summary.

## Table 13.1. Effect of coupling agent/filler interaction

| Filler | Polymer | Coupling | Effect | Refs |
|---|---|---|---|---|
| Al borate | PC/ABS | silane | improved adhesion and orientation change due to response to flow | 43 |
| $Al(OH)_3$ | HDPE<br>PVC<br>silicone | OMCTS<br>silane<br>PDMS | increased adhesion to matrix depends on concentration of modifier<br>crosslinking decreases thermal stability, elongation and tensile str.<br>surface hydrophobization; resistance to solvent extraction and water | 29<br>46<br>37 |
| $CaCO_3$ | EPR<br><br>rubber<br>rubber<br>silicone | maleates<br><br>fatty acid<br>silanes<br>PDMS | decreased disperse component of the surface energy; filler surface energy approaches surface energy of matrix<br>decreased tensile strength and flexural cracking<br>increased green strength, Mooney viscosity, and tensile properties<br>surface hydrophobization; resistance to solvent extraction and water | 21<br><br>49<br>49<br>37 |
| Carbon black | rubber<br>rubber | ammonia<br>APTS | slightly reduced bound rubber (7%)<br>crosslinking through reaction between $-NH_2$ and COOH from rubber | 35<br>36 |
| Carbon fiber | PEEK | polyimide | no improvement of mechanical properties | 52 |
| $Fe_2O_3$ | silicone | PDMS | surface hydrophobization; resistance to solvent extraction and water | 37 |
| Fumed silica | PEG<br>silicone | silane<br>silane | improved mechanical properties and electric conductivity<br>specific interactions decreased | 39<br>40 |
| Glass beads | PET<br><br>Epoxy | APES<br><br>silane | increases interfacial adhesion but hinders crystallization of PET by reducing nucleating activity of filler<br>fracture resistance increased but no improvement of $H_2O$ resistance | 23<br><br>33 |
| Glass fiber | PP<br>PP | MA<br>MA | flex strength increased, notch Izod decreased with MA increasing<br>increased tensile and flexural properties | 26<br>31 |
| Gold - nanosize | PMMA | TOAB | nanoparticles obtained by polymerization in the presence of Au | 30 |
| Hydroxyapatite | | silanes | stable to water coating; stability depends on silane functional group; coating is permeable to ionic transport in body fluids | 24 |
| Jute fibers | Epoxy | ES | increased ; increase tensile retention in $H_2O$ presence | 27 |
| Kaolin | HDPE<br>PA-66 | silane<br>silanes | increased stiffness and fracture resistance<br>improved wet strength and mechanical properties | 35<br>45 |
| $Mg(OH)_2$ | PP<br>PP<br>PP | silane<br>titanate<br>fatty acids | slight increase in flexural and impact strength<br>slight increase in impact strength<br>the most effective surface treatment improving toughness | 32<br>32<br>32 |
| Mica | PP<br>PP<br>PP<br>PA-66 | MA<br>MA<br>MA<br>silanes | increased tensile and flexural properties<br>all mechanical properties increased except impact strength<br>all mechanical properties increased except impact strength<br>improved wet strength and mechanical properties | 31<br>38<br>44<br>45 |
| Quartz | Polyester | silane | increased mechanical and water resistance, filler particle splitting | 50 |
| Silica | EPM<br>SBR<br>SBR<br>rubber | alcohol<br>TESPT<br>TESPT<br>ammonia | reduced interaction with polymer and itself<br>Mooney viscosity decreased; larger pore size the later silica is added<br>Mooney viscosity decreased; reduced cure rate<br>substantially reduced bound rubber (85%) | 48<br>28<br>34<br>34 |
| Talc | PP<br>PP<br>PA-66 | MA<br>MA<br>silanes | increased tensile and flexural properties<br>properties of system can be tailored to application<br>improved wet strength and mechanical properties | 31<br>54<br>45 |
| Wollastonite | PA-66 | silanes | improved wet strength and mechanical properties | 45 |
| Zirconium silicate | HDPE | silane | improved impact strength in comparison to untreated filler | 41 |
| ZnO | PC<br>PET | PDMS<br>PDMS | reduced catalytic activity of pigment in photodegradation<br>reduced catalytic activity of pigment in photodegradation | 42<br>42 |

APES - γ-aminopropyltriethoxysilane; APTS - 3-aminopropyltriethoxysilane; ES - epoxy silane; MA- maleic anhydride; OMCTS - octamethylcycloethoxysiloxane; PDMS - poly(dimethyl siloxane); TESPT - (bis)(triethoxysilylpropyl)- tetrasulfone; TOAB - tetraoctylammonium bromide with $HAuCl_4$

## 13.7 DISPERSING AGENTS AND SURFACE ACTIVE AGENTS

Filler dispersion is a mostly mechanical process which requires sufficient shear for dispersion but two stages of the dispersion process sometimes require the help of additives. These are the stages of surface wetting and stabilization of the dispersion. In both stages, dispersing agents and surface active agents are frequently involved. Their effect on pigment dispersion is discussed here.[55-61] The surface wetting of filler by the matrix is frequently difficult because of high viscosities of the matrix or the lack of compatibility between filler and matrix. Here, dispersing agents can be of help. In the case of systems of high viscosity such as melts or dispersion in high viscosity polymers, the dispersion obtained is usually stable; therefore, no further action is required. But this is not the case in dispersions of filler in low viscosity liquids, where a large difference in density and a tendency of some fillers to flocculate or aggregate causes a good dispersion to be reversed. In such cases, surface active agents help to stabilize the suspensions.

The mechanisms of both processes are essentially similar, even though they are employed in different cases. Both dispersing agents and surface active agents have polar molecules composed of groups which easily interact with filler and matrix. Between these groups there is frequently a chain which is used to create some space between filler and matrix. From this description, it is clear that a filler modified by a surface active agent has modified surface groups, which changes its behavior in the system.

The materials used for modification can be of low molecular weight but can also be small molecular weight polymers such as polyethylene wax, which was used to improve the dispersion of carbon black in polyethylene.[56] This wax helps in breaking carbon black agglomerates, resulting in reduced viscosity and torque and substantially improved dispersion. Also, comb-like grafted block copolymers are used as dispersing aid for polyethylene, resulting in a more efficient use of pigments due to a better dispersion.[55]

Figure 13.4 shows the effect of pH of carbon black on the amount of dispersing agent required.[57] The suspension of carbon black in water was made using the sodium salt of a polymeric carboxylic acid, which is an anionic surfactant. Depending on pH, substantially different amounts of dispersing agent were needed. It is noticeable that the relationship is very steep, which indicates a very small tolerance to the differences in pH. A small difference in pH may completely alter dispersion. Also, difference in the specific surface area and surface treatment of carbon black has a strong influence on the quality of dispersion.

Figure 13.5 is an example of a collaborate activity of surfactant and filler in obtaining small and uniform cell sizes. The control foam was obtained with surfactant alone. The addition of silica increases nucleation and results in a very uniform polyurethane foam. This foam was developed to increase the insulation rating of refrigerator foams.[61]

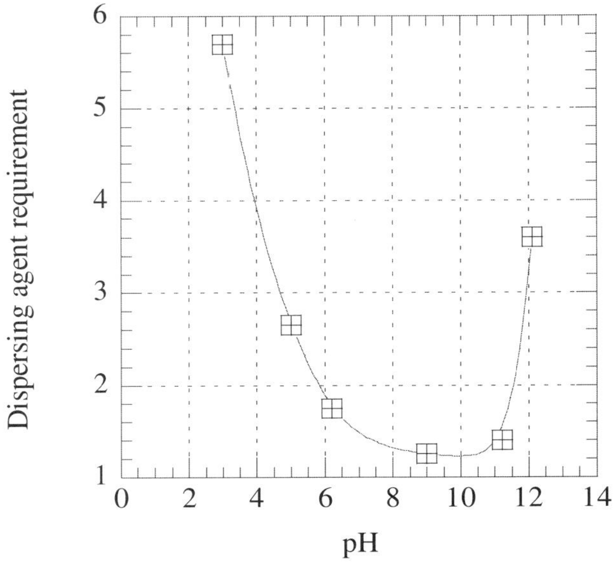


Figure 13.4. Dispersion agent requirement vs. pH in water dispersions of carbon black. [Adapted, by permission, from Foster J K, Sims, E S, *Polym. Paint Col. J.*, **184**, 1994, 312.]

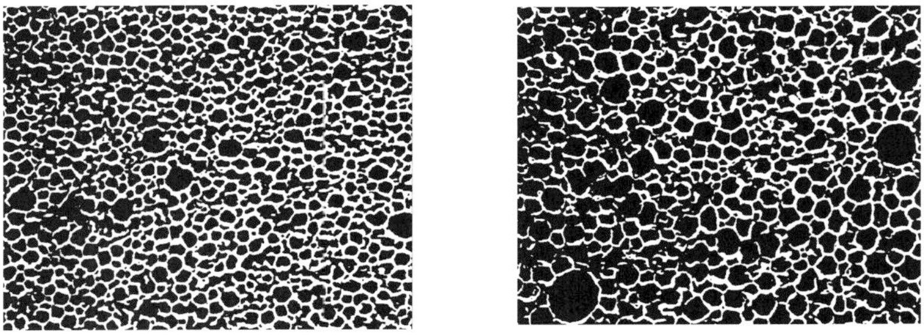

Figure 13.5. Micrographs of rigid polyurethane foams. left - surfactant/silica foam, right - control. Magnification 30x. [Adapted, by permission, from Okoroafor M O, Wang A, Bhattacharjee D, Cikut L, Haworth G J, Polyurethanes '95. Conference Proceedings, Chicago, Il., 26th-29th Sept.1995, 303-9.]

A surface active agent may also help in filler synthesis. Various surfactants can be used to aid the dispersion of precipitating filler. Addition of surfactant produces a more uniform particle size distribution. At the same time, the surface of filler is converted by reaction with surfactant. For example, alcohols and phenols may react with hydroxyl groups, changing surface properties and surface active groups.[59] In interpenetrating networks, such a surface change, caused by the interaction of $Fe_2O_3$ with surfactant, increased microphase separation in comparison with filler used without surfactant.[60]

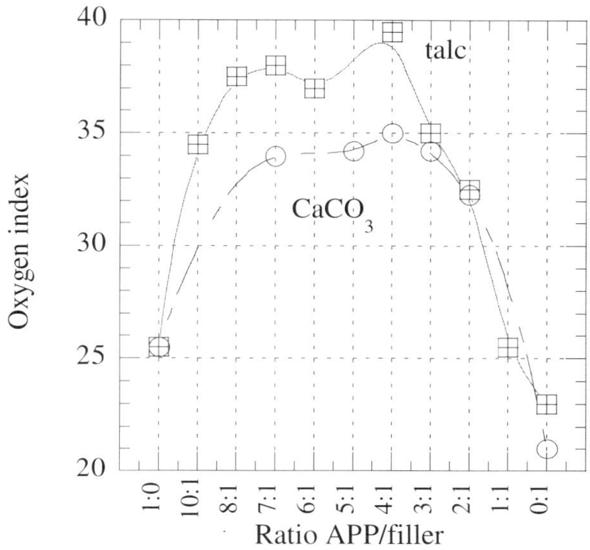


Figure 13.6. Oxygen index vs. ratio of ammonium polyphosphate/filler. [Data from Levchik G F, Levchik S V, Lesnikovich A I, *Polym. Degradat. Stabil.*, **54**, Nos 2-3, 1996, 361-3.]

Surface active agents were also employed in processes of filler recovery from waste paper.[58] An increased concentration of surfactant increased the recovery of kaolin and calcium carbonate from waste paper.

## 13.8 FLAME RETARDANTS

Combinations of inorganic and organic flame retardants are discussed here.[62-68] Figure 13.6 shows than the addition of regular fillers, such as talc and $CaCO_3$, to ammonium polyphosphate increased the fire resistance of PA-6.[62] The function of filler in these combinations is to increase char yield and increase insulation properties of char. On the other hand, ammonium polyphosphate protects char from oxidation and hinders diffusion of combustible gases to the flame.

A more complex fire retarding system was used in propylene-ethylene copolymer.[63] This system consists of huntite/hydromagnesite/antimony trioxide/decabromodiphenyl oxide. Fillers act as effective flame retardants and, in addition, reinforce the copolymer. At the same time, an agent acting in the gaseous phase was needed to reduce the height of flames. A combination of antimony trioxide and decabromodiphenyl oxide plays this role in the system.

Figure 13.7 shows the effect of two combinations on smoke release. Addition of zinc stannate or ammonium octamolybdate, AOM, gives better performance than antimony trioxide alone.[64] Antimony trioxide performance is also improved by brominated flame retardants.[65] Chlorinated hydrocarbons, antimony trioxide, and novel silicone powders form good combinations, reducing the heat release rate and smoke in thermoplastics.[66]

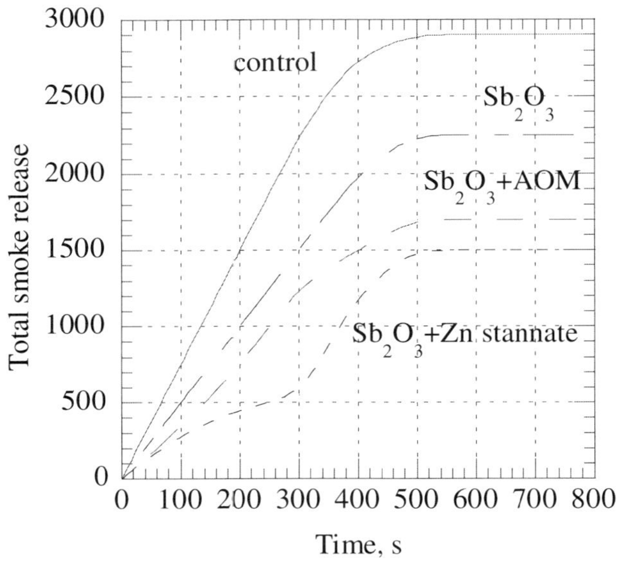


Figure 13.7. Total smoke release vs. time. [Data from Herbert M J, Flame Retardants '96. Conference proceedings, London, 17th-18th Jan.1996, 157-72.]

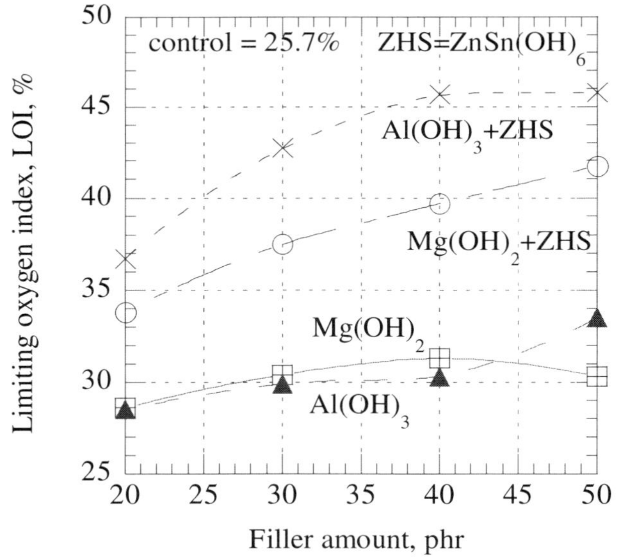


Figure 13.8. The effect of a zinc hydroxystannate coating on limiting oxygen index. [Data from Baggaley R G, Hornsby P R, Yahya R, Cussak P A, Monk A W, *Fire Mater.*, **21**, 1997, 179-85.]

Zinc hydroxystannate was found to enhance the performance of $Mg(OH)_2$ and $Al(OH)_3$. Figure 13.8 shows the effect of filler coated with zinc hydroxystannate on

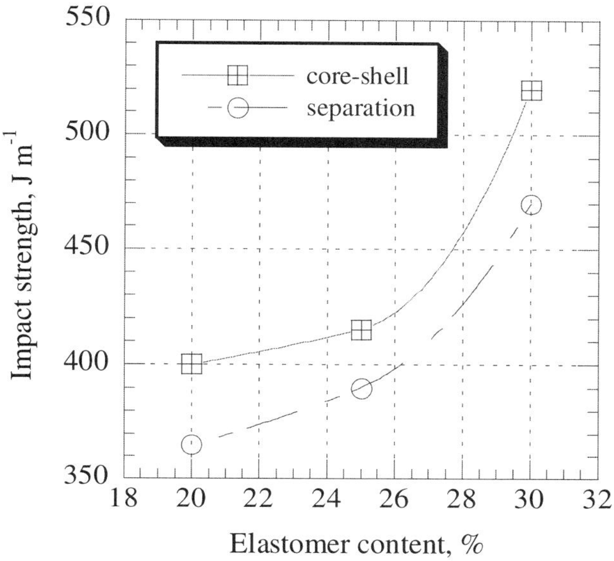


Figure 13.9. Elastomer content vs. impact strength of polypropylene containing 15% talc and maleic anhydride modified ethylene-propylene copolymer. [Data from Yu Long, Shanks R A, *J. Appl. Polym. Sci.*, **61**, No.11, 1996, 1877-85.]

the limiting oxygen index. Substantial improvements result.[68] Similar improvement of ignition time and heat release rate also occurs.

In intumescent fire retardant application, clay was combined with ammonium polyphosphate and pentaerythritol in the polypropylene matrix.[67] The fire retarding properties depended on the composition of clay. Ammonium polyphosphate played the role of carbonization catalyst and pentaerythritol the role of carbonization agent.

## 13.9 IMPACT MODIFIERS

Impact modifiers are frequently used in filled systems.[69-78] The effect of the presence of both filler and impact modifier on mechanical properties depends on the microstructure. Figure 8.33 shows structure combinations.[69] If particles of filler and impact modifier occupy separate sites in the matrix, they form separated structures. If filler particles are embedded in the impact modifier resin, together they form a core-shell morphology. In a core-shell morphology, particles of filler become more elastic and may enhance adhesion to the matrix. These two structures have implications on mechanical properties. Typically, with a core-shell structure, the elastic modulus decreases, elongation increases, and impact strength increases. Figure 13.9 shows that some gains in impact strength were obtained due to a core-shell microstructure.

The above relationships between structure and basic mechanical properties are obtained from the analysis of experimental compounds as well as theoretical analysis. In studies of PVC containing $CaCO_3$ and chlorinated polyethylene as impact modifier, impact strength increased on addition of filler, which formed core-shell structure with the filler.[78] Impact strength also increased with the amount of impact modifier (SEBS and SEBS modified with maleic anhydride) in glass spheres filled PP.[75] Both results are in agreement with the theoretical analysis.

In ultrafine talc-filled rigid PVC modified with acrylic modifier, impact strength consistently decreased with addition of filler regardless of the concentration of acrylic modifier.[73,74] An increase in calcium carbonate caused a substantial increase in impact strength. Similar results were obtained for HDPE/EPDM/CB and HDPE/CB systems.[70] There were high increases in tensile strength due to increased addition of carbon black, but no substantial influence of impact modifier was noted. Impact strength always decreased with increasing carbon black concentration and only impact modifier was instrumental in increasing impact strength.

It can be thus summarized[76,77] that a range of mechanical behavior can be generated by the appropriate choice of processing conditions, filler size, shape and concentration, and elastomer content and its distribution between the matrix and filler. These choices should lead to results predicted by theory.[69]

## 13.10 UV STABILIZERS

There is a growing evidence that fillers may adversely affect the performance of UV stabilizers, including practical immobilization and loss of their effect on stabilized systems.[79-83] These data coming from different sources indicate a growing concern that UV stabilizers, which are expensive additives, may not perform up to expectations. Table 13.2 shows the amounts of various stabilizers absorbed on carbon black. The absorbed quantities are substantial to raise further questions regarding the effect of this absorption on the performance of stabilizers, considering that carbon black is present in many formulations either as filler or pigment. Table 13.3 shows that carbon black is not the only filler which absorbs UV stabilizers.[79] Fillers used in this experiment, conducted in a cyclohexane suspension, were at low concentrations. The concentration of HALS was typical of many formulations. Results show that tertiary HALS is superior to secondary

**Table 13.2. Absorption of UV stabilizers on 30% Monarch 1300 suspended in xylene[83]**

| UV stabilizer | Initial concentration, % | Absorption % |
|---|---|---|
| Tinuvin 384 | 3 | 27 |
| Tinuvin 770 | 1.5/3 | 100 |
| Tinuvin 292 | 1.5/3 | 83/100 |
| Tinuvin 1130 | 3 | 81 |

**Table 13.3. Absorption of secondary and tertiary HALS on different fillers[79]**

| Filler | Concentration, % | | HALS absorption, % | |
|---|---|---|---|---|
| | filler | HALS | secondary | tertiary |
| Talc | 20 | 0.6 | 25 | 6 |
| $TiO_2$ | 4 | 0.6 | 17 | 7 |
| $CaCO_3$ | 4 | 0.6 | 75 | 40 |
| Carbon black | 8 | 0.6 | 20 | 7 |

HALS but that absorption on fillers occurs in both cases. Calcium carbonate is the most absorbing filler in this experiment.

The molecular mobility of various HALS was studied by the spin probe TEMPO, and ESR spectra were recorded.[80] In polyethylene filled with silica having a surface area of 139 $m^2/g$, the concentration of silica was in the range from 0 to 25.9%. It was found that hydrogen bonds are formed between nitroxyl radicals and hydroxyl groups of filler. The formation of hydroxyl groups leads to changes in $A_{ZZ}$ which is the peak-to-peak separation of the rigid limit spectrum. For isolated nitroxyl radicals (no hydrogen bonding) $A_{ZZ} = 3.33$ mT. For hydrogen bonded radicals this value changes to 3.83 mT. The measured values were from 3.4 to 4 mT depending on silica concentration. This suggests that addition of silica immobilizes HALS.

The above presented data may constitute evidence that under model conditions (without presence of polymer) UV stabilizers are adsorbed. It is conceivable that the absorption on filler may potentially play a positive role: it may slow down loss of stabilizers due to their migration and volatility. Figure 13.10 gives data on changes in carbonyl absorption during PP photooxidation. Both calcium carbonate and talc reduce the performance of HALS even though these fillers are only used in a very moderate concentration (10%), much lower than in typical products.

Table 13.4 shows the effect of the presence and absence of 1% silica on the performance of various UV stabilizers in stabilized polypropylene. The performance of all stabilizers was substantially retarded in the presence of silica.[82] The addition of silica without the presence of UV stabilizer improves UV stability of the polymer. The data shows that the effect of filler is not restricted to HALS but extends to a broad spectrum of UV stabilizers.

**Table 13.4. Time to 0.06 carbonyl units for PP stabilized with various stabilizers in concentration of 0.1%[82]**

| Stabilizer | Time, h | |
|---|---|---|
| | without silica | with silica |
| Control | 384 | 432 |
| Irganox 1010 | 1032 | 144 |
| Cyanox LTDP | 912 | 192 |
| Irgafos 168 | 408 | 144 |
| Chimassorb 944 | 1512 | 540 |
| Cyasorb UV 3346 | 1512 | 720 |
| Tinuvin 622 | 2004 | 456 |

It can be postulated that three factors may play a role in stabilization in the presence of fillers. One factor is the discussed absorption which immobilizes stabilizer to the extent that it cannot perform its function. The second is its desorption capability, which, if it exists, may enhance the performance of stabilizer due to its better retention. However, it should be mentioned that photochemical changes occur at very short time scales; therefore, there must always be a sufficient concentration of "free" (not absorbed) stabilizer to react with radicals. The third factor is the effect of filler on structure and related stability of stabilizer. The reduced stability of polypropylene in the presence of some stabilizers in Table 13.4 can be explained only by the formation of degradation products. It

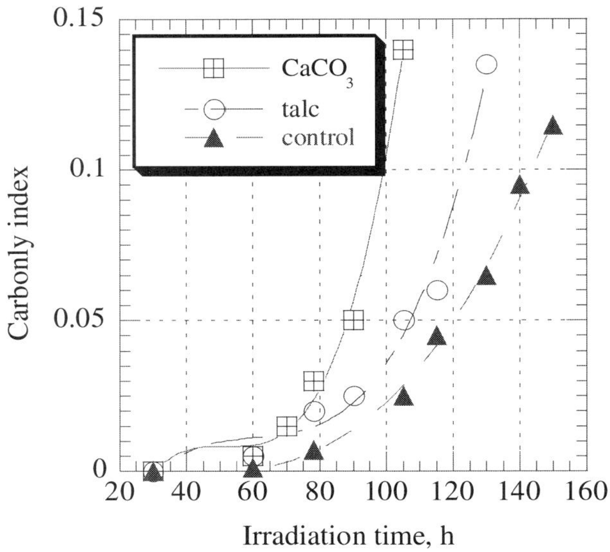


Figure 13.10. Carbonyl absorption during photooxidation of PP containing 0.3% HALS and fillers. [Adapted, by permission, from Hu X, Xu H, Zhang Z, *Polym. Degradat. Stabil.*, **43**, No.2, 1994, 225-8.]

is well established that adsorption of any chemical compound on filler may change its molecular structure, because the molecule must change its geometry to be adsorbed. Perhaps this change makes some products to degrade.

In summary, the data presented above show that there is a concern regarding performance of UV stabilizers in real systems, which typically include fillers. It is known from the automotive industry that the only method by which durability of automotive coatings could be increased was by the use of highly stabilized clear coat. It is also known from this field that a substantial amount of UV stabilizer from clear coat is absorbed on the pigmented coat, by which clear coat stability is reduced. There is also a growing tendency to protect materials with various surface coatings rather than adding UV stabilizers into the material bulk. This trend may reflect the reality of little progress made in stabilizing compounds during the last decade, which followed growing interest in stabilization due to arrival of HALS. The new developments in some pigments such as ultrafine $TiO_2$ should be closely monitored, since some pigments already offer better protection than organic stabilizers and at a fraction of the cost.

## 13.11 OTHER ADDITIVES

Many other additives affect performance of fillers, or their performance is affected by fillers. Some findings are discussed below.[84-97]

Combinations of particulate fillers and fibers give some unique properties. A combination of calcium carbonate and polyamide fibers gave the best dielectric properties in natural rubber. This composition was also the most resistant to natural aging.[87] Addition of mica and zirconia to glass fiber reinforced composites reduces wear and the friction coefficient.[88] In sealants and adhesives, rheological properties depend on fillers. Good non-sag properties are the easiest to obtain by using combinations of spherical and elongated particles.

Properties of silica-filled LDPE depend on the crystallization rate, which depends on nucleation by filler particles. The crosslinking initiated by thermal decomposition of peroxide retards crystallization. The properties of the resultant product in the reactive system depend on the timing of both processes.[89,90]

Friction and wear of metal oxide filled PTFE can be improved by liquid paraffin lubrication, and the friction coefficient decreases by one order of magnitude. On the other hand, the interaction between liquid paraffin oil and metal oxide filled PTFE reduces load-carrying capacity because of absorption on the surface layers of composite.[91]

Diamine salts of fatty acids are used as multifunctional additives in natural rubber compounds filled with carbon black.[92,93] They affect the elastomer-carbon black interface. With an increased concentration of multifunctional additive, the concentration of bound rubber decreases but dispersion of carbon black is improved. In silica filled rubber, multifunctional additive also improves the dispersion of silica, but in addition, it decreases the negative influence of silica filler on vulcanization rate.

The rheological properties of suspensions of montmorillonite depend on the presence of electrolytes and polyelectrolytes.[94] The face-to-face interactions (coagulation) occur in the presence of polyvalent cations. The edge-to-face interactions are strongest at low pH and in the presence of high concentrations of electrolytes.

A combination of calcium hydroxide and vegetable oil in polyethylene gives a cost effective method of filling. Salts of fatty acids are formed during compounding, resulting in improved interaction between polymer and filler. In addition, there is a potential to use this system for waste recycling, where plastic waste is typically contaminated with vegetable oil.[96]

Proprietary additives based on copolymers containing carboxyl groups are rheological modifiers used to lower the viscosity of systems filled with $Al(OH)_3$.[97]

The examples included in this chapter show that addition of filler to any composition must be evaluated for various interactions and interferences in properties other than the reasons for filler inclusion. Simple addition of filler without considering the possible effects may cause essential changes in the material.

## REFERENCES


1 Cochet P, Bomal Y, *Kaut. u. Gummi Kunst.*, **48**, No.4, 1995, 270-5.
2 Suzuki N, Ishida H, *Macromol. Symp.*, **108**, 1996, 19-53.
3 Marciniec B, Gulinski J, *Int. Polym. Sci. Technol.*, **22**, No.2, 1995, T/83-7.
4 Szalinska A, *Int. Polym. Sci. Technol.*, **20**, No.9, 1993, T/101-5.
5 Sherman L M, *Plast. Technol.*, **41**, No.3, 1995, 42-5.
6 Alpern V, Shutov F, *Prog. Rubb. Plast. Technol.*, **11**, No.4, 1995, 268-83.

7 Guriya K C, Tripathy D K, *J. Appl. Polym. Sci.*, **62**, No.1, 1996, 117-27.
8 Malanda L M, Park C B, Balatinecz J J, Antec '96. Volume II. Conference proceedings, Indianapolis, 5th-10th May 1996, p.1900-7.
9 Bomo F, Meeting of the Rubber Division, ACS, Montreal, May 5-8, 1996, paper E.
10 Hindryckx F, Dubois P, Jerome R, Teyssie P, Marti M G, *J. Appl. Polym. Sci.*, **64**, 1997, 439-54.
11 Zhang C-Z, Liu W-M, Xue Q-J, Shen W-C, *J. Appl. Polym. Sci.*, **66**, 1997, 85-93.
12 Jeffs D, *Reinf. Plast.*, **38**, No.6, 1994, 32-5.
13 Benedikt G M, Goodall B L, Eds., **Metallocene-Catalyzed Polymers**, *Plastic Design Library*, Norwich, 1998.
14 Chiang W Y, Yang W D, Pukanszky B, *Polym. Engng. Sci.*, **34**, No.6, 1994, 485-92.
15 Benderly D, Siegmann A, Narkis M, Antec '95. Vol. II. Conference Proceedings, Boston, Ma., 7th-11th May 1995, 3168-71.
16 Herrero C R, Morales E, Acosta J L, *J. Appl. Polym. Sci.*, **51**, No.7, 1994, 1189-97.
17 Roesch J, Barghoorn P, Muelhaupt R, *Makromol. Chem. Rapid Commun.*, **15**, No.9, 1994, 691-6.
18 Zhu J, Ou Y-C, Feng Y-P, *Polym. Int.*, **37**, No.2, 1995, 105-11.
19 Xanthos M, Grenci J, Dagli S S, Antec '95. Vol.II. Conference Proceedings, Boston, Ma., 7th-11th May 1995, 3194-200.
20 Yeh J T, Yang H M, Huang S S, *Polym. Degradat. Stabil.*, **50**, No.2, 1995, 229-34.
21 Zaborski M, Slusarski L, *Composite Interfaces*, **3**, No.1, 1995, 9-22.
22 Evans L R, Meeting of the Rubber Division, ACS, Montreal, May 5-8, 1996, paper D.
23 Ou Y, Yu Z, Zhu J, Li G, Zhu S, *Chinese J. Polym. Sci.*, **14**, No.2, 1996, 172-82.
24 Dupraz A M P, de Wijn J R, v. d. Meer S A T, de Groot K, *J. Biomed. Mat. Res.*, **30**, No.2, 1996, 231-8.
25 Cohen Addad J P, Euradh '94. Conference Proceedings, Mulhouse, 12th-15th Sept.1994, 25-30.
26 Montsinger L V,Antec '96. Volume II. Conference proceedings, Indianapolis, 5th-10th May 1996, 2546-9.
27 Gassan J, Bledzki A K, Antec '96. Volume II. Conference proceedings, Indianapolis, 5th-10th May 1996, p.2552-7.
28 Gorl U, Rausch R, Esch H, Kuhlmann R, *Int. Polym. Sci. Technol.*, **23**, No.7, 1996, T/81-7.
29 Dubnikova I L, Gorokhova E V, Gorenberg A Y, Topolkaraev V A, *Polym. Sci., Ser. A*, **37**, No.9, 1995, 951-8.
30 Gonsalves K E, Carlson G, Chen X, Kumar J, Aranda F, Perez R, Jose-Yacaman M, *J. Mat. Sci. Lett.*, **15**, No.11, 1996, 948-51.
31 Hojabr S, Boocock J R B, Antec 95. Volume III. Conference proceedings, Boston, Ma., 7th-11th May 1995, 3620-7.
32 Hornsby P R, Watson C L, *J. Mat. Sci.*, **30**, No.21, 1995, 5347-55.
33 Lekatou A, Faidi S E, Lyon S B, Newman R C, *J. Mat. Res.*, **11**, No.5, 1996, 1293-304.
34 Wolff S, *Rubb.Chem.Technol.*, **69**, No.3, 1996, 325-46.
35 Savadori A, Scapin M, Walter R, *Macromol. Symp.*, **108**, 1996, 183-202.
36 Bandyopadhyay S, De P P, Tripathy D K, De S K, *J. Appl. Polym. Sci.*, **61**, No.10, 1996, 1813-20.
37 Soares R F, Leite C A P, Botter W, Galembeck F, *J. Appl. Polym. Sci.*, **60**, No.11, 1996, 2001-6.
38 Borden K A, Wei R C, Manganaro C R, *Plast. Compounding*, **16**, No.5, 1993, 51-5.
39 Khan S A, Baker G L, Colson S, *Chem. of Mat.*, **6**, No.12, 1994, 2359-63.
40 Zumbrum M A, *J. Adhesion*, **46**, Nos.1-4, 1994, 181-96.
41 Levering A W, Te Nijenhuis K, *J. Adhesion*, **45**, No.1-4, 1994, 137-48.
42 Tanaka T, Waki Y, Hamamoto A, Nogami N, Antec '97. Conference proceedings, Toronto, April 1997, 3054-8.
43 Nichols K, Solc J, Shieu F, Antec '93. Conference Proceedings, New Orleans, La., 9th-13th May 1993, Vol. II, 1938-42.
44 Borden K A, Weil R C, Manganaro C R, Antec '93. Conference Proceedings, New Orleans, La., 9th-13th May 1993, Vol. II, 2167-70.
45 Zolotnitsky M, Steinmetz J R, Antec '94. Conference Proceedings, San Francisco, Ca., 1st-5th May 1994, Vol. III, 2756-60.
46 Liptak P, *Int. Polym. Sci. Technol.*, **21**, No.8, 1994, T/50-3.
47 Hill R, Okoroafor E U, *Composites*, **25**, No.10, 1994, 913-6.
48 Zaborski M, Slusarski L, Vidal A, *Int. Polym. Sci. Technol.*, **20**, No.11, 1993, T/99-104.
49 Shiroki Y, *Int. Polym. Sci. Technol.*, **20**, No.6, 1993, T/12-21.
50 Kominar V, Narkis M, Siegmann A, Breuer O, *Sci. & Engng. Composite Materials*, **3**, No.1, 1994, 61-6.

51 Grady B P, Antec '97. Conference proceedings, Toronto, April 1997, 2490-3.
52 Davis R M, Gardner S H, Marand E, Laot C, Reifsnider K, DeSmidt H, McGrath J, Tan B, Antec '97. Conference proceedings, Toronto, April 1997, 2494-9.
53 Reis M J, Do Rego A M B, Da Silva J D L, *J. Mat. Sci.*, **30**, No.1, 1995, 118-26.
54 Borden K A, Weil R C, Manganaro C R, Antec '94. Conference Proceedings, San Francisco, Ca., 1st-5th May 1994, Vol. III, 2761-5.
55 Hahn G E, Conference proceedings, Addcon '96, Brussels, 1996.
56 Hollis R D, Hyche K W, Color and Appearance Retec: Effects in Plastics. Conference proceedings, Oak Brook, Il., 20th-22nd Sept.1994, 94-105.
57 Foster J K, Sims E S, Venable S W, *Paint & Ink Int.*, **8**, No.3, 1995, 18-21.
58 Liphard M, Von Rybinski W, Schreck B, *Prog. Coll. & Polym. Sci.*, **95**, 1994, 168-74.
59 Jesionowski T, Krysztafkiewicz A, *Pigment Resin Technol.*, **25**, No.3, 1996, 4-14.
60 Sergeeva L M, Slinchenko E A, Brovko A A, Fainleib A M, Nedashkovskaya N S, *Polym. Sci., Ser. B*, **38**, Nos.5/6, 1996, 225-30.
61 Okoroafor M O, Wang A, Bhattacharjee D, Cikut L, Haworth G J, Polyurethanes '95. Conference Proceedings, Chicago, Il., 26th-29th Sept.1995, 303-9.
62 Levchik G F, Levchik S V, Lesnikovich A I, *Polym. Degradat. Stabil.*, **54**, Nos 2-3, 1996, 361-3.
63 Toure B, Lopez Cuesta J M, Gaudon P, Benhassaine A, Crespy A, *Polym. Degradat. Stabil.*, **53**, No.3, 1996, 371-9.
64 Herbert M J, Flame Retardants '96. Conference proceedings, London, 17th-18th Jan.1996, 157-72.
65 Smith R, Utevskii L, Muskatel M, Finberg I, Scheinart Y, Georlette P, Flame Retardants '96. Conference proceedings, London, 17th-18th Jan.1996, 79-90.
66 Pape P G, Romenesko D J, Antec '97. Conference proceedings, Toronto, April 1997, 2941-52.
67 Le Bras M, Bourbigot S, *Fire & Mat.*, **20**, No.1, 1996, 39-49.
68 Baggaley R G, Hornsby P R, Yahya R, Cussak P A, Monk A W, *Fire Mater.*, **21**, 1997, 179-85.
69 Yu Long, Shanks R A, *J. Appl. Polym. Sci.*, **61**, No.11, 1996, 1877-85.
70 Ou Y C, Zhu J, Feng Y P, *J. Appl. Polym. Sci.*, **59**, No.2, 1996, 287-94.
71 Pukanszky B, Maurer F H J, Boode J W, *Polym. Engng. Sci.*, **35**, No.24, 1995, 1962-71.
72 Yu T C, Antec '95. Vol. II. Conference Proceedings, Boston, Ma., 7th-11th May 1995, 2358-68.
73 Wiebking H E, Antec 95. Volume III. Conference proceedings, Boston, Ma., 7th-11th May 1995, 4112-6.
74 Wiebking H E, *J. Vinyl and Additive Technol.*, **2**, No.3, 1996, 187-9.
75 Stricker F, Muelhaupt R, *J. Appl. Polym. Sci.*, **62**, No.11, 1996, 1799-806.
76 Jancar J, Dibenedetto A T, *J. Mat. Sci.*, **30**, No.9, 1995, 2438-45.
77 Jancar J, Dibenedetto A T, *J. Mat. Sci.*, **30**, No.6, 1995, 1601-8.
78 Ventresca D A, Berard M T, Antec '97. Conference proceedings, Toronto, April 1997, 3574-9.
79 Kikkawa K, *Polym. Degradat. Stabil.*, **49**, No.1, 1995, 135-43.
80 Tino J, Mach P, Hlouskova Z, Chodak I, *J. Macromol. Sci. A*, **A31**, No.10, 1994, 1481-7.
81 Hu X, Xu H, Zhang Z, *Polym. Degradat. Stabil.*, **43**, No.2, 1994, 225-8.
82 Allen N S, Edge M, Corrales T, Childs A, Liauw C, Catalina F, Peinado C, Minihan A, *Polym. Degradat. Stabil.*, **56**, 1997, 125-39.
83 Haacke G, Polymer Stabilizers and Modifiers Conference '97, Hilton Head Island, 1997.
84 Bataille P, Mahlous M, Schreiber H P, *Polym. Engng. Sci.*, **34**, No.12, 1994, 981-5.
85 Suri A, Min K, Antec '97. Conference proceedings, Toronto, April 1997, 1487-91.
86 Tang L-G, Kardos J L, *Polym. Composites*, **18**, No.1, 1997, 100-13.
87 Saad A L G, Younan A F, *Polym. Degradat. Stabil.*, **50**, No.2, 1995, 133-40.
88 Srivastava V K, Pathak J P, *Polym. & Polym. Composites*, **3**, No.6, 1995, 411-4.
89 Janigova I, Chodak I, *Eur. Polym. J.*, **31**, No.3, 1995, 271-4.
90 Janigova I, Chodak I, *Eur. Polym. J.*, **30**, No.10, 1994, 1105-10.
91 Zhang C-Z, Liu W-M, Xue Q-J, Shen W-C, *J. Appl. Polym. Sci.*, **66**, 1997, 85-93.
92 Ismail H, Freakley P K, Sutherland I, Sheng E, *Eur. Polym. J.*, **31**, No.11, 1995, 1109-17.
93 Ismail H, Freakley P K, Sheng E, *Eur. Polym. J.*, **31**, No.11, 1995, 1049-56.
94 Miano F, Rabaioli M R, *Coll. & Surfaces*, **84**, Nos.2/3, 1994, 229-37.
95 Cochet P, Barruel P, Barriquand L, Grobert J, Bomal Y, Prat E, IRC '93/144th Meeting, Fall 1993. Conference Proceedings, Orlando, Fl., 26th-29th Oct.1993, Paper 162.
96 Mlecnik E, La Mantia F P, *J. Appl. Polym. Sci.*, **65**, 1997, 2761-72.
97 Brown N, Linnert E, *Reinf. Plast.*, **39**, No.11, 1995, 34-7.

# 14

# Testing Methods in Filled Systems

This chapter contains a discussion of the following subjects related to the analysis of filled systems:

- The reasons for the use of a particular testing method
- The procedures used for testing
- Available standard methods
- Major findings by various analytical techniques.

The testing methods are divided into two groups: physical methods and chemical and instrumental analysis. Within the group, the testing methods are discussed in alphabetic order according to the same pattern.

## 14.1 PHYSICAL METHODS

### 14.1.1 ATOMIC FORCE MICROSCOPY[1-3]

*Applications*: Atomic force microscopy is an efficient tool for the analysis of surface topography and also surface mechanical properties such as friction and elasticity. It can be used to reveal molecular or atomic order with spatial resolution higher than that of electron microscopy. Hard and flat surfaces can be observed without difficulties. Soft and rough samples may produce image artifacts. The following uses are known in filled systems: determination of surface morphology and nanostructure of high modulus polyethylene fibers,[1] structure and morphology of small diameter (13 μm) aramid fibers,[2] and carbon black aggregates.[3]

*Testing procedure*: Probing repulsive force variations between a very small probe (usually apex radius smaller than 10 nm) and a sample, one can observe surface morphology. Nanoscope from Digital Instruments, Inc., Santa Barbara, California is an instrument chosen in reported applications. The microscope stage allows for sample rotation. Several different probes are used (probes for high magnification of atomic order differ from the probes for macroscopic examinations). Also, different software is available. Samples can be observed in air, but spatial resolution can be enhanced and surface damage avoided by observing samples under water. Surface damage can also be limited by scanning in constant force mode. Sample preparation is very important. A microscope cover slip glass is a frequent substrate used. This glass is cut to smaller sections, for example, 3 × 3 mm. The sample is deposited on the wet glass piece and dried under vacuum at room temperature.[2] A more complex procedure was used in preparation of carbon black aggregates for observation.[3] Carbon black (20 mg) was ultrasonically dispersed in chloroform (40 ml), then a few microliters deposited on glass pieces. In order to obtain a uniform distribution and density of aggregates, the suspension was also deposited by two or three consecutive deposits in the same location. It was also noticed that the surface of the glass support has an important effect on uniformity of distribution. A gold-coated surface gave a more uniform distribution than straight glass and was the substrate of choice for carbon black samples.[3]

*Standard methods*: none

*Major results*: The smallest nanofibrils detected had a width of 10-15 nm (even smaller fibers of 5-8 nm can be seen).[1] The vertical resolution of atomic force microscopy is superior to scanning electron microscopy; therefore, peculiarities of fiber morphology can be easily observed. The results of fiber diameter measurement are in agreement with X-ray diffraction experiments. Roughness parameters of aramid fibers can be determined, which help in improving process conditions.[2] Carbon black aggregates had heights of 50 to 150 nm, which are higher than detected by TEM (20 nm).[3] It is postulated that the understanding of structure of carbon black should be reviewed after broader data on carbon black become available.[3] It is possible that the present understanding is based on artifacts related to the previously used methods such as SEM and TEM.

### 14.1.2 AUTOIGNITION TEST[4]

*Applications*: Determination of specimen autoignition.

*Testing procedure*: Temperature of specimen is kept constant at 430°C in the presence of air, and time to autoignition is measured. In another version, the oxygen atmosphere is modified, and the temperature of autoignition is recorded.

*Standard methods*: ASTM D 1929

*Major results*: Figure 12.3 shows that talc and $Mg(OH)_2$ increase ignition time with increasing concentration of filler.

### 14.1.3 BOUND RUBBER[5]

*Applications*: Bound rubber is a measure of filler surface activity to the matrix, and it is considered as a factor in the estimation of filler reinforcement.

*Testing procedure*: Small pieces of uncured rubber are immersed for several days at room temperature in a large excess of good solvent such as toluene. The sample in contact with solvent becomes divided into three parts: polymer solution, $m_{pI}$, solvent-dispersed filler particles with absorbed polymer chains, $m_{pII}$, and solvent-swollen gel of filler particles connected through polymer chains, $m_{pIII}$. The fraction of polymer bound to filler is determined from the equation: $B = \left(m_{pII} + m_{pIII}\right) / m_p$ where $m_p$ is total mass of polymer. The fraction of polymer not dispersed by solvent is given by the following equation: $G = m_{pIII} / m_p$.

*Standard methods*: NF T 45-114

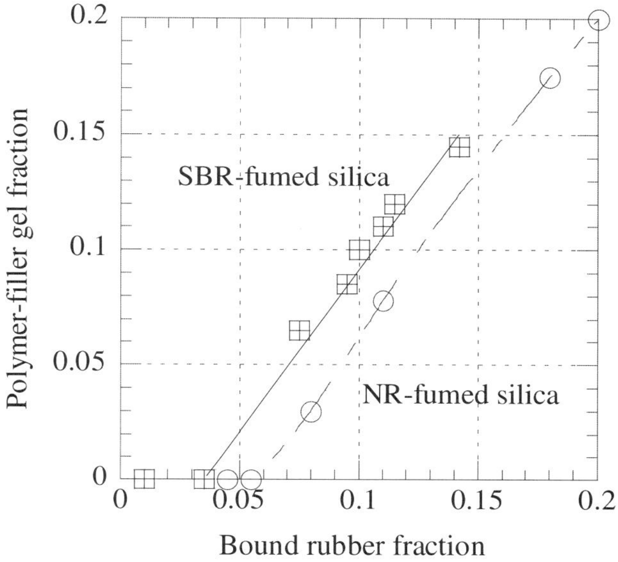


Figure 14.1. Polymer-filler gel vs. bound rubber fraction. [Adapted, by permission, from Karasek L, Sumita M, *J. Mat. Sci.*, **31**, No.2, 1996, 281-9.]

*Major results*: Figure 14.1 shows that at smaller concentrations of filler, there is only absorption of polymer on solvent-dispersed filler. When concentration increases, the gel is formed. Higher molecular weight polymer is bound preferentially. There are many factors influencing bound rubber measurement.[5] With regard to filler, these include concentration, size of aggregates, surface area, and surface activity (functional groups, free radicals from aggregate breakdown, graphitization, coupling agents, hydrophobic agents, surface active agents). Elastomer affects bound rubber determination depending on chemical composition, unsaturations, stability (thermal, mechano-chemical, and oxidative). Several additives affect the determination of bound rubber. These include free radical terminators, promoters, coupling agents, and surface active agents. Other parameters include mixing variables and temperature of extraction. With this variability, results are frequently questionable.

### 14.1.4 CHAR FORMATION[6]

*Applications*: The method is used to determine the effect of formulation variables on char formation and the stability of char to further pyrolysis.

*Testing procedure*: A tube furnace is 1 m long and has a 7.2 cm diameter. At one end of the tube, a forced air flow is supplied. The furnace temperature is measured by a thermocouple located 30 cm from the end of air supply. The temperature is constant within ±2°C and is kept at 600°C. The sample is placed in a holder made from a ceramic material with a tungsten foil lining. In the oxygen-free mode, no air flows in the furnace, but a dry air is supplied at the rate of 1 $m^3$/min in the oxygen-

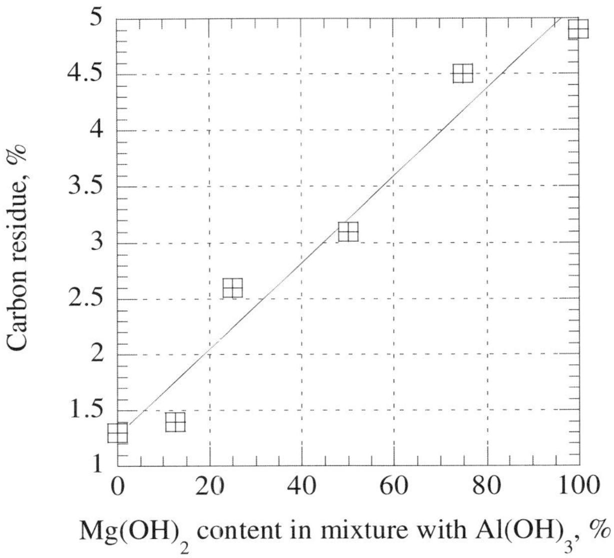


Figure 14.2. Carbon residue remaining from the decomposition of EPDM at 600°C for 5 min vs. percentage of $Mg(OH)_2$ in the mixture with $Al(OH)_3$. [Data from Levesque J L, Fraval J T, Antec '93. Conference Proceedings, New Orleans, La., 9th-13th May 1993, Vol. II, 1957-64.]

ated mode. Typically, the oxygen-free mode was used because of better repeatability.[6] The samples (1 × 4 × 0.2 cm) are placed on wire grid, left in the oven for a specific period of time, removed, immersed in liquid nitrogen to stop the combustion process, and carbon residue determined.

*Standard methods*: AS 2122

*Major results*: Figure 14.2 shows the results of determination of char formation in EPDM filled with a mixture of $Al(OH)_3$ and $Mg(OH)_2$. $Mg(OH)_2$ is more efficient in char formation.

### 14.1.5 CONE CALORIMETRY[4,7-11]

*Applications*: A cone calorimeter, named after a truncated cone shape of the furnace, is a heat release rate calorimeter which permits the determination of heat release under controlled conditions. It determines the critical fire parameters required for a range of natural and synthetic materials using small samples (100 $mm^2$), and simple materials, and is used for composites and combinations of different materials. This apparatus allows simultaneous and continuous determinations of heat release rate, smoke production rate, mass loss rate, concentration of the various combustion gases formed, ignitability, heat of combustion and soot production data for the materials tested.

*Testing procedure*: The equipment was developed by NIST and is produced by PL Thermal Sciences, 300 Washington Blvd., Mundelein, IL 60060, USA. Testing according to ISO 5660 requires a heat flux of 25 $kW/m^2$ and an air flow 24 l/s. The heat released is calculated based on oxygen depletion due to combustion. The mass of the specimen, the rate of mass loss, the time to ignition, the rate of heat release and its maximum value are calculated for the first 3 min of burning by a supplied software. The instrument also calculates specific extinction area, which is the increase of extinction area of the smoke divided by the unit mass of material decomposing to volatile products.

*Standard methods*: ASTM E 1740, BS 476:Part 1, ISO 5660

*Major results*: Due to the computerized determination of many parameters of material combustion and the well controlled process of combustion, this instrument is frequently used in studies of materials containing fillers.

### 14.1.6 CONTACT ANGLE[12-25]

*Applications*: The following uses of contact angle were reported in the literature: surface energy of different sizes for fibers,[12] correlation between contact angle of fiber and interlaminar shear strength of composite,[12] effect of surface treatment of fillers for paints,[13] the matrix-filler adhesion parameter for PS filled with $CaCO_3$,[14] dispersion stability of PEO-modified kaolin particles,[16] determination of contact angle of carbon fibers and its dependence on treatment,[17] wettability of fiber surfaces,[20] effect of fillers on surface free energy of paper coatings,[16] cleanliness of fibers,[22] wetting of filler by polymer,[23] and adhesion between layers.[25]

*Testing procedure*: Several methods are used to determine contact angle. The method of choice depends on the sample type. The most common method used for larger, flat samples is a goniometer method. A water droplet is placed on the material surface and contact angle determined either by optical methods (older instruments) or by the use of software which analyzes the image and calculates the contact angle. The precision of determination depends on sample preparation and time of measurement. Samples with surface roughness and imperfections give confusing results. Also, materials which are rapidly wetted by the liquid used give results of very low precision. The availability of a great number of instruments of different levels of computerization makes a precise determination of contact angle possible. Particulate fillers and fibers are difficult to measure because of their small dimensions. In the case of fibers, the tensiometer method gives precise measurements.[12] One end of the fiber is hung vertically from the arm of an electromicrobalance, and the other end is immersed into the liquid. From the Fowkes equation, interfacial free energies can be calculated. Goniometer was also successfully used to determine the contact angle of the fiber.[12] A precise method was developed based on photographic images taken under the microscope.[17] The method eliminates errors due to the influence of surface roughness of fiber. In the case of particulate fillers, three options are available: determination of the surface energy by inverse gas chromatography,[18,19] the captive droplet technique,[16] and the filler column method.[14] Inverse gas chromatography is discussed in a separate section below. In the captive droplet technique, a smooth surface is obtained by press-

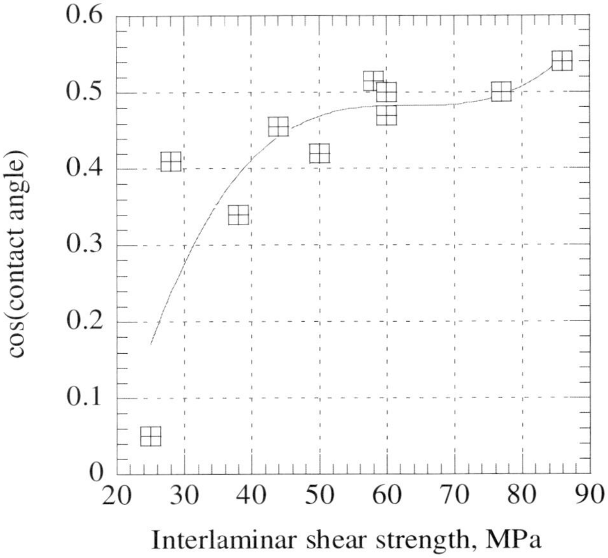


Figure 14.3. Contact angle of carbon fibers vs. interlaminar shear strength of epoxy composites. [Adapted, by permission from Tang L-G, Kardos J L, *Polym. Composites*, **18**, No.1, 1997, 100-13.]

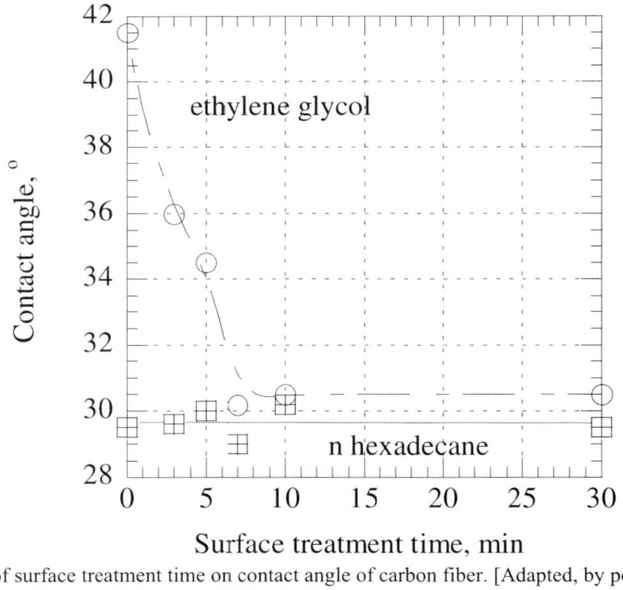


Figure 14.4. Effect of surface treatment time on contact angle of carbon fiber. [Adapted, by permission, from Ogawa T, Ikeda M, *J. Adhesion*, **43**, Nos.1-2, 1993, 69-78.]

ing a ground sample with pressure of 8 tones in a KBr die. A microsyringe deposits a droplet of water on the surface of a sample placed under optical microscope. The

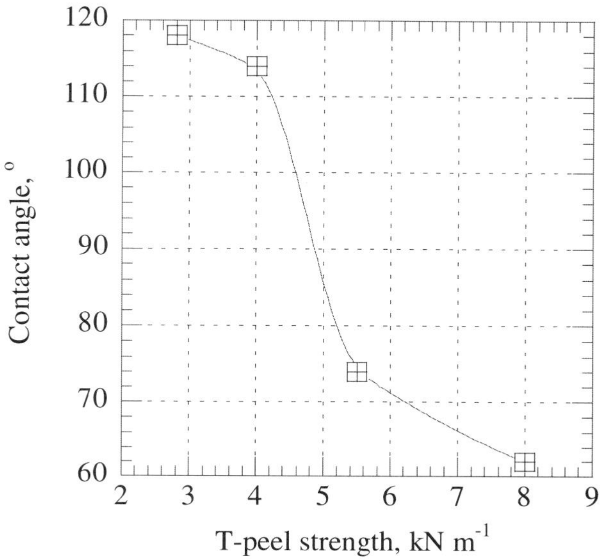


Figure 14.5 Contact angle vs. peel strength of SBR/polyurethane joints. [Adapted, by permission, from Torro-Palau A, Fernandez-Garcia J C, Orgiles-Barcelo A C, Martin-Martinez J M, *J. Adhesion*, **57**, Nos.1-4, 1996, 203-25.]

photograph of the droplet is taken and analyzed.[16] In the filler column method, contact angle is determined from liquid capillary rise.[14]

*Standard methods*: none

*Major results*: Figure 14.3 shows the effect of contact angle of carbon fibers on interlaminar shear strength of epoxy composites. An increase in contact angle of carbon fiber (e.g., increase in functional groups) results in an increase of interlaminar shear strength.[12] Figure 14.4 shows that the disperse component (relative to n-hexadecane) does not change with treatment time, whereas the polar component (relative to ethylene glycol) changes with time. The diminishing contact angle indicates that surface free energy increases, which is expected from the oxidation of carbon fiber.[17] Figure 14.5 shows that with a decrease of contact angle, peel strength of rubber polyurethane joints increases. The samples of rubber differ in silica content. The higher the silica concentration, the lower the contact angle and the higher the adhesion.[25] This is in correlation with SEM studies showing increasing surface roughening with increased addition of silica. The increase in surface roughness increases mechanical adhesion and thus improved peel strength.

### 14.1.7 DISPERSING AGENT REQUIREMENT[26]

*Applications*: The dispersing agent requirement is the minimum amount of dispersing agent required to produce a fluid dispersion of a particular carbon black loading. It is expressed as a weight percent of dispersant per total weight of carbon black. The dispersing agent requirement is an important parameter because a reduced amount of dispersing agent may cause instability of dispersion, whereas too large an amount of dispersing agent may restrict carbon black loading level and reduce weather stability.[26]

*Testing procedure*: Water is added to a blender, followed by ammonium hydroxide to adjust pH to the required level. Carbon black is added in 30 s with low speed mixing. Dispersing agent is added in portions from a burette under high speed dispersion. The endpoint is determined as 5 min of uninterrupted fluidity under high speed mixing. The total time of determination should not exceed 10 min.

*Standard methods*: The method is described in Cabot Corporation Technical Report S-131.

### 14.1.8 DISPERSION TESTS[27-30]

*Applications*: Considering that many properties of filled materials depend on the quality of dispersion, the reasons for conducting dispersion tests are self explanatory.

*Testing procedure*: There is no standard procedure applicable to all cases of filler dispersion because so many methods of processing are used for very diverse groups of materials. Some examples of tests used are given below. Carbon black is added to polymers processed by melting techniques for various reasons such as reinforcement, pigment, UV absorber, conducting additive, etc. Two tests are applicable to monitoring carbon black dispersion in these applications.[27] In one method, a pelletized masterbatch was processed in a Brabender Plasticorder PL-2000 single screw extruder equipped with a set of two screens ($120 \times 400$ mesh which has a 45 μm rating and a second supporting screen of 24 mesh). A pressure transducer measured pressure versus time. The endpoint was either 400 s or a pressure exceeding 2000 psi. The results of the test were compared to a visual method in which a film extruded through a ribbon die, having dimensions of $75 \times 0.24$ mm, was used. The number of imperfections were correlated with the Plasticorder test. The distribution of fibers in a composite is measured by this method.[28] The radiation passes through the sample and a radiograph records the intensity of radiation transmitted through the object. The radiograph is then digitized to measure the intensity of each pixel, from which the distribution of each gray level within the image is determined. Based on the distribution of gray levels, it is possible to measure the local and average glass content. Other fillers such as carbon fiber were used with good results. Surface mapping was used to determine the degree of mixing of aluminum particles in polybutadiene.[29] The measurements were done by electron microscopy coupled with energy dispersive X-ray. In addition to filler dispersion, the preferred orientation was also determined.[29] In carbon black dispersions, four main methods are used: optical, microwave, electric conductivity, and dark field cut surface.[30] In the optical method, the image reflected from the cut surface is compared to standards and the nearest standard selected. The Dielecmeter (Sairem, France) is used in the microwave energy absorption method. The value of the dielectric constant depends

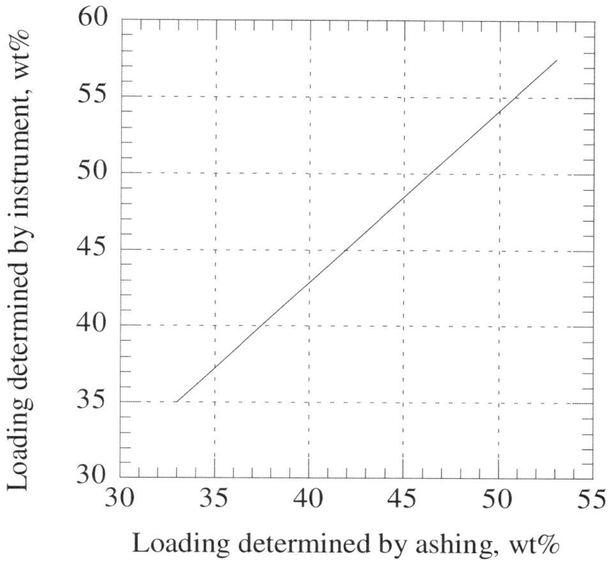


Figure 14.6. Glass loading determined by two methods. [Adapted, by permission, from Scott D M, *Composites, Part A*, **28A**, 1997, 703-7.]

on carbon black dispersion. Used in the electric conductivity method the Tangent Electroscanner (Tangent Scientific Instruments, Dublin, Ireland) has a measuring transducer with electrodes which induce alternating current in uncured rubber. The electrodes scan the surface of samples. The dark field cut surface method measures the light scattered from a cut surface, which depends on carbon black dispersion. Each method used for carbon black has some interference either related to the matrix in which carbon black is dispersed or to the conditions of running the experiment. This makes data difficult to compare.

*Standard methods*: ASTM D 2663, ISO 8780, ISO 8781, ISO 11420, ISO 13549

*Major results*: Figure 14.6 compares results of radiographic imaging with the ashing method. The data indicate that the method offers precise results.[28]

### 14.1.9 DRIPPING TEST[31]

*Applications*: The method is used to determine burning drops produced during material decomposition.

*Testing procedure*: The ignition source is a horizontal electric radiator producing a radiation intensity of 3 W/cm$^2$, positioned 30 mm above the specimen placed on a wire grid. 300 mm below the specimen, cotton wool is placed on receptacle for drops. If the specimen ignites during the first 5 min, the radiator is removed 3 s after ignition. During the second 5 min, the radiator is kept on regardless of whether the specimen burns or not. There are four classes - M.1 to M.4. If during the duration of test (10 min) the cotton wool ignites, the material is classified in class M.4. The assignment to the other classes depends on the amount of residues.

*Standard methods*: French specification NF P 92505
*Major results*: Huntite and hydromagnesite mixture at loading of minimum 25% in ethylene-propylene copolymers eliminates inflammation.[31]

### 14.1.10 DYNAMIC MECHANICAL ANALYSIS[32-34]

*Applications*: General method also used in filler applications. The results give information on mobility of molecules in the presence of filler, changes in the structure of the matrix due to interaction with filler, the effect of fillers on matrix degradation, microphase separation, and other related phenomena.
*Testing procedure*: DMA is mostly used as a scientific tool; therefore the testing procedure is selected depending on the requirements of the studies conducted.
*Standard methods*: none

### 14.1.11 ELECTRIC CONSTANTS DETERMINATION[35-42]

*Applications*: Fillers are essential components of formulations of materials which must have either electric conductivity, or high electric resistance, or EMI shielding capability. The addition of fillers to these compounds requires adequate methods of control.
*Testing procedures*:
ELECTRIC PROPERTIES
Dielectric properties. Dielectric constant and dissipation factor can be measured using a HP 4284A LCR meter with a HP 1645B test fixture.[37] The frequency range is 20 Hz to 1 MHZ. The dissipation factor is a measure of energy lost during the reversal of electric polarization. It is expressed as a fractional energy loss. The impedance bridge method is used to measure the dissipation factor, $\tan\delta$, dielectric constant (permittivity), $\varepsilon'$, and dielectric loss factor, $\varepsilon''$. The configuration of electrodes and specimen painting with conductive paint is similar to the measurement of surface resistivity discussed below. The DC transient current method is an alternative method for the measurement of the dissipation factor and dielectric constants. In this method, a sample is charged with a constant voltage (e.g., 2 kV) for a long duration (e.g., 170 min). The specimen is then discharged. The charging and discharging currents are measured in short intervals (e.g., 5 s), and required constants are calculated.[41]
Electric resistivity. The electric resistance measurement is the same as discussed below under volume resistivity, to which this measurement is temporary adapted. For flexible materials, special electrode systems are developed to clamp sample and electric wires.[35] The measuring equipment is based on a Wheatstone bridge circuit. The conductivity of metal powder-containing epoxy was measured in special dies equipped with built-in brass electrodes inserted to the die. The material was cured in the die to assure good contact with electrodes.[36] Special sample holders and clamping devices are used for precise determination of rubber compounds containing carbon black.[39]

ELECTROSTATIC APPLICATIONS

Volume resistivity, surface resistivity, and charge decay time are major characteristics of electrostatic properties of materials. Volume resistivity, expressed in ohm-cm, is the resistivity of material measured on opposite ends of a material which is 1 cm thick. Surface resistivity, expressed in ohm, is a resistance between two electrodes placed along the same surface of the specimen. The charge decay time, expressed in s, is defined as the time needed to dissipate a certain percentage of charge induced on the surface of material. Other terms used include shielding effectiveness, decay half-time, peak voltage, and a percentage of charge retained. Shielding effectiveness, expressed in decibels, is a measure of attenuation of electromagnetic interference (EMI) by internal reflection, absorption, and partial reflection. Decay half-time, expressed in s, is the time to dissipate half the charge induced. Peak voltage, expressed in volts, is the maximum induced voltage in the charge decay test. The percentage of charge retained is a percentage of charge remaining in charge decay test after 500 s. The most frequently used methods of determining these quantities are characterized below. The respective standards are given in the next section.

Charge decay time. A specimen is placed in Faraday cage with electrodes on each side of the specimen. One electrode induces the charge, the other electrode measures changes in electric field. In this measurement, charge decay time, decay half-time, peak voltage, and the percentage of voltage remaining after 500 s are determined.

Surface resistivity. One side of the specimen is coated with a circle of silver paint surrounded by a ring of silver paint. The uncoated distance between the circle and the ring is an effective length on which surface resistivity is measured. The other surface of the specimen is fully coated with silver paint. Current and voltage are measured and surface resistivity calculated. If samples contain internal or external antistatics, the measurement is performed under a controlled atmosphere to eliminate the influence of temperature and relative humidity. Also, specimen conditioning is used to account for migration of the antistatic to the surface. The surface of specimen containing antistatics is not coated with silver paint, but electrodes are pressed to the surface. The resistivity of conductive pipes is determined in a special test arrangement.[38]

Volume resistivity. There are some differences between standard methods but the general principle is similar. A specimen of standard size is coated with a silver paint on the opposite surfaces to assure good surface conductivity. Two electrodes are attached to both surfaces in the manner minimizing contact resistance. The voltage applied depends on the expected resistance of the specimen and is in the range of 0.1 to 1000 v/mm thickness. Current and voltage are measured between the faces and volume resistance calculated. The edge effect can be minimized by means of guard electrodes.[37]

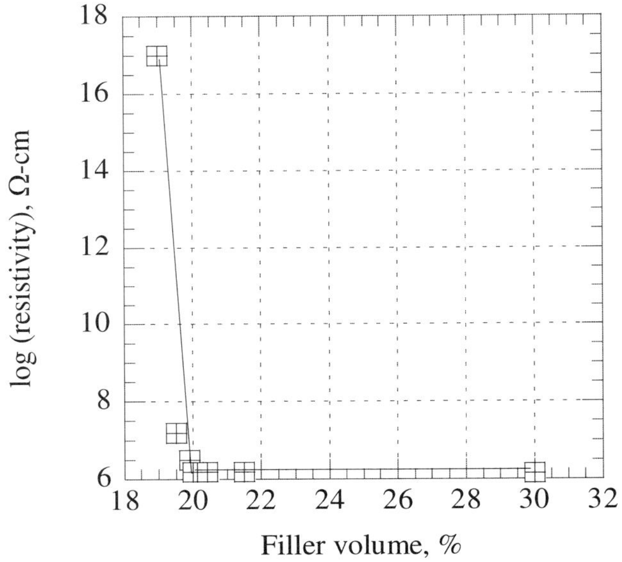


Figure 14.7. Resistivity of aluminum powder filled PMMA. [Adapted, by permission from Lei Yang, Schruben D L, *Polym. Engng. Sci.*, **34**, No.14, 1994, 1109-14.]

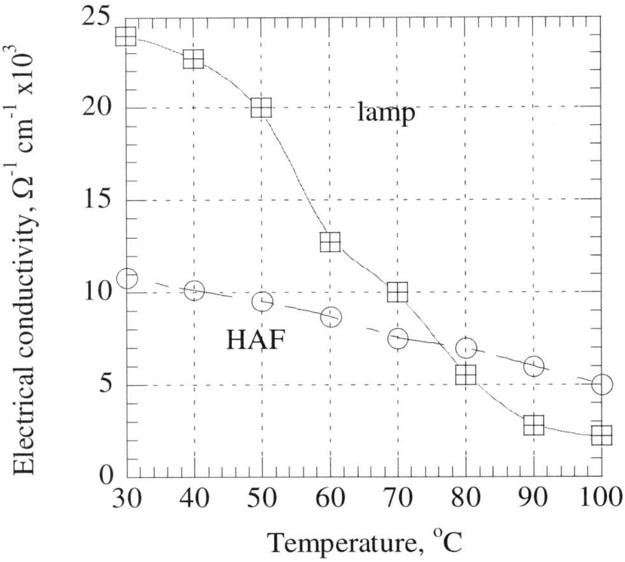


Figure 14.8. Conductivity of butyl rubber filled with carbon black vs. temperature. [Data from Nasr G M, Badawy M M, Gwaily S E, Shash N M, Hassan H H, *Polym. Degradat. Stabil.*, **48**, No.2, 1995, 237-41.]

*Standard methods*: electric resistance - ASTM D 257, shielding effectiveness - ASTM ES 7, BS 6667 (part 1 and 2), static decay - ASTM F 365, BS 2783 (part 2), surface resistivity - ASTM D 25 (specimens containing antistatics), ASTM D 257, ASTM F 1173 (pipe resistivity), volume resistivity - ASTM D 257, DIN 53596, ISO 3915

*Major results*: Figure 14.7 shows that the resistivity of aluminum-filled PMMA changes abruptly. Smaller volumes of filler contribute a little to resistivity but, after certain threshold value of filler concentration, further additions have little contribution. A similar relationship was obtained for nickel powder; the only difference is in the final value of resistivity, which was lower for nickel due to its higher conductivity.[35] The same conclusions can be obtained from conductivity determinations of epoxy resins filled with copper and nickel.[36] Figure 14.8 shows the effect of temperature on the electric conductivity of butyl rubber filled with different grades of carbon black. In both cases, conductivity decreases with temperature, but lamp black is substantially more sensitive to temperature changes.[39] Even more pronounced changes with temperature were detected for the dielectric loss factor and dissipation factor for mineral filled epoxy.[41]

### 14.1.12 ELECTRON MICROSCOPY[43-59]

*Applications*: Electron microscopy as a general tool can be used in numerous applications in filled systems. The most frequent applications include: estimation of adhesion of fibers to the matrix in composites,[44] fracture mechanism of bone cements,[46] histological changes of bone cements and evaluation of the bone cement interface with glass fiber,[47] morphology of filler depleted layers of paint near the substrate,[48] morphology of fillers,[49] filler dispersion and distribution in the matrix,[51,53] filler aggregation after dispersion,[52,53,57] the effect of surface treatment on filler adhesion,[54] the effect of processing methods and flow patterns on filler distribution,[55] the effect of morphology and filler distribution on pyrolysis,[58] and structure of nanocomposites.[59] This broad list of applications shows the capabilities of the method in offering essential information on filled systems.

*Testing procedure*: SEM specimens are usually prepared for examination by prior sputtering with Au or Au-Pd alloys.[44,49] To enhance information, X-ray microanalyzers are used in combination to determine intensities of silicon, titanium, zinc, copper, calcium, nitrogen and phosphorus.[47,57,59] In TEM studies, electron energy loss spectra and element spectroscopic images are obtained, and the 3-window method is used for elemental studies.[57] Specimens are microtomed in a manner that does not distort the actual distribution of sample components. For example, fractured samples are microtomed below the fractured surface, left to relax, and then sputter-coated. Materials are frequently fractured under liquid nitrogen to obtain surfaces for filler distribution observation. In TEM studies, relatively high acceleration voltage is used (80 KeV), but samples are viewed at 120K to prevent degradation.[57] The highest magnification was used in TEM studies of nanocomposites where the smallest particle size of 2.9 nm was detected.[59] This analysis shows that SEM and TEM are used in a fairly conventional manner for the analysis of filled specimens.

*Standard methods*: not applicable

*Major results*: The most interesting interpretation of SEM data regards technology of blend preparation, where electron microscopy permits monitoring of the stages of the process, and the effect of fillers on blend morphology, as well as following the effect of order of component addition on the properties of resultant material.[53] Application of SEM to industrial processing of filled plastics by different methods allows the understanding of the material flow pattern due to the orientation of particles of different sizes.[55] Combination of SEM or TEM with XPS is a very powerful technique. In paints, the cross-sectional distribution of filler in the proximity of coated substrate can be determined.[48] In nanocomposites, not only can particle sizes be measured, but also the chemical structure and morphology of these small particles obtained from mixed components can be determined.[59]

## 14.1.13 FIBER ORIENTATION[60-62]

*Applications*: Fiber orientation in composites or materials processed by other methods depends on processing parameters, characteristics of the fibers, and external stress applied to formed materials. Fiber orientation, on the other hand, affects mechanical performance and electric and thermal conductivity of fiber-containing material. This outlines the reasons for interest in fiber orientation determination.

*Testing procedure*: Microradiography is one method which allows the determination of fiber orientation.[60] Sample with a frozen-in fiber orientation is ground with abrasive paper of progressively finer grit sizes to 0.1-0.25 mm thickness. Soft x-ray white radiation was used to take images of microradiographs. These images were then processed by graphic software such as Adobe Photoshop and orientation determined by digitizing, followed by the determination of orientation function by numerical integration. In another study, orientation distribution function was determined according to a procedure discussed elsewhere.[62]

*Standard methods*: none

*Major results*: Figure 14.9 shows the fiber distribution angle for samples of glass fiber filled polyamide subjected to different levels of strain.[60] Micrographs (not included here) show that within this range of strains, fibers assumed orientations from the totally random (at $\varepsilon = 0$) to perfectly oriented (at $\varepsilon = 2.75$), which is very well reflected by the results of fiber orientation obtained from microradiographic studies.

## 14.1.14 FLAME PROPAGATION TEST[31,63]

*Applications*: The flame propagation test is used to classify materials into four categories from M.1 (nonflammable) to M.4 (highly flammable).[31] In aerospace applications, NASA uses the upward flame propagation test. This test simulates the beginning of a fire with a medium incident heat flux.[63]

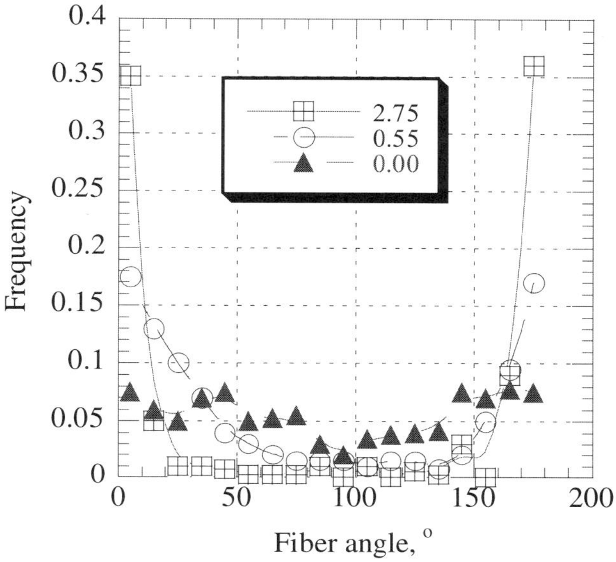


Figure 14.9. Fiber distribution at different strains. [Adapted, by permission, from Wagner A H, Kalyon D M, Yazici R, Fiske T J, Antec '97. Conference proceedings, Toronto, April 1997, 996-1000.]

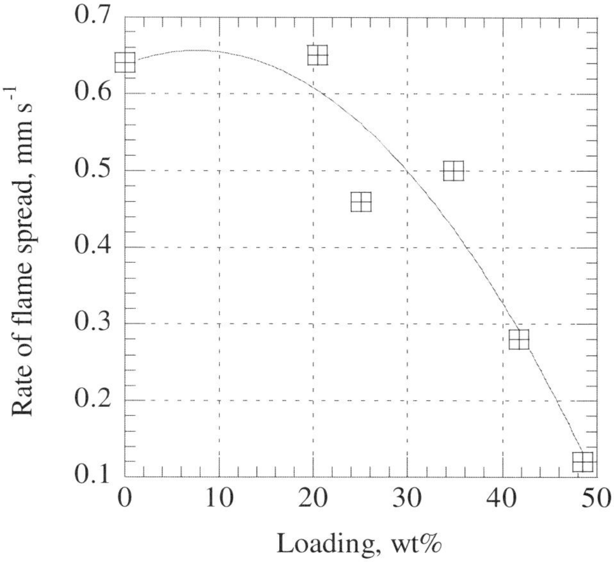


Figure 14.10. Rate of flame spread vs. loading of huntite/hydromagnesite filler in ethylene-propylene copolymer. [Data from Toure B, Lopez-Cuesta J, Benhassaine A, Crespy A, *Int. J. Polym. Analysis and Characterization*, , No.3, 1996, 193-202.]

*Testing procedure*: In the flame propagation test, the sample is clamped horizontally. Two marks are placed on the sample at a distance of 250 mm. A 50 mm space

is provided at the end of the sample for initiation of burning by a Bunsen burner. A time required to reach the second mark is determined, and from this time the rate of flame spread is calculated, expressed in mm/min.[31] The NASA upward propagation test is conducted in an atmosphere containing 30% oxygen. A sample 30.5 cm long is mounted vertically in a spring-clamp fixture. A chemical igniter is geometrically centered 0.64 cm below the front leading edge of the sample. The energy provided by a chemical igniter is 3128 joules and the total burn time of igniter is 25 s, with a flame height of 6.4 cm. The average flame spread rate is calculated by dividing the flame spread length by the flame spread time. The sample fails the test when the flame spread is greater than half the length of the sample.[63]
*Standard methods*: French specification - NF P 92-504; NASA Handbook 8060.1C test 1.
*Major results*: Figure 14.10 shows the effect of huntite/hydromagnesite loading on the rate of flame spread in ethylene-propylene copolymer.[31] A substantial decrease in the rate of flame spread is found when loading approached 50 wt%.

### 14.1.15 GLOW WIRE TEST[4]

*Applications*: Determination of minimum ignition temperature.
*Testing procedure*: The specimen is pressed against a tip consisting of a U-shaped wire coil heated by an electric current. The minimum ignition temperature is measured.
*Standard methods*: AS/NZS 4695, ASTM D 6194, BS EN 60695

### 14.1.16 IMAGE ANALYSIS[64-68]

*Applications*: Image analysis is a useful tool for many purposes of filler testing. Several examples of application exist in the current literature: the study of solid rocket propellant and liners by magnetic resonance imaging,[64] analysis of the behavior of polymers under extension,[65] surface deformation of polymer composites,[66] granulometry of short glass fiber in relationship to processing conditions,[67] and morphology of calcium carbonate filled polypropylene.[68] These and other applications of image analysis permit the transformation of image to numerical parameters, which can then be correlated to other factors of performance.
*Testing procedure*: There is no standard method used in these investigations, since requirements of image analysis must be synchronized with the method and geometry of material testing. At the same time, essential steps are similar, involving image acquisition, conversion of image to digital form, and analysis of the results. The image analysis becomes more complex if kinetic data must be obtained because of the rate with which the image must be captured, in order to free memory for the next image acquisition, and because of the method of further data processing. In magnetic resonance imaging, spin-echo and multislice pulse sequences were used for data acquisition. The resolution of imaging was $70 \times 70\,\mu m$ and image size was $128 \times 128$ pixels. The data for each experiment were based on 48 scans giving a total acquisition time of 1.7 h for spin-echo and 0.85 h for multislice.[64] The method allows

the measurement of loading of filler and its distribution in the material, as well as the detection of voids and imperfections and the visualization of bubbles in the sample. The measuring system for the image analysis of behavior of polymers under large strains includes the following essential elements: camera, video recorder, frame grabber, array processor, time, and PC with large magneto-optical disk.[65] The array processor translates the image from analog to digital. The camera is initially activated by the signal from the tensile testing machine (first image), then subsequently by an electronic trigger which obtains the signal from the timer. A software program written in C language is capable of performing calculations of stress-strain-strain-rate behavior. The specimen for testing has a grid silk-printed on the surface to follow relative displacements of the grid and compare displacements with total draw force acquired from the load cell. This method consists of a measuring system capable of performing the determination without a physical contact with the sample. The use of image analysis in the measurement of scratch resistance is another important application.[66] The measuring system is used to analyze samples which were subjected to scratch testing prior to the optical analysis. The scratch surface is observed under crossed polarizers of an optical microscope at 100× magnification after placing them on a 360° rotating optical stage. The linearly polarized light reflected off the sample surface and a scattered light collected after passing through the crossed polarized analyzer is captured by a camera mounted on the top of the optical tube of the microscope and sent to an analog-to-digital converter. The digitized image for each pixel has gray scale values from 0 (black) to 255 (white) assigned. The result is then used to evaluate surface whitening of materials having different compositions (e.g., concentration of filler). The granulometric characterization of glass fibers for reinforced polypropylene was done by image analysis, which included image acquisition, image treatment, and quantitative analysis.[67] The fibers were dispersed in solvent between two glass slides (about 800 fibers) and observed with a polarizing microscope. Each selected fiber was labeled and captured images digitized. Using morphological tools, the image was filtered to improve quality and transferred into binary images. Software was used to calculate a size factor for each fiber which was then used to determine length distribution expressed by average number or average weight. In a similar application for particulate fillers, SEM images were digitized by an analog-to-digital converter.[68] Using Vidas software, the diameter, shape and orientation of particles were determined. From this review, it is concluded that image analysis methods are a very powerful (and easy to adapt) technology for gathering information on filler morphology in polymer systems. Recent developments in hardware will further contribute to the development of these methods and the understanding of real performance characteristics of fillers.

*Standard methods*: none

*Major results*: Figure 8.40 shows that the void density or whitening of talc filled polypropylene (relative to average scattering and scattering difference) increases

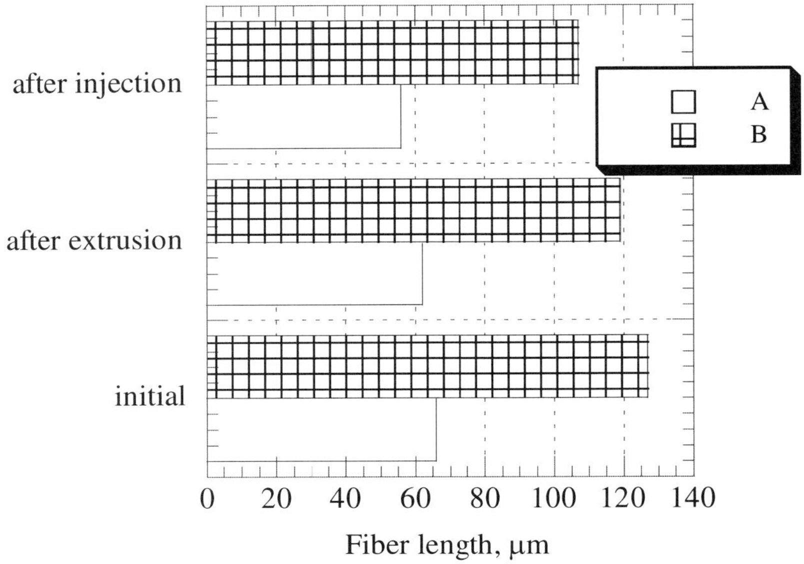


Figure 14.11. The average length of fiber prior to and after processing in glass fiber filled polypropylene. [Data from Averous L, Quantin J C, Lafon D, Crespy A, *Int. J. Polym. Analysis and Characterization*, **1**, No.4, 1995, 339-47.]

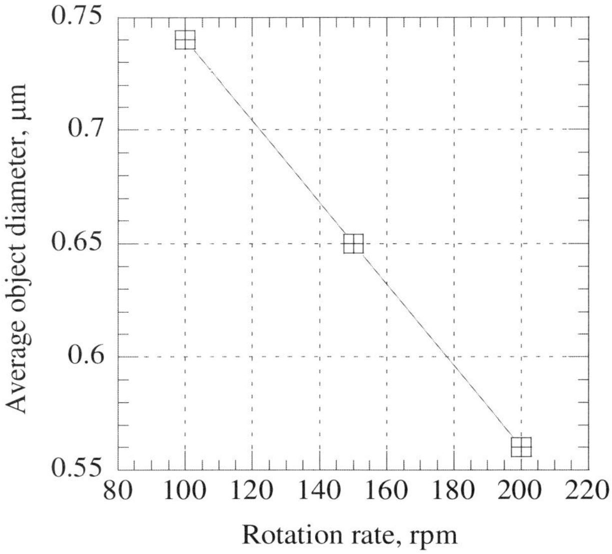


Figure 14.12. Average object diameter vs. rpm of Banbury mixer during incorporation of 10% calcium carbonate in polypropylene. [Data from Herzig R, Baker W E, *J. Mat. Sci.*, **28**, No.24, 1993, 6531-9.]

(and scratch resistance decreases) with talc concentration increasing.[66] Fibers of different length distribution were used for extrusion and injection molding of glass

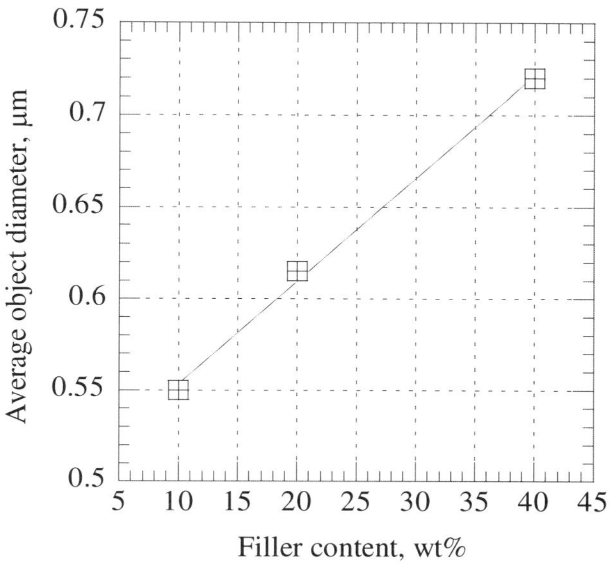


Figure 14.13. Average object diameter vs. $CaCO_3$ concentration after mixing in Banbury mixer with polypropylene. [Data from Herzig R, Baker W E, *J. Mat. Sci.*, **28**, No.24, 1993, 6531-9.]

fiber reinforced polypropylene. The differences of average length of fibers after processing are given in Figure 14.11. Injection molding was more likely to cause changes in fiber length, but the loss of dimension was not large.[67] Figures 14.12 and 14.13 show the effect of mixing in a Banbury mixer on the object diameter (average particle size) of calcium carbonate in polypropylene vs. rpm and filler concentration, respectively.[68] Increased intensity of mixing contributed to particle size reduction whereas the increased concentration of filler caused particle size to increase.

### 14.1.17 LIMITING OXYGEN INDEX[4,31,63,69]

*Applications*: The test illustrates the relative flammability of materials by measuring the minimum concentration of oxygen in atmosphere required to initiate and support flame for more than 3 min.

*Testing procedure*: The minimum concentration of oxygen is measured using a vertically suspended sample in an instrument capable of modifying the atmosphere by mixing oxygen and nitrogen in required proportions (e.g., Stanton Redcroft instrument). The result of determination is presented in a form of % $O_2$.

*Standard methods*: ASTM D 2863, BS/ISO 4589, ISO 4589

*Major results*: Figure 14.14 shows that special fillers are required to increase the limiting oxygen index.[4] General fillers, such as talc, do not affect the limiting oxygen index in any other way than by dilution of the burning component (matrix). $Mg(OH)_2$, fire retarding filler, increases the limiting oxygen index along with a filler concentration increase.

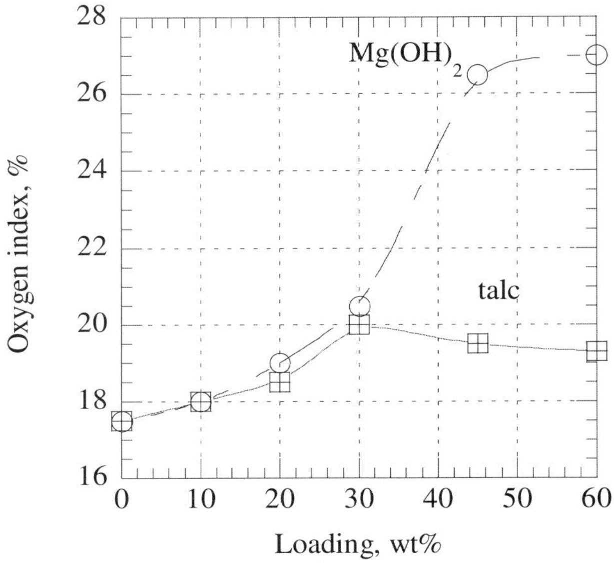


Figure 14.14. Limiting oxygen index vs. filler loading in PP. [Adapted, by permission, from Costa L, Camino G, Bertelli G, Borsini G, *Fire & Mat.*, **19**, No.3, 1995, 133-42.]

## 14.1.18 MAGNETIC PROPERTIES[70,71]

*Applications:* Polymers are nonmagnetic materials but they can be modified by fillers. Plastic magnets, first introduced in 1955, are inferior to cast and sintered magnets but have many desirable properties such as low cost, ease of production, better uniformity and reproducibility.[70] Plastic magnets are used in electronic instruments, communication, household utensils, and audio equipment.

*Testing procedure*: The magnetic permeability measurement is performed on the samples formed to the shape of toroids wrapped uniformly with two sets of wire windings. The primary coil is excited with a sine wave from a function generator. The voltage induced in the secondary coil is measured with a lock amplifier at a frequency range of 0.5 Hz to 120 kHz.[70]

*Standard methods*: AS/DR 96526, ASTM A 772

*Major results*: In non-interacting spheres, the effective permittivity, $\varepsilon_{eff} = \varepsilon(1 + 3\phi)$, where $\varepsilon$ is polymer permittivity and $\phi$ is filler fraction. Figure 14.15 shows that the use of ferromagnetic in LDPE gives substantially better performance.[70] The ferromagnetic filler, HyMu used in this study is composed of nickel, molybdenum, manganese, iron, and carbon. Nanocomposites containing magnetic fillers exhibit superparamagnetism and superferromagnetism. The nanocomposite materials can be changed by the choice of filler concentration because the interparticle distance controls the particle magnetization vector. Therefore, by changing interparticle distance (concentration), one may change properties from superferromagnetic to superparamagnetic at a given temperature.[71]

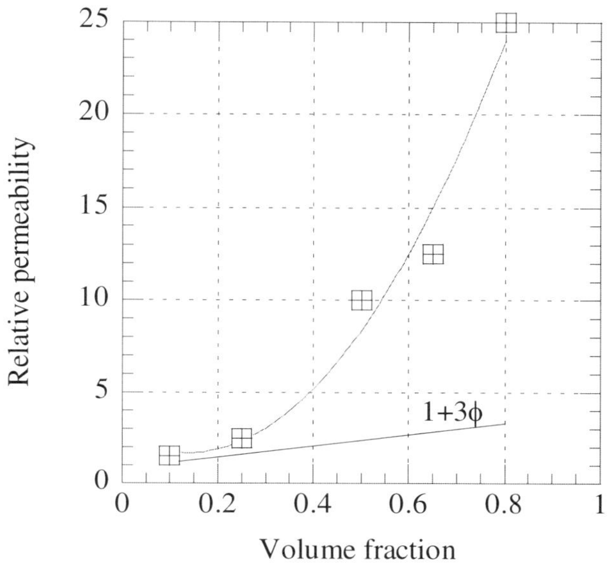


Figure 14.15. Magnetic permeability of LDPE filled with HyMu. [Data from Fiske T J, Gokturk H S, Kalyon D M, Antec '93. Conference Proceedings, New Orleans, La., 9th-13th May 1993, Vol. I, 614-7.]

### 14.1.19 OPTICAL MICROSCOPY[50,53]

*Applications*: Optical microscopy finds several important applications in filled systems, including observation of crystallization and formation of spherulites[50] and phase morphology of polymer blends.[53] In the first case, important information can be obtained on the effect of filler on matrix crystallization. In polymer blends, fillers may affect phase separation or may be preferentially located in one phase, affecting many physical properties such as conductivity (both thermal and electrical) and mechanical performance.

*Testing procedure*: In crystallization studies, samples are placed on a hot stage of a polarizing microscope and crystallization regime (cooling) is strictly controlled. The photomicrographs show the stages of the crystallization and the morphology of spherulites.[50] The effect of temperature, up to the melting point, and the cooling rate on blend morphology can also be observed using a hot stage.[53]

*Standard methods*: not applicable

*Major results*: Carbon black has a better affinity to polyamide than polypropylene. Even if carbon black was added to polypropylene, it was preferentially transferred to the polyamide phase during mixing.[53] In a polymer blend of two polymers, polyamide formed the minor phase and, due to preferential location of carbon black in its phase, the carbon black concentration in the polyamide phase was much higher than expected from the amount of carbon black added. This high concentration of carbon black in the minor phase resulted in substantially increased conductivity of the blend.[53]

### 14.1.20 PARTICLE SIZE ANALYSIS[31,72,73]

*Applications*: The particle size distribution of filler is one essential parameter controlling the performance of filled materials and as such it is a control parameter of commercial fillers.
*Testing procedure*: The most crude method of analysis of particle size distribution is based on sieve analysis. Considering that most fillers contain very small particles, this method is usually inadequate for practical purposes. A Coulter counter is a common instrument used to determine particle size distribution.[31] Particles of filler suspended in water are measured by laser diffraction and recalculated to distribution by volume or weight. Various surface phenomena may affect the readings but still this is one of the most suitable methods of measurement. The measurement of particle size of carbon black is very difficult. The best results are obtained by TEM,[72] but here also very high irregularity of the carbon black shape makes the measurement not very reliable. The particle size of carbon black is frequently correlated with a specific surface area measured by BET method. This again has many other influences considering that carbon black is energetically heterogeneous, and adsorption of gas or liquid depends on functional group concentration.[73] Studies of oxidized carbon black indicated that the specific surface area of carbon black changes on oxidation whereas TEM determined size does not. Determination of the structure of carbon black is even more difficult and results are confusing. It can be summarized that, in spite of the fact that particle size analysis is very important, there is a great deficiency in techniques available to measure it.
*Standard methods*: AS 2879, ASTM D 3360, ASTM D 3849, DIN 66111, ISO 4497

### 14.1.21 RADIANT PANEL TEST[69]

*Applications*: In this test, the flammability of materials is considered as a function of the heat release rate and critical ignition energy. Flammability is inversely proportional to ignition energy and directly proportional to heat released.[69]
*Testing procedure*: The test sample is exposed to heat from a radiant panel at a 30° angle, meaning that the upper portion is severely exposed. The time progress of ignition down the specimen is the so-called flame spread factor. The thermocouples placed above the specimen serve as a heat-flux measuring device to monitor the rate of heat release, called a heat evolution factor. The two measurements multiplied by each other give a flame spread index. Two materials are used as calibrating specimens. The mineral hardboard with an index 0 and red oak with index 100 serve as calibrating specimens.[69]
*Standard methods*: ASTM E 162, ISO/DIS 13927

### 14.1.22 RATE OF COMBUSTION[4]

*Applications*: Flammability of filled systems.
*Testing procedure*: Horizontally held specimen is ignited at one end. The rate of combustion is calculated from the time required for the flame to pass a 25 mm

length. If the flame is extinguished before the 25 mm mark, then an arbitrary zero rate is given to the specimen.
*Standard methods*: AS 2122, ASTM D 635
*Major results*: Figure 12.8 shows that talc increases the rate of PP combustion whereas $Mg(OH)_2$ used in sufficient concentration decreases the combustion rate.[4]

### 14.1.23 SCANNING ACOUSTIC MICROSCOPY[74]

*Applications*: Nondestructive method of determination of carbon fiber reinforced composites. Damage of woven fiber reinforced composites, distribution of filler due to flow in molding techniques, distribution of fiber in composite, and dispersion of carbon black are examples characterizing potential applications of the method.[74]
*Testing procedure*: A sapphire rod with a concave spherical surface at one end and an epitaxially grown piezoelectric transducer on the other end form the acoustic transducer which is the most essential element of the acoustic microscope. A transmitter supplies the RF signal to the transducer. The signal is converted into acoustic waves which propagate through the rod. The sample is scanned and the acoustic wave at each point is reflected back to the transducer acting as receiver. The transducer converts the acoustic wave to an RF signal, which is then translated to the electrical voltage corresponding to the amplitude measurement. This is then stored as a digital value in memory. When scanning is completed, the acoustic properties of the specimen are converted to a gray scale and displayed on a monitor. Samples are examined either by pulse or burst modes. The pulse mode, operating at a lower frequency, is suited for in-depth examination of materials (up to several mm). The instrument responds to either discontinuity in a sample (cracks, voids, delaminations) or changes in properties (inclusions, fibers, crystalline structures). The burst mode operates at a high frequency range which allows for much higher resolution. The method is capable of generating an image of the subsurface and operates about 500 μm below the surface.
*Standard methods*: not applicable
*Major results*: The method is useful for control of expensive materials in responsible applications, such as, for example, materials used in aeronautics. Material control or inspection can be conducted (pulse method) without damage caused to the inspected material. The other essential advantage of the method is related to the fact that fillers can be observed within the filled material, which is not possible by any other technique. In addition, clarity of the micrograph is improved compared with optical microscopy. This method, although unique, has many applications in filled materials and hopefully more data will be known in the future in order to facilitate better understanding of filled materials.

### 14.1.24 SMOKE CHAMBER[4,69]

*Applications*: Smoke evolution of commercial materials and the effect of various additives including fillers on smoke production rate.

*Testing procedure*: An NBS smoke chamber is equipped with six flamelet burners and a radiant heat source. Non-flaming samples are tested using the heater alone. The sample is placed vertically in a front of the heater (2.5 W/cm$^2$). The results are obtained by measuring the percent transmission of a light beam which travels from the bottom of the chamber, through the smoke, to the photomultiplier tube at the top. The results are recorded as a function of time. The transmittance measured is converted to the specific optical density from the following equation: $Ds = V[\log(100/T)]/LA$, where V is chamber volume, L is the light beam path length, A is the sample area, and T is the transmittance.
*Standard methods*: ASTM E 662
*Major results*: General fillers do not affect smoke formation by any means other than simple dilution. Fire retardant fillers such as $Mg(OH)_2$ decrease smoke formation only at high concentrations.[4] Materials which are known catalysts of degradation (e.g., copper) increase smoke formation.[69]

### 14.1.25 SONIC METHODS[75-78]

*Applications*: The physical principle of measurement is similar to the scanning acoustic microscopy discussed in the Section 14.23, but applications and the method of data processing are essentially different. Sonic methods were used in the following applications to filled materials: the effect of particle size and surface treatment on acoustic emission of filled epoxy,[75] longitudinal velocity measurement of tungsten filled epoxy,[76] and in-line ultrasonic measurement of fillers during extrusion.[77] Numerous parameters related to fillers can be characterized by this non-destructive method.
*Testing procedure*: A sample in the form of dumbbell was tested in a tensile testing machine with an acoustic emission transducer attached to its surface using silicone grease and vinyl tape. The acoustic emission generated during tensile testing was analyzed.[75] The velocity of ultrasonic waves traveling in the material under test was used for the determination of longitudinal and shear moduli.[76] The experimental setup included a sample, transmitting transducer, and receiving transducer all immersed in water to improve propagation of acoustic waves. The transmitting transducer was connected to the acoustic wave generator and the receiving transducer to an oscilloscope. The bulk modulus and shear and Young's moduli were calculated from longitudinal velocity.[76] In extruder experiments,[77] the emitting and receiving transducers were installed on the die wall so as to not interfere with the flow of polymer. The attenuation was measured in this experiment because ultrasonic velocity is temperature and pressure dependent.
*Standard methods*: not applicable
*Major results*: Figure 14.16 shows the effect of mean particle size of spherical silica (flame-fused synthetic silica) on acoustic emission. The emission increases with the particle size of filler increasing.[75] The source of this increase in acoustic emission is thought to be related to the fracture of particles, debonding of particles from

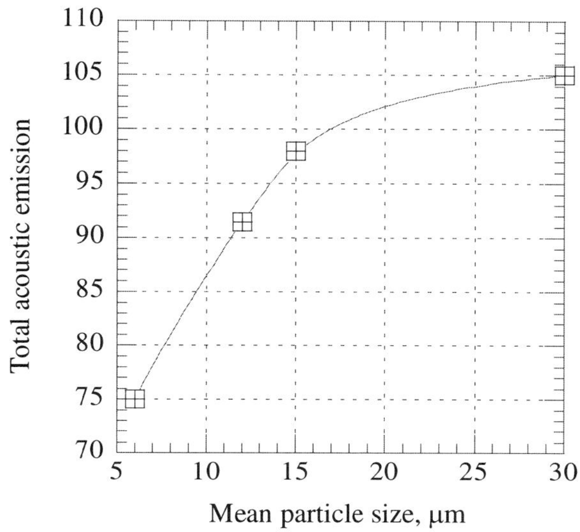


Figure 14.16. The effect of mean particle size of silica on total acoustic emission of epoxy filled with 70% silica. [Data from Ohta M, Nakamura Y, Hamada H, Maekawa Z, *Polym. & Polym. Composites*, **2**, No.4, 1994, 215-21.]

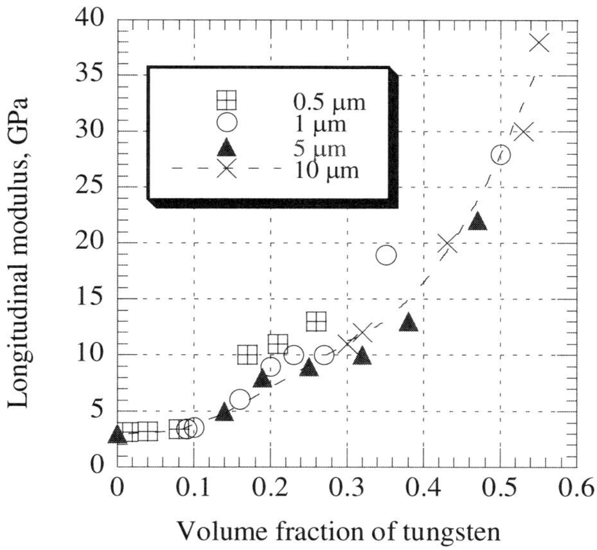


Figure 14.17. Longitudinal modulus vs. volume fraction of tungsten in epoxy. Tungsten having particle sizes from 0.5, 1, 5, and 10 μm is plotted on the graph. [Adapted, by permission, from Nguyen T N, Lethiecq M, Levassort F, Patat F, *Int. J. Polym. Analysis and Characterization*, **1**, No.4, 1995, 277-87.]

the matrix, and cracks formed in the matrix. The tensile strength of the composite decreased with increasing particle size which is in agreement with interpretation. Figure 14.17 shows that the longitudinal modulus increases with addition of tung-

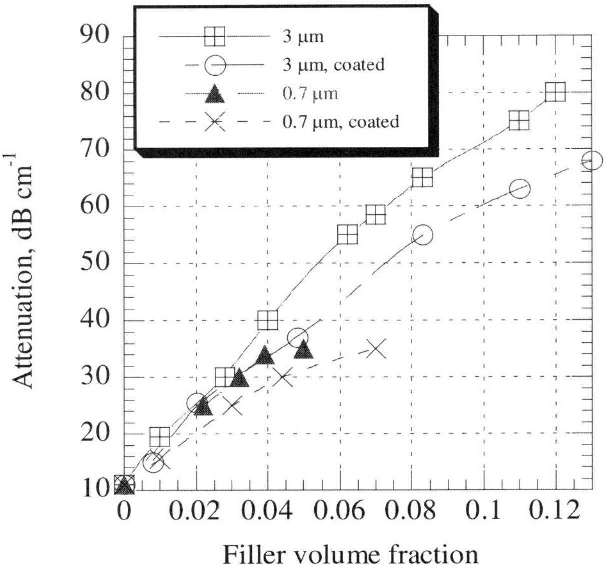


Figure 14.18. Attenuation vs. volume fraction of different fillers. [Adapted, by permission, from Gendron R, Daigneault L E, Tatibouet J, Dumoulin M M, Antec '94. Conference Proceedings, San Francisco, Ca., 1st-5th May 1994, Vol. I, 167-71.]

sten increasing.[76] The particle size in the measuring range has very little effect on modulus. In extrusion experiments, acoustic measurements were used to measure the distribution of residence time in the extruder.[77] This method is an alternative solution to the normal practice of such measurements, which consists of the addition of dye to the extruded compound to evaluate the quality of mixing. A filler ($CaCO_3$ in this experiment) can be used as tracer. Figure 14.18 shows the effect of filler grade on attenuation. Camel-Wite has an average particle size of 3 μm and Camel-Cal of 0.7 μm. Both grades were tested with and without stearic coating. It is noticeable that larger particles give stronger signals (more imperfections) and uncoated grades also give stronger signals (more difficult to disperse). Increasing volume fraction of filler contributes to the increase of attenuation. $CaCO_3$ was found to be a very suitable choice for application (as a tracer to measure residence time required).[77]

### 14.1.26 SPECIFIC SURFACE AREA[78]

*Applications*: Determination of specific surface area of filler which is used as a factor relative to particle size, porosity, and surface activity of filler.

*Testing procedure*: The automated instruments (e.g., Quantachrome NOVA 1000 or Micrometrics Gemini 2360) are used for rapid surface determination. The physical adsorption of nitrogen or other inert gas should be conducted at a pressure which is within the range of linearity.

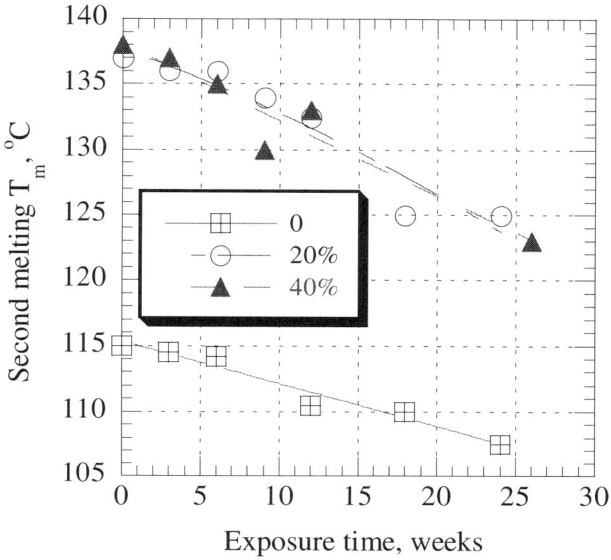


Figure 14.19. Crystallization temperature of talc filled PP vs. UV exposure time. [Adapted, by permission, from Rabello M S, White J R, *Polym. Composites*, **17**, No.5, 1996, 691-704.]

*Standard methods*: ASTM C 1069, ASTM D 1993, ASTM D 5604, BS 4359: Part 1, ISO 9277

### 14.1.27 THERMAL ANALYSIS[50,52,79-98]

*Applications*: Thermal analysis found numerous applications in filled systems. The following studies were conducted: the effect of filler on degradation rate of HDPE,[50] mechanism of degradation of PU foams filled with inorganic flame retardants,[82] thermal degradation of ethyl acrylate copolymer in the presence of fillers,[87] photodegradation of filled polypropylene,[89] the effect of nucleating filler on the photodegradation of polypropylene,[85] the effect of cooling rate on crystallization of HDPE in the presence of fillers,[50] dynamic mechanical analysis of nanocomposites,[52] determination of glass fiber content in a composite,[70] cure kinetics in the presence of fillers,[80] activation energy of curing in the presence of fillers,[90] the effect of filler on the curing kinetics of a composite,[91] glass transition temperature at different filler contents,[81] glass transition temperature of nanocomposites containing whiskers,[88] thermal degradation of polymer blends containing fillers,[92] thermal decomposition of fillers,[94] the effect of filler loading on DTA of filled polymer,[95,96] and formation of an interpenetrating network in the presence of filler.[97]

*Testing procedure*: Testing of thermal properties is a fairly standard procedure which is not discussed here. DSC was used for the studies on the effect of cooling rate on formation of spherulites in the presence of fillers.[50] In order to assure the repeatability of conditions of the experiment, the instrument was calibrated with indium and tin standards.[50] The TGA/DSC instrument was coupled with a mass

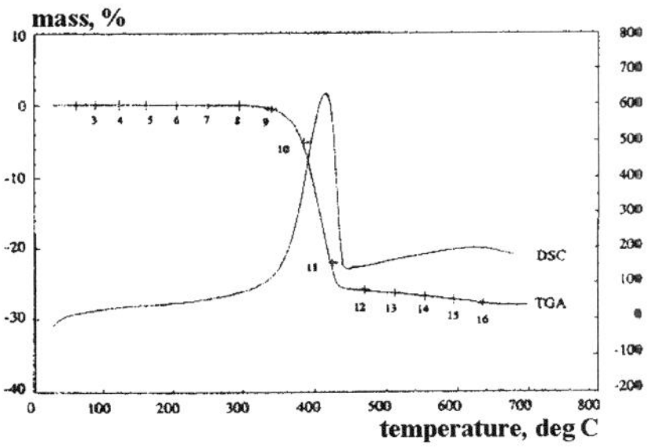


Figure 14.20. TGA curve for $Mg(OH)_2$. [Adapted, by permission, from Hornsby P R, Wang J, Rothon R, Jackson G, Wilkinson G, Cossick K, *Polym. Degradat. Stabil.*, **51**, No.3, 1996, 235-49.]

spectrometer and FTIR and gases released from thermal decomposition analyzed for composition.[94] In most cases, the thermal analysis was used in a standard way.
*Standard methods*: not applicable
*Major results*: Figure 14.19 shows that the crystallization temperature of filled PP is substantially higher than for unfilled polymer.[89] There is no difference in crystallization temperature between 20 and 40% talc. It was reported in the literature that only a small addition of talc increases the crystallization temperature due to nucleation. A further increase in talc concentration does not change crystallization temperature. UV exposure of talc filled PP changes the affinity of PP to talc causing a gradual drop in crystallization temperature. Figure 14.20 shows the thermal decomposition of $Mg(OH)_2$.[94] The water release is accomplished within a temperature range of 320 and 440°C. It was determined by DSC that the reaction is strongly endothermic.

## 14.2 CHEMICAL AND INSTRUMENTAL ANALYSIS

### 14.2.1 ELECTRON SPIN RESONANCE[99-101]

*Applications*: ESR spectroscopy was used to monitor the orientation and distribution of filler particles in polymer composites,[99-101] and molecular movement in filled, crosslinked material was studied.[100]
*Testing procedure*: Free radicals were generated by γ-irradiation from a cobalt source at -75°C. The radical decay study was performed at 40°C.[100] Most inorganic fillers contain paramagnetic impurities and defect centers in their crystalline structure. The orientation of the magnetic axis of these paramagnetic centers can be characterized by ESR.[99] By measuring the magnetic anisotropy of naturally occurring Mn(II) centers in calcium carbonate and in talc the change in orientation of the filler can be followed.[101] The amplitude ratio of characteristic ESR bands gave the order parameter.
*Standard methods*: not applicable
*Major results*: Figure 14.21 shows the effect of the concentration of silica filler on radical decay in a crosslinked and uncrosslinked system.[100] Both fillers and crosslinks restrict molecular mobility. A small addition of filler has a very large effect on radical decay (and molecular mobility). Further addition of filler has a decreased effect on the rate of radical decay.[100] The orientation of the particles changed as a function of composition, depending on processing technology (the type of molding, i.e., injection versus compression molding), and had a specific spatial distribution in the cross-section of the injection molded specimen. A

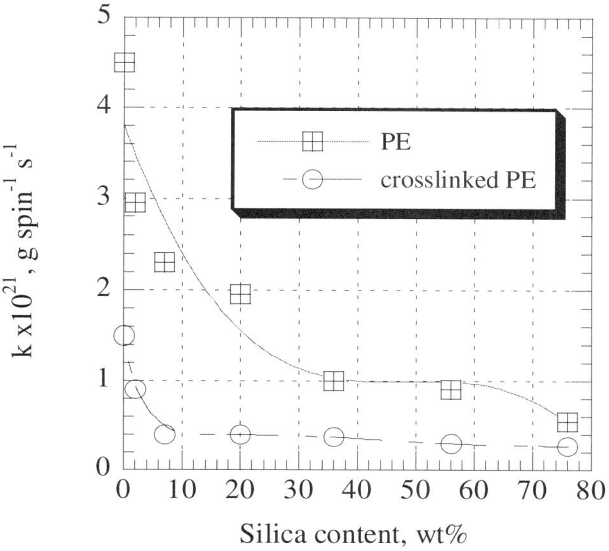


Figure 14.21. Rate constant of free radical decay at 40°C vs. content of silica in PE. [Adapted, by permission, from Szocs F, Klimova M, Chodak I, Chorvath I, *Eur. Polym. J.*, **32**, No.3, 1996, 401-2.]

correlation was found between average orientation of anisotropic particles and the mechanical properties of various composites.

### 14.2.2 ELECTRON SPECTROSCOPY FOR CHEMICAL ANALYSIS[102-106]

*Applications*: ESCA was used in studies of polyurethane containing metal salts,[102] to determine the distribution of silica in a PVA matrix,[103] to analyze surface groups of carbon black,[104,105] and to observe the effect of surface modification on the surface composition of composites.[106]

*Testing procedure*: Standard methods of sample preparation and instrument operation are used.

*Standard methods*: not applicable

*Major results*: The analysis of ultrafine particles in polyurethane shows that $Ni^{2+}$ was reduced in the process to metallic Ni. After 15 days, some Ni atoms on the surface were oxidized, but the Ni atoms residing in the bulk were not affected by oxidation.[102] The differences in concentration between the front and the back of the sample of PVA containing silica were -4, +2, and +2 for C, O, and Si, respectively.[103] This shows good intermixing between polymer and filler. At the same time, the values for all three atoms do not match the theoretical values, which is attributed to the reaction between unreacted ethoxide groups and silica forming covalent bonds.[103] Substantial differences between furnace carbon black and carbon black obtained from the pyrolysis of tires were detected by ESCA.[104] Furnace carbon black had one strong graphitic carbon peak and very small peaks of carbon-oxygen and carbon in aliphatic or aromatic compounds. Carbon black from pyrolized tires had strong peaks of all compounds.[104] Treatment of carbon black by plasma changes the concentration of atoms on the surface of carbon black, but heating these treated samples at 900°C returns them to the initial composition.[105] The concentration of several groups was studied, including C-C, C-O, C=O, COO, keto-enol, and $C-NH_2$.[105]

### 14.2.3 INVERSE GAS CHROMATOGRAPHY[15,18,19,72,93,105,107-116]

*Applications*: Inverse gas chromatography became one of the most important tools used for studying the surface properties of fillers. The following information can be obtained from measurements: isotherms of adsorption of various probes, energy distribution of adsorption, heat of adsorption, surface free energy, dispersive (nonpolar) component of surface energy, polar contribution to surface energy, specific components of the free energy of adsorption, enthalpy and entropy of adsorption, acid-base forces (pair interaction parameter), work of adhesion between polymer and filler. The fillers for the following applications were studied by inverse gas chromatography: the effect of plasma treatment of carbon black,[105] the effect of thermal treatment of carbon black on rubber reinforcement,[18] heats of adsorption of various probes on carbon black,[72] the effect of compression on surface activity of carbon black,[19] the effect of extraction on the properties of carbon black,[110] properties of commercial carbon blacks,[111] the effect of filler surface energy on its performance in paints,[15,108] acid/base properties of fumed silica used in silicone elastomers and silica modification by thermal treatment,[109] surface properties of ZnO and their effect on the reinforcement of elastomers,[112] acid/base properties of $CaCO_3$,[113] characterization of Kevlar fibers,[114] heterogeneity of the filler surface,[115] and the effect of fillers in polymer blends.[116] Reversed-phase liquid chromatography is based on a similar approach but it is not as broadly applied as inverse gas chromatography.[107]

*Testing procedure*: The principle of measurement by inverse gas chromatography is similar to gas chromatography, in which the instrument contains a column placed inside the thermostated chamber, an injection port and a detector. The injected sample, carried through the chromatographic column by a carrier gas, is separated on a stationary phase of column, and measured by detector. In gas chromatography, the stationary solid phase is selected to give the best separation of mobile vapor phase. In inverse gas chromatography, the mobile vapor phase is a probe selected from numerous polar and nonpolar liquids and the stationary phase is composed of filler (or polymer) under testing. The method can be used to characterize independently filler and polymer particles and to analyze polymer-filler interaction. A large body of theoretical treatment of data exists which allows the calculation of numerous parameters listed in the *Applications*. The peculiarity of equipment operation in inverse gas chromatography in comparison to standard gas chromatographic measurements is related to the preparation of the column, which may introduce

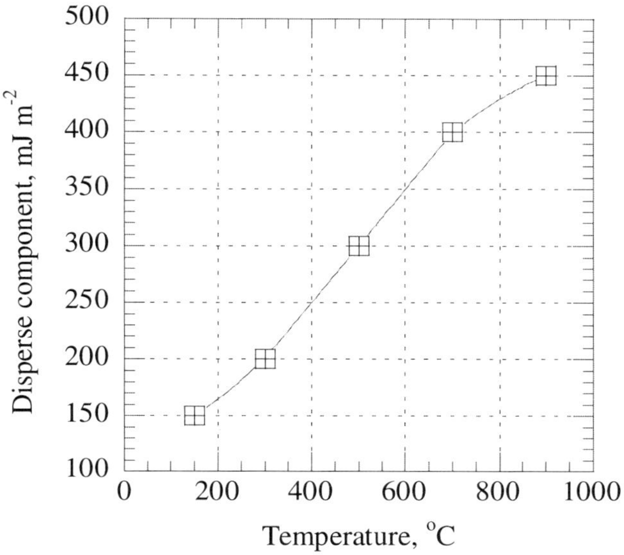


Figure 14.22. Disperse component of carbon black vs. temperature of thermal treatment under nitrogen. [Adapted, by permission, from Donnet J B, Wang W, Vidal A, Wang M J, *Kaut. u. Gummi Kunst.*, **46**, No.11, Nov.1993, 866-71.]

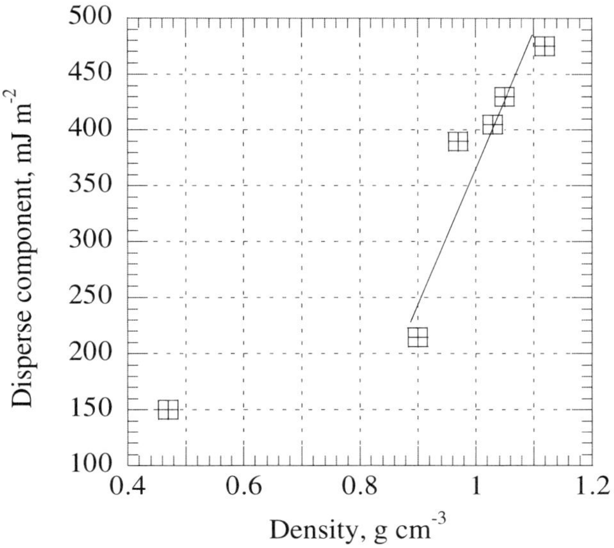


Figure 14.23. Disperse component vs. density of carbon black on compression. [Adapted, by permission, from Wang W D, Haidar B, Vidal A, Donnet J B, *Kaut. u. Gummi Kunst.*, **47**, No.4, 1994, 238-41.]

measurement error. In gas chromatographic measurements, the stationary phase is very well defined if purchased from a reputable supplier. In inverse gas chromatography, the column is prepared by the user with material which was not optimized

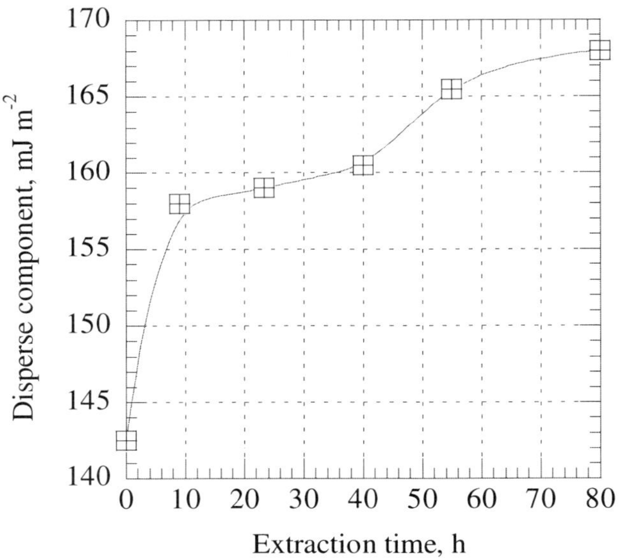


Figure 14.24. Disperse component vs. extraction time of carbon black by boiling toluene. [Adapted, by permission, from Vidal A, Wang W, Donnet J B, *Kaut. u. Gummi Kunst.*, **46**, No.10, 1993, 770-8.]

for unrestricted flow. The material for analysis has to be prepared in such a manner that a negligible flow rate drop occurs when carrier gas passes through the column. In one study,[109] fumed silica powders were agglomerated by dispersing in hexane, allowed to dry under nitrogen and sieved through a wire mesh to obtain a narrow distribution of particle sizes. Similar procedures are used in carbon black studies.[110] It should be pointed out that the method of preparation of material should also not alter surface properties of filler, since it can change the results of measurements. It is also essential to condition a column especially if surface-modified fillers are used, to remove a chemical load which otherwise will affect the conditions of adsorption and detection. Infinite dilution is one essential modification of measurement. In this case, very small samples of probes (close to the detection limit of detector) are injected.[112] Dead volume of a column is minimized by close packing[114] and measured by injecting methane.[112]

*Standard methods*: none

*Major results*: Many examples are available in the current literature concerning how this method can enhance understanding of the use of fillers. Figure 14.22 shows that the dispersive component of carbon black increases with temperature of treatment under nitrogen.[18] About half of the acid groups disappear at 500°C, and they all disappear at 900°C.[18] At around 700°C, the dispersive component begins to reach a plateau. This suggests that the surface chemical groups at the lowest oxygen containing positions do not play a positive role in the increase of surface energy and explains why polymer-carbon black interaction decreases upon oxidation of filler.

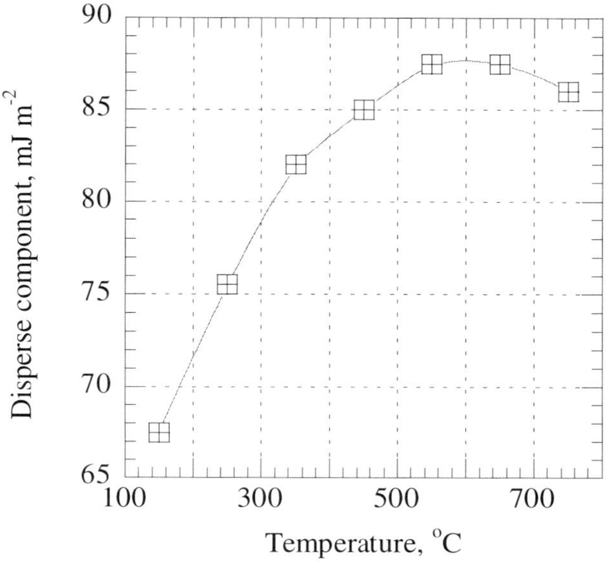


Figure 14.25. Disperse component of fumed silica vs. temperature of thermal treatment. [Adapted, by permission, from Zumbrum M A, *J. Adhesion*, **46**, Nos.1-4, 1994, 181-96.]

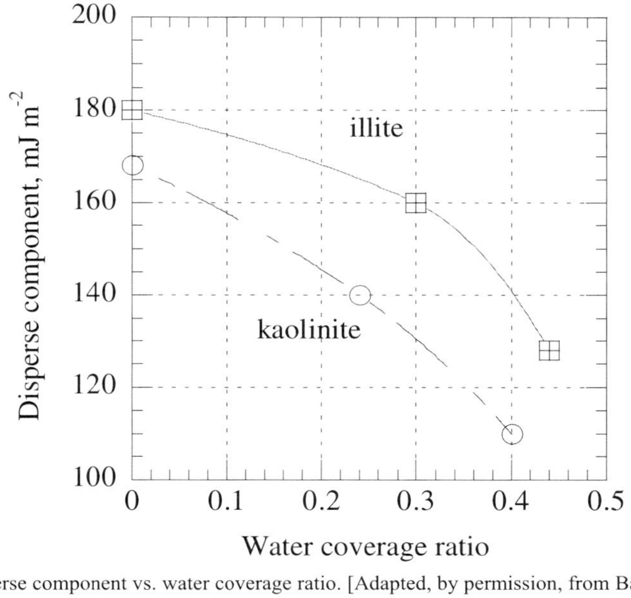


Figure 14.26. Disperse component vs. water coverage ratio. [Adapted, by permission, from Balard H, Saada A, Hartmann J, Aouadj O, Papirer E, *Macromol. Symp.*, **108**, 1996, 63-80.]

Densification of carbon black by compression increases the dispersive component of surface free energy.[19] This process is initially not proportional to density, but after some threshold value at around 0.7 g/cm$^3$ the dispersive component has a linear relationship with density(Figure 14.23). Also, extraction of carbon black with solvents increases dispersive component due to elimination of certain functional groups from the surface (Figure 14.24).[110] The thermal treatment of fumed silica also results in an increase of the dispersive free energy due to condensation of adjacent silanol groups (Figure 14.25).[109] Figure 11.2 shows the effect of acidic and basic calcium carbonates on elongation.[113] Water adsorbed on the filler surface decreases the dispersive component due to shielding of the sites having the highest adsorption energy (Figure 14.26).[115]

### 14.2.4 GAS CHROMATOGRAPHY[92,117,118]

*Applications*: Gas chromatography has limited applications to filled systems. It was used for characterization of the various degradation products of ethylene ethyl acrylate copolymer filled with calcium carbonate by GC-MS,[92] evaluation of ecotoxicological properties of materials containing flame retardants,[117] and determination of carbon black content by pyrolysis gas chromatography.[118]
*Testing procedure*: Volatile products of degradation were absorbed in Tenax cartridges and the ethanolic eluate of the Tenax cartridge was analyzed by GC-MS.[117] The furnace type pyrolyzer was connected to the gas chromatograph, and pyrolysis was conducted under helium. Volatiles were used for the identification of polymer and residue to determine the amount of carbon black.[118] The standard error of determination of carbon black was in the range of 0.2 to 3.4%.
*Standard methods*: not applicable
*Major results*: Calcium carbonate prevents formation of acids by interacting with acid groups.[92] Products of higher molecular weight are produced in the presence of filler.

### 14.2.5 GEL CONTENT[119,120]

*Applications*: Determination of the insoluble fraction of polymer in compounded and processed materials.
*Testing procedure*: A standardized procedure which requires choice of solvent for extraction. In PE determinations, xylene was used as a solvent.[119] A more complex procedure was used to determine the gel content in radiation crosslinked PVC filled with calcium carbonate.[120] The compound was extracted with tetrahydrofuran, and non-dissolved residue was determined. This residue was then used for determination of chlorine by the Schoniger method. From the amount of chlorine, the concentration of polymer was established. The remainder of the gel content was a filler embedded by gel.[120]
*Standard methods*: ASTM D 2765, ISO 19147
*Major results*: The total gel content in radiation crosslinked PVC was increased by the addition of calcium carbonate but only due to the inclusion of filler in gel. When filler content in gel was subtracted, the amount of polymer gel formed decreased as the addition of filler increased due to the stabilizing activity of the filler.[120]

### 14.2.6 INFRARED AND RAMAN SPECTROSCOPY[48,7985,89,94,106,109,121-138]

*Applications*: The following use was made of infrared and Raman spectroscopy: identification of surface groups on treated and untreated fumed silica,[109] identification of silica functional groups and coatings by Raman spectroscopy,[122] silane deposition on various fillers,[127] surface grafting of barium sulfate by acryloamide grafting of wood fiber,[106] spectral absorption of filler,[89,126] crystallization of polymer in the presence of filler,[79] absorption of polyacrylate on alumina,[121] deformation of fibers in composites by Raman spectroscopy,[123] the effect of fillers on UV cure,[124] the effect of zeolite on PVDF crystalline structure,[125] kaolin modification by PEG,[126] characterization of carbon fibers by Raman spectroscopy,[130] the effect of strain on the Raman spectrum of carbon fiber,[134] characterization of carbon fiber by IR,[135] interaction of filler with ZnO,[131] interaction of talc with LDPE,[132] formation of zinc carboxylates during UV irradiation of PE,[138] and photodegradation of polychloroprene in the presence of carbon black.[137]

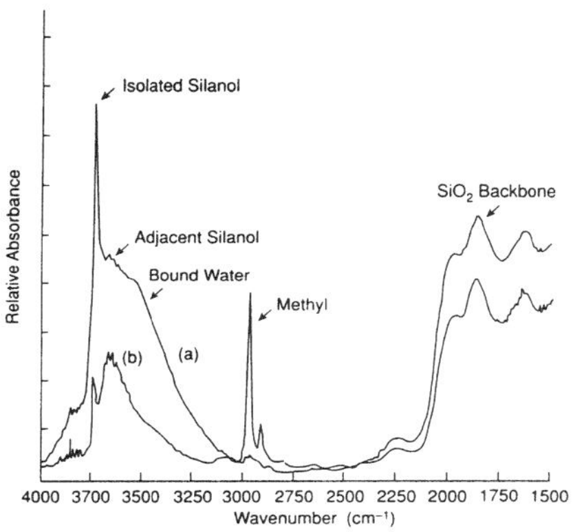


Figure 14.27. Infrared spectra of fumed silica. [Adapted, by permission, from Zumbrum M A, *J. Adhesion*, **46**, Nos.1-4, 1994, 181-96.]

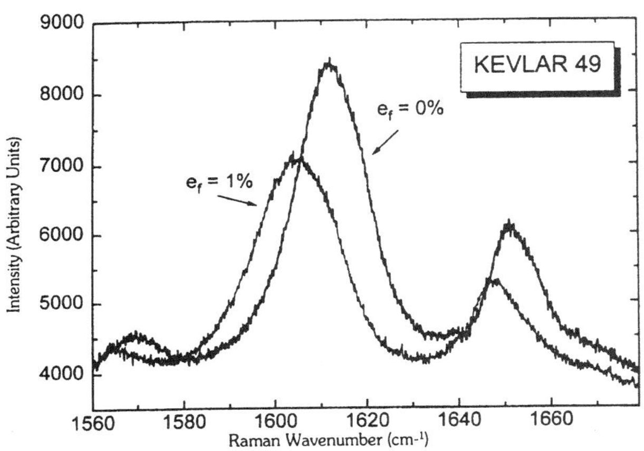


Figure 14.28. Strain induced shift of Raman spectra for a single filament Kevlar. [Adapted, by permission, from Young R J, *Prog. Rubb. Plast. Technol.*, **11**, No.2, 1995, 124-36.]

*Testing procedure*: In general, standard methods were used, with some improvements to obtain better resolution. A photoacoustic detector was used to obtain spectra of fumed silica.[109] A carbon black background was used. In studies of adhesion of coatings on metal substrates, a gold coated background was used as the reference spectrum.[48] Diffuse reflectance spectra, DRIFT, were used to characterize surface species on alumina[121] and kaolin.[126] A Raman microprobe was capable of obtaining spectra from a very small area (2 μm

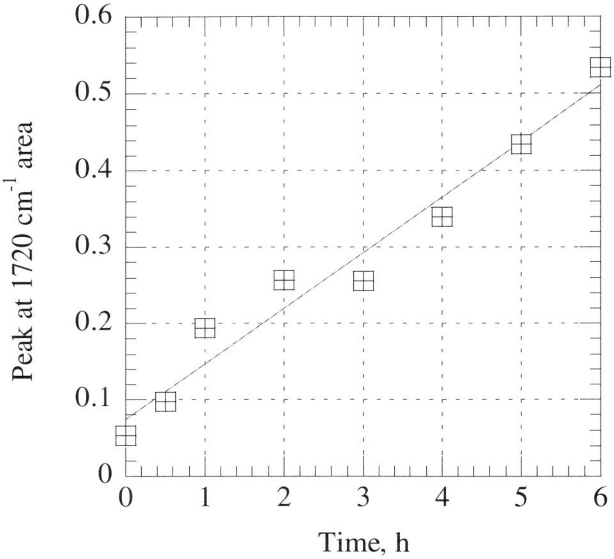


Figure 14.29. Peak surface areas at 1720 $cm^{-1}$ vs. treatment time of carbon fibers. [Data from Ohwaki T, Ishida H, *J. Adhesion*, **52**, Nos.1-4, 1995, 167-86.]

in diameter).[123] A Raman frequency shift was used to study the effect of strain on carbon fiber.[134]

*Standard methods*: not applicable

*Major results*: Figure 14.27 compares the IR spectra of silane-treated and untreated silica. The major difference is in the concentration of functional groups, which are numerous in untreated silica and few in the treated version. Raman spectroscopy was found very useful in determination of functional groups on silica and other fillers.[122] Figure 14.28 demonstrates the effect of strain on the shift in a Raman band of Kevlar fiber.[123] Two main bands at 1610 and 1645 $cm^{-1}$ shift significantly to a lower wavenumber on application of strain. Further studies revealed that there is a very good linear correlation between strain and the 1610 $cm^{-1}$ band shift. Group assignment of kaolin and modified kaolin using diffuse reflectance IR was published.[126] Talc presence in a formulation can be easily recognized on an IR spectrum.[89,132] Oxidation of carbon fiber can be monitored by FTIR measurements (Figure 14.29). Two peaks on the IR spectrum (1720 and 1580 $cm^{-1}$) have a linear correlation with treatment time.[135]

## 14.2.7 NUCLEAR MAGNETIC RESONANCE SPECTROSCOPY[64,124,126,139-153]

*Applications*: NMR has found the following applications in filled systems: carbon black adsorption of SBR,[141] the effect of carbon black loading on cure rate of natural rubber,[148] gel-like behavior of polybutadiene/carbon black mixtures,[150] structure and dynamics of carbon black filled rubber vulcanizates,[153] interaction of

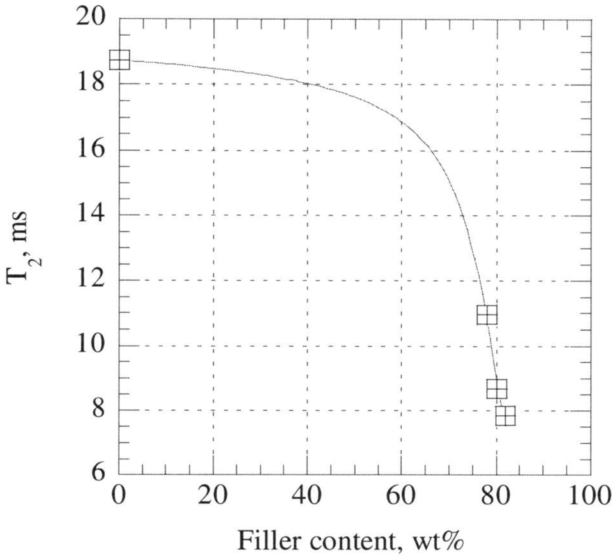


Figure 14.30 Relaxation time, $T_2$, vs. filler concentration in solid racket propellant. [Data from Merwin L H, Nissan R A, Stephens T S, Wallner A S, *J. Appl. Polym. Sci.*, **62**, No.2, 1996, 341-8.]

kaolin, $BaSO_4$, lithopone and ZnS with EPDM,[152] determination of the effect of filler on $T_2$ relaxation times and MRI imaging (see Section 14.1.16),[64] comparison of filled and unfilled UV-curable systems containing $Al(OH)_3$,[124] determination of chemical shifts on modified kaolin,[126] structure comparison of PMMA filled with $Al(OH)_3$,[139] curing of duroplasts in the presence of fillers,[140] structure and dynamics of PMMA/colloidal silica composite,[147] crystallization of PDMS in the presence of silica,[142] stress distribution in filled PDMS by NMR imaging,[143] interaction between silica and rubbers,[145] oxidative aging of carbon black-filled SBR by NMR imaging,[144] crosslink density of filled natural rubber,[146] the effect of carbon black on $T_2$,[146] characterization of absorbed water in aramid fiber,[149] and mobility of hydrocarbon chains in barium sulphate containing nanocomposite.[151] Numerous applications show the versatility of NMR applications in filled systems, which is due to the possibility of use of solid samples and evaluation of molecular mobility.
*Testing procedure*: Several variations of NMR were used as follows: MRI (magnetic resonance imaging),[64,144] solid-state $^{13}C$ NMR,[124,126,139,140,146,147,148,153] $^{1}H$ NMR,[141,143,146,149,152] $^{29}Si$ NMR,[147] $^{23}Na$ NMR,[149] magic angle,[141,146,147,148,149]
*Standard methods*: not applicable
*Major results*: Figure 14.30 shows the relationship between relaxation time, $T_2$, and amount of filler in solid racket propellant.[64] Relaxation time decreases with filler load because the material becomes more rigid with increasing filler loading. Urethane carbonyl, and aromatic carbons were found among others in kaolin modified by urethane.[126] Conformational changes in PMMA were found in a strongly inter-

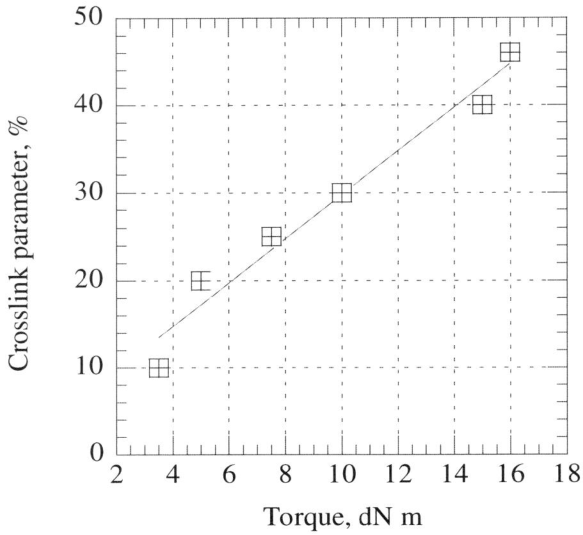


Figure 14.31. Crosslink parameter vs. torque for carbon black filled SBR. [Adapted, by permission, from Fuelber C, Bluemich B, Unseld K, Herrmann V, *Kaut. u. Gummi Kunst.*, **48**, No.4, 1995, 254-9.]

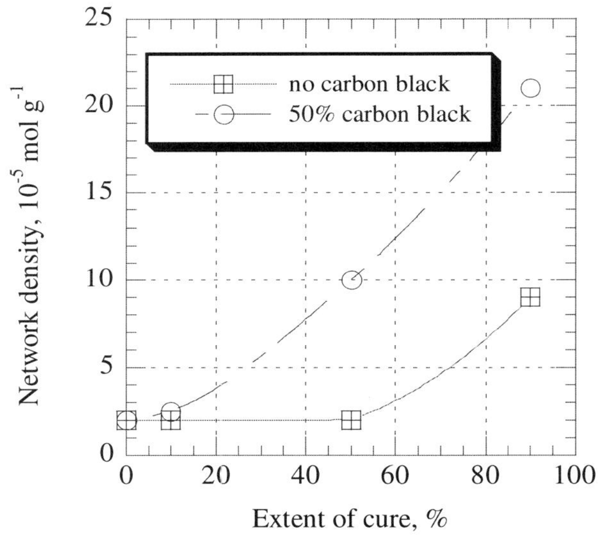


Figure 14.32. The effect of carbon black on network density of natural rubber. [Data from Mori M, Koenig J L, *Rubb. Chem. Technol.,* **68**, No.4, 1995, 551-62.]

acting system with alumina.[139] Figure 14.31 shows correlation of the NMR crosslink parameter, δ, with vulcameter torque.[144] This results are in agreement with theory and show the precision of measurement which can be attained. Figure 14.32 shows that carbon black increases the network density of vulcanized rubber.[148]

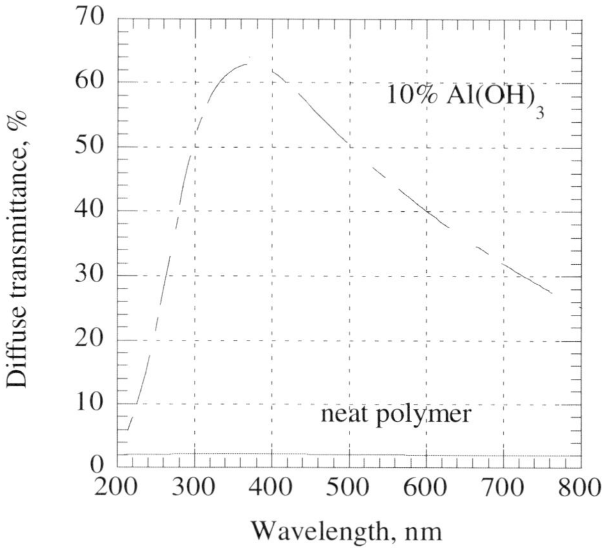


Figure 14.33. Relative diffuse UV/VIS transmittance spectra of PU after UV cure. [Data from Parker A A, Martin E S, Clever T R, *J. Coatings Technol.*, **66**, No.829, 1994, 39-46.]

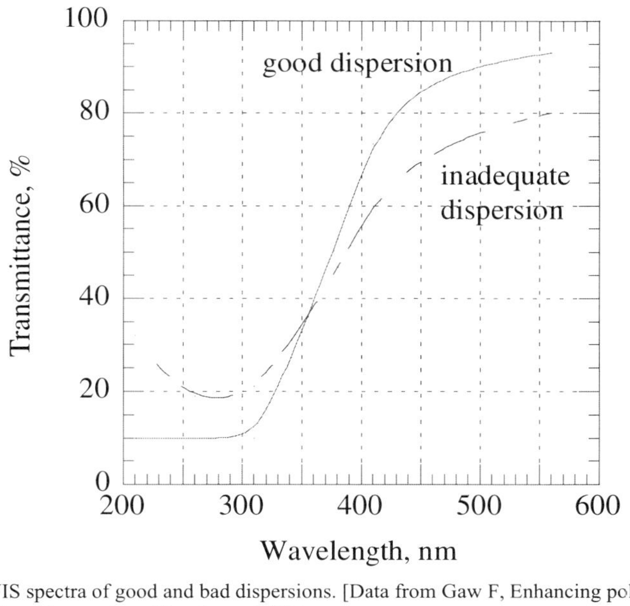


Figure 14.34. UV/VIS spectra of good and bad dispersions. [Data from Gaw F, Enhancing polymers using additives and modifiers, Symposium, Shawbury, 1993.]

### 14.2.8 UV AND VISIBLE SPECTROPHOTOMETRY[40,59,124,154,155]

*Applications*: Due to the nature of fillers, UV and visible spectroscopy is not the most popular method of testing, but there is some useful information which can be obtained from these methods. The following information can be found in the current literature: the effect of fillers on UV cure of coatings,[124] UV spectrum of a nanocomposite,[59] determination of conjugated double bonds in PE,[40] the effect of dispersion on the UV-visible spectrum of $TiO_2$ filled materials,[154] the effect of crystal diameter on the relative opacity of $TiO_2$,[155] and the effect of ultrafine $TiO_2$ on the UV-visible spectrum of PP.[155]

*Major results*: Figure 14.33 shows relative diffuse UV/VIS transmittance spectra for PU films filled with $Al(OH)_3$. This finding corresponds with a higher rate of UV cure of system containing filler.[124] Poor dispersion of $TiO_2$ reduces absorption of UV and imparts whiteness due to the scattering of visible light (Figure 14.34).[154]

### 14.2.9 X-RAY ANALYSIS[55,59,85,89,98,103,156-165]

*Applications*: Numerous uses of x-ray analysis were reported for filled systems. They include: orientation of talc particles in extruded thermoplastics,[55,89,163,165] particle size determination in nanocomposites,[59] crystallinity of talc nucleated PP,[85] crystallinity of polymerization filled PE,[98] diffraction pattern of filled PVA,[103] structure of nanocomposites based on montmorillonite,[156] degree of filler mixing,[157] structural characteristics of fillers,[158,159] structure of carbon black filled rubber,[160] the effect of apatite concentration on the structure of wood pulp,[161] and graphite as template.[164] This list shows the versatility of the method in applications to filled systems.

*Testing procedure*: The testing procedures are adjusted to the experiment, but in the majority of cases typical WAXS measurements are used.

*Standard methods*: not applicable

*Major results*: The results support other studies which is why no examples are given for this method.

### 14.2.10 X-RAY PHOTOELECTRON SPECTROSCOPY[12,59,102,128,135,166-174]

*Applications*: XPS was used for the following purposes: determination of elemental composition of nanocomposites,[59] the effect of oxidation and reduction of carbon fibers by monitoring the O/C ratio,[12,135,174] concentration of functional groups on the surface of carbon fibers,[12,166,174] elemental composition of the surface of carbon fibers,[166,168,171] the effect of surface coating on the surface composition of carbon fiber,[171] chemical degradation of carbon fibers,[171,172] fiber/matrix interface of sized carbon fiber,[173] the effect of fiber sizing technology on surface composition of carbon fibers,[173] surface analysis of barium sulfate modified by 12-hydroxystearate,[128] X-ray radiography of nickel-coated fibers,[167] surface atoms of hydroxyapatite,[169] coating analysis of silane treated hydroxyapatite,[169] and composition of the failure area of an adhesive joint between rubber and metal.[170] This review of applications shows that carbon fibers are the most frequently tested material by XPS.

*Testing procedure*: The methods of testing used were fairly standard techniques of equipment operation. It was mentioned[169] that the sampling depth of XPS (~50-100 Å) is sensitive enough to detect silane having a thickness of 5-10 monolayers.

*Standard methods*: ASTM E 902 (instrument calibration).

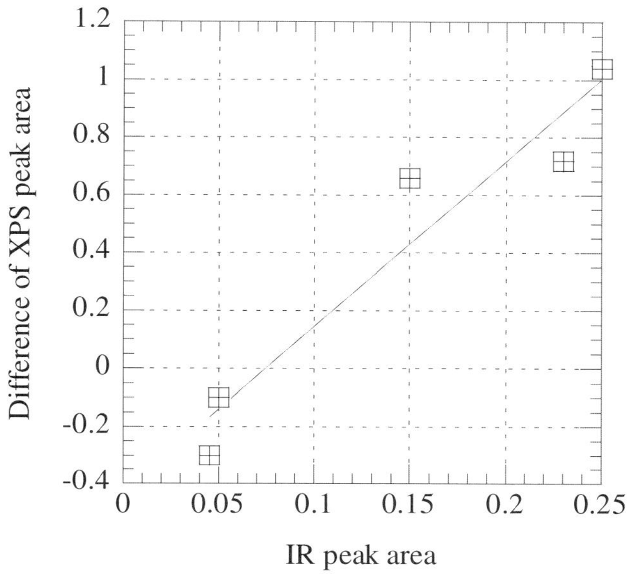


Figure 14.35. Plot of IR peak area vs. difference in XPS area. [Adapted, by permission, from Ohwaki T, Ishida H, *J. Adhesion*, **52**, Nos.1-4, 1995, 167-86.]

*Major results*: Figure 14.35 shows the correlation between IR peak area and difference of XPS peak area.[135] The IR peak area is at 1720 $cm^{-1}$, and the XPS peak area difference is at 288.4 eV. Both peaks were assigned to C=O. The results obtained by the two methods show a good correlation. Figure 6.7 shows the effect of treatment time by oxygen plasma of carbon fiber on the O/C elemental ratio vs. treatment time. A plateau is reached after a short period of treatment.

## REFERENCES


1 Wawkuschewski A, Cantow H J, Magonov S N, *Polym. Bull.*, **32**, No.2, 1994, 235-40.
2 Srinivasan G, Reneker D H, *Polym. Int.*, **36**, No.2, 1995, 195-201.
3 Donnet J B, Custodero E, Wang T K, *Kaut. u. Gummi Kunst.*, **49**, No.4, 1996, 274-9.
4 Costa L, Camino G, Bertelli G, Borsini G, *Fire & Mat.*, **19**, No.3, 1995, 133-42.
5 Karasek L, Sumita M, *J. Mat. Sci.*, **31**, No.2, 1996, 281-9.
6 Levesque J L, Fraval J T, Antec '93. Conference Proceedings, New Orleans, La., 9th-13th May 1993, Vol. II, 1957-64.
7 Hirschler M, Flame Retardants '96. Conference proceedings, London, 17th-18th Jan.1996, 199-214.
8 Cusack P, Flame Retardants '96. Conference proceedings, London, 17th-18th Jan.1996, 57-69.
9 Ferm D J, Shen K K, Antec '94. Conference Proceedings, San Francisco, Ca., 1st-5th May 1994, Vol.III, 3522-6.
10 Whiteley R H, Elliot P J, Staggs J E, Flame Retardants '96. Conference proceedings, London, 17th-18th Jan.1996, 70-8.
11 Pape P G, Romenesko D J, Antec '97. Conference proceedings, Toronto, April 1997, 2941-52.
12 Tang L-G, Kardos J L, *Polym. Composites*, **18**, No.1, 1997, 100-13.
13 Balard H, Papirer E, *Prog. Org. Coatings*, **22**, No.1-4, 1993, 1-17.
14 Godard P, Bomal Y, Biebuyck J J, *J. Mat. Sci.*, **28**, No.24, 1993, 6605-10.
15 Hegedus C R, Kamel I L, J. *Coatings Technol.*, **65**, No.820, 1993, 23-30.
16 Molphy M, Mainwaring D E, Rizzardo E, Gunatillake P A, Laslett R L, *Polym. Int.*, **37**, No.1,

1995, 53-61.
17 OgawaT, Ikeda M, *J. Adhesion*, **43**, Nos.1-2, 1993, 69-78.
18 Donnet J B, Wang W, Vidal A, Wang M J, *Kaut. u. Gummi Kunst.*, **46**, No.11, Nov.1993, 866-71.
19 Wang W D, Haidar B, Vidal A, Donnet J B, *Kaut. u. Gummi Kunst.*, **47**, No.4, 1994, 238-41.
20 Zaborskaya L V, Dovgualo V A, Yurkevich O R, *J. Adhesion Sci. Technol.*, **9**, No.1, 1995, 61-71.
21 Al-Turaif H, Unertl W N, Lepoutre P, *J. Adhesion Sci. Technol.*, **9**, No.7, 1995, 801-11.
22 Lee I, *J. Mat. Sci.*, **30**, No.23, 1995, 6019-22.
23 Bomal Y, Godard P, *Polym. Engng. Sci.*, **36**, No.2, 1996, 237-43.
24 Gatenhom P, Hedenberg P, Karlsson J, Felix J, Antec '96. Volume II. Conference proceedings, Indianapolis, 5th-10th May 1996, 2302-4.
25 Torro-Palau A, Fernandez-Garcia J C, Orgiles-Barcelo A C, Martin-Martinez J M, *J. Adhesion*, **57**, Nos.1-4, 1996, 203-25.
26 Foster J K, Sims E S, Venable S W, *Paint & Ink Int.*, **8**, No.3, 1995, 18-21.
27 Hollis R D, Hyche K W, Color and Appearance Retec: Effects in Plastics. Conference proceedings, Oak Brook, Il., 20th-22nd Sept.1994, 94-105.
28 Scott D M, *Composites, Part A*, **28A**, 1997, 703-7.
29 Yazici R, Kalyon D M, Antec '93. Conference Proceedings, New Orleans, La., 9th-13th May 1993, Vol. III, 2845-50.
30 Guerbe L, Freakley P K, *Kaut. u. Gummi Kunst.*, **48**, No.4, 1995, 260-9.
31 Toure B, Lopez-Cuesta J, Benhassaine A, Crespy A, *Int. J. Polym. Analysis and Characterization*, **2**, No.3, 1996, 193-202.
32 Frounchi M, Burford R P, Chaplin R P, *Polym. & Polym. Composites*, **2**, No.2, 1994, 77-82.
33 Dole P, Chauchard J, *Polym. Degradat. Stabil.*, **47**, No.3, 1995, 441-8.
34 Brovkop O O, Sergeeva L M, Slinchenko O A, Fainleib O M, *Polym. Int.*, **40**, No.4, 1996, 299-305.
35 Lei Yang, Schruben D L, *Polym. Engng. Sci.*, **34**, No.14, 1994, 1109-14.
36 Mamunya E P, Davidenko V V, Lebedev E V, *Polym. Composites*, **16**, No.4, 1995, 319-24.
37 Gokturk H S, Fiske T J, Kalyon D M, Antec '93. Conference Proceedings, New Orleans, La., 9th-13th May 1993, Vol. I, 605-8.
38 Van Beek G A, Pang S S, Lea R H, Antec '93. Conference Proceedings, New Orleans, La., 9th-13th May 1993, Vol. I, 602-4.
39 Nasr G M, Badawy M M, Gwaily S E, Shash N M, Hassan H H, *Polym. Degradat. Stabil.*, **48**, No.2, 1995, 237-41.
40 Svorcik V, Micek I, Jankovskij O, Rybka V, Hnatowicz V, Wang L, Angert N, *Polym. Degradat. Stabil.*, **55**, 1997, 115-21.
41 Wu S-L, Tung I-C, *Polym. Composites*, **16**, No.3, 1995, 233-9.
42 Saad A L G, Younan A F, *Polym. Degradat. Stabil.*, **50**, No.2, 1995, 133-40.
43 Meng-Jiao Wang, Wolff S, Freund B, *Rubb. Chem. Technol.*, **67**, No.1, 1994, 27-41.
44 Zihlif A M, Di Liello V, Martuscelli E, Ragosta G, *Int. J. Polym. Mat.*, **29**, Nos.3-4, 1995, 211-20.
45 Maas S, Gronski W, *Kaut. u. Gummi Kunst.*, **47**, No.6, 1994, 409-15.
46 Topoleski L D T, Ducheyne P, Cuckler J M, *J. Biomed. Mat. Res.*, **29**, No.3, 1995, 299-307.
47 Tamura J, Kawanabe K, Yamamuro T, Nakamura T, Kokubo T, Yoshihara S, Shibuya T, *J. Biomed. Mat. Res.*, **29**, No.5, 1995, 551-9.
48 Roche A A, Dole P, Bouzziri M, *J. Adhesion Sci. Technol.*, **8**, No.6, 1994, 587-609.
49 Furtado C R G, Nunes R C R, de Siqueira Filho A S, *Polym. Bull.*, **34**, No.5/6, 1995, 627-33.
50 Minkova L, Magagnini P L, *Polym. Degradat. Stabil.*, **42**, No.1, 1993, 107-15.
51 Alkan C, Arslan M, Cici M, Kaya M, Aksoy M, *Resources Conserv. & Recycling*, **13**, Nos.3-4, 1995, 147-54.
52 Helbert W, Cavaille J Y, Dufresne A, *Polym. Composites*, **17**, No.4, 1996, 604-11.
53 Tchoudakov R, Breuer O, Narkis M, Siegmann A, *Polym. Engng. Sci.*, **36**, No.10, 1996, 1336-46.
54 Casenave S, Ait-Kadi A, Brahimi B, Riedl B, Antec '95. Vol. II. Conference Proceedings, Boston, Ma., 7th-11th May 1995, 1438-42.
55 Kim K J, White J L, *J. of Non-Newtonian Fluid Mechanics*, **66**, Nos.2/3, 1996, 257-70.
56 Niedermeier W, Raab H, Maier P, Kreitmeier S, Goeritz D, *Kaut. u. Gummi Kunst.*, **48**, No.9, 1995, 611-6.
57 Eisenbach C D, Ribbe A, Goeldel A, *Kaut. u. Gummi Kunst.*, **49**, No.6, 1996, 406-10.
58 Zhao W, Hasegawa S, Fujita J, Yoshii F, Sasaki T, Makuuchi K, Sun J, Nishimoto S, *Polym. Degradat. Stabil.*, **53**, No.2, 1996, 199-206.
59 Jinman Huang, Yi Yang, Bai Yang, Shiyong Liu, Jiacong Shen, *Polym. Bull.*, **37**, No.5, 1996, 679-82.

60 Wagner A H, Kalyon D M, Yazici R, Fiske T J, Antec '97. Conference proceedings, Toronto, April 1997, 996-1000.
61 Gerard P, Raine J, Pabiot J, Antec '97. Conference proceedings, Toronto, April 1997, 526-31.
62 Bay R S, Tucker C L, PhD thesis, University of Illinois, Urbana, 1991.
63 Hshieh F Y, Beson H D, *Fire Mater.*, **21**, 1997, 41-9.
64 Merwin L H, Nissan R A, Stephens T S, Wallner A S, *J. Appl. Polym. Sci.*, **62**, No.2, 1996, 341-8.
65 Haynes A R, Coates P D, *J. Mat. Sci.*, **31**, No.7, 1996, 1843-55.
66 Kody R S, Martin D C, *Polym. Engng. Sci.*, **36**, No.2, 1996, 298-304.
67 Averous L, Quantin J C, Lafon D, Crespy A, *Int. J. Polym. Analysis and Characterization*, **1**, No.4, 1995, 339-47.
68 Herzig R, Baker W E, *J. Mat. Sci.*, **28**, No.24, 1993, 6531-9.
69 Benrashid R, Nelson G L, *J. Fire Sci.*, **11**, No.5, 1993, 371-93.
70 Fiske T J, Gokturk H S, Kalyon D M, Antec '93. Conference Proceedings, New Orleans, La., 9th-13th May 1993, Vol. I, 614-7.
71 Godovsky D Yu, *Adv. Polym. Sci.*, **119**, 1995, 81-122.
72 Herd C R, Meeting of the Rubber Division, ACS, Montreal, May 5-8, 1996, paper B.
73 Gerspacher M, O'Farrell C P, Yang H H, Wampler W A, Meeting of the Rubber Division, ACS, Montreal, May 5-8, 1996, paper C.
74 Lisy F, Hiltner A, Baer E, Katz J L, Meunier A, *J. Appl. Polym. Sci.*, **52**, No.2, 1994, 329-52.
75 Ohta M, Nakamura Y, Hamada H, Maekawa Z, *Polym. & Polym. Composites*, **2**, No.4, 1994, 215-21.
76 Nguyen T N, Lethiecq M, Levassort F, Patat F, *Int. J. Polym. Analysis and Characterization*, **1**, No.4, 1995, 277-87.
77 Gendron R, Daigneault L E, Tatibouet J, Dumoulin M M, Antec '94. Conference Proceedings, San Francisco, Ca., 1st-5th May 1994, Vol. I, 167-71.
78 Maeda S, Armes S P, *Synthetic Metals*, **73**, No.2, 1995, 151-5.
79 Quintanilla L, Pastor J M, *Polymer*, **35**, No.24, 1994, 5241-6.
80 Peng W, Riedl B, *Polymer*, **35**, No.6, 1994, 1280-6.
81 del Rio C, Acosta J L, *Polymer*, **35**, No.17, 1994, 3752-7.
82 Modesti M, Simioni F, Albertin P, *Cell. Polym.*, **13**, No.2, 1994, 113-24.
83 Privalko V P, Novikov V V, *Adv. Polym. Sci.*, **119**, 1995, 31-77.
84 Bigg D M, **Thermal and Electrical Conductivity of Polymers Materials**, Eds. Godovsky Y K, Privalko V P, *Springer*, Berlin 1995.
85 Rabello M S, White J R, Antec '97. Conference proceedings, Toronto, April 1997, 2991-5.
86 Albertsson A-C, Barenstedt C, Karlsson S, *J. Environmental Polym. Degradat.*, **1**, No.4, 1993, 241-5.
87 McNeill I C, Mohammed M H, *Polym. Degradat. Stabil.*, **48**, No.1, 1995, 189-95.
88 Hajji P, Cavaille J Y, Favier V, Gauthier C, Vigier G, *Polym. Composites*, **17**, No.4, 1996, 612-9.
89 Rabello M S, White J R, *Polym. Composites*, **17**, No.5, 1996, 691-704.
90 Jang J, Yi J, *Polym. Engng. Sci.*, **35**, No.20, 1995, 1583-91.
91 Kenny J M, Opalicki M, *Composites Part A: Applied Science and Manufacturing*, **27A**, No.3, 1996, 229-40.
92 McNeill I C, Mohammed M H, *Polym. Degradat. Stabil.*, **49**, No.2, 1995, 263-73.
93 Schreiber H P, Antec '95. Vol. II. Conference Proceedings, Boston, Ma., 7th-11th May 1995, 2446-51.
94 Hornsby P R, Wang J, Rothon R, Jackson G, Wilkinson G, Cossick K, *Polym. Degradat. Stabil.*, **51**, No.3, 1996, 235-49.
95 Toure B, Lopez Cuesta J M, Gaudon P, Benhassaine A, Crespy A, *Polym. Degradat. Stabil.*, **53**, No.3, 1996, 371-9.
96 Toure B, Lopez Cuesta J-M, Longerey M, Crespy A, *Polym. Degradat. Stabil.*, **54**, Nos 2-3, 1996, 345-52.
97 Sergeeva L M, Slinchenko E A, Brovko A A, Fainleib A M, Nedashkovskaya N S, *Polym. Sci., Ser. B*, **38**, Nos.5/6, 1996, 225-30.
98 Beloshenko V A, Kozlov G V, Slobodina V G, Prut E V, Grinev V G, *Polym. Sci., Ser. B*, **37**, Nos.5-6, 1995, 316-8.
99 Pukanszky B, Belina K, Rockenbauer A, Maurer F H J, *Composites*, **25**, No.3, 1994, 205-14.
100 Szocs F, Klimova M, Chodak I, Chorvath I, *Eur. Polym. J.*, **32**, No.3, 1996, 401-2.
101 Rockenbauer A, Korecz L, Pukanszky B, *Polym. Bull.*, **33**, No.5, 1994, 585-9.
102 Chen L, Liu K, Yang C Z, *Polym. Bull.*, **37**, No.3, 1996, 377-83.
103 Suzuki F, Nakane K, Piao J-S, *J. Mat. Sci.*, **31**, No.5, 1996, 1335-40.
104 Darmstadt H, Roy C, Kaliaguine S, Sahouli B, Blacher S, Pirard R, Brouers F, *Rubb. Chem. Technol.*,

68, No.2, 1995, 330-41.
105 Wang W, Vidal A, Donnet J-B, Wang M-J, *Kaut. u. Gummi Kunst.*, **46**, No.12, Dec.1993, 933-40.
106 Sain M M, Kokta B V, Antec '93. Conference Proceedings, New Orleans, La., 9th-13th May 1993, Vol. I, 320-4.
107 Garbow J R, Asrar J, Hardiman C J, *Chem. of Mat.*, **5**, No.6, 1993, 869-75.
108 Hegedus C R, Kamel I L, *J. Coatings Technol.*, **65**, No.820, 1993, 31-43
109 Zumbrum M A, *J. Adhesion*, **46**, Nos.1-4, 1994, 181-96.
110 Vidal A, Wang W, Donnet J B, *Kaut. u. Gummi Kunst.*, **46**, No.10, 1993, 770-8.
111 Donnet J B, *Kaut. u. Gummi Kunst.*, **47**, No.9, 1994, 628-32.
112 Zaborski M, Slusarski L, Donnet J B, Papirer E, *Kaut. u. Gummi Kunst.*, **47**, No.10, 1994, 730-8.
113 Ulkem I, Bataille P, Schreiber H P, *J. Macromol. Sci. A*, **31**, No.3, 1994, 291-303.
114 Rebouillat S, Escoubes M, Gauthier R, Vigier A, *J. Appl. Polym. Sci.*, 58, No.8, 1995, 1305-15.
115 Balard H, Saada A, Hartmann J, Aouadj O, Papirer E, *Macromol. Symp.*, **108**, 1996, 63-80.
116 Persson A L, Bertilsson H, *Composite Interfaces*, **3**, No.4, 1996, 321-32.
117 Kettrup A A, Lenoir D, Thumm W, Kampke-Thiel K, Beck B, *Polym. Degradat. Stabil.*, **54**, Nos 2-3, 1996, 175-80.
118 Ryabikova V M, Zigel A N, Sverdlova S I, Vinogradova G A, *Int. Polym. Sci. Technol.*, **23**, No.1, 1996, T/89-90.
119 Yeh J T, Yang H M, Huang S S, *Polym. Degradat. Stabil.*, **50**, No.2, 1995, 229-34.
120 Chudinova V V, Guzeev V V, Mozzhukhin V B, Pomerantseva E G, Nozrina F D, Zhil'tsov V V, Zubov V P, *Int. Polym. Sci. Technol.*, **21**, No.10, 1994, T/102-4.
121 Lee D H, Condrate R A, Reed J S, *J. Mater. Sci.*, **32**, 1997, 471-8.
122 Gailliez-Degremont E, Bacquet M, Laureyns J, Morcellet M, *J. Appl. Polym. Sci.*, **65**, 1997, 871-82.
123 Young R J, *Prog. Rubb. Plast. Technol.*, **11**, No.2, 1995, 124-36.
124 Parker A A, Martin E S, Clever T R, *J. Coatings Technol.*, **66**, No.829, 1994, 39-46.
125 Abramova N A, Diikova E U, Lyakhovskii Yu Z, *Polym. Sci.*, **36**, No.9, 1994, 1308-9.
126 Molphy M, Laslett R L, Gunatillake P A, Rizzardo E, Mainwaring D E, *Polym. Int.*, **34**, No.4, 1994, 425-31.
127 Zolotnitsky M, Steinmetz J R, *J. Vinyl and Additive Technol.*, **1**, No.2, 1995, 109-13.
128 Tsubokawa N, Seno K, *J. Macromol. Sci. A*, **31**, No.9, 1994, 1135-45.
129 Gerspacher M, O'Farrel C P, Wampler W A, *Rubb. World*, **212**, No.3, 1995, 26-9.
130 Melanitis N, Tetlow P L, Galiotis C, *J. Mat. Sci.*, **31**, No.4, 1996, 851-60.
131 Datta S, Bhattacharya A K, De S K, Kontos E G, Wefer J M, *Polymer*, **37**, No.12, 1996, 2581-5.
132 Singhal A, Fina L J, *Polymer*, **37**, No.12, 1996, 2335-43.
133 Caillaud J L, Deguillaume S, Vincent M, Giannotta J C, Widmaier J M, *Polym. Int.*, **40**, No.1, 1996, 1-7.
134 Leveque D, Auvray M H, *Composites Sci. & Technol.*, **56**, No.7, 1996, 749-54.
135 Ohwaki T, Ishida H, *J. Adhesion*, **52**, Nos.1-4, 1995, 167-86.
136 Sanchez-Solis A, Estrada M R, *Polym. Degradat. Stabil.*, **52**, No.3, 1996, 305-9.
137 Delor F, Lacoste J, Lemaire J, Barrois-Oudin N, Cardinet C, *Polym. Degradat. Stabil.*, **53**, No.3, 1996, 361-9.
138 Gordienko V P, Dmitriev Y A, *Polym. Sci., Ser. B*, **37**, Nos.5-6, 1995, 249-50.
139 Grohens Y, Schultz J, *Int. J. Adhesion Adhesives*, **17**, 1997, 163-7.
140 Domke W D, Halmheu F, Schneider S, *J. Appl. Polym. Sci.*, **54**, No.1, 1994, 83-90.
141 Dutta N K, Choudhury N R, Haidar B, Vidal A, Donnet J B, Delmotte L, Chezeau J M, *Polymer*, **35**, No.20, 1994, 4293-9.
142 Ebengou R H, Cohen-Addad J P, *Polymer*, **35**, No.14, 1994, 2962-9.
143 Bluemler P, Bluemich B, *Acta Polymerica*, **44**, No.3, 1993, 125-31.
144 Fuelber C, Bluemich B, Unseld K, Herrmann V, *Kaut. u. Gummi Kunst.*, **48**, No.4, 1995, 254-9.
145 Ou Y C, Yu Z Z, Vidal A, Donnet J B, *J. Appl. Polym. Sci.*, **59**, No.8, 1996, 1321-8.
146 Legrand A P, *Macromol. Symp.*, **108**, 1996, 81-96.
147 Joseph R, Zhang S, Ford W T, *Macromolecules*, **29**, No.4, 1996, 1305-12.
148 Mori M, Koenig J L, *Rubb. Chem. Technol.*, **68**, No.4, 1995, 551-62.
149 Connor C, Chadwick M M, *J. Mat. Sci.*, **31**, No.14, 1996, 3871-7.
150 Addad J P C, Frebourg P, *Polymer*, **37**, No.19, 1996, 4235-42.
151 Erofeev L N, Raevskii A V, Pisarenko T I, Grishin B S, *Int. Polym. Sci. Technol.*, **23**, No.5, 1996, T/12-4.
152 Hess M, Veeman W, Magusin P, Antec '96. Volume III. Conference proceedings, Indianapolis,

5th-10th May 1996, p.3682-6.
153 Heuert U, Knorgen M, Menge H, Scheler G, Schneider H, *Polym.Bull.*, **37**, No.4, Oct.1996, 489-96.
154 Gaw F M, Enhancing Polymers Using Additives and Modifiers, Rapra, Shawbury, 1993.
155 Robertson D R, Gaw F, AddCon '95, Basel, 1995.
156 Wu J, Lerner M M, *Chem. of Mat.*, **5**, No.6, 1993, 835-8.
157 Yazici R, Kalyon D M, Antec '97. Conference proceedings, Toronto, April 1997, 2076-80.
158 Kovacevic V, Lucic S, Hace D, Glasnovic A, Smit I, Bravar M, *J. Adhesion*, **47**, No.1-3, 1994, 201-15.
159 Anantharaman M R, Kurian P, Banerjee B, Mohamed E M, George M, *Kaut. u. Gummi Kunst.*, **49**, No.6, 1996, 424-6.
160 Strauss M, Pieper T, Peng W, Kilian H G, *Makromol. Chem., Macromol. Symp.*, **76**, 1993, 131-6.
161 Nagieb Z A, El-Sakr N S, *Polym. Degradat. Stabil.*, **57**, 1997, 205-9.
162 Le Bras M, Bourbigot, Le Tallec Y, Laureyns J., *Polym. Degradat. Stabil.*, **56**, 1997, 11-21.
163 Suh C H, White J L, *Polym. Engng. Sci.*, **36**, No.17, 1996 2188-97.
164 Isayama M, Nomiyama K, Kunitake T, *Adv. Mat.*, **8**, No.8, 1996, 641-4.
165 Suh C H, White J L, Antec '96. Vol.I. Conference Proceedings, Indianapolis, 5th-10th May 1996, p.958-62.
166 Wang P H, Hong K L, Zhu Q R, *J. Appl. Polym. Sci.*, **62**, No.12, 1996, 1987-91.
167 Guanghong Lu, Xiaotian Li, Hancheng Jiang, *Composites Sci. & Technol.*, **56**, No.2, 1996, 193-200.
168 Tsutsumi K, Ban K, Shibata K, Okazaki S, Kogoma M, *J. Adhesion*, **57**, Nos.1-4, 1996, 45-53.
169 Dupraz A M P, de Wijn J R, v. d. Meer S A T, de Groot K, *J. Biomed. Mat. Res.*, **30**, No.2, 1996, 231-8.
170 Jacks J, *Plast. Rubb.* Tiejun Wang, Sherwood P M A, *Chem. of Mat.*, **7**, No.5, 1995, 1031-40. Wkly., No.1667, 1996, 7.
171 Tiejun Wang, Sherwood P M A, *Chem. of Mat.*, **7**, No.5, 1995, 1031-40.
172 Wang T, Sherwood P M A, *Chem. of Mat.*, **6**, No.6, 1994, 788-95.
173 Reis M J, Do Rego A M B, Da Silva J D L, *J. Mat. Sci.*, **30**, No.1, 1995, 118-26.
174 Byung Suk Jin, Kwang Hee Lee, Chul Rim Choe, *Polym. Int.*, **34**, No.2, 1994, 181-5.

# 15

# Fillers in Commercial Polymers

Several aspects of this discussion are common to all polymers. These include the major applications of polymers, the most common methods of polymer processing, the most frequently used fillers and their typical concentrations, additives used to incorporate fillers, special methods of filler incorporation, fillers pre-treatment, and special considerations affecting the selection of a filler. These data found in the most recent literature are recorded in a tabular form for clarity and ease of comparison. This is followed by some examples of current use of fillers in particular polymers. The examples can be used to develop numerous new applications for fillers in material improvement. They elaborate on the most recent developments in filler and filled material improvements.

The application of polymer affects choice of filler. For example, to prepare conductive materials, special fillers must be used to obtain the required properties. Also, the method of processing imposes certain constraints on the choice and treatment of the filler before its use. For example, polymers processed at high temperature require fillers which do not contain moisture. This affects both the choice of the filler and/or its pretreatment. The choice of additives used to improve the incorporation of the filler depends on the application and the properties required from a product but it is also determined by the processing method. For example, the viscosity of a melt is reduced by special lubricating agents whereas the viscosity of filler dispersions is controlled by the surface treatment of filler. In some cases, the order of addition is important or a special filler pretreatment is used to achieve the desired results. These methods are discussed in special section in the table. Some fillers simply cannot be used with some polymers. In other cases, special care must be taken to ensure polymer stability or filler may interact with some vital components of the formulation. This subject is discussed in special considerations of filler choice.

## 15.1 ACRYLICS[1-2]

| | |
|---|---|
| **Major polymer applications** | paints, coatings, roofing, sealants, elastomers, fibers, industrial filters, battery separators, carbon and graphite fibers, ceramics |
| **Important processing methods** | mixing/compounding, extrusion, calendering, Banbury mixer, vulcanization, wet and dry spinning |
| **Typical fillers** | calcium carbonate, titanium dioxide, fumed silica, zinc oxide, carbon black, aluminum silicate, graphite, ceramic microspheres |
| **Typical concentration range** | 20-70 wt% |
| **Auxiliary agents** | silane, mostly amine-functional |
| **Special methods of incorporation** | in ceramics manufacture, the presence of sodium ions on the alumina surface could inhibit the formation of covalent bonding with acrylate polymers, the surface adsorption can be regulated by adjusting pH[1] |
| **Methods of filler pretreatment** | neutralization in ceramic applications |
| **Special considerations** | titanium dioxide accelerates UV degradation; copper compounds catalyze thermal and UV degradation; neutral and basic fillers are recommended because acidic fillers and pigments retard cure |

Acrylic polymers also include water emulsions of acrylic resins, acrylate resins used in ceramic applications, and the precursor of carbon fiber, namely acrylonitrile. The table includes also some information on acrylic elastomers. Polymethylmethacrylate is discussed under a separate subsection.

Water emulsions used in paints, coatings, and sealants contain substantial amounts of fillers which are incorporated by the conventional methods of dispersion. Frequently, grinding is used in the paint industry. In sealants, fillers are used for reinforcement, rheology, and crosslinking. The rheology of a sealant may be controlled by the incorporation of fumed silica in quantities around 3 wt%. The non-sag properties of sealant are partially due to fumed silica but are also regulated by pH adjustment in the presence of other additives such as special acrylic resins and polyurethane thickeners. The combination of both effects gives the sealant its final properties. The reinforcement of the sealant is produced through a combination of two processes: the interaction of silica particles and crosslinking through zinc oxide.

Fillers in a sealant are added in limited quantities because high loadings affect the elastomeric properties of acrylic resin. In coatings, larger quantities of fillers and pigments can be used because the coating is not required, in most cases, to tolerate large elongations. Exterior coatings, which require crack bridging capabilities, are an exception. In this case, elongation in excess of 1000 % is required. Here, the filler load is substantially reduced. Some exterior textured coatings or stuccos, require unusual fillers such as silica flours, glass beads, and ceramic microspheres. These fillers are used to obtain different decorative effects. Silica flour added in large quantities plays the role of the classical filler added to reduce price. It often is added in large quantities, sometimes as high as 80 wt%. Silica flour of various particle size is mixed to achieve increased packing density and imitates the effect of the old cementitious stucco. In some finishes, ceramic microspheres of large sizes (several millimeters in diameter) are added to produce surface imperfections. During tooling they form grooves and holes of different sizes which imitate the handmade finishes used throughout the world but, most notably, in Italy. Solid, colored glass beads 1 mm in diameter and larger are used at maximum packing density to obtain a type of finish which has a color similar to the color of the glass beads (or their mixtures). The acrylic resin in this application acts as a binder. Several different colors can be mixed together to obtain its required effect which is usually named after some traditional well-known name, e.g., *pfefer and saltz* (pepper and salt).

Work on ceramics shows the effects of polymer adsorption on the filler surface (Figure 15.1). The adsorption is a reversible phenomenon controlled by the pH of an alumina slurry. Coiled chains are formed at low pH and, due to ionization, stretched out chains form at a higher pH.

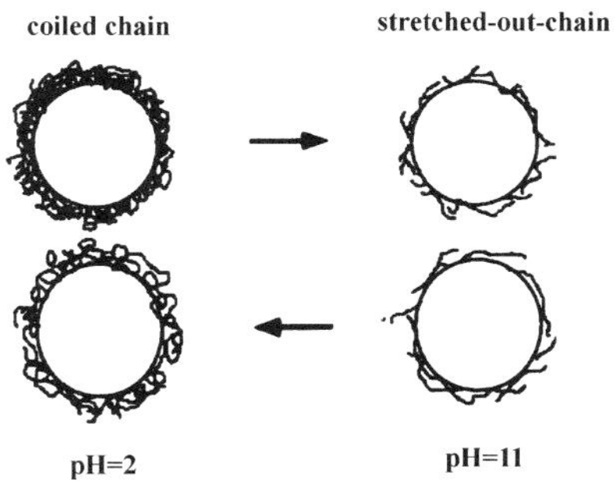


Figure 15.1. Configuration of polyacrylate adsorbed at different pH. [Adapted, by permission, from Lee D H, Condrate R A, Reed J S, *J. Mater. Sci.*, **32**, 1997, 471-8.]

## 15.2 ACRYLONITRILE-BUTADIENE-STYRENE COPOLYMER (ABS)[3-6]

| | |
|---|---|
| **Major polymer applications** | appliance (refrigerator liners, kitchen appliance housings, vacuum cleaners, power tools), automotive (instrument panels, consoles, door parts, knobs, trim, wheel covers, mirror and headlight housing, front radiator grilles), business machines (computers, discs, phones), packaging, pipes and fittings, recreation (snowmobiles, boats, vehicles), toys, luggage, lunch and tool boxes, medical applications, lawn and garden equipment, furniture, hot tubs, military |
| **Important processing methods** | injection molding, extrusion, film lamination, calendering, blow molding, thermoforming |
| **Typical fillers** | talc, magnesium hydroxide, antimony oxide, carbon black, nickel or copper coated carbon fibers, glass beads |
| **Typical concentration range** | carbon black for color and UV protection - 0.5-3 wt%, general and flame retarding fillers - 30-60 wt%; nickel coated fibers for EMI shielding - 10 wt% (about half compared with the required concentration of carbon fiber), stainless steel fiber - 0.5-3 wt% |
| **Auxiliary agents** | premixing of nickel coated fibers in polymer prior to injection molding increased the shielding efficiency of fibers[5] |
| **Special methods of incorporation** | color concentrate of carbon black gave 3 times better jetness compared with mixed-in powdered carbon black[3] |
| **Methods of filler pretreatment** | drying |
| **Special considerations** | moisture in excess of 0.1% will cause bubbling; iron, copper, manganese, cobalt impurities catalyze oxidation |

Filler mixing technology is important in ABS processing. Carbon black must be well dispersed to obtain good jetness and impact strength. High jetness is relatively easy to obtain by the use of high surface area carbon black and by adjusting its concentration to requirements. As the carbon black content is increased, high impact strength becomes more difficult to maintain because impact strength decreases as undispersed surface area increases. The best impact retention is achieved when a lower concentration of carbon black was subjected to two stage mixing. First carbon black is dispersed in ABS and then the granulate obtained after solvent evaporation is used as a color concentrate.[3] The impact retention of carbon black containing ABS dispersed in this way is independent of the surface area of carbon black but decreases as the concentration of carbon black increases.

Dispersion of conductive fillers is even more critical. Here,[5] two methods have been used. In one, the nickel coated carbon fibers were added directly to the ABS. in the other, the coated fibers were pre-dispersed with a solvent in ABS, then the solvent was evaporated to form a granulate. Figure 15.2 shows the results. A two stage dispersion is clearly critical for obtaining good shielding effectiveness. Figure 15.3 shows the effect of nickel coating on carbon fiber performance. A coating

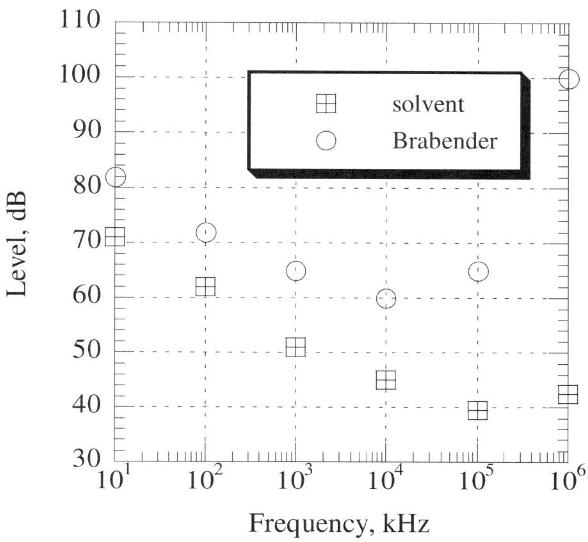


Figure 15.2. Shielding effectiveness of nickel-coated carbon fiber in ABS depending on method of mixing. [Adapted, by permission, from Guanghong Lu, Xiaotian Li, Hancheng Jiang, *Composites Sci. & Technol.*, **56**, No.2, 1996, 193-200.]

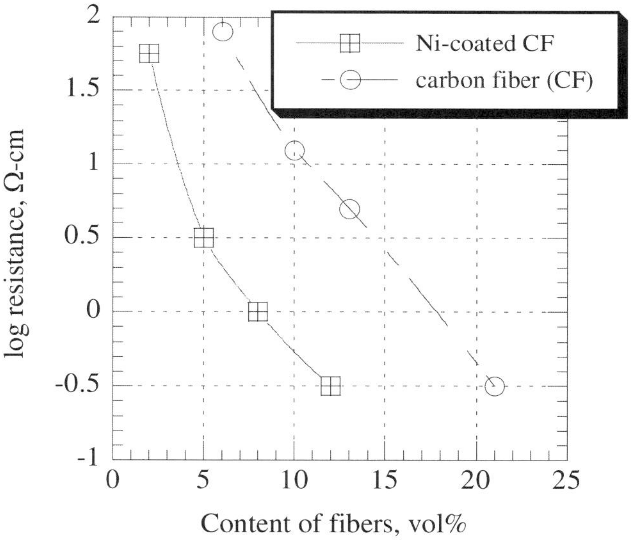


Figure 15.3. Resistivity of ABS composite vs. content of fibers. [Adapted, by permission, from Guanghong Lu, Xiaotian Li, Hancheng Jiang, *Composites Sci. & Technol.*, **56**, No.2, 1996, 193-200.]

with an optimal thickness of 0.2-0.5 μm allows for a substantial reduction in carbon fiber (as much as by one half) needed to obtain the conductance required. Nickel coating makes the fiber more conductive and gives better mechanical properties to the fiber which helps it to withstand processing conditions.

## 15.3 ACRYLONITRILE-STYRENE-ACRYLATE (ASA)[7]

| | |
|---|---|
| **Major polymer applications** | mirrors for personal watercraft, recreational vehicle antennas, pool accessories, exterior cable enclosures, large screen displays, sheet outdoor furnishings, profiles, spas, marine applications, skylights, ski bindings |
| **Important processing methods** | injection molding, extrusion, thermoforming |
| **Typical fillers** | carbon black, glass beads |
| **Typical concentration range** | 5-20 wt% |
| **Auxiliary agents** | silane for treatment of glass beads |
| **Special methods of incorporation** | not reported |
| **Methods of filler pretreatment** | silane coupling agent |
| **Special considerations** | none reported |

The addition of glass beads to ASA resulted in the decrease of tensile and flexural yield strength, notched and unnotched impact strength, and resistance to steady crack propagation. Flexural modulus was the only parameter which was improved.[7]

## 15.4 ALIPHATIC POLYKETONE[8]

| | |
|---|---|
| **Major polymer applications** | automotive (fuel lines and connectors, fuel pump components, fuel tanks, filters, injection rails, air inlet, manifolds, gears, wheel covers), gears for business machines, liners for flexible fuel hoses, automotive fuel system components, industrial molded parts, film fibers |
| **Important processing methods** | injection molding, blow molding, extrusion, powder coating, spinning |
| **Typical fillers** | glass fiber, mica, wollastonite, calcium carbonate |
| **Typical concentration range** | 15-30 wt% |
| **Auxiliary agents** | not reported |
| **Special methods of incorporation** | not reported |
| **Methods of filler pretreatment** | not reported |
| **Special considerations** | not reported |

Several fillers (glass fiber, mica, wollastonite, and calcium carbonate) were compared in this study.[8] Flexural modulus increased as the concentration of filler increased, but the highest rate of increase was obtained first with glass fiber then next with mica. Calcium carbonate and wollastonite caused a much lower change in flexural modulus. Flexural strength was substantially improved by glass fiber but all of the other fillers had practically no effect on flexural strength. The deflection temperature under load was increased by all fillers but the gains obtained with glass fiber were spectacular. Only 5 wt% glass fiber was needed to double the heat deflection temperature. A further increase in glass fiber concentration did not produce any further improvement. Izod impact strength is increased by glass fiber. All other fillers decrease impact strength. This study shows that although all of the fillers tested can be used with polyketone but only one – glass fiber improves all of its mechanical properties.

## 15.5 ALKYD RESINS[9]

| | |
|---|---|
| **Major polymer applications** | coatings, paints, varnishes, electrical applications, pavement marking, artist's paints, putties, printing inks |
| **Important processing methods** | compounding/mixing, molding |
| **Typical fillers** | calcium carbonate, titanium dioxide, glass fiber, silica, iron oxides, clay, mica, zinc oxide, lithopone |
| **Typical concentration range** | 30-60 wt% |
| **Auxiliary agents** | flow control additives, wetting and dispersing additives, antifloating additives |
| **Special methods of incorporation** | grinding, sand milling |
| **Methods of filler pretreatment** | typically no treatment |
| **Special considerations** | filler porosity is important consideration in silica matting agents; some silica matting agents affect thixotropy of alkyd paints |

Alkyd resins are used in many traditional applications which have a well established technology. New studies are infrequently reported. The inherent high gloss of alkyd paints requires special materials which can produce matt surfaces finish. Several products (different grades of Gasil) have been developed recently for this purpose. Rheological studies have shown that some types of silica disrupt the rheological networks of paints. Silica which did this had pore size of 100 to 400 Å. Special types of alkyd resins are required to take advantage of these rheological modifiers. If the pore size of silica is large enough to capture the structure forming components of the alkyd resin, silica affects the performance of alkyd resin. Wax coating does not affect this process of rheological interference. The matting effect depends on the number of particles and the number of which protrude from the surface. These properties, combined with smaller pore sizes, make some silica good matting agents. An added advantage is that they can be stirred into the paint.[9]

## 15.6 ELASTOMERS, TPO

| | |
|---|---|
| **Major polymer applications** | automotive, car bumper, wire insulation, hose tube, sheet |
| **Important processing methods** | injection molding, blow molding, extrusion |
| **Typical fillers** | calcium carbonate, talc, carbon black, wollastonite |
| **Typical concentration range** | carbon black - 1-2 wt%, general filler - 10-30 wt%; proprietary formulations |
| **Auxiliary agents** | not reported |
| **Special methods of incorporation** | not reported |
| **Methods of filler pretreatment** | components must be dry before processing |
| **Special considerations** | talc reduces mold shrinkage |

## 15.7 EPOXY RESINS[10-57]

| | |
|---|---|
| **Major polymer applications** | surface protective coatings (protective and decorative - automotive, metal cans, industrial flooring, anticorrosive paints), electrical/electronics (printed circuit panels, conductive adhesives), composites (building/construction, marine, electrical/electronics, aircraft, communication satellites, automotive, pipes, consumer products), bonding and adhesives, flooring, tooling and casting, biosensors |
| **Important processing methods** | casting, lamination, molding, compounding |
| **Typical fillers** | calcium carbonate, barium sulfate, talc, kaolin, mica, quartz, sand, glass spheres, silica, titanium dioxide, aluminum hydroxide, carbon fiber, glass fiber, aramid fiber, aluminum, copper, silver, iron, graphite, molybdenum disulfide, zirconium silicate, lithium aluminum silicate, vermiculite, slate powder, titanium boride, ground rubber, iron oxide, microvoids |
| **Typical concentration range** | up to 90 wt% (sand); in situ formed silica - 2-43 wt%, alumina - 5-20 wt%, fractal approach to the critical filler volume fraction in conductive composites determines the required concentration of filler[45] |
| **Auxiliary agents** | aminopropyltriethoxy silane,[18,43] epoxy silane,[28] modified polyethyleneimide,[18] epoxy resin[12,46] |
| **Special methods of incorporation** | *in situ* polymerization of metal alkoxides;[11] nanocomposite synthesis[49, 50] |
| **Methods of filler pretreatment** | Kevlar fibers for applications in composites were treated by coating with epoxy resin (1% epoxy resin on the fiber surface) or microwave plasma in an atmosphere of ammonia which grafts amine groups on the fiber surface. Fiber coating with epoxy reduced water absorption of composite whereas ammonia treated fiber behave in a manner similar to untreated fiber; rubber-epoxy interface was modified by carbonyl terminated copolymers of butadiene-acrylonitrile;[12,43] plasma modification of carbon fiber;[21] glass beads were encapsulated by grafted polybutyl acrylate and polystyrene;[30] admicellar polymerization of styrene monomer on the surface of glass fiber improved properties of composite;[36] epoxy sizing of carbon fibers;[46] plasma treatment of carbon fibers was found to produce graphitic, phenolic, carbonyl, and carboxyl groups;[46] vapor-grown carbon fibers are produced by decomposing of gas-phase hydrocarbon in the presence of ultrafine iron catalyst;[47] |
| **Special considerations** | fillers absorb part of the heat generated during the curing;[15] additions of silica (35-52 wt%) to UV-curable epoxy did not have a significant effect on cure rate;[17] the use of microvoids (see Figure 8.35) for epoxy toughening gives the same results as rubber toughening;[24]glass A beads filled epoxy degrades rapidly in a marine environment when glass concentration increases above 12 wt%;[27] $Fe_2O_3$ catalyzes the curing reactions of epoxy resins[29] |

Epoxy mortars for patching applications are the most highly filled materials. With the proper selection of silica sand mixture as the filler, the filler concentration can be as high as 95 wt%. This, and similar materials produced from a reactive polyurethane systems are probably the most highly filled plastics. Another example of highly filled epoxies can be found in the liquid metal compounds which are epoxies highly filled with metals such as steel, aluminum, titanium, bronze, and copper. The loading of metal powder in these materials can reach 70 wt%. This level of iron powder was used in pipeline sealants.[19] The inclusion of metal powders has a bearing on the corrosion protection of epoxy coating.[44] The addition of iron powder improves corrosion protection whereas addition of copper or nickel reduces the protective capabilities of epoxy coatings.[44] Barrier properties of epoxy coatings containing adhesion promoter and glass flakes have been evaluated by electrochemical impedance spectroscopy.[48]

Titanium boride was used in an epoxy system to induce electrical conductivity.[12] A high filler loading of 46 vol% was selected to study the effect of thermal expansion and contraction on electric conductivity. The filler chosen was composed of rigid particles which cannot form particle-particle connections other than through direct contact. Figure 15.4 shows the relationships of relative thermal expansion and specific electric resistivity. The resistivity increases slowly with temperature until the contacts between particles of filler are broken which results in a rapid increase in resistance.[12]

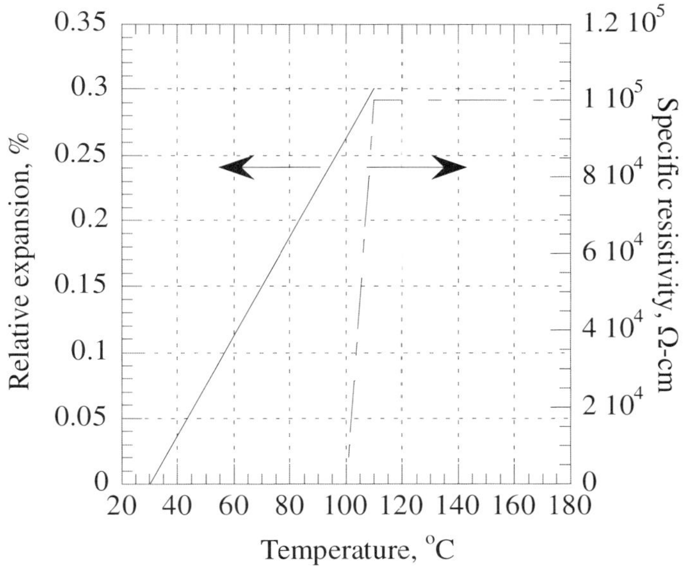


Figure 15.4. Thermal expansion and electric conductivity vs. temperature for epoxy filled with 46 vol% titanium boride. [Data from Strumpler R, Maidorn G, Garbin A, Ritzer L, Greuter F, *Polym. & Polym. Composites*, **4**, No.5, 1996, 299-304.]

The resistivity of epoxy resins depends on the loading of fillers and on their chemical composition (Figure 15.5). When good filler dispersion was the criterion for determining maximum filler loading there was a considerable difference in the

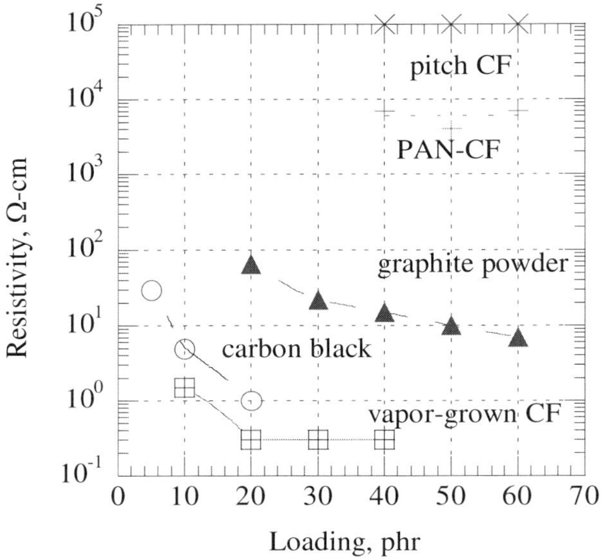


Figure 15.5. Resistivity of epoxy resin vs. filler load. [Adapted, by permission, from Katsumata M, Endo M, Ushijima H, Yamanishi H, *J. Mat. Res.*, **9**, No.4, 1994, 841-3.]

loading achieved by various types of fillers. Only 20 phr of carbon black could be incorporated. Graphite fiber, pitch and PAN carbon fibers could be used up to a concentration of 60 phr. The vapor-grown carbon fiber could be used up to only 40 phr but it was the one which gave the best performance in reduction of resistivity.[47]

Thermal conductivity and expansion are important properties of adhesives used in electronics. Both properties influence the performance of computer chips. Generally, the chip has a protective cover which is attached by an adhesive. The adhesive bond must be maintained during thermally induced movement in the chip. The chip is bonded to its base with an adhesive which must also take thermal movement and, in addition, transfer heat from the chip. Two epoxy adhesives were used in the study: silica filled epoxy (65 and 75 wt% $SiO_2$ epoxy) and epoxy containing 70 wt% Ag.[26] Figure 15.6 shows their thermal conductivities. The behavior of both adhesives is completely different. The silver filled adhesive had a maximum conductivity at about 60°C whereas the maximum for $SiO_2$ filled adhesive was 120°C. The $T_g$ of both adhesives was 50 and 160°C, respectively. Below its $T_g$, the thermal conductivity of the adhesive increases at the expense of increased segmental motions in the chain molecules. Above the $T_g$ the velocity of photons rapidly decreases with increasing temperature and the thermal conductivity also decreases rapidly.

Microwave propagation in carbon black/epoxy resin composites shows that for small particle size inclusions, magnetic wave propagation increases with filler concentration but for large particles the propagation of magnetic waves does not depend on the concentration of the inclusions.[25]

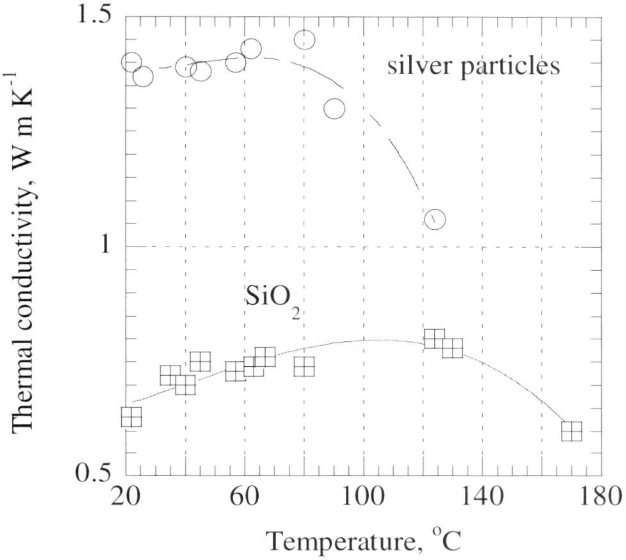


Figure 15.6. Thermal conductivity of filled epoxies vs. temperature. [Data from Nyilas A, Rehme R, Wyrwich C, Springer H, Hinrichsen G, *J. Mat. Sci. Lett.*, **15**, No.16, 1996, 1457-9.]

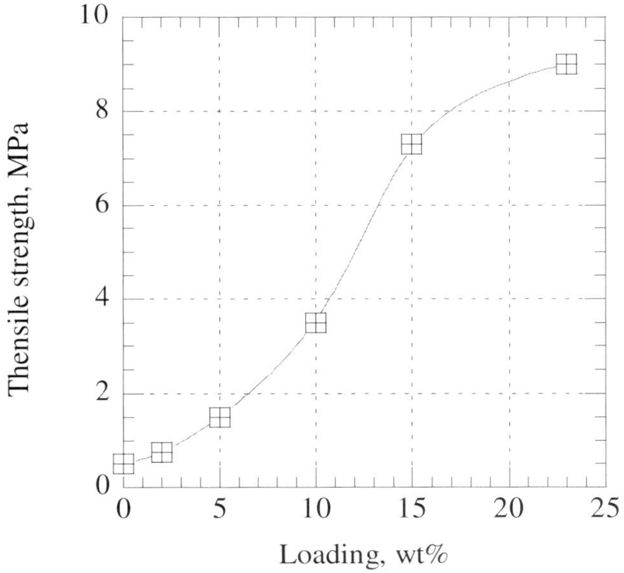


Figure 15.7. Tensile strength of epoxy/montmorillonite nanocomposite vs. filler concentration. [Adapted, by permission, from Lan T, Pinnavaia T J, *Chem. of Mat.*, **6**, No.12, 1994, 2216-9.]

Treatment of glass beads with epoxy silane improved the adhesion of glass beads/epoxy composite which resulted in a substantially lower water uptake and better retention of properties after water immersion.[28] A similar improvement in composite properties was obtained in a jute/epoxy system.[54] Glass beads mixed with rubber particles improved the fatigue resistance of an epoxy composite reducing the localized stress at the crack tip.[32] The methods of improvement of adhesion between the carbon fiber and the matrix are reviewed elsewhere.[53] The compressive strength of an epoxy composite can be improved by glass beads, quartz, and calcium carbonate (see Figure 8.25).[34] The shape of the particle has an effect on fracture behavior of filled epoxy composite.[35] The best properties are obtained with spherical silica particles. Fracture behavior of glass fabric/epoxy laminates can be improved by the addition of alumina.[37] Mechanical properties can be dramatically increased (tensile strength is increased 9 times) when the clay used as a filler is in the nanocomposite form (Figure 15.7).

Filled epoxy resins wastes can be ground and used as a filler for various materials.[39,40] Recycled rubber particles are sometimes used for toughening of epoxy resins.[52]

## 15.8 ETHYLENE VINYL ACETATE COPOLYMER, EVA[58-62]

| | |
|---|---|
| **Major polymer applications** | hot-melt coatings, hot-melt adhesives, wall covering adhesives, paints, tubing, sporting goods, footwear, baby products, controlled release devices, wire and cable (semiconductor shields, automotive wire, automotive ignition, low-smoke cable), asphalt modification, slow burning candles, cap liners |
| **Important processing methods** | mixing/compounding, injection molding, extrusion, Banbury mixer, two-roll mills, cold feed extruders, reaction injection molding |
| **Typical fillers** | calcium carbonate, clay, aluminum hydroxide, magnesium hydroxide, zinc oxide, silica, quartz, red phosphorus |
| **Typical concentration range** | up to 45 wt% (magnesium hydroxide up to 60 wt%)[62] aluminum hydroxide up to 80 wt% |
| **Auxiliary agents** | siloxane,[58] silanes, fatty acids, unsaturated polymeric acids |
| **Special methods of incorporation** | stearic acid is used to avoid sticking during the mixing; red phosphorus is usually added in the form of a masterbatch containing 50 wt% red phosphorus |
| **Methods of filler pretreatment** | thermal treatment,[58] trimethylchlorosiloxane treatment[58] |
| **Special considerations** | red phosphorus is an excellent flame retardant for EVA, used up to 8 wt% it gives V-0 rating |

The dynamic viscosity and shear modulus of silica filled EVA were related to the work of adhesion of filler particles.[58] The work of adhesion depends on the particle size distribution and load of filler. The increase in particle size of filler causes an increase in work of adhesion.

## 15.9 ETHYLENE-ETHYL ACETATE COPOLYMER, EEA[63-66]

| | |
|---|---|
| **Major polymer applications** | hose, tubing, additive to other polymers |
| **Important processing methods** | extrusion/compounding, Banbury mixer, extrusion, calendering |
| **Typical fillers** | calcium carbonate, titanium dioxide, zinc oxide, carbon black, aluminum silicate, graphite, magnesium hydroxide, aluminum hydroxide |
| **Typical concentration range** | carbon black 5-20 wt%, calcium carbonate and many other fillers up to 50 wt% |
| **Auxiliary agents** | silane, mostly amine-functional |
| **Special methods of incorporation** | roll milling |
| **Methods of filler pretreatment** | none reported |
| **Special considerations** | metal hydroxides catalyze the oxidation of char to convert carboxyl groups to carbon dioxide[65] |

Calcium carbonate stabilizes EEA at elevated temperatures. The stabilizing effect depends on particle size, type of calcium carbonate and coating. Calcite gives better stabilization than whiting or precipitated calcium carbonate. Also, a stearate coated grade improves thermal stability. The filler prevents formation of acetic and propanoic acids by interacting with carboxyl groups. Metal hydroxides (Mg and Al) also stabilize EEA as they decompose endothermally. For this mechanism to work, the metal hydroxide must decompose at a much lower temperature than the degradation temperature of the copolymer. Otherwise, the hydroxide will interact with the reactive groups of the copolymer and change the mechanism of the degradation of the polymer.[64,65]

## 15.10 ETHYLENE-PROPYLENE COPOLYMERS, EPR & EPDM[67-84]

| | |
|---|---|
| **Major polymer applications** | automotive radiator hose, garden hose, wire and cable, tires, roofing, gaskets, conveyor belts |
| **Important processing methods** | extrusion, molding, calendering, coating |
| **Typical fillers** | calcium carbonate, calcinated clay, aluminum hydroxide, magnesium carbonate, magnesium hydroxide, antimony trioxide, calcium borate, huntite, hydromagnesite, zinc oxide, talc, silica |
| **Typical concentration range** | carbon black 20-40 wt%, most others 30-65 wt% |
| **Auxiliary agents** | organic silane,[71] aliphatic alcohols[80] |
| **Special methods of incorporation** | processing history has an essential effect on conductivity; shear imposed during mixing causes a fracture of secondary carbon aggregates; increased temperature during mixing may preferentially form rubber-carbon bonds rather than the carbon-carbon bonds required for conductivity; vulcanization temperature may affect recovery of broken connections between carbon-carbon bonds; talc reduces melt viscosity which results in a smooth surface of extruded and calendered products as well as reduced wear of equipment[71] |
| **Methods of filler pretreatment** | silane treated talc increases modulus and reduces compression set;[77] maleated or sulfonated EPDM interacts with zinc oxide;[75,76] esterification of precipitated silica with methanol, propanol, and hexadecanol;[80] modification of precipitated silica with methacrylic and vinyl silane;[80] calcium carbonate was modified with monoallyl and monodecyl maleate[81] |
| **Special considerations** | in conductive applications special conductive carbon blacks must be employed; fire resistant cables have good balance of properties when 180/20 phr alumina/magnesium carbonate is used; for high LOI 400 phr alumina is used; zinc oxide is used as an ionic crosslinker;[72,74,75] carbon black was found to be a very efficient UV stabilizer for a system having a high electrical conductivity;[69] there is an interaction between ionic crosslinks with zinc oxide and hydroxyl groups of carbon black through hydrogen bonding[73] |

In EPR formulations, calcium borate was found to be a good replacement for the combination of antimony trioxide with an organic flame retardant.[67] Calcium borate, in addition to affecting flame retardation, also reinforces the polymer. Another alternative is based on huntite/hydromagnesite filler. Here, some antimony trioxide and organic flame retardant combination must be added. The huntite/magnesite filler combination cannot, by itself, halt flame spread.[68,70]

In carbon black filled EPDM, the production of foamed materials is affected by filler.[78] Cell density decreases with the amount of carbon black but increases with the amount of blowing agent. The size of cells decreases in the presence of carbon black because of the alkaline surface of carbon black.[78]

## 15.11 IONOMERS

| | |
|---|---|
| **Major polymer applications** | membranes, adhesives, elastomeric applications |
| **Important processing methods** | injection molding, extrusion, vulcanization, molding, calendering, coating |
| **Typical fillers** | calcium carbonate, aluminum hydroxide, magnesium carbonate, magnesium hydroxide, zinc oxide, talc, silica |
| **Typical concentration range** | 20-40 wt% |
| **Auxiliary agents** | silanes |
| **Special methods of incorporation** | not reported |
| **Methods of filler pretreatment** | not reported |
| **Special considerations** | not reported |

## 15.12 LIQUID CRYSTALLINE POLYMERS, LCP[85-89]

| | |
|---|---|
| **Major polymer applications** | microwave cookware, fiber optic connectors, capsules for electronic devices, watches, cameras, audiovisual equipment, under-bonnet automotive components, aerospace structures |
| **Important processing methods** | injection molding, fiber spinning, extrusion |
| **Typical fillers** | glass fiber, wollastonite, carbon black, magnesium carbonate |
| **Typical concentration range** | 30-70 wt% |
| **Auxiliary agents** | not reported |
| **Special methods of incorporation** | materials have skin/core morphology in which the relative amounts of skin and core vary with processing parameters as does the distribution of reinforcement[88] |
| **Methods of filler pretreatment** | drying is very important to prevent hydrolysis[89] |
| **Special considerations** | processing result strongly depends on previous thermal and pre-shear treatment which may cause thermal instability;[85] glass and mineral fibers are known to have weak adhesion to LCP therefore decrease its toughness;[87] small additions of carbon black (1%) reduce melt viscosity[89] |

Thermotropic liquid crystalline polymers can be formulated with high concentration of glass fiber to withstand working temperatures in excess of 300°C. In processing LCP, one problem arises. LCP orients itself in the direction of shear or flow – the process which benefits many materials but makes products from pure LCP excessively anisotropic. To balance mechanical properties it has been suggested[87] that some quantities of short glass or mineral fibers be added to LCP.

Figure 9.13 shows the effect of filler concentration on torque. The smallest increase was due to magnesium carbonate and the largest due to the presence of glass fibers.[89] The mechanical properties of filled composites are substantially improved by additions of magnesium carbonate, wollastonite and glass fiber. The most important improvement is in creep resistance (Figures 6.68 and 8.69).

## 15.13 PERFLUOROALKOXY RESIN, PFA[90]

| | |
|---|---|
| **Major polymer applications** | silicon wafer carriers, pump, pipes, fittings, filtration, tubing, column packing, marine coatings, wear resistant products, coating for hostile environments, automotive weather seals for doors and windows |
| **Important processing methods** | injection molding, extrusion, coating |
| **Typical fillers** | magnesium oxide, calcium hydroxide, PTFE, graphite, molybdenum disulfide, carbon black, metal particles |
| **Typical concentration range** | 3-30 wt% |
| **Auxiliary agents** | not reported |
| **Special methods of incorporation** | not reported |
| **Methods of filler pretreatment** | not reported |
| **Special considerations** | metal oxides play the role of acid acceptors |

## 15.14 PHENOLIC RESINS[55,57,91-98]

| | |
|---|---|
| **Major polymer applications** | aircraft interiors, automotive (pump housings, transmission reactors, timing pulleys), marine, construction, coatings, adhesives, carbonless copy paper, abrasives, friction materials, laminates, foundry resins, battery separators, wood bonding, composites, foam, hollow spheres |
| **Important processing methods** | lamination, molding, coating, compounding |
| **Typical fillers** | wood flour, glass fiber, carbon fiber, mica, wollastonite, mineral wool, talc, magnesium hydroxide, graphite, molybdenum sulfide, carbon black, cashew shell particles, alumina, chromium oxide, brass and copper powder, iron particles, steel fiber, ceramic powder, rubber particles, aramid, wollastonite, cellulosic fiber, lignin |
| **Typical concentration range** | 30-60 wt% |
| **Auxiliary agents** | stearates, fluoropolymers, carboxylic groups-containing copolymers which reduce viscosity of filled polyester (BYK-W 995)[92] |
| **Special methods of incorporation** | not reported |
| **Methods of filler pretreatment** | lignin treated by methylolation decreases the rate of cure of phenolic adhesives;[94] carbon fiber was anodically oxidized and subjected to various treatments with coupling agents[97] to improve interfacial interaction with phenolic resins and oxidative stability of carbon fibers; titanate coupling of oxidized fibers resulted in improved adhesion to matrix and enhanced thermal stability of fibers[98] |
| **Special considerations** | cobalt salts reduce UV and thermal stability; steel fiber and acrylic fiber give the best wear retention to brake pads;[55] composites containing graphite have much better flame retarding properties than composites containing aramid or glass fiber;[57] glass fiber slows down the cure rates of novolac resins[96] |

Figure 15.8 shows the effect of aramid fibers on the friction coefficient and the specific wear rate of brake pads. Additions of up to 15 vol% aramid fiber are economical to reduce the coefficient of friction decrease which remains constant up to 40 vol%. At the same time, the specific wear rate decreases steadily as fiber concentration increases. This suggests that wear rate is improved by the material reinforcement.[55]

Fire resistance is an important property of phenolic resins. The combination of phenolic resin with Expancel™ expandable microspheres leads to many useful products. Composites for high speed train interiors take advantage of the light weight, excellent fire rating, and very low thermal conductivity.[91] Polyester filled with aluminum hydroxide is an alternative solution for train interior materials. The resin and filler can be easily processed when viscosity regulating additives are added.[92]

The properties of novolac laminates can be improved by the addition of fillers.[93] Corrosion protective materials suffered from delamination because of varia-

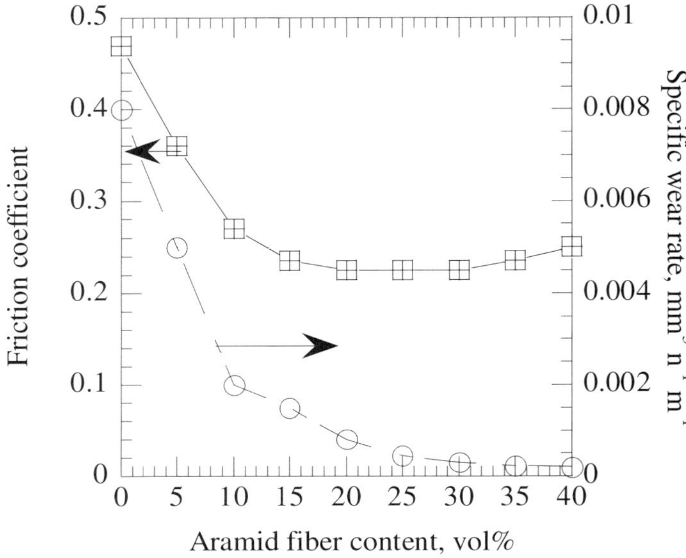


Figure 15.8. Friction coefficient and specific wear rate of phenolic brake pads containing varying concentrations of aramid fiber. [Adapted, by permission, from Bijwe J, *Polym. Composites*, **18**, No.3, 1997, 378-96.]

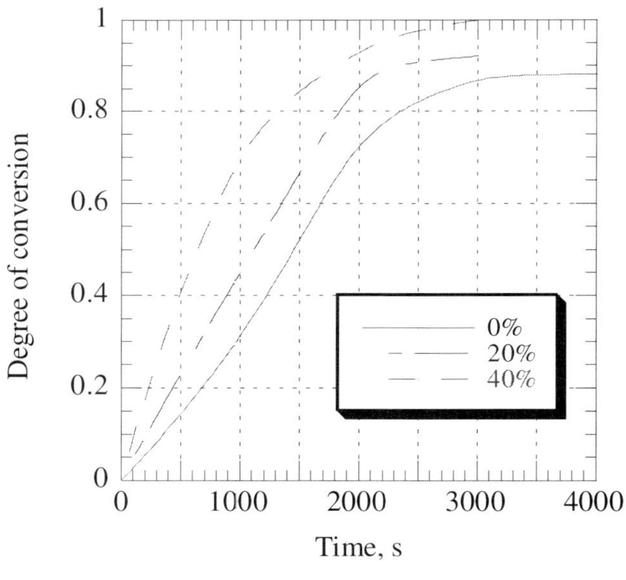


Figure 15.9. Degree of conversion vs. cure time of novolac resin. [Adapted, by permission, from Murayama H, Min K, Antec '97. Conference proceedings, Toronto, April 1997, 759-65.]

tions in thermal expansion rates. When the laminate was made from two layers – one containing carbon fiber and the other filled with 15-20% graphite powder, the heat flow and temperature distribution across the laminate was improved and delamination was eliminated.[93]

Figure 15.9 shows the effect of glass fiber on the rate of conversion of novolac resin.[96] The addition of glass fiber decreases rate of curing and degree of conversion. The chemical mechanism of this process is not discussed.

## 15.15 POLY(ACRYLIC ACID), PAA[99]

| | |
|---|---|
| **Major polymer applications** | dispersants for pigments and fillers, thickeners, toothpaste, hydraulic fluids, ion exchange resins, binder for ceramic, dental cements, polyelectrolytes |
| **Important processing methods** | compounding |
| **Typical fillers** | metal oxide, kaolin, clay |
| **Typical concentration range** | as required for application |
| **Auxiliary agents** | not applicable |
| **Special methods of incorporation** | none reported |
| **Methods of filler pretreatment** | none reported |
| **Special considerations** | poly(acrylic acid) has acid properties therefore it will interact readily with basic fillers such as for example alumina or magnesium hydroxide[99] |

## 15.16 POLYAMIDES, PA[4,100-126]

| | |
|---|---|
| **Major polymer applications** | automotive industry (radiator end tanks, inlet manifolds, rocker covers), electrical components (connectors, switches, motor frames), bearing cages, mechanical handling components, fibers, carpets, tire reinforcement, many other applications |
| **Important processing methods** | melt spinning, injection molding, extrusion |
| **Typical fillers** | glass fiber, carbon fiber, aramid, antimony trioxide, zinc borate, stainless steel fiber, graphite, nickel coated graphite, aluminum flakes, metallized glass |
| **Typical concentration range** | stainless steel fiber - 1-7 wt%, 30 wt% or more graphite, 15-50 wt% carbon black for conductive materials, glass spheres and fibers up to 70 wt%, carbon fiber - 20 wt%, silicon oxide and silicates up to 40 wt%, LCP up to 30 wt%, wollastonite - 40 wt% (can be used in conjunction with glass fiber,[103] copper/polyamide-11 composite was made with 90 wt% spherical copper powder[109] |
| **Auxiliary agents** | silanes,[125] waterborne silanes,[119] compatibilizers in polymer blends |
| **Special methods of incorporation** | for delustering, titanium dioxide is added to polymer at 210°C to avoid excessive agglomeration; the order of addition of glass fiber to PP/PA-6 blend affects blend mechanical performance, glass fiber must be added to already compatibilized blend to avoid filler encapsulation;[107] the use of vacuum hopper and premixing of polymer with copper spheres causes a reduction in porosity of highly filled polyamide[109] |
| **Methods of filler pretreatment** | silane treatment of wollastonite;[103] polyamide has ability to wet carbon fiber, polyamide behaves like a melt at 180°C even though its melting temperature is 225°C[115] |
| **Special considerations** | copper compounds catalyze thermal and UV degradation; titanium dioxide lowers UV stability, titanium dioxide is used as agent; red phosphorus in combination with zinc borate gives V-0 or V-1 rating with halogen-free system and inhibits corrosion because it can trap trace amounts of phosphine produced from red phosphorus |

In applications which require electric conductivity, polyamides are processed either with carbon fiber or with graphite. These applications include business machines (copying machines, computer printers), electronic packaging, carpet fiber, and EMI shielding. Other fillers, such as nickel coated graphite, stainless steel fiber, aluminum flakes and metallized glass are used less often. Polyamide is one of the best EMI shielding materials. When compounded with only 15% nickel coated glass fiber, it gives an attenuation of 50 db. By comparison, polyamide compounded with 30% graphite fiber gives an attenuation of only 30 db. Figure 15.10 illustrates the affect of particle size on resistivity.[110] The general rule for filled materials is that the lower the particle size of the conductive particle the higher the conductivity of the resultant material. In conductive materials filled with fibers, this relationship is more complex and more dependent on filler type.

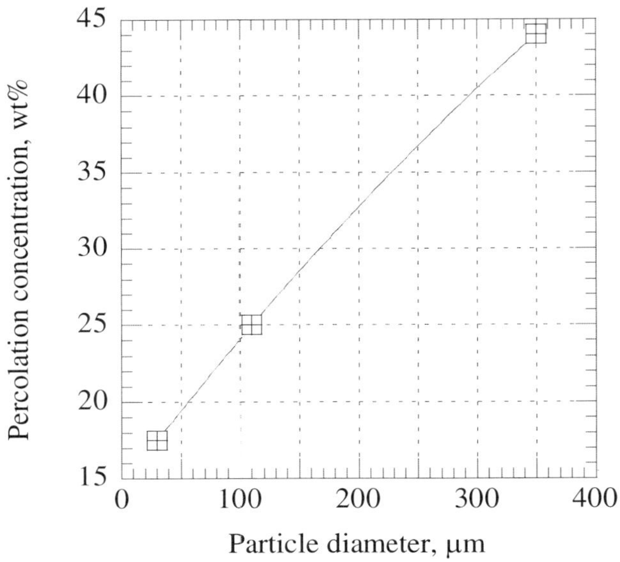


Figure 15.10. Weight fraction of copper for lower percolation concentration vs. particle diameter of copper particles in polyamide. [Adapted, by permission, from Larena A, Pinto G, *Polym. Composites*, **16**, No.6, 1995, 536-41.]

Less than 10% of the polyamide produced is made in a flame retardant version. The best system is composed of a combination of red phosphorus and zinc borate (see table above). The only drawback of this system is its color which is restricted to brick red or black. If other colors are required, ammonium polyphosphate is used either in combination with organic flame retardants or with antimony trioxide. It is possible to manufacture a very wide range of colors in the halogen free system. Some systems make use of the addition of novolac or melamine resins. For intumescent applications, ammonium polyphosphate, in combination with other components, is the most frequently used additive. Figure 13.6 shows that fillers such as calcium carbonate and talc (at certain range of concentrations) improve the effectiveness of ammonium polyphosphate. This is both unusual and important.[100] It is unusual because, in most polymers, the addition of fillers has an opposite influence on the efficiency of ammonium polyphosphate and it is important because ammonium polyphosphate must be used in large concentrations (minimum 20%, typical 30%) in order to perform as a flame retardant.

The use of magnesium hydroxide in polyamides is restricted by the low degradation temperature and the low hydrolytic stability of polyamides. Polyamide 6 and 6.6 begin to degrade at around 350°C whereas magnesium hydroxide releases water between 320 and 440°C. *In situ* production of water lowers the thermal stability of polyamides.[101] The addition of 60 wt% magnesium hydroxide produced a flame

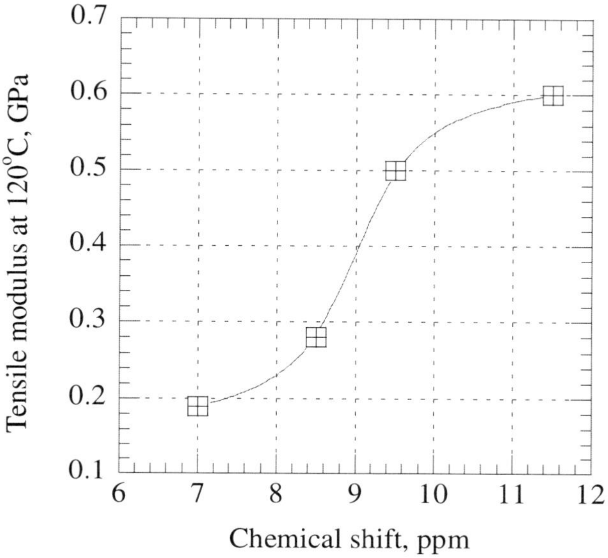


Figure 15.11. Tensile modulus of composites made from nylon and different fillers (montmorillonite, saponite, hectorite and mica) vs. $^{15}$N-NMR chemical shifts of model compounds of fillers. [Adapted, by permission, from Usuki A, Koiwai A, Kojima Y, Kawasumi M, Okada A, Kurauchi T, Kamigaito O, *J. Appl. Polym. Sci.*, **55**, No.1, 1995, 119-23.]

retardant polyamide 6 with its degradation temperature overlapping with that of the filler.[111] A similar method was not successful with polyamide 6.6.

Fillers play an important role in powder coating of polyamide to form articles with a metal-like look (e.g., handles, mountings for radiators and pipes). For a material to be powder coated, it must withstand the stoving temperatures (170°C or more). It must also be electrically conductive, be chargeable and its reactive groups must be able to link with the coating system.[104] Fillers such as metal and metal coated ceramic spheres and carbon fibers are added to polyamide for its strength and paintability.

When aluminum borate whiskers were incorporated in SAN/PA-6 blends, the whisker, due to acid/based interaction had a better affinity to PA-6 than to SAN and for this reason they ended up in residence in the polyamide phase.[105] A similar principle was used to obtain a conductive blend by compounding PA, PP and carbon black.[106] Carbon black has a better affinity to polyamide and thus it prefers to reside in the polyamide phase or in the interphase formed between two immiscible polymers. Even when carbon black is added initially to polypropylene, it still transfers to the polyamide phase. This migration and preferential location of carbon black is not only an interesting scientific principle but it also has important practical implications. For the purpose of conductivity, a certain threshold concentration of carbon black is required to enable percolation and this threshold concentration is substantially lower if the carbon black settles in only one phase. This phase becomes richer in carbon black than the average concentration in blend. This principle allows materials to be formulated in more efficient manner.

Fiber reinforcement plays important role in polyamide processing. The orientation of fibers is not only influenced by the method of processing but also by external strain. Figures 14.9 and 9.29 show the effect of small strains on glass fiber orientation in polyamide 6.[114] The reinforcement also depends on interaction between the filler and the matrix polymer. Figure 15.11 shows the relation between NMR chemical shifts of model compounds based on five different fillers and their tensile moduli. The compounds with a higher positive charge density on the nitrogen atoms in the polyamide molecule form stronger composites because the interaction between filler and matrix has an ionic character.[124]

## 15.17 POLYAMIDEIMIDE, PAI

| | |
|---|---|
| **Major polymer applications** | metal compressors in aerospace applications, pump housings, compressor valve plates, bushings, bearings, wear pads, piston rings and seals, gears, fasteners, plastic engine |
| **Important processing methods** | injection molding, extrusion, compression molding |
| **Typical fillers** | carbon fiber, glass fiber, graphite, fluorocarbon, PTFE |
| **Typical concentration range** | carbon fiber up to 30 wt%, glass fiber 30-40 wt%, graphite up to 20 wt%, PTFE fiber 1-2 wt% |
| **Auxiliary agents** | not reported |
| **Special methods of incorporation** | drying (polyamideimide may blister if it contains moisture and its temperature is increased rapidly) |
| **Methods of filler pretreatment** | drying to prevent degradation and blistering |
| **Special considerations** | post cure which takes several days to obtain peak mechanical properties |

## 15.18 POLYAMINES

| | |
|---|---|
| **Major polymer applications** | flocculation of particulate matter, pigment retention aids in paper (e.g. $TiO_2$), filtration aids, cosmetics |
| **Important processing methods** | compounding |
| **Typical fillers** | titanium dioxide |
| **Typical concentration range** | as required by application |
| **Auxiliary agents** | not reported |
| **Special methods of incorporation** | not reported |
| **Methods of filler pretreatment** | not reported |
| **Special considerations** | not reported |

## 15.19 POLYANILINE, PANI[127]

| | |
|---|---|
| **Major polymer applications** | materials having electric conductivity, protection against static electricity, EMI shielding, corrosion protection |
| **Important processing methods** | compounding |
| **Typical fillers** | silica |
| **Typical concentration range** | up to 40 wt% |
| **Auxiliary agents** | silica can be used in combination with colloid forming polymers such as PVAl or PVP |
| **Special methods of incorporation** | polymerization in the presence of filler[127] |
| **Methods of filler pretreatment** | none reported but the type of filler plays essential role in obtaining stable polyaniline dispersion |
| **Special considerations** | a certain level of fillers is required to obtain stable dispersion; the dispersed polymer or filler must be chosen such that it does not interfere with the conductivity of polyaniline |

## 15.20 POLYARYLETHERKETONE, PAEK[128]

| | |
|---|---|
| **Major polymer applications** | valve seats, pump impellers, valve linings, oil well data logging tools, bearing cages, aerospace, cryogenic propellant tank for supersonic aircrafts, nuclear power plants, satellites, fuel valves, bolts and nuts, heat-resistant gears, vacuum pump blades, butterfly valve seatings, piston rings, chemically resistant bearings and cams, machine tools, horizontal stabilizers for helicopters, ducting, semiconductor wafer carriers, belts, tennis racket strings, surgical instruments, sterilization equipment for medical and dental applications, bone screws, implants, fracture fixation plates, high performance conveyors, hot melt adhesives |
| **Important processing methods** | injection molding, extrusion, wire coating, mixing, melt spinning |
| **Typical fillers** | glass fiber, carbon fiber, graphite, PTFE |
| **Typical concentration range** | 20-30 wt% |
| **Auxiliary agents** | not reported |
| **Special methods of incorporation** | not reported |
| **Methods of filler pretreatment** | heat treatment of carbon fibers increases their resistance to oxidation but their adhesion decreases |
| **Special considerations** | carbon fiber requires treatment to increase adhesion by, for example, oxidation of its surface |

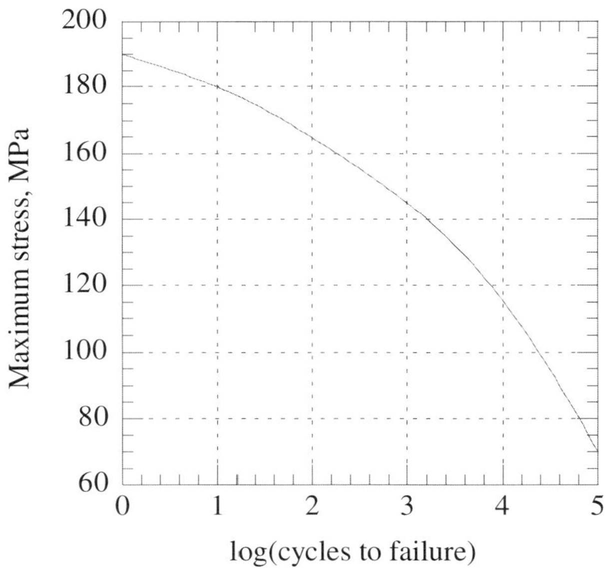


Figure 15.12. Maximum stress vs. cycles to failure for composites based poly(phenylene ether ketone). [Data from Zhou J, Li G, Li B, He T, *J. Appl. Polym. Sci.*, **65**, 1997, 1857-64.]

Figure 15.12 shows the fatigue resistance of carbon and glass fiber filled poly(phenylene ether ketone).[128] The flexural fatigue depends on tensile properties of the composite. The yield strength of the matrix and the quality of the interface affect the fatigue properties of composites.

## 15.21 POLY(BUTYLENE TEREPHTHALATE), PBT[102,103,116,129]

| | |
|---|---|
| **Major polymer applications** | composites, textiles, brush bristles, tire cords, electrical and electronics (connectors, circuit breakers, capacitor housings), automotive (distributor caps, mirror housings, door knobs), housewares, lighting, power tools, sporting goods, plumbing |
| **Important processing methods** | injection molding, extrusion, monofilament extrusion |
| **Typical fillers** | carbon fiber, glass fiber, aramid, mica, talc, calcinated kaolin, antimony trioxide, carbon black, zinc borate, glass spheres |
| **Typical concentration range** | 20-40 wt% |
| **Auxiliary agents** | coupling agents used in composites |
| **Special methods of incorporation** | the process design should account for the much faster crystallization rates of filled materials |
| **Methods of filler pretreatment** | drying, silane coupling |
| **Special considerations** | material must be completely dry before processing to avoid hydrolysis; the concentration of moisture in material ready for processing should be below 0.05% possibly below 0.025%; carbon black is the most effective UV stabilizer followed by $TiO_2$; ferrocene and cobalt salts accelerate UV degradation; zinc borate improves laser marking of glass filled composite; addition of glass spheres to glass fiber filled PBT improves melt flow, shrinkage and warpage;[102] these same improvements cannot be achieved by using talc/glass fiber mixtures;[102] mica controls warpage and increases flexural modulus and strength, heat distortion temperature, and dielectric strength;[103,116] increase in particulate filler content decreases the ductility of weld (the residual strength 50% of the strength of the matrix without filler) but addition of glass fiber only slightly reduces weld strength (10% for 30 wt% glass filled composite)[129] |

## 15.22 POLYCARBONATE, PC[106,108,130-135]

| | |
|---|---|
| **Major polymer applications** | compact disks, optical lenses, camera components, goggles, safety glasses, windows, laminated walls, skylights, copying machines, computer printers, gear, bearings, guide pins, rollers, speedometer needles, windscreens, instrument panels, head lamp covers and housings, tool boxes, dental applications, blood collector containers, disposable syringes, medical tubing, pacemaker components, dinnerware, drinking cups, toys |
| **Important processing methods** | extrusion, calendering, blow molding, injection molding, gas-assisted injection molding, thermoforming, solution casting |
| **Typical fillers** | glass fiber, wollastonite, titanium dioxide, boric oxide, carbon black and graphite fibers for EMI shielding, molybdenum sulfide, graphite, PTFE |
| **Typical concentration range** | 10-40 wt%; carbon black 5-20 wt% (tear strength maximum at 5 wt%, tensile strength maximum at 15 wt%) |
| **Auxiliary agents** | poly(methyl siloxane) and glycidoxy-propyltrimethoxy silane[134] |
| **Special methods of incorporation** | none reported |
| **Methods of filler pretreatment** | moisture content of composite should be below 0.02% to avoid hydrolytic changes as well as blistering |
| **Special considerations** | some inorganic pigments accelerate UV degradation; in PP/PC blend carbon black has higher affinity to PC and it is preferentially located in PC phase (compare with polyamide above);[106] glass fiber reinforced PC has very good retention of mechanical properties on exposure to γ-radiation (see Figure 11.6);[108] thermal stability of PC/carbon fiber composite is also good (see Figure 11.15)[130] |

Flame retardant materials are produced from polycarbonate. Several options are available to produce such materials. A combination of PTFE fibers (2 wt%) and boric oxide (1 wt%) gives V-1 rating. Good results are also obtained by using a combination of alumina and silica or a blend of magnesium carbonate, calcium carbonate with zinc borate. Halogen-free flame retardant grades are readily produced. The addition of zinc borate to polycarbonate substantially reduces the heat release and smoke generation from the compound.[132]

The thermal properties of polycarbonate are not outstanding and the opportunities to improve it are remote. Addition of 30 wt% of carbon or glass fibers increases deflection temperature under load only by about 20°C to the maximum attainable value of 150°C but coefficient of the thermal expansion is drastically reduced which makes it suitable for many of its potential applications.

Polycarbonate is used in applications which require EMI shielding and static control. EMI shielding requires large quantities of conductive fillers. For example, 40 wt% aluminum flake gives an attenuation of 32 dB, 30 wt% graphite fibers gives 42 dB, and 15 wt% nickel coated glass fiber gives 45 dB.[135]

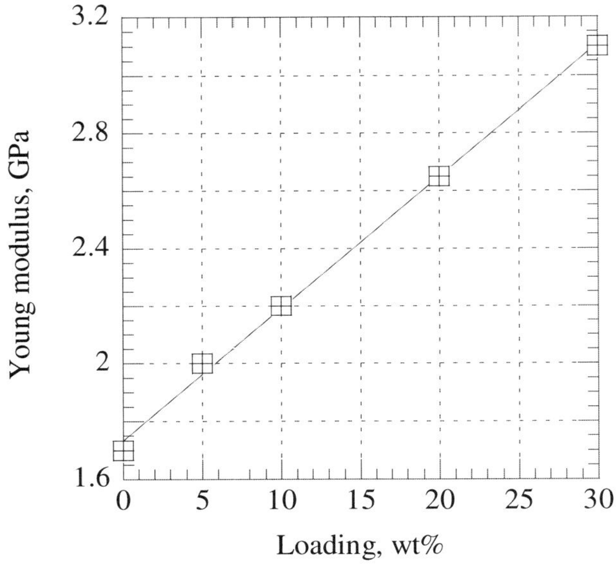


Figure 15.13. Young modulus of polycarbonate vs. amount of carbon fiber. [Adapted, by permission, from Zihlif A M, Di Liello V, Martuscelli E, Ragosta G, *Int. J. Polym. Mat.*, **29**, Nos.3-4, 1995, 211-20.]

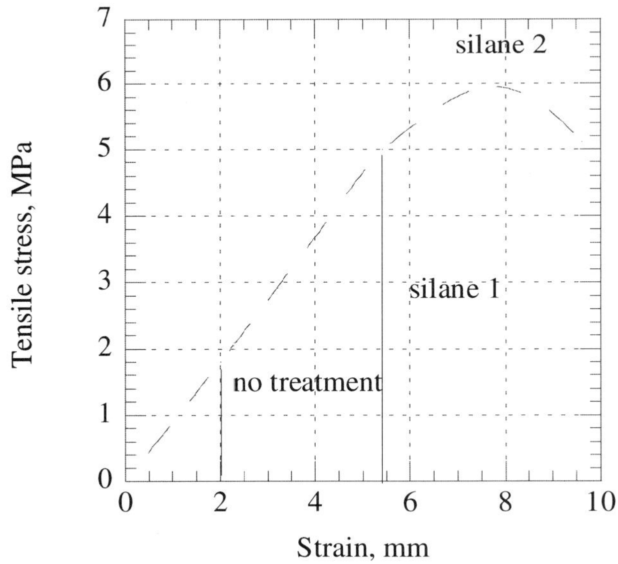


Figure 15.14. Strain-stress curve of 10% zinc oxide filled polycarbonate. [Adapted, by permission, from Tanaka T, Waki Y, Hamamoto A, Nogami N, Antec '97. Conference proceedings, Toronto, April 1997, 3054-8.]

Fiber reinforcement improves mechanical properties by amounts relative to the amount of fiber used. Figure 15.13 gives the relationship between Young modulus and the load of carbon fibers.[133] The effect of carbon fibers on tensile yield stress can be determined from Figure 8.7. Morphological observations indicate that there is good interfacial bonding. Figure 15.14 vividly illustrates the gains in improved mechanical properties that a coupling agent can provide. Two to six percent of coupling agent was used to treat ZnO for a 3-fold improvement in tensile stress.[134]

## 15.23 POLYETHERETHERKETONE, PEEK[85,136-139]

| | |
|---|---|
| **Major polymer applications** | valve seats, pump impellers, valve linings, oil well data logging tools, bearing cages, aerospace, cryogenic propellant tank for supersonic aircrafts, nuclear power plants, satellites, fuel valves, bolts and nuts, heat-resistant gears, vacuum pump blades, butterfly valve seatings, piston rings, chemically resistant bearings and cams, machine tools, horizontal stabilizers for helicopters, ducting, semiconductor wafer carriers, belts, tennis racket strings, surgical instruments, sterilization equipment for medical and dental applications, bone screws, implants, fracture fixation plates, high performance conveyors, hot melt adhesives |
| **Important processing methods** | injection molding, extrusion, wire coating, mixing |
| **Typical fillers** | glass fiber, carbon fiber, graphite, PTFE |
| **Typical concentration range** | typically 30 wt% but high glass loading can go up to 75 wt%[85] |
| **Auxiliary agents** | binders are needed in prepregging step (polyimide is an example of such binder)[137] |
| **Special methods of incorporation** | drying is only necessary to prevent molding defects because the polymer has very high resistance to hydrolysis |
| **Methods of filler pretreatment** | heat treatment of carbon fibers increases their resistance to oxidation but their adhesion decreases |
| **Special considerations** | high temperature treated carbon fiber requires treatment to increase adhesion by, for example, oxidation; wear and friction grades usually contain combination of graphite, carbon fiber, and PTFE |

The introduction of fillers to PEEK creates a higher nucleation rate. The surface of carbon fibers and nuclei within the PEEK matrix compete for crystallization growth. Epitaxial transcrystalline growth was frequently observed on the fiber surface in carbon fiber reinforced PEEK composites.

Reinforcement with glass fiber or carbon fiber doubles the tensile strength and modulus and, at the same time, the impact strength is also increased. Carbon fibers improve properties by at least 50% over PEEK filled with glass fibers. Thermal properties are also improved in this reinforcement, again especially by carbon fiber. Heat deflection temperature more than doubles due to reinforcement to values over 300°C. Solvent resistance is usually determined by the degree of fiber reinforcement and by the adhesion between fiber and matrix. PEEK/carbon fiber composite has an excellent resistance to water (Figure 15.15)[139] and to high temperature.

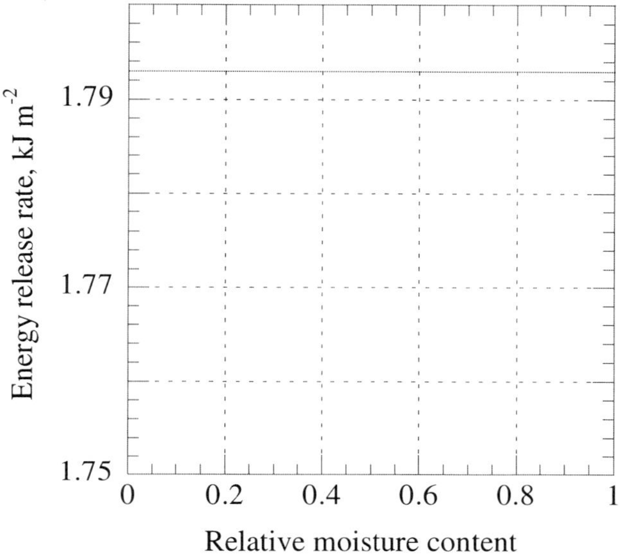


Figure 15.15. Energy release rate, $G_{Ic}$, vs. moisture content. [Data from Selzer R, Friedrich K, *Composites, Part A*, **28A**, 1997, 595-604.]

## 15.24 POLYETHERIMIDE, PEI

| | |
|---|---|
| **Major polymer applications** | microwaveable cookware, electronic connectors, automotive engine sensors, bulb sockets, vacuum pump vanes, aircraft interiors, steam sterilizable surgical components |
| **Important processing methods** | injection molding, blow molding, extrusion |
| **Typical fillers** | glass fiber, carbon fiber |
| **Typical concentration range** | 20-40 wt% |
| **Auxiliary agents** | not reported |
| **Special methods of incorporation** | not reported |
| **Methods of filler pretreatment** | not reported |
| **Special considerations** | PEI has very good fire resistance (LOI 47%) compounding with 30 wt% glass fiber reduces LOI to 32; compounding of PEI with glass fiber does not affect heat deflection temperature |

## 15.25 POLYETHER SULFONE, PES

| | |
|---|---|
| **Major polymer applications** | aircraft interiors, multipin connectors, coil bobbins, integrated circuits sockets, fiber optics connectors, dip switches, automotive fuses, printed circuit boards, transformer wire coatings, microwave cook-ware, sight glasses, membranes, medical applications (due to the resistance to different methods of sterilization), coatings |
| **Important processing methods** | injection molding, extrusion, blow molding, compression molding, thermoforming, casting |
| **Typical fillers** | glass fiber, carbon fiber |
| **Typical concentration range** | 30 wt% |
| **Auxiliary agents** | not reported |
| **Special methods of incorporation** | not reported |
| **Methods of filler pretreatment** | drying to below 0.05 wt% |
| **Special considerations** | moisture causes bubbles, streaking, and splay marks; compounding with glass fiber has very little effect on PES heat deflection temperature and marginally increases LOI but tensile strength is improved in a broad spectrum of temperatures |

# 15.26 POLYETHYLENE, PE[38,45,59,64,83,103,140-195]

| | |
|---|---|
| **Major polymer applications** | packaging and film are the major applications, many others |
| **Important processing methods** | blow film extrusion, molding, cast film extrusion, extrusion, injection molding, rotational molding |
| **Typical fillers** | barium sulfate, calcium carbonate, carbon black, calcium sulfate whiskers, diatomaceous earth, glass fiber, glass spheres, hollow silicates, kaolin, mica, talc, wollastonite, silica, magnesium hydroxide, hydrotalcite, red mud, ground tire rubber, ferromagnetic powder, nickel fibers, wood flour, zirconium silicate, starch, soot, marble, aluminum, lignin, sand |
| **Typical concentration range** | generally ≥40 wt%; calcium carbonate, calcinated kaolin, talc - 20-40 wt%; carbon fiber 5-30% (depending on aspect ratio);[141] titanium dioxide - 1 wt% (occasionally concentrations of 10 wt% are found in thin sections);[143] ferromagnetic powder - up to 90 wt%;[153] stainless steel fiber - 2-3 wt%;[154] nickel fibers for magnetic properties - 10-30%;[155] starch in biodegradable products - 4-8 wt%[165] |
| **Auxiliary agents** | polymer grafting by maleic anhydride or unsaturated diacid anhydride, silanes, titanates, acryloamide was grafted on the surface of calcium carbonate to contain 0.2-1.8% amide groups which were then used for interaction with compatibilizer containing carboxyl groups;[140] aminosilane treatment of kaolin;[142] phosphate treatment of calcium carbonate increases adhesion of mechanical properties of material;[146] low molecular weight polyethylene wax was used as dispersing agent for carbon black[189] |
| **Special methods of incorporation** | usually surface modification of filler is performed prior to filler incorporation into the polymer; high concentration of calcium carbonate (40 wt%) provides a film which has properties similar to paper but surpasses paper in resistance to moisture; chlorinated PE containing carboxyl groups was used as compatibilizer with calcium carbonate;[140] masterbatches of titanium dioxide contain 40-50 wt% pigment;[143] rubber modification to incorporate zirconium silicate without a loss of mechanical strength[159] |
| **Methods of filler pretreatment** | moisture removal by drying, preparation of masterbatches; moisture pickup by titanium dioxide and its concentrates was found to cause lacing which causes film defects, drying of the filler is a simple remedy[150] |
| **Special considerations** | hydrotalcite is used as acid neutralizer with various stabilizing packages; anatase titanium dioxide decreases UV stability; presence of transition metals (Ni, Zn, Fe, Co) affects thermal and UV stability; addition of 15 wt% calcium carbonate increases coating-to-substrate adhesion of polyethylene materials;[149] carbon black in concentration above 2 wt% in conjunction with crosslinking improves UV stability of polyethylene films;[158] carbon black is better antioxidant than many commercial antioxidants;[164] improved thermal conductivity, more stable bubble with calcium carbonate improves output of blown film;[167] talc was used to prevent melt sag in blow molding process;[180,184,186] talc can also be used to reduce gas permeability in film for food packaging industry;[183,188] sand was found to reduce photodegradation of PE;[191] small additions (1-2 wt%) of some metal oxides also give some UV protection whereas other metal oxides increase degradation rate;[192] HALS was immobilized on the surface of silica incorporated into PE which reduced its activity as UV stabilizer[195] |

Infrared studies give us information on filler distribution in PE matrix. Figure 10.12 shows the schematic diagram of the perceived distribution of talc particles in relationship to crystalline formations in PE. The diagram shows that talc is in association with crystallites. This is due to the fact that talc particles affect nucleation through intimate contact and chain alignments on the surface of the filler. The addition of a small amount of titanium dioxide has a strong nucleating action which results in smaller spherulites formed due to the competition of the increased number of growing sites.[143] Nucleation of LDPE was also increased by particulate silica. Adsorption of polymer chains by the filler affects crystallization.[147,148]

Mechanical properties depend on filler-matrix interaction but there are some characteristics of fillers which influence the mechanical behavior of composite. Most fillers increase tensile strength but calcinated kaolin increases tensile strength about 3 times more than calcium carbonate or talc. Impact strength is improved by calcinated kaolin. It is lowered by the addition of either talc or calcium carbonate.[103] This effects can be changed by tailoring the interface between the matrix and the filler. In one example, chlorinated PE containing carboxyl groups was used as compatibilizer.[140] Also, calcium carbonate was modified by grafting acryloamide with 0.2-1.8% amide groups onto its surface. This increased its degree of interaction. The tensile strength of this composite was increased by over 50% and its impact strength by about 120% compared with neat resin (when calcium carbonate was added without compatibilizer, the impact strength of the composite dropped to below 25% of that of neat resin).[140] Similarly, the effect of kaolin was improved by coating it with maleic grafted polyethylene which increased the impact energy of the filled PE by a factor of 4 compared with neat resin.[142] In blown films, calcium carbonate, in the concentration range of 5-20 wt%, was found to increase dart impact strength. But, the tensile strength and elongation decreased as the concentration increased. Additions of talc in the same concentration range decreased all mechanical properties.[152]

The most common method to make PE fire retardant is through the use of phosphoric esters of polyols. An alternate method uses magnesium hydroxide. Figure 15.16 shows the effect of varying amounts of aluminum hydroxide and magnesium hydroxide on the limiting oxygen index (LOI).[190] Magnesium oxide gives a marginally better performance but in both cases a large amount of metal oxide is needed. If such large quantities are used, the impact resistance of the material is substantially reduced. The impact resistance can be improved in these compositions by additions of silane-crosslinkable polyethylene.

The electric conductivity of polyethylene can be improved by addition of many fillers. Figure 15.17 shows the effect of aspect ratio of the fiber on the electric conductivity of polyethylene filled with carbon fiber.[141] Depending on the aspect ratio, different levels of carbon fiber are needed to obtain the same effect. When a very high aspect ratio is used, 2 vol% carbon fiber gives the same effect as can be obtained with 30 vol% of carbon fiber having aspect ratio of 1. Compared with

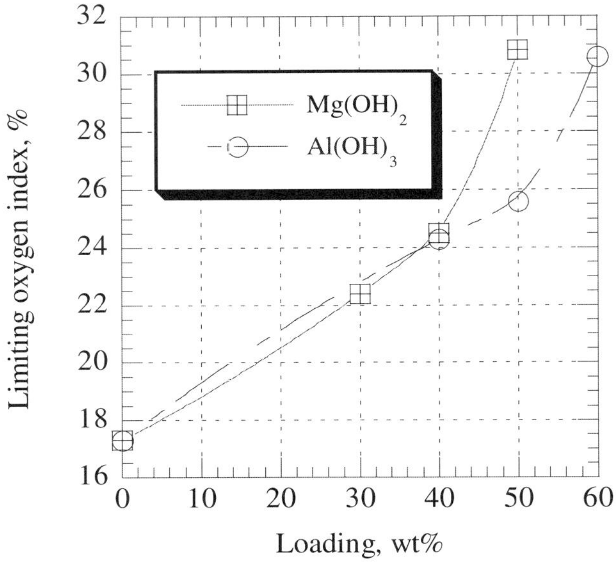


Figure 15.16. Limiting oxygen index of polyethylene vs. concentration of filler. [Adapted, by permission, from Yeh J T, Yang H M, Huang S S, *Polym. Degradat. Stabil.*, **50**, No.2, 1995, 229-34.]

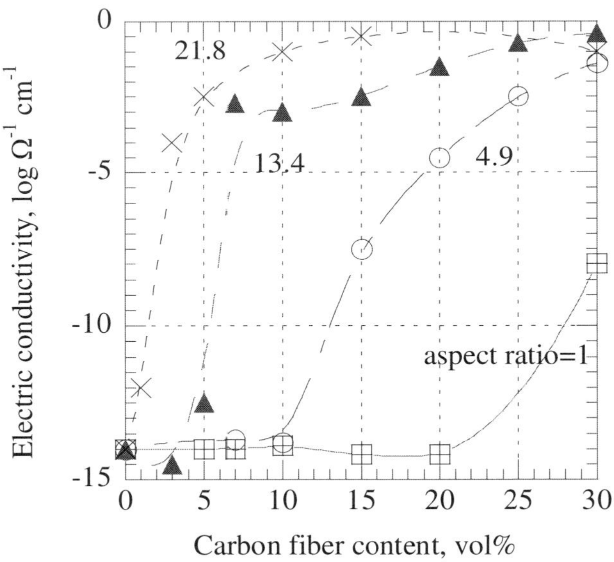


Figure 15.17. Electric conductivity of polyethylene filled with carbon fibers of different aspect ratio vs. volume content. [Adapted, by permission, from Agari Y, Ueda A, Nagai S, *J. Appl. Polym. Sci.*, **52**, No.9, 1994, 1223-31.]

polypropylene, polyethylene required about 50% more carbon black to develop the same conductivity.[145] A better wetting of the filler in polyethylene results in higher

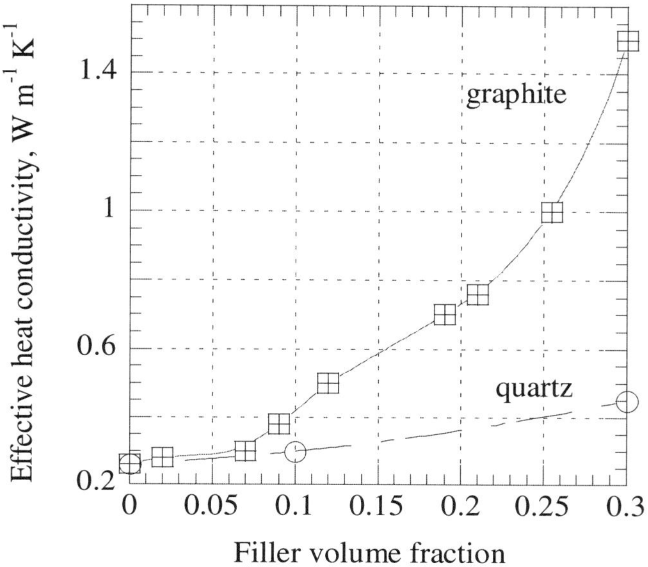


Figure 15.18. Effective heat conductivity of polyethylene vs. filler volume fraction. [Data from Privalko V P, Novikov V V, *Adv. Polym. Sci.*, **119**, 1995, 31-77.]

colloidal stability but also in a higher percolation threshold. This affects the amount of carbon black needed to obtain a certain level of conductivity. For polyethylene to be conductive it must be filled above a certain threshold concentration with conductive filler. Below this threshold level, conductivity remains quite constant and it is largely independent of filler concentration. In this range, conductive filler particles are not in proper contact. When a certain threshold value is reached, the conductivity increases very rapidly as more carbon black is added. Eventually, a plateau of ultimate conductivity is reached which depends on the matrix and on the type of carbon black used.[145] Large concentrations (80-90 wt%) of ferromagnetic materials are required to reach threshold concentration (see Figure 14.15).[153] Conductivity changes rapidly when 3 wt% stainless steel fiber is added to polyethylene.[154] Similarly, magnetic properties can be changed with additions from 10 to 30 wt% of nickel fibers. The fibers must be in the proper orientation to develop optimum magnetic properties.[155] One study[174] sought to obtain a material with low resistivity at room temperature and high resistivity at elevated temperatures. It also attempted to obtain material in which such change occurs within a few degrees Celsius. The composites which were developed can switch rapidly from a low to a high resistivity. These composites are used in devices which can limit electric fault currents. The materials were developed by selecting an appropriate process of incorporation (quality of mixing) and by the choice of carbon black. A composition which combined coarse and fine carbon blacks gave the required performance.

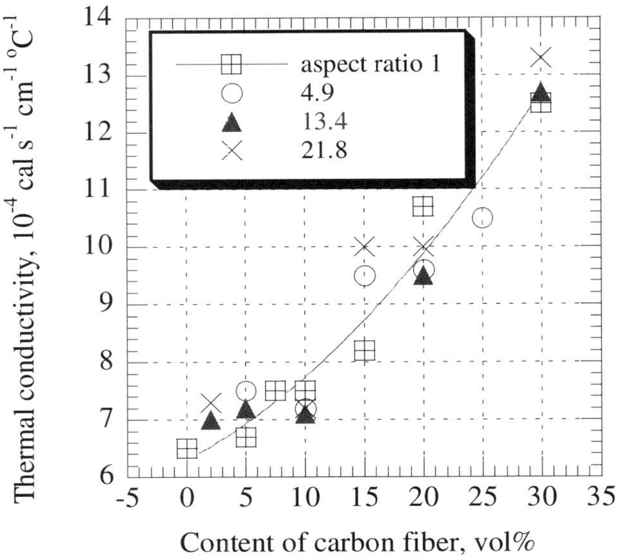


Figure 15.19. Thermal conductivity of polyethylene vs. volume content of carbon fiber. [Adapted, by permission, from Agari Y, Ueda A, Nagai S, *J. Appl. Polym. Sci.*, **52**, No.9, 1994, 1223-31.]

Heat conductivity of composite materials are severely and adversely affected by structural defects in the material. These defects are due to voids, uneven distribution of filler, agglomerates of some materials, unwetted particles, etc. Figure 15.18 shows the effect of filler concentration on thermal conductivity of polyethylene. Graphite, which is a heat conductive material, increases conductivity at a substantially lower concentration than does quartz. These data agree with the theoretical predictions of model.[38] Figure 15.19 shows the effect of volume content and aspect ratio of carbon fiber on thermal conductivity.[141] This figure should be compared with Figure 15.17 to see that, unlike electric conductivity which does depend on the aspect ratio of the carbon fiber, the thermal conductivity is only dependent on fiber concentration and increases as it increases.

## 15.27 POLYETHYLENE, CHLORINATED, CPE[82,196-199]

| | |
|---|---|
| **Major polymer applications** | wire and cable, autoignition wire, roofing membranes, technical hoses, power steering hose, transmission oil cooler hose, car axle boots, automotive air ducts and hoses, impact modification for PVC in pipe, vinyl siding, window profiles and FR ABS |
| **Important processing methods** | peroxide vulcanization, molding, mixing, extrusion |
| **Typical fillers** | clay, carbon black 3-10 wt%, titanium dioxide 1-2 wt%, magnesium oxide is used as thermal stabilizer (typically 5-10 phr) |
| **Typical concentration range** | 20-40 wt% |
| **Auxiliary agents** | not reported |
| **Special methods of incorporation** | CPE was used as PVC modifier to increase filler concentration[199] |
| **Methods of filler pretreatment** | aluminum filler was prepared from the reaction of aluminum chloride and carboxylic acid such as hydroxybenzoic, aminobenzoic, anthranilic acids to form an active filler which improves the mechanical properties of CPE[198] |
| **Special considerations** | presence of zinc, copper, iron and nickel compounds accelerated dehydrochlorination; combination of basic magnesium carbonate and aluminum hydroxide is used as flame retardant and smoke supressant;[82] chlorinated polyethylene adsorbs on the surface of titanium dioxide forming a layer 1-20 nm thick depending on the acid/base interaction parameter of titanium dioxide[197] |

# 15.28 POLYETHYLENE, CHLOROSULFONATED, CSM[82,196,200]

| | |
|---|---|
| **Major polymer applications** | coated fabrics, inflatable boats, roofing, pool liners, industrial effluent pit liners, radiator and heater hoses, wire and cable, adhesives, automotive components (high-temperature timing belts, power steering pressure hose, gaskets, spark plugs), boots, industrial products (hose, rolls, seals, gaskets, diaphragms), and lining for chemical processing equipment |
| **Important processing methods** | vulcanization, coating, extrusion, compounding, injection molding |
| **Typical fillers** | silica, calcium carbonate, carbon black, magnesium hydroxide, basic magnesium carbonate, metal oxides (typically MgO) are used as curing agents |
| **Typical concentration range** | 20-30 wt% |
| **Auxiliary agents** | none reported |
| **Special methods of incorporation** | none reported |
| **Methods of filler pretreatment** | thermal treatment at 800°C and hexadecanol treatment were performed to study the effect that hydroxyl groups have on silica reinforcement[200] |
| **Special considerations** | magnesium hydroxide and basic magnesium carbonate are used as flame and smoke retarding additives[82] |

## 15.29 POLY(ETHYLENE OXIDE), PEO & PEG[201-205]

| | |
|---|---|
| **Major polymer applications** | pharmaceutical applications, controlled release drugs, , polyester fibers, unsaturated polyester resins, oil exploration, polyols, surfactants, haircare, switching elements, polymer electrolytes, lithium batteries, nanocomposites |
| **Important processing methods** | compounding, reacting with other chemicals |
| **Typical fillers** | graphite, fumed silica, molybdenum disulfide, vanadium oxide |
| **Typical concentration range** | graphite - 10-40 wt%; fumed silica - 10 wt% |
| **Auxiliary agents** | not reported |
| **Special methods of incorporation** | not reported |
| **Methods of filler pretreatment** | not reported |
| **Special considerations** | fumed silica used as rheological additive[204] addition of salts decreases solubility of resin in water |

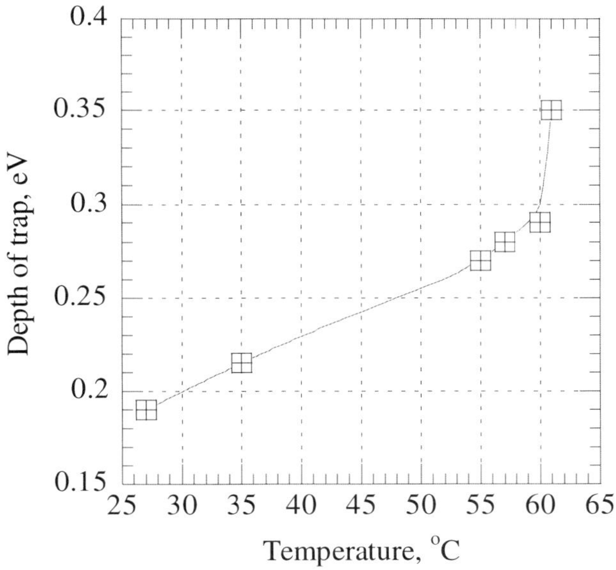


Figure 15.20. The depth of trap vs. temperature. [Adapted, by permission, from Kimura T, Asano Y, Yasuda S, *Polymer*, **37**, No.14, 1996, 2981-7.]

The poly(ethylene glycol)/graphite system acts as a switching element. A switching element is a polymer composite with dispersed conductive particles such as carbon black, graphite or metal particles. The switching element has a low resistance at low temperature (switch-on) and a high resistance at high temperature (switch-off). The principle of action of such system is explained by Figure 15.20. A small difference in temperature causes a rapid change in resistance. The explanations of this phenomenon are two: the classical explanation is based on the premises that with temperature increasing conductive particles are too distant to transfer electrons; the more recent explanation attributes the sudden change in resistivity to the change of dielectric constant of poly(ethylene glycol).[201] A similar system was developed based on polyethylene.

Poly(ethylene oxide) is also used in nanocomposites which contain molybdenum disulfide or vanadium oxide. The inorganic filler and the organic matrix interact at a molecular level forming xerogels, which are nanocomposites with controlled ion mobility.[202,203,205]

## 15.30 POLY(ETHYLENE TEREPHTHALATE), PET[103,206-209]

| | |
|---|---|
| **Major polymer applications** | packaging, bottles, film, fiber, textiles, brush bristles, composites, electrical, automotive, housewares, lighting, power tools, sporting goods, plumbing |
| **Important processing methods** | injection blow molding, extrusion, blow molding, injection molding, monofilament extrusion |
| **Typical fillers** | carbon fiber, glass fiber, aramid, mica, glass spheres, talc, clays, wollastonite, fly ash |
| **Typical concentration range** | glass fiber composites - 30-55 wt%, general purpose fillers - 20-40 wt% |
| **Auxiliary agents** | coupling agents used in composites; γ-aminopropyltrimethoxy silane;[206] maleic anhydride modification[208] |
| **Special methods of incorporation** | filler and interfacial modifier were premixed before extrusion[206] |
| **Methods of filler pretreatment** | drying |
| **Special considerations** | material must be dry before processing to avoid hydrolysis; warping can be reduced by the use of mixed fillers (e.g., glass beads and mica) also, talc and clays; ferrocene and cobalt salts accelerate UV degradation |

The treatment of glass beads with silane improves their adhesion. This benefits most properties of the composite but its tensile strength is slightly lower than that of neat resin. The crystallization rate of PET is higher in the presence of filler. Glass beads cause a heterogeneous nucleating effect. Introduction of a modifier lowers the nucleation rate and acts as a compatibilizer. The combination of glass beads and the modifier increases tensile properties and the impact strength of the composite.[206] Figure 10.7 shows the effect of mica on the crystallization rate of PET.[209]

## 15.31 POLYIMIDE, PI[137,210-212]

| | |
|---|---|
| **Major polymer applications** | aerospace, electronics (mostly films and coatings), photosensitive materials for positive imaging, solar cells, hollow fiber membranes, composites, nuclear power plants, space shuttle, microprocessor chip carriers, structural adhesives |
| **Important processing methods** | vapor phase deposition, spin coating, injection molding, casting, extrusion, drawing of oriented films, compression molding, sintering |
| **Typical fillers** | carbon fiber, glass fiber, graphite lubricant for wear resistant applications; molybdenum sulfide, PTFE, antimony trioxide, barium titanate, clay, silica, aluminum nitride, smectite |
| **Typical concentration range** | in general 2-40 wt%, graphite - 15-40 wt%, PTFE - 10 wt%; clay - 5-15 wt% |
| **Auxiliary agents** | aminosilane, methylpyrrolidone (deagglomeration and stabilization of filler suspension)[210] |
| **Special methods of incorporation** | moisture increases decomposition rate therefore processed materials must be dry; polymerization of PI in the presence of deagglomerated suspension of aluminum nitride[210] |
| **Methods of filler pretreatment** | drying, deagglomeration[210] |
| **Special considerations** | trace metals such as Co, Cu, Ni radically reduce thermal stability; some carbon fibers cause degradation due to the surface impurities; some types of glass reduce stability of polyimide; polyimide is used as modifier of PEEK/carbon fiber composite[137] |

Polyimide was used as a model material in studies of polymer metal interfaces where metal layers were formed by metallization, plasma deposition, chemical vapor deposition, electrochemical deposition, etc.[211] In most of the cases studied, the interpenetration of metal was so good that the metal layer could not be removed by any other means but abrasion. An investigation of interface, determined that the metal particles were found in the surface layers in diminishing quantities perpendicular to the surface and not, as expected, in the form of a sharp borderline between the metal and polymer. Some difficulties exist when metallized polyimides are used for chip production. These diffuse layers of metals complicate design and performance due to the gradients of conductivity which they produce.

Figure 15.21 shows the effect of the method of composite preparation on basal spacing of $CH_3(CH_2)_{n-1}NH_3^+$ montmorillonites and the number of carbon atoms.[212] The processing at 300°C produces more rigid polyimide with a smaller basal spacing. The material prepared by high temperature curing gives the $CO_2$ permeability shown in Figure 15.22. Gas permeability is reduced which is attributed to the layered structure (~10 Å spacing), high aspect ratio, and staircase-like arrangement of layers which all increase the effect of filler. In the simple mixtures of filler and polymer, permeability has the linear relationship with the filler concentration.

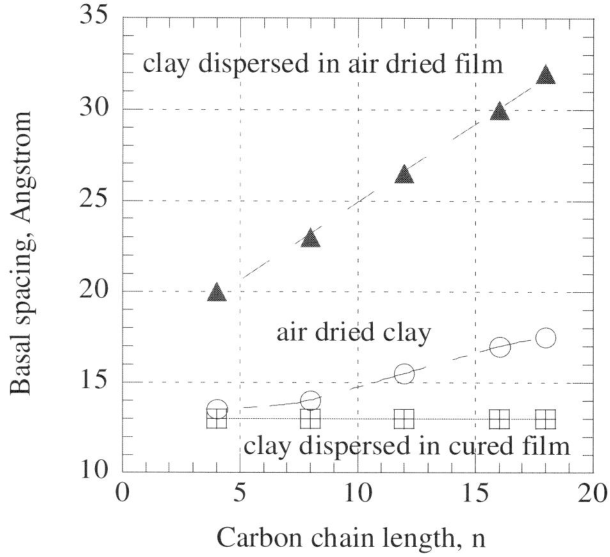


Figure 15.21. Basal spacing of $CH_3(CH_2)_{n-1}NH_3^+$ montmorillonites vs. the number of carbon atoms. [Adapted, by permission, from Lan T, Kaviratna D, Pinnavaia T J, *Chem. of Mat.*, **6**, No.5, 1994, 573-5.]

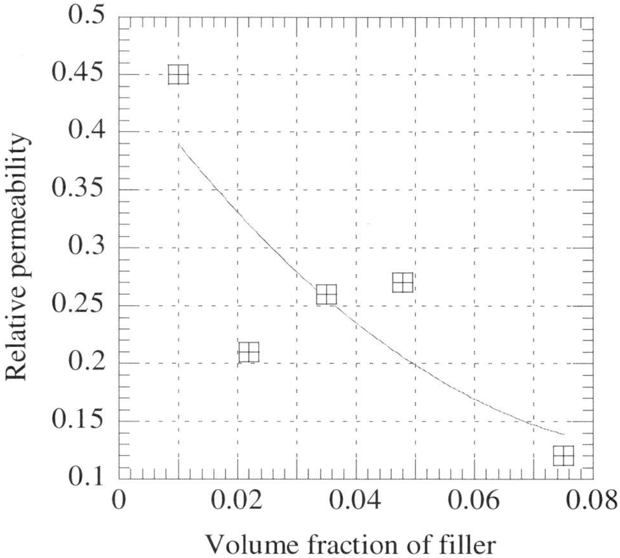


Figure 15.22. $CO_2$ permeability of cured polyimide films containing $CH_3(CH_2)_{17}NH_3^+$ montmorillonite vs. volume fraction of filler. [Data from Lan T, Kaviratna D, Pinnavaia T J, *Chem. of Mat.*, **6**, No.5, 1994, 573-5.]

## 15.32 POLYMETHYLMETHACRYLATE, PMMA[99,197,213-223]

| | |
|---|---|
| **Major polymer applications** | optical fibers, dials, optical components, household items, car rear lights, artificial stones (filled products) for injection molded bath sinks, and kitchen worktops, bone cement, composites, medical applications (e.g. bone cement) |
| **Important processing methods** | casting, injection molding, compression molding |
| **Typical fillers** | aluminum hydroxide, silica, titanium dioxide, glass fiber, mica, barium sulfate, titanium fiber, nickel, aluminum |
| **Typical concentration range** | generally - 20-30 wt%; carbon black - 5-30 wt%; glass powder in bone cement - 30-80 wt%; titanium fiber - 1.5%, aluminum or nickel for conductive applications - 20 vol% |
| **Auxiliary agents** | silanes[221] |
| **Special methods of incorporation** | fillers are most frequently dispersed in monomer in the presence of catalyst and monomer is then polymerized |
| **Methods of filler pretreatment** | carbon black was oxidized prior to incorporation[213] |
| **Special considerations** | according to inverse gas chromatography, PMMA is considered acidic, it therefore interacts better with fillers which have basic character;[99] PMMA was found to form thick layers adsorbed on the surface of glass, titanium dioxide, silica and mica (1400, 51-70, 17, and 110 nm thick, respectively);[197] very strong interaction between polymer and titanium dioxide caused formation of brittle coatings which failed prematurely;[220] formation of clay/$K_2S_2O_8$ complex is a reason for catalytic effect on polymerization, the composites formed have better thermal stability, hardness, and compression strength[222] |

NMR studies indicate that hydroxyl groups on the surface of silica are consumed during polymerization of PMMA. These groups are utilized in a reaction with polymer.[215] Polymer adsorbed on the surface of alumina changes conformation.[223]

Glass powder was used to fill bone cement. Figure 15.23 shows that the affinity index of bone cement increases as glass powder concentration increases. Synthetic material is more readily accepted by the body when more inorganic filler is present.[216] The fact that calcium and silicate ions are consumed from the cement indicates that silane coupling, although advantageous for adhesion improvement, would detract from the bioactive qualities of the cement. The addition of a small amount of titanium fiber (1.5 wt%) to another formulation of bone cement substantially increased its resistance of crack propagation.[217]

Phosphorus-containing PMMA is frequently used for flame resistant applications.

Figure 14.7 shows resistivity of aluminum filled PMMA. The resistance rapidly drops when the concentration of aluminum exceeds 20 vol%. Slightly less (about 18-20 vol%) nickel is needed to obtain the same resistance.[218]

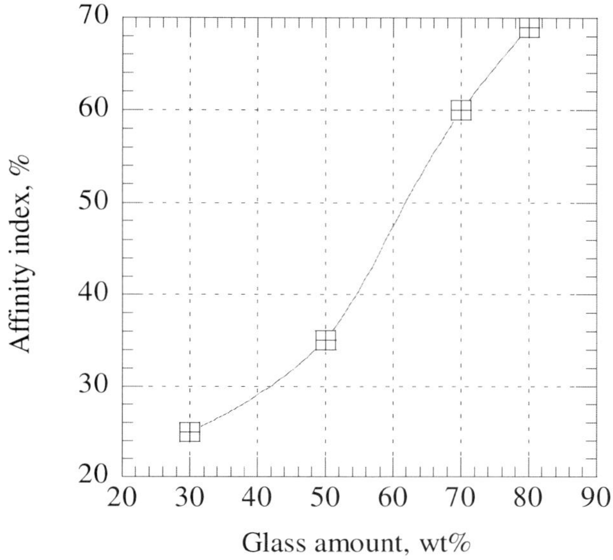


Figure 15.23. Affinity index of bone cement based on PMMA vs. concentration of glass powder. [Adapted, by permission, from Tamura J, Kawanabe K, Yamamuro T, Nakamura T, Kokubo T, Yoshihara S, Shibuya T, *J. Biomed. Mat. Res.*, **29**, No.5, 1995, 551-9.]

## 15.33 POLYOXYMETHYLENE, POM[108,224,225]

| | |
|---|---|
| **Major polymer applications** | appliances, automotive (door handles, window winders, tank filler necks and caps, carburetor, screw caps for cooling system expansion tanks. fuel pumps) phones (dialing units and slider guideways), pneumatic components, parts of textile machines, shower parts, home electronics and hardware, bearings, cams, containers, pump impellers, rollers, springs, clips, and many other applications |
| **Important processing methods** | injection molding, blow molding, rotational molding, extrusion, foam molding, compression molding, transfer molding |
| **Typical fillers** | glass fiber, glass beads, carbon fiber, aramid fiber, carbon black, metal flakes, zinc whisker, talc, calcium carbonate, PTFE fiber |
| **Typical concentration range** | generally - 20-40 wt%; PTFE, aramid fiber - 2-10 wt%, glass fiber 20-30 wt%, glass microspheres - 10-30 wt% |
| **Auxiliary agents** | not reported |
| **Special methods of incorporation** | not reported |
| **Methods of filler pretreatment** | drying |
| **Special considerations** | carbon black is the best UV stabilizer; printability of POM is obtained by addition of talc or calcium carbonate; 0.2-0.3% moisture may reduce thermal stability by 20-30°C; (γ-sterilization degrades POM rapidly) |

The addition of glass beads reduced tensile strength, fracture toughness, and strain energy release rate but improved flexural modulus.[224] The tensile strength is inversely proportional to the square root of the glass sphere diameter. Reinforcing POM with glass fiber improved all its mechanical properties.[225]

## 15.34 POLY(PHENYLENE ETHER), PPO[135,226]

| | |
|---|---|
| **Major polymer applications** | electronics (computer and television housings, keyboard frames, interface boxes), automotive (instrument panels, interior and exterior trim, glove compartments, wheel covers, electric connectors, fuse boxes), air conditioner housings, hospital and office furniture, production of blends |
| **Important processing methods** | injection molding, blow molding, extrusion, thermoforming |
| **Typical fillers** | calcium carbonate, glass fiber, carbon fiber, zinc borate, PTFE, aluminum flake, graphite fiber, nickel coated graphite fiber |
| **Typical concentration range** | glass fiber - 10-40 wt%, carbon fiber - 10-20 wt%, also filled with a combination of mineral filler and glass fiber - 40 wt%, PTFE - 2-3 wt% |
| **Auxiliary agents** | silicone powder[226] |
| **Special methods of incorporation** | PPO is processed with PS which acts as plasticizer |
| **Methods of filler pretreatment** | not reported |
| **Special considerations** | zinc borate, especially in combination with red phosphorus or organic compounds of phosphorus, gives good fire retarding and smoke depressing system; the following are the effects on EMI shielding on addition of conductive fillers: aluminum flake, 40 wt% - 30 dB, graphite fiber, 30 wt% - 40 dB, nickel coated graphite fiber - 42 dB |

# 15.35 POLY(PHENYLENE SULFIDE), PPS[181,227,228]

| | |
|---|---|
| **Major polymer applications** | automotive lighting, ignition and braking systems, carburetor parts, fuel components, chip carriers, phone jacks, IC card connectors, transistor encapsulation, tape recorder head mounts, relay components, motor fans, coil bobbins, sockets, relay units, food choppers, steam hair drier parts, lamp sockets, microwave oven components, pump housings, impeller diffusers, oil well valves, halogen lamp sockets |
| **Important processing methods** | coating, injection molding, blending, compression molding, lamination, thermoforming |
| **Typical fillers** | calcium carbonate, talc, glass fiber, carbon fiber, PTFE, aramid fiber |
| **Typical concentration range** | glass fiber - 20-60 wt%, carbon fiber - 20-30 wt%, PTFE - 10-20 wt%, aramid fiber - 10-15 wt%, general fillers (talc, calcium carbonate) up to 65 wt% |
| **Auxiliary agents** | none reported |
| **Special methods of incorporation** | not reported |
| **Methods of filler pretreatment** | not reported |
| **Special considerations** | glass composites have very high LOI= 47%; glass fiber reinforcement increases heat deflection temperature by more than 150°C to over 260°C at 40 wt% glass fiber |

## 15.36 POLYPROPYLENE, PP[4,21,67,77,103,106,118,136,143,145,157,162,177,180-1,184-6,229-308]

| | |
|---|---|
| **Major polymer applications** | automotive, packaging, furniture, electrical components, fibers, tapes, many others |
| **Important processing methods** | extrusion, blow molding, injection molding, thermoforming |
| **Typical fillers** | calcium carbonate, talc, glass fiber, glass beads, glass flakes, silica flour, wollastonite, mica, sepiolite, magnesium hydroxide, carbon black, clay, metal powders (aluminum, iron, nickel), steel fiber, silicium carbide, phenolic microspheres, wood fiber and flour, antimony trioxide, hydrotalcite, zinc borate, bismuth carbonate, red phosphorus, potassium-magnesium aluminosilicate, fly ash, hydromagnesite-huntite |
| **Typical concentration range** | general range - 20-50 wt%; with some fillers mechanical properties decrease even at low loadings (10%); calcium carbonate - 10-60 wt%, kaolin - 20-40 wt%;[103,270] talc - 20-40 wt%;[184,270] glass beads - 20-50 vol%;[118] carbon black - 10-30 wt%;[77] glass fiber - 1-60 vol%;[157,230] magnesium hydroxide - 60-65 wt% (for V-0 classification); antimony trioxide - 10 wt% (for V-0 classification) |
| **Auxiliary agents** | silanes,[103,107,247,281] titanates;[281] fatty acids;[281] polymer grafting with maleic anhydride to increase interaction,[107,242,243,244,246,247,269] stearic acid, nucleating agents; octamethylcyclotetrasiloxane;[177] amino silane;[231] dimeric aluminates[239] acrylic acid[260] |
| **Special methods of incorporation** | slurry process in which polymer powder and fibers are suspended in water followed by dewatering and wet sheet formation similar to paper manufacture technology; melt impregnation of fiber bundles in equipment containing fluidized bed zone and heating zone followed by extrusion through die; filler encapsulation is faster than blend compatibilization therefore filler must be added to compatibilized blend;[107] compatibilizers were used with glass beads to improve mechanical properties of composite[241] |
| **Methods of filler pretreatment** | if coupling agent is required, filler is usually pretreated before incorporation; fluidized bed glass fiber impregnation in sheet forming; surface modification of chalk and carbon fiber by acetylene gas plasma;[21] stearic and oleic acids;[264] phosphate coating for talc[279] |
| **Special considerations** | chemical composition of filler surface affects nucleation of filler; traces of heavy metals decrease thermal stability and cause discoloration; surface free energy of fillers determines interaction; large difference in thermal properties of fillers and polymer may cause stress; hydrotalcite is used as acid neutralizer with stabilizing packages; anatase titanium dioxide decreases UV stability; presence of transition metals (Ni, Zn, Fe, Co) affects thermal and UV stability; calcium carbonate and talc were found to immobilize HALS stabilizers in PP;[268] with organic masterbatches such as ethylene diamine phosphate V-0 classification can be obtained with 20-25 wt%, at the same time tensile strength and impact strength are substantially reduced |

Fillers affect the nucleation rate as polypropylene crystalizes. The addition of 2.5 wt% titanium dioxide reduces the size of spherulite by a factor of 3 due to an increased number of sites which create competition between growing spherulites.[143] Spherulites did not grow in the surface skin. Even an addition of 40 wt% titanium dioxide did not cause spherulites to grow in the skin. An increased addition of titanium dioxide affected dispersion. On average 2.5 crystals per cluster were found at 10 wt% titanium dioxide and 5 crystals per cluster were detected at 40 wt% addition. Transcrystallinity is the other phenomenon observed in polypropylene. Transcrystallinity develops either on the surface of air bubbles or on the surface of fibers such as glass fibers. Transcrystalline structures are more effectively developed when there is any degree of mechanical stress created around the fiber.[136] The effect of several fibers, such as carbon fiber, E-glass fiber, and Twaron were evaluated using polarizing light microscopy.[234] Both the shear gradient at the interface and the temperature gradient were found to influence transcrystallinity. Figures 7.18 and 10.13 show transcrystallinity on the surface of bamboo fibers.[235] Filler particles oriented during material flow affect the orientation of polymer chains and crystallites because they grow on the surface of the filler particles. A similar orientation of talc particles and polymer chains was found in thermoformed and blow molded polypropylene.[180,181,184,186] The orientation of short glass fibers and related molecular orientation of the matrix can be regulated by process parameters of injection molding.[259] Electron spin resonance studies of calcium carbonate and talc filled polypropylene indicate that filler orientation during injection molding depended on the filler load.[267] The best orientation was obtained at 15 vol% filler.

Mechanical properties of filled polypropylene depend on several factors which are discussed below. The Izod impact strength of carbon black filled polypropylene decreases marginally as the concentration of carbon black decreases.[77] The addition of 20 wt% carbon black to polypropylene produced a substantial (400%) increase in flexural modulus and a 40% increase in flexural strength. Tensile yield strength is a more complex property. Small additions of carbon black (up to 10 wt%) increase tensile yield strength but increasingly higher concentrations eventually decreases tensile yield strength until it drops below that of neat resin. Talc and kaolin in concentrations up to 30 wt% did not change the tensile strength of polypropylene, but improved the flexural modulus by a factor of 2. Talc alone substantially reduced impact strength.[103] Calcium carbonate filled polypropylene had poorer mechanical properties and was more difficult to process than the neat polymer.[255] However, when calcium carbonate was surface coated by stearates, elongation and impact strength were maintained and whiteness and processing characteristics were improved.[255] The addition of glass beads containing rubbery inclusions brought improved toughness to polypropylene.[229] The particles changed the crack growth mechanism by cavitation, shear yielding, and particle matrix debonding. Failure by debonding from the surface of glass beads treated with amine silane occurred at strain of 0.7%.[231] Figures 8.51 and 8.52 show the effect of temperature on the influence of yielding, cavitation, and debonding which are the major mechanisms of composite failure.[232] Figure 15.24 compares the tensile

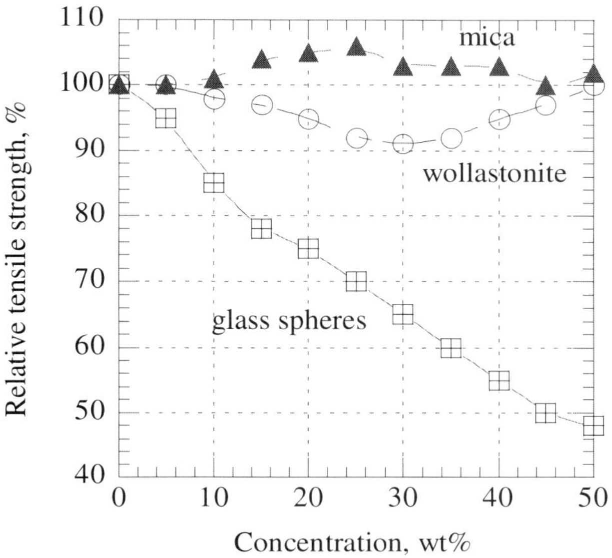


Figure 15.24. Tensile strength of polypropylene vs. concentration of three fillers. [Adapted, by permission, from Jarvela P A, Jarvela P K, *J. Mat. Sci.*, **31**, No.14, 1996, 3853-60.]

strength of polypropylene compounded with selected fillers.[283] Mica and wollastonite do not affect properties of neat resin whereas glass beads decrease tensile strength. The technology of production of microporous propylene sheets took advantage of filler debonding.[237] A sheet of polypropylene which contains 65.8 wt% calcium carbonate is extruded. The base sheets are stretched biaxially in a stretching machine to 500 to 1500%. This causes debonding of the filler which results in a soft microporous membrane with controlled gas and water vapor permeability.[237,278] Microporous polypropylene hollow fibers were prepared with a similar process.[238,276] Glass fibers affect the mechanical properties of polypropylene depending on the concentration, their length and their adhesion to the matrix. Figure 15.25 shows the effect of fiber length on tensile modulus, strength, and Charpy impact.[230] All characteristics of mechanical performance increase when a fiber length increases. Filler mixtures can be selected to optimize the properties of composite. Calcium carbonate and mica, when combined with maleic anhydride, enable the product to meet required properties.[242,251] The proportions of glass fiber and mica may be varied to regulate properties. For example, tensile and flexural strength depend on glass fiber content, and warpage and shrinkage can be regulated by mica.[258] Combinations of fillers with different shapes (platelet, spherical, elongated) give a better performance than would be obtained from a single filler.[283] A novel product hydrated potassium-magnesium aluminosilicate improves several properties of a polypropylene composite, including tensile strength and modulus, weld strength, UV stability, fire retardation and smoke suppression. It has no effect

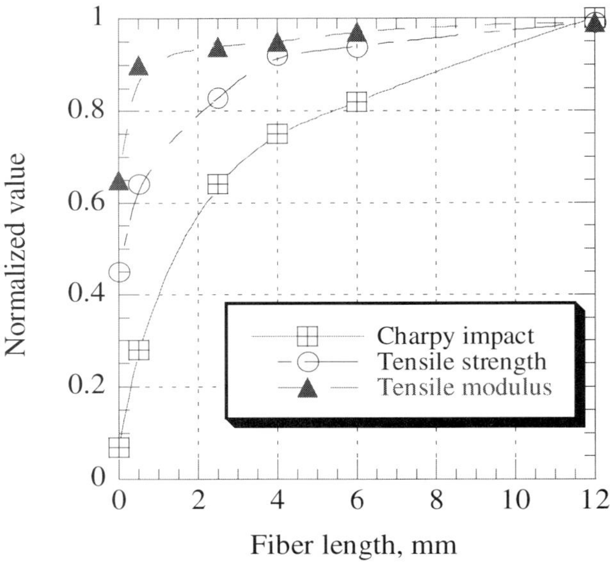


Figure 15.25. Normalized composite properties vs. fiber length. [Adapted, by permission, from Thomason J L, Vlug M A, *Composites, Part A*, **28A**, 1997, 277-88.]

on rheological properties.[249] Smectite was used to prepare nanocomposite with an average interlayer distance of 2.3 nm.[284,285]

Figures 12.3, 12.4, 12.8 and 14.14 contain data comparing the effect of talc and magnesium hydroxide on the burning behavior of polypropylene.[4] Ignition time, ignition temperature, rate of combustion, and limiting oxygen index are substantially improved by the addition of magnesium hydroxide compared to the addition of talc. A substantial improvement is obtained when a large quantity (60 wt%) of magnesium hydroxide is used. The effect of magnesium hydroxide is illustrated by data from cone calorimetry (Figure 15.26).[62] Glass beads reduce heat release by dilution (but heat released is still greater than theoretical calculations). Magnesium hydroxide substantially reduces heat release due to its endothermic degradation which releases water.[62] Magnesium hydroxide is a suitable additive for polypropylene because polypropylene is typically processed at 200°C which is below the temperature at which magnesium hydroxide degrades. Magnesium hydroxide also reduces smoke release.[62] The major drawback is the large amount which must be used to produce these results. Several factors contribute to the performance obtained from magnesium hydroxide including endothermic decomposition, release of water into a vapor phase, dilution of burning components, insulating effect of the oxide char residue. Different grades of magnesium hydroxide produce different results. The reason why is not yet known.[256] A substantial decrease of mechanical properties is observed when magnesium hydroxide is used in the quantity required to obtain fire rating (~60 wt%).[272] Coating the filler and an addition of rubbery par-

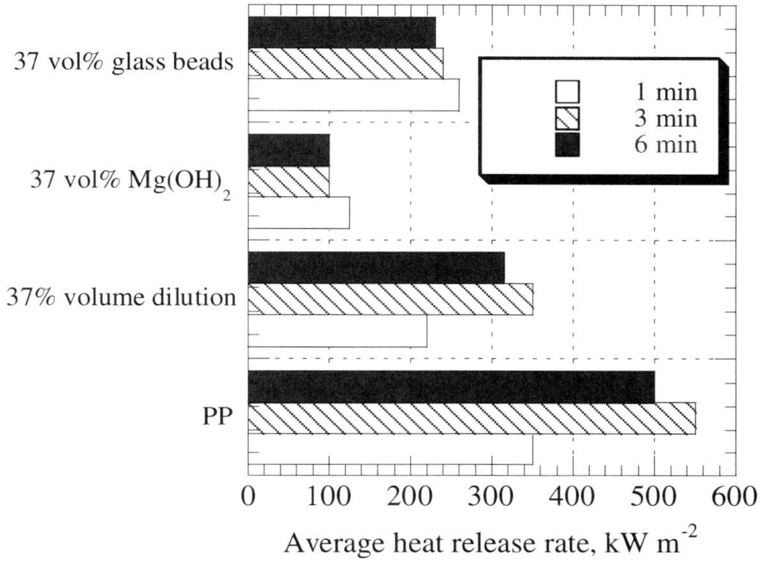


Figure 15.26. The effect of magnesium hydroxide on heat release from polypropylene. [Data from Rothon R N; Hornsby P R, *Polym. Degradat. Stabil.*, **54**, Nos 2-3, 1996, 383-5.]

ticles improves the mechanical performance of the material. Magnesium hydroxide coating materials were evaluated in an extensive study.[281] The objective was to improve the mechanical properties of polypropylene. Of the materials studied, fatty acid derivatives made the best coating.[281] The amount of coating was also very important. For the type of magnesium hydroxide used, the monolayer coverage required a coating of 7 wt% fatty acid derivative.[281]

A carbon black addition above a percolation threshold of 5 vol% increases the conductivity until a plateau is reached at 20 vol%.[145] As the level increases above 10 vol%, the viscosity of the filled polypropylene increases rapidly (see Figure 9.9). As with polyethylene, carbon black is preferentially contained in one phase of a two phase blend.[106] This phenomenon is used in practice to lower the concentration of carbon black required for a certain level of conductivity. Here, again, carbon black is concentrated in the preferred location. Carbon black and copper powder were used to improve connectivity of YBaCuO in ceramic superconductors.[240] Dispersion of copper particles and the related changes in conductivity were enhanced by the presence of acrylic acid modifier.[260]

## 15.37 POLYPYRROLE[127,308-310]

| | |
|---|---|
| **Major polymer applications** | nonmetallic conductors, EMI shielding, battery electrodes, sensors, electronic displays, optoelectronic systems, capacitors, controlled release agents for other components |
| **Important processing methods** | Langmuir-Blodgen technique of monolayer production, solution polymerization over the substrate, electrochemical anodic polymerization, chemical oxidation of pyrrole in carbon black suspension |
| **Typical fillers** | silica, tin oxide, carbon black |
| **Typical concentration range** | carbon black - 10-85 wt%, silica - 1-3wt% |
| **Auxiliary agents** | support materials such as PMMA; chemical oxidant determines the size of particles in polypyrrole/silica nanocomposites[309] |
| **Special methods of incorporation** | polymerization is conducted in a dispersion of filler in monomer[308] |
| **Methods of filler pretreatment** | preparation of a very uniform colloidal suspension of filler especially for chemical synthesis of nanocomposites; the quality of the suspension determines the particle size of primary particles, the thickness of the coating and the uniformity of the material |
| **Special considerations** | EMI shielding without fillers is 45 dB; fillers are added to form stable gels[127] |

In suspensions of carbon black in pyrrole, anodic polymerization takes advantage of the fact that carbon black particles are negatively charged on their surface which makes it possible for them to migrate to a positively charged anode where they become embedded within a growing polypyrrole matrix.[308] This production method is suitable for production of materials for sensors, supercapacitors, fuel cells, etc. The effect of carbon black on the chemical oxidation of pyrrole in carbon black suspensions is shown in Figure 6.26.[308]

## 15.38 POLYSTYRENE & HIGH IMPACT, PS & HIPS[30,226,311-323]

| | |
|---|---|
| **Major polymer applications** | packaging, electrotechnical components, insulating film, household items, toys, blending with other polymers, and numerous other applications |
| **Important processing methods** | injection molding, extrusion, blow molding, thermoforming |
| **Typical fillers** | calcium carbonate, glass beads, barium sulfate, mica, kaolin, talc, glass fibers, silica, montmorillonite, zeolites; PTFE, zinc borate, titanium dioxide, red phosphorus, copper |
| **Typical concentration range** | generally - 10-30 wt%; PTFE - 5-15 wt%; zinc borate - 4-6 wt%; red phosphorus - 7-15 wt% |
| **Auxiliary agents** | silanes (most frequently epoxy and methacryloxy), dispersing agents such as alkanes and stearic acid; silicone powder with methacrylic functional groups for flame retardant applications with magnesium hydroxide[226] |
| **Special methods of incorporation** | for high impact polystyrene, the filler is dispersed in rubber and then incorporated into the matrix polymer; encapsulation of glass beads by grafted polybutylacrylate and polystyrene;[30] free radical grafting of polystyrene onto montmorillonite interlayers;[312] in pan-milling of titanium dioxide with polystyrene, the improvement of impact strength results from the mutual influence on particle size reduction and the creation of new interacting surfaces;[315] melt grafting of glass beads can be conducted by modifying the bead surface with epoxy silane and reaction of epoxy groups with poly(styrene-co-maleic anhydride) through diamine spacer[319] |
| **Methods of filler pretreatment** | drying, especially for extrusion applications |
| **Special considerations** | dewetting angles can be calculated which represent filler-matrix adhesion;[320] carbon black is a compatibilizer of PVDF/PS blends;[321] the thickness of a silane layer coating was estimated to be ~16 nm[322] |

Filler/matrix adhesion determines the rate of particle debonding (Figure 15.27). The debonding of glass beads either treated and untreated with silane begins at the same low strain but the rates and the extent of debonding are different.[311,318] The material properties depend on filler distribution throughout the sample. Figure 15.28 shows that distribution of glass beads in an injection molded article depends on the particle size of the beads. The large and small beads formed a similar core/shell structure but larger particles tended to be located more in the core (about 70%) than smaller particles (about 50%).[316] Also, the distance from gate affects particle distribution. Particles tent to accumulate close to a free surface causing almost double the concentration of beads at the free surface compared with their concentration in the feed. Orientation of talc particles parallel to the wall of the mold or the dies was determined for polystyrene filled with talc.[317] The addition of small amounts (<1 wt%) of rigid particles (barium sulfate and crosslinked polystyrene beads were used in the experiment) dramatically improves the impact strength of

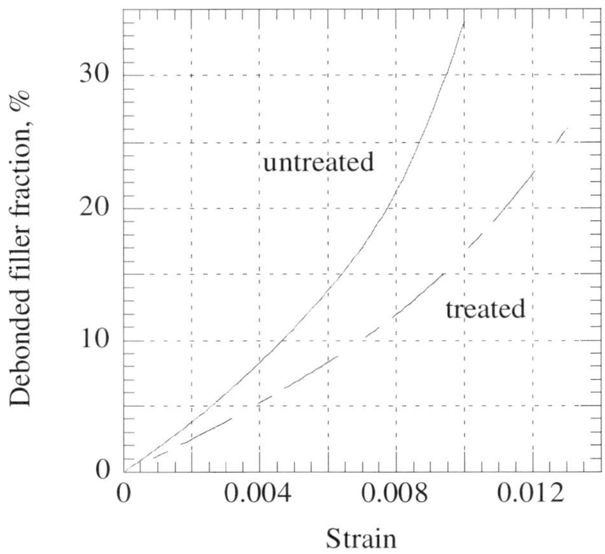


Figure 15.27. Debonded glass bead fraction, $\phi_d$, vs. strain for polystyrene composites. [Adapted, by permission, from Meddad A, Fisa B, *Macromol. Symp.*, **108**, 1996, 173-82.]

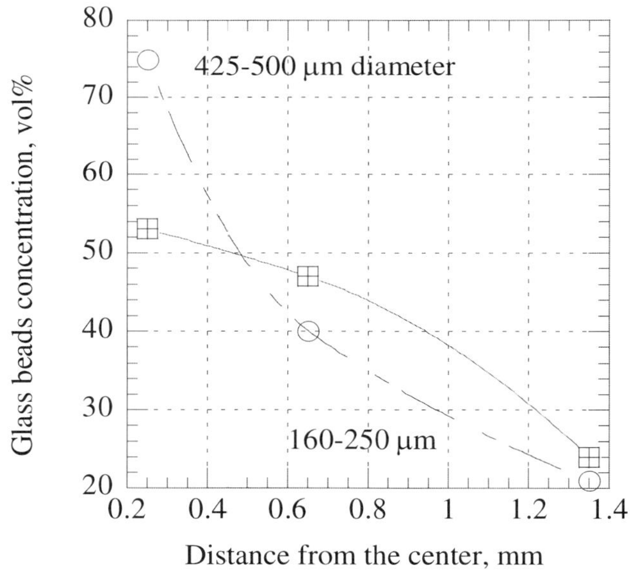


Figure 15.28. Glass concentration across the thickness of the sample. [Data from Ogadhoh S O, Papathanasiou T D, *Composites Part A: Applied Science and Manufacturing*, **27A**, No.1, 1996, 57-63.]

polystyrene. A specific particle diameter must be determined and selected to get the optimum improvement (see Figure 8.19).[314] The mechanism of action is related to the formation of voids and extension of crazes from these voids.

In fire retardant applications, a combination of zinc borate with ammonium polyphosphate gives V-0 rating. The use of zinc borate permits a reduction in the amount of ammonium polyphosphate. Red phosphorus alone or in combination with ammonium polyphosphate or melamine phosphate also produced a V-0 rating. The heat release rate can be effectively improved by small additions (1-2 wt%) of silicone powder in combination with other flame retarding additives or at higher concentration (15 wt%) when used by itself.[226]

## 15.39 POLYSULFIDES

| | |
|---|---|
| **Major polymer applications** | sealants, rocket propellant binders, electrical potting compounds, additives to epoxy, fuel hoses and tubing, insulating glass, fuel-contact sealants |
| **Important processing methods** | vulcanization, moisture or chemical curing of premixed compounds |
| **Typical fillers** | calcium carbonate, carbon black, zinc oxide, calcium oxide and hydroxide |
| **Typical concentration range** | 30-50 wt% |
| **Auxiliary agents** | not reported |
| **Special methods of incorporation** | not reported |
| **Methods of filler pretreatment** | drying for moisture cured systems |
| **Special considerations** | calcium oxide is used as acid scavenger and desiccant; zinc oxide is a curing agent in vulcanization processes; calcium oxide and hydroxide are used as acid scavengers |

## 15.40 POLYSULFONE, PSO

| | |
|---|---|
| **Major polymer applications** | tubing, in medical applications which require resistance to hot water and sterilization, microwave cookware, printed circuit boards |
| **Important processing methods** | injection molding, blow molding, extrusion, thermoforming |
| **Typical fillers** | glass fiber, carbon fiber, aramid fiber, PTFE |
| **Typical concentration range** | glass fiber - 20-30 wt%, PTFE - 8-15 wt% |
| **Auxiliary agents** | not reported |
| **Special methods of incorporation** | not reported |
| **Methods of filler pretreatment** | drying |
| **Special considerations** | material should have a moisture content below 0.05% to prevent bubbling, surface streaks, etc. |

## 15.41 POLYTETRAFLUOROETHYLENE, PTFE[324-327]

| | |
|---|---|
| **Major polymer applications** | wear reduction, friction reduction, film, tubes, gaskets, valve and pump parts, tank lining, laboratory equipment, filtration membranes, bearings, piston rings, seals, non stick coating, electric insulation applications, Gore-Tex™ membranes |
| **Important processing methods** | solid phase forming, sintering, ram extrusion, compression molding, paste extrusion (mixed with lubricants and forced through cold die followed by lubricant evaporation and sintering), spraying, flow coating, dip, coating, film coating, fiber spinning |
| **Typical fillers** | glass fiber, carbon fiber, graphite, metal powders (bronze), molybdenum sulfide, boron nitride, carbon black, Ni-Zn ferrite |
| **Typical concentration range** | glass fiber - 15-25 wt%; graphite - 20-30 wt% |
| **Auxiliary agents** | not reported |
| **Special methods of incorporation** | fillers are premixed with powdered resins and then molded |
| **Methods of filler pretreatment** | not reported |
| **Special considerations** | high temperature deflection decreases with glass fiber loading and after heating to 200°C the effect of reinforcement disappears |

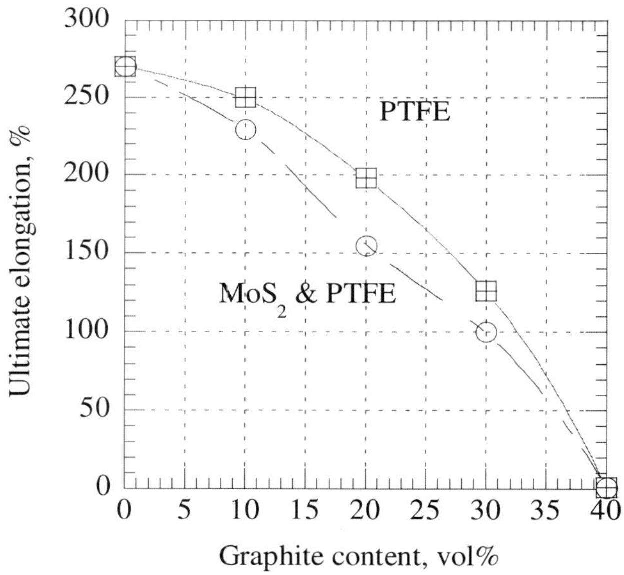


Figure 15.29. Elongation vs. filler content. [Data from Fengyuan Yan, Qunji Xue, Shengrong Yang, *J. Appl. Polym. Sci.*, **61**, No.7, 1996, 1223-9.]

Neat PTFE has several properties which eliminate its need for additives. A very high limiting oxygen index (>95%) means that resin is nonflammable. Its thermal and electric insulation properties and its low friction coefficient means that further improvements are rarely needed. An improvement in mechanical properties may sometimes be needed but this is usually obtained at the expense of chemical resistance because most fillers detract from the chemical resistance of neat PTFE. The addition of NiZn ferrite powder is used to impart ferromagnetic properties to the polymer for the electronic industry.[324] The addition of graphite, molybdenum disulfide, boron nitride, metal powders, and glass fiber increases wear resistance and thermal conductivity. The addition of graphite above 40 wt% increases the porosity of the composite because it generates a change in morphology.[325] Figure 8.37 shows that the wear rate of graphite/PTFE and molybdenum sulfide/PTFE composites changes above 40 wt% of filler.[326] Addition of particulate fillers usually results in deterioration of mechanical properties as Figure 15.29 shows. The principles of operation under high wear conditions with lubrication were studied.[327]

## 15.42 POLYURETHANES, PU & TPU[328-350]

| | |
|---|---|
| **Major polymer applications** | TPU: coatings, footwear, automotive, wire and cable, hose and tubes, film and sheet<br>PU: coatings, sealants, adhesives, foams, primers, mortars, and numerous other products |
| **Important processing methods** | TPU: injection molding, extrusion, blow molding, solution coating of fabrics and calendering coating<br>PU: compounding, mixing, chemical or moisture cure |
| **Typical fillers** | calcium carbonate, calcium sulfate, silica, organic fibers, graphite, mica, bentonites, sand, aluminum hydroxide, sepiolite, rubber particles |
| **Typical concentration range** | calcium carbonate - 30-60 wt%; calcium sulfate - 5-10 wt%; sand up to 95 wt%; glass beads - 30-40 wt%; sepiolite - 40 wt% |
| **Auxiliary agents** | drying agents, process oils, surfactant for silica modified foam with good insulation properties[337] |
| **Special methods of incorporation** | fillers are frequently pre-dispersed in polyol (for better mechanical properties) or plasticizers (to dry while dispersing); hydroxyapatite was modified by reaction with hexamethylene diisocyanate[342] |
| **Methods of filler pretreatment** | TPU: Total moisture of the system should not exceed 0.1% to prevent hydrolysis and blistering<br>PU: depends on application but typically moisture below 0.05% is required to prevent reaction with water leading to property loss and bubbling |
| **Special considerations** | sepiolite treatment below 550°C does not alter the properties which modify the rheological properties of adhesives;[332] filler presence may affect adhesion (see Figure 8.54);[338] rubber particles are hygroscopic and can absorb 1.4% moisture with the rate of 0.01% per minute (similar to carbon black)[340] |

Polyurethane filled with glass beads was studied to determine the effect of concentration, particle size, and coupling (Figures 15.30, 15.31, and 15.32). The value of strain at failure decreases as concentration increases. The mode of failure also changes as the concentration of glass beads increases. At lower concentrations (12 and 24 vol%), the material shows stress softening before the failure. At the highest concentration, simple brittle failure occurs. The material with smaller particles has a higher modulus and withstands higher strain before it fails but the characters of both strain-stress curves are similar. The stress-strain relationship of polyurethane filled with treated glass beads is linear and failure occurs at a higher stress than with untreated beads. The material containing untreated beads has lower tensile properties but higher elongation.[330]

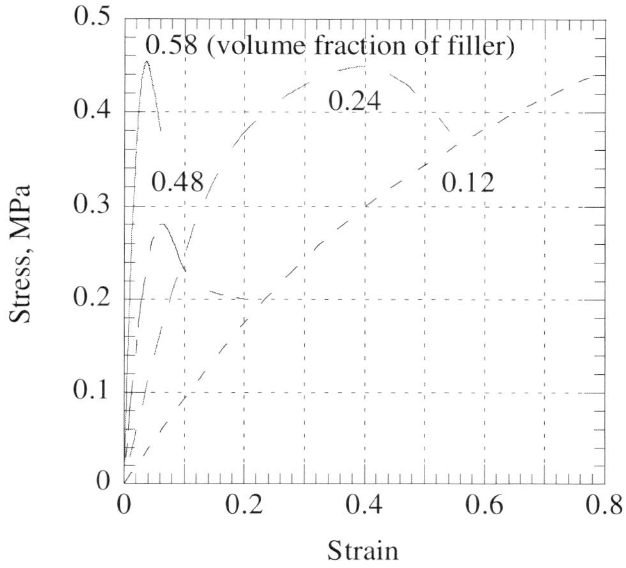


Figure 15.30. Stress-strain curves for glass bead filled polyurethane with different concentrations of glass beads. [Data from Vratsanos L A, Farris R J, *Polym. Engng. Sci.*, **33**, No.22, 1993, 1458-65.]

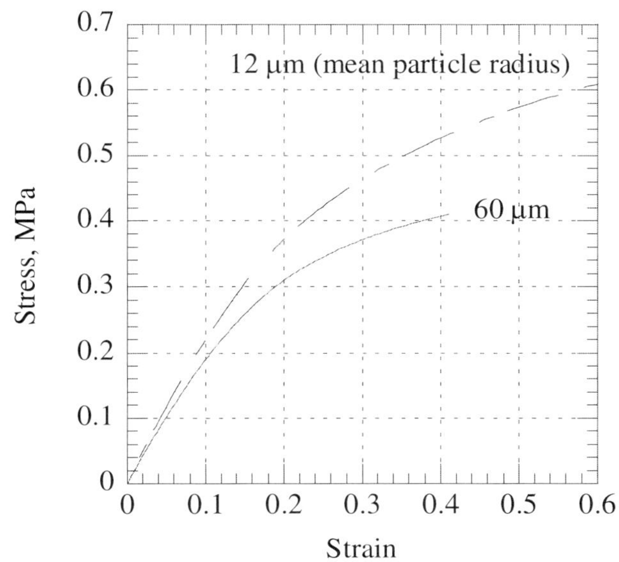


Figure 15.31. Stress-strain curves for glass bead filled polyurethane with different diameter of glass beads. [Data from Vratsanos L A, Farris R J, *Polym. Engng. Sci.*, **33**, No.22, 1993, 1458-65.]

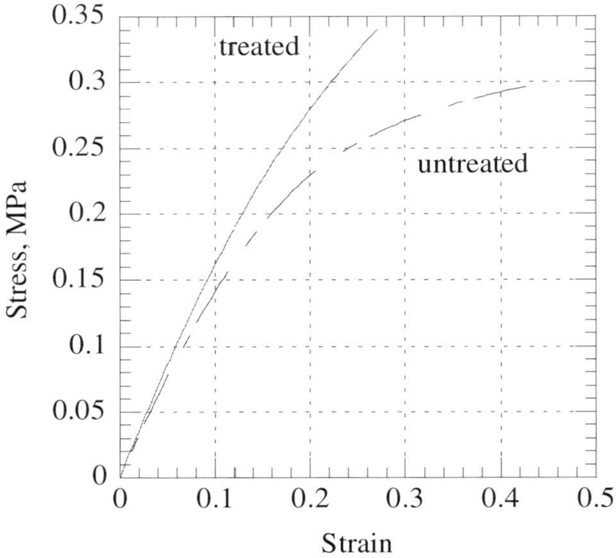


Figure 15.32. Stress-strain curves for glass bead filled polyurethane with coated and uncoated glass beads. [Data from Vratsanos L A, Farris R J, *Polym. Engng. Sci.*, **33**, No.22, 1993, 1458-65.]

Flame retardant polyurethanes are mostly manufactured with compounds of phosphorus, such as ammonium phosphate or polyphosphate.[329] Aluminum hydroxide alone or in combination with melamine is an alternate approach. In intumescent applications, graphite is frequently used. Calcium carbonate is useful as a flame retarding additive, in combination with other flame retarding materials, because of its large endothermic peak found in DTA curves.[331]

## 15.43 POLY(VINYL ACETATE), PVAc[351-356]

| | |
|---|---|
| **Major polymer applications** | production of poly(vinyl alcohol), adhesives, paints |
| **Important processing methods** | mixing/compounding |
| **Typical fillers** | calcium carbonate, clay, mica, talc, aluminosilicate |
| **Typical concentration range** | 20-30 wt% |
| **Auxiliary agents** | thickeners |
| **Special methods of incorporation** | silanes were polymerized in the presence of PVAc to form reinforced composite[354,365] |
| **Methods of filler pretreatment** | not reported |
| **Special considerations** | calcium carbonate (10-20 wt%) enhances rheology; stearate treated calcium carbonate does not interact with polymer and gives composite with a lower mechanical performance;[351] fillers used in adhesives should have limited interaction with matrix otherwise they decrease elasticity of the adhesive joint[355] |

## 15.44 POLY(VINYL ALCOHOL), PVAl[357]

| | |
|---|---|
| **Major polymer applications** | sizing agents, binders, protective colloids, photographic papers, toners, film, water-soluble laundry bags, seed tapes, sanitary pads, belts, printing rolls, controlled drug delivery, membranes |
| **Important processing methods** | casting, extrusion |
| **Typical fillers** | carbon black, silica, calcium carbonate, clay, zinc oxide, titanium dioxide, sand, aluminum oxide, magnesium oxide, zirconia, ferrite, graphite |
| **Typical concentration range** | 20-30 wt%; in some applications up to 98 wt% |
| **Auxiliary agents** | calcium stearate |
| **Special methods of incorporation** | *in situ* precipitation of filler[357] |
| **Methods of filler pretreatment** | not reported |
| **Special considerations** | not reported |

## 15.45 POLY(VINYL BUTYRAL), PVB[358]

| | |
|---|---|
| **Major polymer applications** | safety glass interlayer (automotive windshields), control of light, heat and sound in construction glass, bulletproof glass, adhesives and sealants, binders for rocket propellant, photoconductive papers, magnetic tapes, powder coating, wood sealers and primers, inks, ceramic binders, dry toners, wash primers, composite fiber binders |
| **Important processing methods** | compounding, powder coating, extrusion |
| **Typical fillers** | calcium carbonate, aluminum hydroxide, zinc oxide, rust protective fillers |
| **Typical concentration range** | 20-30 wt% |
| **Auxiliary agents** | not reported |
| **Special methods of incorporation** | not reported |
| **Methods of filler pretreatment** | not reported |
| **Special considerations** | not reported |

## 15.46 POLY(VINYL CHLORIDE), PVC[82,199,358-372]

| | |
|---|---|
| **Major polymer applications** | profiles, cables, siding, windows, pipes, fittings, flooring, footwear, film and sheet, coated fabrics, tubing, gutters, drain pipes, packaging, furniture trim, bottles, gloves, wallpaper, foam backings of carpets, domestic appliances, office equipment, toys, protective clothing, metal protection in automotive and many more |
| **Important processing methods** | extrusion, plastisol coating, thermoforming, calendering, blow molding, rotational molding, injection molding |
| **Typical fillers** | calcium carbonate, clay, talc, silica, antimony trioxide, aluminum hydroxide, magnesium hydroxide, carbon fiber, aluminum fiber, titanium dioxide, carbon black, sand, wood fiber |
| **Typical concentration range** | calcium carbonate - 20-30 wt% (rigid) 30-40 wt% (flexible), talc 5-25 wt%, antimony trioxide - 3-6 wt%, aluminum hydroxide, magnesium hydroxide - 20-40 wt%, sand 40-60 wt% |
| **Auxiliary agents** | lubricants, dispersing agents, compatibilizers (chlorinated polyethylene),[199] acrylic impact modifier to improve impact with talc[360] |
| **Special methods of incorporation** | dry blending including preparation of initial premix, compounding and pelletization |
| **Methods of filler pretreatment** | seldom used |
| **Special considerations** | zinc oxide decreases thermal stability; combination of zinc oxide with carbon black reduces UV stability; high concentrations of $TiO_2$ or carbon black are needed to improve UV stability, at lower concentrations these fillers will reduce UV stability; copper-containing compounds, iron salts, cadmium, cobalt, manganese, lead salts reduce thermal stability of PVC (also fillers containing these metals); sand was found to protect PVC from UV degradation;[359] calcium carbonate interferes with radiation crosslinking of PVC[363,366,370] |

Addition of fillers impacts mechanical properties of PVC. For example, calcium carbonate decreases yield stress and talc decreases impact strength. The aim of research on PVC is to obtain compositions with balanced properties. Ultrafine talc is used in application where impact and stiffness must be simultaneously improved.[360] Flexural modulus is improved by addition of ultrafine and general purpose talcs. But addition of either type of talc rapidly decreases impact strength. If ultrafine talc is incorporated with an acrylic modifier both stiffness and impact are improved (depending on the amounts of both talc and modifier, the flexural modulus can be improved by up to 60% and the impact strength by 1500%).[360] The surface of fillers plays a critical role. Although surface modification of filler can bring advantages, the inherent properties of the untreated filler are equally important. The extensive use of calcium carbonate in PVC is, among other factors (mostly economic), related to its basicity which causes it to interact well with PVC (which is acidic). If surface modification changes this interaction, the performance of the composite may be affected. Fillers are added to PVC to provide thermal and UV stabilization. For example, calcium stearate may play the role of an associate thermal stabilizer when used in a system with calcium salts of fatty acids. These stabilizers use combinations of two or more metals – one of which (e.g. zinc) produces metal chlorides which accelerate PVC degradation. The presence of large amount of calcium salts helps to convert this chloride to calcium chloride which does not increase the degradation rate of PVC. Also, calcium carbonate can react with hydrogen chloride which is produced as PVC degrades. On the other hand, inclusion of fillers which contain admixtures of metals such as iron, nickel, copper, etc. reduces PVC thermal stability. Fillers also affect UV stabilization by adsorption of HALS stabilizers which immobilizes them and prevents them from performing as radical scavengers.

Antimony trioxide is a suitable flame retardant for PVC considering that it requires chlorine to perform. Small additions (3-6 wt%) are sufficient for rigid and semi-rigid compositions. Making plasticized PVC flame retardant requires higher concentrations (15-20 wt%). Antimony trioxide is a white pigment it influences the color. Therefore, small particle size antimony pentoxide from a spray drying process is used to give dark colors as well as the transparent materials. Aluminum hydroxide is another additive used, although, its wide spread application is hampered by a need for larger concentrations at which mechanical properties are affected. Magnesium hydroxide has been used as a fire retardant with an aim to limiting interference with stabilizers.[82] Magnesium hydroxide is marginally better than aluminum hydroxide for smoke suppression.[364] Combinations of aluminum hydroxide and magnesium hydroxide with zinc borate were also studied as a potential replacement for antimony oxide.[367] These combinations provided improvement over antimony oxide in the flame retarding properties of heat release, specific extinction area, smoke, and CO emission. The best performing formulations required higher concentration of inorganic flame retardants (concentration increased by ~250%) but even at the same level of loading, formulations containing zinc borate in combination with aluminum hydroxide gave better performance than combinations of aluminum hydroxide with antimony oxide.[367] Figure 13.8 shows that the performance of aluminum and magnesium hydroxides can be enhanced by coating the powders with zinc hydroxystannate.[372] Smoke reduction can be obtained by various iron, copper, nickel and vanadium compounds which, although reduce the smoke but do affect thermal stability and, frequently, the UV stability of PVC. Other materials used include molybdenum and boron compounds.

In the area of conductive plastics, PVC is used in static control applications and EMI shielding. Static control, involving tiling and sheeting in industrial applications, is achieved by addition of carbon black. EMI shielding is a relatively new application for PVC in which metal and carbon fibers are used.

## 15.47 RUBBERS[71,341,345,346,373-397]

| | |
|---|---|
| **Major polymer applications** | applications are given for specific rubbers in the tables which follow |
| **Important processing methods** | see under individual materials |
| **Typical fillers** | carbon black, talc, in EMI shielding field: silver plated aluminum, silver plated nickel, silver coated glass spheres, silver plated copper, silver, nickel and carbon black |
| **Typical concentration range** | general guidelines: carbon black - 10-40 wt%, clay, talc - 5-60 wt%, calcium silicate - 1-8 wt%, aluminum silicate - 5-40 wt%, magnesium aluminum silicate - 5-30 wt% |
| **Auxiliary agents** | product specific |
| **Special methods of incorporation** | processing history has an essential effect on conductivity; amount of shear imposed and mixing causes the fracture of secondary carbon aggregates; increased temperature during mixing may preferentially form rubber-carbon bonds rather than the carbon-carbon bonds required for conductivity; vulcanization temperature may affect recovery of broken connections between carbon-carbon bonds |
| **Methods of filler pretreatment** | silanes including amino- and mercaptosilanes; treatment is more common with silica and clays |
| **Special considerations** | in conductive applications, special conductive blacks must be employed; in silver containing gaskets, galvanic corrosion is a problem, attention should be given to material with which shield is connected (potential difference), with zinc or aluminum casing nickel filled materials are preferred |

In addition to the general information included here, the specific information is discussed below for different rubber types. Dispersion of fillers is an important part of rubber processing technology. A test was developed to measure dispersion based on viscosity increase.[374] Atomic force microscopy combined with image analysis was used to quantify a dispersion of carbon black.[378] Aggregate size distribution and interaggregate distances can be measured by this method. Good dispersion of filler increases tensile strength, elongation and cut growth and decreases abrasion loss.[381] Rubber mixing with additives including fillers was analyzed elsewhere.[384] The effect of carbon black dispersion on rheological properties of rubber compounds and their morphology can be found in the literature.[385,386] A model of chemical and physical crosslinks was developed[388] and particles debonding conditions were analyzed.[389]

## 15.47.1 NATURAL RUBBER, NR[84, 398]

| | |
|---|---|
| **Major polymer applications** | pumps, valves, piping, hoses, machined components, tubes, boots, waterproof clothing, and bathing apparel, wire and cables, instrument panels, electrician's gloves, tires, heels and soles, conveyor belts, vibration dampers, shock absorbers, latex foams, tire cord impregnation, switchboard panels, plugs, sockets, telephone receivers, storage-battery cases, toys, gloves, pumps, pipes, valves |
| **Important processing methods** | vulcanization, coating, Banbury mixer, Gordon plasticator, skim coating, sheeting, calendering, tubing |
| **Typical fillers** | carbon black, calcium carbonate, dolomite, clays, calcinated clays, talc, soapstone, zinc oxide, fumed silica, borates, iron oxide, zinc oxide, magnesium carbonate, pulverized polyurethane foam, barium and strontium ferrites, magnesium aluminum silicate, nylon fibers, quartz; in EMI shielding field: silver plated aluminum, silver plated nickel, silver coated glass spheres, silver plated copper, silver, nickel and carbon black |
| **Typical concentration range** | carbon black - 20-30 wt%, calcium carbonate, quartz, talc - 15-25 wt%, silica - 15-30 wt%, titanium dioxide - 5 wt%, zinc oxide - 3-5 wt%, magnesium aluminum silicate - 20-40 wt%, barium or strontium ferrite (magnetic fillers) - 15-35 wt%, pulverized polyurethane foam - 15-30 wt% |
| **Auxiliary agents** | fatty acids, amines, silanes, multipurpose additive |
| **Special methods of incorporation** | processing history has an essential effect on conductivity; amount of shear imposed and mixing causes the fracture of secondary carbon aggregates; increased temperature during mixing may preferentially form rubber-carbon bonds rather than carbon-carbon bonds required for conductivity; vulcanization temperature may affect recovery of broken connections between carbon-carbon bonds |
| **Methods of filler pretreatment** | not reported |
| **Special considerations** | in conductive applications special conductive blacks must be employed; in silver containing, gaskets galvanic corrosion is a problem, attention should be given to material with which shield is connected (potential difference), with zinc or aluminum casing nickel filled materials are preferred |

Flexible magnets were prepared by incorporation of barium and strontium ferrites.[377] The permissivity and dielectric loss were reduced by use of calcium carbonate, talc, kaolin, and quartz.[390] Quartz gave the best dielectric properties.

Figure 15.33 shows benzene uptake by natural rubber samples.[398] Filled samples absorb less solvent (lower swelling). The carbon black containing sample had a lower benzene uptake than the silica filled sample. The lower swelling of the carbon black containing sample is due to high bound rubber content, the crosslink density of the black filled vulcanizate, and a strong rubber-filler interaction.

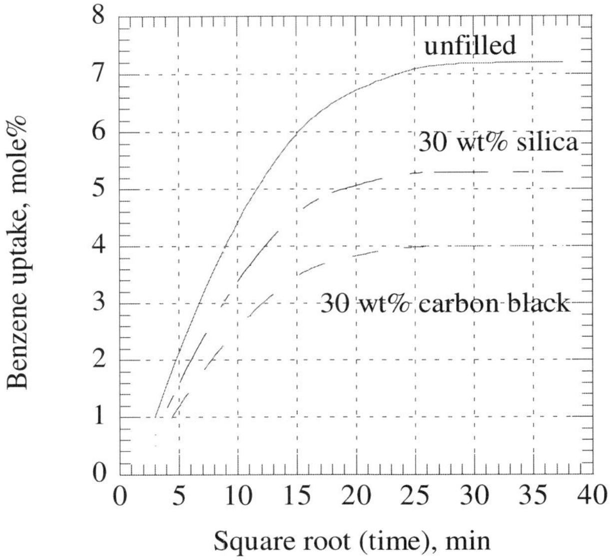


Figure 15.33. Benzene uptake for natural rubber. [Adapted, by permission, from Unnikrishnan G, Thomas S, Varghese S, *Polymer*, **37**, No.13, 1996, 2687-93.]

## 15.47.2 NITRILE RUBBER, NBR[362,399-405]

| | |
|---|---|
| **Major polymer applications** | gaskets, packing, automotive hoses, seals, industrial hoses, printing rolls, belt covers, footwear, hose jackets, polymer modification, tires |
| **Important processing methods** | vulcanization, coating, molding |
| **Typical fillers** | calcium carbonate, kaolin, carbon black, talc, zinc oxide, cellulose fibers |
| **Typical concentration range** | carbon black - 25-50 wt%, calcium carbonate - 20-50 wt%, kaolin - 2-40 wt%, talc 30-40 wt%, cellulose fiber 5-15 wt%, zinc oxide - 2-6 wt% |
| **Auxiliary agents** | stearic acid, aminopropylsilane |
| **Special methods of incorporation** | processing history has an essential effect on conductivity; amount of shear imposed and mixing causes the fracture of secondary carbon aggregates; increased temperature during mixing may preferentially form rubber-carbon bonds rather than the carbon-carbon bonds required for conductivity; vulcanization temperature may affect recovery of broken connections between carbon-carbon bonds; reaction between carboxyl groups of rubber and hydroxyl groups of carbon black occurs during molding at high temperature (190°C), in the case of sulfur-vulcanized systems, low molding temperature (150°C) favors weak physical bonds between rubber and filler[392] |
| **Methods of filler pretreatment** | silane treatment of carbon black during mixing with rubber;[402] co-precipitation of cellulose xanthate and NBR latex[405] |
| **Special considerations** | in conductive applications special conductive blacks must be employed; in silver containing gaskets galvanic corrosion is a problem, attention should be given to material with which shield is connected (potential difference), with zinc or aluminum casing nickel filled materials are preferred |

Figure 15.34 shows the relationship between bound rubber and oxygen content in carbon black. Figure 15.35 gives the relationship between the oxygen content and storage modulus. Figure 15.36 compares oxygen content with interaction parameter. Oxygen containing groups on the carbon black surface can react with NBR carboxyl groups. This reaction controls the concentration of bound rubber. The interaction parameter ($I = \sigma / \eta$) is a ratio of the slope of the stress-strain curve, $\sigma$, to the carbon-carbon networking factor, $\eta$, calculated from the storage modulus. Interaction parameter increases linearly with the concentration of functional groups on the filler surface. The storage modulus depends on reinforcement and chemical bonding between filler and rubber.

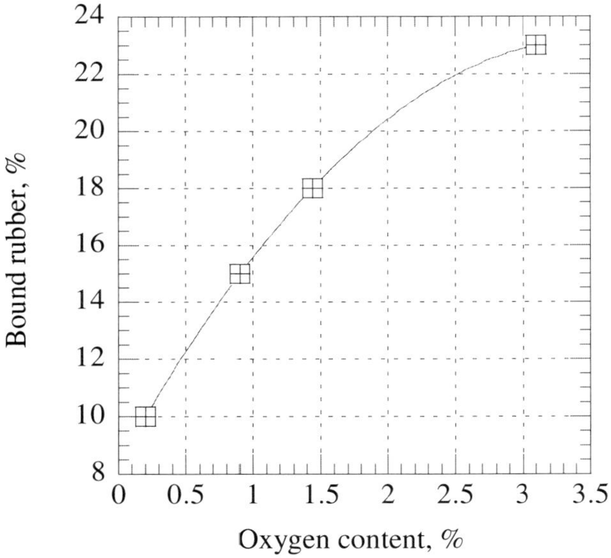


Figure 15.34. Bound rubber vs. oxygen content in carbon black determined for NBR compositions. [Data from Bandyopadhyay S, De P P, Tripathy D K, De S K, *J. Appl. Polym. Sci.*, **58**, No.4, 1995, 719-27.]

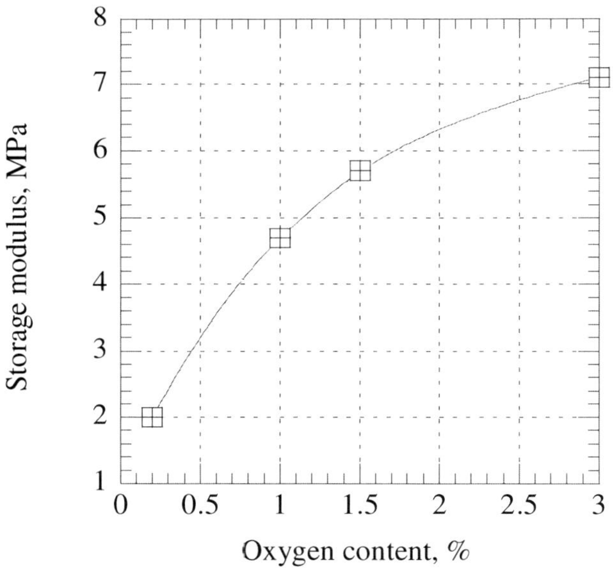


Figure 15.35. Storage modulus vs. oxygen content in carbon black determined for NBR compositions. [Data from Bandyopadhyay S, De P P, Tripathy D K, De S K, *J. Appl. Polym. Sci.*, **58**, No.4, 1995, 719-27.]

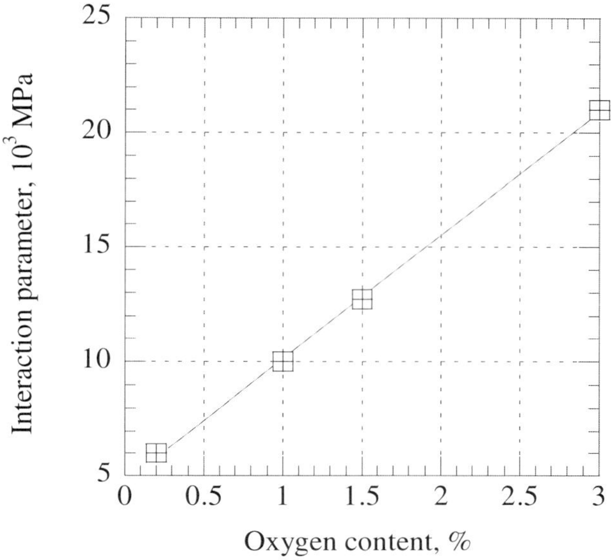


Figure 15.36. Interaction parameter vs. oxygen content in carbon black determined for NBR compositions. [Adapted, by permission, from Bandyopadhyay S, De P P, Tripathy D K, De S K, *J. Appl. Polym. Sci.*, **58**, No.4, 1995, 719-27.]

## 15.47.3 POLYBUTADIENE RUBBER, BR[84,406-408]

| | |
|---|---|
| **Major polymer applications** | modification of other polymers (e.g., HIPS and ABS), golf balls, tires, conveyor belts, hoses, seals and gaskets, rubberized cloth |
| **Important processing methods** | mixing, vulcanization, molding, extrusion, blow molding, injection molding |
| **Typical fillers** | carbon black, zinc oxide |
| **Typical concentration range** | carbon black - 30 wt%, zinc oxide - 3 wt % |
| **Auxiliary agents** | not reported |
| **Special methods of incorporation** | not reported |
| **Methods of filler pretreatment** | not reported |
| **Special considerations** | not reported |

## 15.47.4 POLYBUTYL RUBBER, HR

| | |
|---|---|
| **Major polymer applications** | belting, steam hose, curing bladders, O-rings, shock and vibration products, structural caulks and sealants, water-barrier applications, roof coatings, and gas-metering diaphragms |
| **Important processing methods** | mixing, vulcanization, molding |
| **Typical fillers** | carbon black, zinc oxide, clay; barium titanate, in EMI shielding field: silver plated aluminum, silver plated nickel, silver coated glass spheres, silver plated copper, silver, nickel and carbon black |
| **Typical concentration range** | carbon black - 20-50 wt%, zinc oxide - 2 wt% |
| **Auxiliary agents** | process oil, fatty acids |
| **Special methods of incorporation** | processing history has an essential effect on conductivity; amount of shear imposed and mixing causes the fracture of secondary carbon aggregates; increased temperature during mixing may preferentially form rubber-carbon bonds rather than the carbon-carbon bonds required for conductivity; vulcanization temperature may affect recovery of broken connections between carbon-carbon bonds |
| **Methods of filler pretreatment** | not reported |
| **Special considerations** | in silver containing gaskets galvanic corrosion is a problem, the attention should be given to material with which shield is connected (potential difference), with zinc or aluminum casing nickel filled materials are preferred |

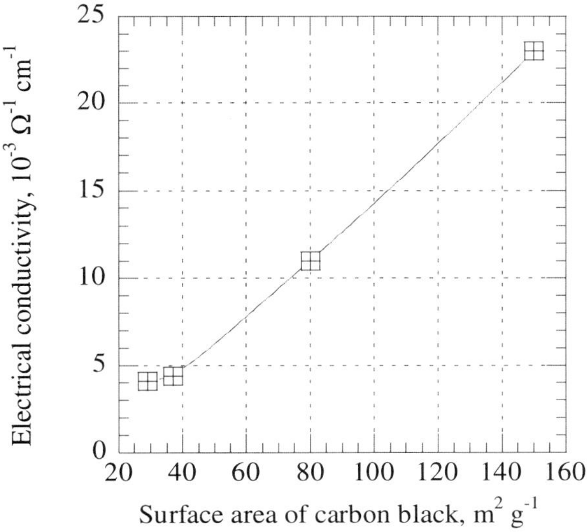


Figure 15.37. Electric conductivity vs. surface area of carbon black incorporated into butyl rubber. [Adapted, by permission, from Nasr G M, Badawy M M, Gwaily S E, Shash N M, Hassan H H, *Polym. Degradat. Stabil.*, **48**, No.2, 1995, 237-41.]

Figure 15.37 shows that conductivity of butyl rubber increases with surface area of carbon black increasing.[409]

## 15.47.5 POLYCHLOROPRENE, CR[198,410-412]

| | |
|---|---|
| **Major polymer applications** | wire and cable jacketing, hose, tubes and covers (auto and industrial), automotive gaskets, seals, CVJ boots and air springs, power transmission belts, molded and extruded goods, cellular products, adhesives, sealants, and protective coatings, foamed wet suits, latex dipped goods (gloves, weather balloons, automotive), paper, and industrial binders (shoe board), construction applications (bridge pads/seals, soil pipe gaskets, waterproof membranes, asphalt modification) |
| **Important processing methods** | vulcanization, dip coating, coating, sheeting, calendering, extrusion |
| **Typical fillers** | carbon black, zinc oxide, magnesium oxide; in EMI shielding field: silver plated aluminum, silver plated nickel, silver coated glass spheres, silver plated copper, silver, nickel and carbon black |
| **Typical concentration range** | carbon black - 20-40 wt%, zinc oxide - 3-4 wt%, magnesium oxide - 2-3 wt%, calcium carbonate, clay silica - 10-70 wt% |
| **Auxiliary agents** | lubricants, surfactants, waxes, oils |
| **Special methods of incorporation** | processing history has an essential effect on conductivity; amount of shear imposed and mixing causes the fracture of secondary carbon aggregates; increased temperature during mixing may preferentially form rubber-carbon bonds rather than the carbon-carbon bonds required for conductivity; vulcanization temperature may affect recovery of broken connections between carbon-carbon bonds |
| **Methods of filler pretreatment** | not reported |
| **Special considerations** | in conductive applications special conductive blacks must be employed; in silver containing gaskets galvanic corrosion is a problem, the attention should be given to material with which shield is connected (potential difference), with zinc or aluminum casing nickel filled materials are preferred; aluminum chloride is an important initiator of degradation; metal oxides are curing agents; carbon black is efficient CR stabilizer[411] |

Fast extrusion furnace black with a particle size of 360 Å, was used to verify different theoretical concepts of percolation which by definition predicts a rapid change in conductance when volume fraction of conductive particles attains a critical value. Figure 15.38 shows the effect of a carbon black addition to polychloroprene. Up to 30 phr carbon black, the conductivity of polychloroprene is almost constant and then it increases linearly as concentration of carbon black increases. The following equation applies: $\sigma = \sigma_0 (P - P_c)^\beta$ where: $\sigma_0$ is constant, P is concentration of conducting particles, $P_c$ is percolation threshold, and $\beta$ is exponent which accounts for cluster size.[410] When data from the Figure 15.38 are replotted as in Figure 15.39 it is evident that the percolation law is valid.

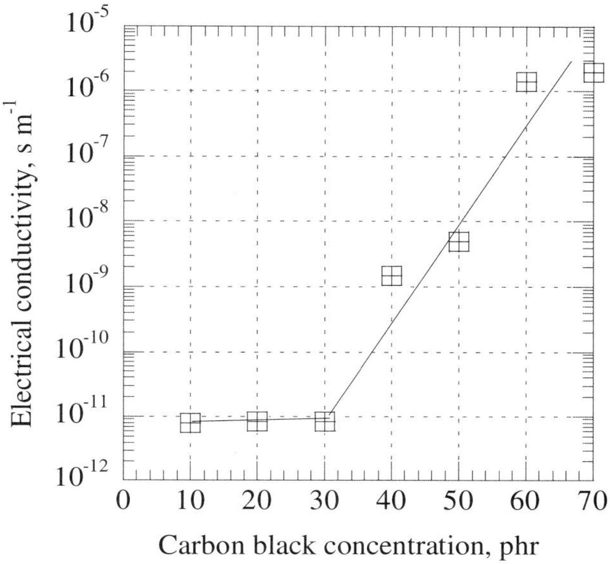


Figure 15.38. Electric conductivity of polychloroprene vs. concentration of carbon black. [Adapted, by permission, from Ali M H, Abo-Hashem A, *Plast. Rubb. Comp. Process. Appln.*, **24**, No.1, 1995, 47-51.]

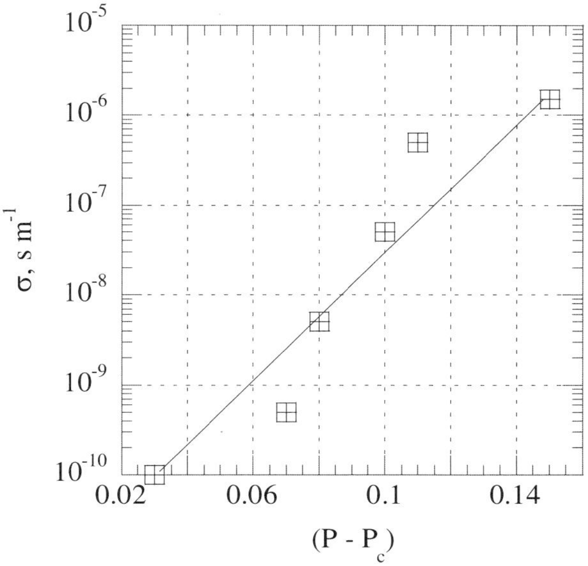


Figure 15.39. Plot of σ vs. $(P - P_c)$ in a logarithmic scale. [Adapted, by permission, from Ali M H, Abo-Hashem A, *Plast. Rubb. Comp. Process. Appln.*, **24**, No.1, 1995, 47-51.]

## 15.47.6 POLYISOBUTYLENE, PIB[399,413]

| | |
|---|---|
| **Major polymer applications** | sealants, roofing membranes |
| **Important processing methods** | compounding, vulcanization, coating, sheeting |
| **Typical fillers** | carbon black, calcium carbonate, kaolin, zinc oxide, clay |
| **Typical concentration range** | carbon black - 20-30 wt%, calcium carbonate - 30-50 wt%, zinc oxide 2-3 wt% |
| **Auxiliary agents** | not reported |
| **Special methods of incorporation** | not reported |
| **Methods of filler pretreatment** | not reported |
| **Special considerations** | not reported |

## 15.47.7 POLYISOPRENE, IR[84,414-416]

| | |
|---|---|
| **Major polymer applications** | pressure-sensitive adhesives, ablatives |
| **Important processing methods** | mixing, vulcanization, extrusion, calendering, molding |
| **Typical fillers** | carbon black, zinc oxide, kaolin, calcium carbonate, silicates, titanium dioxide |
| **Typical concentration range** | carbon black - 20-40 wt% |
| **Auxiliary agents** | stearic acid |
| **Special methods of incorporation** | not reported |
| **Methods of filler pretreatment** | not reported |
| **Special considerations** | zinc oxide is a curing agent |

## 15.47.8 STYRENE-BUTADIENE RUBBER, SBR[207,417-435]

| | |
|---|---|
| **Major polymer applications** | tires, flooring, conveyor belts, shoe products, sheet, tubing, tank and caterpillar tracks, sporting goods, toys, coated fabrics, automotive mechanical goods |
| **Important processing methods** | mixing, compression molding, calendering, vulcanization, coating |
| **Typical fillers** | carbon black, silica, lead oxide (γ-radiation shields), sodium aluminum silicate, clay, mica, kaolin, carbon fiber; crosslinked PS beads |
| **Typical concentration range** | carbon black - 20-50 wt%, precipitated silica - 25-60 wt%, calcium carbonate - 40-70 wt%, lead oxide - 88 wt%, zinc oxide - 1-2 wt%, clay - 20-80 wt%, mica - 20-30 wt%, kaolin - 20-35 wt%, carbon fiber - 15-30 wt% |
| **Auxiliary agents** | silane modification of silica,[420] process oils, lubricants |
| **Special methods of incorporation** | carbon fiber was added at the end of mixing to limit breakage;[427] silane addition stage was varied to improve properties of silica-filled composite[430] |
| **Methods of filler pretreatment** | thermal treatment of carbon black to increase adhesion and amount of bound rubber[420] |
| **Special considerations** | zinc oxide is crosslinking agent;[374] carbon black is a more efficient UV stabilizer in sulfur-cured SBR than in radiation or peroxide cured;[417] vulcanization conditions affects electric conductivity of carbon black filled SBR;[432] solvent diffusion and swelling rate decrease as concentration of carbon black increases[433] |

Carbon black loading and its surface area are the factors which determine mechanical properties of the filled rubber. Figure 15.40 shows the effect of nitrogen surface area and carbon black loading on the tensile strength of a tire compound. An increased surface area of carbon black contributes to an increase in tensile strength. There is a maximum performance which is attainable with each type of carbon black with the general tendency being that maximum performance is obtained at higher concentrations as the surface area decreases.[379] Figure 15.41 shows the effect of the surface area of carbon black on the essential parameters of tire performance.[379] The tensile and tear strength increases as the surface area of carbon black increases but the fatigue life and cut growth decreases when the surface area of carbon black increases.

The addition of kaolin increased the tensile strength of SBR by a factor of up to 4, modulus by a factor of up to 5 and swelling was reduced by 60%.[424] Crosslinked polystyrene beads increase the tensile strength, modulus, and elongation of SBR.[435]

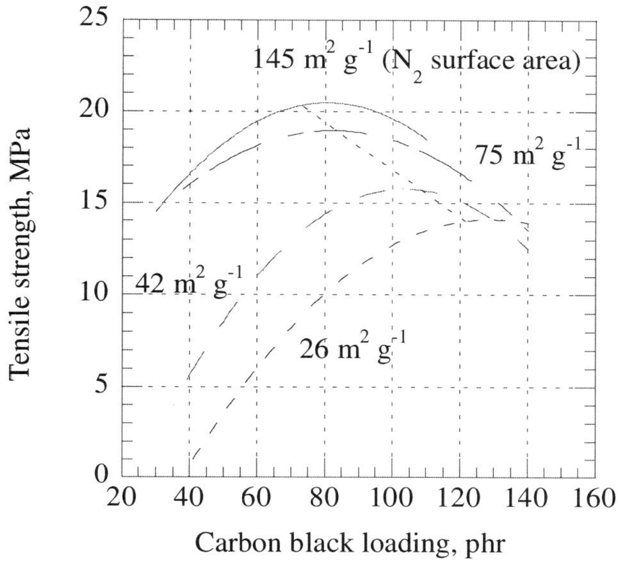


Figure 15.40. Tensile strength of SBR vs. carbon black loading. [Adapted, by permission, from Byers J T, Meeting of the Rubber Division, ACS, Cleveland, October 17-20, 1995, paper B.]

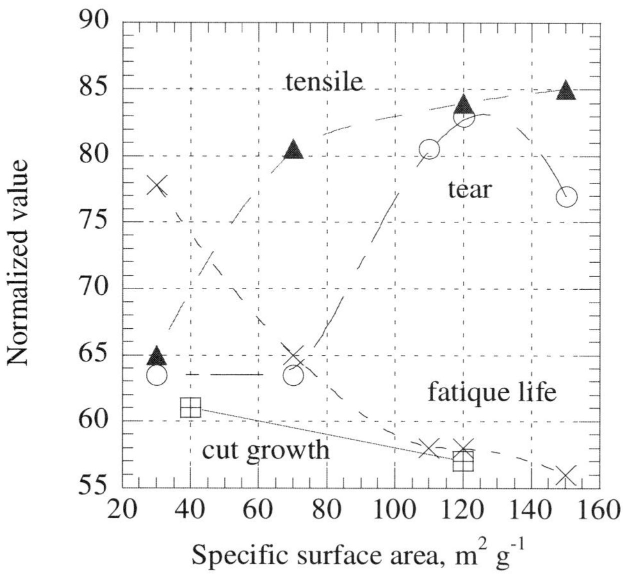


Figure 15.41. Properties of SBR vs. specific surface area of carbon black. [Adapted, by permission, from Byers J T, Meeting of the Rubber Division, ACS, Cleveland, October 17-20, 1995, paper B.]

## 15.48 SILICONES, SI[90,436-452]

| | |
|---|---|
| **Major polymer applications** | automotive (shaft sealing rings, spark plug caps, o-rings, gaskets, ignition cables, coolant and heater hose), general tubing, transfusion and dialysis tubing, door and windows seal, caulking and sealants |
| **Important processing methods** | injection molding, extrusion, room temperature, moisture or chemical cure of premixed compounds, vulcanization, casting |
| **Typical fillers** | fumed silica, calcium carbonate, carbon black, silver, glass beads, metal powders, precipitated silica, aluminum oxide, montmorillonite, mica, zinc oxide |
| **Typical concentration range** | fumed silica - 3-5 wt%, calcium carbonate - 20-50 wt%, glass beads - 5-20 wt%, aluminum oxide - 30-50 wt%, zinc oxide - up to 60% |
| **Auxiliary agents** | silanes |
| **Special methods of incorporation** | alkyl peroxide bridging to vinyl containing siloxane polymers used for conductive applications improved curing characteristics decreased by the presence of carbon black; crosslinking based on platinum catalyzed polyvinylmethyl siloxane and polymethyl hydrogen siloxane is not inhibited by carbon black; *in situ* silica formation;[437] in formation of PDMS nanocomposite, montmorillonite was delaminated in polymer prior to crosslinking[440] |
| **Methods of filler pretreatment** | choice of carbon black for conductive applications is crucial because impurities on carbon black may have an adverse effect on mechanical properties; aluminum oxide and calcium carbonate were coated by a hydrophobic layer of PDMS;[443] heat treatment of fumed silica reduces its ability to reinforce polymer, especially in temperatures above 200°C[444] |
| **Special considerations** | in conductive applications, special conductive blacks must be employed; fillers influence chemical degradation reactions in siloxanes;[438,447,451,452] zinc oxide was found to increase thermal resistance of PDMS[451,452] |

Silicone mechanical properties are inherently poor therefore reinforcement is an essential part of the product development process. A fumed silica addition is one method of silicone improvement by compounding in conjunction with other fillers. The example below shows the results obtained during the development of a nanocomposite material with a low degree of solvent uptake (Figure 15.42). Montmorillonite was delaminated prior to the silicone being cured. The results are compared with the effect of carbon black in SBR. The solvent uptake is inversely proportional to the filler reinforcing strength. Carbon black has strong reinforcing effect on SBR due to molecular interactions but the reinforcement of carbon black still does not match the effect of montmorillonite in the silicone nanocomposite. The advantage of the nanocomposite comes from the differences in surface areas of the fillers. Carbon black has typical surface area in a range from 20 to 100 $m^2/g$ whereas delaminated silicate has a surface area of 750 $m^2/g$. The larger surface area

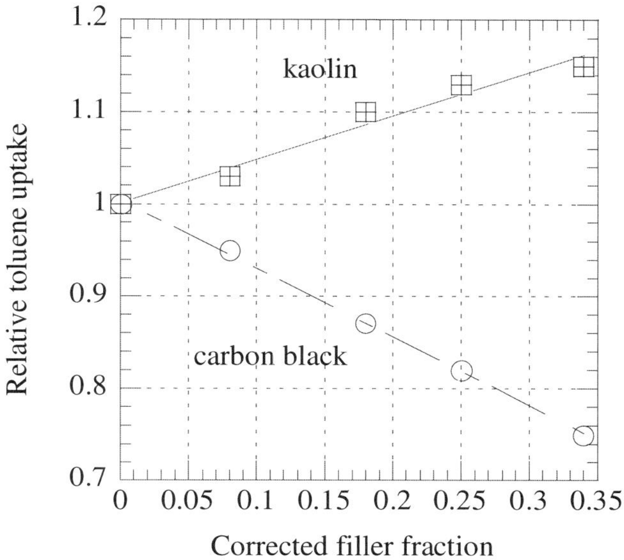


Figure 15.42. Toluene uptake vs. filler concentration. [Data from Burnside S D, Giannelis E P, *Chem. of Mat.*, **7**, No.9, 1995, 1597-600.]

increases the probability of interaction and thus results in the reinforcement and lower solvent uptake shown in Figure 15.42.[440] In another method, small particle sized filler can be incorporated by an *in situ* synthesis.[442] The results obtained show that degradation temperature can be substantially increased by the selection of composition of metal oxide particles which were formed *in situ*.

## 15.49 STYRENE-ACRYLONITRILE COPOLYMER, SAN[105]

| | |
|---|---|
| **Major polymer applications** | housings for electronic and electrical applications, instrument lenses packaging for high barrier properties, bottles, appliances (housings, air conditioner parts, refrigerator shelves, blenders, lenses), housewares (eating utensils, beverage/food containers, display boxes), automotive (dashboard, battery cases) |
| **Important processing methods** | injection molding, extrusion, blow molding, thermoforming, casting |
| **Typical fillers** | glass fiber, PTFE, aluminum borate whiskers |
| **Typical concentration range** | glass fiber - 10-40 wt%, PTFE - 10-15 wt% |
| **Auxiliary agents** | not reported |
| **Special methods of incorporation** | not reported |
| **Methods of filler pretreatment** | drying especially for extrusion; the surface of the injection molded part is affected by moisture |
| **Special considerations** | reprocessing affects color and melt flow index |

## 15.50 TETRAFLUOROETHYLENE-PERFLUOROPROPYLENE, FEP[90]

| | |
|---|---|
| **Major polymer applications** | wire coating, data transmission cable, lined pipes, components for valves and pumps |
| **Important processing methods** | injection molding, wire coating, extrusion |
| **Typical fillers** | graphite, glass fiber, bronze |
| **Typical concentration range** | 10-30 wt% |
| **Auxiliary agents** | not reported |
| **Special methods of incorporation** | not reported |
| **Methods of filler pretreatment** | not reported |
| **Special considerations** | creep and wear resistance are minimized by addition of fillers |

## 15.51 UNSATURATED POLYESTERS[453-456]

| | |
|---|---|
| **Major polymer applications** | composites, corrosion protection, surfboards, cultured stones, composites, bathroom sinks and vanity tops, countertops |
| **Important processing methods** | injection molding, compression molding, resin transfer molding, pultrusion, casting, encapsulation |
| **Typical fillers** | calcium carbonate, aluminum hydroxide, glass fiber, crashed marble, glass fiber, antimony trioxide, carbon black, quartz, saw dust |
| **Typical concentration range** | aluminum hydroxide - 30-80 wt%, quartz - up to 90 wt%, saw dust - 20-50 wt%, calcium carbonate - 50-74 wt%; glass fibers - 20 wt% |
| **Auxiliary agents** | silane;[453] maleic anhydride treatment of saw dust[454] |
| **Special methods of incorporation** | degassing of mold after filling[454] |
| **Methods of filler pretreatment** | coating with thermoplastic polymer; zinc hydroxystannate coating of fire retardant fillers increases their performance |
| **Special considerations** | coated aluminum hydroxide enhances esthetics of coatings |

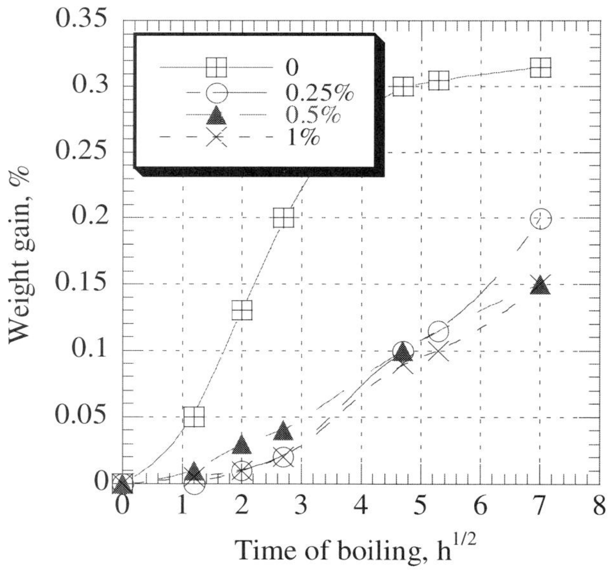


Figure 15.43. Weight gain by unsaturated polyester filled with quartz vs. time in boiling water. [Adapted, by permission, from Kominar V, Narkis M, Siegmann A, Breuer O, *Sci. & Engng. Composite Materials*, **3**, No.1, 1994, 61-6.]

Figure 15.43 shows the effect of boiling in water on weight gain by unsaturated polyester/quartz composite.[453] The addition of 0.5% silane was sufficient to reduce water uptake. A further increase in silane did not contribute to improvement. A sim-

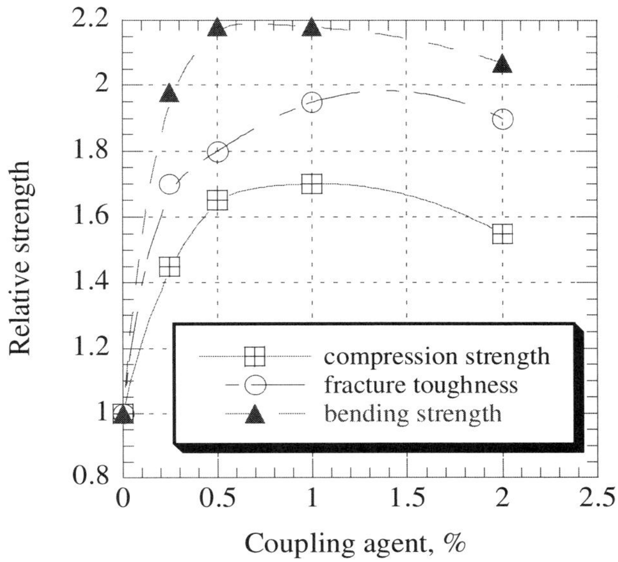


Figure 15.44. Coupling agent effect on mechanical performance of unsaturated polyester filled with quartz. [Adapted, by permission, from Kominar V, Narkis M, Siegmann A, Breuer O, *Sci. & Engng. Composite Materials*, **3**, No.1, 1994, 61-6.]

ilar amount of silane is sufficient to improve bending and compression strength and toughness (Figure 15.44). An inexpensive composite of unsaturated polyester was made with saw dust.[454]

## 15.52 VINYLIDENE-FLUORIDE TERPOLYMERS, PVDF[321,457-460]

| | |
|---|---|
| **Major polymer applications** | membranes, cables, valves, acid storage tanks, tubing, filtration |
| **Important processing methods** | extrusion, molding |
| **Typical fillers** | carbon black, silica, barium titanate, lead zirconium titanate, zeolite, copper powder |
| **Typical concentration range** | carbon black - 5-15 wt%, ceramic filler up to 40 wt%, silica - 20-50 wt%, zeolite - 0.1-1 wt%, copper 10-30 wt% |
| **Auxiliary agents** | tetraethoxysilane[458] |
| **Special methods of incorporation** | pre-dispersion of filler in PVDF solution followed by removal of solvent;[457] *in situ* formation of silica[458] |
| **Methods of filler pretreatment** | not reported |
| **Special considerations** | filling with carbon black can lead to the development of material with switching properties (10 fold increase in resistance at certain temperature)[459] crystallization history has effect on performance of switching material[459] |

## REFERENCES


1 Lee D H, Condrate R A, Reed J S, *J. Mater. Sci.*, **32**, 1997, 471-8.
2 Magrupov M A, Umarov A V, Saidkhodzhaeva K S, Kasimov G A, *Int. Polym. Sci. Technol.*, **23**, No.1, 1996, T/77-9.
3 Yu M C, Bissell M A, Whitehouse R S, Antec 95. Volume III. Conference proceedings, Boston, Ma., 7th-11th May 1995, 3246-50.
4 Costa L, Camino G, Bertelli G, Borsini G, *Fire & Mat.*, **19**, No.3, 1995, 133-42.
5 Guanghong Lu, Xiaotian Li, Hancheng Jiang, *Composites Sci. & Technol.*, **56**, No.2, 1996, 193-200.
6 Tsukuda R, Sumimoto S, Ozawa T, *J. Appl. Polym. Sci.*, **59**, No.6, 1996, 1043-6.
7 Nabi Z U, Hashemi S, *J. Mat. Sci.*, **31**, No.21, 1996, 5593-601.
8 Gingrich R P, Machado J M, Londa M, Proctor M G, Antec '95. Vol. II. Conference Proceedings, Boston, Ma., 7th-11th May 1995, 2345-50.
9 Aldcroft D, *Polym. Paint Col. J.*, **184**, No.4366, 1994, 423-5.
10 Kettrup A A, Lenoir D, Thumm W, Kampke-Thiel K, Beck B, *Polym. Degradat. Stabil.*, **54**, Nos 2-3, 1996, 175-80.
11 Da-Wei Liu, Jackson C L, Barnes J D, *Polym. Adv. Technol.*, **7**, No.4, 1996, 333-9.
12 Strumpler R, Maidorn G, Garbin A, Ritzer L, Greuter F, *Polym. & Polym. Composites*, **4**, No.5, 1996, 299-304.
13 Yang Q, Pritchard G, Phipps M A, Rose R G, *Polym. & Polym. Composites*, **4**, No.4, 1996, 239-46.
14 Srivastava V K, Pathak J P, *Polym. & Polym. Composites*, **3**, No.6, 1995, 411-4.
15 Rebouillat S, Escoubes M, Gauthier R, *J. Adhesion Sci. Technol.*, **10**, No.7, 1996, 635-50.
16 Pearson R A, Azimi H R, Bagheri R, Qian J Y, Antec '95. Vol. II. Conference Proceedings, Boston, Ma., 7th-11th May 1995, 2656-61.
17 Bittmann E, Ehrenstein G W, Antec '96. Volume II. Conference proceedings, Indianapolis, 5th-10th May 1996, 1465-70.
18 Pritykin L M, Razumova O V, Sokolova Y A, Antonov S M, Bolshakov V I, *Int. Polym. Sci. Technol.*, **23**, No.3, 1996, T/80-1.
19 Kushchevskaya N F, Shvets T M, *Int. Polym. Sci. Technol.*, **22**, No.12, 1995, T/21-2.
20 Shin Jen Shiao, Te Zei Wang, *Composites*, **27B**, No.5, 1996, 459-65.
21 Akovali G, Dilsiz N, *Polym. Engng. Sci.*, **36**, No.8, 1996, 1081-6.
22 Wu S-L, Tung I-C, *Polym. Composites*, **16**, No.3, 1995, 233-9.
23 Lambert C, Chauvet H, Larroque M, *Polymer*, **37**, No.15, 1996, 3441-5.
24 Bagheri R, Pearson R A, *Polymer*, **36**, No.25, 1995, 4883-5.
25 Achour M E, Miane J L, Lahjomri F, El Malhi M, Carmona F, *J. Mat. Sci. Lett.*, **14**, No.20, 1995, 1425-9.
26 Nyilas A, Rehme R, Wyrwich C, Springer H, Hinrichsen G, *J. Mat. Sci. Lett.*, **15**, No.16, 1996, 1457-9.
27 Lekatou A, Faidi S E, Lyon S B, Newman R C, *J. Mat. Res.*, **11**, No.5, 1996, 1293-304.
28 Wang J Y, Ploehn H J, *J. Appl. Polym. Sci.*, **59**, No.2, 1996, 345-57.
29 Hong S G, Lin J J,Yuan Ze, *J. Appl. Polym. Sci.*, **59**, No.10, 1996, 1597-605.
30 Luzinov I, Voronov A, Minko S, Kraus R, Wilke W, Zhuk A, *J. Appl. Polym. Sci.*, **61**, No.7, 1996, 1101-9.
31 Achour M E, El Malhi M, Miane J L, Carmona F, *J. Appl. Polym. Sci.*, **61**, No.11, 1996, 2009-13.
32 Azimi H R, Pearson R A, Hertzberg R W, *J. Appl. Polym. Sci.*, **58**, No.2, 1995, 449-63.
33 Palumbo M, Donzella G, Tempesti E, Ferruti P, *J. Appl. Polym. Sci.*, **60**, No.1, 1996, 47-53.
34 Yang Q, Pritchard G, *Polym. & Polym. Composites*, **2**, No.4, 1994, 233-9.
35 Ohta M, Nakamura Y, Hamada H, Maekawa Z, *Polym. & Polym. Composites*, **2**, No.4, 1994, 215-21.
36 Grady B P, Antec '97. Conference proceedings, Toronto, April 1997, 2490-3.
37 Kumar P, Gawahale A R, Rai B, *Adv. Composite Materials*, **4**, No.4, 1995, 279-85.
38 Privalko V P, Novikov V V, *Adv. Polym. Sci.*, **119**, 1995, 31-77.
39 Whalen J P, Poston D D, Antec '94. Conference Proceedings, San Francisco, Ca., 1st-5th May 1994, Vol. I, 924-8.
40 Dzeskiewicz L A, Ralston J D, Antec '94. Conference Proceedings, San Francisco, Ca., 1st-5th May 1994, Vol. I, 931-4.
41 Mamunya E P, Davidenko V V, Lebedev E V, *Polym. Composites*, **16**, No.4, 1995, 319-24.
42 Meijerink J I, Eguchi S, Ogata M, Ishii T, Amagi S, Numata S, Sashima H, *Polymer*, **35**, No.1, 1994, 179-86.
43 Dufresne A, Lacabanne C, *Polymer*, **34**, No. 15, 1993, 3173-8.

44 Kouloumbi N, Tsangaris G M, Kyvelidis S T, *J. Coatings Technol.*, **66**, No.839, 1994, 83-8.
45 Ming Qiu Zhang, Jia Rui Xu, Han Ming Zeng, Qun Huo, Zhi Yi Zhang, Feng Chun Yun, Friedrich K, *J. Mat. Sci.*, **30**, No.17, 1995, 4226-32.
46 Bogoeva-Gaceva G, Burevski D, Dekanski A, Janevski A, *J. Mat. Sci.*, **30**, No.13, 1995, 3543-6.
47 Katsumata M, Endo M, Ushijima H, Yamanishi H, *J. Mat. Res.*, **9**, No.4, 1994, 841-3.
48 Miszczyk A, Szalinska H, *Prog. Org.Coatings*, **25**, No.4, 1995, 357-63.
49 Messersmith P B, Giannelis E P, *Chem. of Mat.*, **6**, No.10, 1994, 1719-25.
50 Lan T, Pinnavaia T J, *Chem. of Mat.*, **6**, No.12, 1994, 2216-9.
51 Konstantellos B, Sideridis E, *Coll. Polym. Sci.*, **273**, No.4, 1995, 307-16.
52 Bagheri R, Williams M A, Pearson R A, *Polym. Eng. Sci.*, **37**, No.2, 1997, 245-51.
53 Tang L-G, Kardos J L, *Polym. Composites*, **18**, No.1, 1997, 100-13.
54 Gassan J, Bledzki A K, *Polym. Composites*, **18**, No.2, 1997, 179-84.
55 Bijwe J, *Polym. Composites*, **18**, No.3, 1997, 378-96.
56 Akay M, Mun S K A, Stanley A, *Composites Sci. Technol*, **57**, 1997, 565-71.
57 Hshieh F Y, Beson H D, *Fire Mater.*, **21**, 1997, 41-9.
58 Shang S W, Williams J W, Soderholm K J M, *J. Mat. Sci.*, **30**, No.17, 1995, 4323-34.
59 Singhal A, Fina L J, *Polymer*, **37**, No.12, 1996, 2335-43.
60 Kerber M L, Ponomarev I N, Lapshova O A, Grinenko E S, Sabsai O Y, Dubinskii M B, Burtseva I V, *Polym. Sci. Ser. A*, **38**, No.8, 1996, 867-74.
61 Haagh G A A V, Peters G W M, Meijer H E H, *Polym. Engng. Sci.*, **36**, No.20, 1996, 2579-88.
62 Rothon R N; Hornsby P R, *Polym. Degradat. Stabil.*, **54**, Nos 2-3, 1996, 383-5.
63 Tang H, Chen X, Luo Y, *Eur. Polym. J.*, **32**, No.8, 1996, 963-6.
64 McNeill I C, Mohammed M H, *Polym. Degradat. Stabil.*, **49**, No.2, 1995, 263-73.
65 McNeill I C, Mohammed M H, *Polym. Degradat. Stabil.*, **48**, No.1, 1995, 189-95.
66 Dole P, Chauchard J, *Polym. Degradat. Stabil.*, **47**, No.3, 1995, 441-8.
67 Toure B, Lopez Cuesta J-M, Longerey M, Crespy A, *Polym. Degradat. Stabil.*, **54**, Nos 2-3, 1996, 345-52.
68 Toure B, Lopez Cuesta J M, Gaudon P, Benhassaine A, Crespy A, *Polym. Degradat. Stabil.*, **53**, No.3, 1996, 371-9.
69 Younan A F, Choneim A M, Tawfik A A A, Abd-El-Nour K N, *Polym. Degradat. Stabil.*, **49**, No.2, 1995, 215-22.
70 Toure B, Lopez-Cuesta J, Benhassaine A, Crespy A, *Int. J. Polym. Analysis and Characterization*, **2**, No.3, 1996, 193-202.
71 *Int. Polym. Sci. Technol.*, **23**, No.7, 1996, T/1-3.
72 Kurian T, Khatgir D, De P P, Tripathy D K, De S K, Peiffer D G, *Polymer*, **37**, No.25, 1996, 5597-605.
73 Datta S, De S K, Kontos E G, Wefer J M, Wagner P, Vidal A, *Polymer*, **37**, No.15, 1996, 3431-5.
74 Datta S, Bhattacharya A K, De S K, Kontos E G, Wefer J M, *Polymer*, **37**, No.12, 1996, 2581-5.
75 Kurian T, De P P, Khastgir D, Tripathy D K, De S K, Peiffer D G, *Polymer*, **36**, No.20, 1995, 3875-84.
76 Datta S, De S K, Kontos E G, Wefer J M, *J. Appl. Polym. Sci.*, **61**, No.1, 1996, 177-86.
77 Chiu H-T, Chiu W-M, *J. Appl. Polym. Sci.*, **61**, No.4, 1996, 607-12.
78 Guriya K C, Tripathy D K, *J. Appl. Polym. Sci.*, **62**, No.1, 1996, 117-27.
79 Kurian T, De P P, Tripathy D K, De S K, Peiffer D G, *J. Appl. Polym. Sci.*, **62**, No.10, 1996, 1729-34.
80 Zaborski M, Slusarski L, Vidal A, *Polim. Tworz. Wielk.*, **38**, No.6, 1993, 263-8.
81 Zaborski M, Slusarski L, *Composite Interfaces*, **3**, No.1, 1995, 9-22.
82 Levesque J L, Fraval J T, Antec '93. Conference Proceedings, New Orleans, La., 9th-13th May 1993, Vol. II, 1957-64.
83 Zhu J, Ou Y-C, Feng Y-P, *Polym. Int.*, **37**, No.2, 1995, 105-11.
84 Coran A Y, Ignatz-Hoover F, Smakula P C, *Rubb. Chem. Technol.*, **67**, No.2, 1994, 237-51.
85 Otaigbe J U, Quinn C J, Beall G H, Antec '97. Conference proceedings, Toronto, April 1997, 1826-30.
86 Zyuzina G F, Vinogradova N K, Gribova I A, Krasnov A P, *Polym. Sci.*, **36**, No.9, 1994, 1205-8.
87 Plummer C J G, Wu Y, Gola M M, Kausch H H, *Polym. Bull.*, **30**, No.5, 1993, 587-94.
88 Liu C, Manzione L T, *Polym. Engng. Sci.*, **36**, No.1, 1996, 10-4.
89 Scaffaro R, Pedretti U, La Mantia F P, *Eur. Polym. J.*, **32**, No.7, 1996, 869-75.
90 Knowles J, *Polym. Paint Col. J.*, **185**, No.4366, 1995, 26-7.
91 Elfving K, Soderberg B, *Reinf. Plast.*, **40**, No.6, 1996, 64-5.
92 Brown N, Linnert E, *Reinf. Plast.*, **39**, No.11, 1995, 34-7.
93 *Reinf. Plast.*, **40**, No.10, 1996, 66-70.
94 Peng W, Riedl B, *Polymer*, **35**, No.6, 1994, 1280-6.

95 Balchandani N, *Pop. Plast. Packag.*, **40**, No.2, 1995, 41-2.
96 Murayama H, Min K, Antec '97. Conference proceedings, Toronto, April 1997, 759-65.
97 Wang T, Sherwood P M A, *Chem. of Mat.*, **6**, No.6, 1994, 788-95.
98 Tiejun Wang, Sherwood P M A, *Chem. of Mat.*, **7**, No.5, 1995, 1031-40.
99 Hegedus C R, Kamel I L, *J. Coatings Technol.*, **65**, No.820, 1993, 31-43
100 Levchik G F, Levchik S V, Lesnikovich A I, *Polym. Degradat. Stabil.*, **54**, Nos 2-3, 1996, 361-3.
101 Hornsby P R, Wang J, Rothon R, Jackson G, Wilkinson G, Cossick K, *Polym. Degradat. Stabil.*, **51**, No.3, 1996, 235-49.
102 Schleifstein R A, Property Enhancement with Modifiers and Additives. Retec proceedings, New Brunswick, N.J., 18th-19th Oct.1994, 89-101.
103 Turner J D, Property Enhancement with Modifiers and Additives. Retec proceedings, New Brunswick, N.J., 18th-19th Oct.1994, 65-87.
104 Baque T, *Kunststoffe Plast Europe*, **86**, No.8, 1996, 22-3.
105 Persson A L, Bertilsson H, *Composite Interfaces*, **3**, No.4, 1996, 321-32.
106 Narkis M, Tchoudakov R, Breuer O, Siegmann A, Antec '95. Vol. II. Conference Proceedings, Boston, Ma., 7th-11th May 1995, p.1343-6.
107 Benderly D, Siegmann A, Narkis M, Antec '95. Vol. II. Conference Proceedings, Boston, Ma., 7th-11th May 1995, 3168-71.
108 McIlvaine J, Antec 95. Volume III. Conference proceedings, Boston, Ma., 7th-11th May 1995, 3346-9.
109 Belanger B, Sanschagrin B, Fisa B, Antec '96. Volume II. Conference proceedings, Indianapolis, 5th-10th May 1996, 1762-7.
110 Larena A, Pinto G, *Polym. Composites*, **16**, No.6, 1995, 536-41.
111 Hornsby P R, Wang J, Cosstick K, Rothon R, Jackson G, Wilkinson G, Flame Retardants '94. Conference proceedings, London, 27th-28th January 1994, 93-108.
112 Hamada H, Haragushi M, Takahashi K, Machida K, Antec '97. Conference proceedings, Toronto, April 1997, 3326-30.
113 Jia N, Kagan V A, Antec '97. Conference proceedings, Toronto, April 1997, 1844-8.
114 Wagner A H, Kalyon D M, Yazici R, Fiske T J, Antec '97. Conference proceedings, Toronto, April 1997, 996-1000.
115 Rammoorthy M, Muzzy J, Antec '97. Conference proceedings, Toronto, April 1997, 2500-4.
116 Canova L A, Antec '93. Conference Proceedings, New Orleans, La., 9th-13th May 1993, Vol. II, 1943-9.
117 Turcovsky G, Antec '94. Conference Proceedings, San Francisco, Ca., 1st-5th May 1994, Vol. I, 796-800.
118 Meddad A, Fellahi S, Pinard M, Fisa B, Antec '94. Conference Proceedings, San Francisco, Ca., 1st-5th May 1994, Vol. II, 2284-8.
119 Zolotnitsky M, Steinmetz J R, Antec '94. Conference Proceedings, San Francisco, Ca., 1st-5th May 1994, Vol. III, 2756-60.
120 Hornsby P R, Wang J, Jackson G, Rothon R N, Wilkinson G, Cosstick K, Antec '94. Conference Proceedings, San Francisco, Ca., 1st-5th May 1994, Vol. III, 2834-9.
121 Ou Y-C, Yu Z-Z, *Polym. Int.*, **37**, No.2, 1995, 113-7.
122 Quintanilla L, Pastor J M, *Polymer*, **35**, No.24, 1994, 5241-6.
123 Messersmith P B, Giannelis E P, *Chem. of Mat.*, **5**, No.8, 1993, 1064-6.
124 Usuki A, Koiwai A, Kojima Y, Kawasumi M, Okada A, Kurauchi T, Kamigaito O, *J. Appl. Polym. Sci.*, **55**, No.1, 1995, 119-23.
125 Ou Y, Yang F, Chen J, *J. Appl. Polym. Sci.*, **64**, 1997, 2317-22.
126 Horst J J, Spoormaker J L, *J. Mater. Sci.*, **32**, 1997, 3541-51.
127 Stejskal J, Kratochvil P, Armes S P, Lascelles S F, Riede A, Helmstedt M, Prokes J, Krivka I, *Macromolecules*, **29**, No.21, 1996, 6814-9.
128 Zhou J, Li G, Li B, He T, *J. Appl. Polym. Sci.*, **65**, 1997, 1857-64.
129 Stokes V K, Antec '93. Conference Proceedings, New Orleans, La., 9th-13th May 1993, Vol. II, 2067-74.
130 Nofal M M, Zihlif A M, Ragosta G, Martuscelli E, *Polym. Composites*, **17**, No.5, 1996, 705-9.
131 Joo Y L, Lee Y D, Kwack T H, Min T I, Antec '96. Vol. I. Conference Proceedings, Indianapolis, 5th-10th May 1996, 64-8.
132 Benrashid R, Nelson G L, *J. Fire Sci.*, **11**, No.5, 1993, 371-93.
133 Zihlif A M, Di Liello V, Martuscelli E, Ragosta G, *Int. J. Polym. Mat.*, **29**, Nos.3-4, 1995, 211-20.
134 Tanaka T, Waki Y, Hamamoto A, Nogami N, Antec '97. Conference proceedings, Toronto, April 1997,

3054-8.
135 Rowberry P J, Conductive Polymers, Conference papers, Bristol, UK, 1992, paper 13.
136 Xavier S F, *Pop. Plast. Packag.*, **41**, No.4, 1996, 61-64.
137 Davis R M, Gardner S H, Marand E, Laot C, Reifsnider K, DeSmidt H, McGrath J, Tan B, Antec '97. Conference proceedings, Toronto, April 1997, 2494-9.
138 Cantwell W J, Zulkifli R, *J. Mater. Sci. Lett.*, **16**, 1997, 509-11.
139 Selzer R, Friedrich K, *Composites, Part A*, **28A**, 1997, 595-604.
140 Wang Y, Lu J, Wang G, *J. Appl. Polym. Sci.*, **64**, 1997, 1275-81.
141 Agari Y, Ueda A, Nagai S, *J. Appl. Polym. Sci.*, **52**, No.9, 1994, 1223-31.
142 Hindryckx F, Dubois P, Patin M, Jerome R, Teyssie P, Garcia Marti M, *J. Appl. Polym. Sci.*, **56**, No.9, 1995, 1093-105.
143 Burke M, Young R J, Stanford J L, *Plast. Rubb. Comp. Process. Appln.*, **20**, No.3, 1993, 121-35.
144 Hashmi S A R, Chand N, *J. Mat. Sci. Lett.*, **14**, No.5, 1995, 377-9.
145 Mamunya E P, Shumskii V F, Lebedev E V, *Polym. Sci.*, **36**, No.6, 1994, 835-8.
146 Fu Q, Wang G, Liu C, *Polymer*, **36**, No.12, 1995, 2397-401.
147 Janigova I, Chodak I, *Eur. Polym. J.*, **30**, No.10, 1994, 1105-10.
148 Janigova I, Chodak I, *Eur. Polym. J.*, **31**, No.3, 1995, 271-4.
149 Ruiz F A, Polymers, Laminations & Coatings Conference, 1995, 647-51.
150 Hansen H, Polymers, Laminations & Coatings Conference, 1995, 653-8.
151 Mayadunne A, Bhattacharya S N, Kosior E, Boontanjai C, Antec 95. Volume I. Conference proceedings, Boston,Ma., 7th-11th May 1995, 1178-82.
152 Ruiz F A, Antec '94. Conference Proceedings, San Francisco, Ca., 1st-5th May 1994, Vol. III, 2766-9.
153 Fiske T J, Gokturk H S, Kalyon D M, Antec '93. Conference Proceedings, New Orleans, La., 9th-13th May 1993, Vol. I, 614-7.
154 Gokturk H S, Fiske T J, Kalyon D M, Antec '93. Conference Proceedings, New Orleans, La., 9th-13th May 1993, Vol. I, 605-8.
155 Fiske T, Gokturk H S, Yazici R, Kalyon D M, Antec '97. Conference proceedings, Toronto, April 1997, 1482-6.
156 Berenbrok P A, Liles B E, Antec '97. Conference proceedings, Toronto, April 1997, 2931-2.
157 Bush S F, Tonkin J D, Antec '97. Conference proceedings, Toronto, April 1997, 3081-85.
158 Turley R S, Strong A B, *J. Adv. Materials*, **25**, No.3, 1994, 53-9.
159 Levering A W, Te Nijenhuis K, *J. Adhesion*, **45**, No.1-4, 1994, 137-48.
160 Albertsson A-C, Barenstedt C, Karlsson S, *J. Environmental Polym. Degradat.*, **1**, No.4, 1993, 241-5.
161 Mateev M M, Totev D L, *Kaut. u. Gummi Kunst.*, **49**, No.6, 1996, 427-31.
162 Jacobson K E, Eller R, Polyolefins VIII. Conference Proceedings, Houston,Tx., 21st-24th Feb.1993, 494-509.
163 Minkova L, Magagnini P L, *Polym. Degradat. Stabil.*, **42**, No.1, 1993, 107-15.
164 Mwila J, Miraftab M, Horrocks A R, *Polym. Degradat. Stabil.*, **44**, No.3, 1994, 351-6.
165 Albertsson A C, Griffin G J L, Karlson S, Nishimoto K, Watanabe Y, *Polym. Degradat. Stabil.*, **45**, No.2, 1994, 173-8.
166 Griffin G J L, *Polym. Degradat. Stabil.*, **45**, No.2, 1994, 241-7.
167 Callari J J, *Plast. World*, **52**, No.3, 1994, 22-6.
168 Svorcik V, Micek I, Jankovskij O, Rybka V, Hnatowicz V, Wang L, Angert N, *Polym. Degradat. Stabil.*, **55**, 1997, 115-21.
169 Le Bras M, Bourbigot, Le Tallec Y, Laureyns J., *Polym. Degradat. Stabil.*, **56**, 1997, 11-21.
170 Tavman I H, *J. Appl. Polym. Sci.*, **62**, No.12, 1996, 2161-7.
171 Hao Tang, Xinfang Chen, Aoqing Tang, Yunxia Luo, *J. Appl. Polym. Sci.*, **59**, No.3, 1996, 383-7.
172 Ou Y C, Zhu J, Feng Y P, *J. Appl. Polym. Sci.*, **59**, No.2, 1996, 287-94.
173 Savadori A, Scapin M, Walter R, *Macromol. Symp.*, **108**, 1996, 183-202.
174 Modine F A, Duggal A R, Robinson D N, Churnetski E L, Bartkowiak M, Mahan G D, Levinson L M, *J. Mat. Res.*, **11**, No.11, 1996, 2889-94.
175 Svorcik V, Rybka V, Hnatowicz V, Bacakova L, *J. Mat. Sci. Lett.*, **14**, No.24, 1995, 1723-4.
176 Sinien L, Xiaoguang Z, Zhongneng Q, Huan X, *J. Mat. Sci. Lett.*, **14**, No.20, 1995, 1458-60.
177 Dubnikova I L, Gorokhova E V, Gorenberg A Y, Topolkaraev V A, *Polym. Sci., Ser. A*, **37**, No.9, 1995, 951-8.
178 Bomal Y, Godard P, *Polym. Engng. Sci.*, 36, No.2, 1996, 237-43.
179 Kobayashi M, Takahashi T, Takimoto J, Koyama K, *Polymer*, **37**, No.16, 1996, 3745-7.
180 Chang Ho Suh, White J L, *Polym. Engng. Sci.*, **36**, No.11, 1996, 1521-30.

181 Suh C H, White J L, *Polym. Engng. Sci.*, **36**, No.17, 1996 2188-97.
182 Fiske T J, Gokturk H, Kalyon D M, Antec '96. Volume II. Conference proceedings, Indianapolis, 5th-10th May 1996, 1768-72.
183 Gill T S, Xanthos M, Wirsbo Co., Antec '96. Volume II. Conference proceedings, Indianapolis, 5th-10th May 1996, 1757-61.
184 Suh C H, White J L, Antec '96. Vol.I. Conference Proceedings, Indianapolis, 5th-10th May 1996, p.958-62.
185 Casenave S, Ait-Kadi A, Brahimi B, Riedl B, Antec '95. Vol. II. Conference Proceedings, Boston, Ma., 7th-11th May 1995, 1438-42.
186 Chang Ho Suh, White J L, *J. Non-Newtonian Fluid Mechanics*, **62**, Nos.2/3, 1996, 175-206.
187 Kretzschmar B, *Kunststoffe Plast Europe*, **86**, No.4, 1996, 20-2.
188 Gill T S, Xanthos M, *J. Vinyl and Additive Technol.*, **2**, No.3, 1996, 248-52.
189 Hollis R D, Hyche K W, Color and Appearance Retec: Effects in Plastics. Conference proceedings, Oak Brook, Il., 20th-22nd Sept.1994, 94-105.
190 Yeh J T, Yang H M, Huang S S, *Polym. Degradat. Stabil.*, **50**, No.2, 1995, 229-34.
191 Sanchez-Solis A, Estrada M R, *Polym. Degradat. Stabil.*, **52**, No.3, 1996, 305-9.
192 Gordienko V P, Dmitriev Y A, *Polym. Degradat. Stabil.*, **53**, No.1, 1996, 79-87.
193 Gordienko V P, Dmitriev Y A, *Polym. Sci., Ser. B*, **37**, Nos.5-6, 1995, 249-50.
194 Beloshenko V A, Kozlov G V, Slobodina V G, Prut E V, Grinev V G, *Polym. Sci., Ser. B*, **37**, Nos.5-6, 1995, 316-8.
195 Tino J, Mach P, Hlouskova Z, Chodak I, *J. Macromol. Sci. A*, **A31**, No.10, 1994, 1481-7.
196 Nichols M E, Pett R A, *Rubb. Chem. Technol.*, **67**, No.4, 1994, 619-28.
197 Hedgus C R, Kamel I L, *J. Coatings Technol.*, **65**, No.821, June 1993, 49-61
198 Yamada H, Inagaki S, Okamoto H, Furukawa J, *Int. Polym. Sci. Technol.*, **21**, No.6, 1994, T/29-35.
199 Ventresca D A, Berard M T, Antec '97. Conference proceedings, Toronto, April 1997, 3574-9.
200 Roychoudhury A, De P P, Roychoudhury N, Vidal A, *Rubb. Chem. Technol.*, **68**, No.5, 1995, 815-23.
201 Kimura T, Asano Y, Yasuda S, *Polymer*, **37**, No.14, 1996, 2981-7.
202 Liu Y J, Schindler J L, DeGroot D C, Kannewurf C R, Hirpo W, Kanatzidis M G, *Chem. of Mat.*, **8**, No.2, 1996, 525-34.
203 Ruiz-Hitzky E, Aranda P, Casal B, Galvan J C, *Adv. Mat.*, **7**, No.2, 1995, 180-4.
204 Khan S A, Baker G L, Colson S, *Chem. of Mat.*, **6**, No.12, 1994, 2359-63.
205 Lemmon J P, Lerner M M, *Chem. of Mat.*, **6**, No.2, 1994, 207-10.
206 Ou Y, Yu Z, Zhu J, Li G, Zhu S, *Chinese J. Polym. Sci.*, **14**, No.2, 1996, 172-82.
207 Casoli A, Charmeau J Y, Holl Y, d'Allest J F, *J. Adhesion*, **57**, Nos.1-4, 1996, 133-51.
208 Rebeiz K S, Rosett J W, Nesbit S M, Craft A P, *J. Mat. Sci. Lett.*, **15**, No.14, 1996, 1273-5.
209 Gerspacher M, O'Farrell C P, Nikiel L, Yang H H, *Rubb. Plast. News*, **26**, No.2, 1996, 39-40.
210 Chen X, Gonsalves K E, Chow G M, Xiao T D, *Adv. Mat.*, **6**, No.6, 1994, 481-4.
211 Wool R P, Long J M, *Macromolecules*, **26**, No.19, 1993, 5227-39.
212 Lan T, Kaviratna D, Pinnavaia T J, *Chem. of Mat.*, **6**, No.5, 1994, 573-5.
213 Asai S, Sumita M, *J. Macromol. Sci. B*, **34**, No.3, 1995, 283-94.
214 Molino L N, Topoleski L D T, *J. Biomed. Mat. Res.*, **31**, No.1, 1996, 131-7.
215 Joseph R, Zhang S, Ford W T, *Macromolecules*, **29**, No.4, 1996, 1305-12.
216 Tamura J, Kawanabe K, Yamamuro T, Nakamura T, Kokubo T, Yoshihara S, Shibuya T, *J. Biomed. Mat. Res.*, **29**, No.5, 1995, 551-9.
217 Topoleski L D T, Ducheyne P, Cuckler J M, *J. Biomed. Mat. Res.*, **29**, No.3, 1995, 299-307.
218 Lei Yang, Schruben D L, *Polym. Engng. Sci.*, **34**, No.14, 1994, 1109-14.
219 Liqing Sun, Aklonis J J, Salovey R, *Polym. Engng. Sci.*, **33**, No.20, 1993, 1308-19.
220 Hegedus C R, Kamel I L, *J. Coatings Technol.*, **65**, No.822, July 1993, 37-43.
221 Sunkara H B, Jethmalani J M, Ford W T, *Chem. of Mat.*, **6**, No.4, 1994, 362-4.
222 Visser S A, *J. Appl. Polym. Sci.*, **63**, 1997, 1805-20.
223 Grohens Y, Schultz J, *Int. J. Adhesion Adhesives*, **17**, 1997, 163-7.
224 Hashemi S, Din K J, Low P, *Polym. Engng. Sci.*, **36**, No.13, 1996, 1807-20.
225 Hashemi S, Gilbride M T, Hodgkinson J, *J. Mat. Sci.*, **31**, No.19, 1996, 5017-25.
226 Pape P G, Romenesko D J, Antec '97. Conference proceedings, Toronto, April 1997, 2941-52.
227 Okiyokota M, Hamada H, Hiragushi M, Hasegawa T, Antec '97. Conference proceedings, Toronto, April 1997, 3319-21.
228 Dreibelbis G L, Antec 95. Volume III. Conference proceedings, Boston, Ma., 7th-11th May 1995, 4374-6.

229 Hornsby P R, Premphet K, *J. Mater. Sci.*, **32**, 1997, 4767-75.
230 Thomason J L, Vlug M A, *Composites. Part A*, **28A**, 1997, 277-88.
231 Sjogren B A, Berglund L A, *Polym. Composites*, **18**, No.1, 1997, 1-8.
232 Asp L E, Sjogren B A, Berglund L A, *Polym. Composites*, **18**, No.1, 1997, 9-15.
233 Averous L, Quantin J C, Crespy A, *Polym. Eng. Sci.*, **37**, No.2, 1997, 329-37.
234 Cai Y, Petermann J, Wittich H, *J. Appl. Polym. Sci.*, **65**, 1997, 67-75.
235 Mi Y, Chen X, Guo Q, *J. Appl. Polym. Sci.*, **64**, 1997, 1267-73.
236 Gahleitner M, Bernreitner K, Neissl W, *J. Appl. Polym. Sci.*, **53**, No.3, 1994, 283-9.
237 Nakamura S, Kaneko S, Mizutani Y, *J. Appl. Polym. Sci.*, **49**, No.1, 1993, 143-50.
238 Nago S, Mizutani Y, *J. Appl. Polym. Sci.*, **53**, No.12, 1994, 1579-87.
239 Liao K, Chen X, Zheng C, *J. Appl. Polym. Sci.*, **57**, No.10, 1995, 1245-50.
240 Acosta J L, Jurado J R, *J. Appl. Polym. Sci.*, **57**, No.4, 1995, 431-7.
241 Roesch J, Barghoorn P, Muelhaupt R, *Makromol. Chem. Rapid Commun.*, **15**, No.9, 1994, 691-6.
242 Jancar J, DiBenedetto A T, *Sci. & Eng. Composite Materials*, **3**, No. 4, 1994, 217-26.
243 Jancar J, Dibenedetto A T, *J. Mat. Sci.*, **30**, No.6, 1995, 1601-8.
244 Jancar J, Dibenedetto A T, *J. Mat. Sci.*, **30**, No.9, 1995, 2438-45.
245 Baranovskii V M, Tarara A M, Khomik A A, *Int. Polym. Sci. Technol.*, **20**, No.1, 1993, T/98-9.
246 Jancar J, Dibenedetto A T, Dianselmo A, *Polym. Engng. Sci.*, **33**, No.9, 1993, 559-63.
247 Chiang W Y, Yang W D, Pukanszky B, *Polym. Engng. Sci.*, **34**, No.6, 1994, 485-92.
248 Baranovskii V M, Bondarenko S I, Kachanovskaya L D, Zelenev Y V, Makarov V G, Ovcharenko F D, *Int. Polym. Sci. Technol.*, **22**, No.1, 1995, T/91-3.
249 Schott N R, Rahman M, Perez M A, Antec '94. Conference Proceedings, San Francisco, Ca., 1st-5th May 1994, Vol. III, 2846-50.
250 Borden K A, Weil R C, Manganaro C R, Antec '94. Conference Proceedings, San Francisco, Ca., 1st-5th May 1994, Vol. III, 2761-5.
251 Jancar J, DiBenedetto A T, Antec '94. Conference Proceedings, San Francisco, Ca., 1st-5th May 1994, Vol. II, 1710-2.
252 Gendron R, Daigneault L E, Tatibouet J, Dumoulin M M, Antec '94. Conference Proceedings, San Francisco, Ca., 1st-5th May 1994, Vol. I, 167-71.
253 Wong K W Y, Truss R W, *Composites Sci. & Technol.*, **52**, No.3, 1994, 361-8.
254 Borden K A, Weil R C, Manganaro C R, Antec '93. Conference Proceedings, New Orleans, La., 9th-13th May 1993, Vol. II, 2167-70.
255 Skelhorn D A, Antec '93. Conference Proceedings, New Orleans, La., 9th-13th May 1993, Vol. II, 1965-70.
256 Hornsby P R, Mthupha A, Antec '93. Conference Proceedings, New Orleans, La., 9th-13th May 1993, Vol. II, 1954-6.
257 Sain M M, Kokta B V, Antec '93. Conference Proceedings, New Orleans, La., 9th-13th May 1993, Vol. I, 320-4.
258 Canova L A, Fergusson L W, Parrinello L M, Subramanian R, Giles H F, Antec '97. Conference proceedings, Toronto, April 1997, 2112-6.
259 Gerard P, Raine J, Pabiot J, Antec '97. Conference proceedings, Toronto, April 1997, 526-31.
260 Suri A, Min K, Antec '97. Conference proceedings, Toronto, April 1997, 1487-91.
261 Barbosa S E, Kenny J M, Antec '97. Conference proceedings, Toronto, April 1997, 1855-9.
262 Ewing F, Leatherman M, Antec '97. Conference proceedings, Toronto, April 1997, 2924-7.
263 Rabello M S, White J R, Antec '97. Conference proceedings, Toronto, April 1997, 2991-5.
264 Clemens M L, Doyle M D, Lees G C, Briggs C C, Day R C, Flame Retardants '94. Conference proceedings, London, 27th-28th January 1994, 193-202.
265 Schott N R, Rahman M, Perez M A, *J. Vinyl and Additive Technol.*, **1**, No.1, 1995, 36-40.
266 Schneider J P, Myers G E, Clemons C M, English B W, *J. Vinyl and Additive Technol.*, **1**, No.2, 1995, 103-8.
267 Rockenbauer A, Korecz L, Pukanszky B, *Polym. Bull.*, **33**, No.5, 1994, 585-9.
268 Hu X, Xu H, Zhang Z, *Polym. Degradat. Stabil.*, **43**, No.2, 1994, 225-8.
269 Borden K A, Wei R C, Manganaro C R, *Plast. Compounding*, **16**, No.5, 1993, 51-5.
270 Desai S C, *Pop. Plast. Packag.*, **39**, No.7, 1994, 53-60.
271 Kim G M, Michler G H, Gahleitner M, Fiebig J, *J. Appl. Polym. Sci.*, **60**, No.9, 1996, 1391-403.
272 Wang J, Tung J F, Fuad M Y A, Hornsby P R, *J. Appl. Polym. Sci.*, **60**, No.9, 1996, 1425-37.
273 Chunmin Ye, Jingjiang Liu, Zhishen Mo, Gongben Tang, Xiabin Jing, *J. Appl. Polym. Sci.*, **60**, No.11, 1996, 1877-81.

274 Wang Z, *J. Appl. Polym. Sci.*, **60**, No.12, 1996, 2239-43.
275 Long Y, Shanks R A, *J. Appl. Polym. Sci.*, **62**, No.4, 1996, 639-46.
276 Nago S, Mizutani Y, *J. Appl. Polym. Sci.*, **61**, No.13, 1996, 2355-9.
277 Yu Long, Shanks R A,Melbourne, *J. Appl. Polym. Sci.*, **61**, No.11, 1996, 1877-85.
278 Nago S, Mizutani Y,Tokuyama Corp., *J. Appl. Polym. Sci.*, **61**, No.1, 1996, 31-5.
279 Liu Z, Gilbert M, *J. Appl. Polym. Sci.*, **59**, No.7, 1996, 1087-98.
280 Alonso M, Gonzalez A, de Saja J A, *Plast. Rubb. Comp. Process. Appln.*, **24**, No.3, 1995, 131-7.
281 Hornsby P R, Watson C L, *J. Mat. Sci.*, **30**, No.21, 1995, 5347-55.
282 Khare A, Mitra A, Radhakrishnan S, *J. Mat. Sci.*, **31**, No.21, 1996, 5691-5.
283 Jarvela P A, Jarvela P K, *J. Mat. Sci.*, **31**, No.14, 1996, 3853-60.
284 Furuichi N, Kurokawa Y, Fujita K, Oya A, Yasuda H, Kiso M, *J. Mat. Sci.*, **31**, No.16, 1996, 4307-10.
285 Kurokawa Y, Yasuda H, Oya A, *J. Mat. Sci. Lett.*, **15**, No.17, 1996, 1481-3.
286 Benderly D, Siegmann A, Narkis M, *J. Mat. Sci. Lett.*, **15**, No.15, 1996, 1349-52.
287 Greso A J, Phillips P J, *Polymer*, **37**, No.14, 1996, 3165-70.
288 Rabello M S, White J R, *Polym. Composites*, **17**, No.5, 1996, 691-704.
289 Eggers H, Schummer P, *Rubb. Chem. Technol.*, **69**, No.2, 1996, 253-65.
290 Pukanszky B, Maurer F H J, Boode J W, *Polym. Engng. Sci.*, **35**, No.24, 1995, 1962-71.
291 Zaini M J, Fuad M Y A, Ismail Z, Mansor M S, Mustafah J, *Polym. Int.*, **40**, No.1, 1996, 51-5.
292 Stricker F, Muelhaupt R, *High Perform. Polym.*, **8**, No.1, 1996, 97-108.
293 Le Bras M, Bourbigot S, *Fire & Mat.*, **20**, No.1, 1996, 39-49.
294 Johnson K C, Antec '96. Volume III. Conference proceedings, Indianapolis, 5th-10th May 1996, 3545-9.
295 Montsinger L V,Antec '96. Volume II. Conference proceedings, Indianapolis, 5th-10th May 1996, 2546-9.
296 Cheung T, Tjong S C, Li R K Y, Antec '96. Volume II. Conference proceedings, Indianapolis, 5th-10th May 1996, 2256-9.
297 Tiganis B E, Shanks R A, Long Y, Antec '96. Volume II. Conference proceedings, Indianapolis, 5th-10th May 1996, 1744-9.
298 Stockblower D, Antec '96. Volume II. Conference proceedings, Indianapolis, 5th-10th May 1996, 1690-4.
299 Ishibashi J, Kobayashi A, Yoshikawa T, Shinozaki K, Antec '96. Vol. I. Conference Proceedings, Indianapolis, 5th-10th May 1996, 386-90.
300 Hojabr S, Boocock J R B, Antec 95. Volume III. Conference proceedings, Boston, Ma., 7th-11th May 1995, 3620-7.
301 Curtis B J, Malloy R, Jones J, Antec 95. Volume III. Conference proceedings, Boston, Ma., 7th-11th May 1995, 3594-8.
302 Averous L, Quantin J C, Lafon D, Crespy A, *Int. J. Polym. Analysis and Characterization*, **1**, No.4, 1995, 339-47.
303 Hattori K, Morford S, Delany D T, *J. Vinyl and Additive Technol.*, **1**, No.3, 1995. 170-3.
304 Driscoll S B, Nanavaty P, Stockbower D W, Polymer Additives for Injection Moulding and Extrusion Applications. Retec proceedings, White Haven, Pa., 18th-19th Oct.1995, p.161-85.
305 Kikkawa K, *Polym. Degradat. Stabil.*, **49**, No.1, 1995, 135-43.
306 Guerrica-Echevarria G, Eguiazabal J I, Nazabal J, *Polym. Degradat. Stabil.*, **53**, No.1, 1996, 1-8.
307 Zhao W, Hasegawa S, Fujita J, Yoshii F, Sasaki T, Makuuchi K, Sun J, Nishimoto S, *Polym. Degradat. Stabil.*, **53**, No.2, 1996, 199-206.
308 Wampler W A, Rajeshwar K, Pethe R G, Hyer R C, Sharma S C, *J. Mat. Res.*, **10**, No.7, 1995, 1811-22.
309 Maeda S, Armes S P, J. *Mat. Chem.*, **4**, No.6, 1994, 935-42.
310 Faguy P W, Ma W, Lowe J A, Pan W P, Brown T, J. *Mat. Chem.*, **4**, No.5, 1994, 771-2.
311 Meddad A, Fisa B, *Macromol. Symp.*, **108**, 1996, 173-82.
312 Akelah A, Moet A, *J. Mat. Sci.*, **31**, No.13, 1996, 3589-96.
313 Kobayashi M, Takahashi T, Takimoto J, Koyama K, *Polymer*, **36**, No.20, 1995, 3927-33.
314 Mitsui S, Kihara H, Yoshimi S, Okamoto Y, *Polym. Engng. Sci.*, **36**, No.17, 1996, 2241-6.
315 Qi Wang, Jizhuang Cao, Guangjin Li, Xi Xu, *Polym. Int.*, **41**, No. 3, 1996, 245-9.
316 Ogadhoh S O, Papathanasiou T D, *Composites Part A: Applied Science and Manufacturing*, **27A**, No.1, 1996, 57-63.
317 Kim K J, White J L, *J. of Non-Newtonian Fluid Mechanics*, **66**, Nos.2/3, 1996, 257-70.
318 Meddad A, Fisa B, *J. Mater. Sci.*, **32**, 1997, 1177-85.

319 Friedrich C, Scheuchenpflug W, Neuhaeusler S, Roesch J, *J. Appl. Polym. Sci.*, **57**, No.4, 1995, 499-508.
320 Godard P, Bomal Y, Biebuyck J J, *J. Mat. Sci.*, **28**, No.24, 1993, 6605-10.
321 del Rio C, Acosta J L, *Polymer*, **35**, No.17, 1994, 3752-7.
322 Alberola N, Bergeret A, S, *Polym. Composites*, **15**, No.6, 1994, 442-52.
323 Li Shucai, Peng Weijang, Lu Xiuping, *Int. J. Polym. Mat.*, **29**, Nos.1-2, 1995, 37-42.
324 Baranovskii V M, Bondarenko V V, Zadorina E N, Cherenkov A V, Zelenev Y V, *Int. Polym. Sci. Technol.*, **23**, No.6, 1996, T/87-9.
325 Fengyuan Yan, Wenhua Wang, Qunji Xue, Long Wei, *J. Appl. Polym. Sci.*, **61**, No.7, 1996, 1231-6.
326 Fengyuan Yan, Qunji Xue, Shengrong Yang, *J. Appl. Polym. Sci.*, **61**, No.7, 1996, 1223-9.
327 Zhang C-Z, Liu W-M, Xue Q-J, Shen W-C, *J. Appl. Polym. Sci.*, **66**, 1997, 85-93.
328 Tabor R L, Polyurethanes '94. Conference proceedings, Boston, Ma., 9th-12th Oct.1994, 288-94.
329 Modesti M, Simioni F, Albertin P, *Cell. Polym.*, **13**, No.2, 1994, 113-24.
330 Vratsanos L A, Farris R J, *Polym. Engng. Sci.*, **33**, No.22, 1993, 1458-65.
331 Dolui S K, *J. Appl. Polym. Sci.*, **53**, No.4, 1994, 463-5.
332 Torro-Palau A, Fernandez-Garcia J C, Orgiles-Barcelo A C, Pastor-Blas M M, Martin-Martinez J M, *Int. J. Adhesion Adhesives*, **17**, 1997, 111-9.
333 Sergeeva L M, Slinchenko E A, Brovko A A, Fainleib A M, Nedashkovskaya N S, *Polym. Sci., Ser. B*, **38**, Nos.5/6, 1996, 225-30.
334 Smock D, *Plast. World*, **54**, No.12, 1996, 35-9.
335 Porter J R, Groseth C K, Little D W, Polyurethanes '95. Conference Proceedings, Chicago, Il., 26th-29th Sept.1995, 532-7.
336 Sendijarevic A, Sendijarevic V, Wang X, Haidar A, Dutta U, Klempner D, Frisch K C, Polyurethanes '95. Conference Proceedings, Chicago, Il., 26th-29th Sept.1995, 418-26.
337 Okoroafor M O, Wang A, Bhattacharjee D, Cikut L, Haworth G J, Polyurethanes '95. Conference Proceedings, Chicago, Il., 26th-29th Sept.1995, 303-9.
338 Oien H T, Polyurethanes '95. Conference Proceedings, Chicago, Il., 26th-29th Sept.1995, 137-41.
339 Chen L, Liu K, Yang C Z, *Polym. Buil.*, **37**, No.3, 1996, 377-83.
340 Bauman B D, Plastics Finishing: Responding to Tomorrow's Global Requirements. Retec proceedings, Detroit, Mi, 21st-22nd March 1995, 62-73.
341 Eisenbach C D, Ribbe A, Goeldel A, *Kaut. u. Gummi Kunst.*, **49**, No.6, 1996, 406-10.
342 Bos M, Van Dam G W, Jongsma T, Bruin P, Pennings A J, *Composite Interfaces*, **3**, No.2, 1995, 169-76.
343 Wang K J, Sue H J, Antec '95. Vol. II. Conference Proceedings, Boston, Ma., 7th-11th May 1995, 1758-64.
344 Urabe N, Takatsugi H, Ito M, Toko H, Fukui M, *Int. Polym. Sci. Technol.*, **22**, No.5, 1995, T/68-72.
345 Sombatsompop N, Sims G L A, *Cell. Polym.*, **15**, No.5, 1996, 317-34.
346 Sims G L A, Sombatsompop N, *Cell. Polym.*, **15**, No.2, 1996, 90-104.
347 Brovkop O O, Sergeeva L M, Slinchenko O A, Fainleib O M, *Polym. Int.*, **40**, No.4, 1996, 299-305.
348 Caillaud J L, Deguillaume S, Vincent M, Giannotta J C, Widmaier J M, *Polym. Int.*, **40**, No.1, 1996, 1-7.
349 Sergeeva L M, Skiba S I, Karabanova L V, Polym. Int., 39, No.4, April 1996, 317-25.
350 Martin-Martinez J M, Macia-Agullo T G, Fernandez-Garcia J C, Orgiles-Barcelo A C, Torro-Palau A, *Macromol. Symp.,* **108**, 1996, 269-78.
351 Kovacevic V, Lucic S, Hace D, Cerovecki Z, *J. Adhesion Sci. Technol.*, **10**, No.12, 1996, 1273-85.
352 Kovacevic V, Lucic S, Hace D, Glasnovic A, *Polym. Engng. Sci.*, **36**, No.8, 1996, 1134-9.
353 Gerard J F, Chabert B, *Macromol. Symp.*, **108**, 1996, 137-46.
354 Landry C J T, Coltrain B K, *J. Macromol. Sci. A*, **31**, No.12, 1994, 1965-73.
355 Kovacevic V, Lucic S, Hace D, Glasnovic A, Smit I, Bravar M, *J. Adhesion*, **47**, No.1-3, 1994, 201-15.
356 Landry C J T, Coltrain B K, Landry M R, Fitzgerald J J, Long V K, *Macromolecules*, **26**, No.14, 5th July 1993, 3702-12.
357 Suzuki F, Nakane K, Piao J-S, *J. Mat. Sci.*, **31**, No.5, 1996, 1335-40.
358 Mukha B I, Kolupaev B S, Mukha Y B, *Int. Polym. Sci. Technol.*, **23**, No.6, 1996, T/57-9.
359 Sanchez-Solis A, Padilla A, *Polym. Bull.*, **36**, No.6, 1996, 753-58.
360 Wiebking H E, *J. Vinyl and Additive Technol.*, **2**, No.3, 1996, 187-9.
361 Malanda L M, Park C B, Balatinecz J J, Antec '96. Volume II. Conference proceedings, Indianapolis, 5th-10th May 1996, p.1900-7.
362 Karpukhin A A, Ledeneva I N, Aleksandrov V I, *Int. Polym. Sci. Technol.*, **23**, No.4, 1996, T/29-31.

363 Ulkem I, Bataille P, Schreiber H P, *J. Macromol. Sci. A*, **31**, No.3, 1994, 291-303.
364 Molesky F, Schultz R, Midgett S, Green D, *J. Vinyl Additive Technol.*, **1**, No.3, 1995, 159-61.
365 Zolotnitsky M, Steinmetz J R, *J. Vinyl and Additive Technol.*, **1**, No.2, 1995, 109-13.
366 Bataille P, Schreiber H P, Mahlous M, Antec '93. Conference Proceedings, New Orleans, La., 9th-13th May 1993, Vol. II, 1757-9.
367 Ferm D J, Shen K K, Antec '94. Conference Proceedings, San Francisco, Ca., 1st-5th May 1994, Vol.III, 3522-5.
368 Chudinova V V, Guzeev V V, Mozzhukhin V B, Pomerantseva E G, Nozrina F D, Zhil'tsov V V, Zubov V P, *Int. Polym. Sci. Technol.*, **21**, No.10, 1994, T/102-4.
369 Liptak P, *Int. Polym. Sci. Technol.*, **21**, No.8, 1994, T/50-3.
370 Bataille P, Mahlous M, Schreiber H P, *Polym. Engng. Sci.*, **34**, No.12, 1994, 981-5.
371 Liptak P, Zelenak P, *Int. Polym. Sci. Technol.*, **20**, No.9, 1993, T/57-9.
372 Baggaley R G, Hornsby P R, Yahya R, Cussak P A, Monk A W, *Fire Mater.*, **21**, 1997, 179-85.
373 Abdel-Aziz M M, Gwaily S E, *Polym. Degradat. Stabil.*, **55**, 1997, 269-74.
374 Cochet P, Barruel P, Barriquand L, Grobert J, Bomal Y, Prat E, IRC '93/144th Meeting, Fall 1993. Conference Proceedings, Orlando, Fl., 26th-29th Oct.1993, Paper 162.
375 Monthey S, Duddleston B, Podobnik J, *Rubb. World*, **210**, No.3, 1994, 17-9.
376 Pehlergard P, *Rubb. S. Africa*, **10**, No.5, 1995, 8-12.
377 Anantharaman M R, Kurian P, Banerjee B, Mohamed E M, George M, *Kaut. u. Gummi Kunst.*, **49**, No.6, 1996, 424-6.
378 Maas S, Gronski W, *Kaut. u. Gummi Kunst.*, **47**, No.6, 1994, 409-15.
379 Byers J T, Meeting of the Rubber Division, ACS, Cleveland, October 17-20, 1995, paper B.
380 Evans L R, Meeting of the Rubber Division, ACS, Montreal, May 5-8, 1996, paper D.
381 Bomo F, Meeting of the Rubber Division, ACS, Montreal, May 5-8, 1996, paper E.
382 Estrin R I, Pesin O Y, *Int. Polym. Sci. Technol.*, **22**, No.1, 1995, T/12-6.
383 Karsek L, *Int. Polym. Sci. Technol.*, **21**, No.10, 1994, T/35-40.
384 Yoshida T, *Int. Polym. Sci. Technol.*, **20**, No.6, 1993, T/29-39.
385 Dick J S, Pawlowski H A, *J. Elastomers Plast.*, **27**, No.1, 1995, 11-38.
386 Leblanc J L, *Prog. Rubb. Plast. Technol.*, **10**, No.2, 1994, 112-29.
387 Li Y, Wang M J, Zhang T, Zhang F, Fu X, *Rubb. Chem. Technol.*, **67**, No.4, 1994, 693-9.
388 Furukawa J, Yamada E, *J. Appl. Polym. Sci.*, **52**, No.11, 1994, 1587-93.
389 Babich V F, Lipatov Yu S, Todosijchuk T T, *J. Adhesion*, **55**, Nos.3-4, 1996, 317-27.
390 Saad A L G, Younan A F, *Polym. Degradat. Stabil.*, **50**, No.2, 1995, 133-40.
391 Heuert U, Knorgen M, Menge H, Scheler G, Schneider H, *Polym.Bull.*, **37**, No.4, Oct.1996, 489-96.
392 Bandyopadhyay S, De P P, Tripathy D K, De S K, *Kaut. u. Gummi Kunst.*, **49**, No.2, 1996, 115-9.
393 de Candia F, Carotenuto M, Gargani L, Guadagno L, Lauretti E, Renzulli A, *Kaut. u. Gummi Kunst.*, **49**, No.2, 1996, 99-101.
394 Nakajima N, *Int. Polym. Processing*, **11**, No.1, 1996, 3-13.
395 Gownder M, Letton A, Hogan H,Antec '95. Vol. II. Conference Proceedings, Boston, Ma., 7th-11th May 1995, 1983-6.
396 Donnet J B, Wang T K, *Prog. Rubb. Plast. Technol.*, **11**, No.4, 1995, 261-7.
397 Lin C R, Lee Y D, *Macromol. Theory & Simulations*, **5**, No.6, 1996, 1075-104.
398 Unnikrishnan G, Thomas S, Varghese S, *Polymer*, **37**, No.13, 1996, 2687-93.
399 Kiselev V Y, Vnukova V G, *Int. Polym. Sci. Technol.*, **23**, No.5, 1996, T/88-92.
400 Nunes R C R, Mano E B, *Polym. Composites*, **16**, No.5, 1995, 421-3.
401 Bandyopadhyay S, De P P, Tripathy D K, De S K, *J. Appl. Polym. Sci.*, **58**, No.4, 1995, 719-27.
402 Bandyopadhyay S, De P P, Tripathy D K, De S K, *J. Appl. Polym. Sci.*, **61**, No.10, 1996, 1813-20.
403 Bandyopadhyay S, De P P, Tripathy D K, De S K, *Polymer*, **37**, No.2, 1996, 353-7.
404 Mandal U K, Tripathy D K, De S K, *Plast. Rubb. Comp. Process. Appln.*, **24**, No.1, 1995, 19-25.
405 de Sena Affonso J E, Nunes R C R, *Polym. Bull.*, **34**, No.5/6, 1995, 669-75.
406 Cheng J, Bigio D I, Briber R M, Antec '96. Vol. I. Conference Proceedings, Indianapolis, 5th-10th May 1996, 365-9.
407 Addad J P C, Frebourg P, *Polymer*, **37**, No.19, 1996, 4235-42.
408 Garbow J R, Asrar J, Hardiman C J, *Chem. of Mat.*, **5**, No.6, 1993, 869-75.
409 Nasr G M, Badawy M M, Gwaily S E, Shash N M, Hassan H H, *Polym. Degradat. Stabil.*, **48**, No.2, 1995, 237-41.
410 Ali M H, Abo-Hashem A, *Plast. Rubb. Comp. Process. Appln.*, **24**, No.1, 1995, 47-51.
411 Delor F, Lacoste J, Lemaire J, Barrois-Oudin N, Cardinet C, *Polym. Degradat. Stabil.*, **53**, No.3,

1996, 361-9.
412 Lawandy S N, Botros S H, Darwish N A, Mounir A, *Polym. Plast. Technol. Engng.*, **34**, No.6, 1995, 861-74.
413 El-Shall M S, Slack W, *Macromolecules,* **28**, No.24, 1995, 8456-8.
414 Khromov M K, Niazashvili G A, Sakhnovskii N L, Koroleva T A, *Int. Polym. Sci. Technol.*, **23**, No.5, 1996, T/82-4.
415 Khromov M K, Niazashvili G A, *Int. Polym. Sci. Technol.*, **23**, No.5, 1996, T/24-6.
416 Kida N, Ito M, Yatsuyanagi F, Kaido H, *J. Appl. Polym. Sci.*, **61**, No.8, 1996, 1345-50.
417 Basfar A A, Silverman J, *Polym. Degradat. Stabil.*, **46**, No.1, 1994, 1-8.
418 Furtado C R G, Nunes R C R, de Siqueira Filho A S, *Polym. Bull.*, **34**, No.5/6, 1995, 627-33.
419 Fuelber C, Bluemich B, Unseld K, Herrmann V, *Kaut. u. Gummi Kunst.*, **48**, No.4, 1995, 254-9.
420 Wolff S, Wang M J, Tan E H, *Kaut. u. Gummi Kunst.*, **47**, No.12, 1994, 873-84.
421 Zaborski M, Slusarski L, Donnet J B, Papirer E, *Kaut. u. Gummi Kunst.*, **47**, No.10, 1994, 730-8.
422 Donnet J B, Wang W, Vidal A, Wang M J, *Kaut. u. Gummi Kunst.*, **46**, No.11, Nov.1993, 866-71.
423 Dutta N K, Choudhury N R, Haidar B, Vidal A, Donnet J B, Delmotte L, Chezeau J M, *Polymer*, **35**, No.20, 1994, 4293-9.
424 Helaly F M, El-Sawy S M, Abd El-Ghaffar M A, *J. Elastomers Plast.*, **26**, No.4, 1994, 335-46.
425 Wolff S, Wang M-J, Tan E-H, *Rubb. Chem. Technol.*, **66**, No.2, 1993, 163-77.
426 Clarke J, Freakley P K, *Rubb. Chem. Technol.*, **67**, No.4, 1994, 700-15.
427 Abdel-Aziz M M, Youssef H A, El Miligy A A, Yoshii F, Makuuchi K, *Polym. & Polym. Composites*, **4**, No.4, 1996, 259-68.
428 Karasek L, Meissner B, Asai S, Sumita M, *Polym. J. (Jap.)*, **28**, No.2, 1996, 121-6.
429 Sendijarevic A, Sendijarevic V, Wang X, Haidar A, Dutta U, Klempner D, Frisch K C, Polyurethanes '95. Conference Proceedings, Chicago, Il., 26th-29th Sept.1995, 418-26.
430 Gorl U, Rausch R, Esch H, Kuhlmann R, *Int. Polym. Sci. Technol.*, **23**, No.7, 1996, T/81-7.
431 Nasr G M, Badawy M M, Gwaily S E, Attia G, *Polym. Int.*, **38**, No.3, 1995, 249-55.
432 Nasr G M, *Polym. Test.*, **15**, No.6, 1996, 585-91.
433 Nasr G M, Badawy M M, *Polym. Test.*, **15**, No.5, 1996, 477-84.
434 Kovacevic V, Lucic S, Hace D, Glasnovic A, *Polym. Engng. Sci.*, **36**, No.8, 1996, 1134-9.
435 Cha Y J, Choe S, *J. Appl. Polym. Sci.*, **58**, No.1, 1995, 147-57.
436 Chen C-H, Cheng C-H, *J. Composite Mat.*, **30**, No.1, 1996, 69-83.
437 Sharaf M A, Kloczkowski A, Mark J E, *Rubb. Chem. Technol.*, **68**, No.4, 1995, 601-8.
438 Visser S A, Hewitt C E, Binga T D, *J. Polym. Sci., Polym. Phys.*, **34**, No.9, 1996, 1679-89.
439 Yuan Q W, Kloczkowski A, Mark J E, Sharaf M A, *J. Polym. Sci., Polym. Phys.*, **34**, No.9, 1996, 1647-57.
440 Burnside S D, Giannelis E P, *Chem. of Mat.*, **7**, No.9, 1995, 1597-600.
441 Leezenberg P B, Frank C W, *Chem. of Mat.*, **7**, No.10, 1995, 1784-92.
442 Wen J, Mark J E, *J. Appl. Polym. Sci.*, **58**, No.7, 1995, 1135-45.
443 Soares R F, Leite C A P, Botter W, Galembeck F, *J. Appl. Polym. Sci.*, **60**, No.11, 1996, 2001-6.
444 Zumbrum M A, *J. Adhesion*, **46**, Nos 1-4, 1994, 181--96.
445 Kaewpanya R, Meinecke E A, 149th ACS Rubber Division Meeting, Spring 1996. Conference preprints, Montreal, 5th-8th May 1996, paper 32.
446 Wang S Q, Inn Y W, *Polym. Int.*, **37**, No.3, 1995, 153-5.
447 Yang A C M, *Polymer*, **35**, No.15, 1994, 3206-11.
448 Ebengou R H, Cohen-Addad J P, *Polymer*, **35**, No.14, 1994, 2962-9.
449 Cochrane H, Lin C S, *Rubb. Chem. Technol.*, **66**, No.1, 1993, 48-60.
450 Okel T A, Waddell W H, *Rubb. Chem. Technol.*, **68**, No.1, 1995, 59-76.
451 Visser S A, *J. Appl. Polym. Sci.*, **63**, 1997, 1805-20.
452 Visser S A, *J. Appl. Polym. Sci.*, **64**, 1997, 1499-1509.
453 Kominar V, Narkis M, Siegmann A, Breuer O, *Sci. & Engng. Composite Materials*, **3**, No.1, 1994, 61-6.
454 Marcovich N E, Reboredo M M, Aranguren M L, *J. Appl. Polym. Sci.*, **61**, No.1, 1996, 119-24.
455 Kenny J M, Opalicki M, *Composites Part A: Applied Science and Manufacturing*, **27A**, No.3, 1996, 229-40.
456 Kenny J M, Opalicki M, Molina G, Antec '95. Vol. II. Conference Proceedings, Boston, Ma., 7th-11th May 1995, 2782-9.
457 Gregorio R, Cestari M, Bernardino F E, *J. Mat. Sci.*, **31**, No.11, 1996, 2925-30.
458 Yano S, Okubo N, Takahashi K, *Macromol. Symp.*, **108**, 1996, 279-89.

459 Zhang M, Jia W, Chen X, *J. Appl. Polym. Sci.*, **62**, No.5, 1996, 743-7.
460 Abramova N A, Diikova E U, Lyakhovskii Yu Z, *Polym. Sci.*, **36**, No.9, 1994, 1308-9.

# 16

# Filler in Materials Combinations

## 16.1 BLENDS, ALLOYS AND INTERPENETRATING NETWORKS[1-31]

Polymer blends and alloys have more complex behavior in the presence of fillers than the binary mixtures of polymer and filler. The same factors, such as filler distribution, filler-matrix interaction, filler-matrix adhesion, particle orientation, nucleation, chemical reactivity, etc. have influence on properties, but this influence is complicated by the fact that there are two or more polymers present which compete for the same filler particles. These complex interactions result in many interesting phenomena discussed below.

The general idea behind polymer blending is to fulfill expectations that two different polymers, having different sets of properties, may contribute their most advantageous features to form product which surpasses properties of individual polymers. This expectation is rarely fulfilled in relation to mechanical properties because most polymer pairs are immiscible and do not have sufficient adhesion between phases. Two strategies are thus applied to find satisfactory materials: compatibilization and use of filler. In the case of compatibilization, a third component is added (most frequently block copolymer, having building blocks of both polymers, or the product of grafting of one polymer with introduced functional groups reactive towards the other polymer). This addition of the third component gives a relatively simple system in which polymers of the blend are still not mixed but are connected at interfaces through this third component. The comparison of this type of compatibilization to addition of filler compatibilized system is fairly simple. The addition of filler offers combinations given in Figure 16.1 These numerous possibilities of interaction can still be increased by the addition of compatibilizers having different interactions with polymers and fillers or by the use of combinations of fillers or both.

The above description stresses either chemical reactions in these combinations or physical interactions between components. In reality there is still additional effect which may induce changes to structure and thus properties. It is a commonly known effect of fillers on the nucleation of polymers. It can be perceived that filler does not affect nucleation of both polymers with the same intensity. In addition, the availability of polymers at the interface with fillers depends on various parameters such as viscosity, acid/base interaction, etc. If these two are included in the number of combinations, there is a theoretical abundance of possible combinations and thus

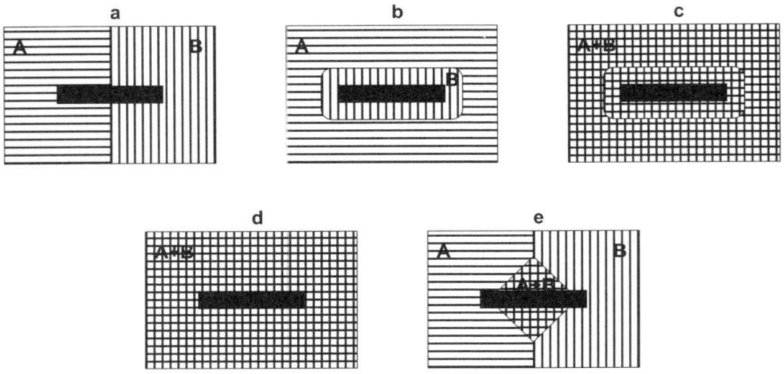


Figure 16.1. Models of interaction of a two component blend with filler. (a) polymers immiscible but both interacting with filler; (b) polymers immiscible but polymer B interacting with filler; (c) polymers miscible but only polymer B interacting with filler; (d) polymers miscible and both interacting with filler; (e) polymers immiscible but both interacting with filler forming a miscible interphase on the surface of filler. [Modified based on Persson A L, Bertilsson H, *Composite Interfaces*, **3**, No.4, 1996, 321-32.]

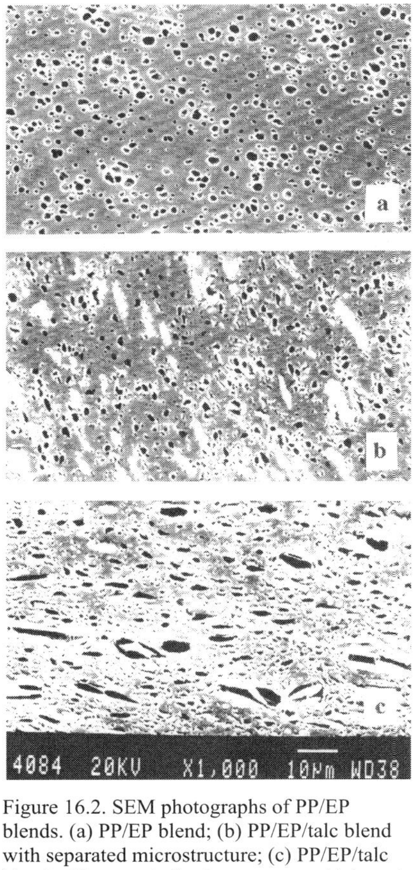


Figure 16.2. SEM photographs of PP/EP blends. (a) PP/EP blend; (b) PP/EP/talc blend with separated microstructure; (c) PP/EP/talc blend with core-shell microstructure. [Adapted, by permission, from Shanks R A, Long Y, *Polym. Networks Blends*, **7**, 1997, 87-92.]

properties with real solutions are always based on combinations of several model patterns. The participation of each idealized component model structure is influenced by the process conditions. Further information on this subject can be found in the expert monographic source.[32]

Figure 16.2 shows examples of morphology of PP/EP blends containing talc.[33] The separated microstructure was obtained by mixing PP, PP grafted with maleic anhydride, and filler, followed by adding EP. In the separated microstructure the dark holes (micrograph b) are EP particles and talc particles are the brightest fragments. The shapes of core-shell particles are random because they depend on the shape of encapsulated particles. Further observations from this study are discussed below.[33]

This abundance of possibilities, combined with still large experimental difficulties in distinguishing between the numerous interactions and morphologies, creates opportunities more for the future than for todays benefit. The existing information is still on the very initial stage of understanding. Inverse gas chromatography was used to explain why, in polyamide-6/SAN blends,

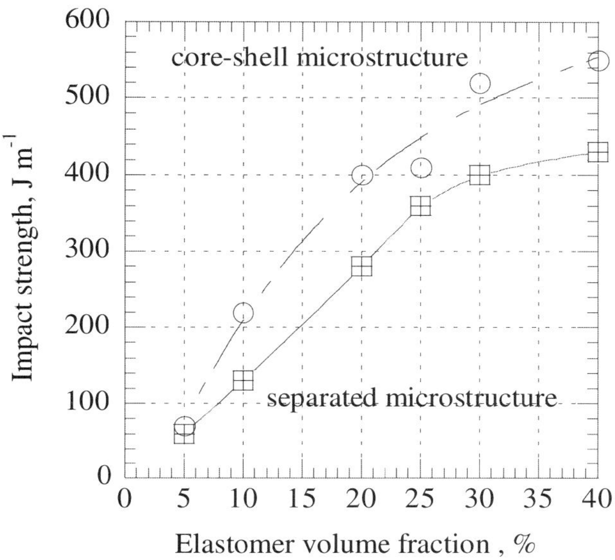


Figure 16.3. Impact strength vs. EP content for PP/EP blend containing 15 wt% talc. [Adapted, by permission, from Shanks R A, Long Y, *Polym. Networks Blends*, **7**, 1997, 87-92.]

polyamide-6 interacts with aluminum borate whiskers. The interaction is believed to be a result of the Lewis acid/base properties of filler and polymers.[25]

Two rubbers – natural and butadiene – were mixed in different proportions and reinforced with cellulose fiber.[26] The results show that the proportion between rubber and fiber content are essential parameters determining tensile strength. The relationships between tensile strength and cellulose content are different for various proportions of both rubbers. The information includes the results of mechanical testing which is not sufficient to uncover the reasons for these different relationships. Factorial experimental design was used to elucidate the effect of compatibilizer, different polymers and fillers. The results of testing about 50 combinations allowed the conclusion that graphite acts as reinforcing agent.[27]

Figure 16.3 shows the effect of core-shell microstructure and separated microstructure on the impact strength of filled blends. Blends containing core-shell structures always give better performance. Figure 16.4 shows the influence of EP on the effect of filler on impact strength. With increasing EP concentration, impact strength increases, but at the same time impact strength becomes increasingly more sensitive to the filler content.[33]

Polymers such as poly(vinylidene fluoride) and polystyrene form incompatible blends.[6] Compounding with copper particles and carbon black affects changes in the morphology of the blends. Both fillers change the crystalline geometry of PVDF. The increase in the amount of copper particles increases the crystallization rate. With carbon black, the crystallization rate initially decreases to increase again at higher loadings, but generally there is no gain compared with unfilled polymer.

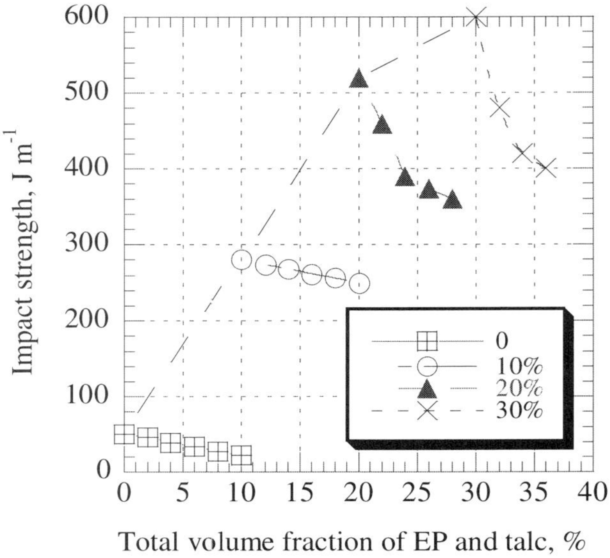


Figure 16.4. Impact strength vs. volume fraction of EP and filler for PP/EP blends. [Adapted, by permission, from Shanks R A, Long Y, *Polym. Networks Blends*, **7**, 1997, 87-92.]

The phase morphology is modified by both fillers which help to micronize and homogenize both phases. Carbon black, in addition, acts as a compatibilizer, which is seen from shifts in the glass transition temperatures of both polymers and in a decrease in melting point of PVDF.[6]

It is not only the blend morphology which is affected by filler presence; filler orientation might be affected by its surface treatment, as was observed in the works on PC/ABS blends containing aluminum borate whisker.[7] When the whisker surface was treated with epoxy silane, its reaction to flow in injection and compression molding processes changed, resulting in a different orientation even though all other parameters of process and composition were the same.[7]

A small addition of talc (8 wt%) changes fatigue resistance. Figure 16.5 shows that crack initiation is delayed (marked in the figure), but crack propagation is the same. The work dissipated has formed the same relationship with crack length for the filled and unfilled samples, but the energy release rate and resistance moment were slightly higher for filled material.[8]

The method of filler incorporation also determines its distribution between the phases.[10] When PE/TPU blend was in the molten state when mixed with carbon black, the island morphology was obtained with carbon black mostly resident in one phase. Mixing at room temperature followed by cold molding (50°C) resulted in the mixture with uniformly distributed carbon black.[10]

The introduction of reactive components during blend formation further increases the complexity of the process and affects results. Polypropylene functional-

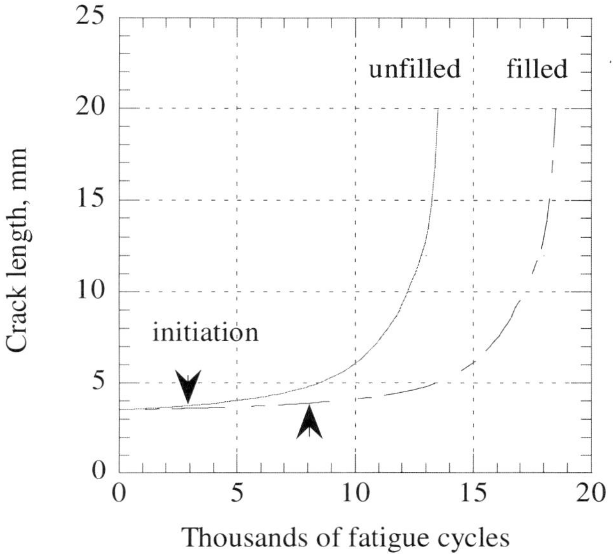


Figure 16.5. Crack length vs. the number of fatigue cycles for PC/ABS blends filled with 8 wt% talc and unfilled. [Adapted, by permission, from Seibel S R, Moet A, Bank D H, Nichols K, Antec '93. Conference Proceedings, New Orleans, La., 9th-13th May 1993, Vol. I, 902-5.]

ized with maleic anhydride was used to compatibilize a blend of polypropylene with polyamide-6 in the presence of glass beads. The kinetics of two processes was important: blend compatibilization and filler encapsulation. The process of blend compatibilization is slower than filler encapsulation. This restricts addition order allowing the blend to compatibilize first, followed by the addition of filler to the already compatibilized blend.

In blends composed of immiscible polymers, amorphous polymer does not affect the crystallization of crystallizable polymer, but if two polymers are miscible, amorphous polymer acts as diluent and affects crystallization of the second polymer. Poly(ε-caprolactone) is a crystallizable component of the blend with poly(vinyl butyral), which is studied in compositions containing carbon black.[2] Typically, blends of these two polymers form very large spherulites, and it is interesting to find out how carbon black affects crystallization and other properties of the blend as well as the distribution of carbon black in relationship to the spherulites. Figure 16.6 shows that spherulite growth rate is independent of carbon black presence (points of carbon black filled and carbon black free blend follow the same relationship). Additional data show that crystallization rate decreases with the amount of PVB increasing.[2] Carbon black aggregates are mainly found in spherulites.

Conducting polymers based on carbon black should contain a small amount of carbon black to facilitate processing. To reduce the amount of carbon black, one must create inhomogeneity in the material either by using the system which rejects carbon black from the crystalline region (opposite to the PCL/PVB blend above) or

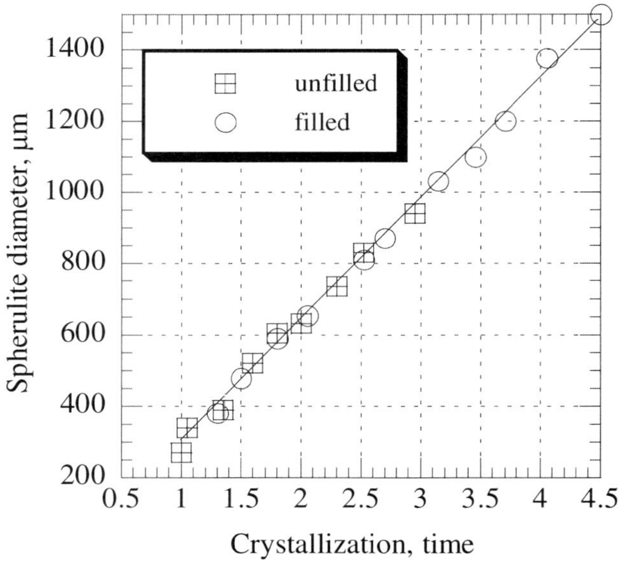


Figure 16.6. Spherulite diameter vs. crystallization time for PCL/PVB blend with (5 wt%) and without carbon black. [Adapted, by permission, from Lee J-C, Nakajima K, Ikehara T, Nishi T, *J. Appl. Polym. Sci.*, **64**, 1997, 797-802.]

by the use of an immiscible blend in which carbon black resides in the minor phase or, better, in the interphase.[3] Two conditions must be fulfilled for a blend to have conductivity at a low concentration of carbon black: heterogeneous distribution of carbon black and adequate phase morphology. Two phase morphologies support this goal: one polymer forms a continuous phase containing carbon black, or both polymers have cocontinuous morphologies and carbon black particles are possibly located in the minor phase. The aim of one study was to use the cocontinuous morphology of a PE/PS blend to improve conductivity.[3] Addition of carbon black was found to increase the continuity fraction from the range of 30 to 45 wt% PE to the range from 50 to 65 wt%. This can only be achieved if carbon black has either interfacial activity or a kinetic effect on the coalescence process. Based on image analysis results, it was concluded that carbon black prevents phase coalescence. The blends containing carbon black were stable during prolonged thermal treatment (24 h at 200°C). Figure 16.7 shows that controlled annealing may help to improve blend conductivity even further since only 0.4 wt% carbon black is needed if either annealing time or temperature are increased. These results suggest that in the region of cocontinuity, carbon black was localized at the interface. Further efforts of this work concentrated on the selective localization of carbon black at the interface by regulating the molding temperature.[11] Figure 18.4 shows the results for a carbon black filled blend of PS/PIP. The results are essentially similar to those given by Figure 16.7, pointing at process temperature as an important factor modifying the distribution of carbon black.

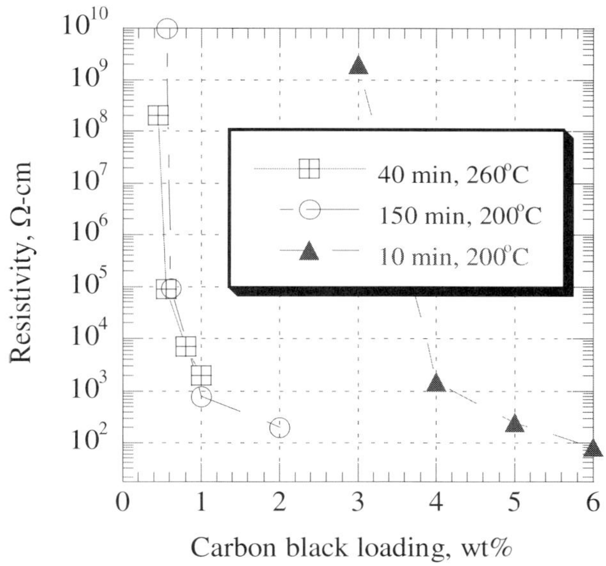


Figure 16.7. Resistivity vs. carbon black content in a PE/PS blend, depending on annealing regime. [Adapted, by permission, from Gubbels F, Blacher S, Vanlathem E, Jerome R, Deltour R, Brouers F, Teyssie Ph, *Macromolecules*, **28**, No.5, 1995, 1559-66.]

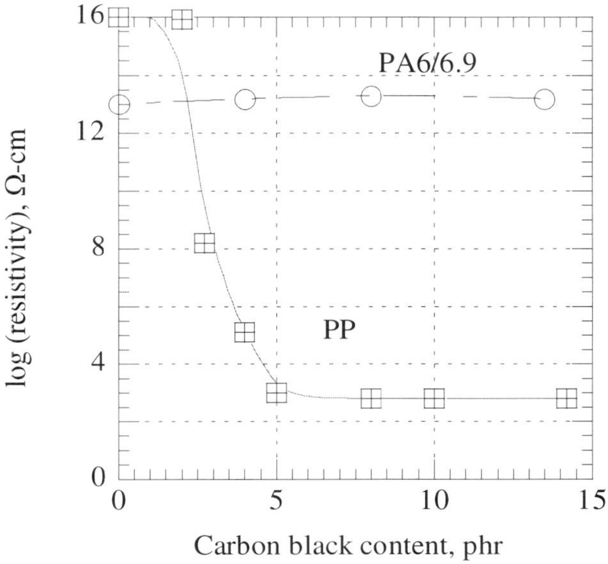


Figure 16.8. Resistivity vs. carbon black content. [Adapted, by permission, from Tchoudakov R, Breuer O, Narkis M, Siegmann A, *Polym. Engng. Sci.*, **36**, No.10, 1996, 1336-46.]

The polymers included in the polymer blend have impact on the development of conductivity behavior. Figure 16.8 shows that carbon black rapidly reduces the resistivity of polypropylene at about 2.5 phr whereas the conductivity of polyamide

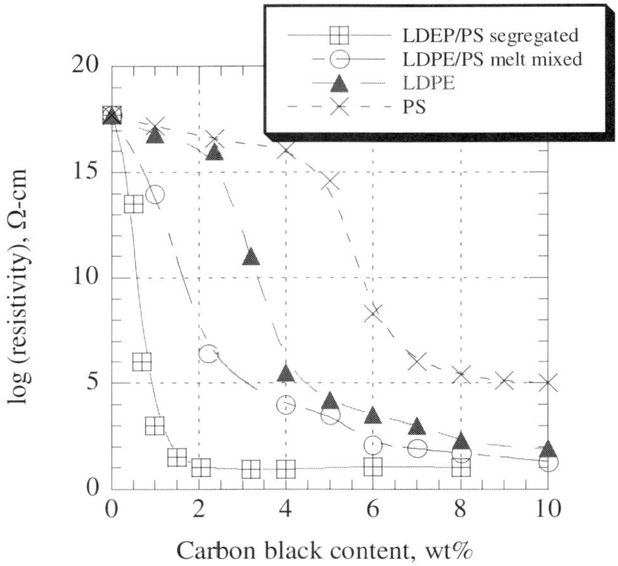


Figure 16.9. Resistivity vs. carbon black content for polymers and blends. [Adapted, by permission, from Kozlowski M, *Polym. Networks Blends*, **5**, 1995, 163-172.]

is not affected in the range of carbon black concentration. At the same time, polyamide has a much higher affinity to carbon black, which is demonstrated by the fact that even if carbon black is first added to polypropylene, carbon black is transferred to polyamide. Mixing of these two polymers with carbon black gives higher resistivity with increasing concentration of polyamide, since carbon black does not influence its conductivity to a great degree.[20] Polycarbonate has a similar characteristic in mixtures with carbon black and polypropylene, but the sudden decrease in conductivity occurs around 5 phr.[24] Polycarbonate, similar to polyamide also has a much stronger affinity to carbon black than polypropylene. Carbon black is thus localized in the polycarbonate phase which is distributed in polypropylene. With a small content of polycarbonate in the blend (up to 30 wt%), carbon black becomes concentrated in polycarbonate phase and resistivity rapidly decreases.[24]

Figure 16.9 compares the effect of carbon black concentration on PS, LDPE, and their blends. The polymer blends have better conductivity than individual polymers. A rapid decrease in conductivity by small additions of carbon black can be obtained in segregated blends, but mechanical properties of such blends are inferior.[34]

The blend morphology containing conductive filler (e.g., carbon black) was simulated by the model based on Cahn's approach. Figure 16.10 shows the two-dimensional cut explaining the localization of carbon black between two incompatible phases, and Figure 16.11 shows the effect of carbon black concentration on the prediction of conductivity.[15] This simple model of interfacial film partitioning

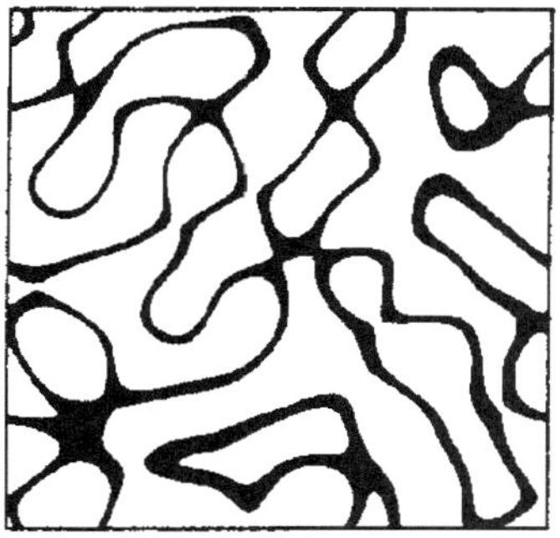

Figure 16.10. Model morphology shown by two-dimensional cut for carbon black filled blend. [Adapted, by permission, from Knackstedt M A, Roberts A P, *Macromolecules*, **29**, No.4, 1996, 1369-71.]

in two polymer phases gives a conductivity exponent in agreement with experiment.[15]

Interpenetrating networks experience different effects of filler incorporation because of the method of their formation and the higher probability of chemical bond formation with fillers due to the reactivity of the system. There are also more regulating factors because of the larger number of components, which affect properties through different sequences of incorporation. One example of improvement is given in Figures 19.23 and 19.24 which show that mica affects damping characteristics of interpenetrating networks.[9] An interpenetrating network composed of polypyrrole and EPDM containing fillers was used to show that conductive materials can be obtained by incorporation of an intrinsically conductive polymer. Fillers such as silica and kaolin were used to improve mechanical properties which are affected by formation of interpenetrating network.[18]

The microphase separation of a semi-interpenetrating network based on polyurethane and poly(epoxy isocyanurate) containing $Fe_2O_3$ was studied by dynamic mechanical spectroscopy. The preferential absorption of polyurethane on

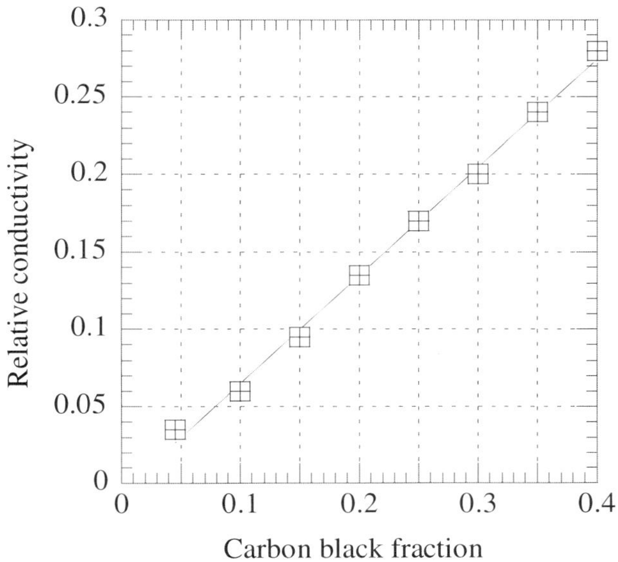


Figure 16.11. Conductivity of simulated blend vs. carbon black fraction. [Adapted, by permission, from Knackstedt M A, Roberts A P, *Macromolecules*, **29**, No.4, 1996, 1369-71.]

filler causes microphase separation.[30] The properties of an epoxy-silica interpenetrating network are discussed.[31] The unique morphological properties can be obtained by network formation, which cannot be obtained by component mixing. Thermal properties such as stability and char yield are substantially improved by network formation.

The above review of current findings shows that polymer blending in the presence of fillers may offer interesting new products. Considerable effort was made in the studies of conductive polymers, and many valuable practical results are available. The effect of fillers on the improvement of mechanical properties is still on the exploratory stage. At the same time, numerous theoretical possibilities exist which should contribute to improved materials in the future.

## 16.2 COMPOSITES[35-62]

One definition describes the composite material as follows: "A material created by the synthetic assembly of two or more materials (a selected filler or reinforcing element and compatible matrix binder) to obtain specific characteristics and properties." Composites began from the idea to replace metals with materials which can match their strength but do not corrode, are lighter, and less expensive. Because these goals were realized in the early days by using reactive glue and fiber, composites are generally thought of as fiber reinforced resins. The meaning today encompasses all fillers, and frequently the term composite is applied to any mixture of polymeric material and filler. With this broad definition, composites are the subject of entire books. In order to avoid repeating information, this discussion includes only some aspects of composites such as

- Mixtures of fibers and particulate materials
- Material design to obtain materials having certain combinations of properties

The remaining aspects of this large and complex group of materials are either found in various parts of this book or can be found in specialized monograph on the subject.[32] The major selection criteria of advanced composites are properties of the fibers or other reinforcing fillers, properties of the matrix, fiber-matrix interface, and processing technology, which combined, results in engineered properties.

There are numerous benefits of mixing the elongated and spherical particles as established in the result of a broad study.[61] Figure 16.12 shows the effect of glass fibers, glass microspheres, and their mixtures on mechanical properties of polyamide-66. The data show that only unnotched Charpy impact test results are lower for the combination of glass fiber and microspheres than for any of the fillers alone. In addition to improvement of mechanical strength parameters, properties can be balanced to the requirement by changing the proportion of both fillers.[61]

The compositions of different fillers are easier to process. Figure 16.13 shows that replacement of part of calcium carbonate by microspheres results in reduction of viscosity.[61] Similar reduction of viscosity can be expected in the mixtures of fi-

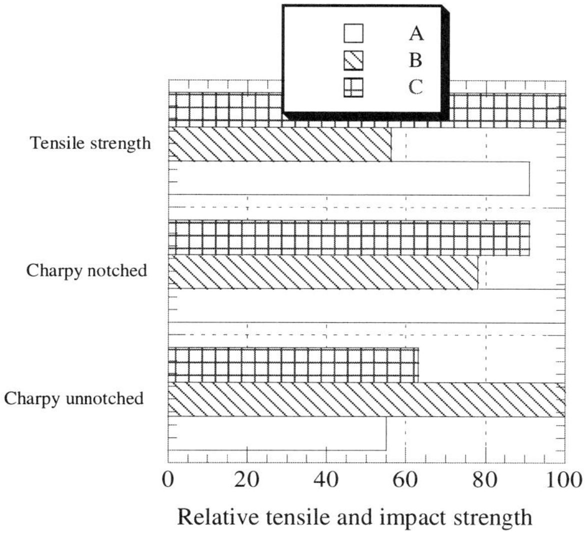


Figure 16.12. Relative values of mechanical performance for polyamide-66 reinforced with 15 wt% glass fiber (A), 15 wt% microspheres (B), and their combination of 15 wt% each (C). [Data courtesy of Abrasivos Y Maquinaria, SA, Barcelona, Spain][61]

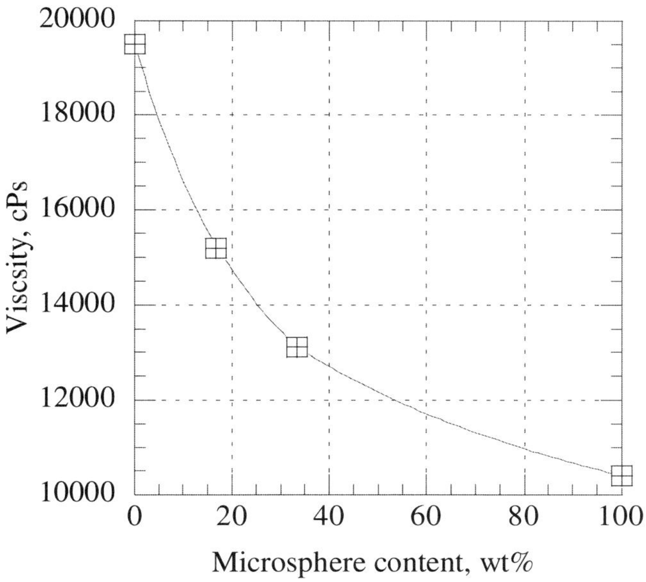


Figure 16.13. Viscosity of polyester resin filled with calcium carbonate and microspheres (total content of filler ($CaCO_3$+microspheres) is 60 wt%). [Data courtesy of Abrasivos Y Maquinaria, SA, Barcelona, Spain][61]

bers and microspheres because of their free flowing character. In injection molding, glass beads have ball-bearing effect which makes flow of material containing fiber much easier which results in lower temperature and shorter cycle.

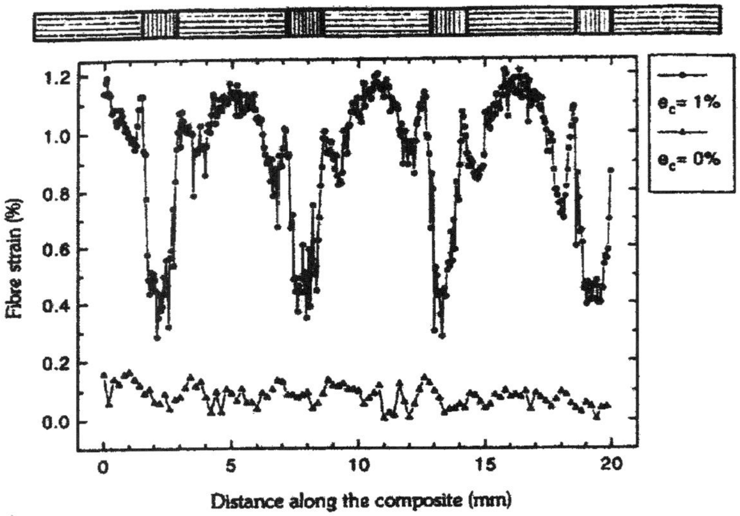


Figure 16.14. Fiber strain vs. distance along composite recalculated from stress-induced Raman band shifts. [Adapted, by permission, from Young R J, *Prog. Rubb. Plast. Technol.*, **11**, No.2, 1995, 124-36.]

Abrasion resistance and impact strength are improved simultaneously when a combination of beads and fibers is used. In these composites, glass beads dissipate energy and fibers protect beads from abrasion.[61]

The deformation of fibers under strain and strain distribution are important for the designer of composites. Figure 14.28 shows the Raman spectra of aramid fiber measured with and without the strain. There is a sufficient shift in spectral peaks to use it for the analysis of complex materials. This method was adapted for evaluation of epoxy resin reinforced with satin aramid fabric. Figure 16.14 shows the results along the warp face.[39] The schematic drawing above the graph shows positions of warp threads which match peaks on the Raman spectrum when the fabric is under strain. The entire strain applied to the composite is transferred to warp threads. This technique has excellent capabilities for mapping stress distribution in composites. Similar information was obtained along the weft face, but in this direction only 20 to 70% strain is transferred.[39]

Combination of several properties is becoming increasingly important in modern industry. One example may be taken from electronics, where in addition to mechanical properties and electric resistance, thermal stability and conductivity are important requirements. It was estimated that the increase of temperature by 10°C reduces time to failure by the factor of two.[40] A finite analysis model was developed which accounts for the following properties of filled composites: microstructure, effect of particle shape, formation of conductive chains, effect of filler aspect ratio, and interfacial thermal resistance. The predictions of the model indicate the most advantageous dispersion of particles in the matrix. Conductive chain formation, merging of conductive chains, branching of conductive chains, and agglomeration at the center give effective thermal conductivities of 1.33, 1.24, 1.16, and 0.73 W/m·K, respectively.[40]

Achieving a combination of several mechanical properties is also a frequent problem. The addition of fillers to composites improves their strength, modulus, and shrinkage, but toughness of the composite is usually reduced. If elastomer is added to the composite, toughness is improved but stiffness decreases. A model was developed in which improvement of all these properties is examined by the application of core-shell fillers. The use of core-shell filler based on carbon black resulted, according to the proposed model, in a composite which had high strength, toughness, and modulus.[49]

Varistors are electric components with variable resistance, depending on the electric current. Varistors should have low breakdown field strength combined with fairly high non-linearity of electrical conductivity. Special varistor powdered filler was prepared to obtain these properties.[53]

Aramid fibers are popular in the reinforcement of various matrix materials. It is frequently the case that low water uptake by a composite is required, which cannot be regulated by the matrix. Figure 16.15 shows that aramid fibers differ in water sorption kinetics. Fiber E gives better performance due to a hydrophobic coating. This lower water sorption of the fiber decreases water sorption of the composite, as was established in the study.[58]

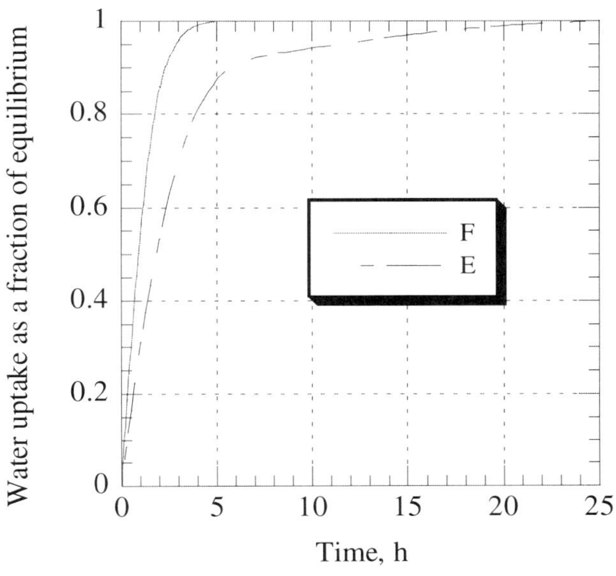


Figure 16.15. Water sorption kinetics of aramid fibers vs. time. [Adapted, by permission, from Rebouillat S, Escoubes M, Gauthier R, *J. Adhesion Sci. Technol.*, **10**, No.7, 1996, 635-50.]

Fire proofing of various composites is an important example of design to introduce material having a certain set of other required properties combined with fire resistance. This is usually a difficult task, considering that fire retarding fillers must be usually used in large quantities, and this deteriorates mechanical properties and makes processing more difficult because of increased viscosity. Such technologies are developed to obtain fire resistant materials which offer good mechanical performance.[59,60]

Considering that properties of the composite depend on filler, matrix, and their adhesion, an interesting proposal was put forward.[56] The so-called matrix or template polymerization was proposed for tailoring the properties of composites. This method allows polymer to grow on organic or inorganic matrices by initiation of polymerization on the surface of the matrix. The polymers produced on matrices frequently have very different properties than polymers obtained without the effect of a template because template changes the configuration and conformation of polymers. The other advantage of template polymerization is in changing reactivity, considering that monomers without template have a different conformation which may hinder the chemical reaction. Introduction of a template may change access to reactive groups and thus change kinetics of the reaction. The study of template polymerization is an interesting field but the findings were not applied in practice because of difficulties in separating the obtained polymer from template. In the area of the composite, this curse may become a blessing since difficulty of separation frequently means excellent adhesion. Some works on nanocomposites apply these methods. More information on template polymerization can be found in the first monograph recently published on this subject.[62]

## 16.3 NANOCOMPOSITES[63-91]

These novel materials receive increasing interest because of their complex structure and interesting performance characteristics. The following points will be discussed here:

- Nanocomposite design and formation
- Resultant particle size
- Potential applications
- Properties

Nanocomposites, in addition to small particle sizes usually are required to have a very uniform composition. These two requirements impose constraints on the methods of synthesis. In addition, the required form of the final product is given consideration in the choice of the method of their synthesis. These constraints resulted in numerous design strategies in methods of nanocomposite manufacture, discussed below.

One method is to produce particulate material of the required size and then to incorporate it in castable matrix, followed by film formation. Figure 16.16 shows the chemistry of the process and the steps from the raw materials to the final film

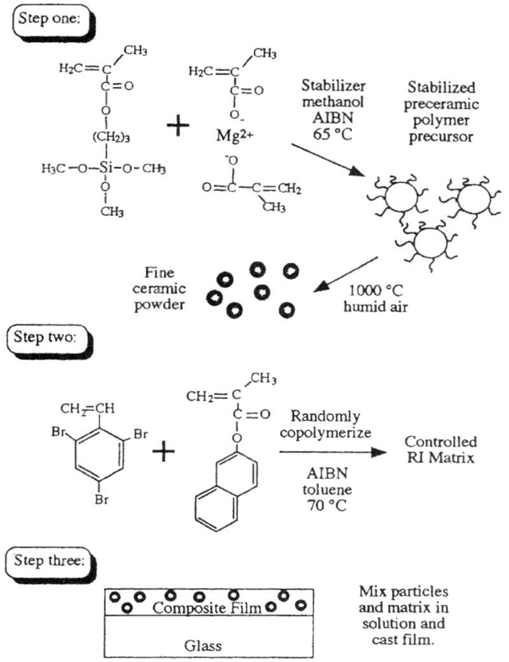


Figure 16.16. *Ex situ* synthesis of optical composite for laser amplifying films. [Adapted, by permission, from Beecroft L L, Ober C K, *Chem. Mater.*, **9**, 1997, 1302-17.]

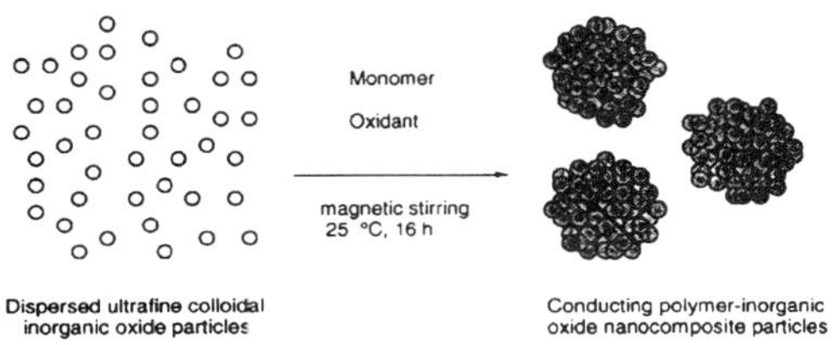


Figure 16.17. Particulate nanocomposite formation. [Adapted, by permission, from Maeda S, Armes S P, *Synthetic Metals*, **73**, 1995, 151-155.]

used for amplification in lasers.[63] In the first stage, a polymer based on silicon and magnesium containing monomers is produced, resulting in beads having sizes of 100-500 nm. These beads are subjected to the treatment at 1000°C, where the organic part is removed and crystalline fosterite is formed, having 100 nm size of beads. In optical applications, if particles are larger than 25 nm, it is very important to select a polymer for the matrix which has a very similar refractive index (refractive index mismatch less than 0.02); otherwise, optical scattering will be experienced. The polymer used in step two was selected for a precise match of refractive index.

Figure 16.17 shows the formation of conductive polymer/inorganic oxide nanocomposite particles. Silica, having a particle size of 20 nm, was dispersed in water, oxidant was added, followed by addition of monomer (pyrrole or aniline), and polymerization was conducted under constant stirring for 16 h at room temperature. "Raspberry" clusters of nanocomposite were obtained.[70]

Silicon nanoparticles in a narrow range of sizes (25-50 nm) were obtained by mechanical attrition.[74] This simple method of production of particles in large quantities may find a broad range of applications.

The development of nanocomposites containing metal particles may be very rewarding since such materials can be useful in catalysts, an optical and electronic devices.[79] Figure 16.18 shows the schematic diagram of the method of synthesis. The polymers used in this material were poly(4-methyl-1-pentene) and poly(tetrafluoroethylene). Dimethyl(cyclooctadiene)platinum(II) was used as the metal precursor. The metal precursor was dissolved in supercritical liquid carbon dioxide

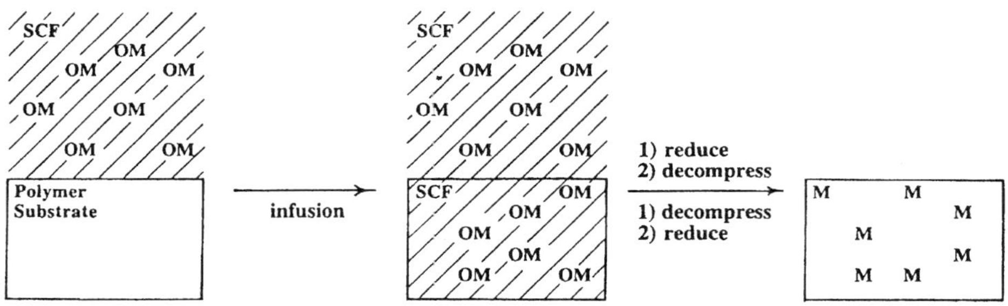


Figure 16.18. Schematic diagram of the synthesis of a polymer/metal nanocomposite. [Adapted, by permission, from Watkins J J, McCarthy T J, *Chem. of Mat.*, **7**, No.11, 1995, 1991-4.]

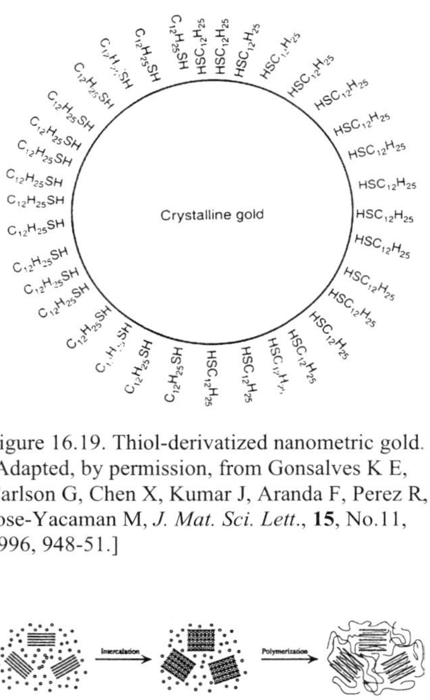


Figure 16.19. Thiol-derivatized nanometric gold. [Adapted, by permission, from Gonsalves K E, Carlson G, Chen X, Kumar J, Aranda F, Perez R, Jose-Yacaman M, *J. Mat. Sci. Lett.*, **15**, No.11, 1996, 948-51.]

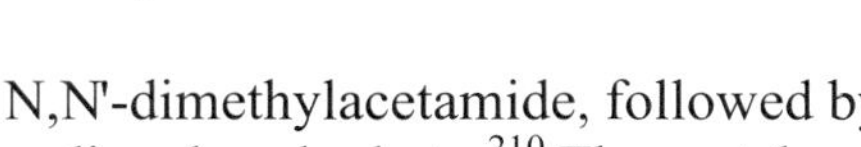

Figure 16.20. Layered composite synthesis. [Adapted, by permission from Messersmith P B, Giannelis E P, *Chem. of Mat.*, **5**, No.8, 1993, 1064-6.]

and infused into solid polymers. In the next step, metal precursor was either chemically or thermally reduced to produce metal domains. The properties of supercritical carbon dioxide are very important here. It has a high permeation rate in all polymers. Neither polymer nor metal precursor need to be soluble in carbon dioxide. This gives a method which can be adapted to any system.[79]

Surface-functionalized gold particles were obtained by a phase-transfer reaction of gold ions with dodecanethiol. Figure 16.19 shows the structure of such particles.[83] These particles are of interest because they have nonlinear optical polarizability. The particles were mixed with MMA and initiator and polymerized. No agglomeration of particles was detected, but the polymerization rate in the presence of nanoparticles was inhibited.[83]

Metal-containing nanocomposites were obtained by dispersion of metal chlorides in polyurethane. Both polyurethane and metal salts were dissolved in N,N'-dimethylacetamide, followed by film casting, and reduction of metal salts by sodium borohydrate.[210] The metal particle size depended on metal salt and concentration.

The layered nanocomposites can be obtained as shown in Figure 16.20. Intercalation of ε-caprolactone into the silicate galleries was done simply by suspending

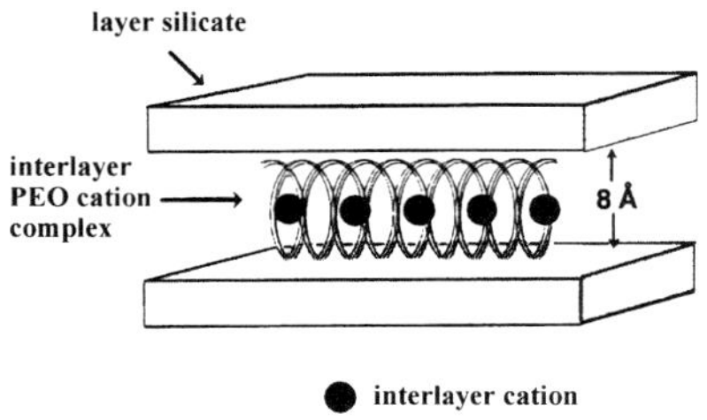


Figure 16.21. Schematic representation of layered nanocomposite with ion mobility. [Adapted, by permission, from Ruiz-Hitzky E, Aranda P, Casal B, Galvan J C, *Adv. Mat.*, **7**, No.2, 1995, 180-4.]

silicate powder in the monomer.[68] In the next step, polymerization was conducted at 100°C, resulting in polymer-intercalated silicate embedded in a poly(ε-caprolactone) matrix. During intercalation, silicate spacing was increased by 12.8 to 14.6 Å. After polymerization, spacing was reduced to 13.7 Å, which is due to silicate layer thickness (9.6 Å) and interchain distance (4 Å). The polymer is strongly adsorbed onto the silicate layers and shows no melting transition.[68] Layered nanocomposites derived from Na-montmorillonite had a regular lattice spacing of 17.7 Å, which is relative to twice the gallery expansion by two layers of polymer.[69]

Figure 16.21 shows structural arrangement of PEO intercalated into phyllosilicate (montmorillonite). Other organic components such as crown ethers can also be used. This is an example that ionic structures can be regularly located within the structure of the composite. These ionic structures can be useful in ion selective sensors.[72]

Vanadium pentoxide xerogels are very reactive layered host materials which can be intercalated by various means such as cation-exchange, acid-base chemistry, or redox reactions.[77] Vanadium pentoxide xerogel was prepared by polymerization of $HVO_3$ after a few days of reaction at room temperature. The resultant xerogel

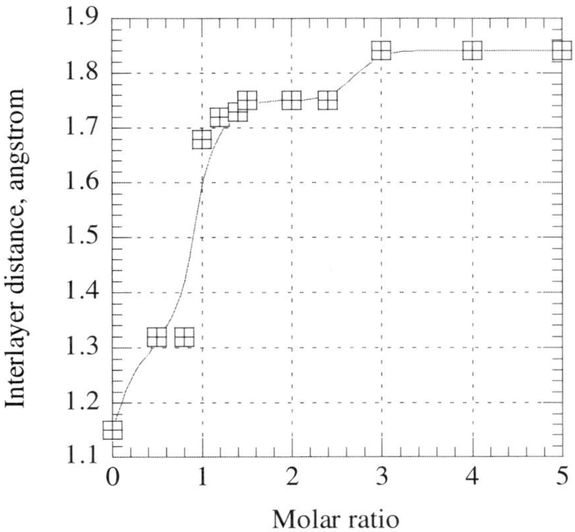


Figure 16.22. Interlayer spacing vs. molar ratio of PEO to $V_2O_5$. [Adapted, by permission, from Liu Y J, Schindler J L, DeGroot D C, Kannewurf C R, Hirpo W, Kanatzidis M G, *Chem. of Mat.*, **8**, No.2, 1996, 525-34.]

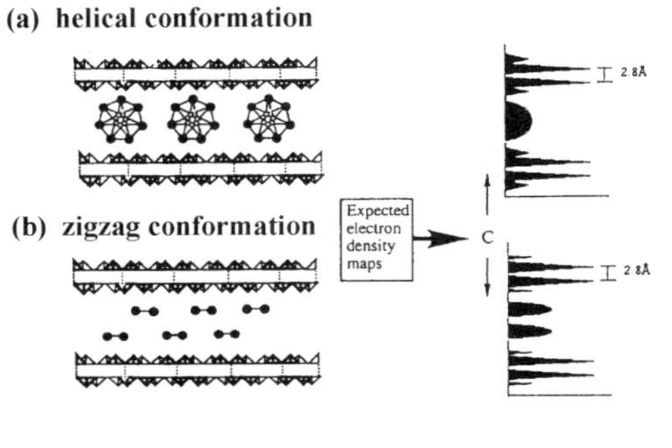


Figure 16.23. Schematic structures of PEO conformation in $V_2O_5$ xerogel and their expected electron density projections along the c axis. (a) coil conformation, (b) zigzag conformation forming bilayer. [Adapted, by permission, from Liu Y J, Schindler J L, DeGroot D C, Kannewurf C R, Hirpo W, Kanatzidis M G, *Chem. of Mat.*, **8**, No.2, 1996, 525-34.]

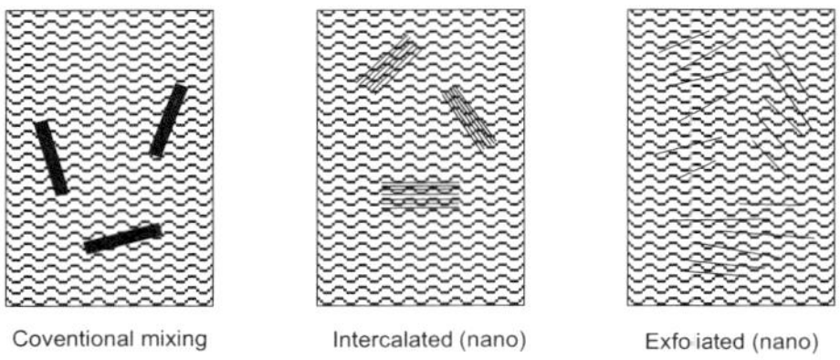


Figure 16.24. Three possible types of polymer/clay composites. [Adapted, by permission, from Lan T, Kaviratna P D, Pinnavaia T J, *Chem. of Mat.*, **7**, No.11, 1995, 2144-50.]

was mixed with different molar ratios of PEO and cast into films after water evaporation at room temperature. The interlayer spacing depended on molar ratio as shown in Figure 16.22. Figure 16.23 shows some possible conformations of PEO chains and their expected electron density projections. It was determined that the bilayer model (b) is correct.

Figure 16.24 shows the schematic representation of dispersed clay particles in a polymer matrix.[78] Conventionally dispersed clay has aggregated layers in face-to-face form. Intercalated clay composites have one or more layers of polymer inserted into the clay host gallery. Exfoliated polymer/clay nanocomposites have low clay content (lower than intercalated clay composites which have clay content ~50%). It was found that 1 wt% exfoliated clay such as hectorite, montmorillonite, or fluorohectorite increases the tensile modulus of epoxy resin by 50-65%.[78]

Montmorillonite was used in a two stage process of nanocomposite formation.[81] In the first step, montmorillonite was intercalated with vinyl monomer and then used in the second step to insert polystyrene by *in situ* polymerization.

Electron behavior, optical properties, catalytic properties, conductivity, and magnetic properties of nanocomposites were discussed in an extensive review paper.[73] Complementary use of electron paramagnetic resonance and nuclear magnetic resonance helped to understand chain mobility in nanocomposites obtained from poly(ethylene oxide) encapped with triethoxysilicon.[80] This nanocomposite is composed of PEO chains attached to silica clusters. It was found that chain fragments close to the silica clusters have hindered mobility due to the reduction of local free volume. The length of this hindered segment is estimated as three ethylene oxide units.[80]

**Table 16.1. Particle size and applications of nanocomposites**

| Filler | Matrix | Particle size, nm | Application | Ref. |
|---|---|---|---|---|
| Fosterite | Tribromostyrene/ naphthyl methacrylate | 100 | optical composite for laser amplifying films | 63 |
| Silica | Polyaniline | 200 | marker particles in diagnostic assays | 70 |
| Aluminum nitride | Polyimide | 20-100 | gas barrier, low thermal expansion | 71 |
| Phyllosilicate | Polyethylene oxide | layered | solid electrolyte, ion-selective materials | 72 |
| Silicon | - | 25-50 | nanoparticles for any application | 74 |
| Vanadium pentoxide | Polyethylene oxide | layered | - | 77 |
| Montmorillonite | Epoxy resin | layered | reinforcement | 78 |
| Platinum | PTFE | 15 | optical, electronics | 79 |
| Montmorillonite | Polystyrene | 150-400 | combination of high strength and high toughness | 81 |
| Smectite | polypropylene | layered | polymer reinforcement | 82 |
| Gold | PMMA | 6-14 | nonlinear optical devices | 83 |
| Cellulose whisker | St-BuAc-copolym. | n/a | polymer reinforcement | 84 |
| Metals (Fe, Co, Ni, Cu) | Polyurethane | 10-150 | | 87 |
| Coated silica | PMA | 150 | polymer reinforcement | 88 |
| Boehmite | | | | 89 |
| Cu2S/CdS/ZnS | Polystyrene | 2.8-7.1 | semiconductor | 90 |

Platy clay mineral reinforced nanocomposites of polyamide, epoxy resins, poly(ethylene oxide), and polystyrene having outstanding mechanical properties have already been commercialized.[82] Spectacular results of reinforcement with cellulose whisker were obtained for styrene butylacrylate copolymer emulsion (6 wt% whisker increased tensile strength by 800-3300% and modulus by 750-16,000%, depending on the method of processing).[84] Because of substantially lower loading with inorganic materials, nanocomposites allow the obtainment of high performance materials at much lower density. Polyamide reinforced with nanoparticles of silicate have improved tensile strength and impact strength and substantially reduced water permeability (Figure 16.25).[86]

Figure 15.42 shows the effect of silica nanoparticles on solvent uptake by the nanocomposite. In addition to much lower solvent uptake, the nanocomposite has increased thermal stability.[91]

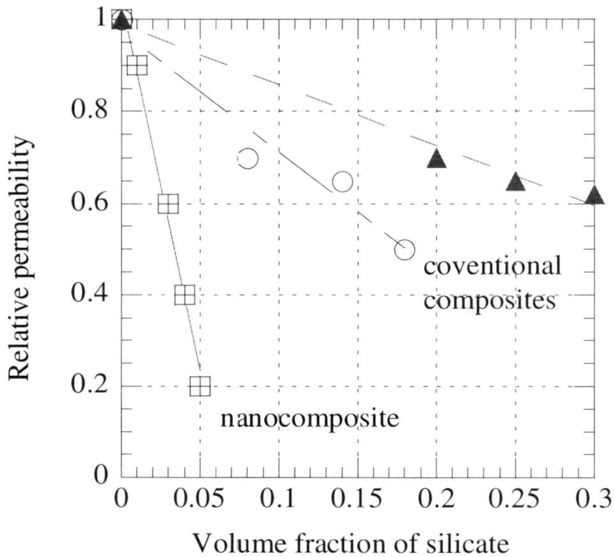


Figure 16.25. Water permeability of nanocompcsite and polyamide composite conventionally filled with silicate. [Adapted, by permission, from Giánnelis E P, *Adv. Mat.*, **8**, No.1, 1996, 29-35.]

## 16.4 LAMINATES[92-96]

The laminates are not the main focus of this book and thus only a few examples from the current literature are given on filler applications in these products.

Waste incineration requires sophisticated scrubbers to contain corrosive volatiles which are in contact with the scrubber wall at elevated temperatures reaching 220°C.[92] Glass mat reinforced novolac epoxy vinyl ester resins have good performance in this application except that they are subjected to thermal stress due to the temperature difference between surface and wall, and they frequently crack. In order to improve performance, a two layer laminate was produced. The base layer is composed of resin reinforced with glass mat. The resin contains 15-20% graphite filler. The surface layer contains 100 g/m$^2$ of carbon fiber. The presence of these two fillers helps to distribute heat more evenly throughout the layer of laminate and prevents cracking even under extreme conditions.[92]

The addition of aluminum trihydrate improved fire resistance of glass epoxy laminates but, because of the high loading required, it decreased the mechanical properties of the laminate. Various components of formulation were studied to improve performance. It was found that the curing agent and impact modifier help to improve the mechanical properties of the laminates. Other fillers were also studied in order to understand the impact of filler on properties. It was found that all fillers (glass beads, quartz, calcium carbonate, mica) reduce mechanical properties of laminates, not just aluminum trihydrate.[93] Aluminum trihydrate was found to be one of better performers in this system.

Aluminum particles were used to reinforce glass fiber/epoxy laminates. The fracture toughness was increased but tensile strength and modulus were decreased.[95]

## REFERENCES


1 Jarvela P A, Shucai L, Jarvela P K, *J. Appl. Polym. Sci.*, **65**, 1997, 2003-11.
2 Lee J-C, Nakajima K, Ikehara T, Nishi T, *J. Appl. Polym. Sci.*, **64**, 1997, 797-802.
3 Gubbels F, Blacher S, Vanlathem E, Jerome R, Deltour R, Brouers F, Teyssie Ph, *Macromolecules*, **28**, No.5, 1995, 1559-66.
4 Tchmutin I A, Ponomarenko A T, Shevchenko V G, Godovski D Y, *Synthetic Metals*, **66**, No.1, 1994, 19-23.
5 Benderly D, Siegmann A, Narkis M, *J. Mat. Sci. Lett.*, **14**, No.2, 1995, 132-4.
6 del Rio C, Acosta J L, *Polymer*, **35**, No.17, 1994, 3752-7.
7 Nichols K, Solc J, Shieu F, Antec '93. Conference Proceedings, New Orleans, La., 9th-13th May 1993, Vol. II, 1938-42.
8 Seibel S R, Moet A, Bank D H, Nichols K, Antec '93. Conference Proceedings, New Orleans, La., 9th-13th May 1993, Vol. I, 902-5.
9 Li Shucai, Peng Weijang, Lu Xiuping, *Int. J. Polym. Mat.*, **29**, Nos.1-2, 1995, 37-42.
10 Zaikin A Y, Nigmatullin V A, *Kaut. u. Gummi Kunst.*, **47**, No.10, 1994, 709-14.
11 Soares B G, Gubbels F, Jerome R, Teyssie P, Vanlathem E, Deltour R, *Polym. Bull.*, **35**, No.1/2, 1995, 223-8.
12 del Rio C, Acosta J L, *J. Appl. Polym. Sci.*, **60**, No.1, 1996, 133-8.
13 Kundu P P, Tripathy D K, Gupta B R, *J. Appl. Polym. Sci.*, **61**, No.11, 1996, 1971-83.
14 Simonsen J, Rials T G, *J. Thermoplast. Composite Mat.*, **9**, No.3, 1996, 292-302.
15 Knackstedt M A, Roberts A P, *Macromolecules*, **29**, No.4, 1996, 1369-71.
16 Hashmi S A R, Majumdar A K, Chand N, *J. Mat. Sci. Lett.*, **15**, No.15, 1996, 1343-5.
17 Ashalatha P V, George K E, Francis D J, *J. Elastomers Plast.*, **28**, No.3, 1996, 223-35.
18 Zoppi R A, De Paoli M A, *Polymer*, **37**, No.10, 1996, 1999-209.
19 Pukanszky B, Maurer F H J, Boode J W, *Polym. Engng. Sci.*, **35**, No.24, 1995, 1962-71.
20 Tchoudakov R, Breuer O, Narkis M, Siegmann A, *Polym. Engng. Sci.*, **36**, No.10, 1996, 1336-46.
21 Seibel S R, Moet A, Bank D H, Sehanobish K, Antec 95. Volume III. Conference proceedings, Boston, Ma., 7th-11th May 1995, 3966-70.
22 Xanthos M, Grenci J, Dagli S S, Antec '95. Vol.II. Conference Proceedings, Boston, Ma., 7th-11th May 1995, 3194-200.
23 Benderly D, Siegmann A, Narkis M, Antec '95. Vol. II. Conference Proceedings, Boston, Ma., 7th-11th May 1995, 3168-71.
24 Narkis M, Tchoudakov R, Breuer O, Siegmann A, Antec '95. Vol. II. Conference Proceedings, Boston, Ma., 7th-11th May 1995, p.1343-6.
25 Persson A L, Bertilsson H, *Composite Interfaces*, **3**, No.4, 1996, 321-32.
26 Vieira A, Nunes R C R, Visconti L L Y, *Polym. Bull.*, **36**, No.6, 1996, 759-66.
27 Simitzis J, Paitontzis C, Economides N, *Polym. & Polym. Composites*, **3**, No.6, 1995, 427-34.
28 McNeill I C, Mohammed M H, *Polym. Degradat. Stabil.*, **49**, No.2, 1995, 263-73.
29 Carre A, *J. Adhesion*, **54**, Nos.1-4, 1995, 167-74.
30 Sergeeva L M, Slinchenko E A, Brovko A A, Fainleib A M, Nedashkovskaya N S, *Polym. Sci., Ser. B*, **38**, Nos.5/6, 1996, 225-30.
31 Li Shucai, Peng Weijiang, Lu Xiuping, *Int. J. Polym. Mater.*, **29**, 1995, 37-42.
32 Lipatov Y S. **Polymer Reinforcement**. *ChemTec Publishing*, Toronto, 1995.
33 Shanks R A, Long Y, *Polym. Networks Blends*, **7**, 1997, 87-92.
34 Kozlowski M, *Polym. Networks Blends*, **5**, 1995, 163-172.
35 Hornsby P R, Premphet K, *J. Mater. Sci.*, **32**, 1997, 4767-75.
36 Bijwe J, *Polym. Composites*, **18**, No.3, 1997, 378-96.
37 Cousin P, Smith P, *J. Polym. Sci., Polym. Phys.*, **32**, No.3, 1994, 459-68.
38 Liauw C M, Hurst S J, Lees G C, Rothon R N, Dobson D C, *Prog. Rubb. Plast. Technol.*, **11**, No.2, 1995, 137-53.
39 Young R J, *Prog. Rubb. Plast. Technol.*, **11**, No.2, 1995, 124-36.
40 Ramani K, Vaidyanathan A, *J. Composite Mat.*, **29**, No.13, 1995, 1725-40.

41 Pukanszky B, Maurer F H J, *Polymer*, **36**, No.8, 1995, 1617-25.
42 Mamunya E P, Davidenko V V, Lebedev E V, *Polym. Composites*, **16**, No.4, 1995, 319-24.
43 Gokturk H S, Fiske T J, Kalyon D M, Antec '93. Conference Proceedings, New Orleans, La., 9th-13th May 1993, Vol. I, 605-8.
44 Davis R M, Gardner S H, Marand E, Laot C, Reifsnider K, DeSmidt H, McGrath J, Tan B, Antec '97. Conference proceedings, Toronto, April 1997, 2494-9.
45 Okiyokota M, Hamada H, Hiragushi M, Hasegawa T, Antec '97. Conference proceedings, Toronto, April 1997, 3319-21.
46 Zihlif A M, Di Liello V, Martuscelli E, Ragosta G, *Int. J. Polym. Mat.*, **29**, Nos.3-4, 1995, 211-20.
47 Zolotnitsky M, Steinmetz J R, *J. Vinyl and Additive Technol.*, **1**, No.2, 1995, 109-13.
48 Carraher C E, *Polym. News*, **19**, No.2, 1994, 50-2.
49 Ou Y C, Zhu J, Feng Y P, *J. Appl. Polym. Sci.*, **59**, No.2, 1996, 287-94.
50 McCabe J F, Wassell R W, *J. Mat. Sci. Mat. In Med.*, **6**, No.11, 1995, 624-9.
51 Liu Q, De Wijn J R, Bakker D, Van Blitterswijk C A, *J. Mat. Sci. Mat. In Med.*, **7**, No.9, 1996, 551-7.
52 Gerard J F, Chabert B, *Macromol. Symp.*, **108**, 1996, 137-46.
53 Glatz-Reichenbach J, Meyer B, Strumpler R, Kluge-Weiss P, *J. Mat. Sci.*, **31**, No.22, 1996, 5941-4.
54 Nofal M M, Zihlif A M, Ragosta G, Martuscelli E, *Polym. Composites*, **17**, No.5, 1996, 705-9.
55 Golubev A I, *Int. Polym. Sci. Technol.*, **23**, No.3, 1996, T/85-7.
56 Kuznezov A, Vasnev V, Gribova I, Krasnov A, Gureeva G, Ignatov V, *Int. J. Polym. Mat.*, **32**, Nos.1-4, 1996. 85-91.
57 Averous L, Quantin J C, Lafon D, Crespy A, *Int. J. Polym. Analysis and Characterization*, **1**, No.4, 1995, 339-47.
58 Rebouillat S, Escoubes M, Gauthier R, *J. Adhesion Sci. Technol.*, **10**, No.7, 1996, 635-50.
59 Brown N, Linnert E, *Reinf. Plast.*, **39**, No.11, 1995, 34-7.
60 Weaver A, *Reinf. Plast.*, **40**, No.11, 1996, 52-3.
61 Delzant M, *A Contribution Towards Studying the Behavior of Composites Containing Microperl® Solid and Microcel® Hollow Glass Microspheres or a Combination of Fibers and Spheres as Fillers.* Abrasivos Y Maquinaria, S A, Barcelona, Spain.
62 Polowinski S, **Template Polymerization**, *ChemTec Publishing*, Toronto, 1997.
63 Beecroft L L, Ober C K, *Chem. Mater.*, **9**, 1997, 1302-17.
64 Cepak V M, Hulteen J C, Che G, Jirage K B, Lakshmi B B, Fisher E R, Martin C R, *Chem. Mater.*, **9**, 1997, 1065-7.
65 Lan T, Pinnavaia T J, *Chem. of Mat.*, **6**, No.12, 1994, 2216-9.
66 Messersmith P B, Giannelis E P, *Chem. of Mat.*, **6**, No.10, 1994, 1719-25.
67 Lemmon J P, Lerner M M, *Chem. of Mat.*, **6**, No.2, 1994, 207-10.
68 Messersmith P B, Giannelis E P, *Chem. of Mat.*, **5**, No.8, 1993, 1064-6.
69 Wu J, Lerner M M, *Chem. of Mat.*, **5**, No.6, 1993, 835-8.
70 Foster J K, Sims E S, Venable S W, *Paint & Ink Int.*, **8**, No.3, 1995, 18-21.
71 Chen X, Gonsalves K E, Chow G M, Xiao T D, *Adv. Mat.*, **6**, No.6, 1994, 481-4.
72 Ruiz-Hitzky E, Aranda P, Casal B, Galvan J C, *Adv. Mat.*, **7**, No.2, 1995, 180-4.
73 Godovsky D Yu, *Adv. Polym. Sci.*, **119**, 1995, 81-122.
74 Bhagwagar D E, Wisniecki P, Papadimitrakopoulos, Antec '97. Conference proceedings, Toronto, April 1997, 1398-1401.
75 Hsieh B R, Melnyk A R, Antec '97. Conference proceedings, Toronto, April 1997, 1394-7.
76 Zaborski M, Slusarski L, Donnet J B, Papirer E, *Kaut. u. Gummi Kunst.*, **47**, No.10, 1994, 730-8.
77 Liu Y J, Schindler J L, DeGroot D C, Kannewurf C R, Hirpo W, Kanatzidis M G, *Chem. of Mat.*, **8**, No.2, 1996, 525-34.
78 Lan T, Kaviratna P D, Pinnavaia T J, *Chem. of Mat.*, **7**, No.11, 1995, 2144-50.
79 Watkins J J, McCarthy T J, *Chem. of Mat.*, **7**, No.11, 1995, 1991-4.
80 Brik M E, Titman J J, Bayle J P, Judeinstein P, *J. Polym. Sci., Polym. Phys.*, **34**, No.15, 1996, 2533-42.
81 Akelah A, Moet A, *J. Mat. Sci.*, **31**, No.13, 1996, 3589-96.
82 Kurokawa Y, Yasuda H, Oya A, *J. Mat. Sci. Lett.*, **15**, No.17, 1996, 1481-3.
83 Gonsalves K E, Carlson G, Chen X, Kumar J, Aranda F, Perez R, Jose-Yacaman M, *J. Mat. Sci. Lett.*, **15**, No.11, 1996, 948-51.
84 Hajji P, Cavaille J Y, Favier V, Gauthier C, Vigier G, *Polym. Composites*, **17**, No.4, 1996, 612-9.
85 Helbert W, Cavaille J Y, Dufresne A, *Polym. Composites*, **17**, No.4, 1996, 604-11.
86 Giannelis E P, *Adv. Mat.*, **8**, No.1, 1996, 29-35.
87 Chen L, Liu K, Yang C Z, *Polym. Bull.*, **37**, No.3, 1996, 377-83.

88 Pu Z, Mark J E, Jethmalani J M, Ford W T, *Polym. Bull.*, **37**, No.4, 1996, 545-51.
89 Carotenuto G, Nicolais L, Kuang X, Zhu Z, *Polym. News*, **21**, No.8, 1996, 280-1.
90 Jinman Huang, Yi Yang, Bai Yang, Shiyong Liu, Jiacong Shen, *Polym. Bull.*, **37**, No.5, 1996, 679-82.
91 Burnside S D, Giannelis E P, *Chem. of Mat.*, **7**, No.9, 1995, 1597-600.
92 *Reinf. Plast.*, **40**, No.10, 1996, 66-70.
93 Yang Q, Pritchard G, Phipps M A, Rose R G, *Polym. & Polym. Composites*, **4**, No.4, 1996, 239-46.
94 Sanchez-Solis A, Padilla A, *Polym. Bull.*, **36**, No.6, 1996, 753-58.
95 Kumar P, Gawahale A R, Rai B, *Adv. Composite Materials*, **4**, No.4, 1995, 279-85.
96 Grady B P, Antec '97. Conference proceedings, Toronto, April 1997, 2490-3.

# 17

# Formulation with Fillers

There are several reasons for formulation development:

- A new product
- Product improvement
- Cost reduction
- Raw material substitution
- Raw material alternatives
- Process engineering

All of the above are components of the product cycle from creation to discontinuation or replacement. These reasons for formulation development may be treated as separate events which become important when urgently needed. Or, they can be a part of a continuous search for better, more economical products. It goes without saying that in many organizations the first option is a frequent choice. These are organizations which are driven by external events, such as

- We do not have product to sell
- Competitor has better product
- We are too expensive
- Raw material production was discontinued and has to be imported from another continent
- Raw material supplier increased price
- Our equipment breaks down every month because material is too viscous to mix
- etc.

It also goes without saying that this is not the type of organization to be in, considering that events should be driven primarily by predictions and planning during product cycles. The methods of formulation have an important influence on the *modus operandi* and *modus vivendi*. The discussion below analyzes components of an organized, planned formulation.

The reasons for formulation development have an impact on the data required to make such a process successful. New product development is always considered a jewel in the crown since it may open new opportunities, and it is usually associated with the conquest of new territory. It is also the most risky undertaking considering the probability of success, patent barriers, health and safety regulations determining the availability of raw materials and rights to produce, and frequently required investment in the new process and equipment. Even this short list of hurdles shows that the development effort is a multifunctional one since it requires not only the input of specialists in all these areas but also the understanding of the market, the competition, product requirements, and all components which convince people to use this product.

In new product development, the general term associated with an excellent solution is *invention*. There is very little guidance as to what this *invention* really means and how one arrives at an inventive solution. One dictionary says that the inventive person is "able to invent things". The archaic definition of invention is "the act of finding". Rather than analyzing various programs created to increase the inventiveness of inventors, let us concentrate on the root of the matter and analyze what this "act of finding" means in the case of formulating with fillers. It is very difficult to imagine that the "act of finding" can be based mostly on imagination without data; rather, the results of one's work suggest that certain properties or functions can be achieved with a particular set of observed properties or that this set of certain properties can be useful to enhance the properties of some product. This stresses a need for data as a stimulant of inventiveness. For example, it would not be possible to design conductive plastics if neither conductive polymers nor conductive additives were available. But the very fact that one can readily get many materials of this kind does not constitute an immediate solution for the problem.

There are examples in this book of fillers which have magnetic properties, but somebody in the act of inventiveness came to the conclusion that such material can be used for removing materials from their solutions and designed a composite which contains magnetizable particles attached to the various materials (e.g., selective adsobent of a particular material). By mixing this composite with a solution of material, material is adsorbed by adsorbent and removed from solution by a simple magnet. A new analytic technique was created which is very useful in pharmaceutical and biochemical research.

It has been known for about a century that the crystallization of material is affected by the concentration of solution and admixtures, but only recently have scientists come to the conclusion that if implants are to be compatible with a particular organism they have to be crystallized in the environment in which they later perform and from which they should not be rejected. This creates a new field of activities dealing with the development of bone materials that are compatible with real conditions of their performance.

Are these two examples inventions or the products of observation of previously generated data? They are both because previous knowledge and data were required to conclude that they can be useful in their applications and there was somebody who made these important conclusions. There are a large number of people involved in research but only some ever contribute new products. This is related to their ability to see ahead to the remedies for the full list of obstacles which separate the beginning of research from its successful conclusion. Applied chemistry is a complex subject involving the interference and interplay of numerous parameters – some sufficiently quantified and some known only by a general description. All of these parameters are important and must be included in an initial experimental plan. It is thus important to consider a sufficient number of these variables in order to increase the probability of success (or better yet to decrease the probability of obstacle which cannot be overcome with a formulation that is almost complete but not sufficient to proceed to the next phase – production). All of this discussion was designed to show the importance of data.

Let us now analyze what kind of data are needed for the various reasons of formulation listed in the beginning of this chapter. We may begin with new product development since it is so crucial to progress, easier life, and business. The development of a new product begins with an idea regardless of how this idea was generated. The idea can be generated by analysis of needs which were not fulfilled so far, or it may be conceived based on association of an observed property of some material with its possible application. It might be the result of development of a new raw material which has properties not available before and which creates new opportunities, or it may simply occur, as it happens sometimes, by coincidence. Some ideas are more conventional than others and thus they will be considered as incremental improvement. A few ideas are simple but so useful that they are considered breakthroughs. In each case there must be

- A need for product
- A cost/performance ratio acceptable in the market place
- A material means for its development
- Work leading to this development
- A successful conclusion – product implementation
- Communication of the success to the potential users.

If any of the above is missing, the undertaking fails to achieve its goals. From the technical point of view, the following data are needed to fulfill the above requirements:

- A detailed list of product properties
- Quantification of product parameters
- A concept of product design
- A list of materials required to formulate product which meets requirements
- Understanding of the roles which components play in fulfilling requirements
- Understanding of interactions between components of the formulation
- The effect of variation in each component on product properties
- The effect of processing parameters on product properties

In addition to these groups of information, many other data are very useful or sometimes very important. They include: health and safety regulations affecting the use of components, production of intermediates and final product, limits of existing equipment which can be potentially used for production, professional level of operators, precision of production, the effect of environmental conditions on product quality (e.g., variations in humidity, temperature, temperature of cooling water, etc.), cost performance ratios of existing solutions, features of the new product which can be easily communicated to customers to get them interested, size of the market, the effect of production scale on the cost of raw materials and production, product patentability, patents which might be infringed on by the new product, and packaging requirements and storage conditions to obtain the required shelf-life. It is important to stress that all components of this list have some price tags which should be closely monitored throughout the process of development; otherwise good product might be too expensive for buyers.

This makes a list which seems impossible to fulfill, but it is rarely realized that behind every product there was a person or a group of people who had sufficient perception of these requirements that the product was introduced into the market and performed for many years. It is also not fully realized that there has to be a person with vision and imagination who early in the development stage understands the interaction of all these parameters which are critical to the end-effect. It is important to consider that a particular raw material (e.g., filler) does not have only one function in the formulation; rather, affects numerous parameters important for product properties. The more these interactions are understood, the more chances exist that useful product is developed, and the shorter the development time.

Let us now consider the next reason for formulation on the list – product improvement. If in the very first stage of product development, sufficient time was spent on

- Understanding of the roles which components play in fulfilling requirements
- Understanding of interactions between components of the formulation
- The effect of variation in each component on product properties
- The effect of processing parameters on product properties

there is available a lot of useful information for reformulation. If the product was developed by a trial and error process, no such background information exists, which makes the task of product improvement similar to that of development of a new product; in addition, it is burdened by the requirement that it should frequently resemble features of existing product so that it would not be recognized by customers as a new product. Such lack of data is quite typical of older products which makes their improvement very costly and difficult.

It is very easy to show that whether it comes down to cost reduction, raw material substitution, raw material alternatives, or process engineering, well organized data become invaluable. Such data not only help to achieve goals quickly, at low cost, and successfully, but also help to build a process which organizes the present and future work and contributes to continuous progress.

The next question arises as to how such data can be generated? There are numerous examples in the literature on how such data are presented and utilized,[1-13] and many books were written on the approaches to the data processing.[14-20] The most common form of information on the performance of additives is the so-called "starting formulation". These formulations are prepared by the manufacturers of raw materials to demonstrate the performance of their additives. This work is done by the raw material manufacturer alone or in the cooperation with some other parties such as, for example, the leading manufacturer of products which make use of the additive. In some instances, such starting formulations are used for the production of real products, but typically they are not suitable for direct application, considering that products should be unique, well engineered, and economical. Some information is given in the format which shows the most important parameters of the raw material (e.g., particle size, shape, color, etc.) vs. the effect of this property on product performance (e.g., with particle size decreasing reinforcement increases). These are useful directions for selection but only as a starting point because in the technological development, these influences have to be quantified. There are patents and research publications which show the results of studies and which usually contain sufficiently detailed descriptions of conditions of material preparation that the experiment can be verified. The literature search, and scientific monographs are invaluable resources of information used to

- Formulate the concept of product design
- List materials required to formulate product
- Understand the roles which components play in fulfilling requirements
- Understand the interactions between components of the formulation
- Understand the effect of processing parameters on product properties

Still, data from the literature is not a substitute for data generated in the development process. Data from the literature save development time by directing effort to meaningful activities and by frequently suggesting inventive solutions.

The ultimate goal of product development is to describe the properties and functions of the product in form of an algorithm – a functional relationship which includes all essential variables and allows one to generate simple relationships of influence of one parameter on another or on the product property (humans are thought to comprehend better the relationship of two variables). Such an ultimate goal is too difficult to reach, but the sets of relationships relevant to the performance of the product are commonplace. And, these form a very good starting point in the quest to describe properties in the quantitative manner, which brings understanding closer to the ultimate algorithm or model.

There are numerous tools which were designed to help formulators cope with the large amount of data. These include relational databases with graphic capabilities, statistical methods of data validation and evaluation, programs to design experiments, process optimization, worst case analysis, safety analysis, quality control analyzing methods and programs, robust design, screening designs, computer aided formulation, multiple goal decision methods, experts systems, and more. These tools are helpful if they are not used to replace human logic, which is paramount to the successful conquest of the complexity of nature.

The so-called experimental design is the most frequently abused tool because of the lack of understanding of chemistry, technology, and mathematics. The most frequent use of this method is by selecting the ranges of concentration of materials, measuring a few properties of the product by crude methods and projecting results of "mathematical analysis" in the form of usually tridimentional graphs which supposedly offer information on how to regulate process or design material.

This book shows that one filler introduced to the formulation affects many different parameters characterizing the product. In typical formulations of finished products 5-30 raw materials are used, each playing some role and potentially interacting with the remaining components. If the entire direction of formulation is predicted based on a crude measurement of the most common product indicators in specification, the entire principle of mathematics is neglected because a sufficient amount of information does not exist to solve equations in a realistic way.

The computerized system which helps most in product development resembles more the so-called expert system, which is a set of relationships quantified by an experiment for the purpose of similar products. Such systems are increasingly more effective with the amount of data (information) increasing. Considering such need, this book will have in the future a companion CD-ROM containing a base of available data which will be periodically updated to build an incremental wealth of information serving two purposes: material selection and data processing for the needs of the formulator.

This chapter, in the view of some readers, may contain information too general to be useful. At the same time, it is very important to stress that the development of high technology products requires adequate effort from the formulators. Fillers produced today are no longer just low cost filling materials but are sophisticated multifunctional additives. They can contribute to a decreased total cost of final products but only if the functions which they were designed for are fully utilized in the product design. This requires sufficient effort and data to produce results.

## REFERENCES


1 Brown N, Linnert E, *Reinf. Plast.*, **39**, No.11, 1995, 34-7.
2 *Eur. Rubb. J.*, **178**, No.7, 1996, 32.
3 Borden K A, Wei R C, Manganaro C R, *Plast. Compounding*, **16**, No.5, 1993, 51-5.
4 Cochet P, Barruel P, Barriquand L, Grobert J, Bomal Y, Prat E, IRC '93/144th Meeting, Fall 1993. Conference Proceedings, Orlando, Fl., 26th-29th Oct.1993, Paper 162.
5 Skelhorn D A, Meeting of the Rubber Division, ACS, Montreal, May 5-8, 1996, paper A.
6 Evans L R, Meeting of the Rubber Division, ACS, Montreal, May 5-8, 1996, paper D.
7 Bataille P, Schreiber H P, Mahlous M, Antec '93. Conference Proceedings, New Orleans, La., 9th-13th May 1993, Vol. II, 1757-9.
8 Cochrane H, Lin C S, *Rubb. Chem. Technol.*, **66**, No.1, 1993, 48-60.
9 Turley R S, Strong A B, *J. Adv. Materials*, **25**, No.3, 1994, 53-9.
10 Tokita N, Shieh C H, Ouyang G B, Patterson W J, *Kaut. u. Gummi Kunst.*, **47**, No.6, 1994, 416-20.
11 Chaturvedi M, Antec '97. Conference proceedings, Toronto, April 1997, 1865-8.

12 Yazici R, Kalyon D M, Antec '97. Conference proceedings, Toronto, April 1997, 2076-80.
13 *Solid Glass Beads and Hollow Glass Microspheres in the Service of Industry. Abrasivos Y Maquinaria SA*, Barcelona, Spain.
14 Bohl A H, Ed., **Computer Aided Formulation**. *VCH*, New York, 1990.
15 Krottmaier J, **Optimizing Engineering Design**. *McGraw-Hill,* London, 1993.
16 Walker N E, **The Design Analysis Handbook,** 2nd Ed. *Newnes*, Boston, 1998.
17 Hancox N L, Mayer R M, **Design Data for Reinforced Plastics**. *Chapman & Hall,* London, 1994.
18 Rhyder R F, **Manufacturing Process Design and Optimization**. *Marcel Dekker, Inc.*, New York, 1997.
19 Wendle B C, **What Every Engineer Should Know About Developing Plastics Products**. *Marcel Dekker*, New York, 1991.
20 Woodbridge E, **Principles of Paint Formulation**. *Blackie*, Glasgow, 1991.

# 18

# Fillers in Different Processing Methods

This chapter contains a discussion of the important changes in the methods of production and in the process parameters which are required to effectively incorporate fillers. The other goal of this discussion is to evaluate the effect of filler incorporation on the properties of final products manufactured by various methods of polymer processing.

## 18.1 BLOW MOLDING[1-4]

Two basic methods are used in this process to deliver material to the processing units. These are extrusion and injection. In the next step, the preformed material is expanded to form parison. There are many commercial variations on this basic technique some of which include continuous-extrusion-blow-molding, coextrusion-and-sequential-blow-molding, and injection-stretch-blow-molding. Both extrusion and injection molding are the subjects of later discussions below, we will concentrate here on the parison formation, its processing, and the related effects.

Two essential material related properties affect parison performance: die swell and sag. The die design and the operation parameters of the extruder or injection equipment are also very important. Die swell is the change of cross-sectional dimensions caused by the elastic recovery from deformations in the die. The sag behavior is a reaction of the material to gravitational forces. Both parameters of the parison determine its size tolerances along its length from the die exit. If the die swell is too low, the thickness of parison will be too small and the part will lack strength and toughness. If the swell is too large, resin is wasted. It is also essential that the parison have uniform thickness along its length from the die. This can be affected either by the changes in die swell vs. filling time or by sag which causes the material to flow by gravity. Figure 18.1 shows that material filled with talc performs better than unfilled material.[3] Die swell is reduced and it becomes uniform with filled material. This figure also shows that a larger addition of talc (40%) produces the same effect as 20% talc. The sag of the extrudate is also improved by talc but here, a larger amount (40%) gives an even better performance than 20% talc.[3] In

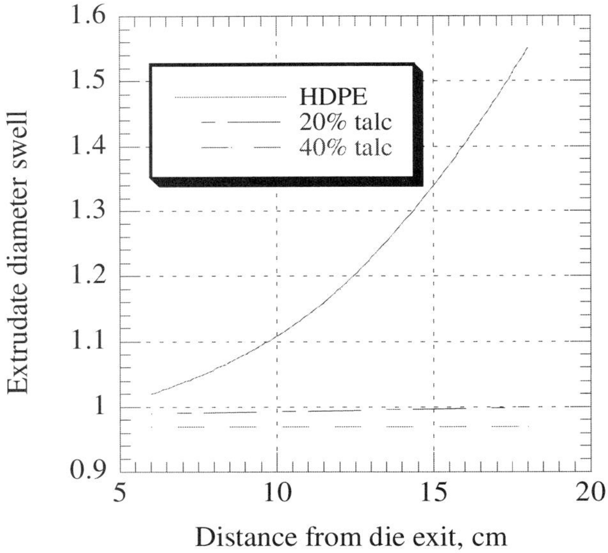


Figure 18.1. Extrudate diameter swell vs. distance from die exit. [Adapted, by permission, from Chang Ho Suh, White J L, *Polym. Engng. Sci.*, **36**, No.11, 1996, 1521-30.]

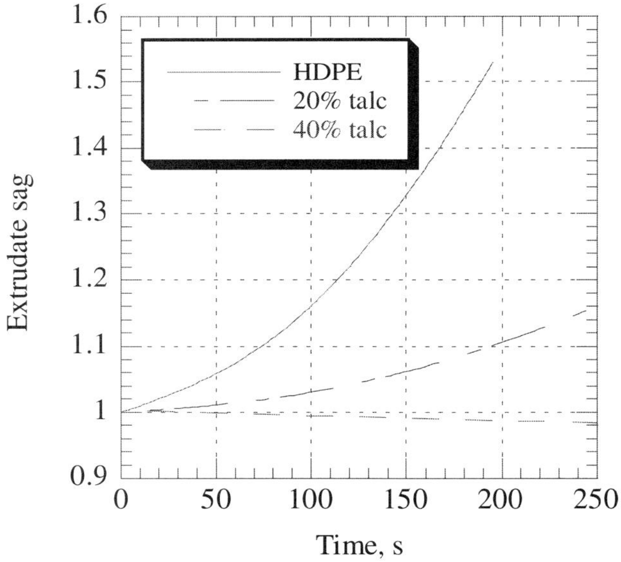


Figure 18.2. Extrudate sag vs. time. [Data from Chang Ho Suh, White J L, *Polym. Engng. Sci.*, **36**, No.11, 1996, 1521-30.]

addition to these two improvements, the material was improved because of orientation of talc particles along the wall.

Similar data come from processing of polyamide-6 with glass fibers. Diameter swell, thickness swell, and weight swell were all substantially decreased by the addition of 12 wt% glass fiber.[1] All of these parameters are almost constant along the length of the parison from the die, giving a uniform parison.

If the parts made in the blow molding process are to be of high quality, the parison must be dimensionally stable. And fillers can play essential roles in this process as pointed out above. During the blowing and cooling stages which follow the parison formation stage, fillers also play essential roles. The blowing stage is associated with large deformations in the material. Particulate materials, especially these with elongated shapes, contribute to the material texture and rheology and allow the material to withstand more severe processing conditions. During the cooling stage, many fillers will increase the crystallization rate which reduces the cooling time.

The addition of fillers to blow molded materials does not require special equipment but frequently the process parameters must be adjusted to compensate for the changes in the rheological properties of the melt. Larger additions of fillers tend especially to introduce non-Newtonian characteristics which require careful consideration. Also, timing of processes requires adjustment to deal with the higher nucleation rate caused by the presence of filler.

## 18.2 CALENDERING AND HOT-MELT COATING[5]

Calendering is a popular method of rubber processing. It is also used in the production of films and in lamination of plastic film to substrates in a continuous web process. Hot-melt coating equipment is another version of calendering equipment used for production of plastic coated textile materials.

Obtaining a completely homogeneous mix and selecting the correct calender processing conditions are essential to obtaining the design performance of the filled coating. The mixing processes are discussed in a separate section later in the chapter.

A calender for processing thermoplastics has several cylinders (usually 3 or 4) which have independent drives and temperature controls. Rubber industry uses simple masticating calenders which usually have only two rolls. In the calender for processing thermoplastics, several parameters must be regulated. These include cylinder speed, the gradient of cylinders speeds between two adjacent cylinders, the gaps between the cylinders, and temperature. In the combination, these parameters affects shear rate, orientation, and stress applied to the material. Even if the material is processed without filler, polymer chains become oriented which results in reinforcement and regulation of the mechanical properties both across the width and lengthwise.

The addition of fillers, especially fibers and platelets improves properties even further, since the filler particles are also oriented. Particle orientation can be produced by relatively small materials strain (a few percent). This is demonstrated in more detail in Chapters 7 and 10. This principle is used in the rubber industry where with simple calenders, fibers are oriented preferentially in one direction. If this material is used for production of belts then these belts have a high rigidity in one direction and flexibility in the other which increases the belt's performance and durability.

More sophisticated calenders used to coat with the thermoplastic polymers give even more control. For example, with the proper choice of conditions, metal flakes can be oriented parallel to the surface which in roofing membrane increases their ability to reflect radiation. The precision of these machines offers very broad range of possible processing conditions.

Incorporation of fillers does not require substantial adjustment in process conditions. In fact, the process is easier in the presence of fillers because the filled material usually has a lower adhesion to the cylinders of calenders which makes it easier to process. Also, cooling times are reduced because fillers increase the crystallization rate.

## 18.3 COMPRESSION MOLDING[6-13]

Compression molding is an example of a simple process which does not require expensive equipment but produces cheaply and in simple shapes. Unlike most other processing methods, compression molding does make materials inexpensive through the use of large quantities of fillers. Compression molding applies unidirectional forces usually perpendicular to the material's surface which may thus affect filler orientation.

Figure 18.3 shows that orientation of talc particles increases slightly with increased filler volume. The negative orientation function parameter indicates that particles are oriented parallel to the mold surface as might be expected from the application of an unidirectional force perpendicular to the surface of the mold.[13] A study compared compression molding with two forms of extrusion and compression molding produced a higher orientation of particles than did the extrusion processes.

In compression molding, molding temperature is an important processing condition, especially in conductive blends. Figure 18.4 shows that, depending on the molding temperature, the percolation threshold of carbon black filled blend can be reduced from 0.8 vol% carbon black at 200°C to 0.2 vol% carbon black at 250°C.[8] Fillers are not distributed equally in the components of a blend but preferentially settle in one component. If all carbon black was preferentially deposited in the polystyrene phase (45 wt% total) then the percolation threshold should be 1.6 vol% and 0.6 vol% because the percolation threshold in carbon black containing polystyrene is 3.6 vol% and 1.3 vol% at 200 and at 250°C, respectively. But data show these thresholds to be lower than predicted. This indicates that carbon black is concentrated at the interphase making it more economical to use blend of polymers in this application.

Pressure during the curing of compression molded item affects its preparation. Figure 18.5 shows that pressure and the type of the binder has an effect on the

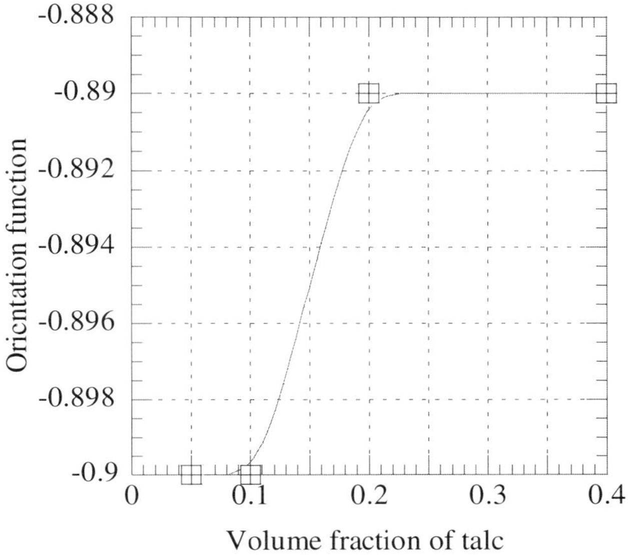


Figure 18.3. Orientation function vs. volume loading of talc in compression molded polystyrene. [Data from Kim K J, White J L, *J. of Non-Newtonian Fluid Mechanics*, **66**, Nos.2/3, 1996, 257-70.]

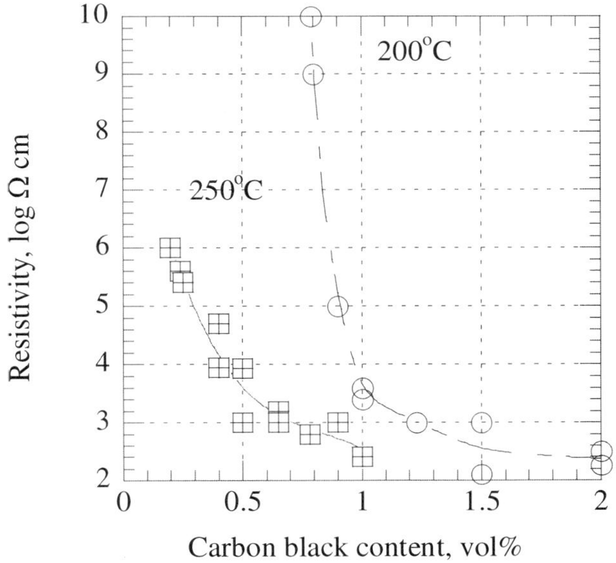


Figure 18.4. Resistivity of PS/PIB blend vs. concentration of carbon blacks. [Adapted, by permission, from Soares B G, Gubbels F, Jerome R, Teyssie P, Vanlathem E, Deltour R, *Polym. Bull.*, **35**, No.1/2, 1995, 223-8.]

resistivity of a carbon black filled composite.[10] The distribution of carbon black in neoprene rubber is affected by pressure but when the same amount of carbon black is used in butyl rubber molding pressure does not influence changes in resistivity.

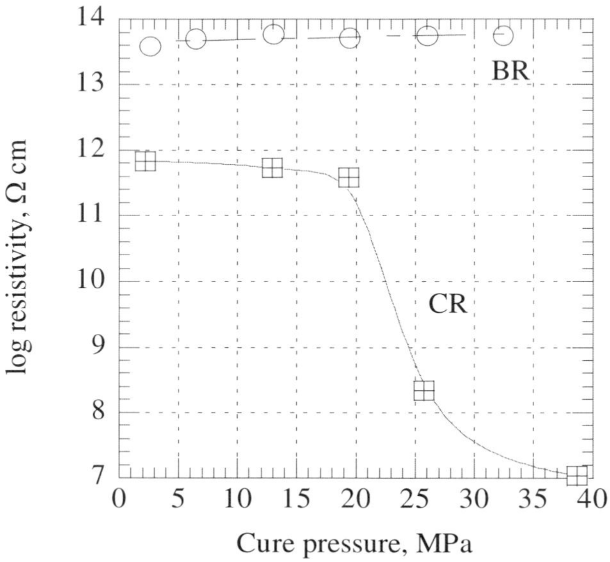


Figure 18.5. Resistivity of neoprene and butyl rubber filled with 40 phr N550 vs. pressure during the cure. [Data from Thompson C M, Allen J C, *Rubb. Chem. Technol.*, **67**, No.1, 1994, 107-18.]

These effects depend on filler concentration and its interaction with rubber. Pressure causes flow in the free rubber which has lower viscosity than rubber which has interacted with carbon black particles. As a result particles may or may not achieve a favorable distribution. If they fail to do so, the resistance is high.

A study of the cure rate[6] of unsaturated polyester containing calcium carbonate and glass fiber indicates that calcium carbonate makes the cure more complete at a rate similar to that of neat resin. The combination of calcium carbonate and glass fiber decreases the rate of cure and lowers the extent of final cure. Small amounts (10 wt%) of talc[11] nucleate crystallizing polypropylene during its compression molding. This level of talc generates the maximum number of crystallization sites. Adding more has no further effect on crystallization.

## 18.4 DIP COATING[5,14]

Dip coating is a process frequently used for the production of gloves. Fillers play a prominent role in dip coating formulations. The formulation must be designed to impart two opposite properties to the coating. It should have a relatively low viscosity to assist dipping process and the viscosity should rapidly increase after the form is withdrawn from the dipping tank.

Rheologically, these formulations are pseudoplastic with a significant yield value. The pseudoplastic characteristic is responsible for the performance of coating and wetting the form in the dipping tank (relatively low viscosity) and for providing a uniform coating of required thickness as form is withdrawn. The yield value causes the material to behave as a solid immediately the form is withdrawn from the tank. If the viscosity is too high an overly thick coating results and surface uniformity may suffer. If the yield value is too low, the material moves by gravity which causes nonuniform thickness of the glove and drips at the ends of fingers. Neither is acceptable. The material should behave as a solid at the elevated temperatures of the next process steps when curing or water evaporation occurs.

Several fillers are employed in these proprietary coatings. In addition to providing the required rheological characteristics, they must contribute to a long shelf-life of the dipping mixtures and give it the stable rheological characteristics which control the performance of coating. The production output is generally too high to regulate process within the batch by adjusting the composition of the liquid mixture.

The mechanical properties of the glove are obviously important. These are strongly influenced by the orientation of filler particles. When they are oriented parallel to the length of the glove, mechanical properties are optimized. It is not desirable that particles become oriented after the glove form is withdrawn from the tank because this causes sagging and dripping problems and variations in thickness. It is important that the proper shear is applied during withdrawal to induce then necessary particle orientation and reinforcement.

## 18.5 DISPERSION[15-37]

Dispersion is the subject of many separate monographs too extensive to be discussed in detail here. It is also a process critical to achieving the best performance of fillers. The properties imparted by fillers depend on the quality of dispersion. In most processes, the aim of dispersion is to decrease particle size and obtain a homogeneous distribution of filler particles. There are also cases in which overmixing may not be desirable. This occurs with fumed silica where overmixing causes irreversible damage to the rheological structure. Another example is in the incorporation of conductive fibers which under harsh mixing conditions are reduced to a size which can no longer give the desired performance characteristics.

Figure 18.6 shows torque vs. time during incorporation of silica in rubber.[23] Initially, the torque is increased because a solid material is being incorporated into the rubber matrix. Following the incorporation, torque decreases rapidly during the crushing phase during which time the agglomerate sizes are reduced but the solid filler has not yet come into proper contact with the matrix. The next stage, when torque increases again, is the wetting stage. During this stage the filler surface is increasingly wetted with rubber. This increases the mix viscosity because, filler interacts with the rubber, its actual concentration in the mixture increases. Now, when viscosity is high and shear rate increases the dispersion process starts causing a size reduction in the filler particles then a distribution of the dispersed particles to form a homogeneous mixture. If this figure is viewed together with Figure 18.7, it becomes clear that several important parameters lead to the attainment of a certain

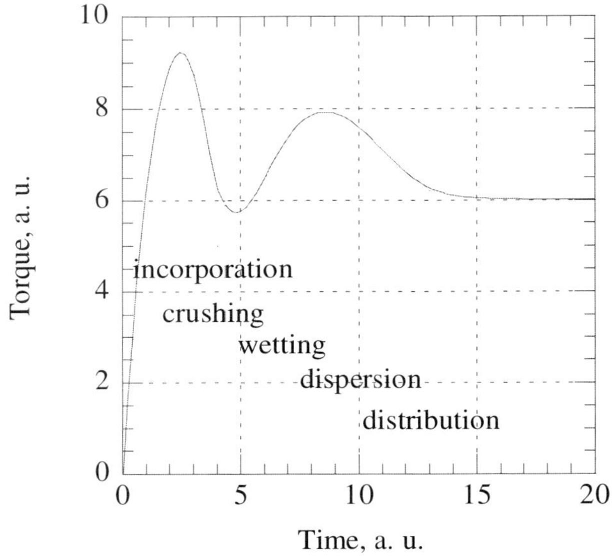


Figure 18.6. Torque vs. mixing time of silica in rubber. [Adapted, by permission, from Bomo F, Meeting of the Rubber Division, ACS, Montreal, May 5-8, 1996, paper E.]

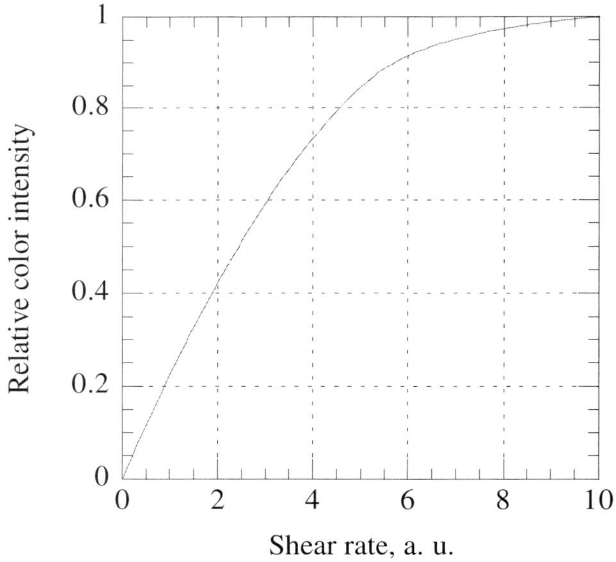


Figure 18.7. Color intensity vs. mixing time and shear rate.

maximum dispersion.[17] The most critical parameters are mixing time and shear rate (torque) (Figure 18.7) but the interaction between the filler and the matrix is also a significant factor. The combination of these three parameters permits an almost unlimited number of permutations making understanding, modelling, and practical

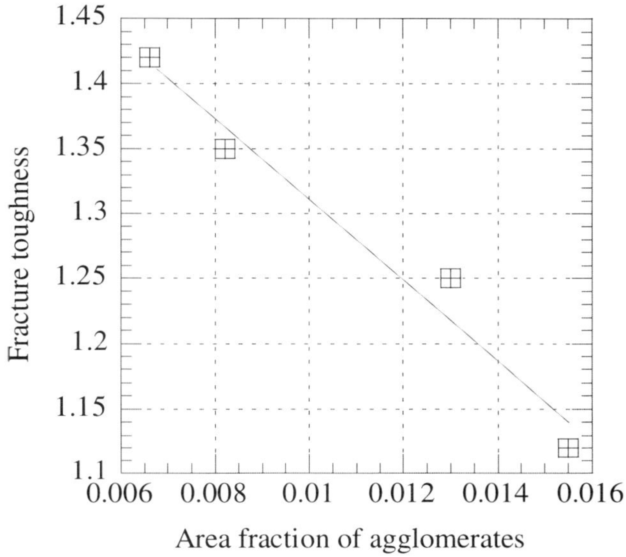


Figure 18.8. Fracture toughness vs. area fraction of agglomerates, $\Phi_a$. [Data from Yeh Wang, Jiang-Shen Huang, *J. Appl. Polym. Sci.*, **60**, No.11, 1996, 1779-91.]

methods of dispersion so difficult that only experimentation can verify assumptions.

Figure 18.8 shows that the fracture toughness of thermoplastic material depends on the dispersion.[26] The degree of dispersion is characterized by the area fraction of agglomerates, $\Phi_a$. The larger the area fraction of agglomerates, the worse is the dispersion. The graph shows that there is a substantial improvement in fracture toughness if dispersion is improved. This is never overstressed considering that we do not attain perfect dispersion but only try to limit imperfect dispersion.

## 18.6 EXTRUSION[13,26,38-54]

The composition of fillers and their degree of dispersion also influence the quality of extruded materials. Extrusion, like many other processes, must disperse fillers, which are usually hydrophilic, in a matrix which is predominantly hydrophobic. This, by itself, is a difficult task. Fillers contain water which may hydrolyze some polymers (e.g., polyamide or polyesters) at the elevated temperatures of processing. Even if water does not contribute to degradative reactions, it still creates the danger of various defects on the surface and in the bulk. Being able to get the best possible dispersion can be difficult, especially with the fine particle sized fillers which contribute to high performance of extruded articles. The smaller the particles the more difficult they are to disperse because the tendency to agglomerate increases as the particle surface area increases.[54] Fillers must be incorporated into a high viscosity melt which makes the process of dispersion more difficult. In spite of these difficulties the extrusion process must be designed for fast, cost effective production of up to several hundred kilograms per hour output from a single extruder.

Many new technologies have been developed to overcome these hurdles. Dispersion of fillers was tackled in many different ways. First, various screw and barrel designs were proposed which were capable of breaking agglomerates. These include the cavity transfer mixer and the pinned barrel. Second, a better technology of melt screening was developed using plate screen changers and continuous rotary disc filters. These filters prevent coarse particles from entering the feed stream. The dispersion and extrusion processes were optimized. It is difficult to optimize the extruder design if the extruder has to perform two functions simultaneously: mixing and dispersing of ingredients and feeding the die for maximum output and quality. Several systems were developed which separate these two functions by using extruders interconnected in series. The mixing extruder or other machine (e.g., ko-kneader) is optimized for dispersion and it feeds the melt to the production extruder. This solution eliminates the need to produce a separate granulate. It also eliminates the uncertainty about the pre-extrusion history of the granulate which affects extrusion results. This process is more efficient and outputs substantially increase.

The adverse effect of water was minimized through the use of venting systems. Traditionally, top of the barrel venting has been used. In a new Werner and Pfleiderer design, side-venting was introduced which allows for more efficient removal of water. The most modern extruders are capable to process PET containing up to 0.4% moisture. Finally, the process economics was improved by direct extrusion and development of continuous mixers which allow the material to be processed in one pass from dry powders. Unlike traditional extrusion which requires the granulate or dry blend to be formed in a separate operation, here, the ingredients are fed to the stream of molten polymer.[52] The line is capable of automatic dosing of all ingredients including fillers and pigments, in-line coloring, and quality control.

These new developments allow materials to be produced in an economical manner into products of high quality. However, the extrusion industry continues to use the traditional technology and, therefore, many of the old problems of filler incorporation remain.

Figure 18.9 shows the effect of filler particle size on extruder throughput for PP filled with talc.[47] Several reasons account for reduction in throughput as the particle size decreases. Increasing the surface area of filler makes mixing more difficult because of agglomerate formation. Smaller particle sized talc has lower bulk density which decreases the conveying efficiency of screw. The relatively large amount of air supplied with the particles decreases the conveying efficiency and increases the time required to extract air.[47] The amount of talc added affects the ratio of throughput, Q, to the screw speed, Ns (Figure 18.10). As the concentration

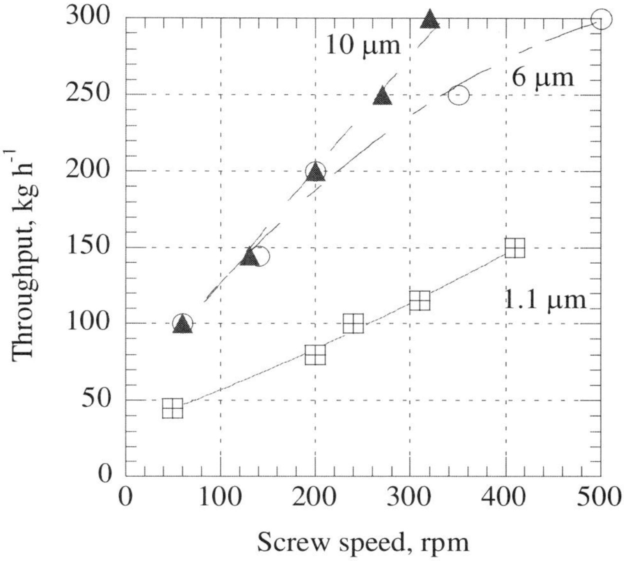


Figure 18.9. The effect of talc particle size on extruder output. [Adapted, by permission, from Ishibashi J, Kobayashi A, Yoshikawa T, Shinozaki K, Antec '96. Vol. I. Conference Proceedings, Indianapolis, 5th-10th May 1996, 386-90.]

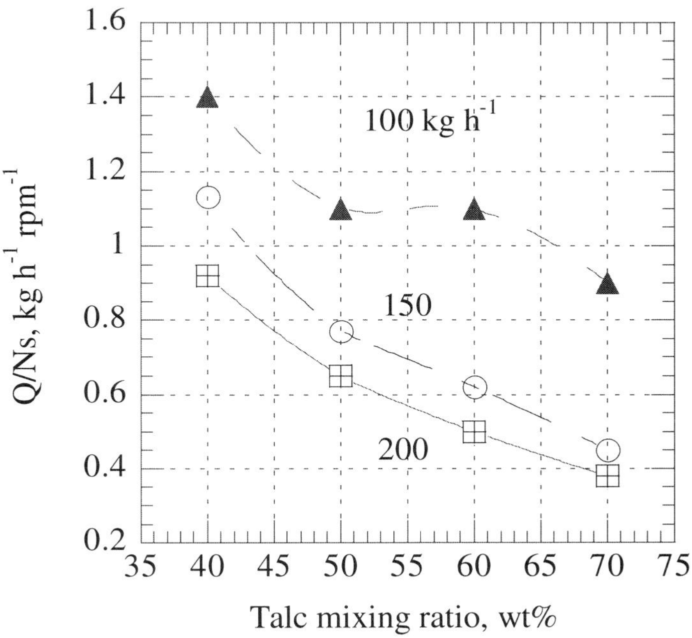


Figure 18.10. Throughput, Q/screw speed, Ns ratio vs. talc loading. [Adapted, by permission, from Ishibashi J, Kobayashi A, Yoshikawa T, Shinozaki K, Antec '96. Vol. I. Conference Proceedings, Indianapolis, 5th-10th May 1996, 386-90.]

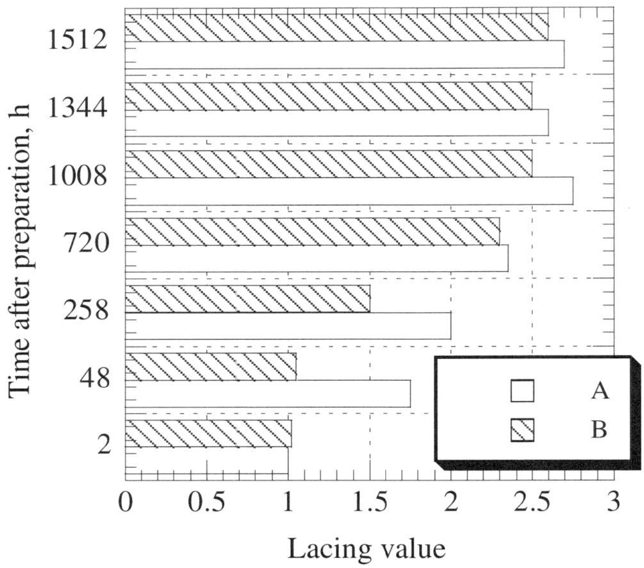


Figure 18.11. Lacing value vs. time of moisture absorption. [Adapted, by permission, from Hansen H, Polymers, Laminations & Coatings Conference, 1995, 653-8.]

of filler increases the Q/Ns ratio decreases which means that Q decreases.[47] The decrease becomes greater as the throughput increases.

Figure 18.11 characterizes the problem with moisture. The addition of titanium dioxide causes lacing in PE films. The lacing is a defect caused by the formation of series of small holes. Investigation has connected lacing to moisture in the titanium dioxide concentrate. Moisture absorption varies depending on the filler or pigment type and the time it is exposed to moisture in ambient air.

## 18.7 FOAMING[55-57]

The production of foamed products is very sensitive to changes in composition and the parameters of processing. The inclusion of fillers complicates the process and careful consideration must be given to the effects that filler incorporation has on material properties. Experimental work must be done to verify process conditions and material performance.

First, a homogeneous suspension of filler must be produced. This suspension should be sedimentation-free because it would influence foam properties. Rheological studies are used to select the appropriate dispersion agents which keep the viscosity close to the viscosity of the unfilled formulations. Complete filler wetting and network formation helps to make the stable suspensions.

From the point of view of chemistry of foaming, the reaction rate in the presence of fillers should be studied to determine if the filler slows or increases the reaction rate. Reaction rate is not only the factor determining the economy of the process. There must be a balance between gas production, viscosity increase, and phase transition. All three must be synchronized to yield the expected results.

Fillers in these systems affect two types of nucleation: nucleation of bubble formation and nucleation of crystallization. Nucleation of bubble formation affects the density of foam. Nucleation of crystallization affects the balance of gas formation and phase transition. The timing of both processes is critical.[55]

Fibers are frequently used in foam reinforcement. Due to the nature of the process, fibers are oriented as the foam rises (strain and movement) but restricted in movement by bubble wall formation (small distances between bubbles do not leave much freedom for fibers to position themselves).

Figure 13.5 shows the effect of a combination of silica and surface active agent on the formation of microporous uniform foam. This structure offers improved thermal insulation capabilities.

## 18.8 INJECTION MOLDING[58-88]

Numerous developments in injection molding machines have contributed to improvements in quality and output but have had only a limited impact on the processing of materials with fillers. But there are a few technological advancements involving fillers. Direct compounding injection molding technology is the most important. Unlike standard methods which use pre-compounded granulates, this method begins from raw materials which are metered from bulk storage facilities and compounded in a two-stage twin rotor continuous compounding mixer. The mixer performs two functions: it melts the matrix material and heats it to the required temperature while dispersing the additives. The compounded material is delivered to an accumulator injection unit which performs the injection functions. This method is a significant advance for the molding industry because it allows inventory to be reduced and a faster changeover. Shear controlled technology is another important development which has an impact on filler processing.[62] Figure 7.3 shows the arrangements of shear units which orient fibers in the required direction. Powder injection molding technology is also related to filler processing technology. Powdery inorganic materials (metals and ceramic powders) are mixed with a sacrificial polymeric binder which allows the required shape to be pre-formed. Once shaped, the binder is removed in a manner which does not distort the initial shape and it is then subjected to a sintering process, yielding a metal or a ceramic part. In normal injection molding good mixing and dispersion are the most basic requirements. But not in the enhanced marbleizing technique which exploits ineffective mixing elements which allow parts to be produced with a marble appearance effect with excellent repeatability. New developments are reported about process aids which allow highly filled materials to be processed with less difficulty through their influence on the flow properties of the material.

The flow of material into the mold cavity affects the way in which particles are distributed in the molded object. Analysis of parts indicates that particles flow by a

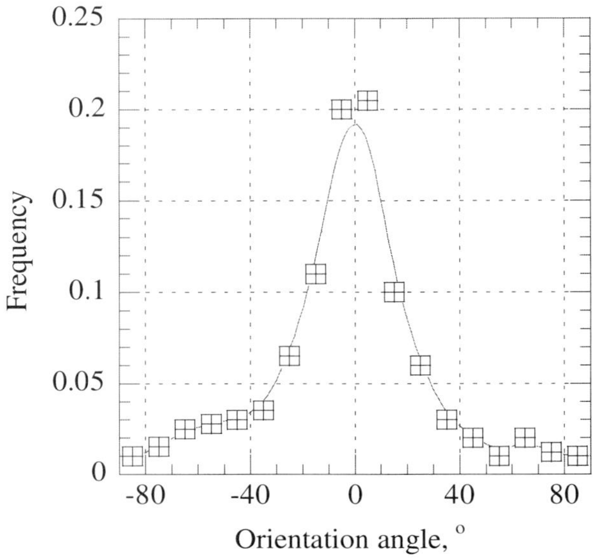


Figure 18.12. Frequency vs. orientation angle for ferromagnetic particles incorporated into a cylindrical part which was injection molded with a magnetic field applied parallel to the direction of flow. [Adapted, by permission, from Fiske T, Gokturk H S, Yazici R, Kalyon D M, *Polym. Eng. Sci.*, **37**, No.5, 1997, 826-37.]

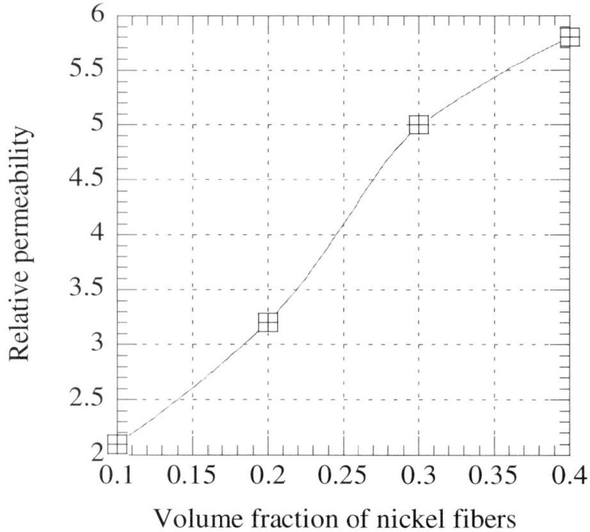


Figure 18.13. Relative permeability vs. concentration of ferromagnetic fibers in PE. [Data from Fiske T, Gokturk H S, Yazici R, Kalyon D M, *Polym. Eng. Sci.*, **37**, No.5, 1997, 826-37.]

fountain flow pattern with a high concentration of particles in the core.[86] During mold filling, the particles concentrate at the surface of the flowing material then are diverted towards the walls by back flow. Fountain flow was also detected in glass

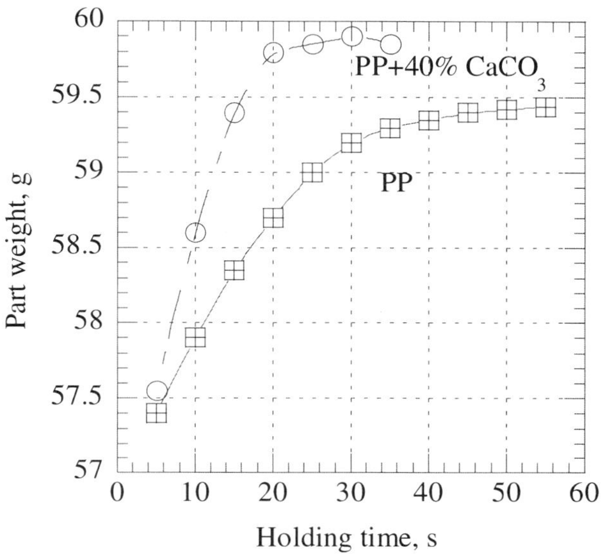


Figure 18.14. Part weight vs. holding time for neat PP and 40% calcium carbonate filled PP. [Adapted, by permission, from Mamat A, Trochu F, Sanschagrin B, *Polym. Engng. Sci.*, **35**, No.19, 1995, 1511-20.]

fiber reinforced liquid crystalline polymer.[79] The lager particles are mostly in the core whereas smaller particles spread outwards from the core.[81]

A magnetic field can be useful in conjunction with material flow to induce the orientation of ferromagnetic particles. Figure 18.12 shows particle orientation in injection molded cylinder with a magnetic field applied in a direction parallel to the flow.[58,70] Particles were well oriented increasing the magnetic permeability of the part. Processing parameters such as temperature and injection rate have a minor influence on particle orientation and permeability. The quantity of fibers had the greatest effect on relative permeability (Figure 18.13).

Figure 18.14 shows that holding time affects the tolerance of part weight. Filled material requires substantially less time to reach equilibrium.[78]

## 18.9 KNIFE COATING[5]

Knife coating is one popular technique for the coating of textile materials. Because the layers of the applied coating are usually very thin, the rheology of the coating is of primary importance. Rheological properties are controlled mostly by the choice of polymer.

In PVC coating formulation fillers play a role. Filler choice mostly depends on the way the filler affect viscosity. The filler should not absorb the plasticizers nor interfere with the pseudoplastic behavior of the paste which is determined by the resin properties and by the choice of plasticizers. Fillers must be completely dispersed, since the gaps between the coated substrate and the knife are very small. There must be no lumps. Fillers should not interfere with deaeration which is critical for all coated layers but most critical for the foamed layer. The presence of air bubbles in the paste affects the uniformity of foam structure.

Silica fillers play a prominent role in the paste rheology of polyurethane and rubber coating mixtures. Silica fillers are also used for surface matting. In polyurethanes, the layer thickness is even thinner than is used with PVC. Dispersion is enhanced by dispersion aids. Fillers are used at relatively low concentrations in PU formulation. But in rubber coatings, large quantities of calcium carbonate are used to decrease cost. Rheology and reinforcement are adjusted with carbon black.

## 18.10 MIXING[89-106]

The mixing process is usually thought of as a progressive process in which two or more components are first blended together to form an inhomogeneous mixture. With additional mixing, the distribution of the components becomes more uniform and finally, more energy is put into the mixture, particles are broken down into progressively smaller and smaller entities and the mixture becomes smooth and uniform. The final stage of the process, the dispersion stage, was discussed in Section 18.5. Here, we will concentrate on the traditional rubber mixing process which has not changed dramatically in 90 years. In spite of its long term use, a full and fundamental understanding of the process is lacking.

The mixing machinery, although now built of better materials and frequently fully computerized and computer designed, is essentially the same as it was at beginning of this century when several excellent designers established the mixing technology. Two types of rotors are generally used: tangential and intermeshing. The tangential rotor first disperses then distributes the materials whereas the intermeshing rotor does it in the opposite sequence. Other differences are related to processing and include fill factor (smaller with intermeshing rotors), speed of ram and rotors (slower with intermeshing rotors), oil incorporation (slower with tangential rotor), etc.

Several mixing processes have been developed throughout the years, such as conventional mixing (rubber added first, followed by powders, plasticizers, oils, curing agents and accelerators), upside down mixing (fillers and oils mixed together before addition of polymer), single stage mixing (convenient because of one step process) and two stage mixing (lowers temperature impact and gives better dispersion). There are also processes which use variable speed mixing and most of them use ram movement. Some more recent technologies such as plate mixers take advantage of applying pressure to material to shorten the mixing time but this technology is not really new since this approach was proposed in the 1920s.

Although different rubber composition requires different equipment and processing, several general principles apply. In many mixers, especially those of older construction, fillers and oils are added in increments to facilitate dispersion. Frequently, in two stage mixing, the first stage is upside down mixing to achieve good dispersion of fillers. This stage is followed, after cooling, by a conventional mixing cycle during which curing additives such as metal oxides are added. The uniform distribution of metal oxides is very important to guarantee uniformity of cure. Many methods are used to aid incorporation of metal oxides. One common approach is to withhold some of the liquid additives (plasticizers, oils) which may interfere with dispersion. Thermal and furnace blacks are more difficult to incorporate and are usually added first. The addition of coarser carbon black tends to help in the incorporation of finer carbon blacks. Fillers may help to disperse other additives such as plasticizers, tackifiers, or lubricants. Filler dosing during mixing can be very critical. If a mineral filler is overloaded it covers the surface of the mixing elements preventing the rubber from sticking to the mixing elements. The rubber then slips over them and no shear is applied. To overcome this, water is sprinkled over the mixture to agglomerate the filler and remove it from the surface of the mixing elements. There are many other methods used by industry to aid mixing. These have been derived from years of practical experience with various compounds, cases, and problems.

Two-roll mixing is another traditional method of mixing which relies on somewhat primitive equipment and on the skills of the operator. It has never been possible to establish standard procedure for this process and batch-to-batch results vary widely.

Problems associated with the general chemistry and physics of mixing are solved based more on practical than scientific principles. It is true that many elements of this process have been evaluated by powerful analytical techniques but knowledge of the kinetics of the process is still based on experience rather than science. The chemical and physical changes are highly complex. When heavy and powerful rotors begin to work on rubber, very high shears and temperatures are involved which contribute to the so-called mastication process. During this process, mechanical forces break bonds in the rubber resulting in free radical formation. This is a detrimental process for rubber because molecular weight decreases but also because reactive sites are formed which may react either with the rubber itself or with additives. The reactions and the numerous directions that these reactions can take are complex and random. The material properties, heat, and shear all influence the results. The process is difficult to control due to high viscosities and random local compositions which cause an inherent inhomogeneity in the system. Traditionally, heat is limited to the requirements of mastication (a state of rheological properties which allows incorporation of ingredients) but there have been suggestions that heat can be used to advantage.[90] A new method of compounding was proposed in which fillers are preheated to temperatures above 100°C. This supposedly increases the efficiency of the mixing process and shortens mixing by up to 30%. In yet another approach, mixers have been fabricated with cooled mixing elements which are designed to remove the heat of mixing and thereby reduce the rate of thermal degradation.

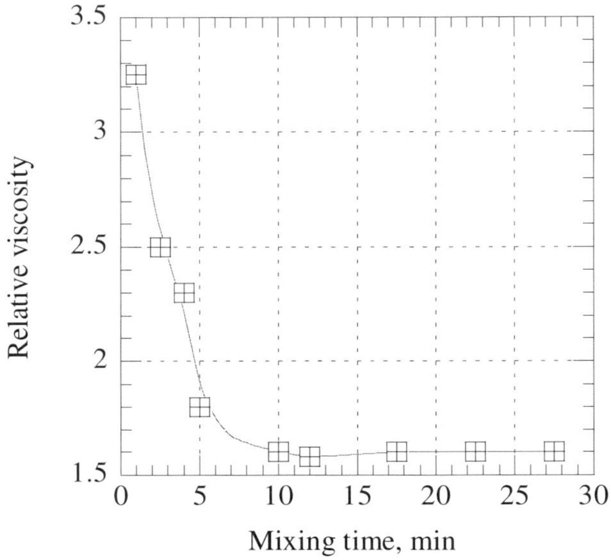


Figure 18.15. Relative viscosity of SBR rubber containing 30 phr carbon black vs. mixing time. [Adapted, by permission, from Clarke J, Freakley P K, *Rubb. Chem. Technol.*, **67**, No.4, 1994, 700-15.]

Several recent findings are reported below to illustrate new observations in the mixing technology. Figure 18.15 shows that the viscosity of SBR rubber containing 30 phr carbon black rapidly decreases at the beginning of the mixing process and then levels out.[91] Viscosity levels out when agglomerates disappear. The power consumption of the mixer, depended on when oil was added relative to carbon black.[93] The shortest mixing time was when oil was added 0.5 min after carbon black. The power consumption had very different characteristics when oil was added together with carbon black but the overall mixing time was as short as before. If oil was added 1.5 min after the carbon black, the mixing time to plateau was substantially increased (about 70%) and the characteristic of power consumption was totally different from both previous cases. For a given set of conditions of mixing only a certain level of dispersion is achieved which cannot be improved by prolonging the mixing time.[97] An improvement in dispersion can be achieved by the choice of mixer or mixing elements or by the selection of mixing condition, or by controlling the addition of components which influence viscosity during the dispersion phase.

Figure 18.16 shows that carbon black incorporation time, BIT, is affected by the addition of ZnO and by the properties of carbon black. If no ZnO is added, the incorporation time decreases as the carbon black structure increases. The opposite is true when ZnO is present during carbon black dispersion.[105] The overall quality of the dispersion improves when ZnO is present and when mixing is carried out at elevated temperature Dispersion quality was assessed by optical measurements.[104]

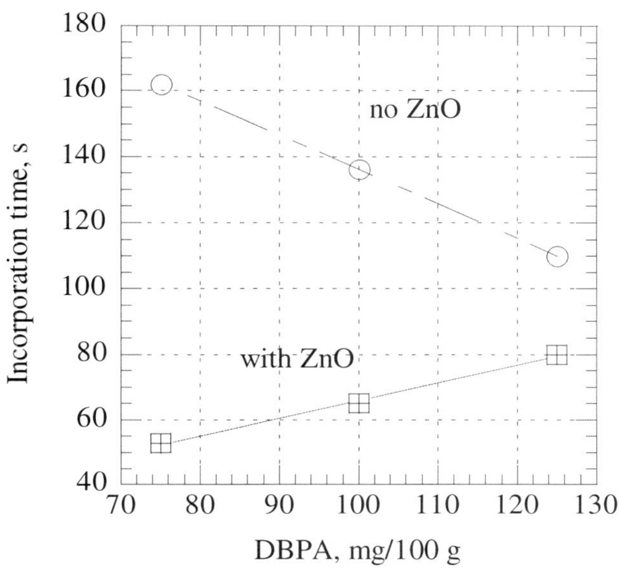


Figure 18.16. Carbon black incorporation time vs. DBPA adsorption. [Adapted, by permission, from Urabe N, Takatsugi H, Ito M, Toko H, Fukui M, *Int. Polym. Sci. Technol.*, **22**, No.5, 1995, T/68-72.]

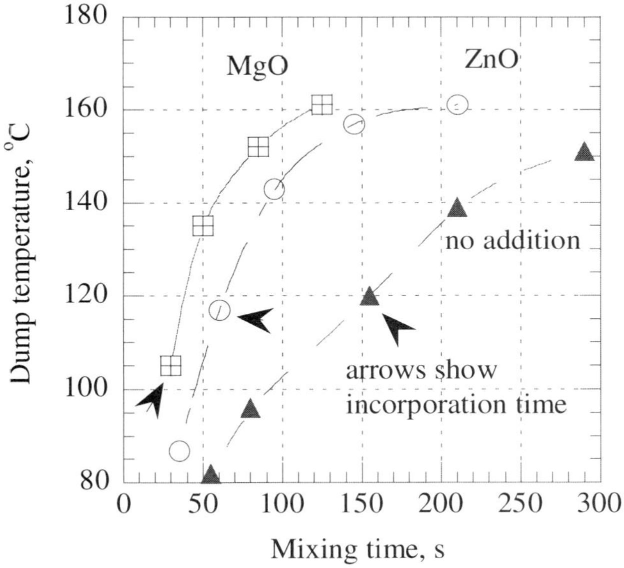


Figure 18.17. Dump temperature vs. mixing time of SBR rubber. [Adapted, by permission, from Urabe N, Takatsugi H, Ito M, Toko H, Fukui M, *Int. Polym. Sci. Technol.*, **22**, No.5, 1995, T/63-7.]

The addition of a crosslinker increases the viscosity of the mixture and the temperature of the batch increases (Figure 18.17). Magnesium oxide, used as a crosslinker, caused the temperature of the mix to rise four times faster than a mix

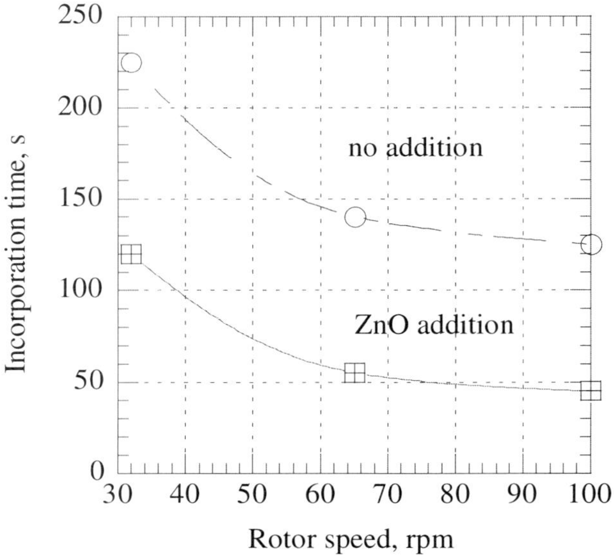


Figure 18.18. Carbon black incorporation time in SBR vs. rotor speed. [Adapted, by permission, from Urabe N, Takatsugi H, Ito M, Toko H, Nakada M, *Int. Polym. Sci. Technol.*, **23**, No.1, 1996, T/29-33.]

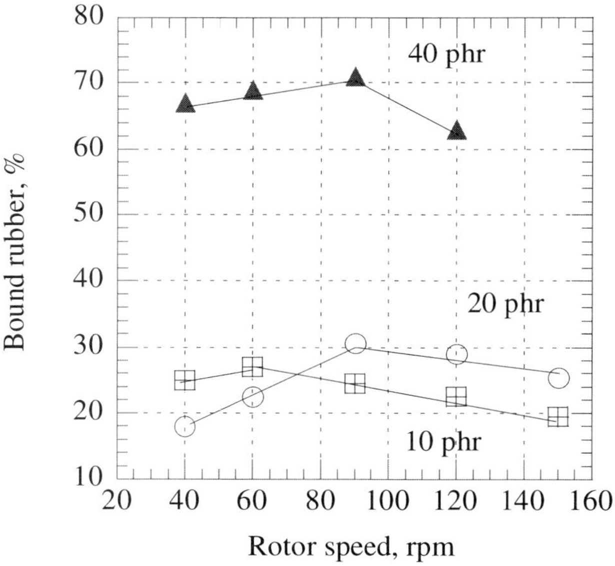


Figure 18.19. Bound rubber vs. mixer rotor speed for natural rubber filled with carbon black. [Adapted, by permission, from Mallick A, Tripathy D K, De S K, *J. Appl. Polym. Sci.*, **53**, No.11, 1994, 1477-90.]

without crosslinker.[104] Zinc oxide also produces a higher rate of temperature increase but its effect is less than that of magnesium oxide. In both cases the dump temperature at the end of the mixing process was much higher than in uncrosslinked rubber batch.

The rotor speed also influences carbon black incorporation time (Figure 18.18). The characteristics of both relationships are similar but longer times are required to disperse rubber with no ZnO added.[105]

The conditions of mixing and the filler concentration affect the bound rubber concentration (Figure 18.19).[89] For a given system, there is a certain critical rotor speed at which the bound rubber concentration reaches a maximum. A further increase in the rotor speed contributes to a decrease in the bound rubber concentration.

## 18.11 PULTRUSION[107-108]

Several types of reinforcements are used in pultrusion. These include: continuous fiber strand, roving, short fiber, and particulate fillers. Aligned fibers provide direct mechanical reinforcement. The functions of particulate fillers are several. This includes reduction of polymerization shrinkage, regulation of rheological properties, aiding in the reduction of fiber agglomeration, and fire retardancy. Many types of particulate fillers are used for these purposes, including most of the major groups of fillers. The industry custom-designs pultruded elements to the requirements of application and customer demands and these wide and varied requirements have lead to the great variety of fillers in traditional use. Because of the wide spread use of pultruded elements in construction, fire retarding properties are very important.

An analysis of the pultruded elements manufactured by the leading producers[107] revealed that the mass content of particulate filler was in the range of 38-50 wt%. This seemingly high range is needed to improve fire retardancy because alumina is often the filler of choice. This concentration of filler substantially reduces the mechanical strength of the element due to the fact that the mass fraction of resin in these elements is low and the resin is responsible for binding the reinforcing components. The performance of the resin depends on its penetration of the structural elements and on the resin's properties which are apparently affected by high concentrations of filler.[107] The particle sizes of filler are also very important. The use of large particle size filler increases fiber agglomeration during the molding process and reduces reinforcement.

Figure 18.20 shows the effect of continuous and long glass fiber concentration on the tensile properties of a pultruded element. The continuous fiber increases tensile more than does long glass fiber. Similarly, flexural modulus and impact strength increase more as the continuous fiber concentration is increased.[108]

## 18.12 REACTION INJECTION MOLDING[94,109-112]

In reaction injection molding, fillers and the reacting mass of matrix may each participate in the reactions, forming systems with a combination of features derived from the high adhesion between components. Numerous variations of this process

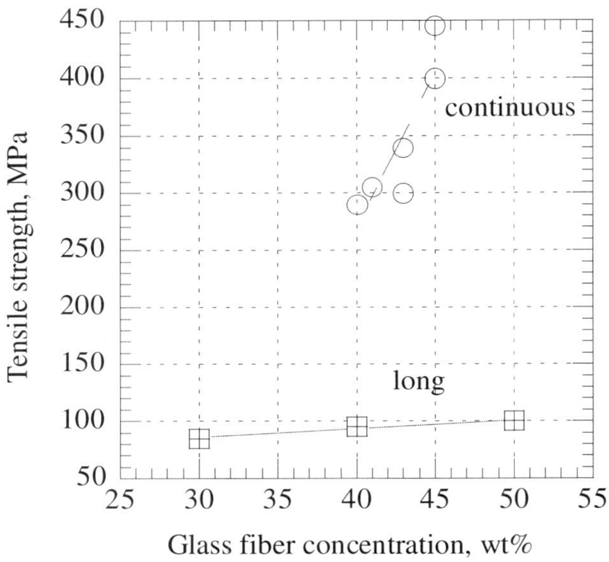


Figure 18.20. Tensile strength vs. glass fiber loading. [Adapted, by permission, from Montsinger L V, Antec '96. Volume II. Conference proceedings, Indianapolis, 5th-10th May 1996, 2546-9.]

have been developed. These include transfer molding, reaction injection molding, reactive extrusion, reinforced reaction injection molding, and structural reaction injection molding. Even pultrusion discussed in the previous section is a variation of reactive processing. All these processes have one common denominator – they use fillers to obtain the required properties. Fillers decrease of warpage, reinforce, reduce thermal expansion, improve dimensional stability at elevated temperatures, and change rheology to improve flow in the mold. Since these systems are reactive, the moisture must be very well controlled. Moisture may interfere with the main curing processes (e.g., polyurethanes). This requires either special grades of fillers or filler pre-drying before incorporation.

In these applications, reinforcement by fibers is complicated by the high probability of generating anisotropic properties because of fiber orientation during the flow. Depending on the geometry and the application of the product, the anisotropy might be advantageous but in most applications efforts are made to obtain better balanced mechanical properties. This is achieved through design of dies or by the use of a micro-mixing processes as presented in Figure 9.7. Increasing the filler loading decreases the effect of micro-mixing.

A combination of reinforcement and polymer grafting gave very good results. Wood fiber was processed with polypropylene grafted with maleic anhydride. Figure 18.21 shows the result.[109] Tensile strength (and also Young modulus) increase with an increased concentration of wood fiber. Similar good results were

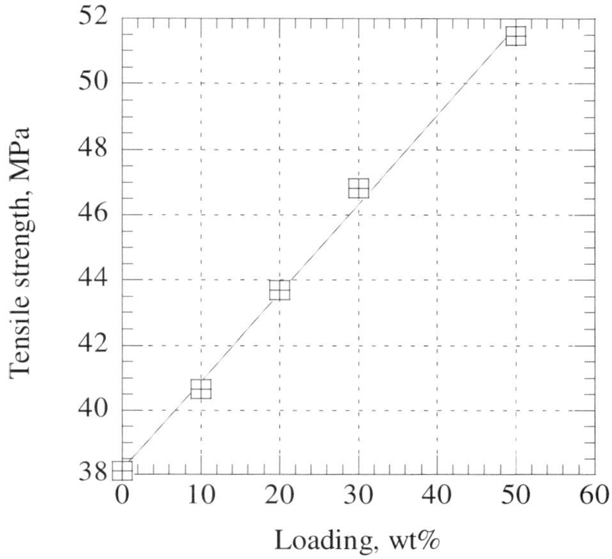


Figure 18.21. Tensile strength of wood filled maleated PP vs. wood fiber loading. [Adapted, by permission, from Collier J R, Lu M, Fahrurrozi M, Collier B J, *J. Appl. Polym. Sci.*, **61**, No.8, 1996, 1423-30.]

obtained with polyurethane reinforced with wood fibers. Good matrix/filler adhesion is important for the improvement of mechanical properties.

The use of filler may adversely affect the surface finish of some products. This is especially true, if high concentrations of filler are used (in some applications materials are filled up to 80 % of maximum packing density of filler). Work is being done to improve the properties of fillers in this application. Grades of wollastonite are available which allow parts with a high gloss to be produced.[111] Another development is in conductive RIM, ready for electrostatic painting without priming.[112]

## 18.13 ROTATIONAL MOLDING[113]

Fillers are not used to any extent in products made by the rotational molding process. Rotational molding is dominated by polyethylene (close to 90% volume) to which even small addition of pigments or fillers (less than 2 wt%) causes a decrease in tensile and impact properties of the products manufactured in this process. Polyethylene is vulnerable to environmental stress cracking which is made worse if fillers are present.

Resin and fillers of different density are separated by rotational forces. Even when a mixture of different particle sized resin is used, the finer particles manage to sift through the larger particles to the mold surface. If a material of higher density such as filler is added, this will, by itself, enhance the movement of heavier particles to the mold surface to cause cross-sectional inhomogeneities in the composition. This has been confirmed by studies involving incorporation of colorants. Two methods are used in coloring: extruded masterbatching and dry blending. If dry blending is used, the mechanical properties (especially impact strength) are reduced on addition of small amount of pigments.

However, carbon black has been used with success as a filler in rotationally molded products. At concentrations of up to 2.5 wt% it improves the weathering resistance of the product. Since dimensional stability and the shape of articles would benefit from the use of fillers and fibers, it is probable that they will start to be used in the future. When this happens, fillers will be introduced in a premixed form to assure homogeneous distribution (forces normal to the surface of product cannot cause the movement of particles towards the mold surface when particles are premixed with polymer).

## 18.14 SHEET MOLDING[114]

Sheet molding uses three groups of fillers for different purposes: metal oxides for curing, fibers for reinforcement, particulate fillers for variety of other purposes such as improvements in mechanical properties, chemical resistance and surface appearance. The technology is very well documented.[114]

The thickening process of the mixture used for sheet molding is one of the most important production steps. It is accomplished by the use of magnesium and calcium oxides or hydroxides. Magnesium oxide is the most frequently used. The process is a two-step reaction. In the first step, magnesium oxide reacts with two carboxyl groups forming a bridge. In the second stage, molecules are aligned by hydrogen bonding through the water produced in the first step, followed by formation of complexes between magnesium and carbonyl groups. These processes increase viscosity in the first step by increasing molecular weight and through the formation of a tri-dimensional network in the second step. These process steps must be well controlled. The rheological properties that they produce are important for product quality. The process of thickening takes about a day and then the viscosity remains stable for several weeks providing the process was correct and raw materials were of good quality. Wet filler and coarse particles in fillers cause problems.

Fillers are used in these products to improve mechanical properties or impart flammability resistance. Fillers are frequently silane-treated to further improve mechanical properties. Fillers must have a low moisture (below 0.1%), a low absorption of resin, and are expected to impart thixotropic properties. There are special cases. For example, if peroxyketals are used as initiators, basic fillers have to be used because acidic fillers interfere with cure times and the shelf-life of the composition. Shape and particle size distribution must be considered in filler selection to impart the desired rheological properties. Calcium carbonate is the most popular filler but aluminum trihydrate, anhydrous calcium sulfate, and silica are also frequently used. Barite is well suited to this application, especially if acid rain resistance is required but its more widespread use is limited because of its high specific gravity. Light weight is usually an essential requirement.

Glass fiber is used for reinforcement. The glass fiber must be properly surface sized to give the best mechanical performance and fire resistance. Good fiber wetting and low moisture content are also required.

Many defects in molded sheet can be related to fillers. Blistering may be caused by moisture but also by insufficient wetting of fibers and entrapped air pockets. Inconsistent cure and thickening is caused by the presence of moisture. Improper fiber orientation may initiate fracture cracks. Flow marks, streaks, warpage, and protruding fibers are also caused by improper fiber orientation.

## 18.15 THERMOFORMING[115-118]

Thermoforming is closely related to sheet molding. Compounding issues are similar to sheet formation. Fillers are added to materials for the same reasons. Again, these are reinforcement, increased dimensional stability, reaction to elevated temperature, fire proofing, and increased stiffness. The processing of a filled sheet is more complex than for unfilled resin because fillers reduce flexibility and elongation. Fiber reinforcement introduces constraints which limit the possible shapes of thermoformed material. Because of a reduced ability to change shape, only simple shapes can be manufactured from reinforced sheets without a substantial increase in the surface area. In order to process reinforced sheets to more complex shapes very large forces are required and results have not been satisfactory due to substantial surface defects.

Fibers and other non-spherical fillers change their orientation during thermoforming. Small strains are sufficient to orient fibers. Experiments have demonstrated that particles of talc orient themselves parallel to the surface of thermoformed parts.[116] The crystallites are oriented in a direction perpendicular to the either the talc or the mold surface. This is because the mechanism of crystallite growth begins on the surface of talc and grows outwards.

## 18.16 WELDING AND MACHINING[119-120]

Welding and machining of filled materials is known to be affected by the presence of fillers because filled materials are harder and their cutting is more difficult.[118] The many processes used for machining and jointing plastic parts, welding is the most affected by fillers in the formulation. Fillers interact with matrix which results in the higher dimensional stability and more resistance to flow in the molten state. It is not clear if the difficulties with welding should be attributed to the presence of fillers or the failure to develop the proper welding process.

Both past and current literature does not provide many answers but there are some preliminary hypotheses. Recent papers on the subject refer to older works.[119,120] These gave some information on welding results. In a hot-tool welding of polypropylene containing 40 wt% calcium carbonate, polypropylene containing

20 wt% short glass fibers, and poly(ether sulfone) containing 30 wt% short glass fibers, the strength of the weld relative to the material was 50-55%. Graphite fiber filled PVDF had a weld strength substantially lower than the material. But, in some instances the weld strength was higher than the tensile strength of unfilled materials. It is difficult to assess these findings because studies made on commercial materials cannot account for the influence of other components of the mixture many of which may also affect welding result. It is also difficult to determine if the conditions for welding of the filled materials were optimized.

One current paper[119] comments on the choice of welding parameters for hot plate welding. When the results of the two studies are compared, they show that the best welding parameters for polypropylene filled with glass fibers were determined by a complex relationship between hot plate temperature, heating time, and welding pressure. Analysis of these parameters with the aid of a model determined the optimal choice of welding conditions for the filled material. Perhaps, these parameters were not optimal for neat resin. Diffusion and geometrical model for analyzing welding results was inadequate for determining optimal processing conditions.

In vibrational welding of poly(butylene terephthalate) filled with mineral filler and glass fiber, the tensile strength of unfilled polymer was compared with the weld strength.[120] For glass fiber filled polymer at all concentrations of glass fiber (15-30 wt%) and for 10 wt% mineral filler, the weld strength was close (90-95%) to the strength of unfilled polymer. Material containing 30 wt% mineral filler had only half the weld strength of neat resin. Remedies must be found in future to prevent the loss of mechanical performance in welding process which offsets results of reinforcement. This may not matter if fillers were added for other purposes than strength. This study also shows that the addition of both mineral filler and glass fibers required adjustment in the conditions of welding. In particular, either the welding time or the pressure applied to the adjoining parts had to be increased to obtain the same result as for the neat resin. More studies are clearly required.

## REFERENCES


1 Wagner A H, Kalyon D M, Antec '96. Vol. I. Conference Proceedings, Indianapolis, 5th-10th May 1996, 953-7.
2 Ruiz F A, Antec '94. Conference Proceedings, San Francisco, Ca., 1st-5th May 1994, Vol. III, 2766-9.
3 Chang Ho Suh, White J L, *Polym. Engng. Sci.*, **36**, No.11, 1996, 1521-30.
4 Suh C H, White J L, Antec '96. Vol.I. Conference Proceedings, Indianapolis, 5th-10th May 1996, p.958-62.
5 Wypych J, **Polymer Modified Textile Materials**. *John Wiley & Sons*, New York, 1988.
6 Kenny J M, Opalicki M, *Composites Part A: Applied Science and Manufacturing*, **27A**, No.3, 1996, 229-40.
7 Kimura T, *Int. Polym. Sci. Technol.*, **22**, No. 4, 1995, T/62-9.
8 Soares B G, Gubbels F, Jerome R, Teyssie P, Vanlathem E, Deltour R, *Polym. Bull.*, **35**, No.1/2, 1995, 223-8.
9 Rockenbauer A, Korecz L, Pukanszky B, *Polym. Bull.*, **33**, No.5, 1994, 585-9.
10 Thompson C M, Allen J C, *Rubb. Chem. Technol.*, **67**, No.1, 1994, 107-18.
11 Alonso M, Gonzalez A, de Saja J A, *Plast. Rubb. Comp. Process. Appln.*, **24**, No.3, 1995, 131-7.

12 Mukha B I, Kolupaev B S, Mukha Y B, *Int. Polym. Sci. Technol.*, **23**, No.6, 1996, T/57-9.
13 Kim K J, White J L, *J. of Non-Newtonian Fluid Mechanics*, **66**, Nos.2/3, 1996, 257-70.
14 Wypych J, **Poly(vinyl chloride) Stabilization**. *Elsevier*, Amsterdam, 1986.
15 Scott D M, *Composites, Part A*, **28A**, 1997, 703-7.
16 Burke M, Young R J, Stanford J L, *Plast. Rubb. Comp. Process. Appln.*, **20**, No.3, 1993, 121-35.
17 Wessling B, *Synthetic Metals*, **57**, No.1, 1993, 3507-13.
18 Coran A Y, Ignatz-Hoover F, Smakula P C, *Rubb. Chem. Technol.*, **67**, No.2, 1994, 237-51.
19 Leblanc J L, *Prog. Rubb. Plast. Technol.*, **10**, No.2, 1994, 112-29.
20 Hegedus C R, Kamel I L, *J. Coatings Technol.*, **65**, No.822, July 1993, 37-43.
21 Liqing Sun, Aklonis J J, Salovey R, *Polym. Engng. Sci.*, **33**, No.20, 1993, 1308-19.
22 Molphy M, Mainwaring D E, Rizzardo E, Gunatillake P A, Laslett R L, *Polym. Int.*, **37**, No.1, 1995, 53-61.
23 Bomo F, Meeting of the Rubber Division, ACS, Montreal, May 5-8, 1996, paper E.
24 Le Bras M, Bourbigot, Le Tallec Y, Laureyns J., *Polym. Degradat. Stabil.*, **56**, 1997, 11-21.
25 Foster J K, Sims E S, Venable S W, *Paint & Ink Int.*, **8**, No.3, 1995, 18-21.
26 Yeh Wang, Jiang-Shen Huang, *J. Appl. Polym. Sci.*, **60**, No.11, 1996, 1779-91.
27 Hao Tang, Xinfang Chen, Aoqing Tang, Yunxia Luo, *J. Appl. Polym. Sci.*, **59**, No.3, 1996, 383-7.
28 Liauw C M, Lees G C, Hurst S J, Rothon R N, Dobson D C, *Plast. Rubb. Comp. Process. Appln.*, **24**, No.4, 1995, 211-9.
29 Bohin F, Feke D L, Manas-Zloczower I, *Rubb. Chem. Technol.*, **69**, No.1, 1996, 1-7.
30 Qi Li, Feke D L, Manas-Zloczower I, *Rubb. Chem. Technol.*, **68**, No.5, 1995, 836-41.
31 Karasek L, Sumita M, *J. Mat. Sci.*, **31**, No.2, 1996, 281-9.
32 Gabrielson L, Edrisinghe M J, *J. Mat. Sci. Lett.*, **15**, No.13, 1996, 1105-7.
33 Ismail H, Freakley P K, Sutherland I, Sheng E, *Eur. Polym. J.*, **31**, No.11, 1995, 1109-17.
34 Shiga S, Oka N, *Int. Polym. Sci. Technol.*, **22**, No.12, 1995, T/43-6.
35 Davis B A, Osswald T A, Antec '96. Vol. I. Conference Proceedings, Indianapolis, 5th-10th May 1996, 54-8.
36 Potente H, Flecke J, Antec '96. Vol. I. Conference Proceedings, Indianapolis, 5th-10th May 1996, 178-82.
37 Yu M C, Bissell M A, Whitehouse R S, Antec 95. Volume III. Conference proceedings, Boston, Ma., 7th-11th May 1995, 3246-50.
38 Lau E, Goodman J, *J. Elastomers Plast.*, **25**, No.4, 1993, 322-42.
39 Ruiz F A, Polymers, Laminations & Coatings Conference, 1995, 647-51.
40 Hansen H, Polymers, Laminations & Coatings Conference, 1995, 653-8.
41 Gendron R, Daigneault L E, Tatibouet J, Dumoulin M M, Antec '94. Conference Proceedings, San Francisco, Ca., 1st-5th May 1994, Vol. I, 167-71.
42 Cheng J, Bigio D I, Briber R M, Antec '97. Conference proceedings, Toronto, April 1997, 162-7.
43 Yazici R, Kalyon D M, Antec '97. Conference proceedings, Toronto, April 1997, 2076-80.
44 Turley R S, Strong A B, *J. Adv. Materials*, **25**, No.3, 1994, 53-9.
45 Nago S, Mizutani Y,Tokuyama Corp., *J. Appl. Polym. Sci.*, **61**, No.1, 1996, 31-5.
46 Gendron R, Daigneault L E, Tatibouet J, Dumoulin M M, *Adv. Polym. Technol.*, **15**, No.2, 1996, 111-25.
47 Ishibashi J, Kobayashi A, Yoshikawa T, Shinozaki K, Antec '96. Vol. I. Conference Proceedings, Indianapolis, 5th-10th May 1996, 386-90.
48 Cheng J, Bigio D I, Briber R M, Antec '96. Vol. I. Conference Proceedings, Indianapolis, 5th-10th May 1996, 365-9.
49 Potente H, Melisch U, Flecke J, Antec '96. Vol. I. Conference Proceedings, Indianapolis, 5th-10th May 1996, 334-7.
50 Huneault M A, Gendron R, Daigneault L E, Antec '96. Vol. I. Conference Proceedings, Indianapolis, 5th-10th May 1996, 329-33.
51 Joo Y L, Lee Y D, Kwack T H, Min T-I, Antec '96. Vol. I. Conference Proceedings, Indianapolis, 5th-10th May 1996, 64-8.
52 Ernst D, *Brit. Plast. Rubb.*, Jan.1997, 4-6.
53 Kretzschmar B, *Kunststoffe Plast Europe*, **86**, No.4, 1996, 20-2.
54 Gale M, Screws for Polymer Processing, Conference Proceedings, Rapra, Shawbury, 1995.
55 Alpern V, Shutov F, *Prog. Rubb. Plast. Technol.*, **11**, No.4, 1995, 268-83.
56 Okoroafor M O, Wang A, Bhattacharjee D, Cikut L, Haworth G J, Polyurethanes '95. Conference Proceedings, Chicago, Il., 26th-29th Sept.1995, 303-9.

57 Tabor R L, Polyurethanes '94. Conference proceedings, Boston, Ma., 9th-12th Oct.1994, 288-94.
58 Fiske T, Gokturk H S, Yazici R, Kalyon D M, *Polym. Eng. Sci.*, **37**, No.5, 1997, 826-37.
59 Averous L, Quantin J C, Crespy A, *Polym. Eng. Sci.*, **37**, No.2, 1997, 329-37.
60 Zhou J, Li G, Li B, He T, *J. Appl. Polym. Sci.*, **65**, 1997, 1857-64.
61 Arakawa K, Iwami E, Kimura K, Nomaguchi K, *J. Reinf. Plast. Comp.*, **13**, No.12, 1994, 1100-15.
62 Allan P S, Bevis M J, *Materials World*, **2**, No.1, 1994, 7-9.
63 Chiang W Y, Yang W D, Pukanszky B, *Polym. Engng. Sci.*, **34**, No.6, 1994, 485-92.
64 Turcovsky G, Antec '94. Conference Proceedings, San Francisco, Ca., 1st-5th May 1994, Vol. I, 796-800.
65 Heberlein D E, Antec '94. Conference Proceedings, San Francisco, Ca., 1st-5th May 1994, Vol. I, 791-5.
66 Canova L A, Fergusson L W, Parrinello L M, Subramanian R, Giles H F, Antec '97. Conference proceedings, Toronto, April 1997, 2112-6.
67 Gerard P, Raine J, Pabiot J, Antec '97. Conference proceedings, Toronto, April 1997, 526-31.
68 Dave P, Brown K, MacIver W, Chundury D, Draucker C, Lightener L, Antec '97. Conference proceedings, Toronto, April 1997, 514-20.
69 Lee S C, Yang D Y, Youn J R, Antec '97. Conference proceedings, Toronto, April 1997, 642-8.
70 Fiske T, Gokturk H S, Yazici R, Kalyon D M, Antec '97. Conference proceedings, Toronto, April 1997, 1482-6.
71 Pechulis M, Vautour D, Antec '97. Conference proceedings, Toronto, April 1997, 1860-4.
72 Portway J, Antec '97. Conference proceedings, Toronto, April 1997, 3050-3
73 Hamada H, Hiragushi M, Takahashi K, Machida K, Antec '97. Conference proceedings, Toronto, April 1997, 3326-30.
74 Okiyokota M, Hamada H, Hiragushi M, Hasegawa T, Antec '97. Conference proceedings, Toronto, April 1997, 3319-21.
75 Plummer C J G, Wu Y, Gola M M, Kausch H H, *Polym. Bull.*, **30**, No.5, 1993, 587-94.
76 Lanteri B, Burlet H, Poitou A, Campion I, *J. Mat. Sci.*, **31**, No.7, 1996, 1751-60.
77 Okamoto M, Shinoda Y, Okuyama T, Yamaguchi A, Sekura T, *J. Mat. Sci. Lett.*, **15**, No.13, 1996, 1178-9.
78 Mamat A, Trochu F, Sanschagrin B, *Polym. Engng. Sci.*, **35**, No.19, 1995, 1511-20.
79 Liu C, Manzione L T, *Polym. Engng. Sci.*, **36**, No.1, 1996, 10-4.
80 Hashemi S, Din K J, Low P, *Polym. Engng. Sci.*, **36**, No.13, 1996, 1807-20.
81 Ogadhoh S O, Papathanasiou T D, *Composites Part A: Applied Science and Manufacturing*, **27A**, No.1, 1996, 57-63.
82 Golubev A I, *Int. Polym. Sci. Technol.*, **22**, No.10, 1995, T/42-3.
83 Belanger B, Sanschagrin B, Fisa B, Antec '96. Volume II. Conference proceedings, Indianapolis, 5th-10th May 1996, 1762-7.
84 Dreibelbis G L, Antec 95. Volume III. Conference proceedings, Boston, Ma., 7th-11th May 1995, 4374-6.
85 Yu T C, Antec '95. Vol. II. Conference Proceedings, Boston, Ma., 7th-11th May 1995, 2358-68.
86 Papathanasiou T D, *Int. Polym. Processing*, **11**, No.3, Sept.1996, 275-83.
87 Luelsdorf P, *Kunststoffe Plast Europe*, **86**, No.6, 1996, 7-9.
88 Golubev A I, *Int. Polym. Sci. Technol.*, **22**, No.10, 1995, T/40-1.
89 Mallick A, Tripathy D K, De S K, *J. Appl. Polym. Sci.*, **53**, No.11, 1994, 1477-90.
90 Naitove M H, Shorr N, *Plast. Technol.*, **40**, No.13, 1994, 16.
91 Clarke J, Freakley P K, *Rubb. Chem. Technol.*, **67**, No.4, 1994, 700-15.
92 Li Y, Wang M J, Zhang T, Zhang F, Fu X, *Rubb. Chem. Technol.*, **67**, No.4, 1994, 693-9.
93 Yoshida T, *Int. Polym. Sci. Technol.*, **20**, No.6, 1993, T/29-39.
94 Agarwal S, Campbell G A, Antec 95. Volume I. Conference proceedings, Boston, Ma., 7th-11th May 1995, 839-42.
95 Yazici R, Kalyon D M, Antec '93. Conference Proceedings, New Orleans, La., 9th-13th May 1993, Vol. III, 2845-50.
96 Rosenov M W K, Bell J A E, Antec '97. Conference proceedings, Toronto, April 1997, 1492-8.
97 White L, *Eur. Rubb. J.*, **176**, No.11, 1994, 22-4.
98 Cochet P, Barruel P, Barriquand L, Grobert J, Bomal Y, Prat E, IRC '93/144th Meeting, Fall 1993. Conference Proceedings, Orlando, Fl., 26-29th Oct.1993, Paper 162.
99 Tan L S, McHugh, *J. Mater. Sci.*, **31**, 1996, 3701-6.
100 Tan L S, McHugh A J, Gulgun M A, Kriven W M, *J. Mat. Res.*, **11**, No.7, 1996, 1739-47.

101 Jarvela P A, Jarvela P K, *J. Mat. Sci.*, **31**, No.14, 1996, 3853-60.
102 Chan C M, *Polym. Engng. Sci.*, **36**, No.4, 1996, 495-500.
103 Urabe N, Takatsugi H, Ito M, Toko H, Fukui M, *Int. Polym. Sci. Technol.*, **22**, No.5, 1995, T/68-72.
104 Urabe N, Takatsugi H, Ito M, Toko H, Fukui M, *Int. Polym. Sci. Technol.*, **22**, No.5, 1995, T/63-7.
105 Urabe N, Takatsugi H, Ito M, Toko H, Nakada M, *Int. Polym. Sci. Technol.*, **23**, No.1, 1996, T/29-33.
106 Nakajima N, *Int. Polym. Processing*, **11**, No.1, 1996, 3-13.
107 Ye B S, Svenson A L, Bank L C, *Composites*, **26**, No.10, 1995, 725-31.
108 Montsinger L V, Antec '96. Volume II. Conference proceedings, Indianapolis, 5th-10th May 1996, 2546-9.
109 Collier J R, Lu M, Fahrurrozi M, Collier B J, *J. Appl. Polym. Sci.*, **61**, No.8, 1996, 1423-30.
110 Haagh G A A V, Peters G W M, Meijer H E H, *Polym. Engng. Sci.*, **36**, No.20, 1996, 2579-88.
111 Turner J D, Property Enhancement with Modifiers and Additives. Retec proceedings, New Brunswick, N.J., 18-19th Oct.1994, 65-87.
112 Porter J R, Groseth C K, Little D W, Polyurethanes '95. Conference Proceedings, Chicago, Il., 26-29th Sept.1995, 532-7.
113 Crawford R J, **Rotational Molding of Plastics**, 2nd Edition, *Research Studies Press Ltd.*, Tauton, England, 1996.
114 Kia H G, **Sheet Molding Compounds. Science and Technology**, *Hanser Verlag*, 1993.
115 Sanchez-Solis A, Padilla A, *Polym. Bull.*, **36**, No.6, 1996, 753-58.
116 Suh C H, White J L, *Polym. Engng. Sci.*, **36**, No.17, 1996 2188-97.
117 Florian J, **Practical Thermoforming. Principles and Applications**. 2nd Ed., *Marcel Dekker*, 1996.
118 Throne J L, **Technology of Thermoforming**. *Hanser*, Munich, 1996.
119 Nonhof C J, *Polym. Engng. Sci.*, **36**, No.9, 1996, 1184-95.
120 Stokes V K, Antec '93. Conference Proceedings, New Orleans, La., 9th-13th May 1993, Vol. II, 2067-74.

# 19

# FILLERS IN DIFFERENT PRODUCTS

In product development, fillers play various roles as outlined in the Chapter 1. In this chapter we will analyze

- Current changes in the market which create new opportunities
- Reasons for filler addition
- Potential improvement expected from filler addition
- Practical results of fillers use

## 19.1 ADHESIVES

Adhesives have very broad range of performance requirements. The performance spectrum ranges from pressure sensitive products where almost minimal adhesion is required, to extremely high performance adhesives with strength equivalent to that of metals. But the scope of the adhesive's performance goes well beyond adhesive strength. Flowability, force to adhere and mechanical, thermal, electrical, barrier, and optical properties as well as chemical and weather resistance and rheological behavior all must be considered in adhesive formulations. These essential parameters are discussed below from the point of view of the influence of fillers.

In the adhesive industry, the balance between tensile properties and adhesion is the most important part of the design process. Structural adhesives are expected to fail cohesively, a convincing argument to the user that the adhesive was designed for a particular substrate. Fillers frequently increase hardness and have reinforcing properties, so the choice of the filler and its concentration are often critical. In addition, adhesion may also be affected by the filler's presence either due to absorption of coupling agents, change in rheological properties (reducing mechanical adhesion), or changing moisture permeability which affects hydrolytic changes at the interphase.

Sepiolite was subjected to a thermal treatment which removes crystallization water at 500°C and constitution water at 850°C.[1] When these changes occur, they are accompanied by crystal folding which introduce substantial changes in the structure and interaction capabilities of sepiolite. Treated and untreated sepiolites were used in the preparation of a polyurethane adhesive to determine if the modification impacts the properties of the adhesive such as its interaction with polymer essential for its performance in adhesive. Figure 19.1 shows the effect of treatment

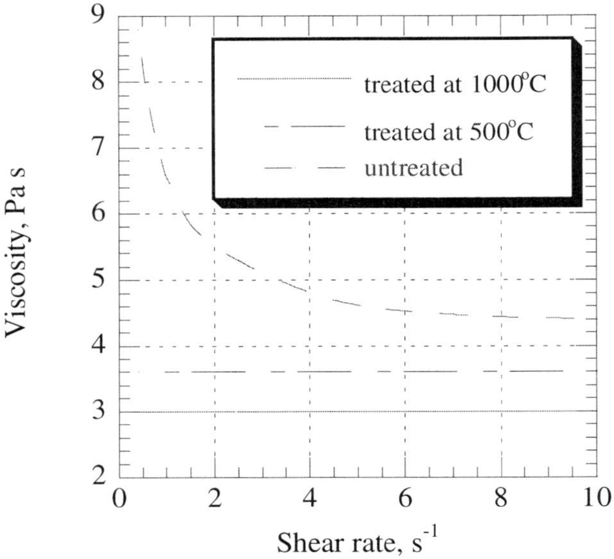


Figure 19.1. Viscosity of PU adhesive containing treated and untreated sepiolite at 10 wt%. [Data from Torro-Palau A, Fernandez-Garcia J C, Orgiles-Barcelo A C, Pastor-Blas M M, Martin-Martinez J M, *Int. J. Adhesion Adhesives*, **17**, 1997, 111-9.]

on viscosity. Treatment and its temperature had an impact on the rheological characteristics of adhesive. Sepiolites treated at 500 and 1000°C imparted Newtonian properties to the adhesive whereas non-treated sepiolite imparted non-Newtonian characteristics (pseudoplastic). Since adhesives are usually expected to have thixotropic behavior (making them easier to apply and eliminating sag after application), the treatment of sepiolite was not beneficial to the adhesive. The tensile strength of the adhesive was also reduced by the thermal treated sepiolite as was elongation. Figure 19.2 shows the effect of the treated sepiolite on green and post-cure peel strength.[1] Green strength is reduced because the treated filler does not have the ability to interact with polymer which otherwise would have formed a network which, in turn, would have contributed to viscoelastic properties of the uncured adhesive. The peel strength of the cured adhesive was improved over the unfilled polyurethane because the filler reduces tensile strength which makes the adhesive less rigid. The green strength of polyurethane adhesive containing fumed silica was increased with small amounts of filler (5-15 wt%) but decreased on a further increase of filler.[4] The peel strength of cured PU adhesive increased as silica concentration from 15 to 50 wt%.[7] These studies indicate that fillers are useful in the regulating and balancing the properties of adhesive to produce increased adhesion.

In a pressure sensitive adhesive, fillers may affect properties such as cohesion, cold flow, and peel adhesion. Most fillers increase cohesion and reduce cold flow. In some formulations, even a small addition of filler dramatically reduces peel ad-

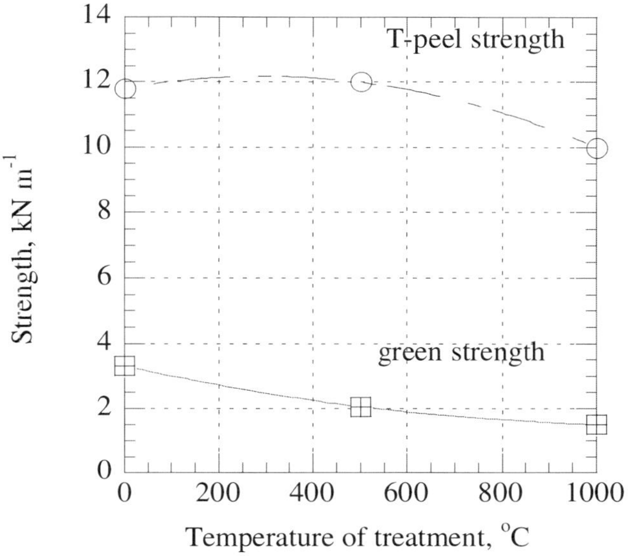


Figure 19.2. Peel and green strength of PU adhesive containing sepiolites treated at different temperatures in an adhesive for roughened rubber. [Data from Torro-Palau A, Fernandez-Garcia J C, Orgiles-Barcelo A C, Pastor-Blas M M, Martin-Martinez J M, *Int. J. Adhesion Adhesives*, **17**, 1997, 111-9.]

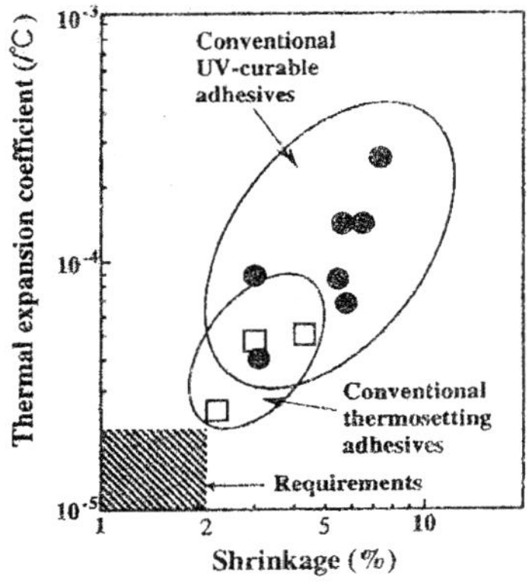


Figure 19.3. Thermal expansion coefficient of epoxy adhesive vs. shrinkage. [Adapted, by permission, from Murata N, Nishi S, Hosono S, *J. Adhesion*, **59**, Nos.1-4, 1996, 39-50.]

hesion either because of interaction with the tackifier or because filler particles at the surface reduce the area of contact between the adhesive and the substrate. Glass beads are used to regulate peel adhesion because of their ability to reduce surface contact.

Studies of UV curable adhesive show how an adhesive may be obtained which has a low thermal expansion coefficient and low shrinkage. Figure 19.3 compares adhesives with performance requirements. The product developed in this study has a shrinkage of 1.2% and a thermal expansion coefficient of less than $2\times10^{-5}$/°C. A spherical quartz filler, surface treated with a silane, was used in this adhesive. The high degree of transparency to UV enabled the quartz to impart these excellent properties.[8]

A model was developed to analyze the conductivity of materials filled with conductive particles.[9] This model was compared with experimental data for four commercial adhesives containing silver flakes. It was discovered that the resistivity is higher in the planar direction (thin films) than in three-dimensional space. Figure 15.5 shows the effect of carbon fibers obtained from different processes on the electrical conductivity of epoxy adhesive.[2] Vapor-grown carbon fibers gave the best increase in conductivity.

## 19.2 AGRICULTURE[10-12]

Greenhouse and mulch films are the most commonly used synthetic film products in agriculture. In both technologies, fillers play an important role. In greenhouse film, calcinated clay is added at up to 10 wt% to change infrared absorption characteristics.[10] The filled film has better insulation characteristics. It transmits less infrared during the day and emits less infrared during the night which makes the temperature in the greenhouse more uniform. The same physical principle is used for laser marking. Unfilled polyethylene film is not suitable for laser marking since light is readily transmitted. In the filled film, a sufficient amount of energy is retained by the film to cause charring.

Mulch film prevents evaporation of moisture from the soil, suppresses weed growth, and prevents soil erosion. In order to perform these functions, mulch film must have a high opacity, weather stability, and proper mechanical characteristics. Carbon black is used to increase UV stability and obtain high opacity. The performance of carbon black in this application depends on the grade, amount, and dispersion. The grade is important because it determines particle size and particle chemistry. Particle size affects the absorption/reflection ratio. Smaller particles tend to absorb more light and coarser particles tend to reflect more light. The surface chemistry of carbon black affects its ability to scavenge radicals which constitutes part of the mechanism by which carbon black works as a UV stabilizer. By choosing the appropriate grade of carbon black, stabilization can be increased by a factor of 4 to 5.[11] The proper choice of grade, concentration, and dispersion method will allow the elongation of the film to be increased by up to 10 times.[11]

The concentration of carbon black is important for both UV protection and opacity (required to prevent the growth of weeds). The amount of carbon black required for a certain level of opacity depends on the film thickness. The same volume of carbon black per unit area must be maintained in the film as the film thickness is reduced in order to retain the same degree of opacity. For a 30 μm thick film this concentration varies between 4 and 9 wt% depending on the type of carbon black used.

Dispersion is also an important factor. Light reflection depends on the surface area of carbon black after dispersion. If large agglomerates are present light reflection (and UV protection and opacity) is reduced. It is a frequent practice to use fully dispersed masterbatches offered either by carbon black manufacturers or compounders to achieve good results.[12]

## 19.3 AEROSPACE

Aerospace uses a large number of plastic components, most of which contain fillers for various reasons. It is difficult to follow the development in these components due to the proprietary nature of the business. Current legislative developments are likely to cause substantial changes in this sector. FAA Regulation FAR25 restricts materials to those which will satisfy fire protection requirements. In the past, regulations allowed many materials which contained organic and inorganic flame retardants for use in aerospace applications. With the new regulations, only certain polymers qualify for the application. These are Nomex and glass fiber laminates based on phenolics and polyketones. These changes reduce the number of flame retarding fillers which can now be used.

In another, relatively new, development, a high purity thermal carbon black was used as a component of an adhesive lining used in the construction of solid-fuel rocket motors and space shuttle motors.

## 19.4 APPLIANCES[13-15]

The use of plastic materials is predicted to grow by 3.5%/year which is higher than the growth predicted for appliance sales (1.8%/year).[13] The major growth with the appliance sector is in thermoplastics (5.4-5.9%/year) followed by styrenics. There is also large change in the type of materials that are being used (replacement of one plastic by another). The largest gains are expected in polyethylene (8.8%/year), polycarbonate (8.7%/year), and polypropylene (6%/year).[13]

Electrostatic applications and fire retarding formulations have yet to become the major focus in these applications. Only in the USA and Canada is a specific fire rating a requirement. In Europe, fire is still mostly prevented by the design of the electrical circuitry rather than the plastic design. Only in some applications, close to the source of possible fire ignition are some flame retarded plastics used.

The major impact on this industry comes from two regulations: elimination of CFCs from foam production and coming requirement in the USA to reduce power consumption by 33%. These two in combination require better insulation materials and constitute the largest opportunity in this sector.

Two studies were reported on the improvement of insulating properties of foams in one case through the use of xerogels[15] and in the other through better dispersion of filler combined with a better uniformity of foam structure.[14] The theoretical calculations supported by the experiment show that decreasing the size of voids increases the k-factor or decreases thermal conductivity.[14] In filled systems, the improvement comes from absorption of radiation and low thermal conductivity of the filler. Studies on filled foam concentrated on the reduction of the radiative losses.[14] Typically, 2 wt% carbon black is added to improve the thermal properties of foam but carbon black is difficult to disperse. A special dispersing agent was developed which resulted in a more uniform structure of foam (see Figure 13.5) which obtained a sufficient decrease in thermal conductivity to fulfill the requirements of the new regulation and decreasing the power consumption by 33%.[14]

## 19.5 AUTOMOTIVE MATERIALS[16-24]

Three major initiatives affect the automotive industry now and in its future. These are fuel conservation (lighter car), ease of assembly and finishing, and post-use recycling. In addition, as always, vehicle components must be cheap, esthetic, and durable. Fillers have obviously many roles to play in fulfilling the demands of this large and powerful industry. In the area of automotive safety, fillers can contribute most to fire retardancy and elimination of static charges. However, these are not yet to priority requirements for the industry. Fire retarding of car interior is a lower priority than the engine compartment where fires may be caused by low performance of wiring and the dimensional instability of plastics. The temperature in some areas of the engine compartment has increased in recent years by about 30°C to 150°C which makes thermal stability and dimensional stability of materials a much more demanding requirement. There are no specific regulations governing the flammability of car interior.

Static electricity also causes safety concerns. Regulations to condense fuel vapors rather than releasing them to atmosphere has created new components in which fires can start. In the US market, conductive plastics such as hoses and filters are in use to minimize static electrical discharges. Either carbon black or steel fibers are used in these products. It can be expected that new parts and materials will become available in the future. Other areas which require conductive plastics include the equipment installed to monitor and control brakes, engine, environment, suspension, etc. These are electronic devices, important in their functions to the safety of car operation, which are sensitive to static charges. All the areas surrounding these devices should be designed to prevent static charges from forming and accumulating.

The major use of fillers is in parts of body and in the car interior. Here, the major goals are as outlined above. The production of lighter but mechanically strong materials is, in most cases, the major requirement. There is one exception. This is the casing for the compact disc and tape players which are required to be heavy to perform their function. In the past this element was manufactured from metals. Introduction of plastic material required that a high density filler be used to bring the density of the part above 3 g/cm$^3$. But in most cases, weight is reduced by metal replacement and reinforcement of plastic material to decrease the thickness without losing mechanical performance. Figure 19.4 shows that gains in mechanical properties of automotive TPOs can be best achieved by optimization of the mixing time. Large gains in properties were obtained when mixing time was increased to improve distribution of polymer phases and dispersion of fillers.[18] In talc concentrations from 0 to 40 wt%, there was no obvious indication that an increased filler content requires longer mixing time. It was found instead that a certain minimum mixing time is required to optimize materials properties. Filler choice and the orientation of the filler particles are the other important determinants of mechanical performance.[19,21]

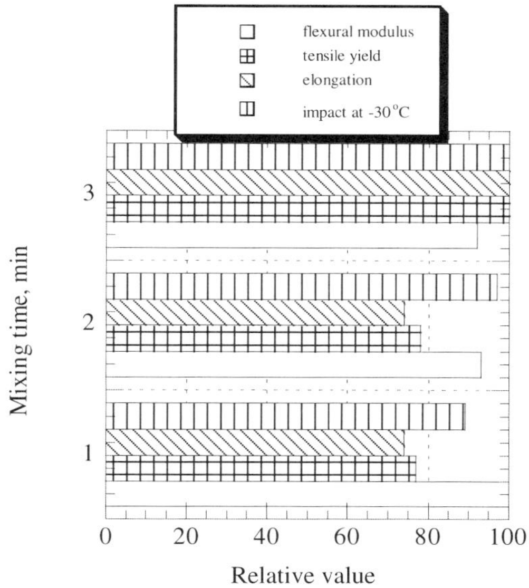


Figure 19.4. Mechanical properties of TPO vs. mixing time. [Data from Lau E, Goodman J, *J. Elastomers Plast.*, **25**, No.4, 1993, 322-42.]

Surface finish of parts is another important source of economic advantage. Surface roughness depends on the orientation of glass fiber. The process was optimized using data from surface smoothness measurements.[19] A conductive RIM was developed to produce material ready for electrostatic painting without the need for further surface preparation.

The choice of plastic material for automotive parts is influenced by its potential recyclability. One recent example[16] shows that the use of recycled material gave the added benefit of an approximately 10% reduction in weight. This involved the use of recycled material in the Chrysler Neon rear spoiler. Major considerations in this respect are thermal and UV stabilities of the reprocessable material combined with low weight and mechanical protection performance characteristics.

## 19.6 BOTTLES AND CONTAINERS[25-27]

Bottles and containers do not use large quantities of fillers because of the predominant need to see the content of these packages. But this is gradually changing because of several requirements, such as protection of the contents from the effect of light, esthetic reasons (color may give the required distinction to the package), the need to use recycled materials, and to enable laser marking.

Many food products lose vitamins and undergo other compositional changes on exposure to light. Fillers or pigments with sufficiently lower light transmission may be added in small amounts to help preserve the contents. A better understanding of filler technology together with the availability of more sophisticated processing equipment enables containers to be made with filler particles oriented to give the best light barrier and improved mechanical properties.[25]

The growing use of recycled materials, especially these which contain mixed plastics requires fillers that will compatibilize these mixed polymers so that mechanical properties can be retained or improved.[26-27]

For laser marking to be effective, the energy of the laser beam must be retained in the material to cause local charring. Properly selected fillers are sufficiently opaque to absorb radiation and aid the process of marking.[10]

## 19.7 BUILDING COMPONENTS[28-30]

Several sections of this chapter discuss building materials (hoses and pipes, pavement, roofing, sealants, siding, and waterproofing). Here, we focus on wall materials and insulation in various extruded and molded profiles. Numerous polymers are used for these two applications. They include polystyrene, phenolic resin, polyvinylchloride, and polyurethanes for insulation purposes and polyvinylchloride, polyurethanes, and polyesters for wall systems and structural elements. The major methods of production include molding, extrusion, and pultrusion.

The major roles of fillers in these applications is to provide reinforcement, fire retarding properties and to lower the cost. Development work on plastic materials is ongoing and the goal is to develop technology which is inexpensive but can perform under adverse conditions. These studies have intensified recently given the increased worldwide demand for new houses. China is an extreme example. It needs 100,000,000 houses to be built in the next five years.

There is a huge market for fillers in building applications. Studies seem to indicate that the primary focus will be on local materials because low cost remains an overriding requirement. Even such a simple material as sand can be used advantageously.[30] Polyvinylchloride has improved resistance to ultraviolet light when sand is used as a filler. The type of sand is important therefore local materials must be studied for the desired applications.

Wood products such as wood fiber and flour as well as other ground natural materials (corn combs, barks, etc.) have also been tested for these applications. They are locally available, inexpensive and composed of material similar to that which is being replaced. Studies show that these materials, rather than being considered scrap, can be useful as a component of building materials.

## 19.8 BUSINESS MACHINES

This sector is a substantial consumer of plastics and fillers. The major requirements include low weight, rigidity, impact strength, aesthetic appearance, dimensional stability, high heat distortion temperature, UV resistance, and flame retardancy. Again, fillers can contribute to the performance of these products. The major polymers used are ABS, POM, PPO, and PS. Lately, new advances in PVC blends have improved the heat distortion temperature and, they too, are finding applications in this equipment. With its inherent flame resistance and good aesthetic appearance, PVC has become an important player in this market.

Several groups of fillers are very important in this sector. They include reinforcing fillers, fillers which may improve heat distortion temperature, and flame retarding additives. Recently, major improvements have been achieved in blending technology and in the incorporation of reinforcing fillers. The low weight of laptop computers is one of the results of these developments. When computer cases were first produced, the thickness of the case wall was 6 mm. Now it barely reaches 2 mm.

There are many applications for fillers in which they are required to impart conductivity to protect sensitive electronic components. Also, the paper handling components of printers, copying machines, etc. require the capability of dissipating electrostatic charges which would otherwise cause paper jams.

## 19.9 CABLES AND WIRES[31-32]

Of the polymers used in the wire and cable industry, three polymers: polyvinylchloride, polyethylene, and chlorinated polyethylene make up to 95% of the usage. Two of three contain chlorine which gives them substantial fire retarding properties. Combination of this intrinsic property with small additions of antimony oxide have provided a system which has been in use for decades. Recently, and partly as a result of the pressures from groups concerned about the environmental impact of chlorine, the emission products from these materials are being evaluated. Especially, these cables which are used in enclosed spaces are under scrutiny and tests are being conducted to evaluate the harmful effect of hydrogen chloride on people and electrical equipment. Highly corrosive gases are evaluated based on their effect on control systems during a fire.

Manufacturers of various fillers continue studies on alternative systems. Most antimony oxide used as a fire retardant can be replaced by a combination of zinc borate without the loss of other properties (in some cases improvements are reported). Another option is to use the same filler systems which are used in polyethylene insulated cables and wires. These are based on magnesium hydroxide and aluminum hydroxide. These systems perform as flame retardants but require a high filler concentration which affects jacket resistance and mechanical performance. Recently, new coated grades have been developed which can be used at up to 65 wt% without the loss of properties or productivity (extrusion rates 2,500 m/min of cable are possible).[31]

Other systems include the mixture of huntite and hydromagnesite which offers relatively good performance. Also, new grades of clay have been developed to improve the resistance of jacketing where large additions of fire retarding fillers affect the electrical insulating properties.

Resolving these issues is important although the current performance is very consistent and satisfactory the environmental and safety issues must be addressed.

There are several polymeric alternatives such as EPDM, EVA, and fluoropolymers which are used for some applications. EVA can be processed with flame retardants free of chlorine. A relatively new product – polyketone – has potential in this application in the future.

## 19.10 COATED FABRICS[33]

Fillers have not played major role in the design of coated fabrics. They have been used to regulate rheology and improve the cost/performance ratio. This is no longer a growth area, so it is unlikely that new developments in high performance fillers will be applied to coated fabrics.

However, three areas: fire retardancy, dirt pick up, and surface tack do exist as outstanding issues in this field and solutions to them exist. Perhaps future work will address these problems.

## 19.11 COATINGS AND PAINTS[34-54]

Paints and coatings are based on traditional technology and there is ample experience in formulating products throughout the industry. Still, recent developments in new fillers and resins can and do contribute to an improved technology of production and better products. Fillers are added to coatings and paints for a variety of reasons, including cost reduction, optical properties (hiding power, color, texture), surface finish (matting additives), electric properties (conductive paints), permeability to water and gases, chemical resistance, UV resistance, rheology (prevention of sag and good application properties). Mechanical properties are seldom addressed by fillers which seems unusual when compared with other products discussed in this chapter.

Still, it is important that fillers interact with the polymer (binder) for various reasons. One is the rheological characteristic of paints. Figure 19.5 shows that many processes may affect how a filler behaves in the system.[35] The simple drying of aluminum hydroxide prior to use contributes to an increased paint viscosity. It should be noted that aluminum hydroxide loses water at 220°C, therefore drying at 80°C may only remove the water adsorbed on the surface of particles. But this is apparently sufficient to increase the interaction with the binder since, when the partially dried filler is added, viscosity almost doubles. Similarly, treatment with 1% triethoxymethacryloylpropylsilane, MPS, contributes to an increased viscosity. This data shows that the same filler can be readily modified to give a variety of different results.

Figure 19.6 shows that interaction involves not only a chemical interaction but can also be physical in nature. Here, different acid/base interactions of different grades of titanium dioxide are involved. The choice of the type of titanium dioxide results in a different thickness in the layer of adsorbed binder. This layer increases the sizes of particles and changes the amount of fillers contributing to the maximum packing density. The calculation of maximum packing density is complicated by

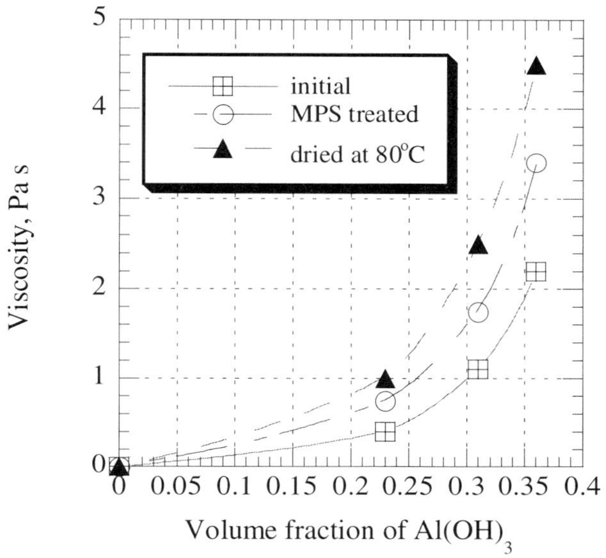


Figure 19.5. Viscosity of polyester coating vs. volume fraction of $Al(OH)_3$ subjected to various treatments. [Adapted, by permission, from Balard H, Papirer E, *Prog. Org. Coatings*, **22**, No.1-4, 1993, 1-17.]

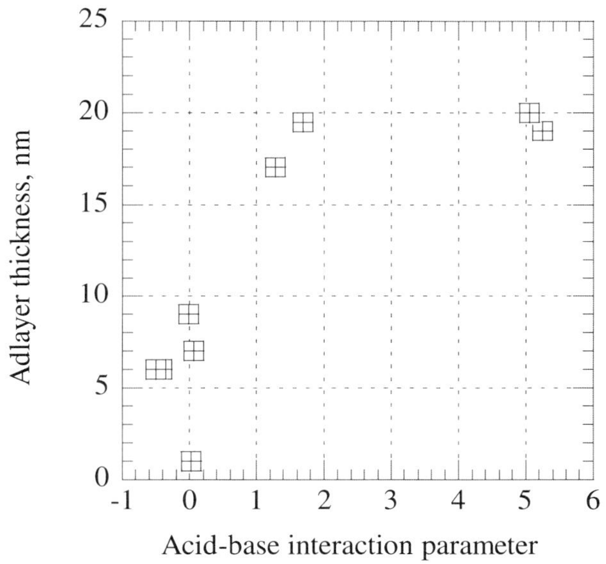


Figure 19.6. The thickness of a layer of adsorbed chlorinated polyethylene on a surface of titanium dioxide vs. the acid/base parameter of the titanium dioxide grade. [Adapted, by permission, from Hedgus C R, Kamel I L, J. Coatings Technol., 65, No.821, June 1993, 49-61.]

the fact that the thickness of the layer is not constant for a given filler but depends on the concentration of filler (Figure 19.7).[36]

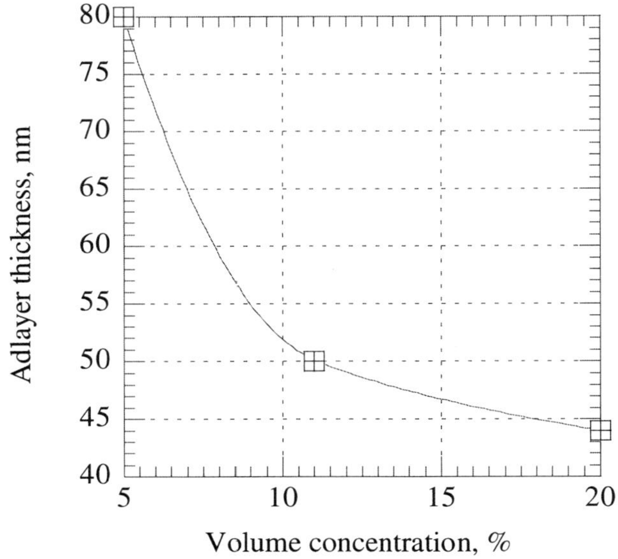


Figure 19.7. The thickness of a layer of adsorbed PVC on titanium dioxide vs. its concentration. [Data from Hedgus C R, Kamel I L, *J. Coatings Technol.*, **65**, No.821, June 1993, 49-61.]

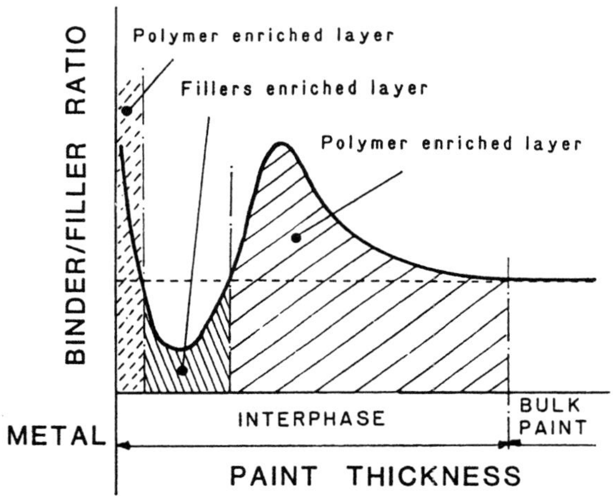


Figure 19.8. Paint/metal interphase model. [Adapted, by permission, from Roche A A, Dole P, Bouzziri M, *J. Adhesion Sci. Technol.*, **8**, No.6, 1994, 587-609.]

The amount of titanium dioxide and its type affects the permeability of the coating. With increased permeability, the corrosion protection of a steel substrate is decreased. The pigment level was increased from 6.4 to 28 vol% and produced a corresponding reduction in corrosion resistance.[37] In this study, the type and quantity of titanium dioxide did not have an effect on the adhesion of the coating to the substrate.[37] Figure 19.8 provides the reason for this lack of effect on adhesion. In this study, a set of coatings was analyzed for their adhesion to metal substrates.[41] Figure 19.8 complements the information given in Figure 7.16 which shows a distribution of filler particles across a cross-section of paint. The layer which is responsible for adhesion is depleted of filler.

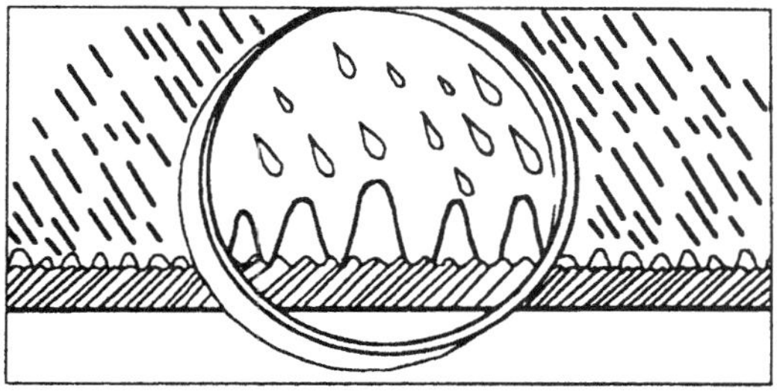

Figure 19.9. Surface spraying with white-hot stainless steel fibers. [Adapted, by permission, from Knowles J, *Polym. Paint Col. J.*, **185**, No.4366, 1995, 26-7.]

Figures 19.9 and 19.10 show two unusual applications of fillers in coatings. Figure 19.9 shows a schematic diagram of the surface of metal (e.g., a pan) on which white-hot particles of stainless steel fibers are sprayed to prepare the surface for a nonstick coating. In the past, these coatings were known to have a limited durability. Durability could be increased by a change to the surface through welding the fibers to the surface. These fibers increase the mechanical adhesion and the durability of coating.[45] The goal of the study illustrated in Figure 19.10 was to determine if ground glass from bottles can be recycled as a paint filler.[50] The paint obtained had a refractive index close to that of both the binder and the ground glass. As long as ground glass was added in quantities below the critical pigment volume, the paint was fully transparent. Above this concentration the paint became opaque.[50]

A primary objective in paints (and paper) formulation is to obtain opacity in the most economical way. In paints, this leads to the search for fillers which may replace some titanium dioxide while still retaining the required degree of opacity.

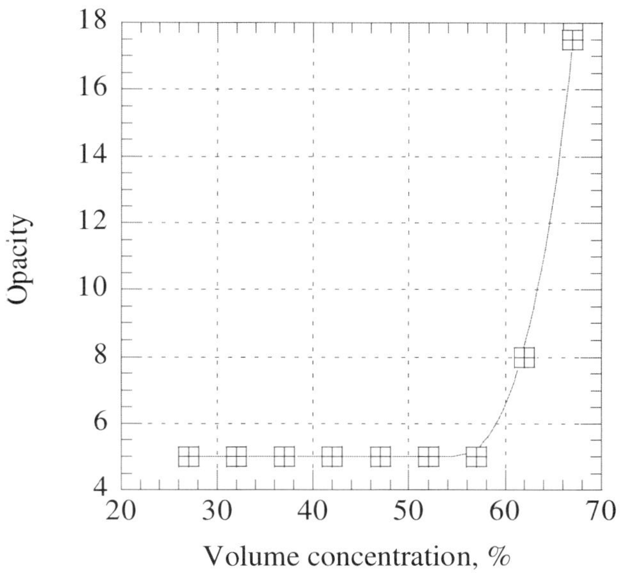


Figure 19.10. Opacity vs. glass volume percent. [Adapted, by permission, from Athey R D, Kirkland T, Lindblom G, Swoboda J, *Eur. Coatings J.*, No.11, 1995, 793-8.]

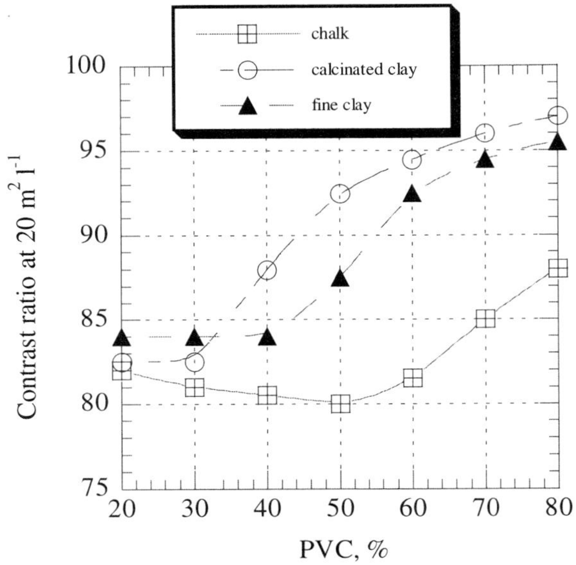


Figure 19.11. Contrast ratio vs. percentage PVC. *Courtesy of ECC International, St Austell, UK* in McGuffog R M, **Clays as Extenders in Decorative Paints**. *ECC International.*

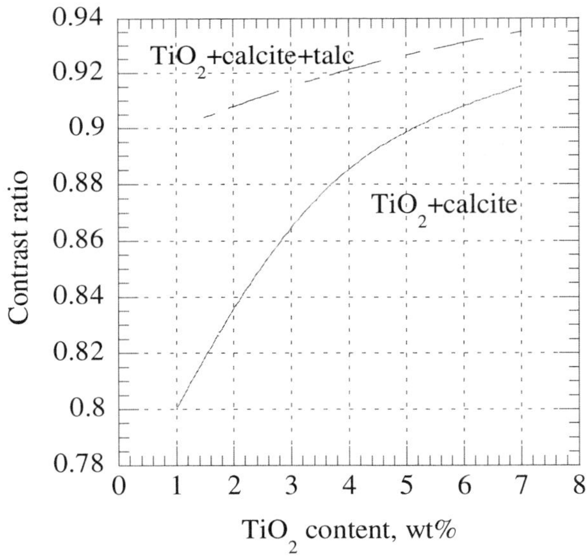


Figure 19.12. Contrast ratio vs. titanium dioxide concentration. 10 wt% calcite (out of 40 wt%) was replaced by talc. *Courtesy of Luzenac, Toulouse, France.*

Figure 19.11 shows one of the candidate products for such replacement.[53] Calcinated clay is used as an extender in paints. Chalk has very limited effect on opacity.

Figure 19.12 shows that the replacement of calcite by the same amount of talc improves opacity and allows a reduction in amount of titanium dioxide.[54]

These works show that there is an interest in finding partial replacements for titanium dioxide which is the major material cost component of paints.

## 19.12 COSMETICS AND PHARMACEUTICAL PRODUCTS[55-56]

Fillers are used by the pharmaceutical industry for three main functions: as colorants, disintegrants, and glidants. Each application demands special properties, as discussed below. Pharmaceutical grade fillers differ from those used by other industries in that they must comply with a high purity standard. The purity of material for pharmaceutical use is not only defined in terms of chemical composition but microbiological contamination is strictly limited.

The pharmaceutical industry uses organic dyes and lakes and inorganic pigments as common colorants. Red, yellow, and black iron oxides, titanium dioxide, calcium carbonate and talc are typical examples of inorganic pigments used in tablet production in order to provide the user with distinctive colors for different products. Inorganic pigments used by the pharmaceutical industry are analyzed for particulate properties (particle size, specific surface area, etc.), refractive index, stability of with respect to heat, UV degradation, and effect of pH. Trace elements are analyzed. These include arsenic, lead, antimony, cadmium, chromium, mercury, copper, zinc, barium, and iron.

The term disintegrant is applied to a substance added to a tablet formulation for the purpose of causing the tablet to break apart in an aqueous environment. Starch is the most commonly used disintegrant but several other materials are also used including inorganic fillers, namely, kaolin and bentonite. These fillers are usually added in concentrations in a range from 5 to 15%. The filler should work in conjunction with the tablet binder and withstand physical forces of compression. A typical disintegrant should, on contact with water, swell, hydrate, change volume or position, or react chemically to produce disruptive changes in the tablet. Kaolin and bentonite swell in contact with water. Their major disadvantage is their off-white color.

Glidants are substances added to cohesive powders in order to improve their flow properties by reducing interparticle friction. The effect produced by glidants depends on their chemical composition, which should be able to form permanent or temporary bonds with cohesive powder. Properties of glidants depend on physical factors such as grain size and shape, moisture content, hygroscopicity, etc. Talc and silica type fillers are typical examples of glidants. These materials, when added to the powdery composition, promote free-flow in the hopper and complete filling of tablet molds.

Theories developed in the study of metal powder compression have been adapted to the compacting of pharmaceutical powders.[55] A mathematical model is used to explain the reasons for defect formation during powder compression because of sticking and capping.[56]

In the cosmetics industry, finely dispersed fillers, are used as abrasives (toothpaste, scrub cosmetics), for their light reflecting properties (sunscreen lotions), for their dehydrating and astringent effect (kaolin in face masks), for their cooling effect (zinc oxide in sunburn lotion), and as cosmetic color additives and extenders (makeup). The many different applications require an extensive range of filler properties.

Within this sector, dentifrice is the most important market for fillers. Traditionally, dentifrice abrasives included dicalcium phosphate dihydrate, calcium carbonate, and insoluble sodium metaphosphate.[21,22] But now, aluminum trihydrate and hydrated silica are the most important fillers in toothpastes. Aluminum trihydrate is used extensively in the European market, whereas hydrated silica dominates the American market. Fillers for toothpaste production are required to have a carefully controlled grain size distribution. It is this property which controls the abrasiveness and the rheology of the toothpaste. Oil absorption depends on grain size distribution and consequently the rheology of the paste is related to the oil absorption of the filler. The abrasiveness of silica is also related to its oil absorption. Since oil absorption increases as grain size distribution decreases, it is not surprising that abrasive RDA (radioactive dentin abrasion) decreases when grain size decreases, with oil absorption increasing. Commercially available pharmaceutical grades of aluminum hydroxide are compatible with humectants, flavoring compounds, detergents, and enzymes. Aluminum hydroxide is also used by pharmaceutical industry (as well as plastics industry) as an antiacid.

Scrub cosmetics use abrasives to remove old horny cells, to massage, to smooth the skin surface, and to remove dirt from skin pores. Natural scrubbing agents are obtained from plant shells, seeds, and oils, and from animal shells and fats. Several inorganic materials are also in use, such as aluminum oxide, silica, kaolin, talc, calcium carbonate, and zirconium dioxide. An inorganic scrub agent should be carefully analyzed for grain size distribution, grain shape and the presence of crystalline forms. Some materials used in scrub agents have a high hardness, and, if they are present in the form of abrasive particles, may cause severe skin damage. Iron oxides, titanium dioxide, and mica are the most frequently used color additives for cosmetics. Zinc oxide and titanium dioxide, in addition to their use as colorants, play the role of UV absorbers, protecting skin from radiation. Talc, kaolin, iron oxides, titanium dioxide, and fumed silica are popular colorants, extenders, and rheology modifiers. Talc is commonly used in formulations where softness and slip are required. A caution: if talc penetrates a wound it may cause talcum granulomae; therefore, in products which may come into contact with wounds, talc should be replaced by aluminum hydrosilicate.

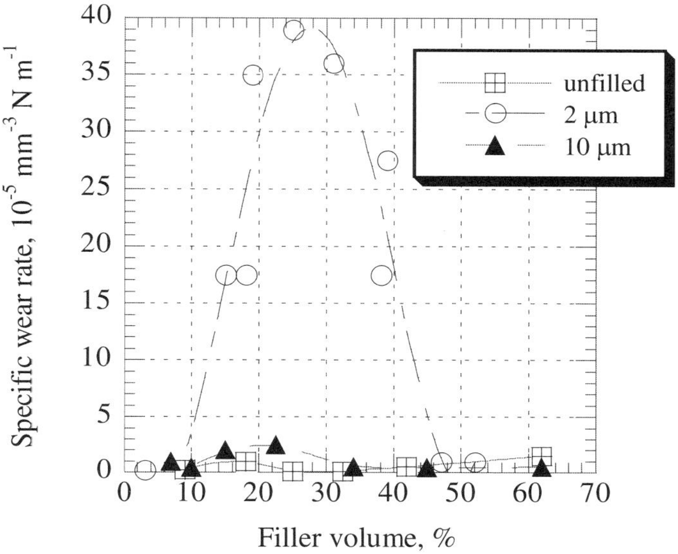


Figure 19.13. Specific wear rate of dental composites vs. filler volume. [Adapted, by permission, from Friedrich K, *J. Mat. Sci. Mat. In Med.*, **4**, No.3, 1993, 266-72.]

## 19.13 DENTAL RESTORATIVE COMPOSITES[57-61]

Adhesion, filler/matrix adhesion, dimensional stability, reinforcement, and wear resistance are the most important concerns in the development of dental composites.[57-61] These requirements are shared with composites used for many other purposes. So much as the methods of testing, mathematical models, methods of interpretation, and remedies developed in other applications may be applied to dental composites.

Figure 19.13 shows that wear rate depends on the concentration of filler and its particle size.[58] During the last decade, the particle size of fillers dropped from 8-30 µm to 0.7-3.6 µm in the present restorative composites.[60] This has increased surface smoothness and decreased plaque retention in unpolished surfaces. The wear rate of the composites increases when fillers are added and small particle size fillers cause more rapid increase of wear rate at a certain range of concentrations. Outside this range, composites have wear rate similar to the unfilled matrix.

The most frequently used fillers are glass powder, lithium aminosilicate, and glass-ceramic. Figure 19.14 shows that properties of dental composites are enhanced by the use of silanes. Treatment with silane also improves water resistance.[61]

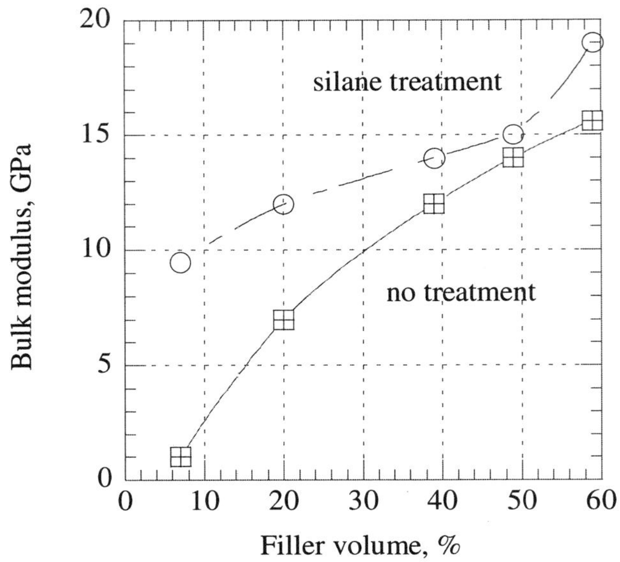


Figure 19.14. Bulk modulus of dental composite vs. filler concentration. [Adapted, by permission, from Jones D W, Rizkalla A S, *J. Biomedical Materials Research (Applied Biomaterials)*, **33**, No.2, 1996, 89-100.]

## 19.14 ELECTRICAL AND ELECTRONIC MATERIALS[62-70]

An extremely large number of products contain components which must meet stringent requirements. Such diversity prevents us from going beyond a general discussion. In the US and Canada products need the Underwriters Laboratories approvals and, in most cases, a V-0 rating is required. When brominated fire retardants were banned in Germany, some polymers such as polyester were affected. This generated a search for alternative fire retardants and different polymers (e.g., polyamide) for fire retardant applications. The opportunity for fire retarding fillers will continue to expand since in other countries (US, Japan, European countries) brominated flame retardants may also be restricted.

Various industries make efforts to introduce static control to work place, products and packaging. As many as 10% of the failures of electronic equipment are related to static electricity. To control static electricity in the work place, many products should be conductive (coatings, mats, bench tops, etc.). Packaging has been developed using conductive fillers. This creates new opportunities for manufacturers of products and fillers.

Electrically insulating and thermally conductive qualities are important in computer chips fabrication. One approach taken is based on boron nitride fillers which offers these two properties. There is also a need to develop materials which are thermally conductive but electrically insulating in high humidity conditions. Polyurethane composites filled with aluminum oxide or carbon fiber can be used for this application. Figure 19.15 shows the effect of the amount of filler on thermal

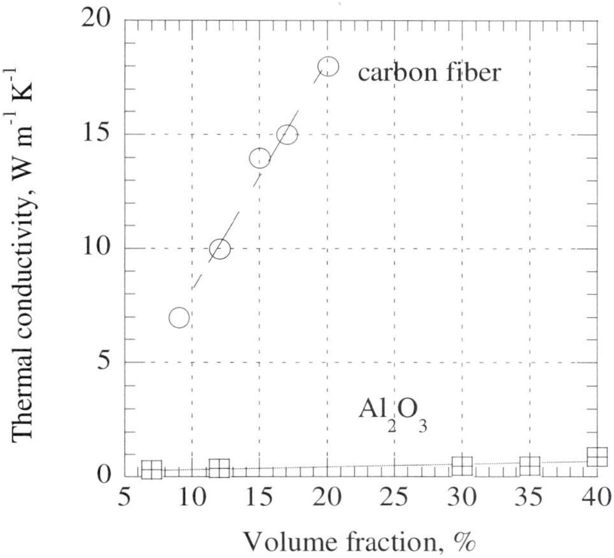


Figure 19.15. Thermal conductivity of PU vs. filler concentration. [Data from Lu X, Xu G, *J. Appl. Polym. Sci.*, **65**, 1997, 2733-8.]

conductivity. A carbon fiber filled material was found to be more resistant to changing humidity.[63]

Z-axis adhesives are a unique class of new products. These adhesives contain conductive particles, which due to their orientation, conduct electricity across their thickness but are non-conductive along their length and width. Several circuit lines can be connected through the same strip of Z-axis adhesive with no current flow between circuit lines.

## 19.15 ELECTROMAGNETIC INTERFERENCE SHIELDING[71-75]

EMI shielding prevents distortion of television, radio, aircraft control signals. Electromagnetic waves have an undesirable effect on people, animals and plants. Because of these interferences, sales of goods which may emit EMI is prohibited. EMI can be controlled by the use of plastic filled with conductive particles, conductive paints, metallization, use of inherently conductive polymers, conductive films, fabrics, and metal shields. Many solutions can thus be offered by fillers. There is a great number of different fillers used in these applications. These include carbon black, carbon fibers, graphite, metal powders, flakes, and fibers, and particles coated with metals. Depending on the type of filler used for EMI shielding, concentrations from 3 to over 40 wt% are required to obtain effective EMI shielding. EMI shielding efficiency is not only dependent on the type and conductivity of material used but it also depends on the particle size shape, its surface finish, its compatibil-

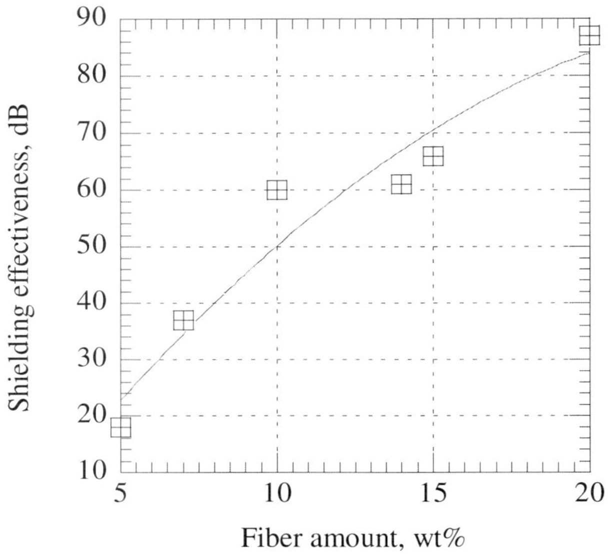


Figure 19.16. Shielding effectiveness vs. concentration of nickel fiber in PC. [Data from Rosenow M W K, Bell J A E, Antec '97. Conference proceedings, Toronto, April 1997, 1492-8.]

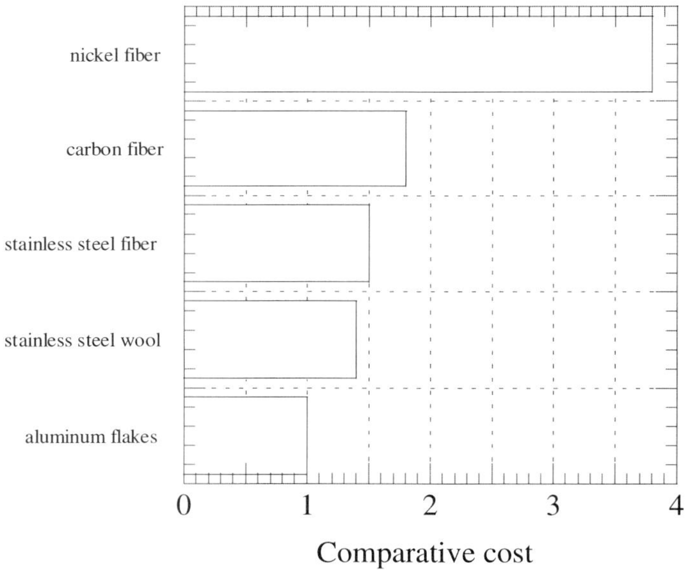


Figure 19.17. Comparative cost of metal filled PES composites. *Courtesy of Transmet Corporation, Columbus, USA.*

ity with matrix, its distribution in matrix, and the incorporation technology used. Shielding effectiveness should be 50 dB.

Figure 19.16 shows how nickel fiber changes EMI shielding effectiveness in polycarbonate. Less than 10 wt% fiber is needed to reach the target value.[71] Figure 15.2 shows that shielding effectiveness depends on the method of incorporation and that the length of fibers can be changed by the process of incorporation.[73]

Figure 19.17 shows comparative cost data which indicate that aluminum flakes and stainless steel fibers are the most cost effective.[76] Carbon black and graphite can, in principle be used in this application, but require a large loading and offer only limited protection. If metal fillers are used, consideration must be given to their oxidation potential and chemical resistance. Surface chemical changes may drastically reduce their performance. Metallization is also an efficient method of producing a conductive material. This was discussed in a recent monograph.[77]

## 19.16 FIBERS[78-79]

Microporous fiber have been produced using the following method.[78] Polypropylene is filled with 25 wt% calcium carbonate. After extrusion, the fiber is stretched at 150°C and calcium carbonate is dissolved by an acid treatment. Pore sizes from 0.01 to 0.02 μm are obtained depending on the type of acid treatment and the stretching ratio. Crystallinity decreases as the stretching ratio decreases and the fiber diameter increases after acid treatment because of shrinkage.

Choosing carbon black as a filler for fibers has many implications. Figure 19.18 shows the effect of carbon black loading on viscosity in PET.[79] Viscosity depends on the type of carbon black. A reduction of 50% viscosity can be attained at the same carbon black concentration simply by change to another grade of carbon black. Moisture absorption, which affects the drying time, can be substantially reduced (by about 50%) by the selection of the appropriate carbon black.[79] Fiber color and tone are affected by the carbon black type and by the method of its dispersion.

## 19.17 FILM[80-85]

Fillers provide films with conductive properties, influence their surface properties, affect their permeability, mechanical and optical properties, and affect their durability against environmental exposure. Various technologies are used to produce conductive films. These include lamination to metal foils (in-plant, using pressure sensitive adhesives), surface coating, and addition of conductive materials. Conductive films are widely used in packaging to limit static electricity.

China clays and calcium carbonates can be used to impart anti-blocking properties to films produced from polyester, and cellulose acetate. Figure 19.19 shows the effect of coated ground calcium carbonate. Even such a small addition as 10 wt% has a substantial effect on blocking. Many other properties such as impact strength, modulus of elasticity, and opacity are improved.[83]

The permeability of HDPE film containing variable quantities of talc is given in Figure 19.20. Talc in smaller quantities reduces oxygen permeability.[79] Talc has

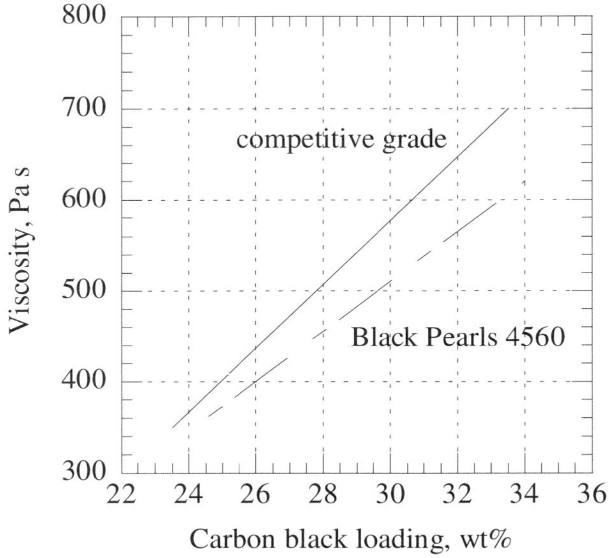


Figure 19.18. Viscosity of PET vs. concentration of carbon black. *Courtesy of Cabot Corporation, Billerica, USA.*

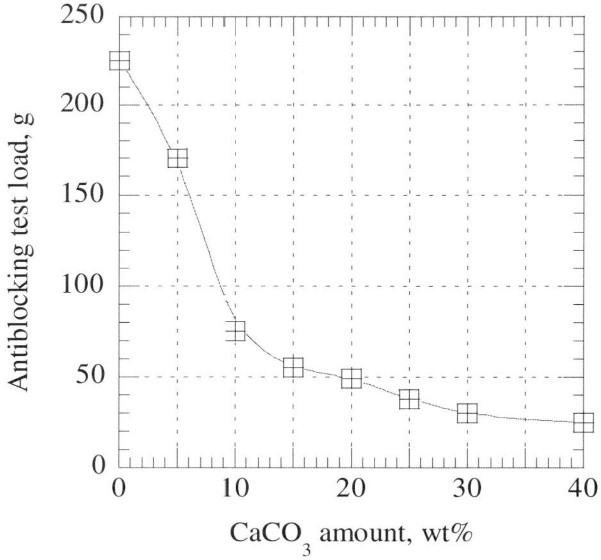


Figure 19.19. Antiblocking properties of LLDPE film containing $CaCO_3$. *Courtesy of ECC International, St Austell, UK* in Johnson S L, Ahsan T, Polymers, Laminations & Coatings Conference, TAPPI Press, Atlanta, 1997.

a flow-induced planar orientation which affects gas permeability. The addition of large amounts reduces permeability but decreases mechanical properties.

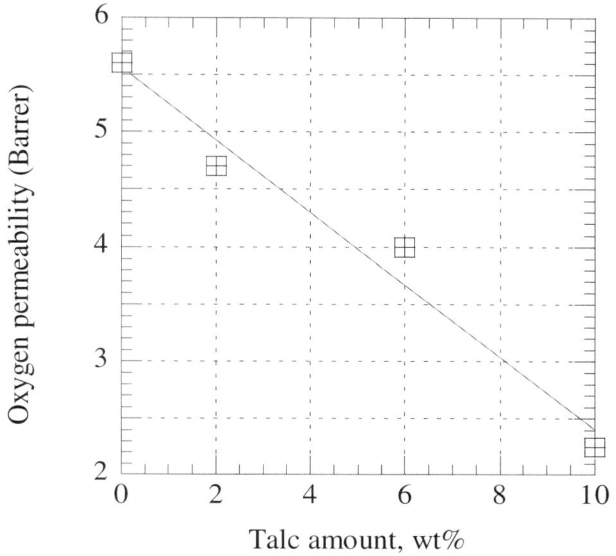


Figure 19.20. Effect of talc on oxygen permeability of HDPE film. [Data from Gill T S, Xanthos M, *J. Vinyl and Additive Technoi.*, **2**, No.3, 1996, 248-52.]

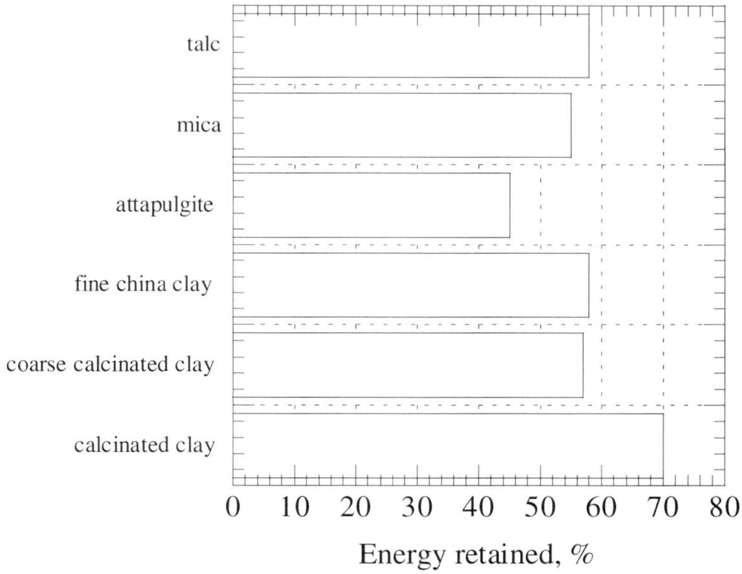


Figure 19.21. The effect of fillers on energy conservation in greenhouse film. *Courtesy of ECC International, St Austell, UK* in Hancock M, *Plasticulture*, **79**, 4, 1988.

As mentioned earlier (section 19.2), the addition of filler reduces the transmission of infrared energy through a greenhouse film. Figure 19.21 shows the effect of various fillers in energy conservation applications.[84] Calcinated clay gives the best results. The reformulation of greenhouse film to include a filler resulted in an increase in crop growth rate and in energy savings.

Fillers improve mechanical properties of films. Calcium carbonate coated with stearic acid improved the impact strength of polyethylene bag films.[80,81] Production output was increased and printability of the film was improved.[81]

Films containing filler may have a substantially improved UV stability due to the addition of carbon black. It is also possible to produce films which will be biodegradable by the combined action of UV degradation and biodegradation.[85] It was confirmed that china clay increases the degradation rate of polyolefins by consuming stabilizers and thus reducing polymer stability.[85]

## 19.18 FOAM[86-91]

Fillers affect foams in several ways: they induce nucleation of gas bubbles and polymer crystallization, they destabilize liquid foam, they interfere with rate of chemical reactions and gas generation processes, they influence the viscosity of liquid premixes, they improve mechanical properties of foam, and foam durability. Care must be used in the selection of fillers for foams. They are used only sparingly in spite of the potential gains in foam performance which might be expected.

A major problem for the foam industry is the recycling of used foam which has now been addressed by the use of pulverized foam as a filler in foam production.[86,89,90] Up to 30 phr of pulverized foam can be added with a positive effect on foam properties. A combination of pulverized foam and carbon black gave good results.[89]

Novel PVC foams are produced with wood fibers.[91] Untreated and silane treated wood fibers were used. It was determined that by the selection of plasticizer level, foaming time, foaming temperature, and silane treatment of wood fiber, the foam of expected properties can be obtained.

## 19.19 FOOD AND FEED[92-93]

There are two main reasons why substances classified as fillers are used in agriculture: feed supplementation and application of pesticides in granular form. The nutritional applications usually involve calcium carbonate, iron salts, and zinc oxide as feed supplements with the required levels of these three metal salts.

Fillers are used in food products in the following functions: colorants, viscosity regulators, nutrients, bleaching agents, and anticaking agents. Some foods include titanium dioxide and alumina or its lakes as food colorants. Calcium sulfate is used as a part of the bleaching system in flour production. Fillers which have high sorptive capacities, like bentonite, are used as viscosity regulators or agents preventing sedimentation of other solids.

Calcium carbonate, calcium sulfate, kaolin, talc, magnesium carbonate, and zinc oxide are frequently used as nutrient supplements or food diluents. Calcium silicate and fumed silica are used as anticaking agents. They are added in concentrations of up to 2% to materials which are hygroscopic. The surface coatings of such material and their ability to absorb moisture make it possible for these materials to transform the hygroscopic materials to free-flowing powders. The type and concentration of fumed silica affects the material's free-flowing properties and the durability of the induced effect. Silane-coated fumed silica performs better because it is hydrophobic.

## 19.20 FRICTION MATERIALS[94]

Asbestos-free fiber-reinforced brake linings are being improved through a continuous development effort. A review of problems and achievements was recently published.[94] Two types of fillers are used in brake pads: particulate fillers and reinforcing fibers. Low cost materials are used as particulate fillers, such as calcium carbonate and barite. Their role is to decrease cost without detracting from the performance of the product.

Fibers are responsible for strength, thermal stability, and frictional properties. 1,200 fibers have been tested to-date for this application. The major groups include aramid, glass, carbon, steel, and cellulose fibers. Each fiber has its own set of problems in the application. This may be price, low melting point, low friction characteristics, corrosion, abrasion of metal elements, low strength, etc. Studies in this field affect the automotive, land transportation, military, and aerospace industries and are being maintained at a high level to further improve the properties of brake materials.

## 19.21 GEOSYNTHETICS[95-96]

Geosynthetics include three groups of products: geotextiles, geogrids, and geomembranes. All materials are required to have UV protection (although the geomembranes are usually covered in application; part of the membrane is exposed and degradation may occur during installation).

Carbon black is the usual UV stabilizer in these products. Weathering studies are reported elsewhere.[97] Several other fillers are employed including titanium dioxide, calcium carbonate, and clay. Clays are used in large volumes in secondary liners where bentonites are used as the absorption media.

## 19.22 HOSES AND PIPES[98-99]

Mechanical performance (tensile, flexural, impact, bursting pressure, and compressive strength), resistance to heat, thermal expansion coefficient, heat distortion temperature, maximum working temperature, burning class, UV stability, and working stress are the most important parameters characterizing performance requirements of pipes and hoses and are used as selection criteria. Fillers can help to fulfill these requirements, but are underutilized. Fillers seem mainly to be used to lower cost.

Carbon black is very frequently used as an UV stabilizer. Pipes which are not normally exposed under their working conditions do need UV stabilization because during storage and transportation pipes are damaged by sun light. The most important parameter of pipe performance – bursting pressure – is severely reduced by the presence of cracks on the pipe surface induced by UV degradation.

For some applications a low thermal expansion coefficient is required and pipes are produced with glass fiber reinforcement which also increases the heat deflection temperature.

Electrically conductive pipes are less well known. These pipes are used to prevent electrostatic hazards in oil tanker applications.[99] The pipe for this purpose was developed using glass fiber for reinforcement and carbon fiber to obtain conductivity. There are also methods of production of such pipe where aluminized glass fiber is mixed with reinforcing glass fiber.

## 19.23 MAGNETIC DEVICES[100-101]

Polymer magnets, although common today, are a quite recent development. The first patent application was filed in France in 1955. Since that time, many ferromagnetic fillers have found their way to market. Plastics magnets do not match the performance of metal magnets but their properties are being systematically improved.

Barium ferrites do not affect the mechanical properties of natural rubber and hence are useful in magnetic applications. Cure rates are increased considerably but ferrites may cause reversion problem.[100] The magnetic properties increase almost linearly with the amount of ferrite added.

A magnetic filler in the form of fiber was added to HDPE and injection molded test products were made.[101] Anisotropic composites were manufactured to induce flow orientation of the fibers. The increased orientation contributed to the generation of a higher permeability magnet. The orientation can be affected by changing the die diameter.

Magnetizable particles were used in the separation of biochemical compounds. Magnetizable particles of iron oxide are produced on different supports capable of absorbing various biological substances which are then removed from suspension by a magnetic field (see more about this application in Chapter 2).

## 19.24 MEDICAL APPLICATIONS[102-110]

The medical sector is the fourth largest consumer of plastics after the packaging, construction, and automotive sectors. It has a relatively high growth rate (8%/year). The growth is in part due to our longer life expectancy but also by new technological advances in medicine.

As is evident from the conducted studies, the majority of problems within this sector are very typical of the plastic industry in general. These include the need to improve reinforcement, durability, fatigue resistance, adhesion, reduce water absorption, etc. In addition to these problems are some issues which are only related specifically to medical applications such as toxicity and biocompatibility. These issues make this research an exclusive field in which the publication of findings is found only in industry specific journals. The medical plastics industry would benefit from greater integration with the mainstream development of plastic materials. It is quite usual to find that although many new materials are available, but only a few are used by medical plastics. The area of prosthesis and dental applications seems dominated by polymethylmethacrylate. The use of fillers is mostly limited to hydroxyapatite and glass. Distressingly studies report numerous fatal failures due to the lack of adhesion between the bone substrate and the implant due to unsatisfactory properties of "Plexiglas" related cements.

Reading these studies one gets a feeling that interdisciplinary effort could be dramatically improved in many strategic ways. However, some pioneering studies are also conducted. One involves the crystallization of inorganic materials in the presence of simulated or specific body fluids. This process affects the structure of the crystal which develops a surface structure similar to that of bone. In this way it becomes more acceptable (or more difficult to recognize as artificial) to living tissues.

Most material studies reported in medical journals are of interest to those involved in mainstream plastic applications. Some medical plastics must perform under constant water immersion. It was reported that absorption of 1% water reduces the fatigue life of PMMA by a factor of four, since bone cement can only be replaced by a surgical operation such a performance is clearly unacceptable.[104] The use of silane to treat the hydroxyapatite filler in this material reduced water uptake. The water uptake increased with increased concentration of hydroxyapatite. In applications, such as dental fillings, increased water uptake is considered helpful since it compensates for the loss of volume due to shrinkage of the filling during curing.

Silanes were also used to treat hydroxyapatite in biodegradable composites which are used for the temporary joining of fractured bones. This material is intended to be gradually replaced by biologically generated material so that a second operation may be avoided. The major problem is that such cements rapidly lose their integrity. Again, a silane coating with suitable functional groups can be used to improve water resistance without affecting the process of gradual biological replacement.[109]

In hard tissue replacement, polyurethane filled with hydroxyapatite is used. The composite must have high strength and stiffness and excellent creep and fatigue resistance. The material in the study met these requirements when hydroxyapatite was pretreated with hexamethylene diisocyanate. The treatment improved the adhesion between the matrix and the filler.[108]

The durability of cemented joint replacements depends on the properties of PMMA bone cement. A titanium fiber was used to reinforce the cement. Before the cement was applied it was centrifuged to remove entrapped air which would cause voids in the connection. The treatment also improved the ductility of the cement. Similar processes are used in many plastic industries to improve elastic properties.

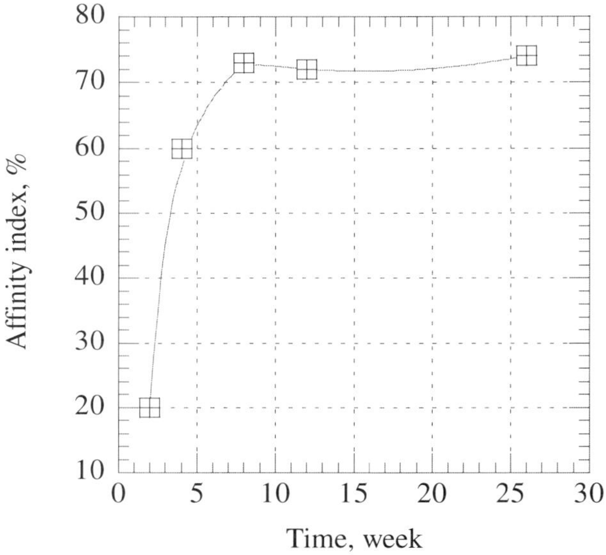


Figure 19.22. Affinity index of bone cement containing 70 wt% glass powder vs. time. [Adapted, by permission, from Tamura J, Kawanabe K, Yamamuro T, Nakamura T, Kokubo T, Yoshihara S, Shibuya T, *J. Biomed. Mat. Res.*, **29**, No.5, 1995, 551-9.]

Figure 15.23 shows that an increased amount of glass filler increases the affinity index of bioactive bone cement. Because fillers are relatively inert they have the potential to improve biocompatibility of artificially produced materials. Figure 19.22 shows that the affinity index increases with time.[106] This suggests that there is an ion exchange between the glass filler and the organism which modifies implant to become more compatible with the surrounding tissue. It was also determined in this study that an increased amounts of filler increased not only the biocompatibility but also the mechanical performance of the composite. Silane treatment was also apart of this study but it is yet to be determined if it will affect the ion exchange process which allows the compatibility to develop.

Fillers are essential in plastics which must be sterilized. Glass fiber in particular is used to withstand sterilization and retain mechanical properties.

Most medical plastics must be kept scrupulously clean and are handled only in a clean room environment. Static charge built up on these parts will attract contaminants therefore it is critical that static charges be dissipated. Fillers perform this function because fillers, unlike organic antistatics, have no tendency to migrate and contaminate the surroundings.

Medical plastics present a large opportunity for filled products but much more work is needed. A review of the literature indicates that efforts are concentrated on materials used for replacement of body tissues. Much less attention is being paid to items produced for everyday use such as packaging, hoses, syringes, etc. which can benefit substantially from incorporation of fillers.

## 19.25 MEMBRANES[111-112]

Two types of membranes may contains fillers: hydrophobic membranes (usually polydimethoxysiloxane filled with zoelites) and heterogeneous ion-exchange membranes which may contain various surface modified inorganic materials.

One important requirement of membranes is that the filler and the matrix must be in good contact which accounts for the widespread use of silane modified materials. An example of such an improvement is reported in work done on silicate used in various polymeric matrices.[112]

Both phases (the matrix and the filler) must be homogeneously distributed and the sizes of particles must be uniform and well controlled. Layered fillers such as montmorillonite are finding growing applications many of which involve the use of nanocomposite technology. One example of intercalation process to produce membrane resulted in the development of nanocomposite with controlled ion mobility.[111]

## 19.26 NOISE DAMPING[113]

Noise pollution can be reduced by controlling the dampening characteristics of a material. The dampening material converts the energy of vibration to heat rather than emitting it to air.[113] Incorporation of fillers gives such characteristics.

Figure 19.23 shows the effect of filler type on vibration dampening. Mica was the most effective filler in this group because of its platelet structure. Figure 19.24 shows the effect of concentration of mica on dampening properties. The dampening characteristic is improved with higher filler. Overloading with filler spoils the effect (dampening properties of material are reduced if the filler concentration is increased beyond a certain level).

## 19.27 OPTICAL DEVICES[114-115]

Fillers are usually considered to be opaque materials but they can play an important role in high technology optical devices. This is possible due to the use of very small particles of controlled size obtained through application of nanocomposite technology.

Optical switches for optical computing devices and hard transparent coatings are two examples of materials which contain such fillers. Very precise mixing technology is required since uniformity of dispersion is critical if acceptable optical properties are to be obtained. Particles must be evenly distributed in the matrix and should have the capability to amplify light by having nonlinear optical properties. Optical scattering should be avoided, particle size and refractive index strongly influence scattering. If particles are smaller than 25 nm, the refractive index mismatch does not matter. If larger particles are used, the difference between the refractive indices of the matrix and the particles should be very small. For example, for particles of 100 nm, the refractive index mismatch must be less than 0.02.[114]

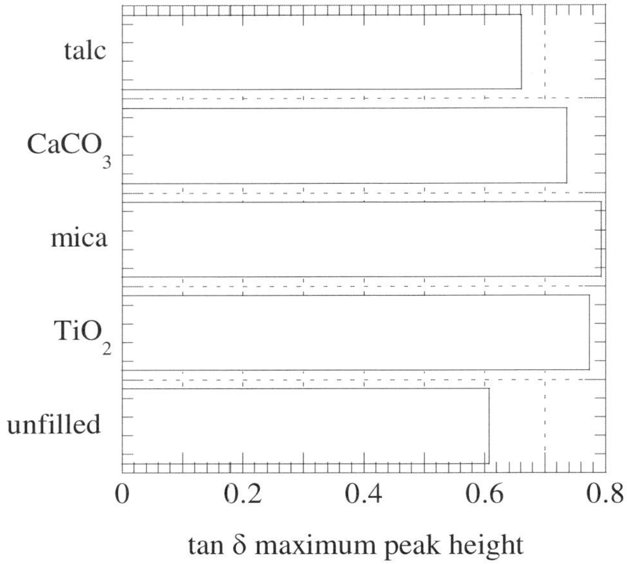


Figure 19.23. The effect of filler type on noise dampening properties of IPN. [Data from Li Shucai, Peng Weijang, Lu Xiuping, *Int. J. Polym. Mat.*, **29**, Nos.1-2, 1995, 37-42.]

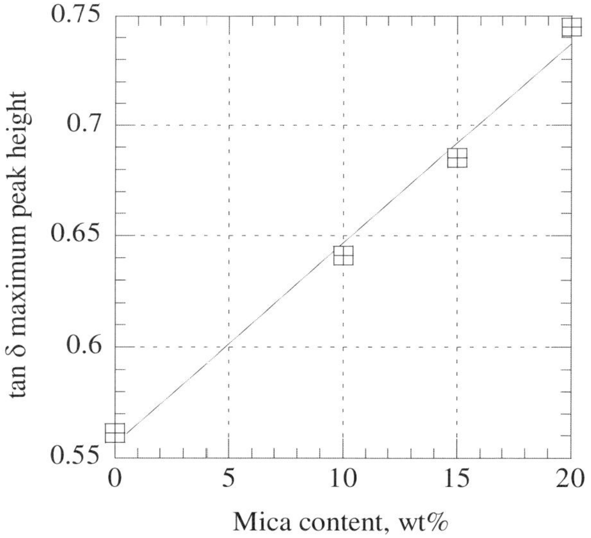


Figure 19.24. The effect of mica concentration on noise dampening properties of IPN. [Data from Li Shucai, Peng Weijang, Lu Xiuping, *Int. J. Polym. Mat.*, **29**, Nos.1-2, 1995, 37-42.]

Several particulate materials are used in these applications, such as CdS, CdSe, silica, $V_2O_5$.

Figure 8.41 shows a reactions scheme by which a nanocomposite was prepared from boehmite.[115] This composite is being commercially used in a hard transparent coating for scratch-resistant glasses.

## 19.28 PAPER[116-120]

The major reason to coat paper with filler-containing mixtures is to improve printability. The secondary requirements include: brightness, opacity, gloss, ink receptivity, and flame resistance. Pigment or filler is the major component of a paper coating. It occupies 70 to 90% of the total dry weight. Pigment selection is based primarily on the expected characteristics of the paper, on the mixing and handling equipment available, on the method of coating, and, perhaps, most importantly on price-performance criteria. Several general principles are considered in filler choice. Fillers can be classified as general purpose fillers used at loading levels greater than 10% of paper weight or specialty fillers used at loading levels less than 10 wt% and more often at less than 5 wt%. The particle size of the general purpose fillers is in a range from 0.5 to 10 μm, whereas specialty fillers (silica, $TiO_2$, etc.) have particles ranging into the tens of nanometers. All fillers have hydrophilic properties, with exception of talc, which is slightly hydrophobic. The ζ-potential of fillers is important to prevent coagulation. A ζ-potential higher than 20 mV is sufficient to prevent flocculation. Interactions between fillers and the other components of paper and the effect of such interactions on paper properties is discussed in detail elsewhere.[120]

A filler in the paper industry is a material which is mixed with fiber to manufacture uncoated stock. If this filler helps to reduce amount of fiber used, it may be called an extender. The particulate used for the coating of paper is called a pigment by the paper industry. Materials of the same chemistry can be used as fillers or pigments but they will frequently have different specification for each application. Based on this, titanium dioxide is a filler if processed together with fiber and a pigment if used to coat paper.

Aluminum trihydrate, barium sulfate, calcium carbonate, clay, amorphous silica, talc, titanium dioxide, and zinc oxide are the major pigments/fillers used by the paper industry. Aluminum trihydrate is mostly used at a level equivalent to 10-20% of the total fillers. It is compatible with other fillers. Aluminum trihydrate improves brightness, opacity, the receptivity of gloss inks, surface smoothness, and flame resistance, it is truly a universal pigment. Used in excessive amounts, it affects the high-shear rheological properties of the coating. Aluminum trihydrate needs more binder than titanium dioxide or clay. Barium sulfate (blanc fixe) is used mostly in photographic paper.

Calcium carbonate is one of the most commonly used fillers in the paper industry. In most cases, it is used in a range from 5 to 50% of the total pigment, but in some processes, at above 70%. Brightening, ink receptivity, and surface smoothness are improved by calcium carbonate. Calcium carbonate is easy to disperse,

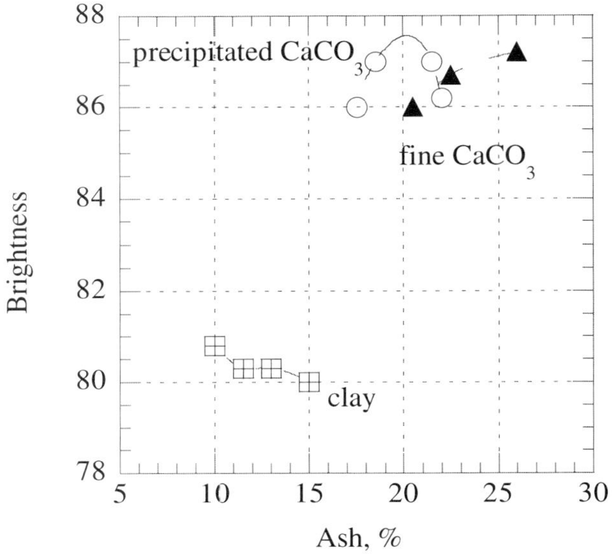


Figure 19.25. Paper brightness vs. ash content. [Data from Anderson T C, Yunko A L, *Pulp Paper*, **53**, 1983, 82.]

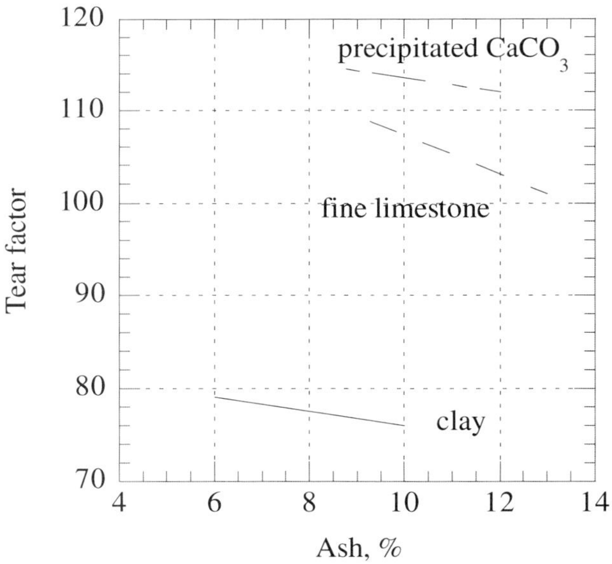


Figure 19.26. Tear factor vs. ash content. [Data from Anderson T C, Yunko A L, *Pulp Paper*, **53**, 1983, 82.]

does not have a tendency to form agglomerates and dispersions containing it are easily stabilized by inorganic dispersants, usually polyphosphate compounds. Calcium carbonate fillers are most beneficial in an alkaline paper process because they

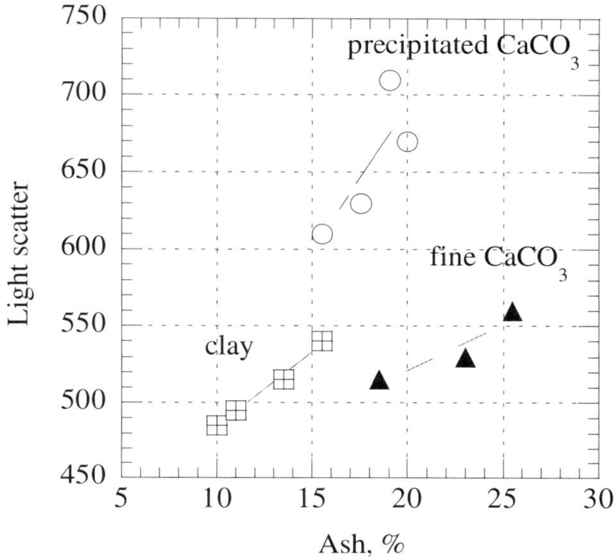


Figure 19.27. Sheet light scatter vs. ash content. [Data from Anderson T C, Yunko A L, *Pulp Paper*, **53**, 1983, 82.]

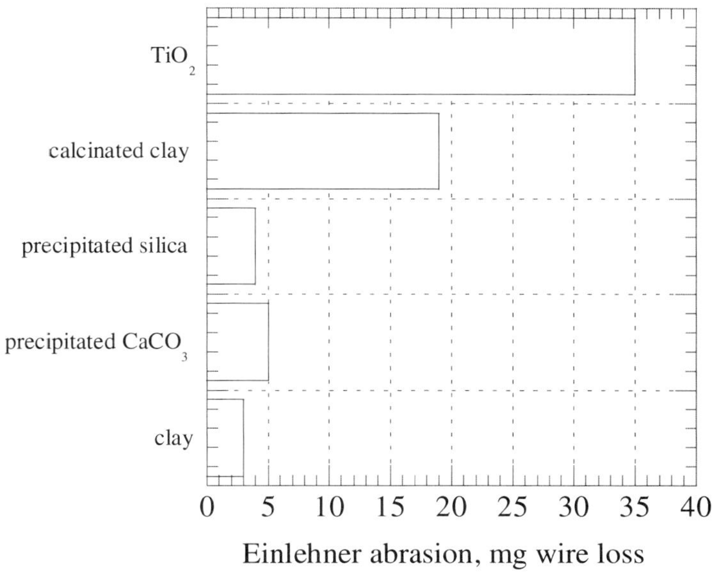


Figure 19.28. Effect of various fillers on Einlehner abrasion. [Data from Anderson T C, Yunko A L, *Pulp Paper*, **53**, 1983, 82.]

provide pH buffering in the required range. There is an ever increasing demand for acid-free paper so the future of calcium carbonate in this application is assured. Figures 19.25 to 19.28 review the results of comparative studies of various pigments on paper properties.[121] Filler fineness does not affect paper brightness, but all other quality factors are related to filler grain size. Pigment distribution in the paper layers as related to brightness and opacity, and the application of retention aids, is discussed in detail elsewhere.[120]

Amorphous silica has become increasingly important in paper technology. It cannot be used alone because it affects coating rheology and adhesive demand. But used in combinations with other fillers, it improves the brightness, opacity, and ink receptivity of paper. Similarly, talc is used as a 10 to 40% replacement for other fillers to improve paper surface smoothness and ink receptivity. Talc is also used in place of clay if clay is not locally available.

Clays and titanium dioxide are the most heavily used fillers and pigments by the paper industry. Clay is easy to disperse because of the hydrophilic character of its surface. It improves brightness, opacity, gloss, ink receptivity, and surface smoothness. It is therefore the universal filler/pigment in paper manufacturing. There are, however, complexities. Fine particle clays decrease gloss but increase surface smoothness. Ink receptivity is generally low with clays, but it also depends on clay particle size and on the calendering process. Excessive adhesive reduces gloss and surface smoothness. It is difficult to achieve all the required properties with a single filler but clay comes close. The technology of clay preparation is exceptionally important because air incorporation into a paper coating affects the brightness and opacity of paper.

Titanium dioxide, with its high refractive index, provides brightness and opacity, which is why it is used in paper manufacturing. Titanium dioxide also contributes to surface smoothness. The rutile form is more opaque than the anatase. Zinc oxide is popular in document duplication papers because of its photoconductivity. Recently, a hollow-sphere polymer pigment has become a useful new material in the paper industry to decrease paper weight.

Chemical composition, particle size, particle shape, specific gravity, surface area, refractive index, brightness, absorptivity, and wettability are the most important criteria used in selecting a filler for the paper industry. For further reading, a recent monograph on the use of fillers in paper industry is an excellent source of practical information.[120]

## 19.29 RADIATION SHIELDS[122]

Filled polymers play a role in primary and secondary protection against $\gamma$-radiation. The photons interact with matter by photoelectric absorption and Rayleigh scattering. For primary partitions which separate an unshielded source from its surroundings, lead bricks or concrete blocks are used. For the secondary partitions which protect personnel from radiation, a protective shield or vest can be made by incorporating metal particles or lead oxide in rubber or plastic. Such shields are used by physicians and dentists or their patients to limit exposure to x-rays. The radiation

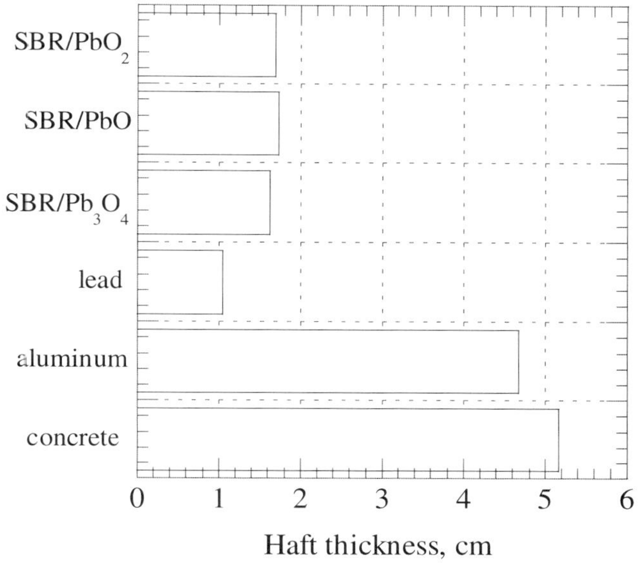


Figure 19.29. Half thickness for $^{60}Co$ $\gamma$-radiation in different materials. [Adapted, by permission, from Abdel-Aziz M M, Gwaily S E, *Polym. Degradat. Stabil.*, **55**, 1997, 269-74.]

shielding is inferior to high density metals but complex shapes available can shield the body effectively.

Figure 19.29 shows the comparative shielding efficiency data for various materials. Rubber filled with lead oxides comes very close in performance to lead and is superior to concrete and aluminum. Exposure of these shields to radiation causes degradation of mechanical properties (hardness, in particular, is increased) but it does not affect shielding efficiency.

## 19.30 RAIL TRANSPORTATION[123]

Polyester composites have been used in subway coaches in the UK.[123] After the King's Cross fire in the London underground a new regulation was introduced dealing with the interior use of plastics which could only be met by phenolics. Polyester parts manufacturers searched for alternatives and, with the cooperation of companies manufacturing additives, they developed a system which surpassed phenolic resins in fire rating. This system contains aluminum hydroxide and an additive to help its incorporation. A large amount (350 phr) of aluminum hydroxide is used. Such a large amount normally increases viscosity and makes production of composites difficult but an additive was found which lowers the viscosity of the composition to below that of the viscosity of the unfilled resin.[123] This system has excellent performance. In addition to its fire retardancy, it has a very low smoke emission and low toxicity.

During the project on the Channel Tunnel, British and French authorities were cooperating on the project and standards of both countries were used with only some differences in the methods of testing. The system described above passed both standards. The German regulations are not as demanding as those of the British and French. An interior plastic panel for passenger rail cars was made 25% lighter by the addition of glass microspheres and it employed the same fire retardancy strategy as described above. This product can pass the German standard but it does not pass French and British regulations.

A primary requirement for plastic composites used in rail car applications is fire and smoke retardancy.

## 19.31 ROOFING

During the last decade many new roofing materials were introduced which are applied in the form of weldable membranes, liquid curable materials, self-adhesive products, and torchable materials. These materials are produced from numerous polymers such as, PVC, chlorinated polyethylene, chlorosulfonated polyethylene, EPDM, acrylics, bitumen, polymer-reinforced bitumen and several other materials. It is beyond the scope of this book to analyze compositional changes in these materials. We will provide a brief overview.

The current market requires a roof which can perform 20 years. Many products are available to do this and more. However, it is becoming more common for warranties only being provided for a 5 year term. Among fillers, only carbon black is important in extending roof durability and it is used widely. Roof durability is related to the plastic behavior of many roofing materials. Efforts have been made to provide better reinforcement to prevent plastic flow and improve dimensional stability.

The second requirement is to conserve energy. This is a factor in summer since winter insulation is done by other means. Many reflective pigments are in use with special emphasis on metal flakes which are becoming increasingly popular either as an additive to the roofing material or as a component of the roof surface coating.

The third requirement is weld durability. Fillers do not appear to be involved in this process but concerns remain that inorganic powder may affect weldability.

## 19.32 TELECOMMUNICATION

Components of switches, relays, and connectors use glass fiber reinforced semicrystalline polymers, such as PA, PPS, PBT, and PET. The following requirements are important in these applications: dimensional stability and precision, low moisture absorption, strength, resistance to creep, electrical insulating properties, and resistance to high working temperatures (~85°C).

Electric insulating properties should be retained at elevated temperatures and in changing relative humidity. The choice of composition is important since the material should not absorb moisture which may corrode metal elements. Usually polyamides 610 and 612 are used since they absorb considerably less moisture than polyamide 6 and 66.

The equipment is designed to operate without failure for more than 20 years. In its design, the flow properties of plastic components are important since a degree of high precision must be maintained in the molded part because metal contacts are frequently very close (less than 1 mm apart). Good flow in the mold is required because some sections are very thin. In the more demanding applications, PPS is used.

## 19.33 TIRES[124-132]

For many decades carbon black has enjoyed a practical monopoly as a filler in the tire industry. It retains this position today, the tire industry consumes 70% of carbon black production. A major breakthrough for non-black fillers came in the early 1970s when a winter tire containing silica in its treading compound was introduced. This resulted in numerous problems with rubber compound processing. The tire compound had different flow and molding characteristics. The development initiated friendly competition (friendly because the major producers of carbon black are also major manufacturers of precipitated silica) which continues to bring improvements to tire performance.

Figure 19.30 shows end use opportunities for non-black fillers.[124] The tire elements show highlights of various opportunities for white fillers. Figure 19.30 omits to show that silica can be used to advantage in treading compound. This is discussed below.

Table 19.1 shows the design criteria for the elements of tire.

**Table 19.1 design requirements of tire elements. Data from Ref. 127.**

| Tire section | Important design criteria |
|---|---|
| White side wall | processability, cure rate, hardness, stress-strain properties, adhesion to adjoining components, resistance to tear, cut/crack propagation resistance, resistance to ozone, oxygen and UV, retention of white color on exposure to environment, low cost |
| Black side wall | resistance to weathering, abrasion, ozone, tear, radial and circumferential cracking, fatigue resistance, protection of adjoining elements of tire |
| Wire coats | good adhesion to brass coated steel wire and to adjoining rubber compounds, tear, fatigue, and age resistance |
| Inner liner | formulated to ensure retention of compressed air in tubeless tire, good air retention, moisture impermeability, flex fatigue resistance, durability |
| Carcass ply coat | promote reinforcement by coating parallel cords |
| Tread | wear resistance, abrasion resistance, traction, speed stability, protection to casing, rolling resistance, ice skid resistance, durability |

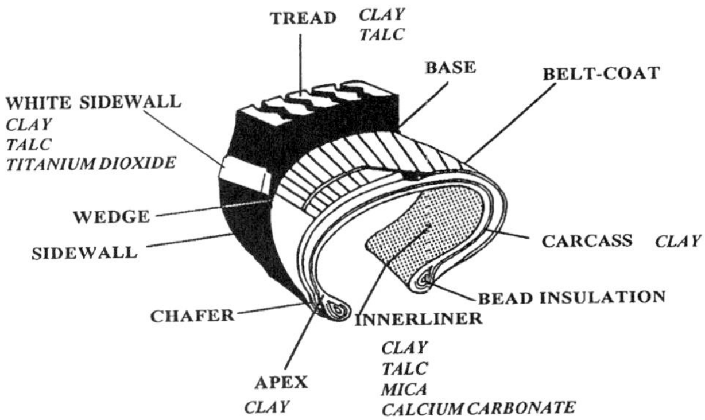


Figure 19.30. Tire cross-section. [Adapted, by permission, from Evans L R, Meeting of the Rubber Division, ACS, Montreal, May 5-8, 1996, paper D.]

The fillers listed in Figure 19.30 are in commercial use performing various functions discussed in detail elsewhere in reference to various tire applications.[124,127]

It was known for a long that the use of silica in tire treads gives better wet and ice skid resistance and better rolling resistance which are important factors in safety and lower consumption of fuel. Real changes in the market occurred recently when reinforcement by silica was improved by the introduction of a new silane. This silane is bistriethoxysilylpropyl tetrasulfone (Si-69). Introduction of this reinforcement removed the barrier of low abrasion resistance and reduced cost of use of silica in comparison with carbon black. The present cost increase due to the use of silica is estimated at $4.75/tire.[128] Part of this cost is raw material ($1) and rest is the cost of additional equipment and processing.[128] But the improvement in tire properties and the possibility of producing "green tire" are offsetting the additional cost and manufacturers are introducing this new technology. One such manufacturer is Michelin.

The technology is still not problem-free. Carbon black is a conductive filler therefore it does not cause an accumulation of static electricity formed because of frictional contact with the road. In a silica reinforced tire, static charge is a problem. Remedial actions have already been taken by the introduction of neoalkoxy zirconates and titanates which are capable of dissipating electrostatic charges. The future of the tire is not yet clear but it is expected that silica's share of this market will be 150,000 tons. This will be supplied to make full silica and silica/carbon black tires.

The carbon black industry is also working on new grades which will offer rolling resistant tires. Two types of carbon black have been tested: super-active carbon blacks and inversion blacks.[132] Both types give promising results.

The reduction of air permeability and the promise of lower cost has increased interest in thermal black use in liners. It is expected that 27,000 tons of thermal black will be needed for this application.[130] The use of thermal black is expected to lower the cost of the liner by 20%.

In the retreading market, new silicas are expected to improve properties of retreaded tires which are currently inferior to new tires.[131] Tread wear resistance and rolling resistance are substantially lower with retreaded tires.

## 19.34 SEALANTS[133-134]

The performance expected from a sealant and the polymer used in its production determine the type of fillers used. From the point of view of performance, sealants can be divided into these having plastic behavior and these with elastomeric behavior. Sealants which have plastic behavior are low-cost and low-performance products which are being gradually eliminated from the market. These sealants use inexpensive fillers to lower cost and regulate non-sag properties. Typical fillers used in these products are calcium carbonate and some fibrous materials which are used as replacement for asbestos which was very popular in the past. This combination of fibrous and spherical particles provided a useful tool to the sealant formulator to regulate non-sag properties which are very important in sealants. Fibers have also been used to reinforce these products because the properties of polymers were poor.

In elastomeric sealants, which now dominate the market, the type of polymer and the application determine filler choice. If the polymer requires reinforcement, reinforcing fillers must be used. Polymers such as silicones and acrylics must be reinforced by fillers. On the other hand, polyurethanes have an excess of tensile strength and in most construction applications this tensile strength has to be decreased by a factor of three to fifteen depending on specific elastomer used. This decrease in tensile properties has to be done carefully to retain good elongation and adhesion to substrates. This requires a combination of fillers and plasticizers which retain all of the other properties required for sealant performance such as rheological characteristics, UV stability, and surface properties which prevent dirt pickup. There are exceptions such as polyurethane automotive sealants which do require reinforcement to give adequate performance.

The mechanism of cure has an important influence on the filler choice. In reactive systems such as polyurethanes, water present in formulation adversely affects sealant stability (shelf-life) which imposes the additional selection criterion of water content. If water is present in the filler it must be removed by additional operations either by an expensive drying process or through the use of chemical moisture scavengers which is also an expensive approach.

Silicone sealant manufacture makes use of fumed silica to regulate rheological properties and reinforce the polymer which, without filler, has too low a tensile and elongation. Precipitated silicas are also used. The use of precipitated silica is regional. The lowest consumption is in Western Europe, followed by USA and the largest is in Japan.[133] Fumed silica may contain a large amount of water depending on the grade. Water in silicone sealants is usually not as critical as it is in polyurethanes because the silicone crosslinker acts as moisture scavenger. Still, control of water input is an important cost and quality factor. In addition to silica, calcium carbonate and titanium dioxide are used in silicone sealants. Silicon polymer is relatively durable on exposure to UV therefore carbon black is used more as a pigment than as a UV stabilizer although, there are grades of silicone sealants which rapidly lose elongation on exposure to UV.

Acrylic sealants have been gaining market share due to their price/performance ratio, high UV durability, and low toxicity. Fillers are used in acrylic sealants for the same reasons as they are used in silicone sealants. Fumed silica is reinforcing filler and plays a role in the rheological system. In addition to fumed silica, zinc oxide is used as crosslinker, calcium carbonate as general purpose filler, and titanium dioxide as a pigment. Acrylic sealants have low tolerance for fillers and high loadings of fillers degrade their properties.

In polyurethane sealants, fillers are used for several purposes. These include rheological control, cost reduction, reinforcement, moisture scavenging, reduction of surface tack, and UV protection. The major requirement is that fillers contain a low level of moisture. A typical specification for fillers is 0.03% moisture which is achieved by filler selection, drying, or moisture scavenging. For cost reduction and rheology regulation, calcium carbonate is the most commonly used filler but many other fillers have been used in these applications. Calcium sulfate, molecular sieves, and some other fillers may be used as moisture scavengers. At high levels, most fillers reduce surface tack and give green strength. However, at these high levels, elongation and fatigue resistance may be adversely affected. The UV protection of sealants is complex. Carbon black provides protection but it also adsorbs UV stabilizers (HALS) therefore use of carbon black is detrimental in most colors. Large quantities of carbon black are used in special applications such as windshield adhesives and sealants. Such high loadings create very high viscosities. The solution lies in the selection of carbon black with appropriate concentration of reactive groups on their surface.

## 19.35 SIDING[135-136]

Vinyl siding is a large consumer of PVC. The most important properties are impact strength and the stability of the color and the material to UV exposure. Relatively low filler content was traditionally used to prevent loss of impact strength and other mechanical properties. Recent findings show that the incorporation of chlorinated polyethylene as an impact modifier allows the amount of calcium carbonate to be increased without affecting tensile strength and elongation and with an increase in impact strength (Figure 19.31).

Titanium dioxide can be used as primary or secondary UV stabilizer in the formulations. Excellent UV stability can be obtained with a high loadings of titanium dioxide.

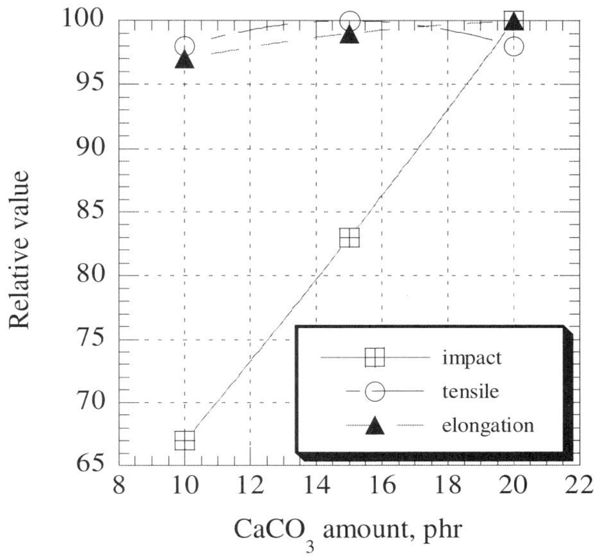


Figure 19.31 Tensile strength, elongation and impact strength of PVC containing 20 phr CPE vs. calcium carbonate content. [Data from Ventresca D A, Berard M T, Antec '97. Conference proceedings, Toronto, April 1997, 3574-9.]

## 19.36 SPORTS EQUIPMENT

Tensile and flexural strength, flexibility, compression strength, abrasion and scratch resistance, coefficient of friction, dimensional tolerance and stability were all improved by incorporating fillers. In sporting goods, such equipment as tennis racquets, skis, ski bindings, skates, etc. were all improved by fillers. High performance fibers such as carbon fiber, aramid , and glass fiber made these advances possible.

## 19.37 WATERPROOFING

Waterproofing has not changed in the last decades as much as other fields. Thus, most significant technological advances has been the introduction of polymer modified membranes. Poor biological resistance of most materials is the factor which has caused delay in its introduction. Bitumen and coal tar are low cost materials which perform well in below grade application because of their low moisture permeability and their excellent biological resistance. Consequently, it is difficult to find cost effective replacements. These traditional products use only a limited amount of fillers because of their adverse effect on mechanical properties. Carbon black is the most widely used additive for regulating rheological properties and reinforcement. Current regulations limiting the use of coal tar and concerns over the toxicity of asphalt may change the situation and force manufacturers to look for less toxic alternative solutions.

By contrast, in above grade applications, where waterproofing coatings are exposed to UV and other degrading environments, many significant changes have occurred. Here, new elastomeric materials have been introduced because crack bridging in a broad range of temperatures is required. These coatings include textured (stucco) and colored acrylic coatings which make the coated substrate waterproof, durable, provide aesthetic values, and low dirt pick up. These new coatings use fumed silica as a rheological additive and for reinforcement, small amounts of calcium carbonate to improve rheological properties and large amounts (textured coatings) of silica sand. Silica sand for this application should have low content of iron to prevent formation of streaks from the iron and its rusts. A variety of glass beads are used for decorative purposes either to develop a certain texture or to impart color.

## 19.38 WINDOWS[137]

A complete window is composed of a frame, gaskets, the insulated glass assembly, seals, and sealants all of which contain fillers. The plastic window market in Europe differs from the North American market. The major difference is in the extruded frame profile and sealing technology. Extruded frame profiles in Germany and other European countries are up to twice as thick as the North American profile. This determines the selection of plastics materials. Also the sealing technology is different. In Germany, polysulfide sealants are used extensively but they are not popular in North America where many other sealing systems are used to increase productivity. PVC containing approximately 10 wt% filler/pigment made up of calcium carbonate and titanium dioxide (about 40% consists of titanium dioxide) is typically the plastic of choice, worldwide. Titanium dioxide plays the additional role of a UV stabilizer. A small number of window frames are produced from glass fiber composite.

The market for composite window materials is expanding because new production facilities are starting up throughout the word in countries which have not traditionally used this type of window. Even in countries where plastic windows are common only about 50% of windows are of this type. They enjoy about 80% of the replacement market but a smaller percentage of the new construction market.

New developments in recycling will reduce consumption of fillers. A technology was developed that recycles the complete window.[137] An automatic line grinds the entire window, then separates the glass, PVC, the metal and the elastomeric seals. PVC is even separated by color and returned for reprocessing. The use of coextrusion permits the use of 80% recycled material. Only the face of the profile is manufactured from a new compound. This technology will not substantially reduce growth because growth is still driven by new markets but eventually demand for new compound should be reduced.

## REFERENCES


1 Torro-Palau A, Fernandez-Garcia J C, Orgiles-Barcelo A C, Pastor-Blas M M, Martin-Martinez J M, *Int. J. Adhesion Adhesives*, **17**, 1997, 111-9.
2 Katsumata M, Endo M, Ushijima H, Yamanishi H, *J. Mat. Res.*, **9**, No.4, 1994, 841-3
3 Xu X X, Crocrombe A D, Smith P A, *Int. J. Fatigue*, **16**, No.7, 1994, 469-77.
4 Martin-Martinez J M, Macia-Agullo T G, Fernandez-Garcia J C, Orgiles-Barcelo A C, Torro-Palau A, *Macromol. Symp.,* **108**, 1996, 269-78.
5 Kushchevskaya N F, Shvets T M, *Int. Polym. Sci. Technol.*, **22**, No.12, 1995, T/21-2.
6 Pritykin L M, Razumova O V, Sokolova Y A, Antonov S M, Bolshakov V I, *Int. Polym. Sci. Technol.*, **23**, No.3, 1996, T/80-1.
7 Torro-Palau A, Fernandez-Garcia J C, Orgiles-Barcelo A C, Martin-Martinez J M, *J. Adhesion*, **57**, Nos.1-4, 1996, 203-25.
8 Murata N, Nishi S, Hosono S, *J. Adhesion*, **59**, Nos.1-4, 1996, 39-50.
9 Wei Y, Sancaktar E, *J. Adhesion Sci. Technol.*, **10**, No.11, 1996, 1199-219.
10 Dufton P W, **Functional Additives for Plastics**, *Rapra*, 1994.
11 **The Role of Carbon Black Morphology in Reducing UV Oxidation of Linear Low Density Polyethylene Films**. Technical Report S-115. *Cabot*.
12 Mathews C., Enhancing Polymers Using Additives and Modifiers II. *Rapra Symposium*, November 1996.
13 *Polym. News*, **19**, 1994, 24-26.
14 Okoroafor M O, Wang A, Bhattacharjee D, Cikut L, Haworth G J, Polyurethanes '95. Conference Proceedings, Chicago, Il., 26th-29th Sept.1995, 303-9.
15 Tabor R L, Polyurethanes '94. Conference proceedings, Boston, Ma., 9th-12th Oct.1994, 288-94.
16 *Reinf. Plast.*, **39**, No.4, 1995, 8.
17 Jia N, Kagan V A, Antec '97. Conference proceedings, Toronto, April 1997, 1844-8.
18 Lau E, Goodman J, *J. Elastomers Plast.*, **25**, No.4, 1993, 322-42.
19 Arakawa K, Iwami E, Kimura K, Nomaguchi K, *J. Reinf. Plast. Comp.*, **13**, No.12, 1994, 1100-15.
20 Yu T C, Antec '95. Vol. II. Conference Proceedings, Boston, Ma., 7th-11th May 1995, 2358-68.
21 Wang K J, Sue H J, Antec '95. Vol. II. Conference Proceedings, Boston, Ma., 7th-11th May 1995, 1758-64.
22 Lancaster G M, Torres A, Brann J E, Polyolefins IX. Conference Proceedings, Houston, Tx., 25th Feb-1st March,1995, 23-30.
23 Porter J R, Groseth C K, Little D W, Polyurethanes '95. Conference Proceedings, Chicago, Il., 26th-29th Sept.1995, 532-7.
24 Elfving K, Soderberg B, *Reinf. Plast.*, **40**, No.6, 1996, 64-5.
25 Chang Ho Suh, White J L, *Polym. Engng. Sci.*, **36**, No.11, 1996, 1521-30.
26 La Mantia F P, **Recycling of Plastic Materials**. *ChemTec Publishing*, Toronto, 1993.
27 La Mantia F P, **Recycling of PVC and Mixed Plastic Waste**. *ChemTec Publishing*, Toronto, 1996.
28 *Plastics in Canada*, **3**, No.2, 1996, 69.
29 Ye B S, Svenson A L, Bank L C, *Composites*, **26**, No.10, 1995, 725-31.
30 Sanchez-Solis A, Padilla A, *Polym. Bull.*, **36**, No.6, 1996, 753-58.
31 Miller B, *Plast. World*, **54**, No.1, 1996, 38-43.
32 Bataille P, Schreiber H P, Mahlous M, Antec '93. Conference Proceedings, New Orleans, La., 9th-13th May 1993, Vol. II, 1757-9.
33 Wypych J, **Polymer Modified Textile Materials**, *John Wiley & Sons*, New York, 1988.
34 Miszczyk A, Szalinska H, *Prog. Org.Coatings*, **25**, No.4, 1995, 357-63.
35 Balard H, Papirer E, *Prog. Org. Coatings*, **22**, No.1-4, 1993, 1-17.
36 Hedgus C R, Kamel I L, *J. Coatings Technol.*, **65**, No.821, June 1993, 49-61.
37 Hegedus C R, Kamel I L, *J. Coatings Technol.*, **65**, No.822, July 1993, 37-43.
38 Hegedus C R, Kamel I L, *J. Coatings Technol.*, **65**, No.820, 1993, 31-43
39 Hegedus C R, Kamel I L, J. *Coatings Technol.*, **65**, No.820, 1993, 23-30.
40 Szalinska A, *Int. Polym. Sci. Technol.*, **20**, No.9, 1993, T/101-5.
41 Roche A A, Dole P, Bouzziri M, *J. Adhesion Sci. Technol.*, **8**, No.6, 1994, 587-609.
42 Robinson S M, *Polym. Paint Col. J.*, **184**, No. 4352, 1994, 311.
43 Foster J K, Sims E S, Venable S W, *Paint & Ink Int.*, **8**, No.3, 1995, 18-21.
44 Aldcroft D, *Polym. Paint Col. J.*, **184**, No.4366, 1994, 423-5.
45 Knowles J, *Polym. Paint Col. J.*, **185**, No.4366, 1995, 26-7.

46 Van Aken L, Hoeck R, *Polym. Paint Col. J.*, **183**, No.4336, 1993, 442-4.
47 Hansen C M, *Prog. Org. Coatings*, **26**, Nos.2-4, 1995, 113-20.
48 Hoy L L, *J. Coatings Technol.*, **68**, No.853, 1996, 33-9.
49 Athey R D, *Eur. Coatings J.*, No.5, 1996, 299-304.
50 Athey R D, Kirkland T, Lindblom G, Swoboda J, *Eur. Coatings J.*, No.11, 1995, 793-8.
51 Hart A, *Polym. Paint Col. J.*, **186**, No.4378, 1996, S3-6.
52 Patel R, *Polym. Paint Col. J.*, **186**, No. 4386, 1996, 23-5.
53 McGuffog R M, **Clays as Extenders in Decorative Paints**. *ECC International.*
54 **Improved Dry Hiding and Matting Effect with Luzenac Talcs**. *Luzenac Group*, 1996.
55 Radebaugh G W, Simonelli A P, *J. Pharm. Sci.*, **74**, 1985, 3.
56 Leuenberger H, Rohera B D, *Pharm. Res.*, **3**, 1986, 65.
57 Akinmade A O, Nicholson J W, *Biomaterials*, **16**, No.2, 1995, 149-54.
58 Friedrich K, *J. Mat. Sci. Mat. In Med.*, **4**, No.3, 1993, 266-72.
59 Anstice M, *Chem. & Ind.*, No.22, 1994, 899-901.
60 Venhoven B A M, De Gee A J, Werner A, Davidson C L, *Biomaterials*, **17**, No.7, 1996, 735-40.
61 Jones D W, Rizkalla A S, *J. Biomedical Materials Research (Applied Biomaterials)*, **33**, No.2, 1996, 89-100.
62 Sethi R S, Goosey M T, London, Chapman & Hall, 1995, p.1-36. Special Polymers for Electronics and Optoelectronics, Chilton J A, Goosey M T, Ed.
63 Lu X, Xu G, *J. Appl. Polym. Sci.*, **65**, 1997, 2733-8.
64 Nalwa H S, Kasai H, Okada S, Oikawa H, Matsuda H, Kakuta A, Mukoh A, Nakanishi H, *Adv. Mat.*, **5**, No.10, 1993, 758-60.
65 Alpern P, Pfannschmidt G, Wagner G, Tilgner R, *Kunststoffe German Plastics*, **83**, No.12, 1993, 18-9.
66 Faguy P W, Ma W, Lowe J A, Pan W P, Brown T, J. *Mat. Chem.*, **4**, No.5, 1994, 771-2.
67 Baranovskii V M, Bondarenko V V, Zadorina E N, Cherenkov A V, Zelenev Y V, *Int. Polym. Sci. Technol.*, **23**, No.6, 1996, T/87-9.
68 Bittmann E, Ehrenstein G W, Antec '96. Volume II. Conference proceedings, Indianapolis, 5th-10th May 1996, 1465-70.
69 Tead S F, Bluem G L, McCormick F B, Plastics for Portable Electronics. Retec Proceedings, Las Vegas, Nv., 5th-6th Jan.1995, 114-30.
70 Kettrup A A, Lenoir D, Thumm W, Kampke-Thiel K, Beck B, *Polym. Degradat. Stabil.*, **54**, Nos 2-3, 1996, 175-80.
71 Rosenow M W K, Bell J A E, Antec '97. Conference proceedings, Toronto, April 1997, 1492-8.
72 Benrashid R, Nelson G L, *J. Fire Sci.*, **11**, No.5, 1993, 371-93.
73 Guanghong Lu, Xiaotian Li, Hancheng Jiang, *Composites Sci. & Technol.*, **56**, No.2, 1996, 193-200.
74 Ardi M S, Dick W, McQueen D H, *Plast. Rubb. Comp. Process. Appln.*, **24**, No.3, 1995, 157-64.
75 Isern-Flecha I, Plastics for Portable Electronics. Retec Proceedings, Las Vegas, Nv., 5th-6th Jan.1995, 132-9.
76 Fillers for EMI Shileding in Composites. Transmet Corporation.
77 Mittal K L, Ed., **Metallized Plastics**. *Marcel Dekker*, New York, 1998.
78 Nago S, Mizutani Y, *J. Appl. Polym. Sci.*, **62**, No.1, 1996, 81-6.
79 Black Pearls® 4560. Premium Performance for Fine-denier Fiber. Cabot
80 Gill T S, Xanthos M, *J. Vinyl and Additive Technol.*, **2**, No.3, 1996, 248-52.
81 Ruiz F A, Antec '94. Conference Proceedings, San Francisco, Ca., 1st-5th May 1994, Vol. III, 2766-9.
82 Callari J J, *Plast. World*, **52**, No.3, 1994, 22-6.
83 Johnson S L, Ahsan T, Polymers, Laminations & Coatings Conference, TAPPI Press, Atlanta, 1997.
84 Hancock M, *Plasticulture*, **79**, 4, 1988.
85 Hancock M, Marsh J E, Lee R L, Mineral Catalyzed Photodegradable Films, ECC International.
86 Sims G L A, Angus M W, Crosley I, Antec '95. Vol. II. Conference Proceedings, Boston, Ma., 7th-11th May 1995, 2166-70.
87 Guriya K C, Tripathy D K, *J. Appl. Polym. Sci.*, **62**, No.1, 1996, 117-27.
88 Alpern V, Shutov F, *Prog. Rubb. Plast. Technol.*, **11**, No.4, 1995, 268-83.
89 Sombatsompop N, Sims G L A, *Cell. Polym.*, **15**, No.5, 1996, 317-34.
90 Sims G L A, Sombatsompop N, *Cell. Polym.*, **15**, No.2, 1996, 90-104.
91 Malanda L M, Park C B, Balatinecz J J, Antec '96. Volume II. Conference proceedings, Indianapolis, 5th-10th May 1996, p.1900-7.
92 Gao Y C, Lelievre J, *Polym. Engng. Sci.*, **34**, No.18, 1994, 1369-76.
93 Lambert C, Chauvet H, Larroque M, *Polymer*, **37**, No.15, 1996, 3441-5.

94 Bijwe J, *Polym. Composites*, **18**, No.3, 1997, 378-96.
95 Thomas R W, Ancelet C R, Brzuskiewicz J E, Antec '93. Conference Proceedings, New Orleans, La., 9th-13th May 1993, Vol. II, 2005-8.
96 Mwila J, Miraftab M, Horrocks A R, *Polym. Degradat. Stabil.*, **44**, No.3, 1994, 351-6.
97 Wypych G, Handbook of Material Weathering, ChemTec Publishing, Toronto, 1996.
98 Hauser R L, Woods D W, Krause-Singh J, Ferry S R, Antec '93. Conference Proceedings, New Orleans, La., 9th-13th May 1993, Vol. I, 341-6.
99 Van Beek G A, Pang S S, Lea R H, Antec '93. Conference Proceedings, New Orleans, La., 9th-13th May 1993, Vol. I, 602-4.
100 Anantharaman M R, Kurian P, Banerjee B, Mohamed E M, George M, *Kaut. u. Gummi Kunst.*, **49**, No.6, 1996, 424-6.
101 Fiske T, Gokturk H S, Yazici R, Kalyon D M, Antec '97. Conference proceedings, Toronto, April 1997, 1482-6.
102 Weng J, Liu Q, Wolke J G C, Zhang D, De Groot K, *J. Mater. Sci. Lett.*, **16**, 1997, 335-7.
103 Jansen J A, de Ruijter J E, Janssen P T M, Paquay Y G C J, *Biomaterials*, **16**, No.11, 1995, 819-27.
104 Deb S, Braden M, Bonfield W, *Biomaterials*, **16**, No.14, 1995, 1095-100.
105 Topoleski L D T, Ducheyne P, Cuckler J M, *J. Biomed. Mat. Res.*, **29**, No.3, 1995, 299-307.
106 Tamura J, Kawanabe K, Yamamuro T, Nakamura T, Kokubo T, Yoshihara S, Shibuya T, *J. Biomed. Mat. Res.*, **29**, No.5, 1995, 551-9.
107 McIlvaine J, Antec 95. Volume III. Conference proceedings, Boston, Ma., 7th-11th May 1995, 3346-9.
108 Bos M, Van Dam G W, Jongsma T, Bruin P, Pennings A J, *Composite Interfaces*, **3**, No.2, 1995, 169-76.
109 Dupraz A M P, de Wijn J R, v. d. Meer S A T, de Groot K, *J. Biomed. Mat. Res.*, **30**, No.2, 1996, 231-8.
110 Molino L N, Topoleski L D T, *J. Biomed. Mat. Res.*, **31**, No.1, 1996, 131-7.
111 Ruiz-Hitzky E, Aranda P, Casal B, Galvan J C, *Adv. Mat.*, **7**, No.2, 1995, 180-4.
112 Duval J M, Kemperman A J B, Folkers B, Mulder M H V, Desgrandchamps G, Smolders C A, *J. Appl. Polym. Sci.*, **54**, No.4, 1994, 409-18.
113 Li Shucai, Peng Weijang, Lu Xiuping, *Int. J. Polym. Mat.*, **29**, Nos.1-2, 1995, 37-42.
114 Beecroft L L, Ober C K, *Chem. Mater.*, **9**, 1997, 1302-17.
115 Schmidt H K, *Macromol. Symp.*, **101**, 1996, 333-42.
116 Al-Turaif H, Unertl W N, Lepoutre P, *J. Adhesion Sci. Technol.*, **9**, No.7, 1995, 801-11.
117 Liphard M, Von Rybinski W, Schreck B, *Prog. Coll. & Polym. Sci.*, **95**, 1994, 168-74.
118 Nagieb Z A, El-Sakr N S, *Polym. Degradat. Stabil.*, **57**, 1997, 205-9.
119 Rybkina E G, Tsybin E V, *Int. Polym. Sci. Technol.*, **23**, No.3, 1996, T/62-5.
120 Hagemeyer R W, Ed., Pigments for Paper, TAPPI Press, Atlanta, 1997.
121 Anderson T C, Yunko A L, *Pulp Paper*, **53**, 1983, 82.
122 Abdel-Aziz M M, Gwaily S E, *Polym. Degradat. Stabil.*, **55**, 1997, 269-74.
123 Brown N, Linnert E, *Reinf. Plast.*, **39**, No.11, 1995, 34-7.
124 Evans L R, Meeting of the Rubber Division, ACS, Montreal, May 5-8, 1996, paper D.
125 Byers J T, Meeting of the Rubber Division, ACS, Cleveland, October 17-20, 1995, paper B.
126 Pehlergard P, *Rubb. S. Africa*, **10**, No.5, 1995, 8-12.
127 Waddell W H, Evans L R, *Rubb. Chem. Technol.*, **69**, No.3, 1996, 377-423.
128 White L, *Eur. Rubb. J.*, **178**, No.8, 1996, 46-52.
129 *Eur. Rubb. J.*, **178**, No.8, 1996, 26.
130 *Eur. Rubb. J.*, **178**, No.7, 1996, 32.
131 Huybrechts F, *Kaut. u. Gummi Kunst.*, **48**, No.10, 1995, 713-7.
132 Forster F, Freund B, *Rubb. S. Africa*, **11**, No.3, 1995, 8-12.
133 EXP 200 for HTV-silicone Rubber. Technical Information TI 1146. Degussa 1995.
134 Nowak R, Schachtely U, Synthetic Silicas for Sealants. Number 63. Degussa
135 Ventresca D A, Berard M T, Antec '97. Conference proceedings, Toronto, April 1997, 3574-9.
136 *Plastics in Canada*, **3**, No.2, 1996, 69.
137 Uhlen H, **Recycling of PVC & Mixed Plastic Waste**, La Mantia F P, Ed., *ChemTec Publishing*, Toronto, 1996.

# 20

# Hazards in Filler Use

Most fillers used today are not hazardous materials but some are, those require special handling and processing to reduce the potential hazard. Table 20.1 gives an overview of properties of fillers which help in the determination of safe handling practices.

**Table 20.1. The properties of fillers**

| Name | CAS # | Health** | Flammability** | Reactivity** | Contact** | Storage color code*** | Toxicity LDLO, mg/kg | Carcinogenicity | TWA, mg/m$^3$ | Silicosis |
|---|---|---|---|---|---|---|---|---|---|---|
| Aluminum | 7429-90-5 | 1 | 4 | 2 | 1 | red | | no | 10 | |
| Aluminum hydroxide | 21645-51-2 | 1 | 0 | 0 | 1 | orange | 150 | no | | |
| Aluminum oxide | 1344-28-1 | 1 | 0 | 0 | 1 | orange | 90 | no | 10 | |
| Aluminum silicate | 1302-76-7 | | | | | | 90 | | 2 | |
| Ammonium phosphate | 7783-28-0 | 1 | 0 | 1 | 1 | orange | | no | 2 | |
| Anthracite | 120-12-7 | 1 | 1 | 0 | 1 | orange | 430 | no | | |
| Antimony pentoxide | 1314-60-9 | | | | | | 4000 | | 0.5 | |
| Antimony trioxide | 1309-64-4 | 3 | 0 | 0 | 2 | blue | 3250 | no | 0.5 | |
| Asbestos | 12001-29-5 | | | | | | 300 | yes | 0.2* | |
| Ash, coal | | 1 | 0 | 0 | 1 | orange | | | | yes* |
| Ash, fly | | | | | | orange | | no | 10 | |
| Barium sulfate | 7727-43-7 | 1 | 0 | 0 | 0 | orange | | no | 10 | |
| Bauxite | 1344-28-1 | | | | | | | no | 10 | |
| Bentonite | 1302-78-9 | | | | | | 1200 | | | |
| Beryllium oxide | 1304-56-9 | 3 | 1 | 0 | | red | | | 0.002 | |
| Boron | 7440-42-8 | | 4 | | | | 310 | | | |
| Calcite | 471-34-1 | 0 | 0 | 0 | 1 | orange | | no | 10 | yes |
| Calcium aluminate | 12042-68-1 | 1 | 1 | 0 | | orange | | | | |
| Calcium borate | 12007-56-6 | 2 | | | | | | | | |

| Name | CAS # | Health** | Flammability** | Reactivity** | Contact** | Storage color code*** | Toxicity LDLO, mg/kg | Carcinogenicity | TWA, mg/m$^3$ | Silicosis |
|---|---|---|---|---|---|---|---|---|---|---|
| Calcium carbonate | 471-34-1 | 0 | 0 | 0 | 1 | orange | | no | 10 | yes* |
| Calcium hydroxide | 1305-62-0 | 1 | 0 | 1 | 2 | orange | 7340 | no | 5 | |
| Calcium silicate | 1344-95-2 | | | | | | | | 10 | |
| Calcium sulfate anh. | 7778-18-9 | 1 | 0 | 0 | 1 | orange | | no | | |
| Carbon black | 1333-86-4 | 1 | 0 | 0 | 1 | orange | | no | 3.5 | |
| Clay | | 1 | 0 | 0 | 0 | orange | | | | yes* |
| Copper spheres | 7440-50-8 | 0 | 0 | 0 | 1 | orange | | no | 1 | |
| Crystobalite | 14464-46-1 | | | | | | | | 0.05 | yes |
| Dawsonite | 12011-76-6 | | | | | | 200 | | | |
| Diatomaceous earth | 61790-53-2 | | | | | | | no | 10 | yes* |
| Dolomite | 471-34-1 | 0 | 0 | 0 | 1 | orange | | no | 10 | yes |
| Ferrite | | 0 | 0 | 0 | 0 | orange | | | | |
| Glass fibers | | 1 | 0 | 0 | 1 | orange | | no | 10 | |
| Gold | 7440-57-5 | | | | | | 58 | | | |
| Graphite | 7782-42-5 | | | | | | | | 2 | yes* |
| Iron | 7439-89-6 | 1 | 3 | 1 | 0 | red | 20 | no | | |
| Kaolin | 1332-58-7 | 0 | 0 | 0 | 1 | orange | | no | 10 | |
| Lead | 7439-92-1 | | 3 | | | | 160 | | 0.15 | |
| Limestone | 471-34-1 | 0 | 0 | 0 | 1 | orange | | no | 10 | yes |
| Lithium aluminum silicate | 1302-66-5 | | | | | | | | 5 | |
| Lithopone | 1345-05-7 | 2 | | | | | | | | |
| Magnesium aluminum silicate | 12174-11-7 | | | | | | | | 10 | yes* |
| Magnesium carbonate | 23389-33-5 | 1 | 0 | 1 | 0 | orange | | no | | |
| Magnesium hydroxide | 1309-42-8 | 1 | 0 | 0 | 1 | orange | | no | | |
| Magnesium oxide | 1309-48-4 | 1 | 0 | 1 | 1 | orange | | no | 10 | |
| Marble | 471-34-1 | 0 | 0 | 0 | 1 | orange | | no | 10 | |
| Mica, muscovite | 12001-26-2 | 1 | 0 | 0 | 1 | orange | | | 3 | yes* |
| Molybdenum | 7439-98-7 | | 3 | | | | 125 | | 10 | |
| Molybdenum disulfide | 1317-33-5 | | | | | | | | | |
| Montmorillonite | 98901-77-4 | | | | | | | no | | |
| Nickel | 7440-02-0 | | | | | | | yes | 0.05 | |
| Palladium | 7440-05-3 | | 3 | | | | | | | |

| Name | CAS # | Health** | Flammability** | Reactivity** | Contact** | Storage color code*** | Toxicity LDLO, mg/kg | Carcinogenicity | TWA, $mg/m^3$ | Silicosis |
|---|---|---|---|---|---|---|---|---|---|---|
| Perlite | 93763-70-3 | 1 | 0 | 0 | 1 | orange | | | 10 | yes* |
| Pyrophyllite | | | | | | | | | 0.14 | |
| Ruthenium dioxide | 12036-10-1 | | | | | | | | | |
| Silica, fumed | 69013-64-2 | | | | | | | | 0.2 | no |
| Silica, fused | 60676-86-0 | | | | | | | | 0.1 | |
| Silica, hydrated | 7631-86-9 | | | | | | | | 10 | |
| Silica, precipitated | 1343-98-2 | 1 | 0 | 0 | 1 | orange | | no | 10 | |
| Silica, quartz | 14808-60-7 | | | | | | | | 0.1 | yes |
| Silica, sand | 7631-86-9 | 1 | 0 | 0 | 1 | orange | | | 80 | |
| Silica, gel | 1343-98-2 | 1 | 0 | 0 | 1 | orange | | no | 10 | |
| Silver | 7440-22-4 | | | | | | | | 0.1 | |
| Smectite | 12199-37-0 | | | | | | | | 2 | yes* |
| Sodium aluminum silicate | 1344-00-9 | 1 | 0 | 0 | 0 | orange | | | | |
| Soot | | 1 | 2 | 0 | 0 | orange | | yes | | |
| Starch | 9005-25-8 | 1 | 2 | 0 | 0 | orange | | | 15 | |
| Syenite | 37244-96-5 | | | | | | | | 10 | no |
| Talc | 14807-96-6 | 1 | 0 | 0 | 1 | orange | | no* | 2 | |
| Titanium boride | 12045-63-5 | 1 | 0 | 0 | 0 | orange | | | | |
| Titanium dioxide | 13463-67-7 | 0 | 0 | 0 | 1 | orange | | no | 10 | |
| Tungsten | 7440-33-7 | | 4 | | | red | | | 5 | |
| Vermiculite | 1318-00-9 | 1 | 0 | 0 | 1 | orange | | no | 10 | |
| Wood flour | 9004-34-6 | | | | | | | | 10 | |
| Wollastonite | 13983-17-0 | 1 | 0 | 0 | 0 | orange | | no | | |
| Zeolite | 1344-00-9 | 1 | 0 | 1 | 1 | orange | | | | |
| Zinc borate | 1332-07-6 | | | | | | | | | |
| Zinc oxide | 1314-13-2 | 2 | 0 | 0 | 0 | orange | 630 | no | 5 | |
| Zinc sulfide | 1314-98-3 | | | | | | | | | |
| Zirconium oxide | 1314-23-4 | 1 | 0 | 0 | 1 | orange | | | 5 | |
| Zirconium silicate | 14940-68-2 | | | | | | | | 5 | |

*in carcinogenicity means - if contains asbestos; in silicosis risk means - if contains crystalline silica; in TWA means - fibers per $cm^3$.
**the following rating is used: 0 - no hazard, 1 - slight, 2 - moderate, 3 - severe, 4 - extreme hazard
***storage area is chosen as follows: orange - general, blue - special, red - hazardous

Several fillers, including asbestos, nickel, and soot, are carcinogens. The health risk due to asbestos exposure is well documented.[1] Exposure to asbestos increases the risk of lung cancer, mesothelioma, asbestosis, pleural thickening, and gastrointestinal cancer. The first four diseases are caused by inhalation of asbestiform fibers, whereas the last is caused by ingestion.

The worst lung damage is caused by asbestos fibers which reach the lower parts of the respiratory tract. Fiber length is also a factor. The fibers longer than 3 μm are the most dangerous. No difference was found[1] between various forms of asbestos. The report[1] also considers other types of fibrous materials both natural and man-made, such as attapulgite, erionite, fibrous glass, mineral wool, and ceramic fiber. These fibers also fulfill the size criteria (diameter, length), with the exception, perhaps, of attapulgite, the fibers of which are generally shorter than1 μm.

People of central Turkey who live in an area rich in erionite frequently suffer from lung cancer and mesothelioma. It is not yet known definitely, if the risk of cancer from exposure to other fibrous materials is considerably lower than that of exposure to asbestos. But the existing data seems to show that the risk of health problems, due to exposure to fibers other than asbestos is less.

Table 20.2 summarizes the data on the effect of asbestos exposure on the death rate caused by various asbestos-related diseases.

**Table 20.2. Mortality data for asbestos related deseases[1]**

| Occupation | # exposed | Total death | Mesothelioma | Respiratory cancer | Gastrointestinal cancer |
|---|---|---|---|---|---|
| Mining & milling | 14,332 | 4,711 | 37 | 372 | 312 |
| Manufacturing | 20,474 | 3,110 | 95 | 547 · | 284 |
| Insulation | 18,594 | 2,871 | 226 | 557 | 150 |
| Shipyards | 8,416 | 3,013 | 31 | 211 | 147 |

These data link asbestos to various forms of cancer. In the last several years, industry has made a big effort to eliminate asbestos from numerous products and asbestos is no longer an active factor.

Some types of talc contain high concentrations of tremolite. Tremolite has two varieties: asbestiform and nonasbestiform. Asbestiform may cause tumors in humans similar to various other forms of asbestos. The nonasbestiform tremolite was recently studied by a group of experts who reviewed previous studies and conducted its own experiments.[2,3] Nonasbestiform tremolite contained in New York state talcs did not produce a carcinogenic response in test animals. It has morphological features different from asbestiform.[2,3] Since talc is commonly used in cosmetic powders these studies are needed to protect the general public from exposure and also prevent the exclusion of materials on any basis other than scientific evaluation. But studies of pharmaceutical talc revealed that some of these talcs contained about 200,000 fibers longer than 5 μm in 1 mg of their mass.[4] A continuing evaluation is needed to segregate safe mineral forms from the dangerous asbestiform variety.

Fillers containing crystalline silica pose dangers. In Table 20.1, there are fourteen groups of fillers which contain some quantities of crystalline silica. In the United States, about 1 million workers are exposed to crystalline silica of these 250 will die from silicosis. The group most exposed are construction workers who use materials containing crystalline silica for sandblasting. The plastics industry was not in the top 10 list of industries but the potential danger in the plastics industry should be understood since the risk can be eliminated with proper protection. Chronic silicosis occurs after 10 or more years of overexposure.[5] But accelerated silicosis from higher exposures can develop in a shorter period of time (5-10 years). The action of crystalline silica on the lungs results in the development of a diffuse, nodular fibrosis involving the parenchyma and the lymphatic system. Fibrosis increases for several years after contact is terminated. Typical complications include cardiac failure and tuberculosis. Fillers containing crystalline silica are included by IRAC in Group 3 which contains materials for which there is insufficient evidence of causing cancer. All fillers which contain more than 0.1% crystalline silica require a warning in MSDS and on the product label.

Various industries use large quantities of fumed and precipitated silica (amorphous silica). These forms of silica do not cause silicosis but protection from exposure to dust must be provided because any excessive dust decreases efficiency of lungs.[6-8]

Other than these, the fillers discussed in this book do not pose special dangers under their normal conditions of use other than when they are accidently exposed to high temperatures. In such a situation, sulfates decompose, emitting toxic fumes; lithopone emits highly toxic fumes containing sulfur oxides and $H_2S$; antimony and zinc oxides emit toxic metal and zinc oxide, respectively.

Thus, with the exception of fibrous fillers, most fillers are relatively harmless materials, although they must be used with reasonable care.

Now information has been developed for carbon black.[9-11] International Agency for Research on Cancer changed the classification of carbon black from Group 3 to Group 2 (possible carcinogen). This was based on a study on rats. This caused one manufacturer, Degussa, to release results of medical studies on 677 employees indicating that carbon black could not be linked to cancer. The results of studies in the USA (2500 workers) and in the UK (1422) came to the same conclusions.[10] How can this conflicting data be assessed? Recent changes in the implementation of new regulations have emphasized cooperation between industry and regulatory agencies.

These new cooperation efforts seem to be effective.[11] A commissioned study was carried out by the University of Birmingham to evaluate occupational exposure to carbon black. The study was conducted in 1987 and again in 1992. During the second study it was discovered that the exposure of workers was reduced by one half compared with the first study. The data from the first study led to the voluntary changes which gave the excellent results.

The potential for dust to explode is the other significant hazard of filler use. Particle size, concentration of particles, and strength of the source of ignition are important criteria for evaluating hazard. Of course, the material forming the dust must also be combustible. Smaller particles are easier to ignite because they have a larger exposed surface area available for ignition and combustion. The lower explosive limit depends on particle size. Also, smaller particles can accumulate greater electrical charges. The lower explosive limit is usually the factor of control. The upper limits are not established. The most violent explosions occur when the amount of available oxygen matches the amount required for complete oxidation of the particulate. The data included in the Table 20.1 shows that metal particulates are the most hazardous explosive dusts. Carbon black is also a hazardous material in this respect. The explosibility index and the ignition sensitivity of carbon blacks is much lower than 0.1.[9] Carbon black presents more of a fire hazard than a dust explosion hazard. None of the 19 samples tested was ignited by an electric spark source.[9] Ignition temperature in air ranges from 445 to 890°C, and in oxygen, from 320 to 450°C. The concentration of oxygen required for ignition was in the range of 35 to 74 percent. The explosibility index of carbon black rose to more than 10 (several orders of magnitude) when carbon black was in a mixture with polystyrene and dioctyl phthalate. Such an increase indicates that mixtures of carbon black with other materials might be even more hazardous.

The moisture level of the particulate increases its ignition temperature. Also, the presence of an inert solid reduces its combustibility. Some particulates (limestone) are used to extinguish fires but mixtures of particulates may increase the explosion hazard. One simple method of reducing explosion hazard is to use an atmosphere of inert gas in processing and transportation. A dust cloud can be ignited by any conventional source but friction sparks, hot surfaces, the heat of friction, and static sparks are the most common sources of ignition.

Another potential hazard in using fillers is the possible formation of toxic materials. Twelve samples of silica fillers were studied with respect to their effect on the formation of N-nitrosamines in rubber compounds.[12] Figures 20.1 and 20.2 provide these data. Filler absorption of $NO_x$ increases as the filler's specific surface area increases. This absorbed $NO_x$ was found to react with amines in rubber and to produce N-nitrosamines. Greater absorption

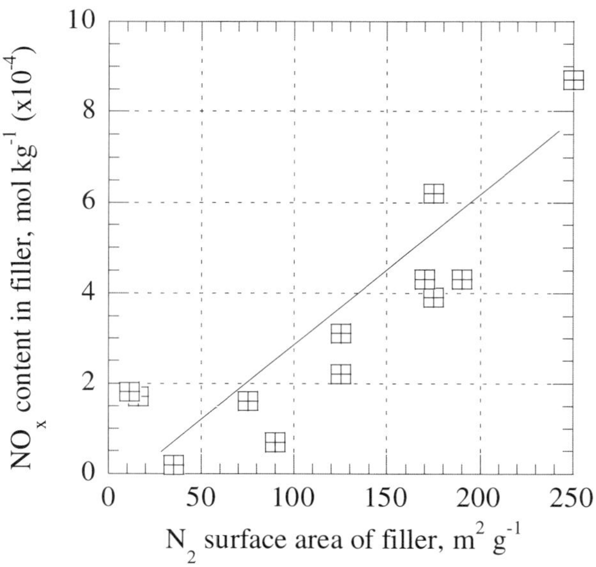


Figure 20.1. $NO_x$ content vs. filler surface area. [Adapted, by permission, from Gorl U, De Kok J J, Bomal Y, Cochet P, Mueller H, *Kaut. u. Gummi Kunst.*, **47**, No.6, June 1994, 430-4.]

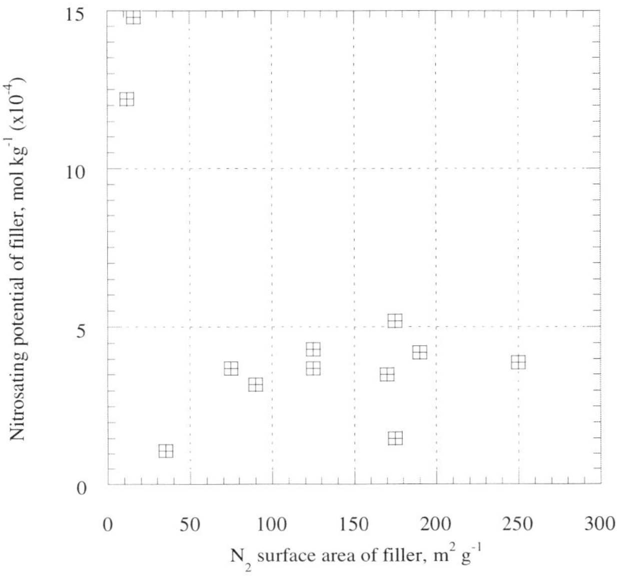


Figure 20.2. Nitrosating potential as a function of filler surface area. [Adapted, by permission, from Gorl U, De Kok J J, Bomal Y, Cochet P, Mueller H, *Kaut. u. Gummi Kunst.*, **47**, No.6, June 1994, 430-4.]

of $NO_x$ did not cause an increase in the formation of nitrosamines. This may be explained by the fact that a precursor is trapped in the porous structure of silica and retained because of its higher chemical activity.

## REFERENCES


1 **Asbestiform Fibers**. Non-occupational Health Risks. *National Academy Press,* Washington, D.C., 1984.
2 Roedelsperger K, Lojewski G H, Brueckel B, Woitowitz H J, *Staub-Reinhalt. Luft*, **44** (2), 62 (1984).
3 **The asbestiform and nonasbestiform mineral growth habit and the relationship to cancer studies**. Report. America Mining Congress and National Stone Association.
4 Wylie A G, Skinner H C W, Marsh J, Snyder H, Garzione C, Hodkinson D, Winters R, Mossman B T, *Toxicology Appl. Pharmacology*, 147, 1997.
5 Sax N I, **Dangerous Properties of Industrial Materials**. *Van Nostrand Reinchold Co.*, New York, 1990.
6 **Hi-Sil Dust Exposure and OSHA Regulations**, *PPG Industries*, Pittsburgh, 1976.
7 **Aerosil. Fumed Silica**. *Degussa*, Frankfurt.
8 **Cab-O-Sil. Properties and Functions**. *Cabot Corp.*, Tuscola, 1985.
9 Nagy J, Dorsett H G, Cooper A R, *US Dept. of the Interior, Bureau of Mines,* TN23.U7, no. 6597.
10 *Eur. Rubb. J.*, **178**, No.8, 1996, 26.
11 Gardiner K, Calvert I A, van Tongeren M J A, Harrington J M, *Ann. Occup. Hyg.*, **40**, No.1, 1996, 65-77.
12 Gorl U, De Kok J J, Bomal Y, Cochet P, Mueller H, *Kaut. u. Gummi Kunst.*, **47**, No.6, June 1994, 430-4.

# INDEX OF ABBREVIATIONS

| | |
|---|---|
| AAS | acrylonitrile, acrylate (ester), styrene copolymer |
| ABA | acrylonitrile, butadiene, acrylate |
| ABR | acrylate, butadiene rubber |
| ABS | acrylonitrile, butadiene, styrene copolymer |
| ACRYLICS | copolymer of methacrylic and acrylic monomers |
| ACS | acrylonitrile, chlorinated polyethylene, styrene terpolymer |
| AMA | acrylate, maleic anhydride terpolymer |
| AMMA | acrylonitrile, methyl methacrylate |
| APO | amorphous polyolefin |
| APES | γ-aminopropyltriethoxysilane |
| APTS | 3-aminopropyltriethoxysilane |
| AS | acrylonitrile, styrene copolymer |
| ASA | acrylonitrile, styrene acrylate |
| BMC | bulk molding compound |
| BMI | bismaleimide |
| CA | cellulose acetate |
| CAB | cellulose acetate butyrate |
| CAP | cellulose acetate propionate |
| CN | cellulose nitrate (celluloid) |
| COC | cycloolefin copolymer |
| COP | copolyester thermoplastic elastomer |
| CPE | chlorinated polyethylene |
| CPVC | chlorinated polyvinyl chloride |
| CR | chloroprene, chlorinated rubber |
| CSM | chlorosulfonated polyethylene |
| CTA | cellulose triacetate |
| CTFE | chlorotrifluoroethylene |
| DAP | diallyl phthalate (thermoset) |
| EAA | ethylene acrylic acid copolymer |
| EC | ethyl cellulose |
| ECTFE | ethylene chlorotrifluoroethylene |
| EEA | ethylene, ethyl acetate copolymer |
| EMAC | ethylene, methyl acrylate copolymer |
| EnBA | ethylene, n-butyl acetate |
| EP | epoxy |
| EPDM | ethylene, propylene diene monomer rubber |
| EPM | ethylene, propylene copolymer rubber |
| EPR | ethylene, propylene rubber |
| EPS | expandable polystyrene |
| ES | epoxy silane |
| ETFE | ethylene tetrafluoroethylene |
| EVA | ethylene, vinyl acetate |
| E/VAC | ethylene, vinyl acetate copolymer |
| EVOH | ethylene, vinyl alcohol |

| | |
|---|---|
| FEP | fluorinated ethylene propylene |
| FRP | fiber reinforced plastic |
| | |
| HDPE | high density polyethylene |
| HIPS | high impact polystyrene |
| HMC | high strength molding compound |
| HMWHDPE | high molecular weight high density polyethylene |
| | |
| I | ionomer |
| IIR | isobutene, isoprene rubber or butyl rubber |
| IPN | interpenetrating polymer network |
| | |
| LCP | liquid crystal polymer |
| LDPE | low density polyethylene |
| LLDPE | linear low density polyethylene |
| LPE | linear polyethylene |
| | |
| MA | maleic anhydride |
| MABS | methyl methacrylate, ABS copolymer |
| MAS | methyl methacrylate, acrylonitrile, styrene copolymer |
| MBS | methyl methacrylate, butadiene, styrene copolymer |
| MDPE | medium density polyethylene |
| MF | melamine formaldehyde |
| MP | melamine phenolic |
| | |
| NBR | nitrile, butadiene rubber |
| NR | natural rubber |
| | |
| OMCTS | octamethylcycloethoxysiloxane |
| OSA | olefin modified styrene acrylonitrile |
| | |
| P | phenolic |
| PA | polyamide (nylon) |
| PAA | polyacrylic acid |
| PAI | polyamide-imide |
| PAEK | polyaryletherketone |
| PAK | polyester alkyd |
| PAL | polyanaline |
| PAN | polyacrylonitrile |
| PANI | polyaniline |
| PAr | polyarylate |
| PARA | polyaryl amide |
| PARS | polyaryloxysiloxane |
| PAS | polyarylsulfone |
| PB | polybutylene |
| PBAN | polybutadiene acrylonitrile |
| PBD | polybutadiene |
| PBG | polybutyleneglycol |
| PBI | polybenzimidazole |
| PBN | polybutylene naphthalate |
| PBS | polybutadiene styrene |

| | |
|---|---|
| PBT | polybutylene terephthalate |
| PC | polycarbonate |
| PC/ABS | polycarbonate/acrylonitrile butadiene styrene blend |
| PCL | polycaprolactone |
| PCT | polycyclohexylene terephthalate |
| PCTFE | polymonochlorotrifluoroethylene |
| PDMS | polydimethylsiloxane |
| PE | polyethylene |
| PEA | polyethylacrylate |
| PEBA | polyether block amide or polyester block amide |
| PEC | polyestercarbonate |
| PEEK | polyetheretherketone |
| PEG | polyethyleneglycol |
| PEI | polyetherimide |
| PEK | polyetherketone |
| PEKK | polyetherketoneketone |
| PEN | polyethylene napthalene |
| PEO | polyethylene oxide |
| PES | polyethersulfone |
| PET | polyethylene terephthalate |
| PET-G | glycol modified polyethylene terephthalate |
| PFA | perfluoroalkoxy |
| PI | polyimide |
| PI | polyisoprene |
| PIB | polyisobutylene |
| PIR | polyisocyanurate |
| PMAN | polymethactylonitrile |
| PMMA | polymethylmethacrylate |
| PMP | polymethylpentene |
| PO | polyolefin |
| POM | polyoxymethylene |
| PP | polypropylene |
| PPA | polyphthalamide |
| PPC | chlorinated polypropylene |
| PPC | polyphthalate carbonate |
| PPE | polyphenylene ether |
| PPI | polymeric polyisocyanate |
| PPO | polyphenylene oxide |
| PPOX | polypropylene oxide |
| PPS | polyphenylene sulfide |
| PPSU | polyphenylene sulfone |
| PPT | polypropylene terephthalate |
| PS | polystyrene |
| PSF | polysulfone, also PSO, PSU, PSUL |
| PSO, PSU | polysulfone |
| PTFE | polytetrafluoroethylene |
| PTMT | polytetramethylene terephthalate |
| PU, PUR | polyurethane |
| PVA | polyvinyl alcohol |
| PVAc | polyvinyl acetate |
| PVB | polyvinyl butyryl |

| | |
|---|---|
| PVC | polyvinyl chloride |
| PVCA | polyvinyl chloride acetate |
| PVDA | polyvinylidene acetate |
| PVDC | polyvinylidene chloride |
| PVDF | polyvinylidene fluoride |
| PVF | polyvinyl fluoride |
| PVK | polyvinyl carbazole |
| PVOH | polyvinyl alcohol |
| PVP | polyvinyl pyrrolidone |
| SAN | styrene, acrylonitrile |
| SB | styrene, butadiene |
| SBR | styrene, butadiene rubber |
| SBS | styrene, butadiene, styrene block copolymer |
| SEBS | styrene, ethylene, butylene, styrene block copolymer |
| SI | silicone |
| SIS | styrene, isoprene, styrene block copolymer |
| SMA | styrene, maleic anhydride copolymer |
| SMC | sheet molding compound |
| SMMA | styrene, methyl methacrylate |
| SVA | styrene, vinyl, acrylonitrile |
| TEO | thermoplastic elastic olefin |
| TESPT | bis-triethoxysilylpropyl-tetrasulfone |
| TOAB | tetraoctylammonium bromide |
| TPE | thermoplastic elastomer |
| TPO | thermoplastic elastomer-olefinic |
| TPE-S | thermoplastic elastomer-styrenic |
| TPU | thermoplastic polyurethane |
| UF | urea formaldehyde |
| UHDPE | ultrahigh density polyethylene |
| ULDPE | ultra low density polyethylene |
| UP, UPE | unsaturated polyester (thermoset) |
| VA | vinyl acetate |
| VAE | vinyl acetate, ethylene |
| VLDPE | very low density polyethylene |

# Directory of Filler Manufacturers and Distributors

| Company name and address | Fillers | Page |
|---|---|---|
| *Abrasivos y Maquinaria, S.A.*<br>Calle Caspe, 79, 2°<br>08013 Barcelona, Spain<br>tel: 34 3 266 10 00<br>fax: 34 3 247 07 21 | glass beads | 87 |
| *Accuratus Ceramic Corporation*<br>14A Brass Castle Road<br>Washington, NJ 07882, USA<br>tel: 1 908 689 0880<br>fax: 1 908 689 8794 | beryllium oxide<br>boron nitride | 45<br>46 |
| *Ace International Inc.*<br>520 North Gold Street<br>Centralia, WA 98531-0885, USA<br>tel: 1 360 736 9999<br>fax: 1 360 736 9797 | wood flour<br>bark flour | 166<br>166 |
| *ACuPowder International, LLC*<br>901 Lehigh Avenue<br>Union, NJ 07083, USA<br>tel: 1 908 851 4500<br>fax: 1 908 851 4597 | copper, brass, and bronze powders | 77 |
| *Advanced Ceramics Corporation*<br>11907 Madison Avenue<br>Lakewood, OH 44107-5026<br>tel: 1 216 529 3900<br>toll free: 800 822 4322 (USA only)<br>fax: 1 216 529 3975 | boron nitride | 46 |
| *Agrashell, Inc.*<br>5934 Keystone Drive<br>Bath, PA 18014, USA<br>tel: 1 610 837 6705<br>fax: 1 610 837 8802 | nut shell flour | 166 |
| *Akzo Nobel Aramid Products Inc.*<br>801 F Blacklawn Road<br>Conyers, GA 30207, USA<br>tel: 1 770 785 2148<br>fax: 1 770 929 8138 | aramid fibers | 178 |

| Company name and address | Fillers | Page |
|---|---|---|
| *Albion Kaolin Company*<br>1 Albion Road<br>Hephzibah, GA 30815, USA<br>tel: 1 706 592 9121<br>fax: 1 706 592 9824 | kaolin | 99 |
| *Albright & Wilson UK Ltd.*<br>P.O. Box 3<br>210-222 Hagley Road West<br>Oldbury, Warley<br>West Midlands B68 0NN, England<br>tel: 44 0121 429 4942<br>fax: 44 0121 420 5151<br><br>*Albright & Wilson Americas*<br>P.O. Box 4439<br>Glen Allen, VA 23058-4439, USA<br>tel: 1 804 968 6300<br>toll free:1 800 446 3700<br>fax: 1 804 968 6385 | red phosphorus | |
| *Alcan Chemicals Europe*<br>Park, Gerrards Cross<br>Buckinghamshire SL9 0QB, England<br>tel: 44 1592 411 000<br>fax: 44 1753 233 444<br><br>*Alcan Aluminio do Brasil Ltda*<br>Av. Americo R. Gianetti, 521<br>35. 400-000-Ouro Preto-MG, Brasil<br>tel: 55 31 551 9140<br>fax: 55 31 551 9260<br><br>*Alcan Chemicals (North America)*<br>Division of Alcan Aluminum Corporation<br>3690 Orange Place<br>Suite 400<br>Cleveland, OH 44122-4438, USA<br>tel: 1 216 765 2550<br>toll free: 800 321 3864 (USA only)<br>fax: 1 216 765 2570<br><br>*Handy Chemical Ltd.*<br>120 de L'Industrie Blvd.<br>Candiac, Quebec JR5 1J2, Canada<br>tel: 1 514 659 9693<br>fax: 1 514 659 6850<br><br>*Indian Aluminium Company, Ltd.*<br>1 Middleton Street<br>Calcutta 700 071, India<br>tel: 91 33 240 2210<br>fax: 91 33 247 3808 | aluminum oxide<br>aluminum trihydroxide<br>zinc borate<br>zinc hydroxystannate<br>zinc stannate | 20<br>22<br>171<br>22<br>175 |

| Company name and address | Fillers | Page |
|---|---|---|
| *American Metal Fibers, Inc.*<br>2889 North Nagel Court<br>Lake Bluff, IL 60044-1460, USA<br>tel: 1 847 295 1200<br>fax: 1 847 295 0520 | steel, copper, and brass fibers | 107 |
| *American Wood Fibers*<br>100 Alderson Street<br>Schofield, WI 54476-0468, USA<br>tel: 1 715 355 1900<br>toll free: 800 642 5448 (USA only)<br>fax: 1 715 355 5721 | wood fibers | 166 |
| *AML Industries, Inc.*<br>P.O. Box 4110<br>Warren, OH 44482, USA<br>tel: 1 330 399 5000<br>toll free: 800 860 5823 (USA only)<br>fax: 1 330 399 5005 | graphite<br>molybdenum disulfide | 92<br>117 |
| *Amoco Performance Products, Inc.*<br>4500 McGinnis Ferry Road<br>Alpharetta, GA 30202-3914, USA<br>tel: 1 770 772 8200<br>toll free: 800 222 2448<br>fax 1 770 772 8753 | carbon fibers (pitch based) | 180 |
| *Amspec Chemical Corporation*<br>751 Water Street<br>Gloucester City, NJ 08030, USA<br>tel: 1 609 456 3930<br>fax: 1 609 456 6704 | aluminum trihydroxide<br>antimony trioxide | 22<br>29 |
| *AMSIB*<br>108 Dover Neck Road<br>Dover, NH 03820-4931, USA<br>tel: 1 603 742 4054<br>fax: 1 603 742 9158 | schungite | |
| *Anthracite Industries, Inc.*<br>P.O. Box 112<br>Sunbury, PA 17801, USA<br>tel: 1 717 286 2176 | anthracite | 25 |
| *Anval, Inc.*<br>301 Route 17 North<br>Suite 800<br>Rutherford, NJ 07070, USA<br>tel: 1 201 939 1065<br>toll free: 1 800 992 6825<br>fax: 1 201 939 1608 | stainless steel powders | 108 |

| Company name and address | Fillers | Page |
|---|---|---|
| *Applied Carbon Technology*<br>953 Route 202 North<br>Somerville, NJ 08876, USA<br>tel: 1 908 707 0807<br>toll free: 1 800672 7979<br>Fax: 1 908 707 0806 | graphite | 92 |
| *Asheville Mica Company*<br>900 Jefferson Avenue<br>Newport News, VA 23607-6120, USA<br>tel: 1 804 244 7311<br>fax: 1 804 245 5236 | mica | 113 |
| *Aspect Minerals*<br>Spruce Pine, NC 28777, USA<br>tel: 800 803 7979 (USA only)<br>fax: 1 704 765 7887 | mica | 113 |
| *Ausimont USA Inc.*<br>P.O. Box 1838<br>Morristown, NJ 07962-1838, USA<br>phone: 1 201 292 6250<br>toll free: 1 800 323 2874<br>fax 1 201 292 0886<br><br>*Ausimont S.p.A.*<br>Via San Pietro 50<br>20021 Bollate<br>Milano, Italy<br>phone: 39 2 6270 6236<br>fax: 39 2 6270 6538<br><br>*Ausimont UK Ltd.*<br>111 Upper Richmond Road<br>Putney<br>London SW15 2TJ, England<br>phone: 44 081 780 0399<br>fax: 44 081 780 2871<br><br>*Montedison Chemicals*<br>Avenida Diagonal 477<br>18 ½ A&B<br>08036 Barcelona, Spain<br>phone: 34 03 419 9200<br>fax: 34 03 4199 9323<br><br>*Montedison France*<br>Defence Avenue 1<br>Rue de Trois Fontanot<br>Immueble MB9<br>92732 Nanterre Cedex, France<br>phone: 33 46 96 34 00<br>Fax: 33 47 25 94 29<br><br>*continued on the next page* | PTFE powder | 122 |

| Company name and address | Fillers | Page |
|---|---|---|
| *Montedison Scandinavia AB*<br>Drakegatan 6<br>41 250 Gothenburg, Sweden<br>tel: 46 031 830 060<br>fax: 46 031 833 975<br><br>*Montedison Deutschland GmbH*<br>Kolner Strasse 3A<br>Postfach 5648<br>D-6236 Eschborn/TS.1, Germany<br>tel: 49 06196 92201<br>fax: 49 06196 482389<br><br>*Montedison Argentina SA*<br>Av. Eduardo Madero 942 piso 8<br>106 Buenos Aires, Argentina<br>tel: 54 1 3 3 011<br>fax: 54 1 112763<br><br>*Montedison do Brazil Ltd.*<br>Avenida Brig. Faria<br>Lima 888 - 5 ½ andar<br>CEP 01452 Jardim Paulistano<br>Sao Paulo, Brazil<br>tel: 55 11 2103325<br>fax: 55 11 8131700<br><br>*Ausimont K.K.*<br>3rd Floor Izumi Akasaka building<br>Minato-ku, Tokyo<br>Japan 107<br>tel: 81 3 3224 7212<br>fax: 81 3 3224 7218<br><br>*Ausimont Singapore Pte. Ltd.*<br>70 Shenton Way<br>#12-03 A/B/C Marina House<br>Singapore 0207<br>tel: 65 223 3581<br>fax: 65 223 2127 | | |
| *Barium & Chemicals, Inc.*<br>P.O. Box 218<br>Steubenville, OH 43952-5218, USA<br>tel: 1 614 282 9776 | barium sulfate | 36 |

| Company name and address | Fillers | Page |
|---|---|---|
| *Bromine Group Dead Sea*<br>P.O. Box 180<br>Beer Sheva 84101 Israel<br>tel: 972 7 629 7608<br>fax: 972 7 629 7819<br><br>*Eurobrom B.V.*<br>The Netherlands<br>tel: 31 70 340 8408<br>fax: 31 70 399 9035<br><br>*Bromine & Chemicals Ltd*<br>United Kingdom<br>tel: 44 171 493 9711<br>fax: 44 171 493 9714<br><br>*Eorobrom B.V. Espana*<br>Spain<br>tel: 34 3 317 9910<br>fax: 34 3 317 6403<br><br>*Ameribrom Inc.*<br>USA<br>tel: 1 800 280 2766<br>fax: 1 212 286 4475<br><br>*Bromisa Ind. E Com. Ltda*<br>Brasil<br>tel: 55 11 258 8288<br>fax: 55 11 259 8228<br><br>*Bromokem Ltd*<br>Japan<br>tel: 81 3 3278 1621<br>fax: 81 3 3278 1625<br><br>*Landkem (Pty.) Ltd.*<br>South Africa<br>tel: 27 11 788 5380<br>fax: 27 11 788 7504 | magnesium hydroxide | 106 |
| *Bekaert Corporation*<br>1395 South Marietta Parkway<br>Building 500, Suite 100<br>Marietta, GA 30067, USA<br>tel: 1 404 421 8520<br>fax: 1 404 426 8107<br><br>*N.V. Bekaer S.A*<br>Bekaer Fiber Technologies<br>Bekaertstraat 2<br>B-8550 Zwevegem, Belgium<br>tel: 32 56 76 61 11<br>fax: 32 56 76 79 66<br>*continued on the next page* | steel fibers | 108 |

| **Company name and address** | **Fillers** | **Page** |
|---|---|---|
| *Bekaer Deutschland GmbH*<br>Postfach 1261<br>D-6380 Bad Homburg v.d. Hohe 1<br>Germany<br>tel: 49 06172 1240<br>fax: 49 06172 35410<br><br>*Bekaer Japan*<br>934 Shin Tokyo Building<br>3-1 Marunouchi 3-Chome<br>Chiyoda-ku, C.P.O. Box 473<br>Tokyo 100-91, Japan<br>tel: 81 3 32142081<br>fax: 81 3 32163570 | | |
| *Buckman Laboratories, Inc.*<br>International Headquarters<br>1256 North McLean Blvd.<br>Memphis, TN 38108, USA<br>tel: 1 901 278 0330<br>toll free: 1 800 BUCKMAN<br>fax: 1 901 276 5443 | barium metaborate | 35 |
| *Burgess Pigment*<br>P.O. Box 349<br>Sandersville, GA 31082<br>tel: 1 912 552 2544<br>toll free: 1 800 841 8999<br>fax: 1 912 552 1772 | kaolin | 99 |
| *Byk Chemie USA*<br>524 South Cherry Street<br>P.O. Box 5670<br>Wallingford, CT 06492-7651, USA<br>tel: 1 203 265 2086<br>fax: 1 203 284 9158 | additives for filler processing | |
| *Cabot Corporation*<br>*Special Blacks Division*<br>157 Concord Road<br>Billerica, MA 01821-7001, USA<br>tel: 1 508 670 7035<br>toll free: 1 800462 2313<br>fax: 1 508 670 7035<br><br>*Cabot Corporation*<br>*Special Blacks Division*<br>200 Raritan Center Parkway<br>P.O. Box 338<br>Edison, NJ 08817, USA<br>tel: 1 908 225 1000<br>toll free: 1 800 648 7993<br>fax: 1 908 225 9644<br>*continued on the next page* | carbon black | 62 |

| Company name and address | Fillers | Page |
|---|---|---|
| *Cabot Performance Materials*<br>P.O. Box 1608<br>County Line Road<br>Boyertown, PA 19512-1608, USA<br>tel: 1 610 369 8500<br>fax: 1 610 367 2068<br><br>*Cabot Canada Ltd.*<br>350 Wilton Street<br>Sarnia, ON N7T 7N4, Canada<br>tel: 1 519 336 2261<br>fax: 1 519 336 8501<br><br>*Cabot Argentina, S.A.I.C.*<br>Larrabure 203<br>2804 Campana, Argentina<br>tel: 54 315 20130<br>fax: 54 315 20178<br><br>*Cabot Brasil*<br>Rua Beira Rio, 57-12° Andar<br>04548-906 Sao Paulo, SP Brazil<br>tel: 55 11 820 2711<br>fax: 55 11 820 9193<br><br>*Cabot Colombiana S.A.*<br>Carretera Mamonal<br>Apartado Aereo 2903<br>Cartagena, Colombia<br>tel: 57 53 685012<br>fax: 57 53 685486<br><br>*Cabot Europa G.I.E.*<br>*Special Blacks Division*<br>Le Nobel<br>2, rue Marcel Monge<br>92158 Suresnes Cedex, France<br>tel: 33 1 46 97 58 00<br>fax: 33 1 47 72 66 47<br><br>*Cabot Carbon Ltd.*<br>Lees Lane<br>Stanlow Ellesmere Port<br>South Wirral<br>Cheshire L65 4HT, England<br>tel: 44 51 355 3677<br>fax: 44 51 356 0712<br><br>*Cabot GmbH*<br>Josef-Bautz-Strasse 15<br>D-63457 Hanau, Germany<br>tel: 49 6181 5050<br>fax: 49 5181 57 1751 | barium titanate | 42 |

*continued on the next page*

| Company name and address | Fillers | Page |
|---|---|---|
| *Cabot Italiana, S.p.A.*<br>Casella Postale 444<br>via Baiona 190<br>48100 Ravenna, Italy<br>tel: 39 544 451612<br>fax: 39 544 451946 | | |
| *Cabot B.V.*<br>P.O. Box 1009<br>Botlekstraat 2<br>3180 AA Rozenburg, ZH<br>The Netherland<br>tel: 31 (0) 1819 45671<br>fax: 31 (0) 1819 45840 | fumed silica | 132 |
| *Cabot Leiden Technical Center*<br>Archimedesweg 15<br>2333 CM Leiden<br>The Netherland<br>tel: 31 71 600300<br>fax: 31 71 227129 | | |
| *Cabot S.A.*<br>Av. Princep d'Asturies<br>66 4rt 1a<br>08012 Barcelona, Spain<br>tel: 34 3 415 50 53<br>fax: 34 3 415 59 15 | | |
| *Cabot Specialty Chemicals, Inc.*<br>3-1-14 Shiba<br>Minato-ku, Tokyo 105. Japan<br>tel: 81 3 3457 7561<br>fax: 81 3 3457 7658 | | |
| *Cabot Australasia Ltd.*<br>P.O. Box 19<br>Altona, Victoria 3018, Australia<br>tel: 61 3 391 1622<br>fax: 61 3 391 9370 | | |
| *Cabot Pacific Area*<br>*Carbon Black Division*<br>c/o Cabot Specialty Chemicals, Inc.<br>6th Floor, RHB 1<br>424 Jalan TunRazak<br>50400 Kuala Lumpur, Malaysia | | |
| *United Carbon India Ltd.*<br>NKM International House<br>178 Babubhai M, Chinai Marg<br>Bombay, India 400 020<br>tel: 91 22 202 1914<br>fax: 91 22 2202 8823 | | |

| Company name and address | Fillers | Page |
|---|---|---|
| *Cancarb Ltd.*<br>1702 Brier Park Crescent N.W.<br>Medicine Hat, AB T1A 7G1, Canada<br>tel: 1 403 527 1121<br>fax: 1 403 529 6093 | carbon black | 62 |
| *Carborundum Corporation*<br>*Boron Nitride Division*<br>168 Creekside Drive<br>Amherst, NY 14228-2027, USA<br>tel: 1 716 691 2051<br>fax: 1 716 691 2090 | boron nitride | 46 |
| *C.E D. Process Minerals Inc.*<br>863 N. Cleveland - Massillon Road<br>Akron, OH 4433-2167, USA<br>tel: 1 330 666 2400<br>fax: 1 330 666 5722 | cristobalite | 78 |
| *Celite Corporation (World Minerals, Inc.)*<br>Headquarters<br>P.O. Box 519<br>Lompoc, CA 93438-0519, USA<br>tel: 1 805 735 7791<br>fax: 1 805 735 5699<br><br>*Celite France*<br>Headquarters<br>9 rue du Colonel-de-Rochebrune<br>B.P. 240<br>92504 Rueil-Malmaison Cedex, France<br>tel: 33 1 47 49 05 60<br>fax: 33 1 47 08 30 25<br><br>*Celite Corporation*<br>295 The West Mall<br>Etobicoke, ON M9C 4Z7, Canada<br>tel: 1 416 626 8175<br>fax: 1 416 626 8235<br><br>*Celite (UK) Ltd.*<br>Livingston Road<br>Hessle<br>North Humberside HU13 0EG<br>England<br>tel: 44 482 64 52 65<br>fax: 44 482 64 11 76<br><br>*Celite Italiana s.r.l.*<br>Viale Pasubio 6<br>20154 Milano, Italy<br>tel: 39 2 65 45 31<br>fax: 39 2 29 00 54 39 | diatomaceous earth<br>hydrous calcium silicate | 80<br>96 |

| Company name and address | Fillers | Page |
|---|---|---|
| *Cellulose Filler Factory Corporation*<br>10200 Worton Road<br>Chestertown, MD 21620, USA<br>tel: 1 410 810 0779<br>toll free: 800 832 4662<br>fax: 1 410 810 0793 | cellulose fibers | 184 |
| *Charis, Inc.*<br>512 Sweet Briar Drive<br>Maryville, TN 37804, USA<br>tel: 1 423 379 1225<br>fax: 1 423 379 1223 | cristobalite | |
| *Charles B. Chrystal Co., Inc.*<br>30 Vesey Street<br>New York, NY 10007, USA<br>tel: (212) 227-2151<br>fax: (212) 233-7916 | production and distribution of over hundred grades of different fillers listed in Chapter 2 | |
| *Chemalloy Company, Inc.*<br>P.O. Box 350<br>Bryn Nawr, PA 19010-0350, USA<br>tel: 1 610 527 3700<br>fax: 1 610 527 3878 | distributor of a broad range of fillers | |
| *CIMBAR Performance Minerals*<br>25 Old River Road S.E.<br>P.O. Box 250<br>Cartersville, GA 30120, USA<br>tel: 1 770 387 0319<br>toll free: 1 800 852 6868<br>fax: 1770 386 6784 | barium sulfate<br>bentonite | 36<br>43 |
| *Climax Molybdenum Company*<br>*Division of Cyprus Amax Company*<br>Centennial Center, Suite 308<br>P.O. Box 0407<br>Ypsilanti, MI 48198-0407, USA<br>tel: 1 734 481 3000<br>fax: 1 734 481 3005<br><br>*Climax Molybdenum Company, Ltd*<br>29 Gresham Street<br>London EC2V 7DA, England<br>tel: 44 1 606 8800 | molybdenum disulfide | 117 |
| *Coal Fillers, Inc*<br>P.O. Box 1063<br>Bluefield, VA 24605, USA<br>tel: 540 322 46 75 | anthracite | 25 |
| *Composite Materials, L L C*<br>700 Waverly Ave.<br>Mamaroneck, NY 10543, USA<br>tel: 1 914 381 4848<br>fax: 1 914 381 4897 | metal coated graphite fibers | 108 |

| Company name and address | Fillers | Page |
|---|---|---|
| *Composite Particles, Inc.*<br>2330 26th Street S.W.<br>Allentown, PA 18103, USA<br>tel: 1 610 791 9900<br>fax: 1 610 791 2486 | polymeric surface activated powders<br>rubber particles surface activated<br>surface activated aramid | 122<br>129<br>178 |
| *Columbian Chemicals Company*<br>Headquarters<br>1600 Parkwood Circle<br>Suite 400<br>Atlanta, GA 30339, USA<br>tel: 1 770 951 5700<br>toll free: 1 800 235 4003<br>fax: 1 770 951 7554<br><br>*Columbian Chemicals Canada, Ltd.*<br>P.O. Box 3398, Station C<br>Hamilton, ON L8H 7M2, Canada<br>tel: 1 905 544 3343<br>fax: 1 905 544 8641<br><br>*Columbian Chemicals, UK*<br>*Sevalco, Ltd.*<br>Severn Road<br>Avonmouth<br>Bristol, BS11 0YL, England<br>tel: 44 117 9235532<br>fax: 44 117 9235333<br><br>*Columbian Carbon International*<br>2 rue de la Couture<br>SILIC 229<br>94528 Rungis Cedex, France<br>tel: 33 146 879241<br>fax: 33 146 879062<br><br>*Columbian Chemicals Europa GmbH*<br>Wiesenauer Strasse 11<br>D-30179 Hannover, Germany<br>tel: 49 511 630 890<br>fax: 49 511 630 8912<br><br>*Columbian Tiszai Carbon Ltd.*<br>P.O. Box 61<br>H-3581 Tiszaujvaros, Hungary<br>tel: 36 49 321 724<br>fax: 36 49 322 003<br><br>*Columbian Carbon Europa S.r.l.*<br>Via S. Cassiano 140<br>P.O. Box 184<br>I-28069 San Martino Di Trecate, Italy<br>tel: 39 321 7981<br>fax: 39 321 798250<br><br>*continued on the next page* | carbon black | 62 |

| **Company name and address** | **Fillers** | **Page** |
|---|---|---|
| *Columbian Carbon Spain SA*<br>Complejo de Santander<br>Gajano (Cantabria)<br>Apartado 23<br>E-39710 Solares, Spain<br>tel: 34 42 503030<br>fax: 34 42 502165<br><br>*Columbian Carbon Japan, Ltd.*<br>8-12 Horidomecho 1-Chome<br>Nihonbaski, Chuɔ-ko<br>Tokyo 103, Japan<br>tel: 81 3 3663 2881<br>fax: 81 3 3667 1569<br><br>*Columbian Carbon Philippines, Inc.*<br>MCC P.O. Box 1969<br>1259 Makati, Metro Manila<br>Philippines<br>tel: 63 2 810 0386<br>fax: 63 2 818 7337 | | |
| *Cortex Biochem, Inc.*<br>1933 Davis Street<br>Suite 321<br>San Leandro, CA 94577, USA<br>tel: 1 510 568 2228<br>fax: 1 510 568 2467 | ferrites | 85 |
| *Crescent Bronze Powder Company, Inc.*<br>3400 North Avondale Ave.<br>Chicago, IL 60618-5432, USA<br>tel: 1 312 539 2441<br>toll free: 1 800 445 6810<br>fax: 1 312 539 1131 | bronze powder | |
| *CSM Industries*<br>21801 Tungsten Road<br>Cleveland, OH 44117, USA<br>tel: 1 216 692 4453<br>fax: 1 216 692 0031 | molybdenum powder | 116 |
| *Degussa AG*<br>P.O. Box 13 45<br>D-63403 Hanau, Germany<br>tel: 49 61 81 59 0<br>fax: 49 61 81 59 30 30<br><br>*Degussa AG*<br>Weissfrauenstrasse 9<br>D-60311 Frankfurt am Main, Germany<br>tel: 49 069 2 18 01<br>fax: 49 069 2 18 32 18 | aluminum oxide<br>carbon black<br>fumed silica<br>precipitated silica<br>titanium dioxide | 20<br>63<br>132<br>139<br>154 |

| Company name and address | Fillers | Page |
|---|---|---|
| *DSM Engineering Plastics Products, Inc.*<br>P.O. Box 14235<br>Reading, PA 19612-4235, USA<br>tel: 1 610 320 6600<br>fax: 1 610 320 6868 | flame retarded grades | |
| *Duke Scientific Corporation*<br>2463 Faber Place<br>Palo Alto, CA 94303, USA<br>tel: 1 650 424 1177<br>toll free: 1 800 334 3883<br>fax: 1 650 424 1158 | glass beads | 87 |
| *DUSLO, a.s.*<br>Drienova ul. 24<br>826 03 Bratislava<br>Slovak Republic<br>tel: 421 7 57 97 385<br>fax: 421 7 230 835 | magnesium hydroxide | 106 |
| *Eagle Picher Minerals, Inc.*<br>6110 Plumas Street<br>Reno, NV 89509, USA<br>tel: 1 702 824 7623<br>fax: 1 702 824 7656 | diatomaceous earth | 80 |
| *ECC International, Ltd.*<br>John Keay House, St. Austell<br>Cornwall PL25 4DJ, England<br>tel: 44 01726 74482<br>fax: 44 01726 623019<br><br>*ECC Americas*<br>5775 Peachtree-Dunwoody Road N.E.<br>Suite 200<br>Atlanta, GA 30342, USA<br>tel: 1 404 843 1551<br>fax: 1 404 843 8872<br><br>*ECC Pacific*<br>101 Thomson Road<br>No. 27 - 05, United Square<br>Singapore 1130<br>tel: 65 2530722<br>fax: 65 2552487 | calcium carbonate<br>ball clay<br>kaolin | 48<br>75<br>100 |

| Company name and address | Fillers | Page |
|---|---|---|
| *Electro Abrasives Corporation*<br>701 Willet Road<br>Buffalo, NY 14218, USA<br>tel: 1 716 822 2500<br>fax: 1 716 822 2858 | aluminum oxide | 20 |
| *EM Corporation*<br>P.O. Box 2400 TR<br>2801 Kent Avenue<br>West Lafayette, IN 47906, USA<br>tel toll free: 1 800 428 7802 | molybdenum disulfide | 117 |
| *Engelhard Corporation*<br>Pigments and Additives Group<br>101 Wood Avenue<br>P.O. Box 770<br>Iselin, NJ 08830-0770, USA<br>Inside North America:<br>tel: 1 732 205 500<br>fax: 1 732 321 0250<br>Outside North America:<br>tel: 1 732 205 7136<br>fax: 1 732 494 9009 | kaolin | 100 |
| *D. J. Enterprises, Inc.*<br>P.O. Box 31366<br>Cleveland, OH 44131, USA<br>tel: 1 216 524 3879 | synthetic carbon | |
| *Evans Clay Company*<br>P.O. Box 595<br>McIntyre, GA 31054, USA<br>tel: 1 912 946 3532<br>fax: 1 912 946 2992 | kaolin<br>glass fibers | 100<br>187 |
| *Engineered Carbons, Inc.*<br>P.O. Box 2831<br>1111 Penn Avenue<br>Borger, TX 79008-2831, USA | carbon filler | |
| *Expancel*<br>Box 13000<br>S-850 13 Sundsvall, Sweden<br>tel: 46 60 13 40 00<br>fax: 46 60 56 95 18<br><br>*Expencel, Inc.*<br>2150-H Northmont Parkway<br>Duluth, GA 30096, USA<br>tel: 1 770 813 9126<br>fax: 1 770 813 8639 | polymeric microspheres | 122 |
| *Fibertec*<br>35 Scotland Boulevard<br>Bridgewater, MA 02324, USA<br>tel: 1 508 697 5100<br>fax: 1 508 697 7140 | various fibers<br>wollastonite | <br>167 |

| Company name and address | Fillers | Page |
|---|---|---|
| *Fiber Sales & Development Corporation*<br>Checkerboard Sq.<br>St. Louis, MO 6364, USA<br>tel: 1 800 325 7108<br>fax: 1 908 968 5117 | cellulose fibers | 184 |
| *Franklin Industrial Minerals*<br>612 Tenth Avenue, North<br>Nashville, TN 37203, USA<br>tel: 1 615 259 4222 | aluminum trihydroxide<br>mica | 22<br>113 |
| *Grefco Minerals, Inc.*<br>P.O. Box 637<br>Lompoc, CA 93438, USA<br>tel: 1 805 736 4501<br>fax: 1 805 735 3735 | diatomaceous earth<br>glass beads<br>perlite | 80<br>88<br>120 |
| *Harwick Standard Distribution Corporation*<br>60 S. Seiberling Street<br>P.O. Box 9360<br>Akron, OH 44305-0360, USA<br>tel: 1 330 798 9300<br>fax: 1 330 798 0214 | distributor of full range of fillers for rubber and plastics<br>fumed silica | <br>132 |
| *Hitox Corporation of America*<br>Headquarteers<br>P.O. Box 2544<br>Corpus Christi, TX 78403-2544, USA<br>tel: 1 512 882 5175<br>fax: 1 512 882 6948 | aluminum trihydroxide<br>barium sulfate<br>titanium dioxide | 22<br>37<br>154 |
| *Huber, J.M. Corporation*<br>*Engineered Minerals Division*<br>One Huber Road<br>Macon, GA 31298, USA<br>tel: 1 912 745 4751<br>fax: 1 912 745 1116 | aluminum trihydroxide<br>barium sulfate<br>calcium carbonate<br>kaolin | 22<br>37<br>49<br>100 |
| *Hyperion Catalysis International*<br>38 Smith Place<br>Cambridge, MA 02138, USA<br>tel: 1 617 354 9678<br>fax: 1 617 354 9691 | carbon fibrils | 180 |
| *I H Polymeric Products, Ltd.*<br>Meopham Triding Estate<br>Meopham, Gravesend<br>Kent DA13 0LT, England<br>tel: 44 (0) 1474 814917<br>fax: 44 (0) 1474 813117 | water absorbing and expanding polymers | |

| Company name and address | Fillers | Page |
|---|---|---|
| *Inco Company*<br>681 Lawlins Road<br>Wyckoff, NJ 07481, USA<br>tel: 1 201 848 1012<br>fax: 1 201 848 1022<br><br>*Inco Europe, Ltd.*<br>5th Floor , Windsor House<br>50 Victoria Street<br>London SW1H 0XB, England<br>tel: 44 181 932 1503<br>fax: 44 181 931 0175 | nickel coated carbon fiber<br>nickel powder and flakes | 108<br>118 |
| *Interfibe Corporation*<br>6001 Cochran Road, A-202<br>Solon, OH 44139, USA<br>tel: 1 216 248 2266<br>toll free: 800 262 3771<br>fax: 1 216 248 2132 | cellulose fibers | 184 |
| *J.B. Company*<br>9 Ginter Street<br>Franklin, NJ<br>USA<br>tel: 1 201 209 8514<br>fax: 1 201 209 7680 | glass beads<br>silver and gold coated glass beads | 88<br>108 |
| *Keith Ceramic Materials, Ltd.*<br>Fisher Way<br>Belvedere<br>Kent DA17 6BN, England<br>tel: 44 181 311 8299<br>fax: 44 181 311 8238 | ceramic powders | |
| *Kentucky-Tennessee Clay Company*<br>1441 Donelson Pike<br>Nashville, TN 37217, USA<br>tel: 1 615 365 0852<br>fax: 1 615 365 0842<br><br>*Kentucky-Tennessee Clay Company*<br>5080 State Route 45 South<br>Mayfield, KY 42066, USA<br>tel: 1 502 247 3061<br>fax: 1 502 247 0293 | ball clay<br>feldspar<br>kaolin | 75<br>86<br>100 |
| *Keystone Filler & Mfg. Company*<br>214 Railroad Street<br>Muncy, PA 17756, USA<br>tel: 1 717 546 3148<br>fax: 1 717 546 7067 | anthracite<br>slate flour | 25<br>149 |

| Company name and address | Fillers | Page |
|---|---|---|
| *Kinetico Inc.*<br>10845 Kinsman Road<br>P.O. Box 193<br>Newbury, OH 44065, USA<br>tel: 1 216 564 9111<br>toll free: 1 800 432 1166<br>fax: 1 216 564 7696 | ceramic beads | 72 |
| *Kronos Canada, Inc.*<br>Suite 206<br>45 Sheppard Ave. East<br>Toronto, Ontario, Canada M2N 5W9<br>tel: 1 416 223 9590<br>fax: 1 416 223 8612 | titanium dioxide | 155 |
| *Laurel Industries, Inc.*<br>30195 Chagrin Boulevard<br>Cleveland, OH 44124-5794, USA<br>tel: 1 216 831 5747<br>toll free: 1 800 221 1304<br>fax: 1 216 831 8479 | sodium antimonate<br>antimony trioxide | 26<br>29 |
| *Luzenac Europe*<br>B.P. 1162<br>31036 Toulouse Cedex 1, France<br>tel: 33 5 61 50 20 20<br>fax: 33 5 61 40 06 23<br><br>Luzenac Naintsch<br>Statteggerstrasse 60<br>Postfach 35<br>8045 Graz, Austria<br>tel: 43 316 69 36 50<br>fax: 43 316 69 36 55<br><br>*Luzenac NV*<br>Scheepzatestraat 2<br>9000 Gent, Belgium<br>tel: 32 9 250 09 11<br>fax: 32 9 251 41 17<br><br>*Luzenac Benelux*<br>Plasky Square 92<br>1030 Brussels, Belgium<br>tel: 32 2 735 10 75<br>fax: 32 2 736 12 90<br><br>*Luzenac Deutschland GmbH*<br>Heltorferstrasse 4<br>40472 Dusseldorf Germany<br>tel: 49 211 471 130<br>fax: 49 211 471 1331<br>*continued on the next page* | talc | 151 |

| Company name and address | Fillers | Page |
|---|---|---|
| *Luzenac America*<br>9000 E. Nichols Ave.<br>Englewood, CO 80112, USA<br>tel: 1 303 643 0400<br>fax: 1 303 643 0444<br><br>*Luzenac Iberica SA*<br>Narcis Monturiol, 2-4-6, 4° dcha<br>08960 Sant Just Desvern, Spain<br>tel: 34 34 733301<br>fax: 34 34 733676<br><br>*Luzenac UK Ltd.*<br>Navigation House, Furness Quay<br>Salford, Manchester M5 2XZ, England<br>tel: 44 161 877 9111<br>fax: 44 161 877 9222<br><br>*Luzenac Europe*<br>Nordic Countries Sales Office<br>Biobalance, Vallensbaekvej 45<br>2605 Brondby, Denmark<br>tel: 45 43 44 33 25<br>fax: 45 43 44 33 21 | | |
| *Malvern Minerals Company*<br>220 Runyon Street<br>P.O. Box 1238<br>Hot Springs National Park, AR 71902, USA<br>tel: 1 501 623 8893<br>fax: 1 501 623 5113 | quartz (tripoli) | 142 |
| *Mica-Tek*<br>A Division of Miller and Company<br>325 North Center Street<br>Suite D<br>Northville, MI 48167-1224, USA<br>tel: 1 248 344 0227<br>fax: 1 248 348 3692<br><br>*Microfine Minerals, Ltd.*<br>Raynesway<br>Derby DE21 7BE, England<br>tel: 44 1332 673131<br>fax: 44 1332 677590 | mica | 113 |
| *Millennium Inorganic Chemicals*<br>200 International Circle<br>Suite 5000<br>Hunt Valley, MD 21030, USA<br>tel: 1 410 229 4400<br>toll free: 1 800 638 3234<br>fax: 1 410 229 4488 | titanium dioxide | 155 |

*continued on the next page*

| Company name and address | Fillers | Page |
|---|---|---|
| Millenium Inorganic Chemicals<br>Australind Headquarters<br>P.O. Box 245<br>Bunbury, WA 6231, Australia<br>tel: 61 08 9780 8333<br>fax: 61 08 9780 8555 | | |
| *Millenium Inorganic Chemicals*<br>China Hong Kong Tower<br>22nd Floor<br>8-12 Hennessy Road<br>Wanchai, Hong Kong<br>tel: 852 2528 4667<br>fax: 852 2865 1246 | | |
| *Millenium Inorganic Chemicals*<br>101 Thomson Road<br>#22-04 United Square<br>Singapore 307591, Singapore<br>tel: 65 255 7255<br>fax: 65 250 0159 | | |
| *Millenium Inorganic Chemicals*<br>P.O. Box 98 950<br>South Auckland Mail Centre<br>Wiri, Auckland, New Zealand<br>tel: 64 09 570 9999<br>fax: 64 09 279 7105 | | |
| *Millenium Inorganic Chemicals*<br>6/F Samdo Building<br>1-170, Soonhwa-Dong<br>Chung-Ku, Seoul, Republic of Korea<br>tel: 82 2 779 4844<br>fax: 82 2 779 4883 | | |
| *SCM Chemicals Europe*<br>P.O. Box 26<br>Grimsby, South Humberside<br>DN37 8DP, England<br>tel: 44 1469 571000<br>fax: 44 1469 571234 | | |
| *SCM Chemicals Europe*<br>Neuer Markt 1<br>D-42781 Haan, Germany<br>tel: 49 2129 93010<br>fax: 49 2129 31584 | | |

*continued on the next page*

| Company name and address | Fillers | Page |
|---|---|---|
| *SCM Chemicals Europe*<br>Centre Paris Pleyel<br>153 bd Anatole France<br>93521 Saint-Denis Cedex 1, France<br>tel: 33 1 49 33 94 70<br>fax: 33 1 48 20 70 11<br><br>*SCM Chemicals Europe*<br>109, Chaussee de Vleurgat<br>B-1050 Brussels, Belgium<br>tel: 32 2 648 7108<br>fax: 32 2 648 5823 | | |
| *Milwhite, Inc.*<br>7050 Portwest Drive<br>Suite 190<br>Houston, TX 77024, USA<br>tel: 1 713 881 1200<br>fax: 1 713 861 7717 | attapulgite<br>barium sulfate<br>barium/strontium sulfate<br>bentonite<br>calcium carbonate<br>talc | 33<br>37<br>41<br>43<br>49<br>151 |
| *MMM Carbon*<br>Avenue Louise 534<br>B-1050 Brussels, Belgium<br>tel: 32 2 627 53 11<br>fax: 32 2 627 53 93 | carbon black | 66 |
| *Morgan Matroc, Ltd.*<br>Bewdley Road Stourport-on Severn<br>Worcestershire DY13 8QR, England<br>tel: 44 0 1299 827000<br>fax: 44 0 1299 827872<br><br>*Morgan Matroc, Ltd.*<br>North American Sales Operation<br>8989 North Port Washington Road<br>Milwaukee, WI 53217, USA<br>tel: 1 888 386 3245<br>fax: 1 888 386 3249 | aluminum oxide | 20 |
| *MO-SCI Corporation*<br>4000 Enterprise Drive<br>P.O. Box 2<br>Rolla, MO 65402-0002, USA<br>tel: 1 573 364 2338<br>fax: 1 573 364 9589 | glass beads<br>glass spheres coated with various metals | 88<br>108 |
| *Nabaltec GmbH*<br>P.O. Box 18 60<br>D-92409 Schwandorf, Germany<br>tel: 49 0 94 31 53 0<br>fax: 49 0 94 31 6 15 57 | aluminum trihydroxide | 23 |
| *Nanophase Technologies Corporation*<br>453 Commerce Street<br>Burr Ridge, IL 60521, USA<br>tel: 1 630 323 1200<br>fax: 1 630 323 1221 | aluminum oxide<br>iron oxide<br>zinc oxide | 20<br>97<br>172 |

| Company name and address | Fillers | Page |
|---|---|---|
| *Non-Metals, Inc.*<br>1870 West Prince Road<br>Suite 67<br>Tucson, AZ 85705, USA<br>tel: 1 520 690 0966<br>fax: 1 520 690 0396<br><br>*China National Non-Metalic Minerals Industry Corporation*<br>No. 142 Xi Wai Avenue<br>Beijing 100044, China<br>tel: 86 10 834 4294<br>fax: 86 10 834 4254<br><br>*CNMIEC Japan Co., Inc.*<br>No. 13-4, 1-Chome, Nihonbashi<br>Muromachi, Chuo-ku<br>Tokyo, Japan<br>tel: 81 3 3242 5260<br>fax: 81 3 3242 5268 | attapulgite<br>bentonite<br>mica<br>pyrophyllite<br>sepiolite<br>talc<br>vermiculite<br>wollastonite | 33<br>43<br>113<br>128<br>130<br>151<br>165<br>167 |
| *Novamet Specialty Products Corporation*<br>681 Lawlins Road<br>Wyckoff, NJ 07481, USA<br>tel: 1 201 891 7976<br>fax: 1 201 891 9467 | silver coated nickel flakes<br>nickel coated graphite<br>nickel flakes and powder | 108<br>108<br>118 |
| *Nyco Minerals, Inc.*<br>124 Mountain View Drive<br>Willsboro, NY 12996-0368, USA<br>tel: 1 518 963 4262<br>fax: 1 518 963 4187<br><br>*Nyco Minerals, Inc. Europe*<br>Ordruvej 24<br>P.O. Box 88<br>DK-2920 Charlottenlund, Denmark<br>tel: 45 39643370<br>fax: 45 39643710 | wollastonite | 167 |
| *Nyacol Products, Inc.*<br>Megunco Road<br>P.O. Box 349<br>Ashland, MA 01721, USA<br>tel: 1 508 881 2220<br>fax: 1 508 881 1855 | antimony pentoxide | 27 |
| *Old Hickory Clay Company*<br>P.O. Box 66<br>Hickory, KY 42051-006, USA<br>tel: 1 502 247 3042<br>fax; 1 502 247 1842 | ball clay | 75 |

| Company name and address | Fillers | Page |
|---|---|---|
| *OMG, Inc.*<br>World Headquarters<br>50 Public Aquare<br>3800 Terminal Tower<br>Cleveland, OH 44113, USA<br>tel: 1 216 781 0083 | | |
| *OMG Americas, Inc.*<br>Headquarters<br>811 Sharon Drive<br>Westlake, OH 44145, USA<br>tel: 1 440 899 2950<br>toll free: 1 800 321 9696<br>fax: 1 440 808 7114 | | |
| *OMG Americas, Inc.*<br>Sales & Production<br>2601 Weck Drive<br>P.O. Box 12166<br>Research Triangle Park, NC 27709-2166, USA<br>tel: 1 919 544 8090<br>fax: 1 919 544 7996 | | |
| *OMG Europe GmbH*<br>Headquarters<br>Morsenbroicher Weg 200<br>P.O. Box 33 03 22<br>D-40435 Dusseldorf, Germany<br>tel: 49 211 961 880<br>fax: 49 211 614 629 | metal powders | |
| *OMG Kokkola Chemicals OY*<br>P.O. Box 286<br>SF-67101 Kokkola, Finland<br>tel: 358 6 82 80111<br>fax: 358 6 82 80307 | | |
| *OMG Vasset SA*<br>59 Chemin de Moisselles<br>95460 Ezanville, France<br>tel: 33 1 39 91 00 20<br>fax: 33 1 39 35 19 65 | | |
| *OMG Asia Pacific Co., Ltd.*<br>Headquarters<br>Suite B, 18F-1, No. 156<br>Min Sheng East Road, secretary. 3<br>Taipei, Taiwan, R.O.C.<br>tel: 886 2 7128566<br>fax: 886 2 7191915 | | |

*continued on the next page*

| Company name and address | Fillers | Page |
|---|---|---|
| *OMG Asia Pacific Co., Ltd.*<br>Ginza Five Building, 5-7<br>Ginza 5-chome, Chuo-ku<br>Tokyo 104, Japan<br>tel: 81 3 5568 5141<br>fax: 81 3 5568 5144<br><br>*OMG Asia Pacific Co., Ltd.*<br>*Singapore PTE, Ltd.*<br>No. 15 Kian Teck Ave.<br>Singapore 628901, Singapore<br>tel 65 266 5788<br>fax; 65 265 0788 | | |
| *OMYA/Plüss-Staufer AG*<br>P.O. Box 32<br>CH-4665 Oftringen, Switzerland<br>tel: 41 62 789 29 29<br>fax: 41 62 789 20 77<br><br>*OMYA S.A.*<br>35 Quai Andre Citroen<br>F-75725 Paris Cedex 15, France<br>tel: 33 1 40 58 44 00<br>fax: 33 1 45 79 73 52<br><br>*OMYA GmbH*<br>Brohler Strasse 11<br>D-5000 Koln 51, Germany<br>tel: 49 221 37 751<br>fax: 49 221 37 18 64<br><br>*OMYA, Inc.*<br>Three PSI Plaza<br>Proctor, VT 05765, USA<br>tel: 1 802 459 3311<br>toll free: 1 800 451 4468<br>fax: 1 802 459 2125<br><br>*Also present in many other locations worldwide* | calcium carbonate<br>dolomite | 49<br>84 |
| *Owens Corning*<br>World Headquarters<br>One Owens Corning Parkway<br>Toledo, OH 43659, USA<br>tel:<br>toll free: 1 800 438 7465<br>fax: 1 419 248 6227<br><br>*N.V. Owens Corning S.A.*<br>178, Chaussee de La Hulpe<br>B-1170 Brussels, Belgium<br><br>*Owens Corning Asia/Pacific*<br>18 Harbour Road<br>Suite 6406-7, Central Plaza<br>Wan Chai, Hong Kong | glass fibers | 187 |

| Company name and address | Fillers | Page |
|---|---|---|
| *Pacific Century, Inc.*<br>P.O. Box 221016<br>Chantilly, VA 20153, USA<br>tel: 1 703 803 3064<br>toll free: 1 800 828 2093<br>fax: 1 703 803 3443 | graphite flakes | |
| *Pierce & Stevens Corporation*<br>710 Ohio Street<br>Buffalo, NY 14203, USA<br>tel: 1 716 856 4941<br>fax: 1 716 856 0942 | polymeric microspheres | 122 |
| *Polar Minerals - Sales*<br>27801 Euclid Avenue<br>Suite 300<br>Cleveland, OH 44132, USA<br>tel: 1 216 731 0011<br>fax: 1 216 731 0009<br><br>*Polar Minerals - Sales*<br>5060 North Royal Atlanta Drive<br>Suite 22<br>Tucker, GA 30084, USA<br>tel: 1 770 934 4411<br>fax: 1 770 934 4376<br><br>*Polar Minerals - Plant*<br>1703 Bluff Rd.<br>Mt. Vernon, IN 47620, USA<br>tel: 1 812 836 5236<br>fax: 1 812 838 4744 | barium sulfate<br>calcium carbonate<br>mica<br>talc | 37<br>49<br>113<br>151 |
| *Piqua Minerals, Inc.*<br>1750 West Statler Road<br>Piqua, OH 45356, USA<br>tel: 1 937 773 4824<br>fax: 1 937 773 0791 | calcium carbonate | 49 |
| *Plastic Methods Co., Inc.*<br>20 West 37th Street<br>New York, NY 10018, USA<br>tel: 1 212 695 0070<br>fax: 1 212 967 5015 | copper coated glass spheres | 108 |
| *Polytechs S.A.*<br>Zone Industrielle de la Gare<br>BP 14<br>76450 Cany Barville, France<br>tel: 33 (0) 2 35 57 81 62<br>fax: 33 (0) 2 35 57 8192 | masterbatches | |

| Company name and address | Fillers | Page |
|---|---|---|
| *Potters Industries, Inc.*<br>Southpoint Corporate Headquarters<br>P.O. Box 840<br>Valley Forge, PA 19482-0840, USA<br>tel: 1 610 651 4700<br>fax: 1 610 408 9723<br><br>*Potters- Ballotini Europe*<br>Pontefract Road<br>Barnsley<br>South Yorkshire S71 1HJ, England<br>tel: 44 (0) 1226 770381<br>Fax: 44 (0) 1226 207615 | glass beads<br>silver coated glass beads<br>silver coated copper powder<br>silver coated aluminum<br>silver coated nickel | 88<br>108<br>108<br>108<br>108 |
| *PPG Industries, Inc.*<br>One PPG Place<br>Pittsburgh, PA 15272, USA<br>tel: 1 412 434 2945<br>toll free: 1 800 243 6745<br>fax: 1 412 434 2197 | precipitated silica<br>glass fibers | 139<br>188 |
| *PQ Corporation*<br>Corporate Headquarters<br>P.O. Box 840<br>Valley Forge, PA 19482-0840, USA<br>tel: 1 423 629 7160<br>toll free: 1 800 777 5780<br>fax: 1 423 698 0614<br><br>*PQ Australia Pty, Ltd.*<br>Melbourne, Victoria<br>tel: 61 3 9791 4066<br>fax: 61 3 9793 6511<br><br>*PQ Hollow Spheres Ltd.*<br>Beverley, Yorkshire, England<br>tel: 44 1482 869385<br>fax: 44 1482 868106 | aluminum oxide<br>antimony pentoxide<br>ceramic beads<br>glass beads<br>silver coated beads<br>silver coated fiber<br>molecular sieves<br>zeolites | 20<br>27<br>72<br>88<br>108<br>108<br>170<br>170 |
| *Quarzwerke GmbH*<br>P.O. Box 1780<br>Kaskadenweg 40<br>D-50226 Frechen, Germany<br>tel: 49 0 22 34 101-0<br>fax: 49 0 22 34 10 1200 | cristobalite<br>fused silica<br>silica sand and flour<br>wollastonite | 78<br>138<br>144<br>167 |
| *Reade Advanced Materials*<br>P.O. Box 15039<br>Riverside, RI 02915-0039, USA<br>tel: 1 401 433 7000<br>fax: 1 401 433 7001 | distributor of full range of fillers | |

| Company name and address | Fillers | Page |
|---|---|---|
| *ReBase Products, Inc.*<br>70 Collier Street<br>Barrie, ON L4M 4Z2, Canada<br>tel: 1 705 734 0400<br>fax: 1 705 734 3403 | calcium hydroxide | 58 |
| *Reheis, Inc.*<br>235 Snyder Avenue<br>P.O. Box 609<br>Berkeley Heights, NJ 07922, USA<br>tel: 1 908 464 1500<br>fax: 1 908 464 7726<br><br>*Reheis Ireland*<br>Kilbarrack Road<br>Dublin 5, Ireland<br>tel: 353 1 322621<br>fax: 353 1 392205 | graphite | |
| *RESCO Products, Inc.*<br>P.O. Box 108<br>Norristown, PA 19404, USA<br>tel: 1 215 292 3500<br>fax: 1 215 279 6070<br><br>*Piedemont Minerals*<br>*Division of RESCO Products, Inc.*<br>P.O. Box 7247<br>Greensboro, NC 27417-0247, USA<br>tel: 1 919 292 0949 | pyrophyllite | |
| *Sachtleben Chemie GmbH*<br>Postfach 17 04 54<br>D-47184 Duisburg, Germany<br>tel: 49 (0) 20 66 22 0<br>fax: 49 (0) 20 66 22 20 00 | barium sulfate<br>kaolin<br>lithopone<br>zinc sulfide | 37<br>100<br>104<br>176 |
| *San Jose Delta Associates, Inc.*<br>482 Sapena Court<br>Santa Clara, CA 95054, USA<br>tel: 1 408 727 1448<br>toll free:1 800 809 4308<br>fax: 1 408 727 6019 | beryllium oxide<br>boron nitride | 45<br>46 |
| *Sekisui Plastics Co., Ltd.*<br>28 F Mitsui Building<br>2-1-1 Nishi-shijuku<br>Shinjuku-ku, Tokyo 163-04, Japan<br>tel: 81 3 3347 9689<br>fax: 81 3 3344 2335 | polymeric spherical fillers | 123 |
| *Shikoku Chemical Corp.* | aluminum borate whiskers | 19 |

| Company name and address | Fillers | Page |
|---|---|---|
| *Sigri Great Lakes Carbon Corporation*<br>2333 Route 22 West<br>Union, NJ 07083, USA<br>tel: 908 851 0060<br>fax: 908 687 4471 | masterbatches | |
| *Silberline Manufacturing Co., Inc.*<br>Lincoln Drive<br>P.O. Box B<br>Tamaqua, PA 18252-0420, USA<br>tel: 1 717 668 6050<br>toll free: 1 800 348 4824<br>fax: 1 717 668 0197 | aluminum flakes and powders | 16 |
| *Silvered Electronic Mica Co., Inc.*<br>P.O. Box 505<br>107 Boston Post Road<br>Willimantic, CT 06226, USA<br>tel: 1 860 456 0831<br>fax: 1 860 423 3506 | mica for electronic applications | |
| *Solvay Alkali GmbH*<br>Postfach 10 13 61<br>47493 Rheinberg, Germany<br>tel: 49 28 43 73 36 07<br>49 28 43 73 25 68<br>fax: 49 28 43 73 37 51<br>49 28 43 73 35 12<br><br>*Solvay-Alkalis*<br>Strategic Business Unit Fillers<br>Via Varesina 2/4<br>21021 Angera (VA), Italy<br>tel: 39 3 31 93 96 00<br>fax: 39 3 31 93 96 09<br><br>*Solvay Austria AG*<br>Parkring 12<br>A-1015 Wien, Austria<br>tel: 43 15 15 8 80<br>fax: 43 15 15 88 90 4<br><br>*Solvay S.A. Benelux*<br>rue du Prince Albert 44<br>B-1050 Bruxelles, Belgium<br>tel: 32 2 5 09 6111<br>fax: 32 2 5 09 6292<br><br>*Solvay-Chem. Spol. S.r.o.*<br>Matechova 3<br>CR-14600 Praha 4, Czech Republic<br>tel: 42 2 42 37 50<br>fax: 42 2 56 00<br>*continued on the next page* | calcium carbonate | 49 |

| Company name and address | Fillers | Page |
|---|---|---|
| *Rode & Rode A/S*<br>Kronprinsensgade 8<br>DK-1114 Kobenhavn K, Denmark<br>tel 45 33 33 07 77<br>fax: 45 33 93 97 93 | | |
| *Solvay Nordic AB*<br>Marknadsvagen 15, Box 1132<br>S-18311 Taby, Sweden<br>tel: 46 8 6 38 00 30 or 32<br>fax: 46 8 6 38 0152 | | |
| *Solvay France S.A.*<br>12, Cours Albert 1er<br>75383 Paris Cedex 08, France<br>tel: 33 1 40 75 80 00<br>fax: 33 1 45 63 28 11 | | |
| *Solvay Chemicals Ltd.*<br>Unit 1, Grovelands Business Centre<br>Boundary Way<br>Hemel Hempstead HP2 7TE, England<br>tel: 44 14 42 23 65 55<br>fax: 44 14 42 23 87 70 | | |
| *Solvay Kemia Kft.*<br>Hungaria korut 114<br>H-1425 Budapest XIV, Hungary<br>tel: 36 1 2 5146 66<br>fax: 36 1 2 51 26 87 | | |
| *Solvay Chemie B.V.*<br>Postbus 8<br>NL-6040 AA Roermond, The Netherlands<br>tel: 31 47 5 38 46 10<br>fax: 31 47 5 38 48 06 | | |
| *Solvay Chemia Sp. Z.o.o.*<br>Ul. Krolowej Marysienki 11/1<br>Pl-02-954 Warszawa, Poland<br>tel: 48 22 64 27 343<br>fax: 48 22 65 17 816 | | |
| *Solvay Slovchem, spol. Sr.o.*<br>Bardosova 33<br>SK-83101 Bratislava, Slovakia<br>tel: 42 7 37 70 46<br>fax: 42 7 37 32 40 | | |

*continued on the next page*

| Company name and address | Fillers | Page |
|---|---|---|
| *Solvay S.A.*<br>Calle Mallorca 269<br>E-08008 Barcelona, Spain<br>tel: 34 3 4 84 74 00<br>fax: 34 3 4 84 75 41<br><br>*Solvay (Schwitzerland) AG*<br>Zurcherstrasse 42<br>CH-5330 Zurzach, Switzerland<br>tel: 41 56 2 69 61 61<br>fax: 41 56 2 69 62 62 | | |
| *SOVITEC (Belgium) S.A.*<br>Zoning Industrial<br>B-6220 Fleurus, Belgium<br>tel: 32 071 81 50 11<br>fax: 32 02 673 41 45<br><br>*SOVITEC (France) S.A.*<br>BP 98<br>F-57190 Florange, France<br>tel: 33 82 52 68 05<br>fax: 33 82 52 91 22<br><br>*SOVITEC Iberica S.A.*<br>Poligono Industrial<br>E-Castellbisbal-Barcelona, Spain<br>tel: 34 3 772 09 01<br>fax: 34 37 72 00 53<br><br>*SOVITEC GmbH*<br>Postfach 695<br>D-6600 Saarbruecken, Germany<br>tel: 49 0681 64010<br>fax: 49 0681 635325 | glass beads | 88 |
| *Sphere Services, Inc.*<br>1055 Commerce Park Drive<br>Suite 100<br>Oak Ridge, TN 37830, USA<br>tel: 1 423 482 7727<br>fax: 1 423 482 3938 | ceramic beads | 72 |
| *Steward*<br>1200 East 36th Street<br>P.O. Box 510<br>Chattanooga, TN 37401-0510, USA<br>tel: 1 423 867 4100<br>toll free: 1 800 634 2673<br>*continued on the next page* | ferrites | 85 |

| Company name and address | Fillers | Page |
|---|---|---|
| *Steward*<br>Brucefield Industrial Estate<br>5 Cochrane Square<br>Livingston EH54 9DR, Scotland, England<br>tel: 44 (0) 1 506 414200<br>fax: 44 (0) 1 506 410694<br><br>*Steward*<br>47 Hill Street, #08-07<br>SCCCI Building<br>Singapore 179365, Singapore<br>tel: 65 337 9667<br>fax: 65 337 9686 | | |
| *St. Lawrence Chemical. Inc.*<br>19201 Clark Graham Ave.<br>Baie D'Urfe, Quebec H9X 3P5, Canada<br>tel: 1 514 457 3628<br>fax: 1 514 457 9773 | distributor of full line of fillers for plastics and rubber | |
| *Strong-Lite Products Corporation*<br>Emmett Sanders Road<br>P.O. Box 8029<br>Pine Bluff, AR 71611, USA<br>tel: 1 501 536 3453<br>toll free: 1 800 255 9057<br>fax: 1 501 536 1033 | perlite<br>vermiculite | 120<br>165 |
| *Struktol Corporation*<br>201 East Steels Corners Road<br>P.O. Box 1649<br>Stow, OH 44224, USA<br>tel: 1 216 928 5188<br>toll free: 1 800 327 8649<br>fax: 1 216 928 8726 | silica/kaolin filler | |
| *Superior Graphite Corporation*<br>120 South Riverside Plaza<br>Chicago, IL 60606, USA<br>tel: 1 312 559 2999<br>toll free: 1 800 325 0337<br>fax: 1 312 559 9064<br><br>*Superior Graphite Corporation*<br>655 Industrial Drive<br>Unit 17<br>Cambridge, ON N3H 4S1, Canada<br>tel: 1 519 650 1608<br>fax: 1 519 650 1803<br><br>*Superior Graphite Corporation*<br>Square de Meaus 25<br>B-1040 Brussels, Belgium<br>tel: 32 2 513 2660<br>fax: 32 2 513 1971<br>*continued on the next page* | graphite | 92 |

| Company name and address | Fillers | Page |
|---|---|---|
| *Superior Graphite Corporation*<br>Box 13000 (Post)<br>Stockviksverken (Courier)<br>850 13 Sundsvall, Sweden<br>tel: 46 60 13 41 18<br>fax: 46 60 13 41 28 | graphite | 92 |
| *Suzorite Mica Products, Inc.*<br>1475 Graham Bell Street<br>Boucherville, Quebec J4B 6A1, Canada<br>tel: 1 514 655 2450<br>fax: 1 514 655 9942<br><br>*Suzorite Mineral Products, Inc.*<br>1040 Crown Pointe Parkway<br>Suite 270<br>Atlanta, GA 30338, USA<br>tel: 1 770 392 8664<br>fax: 1 770 392 8670 | calcium carbonate<br>mica | 49<br>113 |
| *Syncoglas N.V.*<br>Industriepark<br>Drukkerijstraat 9<br>B-9240 Zele, Belgium<br>tel: 32 052 45 76 11<br>fax: 32 052 44 95 02 | glass fiber reinforcement | |
| *Synair Corporation*<br>P.O. Box 5269<br>2003 Amnicola Highway<br>Chattanooga, TN 37406, USA<br>tel: 1 423 697 0400<br>toll free: 1 800 251 7642<br>fax: 1 423 697 0424 | glass fillers | |
| *TAM Ceramics, Inc.*<br>P.O. Box 67<br>4511 Hyde Park Blvd.<br>Niagara Falls, NY 14305-0067, USA<br>tel: 1 716 278 9400<br>fax: 1 716 285 3026 | barium titanate<br>titanium dioxide | 42<br>155 |
| *Technic, Inc.*<br>Engineered Powders Division<br>300 Park East Drive<br>Woonsocket, RI 02895, USA<br>tel: 1 401 769 7000<br>fax: 1 401 769 2472 | gold powder and flakes<br>silver powder and flakes | 91<br>147 |
| *Teledyne Advanced Materials*<br>*An Allegheny Teledyne Company*<br>7300 Highway 20 West<br>Huntsville, AL 35806, USA<br>tel: 1 888 777 6852<br>fax: 1 888 777 9624 | tungsten powder | 164 |

| Company name and address | Fillers | Page |
|---|---|---|
| *TIMCAL, Ltd.*<br>Graphites and Technologies<br>Plant: Bodio<br>CH-5643 Sins, Switzerland<br>tel: 41 41 789 77 00<br>fax: 41 41 789 77 10<br><br>*TIMCAL America, Inc.*<br>29299 Clemens Road 1-L<br>Westlake, OH 44145, USA<br>tel: 1 216 871 7504<br>fax: 1 216 871 6026<br><br>*Changzhou TIMCAL Graphite Corporation, Ltd.*<br>5th Floor, Building C<br>Tianan Industrial Complex<br>Changzhou Hi-Tech Zone<br>Changzhou 213022, P.R.C.<br>tel: 0519 510 0301<br>fax: 0519 510 5118 | graphite | 92 |
| *Tioxide Americas Inc.*<br>901 Warrenville Road<br>Suite 115<br>Lisle, IL 60532, USA<br>tel: 1 708 515 1180<br>fax: 1 708 515 1210<br><br>*Tioxide Canada, Inc.*<br>9999 Cavendish Boulevard<br>Suite 100<br>Ville Saint-Laurent, Quebec H4M 2X5, Canada<br>tel: 1 514 747 9294<br>fax: 1 514 747 6807<br><br>*Tioxide Europe S.A.*<br>1 rue des Garennes<br>BP 89<br>62102 Calais Cedex, France<br>tel: 33 03 21 46 45 00<br>fax: 33 03 21 46 46 34 | titanium dioxide | 156 |
| *Transmet Corporation*<br>4290 Perimeter Drive<br>Columbus, OH 43228, USA<br>tel: 614 276 5522<br>fax: 614 276 3299 | aluminum, copper, brass and zinc particulate materials of various shapes and particle sizes | 16 |
| *Toho Rayon Co., Ltd.*<br>3-9 Nihonbashi 3-chome Chuo-ku<br>Tokyo 103, Japan<br>tel: 81 03 3278 7669<br>fax: 81 03 3278 7734 | carbon fiber<br>nickel coated carbon fiber | 181<br>108 |

| Company name and address | Fillers | Page |
|---|---|---|
| *3M Chemicals*<br>*Specialty Additives*<br>3M Center building 223-6S-04<br>St. Paul, MN 55144-1000, USA<br>tel: 1 612 737 1751<br>toll free: 1 800 367 8905<br>fax: 1 612 736 4133<br><br>*3M Canada Company*<br>P.O. Box 5757<br>London, ON N6A 4T1, Canada<br>tel: 1 800 410 6880 | ceramic beads<br>glass beads | 72<br>87 |
| *United Clays, Inc.*<br>7003 Chadwick Drive<br>Suite 100<br>Brentwood, TN 37027, USA<br>tel: 1 615 370 4500<br>fax: 1 615 370 0802 | ball clay | 75 |
| *United States Antimony Corporation*<br>P.O. Box 643<br>Thompson Falls, MT 59873, USA<br>tel: 1 406 827 3523<br>fax: 1 406 827 3543 | sodium antimonate<br>antimony trioxide | 26<br>29 |
| *U.S. Silica Company*<br>P.O. Box 187<br>Berkeley Springs, WV 25411, USA<br>tel: 1 304 258 2500 | silica sand | 144 |
| *Vanderbilt, R.T. Company, Inc.*<br>30 Winfield Street<br>P.O. Box 5150<br>Norwalk, CT 06856-5150, USA<br>tel: 1 203 853 1400<br>fax: 1 203 853 1452 | kaolin<br>pyrophyllite<br>talc<br>wollastonite | 100<br>128<br>151<br>167 |
| Wacker-Chemie GmbH<br>Hans-Seidel Platz 4<br>D-81737 München, Germany<br>tel: 49 89 62 79 01<br>fax: 49 89 62 79 17 71<br><br>*Wacker-Chemie GmbH*<br>Vienna/Austria<br>tel: 43 222 5 23 45 14<br>Fax: 43 222 5 23 45 14 33<br><br>*N.V. Wacker-Chemie (Belgium) S.A.*<br>tel: 32 2 6 63 13 00<br>fax: 32 2 6 60 25 53<br>*continued on the next page* | fumed silica | 133 |

| Company name and address | Fillers | Page |
|---|---|---|
| *Wacker-Chemie GmbH*<br>Prague/Czech Republic<br>tel: 42 2 2 31 22 77<br>fax: 42 2 2 31 22 62<br><br>*Wacker-Chemie Danmark A/S*<br>Glostrup/Denmark<br>tel: 45 43 43 03 00<br>fax: 45 43 43 03 16<br><br>*Wacker-Chemie Finland Oy*<br>Helsinki/Finland<br>tel: 358 0 6 92 51 55<br>fax: 358 0 6 92 51 45<br><br>*Wacker-Chimie S.A.*<br>Lyon/France<br>tel: 33 72 61 03 00<br>fax: 33 78 95 27 45<br><br>*Wacker-Chemie GmbH*<br>Kallithea-Athens/Greece<br>tel: 30 1 9 23 32 84<br>fax: 30 1 9 22 16 96<br><br>*Wacker-Chemie GmbH*<br>Budapest/Hungary<br>tel: 36 1 2 09 34 52<br>fax: 36 1 2 09 34 58<br><br>*Wacker Chemicals*<br>Dublin/Ireland<br>tel: 353 1 6 62 14 33<br>fax: 353 1 6 62 12 20<br><br>*Wacker-Chemie Italia S.P.A.*<br>Peschiera Borromeo<br>tel: 39 2 51 69 11<br>fax: 39 2 51 69 14 99<br><br>*Wacker-Chemie Nederland B.V.*<br>Krommenie/Netherlands<br>tel: 31 75 6 47 60 00<br>fax: 31 75 6 21 50 61<br><br>*Wacker-Chemie GmbH*<br>Warsaw/Poland<br>tel: 48 2 6 35 33 31<br>fax: 48 2 6 35 65 17<br><br>*continued on the next page* | | |

| Company name and address | Fillers | Page |
|---|---|---|
| *Wacker Quimica Portuguesa, Lda.*<br>Lisbon/Portugal<br>tel: 351 1 3 55 46 55<br>fax: 351 1 3 55 47 85 | | |
| *Wacker Quimica Iberica S.A.*<br>Barcelona/Spain<br>tel: 34 3 2 17 59 00<br>fax: 34 3 2 17 57 66 | | |
| *Wacker-Kemi AB*<br>Stockholm/Sweden<br>tel: 46 8 6 34 40 60<br>fax: 46 8 25 83 30 | | |
| *Wacker-Chemie (Schweiz) AG*<br>Liestal/Switzerland<br>tel: 41 61 9 27 85 85<br>fax: 41 61 9 21 33 15 | | |
| *Wacker-Chemie GmbH*<br>Levent-Istanbul/Turkey<br>tel: 90 212 2 78 46 45<br>fax: 90 212 2 78 44 70 | | |
| *Wacker Chemicals Ltd.*<br>Walton-on-Thames, Surrey/England<br>tel: 44 19 32 24 61 11<br>fax: 44 19 32 23 60 91 | | |
| *Wilh. Willumsen AS*<br>Stabekk/Norway<br>tel: 47 67 59 06 10<br>fax: 47 67 59 19 12 | | |
| *Wacker Chemicals Australia Pty. Ltd.*<br>Melbourne/Australia<br>tel: 61 3 95 25 16 00<br>fax: 61 3 95 21 27 43 | | |
| *Wacker Quimica do Brasil Ltda.*<br>Sao Paulo/Brazil<br>tel: 55 11 5 48 81 33<br>fax: 55 11 2 46 99 75 | | |
| *B.J. Chemicals Ltd.*<br>Concord/Canada<br>tel: 1 416 661 1500<br>fax: 1 416 661 1333 | | |

*continued on the next page*

| Company name and address | Fillers | Page |
|---|---|---|
| *Wacker Chemicals East Asia Ltd.*<br>Guangzhou/China<br>tel: 86 20 3 87 15 84<br>fax: 86 21 65 15 54 65 | | |
| *Wacker Chemicals East Asia Ltd.*<br>Hong Kong<br>tel: 852 25 06 32 28<br>fax: 852 25 06 32 80 | | |
| *Wacker Chemicals (South Asia) Pte. Ltd.*<br>Bombay/India<br>tel: 91 22 6 43 34 76<br>fax: 91 22 6 43 34 75 | | |
| *Wacker Chemicals East Asia Ltd.*<br>Tokyo/Japan<br>tel: 81 3 52 72 31 21<br>fax: 81 3 52 72 31 40 | | |
| *Wacker Chemicals Korea Ltd.*<br>Seoul/Korea<br>tel: 82 2 5 62 68 77<br>fax: 82 2 5 62 67 71 | | |
| *Wacker Mexicana, S.A. De C.V.*<br>Mexico D.F./Mexico<br>tel: 52 5 5 95 75 99<br>fax: 52 5 83 84 34 | | |
| *Wacker Chemicals ( South Asia) Pte. Ltd.*<br>Singapore<br>tel: 27 11 7 89 27 80<br>fax: 27 11 7 89 27 87 | | |
| *Wacker Chemicals ( South Africa) (Pty) Ltd.*<br>Johannesburg/South Africa<br>tel: 886 2 7 18 22 26<br>fax: 886 2 5 45 14 16 | | |
| Wacker Chemicals East Asia Ltd.<br>Taipei/Taiwan<br>tel: 886 2 7 18 22 26<br>fax: 886 2 5 45 14 16 | | |
| Wacker Silicones Corporation<br>3301 Sutton Road<br>Adrian, MI 49221-9397, USA<br>tel: 1 517 264 8500<br>toll free: 1 800 248 0063<br>fax: 1 517 264 8246 | | |

| Company name and address | Fillers | Page |
|---|---|---|
| *WBB Devon Clays Ltd.*<br>Park House, Courtenay Park<br>Newton Abbot<br>Devon TQ12 4PS, England<br>tel: 44 (0) 1626 332345<br>fax: 44 (0) 1626 332344<br><br>*Fuchs'sche Tongruben GmbH & Co KG*<br>Postfach 347<br>D- 56223 Ransbach-Baumbach, Germany<br>tel: 49 (0) 2623 830<br>fax: 49 (0) 2623 8340<br><br>*WBB Pacific Clays Pte. Ltd.*<br>20 Cecil Street<br>#22-01 The Exchange<br>Singapore 049705, Singapore<br><br>*Pornon et Cie SARL*<br>03490 Diou (Allier), France<br>tel: 33 7042 90 05<br>fax: 33 70 42 91 97<br><br>*WBB Italia Srl*<br>Viale Dino Ferrari 75-83<br>I-41053 Maranello MO, Italy<br>tel: 39 536 941082<br>fax: 39 536 940689<br>*WBB China Ltd.*<br>1735 Park-In Commercial Centre<br>56 Dundas Street, Kowloon, Hong Kong<br>tel: 852 2710 9133<br>fax: 852 2332 6077<br><br>*WBB Vingerling*<br>44 Provincialeweg West<br>P.O. Box 70<br>NL-2850 AB Haastrecht, The Netherlands<br>tel: 31 182 50 27 77<br>fax: 31 182 50 30 35 | ball clay | 75 |
| *Wright Industries, Inc.*<br>225 49th Street<br>Brooklyn, NY 11220, USA<br>tel: 1 718 492 5400<br>fax: 1 718 492 5432 | ferrites | 85 |
| *Zemex Industrial Minerals*<br>1040 Crown Point Parkway<br>Suite 270<br>Atlanta, GA 30338<br>tel: 1 770 392 8664<br>toll free: 1 800 765 8997<br>fax: 1 770 392 8670 | barium sulfate<br>mica<br>talc | 37<br>113<br>152 |

| Company name and address | Fillers | Page |
|---|---|---|
| *Zeochem*<br>Chemie Uetikon Subsidiary<br>P.O. Box 35940<br>Louisville, KY 40232, USA<br>tel: 1 502 634 7600<br>fax: 1 502 634 8133 | zeolites (molecular sieves) | 170 |
| *Zinc Corporation of America*<br>300 Frankfort Road<br>Monaca, PA 15061-2295, USA<br>tel: 1 412 774 1020 | zinc oxide | 172 |

# DIRECTORY OF EQUIPMENT MANUFACTURERS

| Company name and address | Equipment type | Page |
|---|---|---|
| *AccuRate Bulk Soilds Metering*<br>*Unit of Schenck AccuRate*<br>746 East Milwaukee Street<br>P.O. Box 208<br>Whitwater, WI 53190, USA<br>tel: 1 414 473 2441<br>toll free: 1 800 558 0184<br>fax: 1 414 473 4384 | solid metering | 220 |
| *Bel-Tyne Products Ltd.*<br>Victoria Works<br>Brewery Street<br>Portwood, Stockport SK1 2BQ, England<br>tel: 44 0161 480 9511<br>fax: 44 0161 429 8706 | bag handling systems | 216 |
| *Chronos Richardson GmbH*<br>Postfach 1155<br>D-53758 Hennef, Germany<br>tel: 49 (0) 2242 8830<br>fax: 49 (0) 2242 883 186<br><br>*Chronos Richardson Ltd.*<br>Arnside Road, Bestwood<br>Nattingham NG5 5HD, England<br>tel: 44 (0) 115 935 1351<br>fax: 44 (0) 115 960 6941<br><br>*Chronos Richardson SA*<br>Via Kennedy 33<br>I-20090 Rodano (MI) Italy<br>tel: 39 (0) 295 328 001<br>fax: 39 (0) 295 328 006<br><br>*Chronos Richardson Inc.*<br>15 Gardner Road<br>Fairfield, NJ 07004, USA<br>tel: 1 201 227 3522<br>fax 1 201 227 8478<br><br>*Chronos Richardson S.A.*<br>2/4 avenue de la Cerisaie<br>Platanes 306<br>F-94266 Fresnes Cedex, France<br>tel: 33 (0) 146 154040<br>fax: 33 (0) 146 683039 | packaging equipment | 203-5 |

| Company name and address | Equipment type | Page |
| --- | --- | --- |
| *Cleveland Vibrator Company*<br>2828 Clinton Avenue<br>Cleveland, OH 44113, USA<br>tel: 1 216 241 7157<br>fax: 1 216 241 3480 | conveying | |
| *Favre & Matthijs SA*<br>Chemin des Fleurettes, 43<br>CH-1007 Lausanne, Switzerland<br>tel: 41 (0)21 616 98 81<br>fax: 41 (0)21 617 57 56 | sampling | 222 |
| *Hapman Conveyors*<br>6002 E. Kilgore Road<br>P.O. Box 2321<br>Kalamazoo, MI 49003, USA<br>tel: 1 616 343 1675<br>toll free: 1 800 968 7722<br>fax: 1 616 349 2477 | in-plant conveying | 210 |
| *Halvor Forberg AS*<br>Hegdal<br>N-3261 Larvik, Norway<br>tel: 47 33 13 34 34<br>fax: 47 33 13 34 35 | drying | 221 |
| *JAYGO, Inc.*<br>675 Rahway Avenue<br>Union, NJ 07083, USA<br>tel: 1 908 688 3600<br>fax: 1 908 688 6060 | blending | |
| *K-Tron*<br>Routes 55 & 553<br>Pitman, NJ 08071, USA<br>Tel: 1 609 589 0500<br>Fax: 1 609 589 8113 | pneumatic conveying<br>semi-bulk unloading<br>feeding | 210<br>215<br>218-9 |
| *Lancaster Products*<br>*Division of Kercher Industries, Inc.*<br>920 Mechanic Street<br>Lebanon, PA 17046, USA<br>tel: 1 717 273 2111<br>fax: 1 717 273 2967 | semi-bulk unloading<br>feeders and mixers | 215 |
| *Littleford Day, Inc.*<br>7451 Empire Drive<br>Florence, KY 41042-2985, USA<br>tel: 1 606 525 7600<br>fax: 1 606 525 1446 | blending<br>drying | 217<br>220 |
| *Niro Inc.*<br>9165 Rumsey Road<br>Columbia, MD 21045-1991, USA<br>tel: 1 410 997 8700<br>fax: 1 410 997 5021 | drying | |

| Company name and address | Equipment type | Page |
|---|---|---|
| *NOVATEC*<br>222 E. Thomas Avenue<br>Baltimore, MD 21225, USA<br>tel: 1 410 789 4811<br>fax: 1 410 789 4638 | blending/precise metering<br>drying | 217<br>221 |
| *Palamatic Handling Systems Ltd.*<br>Cobnar Wood Close<br>Chesterfield Trading Estate<br>Sheepbridge, Chesterfield<br>Derbyshire S41 9RQ, England<br>tel: 44 (0) 1246 452054<br>fax: 44 (0) 1246 451379 | bag handling equipment | 215 |
| *Premier Pneumatics, Inc.*<br>606 North Front St.<br>P O Box 17<br>Salina, KS 67402-0017, USA<br>Tel: 1 913 825 1611<br>Fax: 1 785 825 8759 | rail and vehicle unloading<br>silo and pneumatic systems<br>bag handling equipment | 205-7<br>208-9<br>216 |
| *Spiroflow-Orthos Systems, Inc.*<br>2806 Gray Fox Road<br>Monroe, NC 28110, USA<br>tel: 1 704 291 9595<br>fax: 1 704 291 9594 | conveying systems | 212-4 |

# Index

**I**



**K**



**L**



**M**

**Q**



**R**